THE

Environment

AND

YOU

MW00753465

NORM CHRISTENSEN   LISSA LEEGE

PEARSON

Acquisitions Editor: Alison Rodal
Project Managers: Laura Murray and Mae Lum
Program Manager: Anna Amato
Development Editor: Mary Hill
Editorial Assistants: Alison Cagle and Libby Reiser
Text Permissions Project Manager: William Opaluch
Development Director: Ginnie Simione Jutson
Program Management Team Lead: Michael Early
Project Management Team Lead: David Zielonka
Production Management: Lumina Datamatics
Copyeditor: Lumina Datamatics

Compositor: Lumina Datamatics
Design Manager: Mark Ong
Interior Designer: Tani Hasegawa, TTEye
Cover Designer: Tani Hasegawa, TTEye
Illustrator: International Mapping
Rights & Permissions Project Manager: Donna Kalal
Rights & Permissions Management: Lumina Datamatics
Photo Researcher: Lumina Datamatics
Manufacturing Buyer: Maura Zaldivar-Garcia
Executive Marketing Manager: Lauren Harp

Cover Photo Credit: Volunteers: © Hero Images/Corbis RF (42-46458524); Crop fields: © Ben Bloom/Getty RF (83758293); Solar panels: © Henglein and Steets/Getty RF (127544072); Magoito beach: © CResende/Getty RF (450422919); Namib-Naukluft National Park: © Lucyna Koch/Getty RF (157522186)

Copyright ©2016, 2013 Pearson Education, Inc. All Rights Reserved. Printed in the United States of America. This publication is protected by copyright, and permission should be obtained from the publisher prior to any prohibited reproduction, storage in a retrieval system, or transmission in any form or by any means, electronic, mechanical, photocopying, recording, or otherwise. For information regarding permissions, request forms and the appropriate contacts within the Pearson Education Global Rights & Permissions department, please visit www.pearsoned.com/permissions/.

Acknowledgements of third party content appear on page C-1, which constitutes an extension of this copyright page.

PEARSON, ALWAYS LEARNING, and MasteringBiology are exclusive trademarks in the U.S. and/or other countries owned by Pearson Education, Inc. or its affiliates.

Unless otherwise indicated herein, any third-party trademarks that may appear in this work are the property of their respective owners and any references to third-party trademarks, logos or other trade dress are for demonstrative or descriptive purposes only. Such references are not intended to imply any sponsorship, endorsement, authorization, or promotion of Pearson's products by the owners of such marks, or any relationship between the owner and Pearson Education, Inc. or its affiliates, authors, licensees or distributors.

Library of Congress Cataloging-in-Publication Data
Christensen, Norman L., 1946-
    The environment and you / Norm Christensen, Lissa Leege. — Second edition.
        pages cm
    ISBN 978-0-321-95789-4
    1. Environmentalism.   2. Environmental policy.   3. Environmental protection.   I. Leege, Lissa.   II. Title.
GE195.C575 2015
304.2'8—dc23
                                                                        2014043393

2 3 4 5 6 7 8 9 10—V011—18 17 16 15

ISBN 10: 0-321-95789-X; ISBN 13: 978-0-321-95789-4 (Student edition)
ISBN 10: 0-134-01439-1; ISBN 13: 978-0-134-01439-5 (A La Carte)

About the Authors

Norm Christensen is professor emeritus and founding dean of Duke University's Nicholas School of the Environment. He earned his undergraduate and master's degrees in biology at California State University, Fresno, and his doctorate at the University of California, Santa Barbara. A central theme in Norm's career has been ecosystem change from both natural and human causes. His research includes studies of the causes and consequences of fire in grasslands, shrublands, and forests; of the impacts of human land use and land abandonment on ecosystem change and species conservation; and of the influence of global warming patterns on ecosystem change. Norm has worked on numerous national advisory committees, including the Interagency Taskforce on the Ecological Effects of the 1988 Yellowstone Fire, the Committee on Environmental Issues in Pacific Northwest Forest Management, the Ecological Society of America Committee on the Scientific Basis for Ecosystem Management, and the U.S. Nuclear Waste Technical Review Board. He has served on the boards of The Conservation Fund, Resources for the Future, the Environmental Defense Fund, The Wilderness Society, and the North Carolina Nature Conservancy. He is a fellow in the American Association for the Advancement of Science and past president of the Ecological Society of America.

Undergraduate education, especially at the introductory level, has been an important part of Norm's career at Duke. He has been honored twice by the university with awards for distinguished undergraduate teaching. He was instrumental in the development of Duke's undergraduate program in environmental science and policy, and he has taught the introductory course for this program for over 15 years. This book is very much a product of these efforts and Norm's passion for connecting students with their environment.

Lissa Leege is a professor of biology and the founding director of the Center for Sustainability at Georgia Southern University. She earned her undergraduate degree in biology from St. Olaf College and received her Ph.D. in plant ecology at Michigan State University. Her ecological research concerns threats to rare plants, including the effects of fire and invasive species on endangered plant populations and communities. She has also conducted 20 years of research on the impacts of invasive pines on the sand dunes of Lake Michigan and the subsequent recovery of this system following invasive species removal. Lissa was instrumental in the development of an Interdisciplinary Concentration in Environmental Sustainability for undergraduates at Georgia Southern. Under her direction, the Center for Sustainability engages the campus in an annual *No Impact Week* and reaches the community with an annual GreenFest celebration, as well as a robust sustainability speaker series. Lissa is also involved with the environment on a statewide level as a member of the 2013 Class of the Institute for Georgia Environmental Leadership and a founding member of the Georgia Campus Sustainability Network. She serves on a local tree board and as a board member for Georgia Southern's Botanical Garden.

Lissa has taught non-majors environmental biology for 16 years with an emphasis on how students can contribute to environmental solutions. In 2006, she established an Environmental Service Learning project, through which thousands of environmental biology students have engaged in tens of thousands of hours of environmental service in the local community. Lissa was honored with both college and university service awards and has served as a faculty fellow in Service-Learning. She also teaches principles of biology, biology of plants, and graduate-level sustainability courses, as well as Study Abroad sustainability courses in Italy. Her contributions to this book have been inspired by her passion for engaging students in positive solutions to environmental problems.

Dedication

To Nicholas, Natalie, Noelle, Nicole, Riley, and all other of Earth's children. May we make decisions today that ensure the future beauty, diversity, and health of the environment on which they will depend.

To Micah and Emory, my constant joy and inspiration. I owe you the beautiful world I inherited, and it is my hope that education will motivate all kinds of students to take leadership and action in bringing about a bright and sustainable future.

Preface

It has been said that change is the only constant. For billions of years, Earth's environment and the organisms that inhabit it have been constantly changing. Over tens of millennia we, our species, have constantly changed; each generation's technologies, values, and understanding of its environment have differed from those that preceded it. As a consequence of those technologies and our growing numbers, we have changed Earth's environment more than any other species living now or in the past.

You and the world around you are the current manifestation of this process of inexorable change. The health and well-being of most of Earth's people have markedly improved over the past century but our impacts on Earth's environment have increased significantly. A century ago, our global population was fewer than 2 billion; today there are well over 7 billion of us. What's more, each of us today uses several times more resources and generates several times more waste than our century-ago ancestors. The effects on our environment are alarming. Resources such as water and petroleum are dwindling. Air pollution and water pollution have become commonplace. Rates of extinction among Earth's species are more than 10 times higher than in pre-industrial times, and Earth's climate is warming because of human-caused changes in the chemistry of its atmosphere.

These changes threaten the health of Earth's ecosystems and the well-being of many of its people; they directly affect you. These changes are unsustainable, but they are not inevitable. Sustainability and ecosystems are important themes throughout this book. Sustainable action and change require knowledge and understanding of the ecosystems upon which we depend. Yes, they are complex, but the key elements of ecosystem function and sustainability are beautifully simple. In an increasingly urban and technology-driven world, the connections between Earth's ecosystems and our well-being may seem distant, even irrelevant. But they are at all times immediate and compelling.

We have not downplayed the significant challenges presented by the variety of environmental issues that affect our lives because a balanced view of the challenges is needed. Naïve optimism is not likely to motivate substantial change in our actions and impacts, but neither is pessimism. We can all change the world in directions that are truly sustainable. We are convinced you will be part of that process of change. That confidence and conviction were the motivation for writing this book; hope was the inspiration.

Hallmark Features and New Innovations

A New Author

We welcome Lissa Leege to the author team of *The Environment and You*. Lissa teaches biology and environmental biology at Georgia Southern University and is the director of the university's Center for Sustainability. Lissa's passion for the environment and teaching, as well as her spirit of hope, have added new energy to the second edition.

A Focus on You

A hallmark of the first edition and further reinforced in the second is the importance of humans as agents of environmental change. The effects of those changes on human well-being continue to be a central theme in the second edition. *The Environment and You* emphasizes problem-solving and solutions that will enable you to make more informed choices on actions to support the well-being of humans and the health of the planet.

- *Where You Live* New to the second edition, this feature invites you to use primary data sources to explore environmental principles, issues, and sustainable solutions within the context of your local community. By answering the questions posed, you'll see how concepts and examples from your textbook can be applied to where you live and learn. This will not only satisfy your curiosity but also help you connect local discoveries to central themes of the chapters. Do you know, for example, what biome you live in (Chapter 7) or whether you share your local environment with an endangered species (Chapter 8)? Do you ever think about just how much water you use everyday (Chapter 11)? How about the size of your waste footprint (Chapter 17)? These are just a few of the questions you will explore.

- *Seeing Solutions* Problems need solutions and this feature highlights how individuals and groups around the world are using new approaches to solve environmental problems. Topics include a city that is investing in green space to solve problems associated with transportation, the local economy, and the health of its citizens (Chapter 16); a business that lessens its impact while improving profit and employee–community relations with a focus on the triple bottom line (Chapter 1); a group that supports increased educational opportunity for young women as a means to improve the health and well-being of their communities (Chapter 5); and efforts designed to support underdeveloped countries in dealing with the economic pressures of a changing world (Chapter 8).

- *Agents of Change* This feature showcases the efforts of college students and recent graduates who have taken action to produce sustainable environments and improve human well-being. It is intended to provide guidance and encouragement for any student with a similar drive to make the world a better place. The second edition features seven new inspiring Agents of Change, including Sol Weiner and Tom Clement, Guildford College (Chapter 2); Liz Brajevich, Michigan State University (Chapter 6); and Alex Freid, University of New Hampshire (Chapter 17).

- *Real Questions* We asked students around the country what they wanted to know and they responded with questions such as "Is climate change the reason for increased storms?" Their questions and our answers appear in the margins of every chapter.

Solid Coverage of Environmental Science

Our current understanding of environmental issues is built on a foundation of decades of careful research by generations of scientists. The second edition not only continues to provide many examples to help you understand the role science and scientific data can play in reducing uncertainty surrounding environmental issues but also engages you in the spirit of inquiry scientists use to ask questions and gather evidence to support predictions.

- *Currency* New discoveries are constantly occurring, and our understanding is quickly evolving in all areas of environmental science. Among the many updates to the second edition are recently revised United Nations forecasts for the growth of human populations, the latest information on changes in Earth's climate from the Intergovernmental Panel on Climate Change, and recent innovations in agriculture, energy conservation, and green building practices. This edition provides the most current synthesis of such changes in every environmental field.

- *Motivation* Each chapter opens with an essay about humans and their interaction with or understanding of the environment. From the long-abandoned statues of Easter Island (Chapter 1) to the present-day concerns of the Arctic Inuit (Chapter 9) or the principled attempt by some San Franciscans to return Hetch Hetchy Valley to its original state (Chapter 2), environmental science is full of interesting stories. These stories will help you connect to the scientific concepts introduced in each chapter.

- *Applications and Examples* *The Environment and You* provides numerous explanations of how scientists have found innovative ways to gather the evidence that supports current conclusions and enables informed predictions.

- *Focus on Science* This feature encourages you to think about the process of scientific inquiry and the different methods scientists use to gather evidence by highlighting the work of individual scientists and the contributions they have made. In the second edition, we have emphasized the strategies scientists use to conduct scientific research and added critical thinking questions that will spark class discussion and encourage you to think like a scientist.

- *New Frontiers* New to the second edition, this feature highlights interesting areas of environmental research as well as unique approaches to problem-solving. New Frontier features emphasize the complex interactions between new scientific discovery, ethics, and policy and ask you to consider the implications of the power science has to change the way we live and interact with the environment.

Organized for Learning

The Environment and You is organized to help students understand environmental science.

- Each lesson begins with a big idea so students always have a way to see the forest as well as the trees.
- Manageable amounts of information are organized by key concepts within modules, giving students complete lessons before moving on to the next topic.
- Important concepts are illustrated with clear, purposeful charts and graphs and supported with photographs that capture the essence of the concept being presented.

A new overall chapter organization in the second edition improves the continuity and connectivity between chapters and integrates complex key concepts with relevant environmental issues. Our new organization features the following:

- An integrated approach to ecosystem ecology, where complex biogeochemical cycles appear in context
- A single chapter on the geography of life that includes marine, aquatic, and terrestrial biomes

- A single chapter on water that offers a more cohesive approach, uniting coverage of the physical attributes of water with the discussion of issues related to water quality, conservation, and wastewater management

Supporting All Levels of Students

Students in introductory environmental science classes have vastly different levels of science background. *The Environment and You* is designed and written to serve that diversity.

- **Self-assessment:** Questions at the end of every module allow students to assess whether or not they have truly grasped a topic before they move on. Questions at the end of each chapter are designed to encourage synthesis of concepts and application to real situations.
- **MasteringEnvironmentalScience™:** Used by over a million science students, the Mastering platform is the most effective and widely used online tutorial, homework, and assessment system for the sciences. It motivates students to come to class prepared; provides students with personalized coaching and feedback; quickly monitors and displays student results; easily captures data to demonstrate assessment outcomes; and automatically grades assignments, including concept review activities, 3-D BioFlix® animation activities and quizzes, Graphit! activities, and chapter reading quizzes.

New to the second edition of *Environmental Science and You*, Mastering has an expanded suite of activities designed to help your students practice concepts and develop scientific inquiry skills:

- *Process of Science activities* encourage your students to put scientific inquiry skills into action. These interactive activities guide them through current environmental research and help them understand concepts such as developing a hypothesis, making a prediction, understanding variables and independent variables, and more.
- *Global Connection activities* demonstrate the global relevance of local environmental issues and chapter themes. Your students will be able to draw comparisons between environmental issues in the United States and other countries such as water usage, air pollution, or species habitat loss.
- Expanded *Interpreting Graphs and Data* activities allow students to practice quantitative skills related to graph interpretation and analysis.

Additional *Video Field Trips* have been added to *MasteringEnvironmentalScience*. Assign these videos for use outside of class or use them in class to bring real issues to life. New to the second edition are a visit to a water desalination plant to see how one community is coping with water resource issues and an in-depth look at bee colony decline in the United States.

Acknowledgments

We accept all of the responsibilities of authorship for the second edition of *The Environment and You*, most particularly for any mistakes or flaws. But others deserve much of the credit for its development, organization, presentation, and production. As this project evolved over the course of several years, the Pearson Education publishing team and numerous environmental science colleagues provided much needed guidance and encouragement.

We are especially grateful to Alison Rodal, our acquisitions editor for this edition of *The Environment and You*. She was the catalyst for many of this edition's organizational changes and new features, and her contagious enthusiasm for this project motivated us at every stage. Chalon Bridges was executive editor for the first edition of *The Environment and You*, and many of her ideas for its content, organization, and presentation continue to be important in this edition.

Our development editor, Mary Hill, expertly and cheerfully guided us on this second edition journey, from start to finish. Mary has an exceptional eye for detail on matters ranging from grammar to module organization and layout to connections among chapters. Even more,

we are awed by her nuanced understanding of so many facets of environmental science that informed her suggestions on substance and presentation. Her wonderful sense of humor sustained us throughout this process. Susan Teahan, Melissa Parkin, and Julia Osborne served as development editors for this book's first edition. We remain grateful to each one for important contributions to the creation of *The Environment and You*.

We thank Editor-in-Chief Beth Wilbur and Director of Development Deborah Gale who encouraged and facilitated this project in both its first and second editions, and Executive Editorial Manager Ginnie Simione Jutson for the second edition. In addition, we would not have been able to publish this project without the support from Editorial Director Adam Jaworski and President Paul Corey. Thank you for taking a risk on this project and for your ongoing collective leadership in science education.

Sophie Mitchell and her wonderful team at Dorling Kindersley Education helped craft and execute the original vision for the first edition of this project.

Program Manager Anna Amato very ably managed both the editorial and production processes. She deserves special credit for keeping all of us on track and on time.

Producing a book where text and art are created, designed, and arranged in tandem requires a highly collaborative approach to publishing. We are grateful to our production colleagues for overseeing and orchestrating this effort. David Zielonka led the production team that included Mae Lum and Laura Murray. Mark Ong and Tani Hasegawa were responsible for the page and cover design of this second edition, and Lindsay Bethoney oversaw the compositing of our text files to actual page layouts. We thank Kevin Lear of International Mapping for his leadership in the production of illustrations, graphs, and maps.

Special thanks go to Libby Reiser who was supplements project manager. Libby not only oversaw the production of all second edition supplements but also played a special role in bringing in new Agents of Change for the second edition. We remain grateful to Assistant Editor Rachel Bricker for her leadership in the development of the Agents of Change features in the first edition. We also thank Editorial Assistant Alison Cagle for so skillfully juggling various tasks to support the entire publishing team.

Special thanks to Content Producer Joe Mochnick for overseeing all details on the production of media for the new edition and for MasteringEnvironmentalScience, and to Tania Mlawer and Sarah Jensen for bringing their creativity and expertise to the development of our new MasteringEnvironmentalScience activities. Todd Brown ensured the smooth release of MasteringEnvironmentalScience for the second edition of the text.

We would also like to thank each contributing supplement author for the edition. Jacquelyn Jordan, Clayton State University, did a wonderful job carefully updating the Instructor's Guide. The Test Bank was written and assembled by Tanya Smutka, Inver Hills Community College, and David Serrano, Broward State College. David is also the author of the second edition PowerPoint presentations, carefully updating each chapter presentation to help give instructors a head start in planning each lecture. Justin St Julianna, Ivy Tech Community College, brought his perspective and expertise using media in his own teaching to our new Process of Science coaching activities. Reading Questions were crafted by Nilo Marin, Broward State College. We also thank Erica Kipp, Pace University, for her contribution to the updates in MasteringEnvironmentalScience resources for this edition.

After many years spent creating and crafting this book, there comes a time to pass the torch to marketing and sales. We are grateful to Christy Lesko, Director of Marketing, for her support of this text. Lauren Harp and Amee Mosley brought endless enthusiasm in promoting *The Environment and You*, communicating our vision to instructors all over the country, all with the support of Ami Sampat, Marketing Assistant. We are fortunate to have the support of the many sales representatives who work tirelessly to communicate our vision to faculty and ensure instructors' needs are satisfied. We thank them for their dedication and commitment!

Terrence Bensel, Brian Bovard, Robert Kingsolver, and Lester Rowntree made important contributions in the first edition to chapters on climate change, biodiversity, agriculture, energy, and waste management. Their detailed outlines provided road maps through sometimes unfamiliar territory, and many elements from their drafts of several of these chapters are part of the final product.

We owe much to our students at Duke and Georgia Southern Universities. In many ways, they helped shape the spirit and content of this text. They have been guinea pigs for each of its chapters and volunteered many editorial comments. The book is much the better for their input.

Over the years, each of us has had the benefit of working with wonderful mentors and colleagues, all the while being supported by our families. For each of us, individually, we want to thank those people who are so special to us.

Norm: My undergraduate and master's advisor Bert Tribbey passed along much knowledge and wisdom that appears in these pages, and he has long served as my primary role model for teaching excellence. My Duke colleagues William Chameides, Deborah Gallagher, Prasad Kasibhatla, Emily Klein, Randy Kramer, Marie Lynn Miranda, Joel Meyer, Lincoln Pratson, William Schlesinger, and Dean Urban were key sources of information and constructive criticism.

I am grateful to my family for their patience with me over the life of this project. My wife Portia has been a sounding board for new ideas, an editor of essays and features, and the best friend ever.

Lissa: My Ph.D. advisor Peter Murphy was an excellent role model who always encouraged my love of teaching and ultimately inspired my desire to reach a wider audience. I am grateful to Georgia Southern University and the Department of Biology for providing me the educational leave I needed to pursue this project, and to my museum colleagues for opening my eyes to the exhilaration of teaching beyond my classroom.

I thank my parents for believing in my passion for sustainability and supporting my path. I owe much to my children Micah and Emory for the time they allowed me to dedicate to this book. I hope that they are proud of the outcome when I can finally say yes to the question, "Are you finished with the book yet, Mom?" Finally, I extend my deepest gratitude to my remarkably patient and supportive husband Frank D'Arcangelo, who encouraged me to follow this dream, even though it meant that he would need to be SuperDad even more often.

Second Edition Reviewers

Mark Basinger
Barton College

Terrence Bensel
Allegheny College

Leonard Bernstein
Temple University

Judy Bluemer
Morton College

Scott Brame
Clemson University

James R. Brandle
College of Agriculture and Natural Resources

Meshagae Hunte-Brown
Drexel University

Robert Bruck
North Caroline State University

Kelly Cartwright
College of Lake County

David Charlet
College of Southern Nevada

Peter G. Chege
Black Hawk College

Lu Anne Clark
Lansing Community College

Jacqueline Courteau
University of Michigan

Anthony D. Curtis
Radford University

Andy Dyer
University of South Carolina

Gregory S. Farley
Chesapeake College

Eric G. Haenni
Franciscan University of Steubenville

Jennifer Harper
Bainbridge College

Stephanie Hart
Lansing Community College

Alyssa Haygood
Arizona Western College

Tara Holmberg
Northwestern Connecticut Community College

Barbara Ikalainen
North Shore Community College

Jacqueline Jordan
Clayton State University

Natalie Kee
University of Mount Union

Reuben Keller
Loyola University

Erica Kipp
Pace University

Katherine LaCommare
Lansing Community College

Nilo Marin
Broward College

Carolyn Martsberger
Loyola University Chicago

John McClain
Temple College

Charles McClaugherty
University of Mount Union

Greg O'Mullan
Queens College CUNY

Raymond S. Pacovsky
Palm Beach State College

Barry Perlmutter
Community College of Southern Nevada

Tim Rhoads
Central Virginia Community College

James Salazar
Galveston College

David Serrano
Broward College

Rich Sheibley
Edmonds Community College

Lynnda Skidmore
Wayne County Community College

Justin R. St. Juliana
Ivy Tech Community College

Keith Summerville
Drake University

Claire Todd
Pacific Lutheran University

Brad Turner
McLennan Community College

Daniel Wagner
Eastern Florida State College

Albert Walls
Cape Fear Community College

Jennifer Welch
Madison Community College Kentucky Community & Technical College System

Jennifer Wiatrowski
Pasco-Hernando State College Porter Campus

James R. Yount
Eastern Florida State College.

First Edition Reviewers

David A. Aborn, *University of Tennessee Chattanooga;* Isoken Aighewi, *University of Maryland;* Saleem Ali, *University of Vermont;* John All, *Western Kentucky University;* Mary Allen, *Hartwick College;* Mark W. Anderson, *University of Maine;* Joe Arruda, *Pittsburg State University;* Daphne Babcock, *Collin County Community College;* Narinder S. Bansal, *Ohlone College;* Jon Barbour, *University of Colorado, Denver;* Morgan Barrows, *Saddleback College;* Christy Bazan, *Illinois State University;* Hans Beck, *Aurora University;* Peter Beck, *St. Edwards University;* Diane B. Beechinor, *Northeast Lakeview College;* Terry Bensel, *Allegheny College;* Leonard Bernstein, *Temple University;* William Berry, *University of California, Berkeley;* Lisa K. Bonnaeu, *Metropolitan Community College;* Brian Bovard, *Florida Golf Coast University;* Peter Busher, *Boston University;* Kelly Cartwright, *College of Lake County;* Paul Chandler, *Ball State University;* David Charlet, *College of Southern Nevada;* Marina Chiarappa-Zucca, *De Anza College;* Van Christman, *Brigham Young University, Idaho;* Donna Cohen, *Massachusetts Bay Community College;* John Conoley, *East Carolina University;* Jessica Crowe, *Valdosta State University;* Jean DeSaix, *University of North Carolina Chapel Hill;* Doreen Dewell, *Whatcom Community College;* Dr. Darren Divine, *Community College of Southern Nevada;* Rebecca Dodge, *Midwestern State University;* James English, *Gardner-Webb University;* JodyLee Estrada Duek, *Pima Community College;* Douglas Flournoy, *Indian Hills Community College;* Steven Frankel, *Northeastern University;* Jonathan Frye, *McPherson College;* Karen Gaines, *Eastern Illinois University;* Kurt Haberyan, *Northwest Missouri State;* Anne Hall, *Emory University;* Stephanie Hart, *Lansing Community College;* Harlan Hendricks, *Columbus State University;* Carol Hoban, *Kennesaw State University;* Kelley Hodges, *Gulf Coast Community College;* Tara Holmberg, *Northwestern Connecticut Community College;* Kathryn Hopkins, *McLennan Community College;* Meshagae Hunte-Brown, *Drexel University;* Emmanuel Iyiegbuniwe, *Western Kentucky University;* Tom Jurik, *Iowa State University;* Richard Jurin, *University of Northern Colorado;* Susan W. Karr, *Carson-Newman College;* David K. Kern, *Whatcom Community College;* Kevin King, *Clinton Community College;* Jack Kinworthy, *Concordia University;* Rob Kingsolver, *Bellarmine University;* Cindy Klevickis, *James Madison University;* Steven A. Kolmes, *University of Portland;* Ned Knight, *Linfield College;* Erica Kosal, *North Carolina Wesleyan College;* Janet Kotash, *Moraine Valley Community College;* Robert Kremer, *University of Missouri;* Diana Kropf-Gomez, *Richland College;* James David Kubicki, *The Pennsylvania State University;* Kody Kuehnl, *Franklin University;* Frank Kuserk, *Moravian College;* Troy A. Ladine, *East Texas Baptist University;* Elizabeth Larson-Keagy, *Arizona State University;* Jejung Lee, *University of Missouri;* Lissa M. Leege, *Georgia Southern University;* Kurt Leuschner, *College of the Dessert;* Honqi Li, *Frostburg State University;* Satish Mahajan, *Lane College;* Kenneth Mantai, *State University of New York, Fredonia;* Anthony Marcattilio, *St. Cloud State University;* Heidi Marcum, *Baylor University;* Allan Matthias, *University of Arizona;* Kamau Mbuthia, *Bowling Green State University;* John McClain, *Temple College;* Joseph McCulloch, *Normandale Community College;*

Robert McKay, *Bowling Green State University*; Bram Middeldorp, *Minneapolis Community and Technical College*; Chris Migliaccio, *Miami Dade College*; Kiran Misra, *Edinboro University of Pennsylvania*; James Morris, *University of South Carolina, Columbia*; Sherri Morris, *Bradley University*; Eric Myers, *South Suburban College*; Jason Neff, *University of Colorado, Boulder*; Emily Nekl, *High Point University*; John Olson, *Villanova University*; Bruce Olszewski, *San Jose State University*; Gregory O'Mullan, *Queens College*; Stephen Overmann, *Southeast Missouri State University*; William J. Pegg, *Frostburg State University*; Barry Perlmutter, *Community College of Southern Nevada*; Shana Petermann, *Minnesota State Community and Technical College*; Julie Phillips, *De Anza College*; Frank Phillips, *McNeese State University*; John Pleasants, *Iowa State University*; Brad Reynolds, *University of Tennessee, Chattanooga*; Kayla Rihani, *Northeastern Illinois University*; Carleton Lee Rockett, *Bowling Green State University*; Susan Rolke, *Franklin Pierce University*; Deanne Roquet, *Lake Superior College*; Steven Rudnick, *University of Massachusetts, Boston*; Dork Sahagian, *Lehigh University*; Milton Saier, *University of California, San Diego*; James Salazar, *Galveston College*; Kimberly Schulte, *Georgia Perimeter College*; Michele Schutzenhofer, *McKendree University*; Rebecca Sears, *Western State College of Colorado*; David Serrano, *Broward College*; Garey Simpson, *Kennesaw State University*; Debra Socci, *Seminole Community College*; Ravi Srinivas, *University of St. Thomas*; Craig W. Steele, *Edinboro University*; Michelle Stevens, *California State University, Sacramento*; Robert Strikwerda, *Indiana University, Kokomo*; Keith Summerville, *Drake University*; Jamey Thompson, *Hudson Valley Community College*; Ruthanne Thompson, *University of North Texas*; Bradley Turner, *McLennan Community College*; Lina Urquidi, *New Mexico State University*; Sean Watts, *Santa Clara University*; John Weishampel, *University of Central Florida*; Timothy Welling, *Dutchess Community College*; Kelly Wessell, *Tompkins Cortland Community College*; James Winebrake, *Rochester Institute of Technology*; Chris Winslow, *Bowling Green State University*; Danielle M. Wirth, *Des Moines Area Community College*; Todd Yetter, *University of the Cumberlands*.

Class Test and Interview Participants

Ginny Adams, *University of Central Arkansas*; John All, *Western Kentucky University*; Jeff Anglen, *California State University, Fresno*; Dave Armstrong, *University of Colorado*; Berk Ayranci, *Temple University*; Roy Barnes, *Scottsdale Community College*; Christy Bazan, *Illinois State University*; Sandy Bejarano, *Pima College East Campus*; Leonard Bernstein, *Temple University*; William Berry, *University of California, Berkeley*; Neil Blackstone, *Northern Illinois University*; Christopher Bloch, *Texas Tech University*; Gary M. Booth, *Brigham Young University*; James Brandle, *University of Nebraska, Lincoln*; Robert Bruck, *North Carolina State College*; George Byrns, *Illinois State University*; John Calloway, *University of San Francisco*; Frank Carver, *Forsyth College*; Ken Charters, *Cochise Community College*; Dave Charlet, *Community College of Southern Nevada*; LuAnn Clark, *Lansing Community College*; Jaimee Corbet, *Paradise Valley Community College*; Robert Cromer, *Augusta State University*; Wynn Cudmore, *Chemeketa Community College*; Jane Cundiff, *Radford University*; Lynnette Danzl-Tauer, *Rock Valley College*; James Diana, *University of Michigan, Ann Arbor*; Darren Divine, *Community College of Southern Nevada*; Rebecca Dodge, *Midwestern State University*; David Dolan, *University of Wisconsin, Green Bay*; Michael Draney, *University of Wisconsin, Green Bay*; Renee Dutreaux-Hai, *California State University, Los Angeles*; Johannes Feddema, *University of Kansas*; Richard S. Feldman, *Marist College*; Kevin Fermanich, *University of Wisconsin, Green Bay*; Linda Fitzhugh, *Gulf Coast College*; Laurie Fladd, *Trident Technical University*; Chris Fox, *Catonsville Community College*; Katie Gerber, *Santa Rosa Junior College*; Thaddeus Godish, *Ball State University*; James Goetz, *Kingsborough Community College*; Robert Goodman, *Citrus College*; Larry Gray, *Utah Valley University*; Peggy Green, *Broward Community College, North*; Joshua Grover, *Ball State University*; Kurt Haberyan, *Northwest Missouri State*; George Hagen, *Palo Alto College*; Nigel Hancock, *Long Beach City College*; Wendy Hartman, *Palm Beach Community College*; Kim Hatch, *Long Beach City College*; James Haynes, *State University of New York, Brockport*; Kathi Hopkins, *McClennan Community College*; James J. Horwitz, *Palm Beach Community College*; Joseph Hull, *Seattle Central Community College*; Carolyn Jensen, *Pennsylvania State University, University Park*; David Jones, *North Eastern Illinois University*; Susan Karr, *Carson-Newman College*; Leslie Kanat, *Johnson State College*; Julie Klejeski, *Mesabi Range Community College*; Janet Kotash, *Moraine Valley Community College*; Katherine LaCommare, *Lansing Community College*; John Lendvay, *University of San Francisco*; Paul Lorah, *University of St. Thomas*; Deborah Marr, *Indiana University, South Bend*; Allan Matthias, *University of Arizona*; Shelly Maxfield, *Pima Community College*; John McClain, *Temple Junior College*; Joesph McCulloch, *Normandale Community College*; Rachel McShane, *St. Charles Community College*; Steven J. Meyer, *University of Wisconsin, Green Bay*; Alex Mintzer, *Cypress College*; Jane Moore, *Tarrant County Community College*; James Morris, *University of South Carolina, Columbia*; William Muller, *Temple University*; Hari Pant, *City University of New York, Lehman*; Robert Patterson, *North Carolina State University*; Dan Pavuk, *Bowling Green State University*; Christopher Pennuto, *Buffalo State University*; Barry Perlmutter, *Community College of Southern Nevada*; Julie Phillips, *De Anza College*; Mai Phillips, *University of Wisconsin, Milwaukee*; John Pleasants, *Iowa State University*; Ron Pohala, *Luzerne County Community College*; Juan Carlos Ramirez-Darronsoro, *Ball State University*; Marco Restani, *St. Cloud University*; Brad Reynolds, *University of Tennessee, Chattanooga*; Howard Riessen, *Buffalo State University*; Shamili A. Sandiford, *College of Dupage*; Jodi Shann, *University of Cincinnati*; Loris Sherman, *Somerset Community College*; Brent Sipes, *University of Hawaii, Manoa*; Shobha Sriharan, *Virginia State University*; Edward Standora, *Buffalo State University*; Philip Stevens, *Indiana University, Fort Wayne*; John Suen, *California State University, Fresno*; Jamey Thompson, *Hudson Valley Community College*; Claire Todd, *Pacific Lutheran University*; William Trayler, *California State University, Fresno*; Carl N. Von Endem, *Northern Illinois University*; Zhi Wang, *California State University, Fresno*; Sharon Ward, *Montgomery College*; Jeff Watanabe, *Ohlone College*; Paul W. Webb, *University of Michigan, Ann Arbor*; James W.C. White, *University of Colorado*; Deb Williams, *Johnson County Community College*; Christopher J. Winslow, *Bowling Green State University*; Don Wujek, *Oakland Community College, Auburn Hills*; Lori Zaikowski, *Dowling College*; Carol Zellmer, *California State University, Fresno*; Joseph Zurovchak, *Statue University of New York, Orange Community College*.

Contents

9 Climate Change and Global Warming 266

10 Air Quality 306

11 Water 334

12 Agriculture and the Ecology of Food 374

19 The Environment and You 614

Where YOU LIVE

Stay Focused on the Big Ideas

The Environment and You strives to make navigating, focusing, and learning easier for students.

IMPROVED!
Big Idea Summaries start each lesson so you can easily keep sight of the big picture as well as the supporting details for each module and topic.

Big Idea Statements clearly summarize the learning objective for each topic.

NEW! **Improved chapter organization** strengthens the connections between chapters and integrates key concepts with relevant environmental issues. Highlights include a more cohesive approach to water-related issues, an integrated discussion of ecosystem ecology, and a revised chapter on the geography of life.

11.6 Wastewater Treatment

BIG IDEA Wastewater includes sewage, water from sinks and other household uses, stormwater runoff, and water used by manufacturing facilities and other industries. In the past, wastewater was simply dumped into nearby waterways. Today, in most developed countries, wastewater is treated to protect the environment and prevent the spread of disease before it is returned to streams and rivers to flow back into the natural hydrologic cycle. Unfortunately, in some poor countries, wastewater still goes untreated. The most common methods of managing wastewater are municipal sewage treatment plants and septic systems. Recently, some municipalities have become interested in treating wastewater with methods that mimic the biogeochemical processes of natural ecosystems.

Municipal Wastewater Treatment

■ Wastewater treatment varies widely among developing and developed countries.

Until recently, sewage and other forms of wastewater were simply dumped into nearby waterways. Over time, dilution and natural processes would eventually decompose the sewage and purify the water. In areas with concentrated human populations, however, this approach was problematic because natural processes cannot break down large amounts of waste in a short period of time. Untreated sewage and wastewater harm natural ecosystems, threaten human health, and contaminate surface water, the source of drinking water for the vast majority of Earth's people.

In most developed countries, wastewater is now treated before it is returned to the environment. In less developed countries, the treatment of wastewater is more variable. Many poor countries have virtually no wastewater treatment. In Latin America, only about 15% of the wastewater that is collected is treated in

▼ Figure 11.42 **Wastewater Treatment**
When wastewater enters a modern treatment plant, it is pretreated to remove solids. Primary treatment removes additional solids and produces a homogeneous liquid that is high in organic compounds. In secondary treatment, microorganisms decompose those organic compounds.

some manner. The lack of adequate water treatment is a major cause of the high rates of waterborne illnesses, such as cholera and typhoid fever, in developing countries.

In the United States, most cities and towns manage their wastewater and sewage in **municipal sewage treatment plants (MSTPs)**. These plants use a stepwise process to remove wastes and chemicals from the water (Figure 11.42). When water first enters a sewage treatment plant, it is pretreated to remove large solids, such as rags, feminine hygiene products, sand, and gravel. Insoluble chemicals, such as grease and oils, may also be removed.

Pretreated wastewater then flows into large settling tanks, where it undergoes **primary treatment**. Particles in the wastewater settle to the bottom of the tank, forming a sediment called sludge. The main

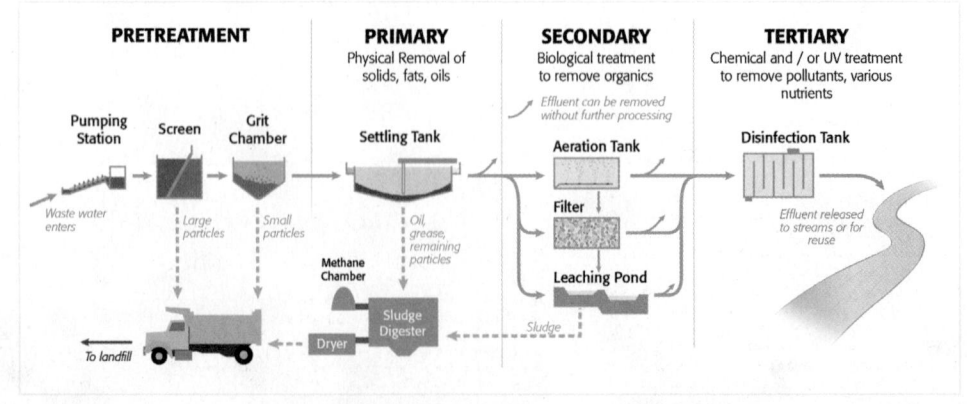

purpose of primary treatment is to produce a relatively homogeneous liquid that can be treated biologically and a sludge that can be processed separately. In many developing countries, municipalities return wastes to the environment after primary treatment.

In **secondary treatment**, bacteria and other microorganisms are used to break down the organic material dissolved in the wastewater. In most treatment plants, secondary treatment takes place in aerated tanks and basins. Some facilities use membranes or gravel filters to further separate solid and liquid wastes (Figure 11.43).

Some treatment plants also use **tertiary treatment** to remove inorganic nutrients from wastewater. In this stage of treatment, wastewater is passed through sand and charcoal filters to remove residual solids and toxins. Next, the water is stored in human-made ponds or lagoons where microorganisms remove significant amounts of dissolved nitrogen and phosphorus. Finally, the water is disinfected with chlorine, ozone, or ultraviolet radiation to reduce the number of microorganisms.

Solid wastes, or sludge, accumulate at each step in this treatment process. Most often, sludge is subjected to digestion by microorganisms, which reduce the volume of organic matter and the number of disease-causing microbes. Sludge is then dried so that it can be transported and disposed of off-site. Usually, it is dumped into landfills or spread onto open land. However, a growing number of treatment plants convert sludge into

pellets that can be used as fertilizers; these pellets are often sold to local gardeners and farmers.

In place of traditional MSTPs, some communities are beginning to use natural or constructed wetlands to purify wastewater that has had primary treatment. Wetlands are very effective at purifying water (see Module 7.7). As water slowly percolates through wetland soils, solid materials are filtered out and microbes decompose the organic matter. Nutrients, such as nitrogen and phosphorus, and contaminants, such as heavy metals, are adsorbed by soil particles or taken up by plants and stored in their tissues.

◀ Figure 11.43 **Your Neighborhood MSTP**
Communities use different components of wastewater treatment depending on their needs. This aerial view of a wastewater treatment plant in Portland, Maine shows primary and secondary treatment.

QUESTIONS 11.6

1. Describe what happens in primary, secondary, and tertiary treatment of wastewater.

2. Explain how an on-site septic system operates.

(MES) For additional review, go to **MasteringEnvironmentalScience**

On-Site Wastewater Treatment

■ If properly maintained, septic systems can isolate waste and protect water supplies.

In less densely populated areas, households often rely on **septic systems** to treat their wastewater and sewage. In these systems, sewage and household wastewater flow to an underground septic tank outside the home. Solids settle to the bottom of the tank where microorganisms begin to break down the waste. Wastewater flows to a series of perforated, underground pipes through which it is released into a **leach field**, where microorganisms in the soil finish breaking down the waste materials (Figure 11.44). Periodically, the solids that settle in the septic tank need to be pumped out and disposed of in a landfill.

Nearly 25% of the households in the United States rely on septic systems to treat their wastewater. When properly maintained, septic systems are an effective means of isolating wastes and protecting water supplies. Maintenance includes monitoring leach fields and occasional pumping of septic tanks. However, about 10% of these systems are not functioning properly. In communities where soil conditions prevent effective leaching, failure rates may exceed 70%. The U.S. EPA reports that failed septic systems are the third most common cause of groundwater contamination.

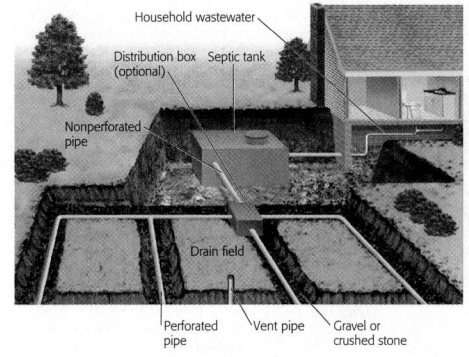

▲ Figure 11.44 **Household Septic System**
Rural wastewater is often treated in on-site septic systems. Wastewater flows into the septic tank, where solid materials are decomposed. Liquid wastes flow out of the tank and into a system of perforated pipes in the leach field, where dissolved organic chemicals are broken down by microbes.

363

Manageable-sized lessons are organized by modules to give you a brief yet complete understanding before moving on to the next topic.

End-of-module questions prompt you to check your understanding at the end of each module.

Find Inspiration in Seeking Solutions to Problems

Co-authors Norm Christensen and Lissa Leege offer a fresh approach by emphasizing problem solving and scalable solutions that inspire students to make more informed choices to support the well-being of humans and the planet.

UPDATED!
Seeing Solutions demonstrates how an organization or community has come together to tackle an environmental problem.

NEW! Where You Live activities invite you to use primary data sources to explore environmental principles, issues, and sustainable solutions in your local community.

Where does your water come from?

Your home has both a street address and a watershed address. Using EPA data, determine in which watershed your home is located.

- What is your watershed called? Where does the water that runs off your street ultimately end up? What upstream rivers lead into your watershed, and what downstream rivers take its water to the ocean?
- Using EPA data for your watershed (impaired water), characterize the water quality in a stream or river near your home.
- What are the most important factors influencing that quality?

SEEING SOLUTIONS

Atlanta's Beltline: Abandoned Railway to Transformative Park Network

How can a blighted railroad corridor dotted with abandoned industrial sites transform into a solution for many common urban problems?

The city of Atlanta, Georgia, holds the auspicious title for the U.S. city with the greatest growth in urban area from 2000 to 2010. Long ago, the city burst through its original boundaries and spread in every direction to occupy a metro area now the size of New Jersey. Atlanta originated as a railroad settlement in the 1830s and was eventually circled by 22 miles of railroad tracks that brought goods to and from the industrial sites located along the outskirts of the city. As the city grew up and its focus shifted to a less industrial economy, the railway waned in importance, and it as well as the industrial sites it served were abandoned. Like many urban systems, Atlanta struggles with sprawl, inadequate public transportation, limited green space, and neighborhoods fragmented by major physical barriers.

Atlanta's award-winning BeltLine began as a thesis project developed by Ryan Gravel in 1999 as he graduated with a master's degree in Architecture and City Planning from the Georgia Institute of Technology. Ryan's vision was to transform the blighted 22-mile railway corridor ringing Atlanta's inner city into a continuous multiuse trail system. The BeltLine would reconnect 45 neighborhoods, revitalize and expand 40 parks, and provide much-needed public transportation via a streetcar system. As an added benefit, Ryan anticipated economic redevelopment of the central city surrounding the BeltLine (Figure 16.26).

Though Ryan's initial vision received accolades, it would take many years, significant and persistent political will, millions of dollars, and a mobilized community to transform it into reality. Several years after graduating Ryan joined an architecture firm, where he discussed his thesis project with colleagues. They put together some concept maps and a letter to send to the mayor, the governor, regional planners, and anyone else who might be able to help. City Councilwoman Cathy Woolard, chair of the transportation committee, gave the project her full support. Together, Gravel and Woolard held meetings in neighborhoods across the city, and Friends of the BeltLine was born. Over the next six years, the plan gained the support of the mayor's office and funding through public-private partnerships.

The first trail opened in 2008. As of January 2014, the BeltLine had four developed trail segments running through 11 miles of new green space. Ultimately, the BeltLine will include 22 miles of pedestrian friendly rail transit, 33 miles of multiuse trails, 1,300 acres of parks, 5,600 units of affordable housing, 1,100 acres of remediated brownfields (industrial wasteland space), public art, and historic preservation (Figure 16.27). The project is expected to be completed over the next two decades.

The economic consequences of the BeltLine are already evident. Property values surrounding the BeltLine were up as much as 30% by 2005, before any part of the project was even complete. In addition, the BeltLine's management group estimates almost $1 billion has been invested in new development surrounding the BeltLine since 2005. Atlanta's BeltLine is now hailed as "the country's best smart growth project"—an engine for new economic development and revitalization of what was once an urban blight. The success of Atlanta's BeltLine shows yet again that what benefits the environment often also benefits the economy.

▼ Figure 16.26 **A Man with a Plan**
Ryan Gravel, the architect whose thesis inspired Atlanta's BeltLine, stands in the foreground of what had once been an abandoned railway line. Eleven miles of trails surrounded by new green space have been created since the project broke ground in 2006, with much more on the way.

Ryan Gravel

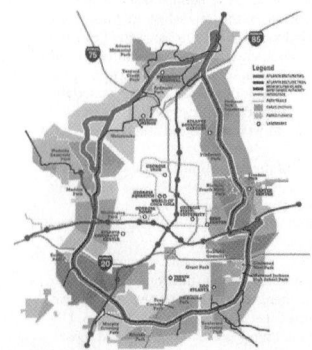

▲ Figure 16.27 **Mapping a Vision**
Atlanta's proposed BeltLine encompasses 1,300 acres of parks, 33 miles of trails, and 22 miles of public transportation. The project is expected to be completed in 20 years.

540

AGENTS OF CHANGE

Water Conservation Competition

Martin Fugueroa is majoring in biology with an emphasis on Human Biology and a minor in sustainability. During his sophomore year at University of California, Merced, Martin Figueroa created the Water Conservation Competition—a month-long water battle between the 14 dorms on campus, which challenges students to reduce their water use. Students can view their daily usage throughout the competition on an online, real-time dashboard that Martin developed with a local technology company. Martin continues to host this competition, which conserves over a million of gallons of water each year.

How did you first get the idea for the UC Merced Water Battle?

UC Merced is located in one of the driest climates in California—the Central Valley, a region also famous for its rich agriculture. When I arrived on campus my freshman year, I noticed a drastic difference in the landscape—parts of the campus had amazing grass and green lands, while other sections were arid and dry. This difference is all due to water resources.

On my drive home to Los Angeles, I observed the large tubes required to pump water to the drier regions of California; just above these were billboards calling for more water resources in the Central Valley. This sparked my desire to learn about where our water originates and how it is distributed.

I then enrolled in a course on sustainability and current environmental issues. This class inspired me to take action on my campus and influence administrators to implement new sustainability standards. My water conservation efforts started out as a campaign, which developed into a competition in the hopes of increasing student participation and awareness. I wanted students to think about water and how our usage impacts the future of our planet.

What steps did you take to create the competition?

The first step was getting approval and support from campus administrators, water stakeholders, and sustainability and housing departments, who would then give me access to facilities and meters needed to track the water usage. I then contacted Aquacue, a water technology company, to help create the water battle dashboard where students can view their water usage in real-time, as well as notify us about leaks. This dashboard greatly increased student participation and allowed students to visually understand their impact.

I also assembled a committee of administrators, stakeholders, and two student groups, Green Campus and Engineers for a Sustainable World. With the help of several classes and professors, we created marketing materials including flyers, short films, commercials, and QR codes, which link students to the water dashboard. Social media provided a great forum for students to encourage their dorm-mates to reduce water consumption, while also fueling competition among the dorms. We kicked off the competition with several tabling events where students could ride our amazing bike blender, drink a free smoothie, and learn about the ways students can conserve water.

370

UPDATED!
Agents of Change showcase inspiring college students and recent graduates who are taking action to develop sustainable environments and improve human well-being. Seven Agents of Change stories are new to the second edition.

AGENTS OF CHANGE

Marisol Becerra, *DePaul University*, Mapping Pollutants in Little Village and Around the World

NEW! Hillary King, *St. Olaf College*, STOGROW

NEW! Eliza Barjbich, *Michigan State University*, Vericomposting

NEW! Alex Fried, *University of New Hamphsire*, Trash 2 Treasure

Rachel Barge, *University of California–Berkeley*, Green Initiative Fund

Will Perez, *Brown University*, Taking Public Health to Rural Haiti

Jessica Franzini, *Stockton College*, Greening Urban Spaces

NEW! Sol Weiner and Tom Clement, *Guildford College*, The Making of *Swine Country*

Varsha Vijay, *Duke University*, Protecting a Unique Biodiversity Hotspot

NEW! Martin Figueroa, *University of California–Merced*, Water Battle

ALSO FEATURED:
Robin Bryan, *University of Winnipeg*, Campaign Against Logging

NEW! Andrew Sartain, University of Oklahoma, Earth Rebirth

NEW! Jen Kelso and Amber White, *Loyola University of Chicago*, Biodiesel Project

Jacob Perritt-Cravey, *University of Florida*, Carbon-Neutral Football Games

Easily Access Current, Accurate Science Information

Our current understanding of environmental issues is built on a foundation of decades of careful research by generations of scientists. This important work is discussed throughout the Second Edition.

NEW! Critical thinking questions can spark class discussion and help you develop inquiry skills and an understanding of scientific discovery.

UPDATED! Focus on Science essays highlight the research of individual scientists and show how scientific research increases understanding of environmental issues.

FOCUS ON SCIENCE

Adapting to Rising Seas

How effective is the conservation of natural habitat as an adaptation strategy to sea level rise and increased storm severity?

Rising sea level and increased storm activity and flooding pose ever-increasing threats to coastal communities around the world. These threats are compounded by rapid population growth and sprawling development in coastal cities. The traditional approach to protect coastal towns has been construction of sea walls and other "hardened" structures. More recently, greater emphasis has been placed on the conservation and restoration of natural habitats such as coral and oyster reefs, sea grass beds, and coastal forests and wetlands that buffer coastlines from waves and storm surges. We know these conservation strategies are effective at particular locations. Katie Arkema and her colleagues at Stanford University Natural Capital Project were interested in determining the value of such conservation practice applied on a large scale, across the entire coast of the United States (Figure 9.45).

Arkema and her team used a combination of *data synthesis* and *ecological models* to address this question. They began by calculating a hazard *index* for each square kilometer of the U.S. coastline based on the physical features that influence water movement, the types of natural coastal habitats at current sea levels, and the likelihood of coastal storms. They then calculated hazard indices based on five sea level scenarios, and for coastlines with or without natural coastal habitat.

Scenario 1 represented current conditions and scenario 2 approximated sea level change expected in a *Sustainable World* future. Scenarios 3, 4, and 5 represented sea level rise with successively greater warming, with 5 corresponding to changes expected under the *Business as Usual* trajectory for global warming. They also mapped data on human populations and property values for each square kilometer of coastline. By overlaying these maps, Arkema was able to convert hazard indices to more direct measures of imperiled human life and property damage.

As expected, the number of people and the amount of property at risk increased with increasing rates of sea level rise (Figure 9.46). The presence of natural coastal habitat diminished those risks by at least 40% in each scenario on a national scale. At least as important, Arkema and her colleagues have produced the first national map indicating where conservation and restoration of reefs, wetlands, and coastal forests have the greatest potential to protect human life and property in coastal communities (Figure 9.47).

Source: Arkema K.K. 2013. Coastal habitats shield people and property from sea-level rise and storms. *Nature: Climate Change* 3: 913–918.

▲ Figure 9.45 **Ecosystem Services** Katie Arkema is interested in finding ways to quantify nature's benefits to people and applying that information to the management of coastal and marine ecosystems.

1. What physical features of a coastline might increase risks associated with sea level rise?

2. Was the effect of habitat protection consistent among the sea level rise scenarios? Explain your conclusion.

3. How might coastal counties use this information to plan future land use?

▲ Figure 9.46 **With and Without Habitat** Bar graphs indicate the number of people and property value at risk nationally. Across all five sea level change scenarios, natural coastal habitats such as reefs, wetlands, and coastal forests substantially diminish risks to life and property in coastal communities.

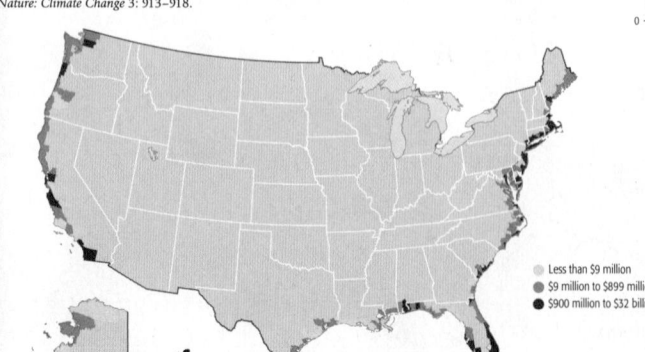

Less than $9 million
$9 million to $899 million
$900 million to $32 billion

◄ Figure 9.47 **Where It Matters Most** The color for each coastal county indicates the total property value for which coastal habitats reduce exposure to storms and sea level rise under sea level rise scenario 4. Coastal habitats protect the greatest value and number of people in New York, Florida, and California.

297

The most current, accurate data and research is presented throughout the text, and citations are provided so you can locate the sources of scientific information.

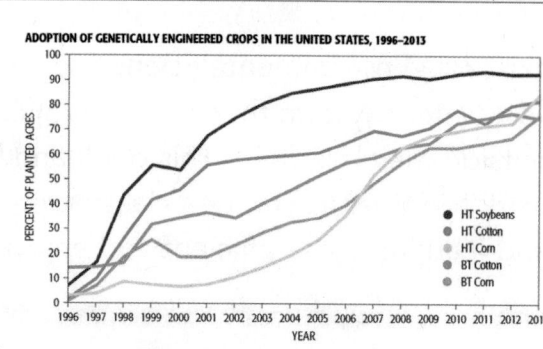

▲ Figure 12.40 **Genetically Modified Crops Crowd the Fields**
GMO corn, cotton, and soybeans have crowded out all other varieties in the United States. More than 90% of these plants are now genetically modified, mostly for resistance to pests and herbicides.

Source: Data from USDA, Economic Research Service and National Agricultural Statistics Service.

9.4 Consequences of Global Warming

BIG IDEA The changes caused by global warming vary from region to region. In some places, increasing temperatures have been accompanied by higher rainfall amounts. In others, they have brought drought. Winters have become milder and shorter in Earth's middle latitudes, and dry seasons have grown longer in some parts of the tropics. Glaciers and ice sheets are melting worldwide. Warming is causing sea levels to rise. Taken together, these changes are having a significant impact on the flora and fauna of many ecosystems.

Drier and Wetter

■ Global warming is producing wetter conditions in some places and drought in others.

The effects of rising temperatures on precipitation vary geographically. Rainfall has increased significantly in eastern North and South America, as well as in most parts of Europe and Asia. In contrast, sub-Saharan Africa, the Mediterranean region, and western North America have been drier. Since 1970, longer and more intense droughts—as measured by decreased precipitation and higher temperatures—have affected wide areas of the tropics and subtropics. At the same time, there has been a worldwide increase in the frequency of rainstorms that result in flooding, even in areas where total annual rainfall has declined (Figure 9.24). Warm air holds more moisture (see Module 3.6).

In regions in which rain is highly seasonal, such as sub-Saharan Africa, global warming appears to be changing the length of wet and dry periods. This is a matter of special concern because food production

depends on the length of wet seasons. Based on current trends and climate models, growing seasons are expected to become shorter over most of sub-Saharan Africa, with the exception of lands very near the equator.

There is evidence that global warming is influencing drought cycles. For example, the El Niño/La Niña/Southern Oscillation is caused by changes in the temperature of surface waters in the equatorial Pacific Ocean. When waters off the west coast of South and Central America are cold, drought is much more common in the southwestern United States. Some climatologists think that since 1970 the length and intensity of El Niño and La Niña events have been outside the range of natural variability. Although climate models predict that such changes will occur, most scientists feel that it is not clear that they are actually underway.

Questions from real students appear throughout the text, along with brief, scientifically accurate responses from authors Norm Christensen and Lissa Leege.

Q *Is climate change the reason for increased storms and global disturbances?*
Ciara Tyce,
Georgia Southern University

Norm: **A** Climate scientists are generally careful not to attribute a particular weather event to global warming. However, increased frequency and intensity of storms and heat waves is consistent with climate model predictions.

New Frontiers discussions emphasizes the complex interactions between new scientific discovery, ethics, and policy.

New FRONTIERS

Revving up Severe Weather?
Ocean temperature is an important factor in the development of tropical storms and hurricanes, and ocean temperatures have increased between 0.25 °C and 0.5 °C (0.45–0.9 °F) over the past century. Warmer sea surface temperatures appear to be associated with the observed increase in the number and strength of tropical storms in the Pacific Ocean. For example, Typhoon Haiyan, which hit the Philippines in 2013, was one of the strongest tropical storms ever recorded. But trends in the Atlantic Ocean are far less clear. The very significant damage from Hurricane Sandy in 2012 was largely a consequence of a combination of sea level rise (see the next section) and poorly managed coastal development (see Module 18.2).

Debate continues regarding the effects of global warming on past and current storm patterns, but there is consensus among scientists that continued sea surface warming will very likely increase the frequency and severity of tropical storms in the future. How much evidence do you believe we need in order to take strong action to mitigate the effects of future strong storms? How much of the risk associated in living in coastal areas should be the responsibility of property owners versus the government?

▲ Figure 9.24 **Deluges and Droughts**
(A) In 2013, torrential rains impacted crop production across much of upstate New York. (B) In 2014, extreme drought in California meant farmers could not grow crops on hundreds of thousands of acres. Global warming may have contributed to both situations.

Learn and Practice with New Online Activities

MasteringEnvironmentalScience® is an online homework, tutorial, and assessment system that helps you quickly master concepts both in and outside the classroom. This book and MasteringEnvironmentalScience work together to create a classroom experience that makes teaching and learning more efficient and enjoyable.

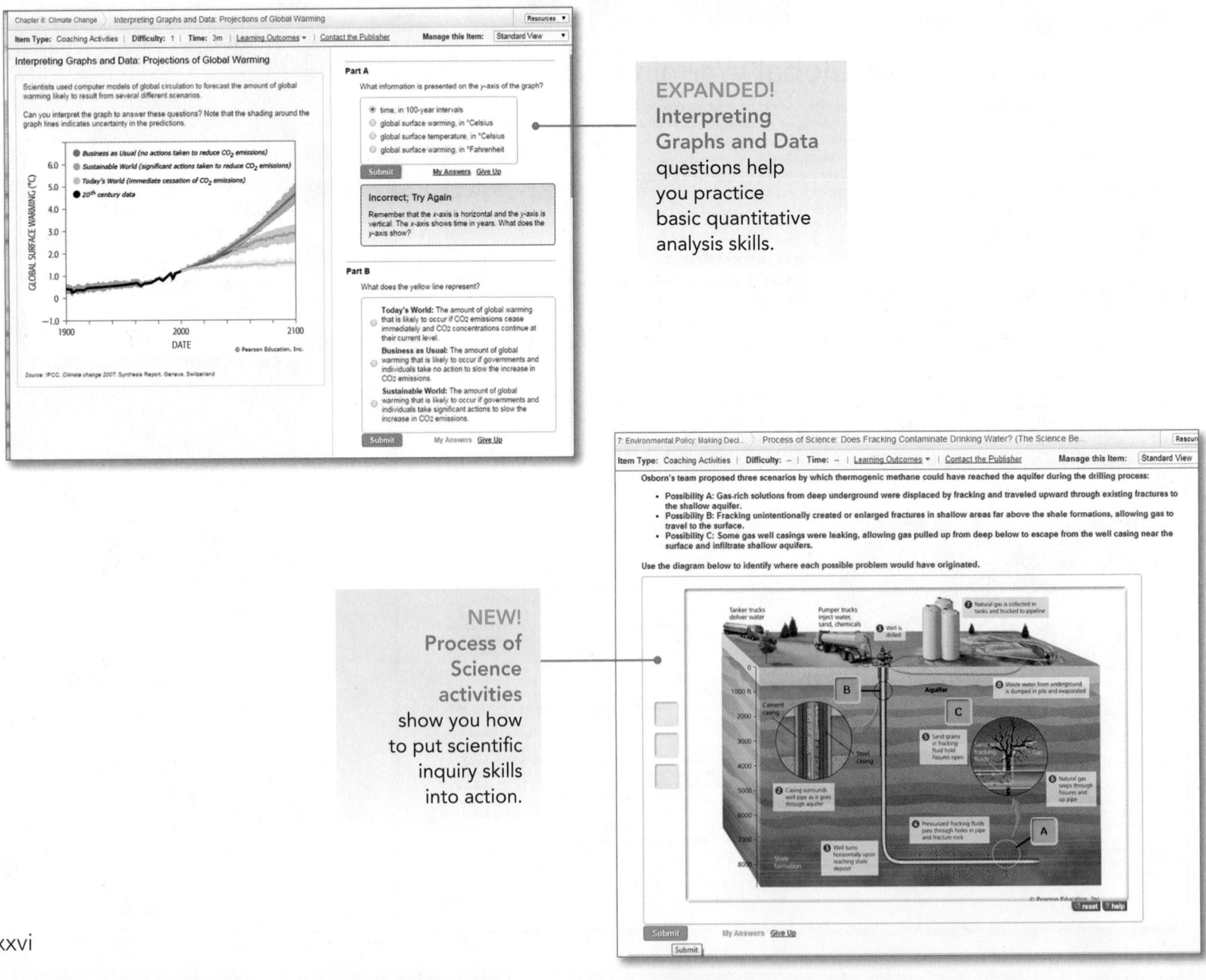

EXPANDED! Interpreting Graphs and Data questions help you practice basic quantitative analysis skills.

NEW! Process of Science activities show you how to put scientific inquiry skills into action.

NEW!
Everyday Environmental Science videos, produced by the BBC, introduce you to connections between environmental science topics and real world issues. Instructors can show film footage in class, during class lectures, or assign as homework in MasteringEnvironmentalScience to engage students in learning about environmental science topics.

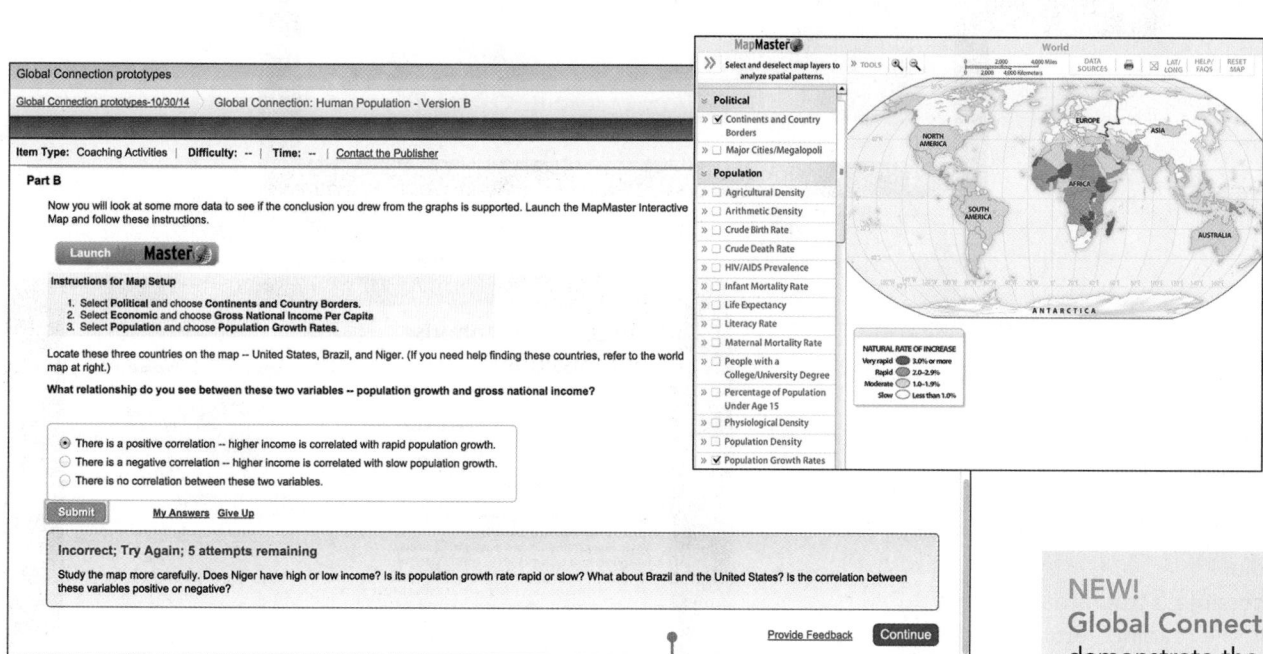

NEW!
Global Connection activities demonstrate the relationship between global and local environmental issues and chapter themes.

Access Study Tools Whenever You Need Them

NEW! Pearson eText 2.0 can be accessed from any web-enabled device, including your computer, tablet, or smartphone.

MasteringEnvironmentalScience®

Using eText 2.0, you can view related videos and animations within the same screen view, reset the type to a larger or smaller size, and read pages in a more comfortable night reading mode.

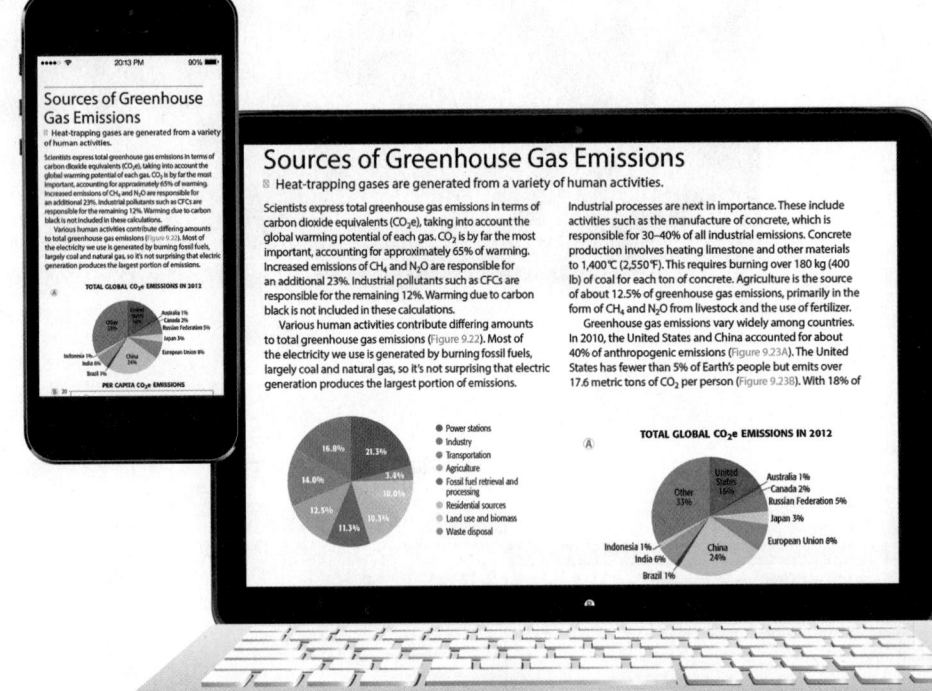

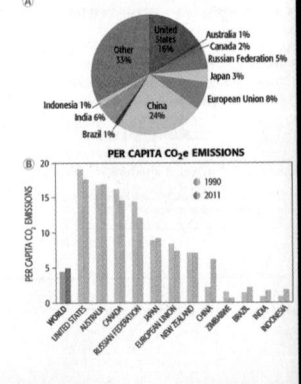

- Now available on smartphones and tablets.
- Seamlessly integrated videos and other rich media.
- Fully accessible (screen-reader ready).
- Configurable reading settings, including resizable type and night-reading mode.
- Instructor and student note-taking, highlighting, bookmarking, and search.

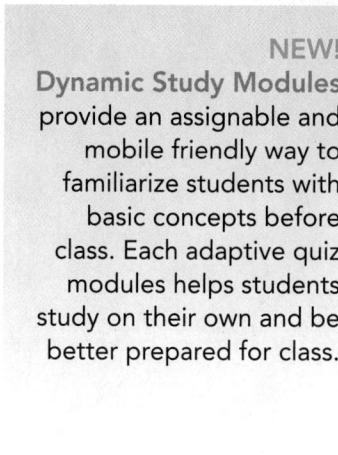

**NEW!
Dynamic Study Modules**
provide an assignable and mobile friendly way to familiarize students with basic concepts before class. Each adaptive quiz modules helps students study on their own and be better prepared for class.

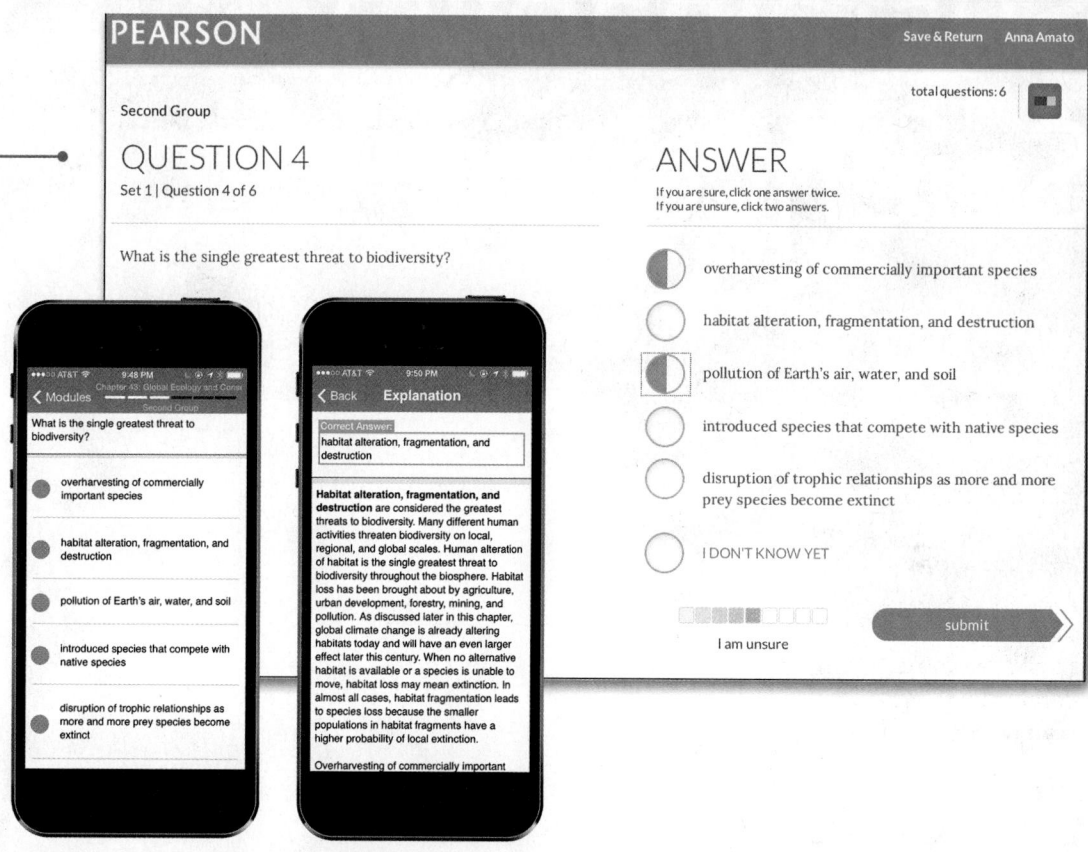

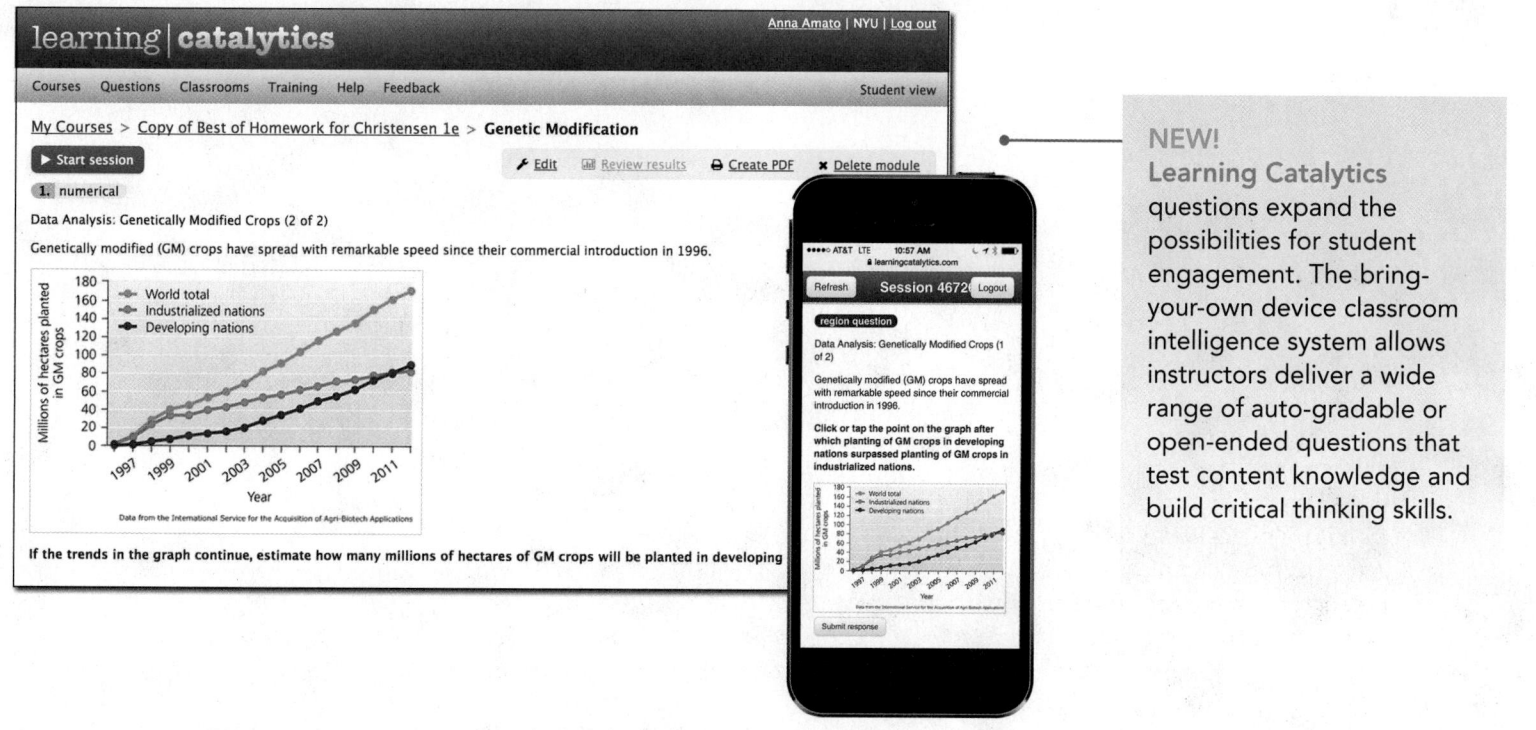

**NEW!
Learning Catalytics**
questions expand the possibilities for student engagement. The bring-your-own device classroom intelligence system allows instructors deliver a wide range of auto-gradable or open-ended questions that test content knowledge and build critical thinking skills.

Environment, Sustainability, and Science

It Takes a Community

Can we collaborate, plan, and act for a sustainable future?

Sometime between A.D. 700 and 1100, people of Polynesian descent colonized a remote island in the southern Pacific Ocean, 3,500 km (2,180 mi) west of Chile. With its diverse forests and productive coastal waters, Rapa Nui or, as it is popularly known, Easter Island must have seemed like paradise to these first inhabitants. In just a few of centuries, these people, also called Rapa Nui, developed an advanced culture with writing and religion, and their population grew to 15,000 to 20,000 (Figure 1.1). But by the time of Dutch explorer Jacob Roggeveen's arrival in 1722, Rapa Nui had been deforested, 21 tree species were extinct, and no tree taller than 10 feet remained. Unable to build seaworthy boats and having decimated the island's land and seabird populations, only 2,000 to 3,000 Rapa Nui remained, eking out a meager living by farming the island's infertile soils.

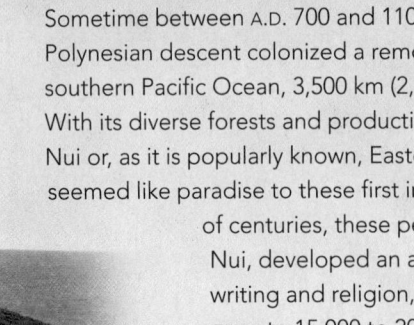

▲ Figure 1.1 **Easter Island**
Hundreds of iconic moai, massive religious statues, testify to a once-robust Rapa Nui community and culture. It is thought that, in the 1600s, the island's few remaining trees were used to transport the moai to the island's perimeter.

The Rapa Nui saga is seen by many historians and ecologists as a classic example of the *tragedy of the commons*. The tragedy, first described in 1968 by philosopher Garret Hardin, is the decline and destruction of the natural resources—forests, wildlife, water, and so on—that are shared in common by members of a community. Cooperation, planning, and regulation among community members could avert this tragedy and sustain such resources indefinitely. But, Hardin argued, cooperative behavior diminishes the net benefit that individual members of the community can obtain by uncontrolled exploitation. In the end, individual greed wins out over community cooperation, and commonly shared resources are overexploited.

We, too, live on an island (Figure 1.2), and possible parallels to the Rapa Nui story are compelling. We share a great many natural resources in common with 7.3 billion other individuals. Supplies of many of those resources are diminishing. Each year, for example, 0.2% of Earth's forests are permanently lost and more than 1% are severely degraded. Nearly a third of marine fish stocks are in decline. The species extinction rate is thought to be 1,000 times greater than in pre-human times. We are polluting critical common resources such as the air we breathe and water we drink. Earth's ecosystems can renew most of these resources, but this would require cooperation, planning, and action among community members.

Are we, too, doomed to the tragedy of the commons? Many environmental scientists believe strongly that we are not. Economist Elinor Ostrom was certainly a vocal champion for this view (Figure 1.3). Although she acknowledged that common resources are often overexploited, she also saw numerous cases where they had been and continue to be sustainably managed. The tragedy of the commons, she argued, is an oversimplification, and community members are not necessarily trapped by their greed. Nor are they unwilling to invest time and energy to agree on sustainable resource management strategies.

Ostrom and her colleagues chronicled numerous examples of community success, including sustainable management of fisheries by the Seri people of northwest Mexico and long-term stewardship of forests by

◄ Figure 1.2 **Home Sweet Home**
It is obvious from this photo of Earth rising over the lifeless surface of our moon that we, too, live on an island. Earth is, so far as we know, the only inhabited and inhabitable place in our solar system. Its resources are finite, but most are renewable if they are used at sustainable rates.

communities in Nepal (Figure 1.4). She also pointed to successful community collaboration on much larger scales such as the Montreal Protocol, an international program that has successfully limited emissions of chemicals that degrade Earth's protective ozone layer.

Ostrom compared these situations to ones where the tragedy of the commons prevailed, and she identified key characteristics of resources, governance, and communities that lead to sustainable management of common resources.

The condition of the resource is very important. When resources become severely exhausted, there are few incentives to manage them sustainably. Therefore, early action is important. Community action is more likely the more important a resource is to the community.

Choice and voice were important governance characteristics shared by sustainable communities. The more options for management, the better, and community members should have equal say in deciding which options to pursue.

▲ Figure 1.3 **A Sustainability Optimist**
Elinor Ostrom was convinced that individuals in communities can sustainably manage their resources through collaborative, collective action. Her innovative studies earned her the 2009 Nobel Prize in Economics.

Sustainable communities share three important characteristics. The first is knowledge; the more communities know and understand about essential resources, the more likely it is that they will collaborate to conserve them. The second is leadership; individuals committed to the future of the community and its resources are essential. The third is what Ostrom called social capital; sustainable communities share a vision for the future, and they share values that shape the means of reaching that future.

Ostrom's key community characteristics appear throughout *The Environment and You*. This includes knowledge and understanding of Earth's ecosystems and our impacts on them. It also includes abundant examples of sustainable actions to reduce or eliminate those impacts. Most important, we demonstrate ways that you as individuals can put these actions into practice. We are certain that you will find many reasons to share Ostrom's belief in the potential for sustainable communities at local, regional, and global levels.

- *What is sustainability?*
- *What are the characteristics of sustainable systems?*
- *What is an ecosystem?*
- *What key characteristics do Earth's ecosystems share?*
- *Why are uncertainty and science important to the sustainable management of systems?*
- *What important questions remain for a sustainable future?*

◀ Figure 1.4 **Sustainable Communities**
Ⓐ The Seri people have sustainably fished and farmed along the coast of the Sea of Cortez for over 2,000 years. Community members share a commitment to the well-being of their community and the resources that support it. Ⓑ For tens of generations, Nepali farmers have sustainably managed their agricultural fields alongside forest ecosystems. These communities understand that intact diverse forests provide fuelwood, timber, and wildlife, and they also control erosion and support pollinators for their crops.

1.1 Environment and Sustainability

BIG IDEA In Earth's long history, no organism has had a greater effect on the environment than have humans. Our ability to appropriate Earth's resources has been a major factor in the rapid growth in our numbers. Over the past century, we have come to understand that our actions have significant consequences for the well-being of the community of all living things and for ourselves in particular. This understanding is the basis for human actions and behaviors that are mindful of essential environmental processes, economically feasible, and fair to all people now and in the future. These are the key prerequisites for a sustainable future.

The Environment and You

■ Environmental science and ecology explore the interactions of humans with the natural environment.

You may use the word *environment* to describe where you are and everything around you. Scientists have a more specific definition: the **environment** is all of the physical, chemical, and biological factors and processes that determine the growth and survival of an organism or a community of organisms. The long list of all the factors that make up your environment would include the gases in the air you breathe and the many life forms that nourish and are nourished by you. **Environmental science** studies all aspects of the environment.

Ecology is the branch of environmental science that focuses on the abundance and distribution of organisms in relation to their environment. Earth's environments have sustained living organisms for at least 3.8 billion years; they have sustained members of our own species for well over 100,000 years. Throughout this time, Earth's environments and the communities of organisms that depend on them have been constantly changing.

▲ Figure 1.5 **The Frontier**
This painting by Albert Bierstadt of seemingly unending wilderness is typical of many 19th-century depictions of American landscapes.

◄ Figure 1.6 **Too Many Uses?**
Rapid population growth in the United States following World War II placed increasing demands on public lands for timber resources and for other ecosystem services such as grazing, recreation, and the protection of water supplies.

Defining Sustainable Actions

■ Our understanding of sustainable behavior has changed through time.

The concept of sustainability is central to environmental science. But what does it mean to be sustainable? Over the last 150 years, we have viewed this concept in different ways.

A century and a half ago, the human use of resources, such as wildlife and fisheries, was determined largely by our needs or perceived needs. Earth's forest resources appeared to be inexhaustible, and wildlife was abundant (Figure 1.5). Rivers, lakes, and coastal waters were teeming with fish. We were fully confident of the environment's capacity to produce an abundance of resources and to absorb and process our wastes. And why not? The world's population was only about 1 billion people, and the population of the United States was less than 75 million. (Today, over 7 billion people inhabit this planet; more than 300 million live in the United States.)

As human populations and their demand for resources grew, supplies of resources began to dwindle. Between 1920 and 1940, the number of scientific studies of the environment increased greatly. With new knowledge, resource managers began to appreciate the need to align the demand for resources with their supply. In forestry,

this included strategies to plant and regrow trees after harvest and to protect them from the threats of pests and fire. In fisheries, this meant establishing catch limits and artificially stocking lakes and streams. These actions helped to sustain the supply of resources.

By 1950, there were 2.5 billion people in the world and 152 million in the United States. As the demand for resources increased, conflicts skyrocketed. In the United States, for example, growth in housing increased the demand for wood from national forests. At the same time, there was increased public interest in using those forests for recreation, for supporting wildlife, and for protecting water supplies. People argued over which use of forest resources should be given priority. To be sustainable, management policies had to recognize the different demands on the environment and its resources. Policies also had to address the conflicts among humans who valued those resources differently. For example, in the management of U.S. national forests there are strongly held values associated with commercial timber management, the provision of clean water, and the conservation of species that have often been in conflict (Figure 1.6).

Where YOU LIVE What resources does your national forest provide?
Find the nearest national forest to your home by searching the web for "U.S. national forest" and the name of your state. What important natural resources and public benefits does your forest provide? What potential conflicts might there be among these uses?

▲ **Figure 1.7 Sustainability as a Commitment to the Future**

The 1987 UN Commission on Sustainability was chaired by Norwegian prime minister Gro Brundtland.

▼ Figure 1.8 **A World of Change**

This is a composite satellite image of Earth at night. Imagine how it might have looked a century ago. How will it change over the next century? The changes in these lights represent changes in Earth's ecosystems, in human values and technologies, and in our environmental impacts.

In its 1987 report, *Our Common Future*, the United Nations Commission on Sustainability declared, "At its most basic level, sustainability means meeting the needs of the present without compromising the ability of future generations to meet their own needs" (Figure 1.7). This declaration added a critical element of time to our concept of sustainability—sustainable management has its eye on the needs of the future.

The UN Commission on Sustainability was concerned with maintaining **human well-being**, a multifaceted concept that includes life's basic necessities, such as food and shelter, as well as good health, social stability, and personal freedom. Their report noted that the factors necessary to ensure human well-being may change from one generation to the next. Thus, sustainability is not a process of maintaining the status quo. Instead, the commission argued that sustainability is maintaining the ability to accommodate three important sources of change.

1. The world is changing.   Earth's environments undergo constant change, sometimes in regular daily and seasonal cycles and sometimes in more complicated patterns. Environmental change is both inevitable and essential.

2. We are changing.   Successive generations of humans have developed and used ever-changing technologies to extract and use resources. Humans also pass their knowledge about their environment and its resources from one generation to the next. As a consequence, our needs for ecosystem goods and services are constantly changing. The value we place on these goods and services changes, too.

3. We are changing the world.   No other organism has ever shaped its environment to the extent that we have. Our use of technologies, such as fire and agriculture, has allowed our numbers to increase and has changed the world. Our increasing numbers have extended our influence (Figure 1.8). Our increasing use of technology has had even greater consequences.

Today we are at a unique moment in history. At no previous time has Earth supported so many humans. At no previous time has overall human well-being been better. We can point with gratitude to actions by previous generations that have helped us meet our current needs, such as the development of agricultural technologies and the establishment of national forests and parks. But we can also identify actions by our ancestors that have impoverished our world. In many regions, poor agricultural practices have permanently diminished soil fertility. Overexploitation has left us with mere snippets of the once vast ancient forests.

Although average well-being among humans is high, variation in well-being is high as well. In developed countries such as the United States, clean water is taken for granted and obesity is all too common. Yet more than one-fifth of Earth's people lack access to clean water and suffer from malnutrition. This disparity has led the United Nations and other world leaders to broaden the definition of sustainability to include the concept of equity. In this context, **sustainability** means meeting the needs of the present *in an equitable and fair fashion* without compromising the ability of future generations to meet their own needs. This is the definition that we will use throughout this text. (It should be noted, however, that scholars and decision makers are not in agreement on what constitutes "equitable and fair.")

Does this definition of sustainability seem too focused on humans? If so, remember that Earth's environments were self-sustaining for billions of years before our time. If we were to disappear tomorrow, they would eventually recover from our impacts.

Our present and future well-being depends heavily on how we treat all of Earth's life and environments. But it also depends on actions that are economically feasible and fair to all people, now and in the future.

Planet, People, and Profit: The Triple Bottom Line

■ Sustainability lies at the intersection of environmental, social, and economic success.

For years, the environmentally minded have insisted on the importance of including the environment in the measurement of profit in business but struggled to develop an accounting system to include these external costs. In 1994, John Elkington, the founder of a British consultancy, SustainAbility, developed the concept of the **triple bottom line (TBL)**. TBL is an accounting framework by which corporations, nonprofit organizations, and even governments can measure three dimensions of performance: environmental, social, and economic. According to TBL thinking, the intersection of these three dimensions is the only sustainable path for the future (Figure 1.9). Also referred to as the "3 P's"—planet, people, and profit—TBL incorporates the dimension of "planet" or environment, as measured by output of pollutants, conservation of endangered species, waste production, energy use, and so on. The social dimension, or "people," includes such measures as human health and well-being, equity (including the difference in salary between top executives and the lowest-paid workers), access to social resources, and benefit to the community. The economic dimension, or "profit," is the only measure that has been traditionally evaluated.

Reporting the TBL presents a challenge for any organization, in that each dimension is measured in different units. How can the value of pollution that was never emitted be measured in dollars? What is the value of a healthy and happy workforce and a community supported by the businesses therein? Despite the challenges of measurement, TBL is gaining popularity among corporations, nonprofits, and governments, so much so that many have shifted their business models to evaluate and improve all three areas. What is motivating this shift? According to MIT/Sloan Management Review's 2011 Sustainability & Innovation Global Executive Study and Research Project, consumer preferences, legislative pressure, and resource scarcity are the most important drivers of this change toward a new business model. In addition, investors are looking more carefully at sustainability practices before supporting corporations.

Of nearly 3,000 global corporations surveyed in 2011, 31% reported that conserving environmental capital for the future and placing value in their communities and workforce has actually increased their profits. For example, Campbell Soup Company invested significantly

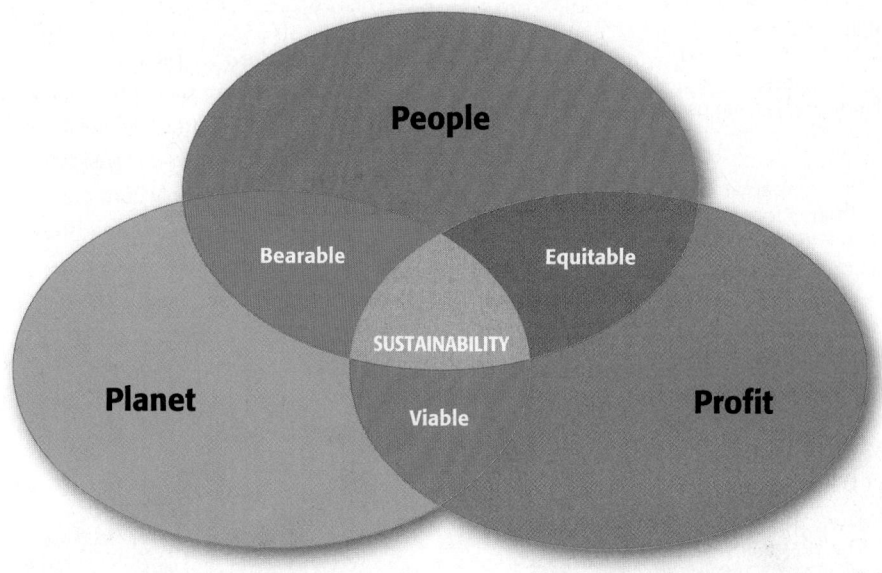

▲ Figure 1.9 **The Triple Bottom Line**
The intersection of these three dimensions is the path to sustainability. Profit without people or planet is short sighted, with no future human or environmental capital to draw from. Environment without people or profit does not support human well-being or allow for economic growth. A focus on the social dimension without accounting for the dimensions of planet or profit does not allow for the preservation of long-term environmental capital or sustaining incomes.

in water efficiency measures, resulting in a savings of 1 billion gallons of water during 2008–2013 and a 15–20% return on investment (money saved due to the improvements). The bottom line is that corporations that incorporate the TBL into their business model also stand to increase their economic profit.

There is much we can learn from the past to help us ensure a sustainable world for the next generation. Yet in many important respects, that generation will be quite different from all earlier generations. By all indications, it will include over 9 billion people. Those people will use technologies and have expectations that we can hardly imagine. If the management of our resources today is to be truly sustainable, we must be very generous in our projection of future needs. The ecosystem concept described next provides a foundation for defining sustainable behavior and actions.

QUESTIONS 1.1

1. Describe how human understanding of sustainable actions has changed over the past century.

2. What kinds of change must be included in our understanding of sustainability?

3. Describe the three dimensions of the TBL. How is each important to sustainability?

(MES) For additional review, go to **MasteringEnvironmentalScience**

DIRTT

Can a company make a profit based on the triple bottom line?

One company that stopped to reconsider its impact and to develop a sustainable approach to business made its intentions clear with its name: DIRTT—Doing It Right This Time. DIRTT creates prefabricated interiors for office buildings, healthcare, and education facilities, and has developed computer design software that drives the entire manufacturing process, creating efficiencies and reducing waste.

DIRTT measures its success by the TBL: planet (environmental), people (social), and profit (economic). For them, the intersection of these three spheres defines a sustainable business model. If the planet is not considered, resources are depleted and the profit will only continue in the short term. If people are not treated equitably, they are not as productive as they might be and the business will fail. If profit is ignored, people will not have jobs and the business cannot continue. All three components play a critical role in the equation.

Planet. DIRTT considers the environment in every part of its business. The product is built on the concept of reuse. Every piece of the interior is interchangeable, reusable, and movable to accommodate the rapidly changing uses for building interiors (**Figure 1.10A**). Like Legos™, new pieces fit with the old, and nothing is ever "orphaned" out. Unlike the modern-day electronics industry, DIRTT does not base its profit on "planned obsolescence" but designs its products to be reused indefinitely.

Construction materials are also selected for their environmental sustainability. For example, the prefab walls are insulated with used denim, deferred from the landfill (**Figure 1.10B**). Shipping materials are reused as well. DIRTT has designed the "cookie"—a small plastic component that effectively separates and protects the prefab materials that are shipped together in stacks to the building site (**Figure 1.10C**). The cookies significantly reduce shipping weight and the amount of wood used in the shipping process. Every shipment includes a "cookie box" into which all the cookies are deposited and sent back to the factory. DIRTT goes so far as to say that they have failed if they must recycle.

People. In the construction industry, seasonal layoffs are common. DIRTT does not hire seasonally and has kept employees on through tough economic times. Because DIRTT guarantees its clients a two-week turnaround from order to shipment, the company always has trained workers ready to build the product at a moment's notice. When business has been slow, workers have been paid to do community service, thus building goodwill in the communities where the factories are located.

In addition, all DIRTT factories have on-site kitchens that offer inexpensive, local food for all employees. This strategy creates more productive workers by providing healthy food options, thereby reducing illness and absenteeism and building a strong sense of community among the workers who get to know each other over their meals. All of these benefits contribute to a stronger team and a more productive work force.

Profit. Finally, DIRTT is also motivated by profit. In 2013, DIRTT went public, which means that shares of the business are now available for purchase by the public. DIRTT leaders recognize that the company cannot be viable if it is not profitable. Although they always consider the needs of shareholders, the company does not resort to short-term gains as a trade-off against long-term environmental benefits or loss of employees. This long-term view, combined with TBL measurements of profit, provides a more realistic evaluation of the company's real success.

▼ Figure 1.10 **A DIRTT-y Interior**
Ⓐ Components of DIRTT's office interiors are designed to be completely movable and interchangeable to accommodate the changing needs of a business.
Ⓑ Insulation made from used denim lines the prefabricated walls and Ⓒ the plastic "cookies" that separate fragile wall components for shipping are sent back to the factory in a "cookie box" for reuse.

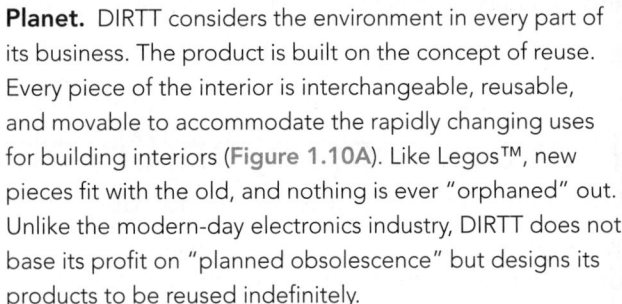

1.2 Ecosystems

BIG IDEA We use the ecosystem concept as a way to organize all of the factors and processes that make up our environment. An ecosystem includes all of the organisms and their physical and chemical environment within a specific area. The flow of energy and matter through ecosystems influences the distribution and abundance of organisms within them. Ecosystems provide resources and services that are essential to our well-being. Today, human activities influence nearly all of Earth's ecosystems, often in ways that diminish the benefits they provide and threaten our well-being.

Ecosystem Function and Integrity

■ The ecosystem concept combines living organisms and their environment.

An **ecosystem** is the combination of a community of organisms and its physical and chemical environment, functioning as an integrated ecological unit. A forest and a lake are familiar examples of ecosystems.

Ecosystems include **biota**, or living organisms, and their **abiotic**, or nonliving, environment. The abiotic characteristics of an ecosystem, such as its climate, determine the distribution and abundance of the organisms that live there. At the same time, the actions of the organisms living in an ecosystem change and shape their environment. For example, trees need deep soil to grow, and as they grow, their decaying leaves increase soil fertility and their root systems penetrate rock fractures and accelerate soil formation.

Ecosystems are connected by the flow of energy and matter. Energy and matter are transformed as they move between living organisms and the abiotic environment. Ecologists refer to the flow of matter and energy and the processes influencing the distribution and abundance of organisms as **ecosystem functions**.

Ecosystem integration. An ecosystem is an *integrated system* made up of living and nonliving parts and the processes that connect them (**Figure 1.11**). **Ecosystem integrity** refers to the web of interactions that regulate ecosystem functions. Ecologists often compare the functional integrity of an ecosystem to that of an individual organism. For example, the tissues and organs of your body work together to carry out the functions necessary for life, such as breathing and digesting food. Similarly, the microbes, plants, and animals in an ecosystem interact with each other and their environment in ways that determine the overall movement of matter and energy through an ecosystem.

Like individual organisms, ecosystems have the capacity to adjust to disturbances and changes in their environment. This capacity depends upon their integrated nature. Severe disturbances, such as hurricanes and fires, may greatly alter the species living in an ecosystem and the way the ecosystem functions. The ecosystem can then undergo a process of change and recovery that ecologists liken to the process of growth and development of an individual organism.

In one very important way, however, the ecosystem-as-organism metaphor is not accurate. The boundaries of a single organism are quite distinct. For example, your skin forms a well-defined boundary between you and your environment. The boundaries of ecosystems are not so clearly delineated. Instead, they are defined by those who study or manage them; often they are quite artificial.

Ecologists study ecosystems of many sizes. The unit of study may be as large as a forest or river basin, an entire continent, or even the entire globe. However, the ecosystem concept is equally applicable at much smaller scales, such as a cup of sour milk or a rotting log.

Ecosystems are often studied and described in the absence of human influences. Ecologists interested in the fundamental workings of ecosystems may choose to study regions of wilderness where human influences are minimal. Today, however, human influences are nearly everywhere. Over the past century, our population has grown from just over 1 billion to over 7 billion people. Changes in culture and technology, such as forms of transportation, energy use, and urban development, have greatly magnified our demand for resources. As a consequence, our impacts on Earth's ecosystems have grown by far more than the sevenfold increase in our numbers.

▼ Figure 1.11 **An Integrated System**
Ecosystems comprise organisms and their nonliving environment, and they are dependent on inflows and outflows of matter and energy.

Inflow: Sunlight and materials like carbon, nitrogen, and water

Transformation of energy and exchange of matter between biota and abiotic environment

Outflow: Heat and materials like carbon, nitrogen, and water

QUESTIONS 1.2

1. Differentiate between ecosystem functions and ecosystem services.

2. Describe four types of ecosystem services.

(MES) For additional review, go to MasteringEnvironmentalScience

Ecosystem Services

■ Ecosystems provide the resources and processes upon which humans depend.

Human well-being depends on **ecosystem services**, the multitude of resources and processes that ecosystems supply to humans. Ecosystem services can be classified into four categories: provisioning, regulating, cultural, and supporting services.

Provisioning services supply us with resources, such as food, water, and the air we breathe. We humans are notable for our ability to modify—often simplify—ecosystems in order to increase their provisioning services. For example, an agricultural field of corn provides a large amount of food to humans but supports far fewer species than the grassland that once grew there.

Regulating services are the ways that ecosystems control important conditions and processes, such as climate, the flow of water, and the absorption of pollutants. These services are a consequence of the integrated, self-regulating nature of ecosystems.

Cultural services are the spiritual and recreational benefits that ecosystems provide. If you have visited the Grand Canyon, Yosemite Valley, or simply taken a quiet walk in the woods near your home, you have benefited from cultural services.

Supporting services are the basic ecosystem processes, such as nutrient cycles and soil formation, that are needed to maintain other services. For example, bees and other pollinating insects provide a supporting service by ensuring that flowers are fertilized and fruits form.

Like all living things, we humans rely on ecosystems for survival as well as shape ecosystems through our actions. But unlike other organisms, human interactions with the environment are mediated by beliefs, knowledge, technologies, and institutions that are passed from one generation to the next. For example, rather than relying on the slow process of biological evolution to adapt to new environments, we dress appropriately for changing weather and invent new technologies to heat our homes. We then pass this knowledge on to our children. Our success as a species, as reflected by our growing numbers, is largely attributable to these unique features (**Figure 1.12**).

In 2006, the United Nations Environment Programme published its *Millennium Ecosystem Assessment*. The work of hundreds of scientists and decision makers, this document evaluated the global impacts of humans on ecosystems and the effects of ecosystem change on human well-being. Three important themes emerged from this assessment. First, many of Earth's ecosystem services are being degraded. Second, there is growing evidence that many ecosystems are becoming more fragile and prone to disturbance. Third, these changes threaten the well-being of all humans, but especially the poor. The report also notes that the effective management of ecosystem services can be a dominant factor in the reduction of poverty and international conflict. The interconnections between ecosystems and human well-being are a major theme of this book.

▼ Figure 1.12 **Humans and Ecosystems** Human well-being is dependent on the services provided by ecosystems. Human activities produce significant changes in the ecosystem functions that deliver those services.

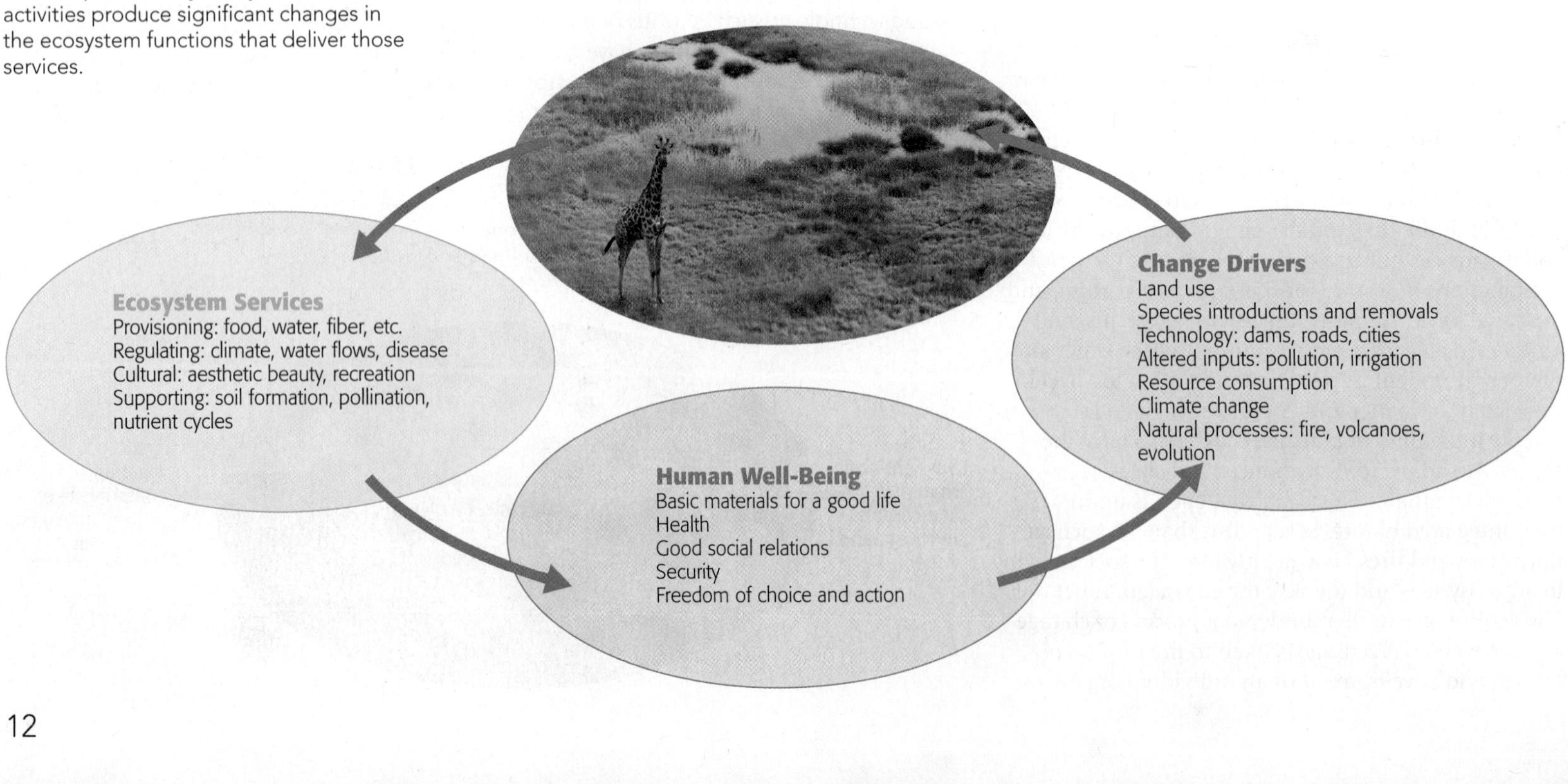

Ecosystem Services
Provisioning: food, water, fiber, etc.
Regulating: climate, water flows, disease
Cultural: aesthetic beauty, recreation
Supporting: soil formation, pollination, nutrient cycles

Change Drivers
Land use
Species introductions and removals
Technology: dams, roads, cities
Altered inputs: pollution, irrigation
Resource consumption
Climate change
Natural processes: fire, volcanoes, evolution

Human Well-Being
Basic materials for a good life
Health
Good social relations
Security
Freedom of choice and action

1.3 Principles of Ecosystem Function

BIG IDEA Ecosystems are integrated systems that can be understood in terms of fundamental scientific principles. Although awesomely complex, ecosystems are not haphazard. Even though living things are continuously consuming and replenishing oxygen, the amount of oxygen in our air remains remarkably constant. The mix of life-forms that surrounds us changes from place to place and season to season in complicated but dependable ways. The organization and function of ecosystems depends on four fundamental principles.

1. Matter and energy are neither created nor destroyed.
2. Ecosystems are always open to gains and losses of matter and energy.
3. Ecosystem processes are self-regulated by interactions among their living and nonliving components.
4. Ecosystem change is inevitable and essential.

Conservation of Matter and Energy

■ Something cannot be created from nothing, and everything goes somewhere.

Energy and matter can be neither created nor destroyed, although in the cores of stars and nuclear reactors matter can be converted into energy. This is known in physics as the law of energy and mass conservation. If you burn 1 g of coal in the presence of oxygen, the ash, smoke, and gases that are released would exactly equal the weight of the coal plus the oxygen. You would also find that the light and heat energy emitted by the fire, plus the energy left in the ash and smoke, precisely equal the chemical energy that was stored in the oxygen and unburned lump of coal. Something—coal—has been used to generate something else—ash, smoke, gases, heat, and light: But no matter or energy has been lost.

The conservation of energy and mass applies to ecosystems, too. By measuring the amounts of matter in the different parts of an ecosystem and the rates at which this matter flows through it, we can characterize ecosystem functions and make predictions about their future behavior. This accounting of amounts and rates is a fundamental tool in ecosystem science and management.

In environmental science, the law of energy and mass conservation is a powerful reminder of two fundamental truths. First, *something cannot be created from nothing*. Energy and matter can be converted from one form to another, but the total amount of any resource is finite. A resource can be exhausted if it is used faster than it is replenished. Second, *everything goes somewhere*. Energy and matter that we put into ecosystems have fates that may influence ecosystem functions (**Figure 1.13**).

▼ **Figure 1.13 Shrinking Lake and Stinking Lake**
Something cannot be created from nothing—Lake Mead near Las Vegas, Nevada, is gradually shrinking as water use exceeds water supply in the Colorado River Ⓐ. Everything goes someplace—China struggles to cope with the numerous chemicals that now pollute Lake Tai, one of its largest lakes. Toxic algae and cyanobacteria are now abundant and create a foul odor that can be detected as far as a mile away Ⓑ.

Ecosystems Are Open

■ Because flows of matter and energy influence ecosystem functions, context matters.

Ecologists often assign arbitrary boundaries to ecosystems based on factors such as type of soil or kind of vegetation. Regardless of the boundaries we choose for them, ecosystems do not have distinct boundaries. Ecosystems are open. Matter and energy can flow into them from their surroundings, and they can also move out. For example, sunlight and rain can enter an ecosystem, and heat and streams can flow out. These gains and losses of matter and energy influence ecosystem functions. An important consequence of this openness is that *context matters*: Functions in an ecosystem are inevitably influenced by the nature of the ecosystems that surround it.

Scientists set the size and boundaries of the ecosystems they study in such a way as to measure inputs and outputs of matter or energy most easily or to monitor key processes within them. For example, ecologists often study the drainage basins of rivers. The boundaries of drainage basins are set by the mountains, hills, and valleys that determine where the water flows. Ecologists can determine the amount of water coming into a drainage basin by measuring rainfall. They can measure the amount leaving by monitoring the flow at the mouth of the river that drains the basin.

To manage ecosystem processes effectively, it would seem logical to do so according to the scale and boundaries by which we can most easily monitor and sustain them. Unfortunately, our political systems often make this difficult to achieve. For example, rivers frequently define the boundaries between counties, states, and countries. These political entities are responsible for managing the water flowing through them. But rivers also divide drainage basins. As a consequence, multiple counties, states, and countries influence and manage water quality and flow (Figure 1.14). This often leads to mismanagement and conflicts among jurisdictions.

Similarly, consider how often straight lines form the boundaries between counties, states, and countries, such as much of the border between the United States and Canada. Such lines have little to do with the movement of the organisms, matter, and energy that shape our environment.

◄ Figure 1.14 **An Arbitrary Boundary**
The Rio Grande separates Mexico from the United States along nearly 70% of the border between the two countries. In this satellite photo, the river flows between El Paso, Texas, on the right and Ciudad Juarez on the left: both cities depend on its resources, contribute pollution to its waters, and argue over its management.

Ecosystem Stability

■ Ecosystem processes are regulated by interactions among biotic and abiotic components.

Ecosystems are constantly changing. Yet key features, such as the total number of species and the processes involving the transformations of matter and energy, remain stable. The stability of ecosystem features and functions derives from the complex interactions among their diverse array of biotic and abiotic components.

Dynamic homeostasis is the process by which systems adjust to changes in ways that minimize how much features or processes vary from their normal values. Your ability to maintain your body temperature around 98.6 °F (37 °C) is an example of dynamic homeostasis. If your temperature cools, even slightly, you put on a sweater or coat to warm yourself. If your temperature rises above this value, you are prone to start shedding layers of clothing.

Examples of dynamic homeostasis abound in ecosystems. Over time, the sizes of plant and animal populations tend to fluctuate above and below average values. The amounts of the chemical elements nitrogen and phosphorus found in lakes or soils vary from time to time, but they maintain long-term average values.

Dynamic homeostasis occurs because of feedback, the ability of a system to adjust based on changes in the system itself. Homeostatic systems "talk and listen to themselves." Again, consider the regulation of your body temperature. A part of your brain called the *hypothalamus* is very sensitive to changes in your body temperature. If your blood temperature drops below 98.6 °F, your hypothalamus sends an "I feel cold" signal to your consciousness. It also sends signals that stimulate involuntary responses, such as shivering or increased metabolism. If your blood temperature rises above 98.6 °F, your conscious brain receives the "I feel warm" signal from the hypothalamus. It also stimulates involuntary responses that cool your body, such as sweating.

The regulation of temperature in your body is an example of a negative feedback system. **Negative feedback** occurs when directional change in a process alters the system in a manner that reverses the direction of that change. A change in your body temperature causes your body to respond in ways that help your body return to its normal temperature. Note that "negative" refers to the *direction* of the change (in the opposite direction), not to the value of the consequences of the change.

Many ecosystem processes are regulated by negative feedback in the interactions between their living and nonliving components. In a pond ecosystem, for example, the rapid growth of algae may deplete the supply of an element in the water that is essential for growth. Consequently, the growth of the algae slows and the supply of that element is again stabilized (Figure 1.15). Negative feedback systems tend to stabilize ecosystems.

As nutrients increase, algal growth increases

High algal populations deplete nutrients, algal growth begins to decline

As algal growth declines, nutrients begin to increase

◄ Figure 1.15 **Negative Feedback**
The growth of algae and the availability of nutrients that control their growth provide an example of negative feedback in which each change is reversed by the other. Increased nutrients increase algal growth; this then diminishes nutrients and subsequently diminishes algal growth. One change is diminished by the other.

Positive feedback occurs when the directional change in a process alters the ecosystem so as to reinforce that change. For example, because tree roots hold soil and limit erosion, the loss of forests on steep mountain slopes accelerates soil erosion. This erosion leads to the loss of even more forest (Figure 1.16). Another example concerns global climate change. As temperatures rise, glaciers melt, decreasing the reflective white cover of Earth. The reduced reflectivity results in greater absorption of sun energy and even more melting ice. If not counteracted, positive feedback can destabilize ecosystems. Note that "positive" does not refer to the value of the outcome of the change: Indeed, the consequences of increased erosion and the loss of glaciers are inevitably negative for the environment. Positive feedback instead refers to feedback *in the same direction* that the process is progressing.

Ecosystem diversity and complexity matter. The dynamic homeostasis of ecosystem functions derives in part from the diversity of species and the complexity of their interactions. Such diversity and complexity provide multiple alternative paths for the movement of matter and energy within an ecosystem. Human actions that diminish diversity often diminish ecosystem stability. This is explored in more detail later in this chapter.

Forest loss initiates erosion

Accelerated erosion accelerates forest loss

▲ Figure 1.16 **Positive Feedback**
Deforestation and erosion can generate positive feedback. On this hillside in Albania, loss of forest stimulates erosion, which then results in loss of more forest and further erosion. One change is reinforced by the other.

Ecosystem Change

■ Change is both inevitable and essential.

Dynamic homeostasis implies change. Ecosystems are constantly changing in response to actions from outside as well as to the processes operating within them. Processes of change in ecosystems have played a significant role in the evolution of the organisms that live in them. As a consequence, species have a variety of characteristics that permit them to adjust to change.

Some changes, such as seasonal changes in temperature, are so regular and predictable that we do not consider them to be disturbances. It is the departures from these regular changes, such as abnormally cold summers or extended winter weather, that we call disturbances. Other disturbances include storms, floods, and fire. Though less predictable, such disturbances are inevitable, and ecosystems must adjust to them.

Ecosystem change in response to disturbance is often homeostatic, controlled by feedbacks that produce regular and predictable patterns. For example, a hurricane may fell many of the trees in a forest (Figure 1.17). This creates an opportunity for new trees to grow. If the climate remains relatively constant, these new trees will produce a very similar forest in 100 to 200 years.

Sometimes processes in ecosystems change because of long-term changes in other processes. For example, a gradual change in the climate of a region can produce new patterns of ecosystem change. Between 5,000 and 10,000 years ago, moist tropical forest grew across the Sahel of Africa, the region south of the modern Sahara Desert. When these tropical forests were disturbed by storms or other events, similar forests soon grew in their place. About 4,000 years ago, the climate of this region became hotter and drier. When tropical forests were disturbed, they were replaced by grasslands and savannas, not forests. The hotter climate had caused a new pattern of changes in the ecosystem.

Over the past 10,000 years, humans have become increasingly important agents of change in most of Earth's ecosystems. Whether or not our actions are sustainable depends on the degree to which we recognize and apply the principles of ecosystem function.

QUESTIONS 1.3

1. What important environmental principles derive from the law of conservation of mass and energy?

2. What is meant by the phrase "ecosystems are always open"?

3. What is the difference between negative and positive feedback?

(MES) For additional review, go to **MasteringEnvironmentalScience**

▼ Figure 1.17 **Ecosystem Disturbance**
This diverse tropical forest ecosystem in the Silver Glen Springs palmetto jungle in the Ocala National Forest, Florida, was disturbed by Hurricane Frances in 2004. It will now undergo change that will result in the regeneration of the forest, a process that will take about a hundred years.

1.4 Acting Sustainably

BIG IDEA Ensuring the well-being of present and future generations is a worthy, motivating vision, but how can we know which of our actions are truly sustainable? For this, we must refer to our understanding of ecosystems. An action is sustainable if it does not impair the functions of ecosystems or their ability to deliver the services upon which we depend. The four fundamental principles of ecosystems are guideposts for assessing whether an action is sustainable.

Managing Resources

■ To be sustainable, actions must conform to the law of mass and energy conservation.

For at least three centuries, scientists have agreed about the unfailing truth of the law of conservation of mass and energy. However, given our overexploitation of natural resources and our dumping of toxic waste into ecosystems, one might conclude that great numbers of us do not share that agreement.

The resources that we take from ecosystems often can be replenished by natural processes or from outside sources. For example, the water taken from the Great Lakes is replenished by rivers that flow into them. Those rivers, in turn, are fed by rain falling in their headwaters. The fish harvested from the Great Lakes are replenished by a complex array of biotic and abiotic interactions that support fish reproduction and population growth. Resources such as water and fish are said to be renewable, but they can be renewed only if we do not use them more quickly than they can be replenished.

Similarly, ecosystems can transform many waste materials and pollutants into less obnoxious or toxic forms. But ecosystems can carry out this function only if the rate of waste flowing in does not exceed the rate at which the ecosystem can transform it. In other words, the law of mass and energy conservation must be followed.

A resource is **nonrenewable** if its amount in an ecosystem declines with virtually any level of use. In the scale of human history, oil is a nonrenewable resource. The world's oil reserves were produced by geologic processes operating over millions of years. Today, we are using these reserves about 1 million times faster than they can be replenished (**Figure 1.18**). Exploitation of nonrenewable resources defies the law of mass and energy conservation.

Sustainable use of resources requires two things: an understanding of their rate of renewal and the ability to manage the rate of their use. For example, many of the factors affecting the flow of water into the Great Lakes are understood, but the causes of decade-to-decade changes in rainfall in the Great Lakes basin are not. The ability to manage the use of water in the basin is complicated by the rapid increase in the number of people living in the region and disputes among states and between the United States and Canada.

▼ Figure 1.18 **A Nonrenewable Resource** It took Earth's ecosystems tens of thousands of years to produce the volume of petroleum consumed by the world's billion-plus automobiles and trucks in a single day.

Understanding Boundaries

■ To be sustainable, actions must acknowledge the importance of boundaries.

Sustainable environmental management requires an understanding of the spatial scales and landscape features that influence ecosystem functions and services. Whether it is a backyard, a national park, or an entire country, arbitrary scales and boundaries limit our ability to monitor and manage ecosystem processes. We cannot manage the quality of water in the lower reaches of a river, for example, if arbitrary boundaries exclude management options upstream.

The scale and boundaries appropriate for managing one process may not align with those appropriate for managing another process or service. For example, river basins are useful units for managing the flow of water, chemicals in the water, and the organisms that live in the water. However, river basins are usually irrelevant to the management of air pollution or populations of large migratory animals.

Sustainable management also requires communication and collaboration across the boundaries that define differences in ownership, management purpose, and nationality. Unfortunately, these human boundaries often represent some of the most significant challenges to sustainable management.

Maintaining Balance and Integration

■ To be sustainable, actions must maintain the homeostatic capacity of ecosystems.

Although our knowledge of ecosystems is far from complete, we must acknowledge that their complexity is key to their integrated character. The many living and nonliving components of an ecosystem play important roles in its functions and the services it provides.

The connection between ecosystem complexity and ecosystem services is similar to the complexity of an automobile. An automobile is made up of thousands of individual parts. To a greater or lesser extent, each part contributes to the automobile's overall function and to the services it provides. The key service you expect from an automobile is to move you from one place to another. Less important services—but ones that might influence which car you purchase—include comfort, climate control, appearance, entertainment, and even navigation. As you drive along, some parts of the car are in use and others are not. If it is summer, you are unlikely to be using the heater. You don't use the brakes until you decide to slow down or stop. Some parts, such as the spare tire and emergency brake, are used only when another part fails.

Now, imagine that you are asked to remove parts from your car sequentially while trying to maintain its most important functions (Figure 1.19). If you have complete knowledge of the relationship between your car's parts and its workings, you can make wise decisions about which services to lose first. For example, you would remove cosmetic features, such as hubcaps and hood ornaments, before giving up a piston or fuel pump. If you were willing to tolerate risk, you might give up the spare tire or emergency brake, since they are only important if other parts fail.

People with a limited understanding of automobiles would approach this disassembly process in a much more random fashion. Unable to associate parts to functions, they would soon disable their cars. They would have a limited ability to set priorities for particular functions or services. They would not understand the risks they might incur by the removal of parts that do not immediately impair a car's function, such as the spare tire or warning lights.

With regard to the manipulation, simplification, and management of ecosystems, we are like children aspiring to be auto mechanics. We can name only a few of the "pieces" in most ecosystems, and we have an even more limited understanding of their importance to particular ecosystem functions and services. We frequently focus on only the most obvious parts and processes or those of most immediate importance to us. Too often, we give little thought to the ecosystem equivalents of spare tires, emergency brakes, and warning lights.

▼ Figure 1.19 **What's Essential?**
Are all these parts necessary for the functioning of your car?

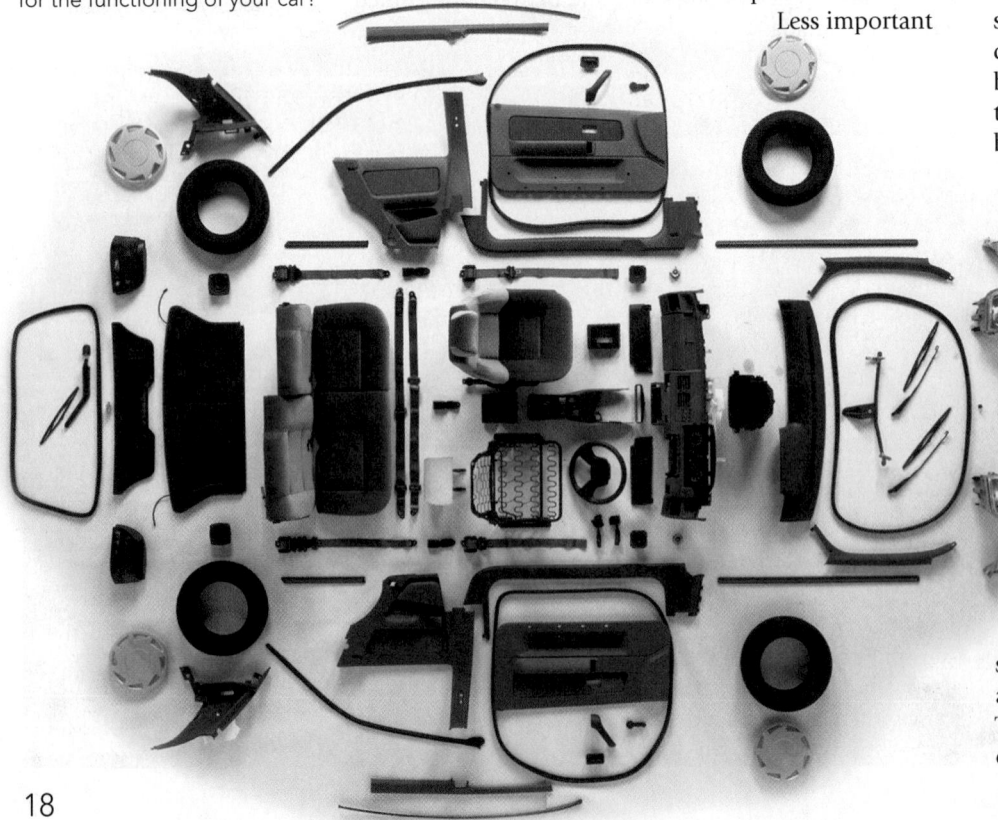

Embracing Change

■ To be sustainable, actions must not ignore the inevitability of change or interfere with ecosystems' capacity to change.

Sustainable environmental management demands that we do more than accept that change is inevitable. It requires that we understand change and explicitly manage it. Past experience and scientific study have taught us much about how to do this successfully. Three lessons are particularly important.

Beware of attempts to halt or alter the tempo of natural change in ecosystems.
Over and over, human attempts to slow or stop normal environmental change have produced surprising and unsustainable consequences. For example, natural factors such as currents and waves are continuously changing the ocean shoreline. Human developers have tried various tactics to halt these changes. Instead of preventing change, these tactics have often damaged coastal ecosystems (Figure 1.20).

Many of our most significant environmental challenges are associated with our limited ability to accommodate inevitable change. In arid regions, plans for managing and allocating water often fail to include contingencies for periods of drought, resulting in much human conflict and hardship. Because agricultural and urban ecosystems lack the homeostatic features of their natural counterparts, they are highly vulnerable to the effects of natural disturbances such as earthquakes and hurricanes.

Copy nature.
For many centuries, humans have been among the most powerful agents of change on Earth. We have altered landscapes, modified the flow of rivers, created and drained lakes, and even changed the chemical composition of air. Indeed, it appears that our activities are now warming Earth's climate. But if change is normal in ecosystems, why should we worry about human-caused change? We must be concerned because the character and tempo of much of human-caused change has no precedent in the evolution of Earth's ecosystems.

Over the past century, we have added hundreds of novel chemicals to the ecosystems. These chemicals, which are not easily broken down by natural processes, are toxic to many organisms. For example, the manufacture of paper and some herbicides and the incineration of commercial and municipal wastes have released large quantities of toxic chemicals called *dioxins* into our environment. In recent years, environmental regulations in the United States have reduced the release of dioxins by 90%. However, because dioxins decompose very slowly, high levels of them remain in many ecosystems.

In other cases, humans have altered the patterns of natural change in ecosystems. For instance, dams that have been installed on rivers to generate electricity also modify the rate of water flow. This causes significant changes in downstream ecosystems, which are adapted to the rates of flow and patterns of flooding that occurred before the dams were built.

Be alert for thresholds of change, or tipping points.
Some human actions can push ecosystems beyond the limits of their normal range of change. This may alter the ecosystems so much that they cannot easily return to their previous state. The point at which an ecosystem changes from one state of homeostasis to another is called a threshold of change, or *tipping point*.

Much of the woodlands in the Sahel region of Africa are at a tipping point. The trees in these woodlands are frequently cut down for fuelwood or to make way for agricultural development. When cut on a small scale, the trees grow back and the woodlands are restored. But when large numbers of trees are repeatedly cut down, the climate becomes drier and soils rapidly erode. When these lands are abandoned, they become deserts rather than woodlands. A threshold of change has been crossed. Human abuse of the land has changed the environment so much that the ecosystem can no longer return to its previous state.

Today, most scientists agree that human activities are contributing to changes in the global climate. With this consensus comes concern that climate change may cause many of Earth's ecosystems to change in ways that cannot be reversed.

▲ Figure 1.20 **Defying Inevitable Change**
Attempts to alter the natural movement of sand along coastal beaches have often produced accelerated erosion such as that along the coast of North Carolina where houses are being washed into the sea.

QUESTIONS 1.4

1. How does ecosystem openness affect the sustainability of human actions?

2. How does maintaining the complexity of ecosystems contribute to their sustainability?

3. Describe three strategies for sustainable human actions in the context of change.

(MES) For additional review, go to **MasteringEnvironmentalScience**

1.5 Uncertainty, Science, and Systems Thinking

BIG IDEA Uncertainty abounds. Part of this uncertainty is the consequence of ignorance—our understanding of Earth's ecosystems is very far from complete. Uncertainty is also a consequence of the complexity of ecosystems, which can produce unpredictable changes.

It's all right to be uncertain; in fact, it is inevitable. **Science** is a process that poses and answers questions objectively in order to increase knowledge and lessen uncertainty. Science depends on careful observation and experimentation to produce results that can be duplicated by others. The results and conclusions of scientific research are open to review and revision by other scientists. New knowledge gained by scientific study is most likely to produce sustainable actions and behaviors when it is applied in ways that recognize the integrated nature of ecosystems.

Uncertainty

■ There is much we do not understand. Why?

It is an understatement to say that there is much that we do not know about ecosystems. Two factors—ignorance and complexity—are especially important sources of uncertainty in our understanding of the environment.

Ignorance. We would be able to meet many environmental challenges more effectively if we simply knew more. Reducing ignorance is the basic justification for most environmental research.

Ignorance presents two different challenges. Sometimes the extent of our ignorance is so overwhelming that it is difficult to know where to start or how to set research priorities. At other times, we don't know what we don't know. Our ignorance leads us to think that we know more than we do, so we fail to ask the right research questions. In his book, *The Discoverers*, Daniel Boorstin wrote, "The great

obstacle to discovering the shape of the earth, the continents and the oceans was not ignorance, but the illusion of knowledge."

Complexity. Although their parts may interact in a predictable manner, complex systems often behave in ways that are practically impossible to predict. It is not that such behavior is random; rather, it is because miniscule changes that are too small to measure in the individual parts of a complex system can lead to very large differences in that system later on. Because ecosystems are so complex, it is virtually impossible for scientists and resource managers to measure and understand all of their interacting components.

The unpredictable behavior produced by the complexity of ecosystems limits our ability to forecast changes. This is similar to the way that the complexity of weather systems limits our ability to predict the weather. Meteorologists can make reasonably accurate predictions about the weather for tomorrow or the next day. The accuracy of weather forecasts diminishes rapidly the further we look into the future. This is because very small variations in the components of weather systems on one day produce significant differences in the future (Figure 1.21).

Uncertainty regarding environmental challenges is often the basis for inaction and conflict. Most famously, this has been the case in the way that various organizations and governments have responded to the challenge of global warming. In recent years, scientific research has shown conclusively that Earth's lower atmosphere is warming, but questions about the magnitude of this change persist. Uncertainties also remain as to whether human activities are responsible for global warming and what actions would be most effective in mitigating this process. These uncertainties even cause some people to argue that we should not take any action at all.

▼ Figure 1.21 **Complexity Yields Uncertainty** Data from several sources allow meteorologists to predict the weather a few days in advance. However, weather predictions for the next month or year are much less certain because of the complexity of Earth's atmospheric systems.

Reducing Uncertainty with Science

■ Science asks questions in a fair and unbiased fashion.

The scientific revolution of the past several centuries has greatly expanded human knowledge and resolved much uncertainty about our world. We are often awed by the technical detail and complicated instrumentation of science. Yet at its core, science is simply the process of asking questions in a way that is as objective, or unbiased, as possible (Figure 1.22).

To find an answer to a question, scientists propose **hypotheses**, or alternative answers that can be tested by careful observation or experimentation. For example, scientists might ask why the population of a particular animal is declining. Hypotheses might be that the animals are suffering from a disease, that too many are being killed by hunters, or that their food supply is diminishing. Scientists then develop **predictions**, "if, then" statements that forecast the outcome of a test of the hypothesis. For example, "if animals are given more food, then their population will increase."

This prediction leads to a test of the hypothesis that food supply is important to population growth. Scientists might perform an experiment in which they provide extra food to half of the population, the **treatment group**, while keeping food levels unchanged for the other half, the **control group**. They would then compare the rates of population growth of the two groups.

The scientific process follows very careful rules regarding hypotheses and the strategies to test them. Most important, a scientific hypothesis must be *falsifiable* by observation or experimentation. This means that it is possible to devise an experiment or observation that could prove the hypothesis wrong. Hypotheses based on an immeasurable "force" or on divine intervention are simply not open to scientific inquiry.

Scientists are particularly concerned about precision and bias in their observations and measurements. In the case of our animal study, **precision** refers to the likelihood that measurements or estimates derived by sampling a subset of the population are representative of the actual values for the entire population. In general, the larger the sample size, the greater the precision. **Bias** refers to the possibility that estimates might be skewed in some particular direction. For example, large individuals in a population are more easily seen and sampled than small individuals. This could lead to bias in estimates of body size. To avoid biased observations, scientists pay special attention to sampling technique.

Experiments generally include one or several treatments relevant to the hypotheses, the results of which are then compared to an untreated control. Furthermore, experiments and observations are run multiple times to ensure that the results can be replicated and conclusions can be confirmed. Finally, scientists subject their results and interpretations to review by other experts before they communicate them in scientific journals.

Although scientists may come to consensus regarding the answer to a particular question, no idea, theory, or hypothesis is immune from further question or revision. No uncertainty ever becomes an absolute certainty.

A set of experiments and observations may lead scientists to agree that a particular hypothesis answers a question or that a specific model explains a phenomenon. However, further study may determine that the model is naïve or misguided and replace it with another, very different model.

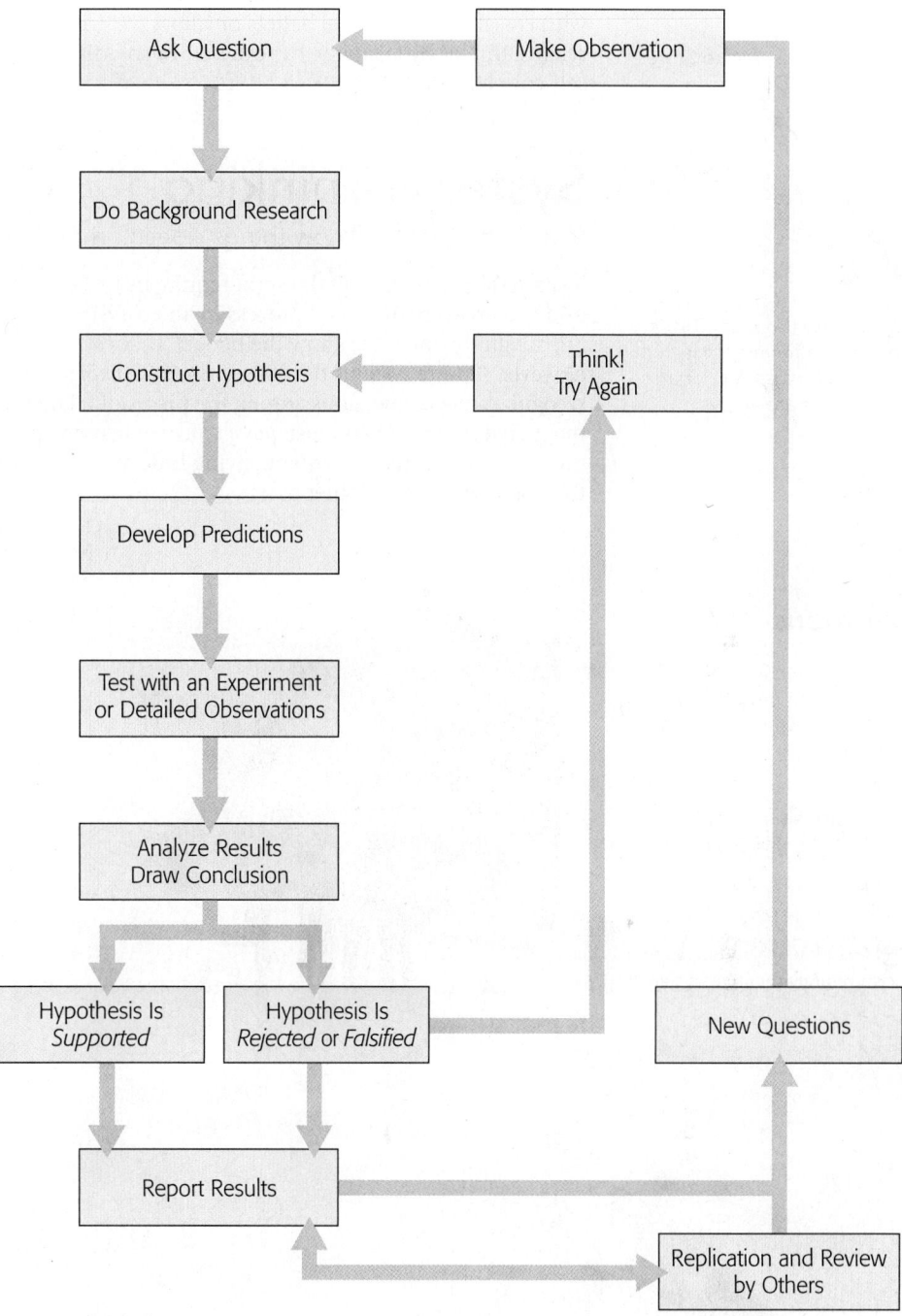

▲ Figure 1.22 **The Scientific Method**

Science is a systematic method for asking questions in an unbiased way. Because results are subject to review and replication by others, science ensures that mistakes are corrected. Inevitably, the process also generates new and interesting questions.

21

Q How can the science community work together to create a common vision for how our environment should be managed?

Patty Adams, Metro State College of Denver

Norm: A *Professional meetings and national workshops provide opportunities for scientists to share new ideas and discuss alternative solutions to environmental challenges. Consensus about appropriate solutions often emerges from these meetings, but debate and dissent are always encouraged.*

As much as they aspire to objectivity, scientists are human. As humans, they may become overly protective of their ideas. Occasionally, they are guilty of advocating for particular answers to questions because they align with closely held personal beliefs or values. However, science thrives by falsifying hypotheses. In so doing, it is self-correcting.

Although scientists often have strong opinions, it is not the role of science to make decisions regarding environmental policies or the management of resources. Rather, science provides a context for assessing and resolving uncertainties regarding specific policy or management alternatives.

Systems Thinking

■ Just because we know things doesn't mean we understand how they fit together.

Sustainable management does not require us to understand every piece and process in an ecosystem (although the more we know the better). It does, however, require systems thinking. *Systems thinking* recognizes the connections among the pieces of a larger, integrated system. In contrast, *piece thinking* focuses on the individual parts of a system, giving little attention to their association with other parts.

Returning to our car metaphor, a piece thinker interested in improving the efficiency of the braking process of a car might focus on the materials that make up the brake pads. Systems thinkers would recognize that the ability to stop is influenced by many interacting features, including brake rotors, tires, and car weight. Thinking across boundaries, systems thinkers would recognize the importance of driver reaction time and road conditions. They would also realize that changes that improve braking efficiency might affect the behavior of other parts of the car.

Many unsustainable actions arise as a consequence of piece thinking. For example, overexploitation of resources such as fishes or forest trees is often a consequence of neglecting the complex ecosystems of which fishes and trees are part. The air pollution we call smog is a result of piece thinking. City planners focused on transportation and industrial development but ignored the ecosystems of their cities, including chemical transformations in the atmosphere and their effects on human health.

Systems thinkers recognize that the parts of systems are actually systems too, with behavior that depends on complex linkages among their parts (**Figure 1.23**). And finally, they recognize that all systems (except, perhaps, the universe) are parts of even larger wholes.

QUESTIONS 1.5

1. Describe two important sources of uncertainty in our understanding of ecosystems.

2. Scientific hypotheses must be falsifiable. What is meant by this statement?

3. How does systems thinking differ from piece thinking?

(MES) For additional review, go to **MasteringEnvironmentalScience**

◀ **Figure 1.23 Systems Within Systems** The longest running predator–prey study involves the interactions of wolf and moose populations on the remote island Isle Royale in Lake Superior. Researchers who study the dynamic of these two species note that the wolf population is as small as it has ever been, while the moose population is on the rise. Their concern is that inbreeding among the wolves has adversely affected their ability to hunt moose, reducing their kill rates. These scientists propose a "genetic rescue" for the wolves, bringing new wolves onto the island, adding to the gene pool. Park officials are concerned about how such a move will affect this isolated ecosystem and so plan to study the environmental impacts of such a decision.

Source: Ecological Studies of Wolves on Isle Royale, Annual Report 2013–14, John A. Vucetich and Rolf O. Peterson, School of Forest Resources and Environmental Science, Michigan Technological University.

Ways of Knowing

What methods did scientists employ to determine the impacts of the insecticide DDT on bird populations?

Skim the pages of a journal like *Science* or *Nature* and you will quickly discover that scientists employ a wide variety of methodologies. These can be classified into four general categories: observation, experimentation, synthesis, and theory.

Observation—Questions about the properties of individuals, populations, or ecosystems, or the changes in those properties through time or from place to place are best answered with careful observations. The properties being measured are called **variables**. For example, ecologists might wish to know how egg production and hatchling survival for a particular bird species varies at different locations. To do this, they could monitor the nests of multiple individuals of this species at a range of sites (**Figure 1.24**). They would ensure precision by sampling as many nests as possible in each place, and they would avoid bias by trying to select sample nests at random from the entire population of nests.

Observations can be used to establish whether there are **correlations** or quantitative relationships between different variables. Positively correlated variables tend to increase and decrease together. Negatively correlated variables tend to change in opposite directions; as one increases, the other decreases. For example, observations of several bird species by ecologists indicated that individual birds that had higher concentrations of DDT-derived chemicals in their tissues produced eggs with thinner shells that were more vulnerable to breakage (**Figure 1.25**). This negative correlation between these variables led to the hypothesis, but it did not prove, that increasing DDT concentrations were the cause of diminished reproductive success in these species.

Experimentation—To firmly establish a causal relationship between variables, an **experiment** is required. For example, a toxicologist might provide birds with food containing different amounts of DDT and examine the effects on their growth or egg production. This sort experiment would likely be carried out using domesticated species such as chickens rather than birds from the wild. The experiment must have a control in which food without DDT is provided in exactly the same way as each of the treatment doses. Control and treatment groups must include several individual birds to ensure precision and avoid bias. This experiment would be run several times to verify that it is repeatable.

▲ **Figure 1.24 Measuring Reproductive Success** This ornithologist is monitoring the number and size of heron nestlings.

Synthesis—Big ideas and grand theories are built on a foundation of facts and information from hundreds, sometimes thousands, of smaller and less grand observations and experiments. Thus, synthesis is a very important scientific methodology. Nowadays, information technologies, including the Internet, greatly facilitate such synthesis. It was just such a synthesis of information from hundreds of studies that led to a general theory to explain why the reproductive success of many bird species was diminished by DDT in their environment.

Theory—Theories are principles devised to explain a group of facts or phenomena. Theories may be formulated as verbal statements about causes and effects. For example, ecologists theorized that concentrations of DDT were especially high in predatory birds like eagles based on many studies establishing that DDT is not broken down or excreted but rather accumulates in the fatty tissues of animals; predators are, therefore, likely to have higher tissue concentrations than the prey that they eat. Scientists are careful to state clearly the facts and assumptions upon which their theories are based.

In this textbook, *Focus on Science* will highlight important research relevant to the study of environmental science. It will generally include questions related to scientific methods and process. You will see how each of these methodologies is applied, individually and in various combinations, to enable you to better understand what is happening around us.

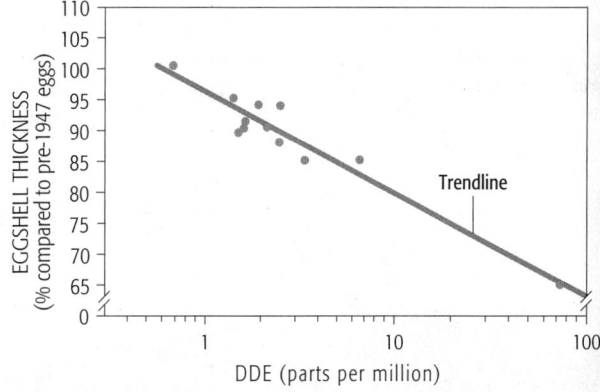

▲ **Figure 1.25 Negative Correlation** In 1969, researchers measured eggshell thickness and concentrations of the DDT residue DDE in parent birds in 10 brown pelican colonies from California, Florida, and South Carolina. Eggshell thickness is expressed as percent of that found in eggs laid prior to 1947, before the widespread use of DDT.

Source: Blus, L.J., et al. 1971. Eggshell thinning in the brown pelican: implication of DDE. *BioScience* 21: 1213–1215.

1.6 Sustainability Science

BIG IDEA In the end, whether or not we choose sustainable actions depends on our understanding of the connections between ecosystems and our well-being. The emerging field of **sustainability science** aims to understand the interactions between ecological systems and social systems, with a particular focus on long-term changes in global systems such as climate and the world economy. Sustainability science is also interested in how stresses, such as pollution and human conflict, affect the homeostasis of ecosystems and social systems. To be successful, sustainability science must embrace a wide range of outlooks and engage scholars from both the natural and social sciences. It is not a laboratory science; rather, it is about the actions we take in the real world. Therefore, scientific exploration and practical application must occur simultaneously. Seven questions are particularly important to the development of sustainability science.

Ecosystem–Social System Research Needs

■ Seven themes are particularly important to the development of sustainability science.

1. Ecosystem–social system connections. Throughout history, we humans have modified ecosystems to increase the provision of the resources we need or want, such as food and water. Increasing these resources has often diminished other ecosystem functions or services. In some cases, human actions have lessened the ability of ecosystems to provide essential services (**Figure 1.26**).

The effects of human activities may not be immediately obvious because the time lag between a particular action and its impact on ecosystem functions may be quite long. Research to understand the physical and biological processes that influence ecosystems from local to global scales must be integrated with studies of the social, political, and economic processes that drive human actions across these scales.

2. Long-term trends. In the past two centuries, humans have achieved a dominion over Earth's ecosystems that is unmatched by any other organism in the 3.8-billion-year history of life. Over 7 billion of us exploit Earth's ecosystems, using technologies that could not have been imagined a century ago. We are apt to see this as a "natural" consequence of our unique human traits such as intelligence, communication, and learning. Yet there is no precedent that assures us that our situation is sustainable. The world a generation from now will be unlike any in the past. Commitment to environmentally sustainable development is a commitment to the well-being of future generations.

3. Ecosystem–social system stability. Human-dominated ecosystems, such as cities and agricultural fields, lack the homeostatic features of their more complex, natural counterparts. When we manage ecosystems to increase their provisioning services, we often sacrifice their regulating or supporting services. As a consequence, they are more vulnerable to disturbances and the effects of long-term change. In other words, ecosystems simplified by humans are more susceptible to drastic changes.

Scientists are learning more about the connections between ecosystem homeostasis and ecosystem complexity. We also need to learn more about how our social systems influence such connections and how our social systems can support them.

4. Thresholds of change. Within a particular range, ecosystems are resilient to change. Outside that range, ecosystems may undergo significant change toward a new stable structure. At that point, it may be difficult or impossible to restore their original structure and function.

We know that such thresholds exist, but we are not yet able to predict their exact ranges. For example, humans have long depended on arid grasslands to support grazing animals. As long as grazing pressure is moderate, such ecosystems are relatively stable and resilient. When rainfall falls below a certain amount, however, such ecosystems are no longer stable, and grasslands quickly become deserts (**Figure 1.27**). At present, we do not know the threshold of rainfall at which this important transition is likely to occur.

▼ Figure 1.26 **Human-dominated Ecosystems** Humans have planted agricultural fields with single crop species on hundreds of millions of acres that once supported diverse and complex prairie ecosystems. These fields provide food that is essential to the growing numbers of us. But they require significant subsidies of fertilizer and water and are more prone to erosion and susceptible to drought than the ecosystems they replaced.

◀ **Figure 1.27 Thresholds of Change**
A goatherd searches for grass for his goats in Niger, Africa. As climates have become warmer and drier, overgrazing has converted grassy woodlands to desert.

5. Human incentives. In general, human actions are influenced by laws and regulations, economic incentives, knowledge, and belief systems. How these factors interact to affect human behavior is not well understood. On a global level, laws and regulations often have limited effects.

Long ago, incentives that influenced human actions were local; their effects on ecosystems were mostly local, too. Today, many human actions are driven by national and global incentives. Globalization of the world's economy makes it difficult to identify incentives for sustainable actions. Incentives that encourage sustainable land stewardship in Bangladesh, where farmers grow their own food, are very different from those for farmers in Kansas who grow food for global markets.

The aspirations of the world's poorest people will be especially important in the years ahead. Improvement in their well-being is essential to reducing conflict among peoples and nations. Identifying incentives for actions that diminish poverty in a sustainable fashion is a particularly important challenge for sustainability science.

6. Monitoring our progress. We have not yet developed a system for monitoring the health or sustainability of the interactions between human social systems and ecosystems. The challenge is a bit like the structure of an automobile's dashboard. We need indicators to show the sustainability of our ecosystems and social systems, similar to the gauges on

a dashboard—the speedometer, oil pressure gauge, and fuel gauge. Local, national, and global communities must agree on what levels or amounts of these indicators are acceptable. For example, similar to setting a speed limit, we must determine what levels of chemicals such as arsenic or mercury can be tolerated in drinking water. We also need to develop protocols for the actions we will take when we approach critical values of these indicators. Finally, we need ways to ensure that these actions are taken, much as we enforce speed limits.

7. Integrating learning and action. Success in sustainability science will require changes in many of our institutions. Scholarly institutions such as universities are organized by disciplines, for example the natural sciences, mathematics, economics, political sciences, and the arts. The challenges of sustainable development cannot be met by the methodologies of any one of these disciplines alone. Instead, it requires interdisciplinary, problem-oriented learning.

Many of the challenges of sustainability are urgent. But the time between the acquisition of new knowledge and its application to real-world problems is often long. It can be accelerated by more effective communication among scholars and those who will apply their scholarship. Research is needed to develop the management systems that unite scientific exploration and application.

QUESTIONS 1.6

1. Give an example of an ecosystem–social system interaction.

2. What is meant by "thresholds of change"?

3. Why is monitoring important for sustainability?

(MES) For additional review, go to
MasteringEnvironmentalScience

The University of California, Berkeley's Green Initiative Fund

Rachel Barge is a graduate of the University of California (UC), Berkeley, where she helped create The Green Initiative Fund (TGIF), which provides approximately $250,000 a year to campus projects that make UC Berkeley more environmentally sustainable. Students began by paying $5 per semester into the fund, which then provided grants students could apply for to champion projects such as energy retrofits, water conservation, and "green" student internships. After graduating, Rachel began her own consulting firm to spread the green fund model to other campuses, raising more than $16,000,000 in the process.

How did you first get the idea for The Green Initiative Fund?

When I arrived at UC Berkeley as a college freshman, the university had just released its first-ever "Campus Sustainability Assessment," a comprehensive analysis of the university's performance in the areas of energy, water, buildings, food, transportation, and purchasing. Although the assessment was a great first step, I soon realized that there was no plan to *implement* most of the great ideas contained in the document. Why? The funding available for such projects was minimal, not even scraping the surface of what needed to be done.

A group of fellow students and I thought we could tackle this problem head-on by raising funds ourselves while also pressuring the administration to put its own money forward. We learned that a green fee program run by the students of UC Santa Barbara was generating more than $150,000 per year for campus sustainability efforts. We thought a similar student fee would be a great way to unleash a significant amount of capital for green projects on our campus.

If anything convinced us of the need to establish this fund, it was learning that our campus had a carbon footprint that was larger than that of the entire nation of Cambodia, a nation of 13 million people. It was an outrageous fact that served as our call to action.

What steps did you take to get started?

About eight students and I formed a core team dedicated to passing the green fee in the spring campus elections. We needed to get the majority of the students to vote in favor of the initiative, which was no easy task. We learned everything about green fees: how to develop their legal structure, how to propose them on the ballot, how to explain their value to fellow students through social media and videos, and how to determine what kinds of projects the fees could pay for.

After nine months of hard work, 69% of the students voted to pass TGIF—we had won our green fee campaign. The university now has approximately $250,000 per year for sustainability initiatives, which is $2,500,000 over 10 years!

What are the biggest challenges you've faced with the project, and how did you solve them?

A fellow student challenged our campaign in court and tried to have The Green Initiative Fund removed from the election ballot. Our opponent found a few small errors in the way we'd set up our campaign. We went to court and fought back hard, winning our legal case and prevailing in the election.

How has your work affected other people and the environment?

It's fun to see the effects of TGIF growing every year. A favorite project of mine is a student-run organic café in Berkeley that supplies students with fresh, local, organic food and serves as a hub for food education and activism. The best part of creating green funds is that you never know what amazing ideas students will have about how to make the campus a better, more sustainable place.

What advice would you give to students who want to replicate your model on their campuses?

Here's my advice for creating a green fund on your campus.

- *Do your research.* Figure out how your campus elections work, how student fees work, and what fees have been created in the past. Every campus is different.
- *Get a core team.* You can't run this campaign alone, so make sure you've got a core team of five to eight people and lots of volunteers for election time.
- *In marketing, focus on the little thing—the size of the fee.* It's only $5, and most students spend that much just to buy breakfast. Isn't $5 a semester a small price to pay for a green campus?

- *Have fun.* These campaigns are your chance to leave a lasting legacy of sustainability on your campus, but make sure to enjoy the process—it goes by fast!

I hope people can learn from my experience: If you have a great idea, go out there and ask people for money to launch your initiative into the world! I had no idea that was possible when I was graduating, but a mentor told me, "You can create a job out of this thing you loved doing in your free time. Put together a proposal and start asking people for money!" It's the best advice I've ever received.

ONE BURRITO HELPS SAVE THE WORLD

For just $5 / semester, TGIF will make our campus more sustainable. That's less than the cost of your lunch.

Summary

Earth's environments comprise all of the physical, chemical, and biological factors that determine the growth and survival of organisms. Humans have altered Earth's environments more than any other single species. These alterations affect our own well-being—our access to food and shelter, health, social stability, and personal freedom. Our actions will be sustainable if they meet the needs of the present in an equitable and fair fashion without compromising the ability of future generations to meet their needs. The intersection of the three dimensions of planet, people, and profit represents a sustainable approach to business. Measuring this (environment, social, and economic) and including it in business planning allows for a more complete accounting of the influence of a corporation or government. Principles of ecosystem function provide basic guidelines for sustainable action. These include a keen awareness of the limits posed by the law of conservation of mass and energy and the openness of ecosystems, conservation of ecosystem complexity and the homeostasis it provides, and the ability to accommodate change. The sustainability of our actions depends especially on our understanding of the connections between ecosystems and human well-being. Studies of ecosystem—social system interactions and stability, long-term trends, thresholds of change, and human incentives for sustainable action—will be key to that understanding.

1.1 Environment and Sustainability

- The environment includes all of the physical, chemical, and biological factors and processes that influence the growth and survival of organisms.
- Sustaining human well-being and the well-being of other organisms depends on human actions and behavior that meet the needs of the present in an equitable and fair fashion without compromising the ability of future generations to meet their needs.
- Sustainability can be evaluated with the triple bottom line approach, considering environmental, social, and economic dimensions of the impact of a corporation, nonprofit, or government on the world.

KEY TERMS

environment, environmental science, ecology, human well-being, sustainability, triple bottom line (TBL)

QUESTIONS

1. How are notions of equity and justice embedded in our current concept of sustainability?
2. Most decision makers agree that equitable treatment and justice among humans must be part of any definition of sustainability. Why is this important?
3. Consider the company, DIRTT, featured in this chapter's Seeing Solutions. How does it address the triple bottom line in its business model?

1.2 Ecosystems

- An ecosystem is a complex community of organisms (biota) and its physical and chemical (abiotic) environment, functioning as an integrated ecological unit.
- The cycling of matter and flow of energy and the processes that influence the growth and survival of organisms are called ecosystem functions.

KEY TERMS

ecosystem, biota, abiotic, ecosystem functions, ecosystem integrity, ecosystem services, provisioning services, regulating services, cultural services, supporting services

Inflow: Sunlight and materials like carbon, nitrogen, and water

Transformation of energy and exchange of matter between biota and abiotic environment

Outflow: Heat and materials like carbon, nitrogen, and water

QUESTIONS

1. Ecologists often liken the functioning of an ecosystem to that of an organism. In what ways does this comparison make sense and in what ways does it not?
2. Over the past 24 hours, you have benefited from many ecosystem services. Name five specific examples and the categories to which they belong.

1.3 Principles of Ecosystem Function

- Although they are complex, ecosystems are highly organized and function in very predictable ways.
- All ecosystems are open to inflows and outflows of matter and energy.
- Ecosystem processes and functions are self-regulated by dynamic homeostasis.
- Ecosystem change is inevitable and essential.

KEY TERMS

dynamic homeostasis, negative feedback, positive feedback

QUESTIONS

1. The boundaries between counties and states are often defined by straight lines. Why might this be a problem for the management of ecosystems?
2. Explain why positive feedbacks often destabilize ecosystems.

1.4 Acting Sustainably

■ Human actions are likely to be unsustainable if they ignore the limited supply of resources, reduce the homeostatic capacity of ecosystems, or interfere with the capacity of ecosystems to change.

KEY TERM

nonrenewable

QUESTIONS

1. The law of conservation of mass and energy is universally accepted. Nevertheless, some human actions indicate the opposite belief. Give two examples of these actions.
2. Much of human management involves replacing complex and diverse ecosystems, such as a forest, with much simpler ecosystems, such as a cornfield. Why are these simpler systems more prone to disturbance?
3. If change is inevitable and normal in ecosystems, why should we be concerned about the changes wrought by humans?

1.5 Uncertainty, Science, and Systems Thinking

■ Acknowledging that there is much that we do not understand about the organization and functioning of ecosystems is an important prerequisite to sustainable action.

■ Sustainable management requires systems thinking, the recognition of the linkages among the parts in the context of a larger, integrated ecosystem.

KEY TERMS

science, hypotheses, predictions, treatment group, control group, precision, bias, observation, variables, correlations, experimentation, experiment, synthesis, theory

QUESTIONS

1. Why is it important to recognize that our knowledge about most environmental issues is incomplete and might change with additional information?
2. Some environmentalists believe that all life is guided by a spiritual force. What role might science have in verifying or refuting that belief?
3. Disagreement among scientists is critical to the scientific process. Why?
4. In many cities rates of garbage production exceed the capacities of processing facilities. Compared to a piece thinker, how would a systems thinker approach this challenge?

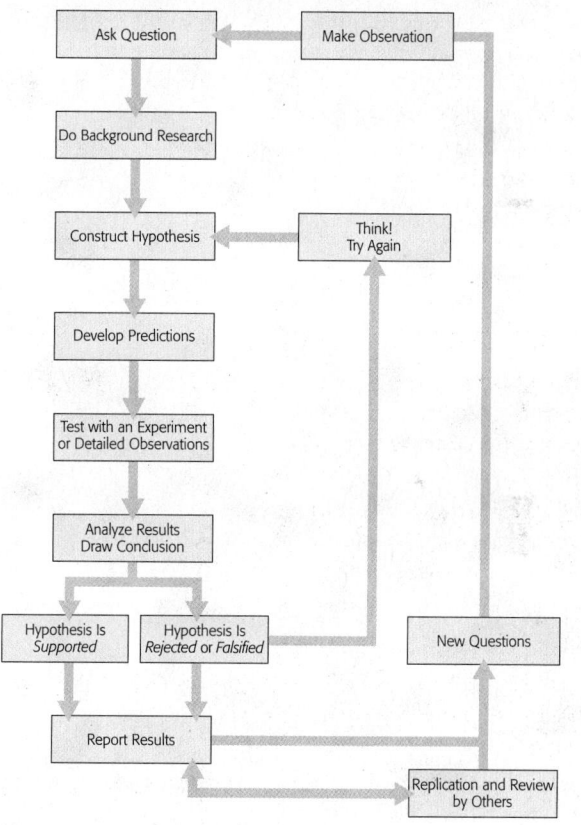

1.6 Sustainability Science

■ Sustainability science is an emerging area of study that focuses on the interactions between ecosystems and human social systems, with a special focus on long-term changes in environment and human populations and society.

KEY TERM

sustainability science

QUESTIONS

1. Research aimed at sustainable solutions to environmental problems must involve many academic disciplines. Using a particular environmental challenge, explain why this is true.
2. Natural disturbances such as hurricanes or droughts may have unexpected consequences in human-dominated ecosystems. Briefly explain why.

MasteringEnvironmentalScience®

Students Go to **MasteringEnvironmentalScience** for assignments, the eText, and the Study Area with practice tests, activities, and more.

Instructors Go to **MasteringEnvironmentalScience** for automatically graded tutorials and questions that you can assign to your students, plus Instructor Resources.

Environmental Ethics, Economics, and Policy

Dam-nation!

Is it right to destroy a wilderness ecosystem in order to benefit human communities?

In winter, deep snow covers Tuolumne Meadows, high in Yosemite National Park. In spring and early summer, the snow melts and the water eventually flows into the Tuolumne River. The water is then captured in the Hetch Hetchy Reservoir and carried by pipeline to San Francisco's public water system, 167 miles away. Since 1934, most of the water consumed by the millions of San Francisco residents has made this journey.

Hetch Hetchy provides the people of San Francisco with exceptionally pure water at a fraction of the cost paid by people in other California cities. Yet in 2012, residents of San Francisco voted on a proposal to require the city to conduct an $8 million study on how the flooded valley could be drained and restored to its former state. Such a restoration would be a formidable undertaking. In addition to the cost of replacing the water that the reservoir currently provides, the costs for removing the dam and restoring the original ecosystem are estimated to be in the billions of dollars.

Why would a city dependent on this water in a state that is notorious for its chronic water shortages even contemplate such an action? What benefits could possibly compensate for these costs? The answers to these questions are found in the history of the Hetch Hetchy project and the complex ethical, economic, and policy dilemmas it represents.

Like Yosemite Valley, Hetch Hetchy Valley is bounded by awesome cliffs and granite domes (**Figure 2.1**). Before the dam was built, lush meadows along the meandering Tuolumne River supported abundant wildlife. For thousands of years, the valley was treasured by all who visited, especially the Paiutes, the indigenous people who hunted and camped here.

As early as 1895, engineers were arguing that the narrow opening and high granite walls of the valley provided an ideal setting for a dam and reservoir. The reservoir would capture the flush of water from melting snows, providing a reliable year-round supply of water. San Francisco's population was growing rapidly, and its mayor, James Duval Phelan, perceived that water supply was the primary limitation for continued growth. Beginning in 1903 and continuing through 1905, Phelan and his proponents repeatedly applied to the U.S.

▲ Figure 2.1 **Before the Dam**
Hetch Hetchy Valley as painted by Albert Bierstadt in 1866.

Department of the Interior for permits to proceed with the project.

Legendary wilderness advocate John Muir and the members of his newly founded Sierra Club opposed the project. They argued that it was contrary to the mission of the newly formed Yosemite National Park and would defile one of nature's wonders. In late 1905, two years after a much-publicized visit to Yosemite with John Muir, President Theodore Roosevelt rejected Phelan's application (**Figure 2.2**). San Francisco formally abandoned its pursuit of Hetch Hetchy water in January 1906.

But then, at 5:16 A.M. on April 18, 1906, a great earthquake struck San Francisco. The earthquake and the fire that followed demolished the city. The rebuilding of the city soon reopened the Hetch Hetchy water debate because by then the city had to pay exorbitant prices to buy water from surrounding communities. The political pressures for developing Hetch Hetchy grew rapidly.

Gifford Pinchot, the chief of the newly created Forest Service and a powerful voice in the Roosevelt administration, was a strong supporter of the project. Pinchot wrote, "I fully sympathize with the desire . . . to protect the Yosemite National Park, but I believe that the highest possible use which could be made of it would be to supply pure water to a great center of population." Muir, on the other hand, passionately fought the project on the

▲ Figure 2.2 **A Memorable Meeting**
President Theodore Roosevelt and John Muir at Glacier Point, Yosemite, in 1903. At this famous meeting, Muir thought he had convinced Roosevelt to preserve Hetch Hetchy Valley.

principle that it was simply wrong to destroy a beautiful wilderness, regardless of its potential human value.

The voices of the Miwok and Paiute Indians, who had lived in Hetch Hetchy for centuries, were not heard in the din of this public debate. Although the development of Hetch Hetchy would cost these people dearly, they were not considered at all. No matter how the valley was developed—with a dam and reservoir or with roads, hotels, and public camps—their access to the valley would be lost forever.

In 1908, Roosevelt succumbed to pressure from Pinchot and others and initiated the permitting process for the dam. But in 1909, Roosevelt's successor, President William Howard Taft, was persuaded by Muir's arguments and put the project on hold. Debate and political wrangling over the project continued until 1913, when the new president, Woodrow Wilson, insisted on its approval. In that same year, Congress passed legislation affirming its support for the project.

Construction of the O'Shaughnessy Dam was completed and the valley filled with water in July 1923 (Figure 2.3). It took 11 more years to complete the aqueduct and water distribution system that would deliver Hetch Hetchy water to thirsty San Franciscans. Few doubt that the city received great benefit from this project. But as California has become more urban, concerns over the continued loss of wilderness have grown. Many influential individuals and organizations now believe that the benefits of a wild Hetch Hetchy Valley exceed the costs of its restoration and replacing its water. Those powerful voices motivated the 2012 ballot proposal in San Francisco. However, given the economic values at stake, it is perhaps no surprise that the proposal was soundly defeated by the citizens of the city.

For over a century, cultural, ethical, and economic values have shaped the debate and policies regarding Hetch Hetchy and myriad other development projects. In this chapter, we explore those values and the ways in which they influence policy and law.

- *How have our attitudes and understanding of environmental issues changed over time?*

- *What is the basis for ethical values about right or wrong actions that affect the environment?*

- *How do economic values influence environmental decisions?*

- *How are attitudes and values translated into policies and laws?*

▼ Figure 2.3 **To Be Continued?**
Hetch Hetchy Valley today with the reservoir that supplies water to the city of San Francisco. The fight continues to see the valley restored, a landscape that Muir described as "one of Nature's rarest and most precious mountain temples."

2.1 Changing Views of Humans and Nature

BIG IDEA What is your connection to the environment? Human attitudes toward the environment are shaped by a variety of factors, including mode of living, cultural history, religious and political beliefs, and knowledge. Connections between humans and the environment were probably obvious to primitive hunter-gatherer communities. But as communities became more agricultural and then developed industrial technologies, the connections became less obvious. During the Industrial Revolution, new knowledge, population growth, and growing environmental impacts raised concerns about the relationship between humans and their environment. Debates raged about whether nature had value in its own right or was valuable only in its service to the needs of humankind. Today, our view of the relationship between humans and the environment continues to evolve in response to the effects of human actions and their consequences for different groups of people.

Pre-Industrial Views

■ Our view of our relationship to nature changed as our dependence on it became less immediate and obvious.

The environment is a central feature in the spiritual traditions of many primitive cultures. For example, many Native American traditions assign sacred status to Mother Earth and Father Sky, whose union, they assert, produced all life. **Animism**, the belief that living and nonliving objects possess a spirit or a soul, is a common feature in the religions of many indigenous people (Figure 2.4). For hunter-gatherers, humanity and nature are unified.

The domestication of plants and animals allowed humans to alter ecosystems for their benefit. For example, farmers in ancient Mesopotamia replaced native grasslands with fields of domesticated wheat. By doing this, they were able to direct the productivity of these landscapes to the support of human populations. Weeding out competing plants and using irrigation allowed farmers to boost crop production even more (see Module 12.2). As agricultural societies turned into market societies, their citizens probably became less aware of their dependence on the natural environment and more involved in social and cultural pursuits.

The monotheistic religions of Judaism, Christianity, and Islam arose during this period of agricultural development. A key feature of these religious traditions is the centrality of humankind among all creation, an *anthropocentric* view evidenced by man having been created in God's image. In Genesis, for example, man is directed to subdue the earth and is given dominion over "every living thing that moves

◀ Figure 2.4 **A Sacred Giraffe**
Archaeologists believe that paintings such as these done about 6,000 years ago by hunter-gatherers in the Tassili n'Ajjer Caves, Algeria, reflect the reverence of these people for the animals they depended on.

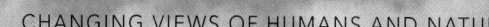

▲ Figure 2.5 **A Sacred Site**
Shinto worshipers enter the Miyajima Temple through the red sea gate first erected in A.D. 1168. The temple is dedicated to the daughters of the Shinto deity of seas and storms and to the brother of the great sun deity. The gate symbolizes the connection between sea and land.

upon the earth." Even so, the relationship of humans to nature is described in different ways in different parts of the Hebrew scriptures, reflecting historical changes in culture and knowledge. Other Judeo-Christian and Islamic writings place greater emphasis on human stewardship of resources.

Eastern religions such as Hinduism, Buddhism, and Shintoism also originated during the time of agricultural development. In general, these religions emphasize the importance of harmony between human actions and nature, and they view humans as inseparable from nature, a *biocentric* view (**Figure 2.5**).

The environment and organisms were common themes in the writings of ancient Greek scholars. None of these scholars had a greater influence on future generations than did Aristotle (384–322 B.C.). Nearly half of his more than 400 works focused on the nature of living things and humankind's relationship to them. He saw the elements of the environment arrayed along a natural "ladder of being" (*scala naturae*) from nonliving matter to plants, insects, and higher animals. On Aristotle's ladder, humans occupy the rung immediately beneath God. For the next two millennia, Aristotle's human-centered view of nature prevailed among Western thinkers.

The Enlightenment and Industrial Revolution

■ New knowledge and ever-growing human impacts on the environment catalyzed new views of the relationship of humans to nature.

By 1700, the scientific method was in wide use, and the Enlightenment was in full bloom. It was widely accepted that Earth was not at the center of the universe, opening up speculation about the position of humankind in the cosmos. Over the next century, philosophers such as Voltaire (1694–1778) would openly question humanity's special relationship to God. In 1809, Jean-Baptiste Lamarck proposed that the human species was simply one of many products of a process of evolution. Lamarck's proposal received little acceptance until 1859, when Charles Darwin's theory of evolution by means of natural selection provided a clear explanation for evolutionary change (see Module 4.5). Throughout this time, the fields of botany and zoology were generating an increased understanding of the connections between organisms and their environment.

Between 1750 and 1850, science spawned technologies such as steam-driven machines, giving birth to the Industrial Revolution. The human impact on the environment grew enormously. Increased commerce and food production generally improved human well-being. But, with the growth of urban industrial centers, fewer people had direct contact with natural environments (Figure 2.6).

As industrialization increased, some people began to look back to the natural world for inspiration. In 1836, Ralph Waldo Emerson published his short book, *Nature*.

▲ Figure 2.6 **Urbanization in the Industrial Age**
As more and more people moved into cities, there were fewer opportunities for people to connect with the natural world. What price have urbanization and industrialization exacted from modern humans and their relationship to nature?

Emerson argued that the obsession of industrial societies with material goods broke the connection between humans and nature. Furthermore, science provides limited insight into nature; humans need to immerse themselves in wild nature if they wish to reunite with its transcendent truths. The publication of *Nature* is often seen as the origin of the American transcendental movement, which included writer-philosophers such as Nathaniel Hawthorne, Walt Whitman, and Henry David Thoreau. For environmentalists, Thoreau's *Walden* is the most enduring of all transcendentalist writings. Many consider Thoreau to be among the founders of the discipline of ecology (Figure 2.7).

Less literary and more scientific, George Perkins Marsh's 1864 book, *Man and Nature*, is viewed by many as the genesis of the modern environmental movement. Marsh argued that nature is essentially "harmonious," but humans have separated themselves from nature and thereby violated its harmonies to the detriment of both nature and humans. Our survival, he argued, requires that we reconcile our activities with the inescapable laws of the natural world.

◄ Figure 2.7 **Walden's Inspiration**
Walden Pond as it appears today. In Thoreau's words, "I went to the woods [to] . . . see if I could not learn what it had to teach and not, when I came to die, discover that I had not lived."

Living in the Modern World: Conservation vs. Preservation

■ Current views of the relationship between humans and their environment have been shaped by debates about how and why we should value nature.

Yellowstone—the world's first national park—was dedicated in 1872. As the government created other parks and forest reserves, their exact mission and value to society was hotly debated. The naturalist John Muir, who was strongly influenced by transcendentalist writings and his own experiences in the Sierra Nevada wilderness, advocated the **preservationist** view that parks and public lands should *preserve* wild nature in its pristine state. Preservationists believed that humans should have access to wilderness parks for their inspiration and beauty but that parks should be protected from consumptive uses such as logging or water diversion. To further these goals, Muir and other preservationists founded one of the earliest nongovernmental environmental advocacy organizations, the Sierra Club.

The leading spokesman for the opposing view was Gifford Pinchot. Pinchot had studied forestry in Europe and was an ardent critic of the destructive forestry practices widely used at that time. Pinchot articulated the **conservationist** view that public resources should be used and managed in a *sustainable* fashion to provide the greatest benefit to the greatest number of people. Conservationists valued nature for the goods and services it provided human beings. Preservationists, in contrast, argued that nature had value in its own right and therefore deserved protection.

Muir and Pinchot, and their respective viewpoints, were at the center of the debate over the Hetch Hetchy water project. To this day, the divide between conservationist and preservationist views continues to define debate on the management of public lands.

Growing up in the midst of the conservation-preservation debates, ecologist and environmentalist Aldo Leopold argued for both the sustainable human use of natural resources and the preservation of wilderness. He articulated a "land ethic" in his book *A Sand County Almanac* (1949): "A thing is right when it tends to preserve the integrity, stability, and beauty of the biotic community. It is wrong when it tends otherwise." Leopold thought that wilderness provides resources that are essential to civilization. He warned, however, that human impacts on ecosystems differ significantly from natural disturbances. They are, therefore, likely to have unintended consequences.

One such unintended consequence resulted from the widespread use of the pesticide DDT. After World War II, this pesticide was sprayed on wetlands across the United States to kill disease-carrying mosquitoes. In addition to killing mosquitoes, DDT destroyed the birds that fed upon them as well as the predators of those birds. In 1962, the American public was alerted to this threat by marine biologist Rachel Carson. Her widely read book,

Silent Spring, pointed out the danger that pesticides posed to natural ecosystems and, ultimately, to humans.

Six years later, Paul Ehrlich's book *The Population Bomb* told the public how the rapid growth of the human population was threatening the environment and human survival. Should trends in population growth continue, Ehrlich warned, our demands would soon exceed Earth's ability to supply us with food and other essential resources.

Today, environmental trends, debates, and warnings dominate news stories and political agendas around the world. Two issues are of particular concern—the widespread extinction of plant and animal species and the warming of Earth's atmosphere. In numerous books and articles, distinguished Harvard ecologist Edward O. Wilson has documented the loss of hundreds of plant and animal species, as well as the threats that these losses pose to our sustainability. In 2007, the Intergovernmental Panel on Climate Change (IPCC) and former vice president Al Gore received the Nobel Peace Prize for their advocacy for action to respond to the threats posed by global warming (Figure 2.8). Their work has helped shape public understanding of the importance of climate change and the need for action to minimize its effects.

▲ Figure 2.8 **Global Recognition, Global Concern** Nobel Peace Prize Laureates Al Gore (left) and Rajendra K. Pachauri (right), Chairman of the Intergovernmental Panel on Climate Change (IPCC), with their Nobel Peace Prize medals and diplomas at the award ceremony in Oslo, Norway, December 10, 2007.

In recent years, theologians representing several different faiths have placed increased emphasis on environmental stewardship. Organizations such as Interfaith Power and Light bring together Jews, evangelical and nonevangelical Christians, and Muslims with common interests in environmental sustainability. These groups advocate care for the environment both as a matter of respect for "God's creation" and as a responsibility to humankind.

Our view of our relationship to nature has evolved considerably through our history. As hunter-gatherers, our dependence on nature was clear. Agriculture, industry, and the development of complex urban societies made our connection to nature less immediate and therefore less obvious. Perhaps this encouraged a sense of dominion over nature. Today, as our numbers near 7.3 billion, we do not doubt our ability to dominate Earth's ecosystems, yet we also have no doubt about our dependence on those ecosystems.

QUESTIONS 2.1

1. Describe how human attitudes were changing around the time that monotheistic religions were emerging.

2. How did transcendental writers, such as Emerson and Thoreau, view nature and our role in it?

3. Describe three major ideas or viewpoints that influence current attitudes toward the environment.

(MES) For additional review, go to **MasteringEnvironmentalScience**

2.2 Environmental Ethics

BIG IDEA The role of humans in the environment has been discussed since ancient times. Yet only in recent decades have we attempted to define appropriate human behavior in the context of the environment. Ethics is the branch of philosophy that explores ways to distinguish between right and wrong conduct or to define moral living. **Environmental ethics** studies the moral relationship of humans to the environment and its nonhuman contents. Systems of environmental ethics vary regarding who or what has value. Nonenvironmental issues, especially fairness and justice between genders or among ethnic groups or races, also influence the determination of whether an action is right or moral.

Doing the Right Thing

■ Different standards for right, or moral, behavior define three ethical traditions.

What makes an action right or wrong? Ethicists have long argued over how to answer this question. The Greek philosopher Plato proposed that *motivation* should be the basis for determining whether an action was good. In his **virtue ethics** system, an action is right if it is motivated by virtues, such as kindness, honesty, loyalty, and justice. Virtue ethics has played a significant role in the development of general ethical frameworks, but because it refers to the rightness of the actions of people toward other people, its language is necessarily anthropocentric, literally human-centered. For example, according to virtue ethics, the neglect of the concerns of the Paiutes in the process that led to building the Hetch Hetchy Reservoir would be considered unkind and, therefore, wrong.

Consequence-based ethics emphasizes the importance of *outcomes*. Right and wrong are defined in terms of pleasure or pain, benefit or harm, and satisfaction or dissatisfaction. An action is right depending on whether or not it delivers pleasure, benefit, or satisfaction. **Utilitarianism** is an example of consequence-based ethics that defines right actions as those that deliver the greatest good to the greatest number. Gifford Pinchot's conservationist approach to environmental decisions was clearly utilitarian. According to consequence-based ethics, building the Hetch Hetchy Reservoir was right because it benefited a far greater number of people than did the preservationist alternative. Note that this same action would be judged unethical if it benefited fewer people than it harmed.

Eighteenth-century philosopher Immanuel Kant advocated **duty-based ethics**—that the rightness or wrongness of actions should be determined by a set of rules or laws. In duty-based ethics, a wrong action is wrong regardless of its outcome. For example, Kant suggested that it is always wrong to lie, even if lying has a positive outcome. Muir's preservationist view of Hetch Hetchy was grounded in duty-based ethics, arguing that destroying a wild place of natural beauty is wrong regardless of how it might benefit humans.

Duty-based ethics presumes an accepted framework, such as the Ten Commandments, for defining whether actions are right. Most such frameworks have been articulated in the context of religious traditions and, like virtue ethics, deal primarily with the actions of humans toward other humans. How do we know what is ethical when it comes to the environment (Figure 2.9)?

► Figure 2.9 **Doing the Right Thing**
In 2012, the final phase of the Keystone Pipeline System, known as Keystone XL, was put before the U.S. government for approval. Three earlier phases were already in operation. XL, the fourth and final phase of the system, was needed to deliver oil from Canada's tar sands straight across the country to refineries in Louisiana. Unlike the three other phases of the pipeline, this one was met with a political firestorm that pitted environmental concerns against economic ones. How would you argue this from an ethical perspective: is it right or wrong? What ethical arguments can you make for each side?

Who or What Matters?

■ Environmental ethics differs regarding who or what has value.

In any ethical framework, we must identify *who* or *what* is to be valued. In consequence-based ethics, we must specify who or what should receive the greatest benefit. In duty-based ethics, we must define to whom or what we have a duty. For example, does the commandment "thou shall not kill" extend to animals?

Systems of environmental ethics also require us to consider whether the value is intrinsic or instrumental. A person, organism, or object valued as an end unto itself is said to have **intrinsic value**. A thing that is valued as a means to some other end has **instrumental value**. Most ethicists agree that every human life has intrinsic value. Some ethicists argue that other living things have value if only as a means to support the well-being of humans. Thus, their value is instrumental.

Anthropocentric ethics assigns intrinsic value only to humans; it defines right actions in terms of outcomes for human beings. Other organisms and objects have instrumental value because they contribute to human well-being. The conservationist view of environmental management is certainly anthropocentric. There are also situations for which the preservationist view could be considered anthropocentric. Things that are not essential to our well-being, such as a place of beauty (Yosemite) or a charismatic animal (panda bear), acquire instrumental value because humans happen to care about them.

Biocentric ethics argues that the value of other living things is equal to the value of humans. Biocentric ethics extends intrinsic value to *individual* organisms beyond human beings; the organisms do not need to benefit humans in order to have value.

There are two distinct schools of biocentric ethics. Some biocentrists argue that for a thing to have intrinsic value, it must be able to experience pleasure or satisfaction. Therefore, it must possess qualities such as self-perception, desires, memory, and a sense of future. In this view, animals such as dogs, chimpanzees, and porpoises have intrinsic value, but plants and animals lacking a complex nervous system do not. This biocentric ethic motivates many in the animal rights movement.

Other biocentric ethicists argue that any individual organism that is the product of natural evolution has intrinsic value. Therefore, an amoeba or a jellyfish has the same value as a porpoise or a chimpanzee (Figure 2.10).

Ecocentric ethics places value on communities of organisms and ecosystems. Ecocentric ethicists believe that *collections* of organisms or critical features in the environment have intrinsic value. Aldo Leopold's assertion that "a thing is right when it tends to preserve the integrity, stability and beauty of the biotic community" is an example of ecocentrism. An ecocentric ethicist might argue that because hunting individual animals improves the health of that species' populations by removing diseased animals, it is a right action. On the other hand, many biocentrists argue that hunting

is unethical because it violates the intrinsic value of the individual organism.

The most expansive form of ecocentrism asserts that intrinsic value derives directly from naturalness. Philosopher Andrew Brennan, for example, argues that a natural entity such as a river has intrinsic value and is not a "mere instrument." Nature does not exist in order to meet our particular needs. Instead, it deserves moral respect in its own right. Because we are the air we breathe, the water we drink, and the food we eat, all elements of the environment have equal intrinsic value. Humans have no right to diminish the diversity and richness of Earth's ecosystems except to meet vital needs. This concept is the basis for what is called the **deep ecology movement**.

▼ Figure 2.10 **What Has Intrinsic Value?**
To which of these organisms do you assign intrinsic value? What is the basis for your decision?

Chrysaora fuscescens
from the phylum Cnidaria

Coronella austriaca
from the phylum Chordata

Pan troglodytes
from the phylum Chordata

Aedes aegypti
from the phylum Athropoda

1. What is the basis for judging right from wrong in virtue, consequence-based, and duty-based ethics?

2. What has intrinsic value in anthropocentric, biocentric, and ecocentric ethical frameworks?

3. What is meant by the phrase *environmental justice*?

(MES) For additional review, go to **MasteringEnvironmentalScience**

Ecofeminism and Environmental Justice

■ The rightness of environmental actions may be influenced by nonenvironmental considerations.

In her 1974 book *Le Feminisme ou la Morte* (*Feminism or Death*), Françoise d'Eaubonne argued that a special connection exists between women and the environment. This environmental ethic, which became known as **ecofeminism**, fostered what she saw as feminine concerns for the interrelationships among humans, nonhuman life, and the environment. She linked the exploitation and domination of women to that of the environment.

Many ecofeminists today point out that women often suffer disproportionately from human destruction of the environment. This connection is made obvious in the plight of women in developing countries who must tend crops and gather fuel wood in badly degraded environments. Ecofeminist and Nobel Prize winner Wangari Maathai saw how deforestation in Kenya deprived women of fuel and led to the erosion of the soil they farmed. In 1977, she established the Green Belt Movement to organize women to plant trees, in effect reforesting the land and thus improving their lives by taking care of the environment that takes care of them (**Figure 2.11**).

Many actions in the environment, such as cutting a forest or locating a waste facility, have different consequences for different people. An action that is considered right according to biocentric or ecocentric criteria may have adverse consequences for some people. With respect to human equity or justice, many would consider such an action to be unethical.

In the United States, for instance, toxic dumps and landfills have historically been located near poor communities, which often have a large racial or ethnic minority population (**Figure 2.12**). In some cases, this placement of waste facilities has been a consequence of direct racial discrimination. In other cases, sites have been selected because land near poor communities is cheap. In either situation, poor communities must bear a large share of the costs and risks associated with such facilities. Situations such as this have given rise to the **environmental justice** movement, which seeks to ensure that in the management

▲ Figure 2.12 **Protesting Injustice**
Residents of rural Williamsburg County, South Carolina protest a plan by state officials to site a large landfill near their homes. About two-thirds of the county's residents are black, and one-third lives below the poverty line. The protest was successful. In 2011, the state formally dropped its plans for the facility.

of natural resources and the environment people are treated fairly regardless of race, gender, or economic status.

Environmental justice is a major consideration in a variety of international negotiations related to global issues. Some ethicists are concerned that the actions of many environmentalists disproportionately benefit the health and affluence of people in developed countries. For example, industrial development in poor countries is often accompanied by increased air and water pollution. The efforts of wealthy nations to impose international regulation of pollutants may be viewed by citizens of developing countries as unjust because such limits thwart their attempts to grow economically and reduce poverty. Justice, they feel, would be for the wealthy nations to provide economic compensation for the costs of conservation and lost opportunities to advance.

In summary, there is no single environmental ethic. The various ethical frameworks focus on different motivating values, such as justice, consequences for humans and ecosystems, and governing principles. Systems of environmental ethics also differ with respect to who or what in the environment has intrinsic value. That said, the UN Environment Programme (UNEP) was established to support what could be described as a growing international environmental ethic that encompasses a variety of actions intended to promote the sustainable development and protection of the global environment.

▲ Figure 2.11 **Empowering Women, Saving the Environment**
The legacy of Wangari Maathai lives on in more than 30 million trees planted by women as part of the Green Belt Movement. There are now more than 6,000 Green Belt nurseries in Kenya alone.

2.3 The Environment and the Marketplace

BIG IDEA Many activities that affect the environment are associated with buying physical goods, such as food and clothing, and intangible services, such as transportation on airplanes or buses. An **economic system** is made up of the institutions and interactions in a society that influence the production, distribution, and consumption of goods and services. In a market economy, supply and demand determine the price and level of production for goods and services. The price that consumers are willing to pay depends on their perceived need for a good or service—a commodity—and the time they will receive it. For several reasons, markets may produce incentives for production or consumption that cannot be sustained without depleting natural resources or damaging the environment.

Economic Systems

■ Economies vary in their complexity and the extent of government control.

One important way in which humans relate to their environment is through economic activity, that is, the production and consumption of goods and services. The most basic economic system is a **subsistence economy**, in which a society meets its needs from its environment without accumulating wealth. Individuals within a subsistence economy may barter to exchange goods and services, but there is no currency. Because people in subsistence economies do not accumulate a surplus of goods, their well-being depends on the continual renewal of resources from their environment. Societies of hunter-gatherers, fishermen, or farmers who live on what they catch or grow are examples of subsistence economies (Figure 2.13).

In **market economies**, the production and consumption of goods and services take place in markets guided by prices based on a system of currency. Individuals and businesses decide on their own what to produce or purchase. Producers are free to determine the price and amount of what they produce. Consumers are free to select what they purchase from among different producers. In a **free market economy**, the government does not influence the marketplace with subsidies, taxation, or regulation.

▶ Figure 2.13 **A Subsistence Economy**
The women at this market in La Paz, Bolivia, often trade one commodity for another.

► Figure 2.14 **It's Not All Planned**
Every day, free-market exchanges occur as small vendors sell their goods and services in a local setting. Even though much of China's economy is planned, this Shanghai street vendor openly sells his oven rolls at a price that works for him.

In a **planned economy**, the government regulates the prices of goods and services and the level of production. A government may simply set production, prices, and salaries by decree. More commonly, planned economies are controlled through a combination of regulation, incentives, subsidies, grants, and taxation.

Today, no country in the world has only one economic system. Although every country has a monetary system, subsistence economies in which people grow or barter for the things they need are found everywhere. In rural communities, for example, farmers may trade services, such as planting or harvesting, with no exchange of money. The United States is often said to have a free market economy, but many sectors of its economy are influenced by government regulations, subsidies, and taxation. China, North Korea, and Cuba claim to have planned economies, but in each of these countries much production and consumption takes place in "black markets" that operate outside the realm of government control (**Figure 2.14**).

Supply and Demand

■ The availability and price of a good or service is determined by the interaction of supply and demand and establishes economic value.

Underlying economic activity is the idea of a value attached to **commodities**, the goods or services that are exchanged. In most cases, money serves as a measure of a commodity's value. In economic terms, what sets that value is the principle of supply and demand, a model that is used to describe the interaction between producers and consumers (**Figure 2.15**). Our understanding of what drives supply and demand in an economy has changed over the centuries. In 1776, Scottish philosopher Adam Smith proposed what he believed to be the basic principles motivating the production and consumption of goods and services in a society.

Smith noted that when a commodity was in short supply, its price increased. As a result, the producer

had a greater **profit**, the difference between the cost to produce a commodity and its price in the marketplace. Smith argued that an increase in price and profit creates incentives for others to produce the same commodity. Supply increases. With a greater supply, competition among producers motivates them to lower prices. If prices for the commodity drop to near the cost of its production, incentives for production decline and supply decrease. Smith referred to this interaction among supply, price, and competition as the "invisible hand" of the economy.

Nowadays, economists accept Smith's assertion that increases in the price of a commodity will increase its supply. But they note that the price and abundance of commodities in a free market are also influenced by consumer demand—the amount of a commodity sought by a community. Consumer demand increases as goods and services become cheaper. **Neoclassical economic theory** focuses on the determination of the price and production of goods and services through the interaction of supply and demand (see Figure 2.15).

Neoclassical economic theory is based on two key assumptions about human behavior. It assumes that those who produce and sell goods and services act to maximize profit. Equally important, it assumes that consumers choose among alternative goods and services based on the satisfaction or benefit they expect to receive from them. In other words, neoclassical economists argue that consumers invariably act to maximize their satisfaction or benefit relative to cost.

▼ Figure 2.15 **Supply and Demand**
As the price of a good or service declines, buyer demand increases, but incentives for producers or sellers decrease. Neoclassical economic theory suggests that this interaction results in an equilibrium between price and production, which is shown here at the intersection of the two lines.

SUPPLY AND DEMAND

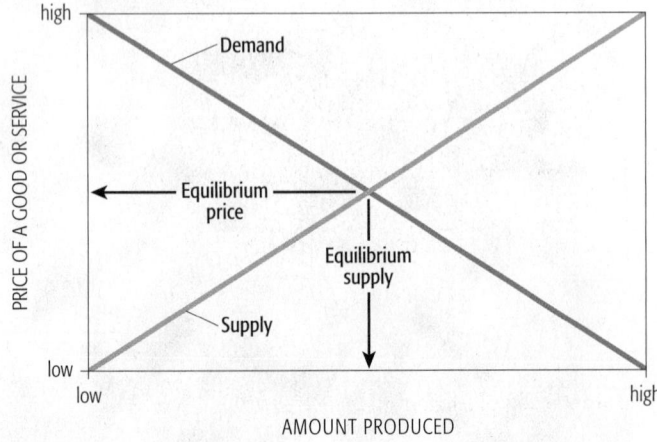

Economic Value

■ The economic value of a good or service is determined both by consumers' need for it and the time when they receive it.

Economists define the **economic value** of a commodity as the price that consumers are willing to pay for it. Typically, that price is largely determined by the benefit that a consumer perceives in a commodity. If you have the means, you might choose to spend a hundred thousand dollars on a luxury car even though at that price you could easily buy four or five cars of good quality. Thus, the economic value of a product is not determined by the cost of production; rather, it is set by what consumers are willing to pay.

The benefit of a good or service—and, therefore, its economic value—is not a fixed quantity. Instead, it depends on what the consumer already possesses and when the consumer will actually receive the good or service.

In general, the benefit provided by a commodity declines as a consumer acquires more of it. For example, the benefit you derive from buying a car may be very high. The benefit you derive from purchasing a second car is much lower. Thus, the benefit derived from a commodity is tightly coupled to the consumer's sense of need. The benefit of a glass of water is far greater to a poor soul wandering in a sun-parched desert than to someone blessed with abundant water (Figure 2.16).

The timing of the delivery of a good or service also affects the perception of its benefit. It is human nature to want something sooner rather than later. Its benefit diminishes if we are not to receive it immediately. Likewise, the economic value of a good or service—a consumer's willingness to pay for it—declines as the time until the consumer will receive it becomes longer.

Consider the effect of delivery time on the value of a natural resource such as fish. The owner of a fish market may be willing to pay a fisherman $10 for 10 pounds of fish delivered today. But the owner would not be willing to pay $10 for 10 pounds of fish that the fisherman promises to deliver a year from now. The rate at which economic value declines with time is called the **discount rate**.

Discount rates are affected by opportunity cost, risk, and consumer need. The **opportunity cost** of buying a thing is equal to difference in the cost of that thing and the cost of the best alternative use of that money. For example, the fishmonger who was asked to pay $10 for fish to be delivered a year from now might be able to invest that $10 in the stock market and earn an additional $3. The opportunity cost for buying the fish in advance would be $3.

Our fishmonger's willingness to pay will be further diminished by risks, or uncertainties about the future. For example, illness or equipment problems may affect the fisherman's ability to deliver fish in the future. Or the actual price of fish a year from now might be lower than the price today.

Finally, the economic value of fish to be delivered at some future time depends heavily on the fishmonger's current needs. If she is desperate to have fish for her shop immediately, she is unlikely to spend $10 for fish that will not be delivered until a year from now. To her, that fish is of little value. In other words, she is likely to assign a very high discount rate to fish delivered in the future.

▲ Figure 2.16 **Abundance Influences Benefit**
In general, the benefit that a consumer or buyer perceives from a unit of a commodity such as water decreases as it becomes more abundant.

Market Complications

■ Markets are never perfect, and prices often do not reflect all costs.

QUESTIONS 2.3

1. Describe three different kinds of economic systems.

2. What does neoclassical economic theory assume about the behavior of sellers and buyers?

3. How and why does the economic value of a commodity change as it becomes less abundant?

4. Give an example of an external environmental cost associated with an economic transaction.

(MES) For additional review, go to MasteringEnvironmentalScience

Neoclassical economics assumes that the rational decisions of producers and consumers serve to maximize profits and consumer satisfaction. But in the complicated markets of the real world, there are often other outcomes. Free markets sometimes produce incentives for actions that deplete resources or damage the environment. There are three reasons why this is so.

Externalized Costs and Benefits. Buyers and sellers rarely bear all of the costs and benefits associated with a product. This is because the production and consumption of products often influence ecosystems or people who are not part of the transaction. **Externalities** are costs and benefits associated with the production of a commodity that affect people other than buyers and sellers. Externalities can be positive or negative.

Positive externalities arise when a third party benefits from an economic transaction between others. For example, a timber company may maintain a healthy forest in order to maximize profits when the timber is harvested. People who don't buy or sell timber can enjoy the recreation, wildlife, and beauty that the healthy forest provides. These benefits are positive externalities.

Negative externalities impart costs to individuals who are neither buyers nor sellers. For example, the timber company might harvest its forest in a manner that pollutes the streams or diminishes the property value of nearby

homes. Here, people who are not involved in the buying and selling of the timber nevertheless pay a cost.

Externalized costs often present significant challenges to human well-being and to sustainable management. For example, oil spills from drilling rigs, tankers, and pipelines are responsible for billions of dollars in costs to communities and the environment that are paid by individuals and society. Those costs are not included in the price of crude oil, nor are they factored into the price of the gasoline you buy at the pump. As a consequence, the marketplace provides no incentives to reduce this negative externality.

Unknown Costs and Benefits. The satisfaction or sense of benefit we realize from any commodity depends in part on what we know about its actual costs and benefits. Yet it is nearly impossible for consumers to have a full understanding of all of the costs and benefits associated with a good or service.

Fifty years ago, for example, ignorance of health risks contributed to the high frequency of cigarette and cigar use among Americans. The perception of the benefits of smoking was enhanced by ad campaigns and Hollywood films, which made smoking appear glamorous and sophisticated (Figure 2.17). Growth in our understanding of the health risks associated with smoking has greatly lessened the perceived benefit of cigarettes. In addition, stiff court fines to tobacco companies and increased tobacco taxes have ensured that some of the health care costs of smoking are now captured in the price of a cigarette.

Limited Resources. Economic models often treat individual commodities as if they are infinitely available or easily interchangeable with other goods and services. Put another way, the availability of commodities like fish or oil are limited mainly by the incentives to producers. This is rarely the case with natural resources and ecosystem services. For example, the Food and Agriculture Organization of the United Nations estimates that over the last four decades nearly 70% of the world's marine fish stocks have been fully exploited or overexploited. During this time, human demand for fish has tripled. Many argue that economic markets have encouraged this exploitation.

Neoclassical economic theory is a powerful framework for understanding human behavior related to the production and consumption of ecosystem goods and services. From the standpoint of sustainability, however, this is only one part of the triple bottom line—environment and social well-being are equally important. Because most resources are finite and their costs and benefits are often externalized or not fully known to consumers, markets can result in depletion of resources or environmental damage. The depletion or overexploitation of a natural resource imparts externalized costs unequally among current and future generations.

► Figure 2.17 **Risky Behavior**
Fifty years ago, magazine ads such as this contributed to the poor understanding of the risks of cigarette smoking among consumers.

2.4 Valuing Ecosystems

BIG IDEA Ecosystem services, such as clean water and beautiful scenery, are not usually bought and sold, but they do have economic value. Their value can be measured by assessing people's willingness to pay for actions that conserve them. Their value can also be assessed by estimating the costs that people must pay if the services are lost.

Economic Valuation of Ecosystem Services

■ Willingness to pay can be used to measure the relative economic value of ecosystem services.

What is the value of unpolluted air or the flood control provided by a forest in a floodplain? Because ecosystem services are not typically bought and sold in markets, how can we assign an economic value to them?

One way to measure their value is by assessing people's willingness to pay the costs of conserving or preserving them. Such assessments do not determine the total value of an ecosystem's services. Rather, they determine the **marginal value**, the difference in people's willingness to pay for one action compared to an alternative.

For example, the marginal value of removing the O'Shaughnessy Dam and restoring Hetch Hetchy could be assessed by asking citizens of San Francisco how much they would be willing to pay for their water if this alternative were pursued. The difference between how much they currently pay and how much they would be willing to pay if the dam were removed is the marginal value of restoring the ecosystem. Notice that the emphasis of marginal valuation is on benefits that are added or lost to humans, not to ecosystems.

What is the economic value of inspiring landscapes and recreational opportunities, such as those provided by Yosemite Valley (or, perhaps, a restored Hetch Hetchy Valley)? **Travel-cost valuation** is the amount of money that people are willing to pay for transportation and lodging to visit such places. Travel-cost estimates may also include the cost of building and maintaining the roads and infrastructure for these destinations. This valuation technique has largely been applied to cultural, aesthetic, and recreational services associated with those places.

In **hedonic valuation**, economic value is determined by the difference in the market price of real estate that is affected by different environmental alternatives. For example, the hedonic value of a public park can be estimated by the difference between the price that people are willing to pay for property near the park compared to the price they will pay for property elsewhere. Similarly, some of the economic costs of a municipal waste facility might be measured by its effect on the value of houses nearby compared to the value of similar houses located far away from the facility (**Figure 2.18**).

Estimating the value of an ecosystem service by surveying people's willingness to pay is called **contingent valuation**. For example, to determine the economic value of an endangered species, respondents might be asked to provide a statement of what they would be willing to pay for actions aimed at its conservation.

The value of the proposed restoration of Hetch Hetchy might be estimated by surveying the willingness of citizens to contribute to its restoration, either directly or through increased taxes.

▼ Figure 2.18 **Hedonic Value**
You might be willing to pay a premium for a home that is located in a natural setting, such as near a lake or a park. If you knew that a sewage treatment facility was located nearby, would that influence the price you would be willing to pay?

Q *Can you put a monetary value on clean air or a beautiful forest? Aren't they priceless?*

Christopher Shad, Cypress College

Norm: A We cannot determine their absolute value; in that sense, they are priceless. We can, however, assess what people are willing to pay for actions that will improve air quality or preserve more forest land. We can also measure economic costs for diminished air quality or the loss of forest cover.

Contingent valuation has been applied to a wide variety of situations; however, the uncertainties associated with this technique make its use controversial. Contingent valuation assumes that respondents' expression of their willingness to pay is an accurate expression of what they would actually pay. This assumption may be untrue. Many studies have shown that expressions of willingness to pay are heavily influenced by how much respondents know about particular ecosystem services and alternative actions. We are likely to assign greater value to things we understand than to those we do not.

Ecological Valuation

■ The value of ecosystem services can be measured by the cost of their possible loss.

In a book called *Natural Capital*, Paul Hawkins, Amory Lovins, and Hunter Lovins argue that all of Earth's resources should be viewed as economic capital. They define **natural capital** as all of Earth's resources that are necessary to produce the ecosystem services on which we depend. Because human actions can improve or degrade these resources, our actions yield consequences that have true monetary value.

The natural capital concept is the basis for **ecological valuation** of ecosystem services. Rather than focusing on willingness to pay, economists determine ecological value by the potential cost of the loss or degradation of an ecosystem service. For example, the 2010 BP oil spill revealed the high ecological value of the services provided by Gulf of Mexico ecosystems. Economists have determined that the spill directly or indirectly affected at least 20 categories of ecosystem services, including commercial fishing, tourism, and hurricane protection by coastal marshes. A recent study suggests that the total value for these services in the Mississippi River Delta alone is $12–$14 billion per year.

Often, the costs or benefits of environmental actions such as deforestation or reforestation are not realized until long after the actions are taken. For this reason, proponents of ecological valuation argue that this approach aligns well with environmental sustainability goals that respect the rights of future generations.

In 1997, a team of scientists headed by University of Vermont professor Robert Costanza used ecological valuation techniques to estimate the annual value of all Earth's ecosystem services. That value in 1997 was $33 trillion. In a 2014 paper, using the same methods but with updated data on costs and changes in the extent of different ecosystem types, Costanza's team revised their estimate upward to $125 trillion/year (**Figure 2.19**). This represents the costs that would be inflicted if these services were suddenly withdrawn. Their valuation demonstrates the significant economic consequences of the degradation of ecosystem services.

Some traditional economists rightly observed that the valuation by Constanza's team did not reflect economic value in the sense of people's willingness to pay. Given that the loss of all ecosystem services would mean our certain death, we humans would likely be willing to pay whatever was necessary to maintain those services. In other words, *the total value of Earth's ecosystems is priceless*.

▶ Figure 2.19
Ecological Value
The orange bars in this graph indicate the percent of Earth's area occupied by each general ecosystem type (biome). The purple bars represent the percent of the total value ($125 trillion/year) of Earth's ecosystem services provided by that ecosystem type. Note that the value of these services from ecosystems such as estuaries, coral reefs, grasslands, and wetlands are quite large in proportion to their areal extent.

Source: Costanza, R., et al. 2014. Changes in the global value of ecosystem services. *Global Environmental Change* 26:152–158.

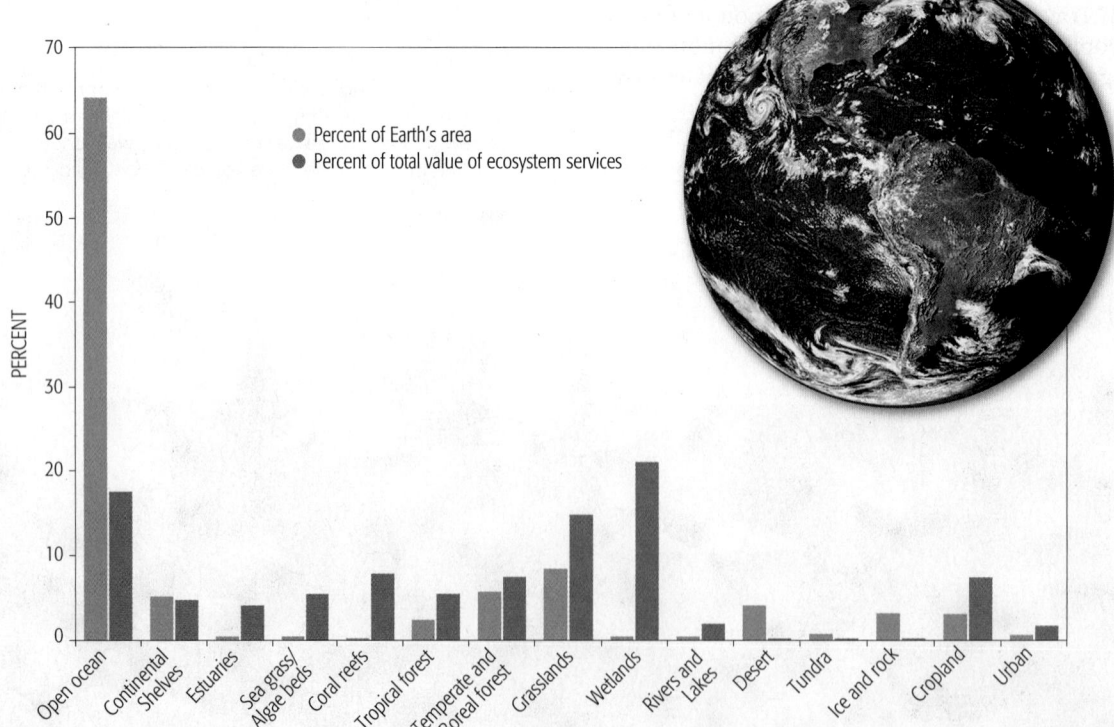

● Percent of Earth's area
● Percent of total value of ecosystem services

PERCENT

BIOME

Open ocean · Continental Shelves · Estuaries · Sea grass/Algae beds · Coral reefs · Tropical forest · Temperate and Boreal forest · Grasslands · Wetlands · Rivers and Lakes · Desert · Tundra · Ice and rock · Cropland · Urban

Calculating Ecological Value

What is the value of a diverse tropical forest to a coffee farmer?

Taylor Ricketts is Professor and Director of the Gund Institute for Ecological Economics at the University of Vermont. Much of his research has focused on the economic value of wild places that are not directly exploited by humans (Figure 2.20). Ricketts has been especially interested in how economic forces can be harnessed to help conserve biodiversity. He could see that natural ecosystems are immensely beneficial to society. But because those benefits are not easily quantified, they are often unappreciated.

More specifically, Ricketts asked the question, What is the economic value of intact tropical forests to coffee plantations? He investigated coffee because it is one of the world's most valuable commodities and it is most often grown in regions where diverse tropical forests are threatened by deforestation.

To produce beans, coffee flowers must be pollinated by insects. Ricketts *hypothesized* that wild bees that live in nearby tropical forests provide this important ecosystem service (Figure 2.21). If those forests were removed, he *predicted*, the number of wild bees visiting coffee flowers would decline, as would the production of coffee beans. The economic value of the pollination provided by the forest could then be calculated by the value of the lost coffee production.

At a very large coffee plantation in Costa Rica, Ricketts established study sites at varying distances from patches of native forest. At study sites within 1 km (0.66 mile) of the forests, he found large numbers of many different bee species visiting the coffee flowers. The number of bees visiting coffee flowers decreased significantly with increasing

▲ Figure 2.20 **Assessing the Economic Value of Pollination** Dr. Taylor Ricketts sampling ripe coffee beans during his Costa Rican studies.

distance from forest. On average, pollination decreased by about 21% at sites more than 1 km from intact forest ecosystems. This difference translated to a monetary loss of over $125 per hectare (over $50 per acre). Ricketts concluded that if the forests near this plantation were cut down, the loss of pollinators would cause economic losses of more than $60,000 each year.

Ricketts found that each hectare of forest was providing pollinating services to the coffee grower worth $385 each year. Costa Rica has a unique Environmental Service Payments Program through which the government provides $42 per hectare annually to landowners who conserve their forests. This compensation, Ricketts noted, is only 11% of the value of a single ecosystem service, pollination.

Source: Ricketts, T.H., G.C. Daily, P.R. Ehrlich, and C.D. Michener. 2004. Economic value of tropical forest to coffee production. *Proceedings of the National Academy of Sciences* 101:12579–12582.

1. What scientific methodology did Ricketts use to establish the relationship between bee visits and the distance of coffee plants from forest?

2. What could happen to the price of a cup of coffee if coffee growers chose to cut the forests surrounding their fields?

3. What implications might this study have for the decline in honeybee populations in North America?

◄ Figure 2.21 **A Key Ecosystem Service** A wild honey bee pollinating coffee flowers.

QUESTIONS 2.4

1. Describe three approaches to estimating the economic value of the ecosystem services provided by conserving a park.

2. How does ecological valuation differ from traditional economic valuation of an ecosystem service?

3. Why is gross domestic product often an unreliable estimate of a nation's wealth?

(MES) For additional review, go to **MasteringEnvironmentalScience**

Measuring the Wealth of Nations

■ Measures of national wealth may externalize the costs of degraded ecosystems and their capacity to generate future wealth.

Economists often estimate the wealth of a nation by its per capita **gross domestic product (GDP)**. GDP is the total value of goods and services produced by the citizens of a country divided by its population size. GDP is a measure of a country's economic standing. Governments and international organizations also use GDP as a basis for assigning financial aid and making loans to nations.

A number of economists have noted that certain actions increase a country's GDP but reduce its human and natural resources, thereby reducing the country's capacity to generate future wealth. In the short term, overharvesting forests or fisheries and mining practices that pollute the ground generate income and add to GDP. But in the long term, these practices diminish the wealth of the country (Figure 2.22). This situation is similar to a shopkeeper who sells his entire inventory and does not restock his shelves. The shopkeeper will make money in the short term, but his capacity to do so in the future is greatly diminished.

To account for the long-term effects of actions, some economists have suggested an alternative measure of national wealth—the **genuine progress indicator (GPI)**. GPI is the GDP plus or minus the economic value of enhancements or degradations to the environment. Most economists agree with the concept of GPI, but there is considerable disagreement on the appropriate methods for measuring it. The World Bank, which provides loans and grants to many of the world's poorest nations, has recently adopted the GPI concept to measure the sustainability of a country's human and environmental resources. Actions such as the destruction of forests and the neglect of waste management systems diminish a country's GPI. Reforestation and investment in health care improve sustainability and increase a country's GPI.

In summary, although many ecosystem services are not typically bought or sold in the marketplace, they do have economic value. That value can be assessed by measuring people's willingness to pay for those services or by calculating the costs that are incurred when the services are impaired. A complete accounting for the wealth of a country considers both the positive and negative impacts of its actions on its human and natural resources. Because GPI includes social and ecological values as well as economic wealth, this approach comes closer to capturing the sustainability triple bottom line principle.

◄ Figure 2.22 **Increasing GDP, Diminishing GPI**

This logging operation in Indonesia increases the country's GDP. However, because it was done in an abusive fashion that reduces future productivity, it diminishes its GPI, or capacity to generate wealth in the future.

2.5 Environmental Policy: Deciding and Acting

BIG IDEA Policies are principles that guide governments and other institutions in setting goals, making decisions, and carrying out activities. **Environmental policies** guide decisions and actions that influence environmental conditions or processes. Policy goals may be achieved through regulation, incentives, partnerships, and volunteerism. The issues that influence environmental policies typically fall into one of eight categories: (1) government versus individual control, (2) competing public values, (3) uncertainty and action, (4) which level of government decides, (5) which government agency has jurisdiction, (6) protection against selfish actions, (7) the best means to an end, and (8) political power relationships.

The Policy Cycle

■ Environmental policy evolves in a cycle and guides several kinds of actions.

Governments typically develop and revise policies through a series of steps called the **policy cycle** (Figure 2.23). The cycle begins with the definition of the problem to be solved. The next step is setting the agenda, the process of determining who will deal with a particular issue and when.

Policy is then formulated through a process of public hearings, debates, and lobbying, in which interested parties present their positions on the issues. Policy formulation culminates in decision making, in which a law may be passed or a management rule developed.

Next, individuals or agencies implement the policy, or carry out the new activities and enforce the laws and rules. The implementation process often requires additional interpretation as the general directions in a law or rule are adjusted to fit the unique situations in the real world. Finally, the laws and the policies that produced them are assessed in a complex political process that includes the news media and the public, as well as elected officials (Figure 2.24). As a consequence of such assessments, policies may be revised or terminated.

Environmental policies may use different methods for guiding actions. **Regulatory mandates** set legal standards for actions. For example, in 1975 the U.S. Congress enacted Corporate Average Fuel Economy (CAFE) regulations, which required automobile companies to meet standards for fuel economy set by the National Highway Traffic Safety Administration and verified by the U.S. Environmental Protection Agency. Other policies are implemented by means of **incentives**, which encourage action by offering something appealing. The Energy Policy Act of 2005, for example, offered consumers tax credits for buying fuel-efficient hybrid vehicles and solar-energy systems. Then, in 2008, it was extended to include small wind-energy systems and geothermal heat pumps. People were not required to do this, but the tax credit was an important incentive to many.

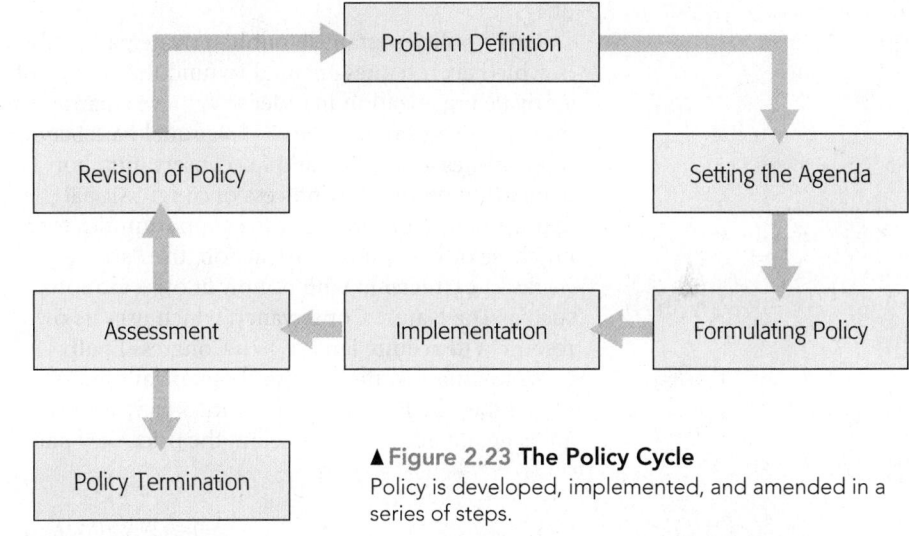

▲ Figure 2.23 **The Policy Cycle**
Policy is developed, implemented, and amended in a series of steps.

▲ Figure 2.24 **Part of the Policy Process**
Over 300,000 people marched through the streets of Manhattan ahead of the U.N. Climate Summit in September 2014. The People's Climate March was organized to persuade delegates to take action to curb global warming.

Market-based policies use economic markets to guide actions. Such policies usually use a combination of regulatory mandates and incentives in the form of credits or rights that are bought and sold. For example, **cap and trade** is a market-based policy in which a regulatory limit, or cap, is set on an action; rights to exceed that limit are traded in markets. The 1990 U.S. Clean Air Act Amendment has been a very successful application of cap and trade. This legislation sets a limit on the total amount of sulfur that can be emitted by all the electrical utilities in the country. From this, performance standards are calculated for each facility. Power plants that emit less sulfur than the standard earn sulfur credits, which can be purchased by power plants whose emissions exceed the standards. These sulfur credits are bought and sold in public markets such as the Chicago Board of Trade.

Other policies establish public-private partnerships, in which tax revenues are used to fund the actions of a private organization in order to achieve a particular outcome. For example, the U.S. National Park Service often wishes to acquire land for conservation but depends upon the slow process of congressional appropriation for funds. Because opportunities for land purchase often require quick action, the Park Service works in partnership with nonprofit organizations such as The Nature Conservancy, which uses its own resources to acquire land. When Congress finally provides funding, the nonprofit organizations resell the land to the Park Service. Private organizations are generally required to resell at the price they paid, but they receive interest on their investment and a commission fee.

Some policies promote **volunteerism**, work performed freely on behalf of a community (Figure 2.25). For example, the Smokey Bear and Woodsy Owl campaigns were developed by the U.S. Forest Service to encourage voluntary behavior that would control destructive wildfires and diminish littering.

▲ Figure 2.25 **Volunteerism**
Volunteers clean up trash along the banks of a stream flowing through their community. A variety of public policies work to encourage such activities.

 Where **What important environmental issues are facing you?**
Do you know how important environmental policies are shaped in your community or on your campus? Select a particular issue recently decided by your county's government, and describe the policy process that led to that decision. Which among Richard Andrews' eight policy issues have played a role in that policy process? Explain.

Policy Decision Framework

■ Environmental policy debates often center on one of eight issues.

A wide variety of issues can affect environmental policy. However, University of North Carolina political scientist Richard Andrews suggests that debates over environmental policy usually fall into one of the following eight categories.

1. Government versus individual control. In the United States, we generally assume that governments should set and enforce standards for environmental hazards that affect public health and well-being, such as chemical pollutants. Decisions related to transportation, energy use, or waste disposal are frequently left to individuals. However, the amount of control given to individuals varies from region to region. For example, some counties and cities leave decisions about recycling metal cans, bottles, and newspapers to the discretion of individuals. Other communities provide recycling services but do not require citizens to use them. Still other communities impose fines on those who violate recycling laws. Who should make these decisions—government or individuals?

2. Competing public values. Decisions about environmental policy often involve trade-offs with other public benefits, most notably economic growth. For example, policies for the management of U.S. National Forests are often influenced by competing values, such as the protection of water and endangered species versus the economic benefits of the jobs provided by the extraction of timber. At an international level, those developing policies to limit the emissions of greenhouse gases must also address concerns about the economic consequences of such policies. Whose values should have priority?

3. Uncertainty and action. In 2000, the European Union endorsed what has come to be known as the **precautionary principle**: When there is reasonable evidence that an action or policy may place human health or the environment at risk, precautionary measures should be taken, even if that evidence is uncertain or inconclusive. In theory, the precautionary principle seems very logical. In practice, policymakers must agree on what constitutes reasonable risk. They must also agree on the trade-offs between different kinds of risk. For example, a policy banning the use of pesticides may lower the risk of certain adverse health effects but increase the risk of lowered crop production. How much do we need to know before we act?

4. Which level of government decides? Many different levels of government are involved in regulating actions related to the environment. In the United States, for example, a federal agency—the U.S. Environmental Protection Agency—sets regulatory standards for air quality. Should individual states have the right to set and enforce their own standards for air quality, so long as they are at least as stringent as federal guidelines? States such as California and New York argue yes, they should be allowed to tailor standards to meet unique situations, such as air quality in major cities. Concerned with having to adjust to a wide range of different requirements, manufacturers and utilities say no. Which level of government should make environmental decisions?

5. Which government agency has jurisdiction? Many different governmental departments and agencies have responsibility for functions that affect the environment. For example, responsibility for controlling the flow of water in rivers resides with both the Bureau of Reclamation in the Department of the Interior and the Army Corps of Engineers in the Department of Defense. If hydroelectric

power generation is involved, the Federal Energy Regulatory Commission of the Department of Energy may also share some responsibility. Shared responsibility may serve the public good if it ensures that competing values are represented in decisions over the management of resources. However, when the responsibilities of agencies overlap or are poorly defined, the result can be confusion, gridlock, or finger pointing. How should decision making be assigned to governmental departments and agencies?

6. Protection against selfish actions. Scientist-philosopher Garrett Hardin argued that conflict between individual interests and the common good is inevitable when people compete for shared, limited resources, such as the fish in a public lake or grazing rights in a public pasture. If individual interests prevail, the common resources are likely to be overexploited. He called this form of overexploitation the **tragedy of the commons** (Figure 2.26). How can environmental policy be formulated to avert tragedies of the commons?

One strategy for avoiding overexploitation of common resources is to assign private ownership to open-access resources. For example, livestock herders may be assigned a specific number of animals that they may graze on a common pasture. They may use their entire quota or sell a portion to others. Alternatively, common resources may be treated as the property of the government and their use regulated by law. Access to our national parks is such an example.

▶ Figure 2.26 **Tragedy of the Commons**

The livestock on this degraded pasture in the African country of Chad are owned by several different families. Because the land is shared with no restrictions, each family has the incentive to put as many animals as it can on the land.

QUESTIONS 2.5

1. Describe the steps in the policy cycle.

2. In what ways might policies guide human actions?

3. List four questions that are important to policy formulation.

(MES) For additional review, go to **MasteringEnvironmentalScience**

Other common resources are more difficult to protect because users are too numerous, geographically separate, or hold conflicting values. The clean water in a river is a common resource. A city or farmer may dump wastes into the river, but the negative effects of their actions may not be realized until far downstream.

Although most people would agree that common resources should be protected, people rarely want to take a large share of the responsibility for such actions. For example, most of us would agree on the need for facilities to manage sewage or municipal waste, but few of us wish to see such operations located near our homes. This "not in my backyard" or NIMBY syndrome has a major influence on a variety of environmental policies (Figure 2.27). What would you do?

7. The best means to an end. In many policy debates, people agree on the desired outcome but disagree about the appropriate strategies for achieving it. Should regulation be enforced with financial or civil penalties? Should incentives take the form of subsidies, tax benefits, or cap-and-trade markets? For example, advocates for policies aimed at reducing emissions of carbon dioxide to Earth's atmosphere disagree on the best way to achieve that end. Some argue that a tax on carbon-based fossil fuels is the best way to reduce emissions. Others argue for setting an overall limit on carbon emissions and allowing individuals and companies to buy and sell the right to emit carbon above that limit. Yet others propose that any measures to reduce carbon dioxide emissions should be voluntary. What kinds of policies are best suited to achieve particular environmental outcomes?

8. Political power relationships. Many policy challenges arise as a consequence of complicated decision-making processes. Collective decisions in a democracy often require compromises, which may produce less than ideal outcomes. The role of power relationships in the U.S. Congress, such as which party holds a majority, is well known. Power relationships can also arise because of the geographical distribution of resources. For example, fossil fuels such as coal and oil are not distributed uniformly among states and countries. This distribution influences the policy positions of these different entities; it is also a major factor in determining the power of individual states or nations to push for their own policy positions. How can power relationships among geographic regions and political constituencies be overcome to develop sound policies?

In summary, the policy cycle generates policies and laws that influence behavior toward the environment by means of regulation, incentives, markets, and volunteerism. Questions related to the role of individuals, the handling of uncertainty, and the nature of governance influence the specific form that environmental policies take.

▼ Figure 2.27 **Not in My Backyard**
We may all agree on the need for sewage treatment plants, but most of us would protest if one was to be located near our home.

2.6 U.S. Environmental Law and Policy

BIG IDEA In the United States, environmental policies and laws are influenced by actions of the different branches of federal, state, and local governments. Provisions in the U.S. Constitution set a framework for the development of many policies, laws, and court decisions. U.S. environmental policy and law have changed in response to changing human values.

Governmental Functions

■ The three branches of the federal government and state and local governments play unique roles in setting environmental policy.

Through the decade following the Declaration of Independence in 1776, the power to legislate, or make laws, resided with each of the 13 states. This included laws related to commerce, land ownership, and individual rights. Each state also had the power to enforce and interpret its own laws. Fearing a return to centralized rule and not wishing to give up their individual powers, the states granted very few powers to the Continental Congress. As a result, there was considerable state-to-state variation in laws, as well as in the interpretation and enforcement of their laws.

In 1787, the U.S. Constitution was adopted by the Constitutional Convention and became the supreme law of the land. The Constitution vested primary legislative, executive, and judicial authority in the federal government and thereby greatly increased the power of the federal government relative to the states. However, the organization of the federal government ensures checks and balances among the activities of the three branches of government (**Figure 2.28**).

All legislative functions are assigned to the Congress, consisting of the House of Representatives and the Senate. As a check on the power of the Congress, legislation is either approved or vetoed by the president. The Congress may override a presidential veto by a two-thirds majority of both houses.

The implementation and enforcement of legislation is the responsibility of the executive branch, which is headed by the president and includes the cabinet and many departments and agencies. The enforcement of particular legislation is generally assigned to one or a few agencies. Agencies then set **regulations**, which are specific rules that establish standards for performance, programs to ensure compliance, and protocols for enforcement.

Wisely, the framers of the Constitution perceived that social change and new knowledge and technology would necessitate continued re-interpretation of law. This interpretive function is vested in the judicial branch, which includes the Supreme Court and lower courts. The various decisions made by the individual courts collectively make up **case law** and establish precedents that influence future court decisions.

A closer look at the law that governs water quality demonstrates how the three branches of the government

work together to create, implement, and interpret environmental policies. In 1972, the Congress passed the Federal Water Pollution Control Act, or Clean Water Act, and President Richard Nixon signed it into law. The act describes broad goals for water quality and assigns the responsibility for setting, monitoring, and enforcing standards of water quality to the Army Corps of Engineers and the U.S. Environmental Protection Agency. These responsibilities are carried out in accordance with hundreds of pages of detailed administrative rules and regulations. In the years since the passage of the Clean Water Act, the courts have ruled on specific matters of interpretation, such as whether activities affecting isolated ponds and bogs are actually covered

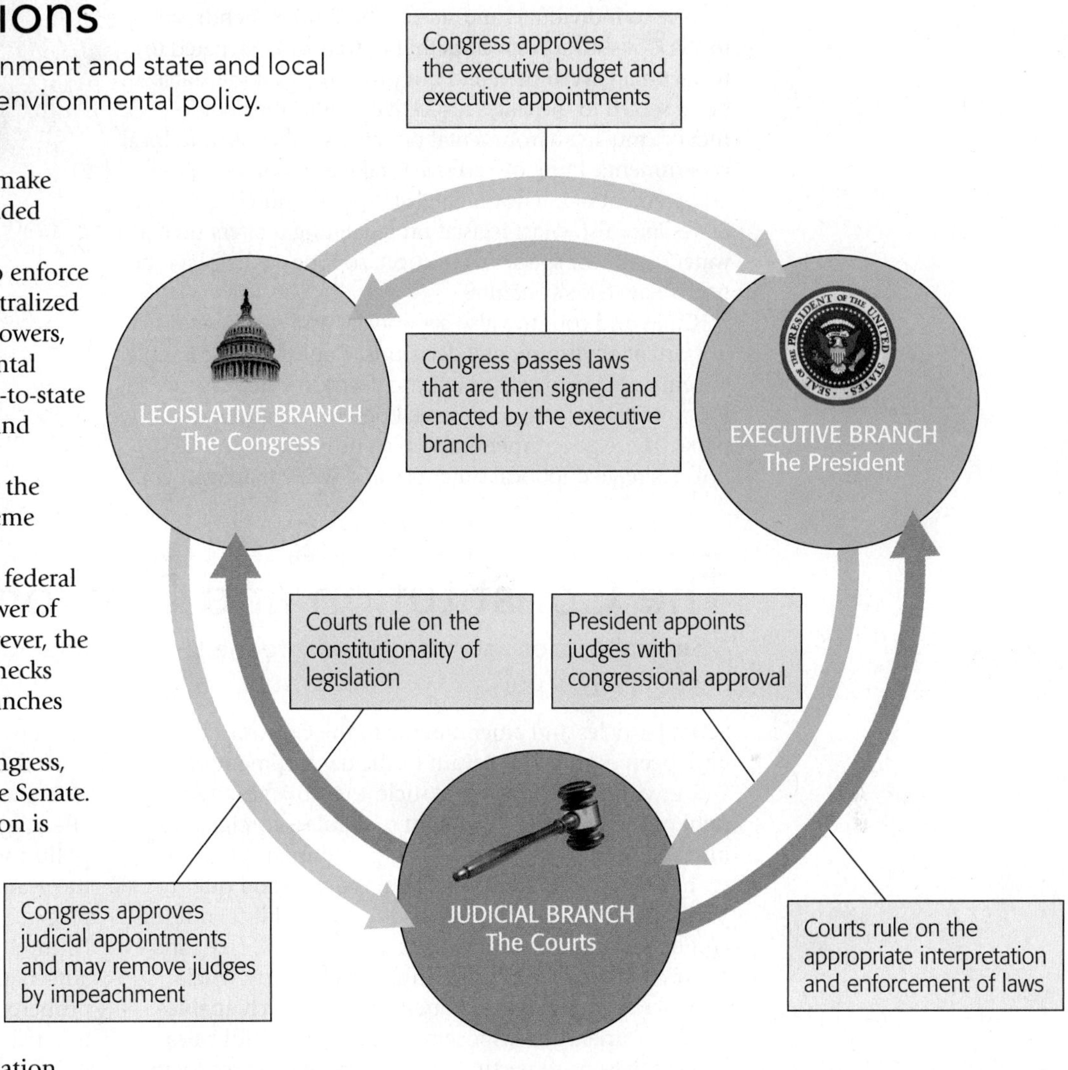

▲ Figure 2.28 **Branches of Government**
The U.S. government is organized to provide checks and balances on the processes of law making, law enforcement, and law interpretation.

Congress approves the executive budget and executive appointments

Congress passes laws that are then signed and enacted by the executive branch

LEGISLATIVE BRANCH The Congress

EXECUTIVE BRANCH The President

Courts rule on the constitutionality of legislation

President appoints judges with congressional approval

Congress approves judicial appointments and may remove judges by impeachment

JUDICIAL BRANCH The Courts

Courts rule on the appropriate interpretation and enforcement of laws

53

by the act (Figure 2.29). In some circumstances, the judicial branch decides where and how laws should be applied. In 2007, for example, the Supreme Court ruled that carbon dioxide (CO_2) is a pollutant and falls within the jurisdiction of the Clean Air Act. They further ruled that the U.S. Environmental Protection Agency has the responsibility to regulate CO_2 emissions from automobiles and industrial sources.

The first 10 amendments to the Constitution, called the Bill of Rights, delineate the power of the government relative to individuals and states. The Tenth Amendment to the Constitution specifies that powers not delegated to the central government and not prohibited for the states be "reserved to the states respectively, or to the people." Indeed, much environmental policy is set by state and local governments. In many cases, federal legislation delegates enforcement of environmental statutes to individual states. States may also enact legislation regulating matters such as water quality or species protection, so long as their laws do not violate federal statutes.

Cities and counties also set policies and make important environmental decisions. Counties, for example, are empowered to make land use decisions, such as the location of landfills, commercial developments, and public parks. These governments also pass ordinances and set policies related to local water use and waste management.

▲ Figure 2.29 **Legislation and Interpretation**
The 1972 Clean Water Act regulates activities affecting the quality of the nation's water. However, the U.S. Supreme Court has interpreted this legislation to apply only to wetlands that drain into rivers and large lakes. It ruled that the act does not apply to actions affecting isolated wetlands such as those often found in the midst of agricultural fields, sometimes referred to as prairie potholes.

The Constitution and Environmental Policy

■ Several articles and amendments to the U.S. Constitution significantly influence environmental policy.

Several articles and amendments to the Constitution have been especially relevant to the development of U.S. environmental policy. Article 1 establishes the right of the federal government to regulate foreign and interstate commerce. The federal government's right to set rules regarding transportation, the flow and quality of navigable waters, commerce affecting wildlife, mining, and transport of natural resources such as oil, coal, and electricity derive from the commerce clause of Article 1.

Article 4 includes the property clause, which enables the government "to dispose of and make needful rules and regulations respecting the territory or other property of the United States." This article is the basis for laws and regulations governing public lands, including the establishment and management of national parks, national forests, and national wildlife refuges. It also is the basis for legislation that allows the government to protect its property by setting restrictions on nearby landowners.

The Fifth Amendment of the Constitution has many provisions that influence environmental policy. Most importantly, it allows the government to take property for public use, but only if it provides just compensation. It is clear that the government must pay a landowner fair market value for land that it acquires for a road, military facility, or national park. However, if the government establishes a regulation that restricts the use of private property, thereby diminishing the value of the property, must the government compensate the property owner? This question is central to environmental policies affecting matters such as land use, regulation of pollution, and protection of endangered species. Over the years, the courts have ruled that if a legitimate public purpose is served, the government need not pay for every "diminution of value." In other words, owners are not automatically entitled to the highest value use of their property. However, recent court decisions have affirmed that the government must provide compensation if a regulation removes most or all of the value of a property.

The Fourteenth Amendment, added to the Constitution soon after the Civil War, requires that the states provide legal due process and "equal protection" to all citizens. Some see this amendment as a constitutional basis for environmental justice. They argue that this amendment requires that environmental services, such as clear air and water, must be provided equally to all citizens, regardless of race or economic status. The courts have supported this interpretation in some situations but not in others. For example, case law requires consideration of equal protection issues if a new waste facility is located in an established community. Such considerations are not required if people knowingly choose to live near an established waste facility.

QUESTIONS 2.6

1. Describe the role of the three branches of government in setting environmental policy.

2. Describe four parts of the U.S. Constitution that influence environmental policy.

(MES) For additional review, go to **MasteringEnvironmentalScience**

2.7 International Environmental Law and Policy

BIG IDEA International environmental policies and laws are established by long-standing norms of behavior, treaties that are ratified by national governments, and decisions and precedents set in international courts. The United Nations, regional consortia, international financial organizations, and nongovernmental organizations all play significant roles in the development of international environmental law and policy. Over the past four decades, global policies and treaties have been developed to restore and conserve the environment while promoting sustainable economic development in poor countries.

Environmental Laws

■ Environmental law derives from established custom, formal treaties, and judicial decisions.

Sovereignty is the concept that a country may behave as it pleases within its borders as long as it does not violate international laws to which it has agreed. These international laws may be in one of three forms: customary, conventional, or judicial.

Customary international laws are accepted norms of behavior or rules that countries follow as a matter of long-standing precedent. The vast majority of the world's governments, including the United States, accept customary international law. For example, it is customary in international law that countries may not use their territory in such a way as to injure the territory of another country and that countries have a duty to warn other countries about environmental emergencies. Therefore, following the major earthquake and tsunami in 2011, Japan was obligated to notify other countries of serious failures at and releases of radioactive materials from its Fukushima Daiichi nuclear facility. The exact time when a principle of action becomes customary law is not clear cut and is a matter of frequent disagreement among countries.

Conventional international laws are established by formal, legally binding conventions or treaties among countries. In 1940, for example, the United States and 17 other countries in the Americas ratified the Convention on Nature Protection and Wildlife Preservation in the Western Hemisphere. By this treaty, these nations agreed to protect migratory birds and to take actions to prevent the extinction of any migratory species.

Judicial international law sets standards for the actions of countries based on the decisions of international courts and tribunals. The International Court of Justice in The Hague, Netherlands, is the most famous of these courts (Figure 2.30). Regional tribunals, such as the European Court of Justice, have been established by treaty among cooperating nations. These courts often have limited authority on environmental issues, but their opinions carry

considerable weight in the interpretation of customary and conventional law. Because so many environmental issues are influenced by global trade, the Dispute Settlement Board of the World Trade Organization (WTO) is also an important source of judicial environmental law.

For several reasons, the enforcement of international environmental law is somewhat limited. First, it is often difficult to verify whether a particular country is complying with the terms of a treaty. Second, countries may enter into an agreement in good faith but lack the capacity to enforce its terms. Third, the international community has few options for punishing countries that violate a treaty.

▼ **Figure 2.30 International Justice**
Environmental disputes among countries are resolved in the International Court of Justice in Peace Palace in The Hague, Netherlands.

Q *Why are international law treaties not fully enforced (or more stringent) in order to protect our air quality and climate?*

Odalys Solares, Florida International University

Norm: **A** International treaties depend not only on the willingness of governments to participate but also on their capacity to fund and enforce treaty agreements. Countries may fail on one or all of these points.

QUESTIONS 2.7

1. What are the three kinds of international laws and how do they come about?

2. What are the three governance bodies of the United Nations? What role does each play in the development of international environmental policy and law?

3. What roles do nongovernmental organizations play in the development and enforcement of environmental laws?

(MES) For additional review, go to MasteringEnvironmentalScience

International Institutions

■ Environmental policies and laws are influenced by a variety of international institutions.

The United Nations. The most important organization shaping international environmental policy and law is the United Nations (Figure 2.31). The 54 member nations of the United Nations Economic and Social Council are charged with promoting international economic and social cooperation and development. As part of its charge, the Economic and Social Council oversees two agencies that play a significant role in international policy. The United Nations Development Programme (UNDP) provides countries with assistance to encourage environmentally sustainable development. The United Nations Environment Programme (UNEP) coordinates environmental programs for the United Nations and has authority over regional and global environmental issues.

Regional Consortia. Countries within different regions have formed organizations to encourage communication and cooperation on matters of particular interest. For example, the Organization of American States represents 35 countries in the Western Hemisphere. Its Department of Sustainable Development coordinates policies and programs aimed at alleviating poverty and promoting economic development in its member nations.

The European Union (EU), created to promote European unity and its influence in the world, is one of the world's strongest regional organizations. It can sign treaties on behalf of its 27 member countries; it also regulates their behavior in many realms, including the environment. The EU European Environment Agency sets standards in a number of areas, including pollution, habitat degradation, fisheries, and waste management. Because these standards apply to many items imported to EU countries, the EU exercises considerable influence on environmental management outside the region.

International Financial Institutions. **Multinational development banks** are institutions that provide financial and technical assistance to countries for economic, social, and environmental development. The largest of these banks is the World Bank, which provides long-term loans and grants to the poorest countries in the world. Regional banks such as the African Development Bank and the Inter-American Development Bank provide similar support. These banks are funded by donations from wealthier countries. In the past, development banks frequently funded projects, such as water diversions and community resettlement, that had positive effects on human well-being but negative environmental consequences. Over the past two decades, development banks have worked hard to incorporate principles of environmental sustainability into their programs.

The WTO was created in 1995 to promote free trade by reducing obstacles to international commerce and enforcing fairness in trading practices. The objectives of sustainable development and environmental protection are included in the introduction to the agreement that created the WTO, and its policies and programs have a significant influence on environmental behavior. WTO agreements and judgments have confirmed the right of governments to protect the environment. However, the globalization of markets facilitated by the WTO may also encourage the unsustainable use of resources in countries with weak environmental laws or countries without the human and financial resources to enforce them. For example, the global market for tropical wood products has encouraged extensive, unsustainable logging in Indonesia. Although Indonesia has laws prohibiting such cutting, they are poorly enforced.

Nongovernmental Organizations. Nongovernmental organizations (NGOs), legally constituted organizations in which there is no governmental participation or representation, have a significant influence on international environmental policy. Some NGOs, such as Greenpeace, advocate for national laws and international treaties to protect the environment. Other NGOs, such as the World Wildlife Fund, work with governments to develop the capacity to implement and enforce laws and treaties (Figure 2.32).

▼ Figure 2.31 **The United Nations**
This is the UN General Assembly Hall in which the member countries consider all matters influencing international peace and security.

◄ Figure 2.32
Nongovernmental Environmental Organizations
The United Nations Environmental Liaison Center International coordinates activities of international NGOs. Since its creation in 1974, the number of member NGOs in the center has grown significantly.

(Bar graph: NUMBER OF NGOs vs YEAR. 1974 ≈ 125, 1993 ≈ 710, 2013 ≈ 850.)

The Global Environmental Facility

Can a development bank promote both economic development and environmental sustainability?

In the early 1980s, sustainable economic development among the world's poorest nations was a high priority for the World Commission on Environment and Development. It envisioned a new era of economic growth based on policies that enhance natural resources and protect humans and ecosystems from pollution. To realize this vision, it argued for greater cooperation among developing and developed countries with regard to many issues, including population growth, alleviation of poverty, energy management, and protection of species and ecosystems.

Following the commission's recommendations, the United Nations established the Global Environmental Facility (GEF) at the 1992 Earth Summit in Rio de Janeiro, Brazil. The GEF is a partnership among 183 countries, development banks, NGOs, and multinational corporations. The GEF uses grants and loans from member countries and development banks to complement funds from private companies. International NGOs often play a significant role in the management of GEF projects.

Since its creation, the GEF has provided over $11 billion in grants and leveraged over $50 billion of additional support from donor countries for 2,700 projects distributed around the world. These projects have been designed to produce environmental benefits, such as the reduction of pollution, the management of water resources, and the support of biodiversity. In addition to producing environmental benefits, GEF projects must encourage sustainable economic development.

The GEF has provided nearly $1 billion to Mexico and the countries of Central America for the development of the Mesoamerican Biological Corridor (MBC). The MBC is not a single place. Rather, it is a group of programs that manage about 600 protected areas and the land surrounding them (Figure 2.33). The MBC unites the goals of species and ecosystem conservation with initiatives for sustainable development that benefit local people (Figure 2.34). Many MBC projects focus on individual countries, but a number of its grants have supported the development of preserves that span international boundaries. Nearly all of these grants have involved cooperation among governments, NGOs, and local communities.

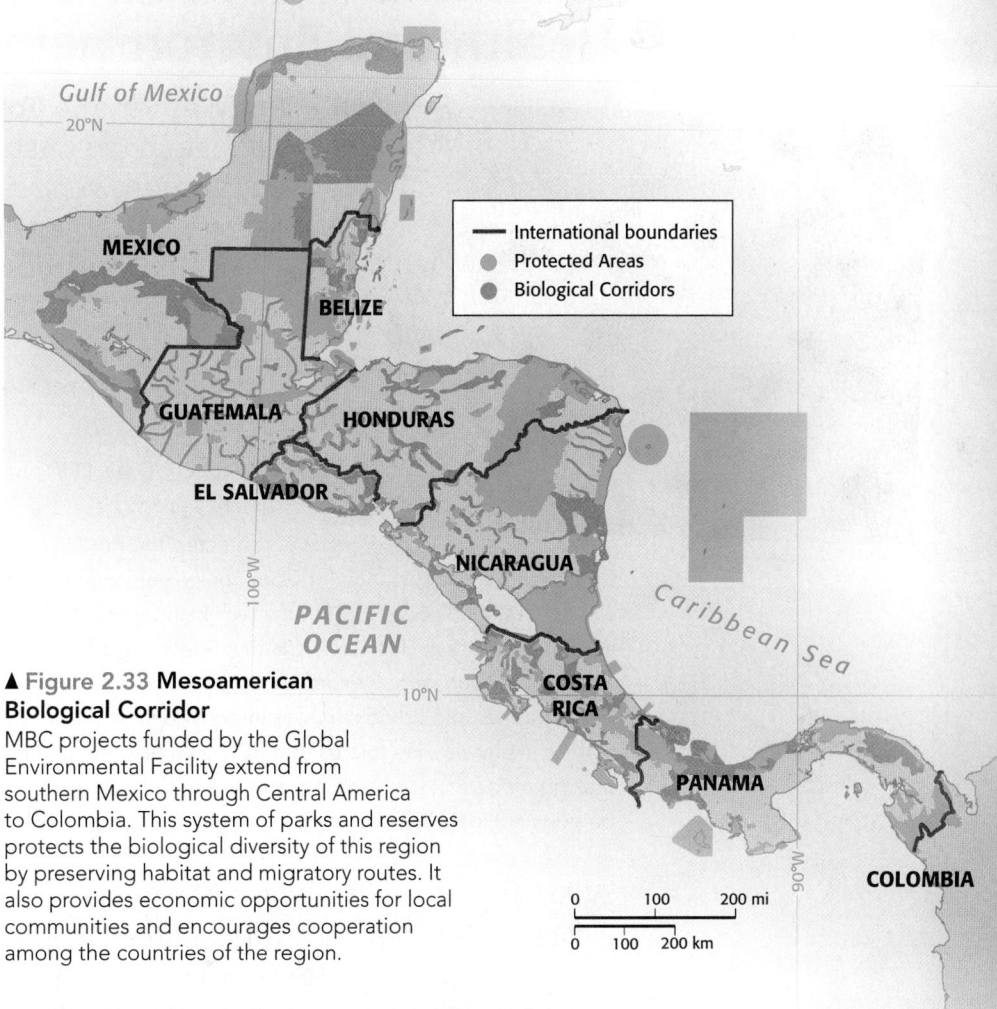

▲ Figure 2.33 **Mesoamerican Biological Corridor**
MBC projects funded by the Global Environmental Facility extend from southern Mexico through Central America to Colombia. This system of parks and reserves protects the biological diversity of this region by preserving habitat and migratory routes. It also provides economic opportunities for local communities and encourages cooperation among the countries of the region.

The MBC is a key reason that nearly 20% of this region is now protected for conservation, helping to save the rich biological diversity of the area. The MBC has also bolstered tourism to every Central American country and improved the economic well-being of their citizens. Just as important, it has encouraged cooperation among countries in a region where, just 20 years ago, border disputes and civil wars were common.

▶ Figure 2.34 **Environmental and Economic Sustainability**
Local naturalists guide ecotourists at the research center in Tortuguero National Park, Costa Rica. The GEF has provided funding for parks such as this and for naturalist training programs.

57

The Making of *Swine Country*: A Film about Public Health and Environmental Justice

Sol Weiner and **Tom Clement** are students at Guildford College in Greensboro, North Carolina. During their sophomore year, in partnership with a rural empowerment association, they created and filmed *Swine Country*, a documentary about the issues surrounding hog farming in Duplin County, North Carolina.

How did you first get the idea for Swine Country?

Sol: In Fall 2012, I contacted the Rural Empowerment Association for Community Help (REACH) with whom I had previously conducted an independent study. I told Devon Hall at REACH that I had a good friend (Tom) who was an experienced photographer and videographer and asked if he was interested in working with us in any way. He told us that REACH was interested in making a documentary about the negative health impacts of industrial livestock operations and it took off from there.

What are the health and environmental issues you witnessed in Duplin County?

Everybody we spoke with pointed to problems with the odor that residents experience from hog operations. The literature shows that odors from livestock waste can raise blood pressure, irritate mucous membrane, and cause stress and embarrassment. Contact with the waste itself can cause resistance to antibiotics, reductions in pulmonary functions, increased exposure to disease-causing pathogens, and increased incidence of asthma. That really is only the tip of the iceberg.

The majority of people who live within close proximity to industrial livestock operations are low-income people of color, living in a vicious cycle of pollution, housing discrimination, and reduced access to safe and equitable forms of economic development. Land values stay low and the cycle continues, a phenomenon that relates to both the environment and public health.

What steps did you take to develop Swine Country?

Sol: The first step was to establish a true collaboration with REACH and develop a mutually beneficial project. Then we forged relationships with people in the community and began interviewing and filming.

Tom: We searched for existing documentation in addition to the new footage we filmed. Historical footage and photographs of Hurricane Floyd, fish kills, etc. were essential to visually explain the magnitude of this problem. Some of the powerful narration came from the Pew Commission on Industrialized Farm Animal Production in 2007 where community members got a chance to explain the problem to industry and legislative representatives. The fact that we were students, working pro-bono and working for social justice made people far more generous with sharing media.

Sol: After film production, we had to cultivate interest on a local level; we showed the trailer at our undergraduate research symposium, held discussions about the issues, and contacted the local press about the premiere.

What are the biggest challenges you've faced with the project, and how did you solve them?

Sol: One challenge with this project was being a minority in an unfamiliar community. Our situation was unique in that we were white college students in an environment where researchers have traditionally exploited marginalized communities. I had to respect that this was not my community, and to follow the lead of the folks at REACH who knew each other and knew the most effective ways to get the work done. I think the best thing to do in these situations is to practice active listening! It's easy for those of us with white privilege to think that we have all the answers, but nothing could be further from the truth. You have to listen to peoples' experiences if you're going to learn anything and be a partner in making change.

Tom: The first couple meetings we attended people seemed very wary of Sol and me, but eventually, people began to trust us. I didn't pull out my camera until our first scheduled interview because it would have felt intrusive. One of the biggest challenges for me was trying to capture the most compelling, accurate story while respecting the privacy of the community members. One of the most

affirming moments of the project was after we showed our first draft to Devon. He beamed at me with astonishment.

How has your work affected other people and the environment?

Within our school community a lot of people have told us that they simply didn't realize pollution from industrial livestock operations was a problem, so just raising that awareness is one important step. Our collective hope is that this documentary will help REACH raise money and awareness. The goal was that this documentary would be told in an authentic voice, narrated mostly by the people who live and work in that environment.

We don't know what impact *Swine Country* will have on the environment, but because we view this work as an extension of what REACH is already doing, we would like to see their work actualized. That would mean stricter environmental regulations and enforcement, a safer

workplace for industrial livestock operation employees, and community involvement and input into what kinds of economic development take place in their communities. If our work makes REACH's work any easier or more effective, we can take pride in that.

What advice would you give to students who want to become involved in the environmental justice movement? How can they replicate your model?

One of the best ways to get involved is to seek out professors who have experience working with communities. Most schools have multicultural education departments and units that specialize in service learning, and both usually have connections in the community. The most important thing you can do is to start asking questions; even if the people you talk to may not have the answers, they may know who does.

Summary

Our attitudes toward the environment are influenced by our mode of living. People who get their provisions directly from the environment usually have a good understanding of their dependence on the natural world. In more urban communities, that dependence is less obvious and often not well understood. Human beliefs about whether actions that influence the environment are right or wrong may be based on motivating virtues, consequences of actions, or rules of behavior. Beliefs are also influenced by ethical decisions about who or what has intrinsic value.

Market forces of supply and demand determine the economic value of natural resources that are harvested or extracted from the environment, such as timber and coal. However, many ecosystem services that are necessary for our well-being, such as clean air and beautiful landscapes, are not bought and sold in markets, but we value them nonetheless. Environmental policies and laws influence human actions through a variety of means. These policies are developed, implemented, enforced, and interpreted by governments and other organizations at the community, national, and international levels.

2.1 Changing Views of Humans and Nature

- Human attitudes toward the environment have changed over time as a consequence of changes in our mode of living, culture, religious and political beliefs, and knowledge.

KEY TERMS

animism, preservationist, conservationist

QUESTIONS

1. How would you expect attitudes toward the environment to differ between hunter-gatherers in a remote wilderness and city dwellers in 19th-century New York?
2. Logging in national forests is very controversial. Describe how preservationist and conservationist views on this issue would likely differ.

2.2 Environmental Ethics

- Environmental ethics is a scholarly discipline that seeks to create frameworks for determining right and wrong human actions with respect to the environment.

KEY TERMS

environmental ethics, virtue ethics, consequence-based ethics, utilitarianism, duty-based ethics, intrinsic value, instrumental value, anthropocentric ethics, biocentric ethics, ecocentric ethics, deep ecology movement, ecofeminism, environmental justice

QUESTIONS

1. A community wishes to build a new elementary school to meet the needs of its growing population. The most convenient site for this new school happens to be one of the few remaining tracts of forest within the city. On what criteria would virtue ethicists, consequence-based ethicists, and duty-based ethicists decide if cutting this forest to locate the school was right?

2. In very dry years or when winters are very cold, the food supply for elk in the Yellowstone wilderness may be limited, and weak or old animals are likely to die. Some argue that park managers should provide supplemental feeding stations to prevent such deaths. What is your view on this matter, and on what ethical framework would you base it?

3. Give an example of an environmental justice issue in your community. How do you feel disputes over such matters should be resolved?

According to...	An act is ethical if...
Virtue Ethics	It is motivated by honesty, justice, or kindness.
Consequence-based Ethics	It results in satisfaction.
Duty-based Ethics	It follows a set of agreed-upon rules.

2.3 The Environment and the Marketplace

- The prices and production of goods and services are determined by their supply and the demand for them.
- Consumers' willingness to pay for a good or service is determined by the satisfaction or benefit that they feel it will bring them.

KEY TERMS

economic system, subsistence economy, market economy, free market economy, planned economy, commodity, profit, neoclassical economic theory, economic value, discount rate, opportunity cost, externality

QUESTIONS

1. The economic system of the United States includes aspects of subsistence, market, and planned economies. Using examples, explain this statement.
2. The cost for producing a particular good or service decreases as the amount produced increases. How is this likely to affect incentives for the producer?

3. Environmental impacts of the production and consumption of commodities are often economic externalities. Give an example of a way that economic systems can capture such external costs in the price of a commodity.

SUPPLY AND DEMAND

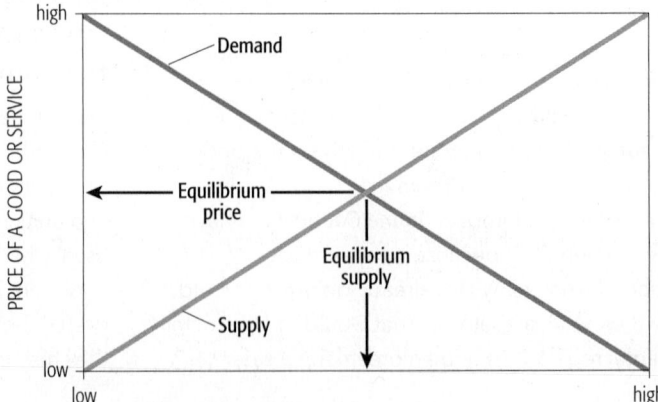

2.4 Valuing Ecosystems

■ The economic value of human actions that either impair or enhance ecosystem services such as clean water or aesthetic beauty can be determined by the difference between people's willingness to pay for the cost of such actions and the cost of alternative actions.

KEY TERMS

marginal value, travel-cost valuation, hedonic valuation, contingent valuation, natural capital, ecological valuation, gross domestic product (GDP), genuine progress indicator (GPI)

QUESTIONS

1. Contingent valuation of economic services is very controversial among economists. Describe two reasons for this controversy.
2. Some economists argue that ecological valuation of ecosystem services does not reflect their true value in an economic system. What is their reasoning?
3. Describe two examples of actions that countries might take that increase their GDP while diminishing their GPI. Give an example of actions that would likely increase both GDP and GPI.

2.5 Environmental Policy: Deciding and Acting

■ Environmental policies are plans that influence human decisions and actions relative to the environment and its natural resources. They do this through mandates, incentives, partnerships, and volunteerism.

KEY TERMS

environmental policy, policy cycle, regulatory mandate, incentive, market-based policy, cap and trade, volunteerism, precautionary principle, tragedy of the commons

QUESTIONS

1. How might variations in supply and demand influence the economic values in a market-based system to manage sulfur emissions from power plants?
2. What factors might influence a country's application of the precautionary principle to a particular environmental problem?
3. Describe two policy alternatives that could be used to reduce degradation of a resource that is shared in common within a community.

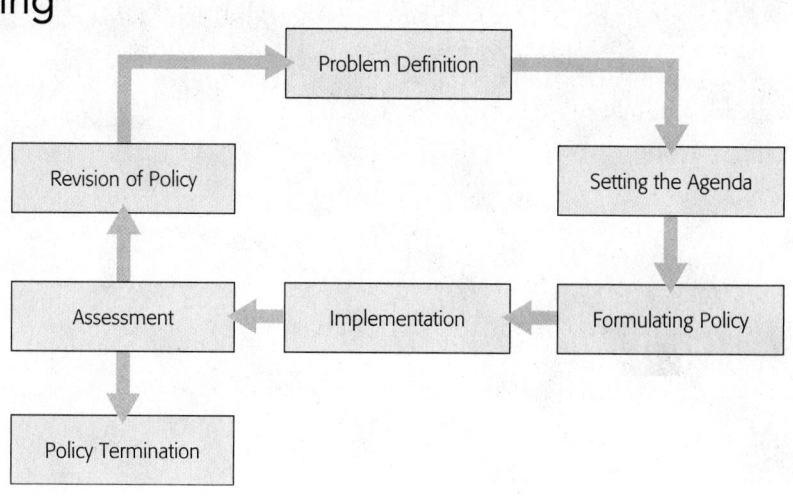

2.6 U.S. Environmental Law and Policy

■ Environmental policies are cast into law by the legislative branch of government, they are implemented and enforced by the executive branch, and they are interpreted through court decisions by the judicial branch.

■ The U.S. Constitution delegates many decision-making powers to governments at the state and local levels.

KEY TERMS

regulations, case law

QUESTIONS

1. In response to public concerns about an environmental issue, a law is passed. Describe the role of each of the branches of government in the development of that law and in its effects on public actions.
2. Government decisions and regulations often influence the economic value of private property. What part of the Constitution determines the government's responsibility to compensate property owners? Under what conditions do property owners have a right to compensation?

2.7 International Environmental Law and Policy

■ International environmental law assumes that each nation may act as it pleases within its boundaries so long as it does not violate international laws to which it has agreed. It is shaped by the United Nations, agreements among nations, and economic and nongovernmental organizations.

KEY TERMS

sovereignty, customary international laws, conventional international laws, judicial international laws, multinational development banks

QUESTIONS

1. Describe ways in which international financial institutions influence international environmental policies.

2. How have the attitudes of individual countries regarding environmental policies been influenced by a country's level of development or wealth compared to other countries?

Customary international laws	Based on long-established precedents and accepted norms of behavior.
Conventional international laws	Based on formal and legally binding treaties.
Judicial international laws	Based on the decisions of international courts.

MasteringEnvironmentalScience®

Students Go to **MasteringEnvironmentalScience** for assignments, the eText, and the Study Area with practice tests, activities, and more.

Instructors Go to **MasteringEnvironmentalScience** for automatically graded tutorials and questions that you can assign to your students, plus Instructor Resources.

The Physical Science
of the Environment

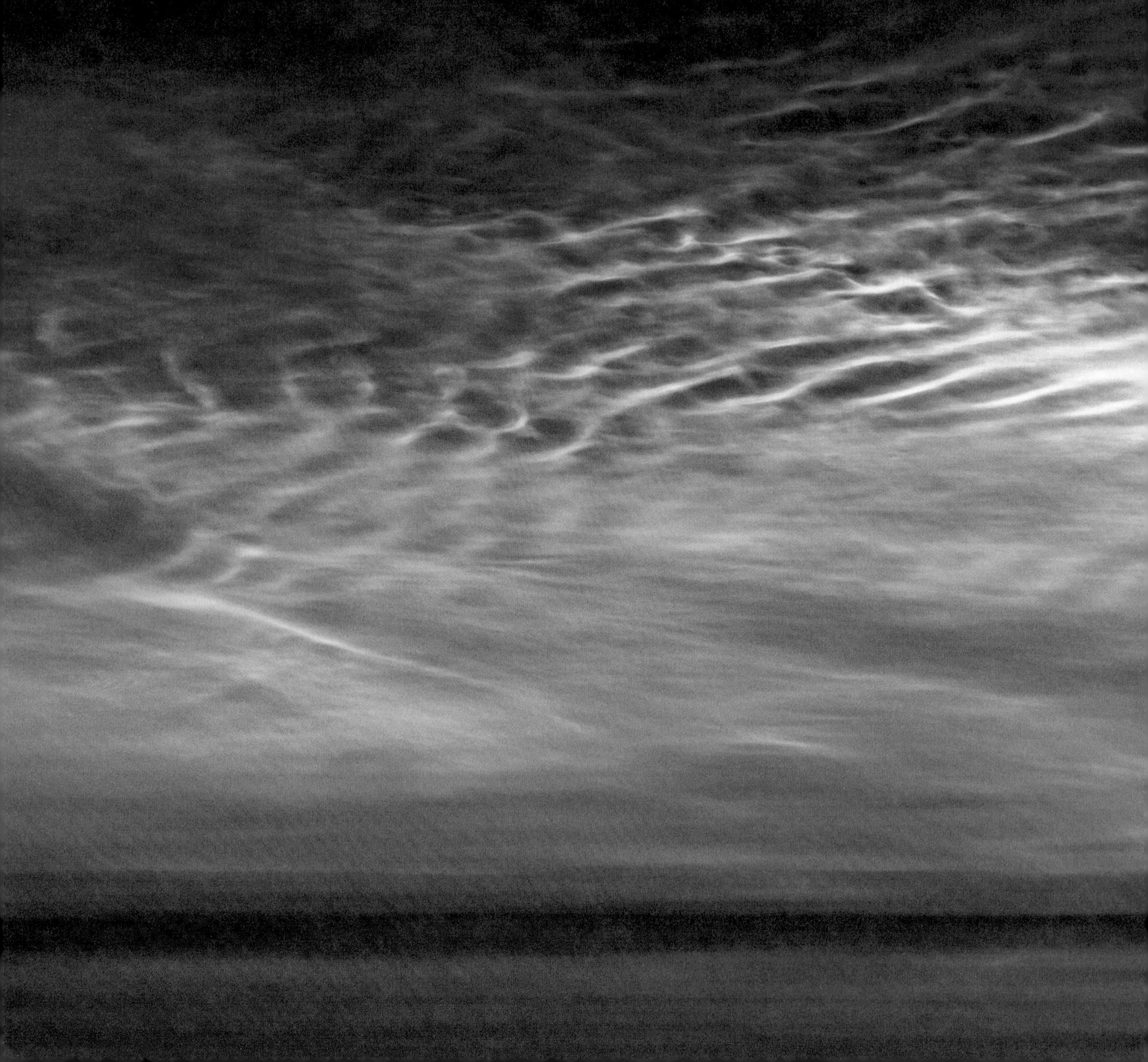

3

Searching for Life Elsewhere

Where is everybody?

The universe is truly vast! Just how immense it really is became obvious in the 1930s, when astronomers discovered that our galaxy, the Milky Way, was only one of more than 100 billion galaxies (Figure 3.1). Each of these galaxies has between 100 and 400 billion stars. A prominent astronomer of that era, Harlow Shapley, argued that somewhere in this vast universe, there must be other forms of life. Even if only one star in a thousand has a solar system, and if only one in a thousand of those solar systems has a planet capable of supporting life, and if life appeared on only one in a thousand of those planets, there would still be several hundred billion planets upon which life would exist. Of course, this leaves us with the question, "Where is everybody?"

Determining whether life exists elsewhere has been a central goal for space exploration. This is particularly true for the many orbiters, probes, and landing craft sent to the planet Mars over the past 40 years. In 1975, two space probes were sent to Mars—*Viking 1* and *2*. Each of these probes had an orbiter and a lander. Photographs taken from the *Viking* orbiters revealed evidence that water had once flowed on the surface of the planet. This evidence of water was encouraging because water is essential to life as we know it.

The *Viking 1* and *2* landers touched down on Mars in 1976. Pictures taken by these landers showed an arid, boulder-strewn landscape. No signs of water or of past or present life were visible (Figure 3.2).

Each *Viking* lander had a robotic arm that was used to sample Martian soil. The soil samples were then subjected to tests designed to determine the presence or absence of life. One of these tests involved mixing the soil with a solution of nutrients similar to those that might have sustained early life on Earth. If this soil mixture released carbon dioxide, it would indicate that nutrients were being processed in a lifelike fashion. To the scientists' great surprise and elation, carbon dioxide was released in abundance!

Elation was soon replaced by confusion as the results of other tests were revealed. For example, Martian soil was heated to see if it would release any of the chemicals that are characteristic of life. None were detected. In another test, soil was exposed to a mixture of gases that might be used by living organisms. There was no evidence that any of these gases were taken up in the soil.

Scientists subsequently determined that intense ultraviolet (UV) light striking the surface of Mars had altered chemicals in Martian soils, making them highly reactive with the nutrient solution. In fact, this UV radiation is so intense that life as we know it cannot survive on Mars' surface. Earth's complex atmosphere shields its ecosystems from this destructive radiation. Evidence from other studies suggest that Mars' atmosphere may at one time have provided similar protection, and subsequent missions have attempted to establish whether life might have existed there in the distant past.

► Figure 3.1 **Worlds Beyond Our Own**
Over a relatively small portion of the sky, the orbiting Hubble Space Telescope captured this image of two colliding galaxies and hundreds of others in the distance.

▼ Figure 3.2 **Is There Life Here?**
The Martian landscape in this Viking image presents no obvious signs of life; however, some chemical reactions in these Martian soils are lifelike.

- What are the basic principles of chemistry and energy transfer that characterize life on Earth?

- How does Earth's environment compare with those of its nearest planetary neighbors?

- What are the physical and chemical characteristics of Earth's interior and its atmosphere and oceans?

- How do these features support climates that sustain abundant life on the one and only location in the universe where we know for certain that life exists?

In 2013, a craft far more sophisticated than any previous lander began its exploration of Mars' surface. The *Curiosity* rover can prowl over tens of miles, and it has instruments to analyze soils and the atmosphere for evidence of conditions that might have supported life in the past (**Figure 3.3**). For example, an instrument called CheMin can analyze soils for evidence of clay minerals. These minerals form in water with very low acidity or alkalinity. These are also the chemical conditions necessary to sustain life.

Exploration of Mars has taught us that determining whether or not life exists on another planet is more complicated than we had hoped. To do so presumes that we understand all the physical and chemical conditions necessary for life, as well as the ways that living things affect their physical environment. However, everything we know about these conditions and effects has come from studies of life on Earth. Our current understanding allows us only to detect life as we know it on Earth. For example, nearly all exploratory sampling and tests have been designed to detect life that is composed of carbon and dependent on water. As those early *Viking* measurements revealed, physical and chemical conditions elsewhere in the universe may produce lifelike responses in the absence of life.

It is also possible that there are other forms of life whose chemistry and behavior are totally unfamiliar to us.

Understanding the physical and chemical conditions necessary to sustain life is an important goal of environmental science and the central theme of this chapter.

◄ Figure 3.3 *Curiosity* This Mars lander is about the size of a small car and carries a very sophisticated array of instruments. In this "selfie," the *Curiosity* is seen traversing soils that appear to have been deposited in the presence of large amounts of liquid water some time in the distant past. The self-portrait is actually a mosaic of a number of images taken from a digital imager that sits atop a robotic arm.

3.1 Chemistry of the Environment

BIG IDEA Matter is the tangible "stuff" of the universe. Despite all the complexity we see in the material universe, especially as we experience it on Earth, all matter is composed of simpler building blocks called **elements**. The physical and chemical properties of elements are attributable to the characteristics of their most basic subunits, **atoms**. Two or more atoms may be held together by chemical bonds to form **molecules**. Water is among the most important molecules in ecosystems. It is a major constituent of all living organisms and is essential to most life functions.

Atoms and Isotopes

■ Atoms are the basic unit of matter.

Nearly all matter consists of one or more of the 92 naturally occurring elements. (We say "naturally occurring" because chemists have artificially synthesized about 20 additional elements.) Chemists give each element a symbol. Hydrogen (H), carbon (C), oxygen (O), and nitrogen (N) are examples of four common elements and their symbols. These four elements play especially important roles in ecosystems.

An atom is the basic subunit of an element that displays the chemical properties of that element. Those properties depend on the numbers of subatomic particles the atom contains. Atoms have a central nucleus that contains positively charged particles called **protons** and electrically neutral particles called **neutrons** (Figure 3.4). The atoms of a particular element always have the same unique number of protons; this number is called the element's atomic number. For example, all hydrogen atoms have a single proton and an atomic number of 1. All carbon atoms have six protons and an atomic number of 6. Note that hydrogen is unique among elements for having a nucleus that does not contain any neutrons.

The nucleus is surrounded by one or more negatively charged particles called **electrons**. The negative charge of a single electron exactly balances the positive charge of a proton. Thus, the overall charge of an atom is neutral when the number of electrons surrounding the nucleus is equal to the number of protons, that is, the atomic number.

Protons, neutrons, and electrons have **mass**, the property responsible for the gravitational attraction of all matter to all other matter. Protons and neutrons account for most of an atom's mass. The mass of each is more than 1,000 times greater than the mass of an electron.

Atoms of an element always have the same number of protons, but they may have different numbers of neutrons. Atoms of an element with different numbers of neutrons are called **isotopes**. For example, carbon atoms always have 6 protons but may have six, seven, or eight neutrons. These isotopes are indicated by the chemical symbol preceded by the total number of neutrons and protons. For example, ^{12}C has six protons and six neutrons; ^{14}C has six protons and eight neutrons (Figure 3.5). Isotopes are important in environmental science because they play important roles in some chemical reactions and energy transformations. Scientists also use them as markers for the passage of time.

The various isotopes of an element generally share the same chemical properties. However, isotopes with more neutrons are slightly more massive than those with fewer neutrons. This difference in mass allows chemists to determine the relative amounts of different isotopes in a sample of a particular element.

Some isotopes, generally the heavier ones, have nuclei that are unstable and can break apart. Such isotopes are **radioactive**, which means they spontaneously "decay," emitting various combinations of high-energy protons, neutrons, electrons, and radiation. This decay may convert them into lighter isotopes of the same or different elements. These new isotopes may be stable or radioactive.

Each radioactive isotope, or radioisotope, decays at a unique rate, which makes them useful markers for the passage of time. That rate is measured by **half-life**, the length of time that it takes for half of a collection of atoms of a radioisotope to decay. Half-lives may be as short as tiny fractions of a second or longer than 1 billion years. For example, ^{14}C has a half-life of 5,730 years; it decays to form ^{14}N. The uranium isotope ^{235}U, which is used in nuclear power plants and weapons, has a half-life of 703 million years; it decays to lead-207 (^{207}Pb).

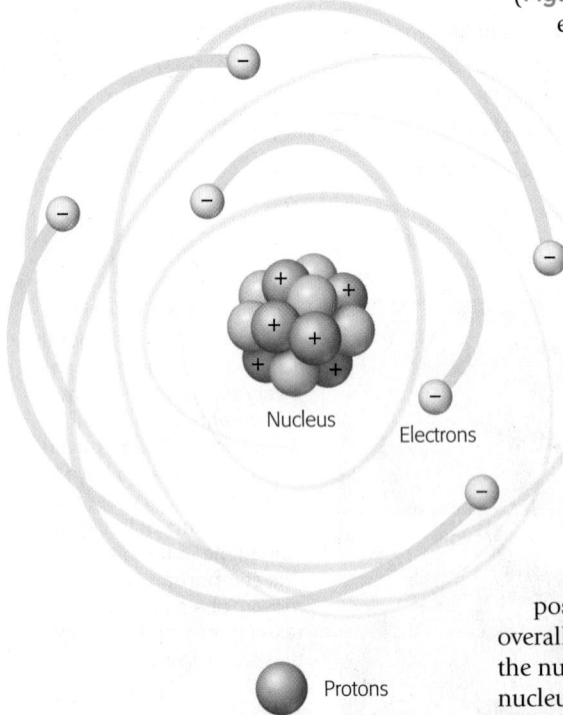

▼ **Figure 3.4 The Atom**
The nucleus of an atom is composed of positively charged protons and uncharged neutrons. Negatively charged electrons swirl in a cloud outside the nucleus. (In this model of a carbon atom, the relative size and spacing of the particles are not to scale.)

Nucleus   Electrons

● Protons

● Neutrons

○ Electrons

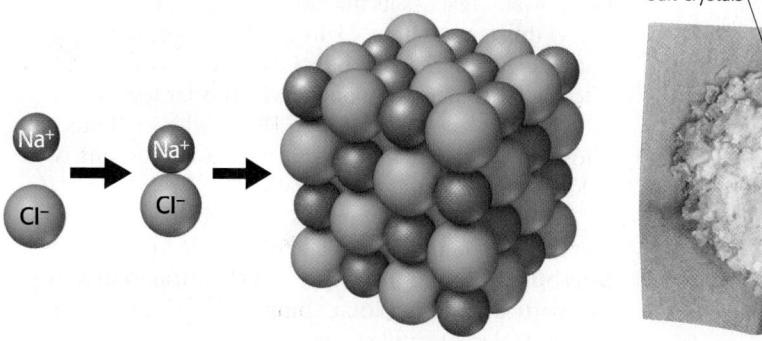

^{12}C ^{13}C ^{14}C

◄ Figure 3.5 **Isotopes of an Atom**
Different isotopes of an element such as carbon have the same number of protons and electrons but different numbers of neutrons.

Molecules and Ionic Compounds

■ Atoms are joined by chemical bonds to form molecules and ionic compounds.

Atoms of an element can combine with other atoms to form myriad kinds of molecules. Some molecules are made of two atoms of the same element, such as the oxygen (O_2) and nitrogen (N_2) in the atmosphere. Molecules that are made of more than one element are called **compounds**. Water (H_2O) and carbon dioxide (CO_2) are two common compounds.

The atoms in a molecule are held together by chemical bonds. The strength of a chemical bond is measured by the energy required to break it. **Covalent bonds** form when atoms share electrons. For example, the two oxygen atoms in oxygen gas (O_2) are held together by a covalent bond: The electrons and their electrical charge are shared equally between the atoms so the bond is nonpolar. In other molecules, atoms share electrons unequally, so part of the molecule is electrically positive and part is electrically negative. For example, in water molecules (H_2O), the oxygen atom attracts electrons a bit more than do the hydrogen atoms. The result is that the end of the molecule with oxygen has a slight negative charge and the ends with hydrogen atoms have a slight positive charge (Figure 3.6).

Chemical bonds can also be formed when one atom transfers one of its electrons to another atom. The atom or molecule donating the electron has a net positive charge, and the atom receiving the electron has a net negative charge. Such electrically charged atoms or molecules are called **ions**, and the electrical attraction between them is called an **ionic bond**. Sodium chloride

▲ Figure 3.7 **Table Salt: An Ionic Compound**
A sodium atom (Na) donates an electron to a chlorine atom (Cl) to form common table salt, NaCl.

(NaCl), also known as table salt, is the most familiar example of an ionic molecule (Figure 3.7).

There are other attractive forces at work, intermolecular forces that bring together atoms and molecules as a consequence of shifts in electrical charge. The strength of these forces depends on the extent of the polarity, that is, the difference in charge that exists between the molecules. One such example is the **hydrogen bond** that forms between water molecules: The hydrogen atom of one molecule is attracted to the oxygen atom of another. Even though relatively weak, these forces can affect the chemical and physical properties of matter. They are also important in a variety of biological functions. One interesting example is that of the gecko and its ability to climb smooth walls because of intermolecular forces that create a bond between the molecules in their toes and the molecules on the surface of the wall (Figure 3.8).

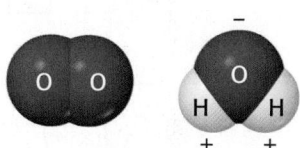

▲ Figure 3.6 **Nonpolar and Polar Molecules**
Electrons and electrical charges are evenly distributed around the nonpolar O_2 molecule. Uneven distribution of electrons between the hydrogen and oxygen of water make it polar.

► Figure 3.8 **Hanging in There**
Animals such as geckos are able to adhere to vertical surfaces because of an intermolecular attraction between the molecules in their feet and the wall's surface.

QUESTIONS 3.1

1. How do isotopes of an element differ?

2. How do covalent, ionic, and hydrogen bonds differ from one another?

3. Why does ice float?

4. What is pH and how does it differ between acidic and alkaline solutions?

(MES) For additional review, go to **MasteringEnvironmentalScience**

The Water Molecule

■ Life depends on the chemistry of water.

The search for life elsewhere in the universe inevitably begins with a search for evidence of water—and that's because every organism on Earth depends on water to live. The cells and tissues of all organisms are largely made of water, and the unique chemical properties of water are critical to their structure and function.

Because of their polarity, water molecules readily bond to one another, forming a network of molecules bound together by hydrogen bonds. Hydrogen bonds are responsible for the cohesive nature of liquid water. As liquid water heats, the hydrogen bonds weaken; at 100 °C (212 °F), virtually all of them are broken. Without hydrogen bonds, water would boil at temperatures well below its actual freezing point (0 °C or 32 °F).

Conversely, as liquid water becomes colder, more hydrogen bonds form, pulling the water molecules closer together. As a result, water becomes denser as it cools. Liquid water reaches its maximum density at 4 °C (39 °F). Below this temperature, hydrogen bonds begin to stabilize. As water cools to 0 °C, its molecules move apart into the lattice structure of ice, which is far less dense than liquid water (Figure 3.9). This is why ice floats. Thus, the winter ice that floats on lakes and oceans is colder than the water underneath.

The polarity of water makes it an excellent solvent—a great many chemicals are able to dissolve in it. **Solubility** refers to the ability of a chemical to dissolve in a particular liquid. Ionic compounds such as NaCl are particularly soluble in water. In contrast, nonpolar molecules, such as those of many oils, are not very soluble in water. Many of the chemical reactions that are essential to the function of living organisms depend on water's unique properties as a solvent.

What makes water such an effective solvent is that the water molecules themselves have a tendency to break apart, or dissociate—into hydrogen ions (H^+) and hydroxide ions (OH^-). Pure water contains equal amounts of these two ions. Chemicals dissolved in water may shift the amount of H^+ ions relative to OH^- ions. The **pH scale** is a quantitative representation of the relative amounts of hydrogen and hydroxide ions in a liquid (Figure 3.10). The pH of pure water is 7, which is said to be neutral. When the concentration of H^+ exceeds that of OH^-, in a solution, the solution is acidic and has a pH value of less than 7. When the concentration of OH^- exceeds that of H^+, the solution is basic, and its pH is greater than 7. Each pH unit reflects a 10-fold change in the amount of each ion. Compared to a neutral solution, a solution of pH 6 has 10 times more H^+ ions than a solution of pH 7, and a solution of pH 5 has 100 (10×10) times more.

The pH of a solution affects many chemical reactions. Variations in the pH of solutions are especially important within organisms and ecosystems. For example, the pH of your blood is slightly alkaline at 7.4: Variations of just 0.1 pH unit can have serious health consequences. As we shall see in later chapters, human activities are increasing the amount of carbon dioxide in Earth's atmosphere. This, in turn, is causing oceans to become more acidic and threatening populations of marine organisms, such as corals and shellfish, that are able to tolerate only very narrow pH ranges.

▼ Figure 3.9 **Hydrogen Bonds**

Ⓐ In liquid water, hydrogen bonds between water molecules are constantly forming and breaking. They become more stable as the temperature decreases. Ⓑ At 0 °C (32 °F), these bonds stabilize and the water molecules form into the crystalline structure we know as ice.

► Figure 3.10 **pH Scale**

In pure water, hydroxide and hydrogen ions are equally abundant, and pH = 7. Lemon juice is acidic because hydrogen ions outnumber hydroxide ions. The reverse is true of milk of magnesia, which is alkaline.

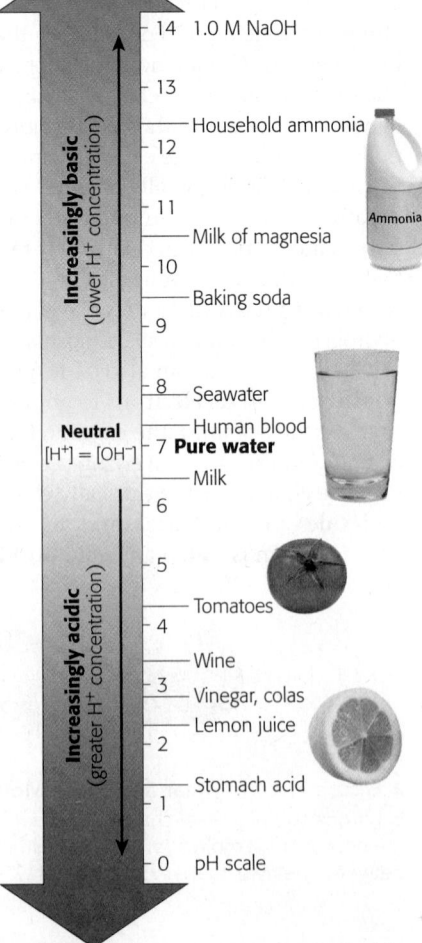

Increasingly basic (lower H^+ concentration)

- 14 — 1.0 M NaOH
- 13
- — Household ammonia
- 12
- 11
- — Milk of magnesia
- 10
- — Baking soda
- 9
- 8 — Seawater
- — Human blood
- Neutral [H^+] = [OH^-] — 7 **Pure water**
- — Milk
- 6

Increasingly acidic (greater H^+ concentration)

- 5
- — Tomatoes
- 4
- — Wine
- 3
- — Vinegar, colas
- — Lemon juice
- 2
- — Stomach acid
- 1
- 0 — pH scale

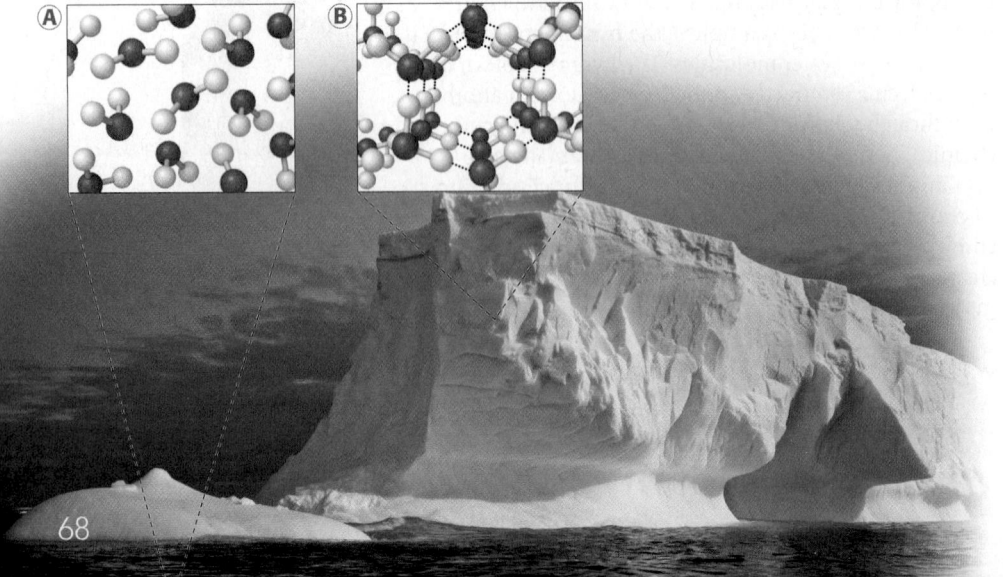

3.2 The Organic Chemistry of Life

BIG IDEA Water is essential for life. However, the primary structural and functional elements of organisms are **organic molecules**, which are made of carbon atoms covalently bonded to hydrogen, oxygen, and other atoms. For this reason, we say that life on Earth is carbon based. Organic molecules range in size from relatively simple hydrocarbons and sugars to large, complex molecules, such as proteins and nucleic acids. Compounds that are not made up of carbon and hydrogen are said to be **inorganic compounds**.

Biological Chemicals

■ Carbon atoms provide the chemical backbone for the complex molecules essential to life.

Life as we know it is often described as carbon-based. Most molecules containing that element are called organic in reference to their central importance to organisms. Carbon atoms are especially versatile because of their ability to bond with up to four other atoms, allowing carbon to form a great number of different compounds.

Hydrocarbons are organic molecules composed entirely of carbon and hydrogen atoms. They are generally not an important constituent in living organisms but may be produced in great quantity as organisms decay. They are an important source of energy. Methane, the primary constituent of natural gas, is the simplest hydrocarbon; its molecules have a single carbon atom and four hydrogen atoms (**Figure 3.11**). Crude oil is the product of the decay of microscopic plants over many eons. It is a mixture of hydrocarbons that range in size from a few to more than 20 carbon atoms.

Most organic molecules we associate with life are large and complex, and for this reason are referred to as **macromolecules**. They are classified into four broad categories: carbohydrates, lipids, proteins and nucleic acids. Among these, carbohydrates, proteins, and nucleic acids are **polymers** that are built up from relatively simpler organic molecules, linked together to form larger chains or networks.

Carbohydrates are an essential building block of life. They are polymers made up of simple **sugars**, an organic molecule that consists of carbon, hydrogen, and oxygen and has the general chemical formula $(CH_2O)n$, where n is between 3 and 7. The sugar glucose, or grape sugar, has the formula $C_6H_{12}O_6$. It is a basic source of energy in most organisms and an ingredient in the synthesis of numerous other organic molecules (**Figure 3.12**). Table sugar is composed of a glucose molecule bonded to a similar sugar called fructose. Ribose ($C_5H_{10}O_5$) and deoxyribose ($C_5H_{10}O_4$) are two very important 5-carbon sugars.

Starch and **cellulose** are carbohydrates that are composed of hundreds of glucose molecules. These large polymers are referred to as **polysaccharides**. Plants use starch to store extra sugar. Cellulose is the primary structural constituent in plant tissues; it is also one of Earth's most abundant organic molecules. The only difference between starch and cellulose is the nature of the bond between adjacent glucose molecules. However, this difference has a significant effect on the ability of organisms to break down the two kinds of molecules. Most organisms can easily break the bonds in starch, releasing glucose molecules to use for food and energy. The bonds between the glucose molecules in cellulose are much more resistant; only a few kinds of microorganisms are able to break those bonds. Being indigestible is a useful property for a structural molecule; it also means that cellulose can serve as a dietary fiber in the human digestive system.

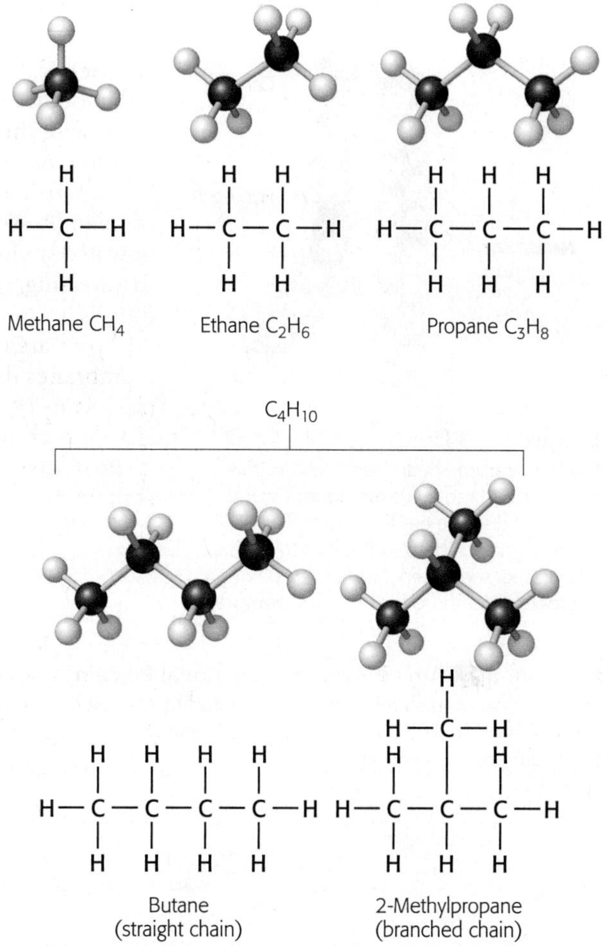

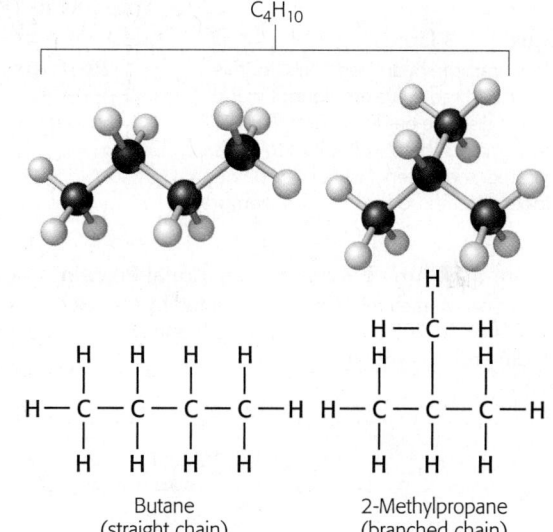

Methane CH_4 Ethane C_2H_6 Propane C_3H_8

C_4H_{10}

Butane (straight chain) 2-Methylpropane (branched chain)

▲ **Figure 3.11 Hydrocarbons** These molecules are among the array of hydrocarbons found in petroleum.

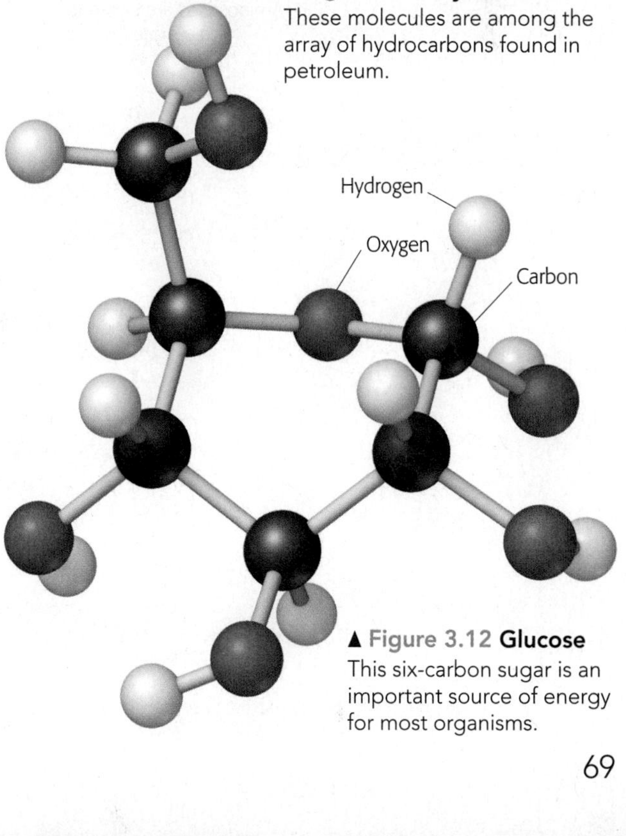

Hydrogen

Oxygen

Carbon

▲ Figure 3.12 **Glucose** This six-carbon sugar is an important source of energy for most organisms.

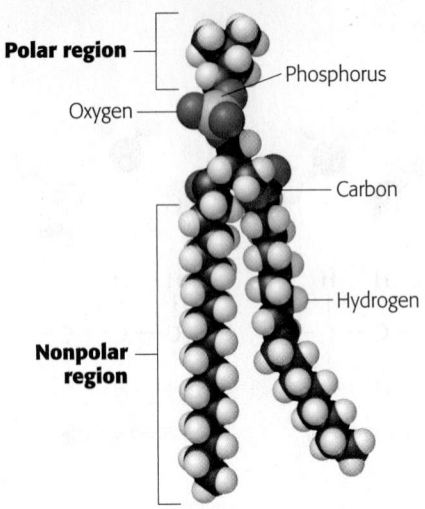

Lipids, which include fats and oils, are organic molecules made of long chains of carbon and hydrogen atoms linked to a smaller "head" containing one to several oxygen atoms. These two regions of a lipid are chemically distinct. The head region is polar, but the rest of the molecule is not, and because of this, most lipids are not soluble in water. As the saying goes, oil and water do not mix. You can see this effect in a bottle of salad dressing: the oil does not dissolve in water or vinegar. Even when you shake the dressing, you can still see two separate liquids.

Lipids are the major constituents of the membranes that surround the cells of all organisms (Figure 3.13). Fats and oils are also important forms of energy storage.

Proteins are polymers made of nitrogen-containing organic molecules called amino acids. The 20 amino acids found in all organisms have different chemical structures, but each has a nitrogen-containing amino group ($-NH_2$) and a carboxylic acid group ($-COOH$). Individual proteins are composed of 100 to over 1,000 individual amino acids (Figure 3.14). The function of each protein is determined in part by its unique sequence of amino acids. The chemical interactions and bonds among the various amino acids in a protein chain cause it to fold into a very complex three-dimensional shape, which is important to its function.

Every organism depends on the actions of thousands of different proteins. Some proteins, such as the keratin that makes up your fingernails and hair, provide structure. Other proteins control cellular activities (hormones, insulin), enable movement (actin, myosin), and protect against disease (antibodies). Many proteins serve as **catalysts**, substances that promote chemical reactions without being consumed in the reaction. Proteins that serve as catalysts are called **enzymes**. Virtually every chemical reaction in an organism is catalyzed by its own unique enzyme.

▲ Figure 3.13 **Lipids**
The long carbon–hydrogen "tails" of this phospholipid molecule are nonpolar and so do not dissolve easily in water. This property makes them well suited to contain and protect the watery inside of a cell when grouped together to form a cell's membrane.

► Figure 3.14 **Amino Acids to Functional Protein**
The function of each protein is determined by the specific sequence of the 20 different amino acids and by its three-dimensional structure.

Typical amino acids

Alanine

Acid group   Amine group

Glutamic acid

Amino acids are linked to form a protein chain

Protein chains assume a three-dimensional structure

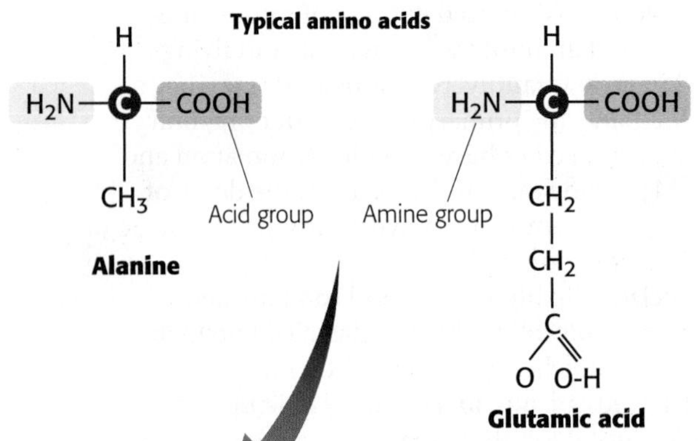

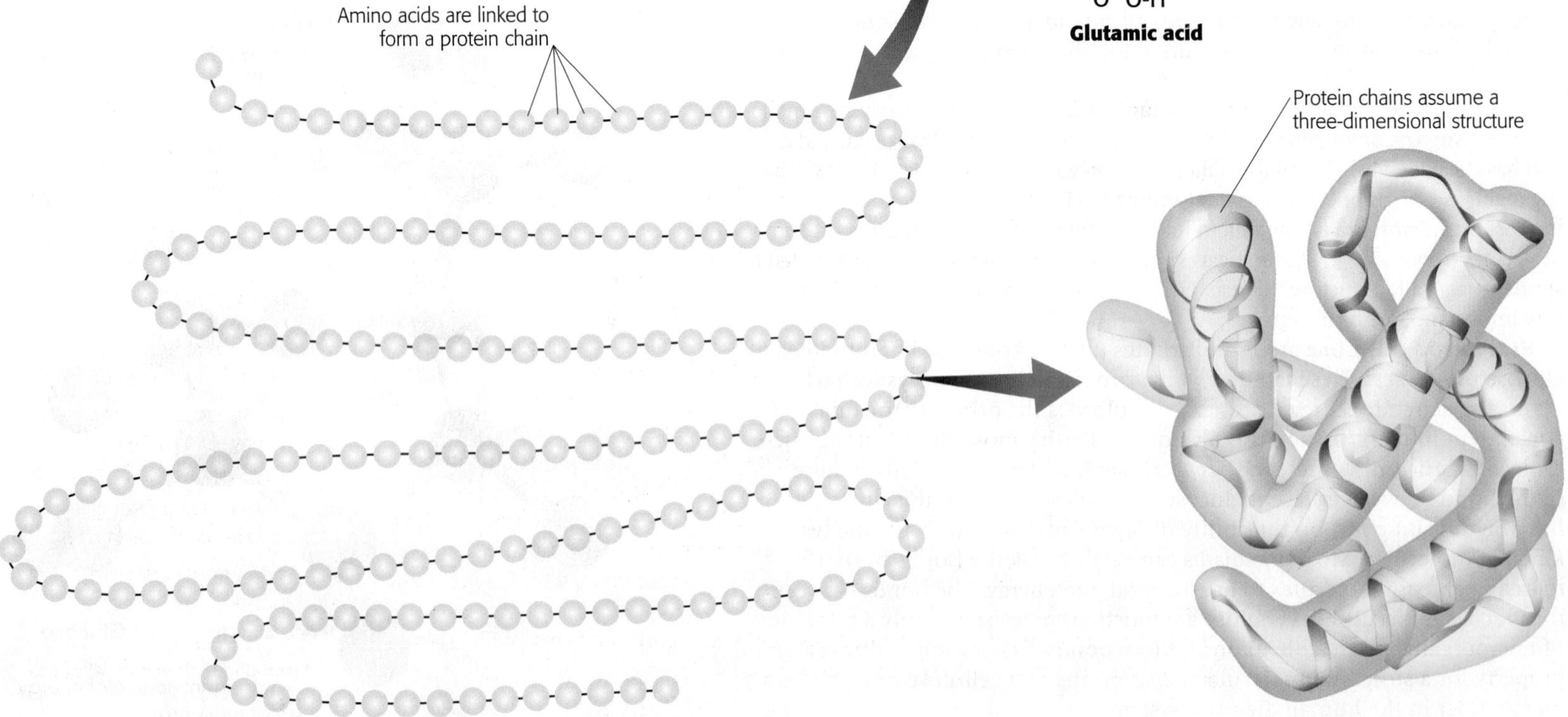

Nucleic acids are polymers of chemical subunits called nucleotides. Each nucleotide is composed of a five-carbon sugar, a phosphate group ($-PO_4$), and a nitrogenous base. The primary function of nucleic acids is to store and transmit information, critical information that is carried in the form of a code built into their basic structure. This information enables every cell in an organism to carry out its functions; it also enables an organism to produce offspring. Two types of nucleic acids are necessary to make this process work: DNA and RNA.

Deoxyribonucleic acid (DNA) forms the hereditary material that is passed from generation to generation in all organisms. Each nucleotide of DNA contains the sugar deoxyribose and one of four nitrogenous bases—adenine (A), thymine (T), cytosine (C), or guanine (G). The hereditary, or genetic, code of each organism is stored in the unique sequence of these nitrogenous bases. The DNA of an individual organism is made of billions of nucleotides; however, the code is broken down into smaller segments of DNA called **genes**. These genes enable a cell to synthesize, or build, the proteins needed to carry out the functions of an organism.

To understand how the genetic code works, you first need to look at the structure of a DNA molecule. Each molecule is made up of two strands. One strand, called the sense strand, carries the code. This is the particular sequence of bases unique to the organism. The second strand carries bases that are complementary to the first strand, forming distinct base pairs (**Figure 3.15**). Adenine (A) pairs only with thymine (T), and cytosine (C) pairs only with guanine (G). When bound together, the two strands form a ladderlike structure. Deoxyribose and phosphate groups form the edges or backbone of this ladder, and paired nucleotides form the rungs. The molecule wraps into a spiral, a form that is described as a double helix.

Ribonucleic acid (RNA) plays a central role in implementing the genetic code by enabling the synthesis of proteins. The structure of RNA is similar to that of DNA, except the nucleotides of RNA contain the sugar ribose rather than deoxyribose and the nitrogenous base uracil (U) rather than thymine (T). Furthermore, RNA typically exists as a single strand rather than a double strand.

Proteins are synthesized in a two-step process. In **transcription**, the DNA code for a particular gene is rewritten as a segment of RNA. This RNA then serves as the template for the assembly of a specific protein in the **translation** process. As you will recall, proteins are made up of a particular sequence of amino acids.

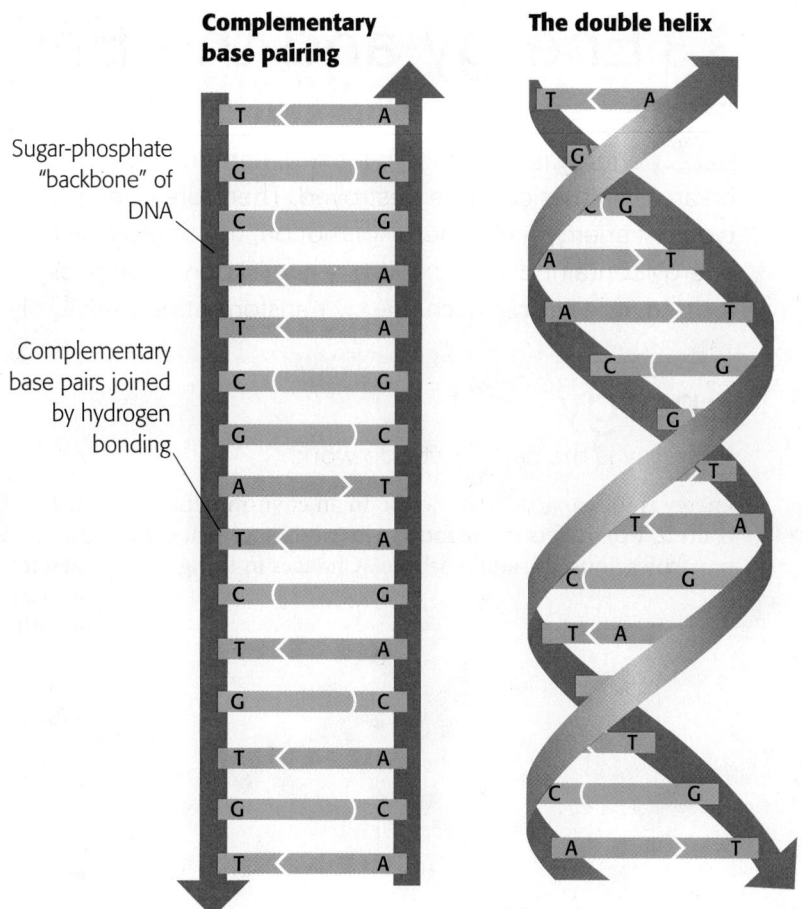

Complementary base pairing

Sugar-phosphate "backbone" of DNA

Complementary base pairs joined by hydrogen bonding

The double helix

▲ Figure 3.15 **DNA Forms the Hereditary Code**
The genetic code that determines the structural and functional attributes of each organism is embedded in the sequence of nitrogenous bases along the DNA strand.

These amino acids are written into the DNA code as a sequence of three distinct bases (TAG, CAA, or GTC, for example). The RNA strand transcribes these triplets using complementary bases (AUC, GUU, or CAG) and this sequence is then translated into a particular sequence of amino acids and a newly synthesized protein.

An organism's complete set of DNA is called its **genome**. With the exception of eggs and sperm, every cell in an organism contains a complete copy of the genome. Individual cell types such as muscle, nerve, and skin differ from one another depending on which portions of the genome are being transcribed and translated in that cell. Individual organisms differ from one another in part because of differences in the sequence of nucleotides in their genomes. They may also differ as a consequence of the effects of differences in their environment that influence how genes are expressed. The role of DNA in the growth, reproduction, and evolution of organisms is discussed in Modules 4.2 and 4.5.

QUESTIONS 3.2

1. How do carbohydrates differ from hydrocarbons?

2. Most lipids are not soluble in water. Why?

3. How does starch differ from cellulose? Why is this difference important?

4. What is an organism's genome?

(MES) For additional review, go to **MasteringEnvironmentalScience**

3.3 Energy and the Environment

BIG IDEA Energy conversions and transformations drive all ecological processes. But, unlike matter, energy is not a tangible thing. Rather, **energy** is the capacity to do work. Energy cannot be created, neither can it be destroyed. Therefore, the amount of energy in the universe is constant. Four types of energy—the energy of motion, the energy in chemical bonds, the energy in light, and the energy contained in matter itself—are important in ecosystems. Energy can be changed from one form to another, but such energy transformations inevitably increase the disorder of the universe.

Energy

■ Energy is the capacity to do work.

Energy is the capacity to do **work**. In an environmental context, work refers to the movement of things, including molecules, muscles, and machines. Changes in things, such as the growth of cells, individual organisms, and populations of organisms also represent work and, therefore, the expenditure of energy. Energy exists in one of two general forms that can be converted one into the other. **Potential energy** is the energy that is stored in a system and available to do work in the future; it is sometimes called the energy of position. **Kinetic energy** is the energy of motion.

For our purposes, we will use the word *system* to describe a group of objects that interact with one another and exchange energy with their environment. For example, consider the system of a ceramic vase about to be pushed from your desk by your pesky cat (**Figure 3.16**). The vase has potential energy that is related to its mass, its height above the floor, and the tug of gravity. As the vase falls, the potential energy in this system decreases, and the kinetic energy of the vase increases. At the instant before the vase crashes into the floor, all of the potential energy of this system has been converted into the kinetic energy of the vase.

Your body movement provides another example of potential and kinetic energy. In this example, potential energy takes the form of the chemical energy stored in the bonds of lipids and carbohydrates taken in from the food you eat. Through a process called cellular respiration, your body can mobilize this energy to coordinate the activities of the different systems in your body. When you decide to move, a whole series of chemical transformations will enable your muscles to contract, converting potential energy to kinetic energy, or motion.

◄ Figure 3.16 **Look Out Below!**
A ceramic vase poised on top of a table has potential energy Ⓐ. As it falls, that potential energy is converted to kinetic energy Ⓑ. When it crashes to the floor, not only has it lost its potential energy, but the system is also much less organized Ⓒ.

Laws of Thermodynamics

■ Energy can be neither created nor destroyed, but it can be transformed.

The potential energy in a system may be transformed into kinetic energy. But all energy transformations conform to two fundamental principles, called the laws of thermodynamics.

The **first law of thermodynamics** states that the total amount of energy in the universe is constant. Energy can be transformed from one form to another, but it can be neither created nor destroyed. This is also called the law of energy conservation.

Although the total energy contained in the universe remains the same, particular systems within the universe may gain or lose energy. Once the vase crashes on the floor, the kinetic energy of its motion will be transformed into the motion of pieces of vase and the floor and the motion of the molecules in the atmosphere through which the vase fell.

The **second law of thermodynamics** states that every energy transformation increases disorder. Disorder, in this case, refers to the fact that the energy in a system that was once organized or "ordered" in such a way as to do work has a tendency to transform into a less usable form as the work is done. This tendency toward a disordered state is described as **entropy**.

A very important consequence of this second law is that in any energy transformation, some energy will be lost from the system. For example, think of the energy contained in the water held at the top of a hydroelectric dam or the chemicals stored in the head of a match. These "systems" have energy available to do work. But once the work is done—water is released to turn a turbine and power a generator or a match lit to provide heat and light—the energy is transformed to a less usable form. It is no longer available to you or the system to do work. In most instances, the "lost" energy has been transformed into **heat**, the kinetic energy associated with the random motion of molecules.

All energy transformations that occur in the universe increase entropy. Although not obvious to you, or your cat, the potential energy in the system of vase and table is transformed into the random motion of molecules as well as debris. Entropy of this system increases. Much of the chemical energy available for moving your arm up to try to catch the vase ends up in the random kinetic motion of molecules in muscles and surrounding tissue. This motion of the molecules makes your body warmer. Entropy increases.

At first, the order and organization of organisms and ecosystems may seem to violate the second law of thermodynamics. It takes a great deal of energy to power the growth, movement, reproduction, and other activities carried out by all organisms. Yet the Earth system appears well ordered almost in defiance of entropy (Figure 3.17). This apparent contradiction disappears when we recognize that ecosystems and their organisms only remain organized because of the constant input of "outside" energy from the sun.

In studying ecosystems, it is important to understand how energy is captured, transformed, and released and the changes this brings to our experience of the material world. There are four types of energy of particular interest: electromagnetic, heat, chemical, and nuclear. We look at those next.

▼ Figure 3.17 **Where Is the Entropy?**
Individual organisms and ecosystems are continuously transforming energy, but they remain highly organized. Their organization depends ultimately on the continuous input of energy from the sun, and it results in an overall increase in the disorder or entropy of the universe.

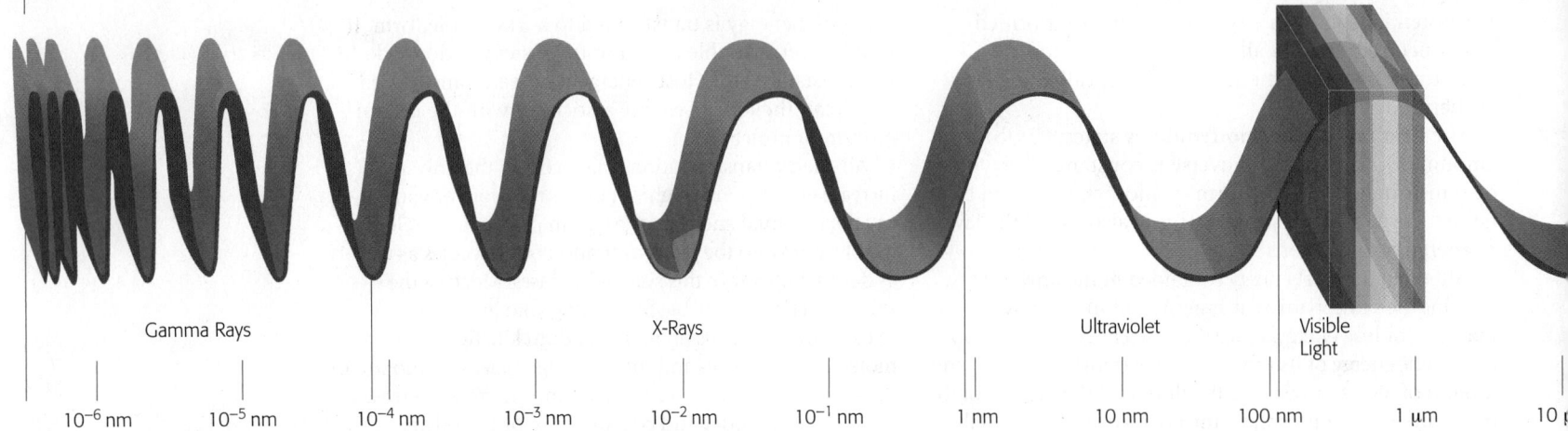

Ionizing radiation—photons carry energy sufficient to ionize molecules and break chemical bonds

Visible light photons can excite electrons—this is what powers photosynthesis and eyesight

Gamma Rays   X-Rays   Ultraviolet   Visible Light

10^{-6} nm 10^{-5} nm 10^{-4} nm 10^{-3} nm 10^{-2} nm 10^{-1} nm 1 nm 10 nm 100 nm 1 μm 10 μm

▲ Figure 3.18 **Electromagnetic Energy**
Light is composed of photons whose energy is determined by their wavelength, the distance from wave crest to wave crest. The shorter a photon's wavelength, the more energy it contains.
Source: Copyright DK LTD

Forms of Energy

■ Four types of energy are especially important in ecosystems.

Electromagnetic radiation. The energy of light is called **electromagnetic radiation**. This energy is transported at a remarkable speed (186,000 miles per second) as "particles" called photons, which have no mass and behave like waves. The amounts of kinetic energy carried by the different forms of electromagnetic radiation are determined by their wavelength, the distance from one wave crest to the next (Figure 3.18). The full range of wavelengths is called the **electromagnetic spectrum**.

Visible light—the light that you are able to see—represents a very small portion of the total electromagnetic spectrum. Different wavelengths of visible light make up all the colors of the rainbow. Photons in this range are sufficiently energetic to elevate the energy level of electrons in organic molecules without breaking their bonds. For example, photons of visible light excite electrons in molecules in the retina of your eyes, providing a signal that your brain interprets as sight. In green plants, photons of visible light excite electrons in chlorophyll, providing the energy for photosynthesis.

Gamma rays, X-rays, and UV light have wavelengths that are shorter than those of visible light. Their photons pack energy sufficient to disrupt the bonds in many

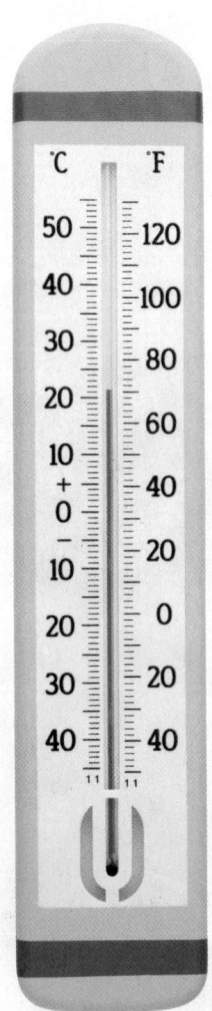

◀ Figure 3.19 **Measuring Kinetic Energy**
Two temperature scales, Fahrenheit and Celsius, are widely used to measure the kinetic energy of collections of molecules.

organic molecules. Thus, these forms of radiation are generally harmful to living systems.

Forms of radiation with wavelengths that are longer than visible light, such as infrared radiation, microwaves, and radio waves, carry less energy than does visible light. Nevertheless, the photons of these wavelengths carry sufficient energy to increase the kinetic energy of molecules. They can also break very weak chemical bonds.

Heat. Heat energy refers to the kinetic energy of molecules. **Temperature** is a measure of the average kinetic energy of a collection of molecules. Over the years, several temperature scales have been devised. In 1727, Daniel Gabriel Fahrenheit proposed a temperature scale in which water freezes at 32° and boils at 212°. Fahrenheit based his scale (°F) on measurements of everyday phenomena, such as very cold days and human body temperature. Fifteen years later, Anders Celsius proposed a temperature scale that set 0° as the freezing point of water and 100° as the boiling point of water (Figure 3.19). Because it is scaled in a simple fashion against the physical property of a common and important chemical, environmental scientists prefer to express temperature in degrees Celsius (°C).

Heat can move in four ways—conduction, convection, radiation, and latent heat transfer (Figure 3.20). **Conduction** is the direct transfer of heat by means of the collisions of molecules. When you touch a hot pot, heat moves from molecules on the pot to the molecules in

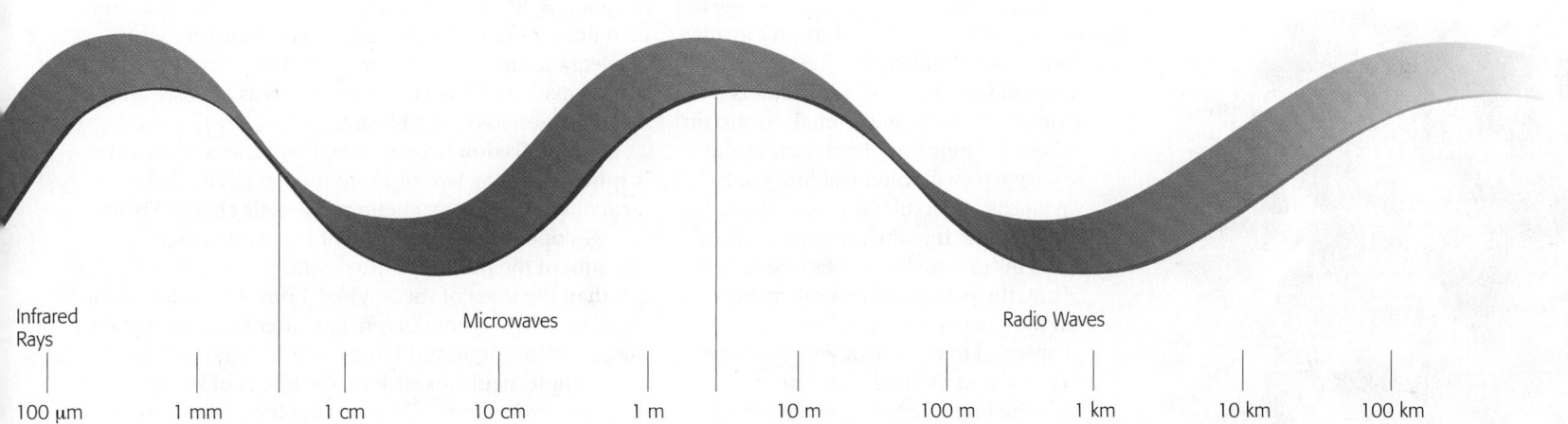

Photons with wavelengths longer than visible light carry sufficient energy to increase the kinetic energy of molecules

Infrared Rays

Microwaves

Radio Waves

100 µm   1 mm   1 cm   10 cm   1 m   10 m   100 m   1 km   10 km   100 km

your fingertips. **Convection** occurs because warm regions in a gas or liquid become less dense and rise, causing the gas or liquid to circulate.

All matter is constantly losing heat by **radiation**, the release of electromagnetic energy. The wavelength of the radiation that is given off decreases with increasing temperature. At 37 °C (98.6 °F), your body radiates heat primarily in the infrared portion of the spectrum, whereas a metal heated to 300 °C (570 °F) glows bright red.

Remember that temperature measures the average kinetic energy of molecules. Within a liquid, some molecules move more rapidly than others. **Latent heat transfer** occurs as the molecules with the highest kinetic energy evaporate. This lowers the average kinetic energy and the temperature of the molecules left behind. You experience latent heat transfer when you sweat and your perspiration evaporates, cooling your skin. Because oceans cover nearly 72% of Earth's surface, the total amount of latent heat transferred to the atmosphere from the oceans is quite significant.

◄ Figure 3.20 **Heat Transfer**
Heat energy can be transferred by Ⓐ conduction, collisions among molecules; Ⓑ convection, circulation due to differences in density; Ⓒ radiation of electromagnetic energy; and Ⓓ latent heat transfer resulting from evaporation of more energetic molecules.

Ⓓ Latent heat transfer
Ⓑ Convection
Ⓒ Radiation
Ⓐ Conduction

Chemical energy. The potential energy associated with the formation or breakage of bonds between atoms is called **chemical energy**. Chemical energy is essential to life. In the process of photosynthesis, green plants use light energy to assemble carbohydrates from carbon dioxide and water (see Module 4.1). The energy in the bonds of these carbohydrates is retrieved by plants and animals to sustain all of life's functions. For example, the energy released from breaking bonds in glucose molecules is captured and transformed into the movement of proteins in muscle cells (Figure 3.21). Thus, the potential chemical energy in the glucose bonds is ultimately converted to the kinetic energy of the contraction of those muscles.

The formation of some chemical bonds releases energy. For example, when hydrogen gas (H_2) bonds with oxygen gas (O_2) to form water, the reaction releases so much energy that it can cause an explosion. The resulting bonds between the hydrogen and oxygen atoms are very strong. The energy required to break a water molecule apart is equal to the energy that was released in its formation.

Nuclear energy. The energy contained in the structure of matter itself is called **nuclear energy**. Albert Einstein's famous equation, $E = mc^2$, states that the energy (E) contained in matter is equal to its mass (m) times the speed of light (c = 299,792,458 meters per second) squared. Thus, even tiny amounts of mass contain enormous amounts of energy. In situations relevant to Earth's ecosystems, the transformation of mass to energy occurs as a consequence of two processes, fission and fusion.

Nuclear fission occurs when the nucleus of an atom is split, producing two or more smaller nuclei and a great deal of electromagnetic and kinetic energy. Fission changes one element into one or more other elements. The sum of the masses of the resulting products is slightly less than the mass of the original atom. The difference in the mass of the system before and after fission is equal to the electromagnetic and kinetic energy that is released. For example, neutrons striking the nuclei of atoms of the uranium isotope ^{235}U can cause the atoms to split into a variety of new atoms and free neutrons, releasing considerable energy. If the ^{235}U is sufficiently abundant, the neutrons released in this reaction may stimulate additional fission events, creating a chain reaction (Figure 3.22). When uncontrolled, such a chain reaction can be explosive, as with an atomic bomb. In the core of a nuclear reactor, the chain reactions are controlled, and the energy that is released is used to boil water and produce steam that is then used to generate electricity.

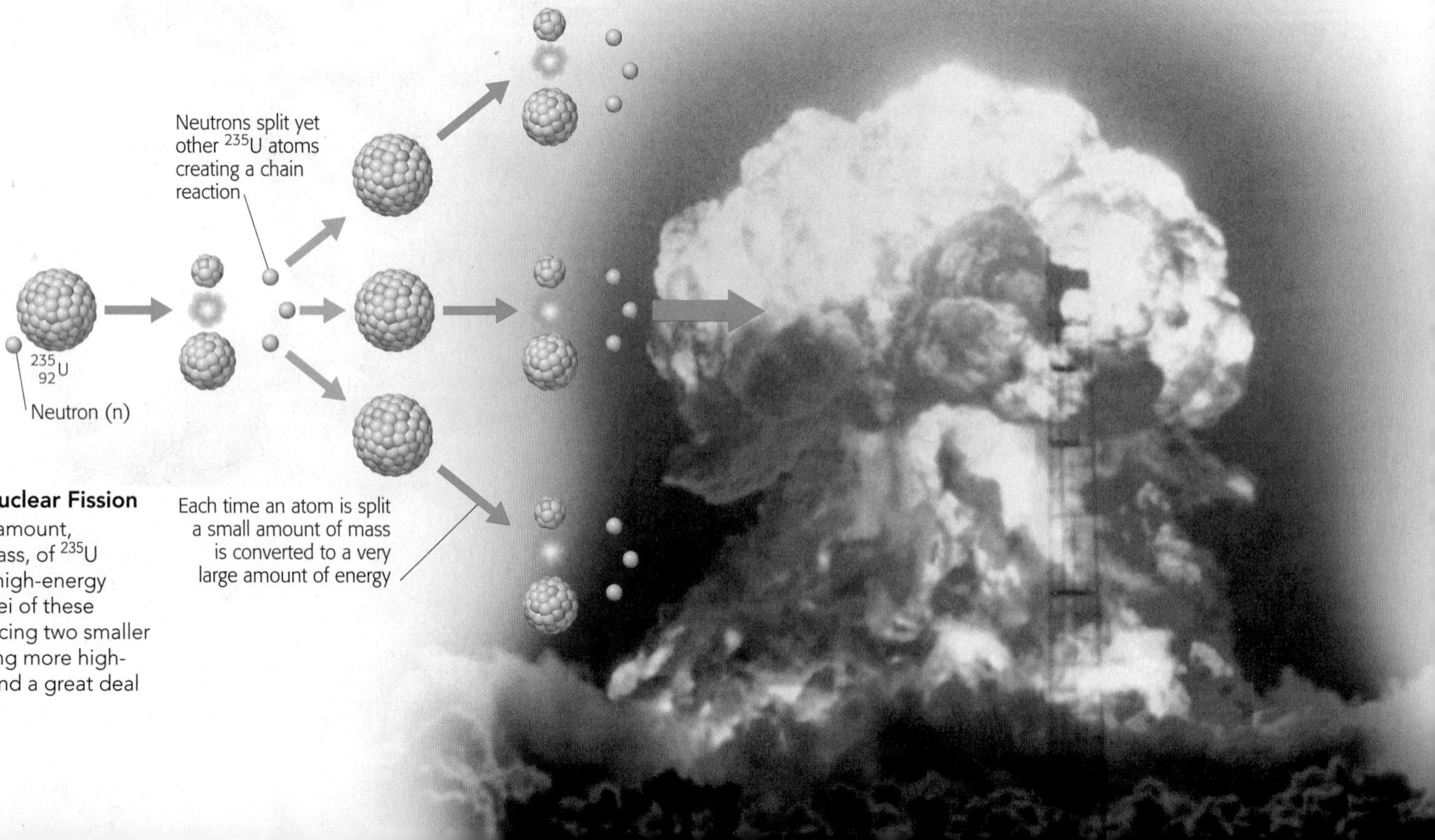

▲ Figure 3.21 **Chemical to Kinetic Energy**
Your body stores a great deal of chemical energy that can be converted to various forms of kinetic energy: your heart beating, your lungs expanding, and your body moving.

Neutrons split yet other ^{235}U atoms creating a chain reaction

$^{235}_{92}U$

Neutron (n)

► Figure 3.22 **Nuclear Fission**
When a sufficient amount, that is, a critical mass, of ^{235}U is bombarded by high-energy neutrons, the nuclei of these atoms split, producing two smaller atoms and releasing more high-energy neutrons and a great deal of energy.

Each time an atom is split a small amount of mass is converted to a very large amount of energy

Nuclear fusion occurs when two atoms collide with so much energy that their nuclei fuse, forming an atom of a new element. The fusion of relatively small atoms, such as hydrogen, carbon, and oxygen, results in a net release of energy as very small amounts of mass are converted to energy. The fusion of massive atoms, such as lead or mercury, requires an input of energy.

By far the most common and most important fusion reaction in the universe occurs when two atoms of hydrogen collide to create an atom of helium (Figure 3.23). The minimum kinetic energy required for this reaction to occur is equivalent to approximately 3,000,000 °C. The electromagnetic radiation and kinetic energy released from this reaction far exceed the energy contained in the colliding hydrogen atoms. This is the energy that powers most stars, including our sun. These stars are so massive that the pull of their gravity provides enough kinetic energy for nuclear fusion to occur. For example, fusion occurs in our sun at temperatures of about 100 million °C.

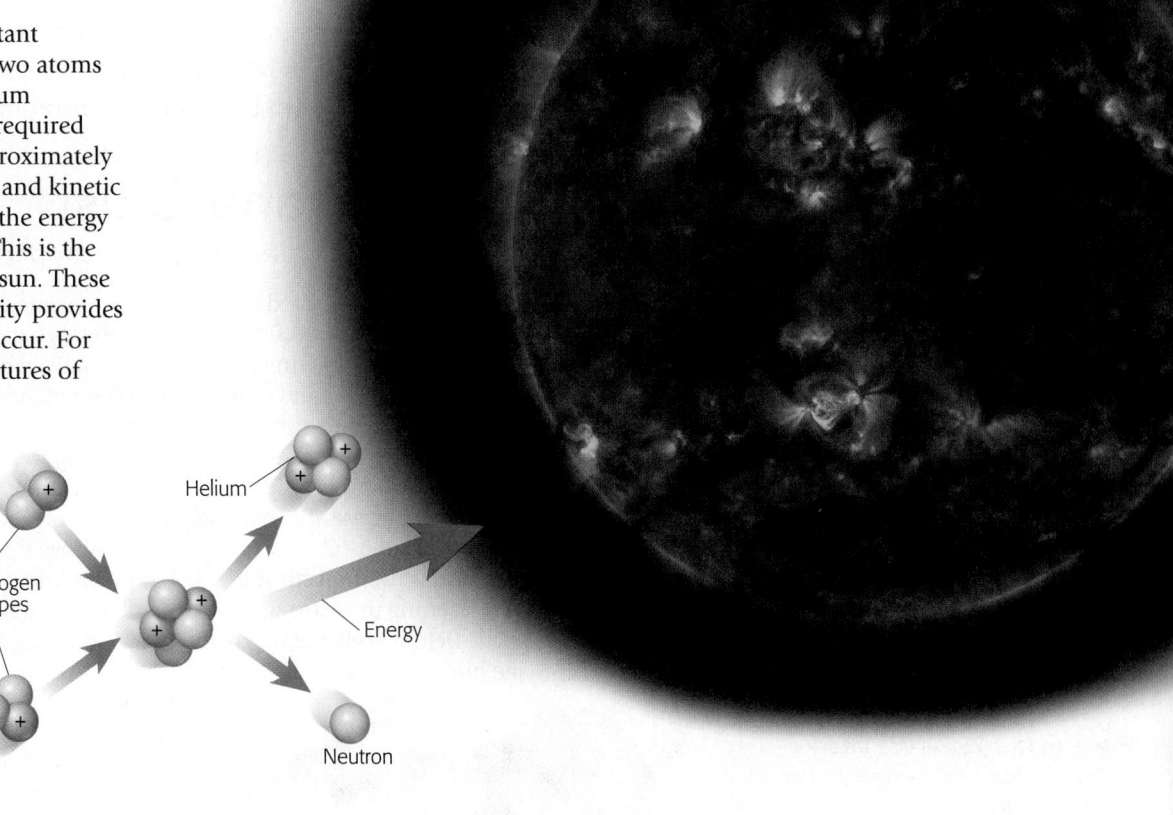

► Figure 3.23 **Solar Power**
High temperatures and pressures within the sun's interior are sufficient to power fusion reactions between isotopes of hydrogen. The molecular model shown here is a much simplified depiction of a very complex reaction.

Helium

Hydrogen isotopes

Energy

Neutron

Energy Units

■ Scientists use several different units to measure energy, depending on the application.

The most fundamental way to measure energy is by the amount of work it does. The joule, named in honor of 19th- century scientist, James Prescott Joule, is such a measure. A **joule (J)** has a very technical definition based on the work done by a force moving an object over a specified distance. In more familiar terms, it is approximately the energy required to heat 1 g of water by 0.24 °C. Scientists and engineers also use other units of energy for particular purposes. All of these units can be converted to joules (Figure 3.24).

Temperature describes the average kinetic energy of the molecules in a system, but the total energy content of a system also depends on the number of molecules within it. A **calorie** (cal) is the energy required to raise the temperature of 1 g of water by 1 °C (1 cal = 4.18 J). Thus, it is sensitive to both the kinetic energy of the molecules and their amount. Biologists and ecologists frequently use calories to measure changes in the energy of organisms and ecosystems. The Calorie (with a capital C) that is commonly used to measure the energy content of food is equal to 1,000 cal, or 1 kcal (small-c calories).

The energy unit most commonly used to measure the everyday use of electricity is the **watt-hour (Wh)**. One kilowatt-hour (kWh) is equal to 1,000 Wh. A watt-hour is the energy expended over an hour at the rate of 1 joule per second. Since there are 3,600 seconds in an hour, 1 Wh is equal to 3,600 J. Thus, a 40-watt lightbulb consumes 144,000 J in the course of an hour.

In summary, energy is neither created nor destroyed; it is only transformed from one form to another. Whenever it is transformed, the entropy in the universe increases. Forms of energy include electromagnetic radiation, heat, chemical energy, and nuclear energy. The units used to measure energy include joules, calories, and watt-hours.

▼ Figure 3.24 **Energy Conversions**

	Joules (J)	Calories (c)	Watt-hours (Wh)
A joule (J) =	1	0.24	0.00028
A calorie (c) =	4.18	1	0.0012
A watt-hour (Wh) =	3,600	861	1

QUESTIONS 3.3

1. Give an example of potential and kinetic energy.

2. What are the laws of thermodynamics?

3. How does the energy content of light relate to its wavelength?

4. Describe four ways in which heat energy can move from Earth's surface to the adjacent atmosphere.

5. Differentiate between nuclear fission and nuclear fusion.

(MES) For additional review, go to **MasteringEnvironmentalScience**

3.4 Earth's Structure

BIG IDEA During the early evolution of Earth, gravity pulled heavier materials into its center, and lighter materials moved toward its surface. As a result, Earth has three distinct layers: a dense core, a less dense mantle, and a light, rocky crust (**Figure 3.25**). The crust is divided into large segments, or plates. Over Earth's history, the movement of these plates has produced significant changes in the geographic configuration of the oceans and continents. Through geologic time, elements within Earth's crust and mantle slowly cycle into the living environment and then back again.

The Core, Mantle, and Crust

■ Solid Earth's three layers have very different physical properties.

At Earth's center is the **core**. The core is composed of a mixture of nickel and iron, with smaller amounts of other heavy elements. The core has two parts: a solid inner core, with a radius of 1,220 km (760 mi), and a liquid outer core that extends 2,260 km (1,400 mi) toward the surface. Both the inner and outer core are under extreme pressure—three million times greater than the atmospheric pressure at Earth's surface. The inner and outer core are also extremely hot, with temperatures in excess of 7,000 °C.

Convection currents and Earth's rotation stir the molten iron of the outer core, setting up a strong magnetic field. This magnetic field orients the magnetic needle of a compass. Some birds and insects also use this magnetic field to navigate over long distances; these animals have magnetized iron crystals in their skin or specialized organs for orientation. More important to the maintenance of life on Earth, this magnetic field deflects the high-energy protons and electrons in the solar wind, causing most of these particles to travel around, rather than through, our atmosphere. Direct exposure to these particles would be lethal to most organisms.

Above the core is a layer of less dense rock called the **mantle**. The mantle, which is 2,900 km (1,800 mi) thick, occupies about 70% of Earth's total volume. The rocks of the mantle are rich in the elements magnesium and silicon, with much smaller amounts of calcium and aluminum. Because they are under great pressure, the solid rocks in the mantle can slowly flow, in a plastic fashion like modeling clay. Heat from the core causes the mantle rock to circulate in huge, slow-moving convection currents. In parts of the upper mantle, rocks may be heated to a liquidlike state called **magma**.

Above the mantle is Earth's **crust**, a thin veneer of solid and relatively light rocks. The crust, which makes up less than 1% of Earth's total volume, varies in depth from 5 to 100 km (3–62 mi). This thin layer is the only part of Earth that directly interacts with living systems.

There are two types of crust, which differ in their chemistry and density. Continental crust makes up the continents and the areas immediately adjacent to them. The many different kinds of rocks in continental crust contain large amounts of lighter elements, especially aluminum and silicon. Ocean crust, which lies beneath most of the deep ocean, is composed of denser and more homogeneous rocks that are similar to the rocks in the mantle. Continental crust varies in thickness from 30 to 100 km (20–62 mi), whereas ocean crust is uniformly about 5 to 10 km (3–6 mi) thick.

Geologists refer to the crust and the upper reaches of the mantle that interact with it as the **lithosphere**. This is the zone of geologic activity that has shaped Earth's surface and continues to do so. Earth's hydrosphere is composed of the liquid water and ice in and on the crust and includes oceans, rivers, and glaciers.

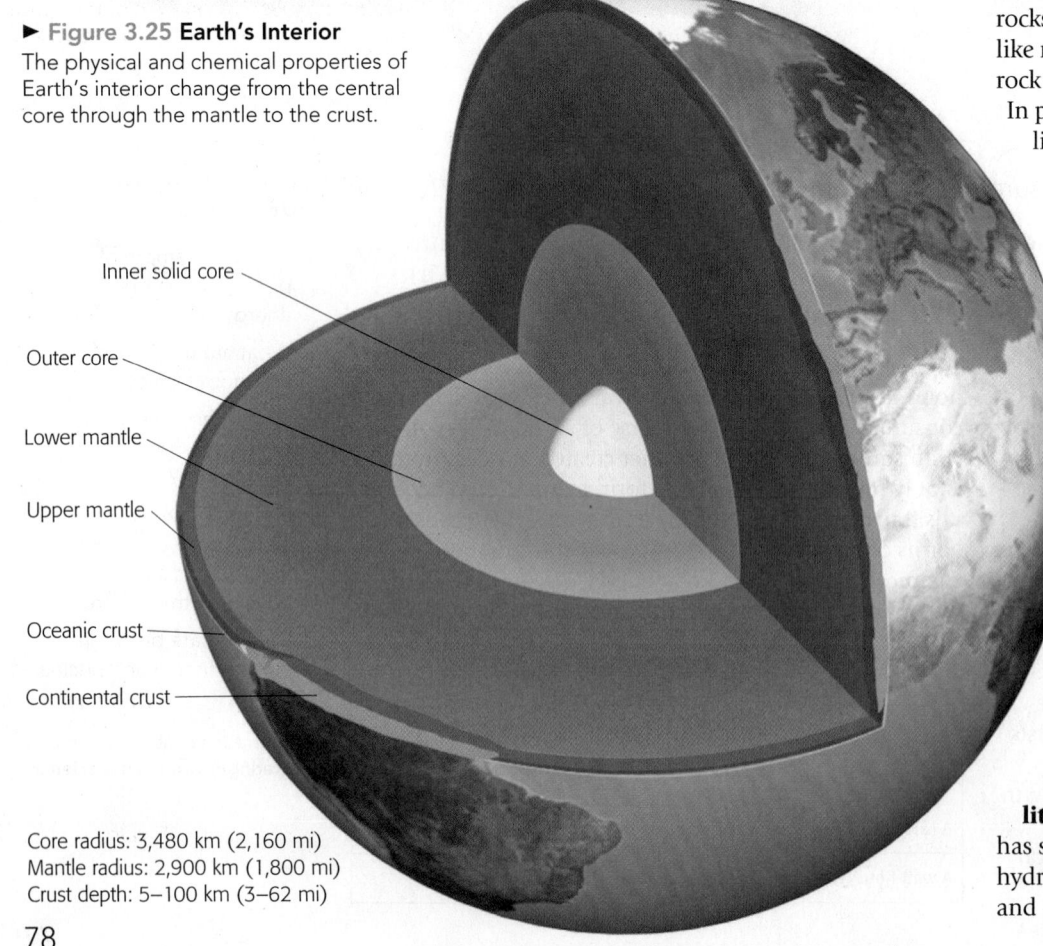

► Figure 3.25 **Earth's Interior**
The physical and chemical properties of Earth's interior change from the central core through the mantle to the crust.

Inner solid core

Outer core

Lower mantle

Upper mantle

Oceanic crust

Continental crust

Core radius: 3,480 km (2,160 mi)
Mantle radius: 2,900 km (1,800 mi)
Crust depth: 5–100 km (3–62 mi)

Building and Moving Continents

■ The geographic configuration and distribution of oceans and continents have changed through time.

Since the existence of relatively accurate world maps, geographers have noticed that the margins of some continents match up with those of other continents. For example, if the Atlantic Ocean separating South America and Africa were removed, these two continents would come together like matching pieces in a puzzle (Figure 3.26). In the 19th and early 20th centuries, geologists observed that sedimentary rocks on separate continents often hold the same kinds of fossils.

In 1911, the German scientist Alfred Wegener attempted to explain these observations by suggesting that in the distant past all of the continents were joined as one large continent. He called this ancient supercontinent Pangea. Wegener proposed that over time Pangea broke apart, and the smaller continents drifted away. However, Wegener could not supply a mechanism that would cause the continents to move. In the absence of such a mechanism, geologists largely dismissed Wegner's hypothesis for the next 50 years.

By the 1960s, evidence was beginning to accumulate that Earth's crust was indeed able to move and that its movements were a consequence of interactions between the crust and the mantle. New techniques allowed geologists to determine when rocks had formed. These techniques revealed that the ages of rocks in the crust vary greatly. Some continental rocks are quite new, whereas

others are nearly 4 billion years old. In contrast, rocks in the ocean crust are relatively young; none are more than 200 million years old. Geologists also observed that the youngest rocks in the ocean crust are found near mid-ocean ridges; the oldest rocks in the ocean crust are found next to the continents.

All of these observations have come to be understood through the theory of plate tectonics. According to this theory, Earth's crust is broken into pieces called **tectonic plates** that float on top of the mantle (Figure 3.27). Continents are embedded in some of these plates. The positions of the plates are not fixed. Instead, they slowly move in relation to each other. What causes tectonic plates to move? As convection currents slowly stir the plastic materials in the mantle, the plates also move. The movement of the mantle is like an enormous conveyor belt, providing the energy that moves the plates along.

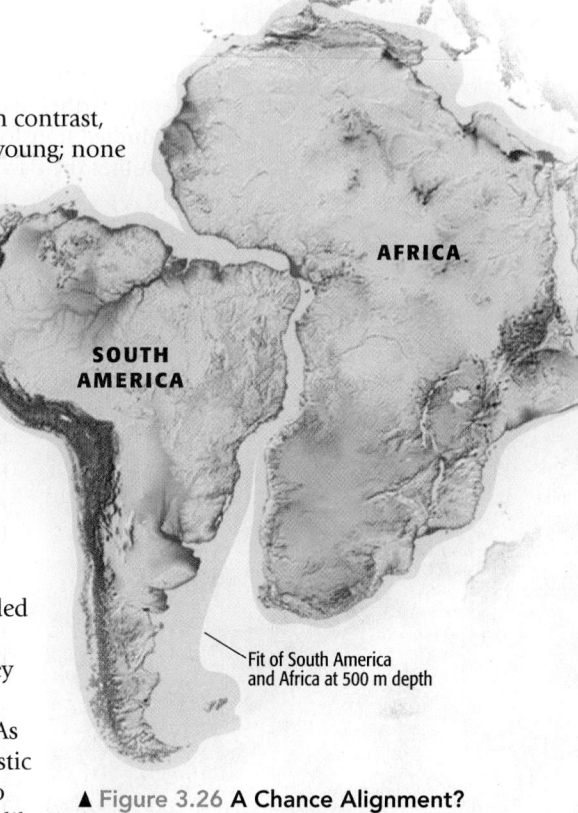

Fit of South America and Africa at 500 m depth

▲ Figure 3.26 **A Chance Alignment?** By the late 19th century, global maps were sufficiently exact that the alignment of the coasts of South America and Africa were quite obvious.

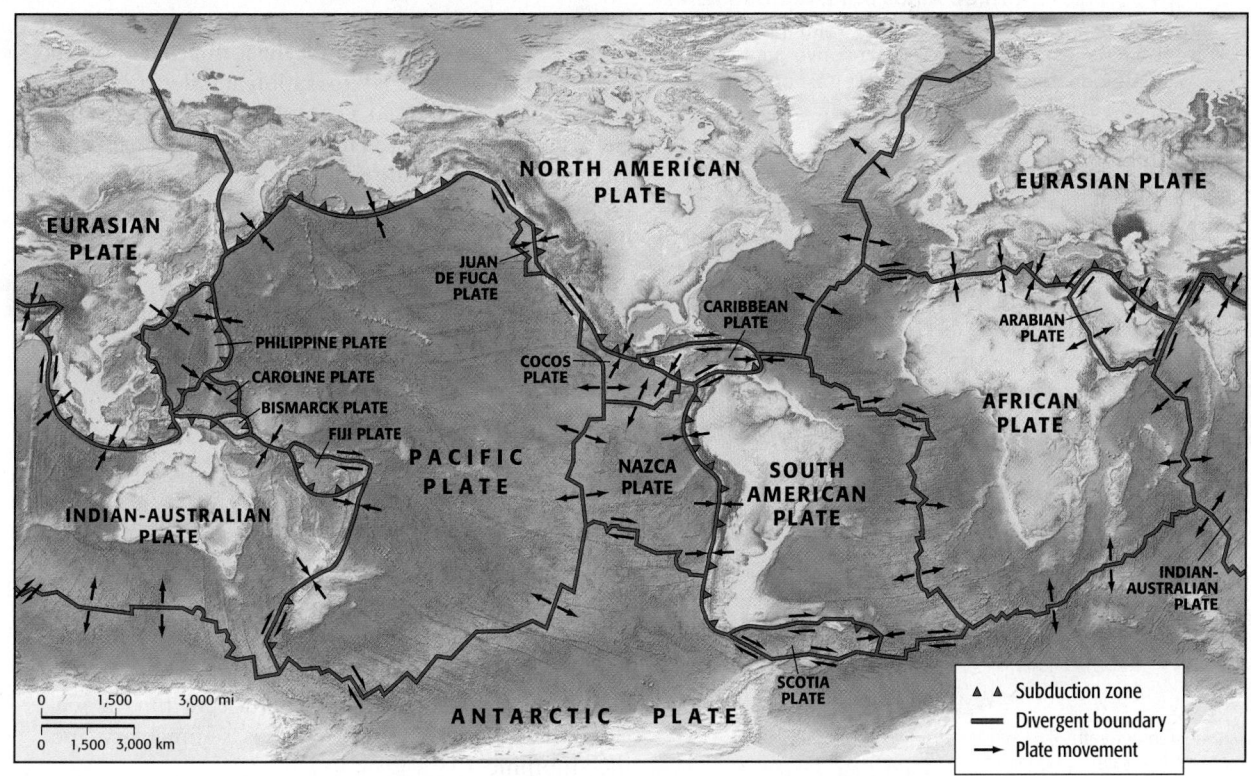

◄ Figure 3.27 **The Fractured and Mobile Crust**

Earth's crust is divided into 17 tectonic plates that move relative to one another.

▲ ▲ Subduction zone
— Divergent boundary
→ Plate movement

▼ Figure 3.28 **A Transform Fault**
A section of the San Andreas Fault in central California. Terrain west of the fault is moving north relative to that to the east.

The places where plates come together are called boundaries; the type of boundary depends upon how the plates are moving in relation to one another. Plates slide past one another along **transform fault boundaries**. As plates move past one another in fits and starts, they may generate earthquakes. The San Andreas Fault is one of the best-studied transform fault boundaries (Figure 3.28). It separates the Pacific Plate, which includes a significant portion of south coastal California, from the western portion of the North American Plate and the rest of California.

Divergent boundaries separate plates that are moving apart. Most such boundaries coincide with ocean ridges. The Mid-Atlantic Ridge is an example of a divergent boundary. It separates the North American Plate from the Eurasian Plate and the South American Plate from the African Plate. At divergent boundaries, hot magma from the mantle rises to the surface, forming new ocean crust.

Tectonic plates collide at **convergent boundaries**. Where ocean and continental plates meet, the dense ocean crust is thrust beneath the lighter continental crust in a process called subduction (Figure 3.29). As it is pushed into the mantle, subducted crust is heated, generating earthquakes and creating an area beneath the continent where magma is abundant. This magma may rise to the surface, forming volcanoes. For most of its length, the west coast of South America forms convergent boundaries with other tectonic plates. The volcanic Andes Mountains are a product of subduction occurring along these boundaries.

Where continental crust collides with continental crust, the land along the boundary is thrust upward, sometimes on a grand scale. For example, the subcontinent of India (part of the Indian–Australian Plate) has been colliding with the Eurasian Plate for nearly 50 million years. Their collision produced the mountains of the Hindu Kush and the Himalayas. The Appalachian Mountains of the United States and Atlas Mountains of Africa were created between 240 and 300 million years ago, when southeastern North America collided with northwest Africa. These two mountain ranges once rose as high as the Himalayas, but have since been eroded to their current stature.

To appreciate how important tectonic activity is to Earth's structure, consider some of Earth's closest neighbors. Mercury, Mars, and our Moon have rigid crusts but lack tectonic plates. Their surfaces bear clear evidence of their distant geologic past. That evidence includes tens of thousands of craters accumulated from impacts with meteors over hundreds of millions of years. Earth, too, has been struck by thousands of meteors, but the subduction, mountain building, and erosion associated with plate tectonics have continuously reworked Earth's surface, erasing all but a few meteor craters.

More importantly, Earth's dynamic crust has been a significant factor in the development of life. The materials that make up the crust contain essential elements in proportions that are helpful for the formation of organic molecules. The continual formation and destruction of the crust has provided a continuous supply of the elements necessary to the function of living organisms and ecosystems. The moving continents have also affected the way in which individual organisms and ecosystems have evolved.

▼ Figure 3.29 **Three Kinds of Plate Boundaries**
Ⓐ Ocean crust is created at divergent plate boundaries. Ⓑ Plates collide at convergent plate boundaries. Ⓒ Plates slide past one another along transform fault boundaries.

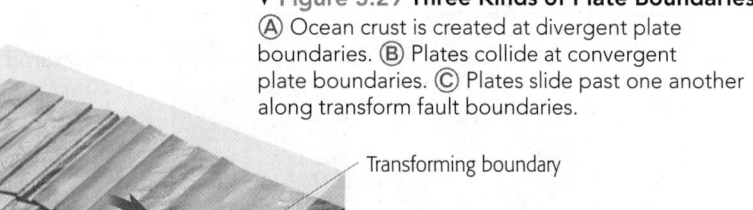

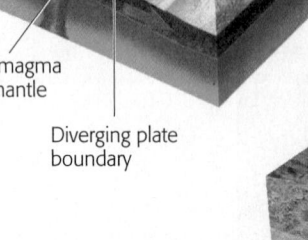

Transforming boundary

Oceanic crust formed when magma cools and solidifies

Rising magma from mantle

Diverging plate boundary

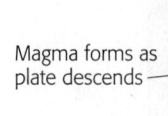

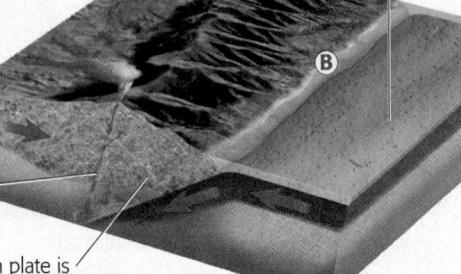

Movement of oceanic plate

Magma forms as plate descends

Ocean plate is subducted beneath continental plate

The Rock Cycle

■ Elements cycle through the lithosphere.

The forces of plate tectonics in combination with weathering, erosion, and deposition cause elements to slowly cycle through Earth's lithosphere as well as its ecosystems. The **rock cycle** is a model that describes a dynamic process in which the physical material that makes up the crust and mantle is slowly transformed from one type of rock to another (Figure 3.30). These geologic processes lead to the wide variety of rocks that make up Earth's crust, an important abiotic factor for any ecosystem. Geologists classify rocks within the rock cycle into three major types: igneous, sedimentary, and metamorphic.

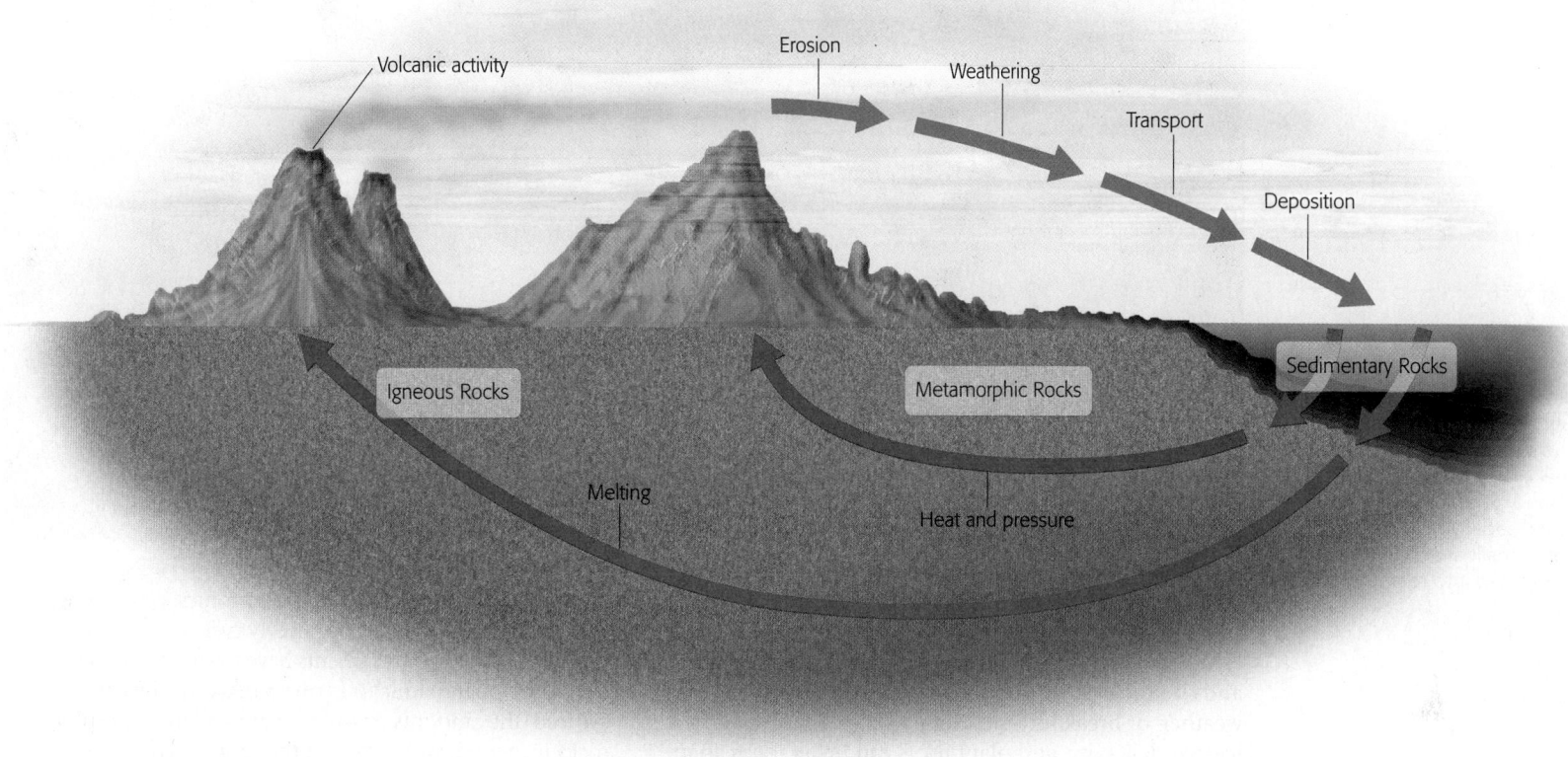

▲ Figure 3.30 The Rock Cycle
Over long periods of time, the matter that makes up one particular rock type can be transformed to create other types of rock. Note that this figure is not drawn to scale.

We start the rock cycle with magma, molten rock. Where temperatures and pressures in the crust or upper mantle are sufficient, all types of rocks melt, forming magma. As magma cools, it solidifies to form **igneous rock**. The characteristics of igneous rock depend on the chemical composition of the magma and the rate at which the magma cools. Magmas formed from material in the mantle solidify into rocks that contain large amounts of iron and magnesium. Magmas formed from continental crust that melted in subduction zones solidify into rocks that contain large amounts of aluminum and silicon.

If magma solidifies beneath Earth's surface, it forms intrusive igneous rock, such as granite (Figure 3.31). Because magma cools very slowly underground, the crystals in the rock have time to grow large, giving intrusive igneous rocks a rough, granular texture. Magma that flows or is ejected onto Earth's surface forms extrusive igneous rock. Basalt, which forms from volcanic lava, is an example of an extrusive igneous rock. Because basalt cools rapidly, it has very small crystals.

► Figure 3.31 Igneous Rock
The granite of Yosemite's Half Dome is an intrusive igneous rock that solidified from magma within Earth's crust and has been exposed by weathering and erosion.

► Figure 3.32 **Sedimentary Rock**
Sedimentary rocks, such as the sandstone and shale that make up the layers of the Grand Canyon, are formed when sediments such as sand or silt are chemically cemented into rock.

Sedimentary rock forms when sediments, such as sand, silt, and the remains of dead organisms, become "glued together" under pressure. A variety of physical and chemical processes on Earth's surface cause rocks to weather, or break into smaller particles. Flowing water, ice, wind, gravity, and plant roots can break rocks apart. As the roots of plants grow into crevices, they force rocks apart and release chemicals that accelerate weathering. Chemicals in decaying organic matter also accelerate rock weathering. The rock particles left by weathering make up the major part of soil. The physical and chemical properties of soils are determined in large part by the nature of the rock from which they develop.

In the process of erosion, weathered rock particles and soil are carried away by gravity, water, wind, or glacial ice. Eventually, these materials are deposited as sediment in locations such as the mouths of rivers and the bottoms of lakes and oceans. Shells, skeletal remains, and undecomposed organic matter from the biosphere also contribute to sediment. Minerals that percolate through sediment glue the particles together, forming rock. As thick layers of sediment accumulate, their weight creates pressure on the sediment layers below, accelerating the process of rock formation.

The physical and chemical characteristics of sedimentary rocks depend on the sediment from which they form (Figure 3.32). For example, sand forms silica-rich sandstones; silt and clay form shale. The calcium-rich shells and skeletons of marine organisms that settle on the ocean floor become limestone. Coal is a sedimentary rock formed from peat and other organic matter that has been transformed by the pressure of the layers of sediment above it.

The very high temperatures and pressures associated with tectonic processes can alter the chemical and physical properties of both sedimentary and igneous rocks, producing **metamorphic rock**. The characteristics of metamorphic rocks depend on the parent rock and the processes of transformation. Under high pressure and temperature, shale becomes slate and limestone becomes marble (Figure 3.33).

Over Earth's long history, plate tectonics, weathering, erosion, and deposition have slowly cycled elements through Earth's crust. Elements have been transported from the upper mantle to Earth's surface. As the continents evolved, the amounts of sedimentary and metamorphic rocks increased. Weathering of the various types of rocks produces soil and releases nutrient elements. These elements can then be taken up in the biosphere or dissolved in the waters of the hydrosphere. By weathering rocks and contributing to the formation of soils and sediments, living organisms play important roles in the rock cycle.

▼ Figure 3.33 **Metamorphic Rock**
The marble in this quarry was formed as sedimentary limestone that was altered chemically under high temperature and pressure.

QUESTIONS 3.4

1. What is the source of Earth's magnetic field?
2. How does continental crust differ from ocean crust?
3. Describe three types of boundaries between tectonic plates.
4. Name and describe the three basic kinds of rock.
5. Describe two roles of plate tectonics in the rock cycle.

 For additional review, go to **MasteringEnvironmentalScience**

3.5 Element Cycles in Earth's Ecosystems

BIG IDEA All of the organisms on Earth and the nonliving environment with which they interact make up the **biosphere**. The biosphere includes the rocks and soils that make up Earth's crust, its fresh water and oceans (the hydrosphere), and its atmosphere. A **biogeochemical cycle** describes the flow of matter through an ecosystem. Most biogeochemical cycles describe the flow of elements, such as carbon or nitrogen, but some describe the flow of compounds, such as water. The distribution and rate of movement of elements in the biosphere is determined by whether or not they are required by organisms, are abundant in the atmosphere, are soluble in water, and are easily immobilized in the lithosphere. Nutrient elements are those that are essential to life's processes.

Biogeochemical Cycles

■ The distribution, abundance, and movement of elements in an ecosystem can be tracked by measuring element pools and fluxes.

A biogeochemical cycle accounts for the flow and chemical transformation of elements as they move among the living and nonliving parts of an ecosystem. Generally, ecosystem scientists describe biogeochemical cycles in terms of pools and fluxes (Figure 3.34). **Pools** are the parts of an ecosystem in which matter may reside, such as the atmosphere, soil, or organisms. **Fluxes** are the rate at which matter moves from one pool to another. By knowing the mass of an element in each pool in an ecosystem and the fluxes of the element through that ecosystem, scientists can account for changes in the abundance of that element within an ecosystem. Ecosystem scientists call this **mass-balance accounting**.

Biogeochemical cycles may be considered for Earth in its entirety or for smaller areas, such as an individual forest or watershed. Ecosystem scientists define pools at various levels of detail, depending on their research interests and management goals. In studies of the global cycling of an element such as carbon or nitrogen, scientists may define only four pools—Earth's crust, waters, atmosphere, and living organisms. More often, scientists divide ecosystems into multiple pools. Thus, a scientist interested in the cycling of an element within a lake would probably designate pools such as lake water, organic sediment, inorganic sediment, and organisms. Because many elements move through food webs, that scientist might further divide the organism pool into primary producers, consumers, and decomposers (see Module 6.4).

▼ Figure 3.34 **Model Biogeochemical Cycle**
Scientists typically represent the element pools as named boxes, and they use arrows to signify the fluxes among different pools.

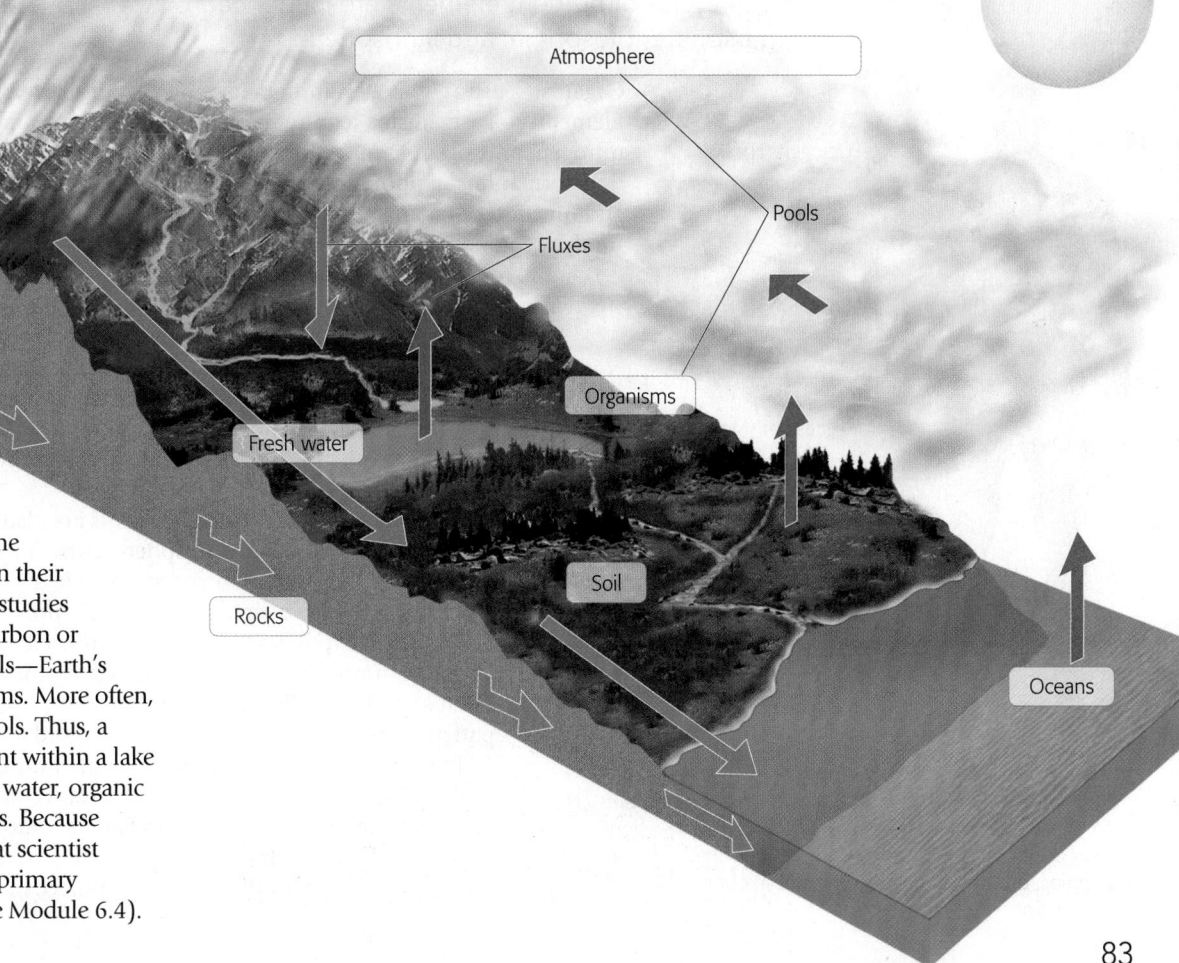

Atmosphere

Pools

Fluxes

Organisms

Fresh water

Soil

Rocks

Oceans

QUESTIONS 3.5

1. What is the difference between a negative and a positive flux?

2. What two factors determine the residence time of an element in a particular pool?

3. What is a nutrient element?

4. Name and describe four factors that influence the global cycling time of nutrient elements.

(MES) For additional review, go to **MasteringEnvironmentalScience**

The **capital** of a particular pool is the total amount (mass) of an element or molecule that it contains. For the entire biosphere, pool sizes are generally measured in teragrams ($1 \text{ Tg} = 10^{12}$ g) or petagrams ($1 \text{ Pg} = 10^{15}$ g). At smaller scales, such as an individual forest or a watershed, pool sizes are measured in grams (g) or kilograms (kg) per unit area, usually per square meter (m^2). The total capital of a particular element in an ecosystem is the sum of the capital in each of the individual pools that compose that ecosystem.

Element fluxes are expressed as the amount of an element that moves from one pool to another in a unit of time. Global fluxes are expressed as teragrams or petagrams per year. Fluxes at smaller scales are described as grams or kilograms per unit area per year (e.g., $kg/m^2/yr$).

Fluxes into a pool are positive, and fluxes out are negative. The net element flux for a pool is the sum of all of the fluxes into and out of it. If the net flux of an element is positive, then the pool's capital of that element is growing. If the net flux is negative, then the pool's capital of that element is shrinking. If the net flux of an element is zero, then the capital of the element in that pool remains constant; such a pool is said to be in **equilibrium**.

Water in a bathtub is a helpful metaphor for these concepts. The flux into the tub is the rate at which water flows from the spigot. The flux out of the tub is the rate at which gravity pulls water down the drain. If you turn on the spigot full force and leave the drain open, the flux of water into the tub may exceed the capacity of the drain to let water out. The net flux will be positive, and the tub will fill up. But if you slowly turn the spigot down, you can find a rate of flow at which the level of water in the tub remains unchanged. Although water is still flowing in and out of the tub, its net flux is zero and the water level is at equilibrium.

Residence time is the average time that an atom of an element or molecule of a compound spends in a pool. Residence time is easily calculated as the size of the equilibrium pool divided by the flux through the pool. Consider the example of the bathtub. If the volume of water at equilibrium is 5 gallons and the spigot is flowing at 1 gallon/minute, then the residence time of a molecule of water in the tub equals 5 minutes. (Note that unlike the capital of elements, quantities of water are measured by volume.) **Cycling time** is the average time that it takes an element or molecule to make its way through an entire biogeochemical cycle.

Nutrients

■ The cycles of nutrient elements are influenced by their abundance and mobility in the atmosphere, biosphere, and lithosphere.

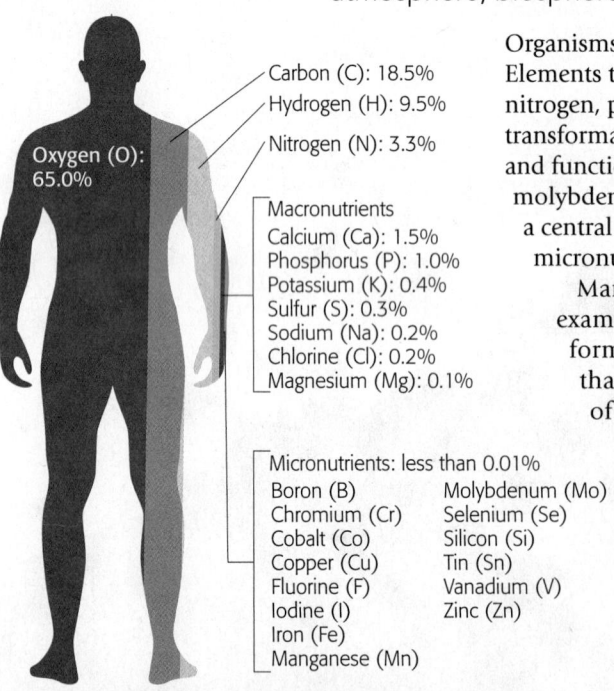

Oxygen (O): 65.0%

Carbon (C): 18.5%
Hydrogen (H): 9.5%
Nitrogen (N): 3.3%

Macronutrients
Calcium (Ca): 1.5%
Phosphorus (P): 1.0%
Potassium (K): 0.4%
Sulfur (S): 0.3%
Sodium (Na): 0.2%
Chlorine (Cl): 0.2%
Magnesium (Mg): 0.1%

Micronutrients: less than 0.01%
Boron (B)   Molybdenum (Mo)
Chromium (Cr)   Selenium (Se)
Cobalt (Co)   Silicon (Si)
Copper (Cu)   Tin (Sn)
Fluorine (F)   Vanadium (V)
Iodine (I)   Zinc (Zn)
Iron (Fe)
Manganese (Mn)

▲ **Figure 3.35 Nutrient Elements**
What elements do you and all life require? Macronutrients are elements we need in large amounts and compose a significant part of our body's mass. Micronutrients are required in smaller quantities, but they play essential roles in our growth and reproduction.

Organisms use at least 25 nutrient elements to carry out their life functions (**Figure 3.35**). Elements that organisms require in comparatively large amounts, such as carbon, oxygen, hydrogen, nitrogen, phosphorus, and sulfur, are **macronutrients**. Macronutrients are important in the energy transformations of photosynthesis and cellular respiration and are also necessary to the structure and function of organisms (see Module 4.2). Organisms also require **micronutrients**, such as molybdenum, manganese, and boron, but only in very small amounts. These elements often play a central role in the catalytic function of enzymes. Different organisms may require different micronutrients.

Many nutrient elements are cycled as organisms eat or are eaten by other organisms. For example, producers such as plants, algae, and cyanobacteria are able to take up inorganic forms of carbon. All other organisms obtain carbon by eating either producers or organisms that eat producers. The rate at which carbon and other nutrients move through the pool of living organisms in Earth's biogeochemical cycles is influenced by the complexity of food webs.

Cycling times for nutrient elements are also related to the size and characteristics of their pools in Earth's atmosphere, crust, and waters. Elements that are abundant in the atmosphere, such as nitrogen, oxygen, and carbon, have short cycling times. Other elements, such as phosphorus, iron, calcium, and magnesium, have chemical forms that are easily tied up in soils and rocks; these elements flow through the biosphere slowly, and their atmospheric pools are typically small. Nutrients that are soluble in water are often abundant in Earth's waters. The cycles of soluble nutrients are greatly influenced by patterns of water movement across the land and in the ocean.

In summary, living organisms require nutrient elements to grow and reproduce. Cycles of these elements are affected by their movement through food webs and by factors that influence their residence times in Earth's atmosphere, crust, and waters. In later chapters, we will explore in detail the global cycling of the macronutrients carbon, nitrogen, and phosphorus and how they interact with ecosystem resources.

3.6 Earth's Atmosphere

BIG IDEA Earth, Venus, and Mars have sufficient gravity to hold an **atmosphere**, the layer of gases above their surfaces (Figure 3.36). The chemical composition of Earth's atmosphere is unlike those of its planetary neighbors; over billions of years, the actions of photosynthetic organisms have decreased the concentration of carbon dioxide and increased the concentration of oxygen. On average, water accounts for about 1% of the molecules in Earth's atmosphere. However, the actual amount of water in the air varies widely from time to time and place to place. Scientists divide Earth's atmosphere into four altitudinal zones, based on changes in temperature and physical characteristics.

Composition of Gases

■ Earth's unique atmosphere sustains life and is a product of life.

The atmospheres of both Venus and Mars are composed of over 96% carbon dioxide (CO_2), with about 3% nitrogen (N_2) and traces of other gases, such as argon (Ar). In contrast, Earth's atmosphere contains 78% nitrogen and 21% oxygen (O_2); CO_2 accounts for a mere 0.039%. Because many organisms depend on oxygen for survival, it is tempting to attribute the presence of life on Earth to the abundance of oxygen in its atmosphere. But it is more correct to say that Earth's unique oxygen-rich atmosphere is a consequence of the presence of life.

It appears that 3 billion years ago the composition of Earth's atmosphere was similar to that of Venus and Mars. When photosynthetic microbes arose, they used atmospheric CO_2 to manufacture carbohydrates, releasing O_2 as a by-product. As a result, the concentration of oxygen began to rise, and the concentration of CO_2 began to decline. Over billions of years, photosynthetic organisms slowly altered the composition of the atmosphere. Thus, we can thank photosynthesis for the oxygen-rich air that we breathe.

Ⓑ Mars

▼ Figure 3.36 **Planetary Atmospheres**
Ⓐ Earth's atmosphere is rich in nitrogen (N_2) and oxygen (O_2), and the clouds indicate lots of water vapor. Ⓑ The Martian atmosphere is nearly 97% carbon dioxide (CO_2), with virtually no water vapor.

Ⓐ Earth

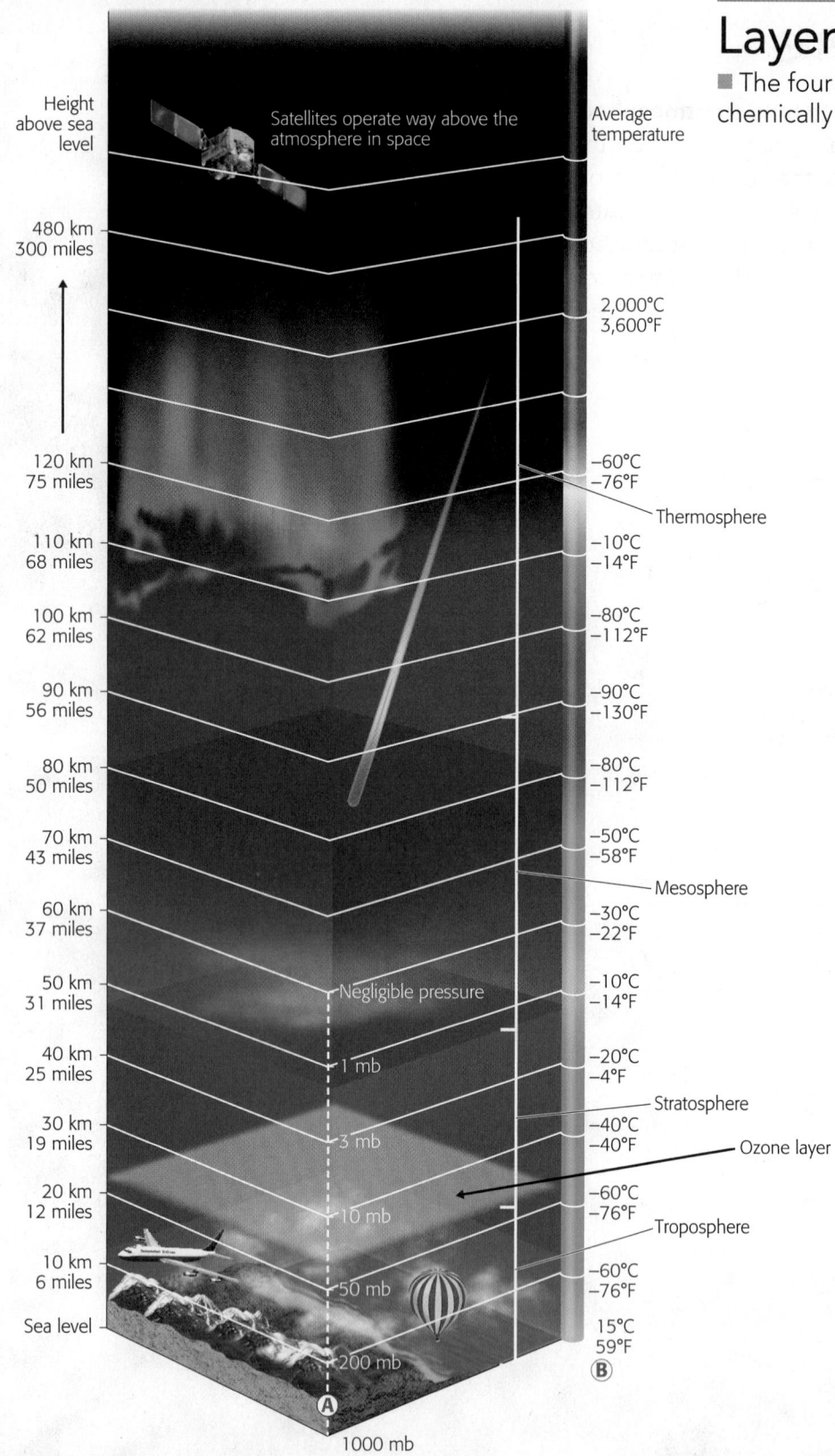

Height above sea level

- 480 km / 300 miles
- 120 km / 75 miles
- 110 km / 68 miles
- 100 km / 62 miles
- 90 km / 56 miles
- 80 km / 50 miles
- 70 km / 43 miles
- 60 km / 37 miles
- 50 km / 31 miles
- 40 km / 25 miles
- 30 km / 19 miles
- 20 km / 12 miles
- 10 km / 6 miles
- Sea level

Satellites operate way above the atmosphere in space

Average temperature

- 2,000°C / 3,600°F
- −60°C / −76°F — Thermosphere
- −10°C / −14°F
- −80°C / −112°F
- −90°C / −130°F
- −80°C / −112°F
- −50°C / −58°F
- −30°C / −22°F — Mesosphere
- −10°C / −14°F
- −20°C / −4°F
- −40°C / −40°F — Stratosphere
- −60°C / −76°F — Ozone layer
- −60°C / −76°F — Troposphere
- 15°C / 59°F

Negligible pressure

1 mb
3 mb
10 mb
50 mb
200 mb
1000 mb

Ⓑ

Ⓐ

Layers of the Atmosphere

■ The four layers of Earth's atmosphere differ from one another chemically and physically.

Earth's atmosphere is approximately 480 km (300 mi) deep. The air at Earth's surface is compressed by the weight of the great column of air above it, affecting the density of atmospheric gases (i.e., the number of molecules per unit volume). The force caused by the pull of gravity on a column of air is **atmospheric pressure**. At sea level, atmospheric pressure is 1,013 millibars (mb), or 14.7 pounds per square inch. With increasing altitude, the weight of the air above decreases, and so does the atmospheric pressure. At an altitude of 5.5 km (18,000 ft), atmospheric pressure is about one-half (500 mb) that at sea level. An airliner at 12.2 km (40,000 ft) is flying through air with an atmospheric pressure of less than 170 mb.

Scientists divide the atmosphere into four layers based on differences in temperature and chemical properties (**Figure 3.37**). The lowest layer, the **troposphere**, provides the air that we breathe. Most of our planet's weather occurs in the troposphere. Air at low altitudes is warmed by heat radiating from Earth's surface. As the air near the ground becomes warmer, it expands, which causes it to rise. As it rises, it cools, and eventually sinks back to the surface. This process of thermal convection keeps the air in the troposphere well mixed. Thus, pollutant gases, such as sulfur dioxide from smokestacks, are quickly dispersed through the troposphere.

If you could travel up through the troposphere, you would find that the air temperature decreases as the altitude increases. This decrease occurs at a rate of 6 °C every kilometer (3.5 °F/1,000 ft). At an altitude of approximately 15 km (9 mi), air reaches a minimum temperature of about −52 °C (−62 °F), and temperature ceases to fall. You have reached the tropopause, the upper boundary of the troposphere.

Above the tropopause is the **stratosphere**, which extends from 15 to about 48 km (9–30 mi) above Earth's surface. Moving upward through the stratosphere, temperature gradually increases. The upper boundary of the stratosphere is the stratopause, where the temperature reaches a maximum of approximately −30 °C (−27 °F)

The chemical composition of the stratosphere is very similar to that of the troposphere, except the stratosphere contains comparatively large amounts of oxygen in the form of ozone (O_3). The concentration of ozone is highest at altitudes between 15 and 35 km (9–22 mi), often called the **ozone layer**. Ozone is particularly effective in absorbing and scattering the abundant UV light that strikes Earth's upper atmosphere. The high amount of energy contained in UV photons can damage many organic molecules, including DNA. Thus, the ozone layer provides important protection to life on Earth.

Unlike the air in the troposphere, the air in the stratosphere undergoes very little convectional mixing. So, when chemical pollutants get into the stratosphere,

▲ Figure 3.37 **Layers of the Atmosphere**
Ⓐ At sea level, atmospheric pressure is 1,013 millibars (mb). With increasing altitude above Earth's surface, the weight of air above, and therefore the atmospheric pressure, diminishes. The atmosphere divides into four unequal layers based on temperature. Ⓑ The temperature and chemical properties of the atmosphere change from the troposphere through the stratosphere, mesosphere, and thermosphere. *Source: Copyright DK LTD*

they are not rapidly dispersed. Rather, they tend to remain there for long periods of time.

Above the stratopause is the **mesosphere**, where the air temperature begins to drop again, reaching lows of –90 °C (–130 °F) at about 90 km (55 mi). Above this is the **thermosphere**, which extends about 480 km (300 mi) into space. The density of gas molecules in the thermosphere is very low indeed. At the boundary between the mesosphere and thermosphere, atmospheric pressure is only 0.0003% of that at sea level. Above about 150 km (95 mi), the gases are so thin that they create virtually no friction with spacecraft. The International Space Station orbits in the thermosphere at an altitude between 320 and 380 km (200–235 mi).

As the name thermosphere suggests, the gases in this layer are heated by direct solar radiation to very hot temperatures, exceeding 2,000 °C (3,600 °F). However, these superheated molecules are so far apart that when astronauts spacewalk in the thermosphere, they do not feel warm. Many gases in the thermosphere are ions, which have electrical charge. These ions interact with the solar wind and Earth's magnetic field to produce the auroras, or northern and southern lights (**Figure 3.38**).

▲ Figure 3.38 **Aurora**
In polar regions, ions from the thermosphere interact with Earth's magnetic field creating a shimmering curtain of glowing particles.

Water in the Atmosphere

■ Water content of the atmosphere varies in a predictable fashion.

The amount of water in the atmosphere varies greatly from place to place and from time to time. On average, water vapor (H_2O) makes up about 1% of the molecules in Earth's atmosphere. At sea level, the **vapor pressure** of water, its relative contribution to total atmospheric pressure, is about 10 mb, or 1% of 1,013 mb. However, the vapor pressure of water in air ranges from as little as 0.01 mb in the driest deserts to over 40 mb in the most humid places.

At any particular temperature, the air can hold only so much water. That amount is the **saturation vapor pressure** for that temperature. Above the saturation vapor pressure, water condenses into a liquid, forming droplets of fog or rain. Saturation vapor pressure—the amount of water the air can potentially hold—increases as the temperature increases (**Figure 3.39**).

Relative humidity is a measure of the extent to which air is saturated, expressed as a percentage. The relative humidity of a mass of air rises and falls with its temperature (**Figure 3.40**). Changes in relative humidity explain the condensation that forms on the outside of a pitcher of iced tea on a hot, humid day. As air circulates near the pitcher, it is cooled. This causes its saturation vapor pressure to drop and its relative humidity to increase. When the air immediately adjacent to the pitcher reaches 100% humidity, droplets of water condense on the glass.

As an air mass becomes cooler, it eventually reaches its **dew point**, the temperature at which the relative humidity is 100%. If the temperature drops to the dew point overnight, water condenses on surfaces, such as your lawn or car. As the temperature warms in the morning, the relative humidity drops and the dew evaporates.

Relative humidity affects the rate at which water evaporates, or changes from a liquid to water vapor. As humidity increases, evaporation occurs more slowly. For example, on a day with low humidity, your perspiration evaporates quickly, cooling your body. On a humid day, your perspiration evaporates more slowly. That's why hot, humid days can seem more uncomfortable than hot, dry days. Similarly, the loss of water from the surfaces of the ocean, lakes, and plant leaves slows down as relative humidity rises.

These transformations of water are especially important in the troposphere, where they influence variations in rainfall and the evaporation across Earth's surface, thereby playing an important role in determining climate. Because of the importance of water in all ecosystems, variations in the amount of water in the atmosphere are of particular interest to environmental scientists.

QUESTIONS 3.6

1. The small amount of CO_2 in Earth's atmosphere is evidence of life. Explain.

2. Describe how temperature changes through each of Earth's four atmospheric layers.

3. What is the relationship between air temperature, saturation vapor pressure, and dew point?

(MES) For additional review, go to **MasteringEnvironmentalScience**

▼ Figure 3.40 **Temperature and Relative Humidity**
Daily change in temperature and relative humidity at Sacramento, California, over three days in May 2014.

Source: U.S. National Oceanographic and Atmospheric Administration, National Weather Service 2014.

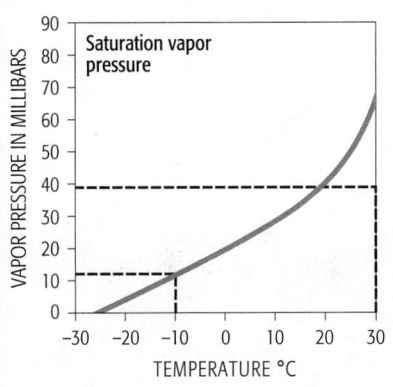

◄ Figure 3.39 **Saturation Pressure**
Water vapor pressure is the amount of atmospheric pressure in millibars that is due to water molecules. The saturation vapor pressure, the amount of water that air can hold, increases with increasing temperature.

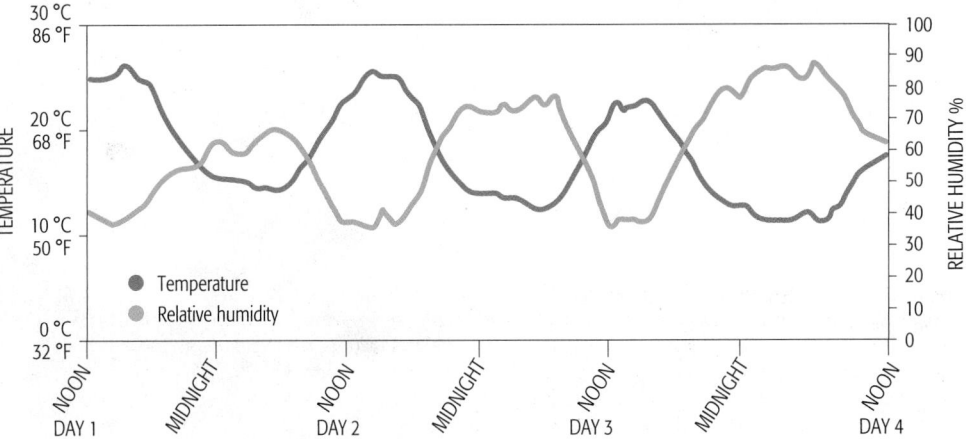

3.7 Earth's Energy Budget, Weather, and Climate

BIG IDEA On average, the amount of energy that our Earth receives from the sun is equal to the amount that it reflects or radiates back into space. However, at different times, energy gain and loss in particular locations may not be in balance. These regional differences produce differences in the temperature of land and water surfaces and the air above them, which influence the circulation of air in the atmosphere and water in the ocean. Together, these patterns affect conditions of weather and climate, such as temperature, relative humidity, and rainfall.

Earth's Energy Budget

■ Earth's energy budget is balanced overall.

On average, solar radiation equivalent to 82,000 joules or 19,500 calories strikes each square meter of Earth's upper atmosphere every minute. The majority of solar radiation arrives as visible light, although there is also considerable UV radiation (Figure 3.41). We know from the first law of thermodynamics that energy cannot be created or destroyed (see Module 3.3). So where does all this energy go?

The fate of this energy is described by an **energy budget**, a system of accounting that measures all the energy entering and leaving the Earth system (Figure 3.42). Of the total radiation striking Earth, 30% is reflected back into space by the upper atmosphere, clouds, and Earth's surface. Clouds are especially important, accounting for two-thirds of the reflected energy. The remaining 70% of solar energy is absorbed by the atmosphere, land, and water, increasing the kinetic energy of the molecules in these substances and making them warmer. Heat gained in this fashion is eventually radiated back into space as infrared radiation.

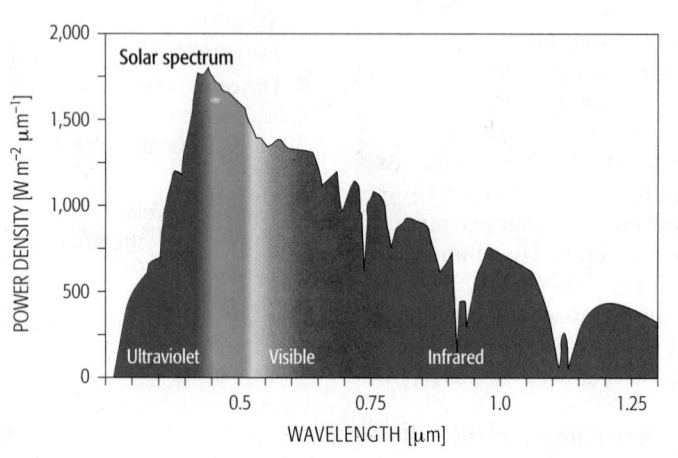

▲ Figure 3.41 **The Solar Spectrum**
Indicated by the rainbow of colors, the solar energy striking the outer reaches of Earth's atmosphere includes significant amounts of light such as ultraviolet and infrared radiation outside the visible spectrum.

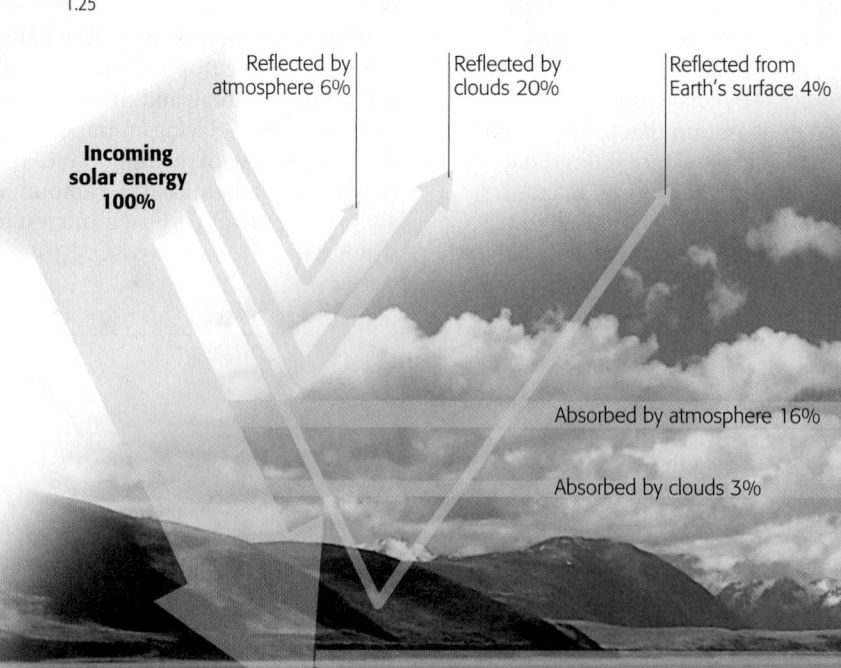

► Figure 3.42 **Earth's Energy Budget**
Over the entire globe and throughout the year, the inflow of energy from the sun equals the amount of energy radiated and reflected to outer space.

Over the course of a year, the energy budget of the entire Earth is balanced—the total amount of energy coming in is equal to the total energy that is reflected or radiated back into space. However, at any particular place or time, the energy budget may be out of balance. The most familiar example of variation in energy balance is the daily temperature cycle.

During the daytime, sunlight warms the atmosphere, land, and water, causing a net heat gain. At night, that heat is lost to space through radiation. Differences in the energy budget from region to region are the primary reason for the variety of climates around the world.

Weather and Climate

■ Local and regional variation in temperatures produces weather and climate.

Climate refers to atmospheric conditions such as temperature, humidity, and rainfall that exist over large regions and relatively long periods of time. In contrast, **weather** refers to short-term variations in local atmospheric conditions. When we say that Los Angeles will experience a cool, foggy morning and hot afternoon, we're talking about weather. When we say that Southern California has mild, moist winters and hot, dry summers, we are describing the climate of this region.

Changes in weather occur as a consequence of changes in the temperature and pressure of air masses. For example, on a sunny summer day, humid air becomes warmer. When this happens, the air expands; because it is less dense, it rises. As the air rises in altitude, it begins to cool. When the air mass cools to its dew point, water condenses, forming a cloud and perhaps a rain shower (**Figure 3.43**). As air cools, it becomes denser, until it eventually begins to sink. As it sinks, it is warmed and its relative humidity decreases.

In areas where air is rising, atmospheric pressure tends to fall. Thus, a drop in air pressure is often associated with an increased likelihood of rain. Conversely, where air is descending, atmospheric pressure rises, usually indicating dry conditions.

Local variations in heat gain and loss produce large areas of relatively warm or cold air. Such air masses may meet along boundaries called fronts. A cold front is a boundary where a mass of cold air is replacing a mass of warm air. Along a warm front, warm air is replacing cold air. In both cases, warm air is pushed up over denser cold air. If the warm air contains sufficient moisture, it may produce rain as it is cooled.

▲ **Figure 3.43 Thunderstorm**
Air warmed over the ocean surface rises in this cloud. As it rises, it cools to the dew point, producing a rain shower.

Q *What role does our ocean play in the weather and weather patterns?*
Christopher Shad, Cypress College

Norm: **A** Oceans affect nearly every aspect of our weather. They influence rainfall patterns because they are the source of most of the water in the atmosphere. Because they absorb solar radiation and circulate heat, they influence temperature patterns worldwide.

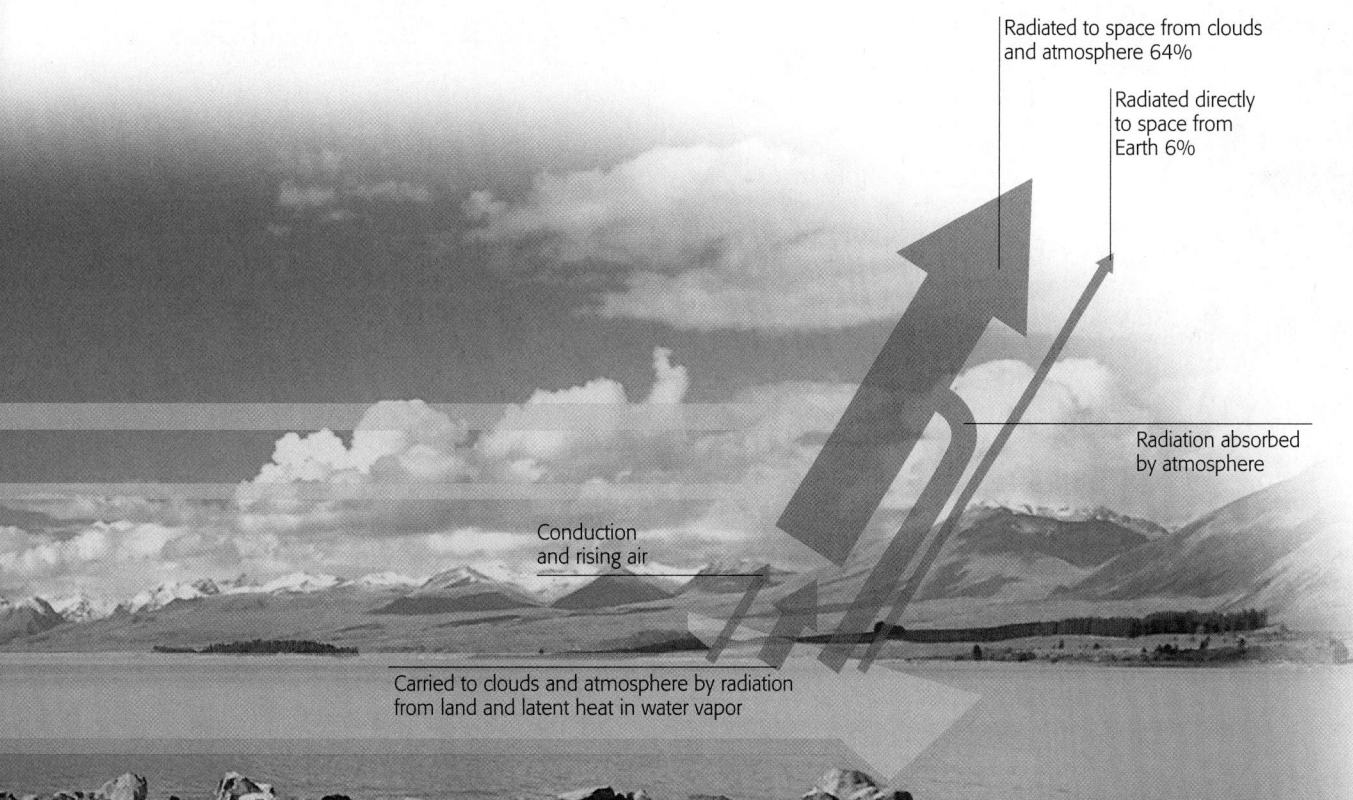

Radiated to space from clouds and atmosphere 64%

Radiated directly to space from Earth 6%

Radiation absorbed by atmosphere

Conduction and rising air

Carried to clouds and atmosphere by radiation from land and latent heat in water vapor

Wind Cells

■ Heat from the equator is transferred toward the poles by convection cells in the troposphere.

Because Earth's surface is curved, the amount of sunlight reaching a specific area on its surface varies with latitude. The amount of radiation striking a square meter of land is highest at the equator and diminishes toward the poles (Figure 3.44). The uneven distribution of solar radiation results in significant differences in air temperature and pressure around the globe. These differences cause air in the troposphere to circulate in large-scale convection currents, called wind cells. The climates of the world are greatly influenced by these wind cells.

At the equator, Earth's surface and the atmosphere immediately above it experience the greatest daily heating. This produces a band of rising air around the equator called the **intertropical convergence zone** (Figure 3.45). As this warm, humid air rises, it cools, and water condenses, forming rain. As a consequence, some of the wettest tropical rainforests are found in this zone. The air then diverges northward or southward, forming two convection cells, called **Hadley cells**, on either side of the equator. The comparatively warm and moist area of Earth beneath the Hadley cells is called the tropical zone. At about 25–30 degrees latitude, the air descends

and is warmed. This produces a zone of low relative humidity; many of the world's driest deserts, such as the Sahara, are found in this zone. The **Ferrel cells** occur at 30–60 degrees latitude in each hemisphere. Here air moves poleward and is warmed along Earth's surface. This region is called the temperate zone and is typified by distinct seasonal changes in temperature. You most likely live in this zone. In the Ferrel cells, air rises in a broad band between 45 and 60 degrees. This zone of rising air is also typified by comparatively high rainfall. The **polar cells** occur between 60 and 90 degrees latitude. Here in the circumpolar zone air moves along Earth's surface toward the equator and temperatures are much colder. The ecosystems associated with each of these zones are described in Chapter 7.

The air circulating in the wind cells interacts with Earth's rotation to cause general wind patterns. As Earth spins on its axis, areas near the equator spin faster than areas near the poles. If you were standing on the equator, you would travel 40,000 km (24,900 mi), Earth's circumference, in a day. As you move toward the pole,

▼ Figure 3.44 **Solar Angle**
The angle at which the sun's radiation strikes Earth increases, and the amount of sunlight received per unit area decreases from the equator to the poles. Solar energy at the equatorial regions is concentrated in a smaller surface area Ⓐ. Solar energy at the polar regions is spread out over a larger surface area Ⓑ.

▼ Figure 3.45 **Wind Cells**
Uneven heating of Earth's surface creates atmospheric convection in the troposphere that creates three distinct wind cells in each hemisphere.

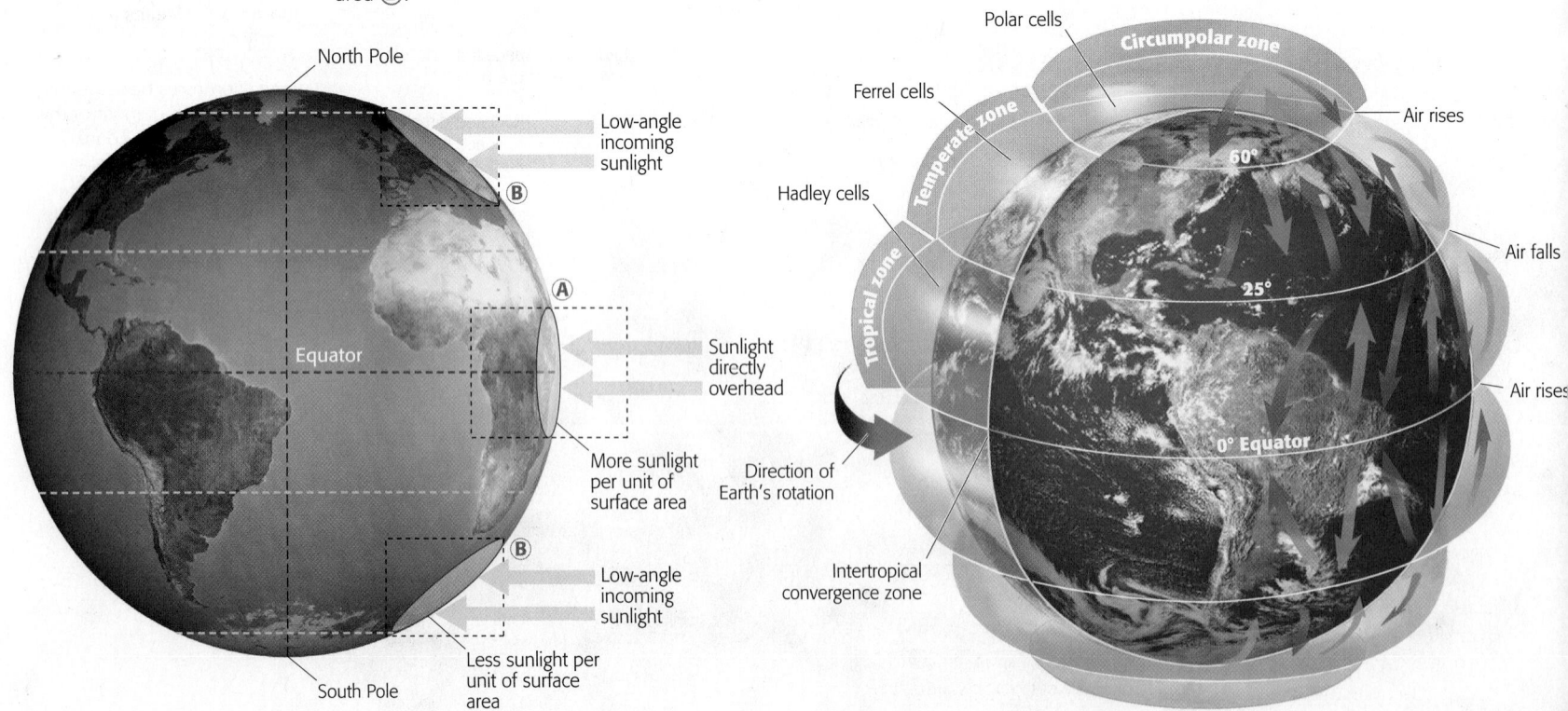

the distance you travel in a single rotation decreases each day, until you reach the pole. Then you would not travel any distance in rotation but simply spin in place. Because the ground is moving at different speeds, air moving over it toward the equator is deflected from east to west. Air moving toward the poles is deflected to the east. This change in wind direction due to Earth's rotation is an example of a **Coriolis effect**.

Because the air near the ground in the Hadley and polar cells is moving toward the equator, the winds seem to be deflected from east to west. Winds that blow from the east are said to be easterly. Thus, regions located within the Hadley and polar cells, the tropical and circumpolar zones, have prevailing easterly winds.

In contrast, the air near the ground in the Ferrel cells is moving toward the poles, so it appears to be deflected from west to east, producing a westerly wind. The temperate zones within the Ferrel cells experience prevailing westerly winds.

In zones where air is ascending or descending, there is no Coriolis effect and no prevailing winds. Winds in these zones are often weak and unpredictable. In the days of sailing ships, the lack of predictable winds in the intertropical convergence zone presented a significant obstacle. Ships attempting to cross the equator were often at a standstill. This region is often referred to as the doldrums.

Ocean Currents

■ Ocean circulation redistributes heat around the Earth.

Oceans contain nearly 98% of the water on Earth. They are the primary source of the water vapor in the atmosphere, and the ultimate destination for water flowing on and below the surface of all land. The circulation of ocean water has important consequences for Earth's climate.

Surface currents. The surface waters of the ocean reach to a depth of about 400 m (1,320 ft). These waters circulate in **ocean currents** that are driven by Earth's rotation, winds, and differences in water temperature. Near the equator, the high input of solar energy warms the surface waters of the Atlantic, Pacific, and Indian oceans. This warm water moves westward in currents propelled by the prevailing easterly winds.

For example, as the equatorial currents approach the east coasts of the Americas and Asia, they are deflected toward the poles. From the poles, water flows back toward the equator along the west coasts of continents. This circular pattern of circulation produces the major ocean **gyres** (Figure 3.46). As a result of Earth's rotation, the circulation of these gyres is clockwise in the Northern Hemisphere and counterclockwise in the Southern Hemisphere.

Warm water loses heat as it moves toward the poles. Water that travels from the poles and back toward the equator is much colder. Thus, at any given latitude, the currents moving along the west coasts of continents are generally much colder than those moving along the east coasts. The temperature of these currents influences the climate of the nearby continents. As the temperature of the ocean surface increases, so does the amount of evaporation to the atmosphere above it. This, in turn, influences the amount of water vapor in air moving onto land. As a result, west coast locations are often drier than east coast locations at the same latitude. For example, rainfall in the U.S. West Coast city of San Francisco is approximately 50 cm (20 in.) per year. Washington, D.C., which sits at nearly the same latitude on the East Coast, receives about 110 cm (42 in.) of rain per year.

If you were to track a typical molecule of water, it would make a complete circuit in a surface gyre in about a year. The velocity of water in ocean surface currents is generally less than 40–65 km (25–40 mi) per day. But some surface currents are much faster. The Gulf Stream, for example, flows out of the Gulf of Mexico between Florida and Cuba at about 160 km (100 mi) per day.

Vertical currents. Ocean waters also circulate vertically, exchanging surface water with water from the depths. This circulation, which is caused by differences in temperature and salinity, is called the **thermohaline circulation**. As water is transported to the North Atlantic, it loses heat and its salinity rises, causing it to become denser. Its greater density causes it to sink.

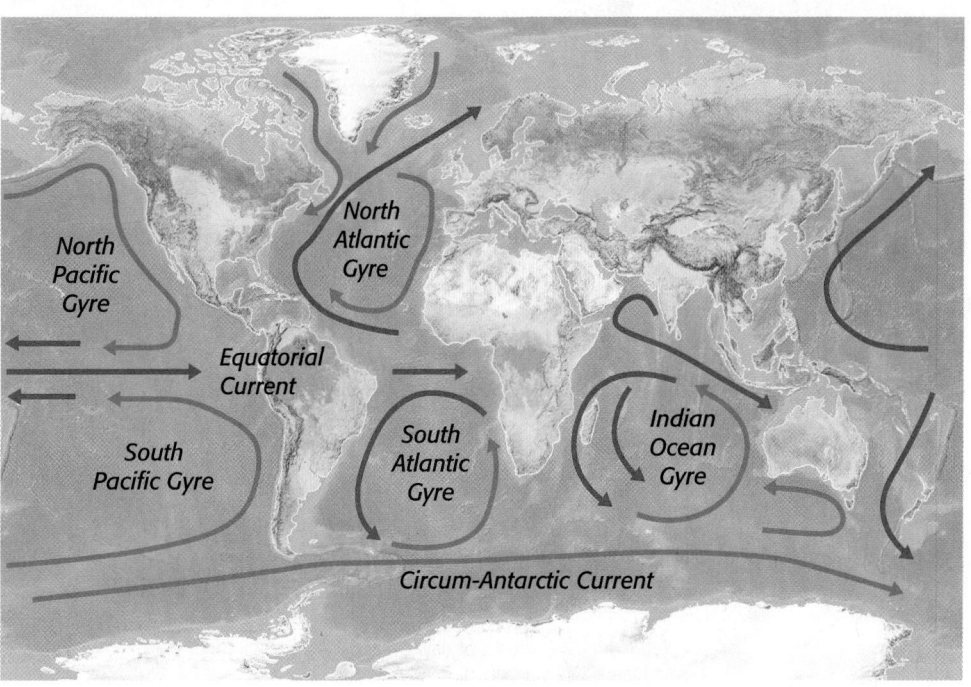

▼ Figure 3.46 **Major Ocean Currents**
The circulation of major ocean gyres brings warm water (red arrows) from equatorial regions along the east coasts of continents and toward the poles. They bring colder water (blue arrows) from the poles along west coasts toward the equator.

North Pacific Gyre

North Atlantic Gyre

Equatorial Current

South Pacific Gyre

South Atlantic Gyre

Indian Ocean Gyre

Circum-Antarctic Current

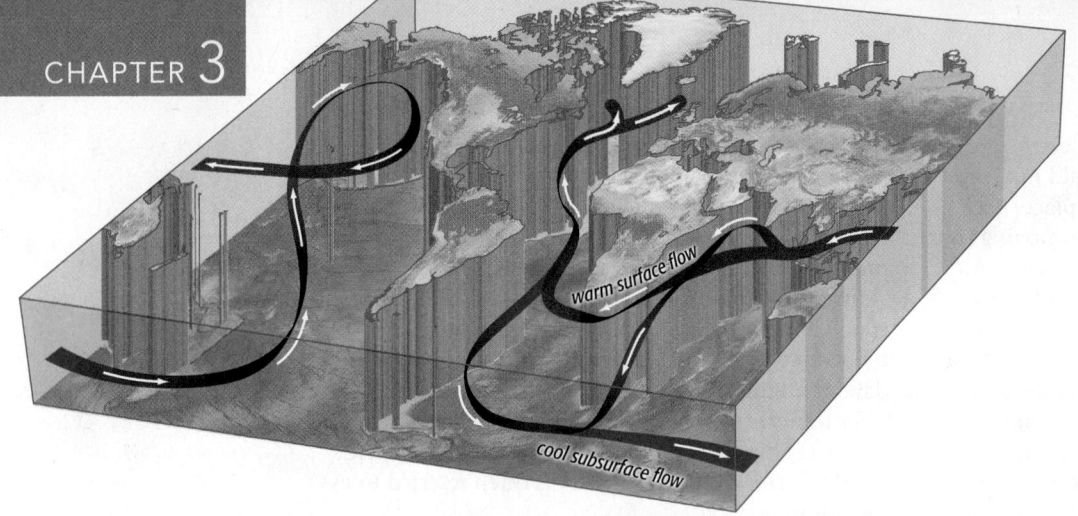

▲ **Figure 3.47 The Great Ocean Current**
Differences in water temperature and salinity affect water density, causing vast amounts of water to move in the thermohaline circulation, which is independent of the surface circulation in ocean gyres. Cold water circulates in the deep ocean, rises in the Indian Ocean and the North Pacific, then slowly travels back to the North Atlantic.

This deep water current then flows south, where it is deflected eastward along the coast of Antarctica. Branches of the circulation finally surface again in the Indian and Pacific oceans (Figure 3.47).

The thermohaline circulation moves very slowly—it takes hundreds of years for a water molecule to make a complete circuit. Yet because it moves vast quantities of water, the thermohaline circulation also moves a great deal of heat. This transfer of heat has a significant effect on global climate. For example, at approximately the same latitudes, the climate of Ireland and Great Britain is warmer than that of southern Alaska because of heat transferred by the thermohaline circulation.

The Seasons

■ Wind cells shift north and south over a year, influencing patterns of seasonal change.

▼ **Figure 3.48 The Seasons**
Through the year, the area of Earth receiving the greatest amount of solar radiation shifts, causing corresponding shifts in wind cells and weather that produce seasonal change.

Seasons occur because Earth is tilted on its axis by 23.5 degrees. Each hemisphere is tilted toward the sun during its summer and away from the sun during its winter (Figure 3.48). Solar energy per unit area is highest in the summer and lowest in the winter.

Seasonal changes in the amounts of radiation reaching different parts of the globe cause the positions of the wind cells to shift north and south. This, in turn, influences the nature of seasonal climate change. Los Angeles, California, for example, is located at about 35° north latitude. During the Northern Hemisphere winter, the city is influenced by the westerly winds of the Ferrel cell, which bring storms and rainfall from off the Pacific Ocean. In summer, the wind cells shift northward. Then the zone of descending air between the Hadley and Ferrel cells is located over Los Angeles, and little rain falls.

Large bodies of water moderate the effects of seasonal changes. Land masses gain and lose heat much faster than do bodies of water. Thus, during summers, land areas tend to gain more heat than the adjacent ocean, and locations in the center of continents become much hotter than locations near the coast. Large areas of hot rising air and low pressure develop over the middle of continents, such as North America and Asia. This low pressure pulls in moist air from adjacent oceans, creating conditions favorable for summer thunderstorms. This effect is greatest near the east coasts of continents where warm waters provide abundant humid air. In the winter, land cools much more rapidly than do adjacent waters. As a result, the coldest winter temperatures are also experienced in locales at the center of continents.

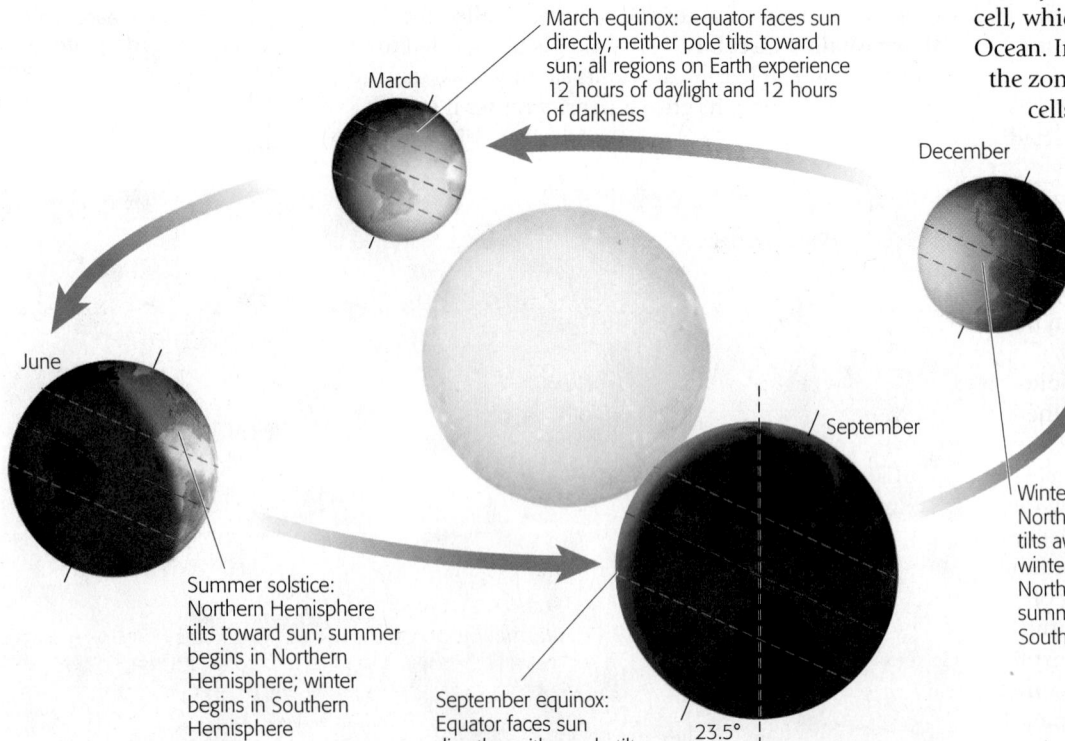

March equinox: equator faces sun directly; neither pole tilts toward sun; all regions on Earth experience 12 hours of daylight and 12 hours of darkness

March

December

June

September

Summer solstice: Northern Hemisphere tilts toward sun; summer begins in Northern Hemisphere; winter begins in Southern Hemisphere

September equinox: Equator faces sun directly; neither pole tilts toward sun; all regions on Earth experience 12 hours of daylight and 12 hours of darkness

23.5°

Winter solstice: Northern Hemisphere tilts away from sun; winter begins in Northern Hemisphere; summer begins in Southern Hemisphere

Depicting Earth's Climate

■ Climatographs integrate seasonal change with temperature and precipitation.

A **climatograph** is a graph that shows the pattern of seasonal changes in temperature and precipitation for a particular location (**Figure 3.49**). The x axis shows the months of the year, and the y axis indicates two factors—the average temperature and the average precipitation for each month. Temperature is shown by a line graph and precipitation by a bar graph.

The availability of moisture is not directly shown on a climatograph, but it can be inferred by comparing the temperature and precipitation curves. Evaporation is directly tied to temperature; higher temperatures produce higher evaporation rates. When evaporation is greater than precipitation, there is a moisture deficit, and the lack of water is likely to limit life functions. When precipitation is greater than evaporation, there is a moisture surplus, and plant and animal growth and

reproduction are limited less by water and more by other factors, such as temperature or soil nutrients.

In the tropical rain forest, temperature and water availability are favorable for plant growth year-round. Climates in most other ecosystems are characterized by fairly distinct growing seasons, when temperature and moisture conditions are favorable for growth, and dormant seasons, when climate limits the activities of plants and animals. In dryer parts of the tropical zone, the dormant season occurs when moisture is in short supply. In the temperate zone, the dormant season typically occurs in winter, when cold temperatures can damage growing tissues. The cold winter dormant season is especially long in the circumpolar zone. The relationships between climatographs and the distribution of different ecosystem types are discussed in detail in Chapter 7.

Where **YOU** LIVE What is your climate like?

Generate a climatograph for your hometown and for another city located at least 1,000 miles away, a place where you might like to live. Average monthly temperature (°C) and precipitation (mm) can be obtained for over 20,000 locations worldwide at the website WorldClimate.

How do seasonal patterns of rainfall and temperature differ at each location? When is moisture likely to be most available for a garden at each location? What are the important factors that are responsible for the differences between these locations?

QUESTIONS 3.7

1. What features of Earth's atmosphere contribute to the energy that it reflects and radiates back to space?

2. What is the difference between climate and weather?

3. Why is rainfall often associated with cold and warm fronts?

4. The intertropical convergence zone shifts north and south throughout the year. Why?

5. Describe the general pattern of circulation in the ocean's surface waters.

6. What is the thermohaline circulation and how does it differ from ocean surface circulation?

(MES) For additional review, go to **MasteringEnvironmentalScience**

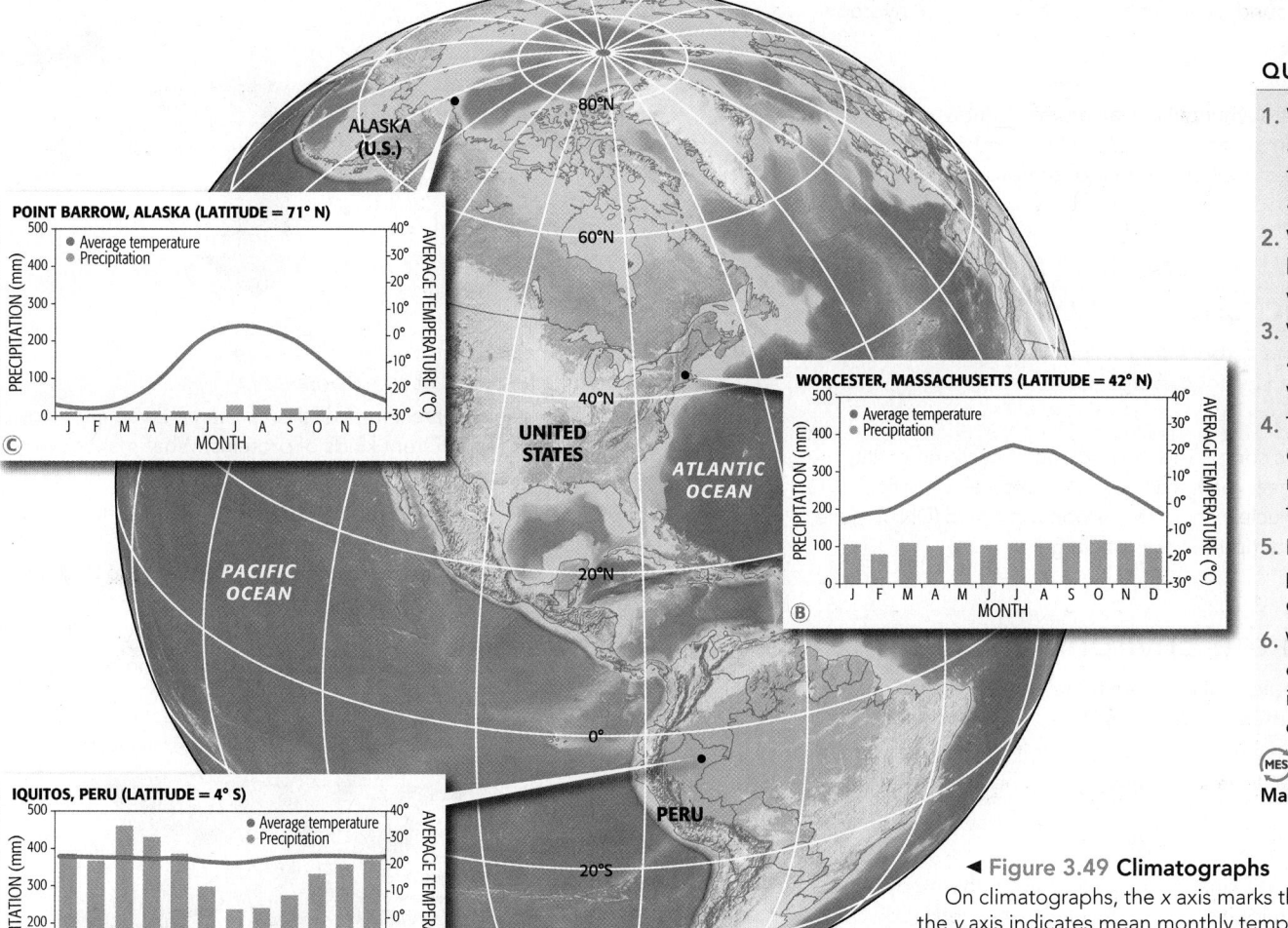

◀ Figure 3.49 **Climatographs**

On climatographs, the x axis marks the months of the year and the y axis indicates mean monthly temperature and precipitation. Ⓐ Warm conditions prevail year-round in tropical zone locations such as Iquitos, Peru. However, rainfall is seasonal, and moisture deficits occur from June through October. Ⓑ The climatograph for Worcester, Massachusetts, located in the temperate zone shows significant seasonal variation in temperature, but rainfall occurs in all seasons. Ⓒ Point Barrow, Alaska, is in the circumpolar zone, and average temperatures above freezing extend only from May through October, and moisture surpluses occur year-round.

Summary

Earth comprises a single large ecosystem. A variety of Earth's features are responsible for the abundant life that occurs here but apparently nowhere else in our solar system. Earth contains the elements necessary for the structure and function of organisms and ecosystems. Unlike other planets in our solar system, Earth has large amounts of water necessary to sustain life. It receives solar energy in amounts sufficient to power the transformation of inorganic molecules to complex organic compounds and to sustain a variety of climates that are favorable to life. Earth's liquid core generates the magnetic field that surrounds it and deflects particles in the solar wind that would otherwise destroy life. Earth's atmosphere filters most of the UV light in the solar spectrum. Its mobile crust recycles rocks and has reshaped and relocated continents, thereby influencing the evolution of Earth's life and ensuring continuous provision of elements necessary to sustain ecosystems. The circulation of heat and water in the atmosphere and oceans is responsible for much of the great variety in Earth's climate. Earth is indeed a wonderful place to live!

3.1 Chemistry of the Environment

- Elements are substances that cannot be broken down or separated into other chemicals by ordinary means. Oxygen, hydrogen, and carbon are 3 of the 92 naturally occurring elements. Molecules are composed of two or more atoms of the same or different elements.

- The unique properties of water are essential to the structure and function of organisms and ecosystems.

KEY TERMS

element, atom, molecule, proton, neutron, electron, mass, isotope, radioactive, half-life, compound, covalent bond, ion, ionic bond, hydrogen bond, solubility, pH scale

QUESTIONS

1. Hydrogen is the simplest element; it has an atomic number of 1. There are three hydrogen isotopes: ^1H, ^2H, and ^3H. Explain how these isotopes differ with respect to their number of protons, neutrons, and electrons.

2. ^3H, the hydrogen isotope called tritium, has a half-life of 12.32 years. What does this mean?

3. Based on the mass of its molecules, water should boil at temperatures well below its freezing point. What accounts for its high boiling point?

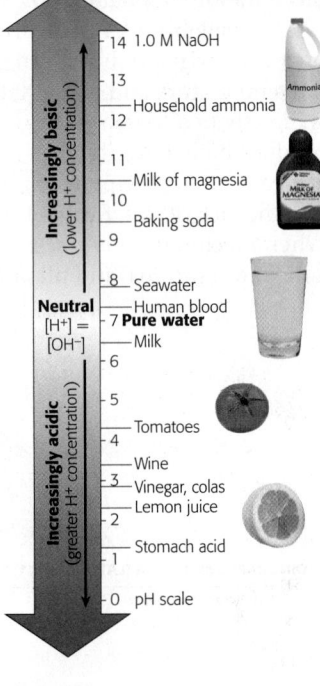

3.2 The Organic Chemistry of Life

- Organic molecules such as carbohydrates, lipids, proteins, and nucleic acids are essential to the functions of organisms.

KEY TERMS

organic molecule, inorganic compound, hydrocarbon, macromolecule, polymer, carbohydrate, sugar, starch, cellulose, polysaccharide, lipid, protein, catalyst, enzyme, nucleic acid, deoxyribonucleic acid (DNA), gene, ribonucleic acid (RNA), transcription, translation, genome

QUESTIONS

1. Describe three different kinds of carbohydrates.

2. The proteins of all organisms are composed of the same 20 amino acids, but there are thousands of different kinds of proteins. What makes the proteins different from one another?

3. What is the general relationship between a gene and the protein that it produces?

4. Describe the role of RNA in the conversion of the DNA code to protein.

3.3 Energy and the Environment

- Energy describes the capacity of a system to do work. Potential energy is stored in a system; kinetic energy is the energy of motion itself.

- Energy can be neither created nor destroyed, but it can be transformed from one form to another. When it is transformed, the entropy of the universe increases.

KEY TERMS

energy, work, potential energy, kinetic energy, first law of thermodynamics, second law of thermodynamics, entropy, heat, electromagnetic radiation, electromagnetic spectrum, temperature, conduction, convection, radiation, latent heat transfer, chemical energy, nuclear energy, nuclear fission, nuclear fusion, joule (J), calorie, watt-hour (Wh)

QUESTIONS

1. You move a book from your desk to a high shelf. Describe the change that occurs in this system in terms of kinetic and potential energy and work.

2. The potential energy represented in the book poised above your desk is considerably less than the energy you expended to put the book up there. What happened to the rest of the energy?

3. Photons of gamma rays, X-rays, and UV light are damaging to living systems. Why?

4. Cats lick themselves to cool off on a hot day. Explain how this works.

5. Gas atoms in the atmosphere are constantly colliding, but these collisions do not cause nuclear fusion to occur. Why not? What would be required to initiate fusion?

3.4 Earth's Structure

- The solid portion of Earth is divided into a dense inner core, a mantle composed of less dense rock, and an even less dense thin outer crust.
- The location and configuration of Earth's continents have changed considerably over its history.

KEY TERMS

core, mantle, magma, crust, lithosphere, tectonic plate, transform fault boundary, divergent boundary, convergent boundary, rock cycle, igneous rock, sedimentary rock, metamorphic rock

QUESTIONS

1. Although Earth's magnetic field is generated deep in its core, it has important consequences for life on the surface. Explain.
2. Some continental rocks are nearly 4 billion years old, but nowhere is ocean crust older than 200 million years. Why is this so?
3. What is the source of the energy that moves Earth's tectonic plates?
4. What does the cratered surface of our Moon tell us about tectonic activity there?

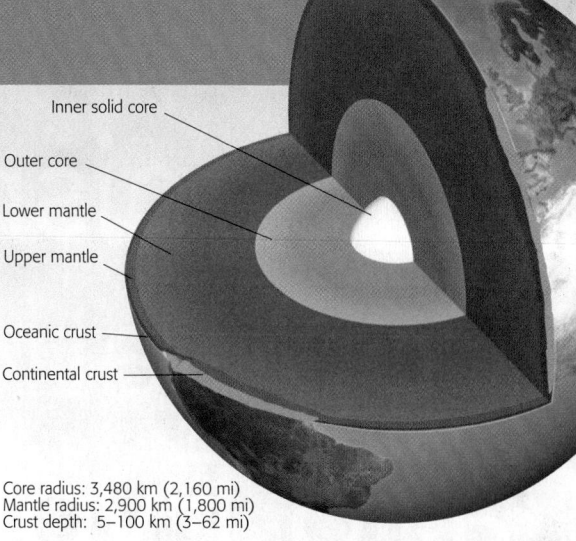

Core radius: 3,480 km (2,160 mi)
Mantle radius: 2,900 km (1,800 mi)
Crust depth: 5–100 km (3–62 mi)

3.5 Element Cycles in Earth's Ecosystems

- The distribution, abundance, and movement of elements in ecosystems can be tracked by measuring element pools and fluxes.
- The cycling of nutrient elements that are required by organisms is influenced by their abundance in the atmosphere, their mobility in the lithosphere, and the complexity of food webs.

KEY TERMS

biosphere, biogeochemical cycle, pool, flux, mass-balance accounting, capital, equilibrium, residence time, cycling time, macronutrient, micronutrient

QUESTIONS

1. If the atmosphere contains 300 Tg of element X, and the flux into and out of the atmosphere of element X is 20 Tg/yr, what is the residence time of element X in the atmosphere?
2. If the net flux of element X into the atmosphere were suddenly to change to +5 Tg/yr, what would you expect the size of the atmospheric pool of element X to be in 10 years?

3.6 Earth's Atmosphere

- Earth's atmosphere is composed mostly of nitrogen (N_2) and oxygen (O_2), with smaller amounts of other gases. It is divided into four distinct layers.
- Vapor pressure is the contribution of water vapor to total atmospheric pressure.

KEY TERMS

atmosphere, atmospheric pressure, troposphere, stratosphere, ozone layer, mesosphere, thermosphere, vapor pressure, saturation vapor pressure, relative humidity, dew point

QUESTIONS

1. Life does not exist on Earth because of its unique atmosphere; rather, its unique atmosphere is a consequence of life. Explain this statement.
2. Why does atmospheric pressure decline with increasing altitude?
3. Gases in the troposphere are constantly mixed, but those in the stratosphere are not as well mixed. Why?
4. On a hot, humid day, water may condense on the outside of windows of an air-conditioned building. Explain this phenomenon using the terms vapor pressure, saturation vapor pressure, relative humidity, and dew point.

3.7 Earth's Energy Budget, Weather, and Climate

- Over a year, Earth's energy budget is balanced; the amount of energy that it receives from the sun equals the amount that is reflected or radiated back into space.
- Climate refers to atmospheric conditions such as temperature, humidity, and rainfall over large regions and long times. Weather refers to short-term variation in those conditions.

KEY TERMS

energy budget, climate, weather, intertropical convergence zone, Hadley cell, Ferrel cell, polar cell, Coriolis effect, ocean current, gyre, thermohaline circulation, climatograph

QUESTIONS

1. In particular places and at certain times of the year, the energy budget may not be balanced. Describe three factors that can produce such imbalances.
2. Many of the world's driest deserts are located at about 30–35° latitude. Why?
3. The "solar equator" is defined as the line around the Earth along which the sun's radiation is most direct. How does this change relative to the actual equator through the year?
4. How does the temperature of the ocean affect patterns of rainfall on adjacent lands?

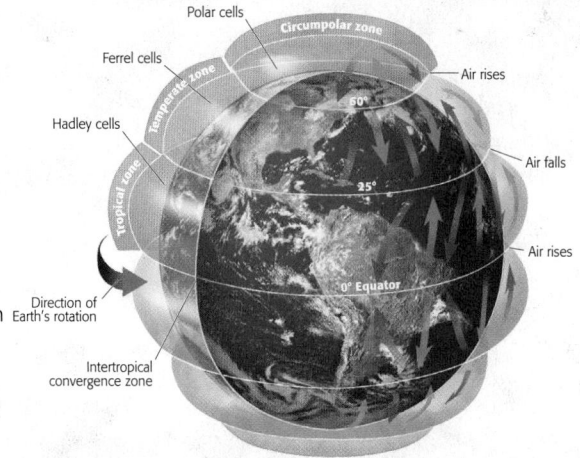

MasteringEnvironmentalScience®

Students Go to **MasteringEnvironmentalScience** for assignments, the eText, and the Study Area with practice tests, activities, and more.

Instructors Go to **MasteringEnvironmentalScience** for automatically graded tutorials and questions that you can assign to your students, plus Instructor Resources.

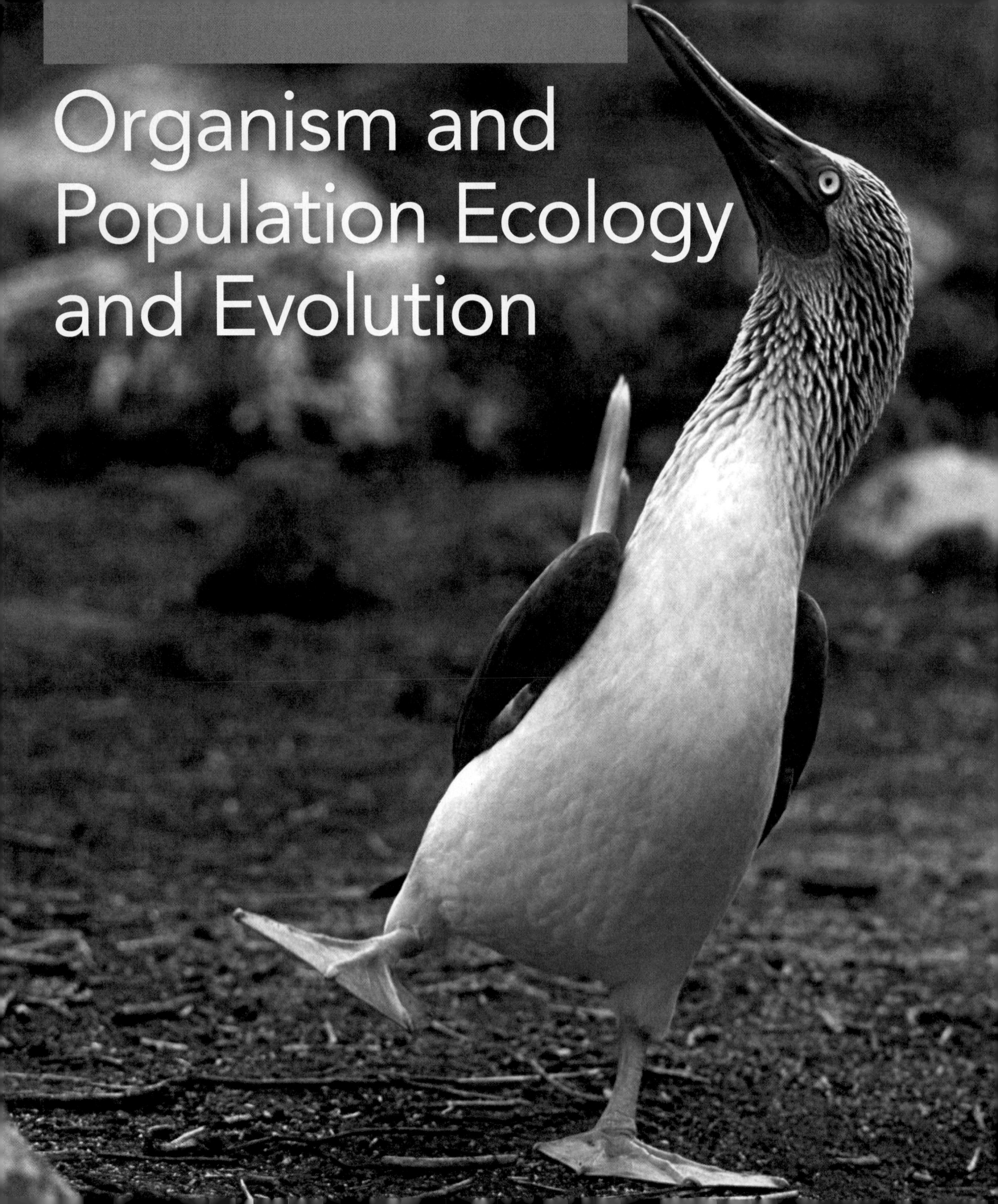

Organism and Population Ecology and Evolution

4

97

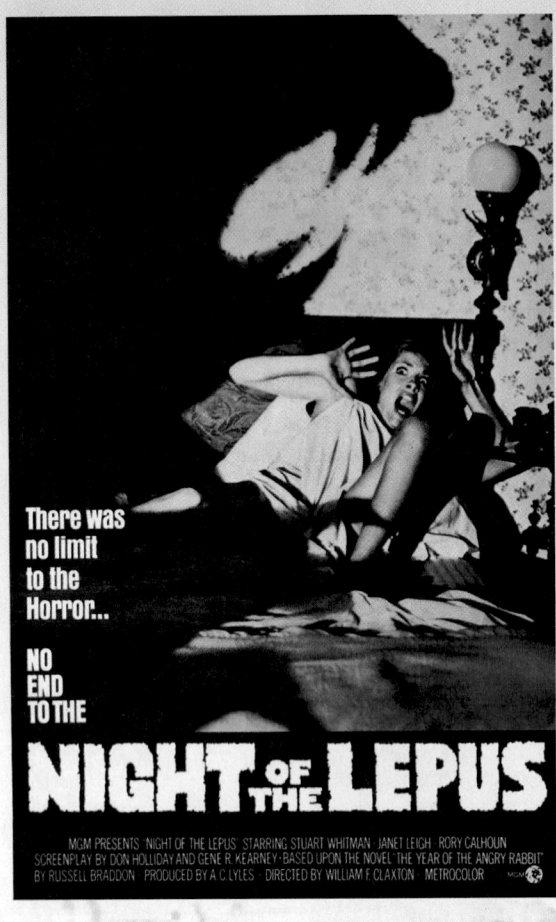

There was
no limit
to the
Horror...

NO
END
TO THE

NIGHT OF THE LEPUS

MGM PRESENTS 'NIGHT OF THE LEPUS' STARRING STUART WHITMAN · JANET LEIGH · RORY CALHOUN
SCREENPLAY BY DON HOLLIDAY AND GENE R. KEARNEY · BASED UPON THE NOVEL 'THE YEAR OF THE ANGRY RABBIT'
BY RUSSELL BRADDON · PRODUCED BY A.C. LYLES · DIRECTED BY WILLIAM F. CLAXTON · METROCOLOR

Genetic Change and Population Growth—Fact and Fiction

Could mutant organisms take over the world?

Mutant creatures run amok is the theme of the 1972 horror film *Night of the Lepus* (Figure 4.1). In it, scientists experimenting with a laboratory rabbit cause a mutation that interrupts its breeding cycle. And wouldn't you know it? The bunny gets away! Much to everyone's horror, the bunny quickly grows to a monstrous size and begins a reproductive frenzy. Matters only get worse when people discover that this mutant prefers the taste of human flesh to carrots!

Fortunately, few of us lie awake at night in fear of ravenous rabbits. But a few pieces of this story have some basis in reality. (Be assured, a humongous, human-hungry hare is not one of those pieces.)

It is true, for example, that rabbit populations are capable of rapid growth when those elements of their natural environment that would keep growth in check are missing. Such was the case when European rabbits were introduced to Australia. In 1859, a shipment of 24 wild rabbits was brought in by a well-meaning European

◀ Figure 4.1 **Horrors Indeed**
Genetic change and explosive population growth are central themes for many horror flicks. *Lepus* is the scientific name for the subgroup of rabbits called hares.

▼ Figure 4.2 **Population Explosions**
Within 10 years of the introduction of a few rabbits to Australia, their populations exceeded millions, devastating the resources upon which they fed.

settler, looking for sport. Others followed with a few rabbits of their own. With abundant food and few natural predators, the rabbit population quickly increased. In just a couple years, there were thousands of rabbits. By 1870, hunters were killing 2 million rabbits each year, but with no apparent effect on their population.

Southern Australia was being overrun; in 1950, Australia's rabbit population was estimated to be 600 million. These rabbits were eating everything they could find, destroying native plants, and leaving little for the native animals to eat (Figure 4.2). To try to control their numbers, the government introduced rabbits infected with a deadly viral disease. The virus spread rapidly, reducing the population by about 60%.

The phenomenon of a change in environment bringing a change to a population is real, as evidenced by the story of Australia's rabbit population. So too is the phenomenon of a genetic change within a population. While the threat of genetic changes depicted in *Night of the Lepus* is overblown, environmental changes such as exposure to novel chemicals or to ultraviolet light can alter the genetic makeup of organisms. The growth and reproduction of individual organisms are determined by genes encoded in their DNA.

More often than not, changes in the DNA code are harmful to the organisms that possess them. For example, in the years since the 2010 BP oil spill in the Gulf of Mexico, fishermen and shrimpers have reported fish and shrimp with a variety of mutations including the absence of eyes and eye sockets (Figure 4.3). Some of these changes appear to be directly linked to changes in the animal's DNA caused by chemicals in the agents used to clean up the oil.

Chemical pollutants may not necessarily be the direct cause of a mutation, but they can still significantly influence the abundance of mutant organisms in a population. For example, the insecticide DDT was first used in 1941 to control populations of the mosquitoes that spread malaria and typhoid; later it was used to control agricultural pests. It was noted at the time that very small numbers of these insects were more resistant to DDT than other members of their populations. This resistance was caused by a genetic mutation that had randomly appeared in populations. However, in the presence of DDT, those DDT-resistant mutants were more likely to survive and reproduce than non-resistant individuals. Over the next two decades, the relative abundance of insects that were resistant to DDT increased dramatically.

To learn the fate of the mythical mutant bunnies and their would-be prey, you will need to see the movie. In this chapter, we will focus on biological processes that take place in real organisms and real populations. We shall see that these processes generally proceed in precise and controlled ways. But, as Australian rabbits, mutant shrimp, and DDT-resistant insects demonstrate, human impacts on the environment can alter populations and the factors that control them.

- *How do organisms acquire energy and how does that energy support their growth and reproduction?*

- *What are the factors that accelerate or limit the growth of populations?*

- *What is the origin of inherited changes, how are they propagated in populations, and what is their role in the evolution of new species?*

- *How does evolutionary change shape our understanding and classification of the diversity of life on Earth?*

▲ Figure 4.3 **Environmental Factors Can Alter DNA**
Environmental factors such as ultraviolet light or chemical pollutants can cause changes in an organism's DNA that produce abnormalities. Scientists suspect that chemical dispersants used to clean up spilled oil may have produced the mutations making these Gulf of Mexico shrimp eyeless.

4.1 The Cell—The Fundamental Unit of Life

BIG IDEA Underlying the great biodiversity found on Earth is the cell, the basic unit of structure and function of all life. The cells of plants, algae, and some bacteria are able to use light energy to synthesize high-energy carbohydrates from carbon dioxide and water. Other kinds of bacteria are able to manufacture carbohydrates by using energy released from inorganic chemical reactions. These synthetic processes are the primary sources of chemical energy for all of Earth's ecosystems. The cells of all organisms—microorganisms, fungi, plants, and animals—depend on carbohydrates to power their functions, including cell growth and division.

▼ **Figure 4.4 Eukaryotic and Prokaryotic Cells** All cells are bounded by a cellular membrane. Eukaryotic cells are distinguished by a nucleus that contains their DNA and cytoplasm that contains different membrane-enclosed organelles. Prokaryotic cells lack such membrane-bound structures. Their DNA is found within the cytoplasm. (DNA is the genetic material common to all living things.)

Prokaryotic cell

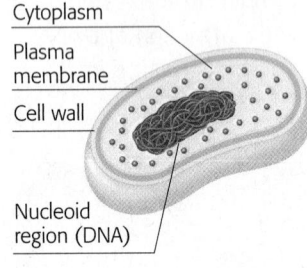

- Cytoplasm
- Plasma membrane
- Cell wall
- Nucleoid region (DNA)

Animal Eukaryotic cell   Plant Eukaryotic cell

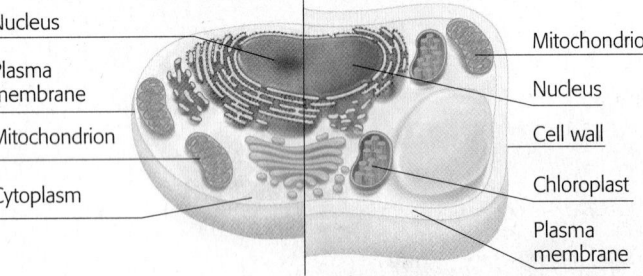

- Nucleus
- Plasma membrane
- Mitochondrion
- Cytoplasm
- Mitochondrion
- Nucleus
- Cell wall
- Chloroplast
- Plasma membrane

Cell Structure

■ Organisms are divided into two groups based on cell structure.

The most ancient evolutionary relationships among organisms can be inferred from the structure of their cells. Biologists use cell structure to broadly classify organisms into one of two groups—prokaryotes and eukaryotes (Figure 4.4). The cells of both groups share a similar chemistry. They are protected by a plasma membrane that regulates the flow of material into and out of the cell. The interior of each is filled with **cytoplasm**, a gel-like substance that includes various structures that do the work of the cell. The two groups of cells do differ in their internal organization.

Prokaryotes, which include the single-celled organisms of the domains Bacteria and Archaea, are thought to resemble the earliest cellular life on Earth. Their cells lack a membrane-enclosed nucleus, and their DNA occurs as a single chromosome in the cytoplasm. There is great diversity among the prokaryotes, with species that perform a multitude of functions in ecosystems, including photosynthesis, the decomposition of wastes, and the cycling of essential elements, such as nitrogen (N) and sulfur (S).

Eukaryotes (domain Eukarya) vary in complexity from single-celled algae and protozoa to plants, fungi, and animals that are made of millions to trillions of cells. The cytoplasm of eukaryotes contains a number of membrane-enclosed **organelles** ("little organs") that perform specialized functions. The most prominent eukaryotic organelle is the nucleus. The nucleus contains most of the cell's DNA, which is organized into distinct chromosomes. Among the other organelles are two that are critical to acquiring and processing energy. The cells of all eukaryotes have mitochondria, organelles that process oxygen and food. The cells of plants and most algae have chloroplasts, organelles that use sunlight, carbon dioxide, and water to produce carbohydrates (food), releasing oxygen as a by-product.

Chemical Functions

■ The cell is the site of life's most important chemical functions.

Plants, some eukaryotic algae, and prokaryotic blue-green algae called *cyanobacteria* nourish themselves through the cellular process of **photosynthesis**. Photosynthesis requires sunlight, carbon dioxide (CO_2), and water (H_2O). Plants and algae absorb CO_2 from the atmosphere and extract H_2O from soil, ocean water, or fresh water.

The chloroplasts in plant cells contain pigments, such as chlorophyll, that are able to capture the energy of light (see Module 3.3). Similar pigments are found in the cytoplasm of algae and cyanobacteria. When sunlight strikes these pigments, they absorb photons of light, increasing the energy level of electrons in the pigments (see Module 3.2). Through a series of chemical reactions, the energy from these electrons is used to combine CO_2 and H_2O to form carbohydrate molecules, such as the sugar glucose ($C_6H_{12}O_6$) (Figure 4.5).

Carbohydrates power the cellular processes of plants and algae and allow these organisms to grow, as well as all the organisms that feed on them. Photosynthesis also produces O_2. Indeed, all of the O_2 in Earth's atmosphere is a by-product of photosynthesis (see Module 3.7).

In some environments where there is no light, certain kinds of bacteria obtain food through the process of **chemosynthesis** (Figure 4.6). For example, in the dark depths of the ocean, chemosynthetic Bacteria and Archaea take in CO_2, O_2, and hydrogen sulfide (H_2S). Through a series of chemical reactions, bacteria convert those molecules to carbohydrates, sulfur (S), and H_2O. The conversion of H_2S to S provides the energy for this process. In such dark environments, chemosynthesis produces the carbohydrates used for the growth and nourishment of all the organisms in the ecosystem.

In **cellular respiration**, a portion of the energy in carbohydrate molecules is used to produce energy-carrying molecules called adenosine triphosphate or ATP. ATP enables a cell to carry out cell functions and facilitate growth (Figure 4.7). Cellular respiration takes place in the cytoplasm of prokaryotes and in the mitochondria of eukaryotes. In cellular respiration, cells break down carbohydrate molecules, leading to the formation of ATP and a release of H_2O, and CO_2. The formation of ATP requires existing molecules to either take on electrons or lose them. This movement of electrons provides the chemical energy for all other cell functions. In the final step of cellular respiration, the O_2 molecules pick up electrons and hydrogen ions to produce H_2O molecules.

Nearly all organisms use the energy from cellular respiration to power their physical, behavioral, and reproductive functions—all of their life processes. Note that plants as well as animals carry out cellular respiration. The many steps in cellular respiration also generate a variety of different chemicals for building the macromolecules—proteins, starch, cellulose, nucleic acids, and lipids—that are essential for growth and reproduction (see Module 3.2). These complex macromolecules may be converted back to simple carbohydrates for use in cellular respiration.

▶ **Figure 4.5 Photosynthesis**
The plant takes in CO_2 from the air and light energy from the sun. CO_2 and light energy react with the H_2O in the chloroplast of the plant cell. The reaction produces carbohydrate, which stores energy.

Cellular respiration requires oxygen. But in some environments there is little or no oxygen; such environments are said to be *anaerobic*. Because they cannot carry out respiration, most plants and animals are unable to survive in anaerobic environments. Many bacteria and fungi, however, are able to survive and grow by using **anaerobic respiration**. In this process, carbohydrates are partially broken down, producing ATP, carbon dioxide, and smaller carbohydrate molecules, such as methanol or ethanol (Figure 4.8). Because the breakdown of carbohydrate molecules is incomplete, anaerobic respiration of carbohydrates yields only about 10–20% of the energy that is released in cellular respiration. The fact that methanol and ethanol can be used as engine fuels is evidence that anaerobic respiration is unable to take advantage of all of the potential chemical energy in carbohydrate molecules.

In summary, nearly all the chemical processes that are necessary to support life and the functioning of ecosystems take place in cells. In all organisms, the energy that is required for these processes is generated by the stepwise breakdown of carbohydrates in respiration. Plants, some algae, and cyanobacteria are able to manufacture carbohydrates from CO_2 and water using light energy (photosynthesis); some bacteria derive carbohydrates from inorganic chemical reactions (chemosynthesis). Animals obtain carbohydrates by eating plants or other animals that eat plants.

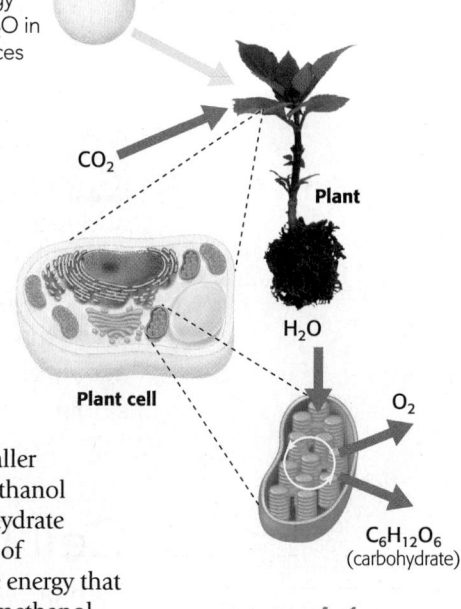

CO_2

Plant

H_2O

Plant cell

O_2

$C_6H_{12}O_6$ (carbohydrate)

QUESTIONS 4.1

1. Describe two differences between the cells of prokaryotes and eukaryotes.

2. Compare the processes of chemosynthesis and photosynthesis.

3. Give two reasons why cellular respiration is essential for all life.

(MES) For additional review, go to **MasteringEnvironmentalScience**

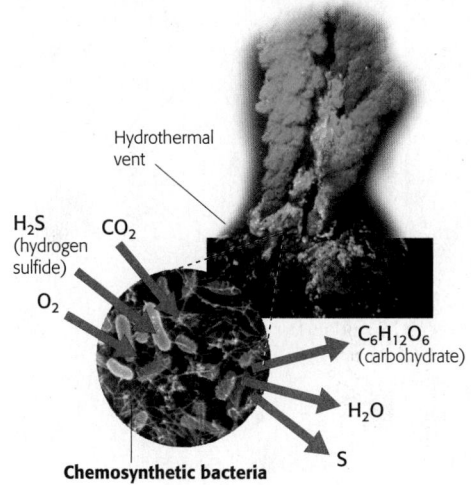

Hydrothermal vent

H_2S (hydrogen sulfide)

CO_2

O_2

$C_6H_{12}O_6$ (carbohydrate)

H_2O

S

Chemosynthetic bacteria

▲ **Figure 4.6 Chemosynthesis**
Bacteria take CO_2 and O_2 from the water and H_2S from the vents. Chemosynthesis produces glucose ($C_6H_{12}O_6$) that is used as energy for the bacteria and water (H_2O) and sulfur (S) that are released into the seawater.

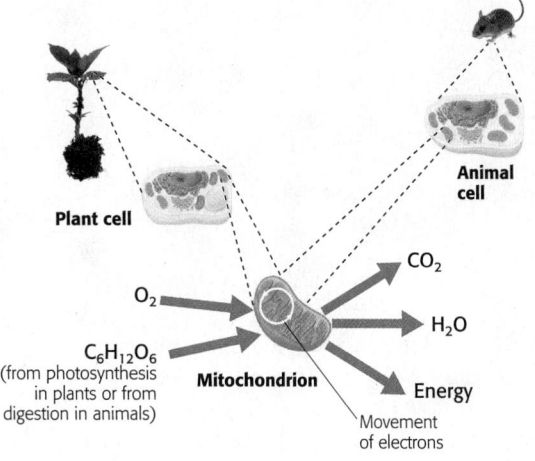

Plant cell

Animal cell

O_2

$C_6H_{12}O_6$ (from photosynthesis in plants or from digestion in animals)

Mitochondrion

CO_2

H_2O

Energy

Movement of electrons

▲ **Figure 4.7 Cellular Respiration**
In cellular respiration, carbohydrate molecules such as glucose are broken down to produce CO_2 and H_2O and release energy needed for organisms to grow and reproduce.

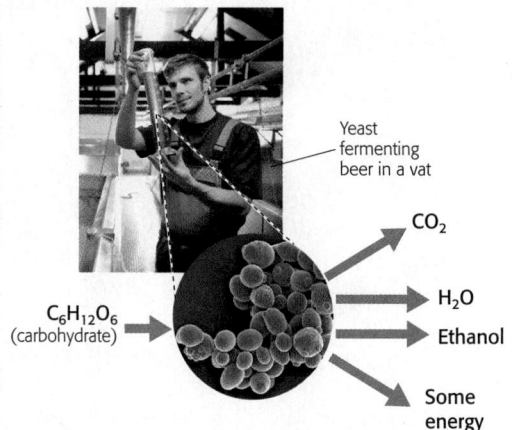

Yeast fermenting beer in a vat

$C_6H_{12}O_6$ (carbohydrate)

CO_2

H_2O

Ethanol

Some energy

▲ **Figure 4.8 Anaerobic Respiration**
The yeast takes in carbohydrate ($C_6H_{12}O_6$). Anaerobic respiration produces energy for the yeast. Carbon dioxide (CO_2), water (H_2O), and ethanol are by-products that are released.

4.2 The Growth and Reproduction of Organisms

BIG IDEA The source of Earth's biodiversity can be found in the cell, in its DNA. Cell growth and reproduction is the basis for the growth of organisms and their populations. How cells grow and reproduce also provides a mechanism through which evolutionary change can occur. Organisms interact with their environment to obtain the material and energy needed for cell growth and division. To reproduce, the cells of organisms divide in a way that allows them to pass their genes on to the next generation. Reproduction may be asexual or sexual. **Asexual reproduction** is accomplished by simple cell division and produces offspring that are genetically identical to their single parent. **Sexual reproduction** involves mating between two parents and produces offspring that are genetically distinct from either parent.

Cell Division and Differentiation

■ An organism grows through the division and differentiation of its cells.

For an organism to grow and develop, it needs a steady supply of carbohydrates. An organism may obtain those carbohydrates through the process of photosynthesis or chemosynthesis, by eating other organisms, or by ingesting the wastes of other organisms. Cells grow larger as they acquire carbohydrates and manufacture the macromolecules that make up cell organelles.

When cells are mature, they may divide to form new cells. The process of cell division ensures that each newly created cell receives a complete set of organelles and genes (see Module 3.2). Before cell division, the DNA in the parent cell is precisely copied, or replicated, so each new cell inherits an exact copy of its parent cell's

DNA (**Figure 4.9**). In eukaryotes, the DNA in a cell is contained in pairs of chromosomes, one of each pair having come from two parent organisms. This form of cell division, where a complete set of DNA is copied and shared between two cells, is known as mitosis.

In complex, multicellular plants and animals, cells differentiate into tissues, such as skin and muscle, that perform specialized functions. Despite the specialization of these cells, all the cells in an organism—with the notable exception of gametes discussed below—contain exactly the same DNA. The tissues differ from one another because the specialized cells use information that is activated from different portions of the organism's DNA.

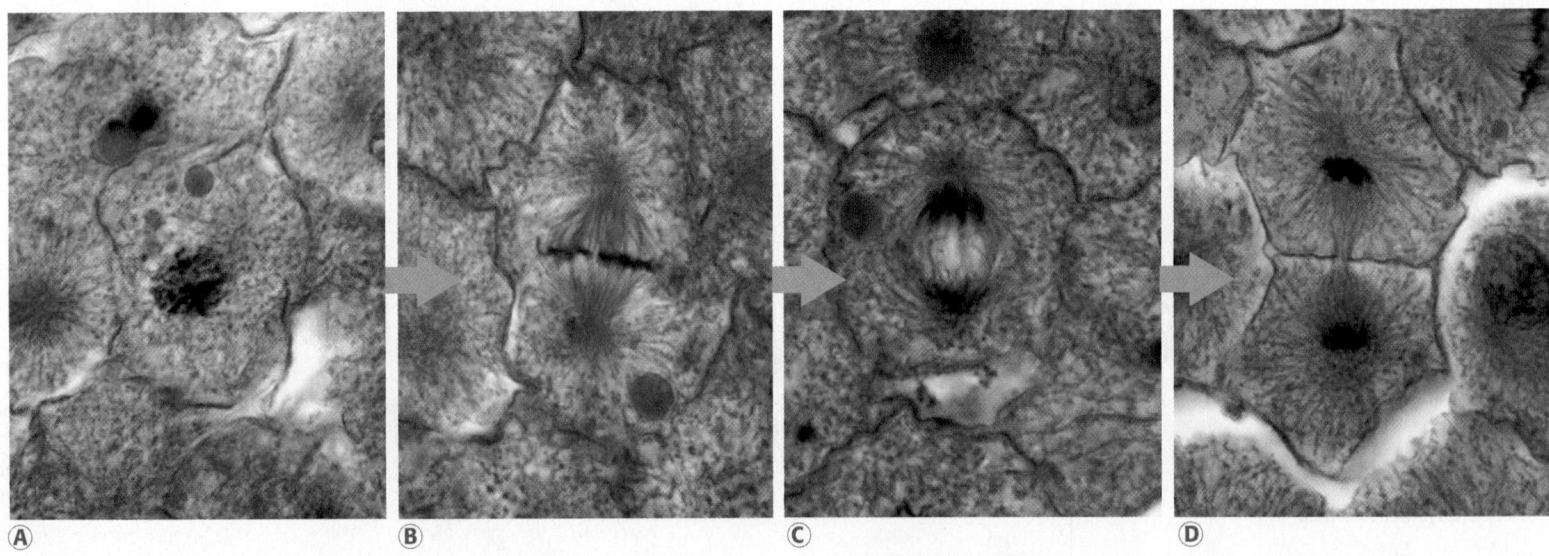

Ⓐ   Ⓑ   Ⓒ   Ⓓ

▲ Figure 4.9 **Cell Division**
Prior to cell division, the cell's DNA is replicated and condenses into distinct chromosomes Ⓐ. Each pair of chromosomes has been duplicated and the duplicates split Ⓑ as one complete set of chromosomes migrates to each of the opposite poles of the cell Ⓒ. In the final stage, a new cell membrane forms between the segregated chromosomes creating two new cells Ⓓ. Each new cell has DNA that is identical to that in the parent cell.

Sexual and Asexual Reproduction

■ The mode of reproduction affects the amount of variation among offspring.

Reproduction allows organisms to multiply, producing offspring that carry some or all of their genes. Some organisms reproduce sexually, and others reproduce asexually. And there are some that use both means of reproduction.

In asexual reproduction, a single parent produces offspring that are exact replicas, or clones, of their parent. In other words, the DNA of the offspring is identical to that of their sole parent (**Figure 4.10**). Most single-celled organisms and organisms with only a few cells reproduce asexually through simple cell division. That includes all bacteria, most protozoa, and some primitive animals. For example, a single-celled amoeba reproduces by dividing into two amoebas. In more complex organisms, cells may divide to form buds that yield clones. Some plants reproduce asexually by means of runners, bulbs, or tubers.

Most plants and animals multiply by sexual reproduction, in which males and females of the same species mate to produce genetically unique offspring. These offspring possess DNA contributed by both parents.

Sexual reproduction is preceded by a special form of cell division that produces sex cells, or **gametes**, with just one set of chromosomes. During this process, referred to as meiosis, the genes from each chromosome pair in the parent cell get redistributed, so that the single chromosomes that result from the division carry a mix of genes from the original chromosome pair. Each gamete has a complete set of single chromosomes, but they are not exact duplicates of the parent cell's chromosomes. Therefore, the DNA of each gamete produced is unique to that cell. Female gametes are called eggs, and male gametes are called sperm.

The union of an egg and a sperm produces a new, genetically unique cell called a **zygote**. The zygote contains two copies of DNA, one contributed by each gamete. The zygote then undergoes simple cell division and differentiation to produce a new, genetically unique organism that shares some characteristics with each of its parents (**Figure 4.11**).

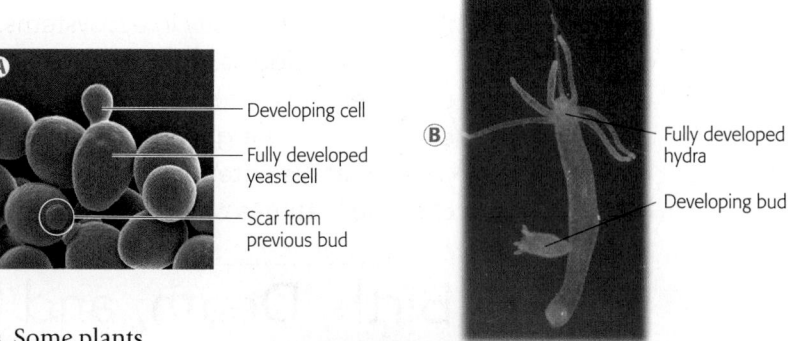

Developing cell

Fully developed yeast cell

Scar from previous bud

Ⓑ Fully developed hydra

Developing bud

Ⓒ

Newly developing strawberry plant

Runner connects budding offspring to parent

Strawberry plant

▲ Figure 4.10 **Asexual Reproduction Produces Clones**
Asexual reproduction produces offspring that are identical to one another. In simple organisms like yeast, this may involve simple cell division Ⓐ. Asexual reproduction also occurs in some animals, such as the sea anemone Ⓑ, and in many plants. For example, when the runners of a strawberry plant take root, a new plant separates from the parent Ⓒ.

QUESTIONS 4.2

1. How does the DNA of a gamete compare to the DNA in the cells of the organism that produced it?

2. How does the DNA of a zygote compare to the DNA of the cells of its mother and father?

(MES) For additional review, go to **MasteringEnvironmentalScience**

◀ Figure 4.11 **Sexual Reproduction Produces Diverse Offspring**
The puppies in this litter share the same parents, but each developed from a separate zygote. Therefore, they are genetically related but also genetically distinct from one another.

4.3 The Growth of Populations

BIG IDEA The growth of populations of individual species is central to the creation and maintenance of biodiversity in ecosystems. Evolutionary change affects populations of organisms and not just individuals. If the organisms in a population with a particular genetic makeup are able to successfully grow and reproduce, the size of the population will increase. In a new, resource-rich environment, the growth of a population is usually exponential rather than arithmetic. A population's growth rate is determined by the rates of birth, death, and migration and how these factors vary over the lifetimes of individuals within the population.

Birth, Death, and Migration

■ Population growth is determined by the rate of birth, death, and migration.

Soon after a few individual organisms move into an environment that is rich in resources, their population will start to grow. In the first few weeks or months, the number of new individuals added to the population is small. But as the size of the population increases, so does the number of reproducing individuals. As a result, the number of individuals added to the population increases with each successive generation. This pattern of steady and continuous growth is said to be exponential. In **exponential growth**, the number of new individuals added to a population in each generation is a multiple of the number present in the previous generation. This is in contrast to **arithmetic growth**, in which the number of new individuals added at each generation is a constant, or set amount, over a given period of time.

The constant by which an exponentially growing population increases is called the **population growth rate**. Population growth rate is typically expressed as the percentage of change over a specific time interval. For example, if a population of 100 individuals grows to 110 individuals over a year, its growth rate is 10% per year.

As a general rule, population growth rates are faster for smaller organisms (**Figure 4.12**). For rapidly reproducing organisms such as bacteria, population growth rates may be calculated for times as brief as an hour. For larger plants and animals, population growth rates are generally calculated for a year. Population growth rate may also be expressed as the **doubling time**, the length of time required for a population to double in size. The doubling time of a population is approximately equal to 70 divided by the population growth rate.

A population's growth rate is a consequence of patterns of birth, death, and migration. Reproduction adds new individuals to the population. The **birth rate** is the number of births in the population per unit of time expressed as a percentage of the population size. Death subtracts individuals from the population. The **death rate** is the number of individuals dying per unit of time expressed as a percentage of the population size. Death rate is also referred to as **mortality rate**.

Migration describes the movement of organisms into and out of a population. The number of organisms

► Figure 4.12 **Population Growth Rate and Body Size**
Population growth rates are expressed here as percent change per year on the left-hand *y*-axis and as doubling time in years on the right-hand *y*-axis. In general, growth rates are low and doubling times quite long for large compared to very small organisms.

moving into an area is calculated in terms of the **immigration rate**, the number of individuals entering the population per unit of time as a percentage of population size. **Emigration rate** is the number of organisms moving out of an area, calculated as a percentage of population size.

Thus, many factors affect the growth of a population. The total population growth rate is calculated as the birth rate plus the immigration rate minus the death rate and the emigration rate.

Population Growth Rate = (Birth Rate + Immigration Rate) − (Death Rate + Emigration Rate)

Exponential Population Growth: A Case Study

■ The introduction of European starlings to America provides a case study for exponential population growth.

In 1890, 60 European starlings were released in New York City for the dubious purpose of introducing to the United States all of the birds mentioned in the works of William Shakespeare. Within a decade, tens of thousands of the birds were foraging in the lawns of the city's parks. By 1920, the total population of European starlings had reached the millions, and the birds were common in all of the New England states. By 1950, hundreds of millions of European starlings were distributed across much of North America (**Figure 4.13**). Rapid, exponential growth in populations like that of the European starling is common when a species is introduced to an unexploited habitat with abundant resources.

Exponential growth can be illustrated by calculating the hypothetical growth of the population of starlings in New York City (**Figure 4.14**). A breeding pair of European starlings typically produces five offspring each year. Assuming that the original population of 60 birds was equally divided between males and females, by 1891 the population of starlings would have increased by 150 to total 210 birds.

30 females × 5 offspring = 150 offspring
150 offspring + 60 original birds = 210 starlings

Making the same assumptions for each successive year, the number of new European starlings added to the population each year would be equal to the number of breeding pairs multiplied by 5. By 1910, this hypothetical European starling population in New York would be nearly 250 million.

Although starlings are very common in New York, their numbers are, thankfully, only a very, very small fraction of this hypothetical prediction. Why are there so many fewer starlings in the real population than in our hypothetical calculation? Our calculation did not consider the rate at which starlings die. In reality, most starlings die during their first year, before they have the opportunity to reproduce. Fewer than 20% of starlings survive beyond two years. Furthermore, each year many birds disperse, or emigrate, to regions outside New York City.

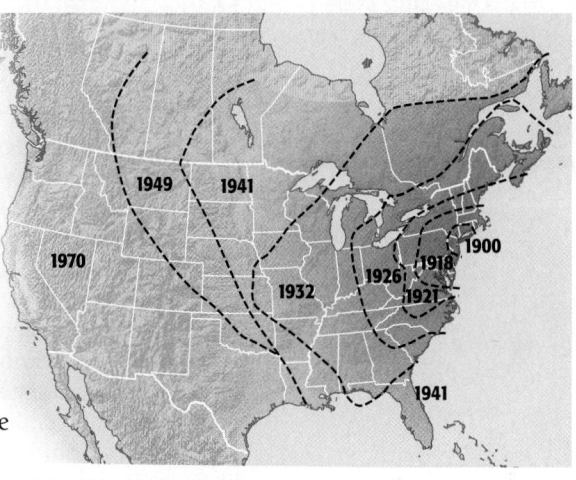

▲ Figure 4.13 **Starlings Invade North America**
The continental population of millions of starlings originated from a source population of 60 individuals released in 1890 in New York's Central Park. The intensity of the shading relates to the relative density of the bird's population.

1895
31,513 birds
15,756 breeding pairs
78,782 offspring

1894
9,003 birds
4,501 breeding pairs
22,509 offspring

1893
2,572 birds
1,286 breeding pairs
6,431 offspring

1892
735 birds
367 breeding pairs
1,837 offspring

1891
210 birds
105 breeding pairs
525 offspring

1890
60 birds
30 breeding pairs
150 offspring

◄ Figure 4.14 **If Population Growth Were Driven by Birth Rate Only**
In just five years, a hypothetical population of 30 male and 30 female starlings will increase to nearly 80,000 birds. This example assumes that each breeding pair produces five offspring each year, that no deaths occur, and that birds do not migrate beyond Central Park.

105

1. Why is the growth of biological populations typically exponential rather than arithmetic?

2. What factors determine population growth rate?

3. What sorts of animals would you expect to have type III survivorship? Which would you expect to have type I survivorship?

4. Why is the actual average number of offspring born to females less than the total fertility rate?

(MES) For additional review, go to **MasteringEnvironmentalScience**

Survivorship and Fertility

■ The likelihood that an organism will survive and reproduce changes over its lifetime.

For European starlings, the chances of dying are relatively constant throughout their life—each year about 60% of the birds in the population die, regardless of their age. By comparison, the death rate for bald eagles is about 35% in the first year of life and less than 10% each year thereafter. Population biologists describe the probability of an organism dying during a particular time interval in terms of **survivorship**. Survivorship curves are constructed by plotting the percentage of individuals in a population that survive to each age (**Figure 4.15**).

Survivorship curves vary widely among species, but the ecological factors that determine survival result in three general patterns of survivorship. Organisms with *type I survivorship* are most likely to die of old age. Type I survivorship is common among large animals and predators that have few enemies. For organisms with *type II survivorship*, the probability of dying is the same at every age. Type II survivorship is observed in organisms such as the European starling, where predators and diseases affect all individuals in the population equally, regardless of their age. In organisms with *type III survivorship*, the very young have the greatest probability of dying because they are most easily taken by predators or are most vulnerable to disease. As organisms with type III survivorship mature, they become more and more likely to survive to old age. Type III survivorship is by far the most common pattern of survivorship observed in nature.

The likelihood of reproduction and the number of offspring produced also changes with age. The rate of reproduction is referred to as the **fertility rate**. For most species, there is an interval between birth and the age of

first reproduction. European starlings, for example, do not begin to reproduce until they are at least one year old; bald eagles do not reproduce until they are at least three years old. Following first reproduction, the **age-specific fertility rate** describes the number of offspring produced by an average female during a particular range of ages. The age-specific fertility rate often increases to an optimum reproductive age and decreases thereafter. In some cases, such as in humans, there is an age beyond which reproduction ceases altogether.

The sum of age-specific fertility rates across all ages is the **total fertility rate**. The total fertility rate is the *potential* number of offspring that an average female in a population can produce if she survives to old age. For a population to grow, total fertility rate must exceed 2.0—the number of offspring needed to replace a female and her mate. If some females die before reaching old age, then the actual average number of births per female is less than the total fertility rate. The difference between the actual average number of offspring per female and total fertility rate is determined by differences in survivorship (**Figure 4.16**).

In general, population growth rate increases as total fertility rate increases. However, population growth rate is also influenced by **generation time**. The generation time of a population is the average difference in age between mothers and their offspring. Among populations with the same total fertility rates and survivorship, those with the shortest generation times will have the highest population growth rates.

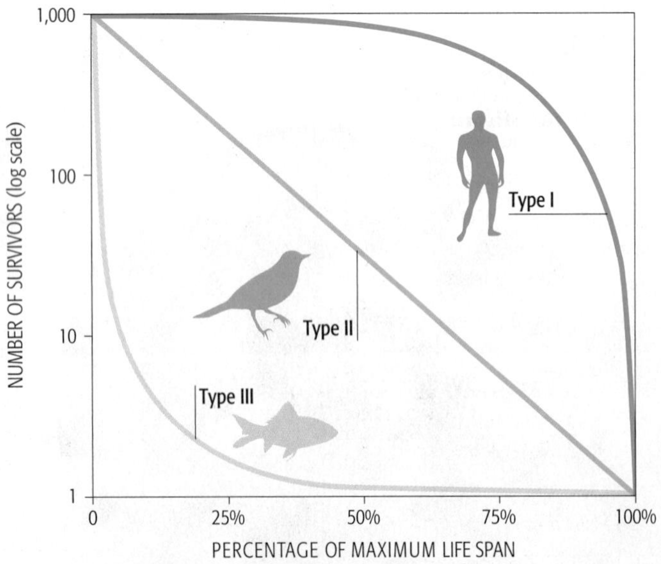

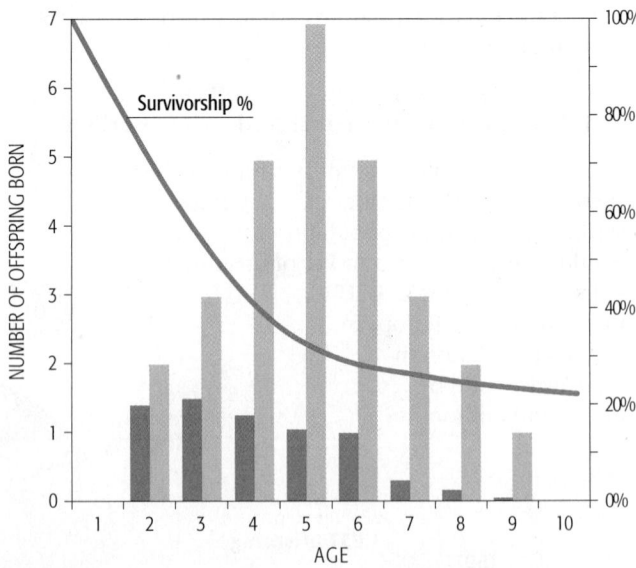

▲ **Figure 4.16 Potential and Actual Fertility Changes with Age**
The green bars represent the average number of births to females in each age class for a hypothetical population. Added together, values of these bars equal the population's total fertility rate. The orange bars represent the extent to which age class actually contributes to the overall birth rate, taking into account survivorship (orange line).

▲ **Figure 4.15 Patterns of Survival**
The risk of death is highest among old individuals in populations with type I survivorship. It is highest among young individuals in populations with type III survivorship. The risk of death is constant, regardless of age, in populations with type II survivorship.

Dinosaur Survivorship

Were the growth and survival of large dinosaurs similar to those of large animals that are alive today?

Is it possible to study the population dynamics of animals that have been extinct for over 63 million years? Such a task might appear hopeless, but Gregory Erickson of Florida State University does just that (**Figure 4.17**). To study long-gone populations, Erickson combines his expertise in the development and evolution of vertebrates with his lifelong fascination with dinosaurs.

Recently, Erikson investigated the fossils of a large carnivorous dinosaur, *Albertosaurus sarcophagus*. (*Albertosaurus* was a close relative of *Tyrannosaurus rex*, made infamous in the *Jurassic Park* movies.) Twenty-two specimens of *Albertosaurus* were unearthed in the Barnum Brown Bone Bed, an aggregation of dinosaur fossils discovered in Alberta, Canada. Independent analysis of the Red Deer Bone Bed indicated that the dinosaurs found there had died over a period of months to years; they were not killed all at once in some sort of catastrophe.

Erickson thought that the *Albertosaurus* fossils might reveal the survivorship of these ancient carnivores. In earlier studies of large lizards and crocodiles, Erickson and other scientists had established that the lines seen in the cross-sections of bones represent yearly increments of growth. The number of lines in a bone shows the age of the animal (**Figure 4.18**). Erickson reasoned that the lines of growth in

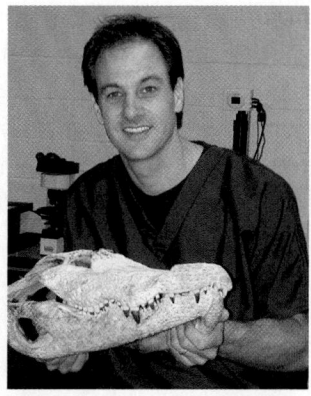

▲ **Figure 4.17 Gregory Erickson**
Florida State University scientist Gregory Erickson has coupled his interest in patterns of bone development in reptiles with his lifelong fascination with the ecology of dinosaurs.

the *Albertosaurus* fossils could also be used to establish the age at which each dinosaur died.

By counting lines in the thighbones of each of the 22 *Albertosaurus* specimens, Erickson found that the age of the dinosaurs ranged from 2 to 28 years. Specimens younger than 2 years are not well preserved in the fossil record and so were unavailable to Erickson. Instead, he referred to other studies of large predators, which indicated a 50–80% death rate during the first years of life. For purposes of comparison, Erickson assumed that 60% of *Albertosaurus* young died before age 2. He then used the data he had amassed to construct a survivorship curve, estimating the percentage of individuals in the population surviving to various ages (**Figure 4.19**).

The results of Erickson's study reveal remarkably high survivorship: 70% of the *Albertosaurus* that were alive at age 2 survived to age 13. This pattern of survivorship is very similar to the type 1 survivorship of existing large animals, such as elephants and rhinoceroses. Erickson attributed the high survivorship of *Albertosaurus* to the fact that by age 2, they were already considerably larger than any other predators that might feed on them.

Erickson found that beyond age 13, the rate of survivorship began to decrease. This decreased survivorship corresponds to the time at which the dinosaurs approached maximum size and are thought to have become sexually mature. At that age, injuries and stress associated with mating and reproduction probably caused mortality to increase. The bones of the oldest (28 years old) specimen show clear signs of deterioration related to old age.

Erickson has recently expanded his analyses to include other members of the tyrannosaur family, including *Tyrannosaurus rex*. His research is an excellent example of how *observations* and *measurements*, even of the fossil remains of long-extinct creatures, can be used to *test hypotheses* about the dynamics of their populations.

1. In what ways did Erickson use modern data to infer patterns of mortality in these ancient fossils?

2. Why was it important that these *Albertosaurus* specimens died over a period of time and not in a single calamity?

3. Studies of fossils often must rely on relatively small sample sizes, in this case 22 individuals. Why might this be important to Erickson's conclusions?

▼ Figure 4.19
***Albertosaurus* Survivorship**
Gregory Erickson used the age of 22 specimens from the Red Deer Bone Bed to construct this survivorship curve for *Albertosaurus*. It is shown here in comparison with survivorship curves of several animals alive today.

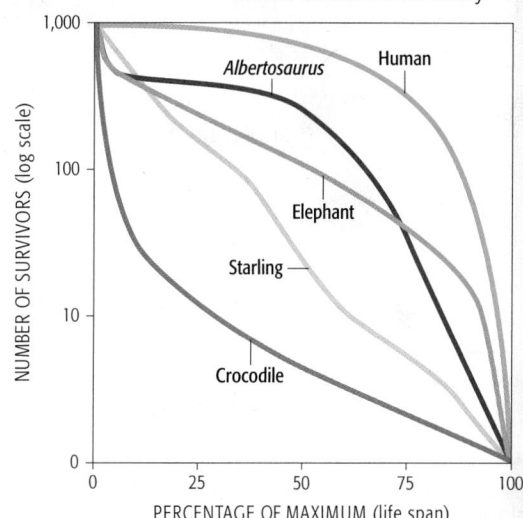

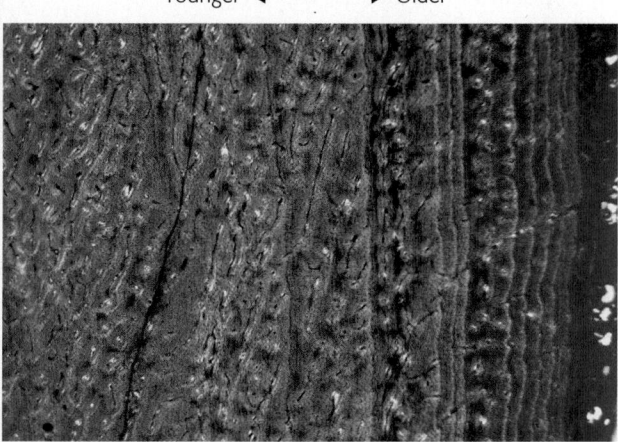

Younger ◄──► Older

▲ **Figure 4.18 Aging a Dinosaur**
Each line in this section of *Albertosaurus* thighbone fossil marks a year in the dinosaur's life and indicates that its growth slowed notably as the animal aged.

Source: Erickson, G.M, P.J. Currie, B.D. Inouye, and A.A. Winn. 2006. Tyrannosaur life tables: An example of Nonavian dinosaur population biology. *Science* 313:213–217.

4.4 Limits on Population Growth

BIG IDEA As a population grows, the amount of unexploited resources available to it decreases. Consequently, competition for those resources increases. As the rate at which a population uses resources approaches the environment's capacity to supply those resources, birth rates decline, death rates increase, and the overall rate of population growth declines. When the rate at which a population uses resources equals the rate at which those resources are supplied, the birth rate equals the death rate, and the population growth rate becomes zero. The population size at which this occurs is called the **carrying capacity**. Thus, the rate at which resources are supplied limits the growth of a population. The growth of a population may also be limited by conditions other than its supply of resources, such as temperature or toxins in the environment. An organism's range of tolerance for many environmental factors determines where it can live.

Environmental Resistance Limits Growth

■ At carrying capacity, the birth rate equals the death rate and the population growth rate stalls at zero.

Population biologists have learned much from monitoring the populations of microbes, flour beetles, fruit flies, and other organisms in the laboratory. In controlled environments, essential resources can be supplied to a population at a constant rate. Factors affecting population growth, including birth rates and death rates, can be monitored precisely. Population growth can be slowed—birth rates may decline and/or death rates may increase—if resources become scarce. The factors like limited resources that slow population growth are often referred to as **environmental resistance**.

Even in a laboratory setting, growth of a population does not continue unchecked. Rather changes in population size follow a predictable and repeatable S-shaped trajectory called a **logistic growth curve** (Figure 4.20). In the beginning of logistic growth, a population grows rapidly, with little environmental resistance slowing it progress. The curve initially takes on the J-shaped trajectory associated with exponential growth. But as resources become increasingly limited, population growth slows until it eventually reaches the carrying capacity of its environment. At carrying capacity, the birth rate equals the death rate, and the population growth rate stalls at 0.0%. Also at carrying capacity, the rate at which resources are supplied to the population equals the rate at which they are used. Environmental resistance, in the form of both biotic and abiotic factors, has slowed the growth rate.

Consider an experiment in which a few bacteria are placed in a flask containing carbohydrates and all the other nutrients the bacteria need to grow; those resources are replenished at a constant rate. Initially, the rate at which the nutrients are supplied far exceeds the rate at which the small population can consume them. Conditions in the flask are favorable for the bacteria to grow and reproduce rapidly. In such ideal conditions, the population growth rate will be maximal and the size of the population will increase exponentially.

As the population of bacteria increases, the rate at which nutrients are consumed increases in direct proportion to the number of bacterial cells. Although the supply of nutrients remains constant, more bacterial cells are using them. The amount available to each individual bacterium declines, and competition among the bacteria for those resources increases. This competition causes a decline in the rate of population growth and an increase in mortality. Eventually, the rate of population growth begins to decline. If the population of bacteria continues to grow, then the rate of resource use will eventually equal the rate at which the resources are supplied. Populations of bacteria usually respond quickly to changes in the supply of resources. As resources diminish, reproduction slows down and bacteria begin to die. At this point, the population growth rate approaches zero, and the population stops growing.

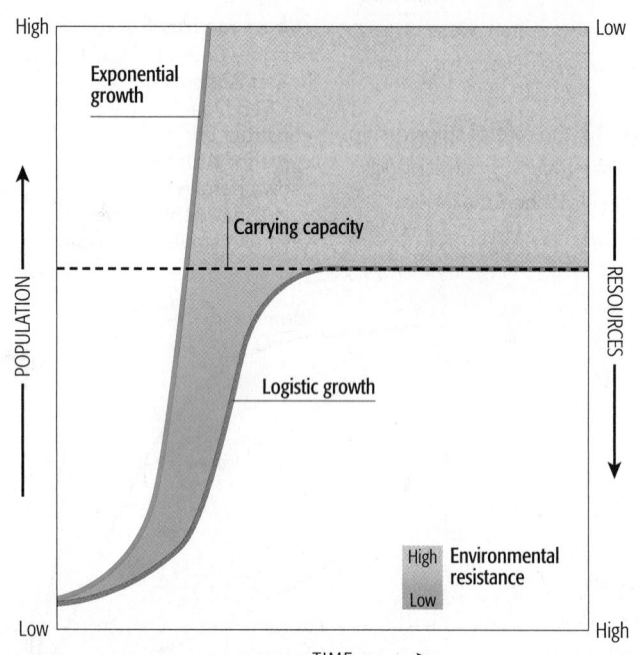

▲ **Figure 4.20 Logistic Growth**
As population size increases, resource limitations eventually slow population growth to zero, producing a logistic growth curve. The carrying capacity is the population size at which resource supply equals population resource use, and the population growth rate is zero.

Alternative Patterns of Population Growth

■ Birth and death rates may not respond immediately to changes in resource supply.

The smooth logistic growth curve shown by laboratory bacteria is typical of organisms with short generation times and simple life cycles. The birth rates and death rates of such organisms are tightly coupled to the supply of resources and change immediately in response to changes in that supply (Figure 4.21).

In organisms that have extended or complex life cycles, the effects of limited resources on birth rate and death rate may be delayed. In such cases, populations may grow beyond their carrying capacity. Subsequent adjustment of birth rate and death rate due to limited resources may cause the populations to fluctuate above and below carrying capacity. In some cases, populations may so far exceed their carrying capacities that resources become very limited. This eventually leads to a population "crash," or negative population growth, or even the disappearance of the population.

When populations exceed their carrying capacity, their use of resources exceeds the rate at which resources are supplied, and resources become scarce. The consequences of scarce resources depend on two factors: the speed with which the population growth rate responds to the diminished supply, and the rate at which the depleted resources are able to recover relative to the continued demands of the population.

Many birds and mammals have population growth rates that respond to diminished resources relatively quickly and depend upon resources that are renewed rapidly. When a population of these organisms overshoots its carrying capacity, the death rate soon exceeds the birth rate, the population growth rate becomes negative, and the population declines . When the population declines, the demand for resources eases, and the supply of resources increases. As resources are renewed, the population's growth rate becomes positive, and the population begins to grow again. In such a scenario, the size of the population may oscillate above and below its carrying capacity.

If a population's response to diminishing resources is excessively delayed, or if the rate at which resources are supplied is much slower than the rate at which they are used, the oscillation in the population size becomes more extreme. Indeed, if resource use continues to outstrip resource supply, the population may crash and disappear.

In nature, crashlike responses are rather common. Plants and animals often invade a new habitat, such as an open field or patch of forest, and increase rapidly by taking advantage of the area's resources. If those resources are supplied at a slow rate or are not renewable, the populations living in that area cannot sustain themselves and soon die out or move away.

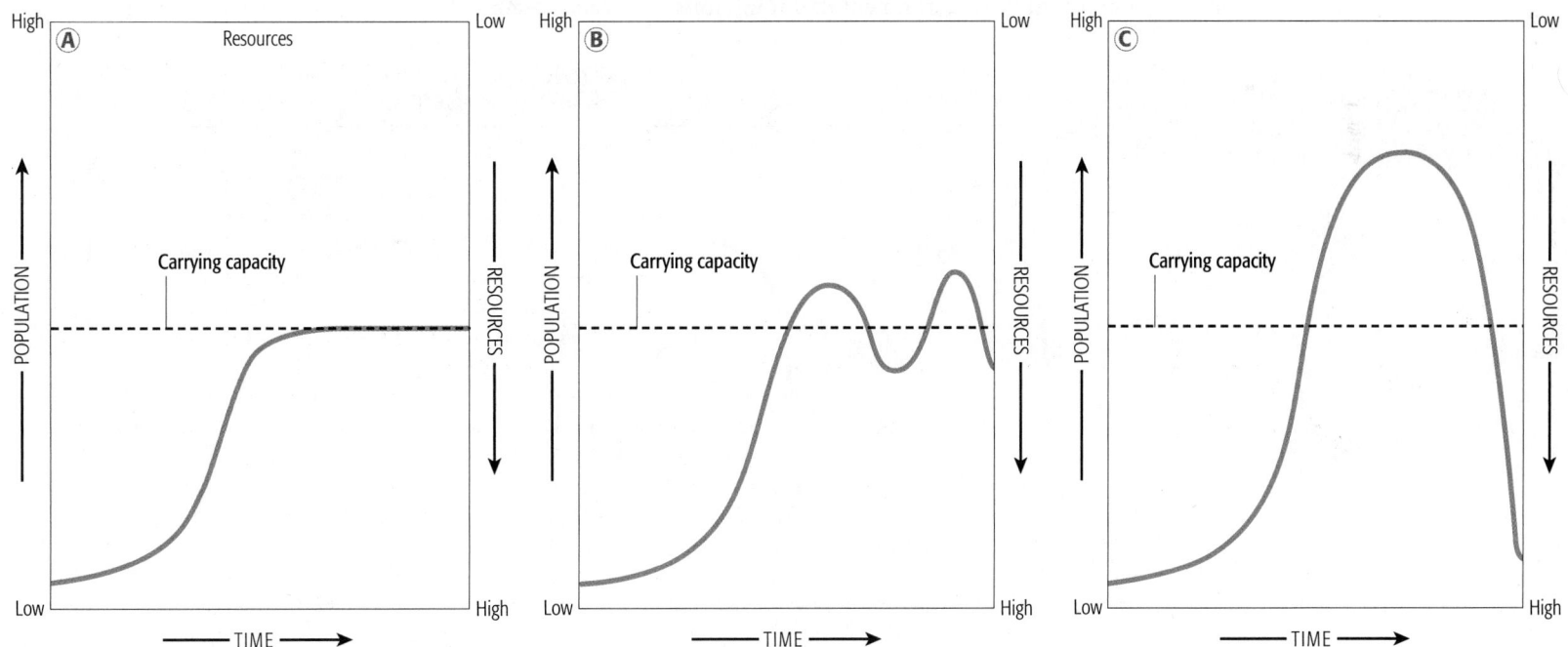

▲ Figure 4.21 **Alternative Patterns of Population Growth**
The graphs here depict hypothetical situations involving resource supply and population growth. Curve Ⓐ represents very tight coupling between population growth rate and resource supply whereas curve Ⓑ indicates a population growth rate that is delayed in response to limited resources but nevertheless recovers when resource supply is renewed. Curve Ⓒ illustrates that the delay between population growth rate response and resource supply is so significant that resource supply cannot recover and the population crashes.

Other Limits on Population Growth

■ Population growth may be influenced by factors other than resource use and supply.

In natural ecosystems, population changes are often erratic and don't seem to follow the growth predicted by theory or observed in the laboratory. Erratic changes occur because populations are influenced by multiple factors of environmental resistance, some of which are not related to resources. In addition, the factors that affect population growth most significantly may change from time to time.

The population growth of a species may be limited by different factors at different seasons of the year. For example, during the breeding season, populations of many migratory songbirds are limited by the availability of nesting sites in temperate forests. During the winter, their populations are limited by the amount of food available in tropical forests.

Temperature is a particularly important factor influencing the growth rate, birth rate, and death rate of all populations. For all living things, there is a minimum temperature below which growth cannot occur and death may be likely. Above the minimum temperature, growth rates generally increase as temperature increases. At an optimum temperature, the population grows most rapidly. Above the optimum temperature, growth rates decline. At some maximum temperature, growth ceases. The span between the minimum and maximum temperatures defines an organism's **range of tolerance**.

An organism's range of tolerance is closely related to the temperatures in the environment that it inhabits.

For example, the range of tolerance for rainbow trout is a water temperature of approximately 10–21 °C (50–70 °F), which closely matches the range of temperatures in the streams where it lives (**Figure 4.22**). The desert pupfish thrives in spring-fed desert pools that vary widely in temperature; the desert pupfish's range of tolerance is very large, 3–40 °C (38–104 °F). In contrast, the ice fish lives in waters near Antarctic glaciers, where temperatures are very cold and do not vary more than a degree or two. The ice fish has a very narrow range of tolerance, between –2 °C and –1 °C (28 °F and 30 °F).

Populations are often sensitive to the presence of toxic chemicals in their environment. Even small amounts of toxins such as cyanide or mercury may limit a population's growth. Sometimes, the organisms themselves produce retarding chemicals. For example, yeast populations may produce ethyl alcohol, which increases in concentration as their populations grow. Depending on the particular genetic strain of yeast, alcohol concentrations between 6% and 14% define the upper limit of their range of tolerance.

Organisms may even have a range of tolerance for resources that are essential to their survival. Plant growth, for example, is often controlled by limited supplies of nitrogen. However, too much nitrogen, such as too much fertilizer applied to a crop, can retard growth or even cause death.

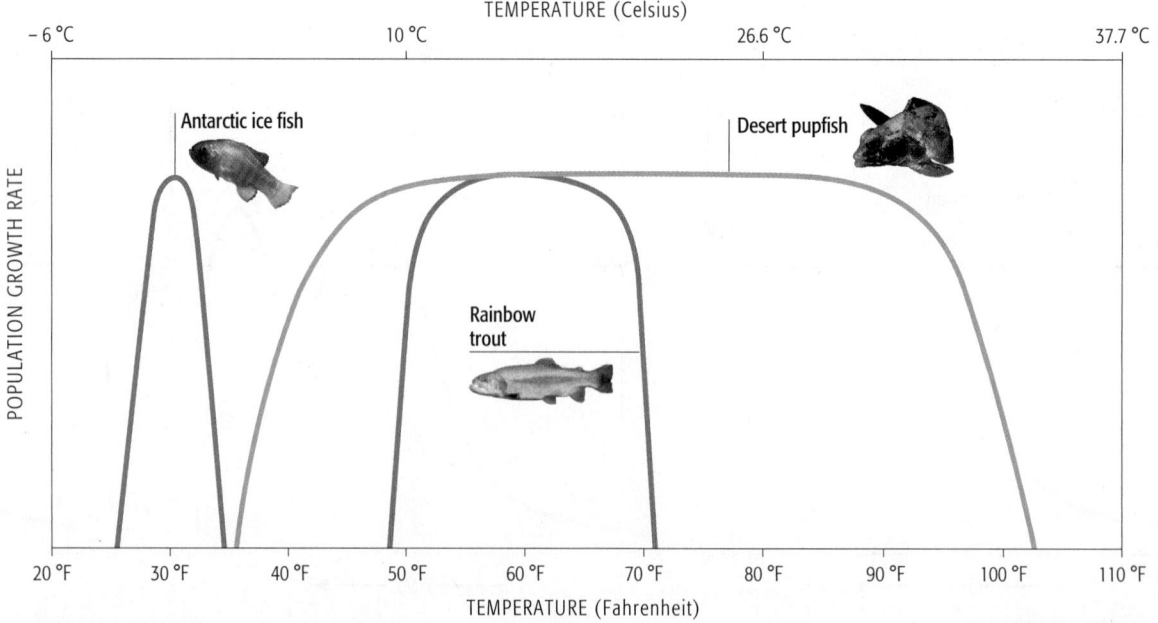

▲ Figure 4.22 **Range of Tolerance Varies among Species**
In general, organism functions and population growth increase with temperature to some optimal level, above which functions rapidly decline. Temperature ranges of tolerance and temperature optimal for different species are related to the environments in which they are found. Which species would be most likely to experience reduced population growth rate if temperatures increased by 3–5 °C?

Habitat and Ecological Niche

■ An organism's range of tolerance for many environmental factors determines its habitat and niche.

At any particular time, each organism must respond to a wide variety of environmental factors. In addition, an organism's range of tolerance for a particular environmental factor is often influenced by the status of other factors. Consider, for example, the interaction of temperature and water. As temperature increases, evaporation also increases. As a result, most terrestrial organisms generally require a greater supply of water when it is hot. These organisms can tolerate higher temperatures when water is plentiful than when it is scarce.

A **habitat** is the complex environment in which an organism is found and upon which it depends for survival. An organism's habitat includes nonliving environmental elements, such as temperature, humidity, and soil, as well as living elements, such as insects and foliage. The size of an organism's habitat depends on the size of the organism and its ecological needs. A salamander may spend its entire life foraging on worms and insects within a few square meters of the forest floor, whereas a mountain lion may hunt for deer over hundreds of square miles of forest, grassland, and shrubland.

Ecologists often describe an organism's habitat as where the organism lives. They call the various activities that define that organism's role in an ecosystem its **ecological niche**. An ecological niche is complex, encompassing every interaction between an organism and its ecosystem—what the organism eats, what eats the organism, and the effects of the organism's activities on the flow of energy and matter through the ecosystem. An organism's habitat and/or ecological niche may change at different times of the year or throughout its lifetime (Figure 4.23 and Figure 4.24). These topics are explored in detail in Chapter 6.

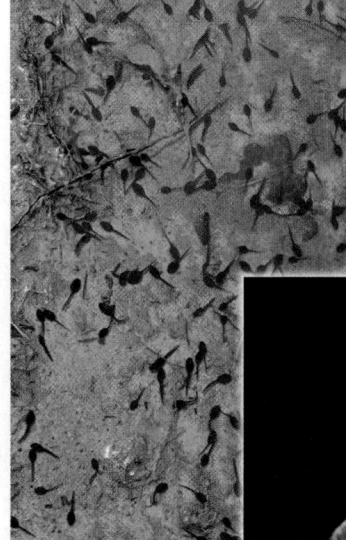

▶ Figure 4.23 **Habitat and Ecological Niche Can Change as Organisms Mature**
The habitat of American toad tadpoles is temporary ponds. The tadpoles feed on plant material and, therefore, occupy the ecological niche of herbivore. The mature toads' habitat is the terrestrial forest floor where their ecological niche is that of a carnivore, feeding on worms and insects.

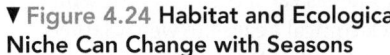

▼ Figure 4.24 **Habitat and Ecological Niche Can Change with Seasons**
The habitat and ecological niche of the Arctic jaeger change during the year. In the winter, it survives by stealing fish from terns and gulls in tropical waters. During the summer, it nests in the Arctic tundra where it feeds on lemmings.

QUESTIONS 4.4

1. Why does environmental resistance increase as population size increases?

2. What factors determine how well a population's growth will match the logistic growth curve?

3. A population's response to a resource or limiting factor may vary from time to time. Why?

 For additional review, go to **MasteringEnvironmentalScience**

111

4.5 Evolution and Natural Selection

BIG IDEA One of the first scientists to recognize the connection between inheritance, variations in behavior and appearance, and a species' ability to persist in an environment was Charles Darwin. In 1859, Darwin published his theory of evolution in a book called *On the Origin of Species by Means of Natural Selection, or the Preservation of Favored Races in the Struggle for Life*. Since his time, the work of many other scientists and the development of the science of genetics have provided ample evidence to support his theory of natural selection as a mechanism for evolution. In this context, the idea of a population takes on new meaning as a species that has the potential to adapt to changes in its environment.

Darwin's Finches

■ Species evolve when populations move into new habitats and discover new ecological niches.

Darwin postulated that the diversity of life is a consequence of **adaptations**, inherited structures, functions, or behaviors that help organisms survive and reproduce. Individuals with adaptations that increase their reproductive success compared to others in the population are said to have greater **fitness**. In the process of **natural selection**, individuals in a population that are most fit survive and leave more offspring, causing their adaptations to become more common. Individuals that are less fit leave fewer offspring, so their features become less common or cease to exist.

Darwin based his theory on a diverse array of observations made over many years. But it was the remarkable diversity of finches on the Galápagos Islands—species found nowhere else—that particularly inspired Darwin's thinking (Figure 4.25). Species belonging to the finch family occur on every continent except Australia. Most finches eat seeds, a diet for which their characteristic thick conical beaks are well adapted. But the 13 species of finches that live on the Galápagos occupy a variety of ecological niches, some of which are decidedly un-finchlike (Figure 4.26). For example, the woodpecker finch (*Cactospiza pallidus*) uses small sticks or cactus spines to prod insects from the crevices and holes in trees (Figure 4.27A). The warbler finch (*Certhidea olivacea*) uses its comparatively fine beak to feed on small insects on the ground (Figure 4.27B). Perhaps most peculiar of all is the sharp-billed ground finch (*Geospiza*

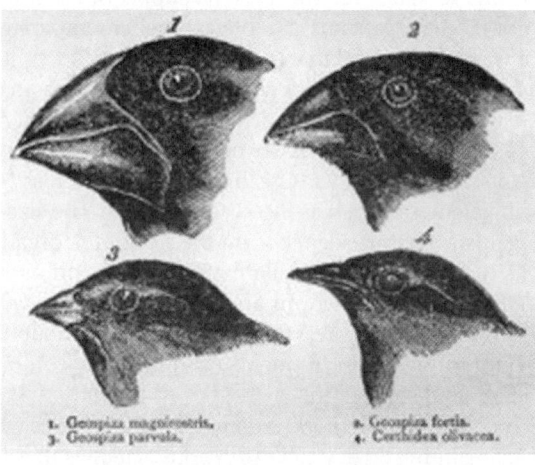

▲ Figure 4.26 **Galápagos Finch Adaptive Radiation**
These drawings depicting very different bill shapes of 4 of the 13 Galápagos finches are from Charles Darwin's research journal. His journal entry reads, "seeing this. . . diversity of structure in one group of birds, one might really fancy that from an original paucity of birds in this archipelago, one species had been taken and modified for different ends."

difficilis), which feeds on the eggs of large seabirds called boobies that nest on the islands (Figure 4.27C).

Because the Galápagos Islands are isolated from South America, Darwin reasoned that relatively few birds from the mainland had successfully migrated to the islands. Even so, he supposed that at some time in the distant past a few finches had found their way to the Galápagos. The finches that made it to the islands found an array of habitats quite different from those on the mainland.

Darwin recognized that individual organisms within each species varied in particular features, such as in their physical structure or behavior. Formal principles of genetics and the genetic basis of mutations were not established until long after Darwin's death. But Darwin did understand that new variations in features appear at random. Most important, Darwin knew that these variations were inherited from one generation to the next.

Darwin reasoned that finches possessing a particular bill shape and feeding behavior were more able to access food than finches with different bill shapes and feeding behaviors. Furthermore, he noted, finches that were better fed were more likely to survive and reproduce healthy offspring. Darwin reasoned that finches with bill shapes

▼ Figure 4.25 **The Galápagos Islands Inspire Charles Darwin**
About 600 miles (1,000 km) west of South America, volcanic activity began producing the Galápagos Islands approximately 5 million years ago. Their isolation from the mainland, austere habitats, and unique flora and fauna catalyzed Charles Darwin's ideas on evolution by means of natural selection.

and feeding behaviors that were more conducive to accessing food would become more abundant in the population than finches with bill shapes and feeding behaviors that were incompatible with finding food.

Based on these observations and this reasoning, Darwin proposed that finches of the Galápagos Islands very gradually evolved into new ecological niches through natural selection. Over a long time, they became separate species. Evolutionary biologists refer to the process by which several species evolve from a single ancestor to occupy new ecological niches as **adaptive radiation**.

◄ Figure 4.27 **Finches Occupy Different Ecological Niches**
Ⓐ The Galápagos woodpecker finch feeds on insects that it prods from crevices in tree trunks and cactus pads using a sharp stick or cactus spine. Ⓑ The warbler finch captures small insects much like the warblers that inhabit temperate forests. Ⓒ The sharp-beaked ground finch feeds on the eggs of large seabirds called boobies that nest on the islands.

Finch Studies Continue

■ Long-term studies of Darwin's finches demonstrate that biological evolution can occur rapidly.

Rosemary Grant and Peter Grant of Princeton University have been studying the Galápagos Islands finches since 1975. The Grants have captured, weighed, measured, and placed leg bands on every finch on the island called Daphne. The data they have gathered include careful measurements of body size, beak size and thickness, and feeding behavior. They also did experiments demonstrating that these traits are genetically inherited.

Between 1977 and 1978, the Galápagos experienced a severe drought, and the Grants recorded the death of thousands of birds. The birds had starved, apparently because their food source, seeds, was in limited supply. Curiously, the deaths did not occur randomly throughout all bird populations. The vast majority of birds that died belonged to certain species, such as the medium ground finch (*Geospiza fortis*).

Normally, medium ground finches are abundant on Daphne. During the drought, however, their population declined from over 1,400 to about 200 birds. The medium ground finch has a comparatively large beak and feeds on a wide range of seeds; it usually avoids very large, tough seeds, such as those of the widespread,

non-native Jamaican fever plant. But during the drought, the only seeds that were available in large numbers were the large tough seeds of the fever plant. Because cracking large seeds requires large jaw muscles, medium ground finches with small beaks and bodies were unable to crack them. Without food, the small-beaked birds died. Only medium ground finches with the largest beaks and bodies were able to survive the drought (Figure 4.28). With natural selection favoring large beaks and bodies, the medium ground finch population underwent significant evolution in both beak and body size within just a few generations.

During the 1980s, rainfall in the Galápagos was plentiful, resulting in abundant and diverse plant growth and the production of a wide variety of seeds. Bountiful food resulted in very successful reproduction among the medium ground finches, and their population flourished. Measurements of birds since then show a return to the predrought range of variation in beak and body size. The Grants' studies suggest that when smaller seeds are abundant, larger beak and body size may be disadvantageous.

Q What is a genetic bottleneck?
Natalie Hollabaugh, Central Michigan University

Norm: Ⓐ A genetic bottleneck occurs when the size of a population becomes so small that significant amounts of genetic variation are lost.

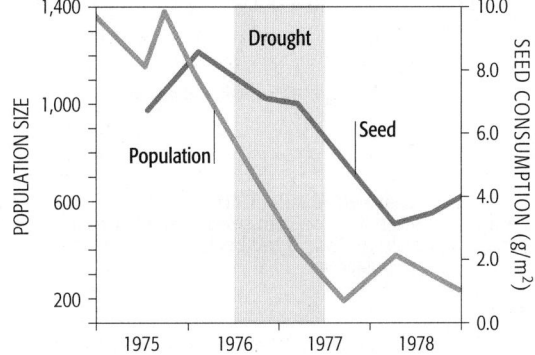

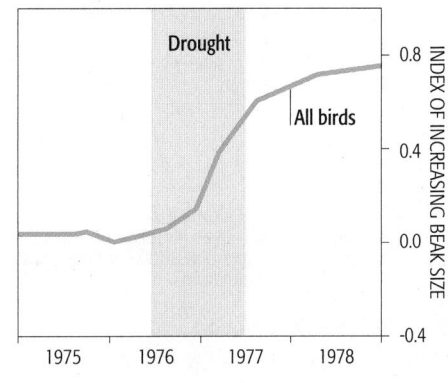

◄ Figure 4.28 **Rapid Evolution**
Between 1976 and 1977, drought on Daphne resulted in very limited seed production by herbs, with the exception of the large-seeded fever plant. During this period, the population of medium ground finches crashed Ⓐ. Those finches with larger beaks were able to survive on the fever plant seeds, and average beak size significantly increased in the population Ⓑ, as measured by an index that tracked the change in the average beak size and body size from one year to the next.

Source: Grant, B.R. and P.R. Grant. 1989. Natural selection in a population of Darwin's finches. *American Naturalist* 133:377–393.

Inherited Variations

■ Mutation is a very important source of new traits that enter a genome.

Darwin's studies and those of many other researchers have demonstrated that populations include considerable individual-to-individual variation for inherited traits. What is the source of that variation? As we saw in Module 4.2, sexual reproduction allows for genetic recombination in gametes and thus variability in offspring. The genes for variation in beak and body size of the Galápagos finches were already present in their genome. More significant to biodiversity is the introduction of new traits through random **mutations**. These mutations, or random changes in the structure of DNA, produce new forms of inherited features. Mutations may arise because of an environmental factor that affects the replication of DNA during cell division. For example, ultraviolet light and chemical pollutants are known to induce mutations. Note that only mutations that are carried in gametes can be passed to the next generation (**Figure 4.29**).

▲ Figure 4.29 **Mutation Is Random**
Albino deer with white coats appear occasionally in populations of otherwise normal deer. Albinism is a consequence of mutation in one of several genes that determine the color of the animal's fur, a mutation that has become part of the deer genome.

Mutation is the source of variation upon which natural selection may act. Most mutations are detrimental and are quickly removed from the population by natural selection. A few mutations have little or no effect on fitness. These may remain in the population and contribute to the range of variation that exists for a feature. For example, a mutation might result in a different plumage color or a slightly larger beak in individual members of a bird species. Changes in the environment might result in natural selection on such variation.

Natural Selection

■ There are three types of natural selection.

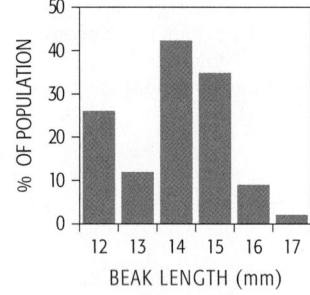

▲ Figure 4.30 **Inherited Features Vary**
Bird beak size is an important attribute that influences feeding behavior. Beak size in the Galápagos Islands cactus finch is influenced by several genes. Differences among individuals for these genes produce such variation.

Source: Grant, B.R., and P.R. Grant. 1989. Natural selection in a population of Darwin's finches. American Naturalist 133:377–393.

As the beaks of Galápagos finches illustrate, there is a range of variation for any inherited feature (**Figure 4.30**). Some genetic variations make an individual organism more fit for its environment, whereas other genetic variations render an individual less fit. Because they can determine an individual's fitness, the range of genetic variations can be directly correlated to natural selection within a population. Evolutionary biologists identify three types of natural selection—directional selection, stabilizing selection, and disruptive selection—based on the relative fitness across the range of genetic variation (**Figure 4.31**).

Directional selection refers to natural selection that favors the survival of individual organisms at one extreme end of the range of variation within a population. The change observed in the average size of finch beaks in response to changes in seed size is an example of directional selection.

Selection that favors traits that fall somewhere in the middle of the range of variation within a population is called **stabilizing selection**. For example, in most species, extremes of body size—very large and very small individuals—are less fit than average-sized individuals. Studies of the effects of major storms and extremely cold weather on bird populations show that very large and very small individuals suffer higher mortality rates than do average-sized individuals. Stabilizing selection is thought to be important for most inherited traits.

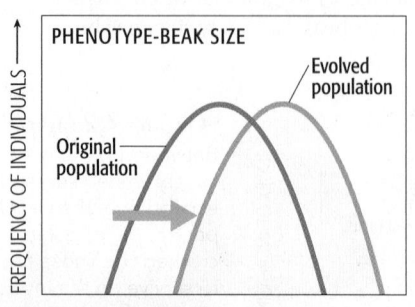

Directional selection
Removes one particular extreme phenotype—here birds with small beaks—from the population

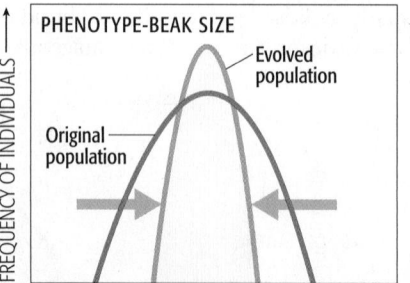

Stabilizing selection
Removes extreme phenotypes—birds with very small and very large beaks—from the population

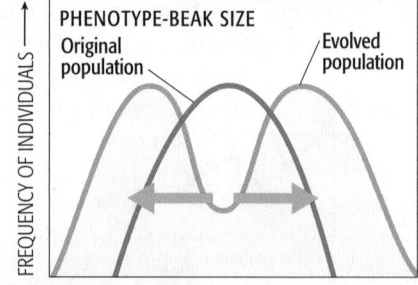

Disruptive selection
Removes medium-sized phenotypes—birds with medium-sized beaks—from the population

▲ Figure 4.31 **Three Types of Natural Selection**
Directional, stabilizing, and disruptive selection differ with respect to which phenotypes in a population are most fit. (A phenotype is the physical expression of traits in an individual based on its genetic makeup.)

Disruptive selection refers to situations in which individuals with genetic traits that fall at the extreme ends of the range are more likely to survive than individuals with average traits. Disruptive selection can occur when a population confronts contrasting environments. For example, if the Galápagos drought had resulted in the survival of plants that produced very small seeds in addition to the large seeds of the Jamaican fever plant, finches with small beaks and finches with large beaks would have been more likely to survive than finches with average-sized beaks.

Genetic Change

■ Chance events produce genetic change in small populations.

Populations of the common Asian ladybird beetle include individuals whose wing covers display a remarkable array of colors and spot patterns. This variation is the consequence of inherited variation in a single gene. In places with small populations of Asian ladybird beetles, researchers have noted that the frequency of colors varies significantly from time to time, with some colors becoming very abundant and other colors disappearing altogether. Researchers have concluded that these changes in the frequency of colors are caused by random events.

Random events may cause the frequency of a trait to increase or decrease simply by chance. Small populations are more susceptible to the effects of random events than large populations. In large populations, the effects of chance events are too small to detect, but as population size decreases, the likelihood that a chance event may cause a noticeable change in a population increases. A change in the frequency of an inherited trait in a population that is brought about by a chance event is called **genetic drift** (Figure 4.32).

The relationship between population size and genetic drift can be understood by comparing two different hypothetical ladybird beetle populations. If a predator were to eat 50 ladybird beetles at random from a population, as a matter of chance the predator might end up eating more beetles of one color than beetles of other colors. In a population of thousands of beetles, this event would have little effect on the overall frequency of the genes for any particular color. But in a population of only 100 beetles, the effect of the random acts of a predator would be significant. For example, if 30 of the beetles in that small population were yellow with no spots, their frequency would be 30%. If the predator ate 50 beetles, 25 of which happened to be yellow, the surviving population of 50 beetles would include only 5 yellow individuals. The frequency of yellow beetles would then be only 10% (5/50 × 100).

Chance can play a particularly important role in determining the genetic makeup of a newly established population. A few individuals from a large population migrating to a new habitat are unlikely to have exactly the same frequencies of characteristics as those found in their parent population. Therefore, the new population they found will be genetically different. Genetic change resulting from the immigration of a small subset of a population is called the *founder effect*.

Some of the best-documented examples of the founder effect are in human populations. Huntington's disease is a severe neurological disorder caused by a variation in a gene that affects the development of brain cells. The disease occurs at an unusually high frequency among the Afrikaner people of South Africa, who are descended from a rather small number of Dutch immigrants who colonized the region in the 18th century; that original population happened to carry an unusually high frequency of the gene for Huntington's disease.

In summary, mutation generates new variations for inherited traits, and sexual reproduction maintains such variations in populations. Evolution occurs when the relative abundance of genetically inherited traits within a population changes. Natural selection is a form of evolution in which factors in the environment favor the survival and reproduction of individuals with particular inherited traits. Genetic drift is a form of evolution that occurs in small populations as a consequence of random events.

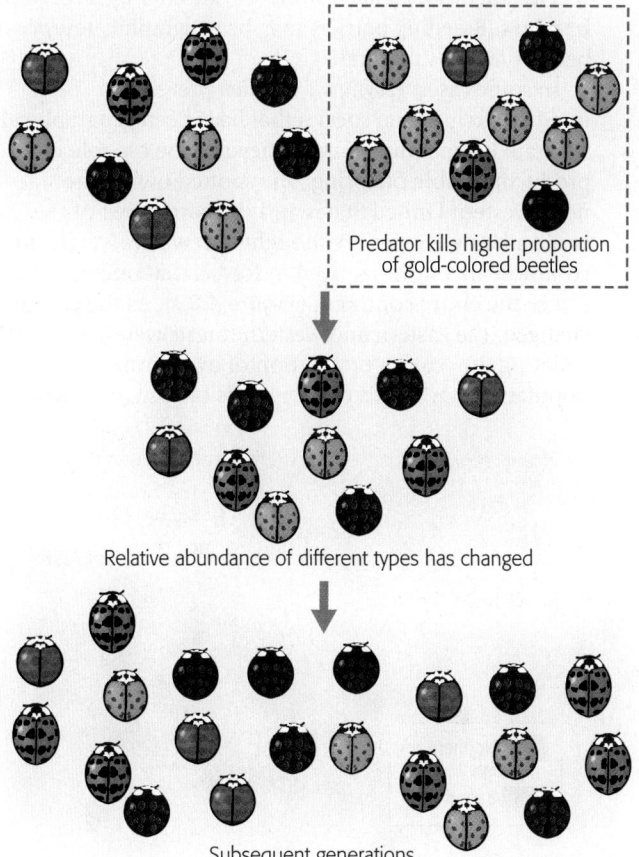

▲ **Figure 4.32 Genetic Drift**
The relationship of population size to genetic drift can be understood by comparing the effects of a hypothetical predator on the frequency of color types in a small ladybird beetle population.

Predator kills higher proportion of gold-colored beetles

Relative abundance of different types has changed

Subsequent generations

QUESTIONS 4.5

1. Based on Darwin's study of the finches on the Galápagos Islands, describe the process of natural selection.

2. Compare directional, stabilizing, and disruptive selection with respect to their effects on inherited variation for a feature.

3. Why doesn't genetic drift occur in large populations?

(MES) For additional review, go to **MasteringEnvironmentalScience**

4.6 The Evolution of Species

BIG IDEA Biodiversity is a measure of the variety of species in the world. A **species** is a group of similar organisms that can potentially mate and produce fertile offspring. Organisms belonging to the same species but different populations can potentially interbreed, whereas organisms belonging to different species cannot. When there are barriers to interbreeding between separate populations, the two populations may undergo evolutionary changes that allow them to occupy separate ecological niches. This is the basis for the evolution of new species.

Preventing Interbreeding

■ Natural selection has resulted in a variety of mechanisms to prevent interbreeding.

In general, each species has a unique set of genetic traits that are not shared with other species. For example, the warbler finch differs from other Galápagos finches in having a fine beak and small body. From an evolutionary standpoint, a more important property of species is that they are reproductively isolated from one another. **Reproductive isolation** means that members of one species do not interbreed with members of other species, even those that are closely related. Closely related species are often isolated from one another by **breeding barriers**. Breeding barriers may be geographic, temporal, behavioral, or structural.

In some cases, *geographic isolation* prevents interbreeding. When species that have been geographically separated come into contact, they may be capable of producing viable offspring. The spotted owl of the northwestern United States and the barred owl of the eastern United States are thought to have evolved from a common ancestor that lived in forests that once extended across the entire continent (Figure 4.33). As the climate changed, the eastern and western forests were separated, isolating the eastern population of owls from the western population. Over tens of thousands of years, the two

populations evolved to have different characteristics. In recent years, barred owls have begun migrating into northwest forests, where they occasionally mate with spotted owls and produce fertile offspring.

Where the geographic ranges of closely related species overlap, unions between species are more likely to occur. Such interbreeding usually results in offspring that are sterile or in some way poorly adapted to their environment. Inherited traits that prevent interbreeding between such species will, therefore, increase their fitness and will be subject to natural selection. Over time, natural selection will favor the development of breeding barriers. Several different kinds of breeding barriers have evolved in plants and animals.

Species may be isolated from one another by reproducing at different times. This is called *temporal isolation* (Figure 4.34). For example, the green frog, the wood frog, and the southern leopard frog inhabit much of the southeastern United States. These three species do not interbreed because their peak breeding periods do not occur at the same time. Their different breeding times are a consequence of each species responding differently to seasonal changes in temperature.

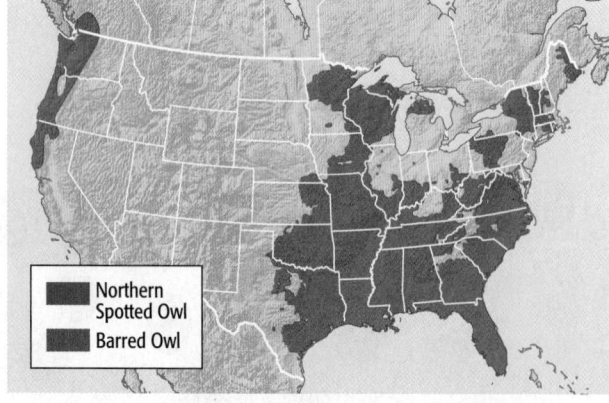

▲ Figure 4.33 **Geographic Reproductive Isolation**
The northern spotted owl inhabits Pacific Northwest forests, and the barred owl is found in forests of the eastern United States. They are reproductively isolated by means of geographic separation. Recent introductions of barred owls to the Pacific Northwest have resulted in successful interbreeding.

Northern Spotted Owl
Barred Owl

GREEN FROG

SOUTHERN LEOPARD FROG

WOOD FROG

FEB   MARCH   APRIL   MAY   JUNE

▲ Figure 4.34 **Temporal Reproductive Isolation**
Different breeding times reproductively isolate the wood frog (*Rana sylvatica*), green frog (*R. clamitans*), and southern leopard frog (*R. sphenocephala*) from one another where they share habitat in the southeastern United States.

Species may also be separated from one another by differences in behavior. This is called *behavioral isolation*. For many animals, mating displays and courtship rituals play a significant role in the choice of mates and can be effective barriers to breeding between species. Such reproductive isolation is especially common among birds that select mates based on mating calls or dances. For example, the mating ritual of the Galápagos Islands' blue-footed boobies includes a complex dance performed first by the male and then by both sexes (Figure 4.35). Neither male nor female boobies will mate with individuals that don't know the dance.

Differences in reproductive structures effectively isolate many closely related species from one another, resulting in *structural isolation*. For example, the shape and color of flowers determines what pollinators are attracted to a plant and, ultimately, whether visiting pollinators will carry the flower's pollen to other flowers for successful pollination. Plants with dark-colored, tubular flowers are especially attractive to hummingbirds, which have long beaks and tongues that are able to reach the nectar at the base of the flowers (Figure 4.36). Yellow and blue flowers with open petals are more typically visited by insects.

Reproductive isolation prevents mating between species. However, on the rare occasions when different species do mate, the genetic differences between them are usually great enough that nonviable, sterile, or partially sterile offspring are produced. For example, no one

questions that tigers (*Panthera tigris*) and lions (*Panthera leo*) are different species. However, on rare occasions where their ranges overlap, they may interbreed. The characteristics of their hybrid offspring depend on the gender of the respective parents. If the father is a lion and the mother a tiger, the offspring are *ligers*, and they resemble a very large lion, but with stripes (Figure 4.37). If dad is a tiger and mom is a lion, the offspring are more like a tiger and are called *tigons*. Male ligers and tigons are completely sterile, whereas their female counterparts are fertile.

In general, interbreeding between genetically distinct populations that are adapted to specific environments produces offspring that are less fit than either of their parents. Therefore, natural selection is likely to favor the evolution of traits or behaviors—breeding barriers—that prevent such mating. This is the primary means by which new species evolve.

QUESTIONS 4.6

1. Two populations of birds may look alike but be considered separate species. On what basis do scientists make such designations?

2. Describe three different kinds of breeding barriers.

(MES) For additional review, go to **MasteringEnvironmentalScience**

▲ Figure 4.36 **Mechanical Reproductive Isolation**
These two closely related species of columbine (*Aquilegia*) are found along streams in the Sierra Nevada. The scarlet columbine (left) grows in sunny habitats, and its red flowers attract hummingbirds. The Sierra columbine (right) grows in shadier locations, and its tubular, cream-colored flowers are visited by large hawk moths.

▼ Figure 4.37 **The Hybrid Liger**
Distinct species may sometimes interbreed. For example, if a male lion breeds with a female tiger, their offspring, ligers, share features of both species. Ligers, however, have much reduced fertility.

▲ Figure 4.35 **Behavioral Reproductive Isolation**
Blue-footed boobies are known for their peculiar mating dances by which males and females recognize each other as members of the same species.

4.7 The Hierarchy of Life

BIG IDEA Finches, frogs, and all Earth's life are the products of at least 3.8 billion years of biological evolution. To understand how biological evolution occurs, scientists consider how species adapt to changing conditions, how they interact with their environment, and how they compare physically, behaviorally, and genetically to other forms of life. **Taxonomy**, the classification of organisms through description, identification, and naming, provides a framework for such comparisons.

Birds

FEAT

Evolutionary Map

■ A hierarchical system of classification reflects the evolutionary relationships among groups.

Domain	Eukarya
A domain is the most fundamental division of Earth's life based on cell structure and basic biochemical characteristics.	The domain Eukarya includes all organisms with eukaryotic cells having a nucleus and membrane bound organelles.
Kingdom	Animala
A kingdom is an overall category containing organisms that share fundamental structures and functions.	The kingdom Animala contains multicellular organisms that obtain energy by eating food. Most have muscles and nerves.
Phylum (Animals) or Division (Plants)	Chordata
A phylum or division is a major subdivision of a kingdom containing one or more classes and their subgroups.	The phylum Chordata includes the vertebrates. Among other shared features they have a nerve cord extending along their backs.
Class	Mammalia
A class is a major subdivision of a phylum or division containing one or more orders and their subgroups.	The class of "warm blooded" vertebrates that have hair or fur and have females that produce milk to nourish their young.
Order	Lagomorpha
An order is a major subdivision of a class containing one or more families and their subgroups.	Rodent-like mammals with four instead of two incisor teeth in the upper jaw.
Family	Leoporidae
A family is a major subdivision of an order containing one or more genera and their subgroups.	This family includes rabbits and hares having short furry tails and elongated ears and hind legs.
Genus	*Lepus*
A genus is a major subdivision of a family containing one or more species.	Rabbit relatives whose young are born with eyes open and a full coat of fur.
Species	*Lepus townsendii*
A species is a group of similar individuals that are able to interbreed with one another in the wild.	The whitefooted jackrabbit is noted for its long ears and well-developed legs.

▲ **Figure 4.38 Classifying** *Lepus*

The hierarchical system of classification defines taxa based on shared characteristics that are inherited from common ancestors.

In taxonomy, each kind of organism is assigned a two-part Latin name. The first part of the name, the **genus**, identifies the organism with its closest relatives. The second part of the name is unique for each species; it may describe a unique characteristic of the species, where the species lives, or the name of a scientist who discovered the species. The two-part Latin name for the white-tailed jackrabbit is *Lepus townsendii*. The genus *Lepus* includes 28 species of large hares whose young are born with their eyes open and a full coat of fur. The species *townsendii*, named in honor of 19th-century naturalist John Kirk Townsend, is distinguished from other members of this genus by its very long ears and hind legs. Its long legs allow this jackrabbit to run as fast as 50 miles per hour.

Taxonomy is a system of hierarchical categories or taxa (singular, taxon), in which each level of classification includes all the levels beneath it (**Figure 4.38**). The lowest level of classification is the species. Closely related species are grouped together in a genus. Closely related genera are grouped into *families*, and families are grouped into *orders*. Groupings continue in this manner up the taxonomic hierarchy to encompass *classes*, *phyla* (or *divisions* for plants), *kingdoms*, and *domains*. At each level of the taxonomic hierarchy, organisms classified into a group are more similar to one another than to organisms belonging to other groups at the same level.

When deciding how to classify organisms, taxonomists are careful to select features that reflect the evolutionary ancestry of organisms. Thus, the taxonomy of life is equivalent to an evolutionary road map.

Protists

Plants

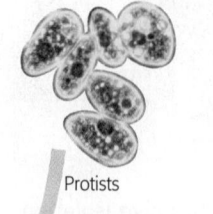

Bacteria

Archaea

Fur

NERVE
MUSC
MOBIL

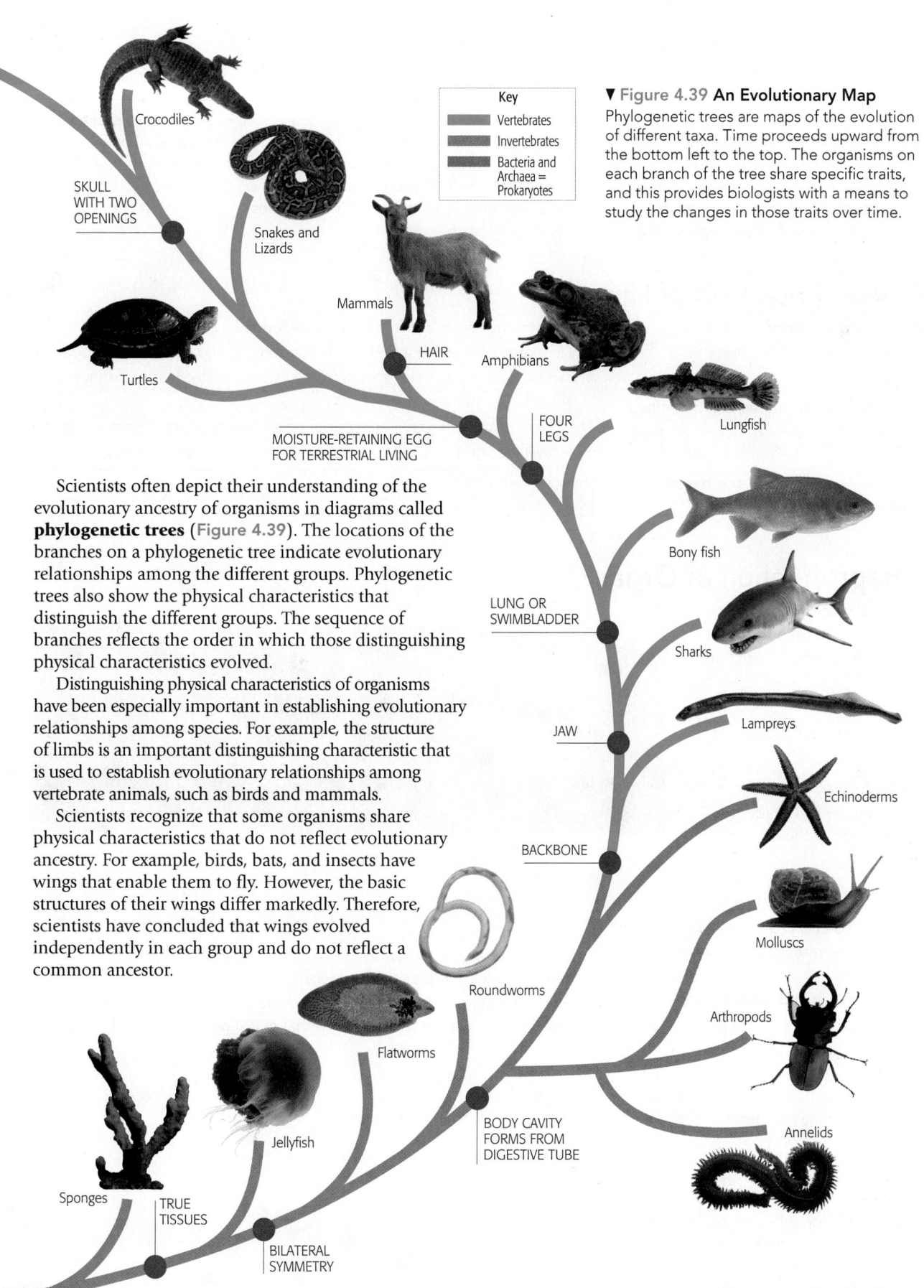

Key
- Vertebrates
- Invertebrates
- Bacteria and Archaea = Prokaryotes

▼ Figure 4.39 **An Evolutionary Map**
Phylogenetic trees are maps of the evolution of different taxa. Time proceeds upward from the bottom left to the top. The organisms on each branch of the tree share specific traits, and this provides biologists with a means to study the changes in those traits over time.

Crocodiles

SKULL WITH TWO OPENINGS

Snakes and Lizards

Mammals

Turtles

HAIR

Amphibians

MOISTURE-RETAINING EGG FOR TERRESTRIAL LIVING

FOUR LEGS

Lungfish

Bony fish

LUNG OR SWIMBLADDER

Sharks

JAW

Lampreys

Echinoderms

BACKBONE

Molluscs

Roundworms

Arthropods

Flatworms

Jellyfish

Sponges

TRUE TISSUES

BILATERAL SYMMETRY

BODY CAVITY FORMS FROM DIGESTIVE TUBE

Annelids

Scientists often depict their understanding of the evolutionary ancestry of organisms in diagrams called **phylogenetic trees** (Figure 4.39). The locations of the branches on a phylogenetic tree indicate evolutionary relationships among the different groups. Phylogenetic trees also show the physical characteristics that distinguish the different groups. The sequence of branches reflects the order in which those distinguishing physical characteristics evolved.

Distinguishing physical characteristics of organisms have been especially important in establishing evolutionary relationships among species. For example, the structure of limbs is an important distinguishing characteristic that is used to establish evolutionary relationships among vertebrate animals, such as birds and mammals.

Scientists recognize that some organisms share physical characteristics that do not reflect evolutionary ancestry. For example, birds, bats, and insects have wings that enable them to fly. However, the basic structures of their wings differ markedly. Therefore, scientists have concluded that wings evolved independently in each group and do not reflect a common ancestor.

QUESTIONS 4.7

1. Explain the evolutionary meaning of the two parts of a species' scientific name.

2. What is implied when two groups of animals are depicted as twigs sprouting from the same branch of a phylogenetic tree?

(MES) For additional review, go to **MasteringEnvironmentalScience**

Summary

Earth's biodiversity and the myriad ecosystem processes that support it ultimately depend on the growth of populations of organisms. The growth of a population is driven by the growth, reproduction, migration, and death of the individual organisms that comprise it. In turn, the growth and reproduction of individual organisms depends on the growth and division of individual cells. The chemical energy needed for these activities ultimately comes from photosynthesis, and organisms access that chemical energy through the process of cellular respiration. The growth of populations may be limited by resources and other factors in the environment that influence the growth and reproduction of the organisms in the population. These environmental factors determine the habitat and ecological niche of each species. Genetic variation causes the individuals within a population to have different traits, which determine whether individuals are able to grow and reproduce successfully. Natural selection acts on that genetic variation to produce evolutionary change and new species.

4.1 The Cell—The Fundamental Unit of Life

- Cell structure is the basis for classifying organisms into two major groups.
- The cell is the site of life's most important chemical functions.

KEY TERMS

cytoplasm, prokaryote, eukaryote, organelle, photosynthesis, chemosynthesis, cellular respiration, anaerobic respiration

QUESTIONS

1. Give examples of prokaryotic and eukaryotic organisms and describe the cellular features that distinguish these two groups.

2. Explain why nearly all life ultimately depends on photosynthesis.
3. What resources are required for photosynthesis, and what are its products?
4. Explain how some organisms produce carbohydrates in the absence of light.

Photosynthesis and respiration involve opposite reactions	Photosynthesis carbon dioxide + water + energy $\longrightarrow$ carbohydrates + oxygen
	Respiration carbohydrates + oxygen $\longrightarrow$ carbon dioxide + water + energy

4.2 The Growth and Reproduction of Organisms

- An organism grows through division and differentiation of its cells.
- The mode of reproduction influences the amount of variation among offspring.

KEY TERMS

asexual reproduction, sexual reproduction, gamete, zygote

QUESTIONS

1. What are clones and how are they produced?
2. All of the cells that make up an organism contain exactly the same DNA. What is the basis for differences in the structure and functions of the cells within an organism?

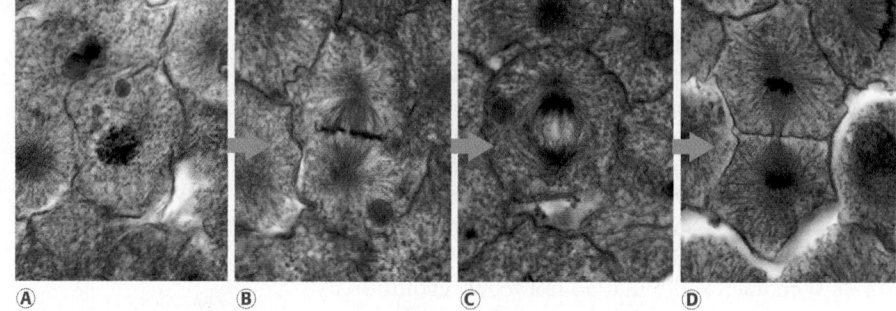

Ⓐ   Ⓑ   Ⓒ   Ⓓ

4.3 The Growth of Populations

- Population growth rate is determined by the rates of birth, death, and migration.
- The likelihood that an organism will reproduce and survive changes over its lifetime.

KEY TERMS

exponential growth, arithmetic growth, population growth rate, doubling time, birth rate, death rate, mortality rate, immigration rate, emigration rate, survivorship, fertility rate, age-specific fertility rate, total fertility rate, generation time

QUESTIONS

1. Differentiate between arithmetic and exponential growth.
2. What is the doubling time for a population with a 1% growth rate? What is the doubling time for a population with a 0.5% growth rate?
3. From a single mating, a female housefly will lay hundreds of eggs. If population growth rate were determined solely by birth rate, Charles Darwin calculated that a pair of flies would produce offspring (children, grandchildren, etc.) equal to the weight of the Earth in just eight generations. Why does this not occur?
4. Give examples of animals that display type I, II, and III survivorship curves and describe ecological conditions that might be responsible for each of these patterns of survival.

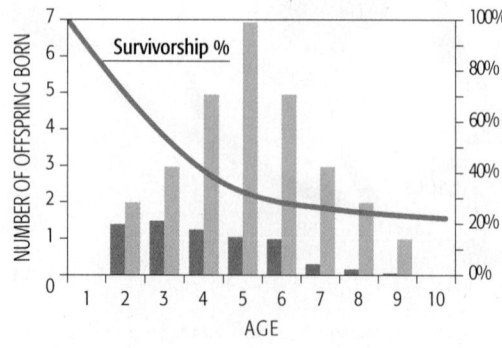

4.4 Limits on Population Growth

■ At carrying capacity, the birth rate equals the death rate and the population growth rate stalls at zero.

KEY TERMS

carrying capacity, environmental resistance, logistic growth curve, range of tolerance, habitat, ecological niche

QUESTIONS

1. As the size of a population nears carrying capacity, how does its use of resources change in relation to the supply of resources?
2. Populations may grow beyond their carrying capacity. Describe and explain two possible consequences of such growth in regard to resource use and resource supply.
3. Explain why an organism's range of tolerance for one factor may depend on the status of other factors in its environment.
4. Using a specific example, differentiate between an organism's habitat and its ecological niche.

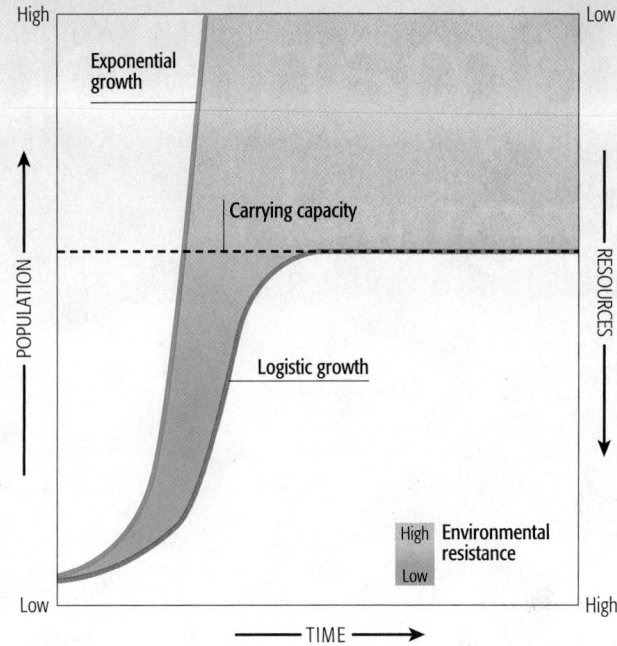

4.5 Evolution and Natural Selection

■ Species evolve when populations move into new habitats and discover new ecological niches.

KEY TERMS

adaptation, fitness, natural selection, adaptive radiation, mutation, directional selection, stabilizing selection, disruptive selection, genetic drift

QUESTIONS

1. Natural selection is sometimes described as survival of the fittest. In addition to survival, what characteristic of an individual organism determines its fitness?

2. As insecticides have become more abundant in the environment, so has the abundance of insects with inherited resistance to them. Explain the likely basis for this change in the relative abundance of insecticide-resistant insects.
3. Explain why genetic drift often occurs in populations recently founded by a few immigrant individuals.

4.6 The Evolution of Species

■ Natural selection has resulted in a variety of mechanisms that prevent interbreeding.

KEY TERMS

species, reproductive isolation, breeding barrier

QUESTIONS

1. Why does natural selection favor the evolution of breeding barriers between species?

2. Describe how two new species might evolve from a parent species, even though they live in the same geographic area.

Temporal isolation	Differences in mating season
Behavioral isolation	Differences in mating calls or dances
Structural isolation	Differences in reproductive anatomy

4.7 The Hierarchy of Life

■ A hierarchical system of classification is an evolutionary map.

KEY TERMS

taxonomy, genus, phylogenetic tree

QUESTIONS

1. The hierarchical system of classification is intended to be a map of evolution. Explain this statement.
2. What is implied about species that occur on the same branch of a phylogenetic tree?

MasteringEnvironmentalScience®

Students Go to **MasteringEnvironmentalScience** for assignments, the eText, and the Study Area with practice tests, activities, and more.

Instructors Go to **MasteringEnvironmentalScience** for automatically graded tutorials and questions that you can assign to your students, plus Instructor Resources.

Human Population Growth

Human Population Growth—By the Numbers

Can we determine how many people Earth can support?

At some moment during the autumn of 2011, the global human population passed the 7 billion mark. We don't know the exact day, much less the hour or minute; United Nations census data and estimates of world population growth are just not that precise. Nevertheless, many noted this milestone in human history. Some people feel that the rapid growth in our numbers is the "mother of all environmental problems." And they have a point. Our

numbers have doubled twice over the past century. If you are 20 years old, Earth's population has grown by nearly 30% (1.6 billion) since your birth. Many argue that the growing number of humans is the primary factor driving our increasing impact on our planet and its life-support systems. From others you will hear, "It's not that simple." Yes, the growth in our numbers is a concern, but the growth in our consumption is even more troubling.

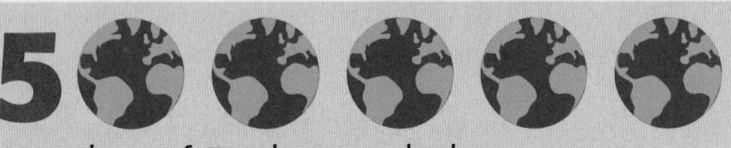

5 Number of Earths needed to support the world's population if everyone lived like **North Americans** do.

55 years Life expectancy of 10 poorest nations

80 years Life expectancy of 10 wealthiest nations

7.03 Average number of children born to a woman in her lifetime in **Niger**

2.06 Average number of children born to a woman during her lifetime in the **United States**

19.7% Percent of married women (or those in a union) that use modern contraception in sub-Saharan Africa

10x Amount of energy & materials used by **Americans** vs people from **Zimbabwe**

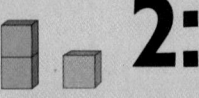 **2:1** Ratio of population of developing to developed countries, **1950**

6:1 Ratio of population of developing to developed countries, projected **2050**

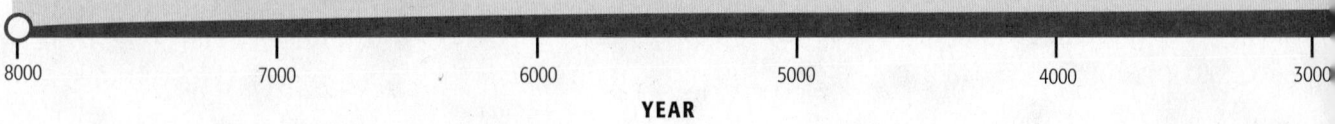 **20,000** Number of girls below the age of 1 that give birth every day in developing countries

8000 B.C. – 500 thousand

8000   7000   6000   5000   4000   3000

YEAR

You will see many numbers included in this chapter. The numbers represent projections, averages, means, percentages, rates, rankings, and ratios, some of which are highlighted in **Figure 5.1**. These numbers are not only important in an academic sense: They will impact your life. The ability of scientists to make accurate predictions using some of these numbers will influence our ability to prepare for their consequences.

Of particular interest are the numbers we use to identify the growth and size of the world population. Just how many people can Earth support? This question has been at the center of much research and argument about human population change. The truth is that no one knows that number. Many feel that we are fast approaching it. Some feel we've already passed it.

- *What are the factors that have influenced human population change through time?*
- *What factors account for the differences that currently exist among countries?*
- *How can we use this information to predict future population trends?*
- *We can quantify the changes in our population; can we do the same for our impact on resources and ecosystems?*
- *How can we slow growth in both numbers and consumption to ensure a sustainable future?*

▼ Figure 5.1 **By the Numbers**
Inspired by Roberts 2011, *Science 333 (29)540–543.*

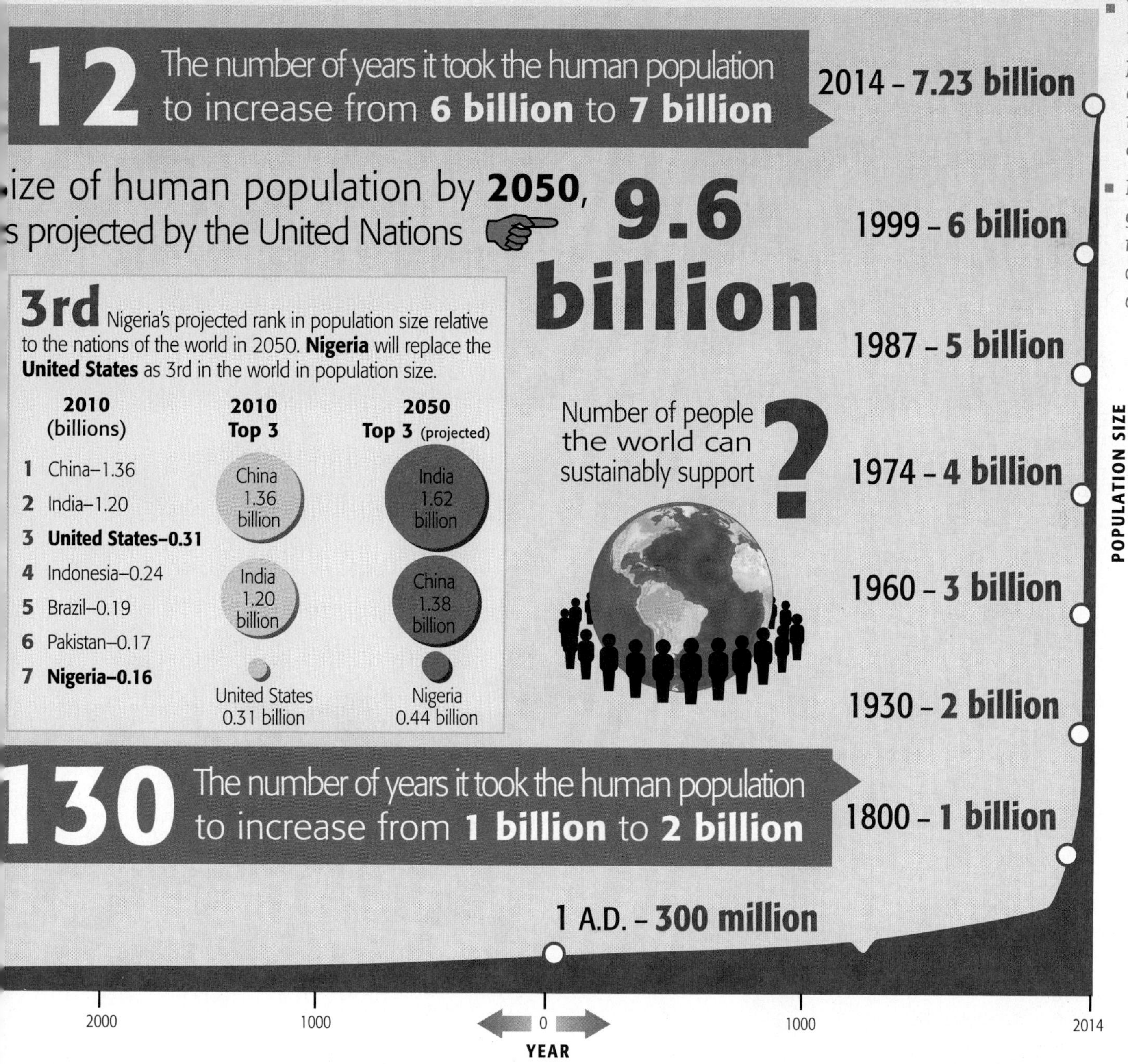

12 The number of years it took the human population to increase from **6 billion** to **7 billion**

ize of human population by **2050**, s projected by the United Nations ☞ **9.6 billion**

3rd Nigeria's projected rank in population size relative to the nations of the world in 2050. **Nigeria** will replace the **United States** as 3rd in the world in population size.

2010 (billions)		2010 Top 3	2050 Top 3 (projected)
1 China–1.36		China 1.36 billion	India 1.62 billion
2 India–1.20			
3 **United States–0.31**		India 1.20 billion	China 1.38 billion
4 Indonesia–0.24			
5 Brazil–0.19			
6 Pakistan–0.17		United States 0.31 billion	Nigeria 0.44 billion
7 **Nigeria–0.16**			

Number of people the world can sustainably support **?**

130 The number of years it took the human population to increase from **1 billion** to **2 billion**

2014 – **7.23 billion**

1999 – **6 billion**

1987 – **5 billion**

1974 – **4 billion**

1960 – **3 billion**

1930 – **2 billion**

1800 – **1 billion**

1 A.D. – **300 million**

POPULATION SIZE

2000   1000   0   1000   2014
YEAR

125

5.1 The History of Human Population Growth

BIG IDEA Over the past 100,000 years, the human population has increased enormously. The earliest humans lived in Africa. As humans moved into new regions, new resources became available and the number of people increased. Eventually, humans migrated throughout the world (Figure 5.2). A number of developments have made it possible for people to live in so many different places on Earth. During the past 10,000 years, innovations in agriculture have increased the food supply, allowing the size and density of human populations to increase. Over the past 300 years, some regions have experienced advances in industry and economic progress that have led to decreasing birth rates, death rates, and rates of population growth (Figure 5.3).

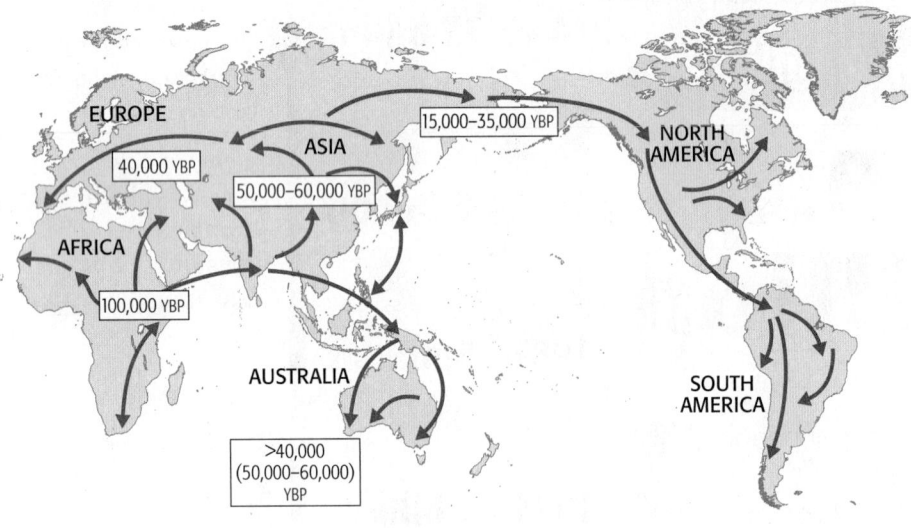

◄ **Figure 5.2 Humans Populate the Globe**
Beginning 100,000 years ago, our species, *Homo sapiens*, began to migrate throughout Africa, followed by another migration into Asia 50,000–60,000 years before present (YBP). Migration from Asia, to Europe and Australia, began some 40,000 years ago. Humans migrated from northern Asia, across the Bering Strait, and into North America 15,000–35,000 years ago.

Source: Cavalli-Sforza, L.L. and M.W. Feldman. 2003. The application of molecular genetic approaches to the study of human evolution. *Nature Genetics* 33:266–275.

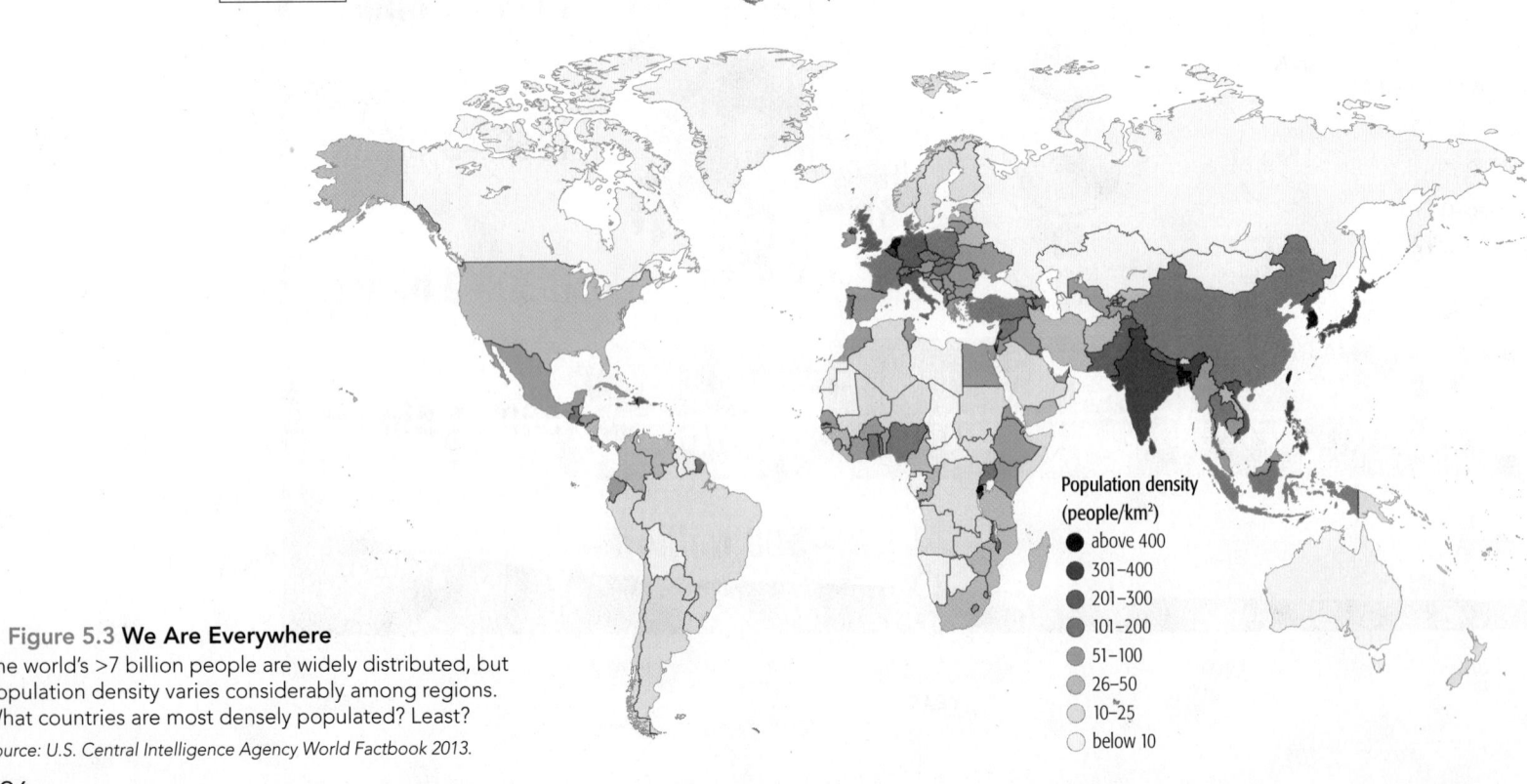

► **Figure 5.3 We Are Everywhere**
The world's >7 billion people are widely distributed, but population density varies considerably among regions. What countries are most densely populated? Least?

Source: U.S. Central Intelligence Agency World Factbook 2013.

Population density (people/km²)
- above 400
- 301–400
- 201–300
- 101–200
- 51–100
- 26–50
- 10–25
- below 10

Three Periods of Growth

■ Human populations experienced three periods of growth.

People who study the size, density, and distribution of human populations are called **demographers**. Demographers identify three periods of human population growth—the pre-agricultural period, the agricultural period, and the industrial period (Figure 5.4). The transitions between these periods occurred gradually, over decades or centuries. Furthermore, these transitions occurred at various times in different parts of the world.

Pre-agricultural period. The pre-agricultural period, which extended over more than 100,000 years, coincided with the development of human culture. The advancement of skills such as tool making, as well as the ability to pass those skills on to successive generations, allowed humans to adapt quickly to new environments. The rate of global population growth during the pre-agricultural period depended on the speed at which humans spread into new regions. It took tens of thousands of years for the human population to double in size. At the end of the pre-agricultural period, the global human population numbered between 5 and 10 million.

Agricultural period. The agricultural period began about 10,000 years ago with the domestication of plants and animals and the development of farming. Breeding and growing livestock and nutrient-rich crops increased the supply of food. New technologies such as irrigation and plowing helped to increase the size of harvests. With more food available, human populations continued to increase. The time required for human populations to double shortened to less than 1,000 years. By the end of the agricultural period, the human population had grown to more than 500 million.

Industrial period. With the advent of the Industrial Revolution around 1800, the industrial period of population growth took hold. Technologies powered by fossil fuels increased food production and allowed agriculture to expand onto land that previously was unfit for farming. With the industrial period came vastly improved sanitation and huge advances in medicine. The rate of population growth accelerated. We are, of course, still in the industrial period, and our population currently exceeds 7 billion.

▼ Figure 5.4 **Three Periods of Human Population Growth**
The first period of population growth was caused by global migration Ⓐ. The second period was initiated by the introduction of agriculture Ⓑ, and the third commenced with the Industrial Revolution Ⓒ. Note that both the time and population scales on the graph are exponential, that is, they use a logarithmic scale. The inset shows human population growth over time on a linear scale. The log scale makes it possible to see increases when the population is still small. Only the large changes in the industrial period are visible at the scale of the linear plot.

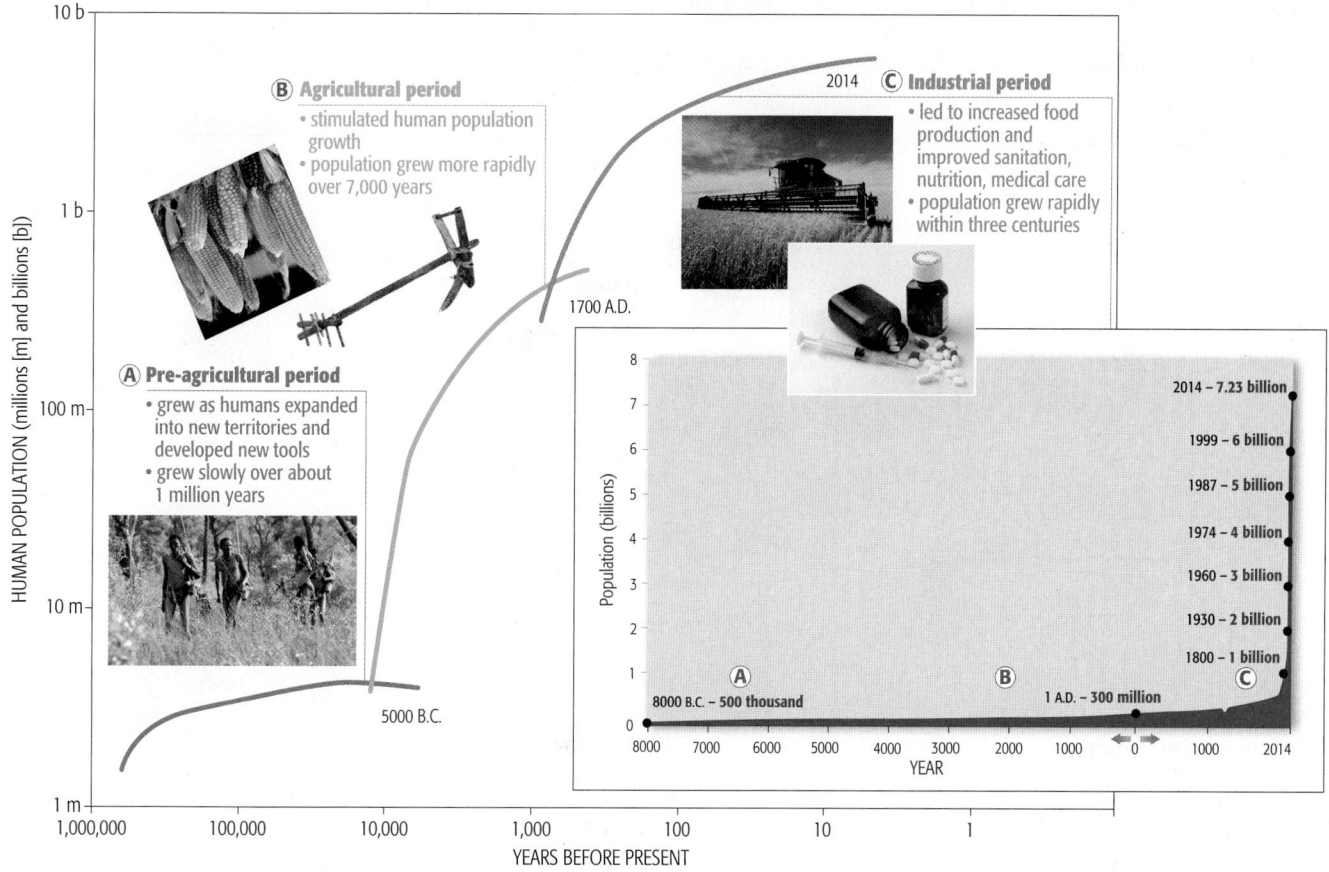

Demographic Transition Model

■ Demographic transition model describes four stages of change in human population growth.

In 1929, the American demographer Warren Thompson published a paper simply titled "Population." In it, Thompson described trends of population growth found in countries that were in the industrial period (see Figure 5.4C). He noted that during the 200 years in which Europe's economy transformed from an agricultural to an urban-industrial base, birth rates on the continent declined by at least 50%. During that same time, the average life expectancy of Europeans doubled, from approximately 35 years to nearly 70 years. But in countries in which industrial development had been slow to progress, birth and death rates remained high. Thompson concluded that economic development improves overall well-being. This improvement results in fewer infant deaths and increased life expectancy. More surprisingly, Thompson found that economic development causes birth rates to decrease.

Thompson's findings were subsequently verified and translated into the **demographic transition model** (**Figure 5.5**). Note that the demographic transition model is descriptive rather than predictive and is based on countries that had access to resources as they were undergoing mortality and fertility transitions. It may be less applicable to resource-limited developing countries today. Notice too that the model describes logistic growth for human populations that pass through all four stages, ultimately leveling out in the final stage (see Module 4.4).

Stage 1.   The first stage of the demographic transition model is the period prior to economic development—*pretransition*. During this stage, human populations are limited by a low availability of food and a high prevalence of disease—factors that negatively affect human well-being. Consequently, death rates are high, particularly among the young. Families compensate for high death rates by having many children, a behavior that is reflected in high birth rates. When viewed over time spans of decades and centuries, birth and death rates equal out, and there is zero population growth, as you can see in the stage 1 green population line in Figure 5.5. Within periods of just a few years, however, birth rates, death rates, and population growth rates may fluctuate considerably. Natural occurrences such as drought or plague, for example, may cause a population to temporarily experience low birth rates and high death rates. One hundred years ago, many of the world's poorest countries were in stage 1.

Stage 2.   The second stage of the demographic transition is called the **mortality transition**. In this stage, improved economic conditions relieve food shortages, produce better living conditions and health care, and expand access to education. Notice in Figure 5.5 how the death rate decreases while the birth rate stays high or even increases. As a result, there is rapid, sustained population growth. Today, very poor countries such as Afghanistan and Niger are in this stage.

Stage 3.   In the third stage of the demographic transition, the **fertility transition**, continued economic development produces social and cultural changes that lead to lower birth rates. People may delay starting families, which lowers total fertility rates, referred to as TFR,

► **Figure 5.5 Demographic Transitions**
In this hypothetical representation, the four stages of the demographic transition are characterized by distinct changes in population birth rates and death rates. The total population (green) rises in stages 2 and 3 as rates of mortality (blue) and then fertility (red) drop, leveling off in stage 4 when births equal deaths.

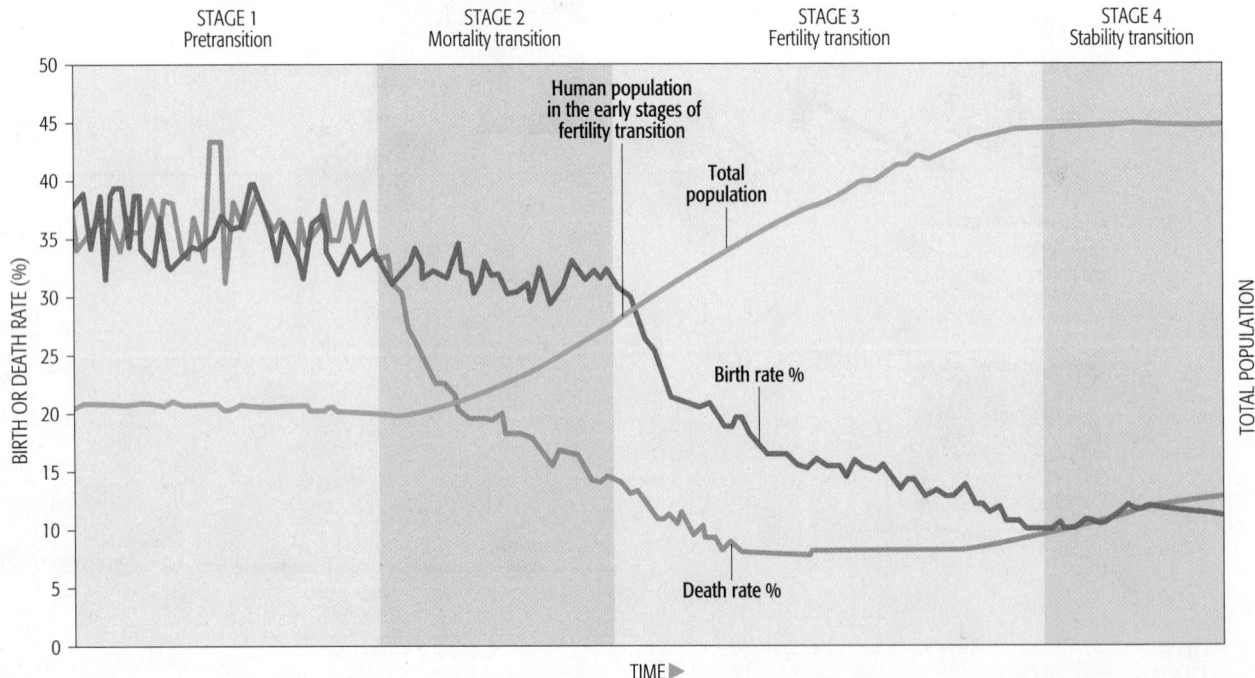

the number of children born to a woman throughout her childbearing years. Families may limit their number of children by using contraceptives or other means of family planning. In the fertility transition, the death rate remains low and the birth rate decreases, so population growth slows. Most of the world's countries are in the fertility transition. However, the United States and many other developed nations are approaching stage 4.

Stage 4. The fourth stage of the demographic transition, sometimes called the **stability transition**, is characterized by a low birth rate and a low death rate. During the stability transition, the birth rate is nearly equal to the death rate. This stage is characterized by zero population growth or even a decline in population. Japan and some western European countries, such as Italy and Spain, are now in stage 4.

A Tale of Two Countries

■ Niger and the Netherlands offer a contrasting look at countries in different stages of the demographic transition.

In 2013, Niger and the Netherlands had one thing in common: a population size of nearly 16.9 million people. By 2050, however, Niger's population is expected to quadruple, while the Netherlands' will increase only slightly. Niger is probably nearing the end of stage 2, the mortality transition, while the Netherlands has low birth and death rates and is transitioning to stage 4 of the model. These projected differences are due in part to the differing birth and death rates of the two countries.

Stage 1. Mortality and fertility were both very high in Niger until 1975 (**Figure 5.6**). The vast majority of families made a living through subsistence farming or herding livestock. Most people were very poor, and many suffered from malnutrition. Disease was widespread and mortality during childbirth common. Reliable population data for the region before 1950 are not available, but life expectancy in Niger in 1950 was 35 years, and 16% of children died before their first birthday.

▶ Figure 5.6 **Demographic Transition in Niger**
In 1950, Niger was pretransitional with high birth and death rates. By 1975 it was entering stage 2. The country is expected to reach stage 3 by the 2020s, but projections still show rapid population growth in 2100. Twice the size of Texas, Niger is located in a landlocked area on the southern edge of the Sahara. Literacy rates are some of the lowest in the world (<30%), and the country suffers from frequent drought. About half the population does not have enough food. Pictured is the capital city of Niamey.

Source: United Nations World Population Projection Revision 2012.

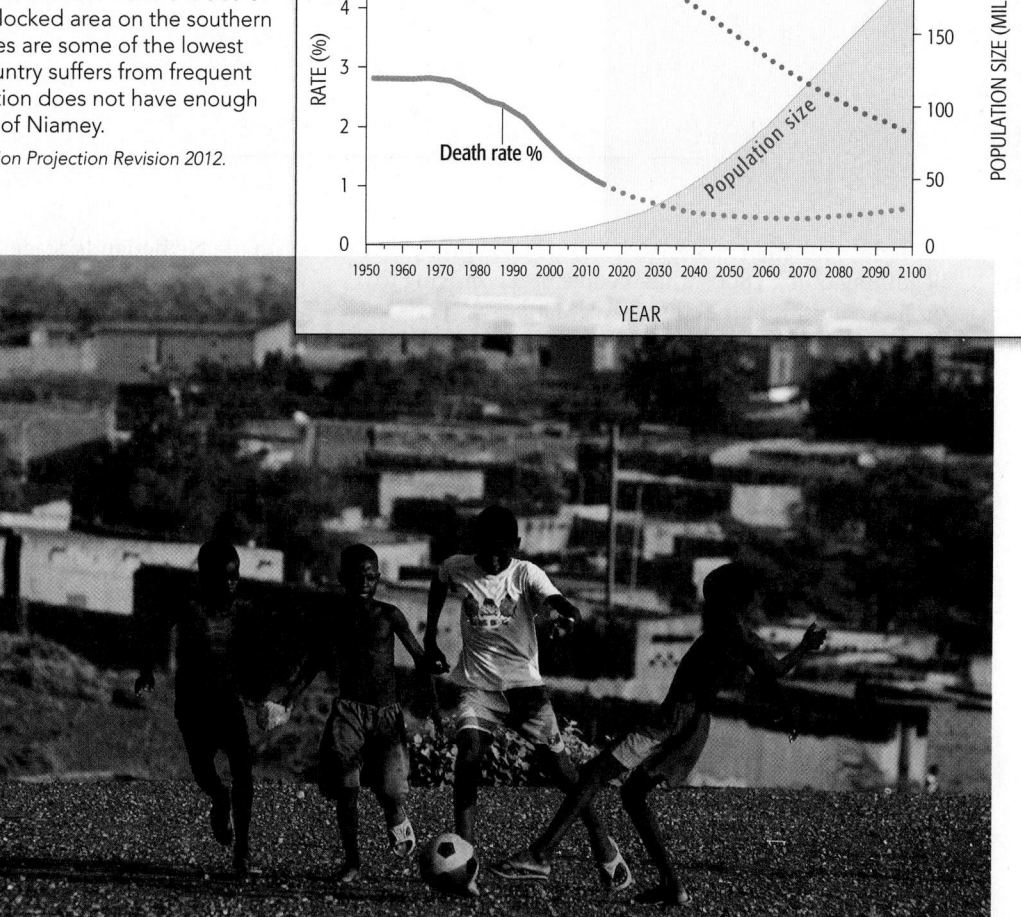

▶ Figure 5.7 **Demographic Transition in the Netherlands**
The Netherlands was in stage 1 before 1800. The mortality transition probably began in the early 1800s. The region's fertility transition began between 1960 and 1970. The United Nations Population Program projections indicate that the Netherlands will enter the stability transition around 2030. Roughly the same size as New Jersey, the Netherlands is located along the North Sea. Today, it has a literacy rate of 99% and a life expectancy of 81 years. It is the 24th wealthiest country in the world as measured by gross domestic product (GDP). Pictured is some of the old and new architecture of The Hague, the seat of the Dutch government.

Source: United Nations World Population Projection Revision 2012.

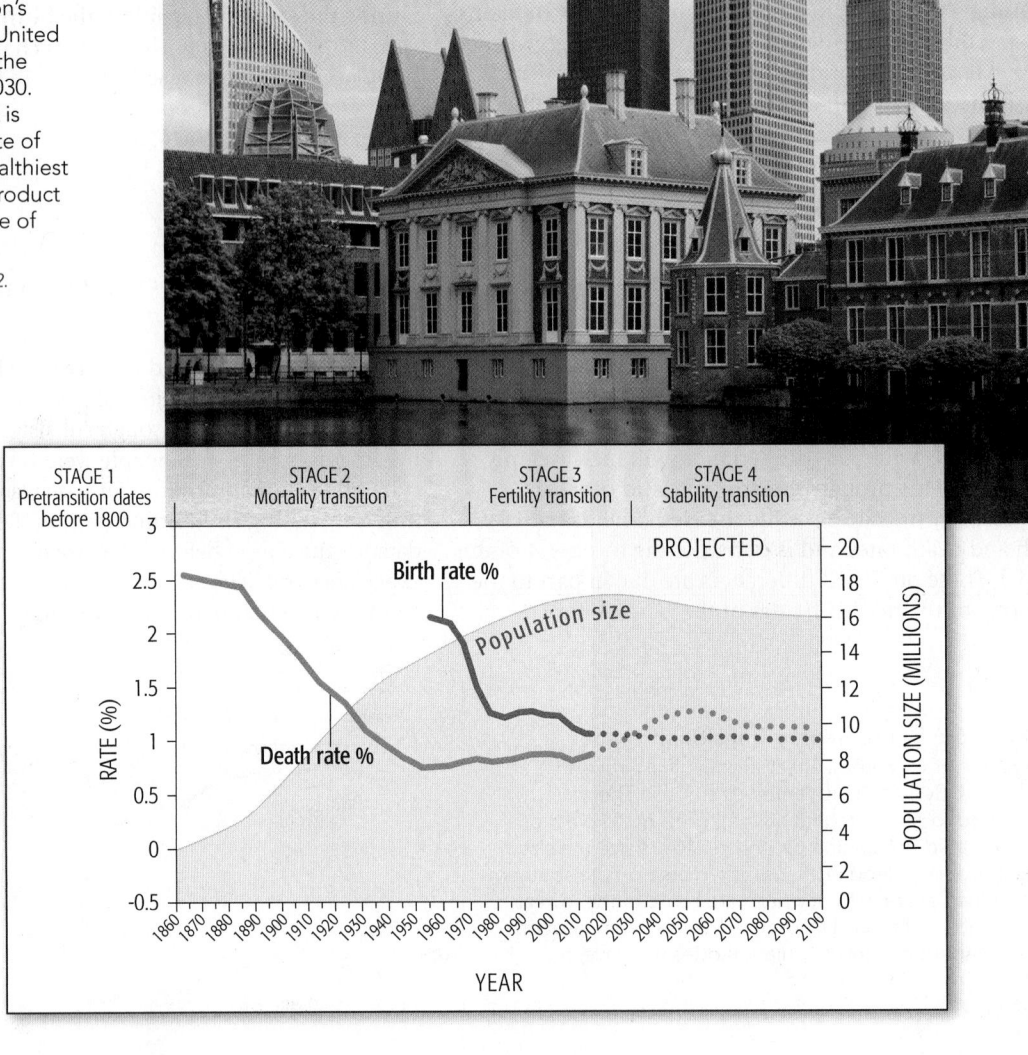

QUESTIONS 5.1

1. What factors contributed to the rapid population growth during the industrial period of human history?

2. During the first stage of the demographic transition, birth and death rates may differ widely from year to year but are nearly equal when calculated over decades. Explain.

3. Why does population growth rate increase at the beginning of the mortality transition?

4. In what stage of the demographic transition is the Netherlands currently? Niger? Why is Niger's population expected to quadruple by 2050 while the Netherlands will remain stable?

(MES) For additional review, go to **MasteringEnvironmentalScience**

In the Netherlands, stage 1 persisted until probably the beginning of the 1800s when people began moving to urban centers from the countryside (**Figure 5.7**).

Stage 2. The mortality transition in Niger probably began around 1970. During the 1970s, the international sale of uranium brought an influx of money to Niger. Sanitation and medical care improved slightly and mortality declined. From 1990 to 2009, Niger tripled the number of children with immunizations for diphtheria, pertussis, and tetanus (DPT3) to 66% and saw slight improvements in access to clean water. The nation continues to struggle with access to medical care, with only 3 hospital beds available for every 10,000 people (2009 data); the rate for the Netherlands is 4.7/1,000. WPP projections predict a continuing decline in mortality through 2055, with a 10% drop in birth rates predicted by 2025. This imbalance between birth and death rates predicts a quadrupling of the population by 2050. Nigerians remain very poor, fewer than 70% of adults are literate, and more than 5% of infants die before their first birthday.

In the Netherlands, stage 2 likely began around 1800. Mortality rates dropped steeply by the 1880s, with the advent of modern medicine, improvements in public health and diet, as well as increases in agricultural production.

Stage 3. In the Netherlands, birth rates began dropping in the late 1960s as access to modern contraception improved. By 2010, birth rates had fallen to half that of the 1950s and birth and death rates were nearly equal. It appears that the Netherlands is on the brink of the stability transition (stage 4), but birth rates still exceed death rates slightly, resulting in a small increase in population each year.

Stage 4. If the current rate of development continues, the United Nations projects that the Netherlands will enter the stability transition near the year 2030. According to current projections, Niger's population will still be growing rapidly in 2100, with no evidence of a shift to Stage 4 in the foreseeable future.

5.2 Global Variation in Human Population Growth

BIG IDEA The human population growth rate—the percent change in the number of people living in an area over a certain time period—is a direct consequence of births, deaths, and migration. Currently, the global population growth rate is approximately 1.09%, but this rate varies greatly among countries and regions. In some regions, the population growth rate is over 3%; in others it is near or below 0%. In general, the population growth rate is highest in the world's poorest countries and lowest in the world's wealthiest countries.

Ⓠ *What number of children per family would produce zero population growth?*
Brandon MacDonald, Citrus Community College

Lissa: Ⓐ *The exact number depends on age-specific mortality rates. As childhood mortality rates diminish, the number of children per family at zero net population growth approaches two.*

Birth Rate

■ Wealth and age influence birth rates.

Worldwide, the birth rate—the number of babies born each year expressed as a percent of the population—is decreasing. Between 1950 and 2013, it dropped from 3.75% to 1.89%. There is, however, considerable variation in birth rates among regions and countries. (See Module 4.3 for an introduction to factors affecting population size.)

Birth rates are highly correlated with the relative wealth of countries, as measured by per capita GDP (see Module 2.4). In the world's poorest countries, birth rates are relatively high: 30–50 of 1,000 women give birth each year. In the wealthiest countries, birth rates are lower; only 8–15 of 1,000 women have children each year (Figure 5.8). To make the comparison, we will use Ethiopia as the example of one of the poorest nations (among the bottom 20) and the United States as one of the wealthiest (among the top 10). In Ethiopia, the birth rate is 4.3%, whereas in the United States it is 1.4%. Because of this difference in birth rates, the number of children born in Ethiopia in 2012 was nearly three-fourths the number born in the United States, even though the total population of Ethiopia is only about one-fourth of the United States.

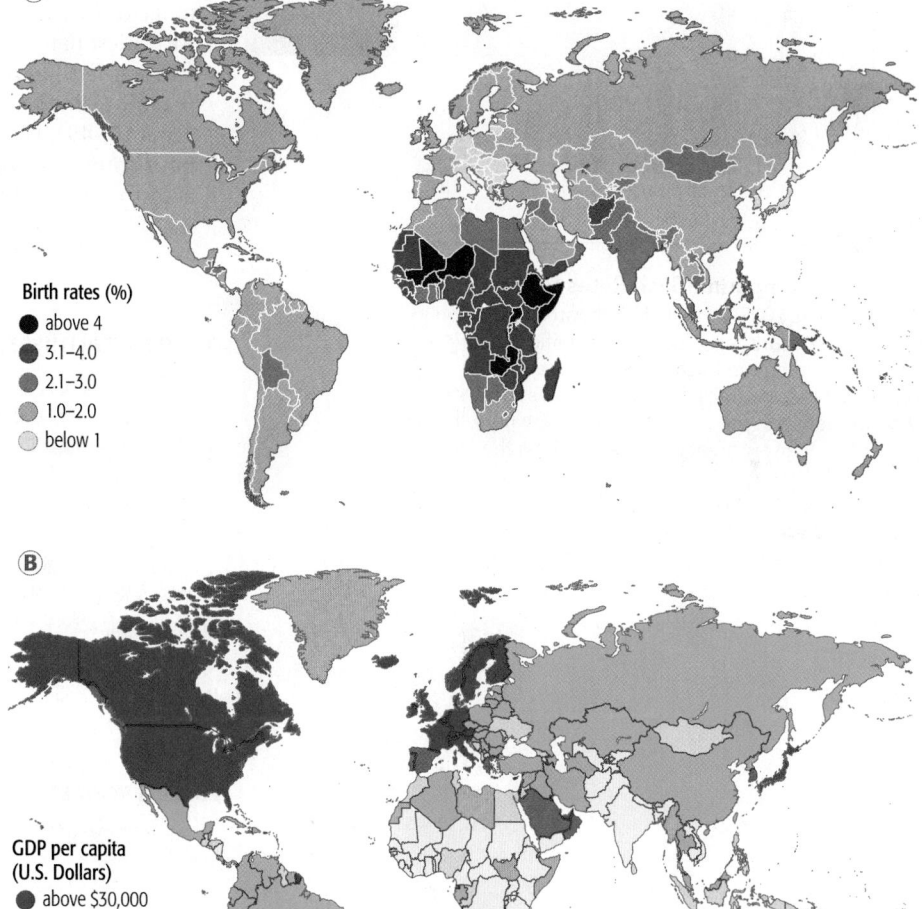

Ⓐ

Birth rates (%)
- above 4
- 3.1–4.0
- 2.1–3.0
- 1.0–2.0
- below 1

Ⓑ

GDP per capita (U.S. Dollars)
- above $30,000
- $20,001–$30,000
- $10,001–$20,000
- $5,001–$10,000
- $2,001–$5,000
- below $2,000
- no data

▶ **Figure 5.8 Birth Rate and Wealth**
These maps compare birth rates and wealth in countries across the globe. In map Ⓐ, notice that the United States has a low birth rate and the countries of sub-Saharan Africa, like Ethiopia, have very high birth rates. Map Ⓑ shows the opposite trend for wealth—GDP is very high in the United States and very low in sub-Saharan Africa.

Source: Data for part A from *CIA Factbook 2013*; data for part B from http://data.worldbank.org.

The difference between the birth rates of poor and wealthy countries can be largely attributed to differences in the age at which women bear children. Women in poorer countries typically have children at a younger age than do women in wealthier nations. **Age-specific birth rates** tell the number of children born in a year per 1,000 women within defined age groups (Figure 5.9). Between 2005 and 2010, for example, nearly 21 of 1,000 15- through 19-year-old girls gave birth in Ethiopia, whereas in the United States only 7 of 1,000 girls in that age range had a child. In many developing countries, high rates of teen motherhood negatively impact young women's health, education, and rights. Additionally, these countries suffer the economic consequences of a reduced labor force.

Higher birth rates among younger women shorten the time between successive generations. Shorter generation times lead to more children growing up and having children, causing a higher population growth rate. The effect of high birth rates among the young is magnified in poor countries, where there are typically more women in younger age classes.

In Ethiopia, women who survive their full childbearing period will bear an average of 4.8 children. Therefore, Ethiopia has a total fertility rate of 4.8. American women who live through their childbearing years will have an average of 2.08 children, resulting in a total fertility rate of 2.08 (see Module 4.3). In the world's poorest countries, total fertility rates range from 4.7 to 7.3, compared to a range of 1.3 to 2.1 in the world's 10 wealthiest nations (Figure 5.10).

Total fertility rates are calculated under the improbable assumption that all women within a population live through their reproductive years (ages 15–49). In wealthy countries, this is close to true. But it is not true in poor countries, where the high incidence of infectious diseases and limited access to medical care cause much higher death rates. In studies completed in 2005, for example, fewer than 3% of American women who lived to age

▲ **Figure 5.9 Comparing Birth Rates by Age**
During the years 2005–2010, the birth rates of women in the 10 poorest countries (blue) were significantly higher across all age groups than the birth rates of women in the 10 wealthiest nations (red). The difference is most striking for women between the ages of 15 and 19, where the birth rate is almost 9 times greater in poor nations than in wealthy ones.

Data from: United Nations World Population Prospects: The 2012 Revision Population Database.

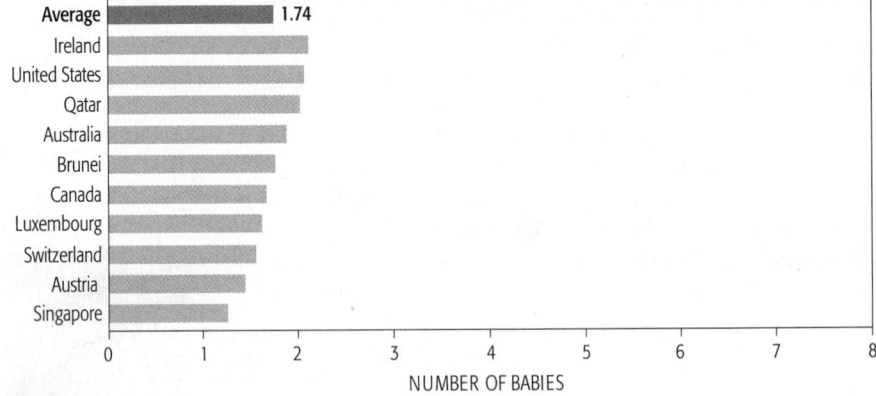

WORLD'S WEALTHIEST NATIONS

▶ **Figure 5.10 Total Fertility Rate and Wealth**
Total fertility rates among the world's poorest countries are approximately three times higher than among the world's wealthiest countries. (The TFR is the sum of all the age-specific fertility rates for a given country.)

Data from: United Nations World Population Prospects: The 2012 Revision Population Database.

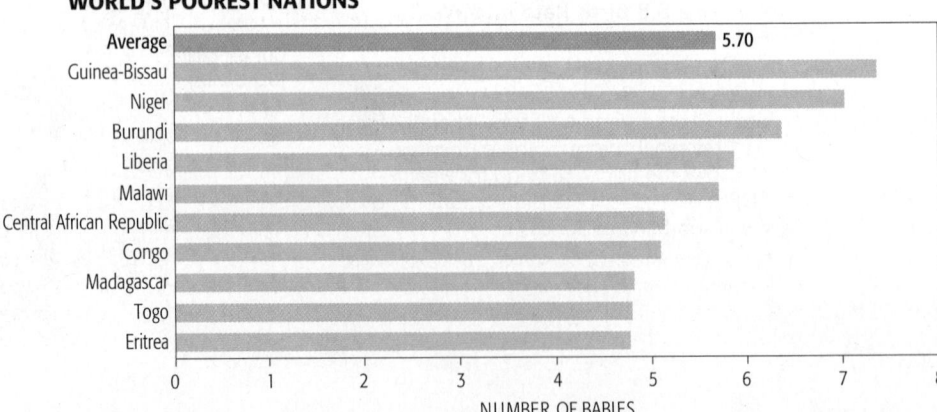

WORLD'S POOREST NATIONS

15 died before reaching 50. In contrast, over 18% of Ethiopian women within that age range died, often from complications associated with pregnancy or childbirth.

Thus, although the total fertility rate for Ethiopia then was 6.1, the average number of children born to women between ages 15 and 50 was actually only 4.8.

Death Rate

■ Wealth and age also influence death rates.

The death rate is the percent of individuals in a population that die each year (see Module 4.3). Like birth rates, death rates are highly correlated with the level of economic development. Over the past 35 years, the worldwide death rate has declined from 1.95% to its current rate of 0.79%. The decline in death rate corresponds to improved nutrition, sanitation, and health care. The healthier we are, the lower our death rate. Because conditions affecting human survival in the world's poorest countries remain well below the standards of developed countries, death rates in the world's poorest countries are nearly twice those of the wealthiest countries (Figure 5.11).

A significant component of the higher death rates in poor countries is infant mortality. **Infant mortality rate** is the percent of infants within a population who die before age 1. Infant mortality is heavily influenced by factors such as sanitation, water quality, and the availability of health care before and after birth. Babies

born in poorer countries are particularly vulnerable to disease and malnutrition during their first year of life. As a result, infant mortality rates in the world's poorest countries are 18 times higher than in the wealthiest countries (Figure 5.12). Of the babies born in Ethiopia in 2013, 5.8% died before age 1; in the United States, only 0.6% of the babies died before their first birthday.

Death rates that deal specifically with age can be expressed in terms of survivorship, the probability of surviving to a particular age (see Module 4.3).

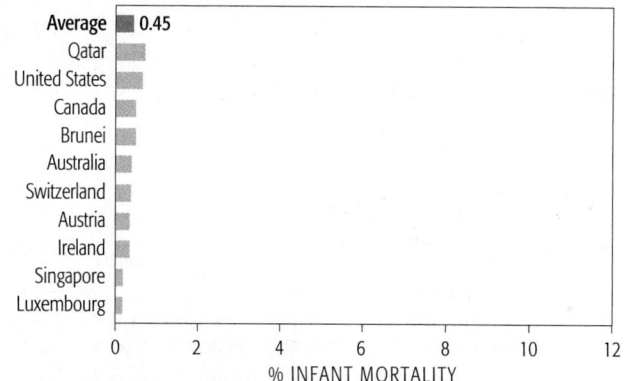
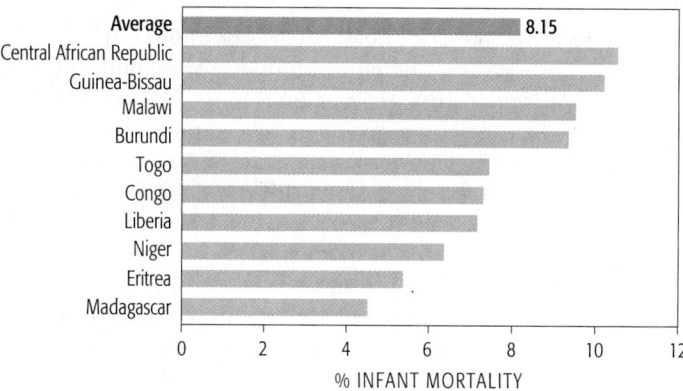

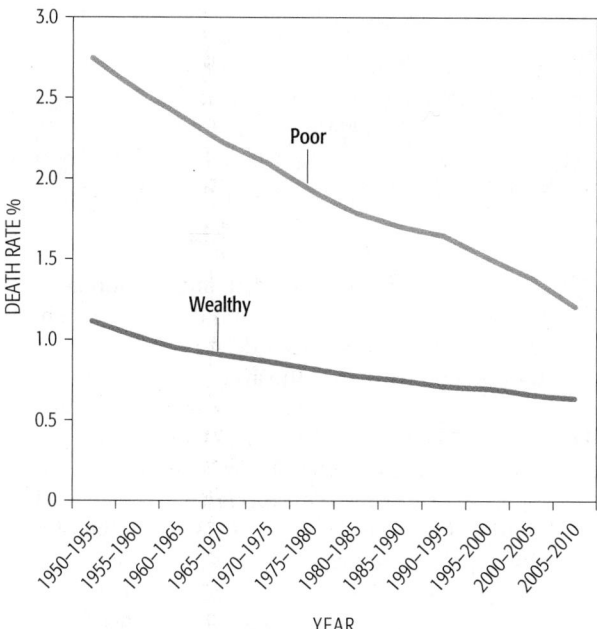

▲ Figure 5.11 **Death Rates Compared in Poor and Wealthy Countries**

In the last 60 years, death rates in the world's poorest countries have declined by more than 40%. But as of 2012, they were still nearly 50% higher than the death rates in the world's wealthiest countries.

Data from: United Nations World Population Prospects: The 2012 Revision Population Database.

▲ Figure 5.12 **Infant Mortality and Wealth**

Infant mortality rates in the world's 10 poorest countries are 10 to 20 times higher than in the 10 wealthiest countries. The variation in these rates among poor countries reflects significant differences in nutrition, sanitation, and access to health care.

Data from: United Nations World Population Prospects: The 2012 Revision Population Database.

▶ Figure 5.13 **Survivorship Curves: United States and Niger**
Survivorship curves show the percentage of people living to a particular age. These curves are for men and women in the United States and Niger. During 2005–2010, over 99% of children born in the United States lived to be at least 5 years old, compared to only 85% in Niger. In both countries, differences in survivorship between males and females are small up to ages 40–50. After age 50, women have better survivorship, especially in the United States.

Data from: United Nations World Population Prospects: The 2012 Revision Population Database.

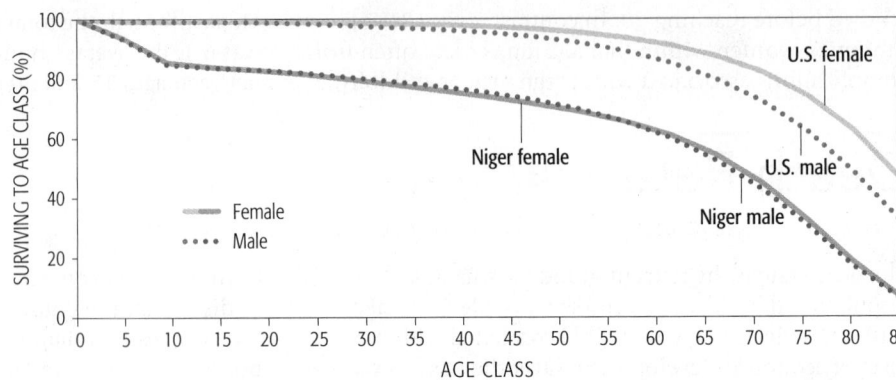

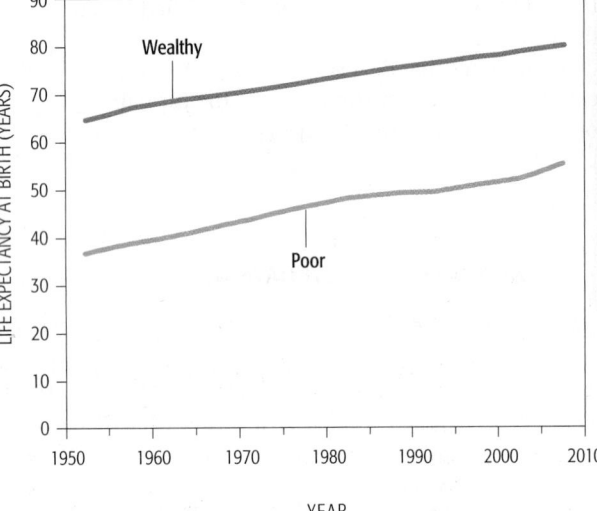

Where YOU LIVE
What do the numbers tell you?
Compare birth rates, death rates, proportion of population under 15, and life expectancy in the United States to a country of interest to you: perhaps someplace you have visited or where you have ancestral roots. Now compare the GDP per capita and secondary-school enrollment of women in both countries. Does this help to explain any differences you see between the countries? What population problems do you predict these countries might face in the next 50 years if any?

See the Datafinder tab on the Population Reference Bureau website to find data.

▲ Figure 5.14 **Life Expectancy and Wealth**
Over the past 60 years, life expectancy has increased in all countries. However, an average baby born in one of the world's 10 poorest countries will live fewer than 55 years. In contrast, a baby born in one of the world's 10 wealthiest countries has a life expectancy of nearly 80 years.

Data from: United Nations World Population Prospects: The 2012 Revision Population Database.

Like birth and death rates, patterns of survivorship differ between wealthy and poor countries. For example, in the African country of Niger, only about 74% of women survive to age 40, compared to 98% in the United States (Figure 5.13). Nutrition, sanitation, and access to health care account for much of this difference.

The survivorship of men and women may differ significantly. Adolescent women often have somewhat higher survivorship than adolescent men, owing to the higher propensity of young men to be involved in accidents and to exhibit violent behavior. In some developing countries, however, poor health care associated with childbearing results in lower survivorship among young women.

Nearly everywhere in the world, the survivorship of women beyond the age of 40 exceeds that of men. The reasons why women live longer than men are not well understood. Men appear to be more susceptible to heart disease, the most significant cause of death in developed countries. Men are also more likely to die in violent

situations, including armed conflict, and to indulge in risky behaviors, such as smoking and drug use, which increases their chances of earlier death.

Life expectancy at birth, the average age to which a baby born at a given time will live, provides a summary of differences in survivorship. As with survivorship, life expectancy is generally higher for women than for men. In 2013, for example, the life expectancy of a baby girl born in the United States was 81.2 years, and that of a baby boy was only 76.2 years.

From a global perspective, life expectancy has increased dramatically over the past six decades, from 46.9 years to 70 years. This increase in life expectancy is largely a result of improvements in the production and distribution of food, cleaner water, and advances in medicine and health care. Sadly, the differences in life expectancy between poor and wealthy countries that existed in 1950 persist today (Figure 5.14). These differences reflect significant variations in human well-being among countries.

Age Structure

■ A population's age structure is a key factor in its future growth.

Survivorship and age-specific birth rates determine a population's age structure, the distribution of a population based on age classes. **Age-structure pyramids**, also known as *age-sex pyramids* or *population pyramids*, illustrate graphically how populations are apportioned according to age and gender. The width of the bars indicates the number of both males (left) and females (right) in a population for each age class shown on the *y*-axis (**Figure 5.15**). Color coding shows approximate pre-reproductive, reproductive, and post-reproductive age categories.

In regions where birth rates are high and survivorship is low, populations have a large proportion of young people. These age-structure pyramids have a wide base, tapering to a narrow tip. For example, in 2014 the median age among Ethiopians was 18 years, meaning that half of the population was 18 years of age or younger (see Figure 5.15A). In regions with low birth rates and high survivorship, age structures are more evenly distributed, indicated by bars of similar width from bottom to top. In 2014 the U.S. median age was 38 years, more than twice that of Ethiopia (see Figure 5.15B).

Age-structure pyramids reflect the history of a country. In Figure 5.15B, the bulge in the 50–64 age categories in the United States represents the post–World War II baby boom. This time-specific increase in population resulted in a secondary bulge in the 20–34 age categories, the offspring of the baby boomers.

The age structure of a population indicates present population growth rates; it also forecasts future growth rates. Broad-based pyramids predict rapid population growth. Where the median age is young and a large proportion of the population is of reproductive age, overall population growth rates are likely to be high. This is because young people have comparatively high birth rates and low death rates (**Figure 5.16**). By comparison, more balanced age structure pyramids predict a stable population size or slow growth, while inverted pyramids predict declining populations.

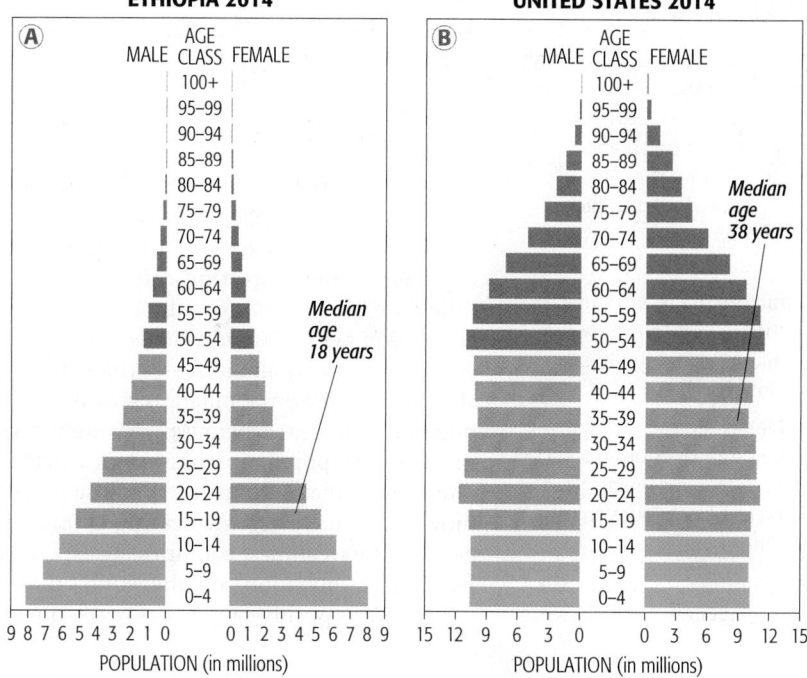

▲ **Figure 5.15 Population Age Structures**
In 2014, the median age in Ethiopia Ⓐ was less than half that of the United States Ⓑ. Notice that a much large proportion of Ethiopia's population falls into the reproductive age range of 15–49 years. (Green bars indicate pre-reproductive age, blue bars are reproductive age, and red are post-reproductive.)

Data from: CIA World Factbook (2014). Retrieved from https://www.cia.gov/library/publications/the-world-factbook/geos.

▼ **Figure 5.16 Age Structure Change**
The current population age structures of Niger and the Netherlands reflect significant differences in their age-dependent fertility rates and survivorship. These differences are projected to result in very different population sizes and age structures by the year 2050.

Data from: CIA World Factbook (2014). Retrieved from https://www.cia.gov/library/publications/the-world-factbook/geos.

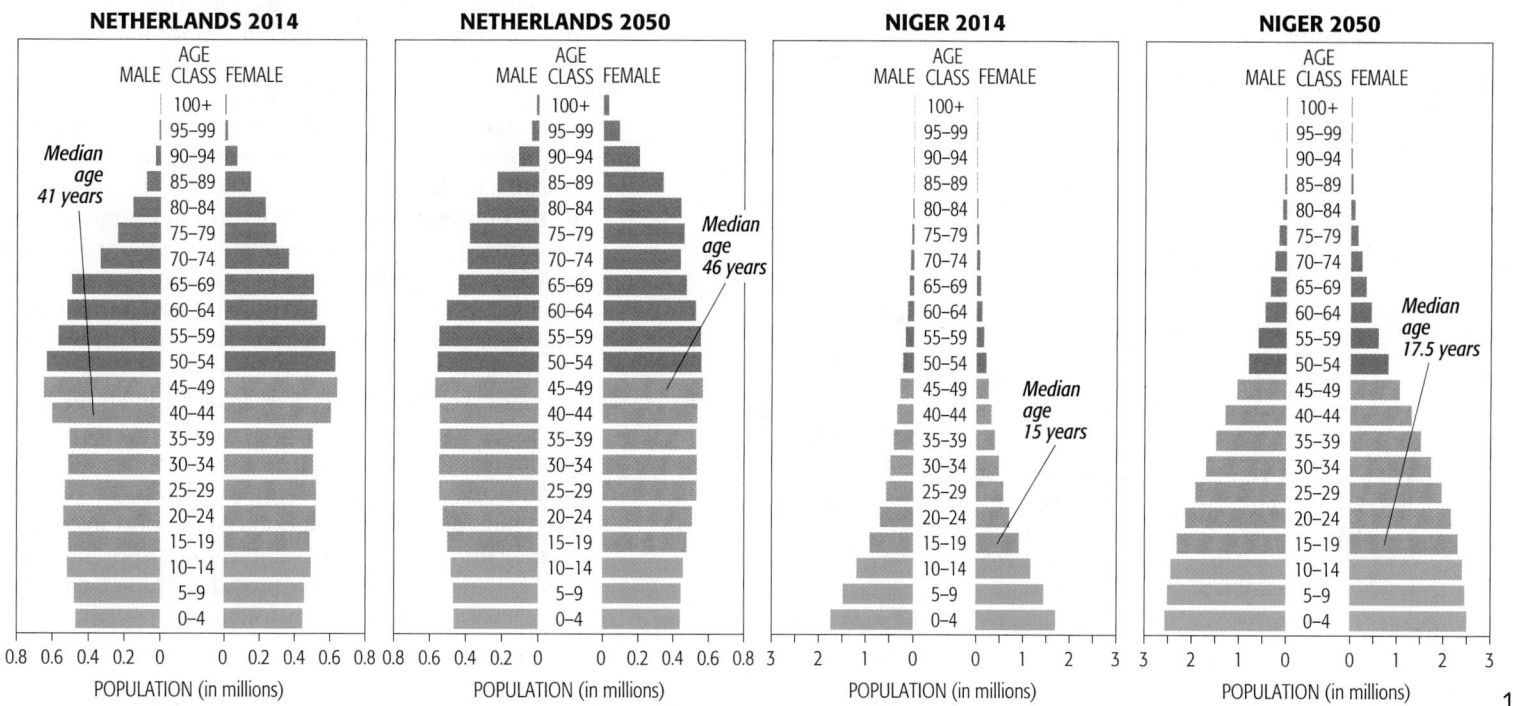

1. Give two reasons why high fertility rates among young women have a greater impact on population growth rate than equally high rates among older women.

2. What are some of the factors that contribute to the relationship between infant mortality rates and wealth?

3. Net migration is positive in some poor countries and negative in others. What factors contribute to this difference?

(MES) For additional review, go to **MasteringEnvironmentalScience**

Migration

■ The relocation of humans is influencing human population growth in many regions.

At any given time, some people are moving into a country (immigrating) and others are moving out (emigrating). Demographers calculate the **net migration rate** as the difference between immigration and emigration per 1,000 individuals in the population. Net migration rates can have a very significant effect on population growth rates. If a population increases because large numbers of people move in, that population has a *positive* net migration rate. If a population decreases because many people move out, it has a *negative* net migration rate. In general, the wealthiest countries are experiencing positive net migration, and the world's poorest countries vary between positive and negative net migration (**Figure 5.17**).

▲ Figure 5.18 **Emigration Push Factors**
Disease, famine, war, and persecution continue to push people to migrate from their home countries. The refugees pictured here have fled drought and famine in Afghanistan.

Among the world's wealthiest nations, positive net migration rates account for 65% of total population growth. In the United States, the rate of population growth in 2013 was 0.9%, and the net migration rate was 0.36%. This rate indicates that immigration accounted for 40% of the total population growth. In Austria, the net migration rate was 0.18%, and the total population growth rate was only 0.02%. The rate of immigration into Austria actually exceeded its population growth rate. Were it not for immigration, Austria's population would actually have decreased by 0.16%.

Net migration is often influenced by situations in neighboring countries. **Push factors** are conditions that force people to emigrate. Push factors include epidemics, drought, famine, war, religious or political persecution, and crowded conditions (**Figure 5.18**). Conditions that encourage people to immigrate into a country are called **pull factors**. Pull factors include opportunities for employment and education, higher living standards, and religious or political freedom.

Migration into poor countries is often due to push factors, such as refugees escaping war in nearby countries. In 2002, both Eritrea and Sierra Leone—two of the world's poorest nations—were surrounded by war-torn countries. That year, the net migration rates into Eritrea and Sierra Leone were three to four times greater than the net migration rate into the United States. Immigration into poor countries complicates and often thwarts efforts for economic and social development.

In the wealthiest countries, aging populations and low rates of population growth have resulted in dwindling workforces. Immigration often provides the additional labor needed to maintain economic output. However, high rates of immigration can lead to political conflicts. In the United States, for example, the cost of meeting immigrants' needs for education, health care, and other public services is frequently a source of contention.

WORLD'S WEALTHIEST NATIONS

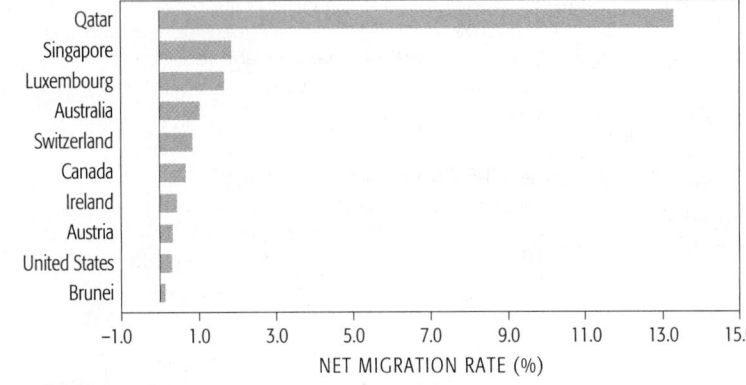

WORLD'S POOREST NATIONS

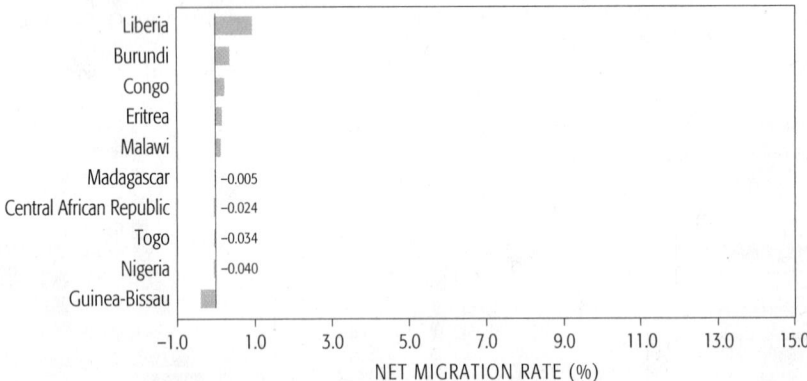

▲ Figure 5.17 **Net Migration Rates and Wealth**
The "pull" of favorable economic and social conditions produces positive net migration rates into wealthy nations. Patterns of migration for poor nations are more complex, being influenced by internal "push" factors as well as conditions in neighboring countries.

Data from: United Nations World Population Prospects: The 2012 Revision Population Database.

5.3 Predicting Human Population Growth

BIG IDEA Predictions about human population growth are usually based on the logistic population model and assumptions about human values and behaviors. English philosopher Thomas R. Malthus was the first to predict limits on the exponential growth of human populations and correlate population growth with human well-being and culture. Since that time, forecasts of human population growth have varied considerably because of the quality of the demographic data available and changing assumptions about human values and behavior.

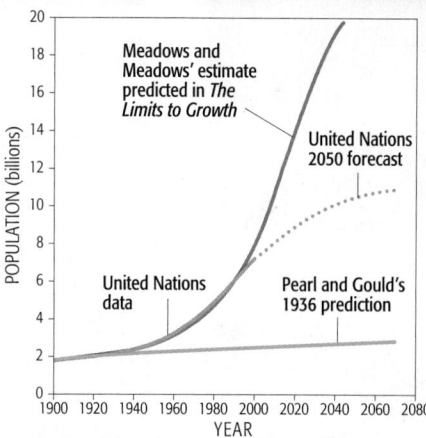

▲ Figure 5.19 **Contrasting Forecasts for Human Population Change**
The forecasts made by Pearl and Gould in 1936 (green) and Meadows and Meadows in 1972 (red) were based on vastly different assumptions about factors affecting future birth and death rates. The solid blue line shows actual census data up to 2010; the dotted blue line shows the United Nations current forecast through the year 2050.

Data from: United Nations World Population Prospects: The 2012 Revision Population Database.

Population Growth Forecasts

■ Forecasts for human population growth rates are based on past growth and assumptions made about future human values and behavior.

In the late 1700s, Malthus explored the effects of limited resources on human populations by comparing data on population growth in the American colonies and on the European continent. In his 1798 *Essay on the Principle of Population*, Malthus observed that in the American colonies, where resources were abundant, human populations were doubling every 25 years (see Module 4.3). In Europe, where resources were much more limited, doubling times were considerably longer.

Malthus noted that human populations grow exponentially, but the resources upon which humans depend typically do not. In fact, Malthus predicted that exponential human population growth would lead to widespread poverty and diminished human well-being. He warned of the effects of starvation upon human survival and reproduction and the suffering it would cause.

By the early 20th century, demographers and researchers were using mathematical models in the attempt to accurately predict human population growth. While the mathematics were sound, assumptions about birth and death rates were subject to change. Remember that the 20th century was a tumultuous one, characterized by economic, social, and political upheaval.

In their 1936 paper, "World Population Growth," statisticians Raymond Pearl and Sophia Gould argued that the rate of global human population growth was slowing. Applying a logistic growth model to the limited data available in the 1930s, they estimated the world's carrying capacity to be 2.6 billion people (see Module 4.4) and predicted that the world population would reach carrying capacity in 2100.

Pearl and Gould based their predictions on data gathered during the period between World Wars I and II. This was a time of widespread economic depression. Demographers of their time assumed that the worldwide poverty would continue. But in fact, the two decades following World War II were characterized by economic progress, high birth rates, and dropping death rates. As a result, population growth accelerated rapidly, and the world population surpassed

Pearl and Gould's estimated carrying capacity in 1953. In his 1968 best-seller, Paul Ehrlich gave a name to this alarming trend, *The Population Bomb*.

Four years later, environmental scientists Donella and Dennis Meadows published *The Limits to Growth*. This highly influential book forecast population growth that far exceeded the Pearl and Gould predictions (Figure 5.19). Meadows and Meadows predicted that the human population would surpass 7.4 billion in the year 2000. They also suggested that as many as 15 billion people might be competing for Earth's limited resources by the year 2050.

Meadows and Meadows assumed that the birth rates and death rates of the 1960s would continue. Around 1970, however, the global population growth rate began to decrease. People in many developing countries began having children later in life, a behavior that significantly slowed the rate at which new individuals were added to the global population. Instead of the 7.4 billion people that Meadows and Meadows had predicted, the world population tipped just over 6 billion in 2000.

Demographers such as Pearl, Gould, Meadows, and Meadows could not foresee how human values and behavior would change with economic development. The latest United Nations forecast on population growth is that the global growth rate will decrease to about 0.45%, and the world population will reach approximately 9.6 billion by the year 2050 (Figure 5.20). Like past predictions, current predictions of moderate population growth are based on assumptions about human values and behaviors. Not only do human values and behavior affect birth and death rates, they also influence the ways in which people use resources.

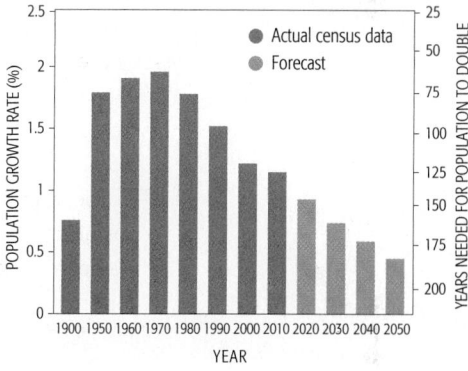

▲ Figure 5.20 **Declining Population Growth Rates**
Worldwide, population growth rates have been decreasing since 1970. The United Nations Department of Economic and Social Affairs forecast is for that trend to continue through the middle of this century.

Data from: United Nations World Population Prospects: The 2012 Revision Population Database.

QUESTIONS 5.3

1. Why did Malthus expect population growth in the 18th century to be faster in the American colonies than in Europe?

2. Forecasts of human population growth depend on assumptions about human values and behavior. Explain.

(MES) For additional review, go to **MasteringEnvironmentalScience**

Forecasting Future Population Trends and Their Uncertainties

Can population models account for uncertainty about the future?

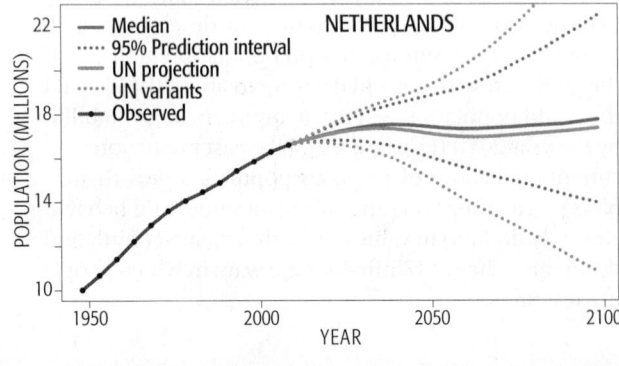

▲ **Figure 5.21 Adrian Raftery**

Adrian Raftery was awarded the Parzen Prize for Statistical Innovation in 2012 for his leadership in the development of models to predict future population trends.

1. How can a researcher determine the accuracy of a model using data from the past?

2. Why might planners and decision makers want to have estimates of the possible error associated with model forecasts?

Forecasts of future population sizes and age distributions for regions, countries, and the world are critically important for accurate planning. They are used by private companies to make business and marketing decisions. They are used by local and national governments, as well as international organizations, to anticipate future resource use. These resource needs include health care services, payments to retirees, and new infrastructure such as schools and transportation. As we have seen, however, historical projections of future population trends have often not matched actual outcomes. Adrian Raftery and his colleagues at the University of Washington have developed models that not only provide more accurate forecasts but also provide planners with estimates of the possible error in those forecasts (**Figure 5.21**).

There are two important problems with traditional population forecast models. First, predictions are *deterministic*. This means that forecasts are strictly determined by specific numerical estimates for births and deaths, as well as immigration and emigration rates, accounting for both men and women in different age categories at a specific moment in time. The United Nations World Population Prospects program estimates possible errors in predictions by postulating low and high values for key rates, but these, too, are deterministic. Second, traditional models do not include important knowledge that could inc rease forecast *accuracy* and also provide an estimate of likely forecast *error*. Such knowledge includes historical trends in demographic rates leading up to a specific moment in time and demographic transition information for regions and countries.

Raftery and his team confronted these problems using the power of modern computers to manage large data sets and make tens of thousands of calculations each second. Using vast historical data sets such as the UN *World Population Prospects*, they determined variation in population outcomes from one time to the next. On the basis of this, they were able to estimate the likely errors in or *confidence intervals* for predictions from one time to the next. In their model,

an estimate of total fertility rate for a future year includes its 95% confidence interval—the range of values within which we can be 95% sure that the future value will actually fall. These models are further refined by incorporating knowledge about trends in regions and countries relative to expected demographic transitions. Changes in important population characteristics are then forecast with models that are *probabilistic* rather than deterministic. This is done by repeating predictive models thousands of times, and each time varying estimates of key demographic rates (birth, death, TFR, etc.) based on likely distributions within their 95% confidence intervals. The expected forecast value is the average of these thousands of model runs, and the variation among the model runs is the basis for calculating the 95% confidence interval for the forecast. Raftery checked the validity of his models by using UN data from 1950 to 1990 to forecast population changes for the period 1995 to 2010. His forecasts were very close to the actual 1995–2010 UN data.

Forecast population trends based on deterministic United Nations projections and Raftery's probabilistic models are compared in **Figure 5.22** for Madagascar and the Netherlands. Raftery's models predict somewhat higher future population sizes for both countries. More important, Raftery's approach also measures the uncertainty in his estimates. Note that, as we attempt to forecast further in the future, both the UN high and low estimates and Raftery's 95% confidence intervals increase significantly. Raftery found, however, that UN high and low estimates greatly underestimate the forecast uncertainties for poor and developing countries such as Madagascar that are undergoing rapid demographic change; they overestimate these uncertainties for wealthy countries in the later stages of demographic transition.

The models developed by Raftery and his colleagues have many applications. They not only provide estimates of likely changes in population size but also provide important forecasts of likely changes in population age structures for individual countries and for the world.

► **Figure 5.22 Population Forecasts and Their Uncertainties**

Actual past and forecast future changes in population size are shown for Madagascar (a developing country) and the Netherlands (a developed country). The 1950–2010 line is actual country data. The solid lines (2010–2100) compare UN (blue) and Raftery's (red) forecasts. The blue dotted lines represent the low and high UN predictions, and the red dotted lines represent Raftery's 95% confidence intervals.

Source: From Raftery, A.E. et al. 2012 Bayesian.

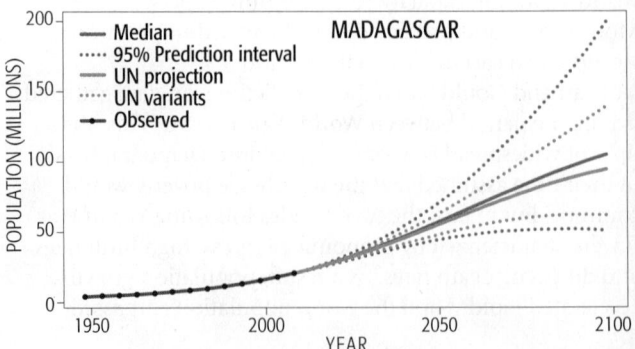

5.4 Resource Use and Population Sustainability

BIG IDEA According to the current United Nations forecast, the world population will reach 9 billion people in approximately 35 years. Can Earth sustain that many people? Whether or not a human population is sustainable depends not only on its size but also on the demands that individuals in the population make on their environment's resources (see Module 1.3). The resources available within countries vary widely, as do the demands for resources made by the citizens of those countries. Generally, the more affluent and technologically advanced a population, the greater its impact on the ecosystems upon which it depends.

Sustainability vs. Carrying Capacity

■ A human population is unsustainable long before it nears its carrying capacity.

How many people can Earth support in a sustainable fashion? Asking such a question is often thought to be the same as asking, What is Earth's carrying capacity? Yet the two questions are not the same.

Carrying capacity refers to the point at which competition for limited resources causes birth rates and death rates to reach equal numbers, resulting in zero population growth. Theoretically, a population can be maintained at carrying capacity as long as its supply of resources remains constant. But at carrying capacity, competition for resources is very intense. The lower birth rates and higher death rates characteristic of populations nearing their carrying capacity indicate that stresses on individuals are extreme (Figure 5.23). The well-being of individuals within a population at carrying capacity is considerably worse than that

experienced by individuals in a population that is well below its carrying capacity.

As a human population approaches its carrying capacity, stress increases. Because human well-being requires more than mere survival, the sustainability of a human population must be below carrying capacity. In fact, sustainability means that enough natural resources are available to guarantee a sufficient level of well-being for generations to come (see Module 1.2).

Thus, two things must be established to answer the question, "How many people can Earth support in a sustainable fashion?" First, we must agree on a standard of human well-being that is sustainable. Second, we must establish the capacity of Earth's ecosystems to provide goods and services to meet that standard.

Q *When do we need to "worry" about population growth?*
Natalie Hollabaugh, Central Michigan University

Lissa: **A** *For the foreseeable future, we should constantly monitor our numbers and act to limit the impact of our population's resource use on Earth's ecosystems and our own well-being.*

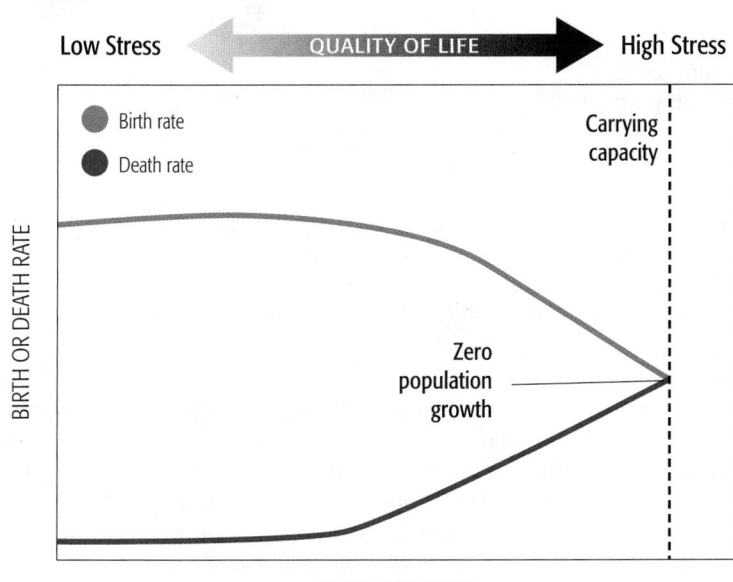

◀ Figure 5.23 **Carrying Capacity and Well-Being**
As a population's size nears carrying capacity, the quality of life of its members, as measured by decreasing birth rates and increasing death rates, declines.

► Figure 5.24 **Ecological Footprints**

An ecological footprint represents the amount of land needed to supply the demand for ecological goods and services. The ecological footprint of an average U.S. citizen is 2.7 times larger than that of an average Mexican citizen and nearly 12 times larger than that of an average Haitian. Footprints are measured in global hectares per capita (gha), a unit of biologically productive land and water area, scaled by the productivity of each land type.

Data from: Ecological Footprint Atlas 2010.

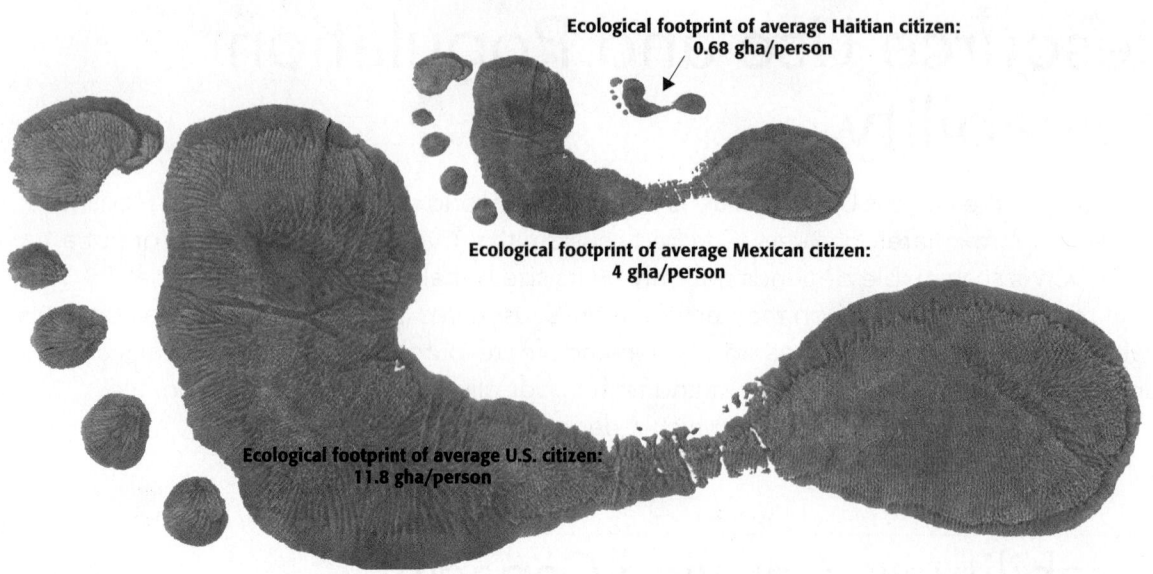

Ecological footprint of average Haitian citizen:
0.68 gha/person

Ecological footprint of average Mexican citizen:
4 gha/person

Ecological footprint of average U.S. citizen:
11.8 gha/person

Human Resource Use

■ Ecosystem goods and services determine the potential growth of human populations.

The impact of a human population on an environment depends on the total number of people living there and on the consumption habits of each individual within that population. If individuals in a population consume increasing amounts of an ecosystem's goods and services, their collective impact on the environment increases, even if the size of the population does not change. Because human consumption varies so dramatically across the globe and over time, estimates of the number of humans that any given environment can sustain are highly variable.

The extent to which humans consume ecosystem goods and services is often described in terms of an **ecological footprint** (Figure 5.24). An ecological footprint is an estimate of the area of land needed to supply human demands for resources. Such demands include shelter, energy, food, and land to absorb wastes. For example, the construction of a house requires timber that grew on a certain amount of land. Housing construction also depends upon the land used to build factories to make concrete for foundations, roofing materials, and many other supplies. In a similar respect, the fossil fuels used to heat homes and to power cars are extracted from land and processed in refineries built on land. Even the carbon dioxide emissions from burning those fossil fuels can be converted into the acres of forest required to absorb those emissions.

Sometimes ecological footprints are calculated as the total land area needed to sustain an entire population. More often, ecological footprints are expressed as the

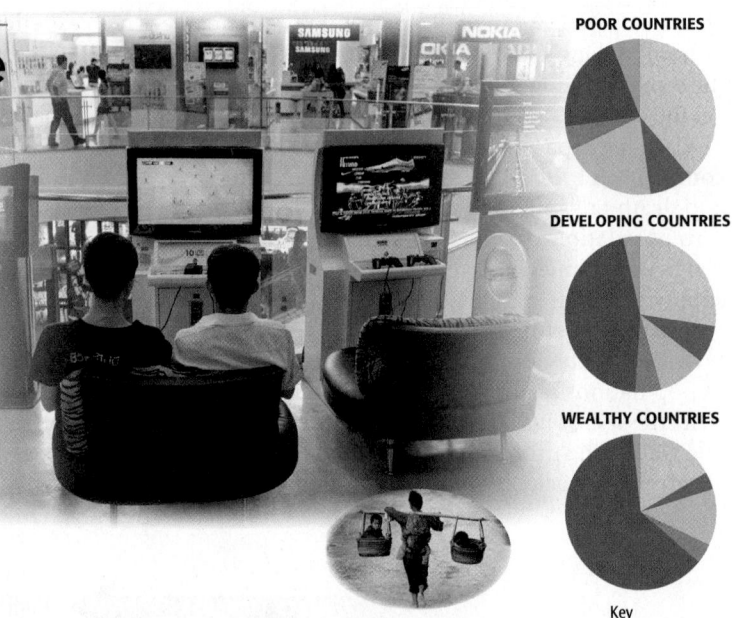

POOR COUNTRIES

DEVELOPING COUNTRIES

WEALTHY COUNTRIES

Key
● Cropland
● Grazing
● Forests
● Fisheries
● Carbon footprint
● Built-up land

▲ Figure 5.25 **Footprint Components and Wealth**

As countries become more affluent, their footprints are increasingly determined by the extraction and burning of fossil fuels and the generation of electric power, indicated by the segment "Carbon footprint." The segment "Built-up land" refers to the area of land covered by human infrastructure, including transportation, buildings, reservoirs for hydropower, and so on.

Data from: Global Footprint Network, www.fooprintnetwork.org.

land area needed to sustain an average person within a population, or the total ecological footprint divided by the population size. The size of ecological footprints varies greatly from country to country. As a country becomes more affluent, its ecological footprint is increasingly defined by its use of fossil fuels, which increase its carbon emissions (Figure 5.25).

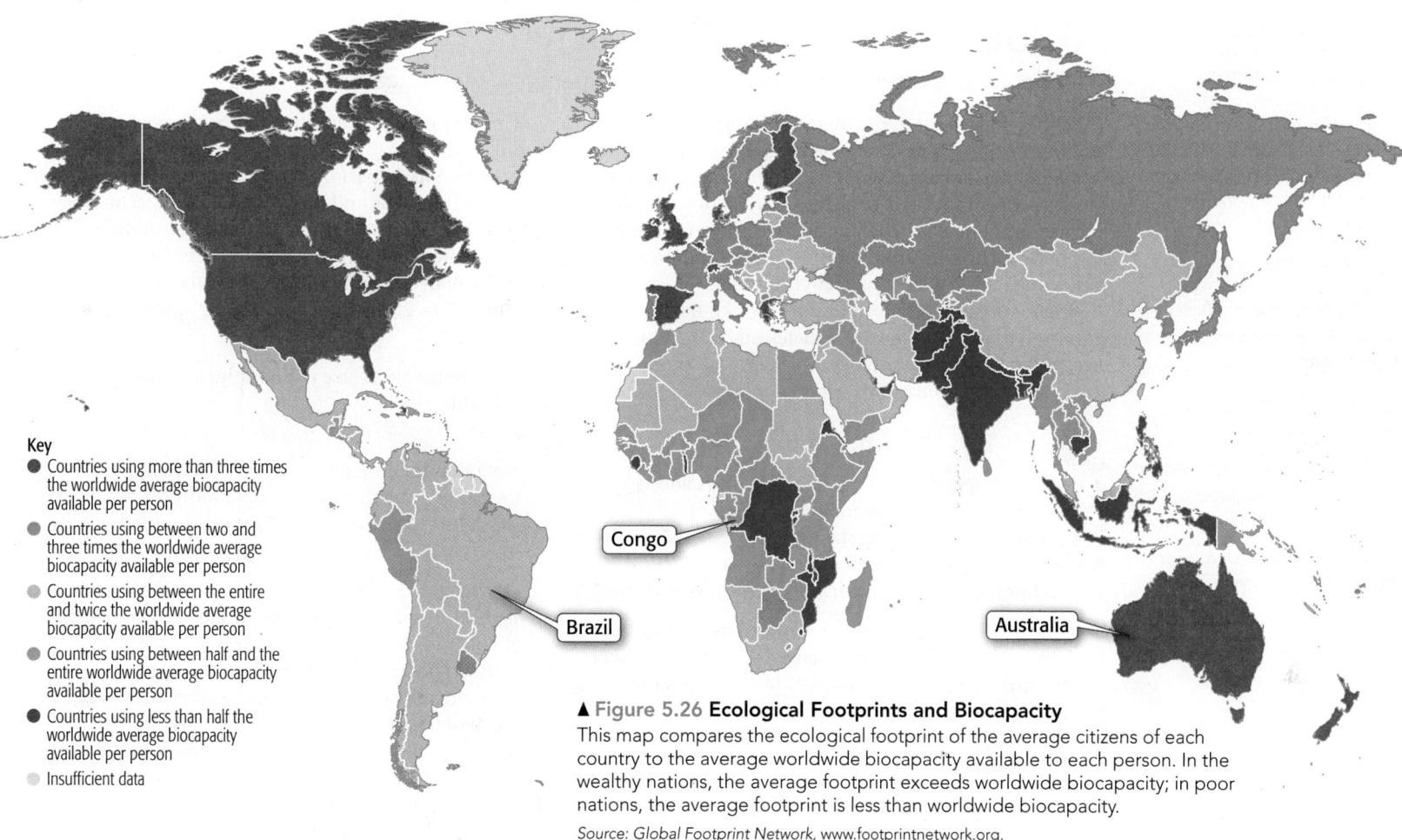

Key

- Countries using more than three times the worldwide average biocapacity available per person
- Countries using between two and three times the worldwide average biocapacity available per person
- Countries using between the entire and twice the worldwide average biocapacity available per person
- Countries using between half and the entire worldwide average biocapacity available per person
- Countries using less than half the worldwide average biocapacity available per person
- Insufficient data

Congo

Brazil

Australia

▲ Figure 5.26 **Ecological Footprints and Biocapacity**
This map compares the ecological footprint of the average citizens of each country to the average worldwide biocapacity available to each person. In the wealthy nations, the average footprint exceeds worldwide biocapacity; in poor nations, the average footprint is less than worldwide biocapacity.

Source: Global Footprint Network, www.footprintnetwork.org.

The ecological footprint of a human population is often compared to its biocapacity (Figure 5.26). **Biocapacity** is a measure of the area and quality of land available to supply a population with resources. Countries with small populations relative to their total land area, such as Australia, Brazil, and Congo, have biocapacities that are larger than their ecological footprints. However, the average footprint of the citizens in these sparsely populated countries may still be larger than the average *worldwide* biocapacity available to each person, as it is for Australia and Brazil.

A population exceeds its biocapacity when it draws a portion of its ecosystem goods and services from outside its geographical boundaries. For example, Haiti has the relatively small ecological footprint of 0.68 global hectares (gha) per person. However, deforestation and environmental degradation have left Haiti with a biocapacity of only 0.31 gha per person; thus, its ecological footprint exceeds its biocapacity. As a result, the people of Haiti must depend on ecosystem goods and services imported from other countries.

The average American has a very large ecological footprint, with each person requiring about 8 gha of land to supply food, housing, transportation, and other needs. The biocapacity available to meet these needs within the United States is 3.9 gha per person. Therefore,

the ecological footprint of Americans extends well beyond the borders of the United States. It can only be accommodated by appropriating the biocapacity of other parts of the world. Countries whose ecological footprints exceed biocapacity are sometimes called *ecological debtors.*

The great majority of the world's countries use more resources than are available within their borders. Some estimates suggest that the combined ecological footprint of the world's countries—each living its current lifestyle— already exceeds global biocapacity by approximately 40%. If all countries adopted the lifestyle of the average North American, the world would exceed its biocapacity five times over. Put another way, it would take five planet Earths to sustain the planet's >7 billion people according to the North American lifestyle. Worldwide, biocapacity per person is decreasing as the human population increases. However, the gap between wealthy and poor nations is widening when it comes to resource use. The footprints of wealthy countries continue to increase (+8% in a 10-year period), but that of poor countries has declined (−11% over that same time period). In other words, the average person in a wealthy country is making ever-greater demands on Earth's resources, while the average person in a poor country is consuming much, much less.

QUESTIONS 5.4

1. What information is needed to determine the size of a sustainable human population rather than the global carrying capacity of the human population?

2. Although Brazil has 11% less total land area than China, its biocapacity is 77% greater. What factors may be contributing to this difference?

3. Even if the populations of many developing countries were to stabilize, their impacts on the environment are likely to continue to increase. Explain why.

(MES) For additional review, go to **MasteringEnvironmentalScience**

Affluence and Technology

■ Affluence and technology impact the environment at a global scale.

The different sizes of the ecological footprints of rich and poor countries illustrate the central role of affluence in the rate of consumption of resources. A great deal of modern consumption is related to the use of technology, such as appliances for cooking and cleaning, cars, and computers. The consumption of resources and the use of technology impact the global environment.

Scientists Paul Ehrlich and John Holdren have proposed that human impact is determined by population size, affluence, and technology. They have summarized this relationship in the **IPAT equation** (Figure 5.27):

$$I = P \times A \times T$$

where

I = environmental impact (loss of resources and degradation of ecosystems)
P = population (including size, growth, and distribution)
A = affluence (consumption per individual)
T = technology (things that demand resources and energy)

Although some technologies improve the environment, the T in the IPAT equation represents the technologies that are destructive to the environment. Technologies that reduce the impact on the environment, such as solar panels and scrubbers that control air pollution, are instead represented as 1/T, modifying the equation to

$$I = (PA) \times 1/T$$

It is worth noting that wealth is not inherently destructive to the environment: It can be used to develop sustainable technologies and to conserve land area, for example. Only when wealth results in greater resource consumption does its environmental impact increase. The IPAT equation is a conceptual model; it is not designed to calculate numerical values.

The IPAT equation emphasizes that the size of a population is not the only factor to consider when determining its impact on the environment. Consumption of ecosystem goods and services and use of technology are also key factors. For the sake of demonstration, you can calculate the environmental impact of an imaginary population by plugging simple numbers into the IPAT equation: $P = 2$, $A = 2$, $T = 2$. The equation becomes:

$$I = 2 \times 2 \times 2 = 8$$

If the population size (P) doubles to 4, its impact (I) also doubles:

$$I = 4 \times 2 \times 2 = 16$$

Note that the impact on the environment would be the same if the size of the population (P) remained constant but its use of technology (T) doubled:

$$I = 2 \times 2 \times 4 = 16$$

Of course, the interaction of the factors is not as simple as the IPAT equation may suggest. A change in one factor invariably causes another factor to change. For example, increases in population may lead to increases in technology. Increases in affluence also tend to increase technology.

Much of the impact of affluence and technology is associated with the use of energy to power motor vehicles and generate electricity. Although people living in developed countries make up only about 16% of the world population, they consume approximately 75% of the total energy used each year. Ehrlich and Holdren estimate that the average citizen of a developed country has 7.5 times more impact on Earth's ecosystems than the average inhabitant of a developing country. Based on energy use alone, Ehrlich and Holdren suggest that the environmental impact of an average American is nearly 30 times greater than that of a citizen of Haiti or Niger.

▼ Figure 5.27 **IPAT**
The IPAT equation proposes that human impact (*I*) is determined by population size (*P*), affluence (*A*), and technology (*T*).

Impact =	Population ×	Affluence ×	Technology
loss of resources and degradation	including size, growth, and distribution	consumption per individual	things that demand resources

As consumption (*A*) increases around the globe, it plays an increasingly significant role in the calculation of Impact (*I*) on Earth (Figure 5.28). The people of developing countries are striving to improve their well-being by acquiring more material goods and better access to technology. As the population and per capita consumption of developing countries increase, their impact on the environment will increase dramatically. For example, between 1980 and 2001, people in developing countries in the Middle East, North Africa, and Asia increased their consumption of electricity by over 370%. During those two decades, the population in the developing countries of Asia grew 1.7%, and the region led the world in pace of economic expansion.

China is the world's most populous country. It is also the second-largest consumer of energy, behind the United States. On average, an American citizen uses over five times more energy than a Chinese citizen. However, between 1990 and 2010, per capita energy use in China grew by over 100% compared to growth of about 12% in the United States.

Because the growth rate of the world's human population is declining, there is hope that we may be able to achieve sustainability. Yet to reach this goal, the rate of consumption will need to change, particularly in the world's richer countries (Figure 5.29). Sustainable development will require increased conservation of ecosystem goods and services. New technologies that help us use those goods and services more efficiently may be the answer. Greater efficiency may provide the means to diminish our collective footprint while sustaining our well-being.

▶ **Figure 5.28 Population vs. Affluence**
Ⓐ An average family from the west African country of Mali is shown atop their home with all their possessions. This man has two wives and eight children, with more likely on the way. *P* is large in the IPAT equation for Mali and other developing countries. Ⓑ For an average American family with all their possessions, *A* is very large in the IPAT equation. This is typical for the United States as well as other developed countries.

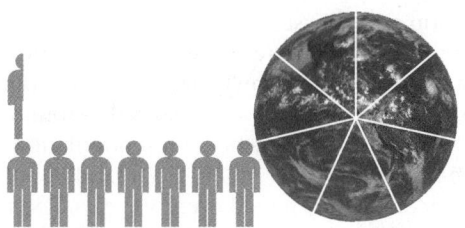

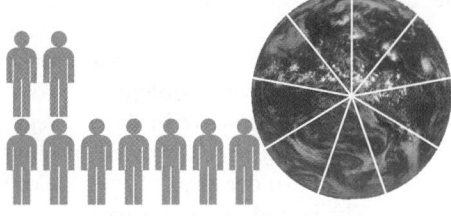

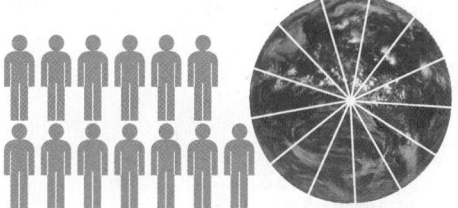

Present: World population >7 billion
With its present population of >7 billion, Earth's natural resources are supporting many millions of people at a high level of expected well-being and many more millions at a low level.

2040: World population 9 billion
If, as expected, the world's human population increases by nearly 50% by 2040, expectations of what is a sufficient level of well-being will have to decrease.

Future: World population 13 billion
If the world's population doubles to over 13 billion, there will have to be a radical shift in expectations with the major consumers modifying their behavior, attitudes, and lifestyles to minimize consumption.

▲ **Figure 5.29 Population Growth Scenarios**
The ability of Earth's environment and natural resources to help future generations meet their needs depends not only on how populations grow but also on how demands for resources change.

Q *In what ways would the government be able to recognize that the population is getting out of hand?*
Megan Reinle,
Northern Illinois University

Lissa: **A** *Symptoms of overpopulation or excessive resource use include increased air and water pollution and shortages of essential resources. They also include measures of decreasing human well-being, such as high infant mortality and diminished life expectancy.*

5.5 Managing Population Growth

BIG IDEA Efforts to manage human population growth began a century before the establishment of demography as a science and the development of the demographic transition model. In the early 1800s, people who were influenced by Malthus began to advocate birth control and limitations on family size. At the advent of the 20th century, providing greater access to family planning emerged as a practical solution to the human population growth problem. As the 20th century came to a close, it was understood that economics, education, the empowerment of women, and family planning must be considered when attempting to manage human population growth. Aging populations, found in many developed nations, also present societal challenges. Nations facing labor shortages and rising health care costs brought on by aging often encourage immigration and increased fertility to reverse these trends.

Family Planning

■ Family planning advances on both national and international fronts.

In 1822, the British politician Francis Place published a book called *Illustrations and Proofs of the Principle of Population*. This book initiated the **neo-Malthusian movement**. Like Malthus, the neo-Malthusians believed that as the human population increased, it would soon outpace the supply of food. Place and his fellow neo-Malthusians were the first to advocate contraception. (Ironically, Malthus himself was firmly against contraception as a means to manage population growth.) At first the ideas of the neo-Malthusians were dismissed, but by the end of the 19th century many scholars and some governments believed that there was a need to manage human population growth and to promote contraception.

In the first half of the 20th century, two world wars and the Great Depression consumed the attention of the United States. Consequently, national efforts to slow population growth gained only modest momentum. In 1916, Margaret Sanger succeeded in opening the first birth control clinic in the United States, where she provided poor women with access to contraception. In 1921, she founded the American Birth Control League, which later became Planned Parenthood. Despite sometimes violent opposition, scores of birth control clinics opened over the next two decades.

The period immediately following World War II produced a **baby boom** within the United States, as returning soldiers and their spouses focused on starting and growing families (Figure 5.30). The 1950s and 1960s were a period of international dialogue and activity on the subject of human population growth. In 1952, representatives of numerous countries, including India, the United Kingdom, and the United States, ratified the constitution of what is now the International Planned Parenthood Federation (IPPF). A nongovernmental organization, the IPPF promoted sexual and reproductive health and the rights of individuals to make their own choices in family planning.

By the 1970s, the IPPF and other nongovernmental organizations such as Family Life and Family Health International were working to slow human population growth. They did so through promotion of universal access to contraceptives, such as birth control pills and condoms, as well as family planning, educational programs, and counseling services.

About the same time as the ratification of the IPPF constitution, the United Nations began to support international family planning organizations and to construct a network for international conferences. In these conferences, representatives from countries around the world met to negotiate solutions to population growth issues. In 1954, the first UN World Population Conference resolved to gather more complete information on population changes in developing countries. The conference also resolved to found regional training centers where local population issues could be addressed. Since then, UN World Population Conferences have been held every 10 years. Recent conferences have focused on the relationships between population growth, economic development, and human rights.

Today, family planning policies encompass a broad range of strategies and priorities. Access to contraceptives remains a high priority. Programs that deal with education on reproductive health, including the prevention, diagnosis, and treatment of HIV/AIDS, have also become important for governmental and nongovernmental organizations.

▼ Figure 5.30 **The Baby Boom**
The economic growth that followed World War II fostered a period of unprecedented population growth and development known as the baby boom.

Development and Population

■ Economic development, education, and empowerment of women decrease human population growth.

The 1974 UN World Population Conference marked a milestone in international negotiations on human population. Developing nations argued that *economic development*, not expanded family planning programs, offered the most promising approach to slowing population growth and alleviating poverty. Falling into line with Thompson's demographic transition model, developing countries adopted the motto, "development is the best contraceptive."

In 1994, the international dialogue on slowing population growth shifted from economic development and the alleviation of poverty to education and the empowerment of women. That year, the International Conference on Population and Development held in Cairo, Egypt, produced the Cairo Consensus.

The Cairo Consensus emphasized improvements in health care, well-being, and rights—particularly for women—as a way to achieve the social, political, economic, and environmental stability necessary to slow population growth. Behind the Cairo Consensus was this reasoning: Once basic educational and health care needs are met, an individual is in a better position to secure employment and to make deliberate decisions about family size. Employment provides greater economic opportunity for both men and women. It also reduces the perceived need for domestic labor, which is often the motivation for having numerous children. Education and employment also give women greater social status than the traditional roles of bearing and rearing children (Figure 5.31).

▲ Figure 5.31 **Education and Population Growth**
Where economic development has improved educational opportunities for women, birth rates and population growth rates have declined.

Aging Populations

■ Populations with slow or negative growth experience economic consequences due to a shrinking workforce and an aging population.

Despite rapid increases in population size, the world population is older than it has ever been. Aging populations, or those that are increasing proportionally in the 65-and-older age class, are especially pronounced in developed countries due to increasing life expectancy and decreasing fertility. European countries, for example, have the highest median age in the world with a projected increase from 14% in the 65+ age class in 2010 to 25% in 2050.

The "demographic time bomb" of aging populations presents significant challenges. These include soaring health care costs, as well as a shrinking labor force and with it the consequent reduction in pension revenue to support the growing segment of elderly. Some European countries have attempted to combat these problems with incentives for others to immigrate into their country as well as encouraging younger citizens who have moved away to return. In addition, some nations have implemented large tax incentives for those who have children.

Balancing human population growth is no simple matter: either too much or too little growth results in societal challenges. Ultimately, both the IPAT model and the ecological footprint concept suggest that consumption is a critical part of the equation when evaluating the impact of a population upon the world's resources.

QUESTIONS 5.5

1. How have the agendas of the UN World Population Conferences changed over the past 50 years?

2. Some countries are considering policies to increase their population growth rates. Which countries and why?

3. Why do many programs focused on managing human population growth encourage the education and empowerment of women?

(MES) For additional review, go to **MasteringEnvironmentalScience**

Q Are policies on having children such as China's "one child per family" rule really necessary?

Jaymee Castillo,
University of California,
Los Angeles

Lissa: A National policies on population growth are essential, but they need not be as restrictive as China's.

Two Approaches to Population Growth

■ China and India offer case studies on different approaches to managing human population growth.

As with all other aspects of human population and well-being, the strategies for managing human population growth and their results vary greatly across regions and countries. The policies of China and India show different approaches to managing population growth in the most populous nations on earth.

◀ Figure 5.32 **China's Changing Demography**
Implemented in 1978, China's one-child policy has had a significant effect on birth rates and population growth rates.

Source: United Nations World Population Prospects: The 2012 Revision Population Database.

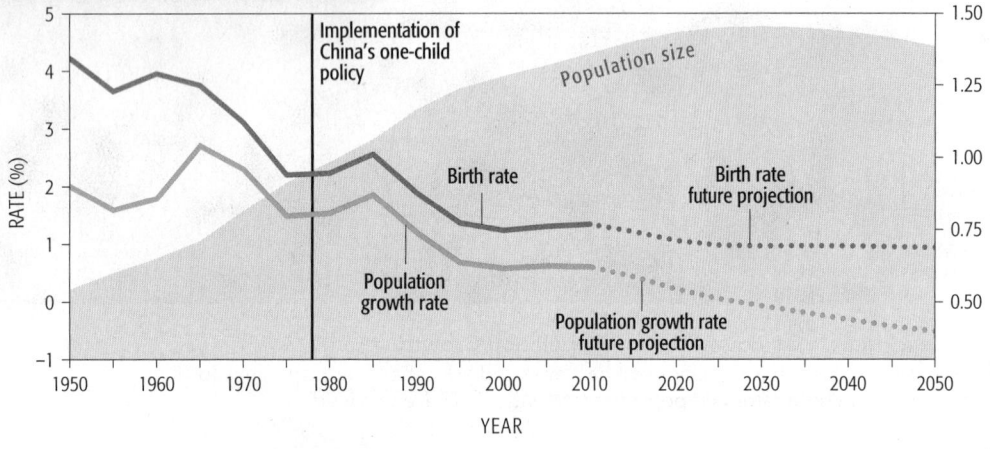

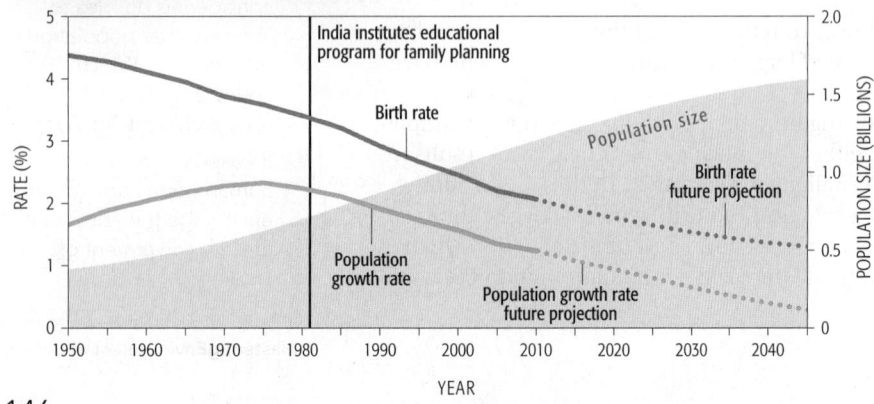

◀ Figure 5.33 **Effect of India's Population Management Practices**
India's educational approach to fertility reduction has slowed population growth and birth rates, but not as quickly as China's one-child policy.

Source: United Nations World Population Prospects: The 2012 Revision Population Database.

China. In the early 1970s, China began a public campaign to reduce its population growth, which culminated in 1978 with the one-child policy. The one-child policy aimed to reduce China's population growth rate from 1.2% to 0% by the year 2000. Through education, financial subsidies, and strict penalties, China encouraged couples to have only one child. China's population growth rate is currently about 0.6%, and its birth rate has dropped from 3.6% to 1.2%. It appears that the one-child policy is working, and the UN forecasts that China's population growth rate will hit 0% by 2030 (Figure 5.32).

China has succeeded in reducing its birth rate and population growth, but not without unintended consequences. The sex ratio of Chinese children is significantly skewed towards males. In 2010, 120 boys were born for every 100 girls. By comparison, a typical sex ratio in the United States is 105 boys to 100 girls. In addition, the Chinese population is now aging rapidly. China reported a 20% drop in primary aged children from 2002 to 2012, causing the closure of more than 13,600 primary schools nationwide. The working age population peaked during 2009 and the elderly are projected to constitute a greater percentage of the population than 0- to 14-year-olds by 2025. China has recently begun to relax its one-child policy and is now contemplating the challenges of an aging population.

India. Unlike China, India's approach to population management is based on education and voluntary participation. In 1966, India created a Department of Family Planning within its Ministry of Health. That program was firmly rooted in the philosophy that family planning should be voluntary and decisions regarding strategies for managing population growth should be delegated to individual states. In 1986, the Department of Family Planning began an educational program to encourage couples to postpone marriage and children. This program was developed in the context of a larger governmental campaign to increase child survival, elevate women's status and employment levels, and increase literacy in the general population.

In 2013, India's population growth rate was 1.3% and its birth rate was 2.1%. Although India's birth rates are declining in accordance with the fertility transition of the demographic transition model, its population growth rate will not drop below 1% until about 2020 (Figure 5.33). Then the country's population will be nearly equal to China's projected 2020 population of 1.35 billion.

Women Deliver!

Can equality, health care, and education for women reduce population growth and increase well-being?

Efforts to curb population growth often attract furious debate. More often than not, these debates center on emotional issues such as immigration policy or access to abortion. Yet in many countries, population growth rates are declining as a consequence of actions that almost everyone agrees are appropriate—education, health care, and improved opportunities for women.

Since 1969, the central mission of the United Nations Fund for Population Activities (UNFPA) has been to "realize a world where every pregnancy is wanted, every birth is safe, every young person is free of HIV, and every girl and woman is treated with dignity and respect." In 2010, UNFPA Executive Director Thoraya Ahmed Obaid said, "change cannot be imposed from the outside; to be lasting it must come from within. Development should be locally owned." This is the spirit in which UNFPA works with countries and communities to reduce population growth while improving the physical and social well-being of women.

UNFPA has three core areas of focus. First, it assists governments in gathering data on the dynamics of their populations. These data are needed to generate the political will for action and to create and maintain sound policies. Second, UNFPA assists governments in delivering reproductive health care to women. This assistance is far ranging, including medical care of mothers and children before and after birth, diagnosis and treatment of infertility and sexually transmitted diseases, and family planning. Third, UNFPA encourages gender equality and the empowerment of women through programs that educate girls, create economic opportunities for women, and encourage women's participation in government.

The education of women in developing countries tends to lower the rate of population growth. In countries in which at least 48% of high school students are women, the average annual birth rate is 2.18%. In countries in which fewer than 45% of high school students are women, the average birth rate is 3.57% (**Figure 5.34**). Girls who attend high school are likely to delay marriage and childbearing until they are older. Married women with a secondary education are more likely to use contraceptives and rear smaller families.

UNFPA works with agencies like the Roma Youth Initiative Be My Friend, in Bosnia Herzegovina, developed by 26-year-old Melina Halilovic. The customs of the Roma people tend to encourage girls to marry early (50% before the age of 18), often bringing an end to their formal education. Ms. Halilovic, a member of the Roma community herself, completed her education and now helps young Roma girls to remain in school. "Success for me is when children from our workshop say to us: 'We are going to school now,'" says Melina. "For so many years we have been talking with parents, telling them they should send their children to the school, even [if] they don't have money. And I'm happy when they come to us and say they are going to school now. I think this is the major success for us. For all of us." Education matters. Only 5.6% of Roma girls with a secondary education bear children as adolescents, as compared to 27% among all Roma women aged 15–19 (UNFPA Dispatch, 30 October 2013 "First School, Then Marriage and Babies").

Improved maternal education and reproductive health also diminish infant mortality. For example, a study in Egypt revealed that nearly 9% of children born to mothers with no education die before the age of 5, compared to 3.8% of children born to mothers who have completed secondary education. It may seem paradoxical, but in the long run improved infant survival results in *decreased* birth rates. This is because the loss of children is often an incentive for women to conceive more children.

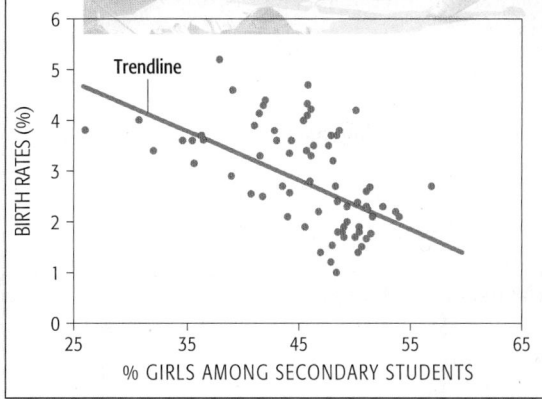

▲ Figure 5.34 **Yes, Education Matters** Students from Macedonia, Bulgaria, Bosnia and Herzegovina, Serbia, Croatia, and Slovenia dance together as part of a graduation march to celebrate their graduation from high school. The graph compares birth rates with girls' representation in high schools (percentage of student body) for 78 countries in which per capita GDP is less than $10,000/year.

Summary

Although the global population growth rate is declining, in many of the world's poorest countries it remains high. In general, where population growth is high, human well-being is comparatively low. Economic development improves nutrition, sanitation, and health care; it also results in lower birth, death, and population growth rates. However, economic development often increases demands on Earth's resources. Our prospects for future sustainability depend on both the continued decline in the human population growth rate and the reduction of our individual demands on the environment.

Demographic measure	Definition/calculation
Birth rate (expressed as %)	# individuals born per year/initial population size
Age specific birth rate (expressed as %)	# individuals born per year to an age category of women/ # women per age category
Age specific fertility rate	average # of children born to a woman in a specific age category
Total fertility rate	sum of age specific fertility rates = average # of children a woman will bear in her lifetime if she lives through her reproductive years
Death rate (expressed as %)	# individuals dying per year/initial population size
Infant mortality rate (expressed as %)	# individuals dying before one year of life/# individuals born during a year
Life expectancy	average # of years of life expected by a hypothetical cohort of people subject to mortality rates expected for a given period
Migration rate (expressed as %)	(# immigrants − # emigrants during a time period)/ initial population size

5.1 The History of Human Population Growth

- Over the past 100,000 years, human populations have experienced three general periods of growth: the pre-agricultural, agricultural, and industrial periods.
- Industrial development and economic progress are correlated with demographic transitions that involve a general decline in birth rates and death rates and an overall decline in population growth rate.

KEY TERMS
demographer, demographic transition model, mortality transition, fertility transition, stability transition

QUESTIONS
1. What factors have caused changes in human population growth rates during the past 100,000 years?
2. Describe the four stages of the demographic transition model of population change during the industrial period.
3. How do population age-structure pyramids change during a demographic transition?

4. Explain why economic development and human well-being result in continuously declining birth rates during the fertility transition stage of the demographic transition.

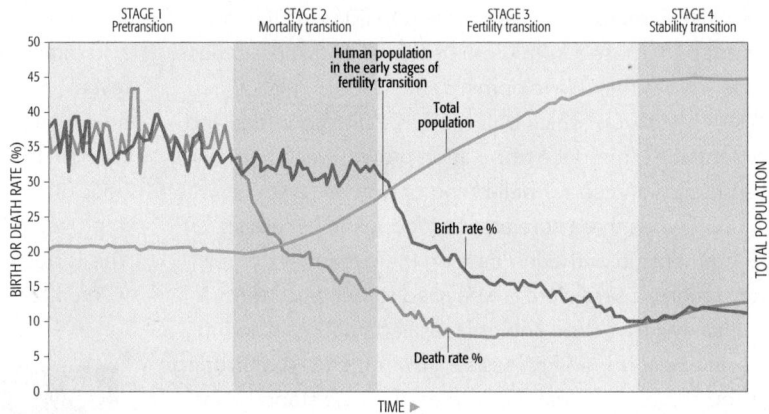

5.2 Global Variation in Human Population Growth

- Since 1970, the world's human population has continued to grow, but the population growth rate has steadily decreased.
- The magnitude of these changes varies among regions and countries and is highly correlated with wealth and well-being.

KEY TERMS
age-specific birth rate, infant mortality rate, life expectancy, age-structure pyramid, net migration rate, push factors, pull factors

QUESTIONS
1. How are birth rate, total fertility rate, and age-specific birth rate related to one another?
2. Total fertility rates overestimate the actual number of children an average woman will bear in her lifetime. Why is this so?
3. What factors account for the global decrease in infant mortality rates and increase in life expectancy over the past 50 years?
4. Two countries with identical age-specific birth rates may have very different total birth rates because they have different age-structure pyramids. Explain.
5. In Germany, the birth rate is lower than the death rate, yet the population growth rate is positive. Why is this so?

	Wealthy Countries	Poor Countries
Total Fertility Rate	Less than 2.0 children	Between 4 and 7 children
Birth Rate	Low	High
Infant Mortality Rate	Low	High
Life Expectancy	High	Low

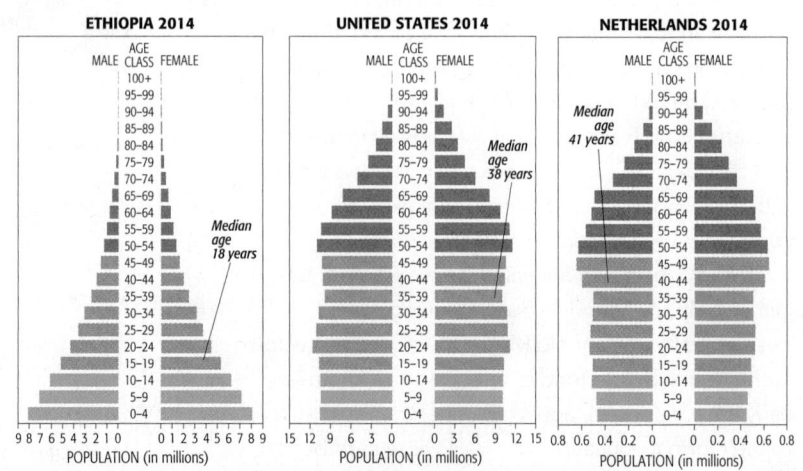

5.3 Predicting Human Population Growth

■ Forecasts of future population growth are dependent on the quality of data used and the assumptions made with regard to human values and behaviors.

■ Current forecasts predict that the global population growth rate will decrease and that the world population will reach approximately 9.6 billion by the year 2050.

QUESTIONS

1. To understand the importance of resources to population growth, Malthus chose to compare 18th-century populations in the American colonies to Europe. Why?

2. Why is forecasting human population growth more difficult than forecasting population growth in other animals?

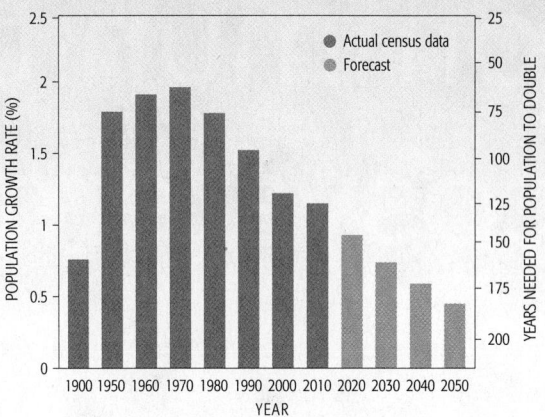

5.4 Resource Use and Population Sustainability

■ The ecological footprint of a population is the area of land necessary to meet human demands for shelter, energy, food, and other resources. Biocapacity is the area of land available to meet those needs.

■ Human impact on Earth's resources is determined by population size, affluence, and technology.

KEY TERMS
ecological footprint, biocapacity, IPAT equation

QUESTIONS

1. Why is the number of people that the world can support in a sustainable fashion considered to be much lower than the global carrying capacity?

2. The ecological footprint for the average person in a country may be small, but the country may still be an ecological debtor. Explain.

3. The world's biocapacity is estimated by some to have declined by more than 12% between 1995 and 2005. What factors may have contributed to this decline?

4. In 2000, China's GDP was ~$1.2 trillion and its population was 1,280,429,000. By 2010, its GDP was $5.9 trillion and its population size was 1,359,821,000. If GDP is used as a measure of affluence, and technology is held constant, how has China's impact changed proportionally during those 10 years? How would this be different if China's GDP had not increased?

5.5 Managing Population Growth

■ Efforts to manage human population growth began in the early 19th century and continue today with an emphasis on family planning, gender equality, and sexual and reproductive health.

■ Aging populations in developed nations present management challenges as the labor force is reduced relative to the elderly.

KEY TERMS
neo-Malthusian movement, baby boom

QUESTIONS

1. What was the neo-Malthusian movement, and why was it important?

2. Why have global conferences and initiatives on human population growth recently begun to focus on the education and empowerment of women and reproductive health?

3. Why do some argue that understanding and closing the gap between the world's poorest and wealthiest individuals, regions, and countries are critical in defining and securing sustainable population growth?

4. What are some of the challenges associated with aging populations?

MasteringEnvironmentalScience®

Students Go to **MasteringEnvironmentalScience** for assignments, the eText, and the Study Area with practice tests, activities, and more.

Instructors Go to **MasteringEnvironmentalScience** for automatically graded tutorials and questions that you can assign to your students, plus Instructor Resources.

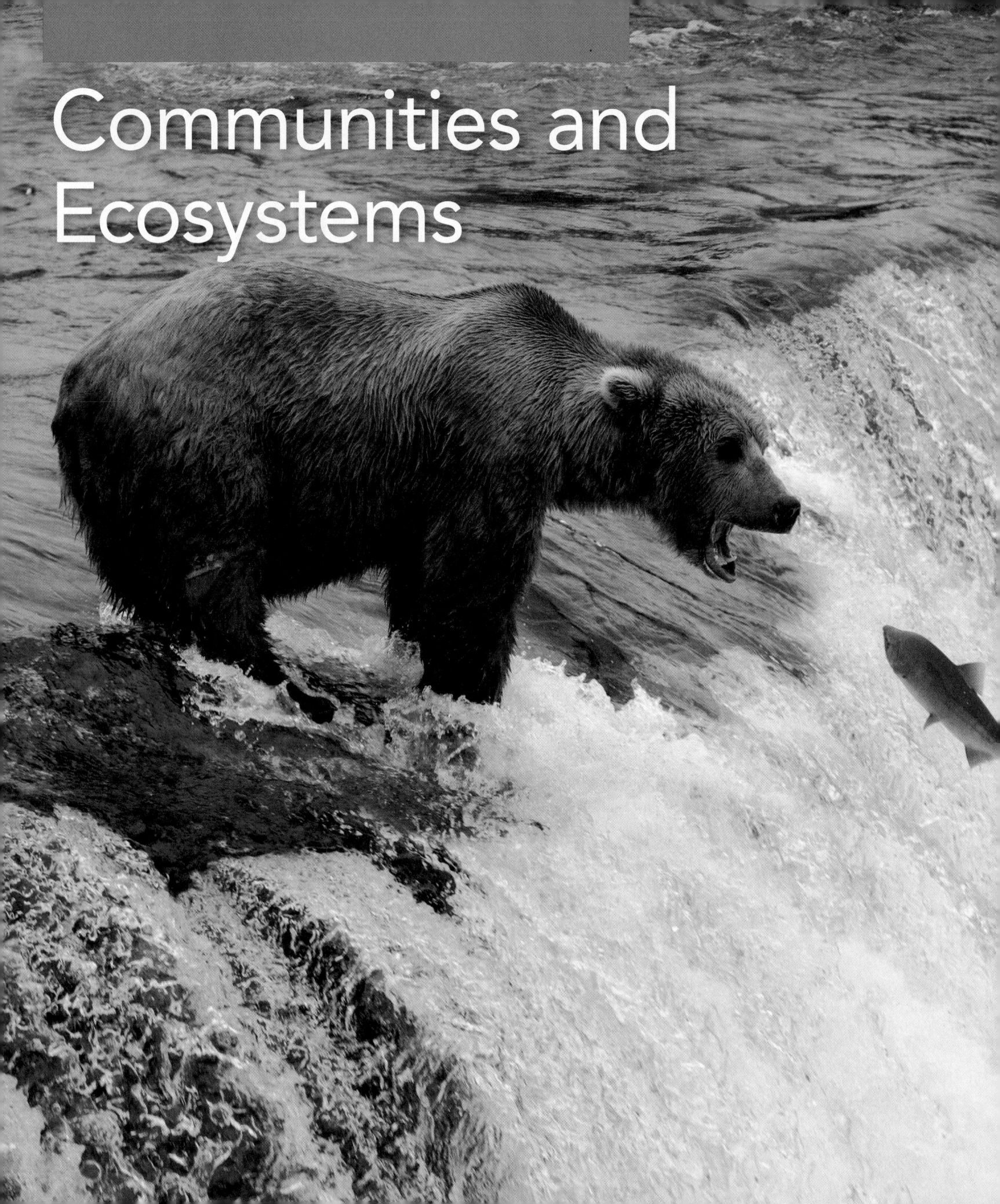

Communities and Ecosystems

6

The Web of Life

What are the interconnections among the many species in an ecosystem?

Images of old-growth forests in the Pacific Northwest often show spotted owls perched on the limbs of towering ancient firs, cedars, and hemlocks (Figure 6.1). But the growth and reproduction of these majestic trees and charismatic animals depend on the activities of a myriad of organisms that are largely unseen and certainly underappreciated.

In the soil, many different species of fungi form intimate associations with the roots of these majestic trees, benefiting both the fungi and the trees. Thin fungal strands grow through the mineral soil and organic litter, extracting nitrogen, phosphorus, and other nutrients and transferring them to the trees. For their part, the trees provide the carbohydrates necessary to support the growth of the fungi.

Fungi reproduce by means of spores. Most fungi form reproductive structures that grow above the ground. The most familiar of these reproductive structures are mushrooms; each kind of mushroom represents a different species of fungus. The spores of mushrooms are usually dispersed by the wind.

However, one group of forest fungi produces its spores inside odd-shaped structures that grow underground. These structures are known to the gourmet as truffles. The spores of truffles are dispersed by animals that feed on them, such as chipmunks and voles. To attract would-be dispersers, truffles release an array of aromatic—and often very tasty—chemicals. When animals eat the truffles, the spores pass through their digestive systems unharmed. The spores are eventually deposited in the animals' feces. In the new location, the feces provide a dose of nutrients that help the spores grow into new fungi.

Among the most important animals that disperse truffle spores in the forests of the Pacific Northwest are northern flying squirrels. These small squirrels nest in tree cavities high above the ground, but they feed on the forest floor. Several tree species depend upon the truffle-forming fungi to obtain the nutrients they need to grow; the distribution of these trees is tightly linked to the activities of flying squirrels (Figure 6.2). This complex connection between the trees and flying squirrels is so strong that if the forest experiences a disturbance, such as

▶ Figure 6.1 **Old-Growth Forest**
The large fir, cedar, and hemlock trees of Oregon's old-growth forests depend on fungi associated with their roots to acquire nitrogen and phosphorus.

fire or logging, the flying squirrel population must be reestablished in order for the forest itself to be reestablished.

Flying squirrels form a significant part of the diet of the spotted owls, foxes, and weasels that live in the old-growth forest. These predators prevent flying squirrels from becoming so abundant that they eat the truffles faster than they are produced. The chemical energy that powers the movements of a flying squirrel, an owl, a fox, or a weasel is provided through a complex web of feeding interactions that begins with photosynthesis in the leaves of the ancient firs, cedars, and hemlocks. Equally, the sustained growth of these trees depends on the interactions among the animals in this web.

In this chapter, we will explore the different kinds of interactions among organisms in ecological communities and their importance to the productivity of ecosystems.

- What is the nature of interactions among community members that depend on shared resources?

- What is the nature of interactions in which one organism is an essential resource to another?

- What is the nature of interactions in which organisms are mutually dependent on one another?

- How does the flow of energy through ecosystems influence the relative abundance of organisms?

- How do these interactions influence the cycling of carbon and the overall productivity of ecosystems?

- How are the complex webs of interactions among organisms affected by disturbances?

▼ Figure 6.2 **Connections in the Forest**
A walnut-sized truffle Ⓐ is shown here with tree roots and associated strands of fungus. Spores of the truffle are dispersed by animals such as the flying squirrels Ⓑ that feed upon these fungi on the forest floor. In turn, flying squirrels are an important food source for several carnivorous animals, including the endangered northern spotted owl Ⓒ.

153

6.1 Competition for Shared Resources

BIG IDEA The species that interact within a specific area do form a web of life, which scientists refer to as an **ecological community**. Interactions among organisms that compete for shared resources are particularly important. In **intraspecific competition**, members of the same species pursue limited resources. In the forests of the Pacific Northwest, flying squirrels compete with each other for truffles, a favored food. If the number of flying squirrels increases, the demand for truffles also increases. Thus, competition among individual flying squirrels becomes more intense.

Other forest animals, such as chipmunks, also eat truffles. The interactions among different species as they compete for shared resources are called **interspecific competition**. Species may compete directly by using and depleting a shared resource. They may also compete by interfering with one another's ability to access a resource. Interspecific competition limits the distribution and population size of most species. Competing species are able to coexist because they use different portions of their shared environment.

Interspecific Competition

■ Interspecific competition occurs when two or more species seek the same limited resource.

Foresters were among the first to describe interspecific competition. They found that when trees of one species were removed from a forest, the remaining trees grew more rapidly. The abundance and diversity of herbs and seedlings growing on the forest floor also increased. Foresters hypothesized that this response was due to decreased competition for water and nutrients in the soil. This hypothesis was tested by digging deep trenches around 3–6 meter (10–20 feet) square plots of land; the trenches cut off all the roots growing into the plots. Conditions in the trenched plots were then compared to those in nearby, untrenched plots (the controls). In all cases, the soil in the trenched plots was found to contain more water and nutrients than the soil in the control plots. The number and growth of plants in the trenched plots were also greater than in the controls. Reduced competition for water and nutrients in the trenched plots had allowed more plants to grow (Figure 6.3).

In the 1930s, Russian biologist G. F. Gause devised a set of elegant laboratory experiments that provide the basis for our formal understanding of interspecific competition. Gause grew two different species of the single-celled *Paramecium*—*P. aurelia* and *P. caudatum*—separately and together (Figure 6.4). Populations of both species always increased more rapidly when they were grown alone (see Module 4.3). When grown together, populations of both species grew more slowly. Eventually, *P. aurelia* totally displaced *P. caudatum*.

The results of his experiments with *Paramecium* species, along with similar experiments performed on other organisms, led Gause to form this postulate: two species that directly compete for essential resources cannot coexist; one species will eventually displace the other. This postulate has come to be known as the **competitive exclusion principle**.

▼ Figure 6.3 **Forest Competition**
Forest trees compete for soil nutrients and water with the herbs and shrubs that grow beneath them.

How Competitors Coexist

■ If the competitive exclusion principle is true, why are ecological communities so rich in species?

An acre of tropical forest may include over 100 species of trees, all of which depend on the same soil, water, and nutrients. Freshwater lakes may have dozens of species of fish, all of which feed on the planktonic algae and animals suspended in the water. Indeed, two or more species of *Paramecium* may be found in the same lake. These and many other examples from ecological communities in nature seem to contradict Gause's principle. If two competing species cannot coexist in the laboratory, how are they able to coexist in natural settings? This question has been the basis for hundreds of ecological studies.

The distinguished ecologist G. Evelyn Hutchinson provided one of the most important explanations for the coexistence of competing organisms. He did this by refining the concept of the ecological niche (see Module 4.4). Hutchinson suggested that each species has a **fundamental niche**, the complete range of environmental conditions needed to support the species. This includes physical aspects such as temperature and

pH and material resources such as food and water. Hutchinson noted, however, that few species actually grow and reproduce in all parts of this theoretical range. Rather, species usually exist only where they are able to compete effectively against other species and avoid the things that eat them. Hutchinson used the term **realized niche** to describe the range of conditions where a species actually occurs given the constraints of competition.

Species whose fundamental niches overlap are potential competitors. Hutchinson suggested that these potential competitors are able to coexist because they divide up the fundamental niche. Hutchinson called this division of resources **niche differentiation**.

Niche differentiation occurs among many different kinds of organisms. In another pivotal study, Robert MacArthur, a student of G. Evelyn Hutchinson, observed the behavior of five different species of warblers, small insectivorous birds that live together in the evergreen forests of New England. During nesting season, the

GROWN SEPARATELY

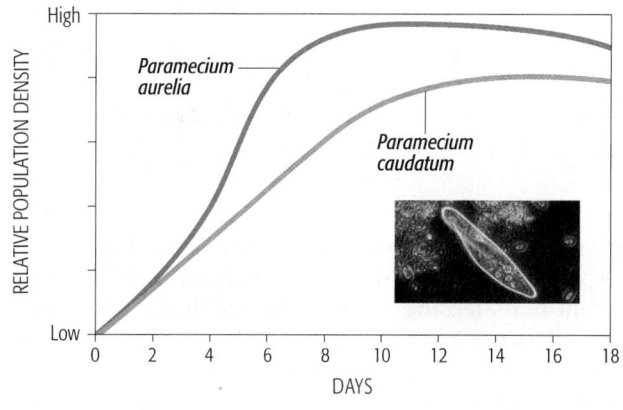

GROWN TOGETHER

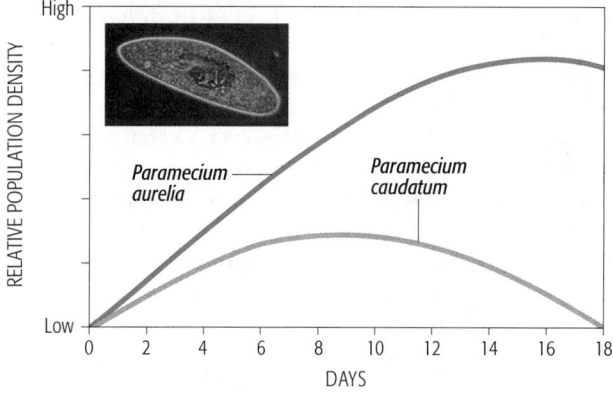

▲ Figure 6.4 **Competitive Exclusion Among *Paramecia***
Not only were the growth rates of both *P. aurelia* and *P. caudatum* lower when grown together than when grown in isolation, but eventually, one species actually excluded the other.

Source: Gause, G.F. 1932. Experimental studies on the struggle for existence. *Journal of Experimental Biology* 9: 389–402.

Where YOU LIVE **How do different bird species coexist where you live?**

Birds are among the most widely studied of all animals and the variation in bird species relates directly to their particular feeding niches. You can easily identify the most common bird species in your backyard and their primary food sources using the Cornell University Ornithology Laboratory's All About Birds website.

What are the six most common bird species in your area? Which ones appear to be competing with one another for food? What differences in their behavior do you think contribute to their ability to coexist?

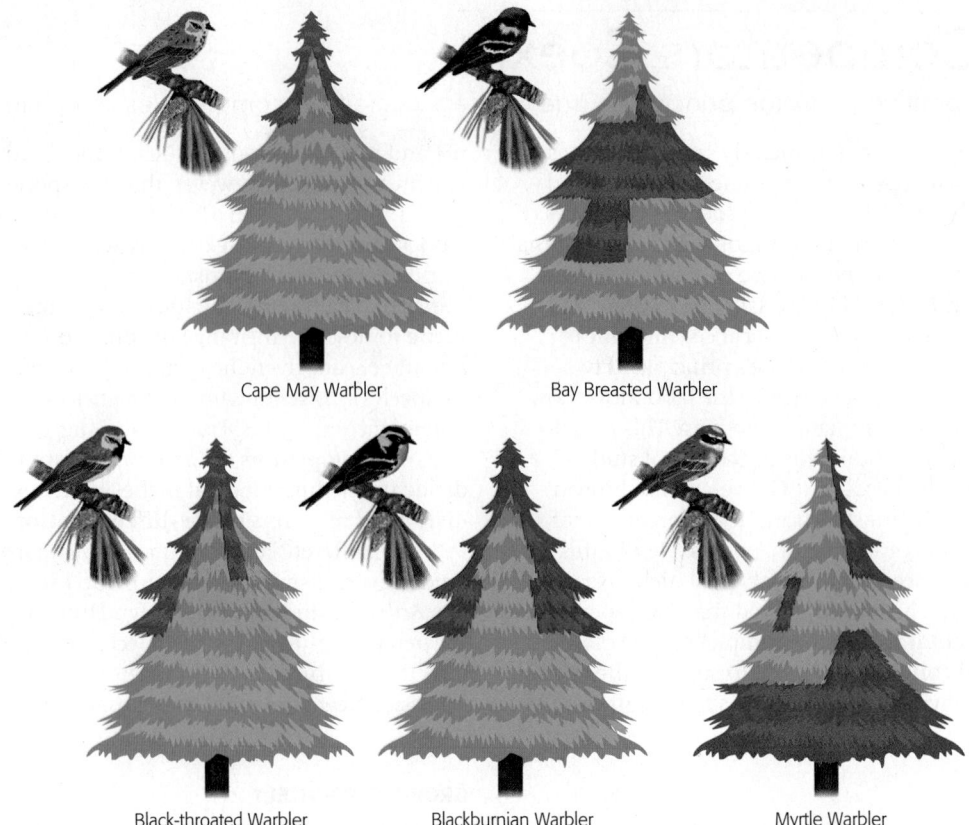

Cape May Warbler   Bay Breasted Warbler

Black-throated Warbler   Blackburnian Warbler   Myrtle Warbler

▲ Figure 6.5 **Coexistence in Competing Warblers**
Although they feed on the same insect larvae, the diverse species of warblers in the spruce and fir forests of New England coexist because they feed in different portions of the forest canopy.

Source: MacArthur, R.H. 1958. Population ecology of some warblers of northeastern coniferous forests. *Ecology* 39: 599–619.

▼ Figure 6.6 **Niche Differentiation Among Prairie Plants**

Many plant species coexist in grassland ecosystems, such as the tallgrass prairie of North America, because they have very different niches below ground.

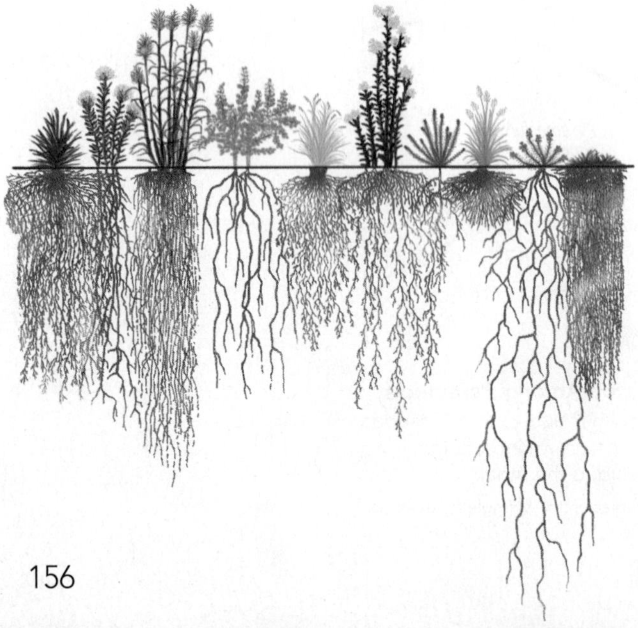

primary food of all the warblers is the caterpillar, especially spruce budworms. MacArthur's careful studies of the birds' feeding behavior revealed that each species competes most effectively in a different part of the forest canopy, and that is where each species can be found (**Figure 6.5**).

The diverse grasses and herbs that grow in native prairies provide another example of niche differentiation. Above ground, these plants appear to be vying for the same space and resources. However, a careful mapping of root systems reveals that different species are adapted to exploiting different portions of the soil (**Figure 6.6**). In addition, some species compete most effectively when growing in bright light, whereas others compete effectively when growing in the shade of taller plants.

Some of Gause's experiments support Hutchinson's niche differentiation hypothesis. Under any specific set of conditions—the same temperature, water availability, food source, and so on—Gause's principle holds true. But if conditions change, competition among species may produce different winners and losers. For example, if waste products are periodically removed from the growth medium, the outcome of the competition between *P. aurelia* and *P. caudatum* is reversed and *P. caudatum* wins. Thus, in a complex environment where waste materials are collected in some places and not in others, these two species could coexist.

Time is required for one species to competitively displace another, and the competitive exclusion principle presumes that environmental conditions remain constant during that time. In nature, however, environments change from season to season and from year to year, so conditions that are favorable to a particular species may not persist. Environments that are constantly changing may allow competing species to coexist.

In estuaries, where rivers flow into the sea, tides and storms are constantly changing the temperature, salinity, and oxygen content of the water. These changing conditions produce rapid changes in the populations of various species of plankton. Particular species become abundant, but soon, changing conditions favor other species.

▲ Figure 6.7 **Exploitation Competition**
Desert shrubs and cacti exploit water and nutrients and thereby prevent other
shrubs from establishing in their vicinity. This often produces regular spacing
among desert plants.

Exploitation and Interference

■ Mechanisms of competition vary.

Organisms compete with one another by two general
mechanisms—exploitation and interference. Imagine
that two diners are served a rather meager meal. Both
are very hungry, and there is not enough food to satisfy
both of them. One diner might outcompete the other
by simply eating more quickly. This is **exploitation
competition**. Alternatively, one diner might get more
food by threatening to stab his competitor with a fork,
thus scaring him from the table. This is **interference
competition**.

In exploitation competition, successful competitors
are able to take up or utilize resources more efficiently.
Therefore, the growth rate of their population per unit
of limiting resource is higher. In deserts, plants compete
fiercely for water. Individual plants deplete the supply of
water near their roots, limiting the ability of other plants to
grow nearby. The very regular spacing of desert shrubs is a
consequence of exploitation competition (Figure 6.7).

Interference competition is common among animals
that fend off would-be competitors with aggressive or
territorial behavior. Such behavior is common among
the various scavengers that compete for the remains of
animals killed by predators (Figure 6.8).

Interference competition is also common among plants
and microbes. These organisms compete by releasing
chemicals that slow the growth of their competitors. For
example, the leaves of garlic mustard release a chemical
that inhibits the development of associations between
fungi and the roots of trees and other herbs, thereby
slowing the growth of the other plants. Garlic mustard

was brought to North America as a culinary herb in about
1860. Soon it escaped into eastern forests, where its
populations grew rapidly. In many places, it is now the
dominant herb, excluding other plant species. Through
interference competition, garlic mustard has significantly
decreased the diversity of the ecological community where
it is now abundant.

▼ Figure 6.8 **Interference Competition**
Scavengers such as hyenas, jackals, and vultures demonstrate
interference competition by aggressively competing with one
another for remains of an animal recently killed by lions.

QUESTIONS 6.1

1. How does competition
affect the growth rates of
competing species?

2. Explain how species that
depend on some of the
same resources may
coexist.

3. What is the difference
between exploitation and
interference competition?

(MES) For additional review, go to
MasteringEnvironmentalScience

157

6.2 Herbivory, Predation, and Parasitism

BIG IDEA A key question about any ecological community is *who is eating whom*? As seen in the example of the old-growth forests of the Pacific Northwest, underlying the complex interactions of organisms in an ecosystem is the need for any organism to get the energy and material it needs to grow and reproduce. In fact, feeding relationships are important in determining the abundance of most organisms. **Consumers** feed on other live organisms. **Herbivores** eat plants, and **predators** hunt and kill animals. **Parasites** live in or on other plants or animals but usually do not kill them directly. The fitness of consumers depends on adaptations that allow them to acquire and utilize their food most effectively. Because consumers directly influence the fitness of the organisms they eat, they exert strong selective pressure on their prey (see Module 4.5).

▼ Figure 6.9 Herbivores Within Herbivores
The complex stomachs of ungulate herbivores harbor a diverse ecosystem of microbes such as this ciliated protist *Ophryoscolex* that can break cellulose down to simple sugars.

Herbivores

■ Herbivores feed on plant materials.

Carbohydrates and all other organic food molecules are ultimately produced by photosynthetic plants and algae, which capture energy from the sun and convert it into chemical energy. Herbivores play a crucial role in transferring the energy stored in those molecules to the rest of the ecosystem (see Module 4.1).

Herbivores use several different feeding strategies to exploit plants. This exploitation often provides benefits to plants, too. Fruit- and seed-eating herbivores often play an important role in dispersing seeds. A diverse array of herbivores, particularly insects, feeds on pollen and nectar. In so doing, they obtain energy for themselves while also facilitating sexual reproduction and ensuring genetic diversity in the plants they pollinate (see Module 4.6).

Many herbivores are grazers, feeding directly on the leaves and young stems of plants. In most cases, grazing animals do not kill the plants, but they certainly do slow their growth. Grazing animals face significant feeding challenges, not the least of which is related to the basic chemistry of the plants upon which they feed. The leaves and stems of plants are supported by cellulose,

a polymer of the sugar glucose (see Module 3.2). The chemical bonds that link glucose molecules together in cellulose are very resistant to the digestive enzymes of animals. Ungulates—cloven-hoofed herbivores such as cattle and goats—are able to digest cellulose because of the activity of microorganisms that live in their multichambered stomachs (Figure 6.9). These microbes produce an enzyme that can break down cellulose. Many other grazing animals, including horses, rodents, and insects, also depend on microbes that live in specialized areas of their digestive tracts.

Plants have evolved a variety of defenses against herbivores, including thorns, irritating hairs, and distasteful or toxic chemicals. Natural selection has favored the evolution of herbivores with adaptations that allow them to thwart, or even take advantage of, these plant defenses. This tit-for-tat evolution of prey and the species that eat them is called **coevolution**.

Coevolution can produce some very surprising results, as in the case of milkweed plants and monarch butterflies. Most species of milkweed produce and store large amounts of chemicals called alkaloids, which are toxic to most animals. Nonetheless, the caterpillars of monarch butterflies feed almost exclusively on milkweed plants (Figure 6.10). Not only are these caterpillars able to tolerate the alkaloids, but they also actually store these chemicals in their own tissues. The alkaloids make the monarchs toxic to birds and other predators that otherwise might eat them.

Given the wide diversity of herbivores, ecologists have wondered why the world is so green. Most ecosystems seem to have abundant plant resources that are not being eaten by herbivores. Although defense mechanisms in plants may play a role, it appears that populations of herbivores are often limited by predators, the things that eat them, rather than by the abundance of the things they eat.

◄ Figure 6.10 Milkweed–Monarch Butterfly Coevolution
The milkweed and the caterpillar that becomes a monarch butterfly present an example of coevolution. Just as the milkweed protects itself from most predators by producing toxic alkaloids, the monarch has adapted so it is able to feed on milkweed in spite of these toxins. By pollinating its flowers, monarchs provide benefit to the milkweed.

Predators

■ Predators hunt, kill, and consume their prey.

Predation is the process by which animals capture, kill, and consume animals of another species, their prey. Predators employ two basic feeding strategies— filter feeding and hunting.

Filter-feeding predators use webs or netlike structures to catch their prey. Spiders use webs to "filter" organisms from their environment. As marine shrimp swim through the water, they trap small organisms in the hairlike setae on their legs and transfer them to their mouths. Blue whales, Earth's largest living animals, are filter feeders (Figure 6.11). Blue whales and their baleen whale relatives use their comblike "teeth" to filter plankton from the water.

Hunting predators actively stalk and capture their prey. Natural selection has favored predators with keen senses of sight and smell, as well as structures such as talons, claws, and sharp teeth that allow predators to seize and kill their prey. Coevolution has played a significant role in predator–prey relationships, and prey organisms have evolved a variety of defenses against capture (Figure 6.12).

Populations of predators are often limited by the availability of prey. Likewise, populations of prey are often limited by their predators. Where predators are dependent on a single species of prey, interactions between the two species may result in synchronized cycles of population growth and decline. This was demonstrated in a classic study of numbers of snowshoe hares and their predator, the Canada lynx (Figure 6.13).

Most predators feed on a number of different prey species. When a particular prey species is abundant, predators focus their attention on it. As predation depletes that species' numbers, the predators switch their attention to other, more abundant species. **Prey switching** provides predators with a steady supply of food; it also ensures that none of the prey species will be totally eliminated.

From the standpoint of an individual prey animal, the consequences of predation are far from positive. However, predation often has beneficial effects on the health of prey populations as a whole. Predators usually do not select their prey at random. Rather, they pick out the youngest, oldest, or sickest individuals. As a consequence, predation can increase the overall health of the prey population.

In many regions of the eastern United States, populations of white-tailed deer have surged, partly due to the absence of natural predators. The incidence of chronic wasting disease, a fatal neurological disease of deer, has also surged. Where predators are present, deer with chronic wasting disease are more likely to be preyed upon than healthy deer. Thus, the incidence and spread of chronic wasting disease is diminished by predation.

▼ Figure 6.11 **Giant Filter Feeders**
Baleen whales, like this blue whale, fill their expandable mouths with plankton-rich water, then keep the plankton in their mouth while forcing the water back out through their comblike upper teeth.

◄ Figure 6.12 **Defending Against Predation**
Natural selection has produced adaptations among prey organisms that allow them a better chance of escaping their predators. Such adaptations include body structures that imitate nonfood parts of the environment, like those of this walking stick.

▲ Figure 6.13 **Predator–Prey Population Cycles**
Numbers of snowshoe hares and Canada lynx trapped by the Hudson Bay Company reveal changes in their populations between 1840 and 1940. As snowshoe hare population size grows, predation by the Canada lynx increases and hare populations then decline. This decline in hare population size may subsequently cause a decline in the lynx population, although this appears to depend on the availability of other prey.

Data from: MacLulich, D. A. 1937. Fluctuations in numbers of the varying hare (*Lepus americanus*). *Univ. Toronto Studies, Biol. Ser. no. 43*: 1–136.

Parasites

■ Parasites depend on living hosts for nourishment.

Parasites live and feed in or on other organisms, their hosts. Usually, a parasite does not kill its host, but it does harm the host and may contribute to its eventual death. Many parasites, such as viruses and tapeworms, live inside their hosts. Other parasites, such as ticks and leeches, are external.

Some parasites have relatively simple life cycles. For example, the common human cold virus is a parasite that multiplies in host cells and is easily transmitted to the next host in droplets of saliva from a sneeze. Other parasites, such as the single-celled protist that causes malaria, have very complex life cycles that require multiple hosts and vectors. **Vectors** are organisms that carry the parasite but are unaffected by it.

Parasites multiply within individual hosts. However, scientists measure the population growth of parasites in terms of their spread, or the spread of the disease they cause, in the host population. The spread of parasitic disease is governed by the following four factors.

Abundance of hosts. As the number of individual hosts increases, the likelihood that a parasite will be transmitted also increases. For example, the devastating outbreaks of bubonic plague in Europe during the 14th century were associated with the movement of a large number of people to cities (Figure 6.14). The spread of the plague was enhanced by the large numbers of another host for the plague bacteria, the rats that lived in the cities.

Accessibility of hosts. The ability of parasites to locate and infect hosts is influenced by the diversity of the ecological community in which they live. In the southeastern United States, for example, the parasitic southern pine bark beetle lays its eggs beneath the bark of pine trees. Its larvae then feed on the trees' living tissues, killing the trees in less than a year. Adult beetles locate new host trees by sensing aromatic chemicals given off by the pine. In plantations where only pines are grown, infestations of pine bark beetles spread very quickly (Figure 6.15). In forests with many species of trees, it is more difficult for beetles to locate new hosts, so the infestation spreads from pine to pine much more slowly.

Transmission rate of parasites. All parasites must have adaptations that allow them to move from one host to another, but rates of transmission vary. For example, two fungal diseases that have had significant effects on the forests of eastern North America—Dutch elm disease and chestnut blight—have different rates of transmission. Both of these fungi were accidentally introduced to U.S. forests on non-native nursery trees imported from Asia. (Dutch elm disease was first identified in the Netherlands, where trees were also accidentally infected by fungi from Asia; hence its name.)

The fungus that causes chestnut blight releases large numbers of wind-dispersed spores, which move from tree to tree very quickly. The blight was first introduced in 1890, and by 1930 virtually all of the adult American chestnut trees in eastern North America were dead. In contrast, the spores of the fungus that causes Dutch elm disease are transmitted by an insect parasite, the elm bark beetle. Although the Dutch elm fungus kills trees very quickly, its insect vector moves from tree to tree comparatively slowly. The disease was first observed in North America in 1928. Although it has killed many trees, it has spread much more slowly than the chestnut blight. Thus far, many elm trees have escaped infection.

▼ Figure 6.14 **Disease Spread and Host Density** Rats harboring plague-infested fleas were abundant in medieval towns and villages, like that depicted in this Bruegel painting. These parasites were easily transferred to dense populations of people.

New **FRONTIERS**

Resurrecting Species

All that remains of once vast populations of American chestnut are occasional sprouts. Only a very small number of these sprouts will survive long enough to produce flowers and seeds. Now, a group of scientists from the American Chestnut Foundation hopes to restore American chestnut to forests in the eastern United States by taking advantage of what we know about genetics and biotechnology to develop a disease resistant version of the American chestnut.

By an arduous process of pollinating flowers of American chestnut sprouts with pollen from the resistant Asian chestnut, these scientists were able to "back-cross" through six generations to produce a population of a few thousand resistant seedlings that they call restoration chestnuts. In a carefully monitored program, the U.S. Forest Service recently began planting small numbers of restoration chestnuts in the wild forests of North Carolina, Virginia, and Tennessee. Some scientists argue that because the restoration chestnut is not the same as the native species and forests have changed in their absence, this reintroduction program should be halted. Are we "playing God" with the American chestnut? How would you frame the ethical questions we would need to consider to regulate such activity? How far is too far in resurrecting a species?

Length of life of an infected host.   The longer an individual host lives with a parasite, the longer it can transmit it. A host that lives a long time provides a great number of opportunities for its parasite to be transmitted to other hosts. Highly virulent parasites that quickly immobilize or kill their hosts limit the opportunities for the parasites to reproduce and spread. Thus, the negative effects of parasites on hosts with which they have coevolved over long periods are generally considerably less than those in new host–parasite interactions.

For example, the chestnut blight fungus is found in nearly all Asian chestnut trees but has very little impact on their growth or reproduction. It is likely that the fungus coevolved with Asian chestnut trees for tens of thousands of years. During that time, Asian chestnut trees developed adaptations that reduced the effects of the fungus. It is also likely that the fungus evolved in ways that prolonged the life of its host. However, when the chestnut blight was accidentally introduced into North America, it came into contact with chestnut trees that had not evolved resistance to it. As a result, the blight rapidly killed almost all of the American chestnut trees.

Clearly, what an organism eats and what it is eaten by affect its ability to grow and reproduce. Herbivory, predation, and parasitism significantly affect the relative abundance and diversity of species in an ecological community. Coevolution produces balance and stability among populations of the eaters and the eaten.

QUESTIONS 6.2

1. Describe two factors that may prevent herbivores from devouring all of the plants upon which they feed.

2. What is prey switching?

3. What four factors influence the spread of a parasitic disease?

(MES) For additional review, go to **MasteringEnvironmentalScience**

▲ Figure 6.15 **Disease Spread and Community Diversity**
Where pines are grown in single-species plantations Ⓐ, parasitic pine bark beetles Ⓑ can spread rapidly from host tree to host tree. However, in complex communities where pines grow interspersed with oaks Ⓒ, beetles have more difficulty locating their host.

6.3 Mutualism and Commensalism

BIG IDEA Intimate interdependencies between species are called **symbioses**. The relationship between a parasite and its host is one form of symbiosis. A symbiotic relationship between two species in which both benefit is called **mutualism**. Mutualistic species are better able to secure resources by living together than by living separately. The symbiosis between truffle fungi and the roots of trees is an example of a mutualism. In addition to the benefits they provide to the mutualistic pair, mutualisms may provide benefits to the larger ecological community. **Commensalism** refers to associations that benefit only one species, leaving the other species unaffected.

Mutualisms and Commensalisms

■ Mutualistic relationships benefit both interacting species; commensal relationships benefit only one species.

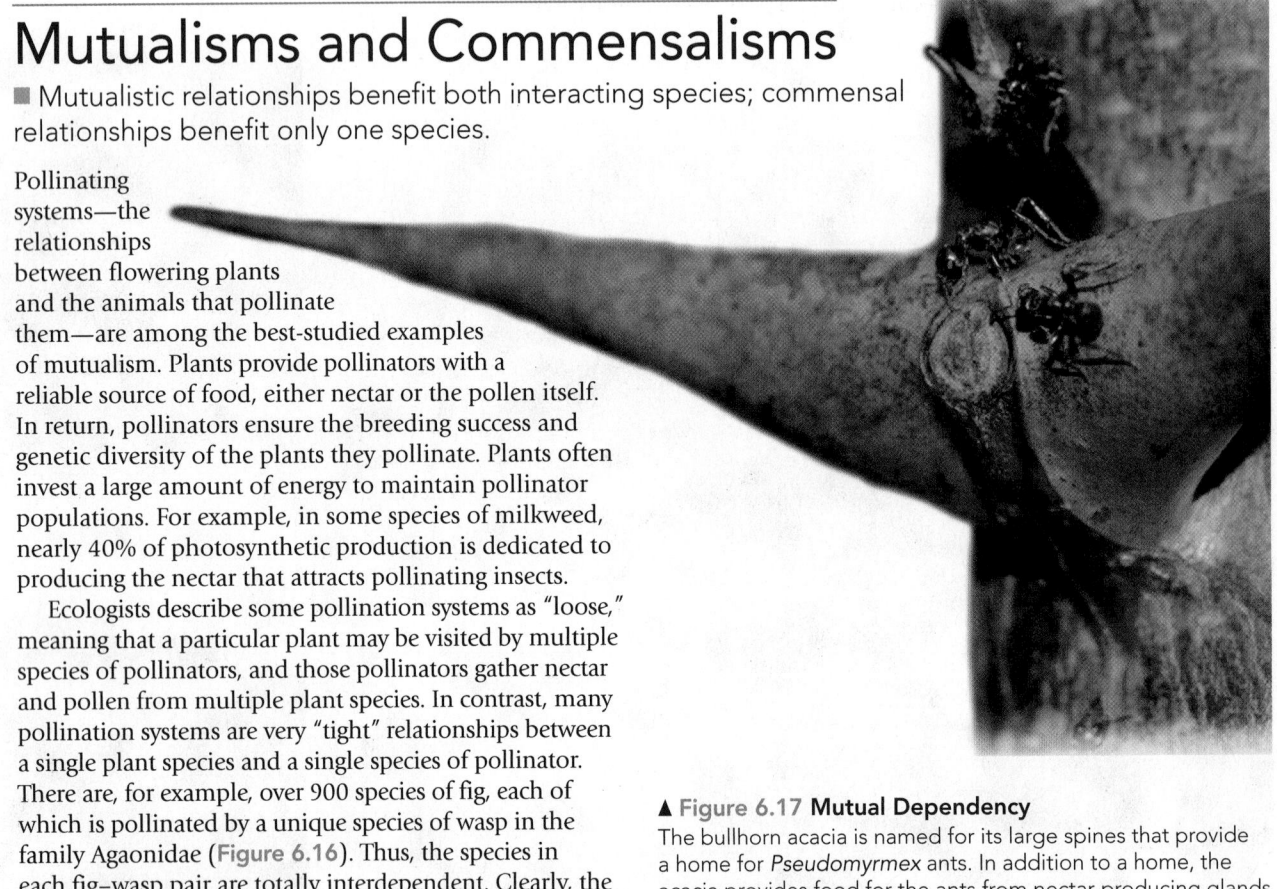

Pollinating systems—the relationships between flowering plants and the animals that pollinate them—are among the best-studied examples of mutualism. Plants provide pollinators with a reliable source of food, either nectar or the pollen itself. In return, pollinators ensure the breeding success and genetic diversity of the plants they pollinate. Plants often invest a large amount of energy to maintain pollinator populations. For example, in some species of milkweed, nearly 40% of photosynthetic production is dedicated to producing the nectar that attracts pollinating insects.

Ecologists describe some pollination systems as "loose," meaning that a particular plant may be visited by multiple species of pollinators, and those pollinators gather nectar and pollen from multiple plant species. In contrast, many pollination systems are very "tight" relationships between a single plant species and a single species of pollinator. There are, for example, over 900 species of fig, each of which is pollinated by a unique species of wasp in the family Agaonidae (Figure 6.16). Thus, the species in each fig–wasp pair are totally interdependent. Clearly, the evolution of species within the fig genus has been tightly coupled to the evolution of agaonid wasps.

▲ Figure 6.17 **Mutual Dependency**
The bullhorn acacia is named for its large spines that provide a home for *Pseudomyrmex* ants. In addition to a home, the acacia provides food for the ants from nectar-producing glands and high-protein nodules. The ant aggressively protects the acacia from other herbivores.

▶ Figure 6.16 **Figs and Wasps**
Each species of fig depends on a unique species of tiny wasp to transfer pollen from male to female flowers. In return, the fig fruit provides habitat for wasp larvae to grow and mature. Here, mature wasps emerge from a sliced-open fig.

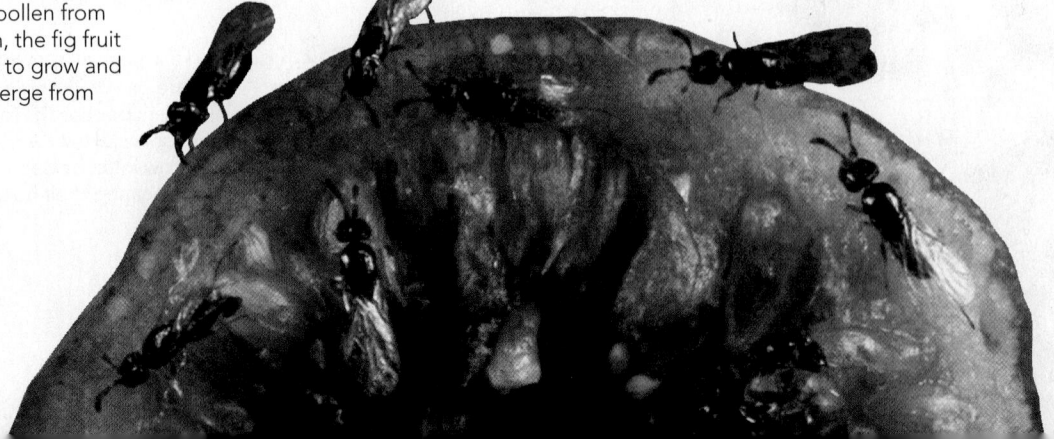

In some mutualisms, one partner provides food to the other partner in exchange for protection from predators or herbivores. The interdependency between acacia ants and the bullhorn acacia trees that grow in the dry tropical forests of Central America is one such mutualism (Figure 6.17). The bullhorn acacia gets its name from its large hollow thorns, in which acacia ants make their nests. In addition to providing a home for the ants, the acacia also provides them with two sources of food—protein-rich nodules that grow on the tips of its leaves and carbohydrate-rich nectar that is secreted from the base of its leaves. For their part, the very aggressive acacia ants ward off the many insect herbivores common in these forests. When acacia ants were experimentally removed from acacias, other insect herbivores increased by nearly 800%, and the acacias were quickly defoliated.

Scientists hypothesize that many mutualistic relationships have developed as a consequence of coevolution between predators or parasites and their hosts. In this scenario, natural selection would favor changes in the behavior of parasites or predators that help them protect their hosts from competing parasites or predators. Natural selection would also favor traits that enhance the host's relationship with its former predator or parasite. Such traits might include the protein-rich nodules and nectar that are produced by the bullhorn acacia.

The effects of mutualism often extend to the whole ecological community. For example, fungus–plant root associations and the interdependency between many herbivores and the cellulose-digesting microbes in their guts are critical to the functioning of entire food webs.

Mutualism plays a particularly important role in the global cycling of nitrogen. Nearly 80% of Earth's atmosphere is nitrogen gas (N_2). Although plants and animals require nitrogen, they are not able to utilize, or fix, atmospheric N_2. Only certain bacteria and cyanobacteria are able to fix nitrogen. Some nitrogen-fixing bacteria form very tight mutualisms with the roots of some plants. In these relationships, the plant provides carbohydrates to the bacteria, and the bacteria provide nitrogen to the plant. Symbiotic nitrogen fixation occurs in many groups of plants but is particularly common among members of the legume family.

In commensalism, one member of a pair of interacting species benefits and the other is unaffected. For example, hermit crabs protect their soft bodies by occupying the empty shells of marine snails (Figure 6.18). Since the snails have long been dead, they are not hurt by the crabs, nor do they derive any benefit from this relationship. Similarly, the branches of tropical trees provide support for a diverse array of epiphytes, plants such as orchids, bromeliads, and cacti that grow on the trees but get their nutrients from the air, rain, and forest debris. The trees are not harmed by this association, nor do they appear to receive any benefit.

Mutualistic and commensal relationships are key to the success of many species in ecological communities. These relationships are the result of coevolution; many probably arose from what were once predator–prey or parasite–host interactions.

QUESTIONS 6.3

1. Differentiate between mutualism and commensalism.

2. Describe the role of mutualism in the cycling of nitrogen in ecosystems.

(MES) For additional review, go to **MasteringEnvironmentalScience**

▶ Figure 6.18 **Commensal Hermit Crab**
In this particular commensal relationship, the hermit crab inhabits the shell of a snail species, in this case a triton. The crab thus depends on snails for their shells, but the snails are unaffected by this relationship.

6.4 The Flow of Energy in Ecological Communities

BIG IDEA Much of the diversity of ecological communities relates to the patterns of **energy flow**—the transfer and transformation of high-energy organic molecules—within them. These patterns are determined by who is feeding on whom, as shown by the relationship of the truffle fungus to the flying squirrel and the fly squirrel to the spotted owl. To understand the flow of energy in communities, ecologists classify organisms by feeding levels, or **trophic levels**, based on their source of food. The feeding interactions among species in a community are best described as a **food web**. The stability or sustainability of a food web may be very sensitive to the abundance of one or a few particular species.

Food Chains

■ The flow of energy through ecological communities is often depicted in simplified food webs or food chains.

► **Figure 6.19 Energy Flow in a Grassland Ecosystem**
Ecologists organize ecological communities by feeding rank or trophic level. The arrows indicate the flow of energy among trophic levels. Dead organic matter and waste are consumed by a diverse array of decomposers.

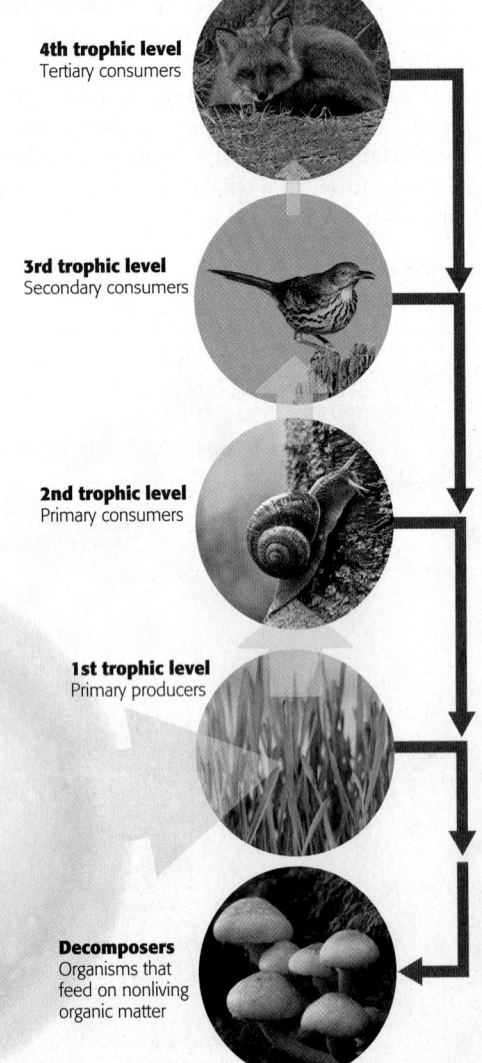

4th trophic level
Tertiary consumers

3rd trophic level
Secondary consumers

2nd trophic level
Primary consumers

1st trophic level
Primary producers

Energy from the sun

Decomposers
Organisms that feed on nonliving organic matter

Feeding relationships among organisms are often depicted as simplified food webs called **food chains**. In a food chain, one species or trophic level is eaten by another, which is eaten by yet another, and so on (Figure 6.19).

In the first, or lowest, trophic level are the primary producers. **Primary producers** transform energy from sunlight or certain inorganic chemicals into high-energy carbohydrates. Green plants, algae, and cyanobacteria, as well as chemosynthetic bacteria found in soils and deep-sea hydrothermal vents, are primary producers (see Module 4.1). Primary producers are the ultimate source of the chemical energy used by all members of ecological communities.

Above the first trophic level are those made up of consumers—organisms that feed on other living organisms. The second trophic level is composed of herbivores, such as grazing mammals and insects; animals that feed on primary producers are also called **primary consumers**. Carnivores such as lions or insect-eating birds that feed directly on herbivores make up the third trophic level; they are called secondary consumers. Tertiary consumers—organisms that feed on secondary consumers—form the fourth trophic level. You might imagine such a feeding chain extending indefinitely. But in fact, communities with more than four trophic levels are quite rare, for reasons that will soon become clear.

Much energy is consumed by **decomposers**, organisms that feed on nonliving organic matter. Scavengers, such as vultures, jackals, and hyenas, feed on the dead bodies of other animals, often the remains left by predators. Termites, wood roaches, and a great variety of other organisms feed on plant litter, such as fallen leaves and dead wood. Microorganisms such as bacteria and fungi obtain their energy from very small bits of organic matter.

Energy and Biomass Pyramids

■ Available energy diminishes at each successive trophic level.

Organisms at each trophic level transform what they eat into **biomass energy**, the food that can be consumed by higher trophic levels. *Biomass* refers to all matter that is derived from living or once-living organisms. To organisms at a particular trophic level, the creatures that make up the trophic level upon which they feed might be considered "machines" that convert inaccessible energy into food energy. Thus, a second-level herbivore transforms leaves and stems into flesh that can be eaten by a third-level carnivore. However, most of the energy that is consumed by any organism is burned in respiration or lost as waste. Only a small fraction becomes biomass. (Imagine how large you would become if you converted all of your food into body mass, pound for pound!)

The fraction of energy that the organisms in one trophic level make available to the next trophic level is called **trophic level efficiency** (Figure 6.20). As a rule of thumb, trophic level efficiencies are assumed to be about 10%. For example, herbivores convert one-tenth of the energy available to them from plants into herbivore biomass. However, trophic level efficiencies vary a great deal and are often much lower than 10%. For example, warm-blooded carnivores use a large proportion of the energy from the food they consume to maintain their body temperature and to stalk and kill their prey. Thus, the trophic level efficiency for tertiary consumers is often only 2–3%.

The total energy stored by organisms in an ecosystem usually decreases significantly from one trophic level to the next. In other words, the amount of energy available to each trophic level is considerably less than the energy available to the trophic level beneath it. Both the numbers and biomass of organisms generally decrease from one trophic level to the next. This produces a pyramid of trophic level energy (Figure 6.21).

It is logical that the biomass at each trophic level would form such a pyramid. Throughout its life, an herbivore must consume many times its weight in plant matter, and a carnivore must consume many times its weight in prey. Food becomes increasingly scarce at each successive trophic level, and organisms that feed at higher trophic levels must expend more energy per unit of biomass to obtain their food. In most ecosystems, there is not enough energy available to support the activities of consumers beyond the third or fourth trophic level.

An exception to the biomass pyramid pattern is found in many lakes and parts of the ocean. There the biomass of the primary producers—planktonic algae, or phytoplankton—may be less than the biomass of the animals that feed on them. This occurs because phytoplankton carry out photosynthesis, grow, and multiply very rapidly but are constantly being grazed by herbivorous zooplankton. This has important consequences for the cycling of carbon in aquatic and marine ecosystems (see Module 6.5).

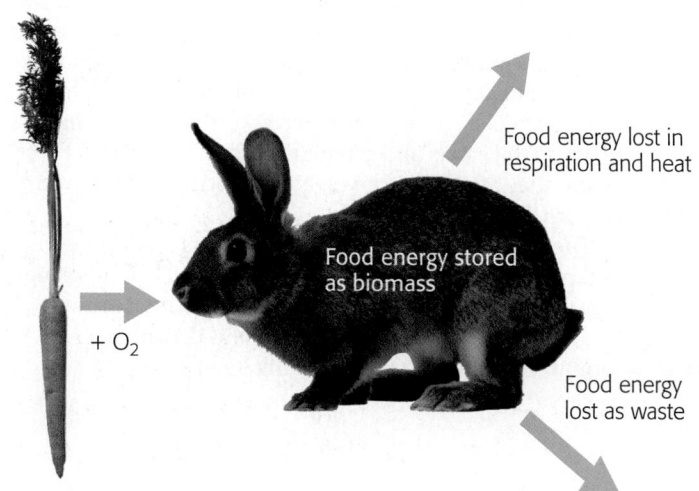

Food energy lost in respiration and heat

Food energy stored as biomass

+ O₂

Food energy lost as waste

▲ Figure 6.20 **An Organism's Energy Budget**
Of the food consumed by an organism, generally less than 10% is converted to organism tissue and is available to the organisms that feed on it. The remainder is respired or passed as waste.

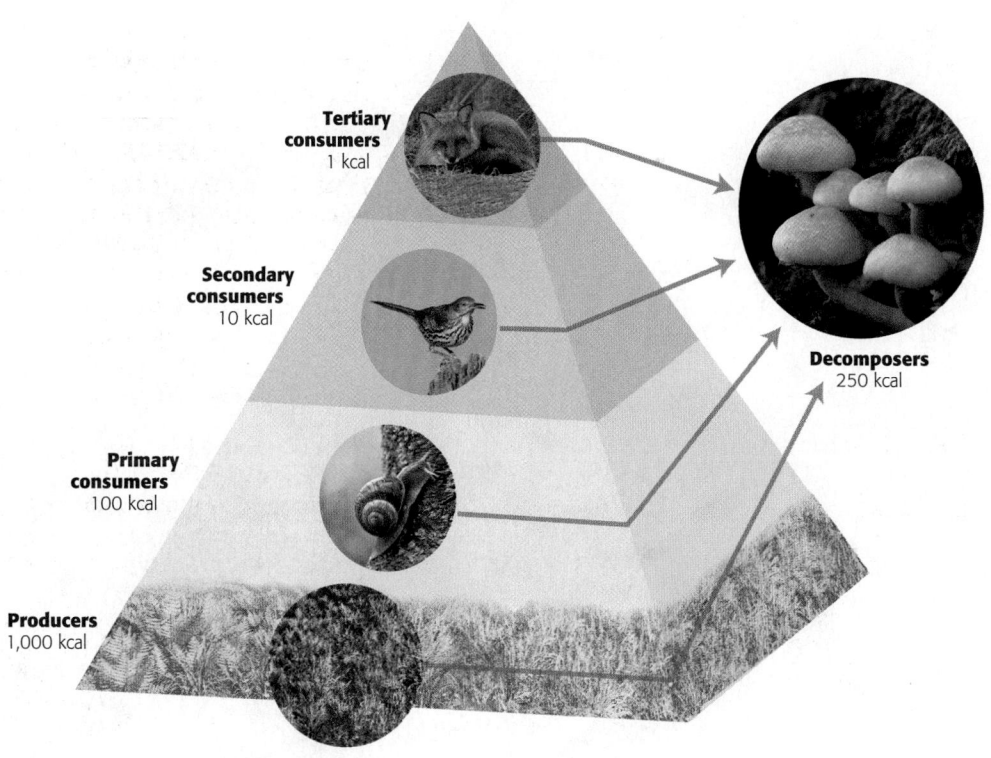

Tertiary consumers
1 kcal

Secondary consumers
10 kcal

Primary consumers
100 kcal

Producers
1,000 kcal

Decomposers
250 kcal

▲ Figure 6.21 **The Pyramid of Energy**
In general, the biomass of organisms and the total amount of energy they store decrease from one trophic level to the next, resulting in pyramids of biomass and energy. These numbers are typical for a grassland ecosystem.

Food Web and Species Diversity

■ Food web complexity supports the stability of ecological communities.

The hierarchy of trophic levels is often described as a food chain. In most ecosystems, however, feeding relationships are more complex. Each trophic level usually includes many species; predators often feed on more than a single prey organism; and many species obtain their food from more than one trophic level. Because the connections between the eaters and eaten rarely form a simple chain, they are more accurately described as a food web (Figure 6.22).

The linkages among primary producers, herbivores, and predators form a consumer food web. Decomposing organisms are often eaten by other organisms, forming decomposer food webs. Consumer and decomposer food webs are nearly always highly interconnected

(Figure 6.23). For example, earthworms are decomposers that feed on decaying organic matter in the soil that comes from many different microorganisms. Earthworms are often eaten by birds. These birds are, in turn, eaten by predators such as hawks. Waste organic matter from these consumers eventually reenters the decomposer food web.

The stability of a community food web is directly related to the diversity of its species and the complexity of their connections. In a relatively simple ecosystem with few interconnections, the loss of a single species may be very disruptive. But if a species disappears from a complex food web, many alternative pathways for energy transformation remain.

Q *Why are carnivores so frequently on the list of endangered species? Is it always a result of hunting?*

Whitney, Old Dominion University

Norm: A *Because carnivore populations are small and they often have very large ranges compared to herbivores, they are more vulnerable to habitat loss, environmental change, and hunting pressures.*

▼ Figure 6.22 **Marine Food Web**
Marine food webs are often quite complex. The thickness of arrows refers to relative amounts of energy transferred among food web members.

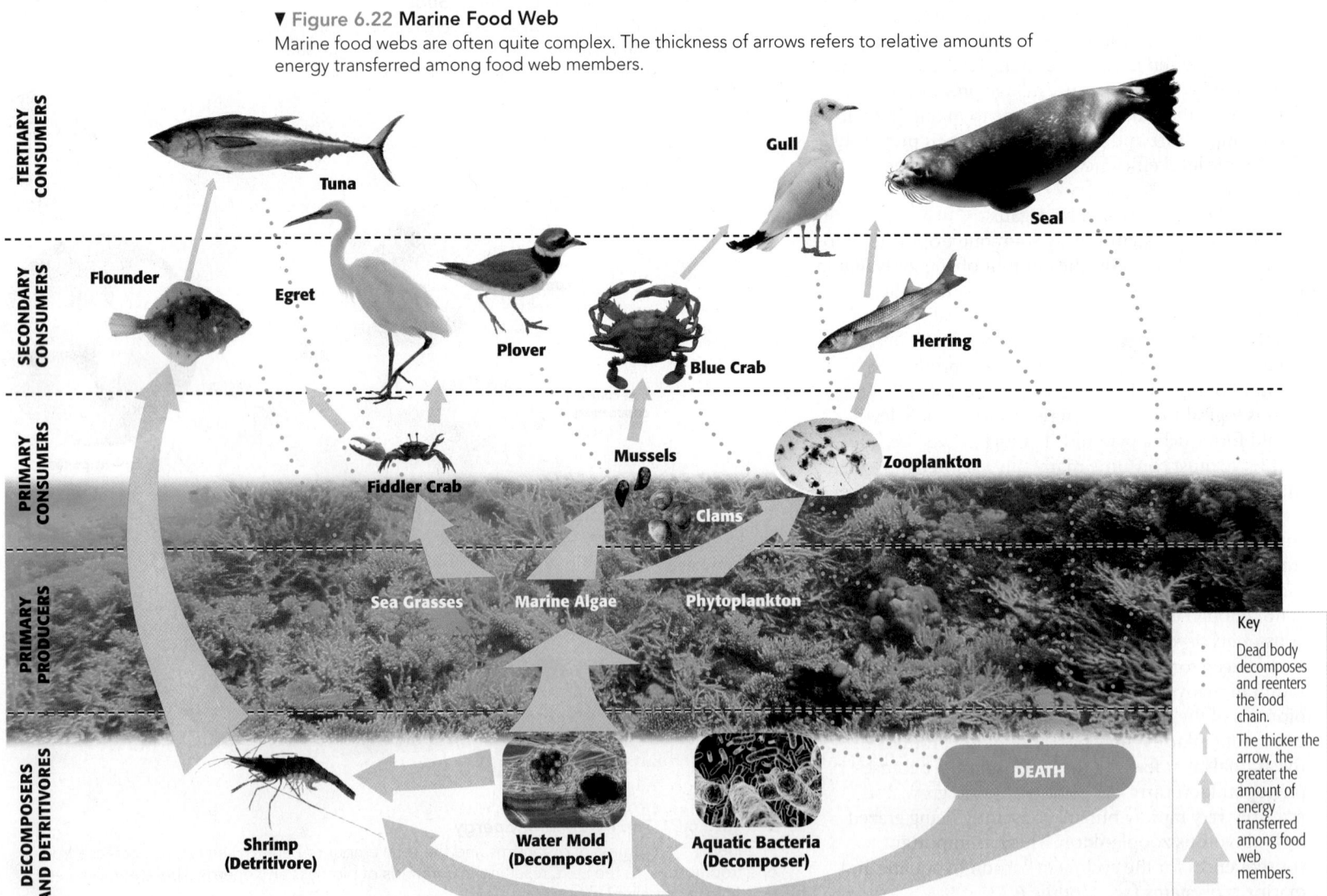

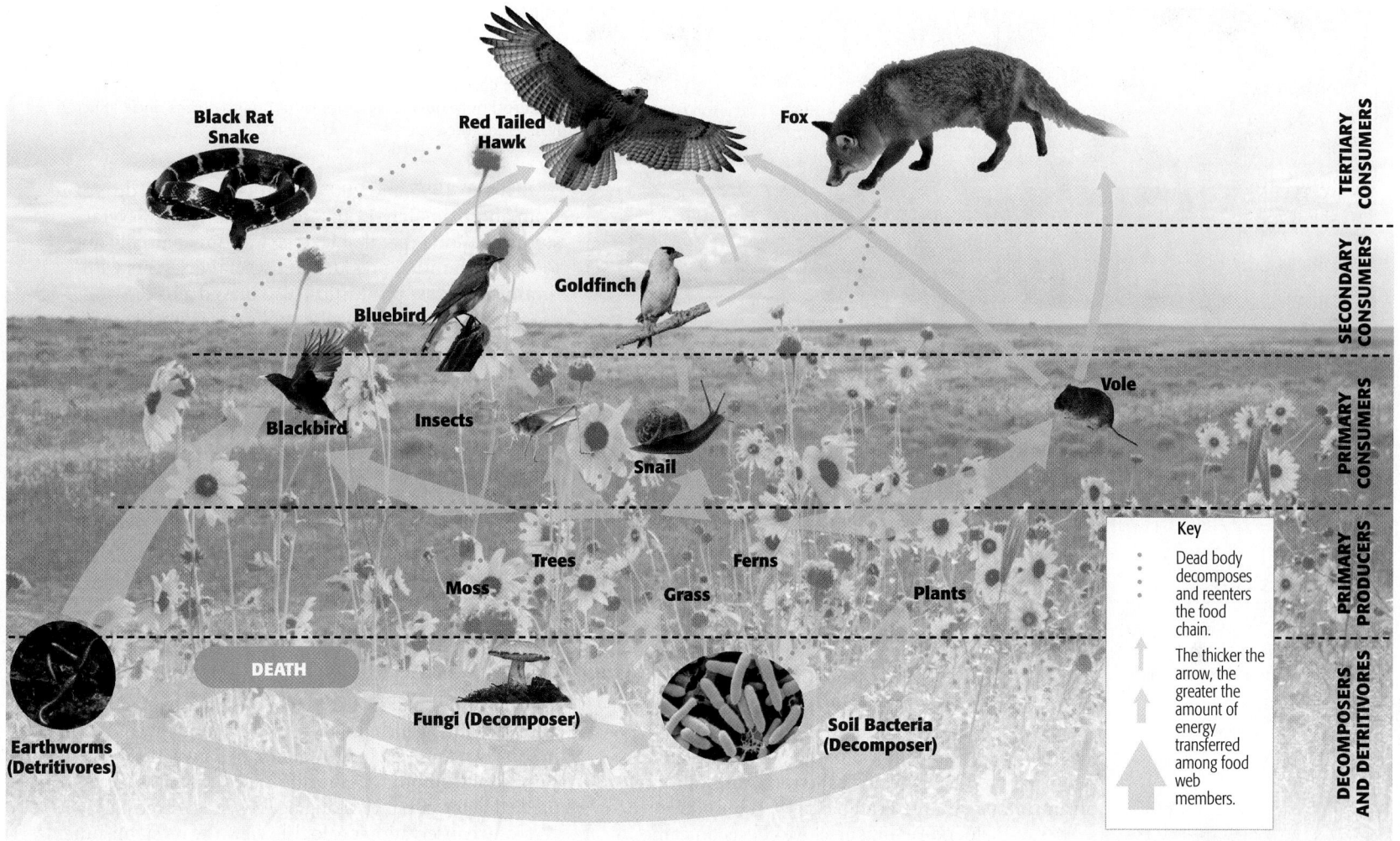

Black Rat Snake

Red Tailed Hawk

Fox

Bluebird

Goldfinch

Blackbird

Insects

Snail

Vole

Trees   Ferns

Moss   Grass   Plants

DEATH

Fungi (Decomposer)

Soil Bacteria (Decomposer)

Earthworms (Detritivores)

TERTIARY CONSUMERS

SECONDARY CONSUMERS

PRIMARY CONSUMERS

PRIMARY PRODUCERS

DECOMPOSERS AND DETRITIVORES

Key

Dead body decomposes and reenters the food chain.

The thicker the arrow, the greater the amount of energy transferred among food web members.

▲ Figure 6.23 **Terrestrial Food Web**
Because plants produce large quantities of wood and cellulose that cannot be digested by herbivores, much energy in terrestrial food webs is processed by decomposing organisms. Like birds feeding on worms, decomposers may then be consumed by predators.

Keystone Species

■ Keystone species are important in the stability of many food webs.

Keystone species, so named in reference to the wedge-shaped keystone that holds an arched structure together, play a particularly important role regarding the abundance of other species. Not surprisingly, plant species that account for a significant portion of total primary production are often keystone species.

Carnivores can also be keystone species, even though they are generally much less abundant than organisms at lower trophic levels. Carnivores often regulate the size of herbivore populations. The loss of carnivores can create a **trophic cascade**, in which the populations of herbivores increase, resulting in the overconsumption of primary producers.

Our increasing understanding of the importance of keystone carnivores is illustrated by the change in attitudes toward the management of wolves in

Yellowstone National Park over the past century. In the early 1900s, managers believed that large predators such as wolves and mountain lions were a threat to deer and elk, which were considered to be more desirable species. In 1914, managers began a program to rid Yellowstone of wolves and mountain lions. This campaign was especially effective against the wolf; the last of Yellowstone's native wolves was killed in 1924. Within a few years, the number of elk feeding on the park's North Range had more than doubled, and their numbers remained high over the succeeding decades.

Beginning in the 1980s, ecologists began to notice the absence of young trees in the many populations of quaking aspen that border the North Range. Data gathered from these populations in the 1920s revealed that aspen were then producing abundant sprouts

◀ Figure 6.24 **Yellowstone Aspen Decline**
Healthy stands of quaking aspen include trees in all different sizes and age classes. In Yellowstone National Park, where a trophic cascade occurred, many aspen stands such as the one pictured here have only older or unhealthy trees, indicating little or no recent tree reproduction.

Between 1995 and 2003, the elk population dropped from approximately 17,000 to 9,000 animals. Several studies indicate that that change was only partially due to wolves; hunting pressure and several dry years with diminished grass production also played a role. Also, the impact of wolves on elk numbers is probably due less to direct predation and more to wolves limiting the amount of range where elk can safely feed. Wolf activity on the North Range is quite variable; where such activity is highest, elk populations are lowest. Where wolves are most active, young sprouts and shoots of aspen have begun to appear in the aspen stands.

Since 2003, elk numbers dropped even further. In 2013, there were fewer than 4,000 animals. The number of wolves has declined to about 50 individuals. This is due in large part to the smaller elk population. Wildlife biologists are uncertain about whether the elk and wolf populations will come into equilibrium, or if the elk population will rebound as wolf predation diminishes. This uncertainty has been further increased by the accidental introduction of parvovirus, a parasitic virus that causes distemper in domestic dogs, which has increased the mortality of pups in some wolf packs. The appearance of parvovirus was a surprise and emphasizes the difficulties ecologists face in predicting the effects of changing ecosystem diversity.

Many conservation ecologists view the reestablishment of wolves on Yellowstone's North Range as a model for how to restore species that were once extinct in an area. However, this project is still in its early stages, and important challenges lie ahead. Nevertheless, it demonstrates how comparatively small numbers of predators can have a significant effect on the diversity of an ecological community through their effects on the flow of energy in complex food webs.

QUESTIONS 6.4

1. Why do ecological communities usually have a maximum of four trophic levels?

2. Why does food web complexity generally increase community stability?

(MES) For additional review, go to **MasteringEnvironmentalScience**

and young trees (see Figure 6.24). In the decade that followed, aspen reproduction ceased. On close examination, similar patterns of change were found in populations of cottonwood and willow. Amid much controversy, several researchers concluded that the decline in the reproduction of these trees was a consequence of overgrazing by the abnormally abundant elk.

Arguing that the removal of the wolf had been a colossal moral and ecological error, conservation organizations mounted campaigns to reintroduce wolves to the park. Fearing negative effects on game and livestock, most hunters and ranchers opposed the reintroduction of wolves. After years of negotiation, however, a plan for wolf reintroduction was approved by the U.S. Fish and Wildlife Service. In 1995 and 1996, 31 wolves were released on the North Range. The reintroduction was quite successful. By 2003, nearly 100 wolves divided among 14 packs were hunting on Yellowstone's North Range (Figure 6.25).

▼ Figure 6.25 **The Wolf–Elk–Aspen Trophic Cascade**
In Yellowstone's North Range, where wolves have become particularly common, elk activity is diminished, allowing young trees to survive and grow in the quaking aspen groves.

Trophic Cascades Across Ecosystems

How could the presence or absence of fish in ponds possibly influence the production of seeds in the plants that grow around them?

Tiffany Knight and her collaborators discovered a remarkable trophic cascade that connects the ecological communities of ponds to the terrestrial communities that surround them (**Figure 6.26**). Knight's team used natural differences among ponds in the University of Florida's Ordway-Swisher Biological Station to create an innovative *field experiment*. They selected eight ponds that were similar in size and depth, but four of the ponds contained fish and four had no fish. Fish are important predators of the insect larvae that live in these ponds. Knight and her collaborators *predicted* that the presence or absence of fish might influence insect populations. They also *hypothesized* that the species that interact with those insects, such as the plants that they pollinate, might be affected.

The aquatic larvae of dragonflies are eaten by several species of fish that live in some of the ponds. Adult dragonflies are voracious predators of other flying insects, including various flies, bees, and butterflies that pollinate plants growing around the ponds (**Figure 6.27**). A shrub called St. John's wort grows abundantly around all the ponds and depends on these insects for pollination (**Figure 6.28**). Knight hypothesized that a variation in the abundance of dragonflies might influence the abundance

▲ Figure 6.26 **Plant Population Biologist**
Tiffany Knight of Washington University, St. Louis, Missouri, is primarily interested in factors influencing the long-term persistence of plant populations.

of pollinators, thereby affecting the reproductive success of plants that depend on those pollinators, such as St. John's wort. Thorough sampling revealed that the larvae of all dragonfly species were two to five times more abundant in ponds without fish than in ponds with fish. Adult dragonflies were also much more abundant around fish-free ponds.

Knight monitored the pollinators visiting St. John's wort. She found that the visitation rates of flies, butterflies, and bees were much higher around ponds with fish and few dragonflies than around ponds with no fish and abundant dragonflies. Knight also demonstrated that seed production in St. John's wort was limited by pollen supply. Because more pollinators visited them, the flowers of shrubs growing around ponds with fish produced many more seeds than did the flowers of shrubs growing near ponds without fish.

The work of Knight and her collaborators is among the first to demonstrate that trophic cascades can operate across the boundaries of very different ecosystems, such as between ponds and the land adjacent to them. Now at Washington University, Knight is continuing to research how interactions among organisms at different trophic levels influence processes such as plant pollination.

1. Describe the important interactions that connect pond fish populations to reproduction in the St. John's wort that grows around ponds.

2. How did Knight and her colleagues use natural differences among ponds to test hypotheses about these interactions?

▲ Figure 6.27 **Dragonflies Eat Pollinating Insects**
Dragonflies feed on a variety of flying insects, including many important pollinators. Here a dragonfly devours a small butterfly.

▲ Figure 6.28 **St. John's Wort**
Seed production in the St. John's wort shrub flowers depends on pollen delivered by a variety of flies, bees, and butterflies.

6.5 The Carbon Cycle and Ecosystem Productivity

BIG IDEA The flow of energy among trophic levels in an ecosystem is directly related to the cycling of carbon. We are composed of numerous elements, but we refer to ourselves as carbon-based life forms. Carbon is the element most closely identified with life and the activities of ecosystems, and it comprises nearly 50% of all ecosystem biomass. Nevertheless, it contributes a miniscule amount (0.032%) to the total mass of Earth's crust and atmosphere; the vast majority of this carbon is found in sedimentary rock with cycling times of hundreds of millions of years. The remaining carbon—just 0.0005%—cycles much more rapidly among Earth's atmosphere and various ecosystem pools (see Module 3.5). The carbon cycle in terrestrial ecosystems differs significantly from that of ecosystems in aquatic and marine environments. Human activities such as deforestation and the burning of fossil fuels have altered carbon cycles in individual ecosystems and globally.

The Carbon Cycle

■ Photosynthesis and cellular respiration are key steps in the movement of carbon through ecosystems.

The flux of carbon between the atmosphere and terrestrial and aquatic ecosystems is tightly coupled to the rate of photosynthesis and the rate at which organic carbon is either respired or burned. Primary producers—plants, algae, and cyanobacteria—use light energy to add hydrogen ions and electrons to CO_2, producing carbohydrates and other organic molecules (see Module 4.1).

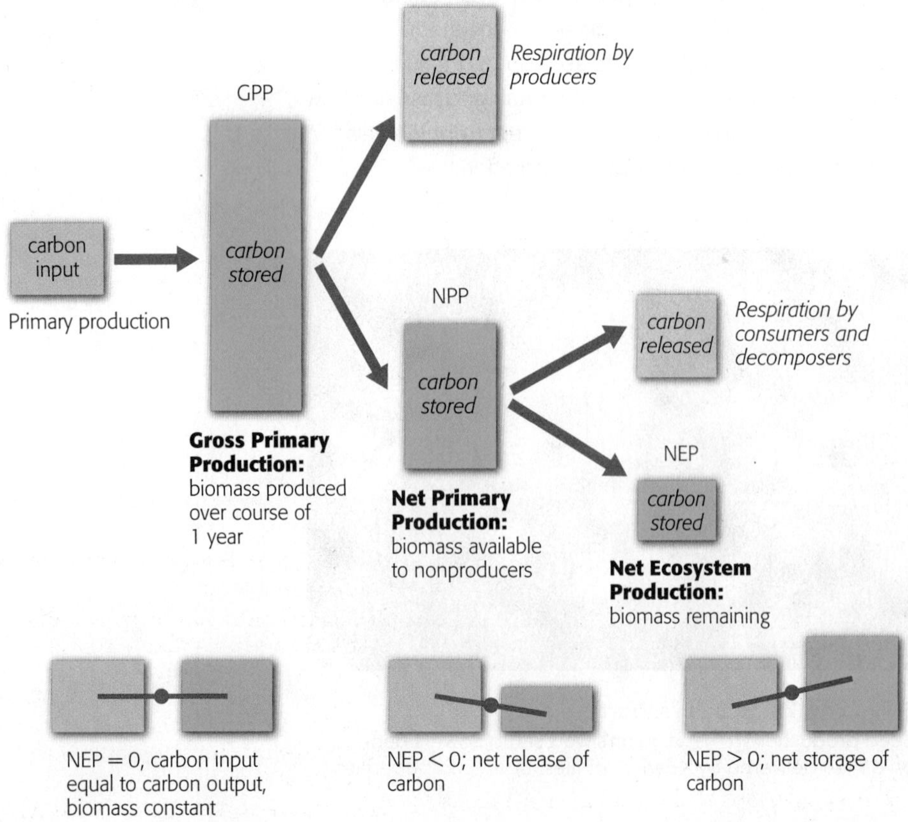

Organic forms of carbon, such as carbohydrates, hydrocarbons, lipids, and proteins, store large amounts of chemical energy; all organisms at every trophic level access this energy through the process of cellular respiration (see Module 4.1). In cellular respiration, hydrogen is removed from organic compounds, oxygen is added, and organic carbon is converted back to CO_2. A similar process occurs when organic compounds are burned, as in a wildfire or in the engine of your car.

The total amount of CO_2 that primary producers convert to organic carbon each year is called **gross primary production (GPP)**. These producers use approximately half of the GPP in cellular respiration to carry out their own life functions. The remaining half is **net primary production (NPP)**. NPP is the amount of organic carbon available to all the nonphotosynthetic organisms, or consumers and decomposers, in an ecosystem. It therefore has important consequences for the number of trophic levels in and complexity of ecosystem food webs.

Net ecosystem production (NEP) is the amount of organic carbon left each year after subtracting the respiration of nonphotosynthetic organisms from NPP. In other words, NEP is the net flux of carbon into an ecosystem—the rate at which carbon moves into Earth's carbon pool (**Figure 6.29**). If the total carbon in the biomass of organisms in an ecosystem remains constant, that is, the same amount of carbon is brought in by photosynthesis as is released by respiration, the NEP is zero. If the biomass decreases, as it does when an ecosystem is burned or degraded by human activities, NEP is negative—more carbon is released than stored. If the biomass of an ecosystem increases, as it does as ecosystems recover from such disturbances, NEP is positive—more carbon is stored than released. NEP is also positive in lake and ocean sediments and bog and swamp soils where poor aeration limits the rate of cellular respiration.

▼ **Figure 6.29 GPP, NPP, and NEP**
Gross primary production is the sum of all photosynthesis in an ecosystem. Net primary production is the energy actually stored in plants and available for consumers. Net ecosystem production is what remains of NPP after respiration by all nonphotosynthetic organisms.

GPP

carbon released — Respiration by producers

carbon input

carbon stored

Primary production

Gross Primary Production: biomass produced over course of 1 year

NPP

carbon stored

Net Primary Production: biomass available to nonproducers

carbon released — Respiration by consumers and decomposers

NEP

carbon stored

Net Ecosystem Production: biomass remaining

NEP = 0, carbon input equal to carbon output, biomass constant

NEP < 0; net release of carbon

NEP > 0; net storage of carbon

Terrestrial Carbon

■ Most of the carbon stored in living biomass is found in terrestrial ecosystems.

Worldwide, approximately 560 Pg of carbon is stored in living biomass on land (**Figure 6.30**). Well over 90% of this biomass is in plants. In any particular place and time, ecosystem biomass may be increasing or decreasing, and NEP may be negative or positive. But scientists generally assume that Earth's total terrestrial biomass pool remains relatively constant—NEP is 0—in the absence of large-scale human impacts like widespread deforestation. This implies that the total amount of carbon released to the atmosphere as CO_2 in respiration by all the organisms on Earth is equal to GPP, the total amount of CO_2 carbon taken up by plants in photosynthesis.

The soils of terrestrial ecosystems contain large amounts of organic carbon—approximately 1,500 Pg. (This is a very rough estimate because it is difficult to measure soil carbon, and soils are poorly studied.) Plant litter, waste, and dead organisms are constantly adding carbon to soils. Some of this carbon is in organic compounds that are quickly consumed by decomposers, such as insects, worms, fungi, and bacteria. Other organic compounds are resistant to decay and may reside in soils for hundreds or even thousands of years.

Because of geographical differences in the length of growing seasons and the availability of water, terrestrial NPP and biomass pools vary widely among ecosystems. In general, NPP and biomass pools are high in warm,

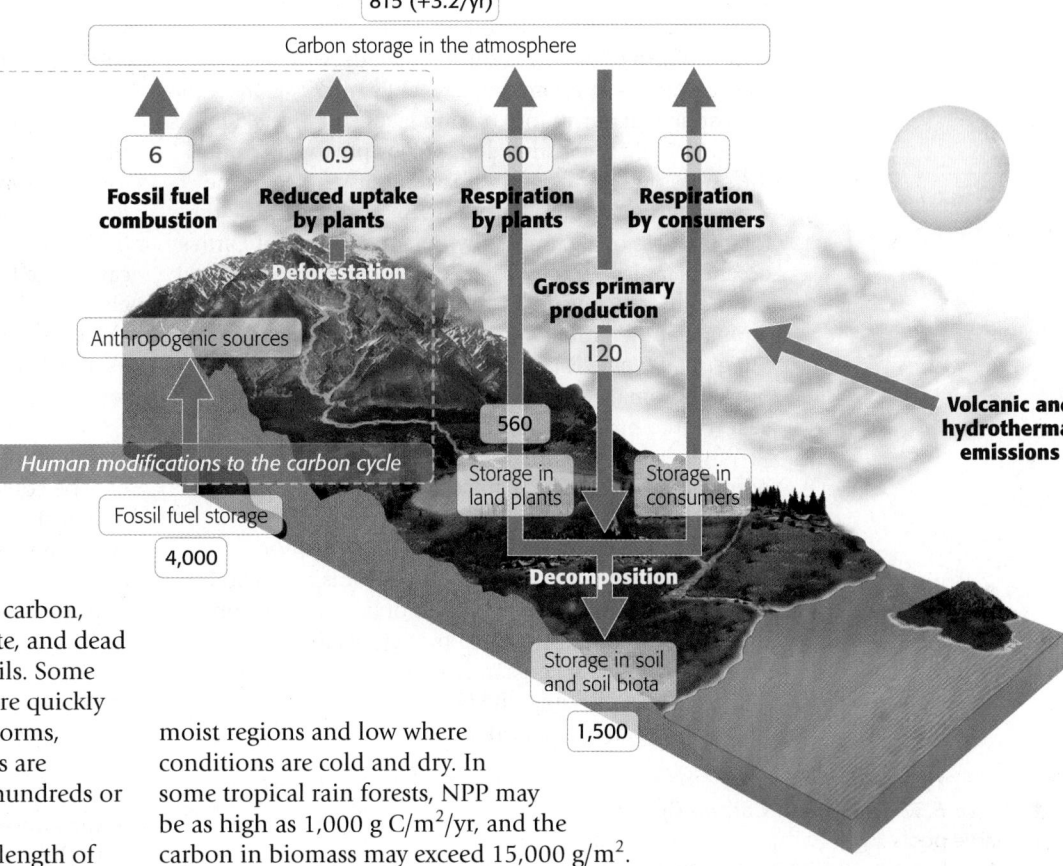

moist regions and low where conditions are cold and dry. In some tropical rain forests, NPP may be as high as 1,000 g C/m²/yr, and the carbon in biomass may exceed 15,000 g/m². In the driest deserts, NPP may be only 40 g C/m²/yr, and the carbon in biomass may be less than 300 g/m² (**Figure 6.31**).

▲ **Figure 6.30 The Terrestrial Carbon Cycle**
The important pools (Pg = 10^{15} g) of the carbon cycle are indicated in black and fluxes (Pg/yr) are indicated in red. The pools and fluxes enclosed by the blue dashed line are those specifically affected by human activities.

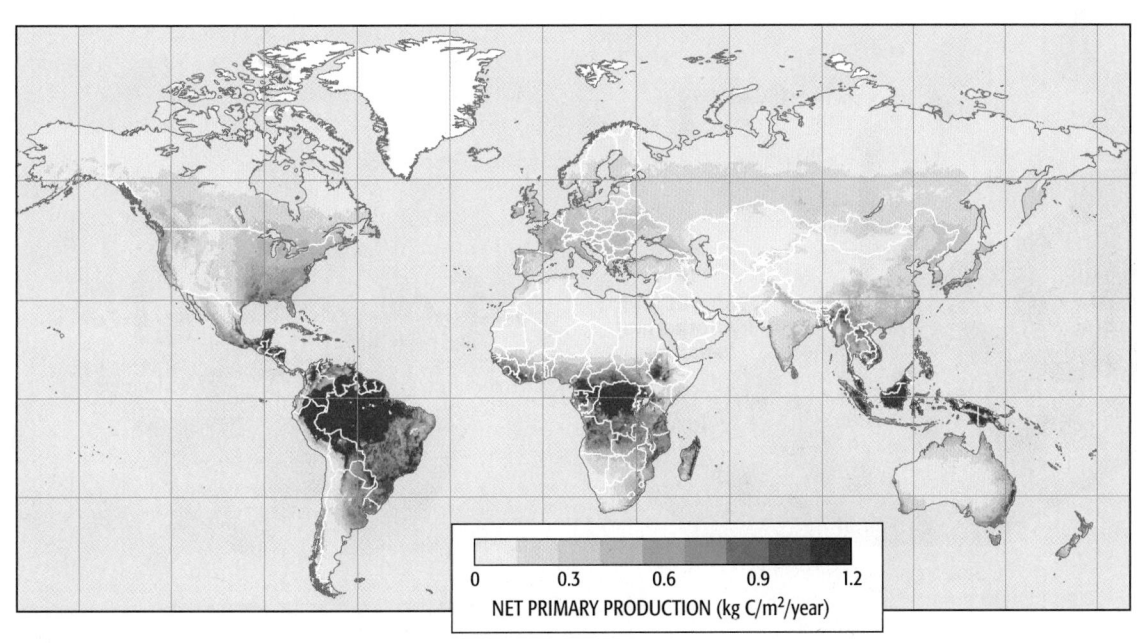

◀ **Figure 6.31 Terrestrial Net Primary Production**
Net primary production in the tropical forests of warm and moist regions, such as the Amazon of South America, central Africa, and southeastern Asia, is 10–20 times that of hot, dry deserts such as those found in the southwest of North America and parts of Africa and Asia.

Source: "Net Primary Productivity" from ATLAS.

171

Aquatic and Marine Carbon

■ Living biomass forms a tiny fraction of total ocean carbon but has a very significant influence on the ocean carbon cycle.

Oceans, lakes, and rivers contain a very significant global pool of carbon, about 38,000 Pg. Much of this carbon is in the form of dissolved carbon dioxide, written as CO_2-C. Each year, approximately 92 Pg of CO_2-C from the atmosphere dissolves into oceans, lakes, and rivers. Of this amount, 51 Pg is consumed in NPP by phytoplankton—the photosynthetic algae and bacteria suspended in the water. This rate of productivity is remarkable because the total biomass of algae and phytoplankton in all of Earth's oceans is far less than 0.1 Pg, a minuscule fraction of total ocean carbon. Considered another way, every gram of biomass in the primary producers in the ocean contributes 510 g of NPP per year. In terrestrial ecosystems, each gram of plant biomass contributes only about 0.12 g of NPP per year (Figure 6.32).

What causes this difference between terrestrial and ocean ecosystems? There are two causes. First, much of the biomass of plants, the primary producers on land, is not green; rather, it is found in long-lived, nonphotosynthetic structures such as roots and wood. Green leaves contain less than 1% of the total carbon in a forest. Second, phytoplankton grow and reproduce very rapidly but are constantly being eaten by marine herbivores. So although the NPP of phytoplankton is high, their total biomass at any moment is comparatively low (see Module 6.4). Consequently, the residence time for carbon in the biomass pool of phytoplankton is much shorter than its residence time in the biomass pool of terrestrial plants.

Much of the CO_2 that dissolves in ocean water reacts to form carbonate (CO_3^{2-}) and bicarbonate (HCO_3^{-}) ions. Dissolved CO_2, carbonate, and bicarbonate account for more than 99% of ocean carbon. Carbonate and bicarbonate combine with calcium to form calcium carbonate ($CaCO_3$), which is the primary structural element in the shells of familiar marine animals such as clams and mussels. However, most calcium carbonate is used to form the skeletons and shells of less well-known microscopic organisms, such as foraminifera.

When marine organisms die, the calcium carbonate in their shells and skeletons may dissolve, and the carbonate may be converted back to CO_2. But often the remains of marine organisms sink to the ocean floor, forming layers of sediments that eventually are transformed into limestone and other sedimentary rocks. This process is an important connection between the carbon cycle and the rock cycle (see Module 3.4).

Over tens of thousands of years, organic matter in ocean sediments may be transformed into oil and natural gas. The rate at which organic carbon is buried in sediments is thought to be far less than 0.01 Pg/yr.

Each year, 90 Pg of carbon from the oceans returns to the atmosphere as CO_2. This is about 2 Pg less than the amount that is absorbed each year. This 2 Pg surplus is much greater than the annual flux of carbon into sediments. Thus, marine scientists believe that the total pool of carbon in the oceans must be increasing. However, given the large size of the ocean carbon pool, this increase is very difficult to detect.

▼ Figure 6.32 **The Ocean Carbon Cycle**
The large pools (Pg = 10^{15} g) of the ocean carbon cycle are indicated in black, and the fluxes (Pg/yr) are indicated in red. The total pool of carbon in living biomass in oceans is less than 1 Pg.

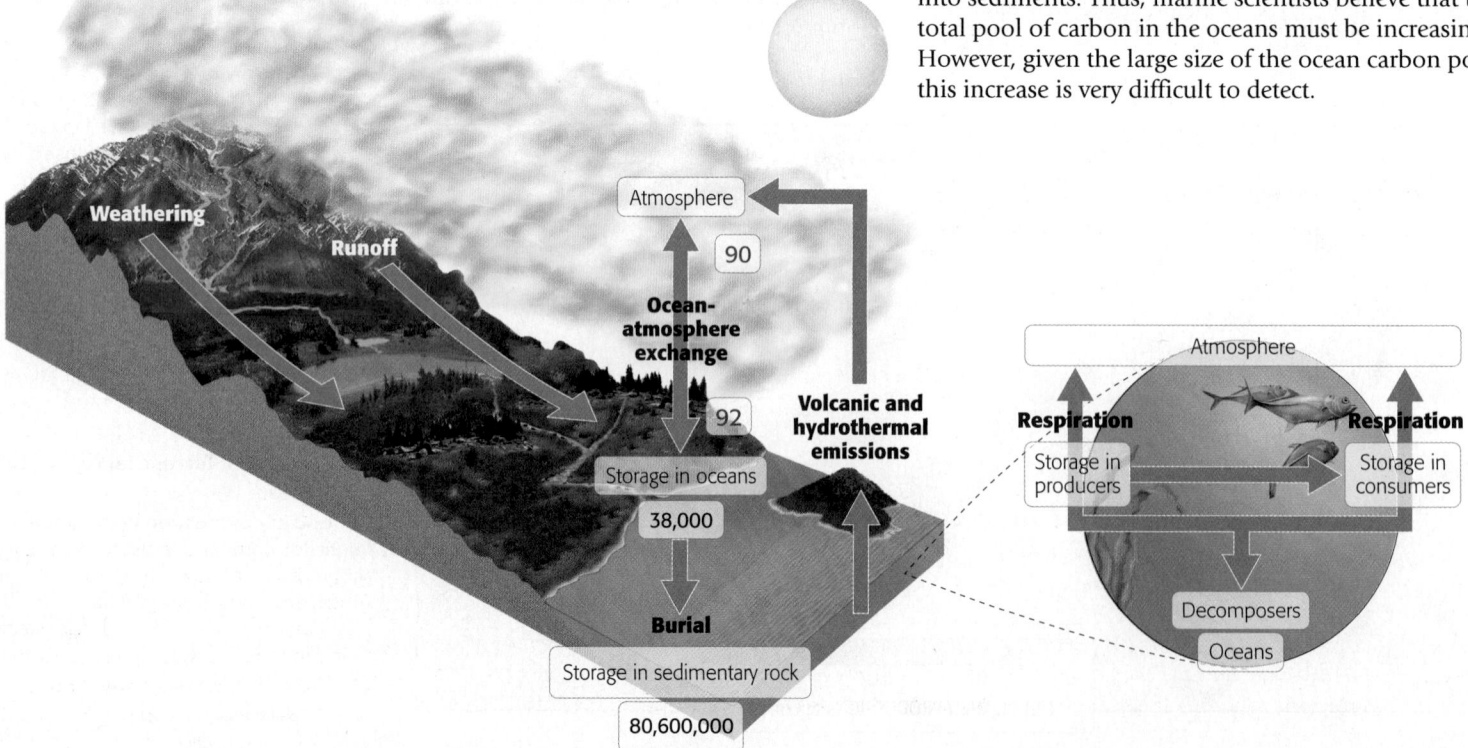

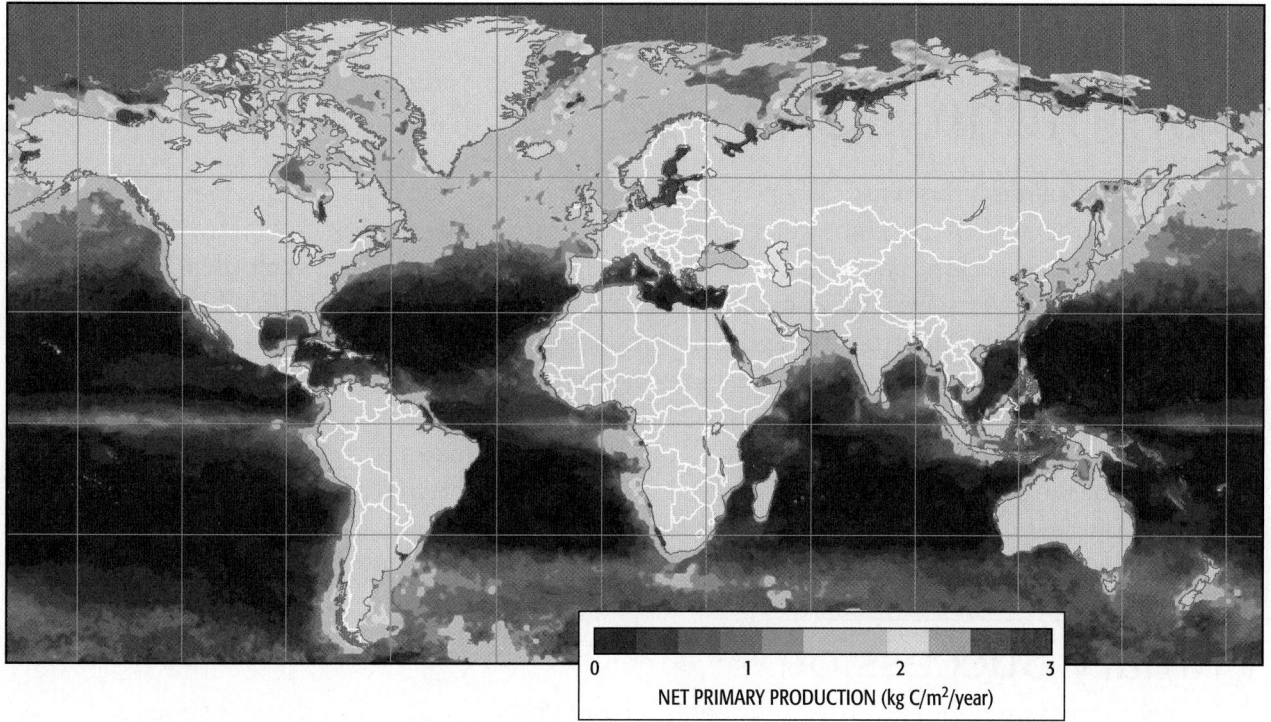

▲ Figure 6.33 Marine Net Primary Production
Except where nutrient inputs are high, such as near coastlines, in polar regions, or where nutrient-rich water from the depths is brought to the surface, net primary production over most of the ocean is less than 150 g C/m²/yr.

The rates of exchange of carbon between the ocean and the atmosphere vary considerably from location to location. Among the factors influencing this variation are temperature and local factors influencing NPP. The solubility of CO_2 gas in water decreases as water temperature increases. Therefore, all other things being equal, the flux of CO_2 from the atmosphere to the ocean decreases with increasing surface ocean temperatures.

Because light can penetrate ocean water to a depth of only 50–100 m, nearly all of the photosynthesis in the ocean takes place near the surface. Actual values of NPP vary greatly from location to location. Over most of the open ocean—more than 90% of the ocean surface—NPP averages 130 g C/m²/yr. But near land where there is an influx of nutrients such as nitrogen and phosphorus, NPP increases by 200–400%. This difference in productivity has significant consequences for local rates of CO_2 exchange with the atmosphere; it also affects the composition and diversity of ocean ecosystems (Figure 6.33).

Human Impacts

■ We are significantly altering Earth's carbon cycle.

Like all consumers, humans obtain their energy from organic compounds that are ultimately derived from the NPP of ecosystems. As such, we compete with all other forms of life for that NPP. The rapid increase in our numbers, particularly over the past century, is powerful evidence that we have been quite successful in that competition (see Module 5.3). Scientists estimate that today humans harvest directly or manage 32–40% of global NPP. By appropriating this large fraction of global production, our impact on the carbon cycle has been profound.

Fossil fuels, the oil, natural gas, and coal that powers our cars and generates much of our electricity, were created eons ago in ecosystems such as bogs, swamps, and ocean sediments with high NEP. Burning those fuels returns that long-stored carbon to the atmosphere as CO_2. These human impacts on Earth's carbon cycle are discussed in much greater detail in Chapter 9.

QUESTIONS 6.5

1. What is the difference between net primary production (NPP) and net ecosystem production (NEP)?

2. How does the biomass of primary producers in terrestrial ecosystems compare with the mass of primary producers in marine ecosystems?

3. What percent of Earth's NPP is directly harvested or managed by humans?

(MES) For additional review, go to **MasteringEnvironmentalScience**

6.6 Disturbance and Community Change

BIG IDEA **Ecological disturbances**, such as fires, hurricanes, or the logging of forests, may result in the loss of many or all of the species in an ecological community. Environmental features that are left behind following a disturbance, such as soil and woody debris, are called **ecological legacies**. As species reestablish themselves, they alter the environment, often making it possible for other species to become established. This process of post-disturbance change in an ecological community is called **succession**. It often leads to diverse and stable ecosystems such as the Pacific Northwest old-growth forests. The nature of successional change is influenced by the severity of the disturbance and the ecological legacies that the disturbance left behind. Succession may also influence the likelihood of future disturbances, producing disturbance cycles. The specific patterns of succession vary from place to place and from time to time, depending on factors such as variations in environmental conditions, climate change, and the character of surrounding ecological communities.

▼ Figure 6.34 **Glacier Retreat and Primary Succession**
The red lines on this map indicate the position of the glaciers at different dates. In 1760, Glacier Bay was completely filled with glacial ice. Since that time, glaciers have retreated over 100 km (62 miles), gradually making rock surfaces available for primary succession.

Primary Succession

■ Primary succession depends on environmental change brought about by pioneer species.

Primary succession occurs where a disturbance has removed virtually all ecological legacies. Because it occurs slowly, often over centuries or even millennia, primary succession is difficult to study directly. Instead, ecologists infer the process of primary succession from relatively short-term studies of landscape patterns. Among the first ecologists to use this approach was William S. Cooper, who studied the recently exposed lands in Glacier Bay, Alaska.

When explorer George Vancouver first mapped Glacier Bay in 1798, a glacier that was thousands of feet thick extended nearly to the mouth of the bay. In the 200 years since then, that glacier has retreated nearly 100 km (60 miles), leaving behind newly exposed rock. Cooper wondered how ecological communities had become established on such barren sites and how they had changed over time. He reasoned that rock surfaces at the mouth of the bay had been exposed for a longer period than those closer to the glacier. Using historical maps, Cooper was able to reconstruct the exact age of the rock surface at each location. By comparing communities at various distances from the glacier, he was able to reconstruct the overall pattern of change (**Figure 6.34**).

Primary succession in Glacier Bay begins when bare rock is first exposed. Within a couple of years, lichens and mosses colonize the rock (**Figure 6.35A**). Ecologists call these earliest colonists **pioneer species**. Pioneer species share three important characteristics. First, they are widely dispersed; lichens and mosses are dispersed by spores. Second, they are able to grow under very harsh and resource-poor conditions. Third, as they grow, they alter their environment in ways that allow other plants to become established. Lichens and mosses, for example, release small amounts of organic acids that accelerate the weathering of nearby rock. This rock weathering, in combination with the organic litter produced by the lichens and mosses, results in the accumulation of small amounts of soil around these plants.

After about 10 years, sufficient soil accumulates in some locations to support the growth of herbs and a few low shrubs. Roots from these plants penetrate cracks in the rock, enlarging them and accelerating the processes of weathering and soil accumulation. Over the next 50 years, the community of herbs and shrubs becomes more diverse and expands across the bare rock. Among these plants, the mountain avens, or *Dryas*, is especially important; symbiotic bacteria associated with its roots fix nitrogen from the air, enriching the soil around it (Figure 6.35B).

Within 60–100 years, birch and alder trees become established, forming a dense forest. Like *Dryas*, the alder has symbiotic nitrogen-fixing bacteria associated with its roots; these bacteria enrich the soil (Figure 6.35C). Eventually, the nitrogen-rich soil allows the seedlings of white spruce and other cone-bearing trees to take hold. After 100 years, saplings of these trees begin to grow through openings in the birch and alder canopy. After 250 years, these trees form a mature white spruce forest that is similar to the forests growing on sites that have been free of glacial ice for a thousand years or more (Figure 6.35D).

NEP changes in a predictable fashion through this process. It is initially low as slow growing mosses and lichens get a foothold, but it increases rapidly through the *Dryas* and alder stages as biomass increases. In mature spruce forests, consumers and decomposers utilize a greater fraction of NPP, leading again to a low NEP, potentially one approaching zero, as the now significant rate of photosynthesis is balanced by a high rate of respiration.

From this and similar studies of primary succession in other regions, ecologists have inferred five important principles.

1. As plant species become established and grow, they may alter the environment around them in ways that make it more habitable for other species. This process, called **facilitation**, is especially important in the early stages of succession.

2. Facilitating species, such as *Dryas* and alder, eventually disappear because they are unable to compete effectively with their successors.

3. The migration of plant species to a site depends on the proximity of seed sources and the ability of the plants to disperse their seeds. Early invading herbs and shrubs often have small wind-dispersed seeds and are abundant in nearby locations.

4. The combination of migration, facilitation, and competition results in populations of different species replacing one another over time.

5. Eventually, a community of plants that is able to reproduce generation after generation may become established; this is the **climax community**. In southern Alaska, the climax community is the group of species associated with a white spruce forest.

These principles apply to succession in many, but not all, situations. For example, we shall see that succession does not always lead to stable climax ecosystems.

◄ Figure 6.35 **Primary Succession**
As glaciers retreat, pioneer species such as lichens and mosses Ⓐ initiate the process of primary succession, which culminates in 200–300 years in mature spruce forest Ⓓ. Early invading species often alter the environment, making it more favorable for their successors Ⓑ and Ⓒ.

Ⓐ Pioneer mosses and lichens—rock weathers and organic matter accumulates

Ⓑ Herbs and low shrubs—diversity increases and soil develops

Ⓒ Invasion by alder, a nitrogen-fixing tree

Ⓓ Invasion by spruce and other conifers

Secondary Succession

■ Secondary succession is often accelerated by pre-disturbance ecological legacies.

Primary succession occurs following major disturbances that leave very few ecological legacies. **Secondary succession** is the process of change following disturbances that leave behind legacies such as soil, woody debris, or plant seeds. Disturbances that initiate secondary succession include ice and wind storms, fire, logging, and various kinds of human land use. The patterns and rate of change that such disturbances initiate are greatly influenced by their legacies.

Old-field succession, the reforestation of abandoned farmland, is one of the most carefully studied examples of secondary succession. During the 1700s, much of the original forest in the southeastern United States was cleared for agriculture. That land was then farmed for a century or more. Over this period, poor farming practices led to extensive soil erosion and nutrient loss. By the time of the Civil War (1860), these lands had lost much of their original productivity. In the next 80 years, loss of soil productivity and economic hard times caused many farmers to abandon their fields. This started the process of old-field succession on millions of acres (Figure 6.36).

In the year immediately following abandonment, old fields are dominated by short-lived grasses and herbs, including several weeds that had previously competed with the farmers' crops. The seeds of longer-lived perennial herbs, such as aster, goldenrod, and broomsedge, also germinate in the first year (Figure 6.37). During the next few years, these perennials outcompete the first-year weeds and dominate the old field. The litter they produce gradually replenishes the organic matter and nutrients in the soil.

During this time, seeds of wind-dispersed trees, such as pines, sweet gum, and tulip tree, also germinate. In three to four years, saplings of these trees begin to emerge above the grasses and herbs. These saplings provide perching sites for birds, which disperse the seeds of trees such as black cherry and dogwood.

In these high-light and rather nutrient-poor environments, pines generally grow faster than the other trees. After 8 to 10 years, the young pine trees form a closed canopy that casts dense shade. Old-field grasses and herbs

▲ Figure 6.36 **Abandoned Farmland Creates Opportunities for Succession**
Poor farming practices and economic hardships led to the abandonment of much badly eroded land in the southeastern United States following the Civil War. This abandonment initiated old-field succession over broad expanses of this region.

are unable to grow in dense shade, so they gradually disappear. In contrast, many of the shrubs and trees that grew in the old field are able to subsist beneath the pines. Soon, other shade-tolerant herbs, shrubs, and trees begin to invade. These new species are often transported by the growing populations of squirrels and other animals for which the pine forest provides suitable habitat.

Pines in this region grow quickly. In just a decade or two, this process converted vast areas of the Southeast from eroded farmland to pine forest. Lands that had once produced cotton and corn became some of the nation's most important sources of timber. The rapidity of this reforestation was facilitated by the legacy of old-field soils, which were enriched by the growth and decay of pioneer grasses and herbs.

► Figure 6.37 **Old-Field Succession**

In secondary succession on abandoned farmland, the legacy of soil accelerates the process of change from old-field pioneers Ⓐ, to pine forest Ⓑ, to mature deciduous forest Ⓒ.

0–10 YEARS

Ⓐ Hardy wind-dispersed herbs and grasses dominate old fields in the first few years following abandonment.

10–15 YEARS

Ⓑ In the high light and poor soil conditions, pines are able to grow more rapidly than other trees and form an even-aged stand in 10–15 years.

150 YEARS

Ⓒ Pines are not able to reproduce in their own shade and are eventually replaced by deciduous, broad-leaved trees like oaks, maples, and hickories.

After 70 to 150 years, individual pine trees begin to die. Pines are well adapted for growing in sunny old fields, but their seedlings are unable to become established in the shade cast by mature trees. As the pines die, they are gradually replaced by broad-leaved trees that are more tolerant of shade, such as oaks, hickories, and maples. Many southeastern forests that have not been harvested for timber are currently in this stage of secondary succession. If left uncut, in about 150 years these forests will be dominated by a climax forest composed of a diverse mixture of broad-leaved deciduous trees.

Because succession produces important changes in habitat, the animal communities also change. Studies in Georgia, for example, revealed significant changes in the abundance and diversity of birds during old-field succession (Figure 6.38). Because many of these birds transport seeds, their presence affects the composition of the plant community.

The character and rate of secondary succession depends on the ecological legacies left by the disturbance. For example, the pattern of change following logging in a 50-year-old pine forest is quite different from that in old-field succession. The soils left by logging contain considerably more organic matter and nutrients than those available in the soil of abandoned fields. Logging also leaves the seeds and saplings of many shrubs and trees. Within as few as three years, a diverse mixture of shrubs and trees forms a dense cover. Unless foresters use fire or herbicides to reduce their growth, broad-leaved deciduous trees soon grow over and eliminate the pines.

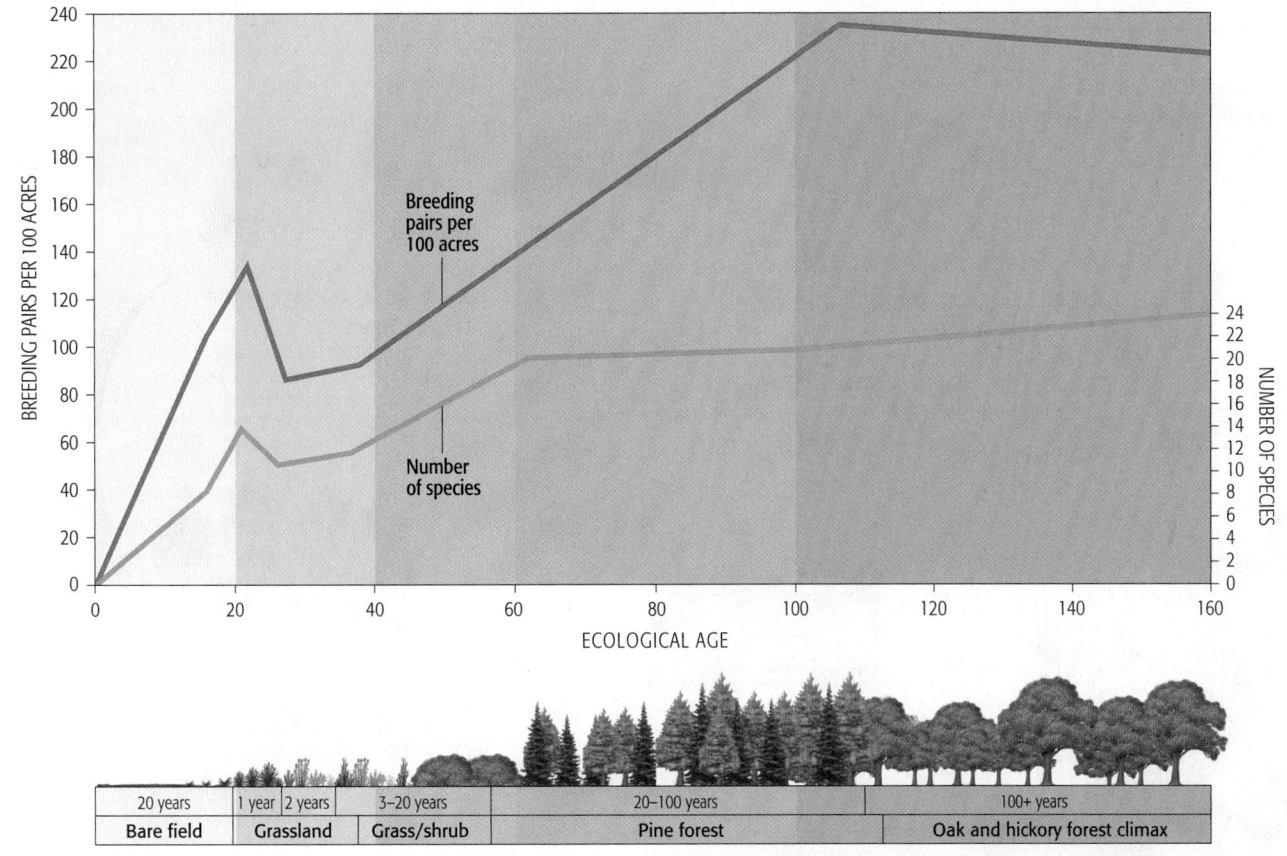

◄ Figure 6.38 **Succession and Bird Diversity**
The number of bird species and the overall abundance of birds increase during old-field succession as more bird habitats become available. For their part, birds accelerate this process by transporting seeds of many late-succession species.

Source: Johnston, D.W. and E.P. Odum. 1956. Breeding bird populations in relation to plant succession on the Georgia Piedmont. *Ecology* 37: 50–62.

Cyclic Succession

■ Succession may increase the likelihood of disturbance, which leads to cycles of change.

Pine cone seeds

Fire causes pine cone to open up and release seeds

▲ **Figure 6.39 Lodgepole Pine Requires Fire to Reproduce**
The seed-bearing cones of the lodgepole pine remain closed until heated by fire. They then release their seeds, which are able to establish themselves in the high-light environment that follows the fire.

Early studies seemed to suggest that ecological communities become increasingly stable as succession proceeds. In some circumstances, ecosystems do appear to become less disturbance-prone over time. However, recent research reveals that succession in some ecosystems may actually increase the likelihood of disturbance. Where this is true, climax communities may not exist. Instead, there is a continuous cycle of disturbance and succession.

One such cycle of change occurs in the lodgepole pine forests of the mountains of western North America. More than 100 years ago, ecologists recognized that the lodgepole pine depends on fire for its survival. Many lodgepole pines produce cones that do not open when they mature. Instead, the scales of their cones are glued shut by resins that keep their seeds inside (Figure 6.39). The heat of forest fires melts these resins, opening the cones and releasing the seeds. Fires usually leave behind bare mineral soils; the seeds of lodgepole pines germinate best in such soils. Fires also burn away underbrush, allowing bright sunlight to reach the ground; the seedlings of the lodgepole pine require bright light to become established. Thus, without fire, the lodgepole pine cannot disperse its seeds, nor can its seeds germinate successfully.

In the years immediately following an intense fire, large numbers of lodgepole pine seedlings become

established and form a dense, shrublike cover. In a few decades they grow into a stand of trees that is 40 to 60 feet tall (Figure 6.40). Lightning or careless humans may start fires in these young forests. Fires typically ignite in the needles and small branches on the ground. Because there is little fuel between the forest floor and the needles and branches in the forest canopy, fires that burn through these young forests tend to remain on the ground. Such fires burn with very low intensity—not enough to damage the trees or open their cones.

After 150 years, the pines mature and individual trees begin to die. This creates openings in the canopy and opportunities for the establishment of shade-tolerant spruce and fir trees. As this process proceeds, woody debris accumulates. The distribution of fine, flammable fuels from the forest floor to the canopy becomes much more continuous. Because the branches of the spruce and fir trees are closer to the ground than those of the pines, these trees create "ladders" that can carry fire from the forest floor into the forest canopy.

Thus, as succession proceeds, the forest becomes much more prone to the very intense fires that kill trees. Intense fires create the conditions that are necessary for the reestablishment of the lodgepole pines. The length of this cycle is generally about 200 to 300 years, although it is affected by factors such as yearly rainfall and the occurrence of lightning.

▼ **Figure 6.40 Lodgepole Pine Forest Fire Cycle**
As lodgepole pine forests undergo succession, wildfire becomes more likely because of changes in the amount and distribution of flammable plant material. The resulting pattern of change is called cyclic succession.

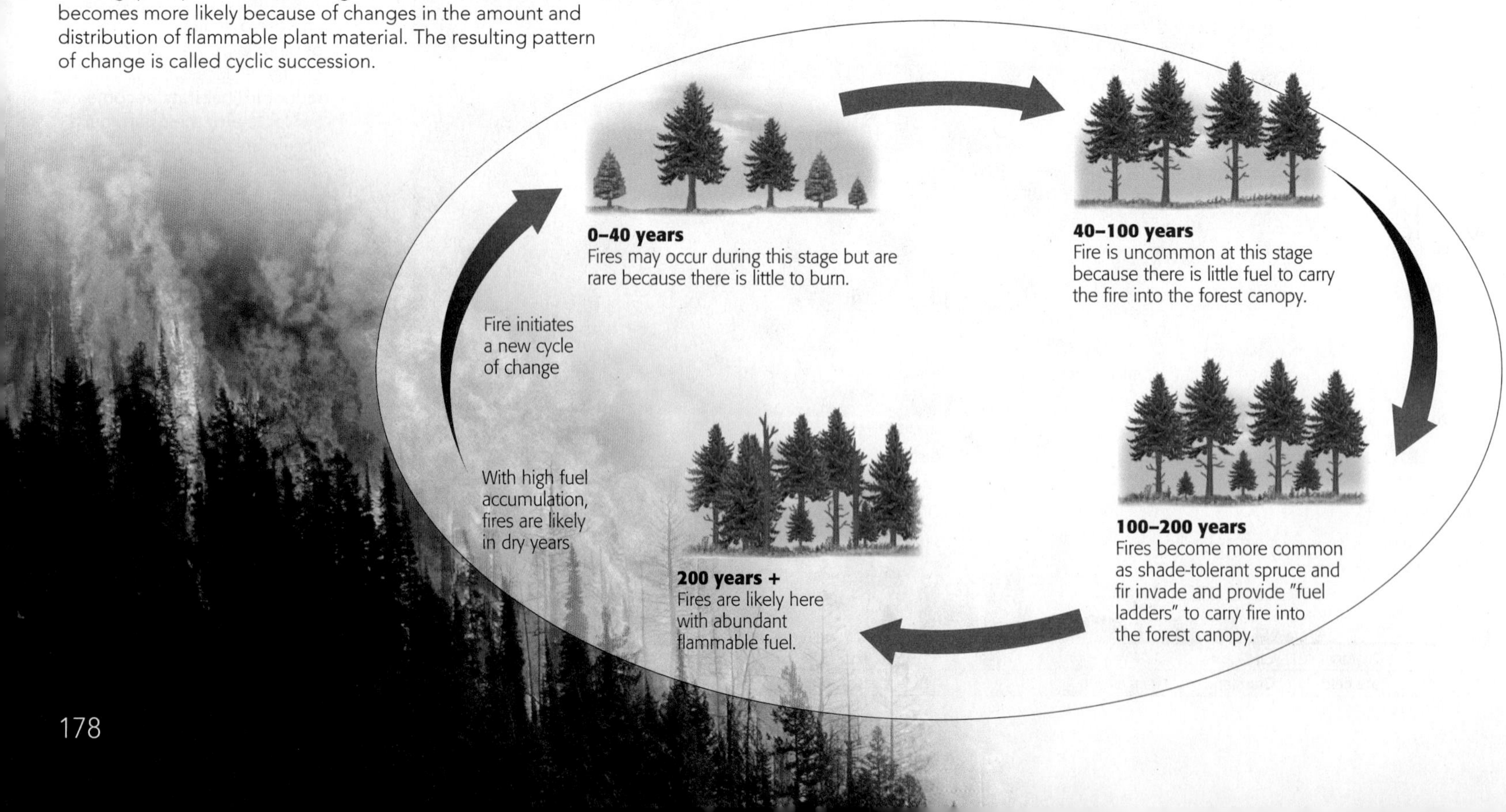

Fire initiates a new cycle of change

With high fuel accumulation, fires are likely in dry years

0–40 years
Fires may occur during this stage but are rare because there is little to burn.

40–100 years
Fire is uncommon at this stage because there is little fuel to carry the fire into the forest canopy.

100–200 years
Fires become more common as shade-tolerant spruce and fir invade and provide "fuel ladders" to carry fire into the forest canopy.

200 years +
Fires are likely here with abundant flammable fuel.

Other kinds of disturbances can also cause cycles of change. In the high-elevation fir forests of northern New England, wind, ice, or insects can damage the canopy, exposing the trees to the prevailing winds. Eventually, these strong, icy winds kill the trees. As trees die, the trees immediately behind them are exposed to the same stresses, increasing the likelihood of their dying. This process initiates a "wave" of tree death that moves across a landscape at a rate of 3 to 10 feet per year. Succession occurs in the "wake" of such waves. The youngest trees grow near the front of the wave, with successively older trees trailing behind. At any given time, several such

waves can be observed moving across many northern mountainsides.

Because they often occur in patches of varying size, cyclic disturbances on large landscapes often produce a mosaic of patches representing different stages of succession. **Return time** is the average time between disturbances at any given place on such a landscape. Return times may be as short as a few years, as in frequently burned grasslands and savannas; many decades long, as in the chaparral of the southwestern United States; or more than a thousand years, as in tropical or temperate rainforests.

The Importance of Place and Time

■ Successional change at a place is affected by the changes occurring in surrounding places.

Ecologists once viewed climax communities as the inevitable end state of succession. Within a particular climatic zone, they thought that succession would eventually produce the same climax assemblage of species regardless of the nature of the disturbance. In southern Alaska, for example, succession would inevitably lead to a white spruce forest. In the southeastern United States, it would lead to a broad-leaved deciduous forest. Many ecologists argued that the climax community was the most stable assemblage of species possible in that environment.

Today, ecologists recognize that multiple factors can affect the process of change following a disturbance. Local variations in terrain can produce conditions that favor the establishment of different assemblages of species. In the Glacier Bay region of Alaska, for example, succession in low, moist areas leads to the formation of peat bogs and forests of black spruce rather than white spruce (Figure 6.41).

Succession in a particular disturbance patch is also influenced by the composition of species in communities immediately surrounding it. After 150 years, an old field surrounded by a forest of oak and hickory trees is more likely to have a large number of oaks and hickories than a similar field in the midst of a pine forest.

Recent studies in North Carolina reveal that patterns of succession in the herbs that grow in pine forests are significantly different from the patterns found in studies conducted between 1930 and 1950. Three factors appear to have caused this difference. First, changes in the landscape associated with human development have altered the patterns of seed dispersal. Second, deer populations have increased 10-fold, and deer grazing has significantly diminished the abundance of many herb species. Third, introduced non-native species, such as garlic mustard, have become more abundant in many locations.

From decade to decade and century to century, changing environments may cause important changes in the process of succession. Ecologists now understand that long-term climate change and human-caused changes, such as widespread deforestation or the introduction of non-native species, can alter the course of succession.

QUESTIONS 6.6

1. Explain why pioneer species may be critical to the recovery of communities after a disturbance.

2. Why do ecological legacies accelerate the process of succession?

3. Describe a situation in which successional change increases the likelihood of disturbance.

4. How does net ecosystem production change during primary succession? Cyclic succession?

(MES) For additional review, go to **MasteringEnvironmentalScience**

▶ Figure 6.41 **Local Environment Alters Succession**
Where drainage is poor around Glacier Bay, succession leads to a so-called muskeg bog with scattered black spruce and an understory of peat moss, which both flourish in moist environments.

Vermicomposting at Michigan State University

Liz Brajevich is a student at Michigan State University in East Lansing, Michigan. During her freshman year, she developed a worm-composting project for campus cafeterias, funded by the "Be Spartan Green" student fund.

Why and how did you first get involved with vermicomposting?

I was browsing a children's outdoor magazine and found an article on an elementary school using worms to break down their cafeteria food waste. I couldn't wait to get started at my own high school but hit a lot of walls with the administration as I tried to implement the program. Live worms seemed like too much of a liability to my school, so I worked classroom by classroom. My own environmental science teacher supported a bin in her room, so I reached out to other teachers in classrooms across my hometown of Los Angeles, California, starting small vermicomposting bins at 20 local elementary schools, middle schools, and high schools.

What steps did you take to get this project started at MSU?

I am a member of RISE, the Residential Initiative on the Study of the Environment. It's a living learning community where students live on the same floor freshman year, and take an environmental issues seminar as a group. In our Environmental Issues seminar, my professors encouraged us to formulate grant proposals for my university's Office of Campus Sustainability "Be Spartan Green" student fund. I worked with three classmates to apply, and we were funded $5,000. This allowed us to build three large vermicomposting bins with a total of 10,000 red wriggler worms. We talked to the cafeteria and set up a system where we pick up food waste from the cafeteria once a week and add it to our worm bins. We use the remainder of our funding to pay students salaries, and also we want to expand the number of students that have small composting bins in their rooms; we currently have 15 students with in room vermicomposting bins, but we'd love to have the whole RISE floor!

What are the biggest challenges you've faced with the project, and how did you solve them?

The biggest challenge for projects is usually funding—however for our project, we are lucky to have more than enough funds to get started. Our big road bump is people's attitudes. Before students know how vermicomposting works, the thought of rotting food and creepy crawlies freaks them out. Vermicomposting is really about an amazing, ecological process that turns waste into a resource, without yucky smells, so it can be done indoors or out. When we wanted to distribute bins for students to compost in their dorm room, residential services initially denied our requests. It took expertly worded emails and proposals full of justifications to get them to support a pilot project of just three students! Patience and persistence are key when you have to win your university's support, but they exist to serve the students, so when there is a student demand that really helps.

What is your major, and how has it helped you in your leadership role?

My double major in Fisheries and Wildlife and Environmental Economics and Policy was key in soliciting support for my project; it was through that program that I found eight freshman in the second year of our project

which has enabled us to create targeted action teams! We have a research team, an outreach team, and a management team. The first year, I had a lot on my plate, juggling permission forms, outreach events, and surveying students about vermicomposting. Now, the research team has presented at the Undergraduate Research and Arts forum on surveys that show student's attitudes of acceptance towards vermicomposting. The outreach team has held booths and workshops, and the management team is making sure the worms get fed! The support of the RISE cohort has been instrumental in the success of the project and my ability to be a leader because of the support I receive.

How has your work affected other students, your surrounding community, and the environment?

Our community has benefited from over 300 kids having our traveling worm bins brought into their elementary schools to teach them about vermicomposting, the life cycle, and the importance of decomposers in taking care of our environment! Our environment has benefited from over 1,000 lbs of food waste being diverted from landfills and turned into nutrient rich fertilizer in just one year. When food waste breaks down in landfills, it releases methane, a potent natural gas. Vermicomposting keeps this methane from entering the atmosphere, plus creates a fertilizer we can use instead of synthetic fertilizers that increases soil's water retention capacity by 20% and adds nitrogen, phosphorus, and potassium to soils (as well as beneficial microorganisms!). Our students at Michigan State have an opportunity to have hands on vermicomposting experience and join one of our ever-growing student action groups focusing on expanding support for vermicomposting.

What advice would you give to students who want to replicate your model on campus?

Vermicomposting requires minimal investment costs—which is great! Even a student fundraiser has the potential to fund the implementation of a small vermicomposting set up. It is really important to have a group of dedicated students, it can be really overwhelming to try and manage on your own! It's a living system, so you have to make sure that at least once a week someone can feed and check the worm bin for moisture, having enough bedding, and so on. We handed out really funny stickers and buttons that said things like "Break it down now y'all" and "I have worms" with a super-worm logo to get students aware of our project and start the discussion about vermicomposting.

Also finding a group to work with, whether it's a class, or a degree program, can really help you get both student and professor support.

What do you hope for vermicomposting at Michigan State University in the coming months? Years?

Looking ahead, it would definitely be amazing to get vermicomposting programs associated with each cafeteria on our huge campus, but first we'll need to build more student support. We do have a lot of students, faculty, and community members asking us how to start their own worm bins, so we will be moving forward with selling small worm bins to those interested and using the funds to further our project. We are currently working out logistics (like how to run the program, file taxes, the list goes on!) but we hope that by sharing with those with an interest in vermicomposting, we can build a greater network of individuals committed to reducing their environmental impact. Our research team also has some great ideas for further studies, including before and after studies on students who participate in our worm bin workshops, economic analysis of vermicomposting, and how residential programs like RISE impact student's participation in programs. Vermicomposting at MSU is going to continue to grow and would love to share that with universities across the United States and even the world. We have lots to learn, and as we learn we can share those victories with other environmentally minded students and schools to have a big impact!

Summary

Ecological communities are clearly more than a haphazard assemblage of species. Coevolution among competitors, predators and prey, parasites and hosts, and mutualistic partners has resulted in a high level of integration and interdependency among community members. This integration results in the efficient processing of energy in food webs. Organisms at each trophic level store a relatively small fraction of the energy that they take in. Therefore, the number of organisms and amount of biomass diminishes from one trophic level to the next. The flow of energy among trophic levels in an ecosystem is directly related to the cycling of carbon and variations in net primary production. The integration among members of ecological communities is also central to their ability to recover from disturbance. This recovery process, called succession, is influenced by the kind of disturbance that initiates it, as well as variations in the physical environment, in the species composition of nearby communities, and in the climate.

6.1 Competition for Shared Resources

■ Competition is the set of interactions among individual organisms that are pursuing a shared but limited resource.

KEY TERMS

ecological community, intraspecific competition, interspecific competition, competitive exclusion principle, fundamental niche, realized niche, niche differentiation, exploitation competition, interference competition

QUESTIONS

1. When two organisms compete, both have lower population growth rates than when they grow separately. Explain why.
2. In the Northwest old-growth forests, three mammals—northern flying squirrels, chipmunks, and voles—eat truffles. By what mechanisms might these three mammals coexist?
3. The diversity of herbs along frequently disturbed roadsides is often high. What factors might account for that diversity?
4. Describe an example of exploitation competition and an example of interference competition.

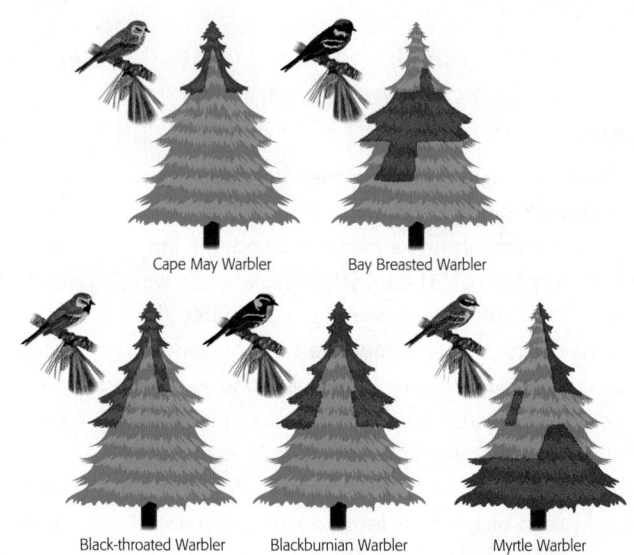

Cape May Warbler   Bay Breasted Warbler

Black-throated Warbler   Blackburnian Warbler   Myrtle Warbler

6.2 Herbivory, Predation, and Parasitism

■ Consumers include herbivores, predators, and parasites; the growth of their populations may be limited by the things they eat or the things that eat them.

KEY TERMS

consumer, herbivore, predator, parasite, coevolution, prey switching, vector

QUESTIONS

1. Why is herbivory an especially challenging way to obtain food?
2. Why are predators important to the well-being of their prey as well as the entire ecological community?
3. The common cold is caused by a viral parasite. New strains of the cold virus spread rapidly, especially in the wintertime. What factors allow new strains to spread so quickly?
4. Newly introduced, non-native predators and parasites often have devastating effects on indigenous host species. Such effects are far more severe than those caused by native predators and parasites. Why might this be so?

6.3 Mutualism and Commensalism

■ Symbioses are intimate interdependencies between different species; mutualism and commensalism are two kinds of symbiosis.

KEY TERMS

symbiosis, mutualism, commensalism

QUESTION

1. What role does coevolution play in the development of mutualistic relationships in ecological communities?

6.4 The Flow of Energy in Ecological Communities

- Food chains and food webs describe the path of energy flow among organisms in an ecological community.
- Within a food chain or web, organisms are assigned to trophic levels based on their food source. Green plants make up the first trophic level; herbivores make up the second trophic level; carnivores that feed on herbivores make up the third trophic level, and so on.

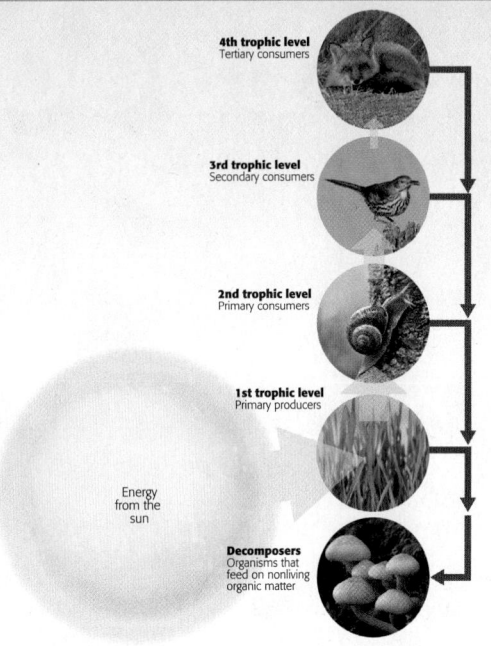

KEY TERMS

energy flow, trophic level, food web, food chain, primary producer, primary consumer, decomposer, biomass energy, trophic level efficiency, keystone species, trophic cascade

QUESTIONS

1. What factors might influence trophic level efficiency?
2. Sharks are considered to be important keystone species in some marine communities. Why might this be the case?
3. Why have conservationists been particularly interested in the restoration of carnivore populations?

6.5 The Carbon Cycle and Ecosystem Productivity

- Photosynthesis and cellular respiration are key steps in the carbon flux between the biosphere and the atmosphere.
- Human activities are increasing the flux of carbon into the atmosphere.

KEY TERMS

gross primary production (GPP), net primary production (NPP), net ecosystem production (NEP)

QUESTIONS

1. Describe a specific situation in which net ecosystem production is positive and one where it is negative.
2. The total pool of carbon in green plants is nearly 5,000 times greater in Earth's terrestrial ecosystems than the pool of carbon in the primary producers in the oceans. However, the total net primary production of Earth's marine ecosystems is nearly equal to that of its terrestrial ecosystems. Explain how this can be true.
3. What factors influence differences in NPP from one place to another in terrestrial ecosystems? How do those factors compare with the factors that influence NPP in marine ecosystems?
4. Describe how human land use is contributing to the increased flux of CO_2 to Earth's atmosphere.
5. Fossil fuels such as oil and coal are the NEP of a distant time. Explain.

6.6 Disturbance and Community Change

- Disturbance such as fires, storms, or land clearing initiate the process of change in ecological communities called succession.
- The processes of primary, secondary, and cyclic succession are influenced by local variations in the physical environment, the composition of species in nearby ecological communities, and long-term changes in climate.

KEY TERMS

ecological disturbance, ecological legacy, succession, primary succession, pioneer species, facilitation, climax community, secondary succession, old-field succession, return time

QUESTIONS

1. Describe the roles of facilitation and competition in successional change.
2. Details of succession such as the rate of change or the sequence of species establishment often vary from location to location within a region. What factors influence such variation?
3. What kinds of natural disturbance occur in ecosystems near where you live?
4. How might they influence the biodiversity of the landscapes within which they occur?

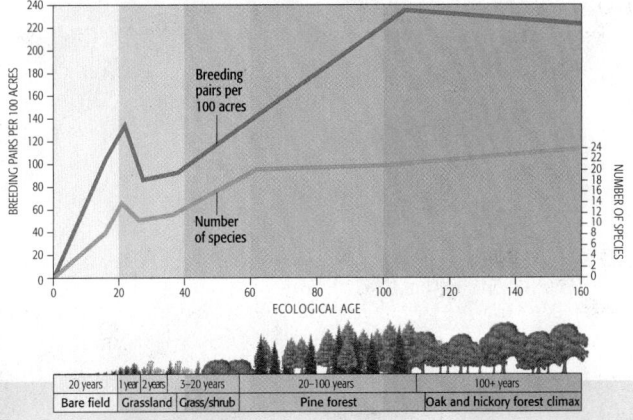

MasteringEnvironmentalScience®

Students Go to **MasteringEnvironmentalScience** for assignments, the eText, and the Study Area with practice tests, activities, and more.

Instructors Go to **MasteringEnvironmentalScience** for automatically graded tutorials and questions that you can assign to your students, plus Instructor Resources.

The Geography of Life

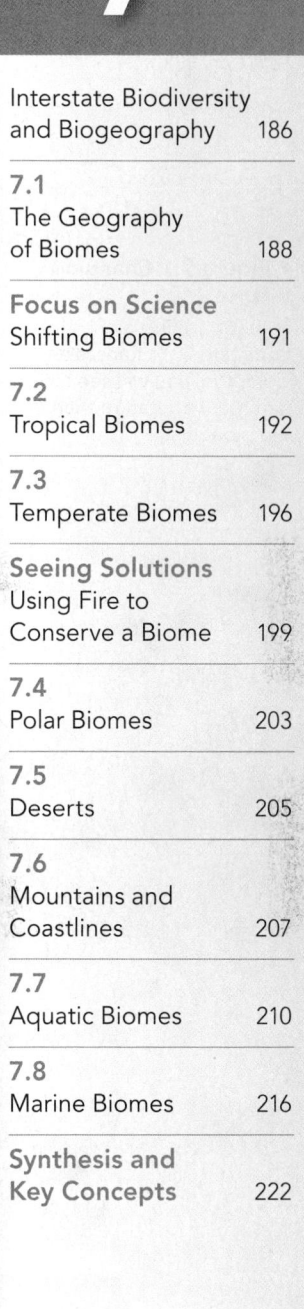

7

185

Interstate Biodiversity and Biogeography

How are human activities changing the kinds of plants and animals found in grasslands?

Have you ever driven cross country on Interstate 70? If so, you know that between Kansas City, Missouri, and Denver, Colorado, lie 600 miles of broad flat countryside (**Figure 7.1**). Although the highway climbs nearly 4,500 feet over this distance, it does so almost imperceptibly. As you travel west, perfectly rectangular fields of soybeans and corn gradually give way to circular plots of irrigated cotton and barley and finally to large expanses of heavily grazed grassland. There is little natural vegetation to be seen.

One hundred and seventy-five years ago, pioneer families traveled this same route for weeks, even months. What they saw on their journey was vastly different from the scenery on your day-long trek.

Covered wagons moved across eastern Kansas through a lush growth of waist-high grasses. The roots of these tall grasses formed a dense mat, or sod. A very diverse array of native sunflowers and other herbs grew among the grasses, and vast herds of bison grazed on the grasses and herbs. Traveling westward through central Kansas, the diversity of grasses and herbs diminished, and they became shorter in response to warmer and drier conditions. Bunchgrasses, species that grow in distinct tufts or tussocks, became increasingly common among the sod-forming species. Onto the high plains of western Kansas and eastern Colorado, wagons maneuvered around ancient clumps of bunchgrass and scattered bits of sagebrush. As the settlers moved west, the herds of bison grew smaller and were joined by pronghorn antelope and elk. It is almost impossible to believe this route is the same one you travel today.

▼ **Figure 7.1 Changing Vistas**
Today, traveling along Interstate 70 in Kansas and Colorado, you will see far different vegetation than did the pioneers.

Short-grass Prairie

Mixed-grass Prairie

Tall-grass Prairie

▲ Figure 7.2 **Similar in Appearance**
Rheas, large flightless birds that can reach 170 cm (67 in.) in height, are seen here feeding in the dry grasslands of Patagonia in southern Chile. These grasslands look remarkably similar to the grasslands of the Colorado high plains; however, the species of grasses and herbs are very different in these two places.

- What factors influence variation in the number and kinds of species from place to place?

- How has evolution among unrelated species of plants and animals produced such similar communities of organisms on different continents?

- Where do the different types of communities occur, and what determines their distribution?

- How have humans altered these patterns, and how are they likely to change in the future?

In the early 1800s, there were places on other continents where you could have traveled a similar distance through seemingly identical vegetation. In eastern Europe, for example, you would have seen a transition from tall grasses to short bunchgrasses if you had traveled from eastern Ukraine to the lower slopes of the Ural Mountains. In South America, you would have seen a similar transition by traveling westward from Argentina's coastal plain to the eastern slopes of the Andes. On close examination, however, you would have found that the species of grasses, herbs, and grazing animals were completely different from one continent to another (**Figure 7.2**). The geographic distribution of different species is typically referred to as *biogeography*.

Today, if you pull off Interstate 70 or travel back roads in eastern Europe or Argentina, you can still find remnants of natural grasslands; those that remain have had to be protected by government agencies or private land conservancies. Some of these grasslands look pretty much as they did to long-ago travelers. But in most places you will discover that human activities and the appearance of invasive species have brought significant changes.

At the University of Kansas Field Station, just 20 miles west of Kansas City, tall-grass prairie has been replaced by dense woodlands of oak trees. Suppression of wildfire and the loss of grazing animals such as bison have allowed these oak trees to invade the grassland. Much of the mixed-grass prairie of the Kirwin National

Wildlife Refuge in central Kansas has been invaded by a non-native grass called smooth brome. Where smooth brome has become dominant, it has reduced the diversity of native grasses and herbs. This, in turn, has diminished the diversity of the animals that feed on the native plants. The U.S. Bureau of Land Management oversees numerous large tracts of bunchgrass-dominated prairie in eastern Colorado. Here, cattle grazing and climate change are encouraging the invasion of desert species, such as sagebrush (**Figure 7.3**).

As you travel along a stretch of road anywhere on Earth, you can recognize vegetation types, such as grassland or evergreen forest, although you may not be able to name the species. On such a journey, think about the landscapes you see and the variety of the biological communities they support.

▼ Figure 7.3 **More Change**
Although protected from the plow, natural areas along I-70 have changed considerably since pioneer times. Ⓐ In the east, former prairie is now woodland. Ⓑ In the mixed-grass region, non-native species such as smooth brome have invaded mixed-grass prairie. Ⓒ Overgrazing and climate change are converting parts of the short-grass prairie to desert.

7.1 The Geography of Biomes

BIG IDEA The process of natural selection has resulted in remarkable similarities among unrelated organisms that grow in similar environments, even on widely separated continents. The major assemblages of similar organisms found in particular environments are called **biomes**. Climate is the most important environmental factor distinguishing terrestrial biomes. Earth's terrestrial biomes occur in one of three general climate zones—the tropical zone, temperate zone, and polar zone. Specific climates within each biome are determined by patterns of temperature, precipitation, and evaporation. Climates within a region are also influenced by the presence of mountains and large bodies of water. Wetland and coastline biomes are found in the transitions between terrestrial environments and environments dominated by liquid water. The assemblages of organisms in Earth's waters occur along a gradient from aqueous or freshwater environments to marine or saltwater environments. The individual biomes in aquatic and marine environments are characterized by differences in water chemistry and patterns of water flow.

Characteristics of Terrestrial Biomes

■ Atmospheric circulation and climate determine the distribution of terrestrial biomes.

If you visit places in North America, Europe, and Asia with warm, wet summers and cold winters, you will find an abundance of broad-leaved deciduous trees. These trees look alike, but they are not the same species. If you visit a desert in the American Southwest or South Africa, you are likely to find plants with thick, waxy leaves and stems that are armed with spines. But on closer examination, you'll discover that although the plants are similar in structure, they belong to entirely different groups of plants (see Module 4.7, Figure 7.4). Why do plants that grow so far apart look so much alike?

Evolutionary biologists have repeatedly observed that similar selective pressures tend to produce remarkably similar traits among species that have no close genetic relationship to one another. The process by which similar environments select for similar features among species that are not closely related is called **convergent evolution**. Because of convergent evolution, plants that grow in similar climates often have very similar adaptations. These include similarities in anatomy, in range of tolerance for variations in temperature and moisture, and in responses to seasonal change.

Biomes are defined by the key features of the plants that grow in them. The most significant of these features are growth form (trees, shrubs, or grasses), leaf type (broadleaf or needle-leaf), and adaptations to changing seasons (deciduous or evergreen leaves).

Biomes can be grouped into three climatic zones, which are defined by latitude and patterns of atmospheric circulation. The **tropical zone** straddles the equator and extends to about 25° latitude north and south, corresponding to the Hadley wind cells in the troposphere (see Module 3.7, Figure 3.45). Tropical rain forest, tropical seasonal forest, and tropical savanna biomes are found in the tropical zone. The **temperate zone** falls between 25° and 60° latitude north and south, which corresponds to the area of prevailing westerly winds in the Ferrel wind cells. Temperate deciduous forest, temperate evergreen forest, chaparral, and grassland biomes occur here. The **polar zone** extends above 60° latitude north and south; its climate is determined by the polar wind cells. Boreal forest and tundra biomes are located in the polar zone. Desert biomes are found in very dry areas of all three climatic zones.

▶ **Figure 7.4 Look Alikes**
The *Euphorbias* growing in the African Namib Desert Ⓐ look very similar to the organ pipe cacti (*Stenocereus*) Ⓑ, which are abundant in North American deserts. However, the two plants are genetically distinct.

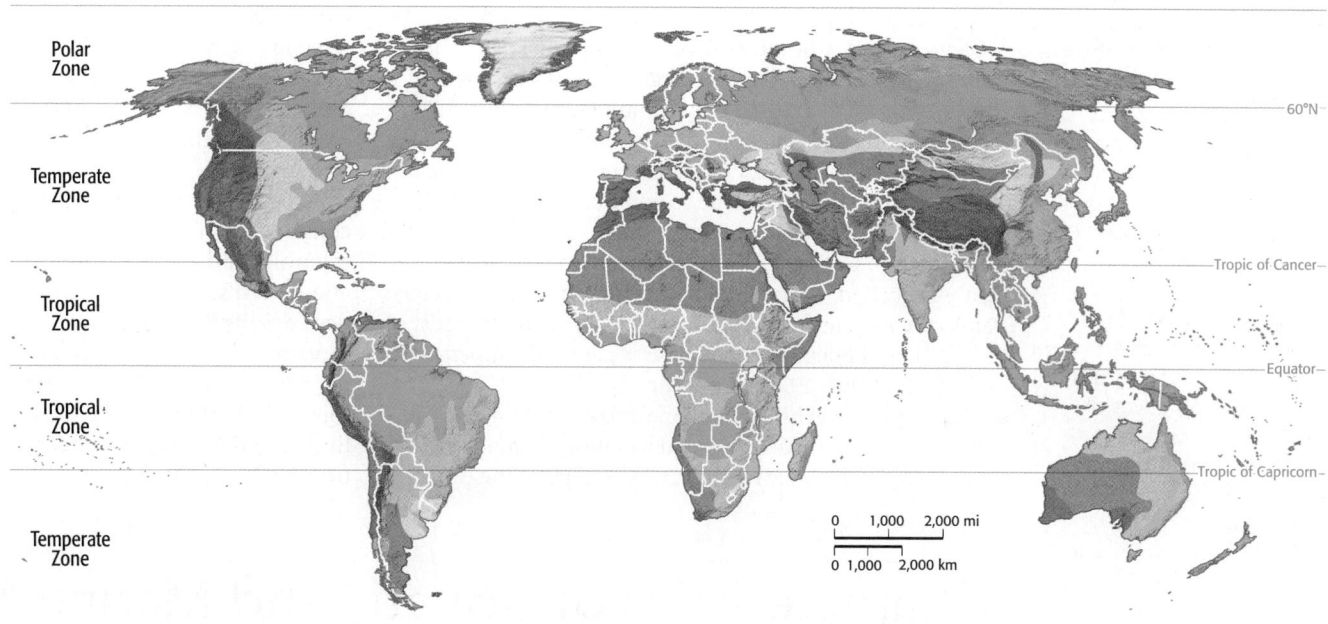

- ● Temperate evergreen forest
- ○ Temperate grassland
- ◐ Temperate deciduous forest
- ● Tropical rain forest
- ● Tropical seasonal forest
- ◐ Savanna
- ● Desert
- ○ Tundra
- ◐ Taiga
- ● Chaparral
- ◐ Mountainous regions
- ○ Ice cap

► **Figure 7.5 Earth's Biomes**
Ten major biomes plus mountainous regions are distributed across Earth's land masses. At this scale irregularities in boundaries and local variation in biome distribution are not represented.

Within these broad climatic zones, the distribution of biomes is influenced by the location of mountain ranges and by the proximity to large bodies of water, such as the ocean. Mountains alter the circulation of air and influence temperature and precipitation in the regions around them.

Proximity to water affects climates because land surfaces heat up and lose their heat more quickly than do bodies of water. As a result, daily and seasonal temperatures vary widely at locations within the interior of continents, away from large bodies of water. Such areas are said to have **continental climates**. Within the temperate zone, most grassland and desert biomes have continental climates. Lands near oceans or large lakes experience much less variation in temperature; such regions have **maritime climates**. The climates of the temperate deciduous forest and temperate evergreen forest biomes range from maritime near coasts to continental inland.

At first glance, the distribution of biomes within a climatic zone and from continent to continent may seem to follow no discernible pattern (**Figure 7.5**). The tropical rain forest biome, for example, extends throughout northern South America, but is found primarily along the western coast and central regions of equatorial Africa. The temperate deciduous forest biome dominates western Europe, but in North America this biome is confined to the east.

A much clearer pattern emerges if you plot the distribution of biomes according to average annual temperatures and precipitation (**Figure 7.6**). All of Earth's biomes fall into the triangular area between cold and dry, warm and dry, and warm and wet. Note that because cold air cannot hold much moisture, there are no regions with low temperatures and high precipitation (see Module 3.6).

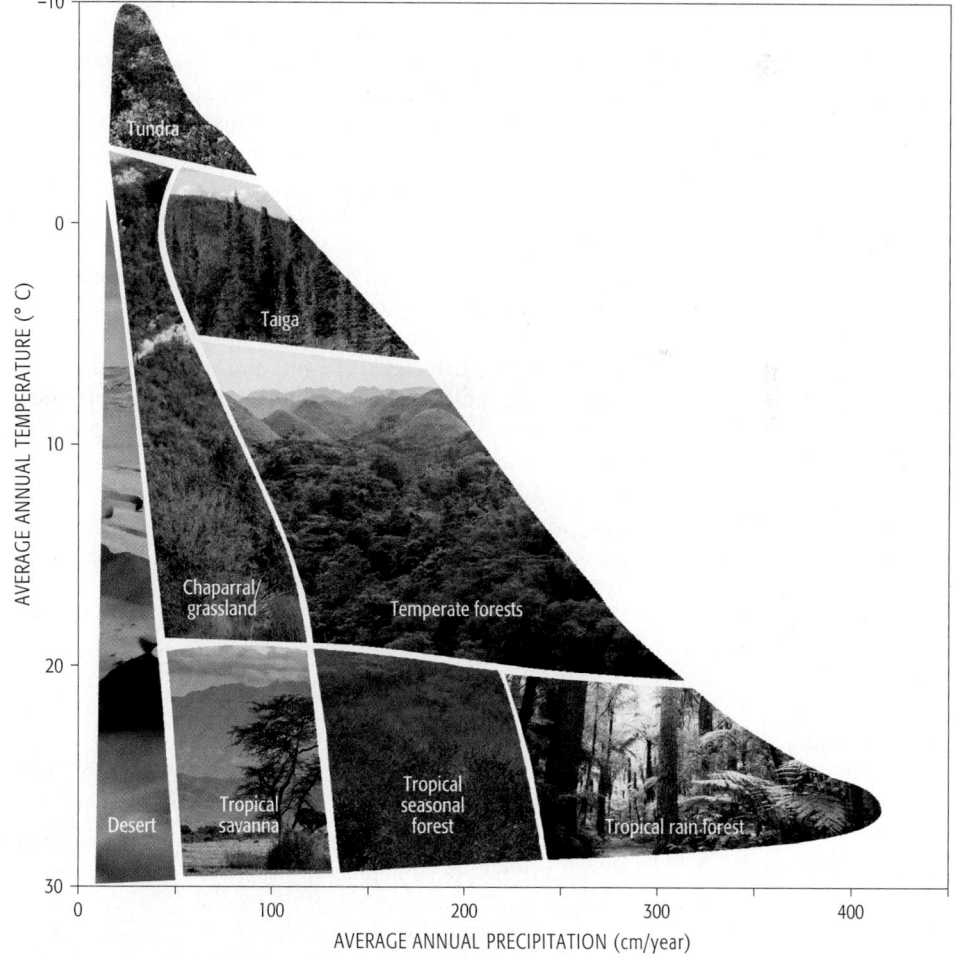

▲ **Figure 7.6 Temperature and Precipitation**
The tiles in this triangular mosaic depict the range of average temperatures and rainfall in which each biome occurs.

Source: Whitaker, R.H., 1975. *Communities and ecosystems.* New York: Macmillan.

189

Where
YOU
LIVE

What terrestrial biomes occur near you?

Scientists and land managers divide the broad biome types into *ecoregions* based on the most important species and unique environment for different locations. You can find a map of North American ecoregions at the EPA website. What ecoregions occur within 100 miles of your home or college? What important plant and animal species typify these ecoregions?

The U.S. Global Change Program provides detailed descriptions of and information about the likely effects of future climate change on U.S. ecoregions on the Explore page of their Global Change website. Are there threatened or endangered species in the ecoregions near you? How is climate change likely to alter these ecoregions?

Total annual precipitation, in the form of rain or snow, is a major factor in the distribution of biomes. But the amount of water available to organisms in a biome is also influenced by the rate of evaporation (see Module 3.7). Evaporation rates generally increase with increasing temperature. This means that if two regions receive the same amount of rainfall, the region that is warmer will have less water available. For example, Tucson, Arizona, and Pueblo, Colorado, both receive an average of about 310 mm (12 in.) of rain each year. Tucson, with an average annual temperature of 20.3 °C (68.5 °F), is in the midst of a desert, and Pueblo, with an average annual temperature of 10.9 °C (51.7 °F), is surrounded by grassland.

Each terrestrial biome is characterized by a unique climatograph depicting seasonal variations in temperature and precipitation (see Module 3.7). Species of the temperate deciduous forest, for example, are able to tolerate cold winters but require warm, wet summers. The species of the temperate evergreen forest, by contrast, can tolerate drier conditions during the warm summer months.

Although climate is the most important factor influencing the distribution of biomes, factors such as soil and fire may also play a role, particularly where one climate region merges, or transitions, into another. For example, in the climatic transition between grassland and desert, sandy, well-drained soils are likely to support desert. Soils with more silt and clay, which retain more moisture, are likely to support grassland. Likewise, in the climatic transition between grassland and deciduous forest, such as in eastern Kansas, grassland occurs where fires are frequent; the absence of fire allows deciduous trees such as oaks to grow.

Characteristics of Aquatic and Marine Biomes

■ Differences in salinity and patterns of flow distinguish the biomes found in Earth's waters.

Over two-thirds of Earth's surface is covered with water, and much of Earth's life is immersed in it. Over 99.7% of the liquid water on Earth's surface is found in its saline oceans; the remainder occurs as fresh water in wetlands, rivers, and lakes. Freshwater ecosystems are said to be **aquatic**, and saline ocean ecosystems are said to be **marine** (Figure 7.7).

Aquatic biomes include wetlands, streams, and lakes. Wetland biomes represent a transition from terrestrial to aquatic and marine environments. They include a variety of ecosystems in which soils are saturated with water for large portions of the year. Gravity moves fresh water continuously down streams and rivers, and the characteristics of ecosystems in this biome are determined by this flow. Where water is impounded in lakes and ponds, water movement is much more subtle. In this biome, ecosystems vary with respect to water depth and proximity to the shoreline.

The estuary biome represents the transition between freshwater rivers and oceans. The plants and animals of this biome are uniquely adapted to changes in salinity related to seasonal variations in stream flow and the daily tidal cycle.

Oceans include several biomes that are influenced by water depth and the proximity to shore. Organisms of the intertidal zone biome must be able to tolerate daily cycles of immersion and exposure, as well as the physical action of the ocean surf. In a manner similar to lakes, ocean biomes below the low tide mark vary with respect to distance from land and water depth.

QUESTIONS 7.1

1. Although they are not genetically related, plants in Asian deserts resemble those found in North American deserts. Explain why?

2. How do Earth's three major climatic zones relate to the distribution of global wind cells?

3. Why are winter and summer temperatures in the coastal city of Santa Monica, California, milder than in San Bernadino, 40 miles inland?

(MES) For additional review, go to **MasteringEnvironmentalScience**

► Figure 7.7 **Water Life**
Aquatic and marine organisms, such as these fish and the plankton upon which they feed, share adaptations that allow them to extract O_2, CO_2, and essential nutrients from their liquid environment.

FOCUS ON SCIENCE

Shifting Biomes

Is global warming changing the distribution of plants and animals on Earth?

Wolfgang Lucht and his coworkers at the Humboldt University in Berlin, Germany, are interested in determining how climate change will affect the worldwide distribution of biomes (**Figure 7.8**). To do this, they are using images taken by *remote sensors* on Earth-orbiting satellites.

Remote sensors are able to measure a variety of characteristics of Earth's surface. For example, the relative amounts of green and red light provide a direct measure of leaf cover. A satellite image is a composite of thousands or even millions of individual sample points, or pixels. Each pixel may represent anywhere from a few square meters to tens of square kilometers of land. By comparing satellite images obtained at different times, scientists can detect changes in the distribution of biomes on continents. This kind of analysis is widely used by scientists to measure changes over time.

The time that Lucht is most interested in is the future. How will the distribution of Earth's biomes change in the different scenarios of predicted greenhouse emissions and associated global warming? He and his associates address this question by using *computer simulation models* that are based on correlations

▲ **Figure 7.8 Biome Change Forecaster**
Wolfgang Lucht and his colleagues use data from satellite images and climate simulation models to predict the effects of global warming on the distribution of biomes.

between past changes in climate and the distribution of biomes and on knowledge of how plants respond to changes in temperature and moisture (see Module 9.5). These simulation models allow Lucht and his team to perform "what if" experiments about the likely redistribution of biomes given different scenarios for reducing greenhouse gas emissions.

Results from Lucht's Berlin research group form the basis for the Intergovernmental Panel on Climate Change's predictions of how global warming will cause biomes to shift (**Figure 7.9**). In North America and Asia, they predict that boreal forests will spread north into areas that are now tundra, while temperate grasslands will shift into areas where boreal forests now grow. The distribution of deserts and tropical biomes are also likely to undergo significant changes. The exact nature of these changes is difficult to predict because it is uncertain how warmer temperatures will affect rainfall.

To resolve these uncertainties, Lucht and his team are now working to incorporate knowledge from recent research into their computer models. They are also investigating how wildfire and human land use may affect future patterns of change.

1. What is the nature of the data that Lucht and his team use in their models?

2. How might the distribution of biomes affect the economic activity of a region?

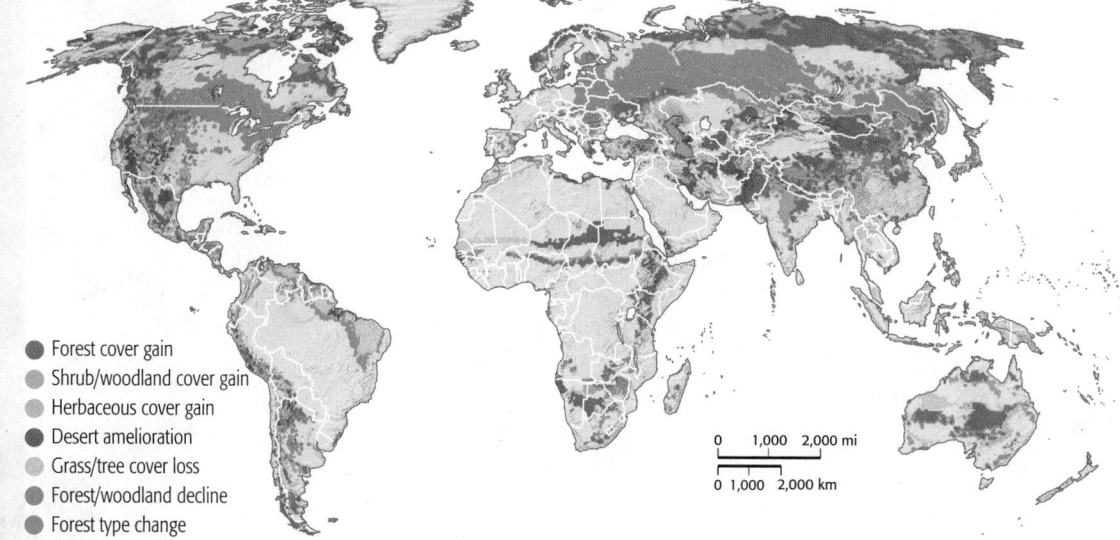

- Forest cover gain
- Shrub/woodland cover gain
- Herbaceous cover gain
- Desert amelioration
- Grass/tree cover loss
- Forest/woodland decline
- Forest type change

0   1,000   2,000 mi
0   1,000   2,000 km

◄ **Figure 7.9 Future Biome Shifts**
This map depicts likely changes over the next century in the distribution of general biome types if current trends in global warming prevail. Note particularly the increase in forest and woodland cover (green) in the Arctic, and forest and woodland decline (orange) in central North America and Asia. The tan regions are where biomes are projected to remain the same.

Source: Fischlin, A., G.F. Midgley, J.T. Price, R. Leemans, B. Gopal, C. Turley, M.D.A. Rounsevell, O.P. Dube, J. Tarazona, A.A. Velichko et al., 2007. Ecosystems, their properties, goods, and services. Climate Change 2007: Impacts, Adaptation and Vulnerability. Contribution of Working Group II to the *Fourth Assessment Report of the Intergovernmental Panel on Climate Change*, M.L. Parry, O.F. Canziani, J.P. Palutikof, P.J. van der Linden, and C.E. Hanson, eds. Cambridge: Cambridge University Press, 211–272.

7.2 Tropical Biomes

BIG IDEA The tropical zone, which extends approximately 25° north and south of the equator, supports some of Earth's most productive and diverse biomes. Throughout the tropics, average monthly temperatures are warm (generally above 20 °C or 68 °F) and vary little throughout the year. Except in high mountains, temperatures do not drop below freezing. The total amount of precipitation and the intensity and duration of the dry season determine whether a tropical location supports a rain forest, a seasonal forest, or a savanna.

Tropical Rain Forest

■ Tropical rain forest is home to more plant and animal species than any other biome.

Tropical rain forest occurs where annual rainfall is greater than 2,000 mm (80 in.) and falls throughout the year (Figure 7.10). The plentiful rainfall and warm climate support an enormous diversity of plants and animals.

If you could stand in the middle of a tropical rain forest, trees would literally tower overhead. This forest has several distinct layers of plants (Figure 7.11). The leafy "ceiling" of the forest, called the canopy, consists of hundreds of species of broad-leaved, evergreen trees that grow 30–40 m (98–131 ft) tall. Occasional emergent trees poke above the rest of the canopy, reaching heights of 55 m (180 ft). Little light filters through the thick canopy to the layers of vegetation in the mid- and understory. However, fallen trees create openings or gaps in the canopy; light that shines through these openings allows saplings, shrubs, and herbs to flourish.

Because of the consistently favorable climate, net primary production in tropical rain forests is much greater than in any other terrestrial biome (see Module 6.5). This high rate of productivity is a bit surprising, given that rain forest soils contain little organic matter and few nutrients. The high productivity depends more on rapid recycling of nutrients than on nutrients stored in soil. Insects, fungi, and bacteria rapidly decompose leaf litter in the warm, moist soil. The nutrients released by this decomposition are quickly taken up by the roots of the fast-growing plants.

The complex layers of vegetation in the rain forest provide favorable habitat for an enormous array of animals. In fact, the tropical rain forest hosts more animal species than any other biome. There are, of course, the species you expect, such as monkeys,

▶ **Figure 7.10 Tropical Rain Forest Climate**
As this climatograph for Iquitos, Peru, in the Amazon Basin indicates, rainfall is persistent and temperatures are continually warm throughout the year in tropical rain forests. These conditions support remarkably diverse forests of broad-leaved evergreen trees such as this.

IQUITOS, PERU (LATITUDE = 4° S)

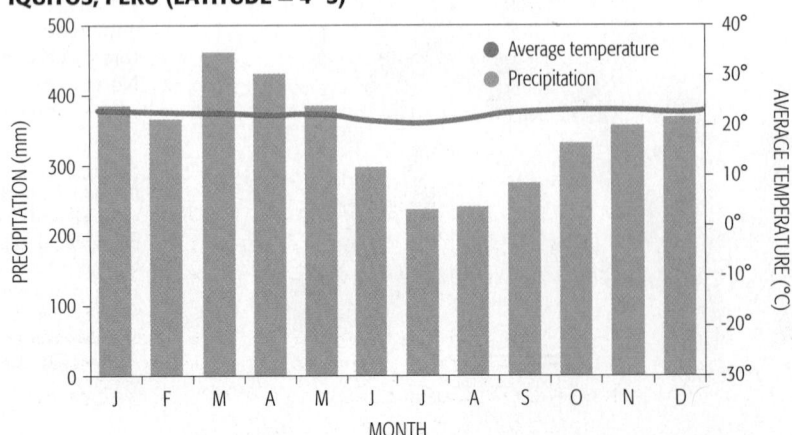

▲ Figure 7.11 **Complex Structure**
The vertical arrangement of foliage of different trees in the tropical rain forest creates a diverse array of habitats for animals. For example, many monkey species spend their entire lives in the forest canopy, whereas other species are confined to the forest floor.

snakes, tree frogs, and a great variety of birds. But the vast majority of the animals here are insects, many of which have yet to be described and named. A single hectare (2.47 acres) of tropical rain forest may support over 2,000 species of beetles.

Much of the diversity in the tropical rain forest is a consequence of evolution that has produced very specialized ecological niches (see Module 6.1). The value of species diversity in tropical rain forests cannot be overstated. The integrated activities of the many species sustain this biome's rapid nutrient cycling and high productivity. This diversity also provides a variety of benefits to humans. Many rain forest plants and animals have been used in the development of medicines and pest-resistant food crops (see Module 8.1). Given that well over half of all tropical species have yet to be discovered, we can conclude that many of these benefits have yet to be realized.

Deforestation, or the clearing of forests, is without question the greatest threat to tropical rain forests (see Module 13.3). Where endemism is high, destruction of forests eliminates a large fraction of habitat for many species, putting them at risk for extinction.

When tropical rain forests are cut and cleared for farming, nutrient cycles are disrupted and the topsoil is easily eroded. Without topsoil, crops fail. But where farming has compacted the soil and caused the loss of nutrients, it is very difficult to reforest the area by planting seeds or young trees. Deforestation also reduces the ability of the rain forest to take up and store carbon, thereby contributing to the increase in greenhouse gases in Earth's atmosphere (see Module 13.4).

Q *What can we do to prevent further loss of tropical forest biomes?*
Christopher Shad, Cypress College

Norm: (A) Poverty in many tropical zone countries encourages deforestation. Many governments and environmental organizations are pursuing strategies to diminish poverty in ways that encourage tropical forest conservation.

Tropical Seasonal Forest

■ Seasonal drought determines the distribution of tropical seasonal forest.

Away from the equator and toward the Tropics of Cancer and Capricorn (23° N and S latitude), annual rainfall diminishes to 1,500–2,500 mm (60–98 in.), and months with little or no rain become more common. Here, **tropical seasonal forest** replaces tropical rain forest. The wettest regions of this biome support forests with canopy trees that are 40 m tall (131 ft); the driest regions support only scrubby woodlands.

In tropical seasonal forests, wet and dry seasons each span about half of the year (Figure 7.12). Dry seasons are longer and more intense in areas that are shielded from the wet easterly trade winds. Such areas include the west side of the Andes and the west sides of many tropical islands, such as Hawaii.

During the rainy season, plants thrive and animals find abundant food, water, and cover. But the organisms that live in this biome must also have adaptations that

allow them to survive drought. When the dry season begins, many trees lose their leaves, a response that reduces their need for water. Such trees are said to be drought deciduous.

During the wet season, large herbivores such as tapirs in South America and the black rhinoceros in Africa roam throughout the forests (Figure 7.13). In the dry season, large animals such as these must migrate to find water. Disturbance and fragmentation of landscapes create challenging barriers to the seasonal migration of the animals of the tropical seasonal forest.

Like tropical rain forests, large expanses of the world's tropical seasonal forests have been cut and burned and the land converted for agriculture. Disturbance and fragmentation of landscapes create challenging barriers to the seasonal migration of the animals of the tropical seasonal forest.

▼ **Figure 7.12 Tropical Seasonal Forest Climate**
A distinct dry season from December through March typifies the climate for the tropical seasonal forest in northern Venezuela. Many of the broad-leaved trees lose their leaves during this dry season.

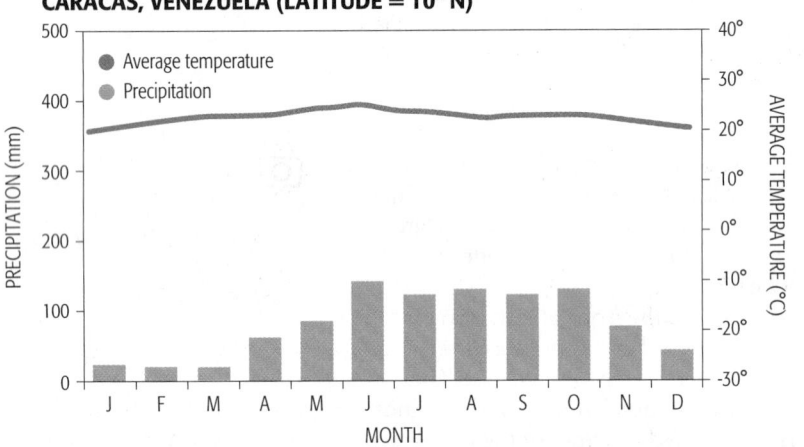

CARACAS, VENEZUELA (LATITUDE = 10° N)

▲ Figure 7.13 **The Brazilian Tapir**
Because of the deforestation of tropical seasonal forest, this once-common horse relative has disappeared over much of its range.

Tropical Savanna

■ Tropical savannas are habitat for large herds of animals that graze on the sea of grass.

Tropical savanna occurs where rainfall is highly seasonal and drought conditions generally persist for more than half of the year (Figure 7.14). This is an ecosystem of open grasslands and a few scattered trees. Like the trees of the tropical seasonal forest, many of the trees that live in the tropical savanna are drought deciduous.

In addition to seasonal moisture deficits, several environmental factors influence the growth and diversity of plants in this biome. Large herds of grazing animals feed on the plants, and frequent fires set the tinder-dry grasslands ablaze. The combination of animal grazing, drought, and fire limits the establishment of trees. As a result, there are sufficient light and soil resources to support the characteristic carpet of grass.

Tropical savanna is a biome of seasonal feast and famine. With the rain, grasslands flourish, supporting large populations of grazing animals. Kangaroos and emu thrive in the tropical savanna of Australia, and wildebeests, zebras, and giraffes roam the tropical savanna of Africa. These herbivores feed a diverse array of predators, including lions and cheetahs. When tropical savanna is dry, large wild herds of grazers and their predators migrate to where water and food are more abundant (Figure 7.15).

The climates where tropical savannas are found overlap considerably with those of tropical seasonal forests. This has led to debate among scientists about the role that humans have played in creating and maintaining tropical savannas. There seems to be

▲ Figure 7.15 **A Seasonal Resource**
The timing of rain varies across the large expanses of tropical savanna in Africa. The large mammals of this biome, such as these elephants, must therefore migrate throughout the year to obtain this precious resource.

growing agreement that, by grazing domestic animals and setting fires, humans have increased the extent of tropical savannas in parts of Africa, South America, and Australia.

The nomadic herders who live in tropical savannas have traditionally moved their cattle in imitation of the migratory patterns of the wild herds. Today, the boundaries of countries and private property limit such migration. This, in turn, promotes overgrazing, the loss of grasses, erosion, and eventually the conversion of savanna to desert. Global warming is accelerating this transition in many regions, especially in sub-Saharan Africa. The growing deserts threaten the well-being of millions of people.

QUESTIONS 7.2

1. Describe two factors that contribute to the very high productivity of the tropical rain forest biome.

2. How does being deciduous help trees survive the dry season?

3. The dry season is longer in tropical savannas near the Tropic of Cancer than in tropical savannas nearer the equator. Why is this so?

(MES) For additional review, go to **MasteringEnvironmentalScience**

▼ Figure 7.14 **Tropical Savanna Climate**
An understory of grasses with scattered trees typifies the tropical savanna of northern Botswana. In this biome, the dry season is longer, and less rain falls during the rainy season than falls in the tropical seasonal forest.

MAUN, BOTSWANA (LATITUDE = 20° S)

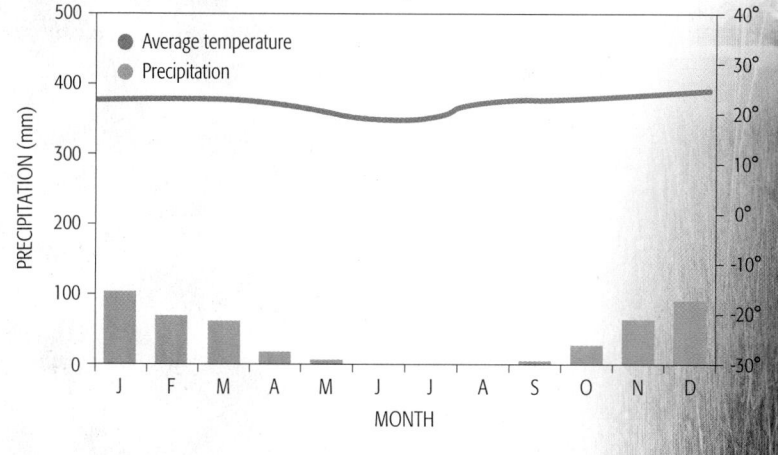

7.3 Temperate Biomes

BIG IDEA Over 60% of Earth's landmass is in the temperate zone, which extends from 25° to 60° latitude north and south of the equator. Most of this land is in the Northern Hemisphere on the continents of Asia, Europe, and North America. In the temperate zone, average annual temperature ranges from 5 °C to 20 °C (41–68 °F), and total annual precipitation ranges from 200 mm (8 in.) to over 2,000 mm (80 in.). The biomes in the temperate zone are temperate deciduous forest, temperate evergreen forest, chaparral, and temperate grassland. Seasonal changes in temperature and rainfall are especially important in determining the distribution of these biomes. Most primary production occurs during the warm growing season, which may be from 4 to 10 months long. Plant and animal activity is typically limited during the winter when temperatures may drop below freezing. Much of Earth's human population resides in the temperate zone, and human impacts on the biomes in this zone have been very significant.

Temperate Deciduous Forest

■ Little of temperate deciduous forest remains undisturbed.

As the name of this biome suggests, the **temperate deciduous forest** is dominated by broad-leaved trees that lose their leaves in the autumn and grow a new set in the spring. Temperate deciduous forest occurs in regions with moderate summers and cold winters. Precipitation is relatively even throughout the year (**Figure 7.16**). Here, the growing season is generally defined as the period between the last hard frost (nighttime temperatures of less than 0 °C) in spring and the first hard frost in autumn. In the northern portion of this biome, the growing season extends from May through September; in Florida and along the Gulf of Mexico, many plants retain their leaves and grow from early March through November.

The spectacular display of autumn leaf colors is one of the most notable features of the temperate deciduous forest. The annual loss and regrowth of leaves has important consequences for the forest ecosystem. The decomposition of autumn leaves returns essential nutrients and adds organic matter to forest soils. The flush of new leaves in the spring provides abundant food for large numbers of plant-eating insects. These insects are, in turn, eaten by animals such as birds and bats. The deep layer of leaf litter provides cover for a wide variety of organisms inhabiting the forest floor; it also supplies the energy that supports very complex food webs of microbes, fungi, invertebrates, birds, and amphibians (see Module 6.4).

▶ **Figure 7.16 Temperate Deciduous Forest Climate** There is reliable rainfall year-round in the temperate deciduous forest biome, represented here by a forest in central Massachusetts. Most of the trees in this biome survive the cold winter by losing their leaves.

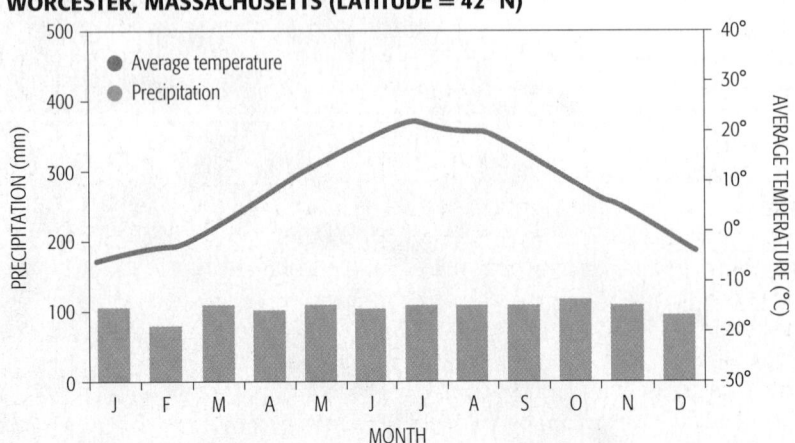

WORCESTER, MASSACHUSETTS (LATITUDE = 42° N)

Many bird species, such as the wood thrush, scarlet tanager, and ruby-throated hummingbird, migrate between tropical and temperate forests. They build nests and breed in the temperate deciduous forests of eastern North America during the spring and summer, when there is ample food. In the fall, they migrate to the tropical forests of Central and South America to feed over the winter months.

Human populations and urban development are particularly dense where temperate deciduous forests occur; as a result, very little of this biome has remained undisturbed. In the eastern United States, for example, from 50 to 70% of the deciduous forest was cleared for agriculture between 1700 and 1900 (Figure 7.17). Most of the remaining forest was significantly altered by livestock grazing and logging for fuel wood. After 1900, farmers abandoned their farms, and much of the cleared land lay fallow. In the ensuing years, trees gradually began to reinvade, creating large expanses of immature forest (see Module 6.6).

Today, urban development and sprawl are causing the fragmentation and loss of temperate deciduous forests in much of the eastern United States, Europe, and China (see Module 13.4). Habitat loss and fragmentation in temperate deciduous forests, combined with deforestation in tropical forests, have caused the populations of many migratory birds to decline. Human disturbance has left the temperate deciduous forest particularly susceptible to invasion by non-native plant and animal species.

▼ Figure 7.17 **Changing Forests**
Temperate deciduous and evergreen forest cover in the United States has changed considerably in the last 400 years.

Source of 1620 and 1920 maps: Greeley, W.B. 1925. The relation of geography to timber supply. *Economic Geography* 1:1–11.

Source of 2010 map: http://fia. fs.fed.us/slides/current-data.pdf

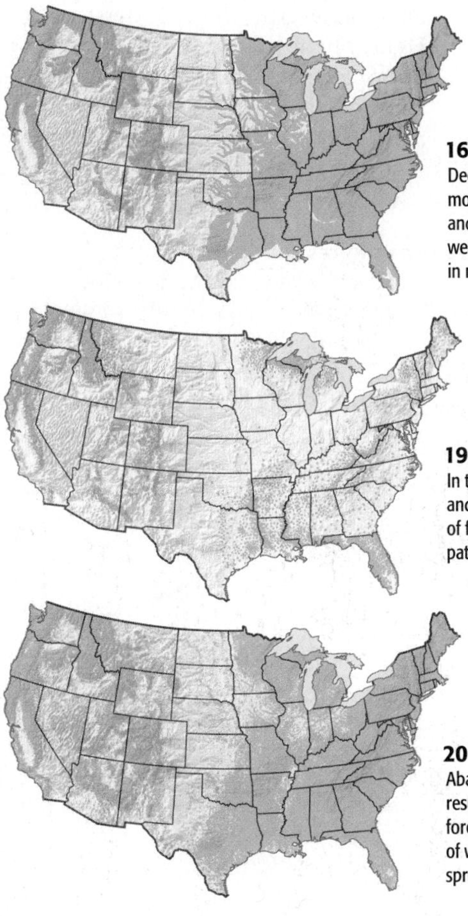

1620
Deciduous forests covered most of the eastern United States and evergreen forest dominated western landscapes, particularly in mountainous areas.

1920
In the 300 years that followed, logging and agriculture diminished the extent of forests in the West and left only scattered patches of forest in the East.

2010
Abandonment of croplands has resulted in a marked increase of forest cover in the East. Suppression of wildfire has contributed to the spread of evergreen forest in the West.

Temperate Evergreen Forest

■ The temperate evergreen biome includes some of the oldest and largest organisms on Earth.

The dominant trees of the **temperate evergreen forest** keep their leaves throughout the year. This biome generally experiences less precipitation and warmer temperatures than the temperate deciduous forest, and summer months are notably drier than winter months (Figure 7.18).

Where heavy winter rains and summer fogs provide ample year-round moisture, evergreen forests are especially lush and diverse, forming **temperate rain forests**. Temperate rain forests grow in coastal regions where temperatures are mild throughout the year. The coastal redwood forests of northern California and the cedar and fir forests of Oregon and Washington are examples of temperate rain forests.

In both the temperate evergreen forest and temperate rain forest, the dominant trees are towering evergreens; these evergreens include some of the largest and oldest organisms that have ever lived. The redwood and giant sequoia of California, the Douglas fir of the Pacific Northwest region of North America, and eucalyptus of Australia can grow to heights of over 90 m (295 ft) and live thousands of years.

In the temperate evergreen forests of the Northern Hemisphere, needle-leaved **conifers**—cone-bearing trees such as pine, spruce, and fir—dominate the canopy. Broad-leaved species, such as dogwood and maple, grow in the understory. In the Southern Hemisphere, the temperate evergreen forest is dominated by broad-leaved trees such as the kauri of New Zealand and eucalyptus trees of Australia.

Mature temperate evergreen forests and rain forests have a diverse array of tree sizes and ages, as well as an accumulation of dead trees, or snags, and fallen logs. Cavities in old trees, snags, and logs provide nesting sites for a variety of animals, including owls, flying squirrels, and bats. The woody litter of the understory provides favorable habitat for animals such as banana slugs and voles.

Logging and the suppression of wildfire are the two most important human impacts on these forests. Logging has diminished the extent of old-growth forests in many areas, threatening the survival of the animals that depend on the habitats found in mature and old-growth forests.

Prior to human intervention, naturally occurring wildfires played an important role in the ecology of the evergreen forests. But for the last century, humans—fearing for life and property and believing that fires were ecologically destructive—have worked to prevent or suppress wildfires. In many locations, the result has been the accumulation of litter and invasion of shade-tolerant trees. This has diminished the quality of habitat for many species, and has actually made these areas more prone to large fires that are difficult to manage (see Module 13.4).

▼ **Figure 7.18 Temperate Evergreen Forest Climate**
This biome occurs in regions with cool, moist winters and comparatively dry summers, as are found in northwestern Oregon. The old-growth evergreen forest has a complex structure with a variety of tree sizes and an understory with abundant shrubs, herbs, and ferns.

PORTLAND, OREGON (LATITUDE = 45° N)

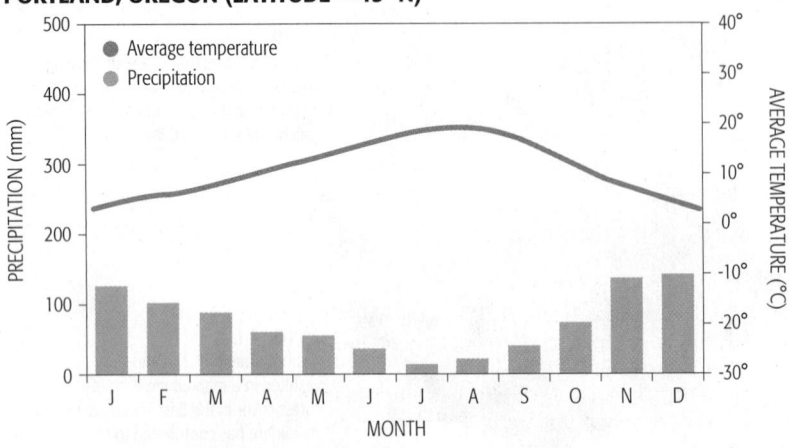

Using Fire to Conserve a Biome

How can fire conserve populations of Earth's largest trees?

The U.S. National Park System is, without doubt, one of the world's most ambitious efforts to conserve the natural diversity of entire biomes. But since its establishment, the management of wildfires within the national parks has been a dilemma. A century ago, the idea of allowing even small natural fires to burn seemed contrary to the goals of park conservation.

By the late 1960s, it had become obvious to many ecologists and firefighters that the suppression of natural fires was having very negative consequences, especially in temperate evergreen forests. Nowhere was this more apparent than in the majestic groves of giant sequoias, a type of temperate evergreen forest restricted to a few areas in the California Sierra Nevada. Prior to protection, low-intensity fires set by lightning or Native Americans burned across the floors of these forests every 5 to 15 years, consuming woody debris that had accumulated since previous fires. Without harming large trees, these fires created the soil conditions necessary for the seeds of the giant sequoia to germinate (Figure 7.19).

Ironically, instead of protecting the giant sequoias, the suppression of fire actually threatened their survival. Because there had been few fires during the past 80 years, large quantities of woody debris had accumulated. Shade-tolerant trees, such as fir and incense cedar, had invaded most sequoia groves. Because their seeds had not germinated, there were virtually no young sequoias. It was especially disconcerting that the accumulated undergrowth increased the likelihood of unnaturally intense fires, which can actually kill the giant sequoias (Figure 7.20).

In 1968, the National Park Service initiated a controversial plan to reintroduce fire to these ecosystems. Low-intensity prescribed fires were introduced. To prevent intense fires and minimize the risk to life and property, smaller trees and brush were first cut down. In setting prescribed fires, managers take special precautions to keep the fires from spreading to nontarget areas and minimize the impacts of smoke on air quality. Following prescribed fires, managers must monitor the spread of invasive non-native plants; sometimes they weed out the invasive species.

Today, the giant sequoia groves have regained their open, parklike structure, and young giant sequoias are once again abundant. Fire management in the giant sequoia groves serves as a model for the design and execution of prescribed fires in other temperate evergreen forests.

▶ Figure 7.19 **Beneficial Fire**
Natural fires in giant sequoia groves primarily burned litter and woody debris on the soil surface. The heat from these fires did not damage the forest canopy.

▲ Figure 7.20 **Effects of Fire Suppression**
When fires are suppressed, litter accumulates and shade-tolerant trees invade. Ⓐ Light surface fire keeps the sequoia forest understory open and provides a suitable habitat for sequoia seedling establishment. Ⓑ The triangular fire scars that are found on nearly all mature sequoia trees are evidence of repeated fires over hundreds of years. This superficial damage to the tree's exterior does kill the tree.

Chaparral

■ Summer drought is an especially prominent feature of this biome.

Chaparral is a biome of evergreen shrublands and low woodlands. It occurs where winters are mild and moist and summers are hot and dry (Figure 7.21). These climatic conditions are often said to be Mediterranean because they are characteristic of the lands surrounding the Mediterranean Sea. Similar climatic conditions occur on the western edges of continents between 30° and 40° latitude in southern California, Chile, South Africa, and southwestern Australia.

Chaparral vegetation includes a diverse array of evergreen shrubs, such as the chamise and manzanita of southern California. Most of these shrubs are **sclerophyllous**, which means "stony leafed." The thick, hard leaves of sclerophyllous shrubs are resistant to water loss and wilting, an adaptation that protects the plants from intense summer drought (Figure 7.22). Sclerophyllous shrubs such as the rosemary and thyme of southern Europe contain fragrant resins and oils that make them valuable as spices.

Fires occur naturally during the hot summer drought. The dense shrubby growth of the chaparral allows fires to burn intensely and spread great distances. Oils and resins in the leaves of many shrubs contribute to this flammability. But the shrubs of this biome are also highly adapted to such fires. Some can sprout from fire-resistant roots within weeks of burning. Others produce seeds that remain inactive in the soil until the heat of a fire stimulates their growth.

Few large mammals are native to the chaparral, probably because the shrubs provide little forage. However, many rodents and birds feed on the seeds and young seedlings beneath the shrub canopy. In chaparral woodlands, grass and acorns provide food for deer and herds of domestic pigs, sheep, and goats.

Humans have been altering the chaparral around the Mediterranean Sea for thousands of years. In Portugal and Spain, for example, much of the chaparral has been transformed into a low-intensity agricultural landscape dominated by cork oak trees. Farmers sustainably harvest the bark of these trees to make wine corks and insulation materials. In many parts of the world, chaparral has been transformed into agricultural uses such as vineyards and olive orchards.

Urban and suburban development has increased the frequency of fires in this biome (see Module 18.2). Development has also fragmented and diminished the extent of the native chaparral and made way for the invasion of alien species of plants.

▼ Figure 7.21 **Chaparral Climate**
This biome, found in places such as southern California, is dominated by shrubs and low trees that are tolerant of severe summer drought.

SANTA BARBARA, CALIFORNIA (LATITUDE = 34° N)

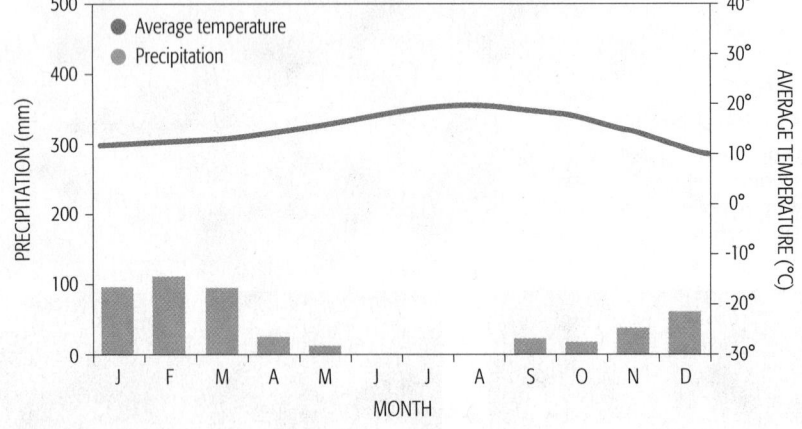

▲ Figure 7.22 **Stonelike Leaves**
The sclerophyllous leaves of manzanita (*Arcostaphylos*) are resistant to the hot, dry summers of California's chaparral.

Temperate Grassland

■ Agriculture and cattle grazing have altered over 90% of Earth's temperate grasslands.

Grasses interspersed with a diverse array of herbs are the dominant vegetation in the **temperate grassland** biome. The climate of this biome is too dry to support forest, woodland, or shrubland but wet enough to prevent the land from becoming desert. In most temperate grasslands, the climate is decidedly continental—winters are often long and cold, but summers can be quite hot (Figure 7.23).

The productivity and diversity of temperate grasslands increase with increasing amounts of rainfall. In the wettest areas of the biome, such as the Pampas region of Uruguay and Argentina and the tall-grass prairies of Illinois and eastern Kansas, grasses may grow as tall as 2 m (7 ft). The roots of these grasses grow together in a dense mat, forming a sod that is very rich in organic matter.

▼ Figure 7.23 **Temperate Grassland Climate**

This biome has cold winters and hot summers and receives only modest rainfall. It is dominated by grasses and herbs whose roots form a dense sod. The Konza Prairie of eastern Kansas is one of the very few temperate grasslands that has been preserved for study through the intervention of The Nature Conservancy and Kansas State University. It remains relatively undisturbed by human activities.

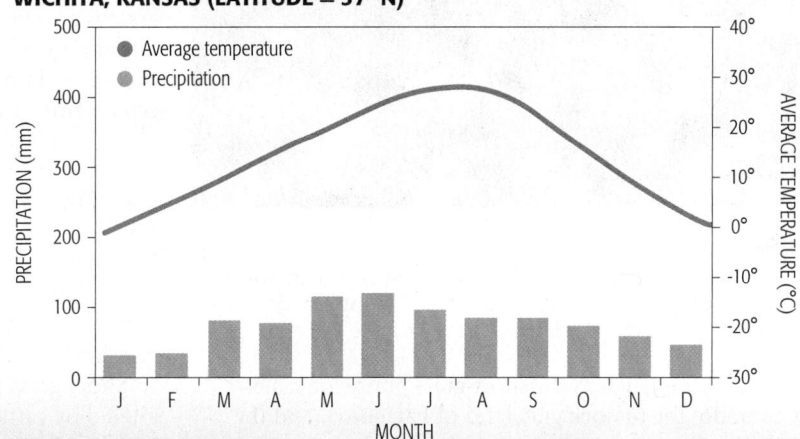

WICHITA, KANSAS (LATITUDE = 37° N)

201

QUESTIONS 7.3

1. The growing season in the temperate deciduous forest biome varies from four months in Maine to over nine months in northern Florida. What factor is most responsible for this variation?

2. Compare the climate of the temperate deciduous forest biome to that of the temperate evergreen forest biome. What is the main distinguishing factor?

(MES) For additional review, go to **MasteringEnvironmentalScience**

▲ **Figure 7.24 Bunch Grasses**
These tussocks of fescue in the arid grasslands of Patagonia in southern Chile are adapted to drought and are able to grow back after being grazed. Individual clumps can survive for centuries.

▼ Figure 7.25 **Much of the Grassland Biome Today**
A century ago, this hay field on the high plains in Alberta, Canada, was a diverse grassland; today, it supports mainly wheat.

In drier areas, such as the short-grass prairie of eastern Colorado, the tussock grassland of Patagonia, and the Russian steppe, grasses are comparatively short, growing to about 0.5 m (1.5 ft). Here, bunchgrasses that grow in distinct clumps or tussocks are common (Figure 7.24). The roots of these grasses typically do not form sod.

In grasslands where summer rainfall is common, most of the grasses and herbs are perennials—plants that survive for several years. Where summers are dry, as in Mediterranean temperate grasslands, the dominant plants are annuals that grow from seeds each year, reproduce, and then die at the end of the growing season.

Much of the biomass of temperate grasslands is located underground in extensive root systems. These roots efficiently extract water and nutrients from the soil and protect the very fertile soil from erosion. As the harsh winter begins, grasses and herbs die back to the ground. In the spring, plants revive quickly, fed by the nutrients stored in their roots. Because of their ability to resprout from roots, grassland plants are resilient to fire and animal grazing.

Prior to the modern expansion of agriculture, the grasses and herbs of this biome supported a diverse array of herbivores, as well as the omnivores and carnivores that preyed upon them. Only remnants of those original ecosystems can be found today.

No biome on Earth has been so extensively transformed by human actions as the temperate grassland. More than half of the world's supply of plant and animal food comes from regions that once supported temperate grasslands. Virtually all (99.9%) of the North American tall-grass prairie has been converted to agriculture (Figure 7.25). Unfortunately, once the sod has been plowed, native grasslands can be restored only with great effort.

Short-grass prairie and steppe regions have been altered by cattle grazing or, where irrigation is available, by farming. In some regions, grazing and farming have proven unsustainable. These disturbances have also made the grasslands much more susceptible to invasion by non-native species. The combination of overgrazing and global warming is causing the driest temperate grasslands to change into desert.

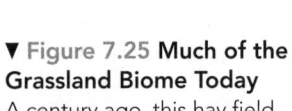

7.4 Polar Biomes

BIG IDEA Polar environments are generally found above 60° north or south latitude, where very cold annual temperatures (less than 5 °C, or 41 °F) and short growing seasons limit the abundance and diversity of life. Two biomes occur here—boreal forest and tundra. Most of these biomes are found in the Northern Hemisphere. Except for ice-covered Antarctica, the Southern Hemisphere has little land in the polar zone.

Boreal Forest

■ Boreal forest trees thrive in cold and wet conditions.

Where the growing season is less than four months and winters are long, dry, and bitterly cold, temperate forests and grasslands give way to the open coniferous forests of the **boreal forest** biome (Figure 7.26). Russians call the vast boreal forest of Siberia **taiga**. Similar forests are found in northern Europe, Canada, and Alaska. Annual precipitation in this biome varies between 500 and 1,500 mm (20 to 59 in./yr), most falling as snow. The cold weather here slows the rate of evaporation, and the relatively flat terrain impedes drainage. As a result, soils usually remain wet or boggy throughout the growing season.

Evergreen conifers such as spruce and fir dominate the boreal forest. Their small, needlelike leaves and drooping branches shed heavy snowfall that otherwise would break their branches. The understory of the forest is composed of low shrubs, herbs, and lichens. Boreal forest trees and understory plants are able to photosynthesize at temperatures very near freezing, completing pollination and seed production within the very short growing season. The harsh climate actually "shelters" the biome from invasive alien species.

Animals such as moose, bear, lynx, snowshoe hare, and spruce grouse live in boreal forests year-round. These animals have well-insulated bodies. Many animals hibernate through the coldest winter months. In summer, birds such as sparrows and warblers breed in the forest, migrating north from temperate and tropical biomes.

Near the transition from the temperate zone, boreal forest trees form a dense forest, reaching heights of over 20 m (66 ft). Farther north, **permafrost**, permanently frozen soil 30–100 cm (11–39 in.) beneath the soil surface, becomes more prevalent. The permafrost, harsher winters, and shorter growing season limit both productivity and species diversity. Here, the dominant black spruce trees grow only 3–5 m (10–17 ft) tall, and the understory is dominated by mosses and lichens.

Lightning-set fires naturally burn over millions of hectares of boreal forest each year. Therefore, the expansive boreal forest landscapes of Alaska, Canada, and Siberia comprise a mosaic of large patches of forest in different stages of succession (see Module 6.6). Global warming appears to be producing drier conditions in many parts of this biome. Consequently, fires have become more frequent and extensive (see Module 9.4).

▼ Figure 7.26 **Boreal Forest Climate**
The Russian word for this biome is *taiga*, which translates loosely as "little sticks." This is an apt description of the northern portions of this biome, as in Canada's Northwest Territory, where short growing seasons and boggy soils limit tree size.

FORT SMITH, CANADA (LATITUDE = 60° N)

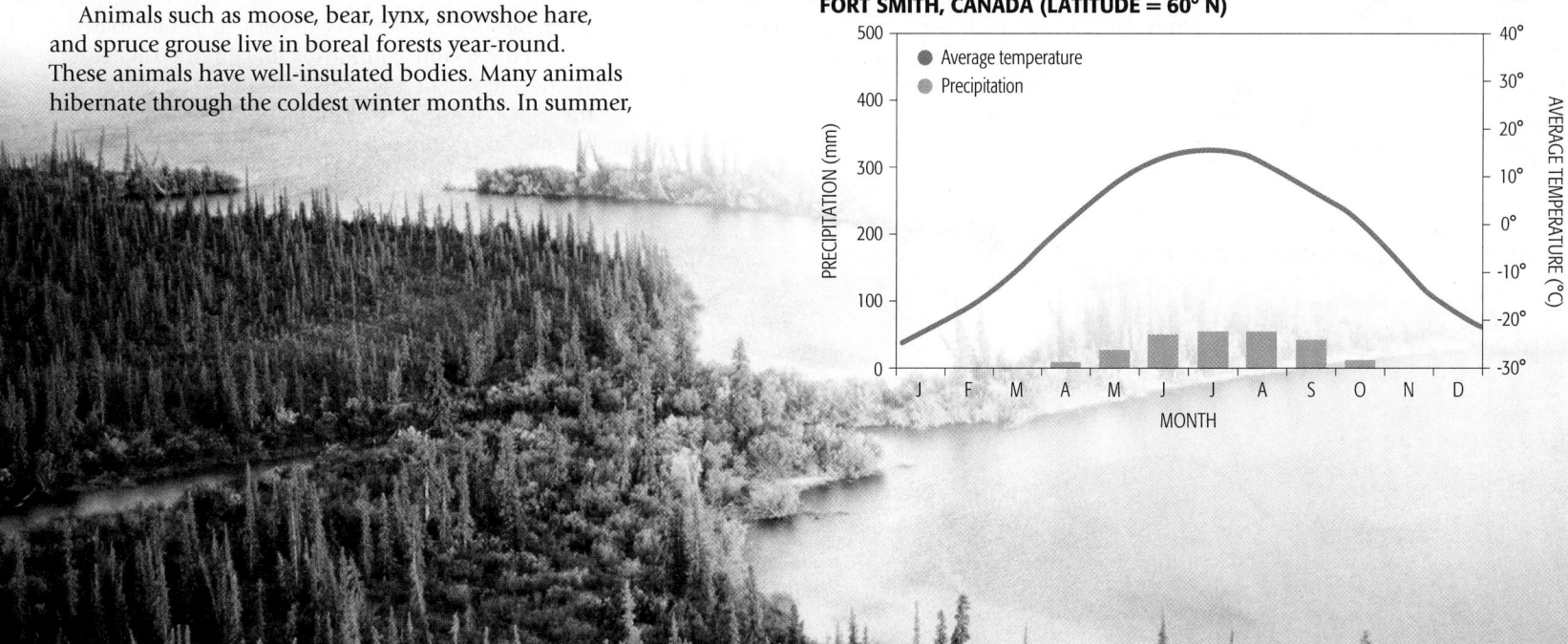

QUESTIONS 7.4

1. The boreal forest extends much farther north in coastal locations than in interior continental locations. Explain this phenomenon.

2. Describe two adaptations that contribute to the ability of tundra plants to survive their harsh environment.

3. What is permafrost and how does its depth differ between boreal forest and tundra biomes?

(MES) For additional review, go to **MasteringEnvironmentalScience**

In recent years, bark beetles have infested large areas of boreal forest in North America and Asia. These infestations are thought to be related to the longer growing seasons caused by global warming. Bark beetles normally complete only a single growth cycle in a year. Longer growing seasons allow them to complete two growth cycles each year, boosting their population.

In the southern boreal forest, logging for timber and paper pulp is the most important human disturbance. In Siberia, for example, over 12 million hectares (30 million acres) of boreal forest are cut each year, and the rate of cutting is increasing. Because of the cold, wet conditions, reforestation in taiga is very slow.

Tundra

■ Tundra exhibits a treeless landscape of subtle beauty and harsh conditions.

In the northernmost polar regions, where growing seasons are shorter than three months and winters are most harsh, boreal forest gives way to **tundra**, treeless landscapes dominated by grasses, herbs, and low shrubs (Figure 7.27). Tundra receives less precipitation than many deserts, only 100–500 mm/yr (4–20 in./yr). Even so, the limited evaporation associated with cold weather causes most tundra soils to remain wet through the summer.

The surface or active layer of tundra soil melts in spring, but permafrost remains 20–50 cm (8–20 in.) below the surface (Figure 7.28). As the active layer thaws, tundra plants rely on nutrients stored in their roots to begin to grow and flower while still surrounded by snow. Freezing temperatures may occur at any time during the summer; however, tundra plants remain resistant to cold and can resume growth immediately when conditions are favorable. Because of global climate warming, the active soil layer is receiving more heat each year. As a result, the depth to the permafrost is increasing in many tundra regions (see Module 9.4).

The summer growth of tundra plants fuels a complex food web. Herbivores include rodents, such as lemmings, and grazers, such as caribou (called reindeer in Eurasia). Dominant carnivores include the arctic fox, wolves, and grizzly bears. The very moist summertime conditions provide extensive breeding habitat for mosquitoes and black flies, both of which prey on mammals.

▲ Figure 7.28 **Permafrost**
The permanently frozen soil is visible beneath the active layer where a stream has eroded into tundra on the North Slope of Alaska.

Several human cultures have developed sustainable lifestyles in the seemingly inhospitable tundra. The North American Inuit sustain themselves as hunter-gatherers, and the Lapps of northern Scandinavia make a living by herding reindeer. More recently, humans have exploited some tundra areas for the mineral resources of oil and gas.

▼ Figure 7.27 **Tundra Climate**
This climatograph from northernmost Alaska depicts harsh winters and very short growing seasons. This climate pattern is similar to that of the tundra ecosystems at the base of Mt. McKinley shown in this photograph.

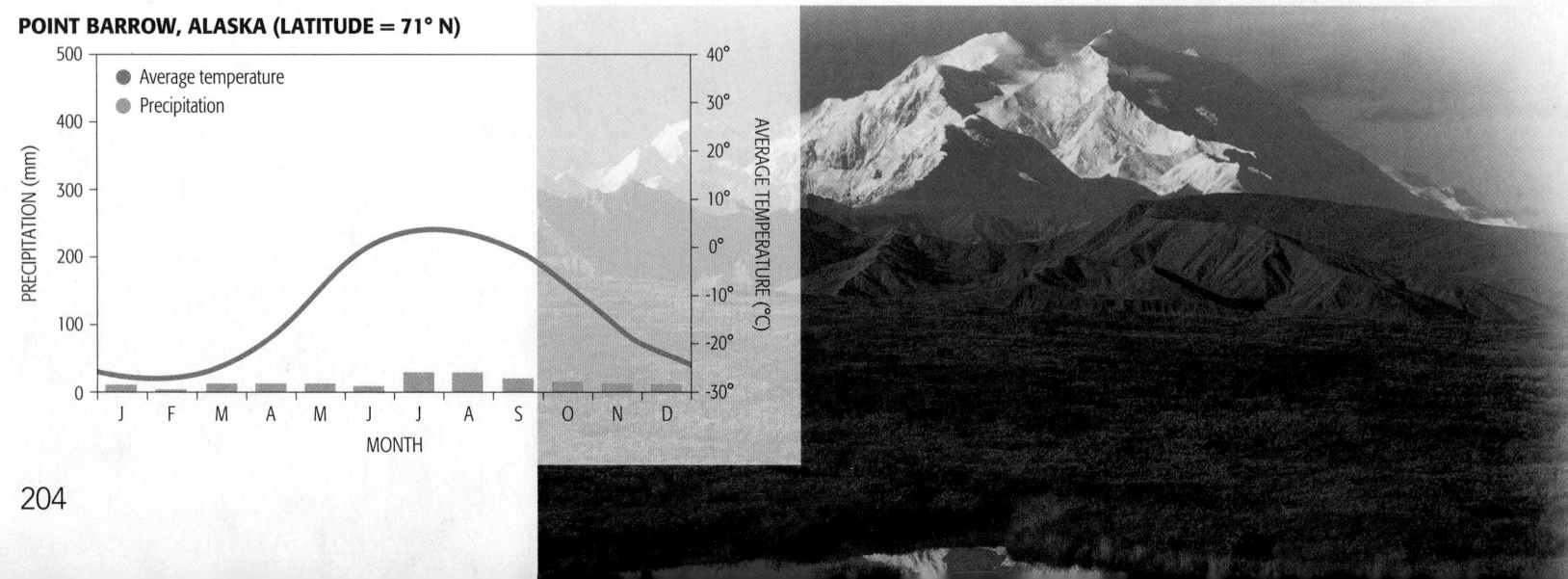

POINT BARROW, ALASKA (LATITUDE = 71° N)

- Average temperature
- Precipitation

7.5 Deserts

BIG IDEA The foremost characteristic of the **desert biome** is a sustained and significant moisture deficit. Compared to other biomes, average annual temperatures in most deserts are high, but polar deserts can be quite cold. Deserts experience some of the widest daily and seasonal temperature fluctuations of any of Earth's biomes. Deserts occur in all three climatic zones but are most extensive in the transition region between the tropics and the temperate zone. Desert plants and animals have a variety of adaptations that minimize water loss. Climate change, livestock grazing, and unsustainable agriculture are causing deserts to expand into other biomes, particularly grasslands and tropical savanna.

Defining Deserts

■ Deserts occur worldwide in very arid environments.

All deserts are dry. Annual rainfall is typically well below 250 mm (10 in.), and annual evapotranspiration, the loss of water vapor from soil and leaf surfaces, is much greater than annual precipitation (Figure 7.29).

You probably think of deserts as being very hot, and most are. But the range of annual average temperatures among the deserts of the world is quite large. For example, the central portion of the Sahara Desert in Africa has an annual average temperature of 27 °C (80 °F), whereas that of the Wright Valley in Antarctica is –20 °C (–4 °F), well below freezing. Yet both of these deserts are extremely dry, receiving far less moisture as precipitation than they lose by evaporation. Many temperate deserts have marked seasonal changes in temperature, with midday conditions varying from far below freezing in winter to over 40 °C (104 °F) in summer.

Deserts are noted for having large fluctuations in daily temperature. In California's Death Valley, for example, it may be below freezing at dawn and hotter than 38 °C (100 °F) by midafternoon. Within hours after sunset, desert temperatures may drop by 15–20 °C (27–36 °F). This is more than twice the usual temperature variation in a tropical rain forest over an entire year.

▼ Figure 7.29 **Desert Climate**
Desert is the driest biome on Earth and occurs in tropical, temperate, and polar zones. Deserts such as the dry valleys of Antarctica Ⓐ, the Asian Gobi Ⓑ, and the North American Sonoran Ⓒ support plant and animal species that are adapted to dry conditions and highly variable temperatures.

TUCSON, ARIZONA (LATITUDE = 32° N)

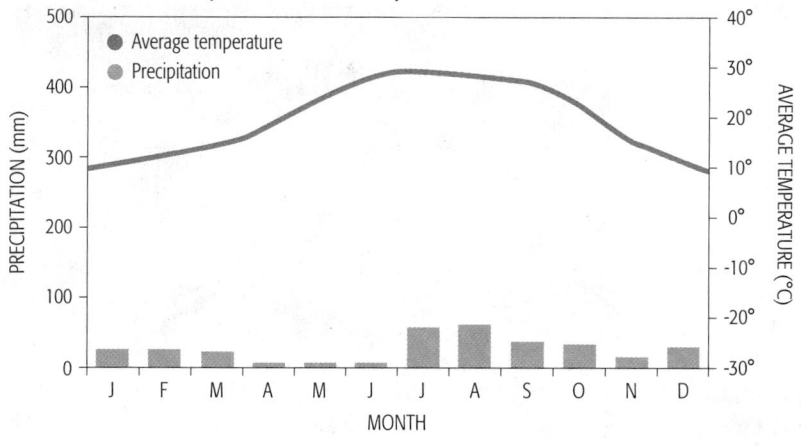

Q *Are the desert biomes getting larger?*
Breeann Sharma,
Arizona State University

Norm: **A** Yes, climate change, overgrazing, and unsustainable agriculture are converting drier parts of the tropical savanna and grassland biomes to desert.

QUESTIONS 7.5

1. In general, plant and animal diversity is highest in deserts where the warmest months correspond to the months of highest rainfall. How would you explain this relationship?

2. Why are some of the world's most extensive deserts found in the transition between the tropical and temperate climatic zones?

(MES) For additional review, go to **MasteringEnvironmentalScience**

Deserts occupy more than one-fourth of Earth's land surface. Extensive deserts such as the Atacama of South America, the Sahara of northern Africa, and the Great Sandy of Australia extend from the tropical zones into the temperate zones. Their distributions correspond to the zone of dry air descending with the equatorial Hadley convection cells (see Module 3.7).

The Sonoran, Mojave, and Great Basin Deserts of North America and the vast Gobi Desert of central Asia are found in temperate zones, within the interior of continents. The aridity of these deserts is accentuated by rain shadows caused by air masses moving over large mountain ranges. (This process is described in upcoming Figure 7.33.) The Gobi Desert, for example, extends to the north of the Himalaya and Hindu Kush Mountains. The southerly monsoon winds dump most of their moisture on the southern slopes of these mountains, so they bring only scant moisture to the Gobi.

The strong cold winds of the polar zone carry very little moisture, therefore creating deserts in some regions. Polar deserts are found on the Queen Elizabeth Islands of northeastern Canada and in the Wright Valley of Antarctica.

The composition of desert vegetation varies greatly, from just a few scattered herbs and lichens growing on barren soil to diverse shrublands with cacti the size of trees. Desert plants display a wide range of adaptations to dry conditions. Many desert plants are **succulent**, meaning they take up large quantities of water and store

it in thick, fleshy stems or leaves. Succulents such as cacti possess abundant spines that protect them from thirsty herbivores. Other desert plants become dormant during dry periods, growing and reproducing only after rain (Figure 7.30). In the driest deserts, herbs, mosses, and lichens form dense, drought-resistant "cushions," which are interspersed between bare ground and rock.

Despite the limited water and severe weather, a great many animals have adaptations that allow them to survive in deserts. The majority of desert animals are nocturnal, active primarily during the cool nighttime hours when water loss to evaporation is low. Some desert animals, such as the Mojave ground squirrel, hibernate in deep burrows during the hottest and driest months; their burrows are much cooler than the surface, and their bodies require far less water in this dormant state. Many desert rodents use water so efficiently that they are able to survive on water they generate from the respiration of their food. Such animals can live their entire lives without drinking a drop of water (Figure 7.31).

Although human activities have affected many desert locations, their impact is generally less extensive than in other biomes. Nomadic herders have lived in many of the world's deserts for thousands of years, moving their animals to find water and food. Until recently, farming in deserts was limited to areas close to springs, wells, or rivers. Today, humans use canals to divert large quantities of water to irrigate agricultural fields and to support rapidly growing cities. Although harsh conditions have limited their invasion, non-native species are increasingly common. Over 110 non-native annual herbs now grow in the California Mojave Desert.

Human actions are causing deserts to spread into transitional areas between other biomes. Overgrazing and poor farming practices in marginal grasslands and savannas allow desert plants to become established. Global warming is encouraging the expansion of deserts in Asia and sub-Saharan Africa.

▼ Figure 7.30 **Enduring Drought**
Some desert plants, such as the ocotillo of the Sonoran Desert, are leafless and dormant during dry periods. When rains come, however, they produce leaves and flowers within days and seeds within just a few weeks.

▼ Figure 7.31 **A True Nondrinker**
From birth to death, the kangaroo rat of North American deserts never drinks a drop of water. It is able to extract all the water it needs from its food.

7.6 Mountains and Coastlines

BIG IDEA Mountains and coastlines are not biomes; instead, they are geographic features that are found in all three climatic zones. Both mountains and coastlines are characterized by gradients of environmental change that support distinctive communities of organisms. On mountains, changes in elevation produce a variety of climates. On coastlines, the most important gradients are those of salinity and water availability. Mountain and coastal communities are particularly vulnerable to human activities and global warming.

Mountains

■ One mountainside may experience as much climatic variation as does all of North America.

Hiking up the San Francisco Peaks in Arizona, you see some remarkable changes. In just a few miles, you travel from the Sonoran desert to a climate like the tundra. The change in plant and animal communities as you travel up the mountain is similar to the changes you might observe traveling from lower to higher latitudes (Figure 7.32). Biomes typical of warm, dry climates occur at lower elevations. At higher elevations, they are replaced with biomes associated with cooler, moister environments.

What causes these differences in climate and vegetation? The key factors are differences in air temperature and prevailing winds. The atmosphere becomes cooler with increasing elevation. Winds move air up a mountainside. (Prevailing winds blow toward the *windward* side of the mountain.) As air moves upward, it becomes cooler and its capacity to hold water vapor decreases (see Module 3.7). Clouds form, releasing rain or snow on the mountainside. On the other side of the mountain (the *leeward* side), the air moves downward, and the process is reversed. As the descending air warms, its capacity to hold moisture increases and it delivers less rainfall (Figure 7.33).

This movement of air results in two very predictable patterns of moisture distribution on mountain slopes. First, precipitation increases with increasing elevation. Second, where there are persistent prevailing winds, the windward sides of mountains are much wetter than the leeward sides. The dry region on the leeward side is called a **rain shadow**. Many of Earth's deserts occur in the rain shadows of high mountains.

The region of transition from boreal forest to tundra-like vegetation on a mountain is called the **tree line**. In general, the tree line occurs at progressively lower elevations in mountains at higher latitudes. On the

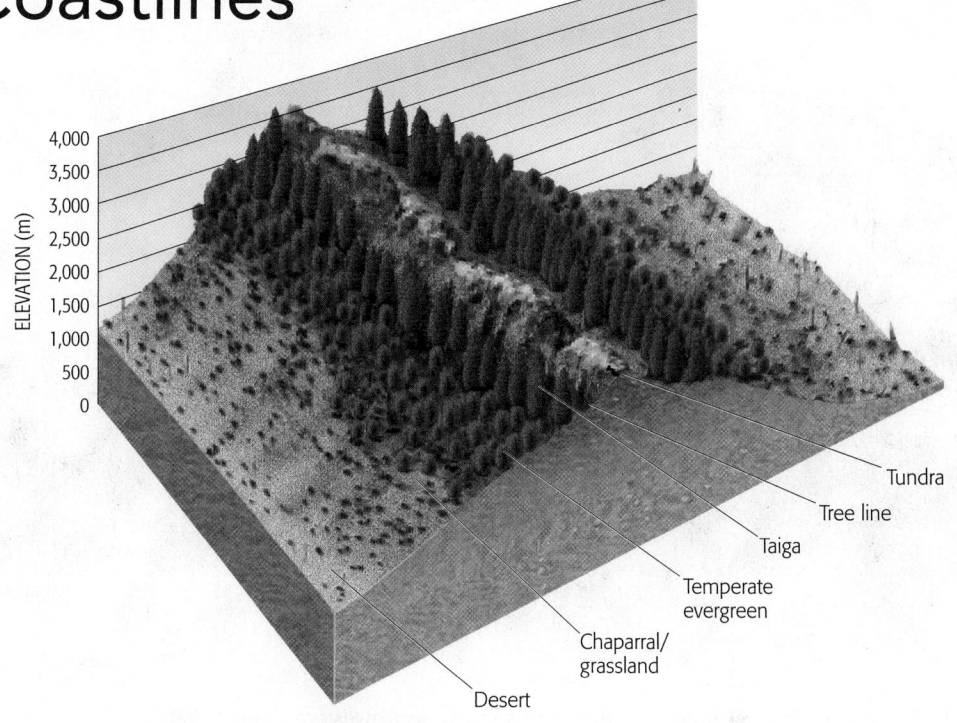

▲ **Figure 7.32 Mountain Biomes**
The distribution of biomes along the slopes of the San Francisco Peaks of Arizona is similar to what you might observe driving from Mexico to northern Canada.

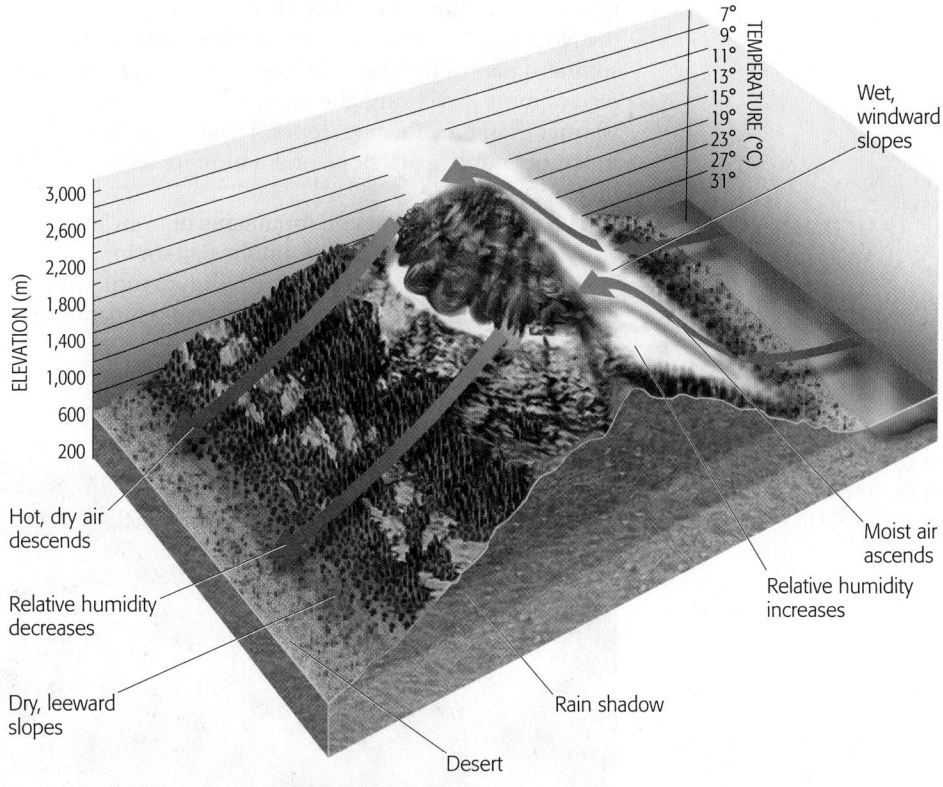

▲ **Figure 7.33 Mountain Climates**
As air rises up a mountainside, it is cooled and produces rainfall. As it descends on the other side, the air is heated and dried.

207

▲ Figure 7.34 **In the Clouds**
This cloud forest in the central mountains of Costa Rica receives ample rain and is often enshrouded in fog. Tree species similar to those found in some temperate forests grow here.

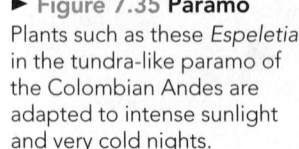

► Figure 7.35 **Paramo**
Plants such as these *Espeletias* in the tundra-like paramo of the Colombian Andes are adapted to intense sunlight and very cold nights.

mountains in the southwestern United States, trees grow as high as 3,650 m (12,000 ft); in the Olympic Mountains of Washington State, the tree line is at 2,400 m (8,000 ft); in the Wrangell Mountains of southern Alaska, the tree line is at about 1,000 m (3,280 ft).

The direction of the sun's radiation is also a factor in the distribution of vegetation on mountains. In the Northern Hemisphere, the sun's path tracks in the southern part of the sky. South-facing slopes, therefore, receive more direct sunlight and are considerably warmer than north-facing slopes. These differences in temperature due to slope orientation increase with increasing latitude.

The pattern of biomes on mountains in tropical regions is distinctly different from the pattern in temperate regions, partly because tropical mountains experience comparatively little variation in temperature from season to season. The warm lower elevations of tropical mountains are generally covered with rain forest. At cooler and wetter elevations, rain forest is replaced by cloud forest, a region resembling the temperate biome (Figure 7.34). Even in the tropics, elevations above 4,000 m (13,200 ft) are too cold to permit tree growth; instead, the land is dominated by a tundra-like collection of grasses and herbs called **paramo**. Because the paramo is often above the clouds, plants that grow there must be adapted to frosty nighttime temperatures and sweltering daytime temperatures (Figure 7.35).

Ecosystems on mountain slopes are especially vulnerable to human activities. Logging, farming, and mining expose the already thin soils to high rates of erosion. Once trees are removed, the reforestation of steep slopes is especially difficult.

Global warming is having a direct effect on the distribution of mountain ecosystems. Long-term studies in the Andes of Colombia and Peru reveal that the ranges of many bird species have shifted to higher elevations over the past 50 years. Similar shifts in plant and animal species have been observed in the mountains of western North America.

For some species, these climatically driven shifts could result in the complete loss of habitat. In the Appalachian Mountains of Virginia, North Carolina, and Tennessee, spruce and fir trees grow at the very top of some peaks. Warmer conditions are allowing broad-leaved deciduous trees to grow at higher and higher elevations, encroaching on the range of the conifers. As a result, spruce, fir, and many of the species associated with them could disappear from these peaks.

Coastlines

■ Salt water and wave action make the conditions in coastal environments especially harsh.

Coastal environments occur in the narrow transition between terrestrial biomes and the ocean. Salt-laden water droplets and the physical action of waves, currents, and storms make coastal environments especially challenging habitats. Because of the environmental stresses they share, the communities of organisms that live on coastlines around the world are remarkably similar. And because ocean currents can carry their seeds and fruits long distances, certain species of plants, such as coconuts and mangroves, grow on coastlines around the world (Figure 7.36).

The sea strand—the place where the ocean meets the land—is a high-salinity environment that is continually being reshaped by the rise and fall of tides and the action of waves and currents. Salt in sea spray is concentrated by evaporation and transported to land by wind off the ocean. This salt is then deposited on plants and soils near the coast, causing water to diffuse from plant tissues and limiting plant growth. Most plants are unable to grow in such a salty environment. But just a few minutes' walk away from the strand, and the salinity and availability of water change greatly. These gradients determine the distribution of grasses, shrubs, and trees.

Only a few annual plants can endure the harsh conditions near the water's edge (Figure 7.37). Once these annuals are established, they stabilize the sand and provide just enough protection from salt spray to allow perennial grasses, such as sea oats, to become established. When perennials gain a foothold, sand accumulates around their bases and sand dunes begin to form. As more and more perennials become established, the sand dunes

grow larger and become more stable. Once stabilized, sand dunes protect plants from wind-borne salt. In fact, a diverse array of plants can grow on the leeward sides of sand dunes. Shrubs and even trees can grow in the most stable and protected coastal environments.

The animals that live in coastal environments are also affected by salinity and the availability of water. Terrestrial ghost crabs forage for small insects and small bits of vegetation amid coastal dunes. To survive in this salty environment, vertebrates such as rabbits and deer depend on the presence of small freshwater streams and ponds.

On some coastal islands there are large populations of domestic horses, pigs, and sheep; these animals were abandoned by sailing vessels over the centuries. Where these species are abundant, they have significantly altered the native vegetation.

Human developments in coastal areas, such as vacation homes and condominiums, have resulted in considerable loss of coastal environments. Human developments also destabilize sand dunes, rendering coastlines more vulnerable to major storms and hurricanes. The human impact on coastal environments is being accentuated by the rising sea levels associated with global warming (see Module 9.4).

(see Module 9.4)

QUESTIONS 7.6

1. What is the cause of mountain rain shadows?

2. The tree line on Mount Shasta, California, is considerably higher on south-facing slopes than on north-facing slopes. Explain why this is the case.

3. Destruction of a sand dune near the edge of the sea often results in the death of plants that grow behind them. Why is this?

4. The same or genetically very similar species of plants are often found on distant continents growing in coastal environments. Why is this so?

(MES) For additional review, go to **MasteringEnvironmentalScience**

▼ Figure 7.36 **Coastline Vegetation**
Plants growing in coastal environments must be able to withstand the effects of salt carried in ocean spray as well as the physical effects of ocean waves.

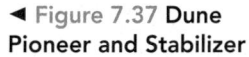

◄ Figure 7.37 **Dune Pioneer and Stabilizer**
Sea rocket Ⓐ is an annual plant that can grow just above the high-tide mark along beaches. Its seeds are dispersed by ocean currents. Sea oat Ⓑ is a common grass on coastal dunes in eastern North America. Its fibrous roots stabilize the sand in the dunes against the forces of wind and moving water.

7.7 Aquatic Biomes

BIG IDEA Aquatic biomes are found at all latitudes and, unlike terrestrial biomes, are characterized more by differences in the chemistry and depth and flow of water than by seasonal patterns of temperature and precipitation. They often occur within the context of terrestrial biomes. Gravity moves fresh water continuously down streams and rivers, and variations in water flow create a variety of unique environments. Water is impounded in lakes and ponds where its movements are far more subtle. These liquid ecosystems each have unique physical and chemical environments that support equally unique and diverse kinds of microbes, plants, and animals. Wetlands are an important transition between terrestrial biomes and aquatic and marine biomes where soils are often or always saturated with water.

Streams

■ Stream biodiversity is influenced by patterns of water flow.

Under the force of gravity, water flows across the surface of the land. Scientists use the word *stream* to describe all natural bodies of flowing water, regardless of their size. Depending on its size and location, a stream may be referred to more specifically as a rill, a brook, a creek, or a river.

The amount of water flowing in a stream varies over time. **Perennial streams** flow all year round. Their flows vary from season to season and year to year, depending on the local patterns of precipitation and evapotranspiration. **Intermittent streams** flow only at certain times. Streams in headwaters are often intermittent, flowing only when it rains. Intermittent streams also occur where patterns of rainfall and evapotranspiration are distinctly seasonal.

A stream's **channel** is the waterway through which it normally flows. Next to the channel is the **floodplain**, the land that experiences periodic flooding. Hydrologists define the extent of a floodplain in terms of time. In headwaters where the slope, or the gradient, of the land is steep, streams have virtually no floodplains. Where the gradient is low, streams often have very broad floodplains. Stream channels in wide floodplains are often bordered by embankments called **levees**. Natural levees form from sediments deposited during floods (Figure 7.38).

Scientists use the word **lotic** to describe ecosystems dominated by flowing fresh water. Lotic ecosystems include the communities of organisms that live in streams as well as the communities of organisms living on lands next to and beneath a stream. The flow of water

▼ Figure 7.38 **Floodplain Features**
A river's floodplain, shown here in cross-section, is shaped by the changing flow of water over thousands of years. Natural levees separate the stream channel from backwater areas, which receive water only during floods.

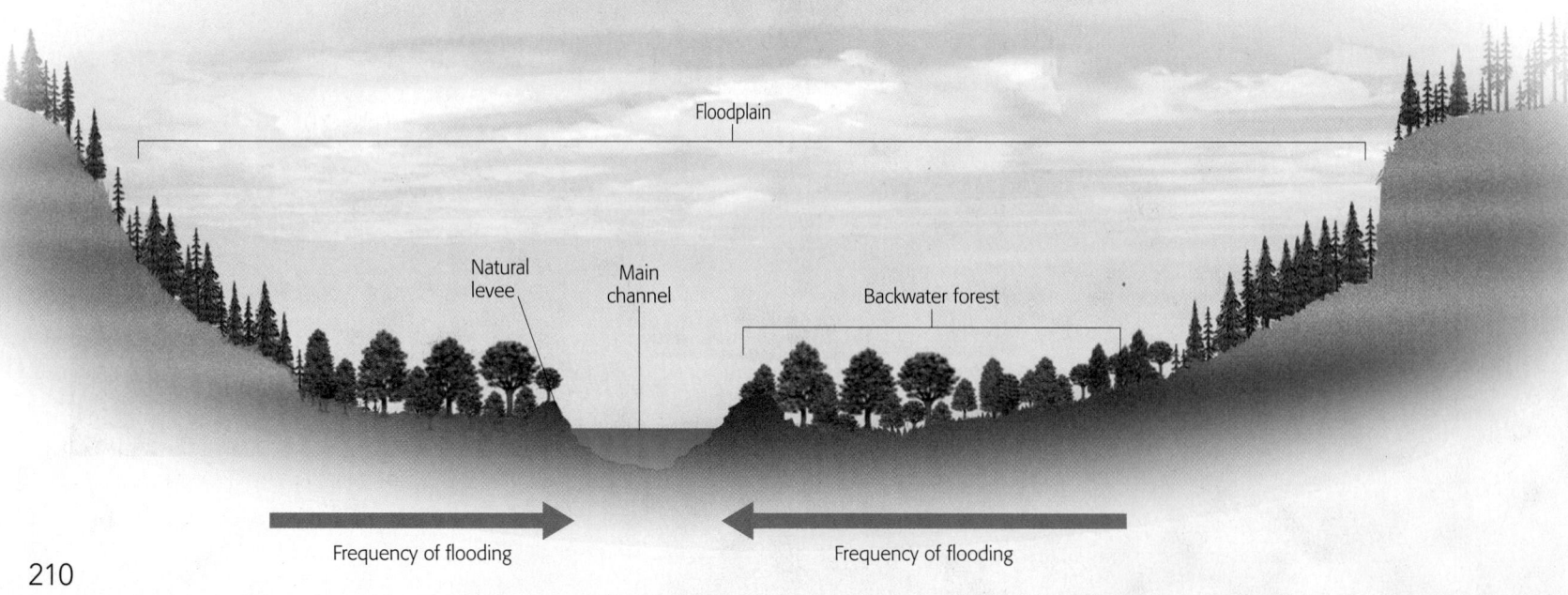

Floodplain

Natural levee

Main channel

Backwater forest

Frequency of flooding

Frequency of flooding

greatly influences biodiversity and the movement of energy and nutrients through lotic ecosystems.

Much of the primary production that supports stream food chains comes from **detritus**, organic matter such as dead leaves and nutrient-rich soil that wash in from adjacent terrestrial ecosystems. Algae attached to rocks or other fixed structures in streams provide additional primary production (Figure 7.39).

In fast-flowing streams, planktonic algae are sparse because the flowing water quickly carries them downstream. Where the flow of water is slower or halted by natural or human-built barriers, planktonic algae are more abundant. Nitrogen and phosphorus in agricultural and municipal runoff also encourage the growth of algae, and high quantities of these elements can lead to eutrophication (see Module 11.4).

The primary consumers in streams have three main ways of feeding (Figure 7.40). Scrapers are animals that use rasping mouthparts to feed on the algae attached to rocks. Snails and some species of insects and fish are scrapers. Shredders feed directly on leaves, small twigs, and other pieces of detritus that fall into streams. These consumers include the larvae of numerous insect species. Filter feeders use gills or netlike structures to collect small bits of detritus carried in the flowing water. Freshwater mussels, worms, and the larvae of blackflies and caddisflies are common stream filter feeders. All of these animals have adaptations that allow them to grasp or attach themselves to rocks and avoid being carried downstream by the flowing water.

Stream herbivores and detritus feeders are very sensitive to changes in water chemistry and sediment load. Thus, their numbers and diversity are commonly used as indicators of stream pollution. When pollution is low, they are generally abundant. When pollution is high, they are scarce or absent.

The various herbivores and detritus feeders are the primary prey for a number of carnivorous animals, including crayfish, beetles, and the larvae of dragonflies and damselflies. They are also the primary source of food for many carnivorous fishes, such as trout. These carnivores depend on their mobility to capture their prey.

The region of transition between a stream and the terrestrial ecosystems that surround it is called the **riparian zone**. The plants growing at the edge of the stream channel must be able to tolerate frequent flooding, saturated soils, and low levels of soil oxygen. Farther from the stream bank, flooding is less frequent and soils contain more oxygen. The farthest extent of the riparian zone is rarely flooded, but many of the plants growing there draw their water from the shallow aquifers associated with rivers.

Although riparian zones represent a relatively small part of the landscape, they include a diverse array of species and provide a variety of ecosystem services. Organic matter from riparian vegetation provides much of the energy that sustains stream food chains. Riparian vegetation and soils filter and cleanse the runoff and groundwater coming into rivers. This is particularly important in agricultural landscapes, where riparian zones absorb excess nutrients running off fields. During storms, the plants and soils in the riparian zone slow down the movement of water, thereby alleviating the effects of flooding.

The region of saturated sediment next to a stream and immediately beneath it is called the **hyporheic zone**. A variety of microorganisms and invertebrate animals live in this zone. Stream water is constantly flowing into and out of the hyporheic zone, filtering out organic matter and sediment as it does. This process facilitates nutrient cycles. Organisms in both the stream channel and the hyporheic zone cycle organic matter through the food web and return dissolved nutrients to the water.

◀ Figure 7.39 **Energy Input**
Leaves and other debris from terrestrial ecosystems provide much of the organic energy for food chains in small streams.

▲ Figure 7.40 **Stream-feeding Strategies**
Freshwater snails Ⓐ use a rasplike tongue or radula to scrape algae from stream rocks. Blackfly larvae Ⓑ shred leaf material with their grinding mouthparts. Freshwater mussels Ⓒ obtain their energy by filtering organic matter from the water.

Lakes and Ponds

■ Habitats vary across lakes and at different depths, most lakes have a short life span.

Lakes and ponds are inland bodies of water that fill topographic basins. Although the water in them may slowly circulate, lakes and ponds have little or no directional flow or current. There are over 300 million inland bodies of water on Earth, but most of them (90%) are smaller than 1 ha (2.5 acres). Bodies of water that are fed by a stream and have a surface area greater than 5 ha (12.5 acres) are generally classified as lakes; smaller bodies of water are called ponds. However, this distinction is somewhat arbitrary, and there is much overlap in the use of these terms, even by scientists.

The life span of most lakes is relatively short, at least in terms of geologic time. Most lakes are considerably less than 1 million years old. Why are they so short-lived? Lakes are constantly receiving sediments from their banks and the rivers that feed them, causing them to fill in. Over time, stream deltas gradually invade their waters. Vegetation grows inward from the edges of lakes, increasing the rate at which mineral and organic sediments accumulate. Eventually sediments fill so much of the lake that it is transformed into a wetland, and other plant species become established.

Most lakes have **open basins**, meaning they are drained by a well-defined stream or river. Lakes with **closed basins** have no outlet stream. They lose water by evaporation and percolation into underlying sediments. As water evaporates, dissolved mineral salts such as sodium chloride are left behind. Thus, where evaporation is the prevailing mode of water loss, as in the Great Salt Lake, the Dead Sea, and the Aral Sea, lake water may be saturated with salts.

Scientists use the word **lentic** to refer to lake ecosystems. They divide lake environments into the **benthic zone**, the waters and sediments on the lake bottom, and the **pelagic zone**, the waters that are not close to the bottom (Figure 7.41). The pelagic zone is further subdivided into the **photic zone**, which receives sufficient sunlight to support photosynthesis, and the **aphotic zone**, deep waters that receive little or no sunlight. The depth of the photic zone depends upon the clarity of the water, varying from 1 to 2 m (3.3–6.6 ft) in very turbid lakes to over 50 m (164 ft) where waters are very clear.

Energy and nutrients are cycled among these zones by sedimentation and the circulation of lake water. In the pelagic zone, the most important primary producers are **phytoplankton**, the various algae suspended in the water. Phytoplankton include cyanobacteria, green algae, and single-celled organisms called diatoms. Phytoplankton use the light energy in the photic zone to carry out photosynthesis—taking up dissolved carbon dioxide and manufacturing carbohydrates. This process also enriches the water in the photic zone with oxygen.

Zooplankton, the diverse array of tiny protists and animals suspended in pelagic waters, feed on phytoplankton as well as bits of organic matter (Figure 7.42). Phytoplankton and zooplankton are eaten by filter-feeding animals, such as insect larvae. Those animals are eaten by small fish, which are themselves prey for larger fish. Phytoplankton and zooplankton are constantly sinking into the aphotic and benthic zones.

▼ Figure 7.41 **Lentic Zones** The open waters of a lake make up the pelagic zone. The bottom and near-bottom waters are the benthic zone. The littoral zone occurs near the shore where enough light passes through the water to support rooted vegetation.

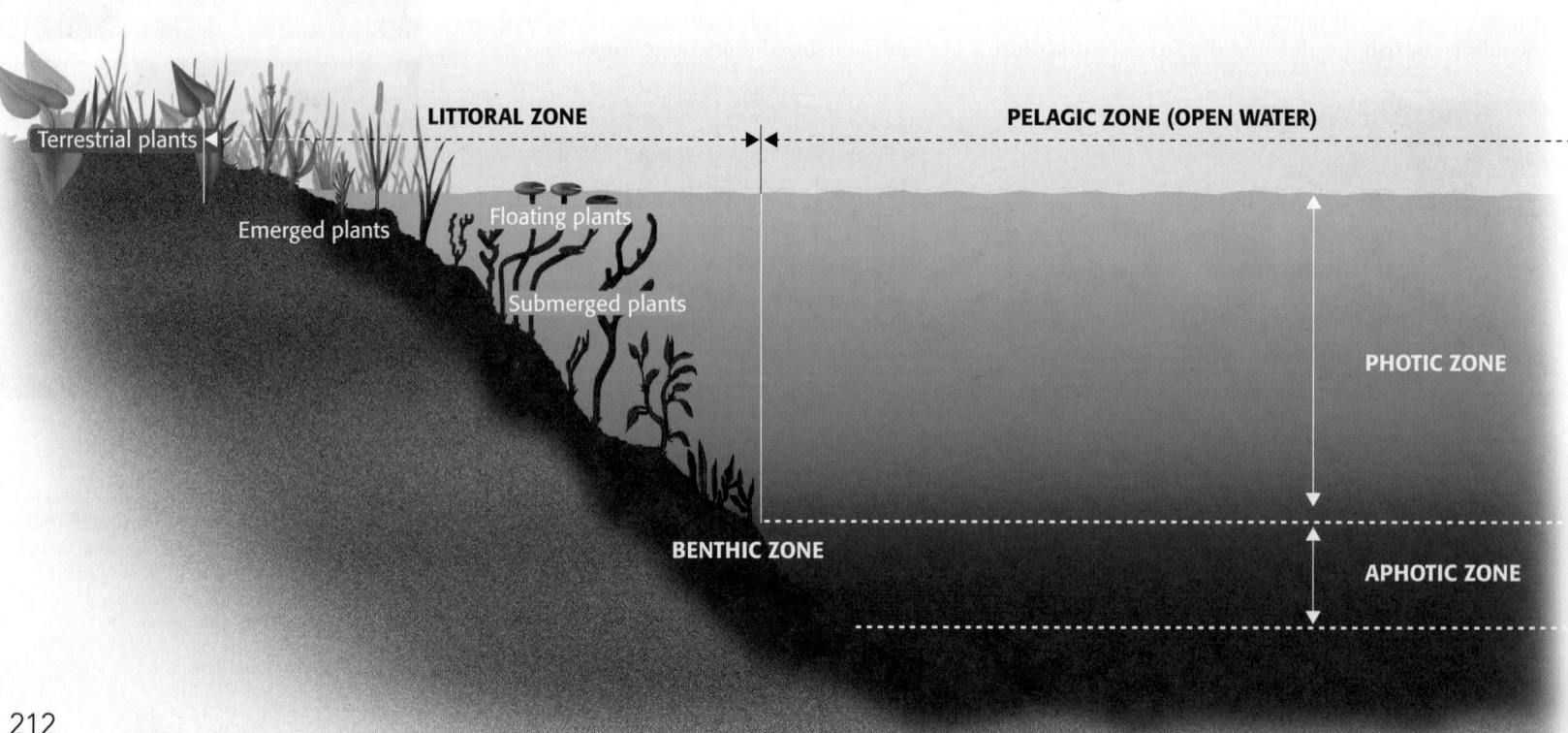

Terrestrial plants

Emerged plants

Floating plants

Submerged plants

LITTORAL ZONE

PELAGIC ZONE (OPEN WATER)

PHOTIC ZONE

BENTHIC ZONE

APHOTIC ZONE

◄ **Figure 7.42 Lake Plankton**
Phytoplankton, such as the green algae in this micrograph, form the base of lake pelagic food webs. These primary producers are consumed by microscopic zooplankton such as water fleas Ⓐ, copepods Ⓑ, water mites Ⓒ, protozoa Ⓓ, and rotifers Ⓔ.

There they may be eaten by small invertebrates and fish. These animals are preyed upon by larger fish.

The total pool of nutrients in a lake has a significant effect on its net primary production. The water and sediments of young lakes often contain low amounts of nitrogen and phosphorus. These conditions support only small populations of phytoplankton, so the waters of young lakes are generally clear. Over time, streams deliver additional nutrients to lake waters and sediment. As a result, primary production increases and the clarity of the water diminishes.

When organisms in the photic zone die, their remains drift to the bottom of the lake, causing nutrients such as nitrogen and phosphorous to accumulate in the sediments of the benthic zone. If lake waters remain unmixed, the pelagic zone gradually becomes depleted of nutrients. However, this does not occur in shallow lakes and ponds because winds continuously mix the water, moving it among the benthic, aphotic, and photic zones.

The waters of deeper lakes are mixed by seasonal changes in temperature and density. In the summer, sunlight heats the surface waters so they are warmer than the deeper waters, and therefore less dense. That difference causes the warm water to stay on the surface. In the fall, the surface waters cool more quickly than the deeper waters. As a result, the denser surface waters sink, forcing the nutrient-rich waters from the bottom upward. That change causes deep water to circulate to the surface. In such lakes, nutrient availability and primary production change from season to season.

The **littoral zone** is the shallow, sloped area relatively close to shore. This zone is often defined by the distribution of rooted vegetation. Thus, its precise depth depends upon how murky the water is, which affects the penetration of sunlight. The aquatic plants in the littoral zone provide much of the primary production. Leaves and other detritus from nearby terrestrial plants are an additional source of food. This abundant organic matter supports a variety of invertebrates, including aquatic worms, snails, insect larvae, and crayfish. These animals are food for fish that live in the littoral zone, such as bass, and other animals, such as turtles, wading birds, and raccoons, that feed at the water's edge (Figure 7.43).

◄ **Figure 7.43 Shoreline Critters**
The primary production of the plants growing in the littoral zone supports complex food webs that include carnivores such as racoons and great blue herons.

Wetlands

■ Wetland species are adapted to saturated, anaerobic soils.

Wetlands are ecosystems in which the soil is saturated with water all year long or for a significant part of the year. Wetlands are like sponges; they soak up large amounts of water during storms and floods, and then release it slowly over time. Plants that live in wetlands have adaptations that allow them to grow and reproduce in such saturated soils. They have to be able to survive in low-oxygen conditions, because oxygen diffuses through saturated soils very slowly. While most plants cannot tolerate such conditions, some species of plants have adaptations that allow them to grow in these water-logged soils; these plants are called **hydrophytes** (literally "water plants"). The roots of most hydrophytes are able to carry out anaerobic respiration, as do many bacteria and fungi. In addition, the stems of many wetland herbs contain specialized tissues that transport air.

The anaerobic conditions in wetlands also slow the decomposition of dead plants and animals, limiting the availability of nitrogen and phosphorus. Wetland shrubs and herbs have root systems that are efficient at extracting limited nutrients from soil. In addition, most hydrophytes are quite frugal in their use of nutrients.

Many aquatic vertebrates and invertebrates thrive in wetlands, providing food for a host of other animals, including a notably wide range of water birds. Wading birds, such as herons and egrets, stalk their prey in the shallow waters, looking for frogs, small fish, and insect larvae. Swans and geese filter food from wetland sediments, and many ducks dive for fish in deeper water. Many reptiles, including freshwater turtles, crocodiles, alligators, and the anaconda (the world's largest snake),

▲ Figure 7.44 **Wetland Mammals**
Botswana's Okavango Delta wetlands support a diverse array of animals, including hippos that feed on submerged wetland vegetation.

are at home in wetlands. Some mammals, such as beavers, muskrats, capybaras, water buffalo, and hippopotamuses are wetland specialists, too (Figure 7.44).

Intact wetlands provide a host of ecosystem services. In addition to providing habitat for fish and wildlife, they store floodwater and protect water quality. Water flowing into wetlands is slowed down considerably by the dense vegetation, peat, and soils.

Wetlands are classified into four categories—marshes, swamps, bogs, and fens—based on the source of their water supply and the kinds of plants that grow in them. These are the landscapes that often feature in the storytelling associated with American writers of the South and 19th century English novels.

Marshes are periodically or continuously flooded wetlands that are dominated by herbaceous plants, including grasses, rushes, reeds, and cattails. When marshes are flooded, water levels are generally shallow. Marshes may be tidal or nontidal.

Tidal marshes are found along seacoasts where the level of water is influenced by the daily and monthly ebb and flow of tides. The most extensive tidal marshes are salt marshes, which occur in salty water near the ocean. Although salt marshes are highly productive ecosystems, their vegetation is often dominated by a single species, cord grass (Figure 7.45). Upstream from salt marshes, but still within the influence of tides, are tidal freshwater marshes. Because their water is not salty, a great many more plant species can coexist in these marshes.

Nontidal marshes are also dominated by grasslike herbs. These freshwater marshes occur on floodplains and in poorly drained depressions. They also occur along

◄ Figure 7.45 **Tidal Marsh**
This salt marsh on the Edisto River near the coast of South Carolina is inundated by tidal waters twice each day. Tidal marshes may extend inland for 10–20 miles.

▲ Figure 7.46 **Swamp Forest**
Bald cypress trees dominate this swamp forest in the shallow waters of Caddo Lake on the border between Louisiana and Texas. The roots of these trees are adapted to the anaerobic conditions that are typical of flooded soils.

▲ Figure 7.47 **Mangroves**
These mangroves in Florida's J.N. "Ding" Darling National Wildlife Refuge are able to tolerate anaerobic and saline conditions associated with regular tidal inundation by seawater.

QUESTIONS 7.7

1. Describe the three different groups of animals that feed on plant matter in streams.
2. Why is the hyporheic zone beneath a stream's channel important to the organisms in the waters above it?
3. Lakes have relatively short lives in geologic time. Why is this so?
4. What factors influence primary production in lakes?
5. Name and describe four types of wetlands.
6. What factors limit plant growth on wetland soils?

(MES) For additional review, go to **MasteringEnvironmentalScience**

the edges of streams and lakes, where they form part of the littoral zone. Nontidal marshes include temporary, or ephemeral, ponds, which contain water only during wet seasons or wet years. In regions with distinct wet and dry seasons, temporary ponds undergo predictable patterns of change as they fill up and dry out.

Swamps are dominated by shrubs or trees and fed by flowing water. Forested swamps are found in the broad floodplains of many of the world's major rivers. They are especially common in the southeastern and south-central United States, where they are dominated by tall trees such as bald cypress and water tupelo (Figure 7.46). Mangrove swamps are found along seacoasts in many tropical and subtropical regions. The shrubs and small trees of mangrove swamps are able to tolerate high levels of salinity and regular tidal inundation (Figure 7.47).

Bogs are wetlands with peat deposits that support a variety of evergreen trees and shrubs. Sphagnum moss often forms a thick, spongy carpet under these woody plants (Figure 7.48). The primary source of water in bogs is rainfall. Because rainfall contains few nutrients, the soils of bogs typically have low levels of nutrients. Bog plants have adaptations that allow them to grow and reproduce despite the limited supplies of nitrogen and phosphorus.

Fens are wetlands that are fed primarily by groundwater. They have peaty soils that typically support a variety of grasses and grasslike rushes, as well as occasional patches of woody vegetation. Fens are quite common in the glaciated regions of North America and Asia, but they also occur in subtropical settings. The Florida Everglades is one of the largest fens in the world (Figure 7.49).

▲ Figure 7.48 **Peat Bog**
The rainwater that feeds this black spruce peat bog in Algonquin Provincial Park, Ontario contains very few nutrients.

► Figure 7.49 **Fen**
The Everglades of south Florida is one of the world's largest fen wetlands. It is fed from waters that drain Lake Okeechobee.

7.8 Marine Biomes

BIG IDEA Marine biomes are characterized by saline waters and they differ from one another with respect to amounts of dissolved salt, the effects of tides, water depth, and proximity to shore. Estuaries are the transition from freshwater streams to the ocean, and estuarine organisms are adapted to variations in salinity related to tides and seasonal changes in stream flow. Ocean waters contain about 3.5% dissolved salts. The shoreline zone between the low and high tide marks is a unique habitat in which organisms must tolerate daily cycles of immersion and exposure. The diversity and productivity of ecosystems of the ocean waters near the edges of continents are enhanced by nutrients exported from terrestrial and aquatic ecosystems. Open ocean ecosystems differ with respect to variations in available light associated with water depth.

Estuaries

■ Estuarine production benefits adjacent terrestrial and marine ecosystems.

Estuaries are partially enclosed bodies of water where fresh water from rivers and streams mixes with ocean water. Estuaries are known by a variety of names, including bays, lagoons, harbors, and sounds, although not all bodies of water called by those names are truly estuaries. True estuaries are regions of transition from freshwater to marine ecosystems. Their defining feature is the mixing of fresh water and salt water. Familiar examples of estuaries include Chesapeake Bay, New York Harbor, San Francisco Bay, and Puget Sound.

The net movement of water in estuaries is from streams and rivers to the ocean. But because estuaries are usually broad and shallow, there is little difference in elevation from the river end to the ocean end. As a result, the flow of water is slow and often very complicated.

Fresh water is less dense than salt water, so the fresh water from rivers generally flows out onto the surface of an estuary. As the surface water moves toward the ocean, it mixes with seawater and becomes saltier. This denser, saltier water then flows back toward the river along the bottom of the estuary.

A number of factors affect the circulation pattern and flushing time of an estuary. Tides produce a regular ebb and flow of ocean water. Winds and storms move surface water in various directions and contribute to the mixing of fresh water and salt water. As a consequence, water levels and salinity at particular locations vary considerably from day to day, month to month, and season to season.

Nearly all estuaries are affected by seasonal variations in the rate of inflow from rivers. For example, the peak flow from the rivers that feed Chesapeake Bay occurs in the spring. By early autumn, their flow is much lower. This seasonal change in flow produces predictable changes in salinity and water movement (**Figure 7.50**).

Water movement and flushing time are also influenced by the size of the estuary, the complexity of its shoreline, and the number and size of the streams feeding into it. Chesapeake Bay and Puget

◄ Figure 7.50 **Seasonal Salinity Change**
In the spring, heavy runoff in the rivers feeding Chesapeake Bay keeps salinity relatively low Ⓐ. Through the summer and into the fall, the rivers flow more slowly, allowing more seawater into the Bay and increasing its salinity Ⓑ.

Source: Chesapeake Bay Foundation. http://www.cbf.org/page.aspx?pid=943. Last accessed 8-9-11

▲ Figure 7.51 **Estuary Filters**
A single oyster filters about 200 L (about 53 gal) of estuary water each day. Where these animals have not been overharvested or depleted by pollution, their populations remain high. Healthy populations can filter the entire volume of an estuary in a week or two.

▼ Figure 7.52 **Estuary Critters**
The American oystercatcher Ⓐ is one of the many species of wading birds that feed on mollusks and crustaceans, such as the fiddler crab Ⓑ.

Sound are fed by numerous rivers, each with its own pattern of flow. Where large rivers enter the sea, the mixing of fresh water and salt water may extend well beyond the coast.

The complex topography and movement of water in estuaries usually results in a wide variety of habitats. Along the shore there may be sandy or stony beaches. At low tide, broad expanses of mudflats may be revealed. Most estuaries have salt marshes with tidal rivers flowing through them, and in warmer climates there are mangrove swamps. In the shallow water there may be beds of eel grass. Each of these habitats supports a distinctive community of plants and animals.

Abundant light and the constant input of nutrient-rich sediments make estuaries among the most productive ecosystems on Earth. The freshwater and saltwater wetlands that fringe most estuaries provide energy to their waters in the form of organic carbon. High rates of photosynthesis in seaweeds and phytoplankton also contribute to the high rate of net primary production. This productivity promotes an abundant and diverse array of other organisms.

The sheltered waters of estuaries provide ideal habitat for a variety of crustaceans, marine worms, and mollusks. Oysters, clams, and many kinds of worms filter organic matter from estuarine water and sediments (Figure 7.51). Crabs and other animals scavenge for larger bits of dead organic matter. These organisms are food for other animals, including a great variety of shorebirds and

mammals such as raccoons and opossums. The abundance and variety of food make estuaries ideal stopping places for many migratory birds (Figure 7.52).

Many species of fish rely on the productive and sheltered waters of estuaries for spawning and early development. For this reason, estuaries are often called the nurseries of the sea. Fish such as flounder and striped bass may complete their entire life cycles within estuaries. Many large marine fish, such as mackerel, tuna, and swordfish, begin their lives in estuaries and then migrate to the open ocean.

Oceans

■ Variations in ocean depth and nutrient availability produce a variety of habitats.

Oceans cover 71% of Earth's surface and contain most of its water. Over millions of years, sodium chloride (NaCl) and other minerals have accumulated in ocean waters, so now their average salinity is about 3.5%. The ocean includes a great number of different habitats, including coastal regions and the open ocean. Like lakes, oceans are divided into zones which each provide habitat to unique kinds of organisms.

Intertidal zone. The area along seacoasts that is submerged at high tides and exposed during low tides is the **intertidal zone**. Intertidal habitats include sandy beaches, mudflats, and rocky outcrops. Organisms in intertidal habitats must be able to tolerate regular cycles of wetting and drying as well as variations in water salinity.

Two factors contribute to the especially diverse array of seaweeds and animals found in rocky intertidal habitats (Figure 7.53). First, patterns of tidal inundation and exposure vary significantly over short distances up along the intertidal rocks. And intertidal species have adapted to very unique niches along this gradient. For example, several barnacle species may co-occur on the same large rock, but each species occupies a particular 0.5–1 m elevation zone

▲ **Figure 7.54 Beach Denizens**
Small flocks of sanderlings dashing in and out of the shallow surf are a familiar site on most beaches. These small birds eat small crustaceans that burrow near the surface of the sand.

relative to mean sea level. Second, rocky intertidal habitats are constantly being disturbed by the crashing surf. In much the same way that intermediate levels of disturbance promote diversity in grasslands, this disturbance promotes the coexistence of many species (see Modules 6.1, 8.4).

Most of us probably consider half-clad members of our own species to be the most important biological community occupying intertidal beaches. However, a closer look reveals a very diverse ecosystem. Most of the species of intertidal beaches are hidden from view beneath the sand surface. Here worms, mollusks, and mole crabs filter organic matter from the saturated sand. These animals support a food web that includes a diverse community of shore birds (Figure 7.54).

▼ **Figure 7.53 Between the Tides**
Rocks in the intertidal zone provide a secure substrate to which a rich variety of seaweeds and animals can attach themselves. However, these organisms must tolerate significant wet and dry periods as the tides move in and out.

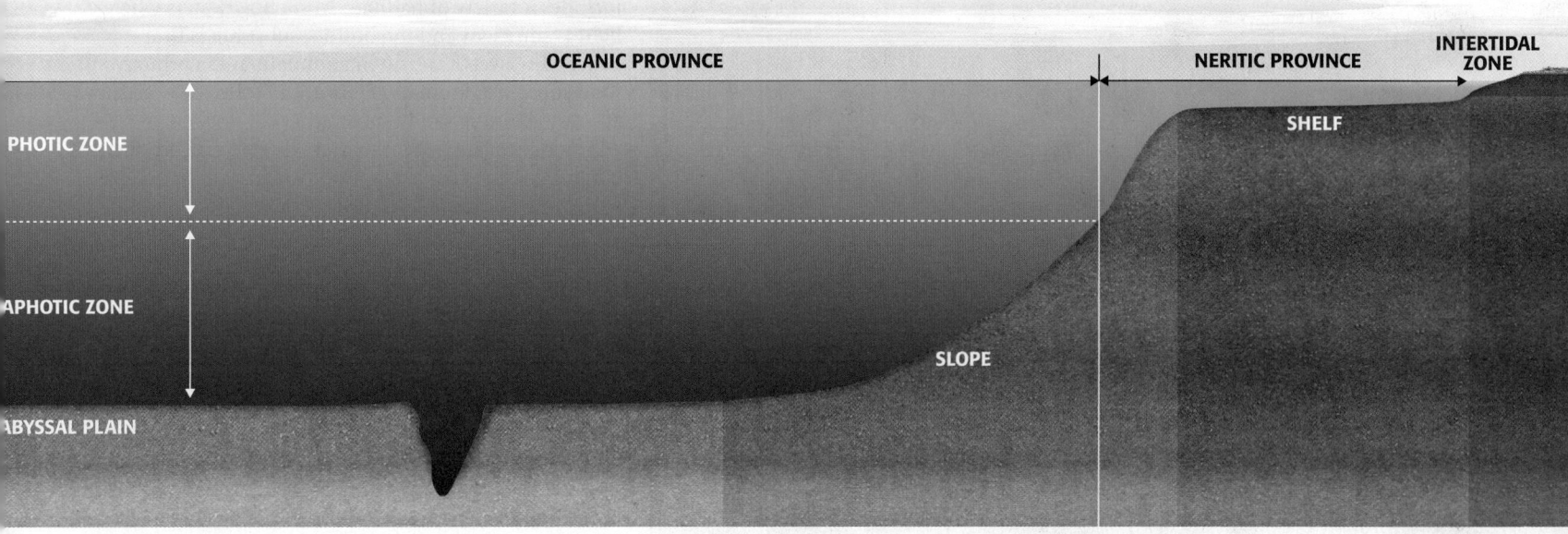

▲ Figure 7.55 Ocean Provinces and Zones
As with lakes, Earth's oceans can be divided into several zones. Photosynthesis occurs in the photic zone. Primary production is highest in the coastal ocean zone because of nutrient inputs from estuaries.

Pelagic zone. The pelagic zone comprises all of the oceans' open waters. It is subdivided into two provinces based on proximity to shore (Figure 7.55). The **neritic province** extends from the low tide mark to the edge of continental shelves. Primary production by phytoplankton is high here, especially near estuaries, because of the steady influx of nutrients (see Module 6.5). The **oceanic province** includes all of the ocean waters beyond the continental shelves. Primary production is much lower in most of this province because essential nutrients like phosphorus and iron are low compared to their neritic zone concentrations. However, production in the oceanic zone can be very high where currents bring deep ocean waters and the nutrients they contain to the surface. One such area is the deep waters of the Gulf of Alaska. Here the ocean's thermohaline circulation brings nutrient-rich water to the surface. These productive waters support a very complex food web with five trophic levels (see Module 6.4).

The pelagic zone is also subdivided into photic and aphotic zones, based on the penetration of light sufficient to support photosynthesis. The depth of the photic zone varies considerably. In neritic zone waters that have large quantities of suspended plankton and sediment, light may only penetrate a few meters. Away from coasts and in deeper ocean waters, plankton are often sparse, and the photic zone may extend to depths greater than 200 m (660 ft). Even so, most of the ocean is in the aphotic zone—94% of ocean waters are deeper than 200 m (660 ft), and nearly half are more than 3,000 m (9,800 ft) deep.

Organic matter (and the nutrients it contains) is constantly settling out of the photic zone into deeper waters. The constant "rain" of this organic matter, or detritus, and the daily migration of zooplankton between photic and aphotic zones support a diverse array of pelagic invertebrates and fishes. The rain of detritus diminishes with depth. At great depths, it is sparse indeed. Organisms here display a wide range of adaptations for acquiring energy (Figure 7.56).

▼ Figure 7.56 Aphotic Zone Predator
This angler fish has a luminescent "lure" projecting from its head that it uses to attract prey in dark deep ocean waters.

Benthic zone. The ocean bottom, or benthic zone, includes a variety of habitats. In shallow waters where light is sufficient and the bottom substrate is firm, attached seaweeds are often abundant and a primary food source for animals (Figure 7.57). Few seaweeds can attach to shifting sandy substrates.

Coral reefs are by far the most diverse and productive benthic marine ecosystems. They occur primarily in shallow tropical waters between the equator and 30° north and south latitude (Figure 7.58). The predominant organisms are stony corals. The calcium carbonate skeletons of these invertebrate animals form the physical structure of the reef. Corals form symbiotic relationships with single-celled algae, which are the most important primary producers in coral reef ecosystems.

Coral reef ecosystems support extraordinary biodiversity, even though they are found in very nutrient-poor water (Figure 7.59). About 25% of all marine animal species are associated with coral reefs. The close symbiotic relationship between the coral and its symbiotic algae ensures that scarce nutrients are efficiently recycled. Furthermore, many of these symbiotic algae are able to fix nitrogen (see Module 12.2).

Coral reefs provide important ecosystem services including fisheries, tourism, and shoreline protection.

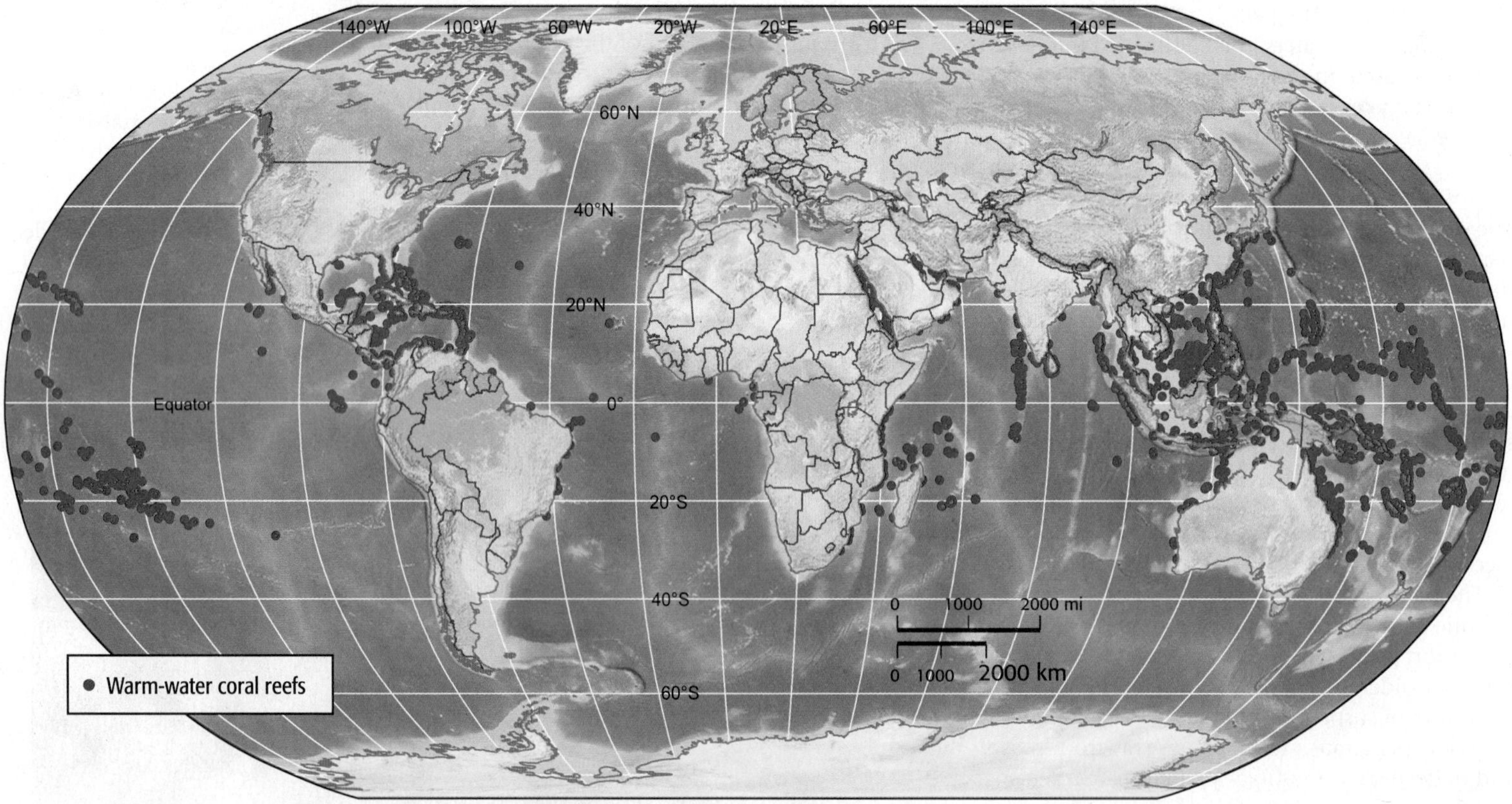

▲ Figure 7.57 **Kelp Forest**
In nutrient-rich coastal waters, giant kelp over 31 m (100 ft) long form a forest that sustains a diverse array of animals.

• Warm-water coral reefs

▲ Figure 7.58 **Coral Reef Distribution**
Nearly all of Earth's coral reefs are confined to warm, very clear tropical waters. However, some species of coral do grow in colder waters of the temperate zone, but they do not form large reefs.
Source: United Nations Environmental Program World Conservation Monitoring Centre. http://data.unep-wcmc.org/datasets/13

▲ Figure 7.59 **Coral Reefs**
These ecosystems owe their remarkably high diversity to primary production by symbiotic algae associated with the coral animals.

QUESTIONS 7.8

1. Explain how water in estuaries moves both toward and away from the sea.

2. Why are ecologists concerned that overharvest of oysters and other shellfish contributes to estuary pollution?

3. In what regions of the ocean is primary production likely to be highest? Why?

4. What is the source of energy for benthic animals in the deep ocean?

(MES) For additional review, go to **MasteringEnvironmentalScience**

But they are fragile and threatened by a variety of human actions. Global warming appears to be a major factor contributing to a phenomenon called coral bleaching, and ocean acidification associated with rising levels of CO_2 in the atmosphere may severely affect coral growth (see Modules 8.5 and 11.4). Urban and agricultural activities are polluting waters around many coral reefs. These ecosystems are also under threat from destructive fishing practices and excessive tourism.

In the deep ocean, the benthic substrate is a fine ooze composed of the skeletons of the planktonic organisms that have drifted down from the waters above. In these deep waters there is no light to support algae. Instead, invertebrates scavenge for bits of organic matter that rain in from the waters above.

The organisms that live on and near deep-sea thermal vents are part of the ocean's most peculiar ecosystem. Here, primary production is carried out by chemosynthetic bacteria, which use the chemical energy in the sulfur released from the vents to produce carbohydrates (see Module 4.1). These bacteria are consumed by filter-feeding clams and worms, which then become prey for other organisms (Figure 7.60).

▼ Figure 7.60 **Thermal Vents**
Chemosynthetic bacteria are the primary source of food for these peculiar segmented worms that live in the waters around deep-ocean thermal vents.

221

Summary

Because of convergent evolution, places on Earth that have similar climates often support plants and animals with similar physical and physiological traits. These similar communities of organisms are called biomes. Most of Earth's terrestrial biomes occur in one of three general climatic zones—the tropical zone, temperate zone, or polar zone. Tropical biomes include tropical rain forest, tropical seasonal forest, and tropical savanna; these biomes experience warm temperatures throughout the year. Temperate zone biomes include temperate deciduous forest, temperate evergreen forest, chaparral, and temperate grassland; they are warm in the summer and generally cold during the winter. Boreal forest and tundra are polar biomes; they have short growing seasons with very cold winters. The desert biome occurs in all three climatic zones where there are significant, sustained water deficits.

Mountains and coastlines are not usually considered to be biomes, but they are widely distributed and support unique ecosystems.

Aquatic biomes include wetlands, streams, lakes, and ponds. Wetland soils are saturated with water all or nearly all of the time. Stream ecosystems include not only flowing waters, but also the floodplain around them and the hyporheic zone beneath the streambed. Lakes fill topographic basins that may be either open or closed. Lake ecosystems are influenced by the depth to which sunlight penetrates the waters and the recycling of nutrients from lake sediments. Marine biomes include estuaries and various ocean ecosystems. Estuaries are partially enclosed bodies of water where rivers and streams meet the sea. They support very diverse ecosystems that are influenced by variations in salinity and tidal flows. Biomes in the ocean differ with respect to the influence of tides, and the availability of nutrients and light.

7.1 The Geography of Biomes

- As a consequence of convergent evolution, organisms that live in similar environments tend to share similar features.
- Biomes are major communities of similar plants and animals that are adapted to a particular environment.
- Climate is the most important factor distinguishing terrestrial biomes.
- Aquatic and marine biomes differ with respect to water salinity.

KEY TERMS

biome, convergent evolution, tropical zone, temperate zone, polar zone, continental climate, maritime climate, aquatic, marine

QUESTIONS

1. Many biomes extend to higher latitudes in near-coastal locations compared to locations in the interior of continents. Why is this the case?
2. The climate–biome model reveals that every biome can exist at higher temperatures when the rainfall is also higher. Explain.

7.2 Tropical Biomes

- Tropical environments are distributed 25° north and south of the equator; tropical rain forest, tropical seasonal forest, and savanna biomes are found in this climate zone.
- Warm temperatures prevail year-round in tropical environments. Rainfall occurs year-round in tropical rain forest, while both tropical seasonal forests and tropical savannas experience seasonal drought.

KEY TERMS

tropical rain forest, tropical seasonal forest, tropical savanna

QUESTIONS

1. The diversity of tropical rain forest insect species is directly correlated with the diversity of trees and shrubs. Explain why.
2. The eastern slopes of mountains in Central America support tropical rain forest whereas the western slopes sustain tropical seasonal forest and savanna. What is the likely cause of this difference?
3. Without fire and grazing, tropical savanna would become woodland. Too much grazing by domestic animals has transformed some tropical savanna into desert. Explain both of these observations.

7.3 Temperate Biomes

- Temperate environments extend north and south of the tropics between about 25° and 60° latitude.
- The distribution of temperate deciduous forest, temperate evergreen forest, chaparral, and temperate grassland biomes is determined by differences in winter temperatures and seasonal rainfall patterns.

KEY TERMS

temperate deciduous forest, temperate evergreen forest, temperate rain forest, conifers, chaparral, sclerophyllous, temperate grassland

QUESTIONS

1. Explain why soil nutrients and insect diversity change considerably throughout the year in temperate deciduous forest.
2. Why did the suppression of wildfires in temperate evergreen forests actually increase the risk of large, intense wildfires?
3. Where summers are hot, such as in Texas, grassland occurs where average annual rainfall exceeds 80 cm (32 in.). In cooler areas, such as Wisconsin, deciduous forest grows with this amount of rain. Why does one climate support grassland and the other support forest?

7.4 Polar Biomes

- Polar biomes of boreal forest and tundra are found above 60° in the Northern Hemisphere.
- In both the boreal forest and tundra, extreme cold and short growing seasons limit the abundance and diversity of life.

KEY TERMS

boreal forest, taiga, permafrost, tundra

QUESTIONS

1. Why is the total precipitation (rain and snow) lower in polar biomes than in most temperate and tropical biomes?
2. Even though precipitation is low, boreal forest and tundra are often wet and boglike. Why?
3. Describe three important effects of global warming on the boreal forest and tundra biomes.

7.5 Deserts

- The desert biome is characterized by arid conditions and widely varying temperatures.

KEY TERMS

desert biome, succulent

QUESTIONS

1. Why do deserts often occur adjacent to major mountain ranges?
2. Explain why deserts are expanding in many places.

7.6 Mountains and Coastlines

- Mountains and coastlines are not actual biomes; they are geographic features with unique communities of plants and animals that occur in all three climatic zones.

KEY TERMS

rain shadow, tree line, paramo

QUESTIONS

1. The eastern slopes of mountains in Central America support tropical rain forest whereas the western slopes sustain tropical seasonal forests and savanna. Explain why this is the case.
2. Describe three ways in which human activities affect coastal ecosystems.

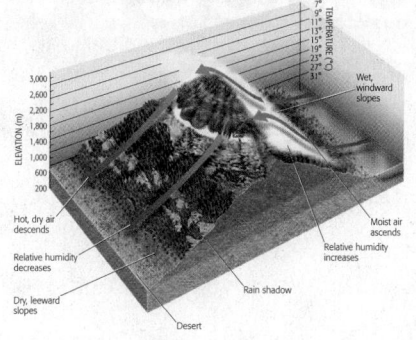

7.7 Aquatic Biomes

- Wetlands differ with respect to the source of water and the nature of water flow.
- The volume and speed of water flowing in streams determines the diversity of ecosystems in their channels and across their floodplains.
- Natural lakes support a variety of ecological communities along their shores and at different depths.

KEY TERMS

perennial stream, intermittent stream, channel, floodplain, levee, lotic, detritus, riparian zone, hyporheic zone, open basin lake, closed basin lake, lentic, benthic zone, pelagic zone, photic zone, aphotic zone, phytoplankton, zooplankton, littoral zone, hydrophyte, marsh, swamp, bog, fen

QUESTIONS

1. Compared to swamps and fens, bog soils contain very low amounts of nutrients. Why?

2. Explain this statement: Stream food chains are often sustained by terrestrial primary production.
3. Why is the salt content generally high in closed basin lakes?
4. Why is primary production in ponds and lakes generally higher following periods of windy weather?

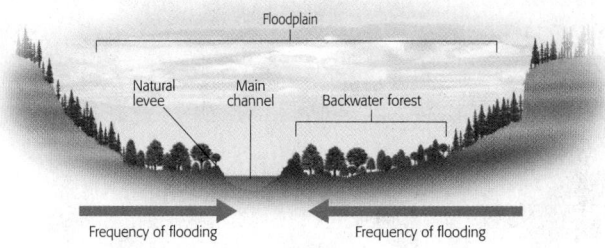

7.8 Marine Biomes

- Estuaries are partly enclosed bodies of water where fresh stream water meets saline ocean water.
- The distribution of different marine ecosystems is influenced by the availability of nitrogen and phosphorus and the penetration of light into ocean waters.

KEY TERMS

estuary, intertidal zone, neritic province, oceanic province

QUESTIONS

1. Why is the salinity of the water in estuaries typically lower in the spring and higher in the summer?
2. Why does the photic zone in near coastal ocean waters generally extend to only 15–25 m (50–80 ft) compared to nearly 100 m (330 ft) farther from shore?

3. Explain why some of the ocean's most complex and diverse ecosystems occur in zones of upwelling.

MasteringEnvironmentalScience®

Students Go to **MasteringEnvironmentalScience** for assignments, the eText, and the Study Area with practice tests, activities, and more.

Instructors Go to **MasteringEnvironmentalScience** for automatically graded tutorials and questions that you can assign to your students, plus Instructor Resources.

Biodiversity
Conservation

225

A Requiem for the Po'ouli

Why are Hawaii's unique bird species disappearing?

In 1973, students from the University of Hawaii discovered a new species of bird on the island of Maui. Biologists estimate that we have identified and named fewer than 15% of Earth's plants and animals. But because birds are often brightly colored and quite vocal, and because they are heavily studied by thousands of amateur bird-watchers and professional ornithologists, biologists think that the vast majority of Earth's birds have already been described. Discovering a new bird is a big deal!

The newly discovered bird was given the common name po'ouli (pronounced "poh-oh-U-lee"). The po'ouli is a member of a subgroup of the finch family called honeycreepers (**Figure 8.1**). Honeycreepers are found on the Hawaiian Islands and nowhere else. Other honeycreepers have long bills and sip nectar from flowers. But the po'ouli has a short bill and eats only small snails that live in the moist tropical forests on the windward side of Maui.

Although this region of Maui had been heavily studied prior to 1973, the po'ouli had escaped detection for two reasons. First, it is cryptic—its coloring blends into its surroundings, and it has a very quiet call. Second, its population was very small. A census in the early 1980s located about 140 birds. Since then, the number of birds has diminished sharply. In 1994, only six birds could be located, and two years later only three were found.

In September 2004, a single po'ouli was captured in an effort to breed more individuals in captivity. Sadly, that bird died of malaria two months later. Despite intensive searches by dozens of professional and amateur birders, no po'ouli have since been found. It is likely that the bird is now extinct.

Unfortunately, the po'ouli's extinction is only one of a great many extinctions on these islands. Prior to human colonization, at least 140 unique bird species lived on the Hawaiian archipelago. When the Polynesians arrived about 800 years ago, many species, such as the flightless Hawaiian goose, were simply too easy prey for human hunters. Other species were displaced by animals that humans brought with them. Rats—those ever-present animals that accompany humans wherever they go—fed on eggs and nestlings, probably causing the disappearance of many species.

By the time the first European, Captain James Cook, set foot on the Hawaiian Islands, at least 43 bird species had already disappeared. Since Cook's landfall in 1778, an additional 18 species have disappeared. Habitat loss associated with urban and agricultural development caused some extinctions. In other cases, the introduction of yet more non-native predators, such as the mongoose, was responsible (**Figure 8.2**).

▼ Figure 8.1 **Alas, the Poor Po'ouli**
Discovered just over 40 years ago, the last po'ouli was seen in the wild in 2004.

- In what ways is the diversity of life important for ecosystems and humans?

- What factors are contributing to the extinction of species?

- What management strategies can we employ to preserve Earth's diversity?

- What national and international policies are in place to conserve Earth's biodiversity?

▲ Figure 8.2 **An Alien Threat**
The Indian mongoose was introduced to Hawaii in 1872 in order to control rats. Because it also feeds on eggs and nestlings, it has had a devastating effect on populations of Hawaii's native birds.

Ornithologists believe the total number of bird species in Hawaii nowadays is approximately the same as it was 300 years ago, but many of these are species that have been introduced from elsewhere. Aggressive non-native birds, such as the common myna and the house finch, compete with native species, causing additional stress to struggling populations.

Today, only 71 of Hawaii's 140-plus unique bird species remain. Of those, 30 are listed as endangered under the U.S. Endangered Species Act (ESA). Fifteen of these are critically endangered, meaning that fewer than 500 individuals survive in the wild (**Figure 8.3**).

Conservation biologists estimate that in the absence of people, one bird species in the world would go extinct every 400 to 500 years. Yet over the past 800 years in Hawaii alone, bird species have gone extinct at the rate of one every 10 to 12 years. Worldwide, bird species are being lost at the rate of about one each year; many fear that rate could increase 10-fold by 2100.

Over the past several decades, efforts to reverse these trends in Hawaii have produced some success. The Hawaiian Endangered Bird Conservation Program is a unique partnership composed of the U.S. Fish and Wildlife Service (USFWS), the State of Hawaii Division of Forestry and Wildlife, and the Zoological Society of San Diego. Since 1994, managers in this program have used a variety of strategies to save the native birds, including habitat protection, control of non-native species, and captive breeding. As a result, the populations of many endangered birds are beginning to increase (**Figure 8.4**).

▲ Figure 8.3 **On the Brink**
The akiapola'au is a honeycreeper found only on the big island of Hawaii. As a consequence of habitat loss and non-native predators, fewer than 500 individuals persist in the wild.

◄ Figure 8.4 **Captive Breeding Success**
In 1995, only about 20 puaiohi, a small thrushlike bird, survived on the island of Kauai. As part of the Hawaiian Endangered Bird Conservation Program, over 200 puaiohi have been hatched and 150 have been reintroduced into the wild. The wild population is now growing.

227

8.1 What Is Biodiversity?

BIG IDEA **Biodiversity** is the variety of life in all its forms and combinations and at all levels of organization. Biodiversity is important within ecological communities, such as forests and lakes, and at larger scales, such as entire regions and biomes (see Module 7.1). In biological communities, biodiversity is measured by the number of different species, their relative abundance, and their spatial distribution. Within populations of a single species, genetic variation among individuals is a measure of biodiversity. Environmental factors such as temperature, availability of water, and patterns of disturbance cause the numbers and kinds of species to differ from place to place.

Landscape Biodiversity

■ Environmental gradients and disturbances produce landscape biodiversity.

Landscape diversity refers to the differences in the variety and abundance of species from place to place. This diversity can be measured by collecting, or sampling, the species in particular locations and comparing those samples to other locations.

You might readily expect variation in species as you move from one place to another, but let's consider a unique situation in which the GPS coordinates in your smart phone barely change, but your altitude does. Imagine hiking a mountain, sampling the numbers and kinds of plants and animals as you go. You'll find that no two places are exactly alike. As you climb, new species appear and others disappear. If you compare the species living in places that are relatively close to each other—perhaps separated by an elevation of 100 m (328 ft)—you will find that they are quite similar. Over longer distances, however, that similarity diminishes. Two thousand meters (6,560 ft) up the mountain, for example, the species in your sample will be totally different from those at the base.

If you hike up different sides of the same mountain, you may find similar patterns in the composition of species as the elevation increases. In the Appalachian Mountains of the southeastern United States, for example, the species of trees growing at low elevations include a diverse array of oaks and hickories. At higher elevations, beech, sugar maple, and buckeye are more prevalent; higher yet, spruce and fir trees predominate. Nevertheless, on each trail you hike you would find significant differences in the composition of trees; on one hike you might find more pines and on another more hemlocks.

After many such hikes, you could analyze your samples and classify them into communities composed of similar species. Ecologists often do this based on the prevalence of common trees and shrubs. You could then produce a map of the types of communities on your mountain. Such a map is a visual depiction of the mountain's landscape diversity (Figure 8.5).

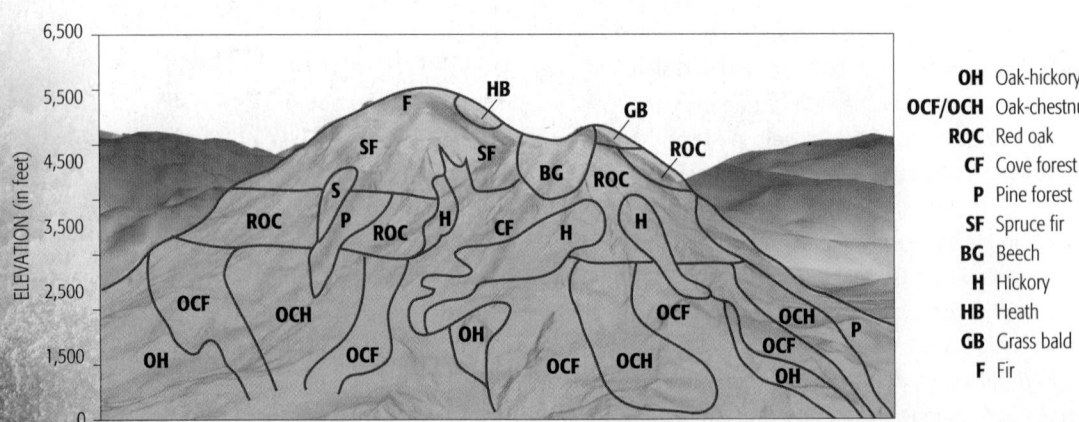

▲ Figure 8.5 **Landscape Diversity**
Ecologist Robert H. Whittaker sampled vegetation in the Great Smoky Mountains and then used his data to produce this generalized map of the forest communities on a typical mountain.

Source: Whittaker, R.H. 1956. The vegetation of the Great Smoky Mountains. *Ecological Monographs* 26: 1–80.

The pattern of landscape diversity is a consequence of two important factors—gradual changes in environmental factors and the history of local disturbances. Many environmental factors change from the base of a mountain to its peak. In general, as elevation increases, temperature and length of growing season decrease and precipitation increases (see Module 7.6). Temperature and length of growing season are also affected by the orientation of the mountain toward the sun, which causes differences in the amount of energy reaching north- and south-facing sides of the mountain. Other environmental factors include proximity to streams and variations in soil. For example, soils on high mountain ridges are often more acidic than those in low valleys. Each species has a specific range of tolerance for these environmental gradients; these tolerances determine its distribution on the mountain.

Unusual climatic conditions and types of soils in a particular landscape often support species that cannot live in more typical environments. In the southern Appalachians, for example, some species of mosses and salamanders are restricted to the very moist conditions along streams; others live only on outcrops of particular rocks, such as granite or serpentine (Figure 8.6). Species that are found only in specific environments or restricted localities are said to be **endemic**. Landscapes with many unique species are said to have a high amount of endemism.

The second factor affecting landscape diversity are disturbances, such as fire and severe winds. Disturbances

▲ Figure 8.6 **An Endemic Species**
The moisture-rich forests of the southern Appalachian Mountains are an ideal home for Jordan's salamander, one of 20 salamander species found only in this location.

create a mosaic of patches undergoing different stages of succession (see Module 6.6). Plant and animal species vary from patch to patch, depending on the kind of disturbance and the length of time since the disturbance occurred (Figure 8.7).

Severe, crown-killing fire

Less severe, crown-scorching fire

Light fire, burned only in the forest understory

◄ Figure 8.7 **Disturbance Mosaic**
In 2012 a wildfire burned across this forested landscape near Colorado's Rocky Mountain National Park creating a complex mosaic of patches. Some patches burned severely, others less so, and others burned hardly at all. The future composition and abundance of plant species in these patches and the transitions between them will vary depending on the severity of the fire.

Community Biodiversity
■ The number of species, their relative abundance, and their arrangement in space are measures of community biodiversity.

At each location along your many mountain hikes, your samples of the flora and fauna provide measures of biodiversity. The simplest measure of biodiversity is **species richness**, the total number of species in each sample. Species richness varies greatly from location to location. In general, it is high in favorable environments, such as warm, moist climates and fertile soils along streams, and lower in less favorable locations, such as high-elevation ridgetops where growing seasons are short and soils are shallow. Fewer species are able to grow and reproduce in these less hospitable habitats.

Diversity can also be measured by **species evenness**, the relative abundance of the different species in a community. Suppose you collect butterflies from two different fields. Your samples might look like those in Figure 8.8. Since both samples have 12 butterflies and

4 species, their species richness is the same. But because most of the butterflies in field A belong to the same species, the butterflies in field A seem less diverse than those in field B. In this example, field B has far greater species evenness than field A.

Among organisms such as trees, songbirds, and butterflies, species evenness is generally low. A few species are common, while the majority of species are represented by far fewer individuals. Within a deciduous forest, birds such as titmice and chickadees may be abundant, while most other species, such as tanagers and owls, are represented by many fewer individuals. At any given time, most of the species in a community are represented by relatively small numbers. Yet over the seasons and from year to year, there may be considerable variation as to which species are more abundant and which are less abundant.

▲ Figure 8.8 **Species Evenness**
The total number of butterflies and butterfly species is the same in these two samples, but species evenness is much lower in field Ⓐ than in field Ⓑ. Which sample appears more diverse to you?

Conservation biologists are interested in both species richness and species evenness. Loss of species richness indicates that some species are disappearing from an area and may indicate a threat to the survival of those species. Species evenness reflects a variety of interactions among species, so changes in this measure may signal important changes in the environment. For example, there is typically high species evenness among the algae in healthy streams. Nitrogen pollution often causes one or two species to grow more rapidly, crowding out the other species. In such cases, a decrease in species evenness indicates stream eutrophication (see Module 11.4).

Community biodiversity is also affected by **structural complexity**, the three-dimensional distribution of species and biological features. Structural complexity is relatively easy to visualize but difficult to quantify. The structure of an old-growth forest with its towering trees, shrub layer, snags, and logs is more complex than the single layer of vegetation in a grassland. The forest provides a much greater abundance of localized habitats such as tree cavities and rotting logs. As a consequence, it has more species than the single layer of grassland plants (Figure 8.9). Similarly, the structure of a coral reef, with its multiple layers extending upward for many meters, is far more complex than that of a flat, sandy ocean bottom (Figure 8.10). So it is not surprising that the reef is home to a great many more species than the sandy ocean bottom.

▲ Figure 8.9 **Complexity Causes Diversity**
A tropical forest Ⓐ has far more structural complexity than a grassland Ⓑ. It is easy to see how the complexity of the forest creates numerous habitats that support a great many species.

▲ Figure 8.10 **Marine Structural Complexity**
The structural complexity of coral reefs Ⓐ supports many more species than nearby sandy ocean bottoms Ⓑ.

Genetic Biodiversity

■ The diversity within populations is measured by genetic variation among individuals.

Within populations of individual species, biodiversity is measured by **genetic diversity**, the genetic variation among individuals. Genetic diversity within our own species is especially obvious in the variation of traits such as height, hair and skin color, and blood types. Genetic diversity in populations provides the variation upon which natural selection acts, allowing the evolution of new adaptations as environments change. Without genetic diversity, species could not compete in the different environmental conditions across their geographic ranges and through time (see Module 4.5).

In most populations of plants and animals, genetic variation is maintained by sexual reproduction. Sexual reproduction ensures that offspring are genetically distinct from one another and from their parents (see Module 4.2). Genetic diversity of populations is maximized by **outbreeding**, or mating between individuals that are not closely related. If the pool of potential mates is small, **inbreeding**, or mating between closely related individuals, is more likely. The more limited genetic diversity within such a population increases the chance of genetic diseases and reproductive failures. Maintaining genetic diversity is especially important in the conservation of threatened and endangered species. As species lose habitat, their populations decrease. The small remaining populations often lack the genetic variation needed for long-term evolutionary change.

Small population size also increases the likelihood that the adverse effects of genetic defects will put the species at even greater risk. Where a population becomes very small, conservation biologists may implement captive breeding programs to restore genetic diversity.

For the Florida panther, conservation biologists chose a different approach. In the early 20th century, there were 30 subspecies of Florida panthers living across the southeastern United States. By 1995 only a single population remained, and it had been so severely reduced in number that they were in danger of extinction. There were fewer than 30 of the big cats living at the southernmost tip of Florida. They suffered from heart defects and reproductive problems that meant shorter life spans and low birth rates. In response, a group of government agencies, including the U.S. Fish and Wildlife Service and the Florida Fish and Wildlife Conservation Commission, agreed to the release of eight female pumas into their territory. The pumas, small mountain lions, had been captured in Brewster County, Texas. This influx of new genetic material into the population has helped to stabilize this population (**Figure 8.11**).

In popular publications, biodiversity is often equated to species richness. However, conservation of biodiversity must take into account the diversity of life at all levels of organization, including genetic diversity within species and the diversity of habitats and ecosystems across landscapes.

QUESTIONS 8.1

1. Describe two factors that influence landscape diversity.

2. What is the difference between species richness and species evenness?

3. Why does the genetic diversity of a population often decline when the populations become smaller?

(MES) For additional review, go to **MasteringEnvironmentalScience**

▼ Figure 8.11 **Genetic Diversity Improves Survival**
Genetic variation within a species reduces the risk of the genetic disorders associated with inbreeding. These panther cubs were produced by crossbreeding endangered Florida panthers with female pumas from Texas. As a result, these cubs do not display some of the traits associated with inbreeding, such as undescended testicles in males and a kinked tail caused by a deformity of the vertebral column.

8.2 Why Biodiversity Matters

BIG IDEA Biodiversity is valuable in many ways. A diverse array of species is required to maintain ecosystem functions and the services they provide to humans. Biodiversity helps ecosystems respond quickly to disturbances and environmental change. In ancient hunting and gathering cultures, wild plants and animals were the source of food and clothing (**Figure 8.12**). Biodiversity has great economic value to modern humans, who depend on many different species of plants and animals for food, fuel, shelter, clothing, medicines, and other purposes. Many people believe that Earth's species are valuable in their own right, independent of their value to humans (see Module 2.3). Therefore, the other creatures sharing our planet deserve our care and concern.

▲ Figure 8.12 **Humans Have Always Depended on Biodiversity**
These cave paintings in Lascaux, France, date from 15,000 to 20,000 years ago. They depict the diversity of animals upon which the painters depended for food and clothing.

Ecosystem Functions and Services

■ Ecosystem functions and services depend on the actions of individual species and interactions among species.

How do different levels of species diversity affect ecosystem functions? For over 20 years, David Tilman and his colleagues at the University of Minnesota have been carrying out experiments to answer this question. In a prairie at the Cedar Creek Natural History Area, Tilman's team planted a number of experimental plots with different numbers of grass and herb species. In each plot, the team measured ecosystem processes such as net primary production, nutrient uptake, and water use (**Figure 8.13**). Their results showed that all of these processes were enhanced as species richness increased.

Similar studies carried out in a variety of ecosystems around the world have reached the same conclusion: Species richness increases ecosystem functions. Scientists do not fully understand *why* biodiversity enhances these functions, although the effects of sampling and complementarity appear to be important.

Species within a community vary with regard to their importance to particular ecosystem processes, such as net primary production. As biodiversity increases, there is a higher likelihood that an ecosystem will include a species that has a big effect on a process. This increased likelihood is known as the **sampling effect**.

Species within a community usually use resources in a complementary fashion. For example, the roots of various prairie herbs and grasses reach different parts of the soil, resulting in a fuller use of soil water and nutrients (see Module 6.1). Because individual species exploit different parts of the environment, the effect is that groups of species use resources more efficiently than would any single species. This is termed the **complementarity effect**.

Biodiversity affects many ecosystem services upon which humans depend. For example, water flowing through wetland ecosystems is purified of chemical wastes. Wetlands also moderate the flow of water, thereby minimizing floods. Over half of the wetlands in the United States have been destroyed or degraded, resulting in the loss of these services. The destruction of wetlands along the Gulf Coast is thought to have contributed to the devastating floods associated with Hurricanes Katrina and Rita in 2005. Where wetland ecosystems are being restored, these services are enhanced by increasing the amount of biodiversity.

▲ Figure 8.13 **Diversity Increases Productivity**
David Tilman and his colleagues set up these experimental plots at the Cedar Creek Natural History Area in Minnesota. Note the differences in the growth and greenness of the plants in the different plots. The greenest and most productive plots also have the highest species richness.

Q *What is the worst case scenario from losing one species of herbivore, for example?*
Breeann Sharma, Arizona State University

Norm: Ⓐ The loss of a single herbivore species can result in the loss of carnivore species that depend on it. If it was once common, its loss can also influence the diversity and abundance of various plant species.

Ecosystem Stability

■ Biodiversity enables ecosystems to resist or recover from environmental changes and disturbances.

Biodiversity in a community buffers the effects of environmental change. During their long-term studies, Tilman's group noted considerable year-to-year variation in the annual rainfall in Minnesota. In general, lower rainfall produced lower net primary production, but greater species diversity lessened this effect (Figure 8.14).

Why were the more diverse communities more resistant to drought? Diverse ecosystems include species with different tolerances to the availability of water. Some plants grow better in wet years, and others grow better in dry years. From year to year, the abundance of the various species changes, but primary production remains relatively constant. However, the loss of just one or a few species from an ecosystem can diminish this drought resistance.

Biodiversity also plays a significant role in the ability of ecological communities to recover from disturbances (see Module 6.6). Because they are widely dispersed and able to tolerate harsh open habitats, pioneer species such as mosses and lichens are able to colonize disturbed areas. In so doing, they stabilize and enrich soils, providing habitat for the establishment of other species. The total process of succession from bare rock to a forest involves many more species than are present at any single stage in that process.

The interactions among species across different food web trophic levels provide a powerful example of the stabilizing role of biodiversity in ecological communities (see Module 6.4). The loss of top carnivore species can cascade through a food web, resulting in explosive growth of herbivore populations, often with devastating effects on the diversity of plants.

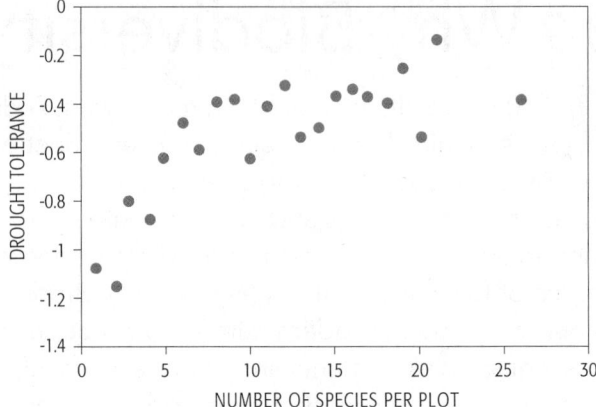

▲ **Figure 8.14 Diversity Increases Drought Tolerance**
In Minnesota prairies, drought diminished net primary production (NPP) less in experimental plots where the species richness of plants was high. Drought tolerance was measured here as the natural logarithm of the ratio of NPP at the height of drought to NPP in a wet year. If NPP was unaffected by drought, this drought tolerance index would equal zero.

Data from: Tilman, D. and J.A. Downing, 1994. Biodiversity and stability in grasslands. *Nature* 367: 363–365.

Historical changes in populations of sea otters in the coastal waters of southern Alaska and British Columbia illustrate the importance of such trophic cascades to the stability of complex ecosystems (Figure 8.15). Sea urchins, which feed on the young growth of seaweeds such as the giant kelp, are among sea otters' favorite food sources. Where sea otters have been eliminated by hunting or predation by killer whales, sea urchin numbers increase dramatically. Overgrazing by sea urchins can result in the near elimination of kelp forests. This trophic cascade from otters to sea urchins to kelp has myriad effects on other species. For example, loss of kelp forest results in a significant decline in a variety of fish species. This, in turn, influences the diets and populations of fish-eating birds such as gulls and bald eagles.

Economic Value

■ Biodiversity provides goods and services that are essential to humans.

Throughout human history, a great many species of plants and animals have been traded in economic markets (see Module 2.3). In modern societies, biodiversity is harvested in activities such as lumbering, fishing, hunting and trapping for meat and furs, and collecting exotic animals and tropical fish for sale as pets. Rare plants are dug for sale to gardeners or collected to make medicines. These direct uses of biodiversity, in effect provisioning ecosystem services, are economically significant. For example, over 100 million metric tons of wild fish are harvested each year, providing roughly 20% of the animal protein consumed by humans around the world. Forest trees provide wood for construction and

▶ **Figure 8.15 Otters Matter**
Where sea otters have disappeared because of hunting or predation by killer whales, explosive growth of sea urchins virtually eliminates giant kelp forests and their associated fish species. This, in turn influences food supplies for predatory birds. Positive and negative effects in these interactions are denoted by plus and minus signs in the diagram.

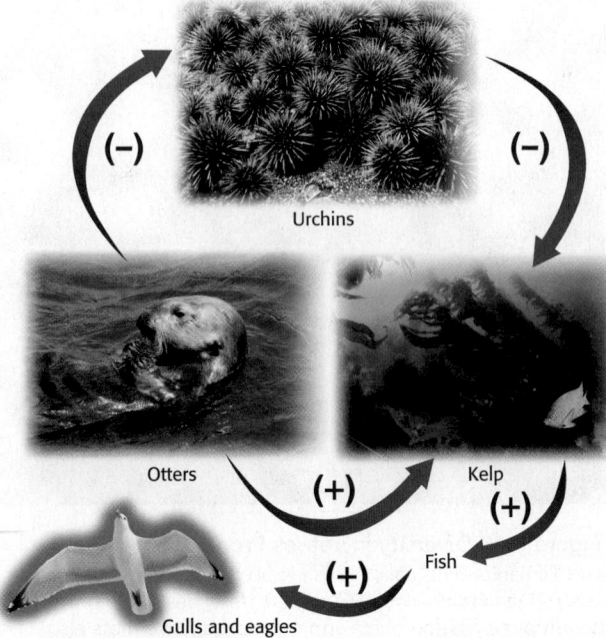

Urchins

Otters

Kelp

Fish

Gulls and eagles

fiber for paper. The many species within forests produce wild nuts and fruits, mushrooms, and understory plants that provide a continuous source of income for people living in forested regions.

Recreational hunting and fishing contribute significantly to the economies of rural communities. Money spent on equipment, fishing lodges, and hunting preserves multiplies the economic value of fish and game as much as 10-fold or even 100-fold. Every year, for example, the Platte River in Nebraska attracts 176,000 people who come to hunt game birds and 247,000 people who come to fish. A study commissioned by the Environmental Protection Agency (EPA) estimates that the combined input of these outdoor enthusiasts contributes about $50 million to Nebraska's annual economy. The economic value of the various game species produces social pressure to save habitats in the Platte River valley, which otherwise might be lost to agriculture or development.

Biodiversity is an important attraction for many tourists, generating hundreds of billions of dollars of economic value annually. For example, Costa Rica derives more than 7% of its total GDP from **ecotourism**, in which visitors come to see the natural beauty and biodiversity of its extensive tropical forests (**Figure 8.16**). After the establishment of Costa Rica's national park system, visits by international tourists rose from 60,000 in 1983 to nearly 2 million in 2010.

Biodiversity-related tourism enhances the economies of industrialized nations, too. A recent study estimated that there are 46 million bird-watchers in the United States. These birders spend a total of $85 billion per year on equipment and trips, creating over 800,000 U.S. jobs.

Biodiversity is also of great value to biotechnology and medicine. About 20% of the pharmaceuticals distributed in the United States come from plants (**Figure 8.17**). Indeed, many of these plants have been used for centuries by native peoples in many locations. And over half of all new drugs developed in the past 20 years come

from chemical compounds derived from plants, animals, or microorganisms collected from ecosystems such as tropical forests, deserts, and coral reefs. In the year 2000, global sales of medicines developed from marine animals surpassed the $100 billion mark. A single product made from chemicals found in a sea sponge and used in treating viral infections is generating millions of dollars in profits each year. These examples suggest that the vast array of species that have not been studied represent an enormous pool of potential economic value.

Without the wealth of species found in the world's many natural ecosystems, biotechnology would have fewer chemical tools to use and limited sources of genetic information for research. Because biodiversity is the fundamental source of biological information, preserving it will further biotechnology and provide new treatments for human diseases.

There are important economic benefits of biodiversity that are not directly tied to traditional markets. Biodiversity loss that results in the loss of ecosystem services can have significant economic consequences (see Module 2.4). For example, the loss of diverse coastal wetlands and mangrove forests has resulted in increased storm surges and flooding during hurricanes and typhoons. The loss of such regulatory and support services has resulted in increased loss of life and property.

▲ Figure 8.16 **Ecotourism and Biodiversity**
This swinging bridge in Costa Rica's Monteverde National Park is a popular spot for bird watchers from all over the world. Tourists interested in biological diversity provide economic support to local communities, which encourages the conservation of the biodiversity that surrounds these communities.

Plant Name	Medicine	Treatment Use
Foxglove (*Digitalis purpurea*)	Digitalin and digitoxin	Heart failure
Goat's rue (*Galega officinalis*)	Metformin	Diabetes
May apple (*Podophyllum peltatum*)	Teniposide	Childhood leukemia
Octea (*Ocotea glaziovii*)	Glasiovine	Depression
Pacific yew (*Taxus brevifolia*)	Taxol	Cancer
Quinine tree (*Cinchona pubescens*)	Quinine	Malaria prevention
Rosy periwinkle (*Catharanthus roseus*)	Vincristine	Childhood leukemia
Sweet wormwood (*Artemisia annua*)	Artemisin	Malaria treatment
Velvet bean (*Mucuna pruriens*)	L-Dopa	Parkinson's disease
Willow (*Salix* spp.)	Aspirin (salicylic acid)	Pain, inflammation

▲ Figure 8.17 **Medicinal Plants**
These are just a few examples of the many plants that provide drugs that improve human well-being.

Existence Value

■ Most people consider human-caused extinction of species to be unethical.

Many people believe that plants and animals have an intrinsic value, regardless of their importance to human beings (see Module 2.3). Therefore other living things deserve respect and protection. Most of the world's ethical and religious traditions recognize a moral obligation to protect other species. The idea of biodiversity's intrinsic value is also well established in the history and philosophy of science. This ethical belief is captured in the United Nations Charter for Nature, which states, "every form of life is unique, warranting respect regardless of its worth to man, and, to accord other organisms such recognition, man must be guided by a moral code of action."

QUESTIONS 8.2

1. In what ways does increased biodiversity increase ecosystem functions?

2. Diverse ecosystems are generally more stable than less diverse ecosystems in the context of year-to-year variations in climate. Why is this so?

 For additional review, go to **MasteringEnvironmentalScience**

8.3 Global Patterns of Biodiversity

BIG IDEA Biologists estimate that Earth supports between 7 and 10 million species of plants, animals, and fungi. This excludes species of microscopic bacteria, archaea, and protists (see Module 4.7), which may also number in the millions. Our knowledge of this biodiversity is sadly incomplete. Only about 1.5 million species of plants and animals—10 to 20% of the total—have been described and assigned a scientific name. The vast majority of species live on land. Of these terrestrial species, over 80% are animals, most of which are insects. In the oceans, almost half of the species are algae. In general, species diversity is greatest near the equator and diminishes toward the poles. On land, tropical rain forest is the most species-rich biome; in the ocean, coral reefs are the ecosystem with the greatest diversity. Across the globe, conservation biologists have identified about 25 areas that possess high biodiversity, have many endemic species, and are particularly threatened by human activities.

Mapping Species Richness

■ The number of species in biomes decreases from the equator toward the poles.

Biological explorers who sailed around the world seeking new and exotic forms of life noticed that species richness was greatest in the tropics and declined as they traveled north or south toward the poles (Figure 8.18). Tropical forests are the most species-rich communities on land, and tropical coral reefs are the most species-rich communities in the ocean. Forests, prairies, and ocean ecosystems in the temperate zone have intermediate species diversity. Communities in polar regions have low numbers of species. Within a particular region, species diversity is also affected by rainfall and elevation, but the underlying global pattern is clear. Over long periods of time, more species have evolved and persisted in the tropical zone compared to temperate and polar zones.

Biologists have proposed several explanations for this latitudinal pattern in the evolution of species richness. Differences in primary production, past disturbances, habitat gradients, and ecosystem complexity among polar, temperate, and tropical regions have received particular attention. There is spirited debate among biologists regarding the relative importance of these various differences. None by itself provides a sufficient explanation. It is likely that all of these factors have contributed to differences in patterns of species evolution from the tropics to temperate and polar regions.

Net primary production. The warm weather and long growing seasons in equatorial ecosystems promote higher net primary production rates than those

▼ **Figure 8.18 Global Species Diversity**
The richness of plant species in ecosystems (number per 10,000 km² or 3,860 mi²) varies widely from region to region. The patterns of variation in species richness are similar for other organisms, such as vertebrates and insects. How does diversity vary among places you have visited?

Source: Bartblott, W. et al. 1999. Terminological and methodological aspects of the mapping and analysis of global biodiversity. *Acta Botanica Fennica* 162: 103–110.

in temperate or polar ecosystems. Higher net primary production means that more energy is captured in photosynthesis and consumed by the entire ecosystem. This greater availability of energy has the potential of supporting more species at each successive trophic level of the food web (see Module 6.4).

Pleistocene climate change. Ecologists once believed that tropical regions were rich in species because their climates had been relatively stable for hundreds of thousands or even millions of years. Thus, tropical biomes were spared the glaciers and shifting climates that probably wiped out many species in Arctic and temperate regions. However, recent evidence suggests that during the last ice age, climate change in tropical regions, such as the Amazon, was much more significant than previously thought. Today, scientists are much less certain about how changes in climate have affected differences in species richness among biomes.

Habitat gradients. In temperate regions, the weather changes significantly through the course of the year, favoring the evolution of organisms that can tolerate a broad range of temperatures. As a result, temperate species are often widely distributed. For example, many species are able to live at a wide range of elevations on mountain slopes. In the tropics, the lack of seasonal temperature changes causes sharper distinctions among habitats, especially in mountainous areas. The tops of a tropical mountain are always cool, mid-elevations are always moderate, and lower elevations are always warm. Some biologists contend that the uniform conditions within elevation zones have favored the evolution of species with narrow ranges of tolerance. As a consequence, tropical species have comparatively specialized ecological niches with respect to elevation.

New FRONTIERS

Counting Species

Knowing the total number of Earth's species is important. Global extinction rates for species of plants and animals appear to have increased 100- to 1,000-fold over the past century, and these losses threaten many of the ecosystem services that we depend on. In the past two decades, published estimates of Earth's species richness range from as few as 3 million to as many as 100 million. Why have estimates of worldwide species richness varied wildly?

As recently as 2001, scientists did not even have a firm estimate of the number of species that had actually been discovered, described, and catalogued. The organization Catalogue of Life was established in that year to provide that information. They listed 1,586,385 species in their online catalogue as of September 2014. By identifying numerical patterns with higher taxonomic levels such as genera, families, and orders (see Module 4.7), researchers have refined their estimate of Earth's species richness to 8.7 million, give or take 1.3 million. About 75% of those species are thought to occur on land; the rest live in Earth's aquatic and marine ecosystems. This is by no means the final word. Improved computer information systems and new approaches in the analysis of species' DNA promise more precise future estimates. Thousands of biologists are working to catalogue and name as many of Earth's species as possible. But some argue that it is not necessary, nor is it worth the time and effort. What do you think and why?

Source: Mora, C., D.P. Tittensor, S. Adl, A.G.B. Simpson, and B. Worm. 2011. How many species are there on Earth and in the Ocean? *PLoS Biology* 9 (8): e1001127 DOI:10.1371/journal.pbio.1001127.

This allows species that are potential competitors to coexist along gradients of elevation.

Ecosystem complexity. Tropical regions favor the evolution of organisms with structures that increase habitat diversity. The multiple layers of a tropical forest provide unique habitats for a diverse array of vines and epiphytes (plants that grow on other plants). These plants, in turn, provide habitats for a diverse array of animals. In the oceans, the complex structures of coral reefs provide distinctive habitats for many different species. Some biologists note that the diverse array of herbivores and predators in tropical ecosystems promotes diversity among the species upon which they feed (see Module 6.4).

Species per 10,000 km²
- Fewer than 100
- 100–500
- 500–2,000
- 2,000–4,000
- More than 4,000

QUESTIONS 8.3

1. What factors likely contribute to the differences in species richness between ecosystems at the equator and those in polar regions?

2. What two features define a biodiversity hotspot?

(MES) For additional review, go to **MasteringEnvironmentalScience**

Biodiversity Hotspots

■ The high biodiversity of certain places on Earth is especially threatened by human activities.

Certain places have an unusually large number of endemic species. Conservation biologists define such regions as **biodiversity hotspots** if the human threats to their habitats are especially high. Conservation International, an organization that uses biodiversity hotspots to prioritize its activities, classifies a region as a hotspot if it contains at least 1,500 species of endemic plants and at least 70% of its original area has been severely altered by human activities.

Using Conservation International's criteria, over 35 biodiversity hotspots have been identified (Figure 8.19). These hotspots represent less than 12% of Earth's total area, but are estimated to contain at least 44% of all plant species and 35% of all vertebrate animal species. As of 2009, over 90% of the habitat in these regions had been destroyed or severely altered.

What makes a hotspot hot? In most hotspots, three factors have allowed the evolution of many endemic species—unique habitats, topographic diversity, and isolation. Two regions with Mediterranean-type climates, California and South Africa, provide excellent examples of the role of these factors (see Module 7.3). Both regions have complicated geologic histories that have produced a diverse array of unique soils. Natural selection has favored the evolution of a great many endemic species that are able to grow in these soils. Both regions also have several mountain ranges. The various elevations and differences between moist and dry slopes produce a great diversity

of mountain habitats. These unique habitats and the topography of the mountains isolate populations from one another. This geographic isolation limits interbreeding among populations within the hotspot, allowing new species to evolve (Figure 8.20; see Module 4.6).

▲ Figure 8.20 **Hotspots Indeed**
In South Africa, shrubby ecosystems called fynbos include a great many endemic species, such as the red aloe in the center of this image. This picture was taken from atop Table Mountain near Cape Town, part of the Cape Floristic Region shown in Figure 8.19.

▼ Figure 8.19 **Biodiversity Hotspots**
The areas labeled here are designated as hotspots because of the high number of endemic species they contain and the large amount of habitat that has been lost from them.

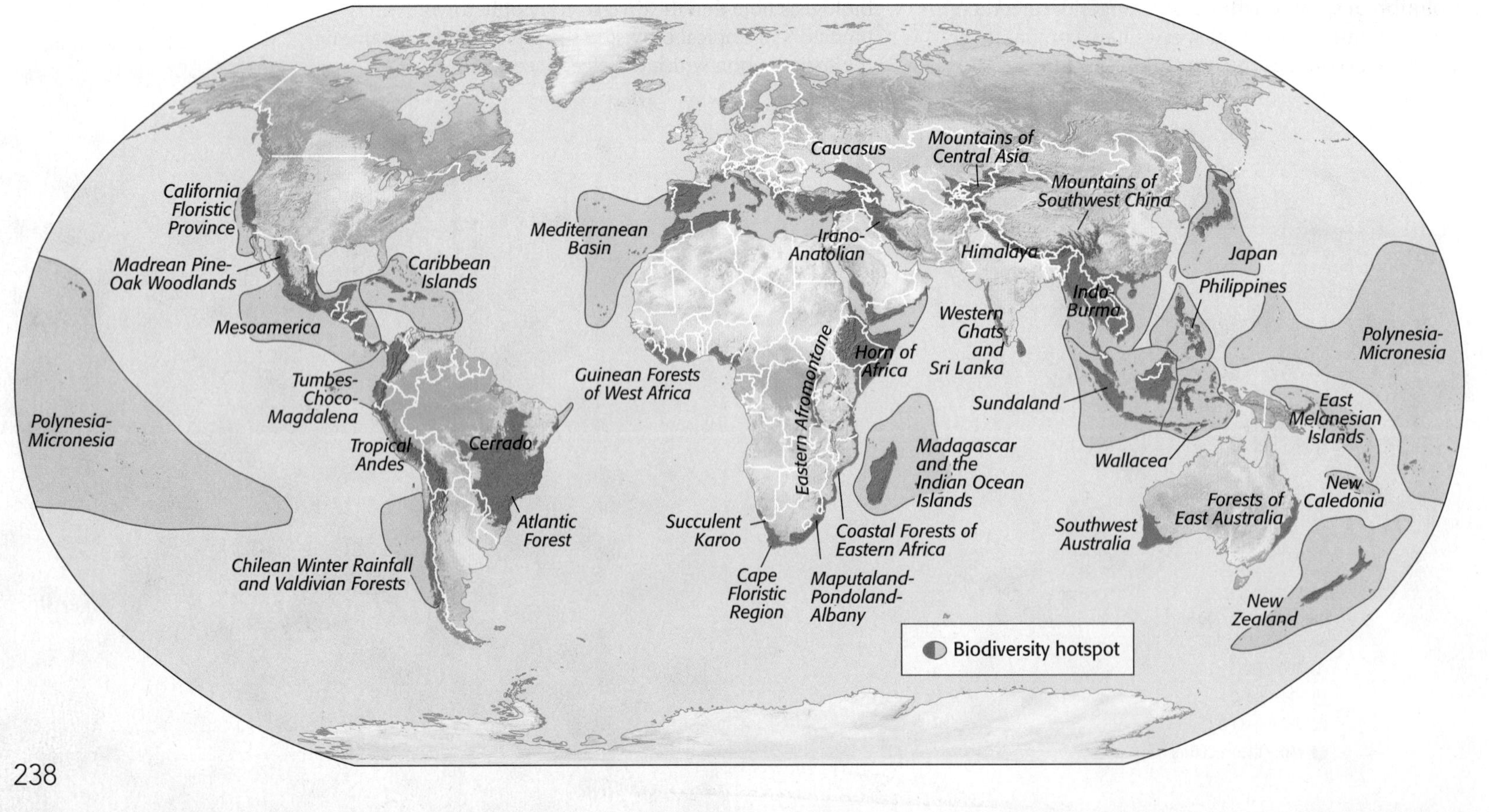

California Floristic Province

Madrean Pine-Oak Woodlands

Caribbean Islands

Mediterranean Basin

Caucasus

Mountains of Central Asia

Mountains of Southwest China

Irano-Anatolian

Himalaya

Japan

Philippines

Mesoamerica

Tumbes-Choco-Magdalena

Polynesia-Micronesia

Tropical Andes

Cerrado

Atlantic Forest

Chilean Winter Rainfall and Valdivian Forests

Guinean Forests of West Africa

Eastern Afromontane

Horn of Africa

Western Ghats and Sri Lanka

Indo-Burma

Sundaland

Madagascar and the Indian Ocean Islands

Wallacea

Polynesia-Micronesia

East Melanesian Islands

Succulent Karoo

Cape Floristic Region

Maputaland-Pondoland-Albany

Coastal Forests of Eastern Africa

Southwest Australia

Forests of East Australia

New Caledonia

New Zealand

Biodiversity hotspot

8.4 Differences in Biodiversity Among Communities

BIG IDEA Over long time spans, variations among biomes in species richness have been shaped by differences in rates and patterns of species evolution. However, from locality to locality and time to time, biodiversity varies significantly among individual ecological communities. Evolution has and continues to play a role in these variations, too. For example, natural selection has produced species that are adapted to the variety of unique habitats created by differences in topography, geology, and ecosystem structure. But non-evolutionary processes and factors are also important. These include interactions among species within communities, such as competition and predation, and natural and human disturbances. In general, the number of species in a community is determined by the equilibrium between the rate at which new species migrate into the area and the rate at which species disappear from that area. Migration rates are influenced most by the degree to which areas are isolated from sources of new migrants. Disappearance rates are influenced most by the size of islands or habitat patches.

Habitat Diversity

■ Topographic and geologic variation fosters diversity.

Within any region, environments with greater diversity of habitats support more species than environments with more uniform habitats. Landscape features such as mountains, valleys, lakes, and rivers create many kinds of habitats. This complexity of habitats increases the number of potential ecological niches, which in turn support a greater diversity of species. Coastal areas that are broken up into chains of islands tend to have more species than coastlines without islands. In marine ecosystems, areas of the ocean with rocks or reefs on the bottom support more species than areas with flat, featureless seabeds. Building an artificial reef on a flat seabed provides places to which organisms such as corals and algae can attach themselves, thereby increasing the number of species in the ecosystem (**Figure 8.21**).

Habitat complexity also affects ecosystems on a smaller scale. A natural stream with its riffles and deep pools, boulders, sunken logs, and sandbars can support a species-rich community of aquatic life. If dredging equipment is used to straighten out the stream and give it uniform depth, the resulting canal has fewer habitats, and species richness declines significantly.

▼ Figure 8.21 **Old Shipwrecks Support Marine Ocean Diversity** This old ship was sunk off the coast of Florida in order to create a complex habitat that will support many marine species.

Species Interactions

■ Competition and predation influence biodiversity.

In most ecological communities, many species compete for common resources, such as food, water, or nesting sites. Competitors are able to coexist because each species occupies a different ecological niche; that is, each species has a different range of tolerance for environmental factors or acquires its resources in a slightly different way (see Module 6.2). Changes in the environment, such as an increase in an essential nutrient, can alter the competitive relationships among species. This, in turn, can cause one or a few species to dominate the others, resulting in a loss of biodiversity. For example, nitrogen pollution in streams often results in the rapid growth of one or two species of algae. Other species of algae, which are no longer able to compete for light or other resources, decrease in number or disappear entirely.

Predators usually increase the biodiversity of an ecosystem. This fact may seem counterintuitive, but it makes sense in the context of competition among the prey species. Predators usually feed most heavily on the species of prey that are most common and require the least energy to capture. As the numbers of a particular prey species decrease, resources become available for other species, which increase in number. Predators then focus their attention on the species that have now become more common, a phenomenon called prey switching. Consequently, no individual prey species is able to become dominant (Figure 8.22). That is why the removal of top predators, such as wolves or large predatory fish, can have devastating effects on the diversity of the prey species upon which they feed (see Module 6.4).

Herbivorous animals appear to play an especially important role in the diversity of tropical rain forest trees.

▲ Figure 8.22 **Predation Increases Diversity**
The diversity of savanna ungulates, such as these zebras and wildebeests, is partly the result of predation by carnivores, such as lions and leopards. No single prey species is able to become dominant enough to exclude its competitors.

Ecologists have long been astounded at the diversity of tree species in these ecosystems, even within very small areas. Even an area as small as a single hectare (2.5 acres) may contain over 100 different tree species. Many different animals feed on the seeds and seedlings of these trees, and many of these herbivores feed only on a specific species of tree.

Experimental studies have demonstrated that the number of herbivores for any one particular tree species is highest in the vicinity of that species' largest trees. As a consequence, the probability that seeds from that species will successfully germinate and grow is lowest near adult trees of that species (Figure 8.23). Other tree species are more likely to germinate and grow there. When extended throughout a forest, this interaction between trees and their herbivores results in far greater diversity than would be found if the seedlings of each tree species grew best near their parent trees.

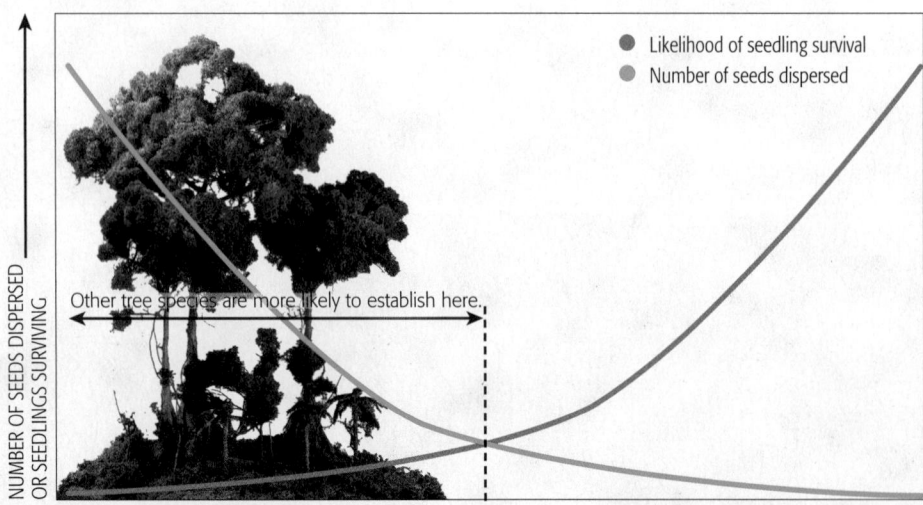

Other tree species are more likely to establish here.

● Likelihood of seedling survival
● Number of seeds dispersed

NUMBER OF SEEDS DISPERSED OR SEEDLINGS SURVIVING

DISTANCE FROM PARENT TREE

◄ Figure 8.23 **Herbivory Increases Tropical Tree Diversity**
Although the seeds of any particular tree species are more likely to fall near their parent tree, predators that feed on those seeds are also likely to be more abundant there. Despite fewer seeds being dispersed to distant locations, the likelihood of survival is higher than nearer the tree. Trees of the same species are therefore much less likely to grow near one another.

Disturbance

■ Biodiversity is highest with intermediate levels of disturbance.

Natural disturbances such as fires or hurricanes occur in most ecosystems. Such disturbances initiate a succession of change in the abundance of different species (see Module 6.6). Certain kinds of disturbances also increase the diversity within individual communities. For example, frequent fires have been shown to maintain the high diversity of herbs and grasses in prairies and savannas (Figure 8.24). In these ecosystems, fire acts like an herbivore, preventing any particular herb or grass species from dominating competing species. Ecologists have shown that biodiversity in grasslands, savannas, and shrublands is highest at intermediate levels of disturbance. If fires are too infrequent, few species flourish. However, many species cannot tolerate fires that are very hot or too frequent, and such conditions also diminish biodiversity.

◄ Figure 8.24 **Disturbance and Diversity**
Like an herbivore, fire in a tall grass prairie limits competitive exclusion and helps increase diversity. These summer wildflowers (inset) are growing in a recently burned Wisconsin prairie.

Local Immigration and Extinction Rates

■ The species richness of communities on islands is determined by the rates at which species arrive and disappear.

It stands to reason that the number of species in a particular place—say birds in a patch of forest or fish around a shipwreck—would be influenced by the rates at which new species appear and others disappear from that place. The rate of appearance is determined by the rate at which species immigrate, or move into an area. The rate of local extinction is determined by the rates at which populations of species are lost from that area.

The importance of immigration and extinction was first clearly articulated by biologists R.H. MacArthur and E.O. Wilson in a theory to explain the variation in the number of species found on different ocean islands. Their **equilibrium theory of island biogeography** states that the diversity of species on an island is largely determined by two factors: the rate at which new species migrate to the island and the rate at which species disappear from the island, or become locally extinct.

MacArthur and Wilson hypothesized that distance from the mainland is the single most important factor influencing the rate at which species migrate to an island. Most plants and animals come to an island from the mainland, and it is easier to reach nearby islands than those that are far away. Therefore, islands close to the mainland have higher rates of immigration than more distant islands.

They also hypothesized that the rate of species loss from an island is most influenced by island size. Small islands support smaller average population sizes than large islands. As populations become smaller, they are more prone to disappear.

MacArthur and Wilson argued that the diversity of species on an island reaches an equilibrium number when the rate of new species immigration is equal to the rate of local species loss. At the equilibrium number, individual species may come and go, but the total number of species remains relatively constant. Islands closer to the mainland have higher immigration rates. The smaller an island, the smaller the size of its plant and animal populations and the higher their rate of species loss. Thus, the theory predicts that large islands close to the mainland will have more species than small, distant islands (Figure 8.25).

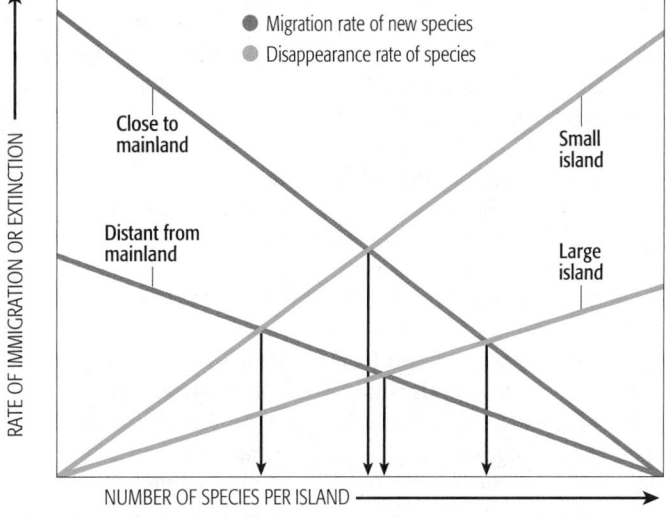

● Migration rate of new species
● Disappearance rate of species

Close to mainland

Distant from mainland

Small island

Large island

RATE OF IMMIGRATION OR EXTINCTION

NUMBER OF SPECIES PER ISLAND

◄ Figure 8.25 **Island Diversity**
The number of species on an island is determined by the balance between its rates of immigration and disappearance. The equilibrium number of species for each island occurs where the two lines cross as indicated by the black arrows. How does the equilibrium number for a large island close to the mainland compare with that of a small island that is distant from the mainland?

241

▲ Figure 8.26 **Sky Islands**
The peaks of the Santa Catalina Mountains near Tucson, Arizona, provide islands of unique habitat separated from one another by a sea of desert.

The equilibrium theory of island biogeography has become a fundamental principle in conservation biology. The importance of island size and distance to mainland to plant and animal biodiversity has been verified in clusters of islands around the world. On the hundreds of islands of Micronesia in the South Pacific, for example, the diversity of birds, mammals, and plants is highest on the largest islands and on islands closest to mainland sources of species.

The principles of this theory can be applied to other kinds of isolated habitat patches. For fish species, a pond is like an island of livable habitat surrounded by inhospitable dry land. As the theory predicts, the number of fish species living in lakes increases with lake size. It also increases with the number of rivers and streams connecting them.

For wildflowers growing on the top of a mountain in Arizona, the surrounding hot, dry lowlands represent a barrier as forbidding as the open sea. In recognition of the isolation of plants growing at high altitudes, mountaintops are sometimes called "sky islands" (Figure 8.26). As the theory predicts, small, isolated mountaintops have fewer species of plants than do larger mountains that are in close proximity to one another.

An important corollary to the theory can be used to help predict the biodiversity of ecosystems on the mainland. A large, continuous ecosystem will support more species than an ecosystem composed of several smaller, disconnected pieces. In other words, the biodiversity of an ecosystem depends not only on the total extent of that ecosystem but also on how fragmented the ecosystem is. Thus the biodiversity of herbs or birds in a forest depends not only on the total area of forest but also on how well the individual forest parcels are connected to one another. Large, interconnected forests have greater biodiversity than forests that are cut into disconnected patches by farms, roads, or housing developments (Figure 8.27).

QUESTIONS 8.4

1. Describe how competition and predation affect the diversity of communities.

2. Why do disturbances such as fire often increase the diversity of communities?

3. Large islands generally have more species than smaller islands, and islands close to the mainland have more species than those farther away. Why?

(MES) For additional review, go to **MasteringEnvironmentalScience**

▲ Figure 8.27 **Fragmented Forests**
Forest once covered this entire landscape. Agriculture, roads, and development now divide it into isolated patches and limit the migration of species between patches. Populations in small patches are more likely to disappear.

8.5 Threats to Biodiversity

BIG IDEA Human activities are causing the global rate of extinction to increase at an alarming rate. Among those activities, the destruction of habitat is most important. Even where the amount of habitat remains high, development and land use are dividing it into smaller and more isolated fragments. As human populations have grown, they have exploited the populations of many plants and animals to extinction or near extinction. Humans have relocated species around the globe, introducing an array of invasive non-native competitors, predators, and diseases to every biome. Air and water pollution, altered patterns of disturbance, and global climate change present other threats to Earth's biodiversity.

Habitat Loss and Degradation

■ Worldwide, the single greatest threat to biodiversity is habitat loss.

The foremost need of every species is abundant habitat in which to grow and reproduce. However, human activities have greatly reduced the extent of many ecosystems and the unique habitats associated with them. In 2011, the International Union for the Conservation of Nature (IUCN) "red listed" 562 of the world's bird species as critically endangered or endangered. Of these species, 82% were listed because of habitat loss.

The extent of habitat loss is particularly alarming in a number of ecosystems. By 1900, over 95% of North America's deciduous forests had been heavily logged or cleared for agriculture. Agriculture has taken an even higher percentage of North American grasslands. Only 3% of North America's mixed and tall-grass prairies remain, and most of these grow in small, isolated patches. Two hundred years ago, longleaf pine savannas covered millions of hectares in the southeastern coastal plain of the United States. Only 1% of this ecosystem is left today. Longleaf pine savannas provided habitat for very diverse endemic flora and fauna. Many of these species are now listed as endangered under the Endangered Species Act (Figure 8.28).

The dwindling tropical forests are of special concern because they contain a disproportionate number of the world's species. These forests were originally vast, covering more than 5 million square miles of Earth's tropical regions. Today about half of these forests are gone, and the clearing of tropical forests continues at a rate of one-half million square miles every decade. The effects on biodiversity are sadly predictable. Biologists

RED-COCKADED WOODPECKER

GOPHER TORTOISE

▶ Figure 8.28 **A Vanishing Ecosystem**
Longleaf pine forests provide important habitat for many species. The loss of this habitat threatens the survival of the endangered red-cockaded woodpecker and gopher tortoise.

243

estimate that the current global extinction rate in tropical forests is 1,000 species per year. And this rate is likely to increase as survivors in small patches of forest suffer the inevitable fate of isolated populations whose numbers become too small to sustain themselves.

Marine ecosystems have also suffered significant habitat loss. Coastal development and boating have destroyed many of the extensive sea grass beds once found in shallow coastal waters. These beds provided habitat for a variety of shellfish, and their grasses were an important source of food for endangered sea turtles, manatees, and dugongs (Figure 8.29). Twenty percent of Earth's coral reefs, the most diverse of marine habitats, have been destroyed, and another 20% have been severely damaged. In some places the picture is much bleaker. In the Philippines, for example, only 10% of the coral reefs remain intact.

◄ Figure 8.29 **Sea Grass Beds**
Sea grass beds in shallow tropical waters are a critical resource for a great many animal species, including the endangered dugong.

Q I understand that there are more and more species going extinct. Is this mostly due to humans or is it just natural selection?

Jaymee Castillo, University of California, Los Angeles

Norm: A It is mostly due to humans. The current rate of species loss is more than 100 times greater than in pre-human times.

Habitat Fragmentation

■ Human development and land use is dividing landscapes into ever smaller and more disconnected habitat islands.

As human activities expand across the globe, large areas of natural habitat are carved up into smaller fragments, which are increasingly separated from one another. Vast landscapes of adjoining habitat have been transformed into disconnected patches, or "islands," of habitat. In many places, urban development and road construction further isolate these patches from one another. Populations within habitat patches are smaller and more vulnerable to local extinction, and recolonization of patches is increasingly difficult. Habitat fragmentation is a major factor in the decline in the diversity of forest songbirds in the eastern United States (Figure 8.30).

When suitable habitat is divided into small patches, the life histories of many organisms are disrupted. Tree seeds, for example, tend to blow into adjacent land that is unsuited for seedling survival. The food and shelter needed by animals may be separated. Small animals tend to wander out of their preferred habitat and die before they find their way back. Larger animals may need breeding territories that are larger than the fragment of habitat they occupy. Migration pathways may be blocked by roads or highways, and the sound of traffic often disturbs or confuses wildlife.

Habitat fragmentation has also played a significant role in the loss of species in some aquatic ecosystems. For example, thousands of dams divide populations of fish that once lived in free-flowing streams. Dams also create insurmountable barriers for shad, salmon, and other fishes that migrate up and down streams as part of their reproductive cycles.

▼ Figure 8.30 **Forest Songbird Decline**
Based on censuses carried out for over a century, the number of songbird species, such as this summer tanager, is declining in forests across the eastern United States. Not only is there less forest habitat, but that habitat is also becoming much more fragmented. To hear as well as see a summer tanager, go to Audubon's website and search for the bird by name.

Overharvesting

■ Species with large bodies, slow population growth rates, and flocking behavior are particularly vulnerable to human exploitation.

The human activities of hunting, fishing, trapping, and collecting endanger many plants and animals. The species in greatest danger are those that humans value as sources of food or medicine and those that are prized by collectors, such as exotic tropical birds and trophy animals.

There are many historical examples of overexploitation of natural populations. Whales once dominated the seas of the world, but several species have been driven near to extinction by fleets of whaling ships seeking their oil and meat. In the 1800s, passenger pigeons were so common that their migrating flocks darkened the sky for days as they passed over American frontier towns. These birds were hunted to extinction because they were easy to shoot, pack in salt, and ship to Europe where they were prized as an exotic meal. Ginseng plants have become rare in the Appalachians because poachers dig and sell the knobby ginseng root, which is especially valued as an herbal medicine in Asia.

The condition of the world's fisheries is a serious problem. The UN Food and Agriculture Organization (FAO) classifies the majority of the world's marine fish species as being fully exploited. For these species, any increase in fishing level will be unsustainable and will cause fish populations to decline. The FAO estimates that over 30% of fish species are being overfished; the stocks of these species are in rapid decline (Figure 8.31).

The physical and behavioral characteristics of some species make them particularly vulnerable to overharvesting. Large animals are generally more vulnerable to human hunting. The populations of large animals typically have slow growth rates, so they are slow to recover if they are diminished. Animals that feed at higher trophic levels tend to have smaller populations and are also more vulnerable to harvesting pressure. Animals that live in herds, flocks, or schools are easier to capture and kill in large numbers than are solitary animals. Certainly, any trait that makes an animal economically desirable puts it at greater risk of overharvest.

The Nassau grouper, which is a highly valued item on seafood menus, unfortunately exhibits all of these traits. As a result, it has been fished nearly to extinction throughout most of its range in the Atlantic and Caribbean oceans (Figure 8.32). The Nassau grouper is a large predator at the top of the food chain of coral reef ecosystems. It takes up to seven years for this fish to reach sexual maturity and decades for it to reach its maximum size.

Nassau groupers are solitary hunters, a feature that might make them difficult to exploit, except for one point in their reproductive cycle. Every winter month on the night of the full moon, groupers from hundreds of miles around gather at a common spawning site. These assemblies occur in the same locations year after year. From long experience, people fishing for grouper know exactly where and when to drop nets and lines to bring in boatloads of fish. The tragic impact of this fishing method is magnified because many of the spawning groupers are caught before they are able to release their eggs. As a result, the spawning ground can be fished out completely, and when it is, no groupers ever return to that site. Only strictly enforced fishing laws limiting the grouper season to nonbreeding months or forbidding all grouper fishing in some marine preserves will be sufficient to save this valuable fish from complete extinction.

▼ Figure 8.32 **Safety in Numbers?**
The Nassau grouper is a solitary reef fish for most of its life Ⓐ, but it is easily captured when hundreds of individuals gather in schools to spawn in the waters near some Caribbean islands Ⓑ. The fish in these images are 2–3 m (6–10 ft) long.

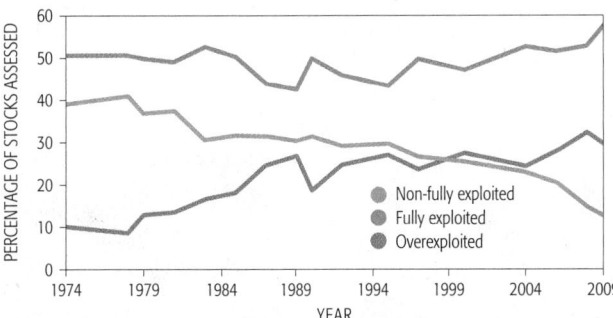

▲ Figure 8.31 **World Fisheries in Danger**
The world's fisheries have historically been treated as if they were inexhaustible. Today, nearly 90% of the world's fish stocks are fully exploited or overexploited. Note that the total value for each year is 100%.

Source: United Nations Food and Agriculture Organization. 2012. *State of the World's Fisheries and Aquaculture Report.* Rome, Italy: FAO.

Non-Native Invasive Species

■ Newly introduced species may be invasive because of their own biological traits or because of the absence of predators or competitors.

Through commerce and travel, humans have redistributed tens of thousands of Earth's species. Some of this redistribution has been intentional. For example, the plants on our farms and in our gardens come from locations around the world. Many other introductions have been unintentional, as when wood-boring beetles arrive in shipments of wood from other continents. But however they have come, non-native species now live in nearly every ecosystem on Earth. In some places they are more abundant than their native counterparts. In all too many ecosystems, the introduction of non-native species has had disastrous consequences for individual species and for biodiversity overall.

When introduced to a new place, most species have difficulty competing with native plants and animals, which are already well adapted to their environment. However, some introduced species have biological traits or behaviors that allow them to overcome such competition and flourish. These traits make them more likely to become invasive. These invasive traits give an introduced species a lopsided advantage in survival and reproduction, allowing it to outcompete native species. As the invader consumes most of the available resources, native species are driven to extinction.

Garlic mustard, for example, is a small biennial plant introduced from Europe that is displacing native wildflowers in woodland communities throughout the eastern United States (Figure 8.33). The secret to this invader's success is natural herbicides secreted by the plant's roots that change the microbial community in the soil and poison neighboring plants, allowing garlic mustard to take over the forest floor.

In other cases, the lack of competitors allows an introduced species to become invasive. For example, the brown tree snake has had a devastating effect on biodiversity, especially on tropical islands of the Pacific (Figure 8.34). Native to Indonesia, this active snake grows up to 8 feet long. It hunts at night in the branches of trees and is a deadly predator on birds and small animals. Brown tree snakes were accidentally introduced to Guam, the northernmost island in the Hawaiian archipelago, during World War II. Guam has no native snakes that might compete with the brown snake. Furthermore, birds on the island had never developed behavioral defenses against this kind of predator. As a result, brown tree snakes devastated bird populations, driving 9 of

Guam's 11 forest-dwelling bird species to extinction in the wild. In addition to birds, the snakes have eliminated small mammals and even lizards on Guam. Thanks to extraordinary efforts by the U.S. Department of Agriculture and the state Division of Forestry and Wildlife, the brown snake has not reached other Hawaiian islands.

Introduced species can also become invasive because they encounter relatively few predators or diseases in the new community. Without natural enemies, the alien population can grow quickly, driving native species to extinction. This certainly seems to have been a factor with the introduction of European rabbits to Australia and their subsequent population explosion. In such cases, the intentional introduction of a predator or disease from the invader's native ecosystem may help control its populations. For example, rabbit populations in Australia have been controlled to some extent by the introduction of a virus from European rabbits.

Wild animals and plants can usually tolerate diseases or parasites that their ancestors have encountered in generations past. Over time, natural selection builds genetically based resistance to common diseases. Just so, European hemlock trees are able to tolerate a European parasitic insect, the hemlock adelgid, which sucks nutrients from their leaves. However, the recent introduction of the hemlock adelgid to eastern North America threatens to eliminate the Canadian and Carolina hemlocks (Figure 8.35).

As a general rule of thumb, conservation biologists estimate that about 10% of the non-native species are able to become established when they enter a foreign location. Of these species, about 10% become invasive. Based on this, we might predict that non-native, invasive species should be rare. But the number of such species is large because humans have relocated tens of thousands of species.

▲ Figure 8.33 **An Alien Plant**
Garlic mustard's native habitat is the forests of Europe and western Asia, but it now dominates the understory of deciduous forests across eastern North America.

▶ Figure 8.34 **An Alien Predator**
The native birds of Guam have few natural defenses against the introduced brown tree snake, which crawls among branches and hunts for prey at night.

▼ Figure 8.35 **Hemlock Decline**
Across the eastern United States, populations of hemlock trees are in severe decline Ⓐ because of the introduction of the European hemlock adelgid. These aphidlike insects, which form white tufts on the undersides of needles and branches Ⓑ, rob trees of nutrients and ultimately kill them.

Pollution

■ Air and water pollution degrade habitats and are directly responsible for the disappearance of some species.

Air and water pollution have directly affected biodiversity in many locations. Lichens, for example, are especially sensitive to air pollution, which has diminished their abundance in many areas, especially near cities. Lichens are symbioses between a fungus and an alga. They are able to survive in some of Earth's most extreme environments, such as on rocks in Antarctica and the driest deserts. Nevertheless, lichens are especially sensitive to chemical and particulate pollutants in the air—so much so that the U.S. Environmental Protection Agency uses their diversity as an index of urban air pollution (**Figure 8.36**).

In many parts of the world, the biodiversity of streams has been diminished by sedimentation and chemical pollution. Many species of crayfish, mussels, and aquatic insects make their living by filtering pieces of dead organic matter from the flowing water. However, even small amounts of sediment can clog these animals' gills, and chemical pollutants quickly concentrate to toxic levels in their bodies. Of the 305 species of freshwater mussels in North America, 157 are now either extinct or in danger of extinction. Of the 330 species of North American crayfish, 111 are either extinct or threatened. The EPA has found that just as the diversity of lichens indicates air quality, the diversity of aquatic insects indicates water quality—their diversity diminishes directly in relation to increased water pollution.

▲ Figure 8.36 **Lichen Diversity and Air Pollution**
The many different species of lichen growing on this tree branch indicate that the tree is growing where air quality is high.

Where **YOU LIVE**   What endangered species live near you?

The U.S. Fish and Wildlife Service maintains a listing of threatened and endangered species that can be accessed by state and county at their website for endangered species. This website includes an interactive map that provides detailed information on many listed species.

■ How many species are listed for your state? For your county?

■ Select three animal and three plant species of particular interest to you. What are the key threats to each of these species and what actions are being taken to recover viable populations of them?

Altered Patterns of Disturbance

■ Changes in the frequency and severity of fires are diminishing biodiversity in some ecosystems.

Some ecosystems have natural cycles of disturbance, such as periodic fires or windfalls from hurricanes. A great many species depend upon these regular disturbances to reproduce and grow. If humans alter these disturbance cycles, the biodiversity of the ecosystems may be diminished.

The jack pine ecosystem is one such example. The jack pine forests of Minnesota, Wisconsin, and Michigan once had frequent fires, which were set by lightning. These fires consumed both overstory trees and understory vegetation and converted leaf litter to mineral ash. Jack pines are well adapted to these ground fires. In fact, they cannot reproduce without fire because their cones do not release their seeds until after they have been subjected to its heat. A few days after a fire, the cones open and their seeds drop onto the burned ground. There the seeds sprout and seedlings thrive, establishing a new jack pine forest.

Other species that live in the jack pine ecosystem are also adapted to the cycles of fire. Most notable of these is a tiny songbird called Kirtland's warbler (**Figure 8.37**). These birds nest only in stands of jack pines that are between 5 and 20 years old; they do not nest in older

stands. Suppression of fire and other timber management procedures have greatly reduced the number of young stands. As a result, Kirtland's warbler is critically endangered; fewer than 500 birds persist in the wild.

▶ Figure 8.37 **Jack Pine and Kirtland's Warblers**
Young stands of jack pine are critical habitat for the endangered Kirtland's warbler. Because humans have limited the number of recent fires, few such stands of jack pines remain.

QUESTIONS 8.5

1. What is the single most important factor contributing to the loss of species worldwide?

2. Why does the fragmentation of landscapes diminish the diversity of species within habitat patches?

3. What factors influence the vulnerability of animal species to overharvesting by humans?

4. What are the factors that influence whether non-native species become invasive or not?

(MES) For additional review, go to **MasteringEnvironmentalScience**

Climate Change

■ Disappearing sea ice endangers polar bears, and warmer seas threaten coral reefs.

There was considerable fanfare in late 2007 when Secretary of the Interior Dirk Kempthorne initiated the legal process of listing the polar bear as threatened under the U.S. Endangered Species Act. Kempthorne's decision was based on the persistent year-to-year decrease in sea ice off the Arctic shores of Canada and Alaska. Polar bears depend on sea ice as a platform from which to hunt for their most important prey, the Arctic seal (**Figure 8.38**). There had been much debate about whether the polar bear should be given legal status, and controversy over the decision continues. For example, critics point out that most polar bear populations elsewhere in the world are either stable or increasing. What is clear, however, is that Arctic climates are warming at a rapid pace, and this warming is having a significant impact on the tundra and the extent of near-shore sea ice (see Module 9.4).

Climate change also affects tropical oceans. A great many coral reefs have experienced a phenomenon called coral bleaching, which many scientists believe is tied to the warming of tropical oceans. Corals have a symbiotic relationship with algae that live inside their bodies. The algae are a critical source of nutrition for the corals; the algae also give the coral animals their color. If the water becomes too warm, the stressed coral animals expel the algae, causing the corals to lose their color. Bleached corals may regain algae, but often they simply die (**Figure 8.39**).

Climate changes due to carbon dioxide and other atmospheric pollutants will challenge living organisms for many years to come (see Module 9.4). Although humans can respond to climate change by planting different crops or moving to more hospitable areas, other species have fewer options. Alpine plants growing on a mountaintop cannot move to a higher mountain if climate change melts the snow and warms their habitat beyond their limits of tolerance. Stressed corals in a warming ocean cannot easily establish new reefs in cooler waters. Regardless of the eventual distribution of biomes such as deserts, savannas, temperate forests, and tundra, whenever their boundaries are shifted by climate change, biodiversity will suffer transitional losses.

▼ Figure 8.38 **Threatened by Climate Change**
The U.S. Fish and Wildlife Service has concluded that most polar bear populations in North America are threatened by the reduction in sea ice caused by global warming. Polar bears hunt seals—their primary prey—from ice floes.

▼ Figure 8.39 **Coral Bleaching and Climate Change**
Warming of tropical oceans appears to be one of the factors contributing to a significant increase in the worldwide bleaching of corals. Bleached corals expel their symbiotic algae. Some corals recover, others do not.

8.6 Strategies for Conserving Biodiversity

BIG IDEA Effective conservation of diversity requires attention not only to the health of populations of individual species but also to the design and management of the places where they live.

Preserves are areas that are protected by various means to maintain biodiversity. Based on the equilibrium theory of island biogeography, preserves should be as large as possible in order to minimize the rate of local extinctions within them. To maximize the migration of species into these areas, they should be connected to or close to other preserves. Lands that surround preserves should be managed to provide additional habitat and to allow species to migrate freely. Patterns of natural disturbances should be maintained. Human activities in preserves must be carefully managed. Healthy populations of wild species depend on the maintenance or restoration of habitat, the management of competitors and predators, and the maintenance of genetic diversity.

Preserves and Protected Areas

■ Conservation of biodiversity in preserves depends on principles of island biogeography.

The preservation of natural areas is a central strategy in the conservation of species and the habitats upon which they depend. Nearly every country has protected some of its lands and coastal waters as preserves, parks, or national forests.

Twelve percent of Earth's land surface—more than 17 million square kilometers (6.53 million square miles)—falls within legally designated preserves. This may seem like a considerable amount of land, but three factors limit the effectiveness of these preserves in conserving Earth's biodiversity. First, countries have tended to preserve areas with the least economic value, such as very arid, rocky, or mountainous lands. Land that is valued for other uses, such as agriculture or urban development, is far less likely to be preserved. As a result, many of the ecosystems that are most threatened by human development lack protected status. Second, even though an area has been designated as a park or preserve, it may not

be protected from human activities. Poor countries may lack the economic or human resources to protect such areas. In other countries, lax administration and corruption allow or even encourage destructive activities within preserves (Figure 8.40). Third, preserves are often designed or managed in ways that do not offer optimal protection to their biodiversity.

► Figure 8.40 **Paper Parks**

It is not enough for countries to simply designate parks; they must be protected and managed, too. Destructive activities in parks are more common than you might think. Here, a government helicopter flies over an illegal logging operation in the middle of a Brazilian national park.

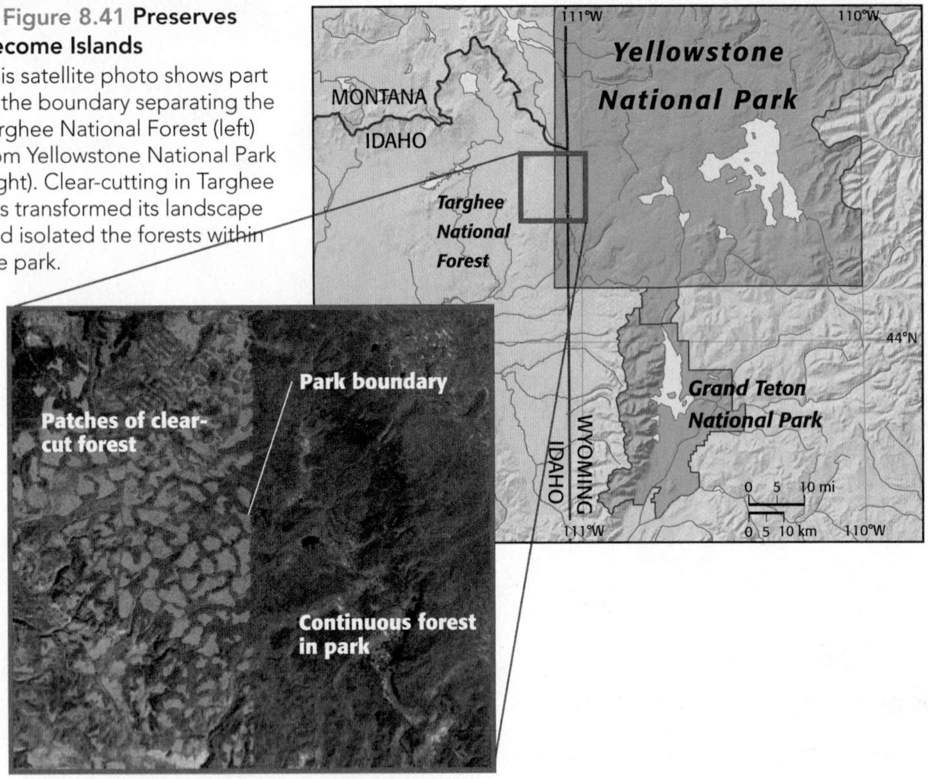

▶ **Figure 8.41 Preserves Become Islands**
This satellite photo shows part of the boundary separating the Targhee National Forest (left) from Yellowstone National Park (right). Clear-cutting in Targhee has transformed its landscape and isolated the forests within the park.

MONTANA
IDAHO
Yellowstone National Park
Targhee National Forest
WYOMING
IDAHO
Grand Teton National Park
44°N
111°W   110°W
0 5 10 mi
0 5 10 km
111°W   110°W

Patches of clear-cut forest

Park boundary

Continuous forest in park

Preserves are islands of sorts. Once their boundaries are delineated, pressure for human activities and development in the areas around them often increases (Figure 8.41). As a consequence, they become surrounded by a "sea" of less habitable, or even hostile, land. As islands, preserves are subject to the patterns of species diversity predicted by the equilibrium theory of island biogeography. Based on this theory, conservation biologists have developed six guidelines for designing and maintaining preserves in order to maximize biodiversity. They are as follows: (1) big is better than small, (2) connected is better than unconnected, (3) near is better than far, (4) buffers matter, (5) accommodate landscape change, and (6) manage people.

Big is better than small. Just as with oceanic islands, larger preserves have a more extensive representation of habitats and support larger populations of species than smaller preserves. Therefore, individual species are less likely to disappear from large preserves. For example, the country of Madagascar is identified as a world biodiversity hotspot because of its many endemic species and the rapid rate at which its interior forests are being cleared. An especially diverse community of lizards, including two-thirds of the world's chameleons, is found here and nowhere else. Exhaustive field studies show that each of these chameleon species has a unique set of habitat preferences. Large preserves that extend over more than one part of Madagascar are needed to conserve habitat for as many species as possible.

Connected is better than unconnected. Over time, the populations of species within a preserve fluctuate. The persistence of many species in an area depends on their ability to recolonize that area. This is especially important in relatively small preserves, where fluctuations are likely to result in the disappearance of a species.

A clever way to overcome the disadvantage of small preserves is to create **migration corridors** that connect several habitat fragments into one large preserve (Figure 8.42). Examples of migration corridors include strips of forest left along streams, greenways planted between city parks, and railroad rights-of-way containing remnants of native prairie vegetation in the American midwest. Animals on the move tend to stay within their preferred habitat, so these strips of suitable vegetation encourage migration between habitat fragments. Animals moving through the corridors often help plants migrate through the corridors by dispersing their seeds. Populations of plants may also become established within the corridors, providing sources of seeds for the recolonization of the habitat fragments. If organisms are able to move among habitat fragments, outbreeding among unrelated individuals is facilitated, and the unfavorable effects of inbreeding are reduced.

▼ **Figure 8.42 Connecting the Landscape**
Hedgerows between these English pastures provide corridors that connect patches of forest. Many species of birds and mammals travel through these corridors.

Near is better than far. When local extinction occurs within a small preserve, reestablishment of the species depends on its migration from other sites. The closer a source population is to the affected preserve, the easier it is for immigrants to fill the void. Just as islands close to the mainland support a higher equilibrium species number, preserves located close to other preserves maintain greater biodiversity. The ability of species to migrate among preserves may play an important role in the survival of many species as climates change in the future.

The United States has formally designated over 109 million acres in over 750 locations in the National Wilderness Preserve System. In the western United States, wilderness areas are often located adjacent to or within tens of miles of one another, facilitating plant and animal migration among them. East of the Mississippi, there are many fewer wilderness areas, and they are often separated from one another by hundreds of miles. The large distances between preserves impede migration and increase the challenges for conserving biodiversity in the East.

Buffers matter. For biological preserves to be effective, their managers must pay attention to activities on the lands that surround them. Land use around the edges of a preserve has a significant effect on the biodiversity within the preserve. A city park bounded by busy streets is truly an island; it has no buffer around its edges to minimize the effects of disturbance from human activities (Figure 8.43). Pollution and noise from the city penetrate the park. Animals wandering across the park boundaries are unlikely to survive long enough to return. The result is a constant drain on species numbers. By contrast, a preserve surrounded by grazing land or forest managed for timber is buffered from many human activities and will support greater biodiversity than one would expect based on its size alone.

Accommodate landscape change.
Disturbances such as fires, ice storms, strong winds, insect infestations, and fungal diseases are a natural part of nearly all ecosystems. These disturbances create a mosaic of landscape patches undergoing various stages of succession. These patches provide diverse habitats that allow many different species to survive (see Module 6.6). Conservation managers should accommodate disturbances that are similar to those that occur naturally and to which native plants and animals

▲ **Figure 8.43 No Buffer to Be Found**
Central Park in New York City provides important habitat to many species, but there is no buffer around the park to protect its biodiversity from the human activities going on immediately adjacent to it.

are adapted. However, disturbances that are significantly more frequent or intense than in the past are likely to diminish biodiversity.

Manage people. Managing the activities of people is one of the most vexing challenges in many parks and preserves. In the United States, for example, national parks are set aside to provide recreation for people as well as to conserve biodiversity. Reconciling the needs of wildlife with the demands of human recreation is a continuing issue for park managers.

In addition to conserving biodiversity, preserves in many countries are set aside to meet the needs of indigenous people, such as hunting, gathering, and subsistence agriculture. In some cases, the potential conflicts between the needs of indigenous people and those of the plants and animals that surround them appear to be well managed. For example, over half of the preserves in the Brazilian Amazon are officially designated for use and management by indigenous people. Illegal logging and poaching are less common in these indigenous preserves than in other types of preserves.

In other parts of the world, conflicts between humans and endangered species are a significant challenge. In India, for example, many preserves have been established to protect tigers and Asian elephants. Unfortunately, these animals often present a threat to people and their homes and farms. When elephants destroy crops or tigers attack livestock or children, people have little sympathy for the conservation of these threatened species (Figure 8.44).

▼ **Figure 8.44 Biodiversity and People in Conflict**
People live in and near many of India's preserves that are set aside for animals such as Asian elephants and tigers. Elephants can be very destructive to farms and villages and are often subject to poaching.

Conservation Corridors

Do connections between habitat patches actually increase their diversity?

1. How did these researchers ensure that increased number of species in the connected corridor treatments were due to the connections, not the added area of the corridor?

2. Why might corridors be more important for herb species that are dispersed by the wind than by herbs that are dispersed by birds?

The theory of island biogeography predicts that species diversity will be higher in preserves and patches of habitat that are connected to one another. For this reason, protected habitat corridors have become a central feature of management plans worldwide. But are these corridors really effective? Establishing and maintaining corridors is costly, so managers would prefer to devote their limited resources elsewhere if corridors are ineffective.

Until 2006, there was no *experimental evidence* addressing this question for whole communities of organisms because it was too difficult to design experiments on such a large scale. Then Nick Haddad and his student Ellen Damschen devised an innovative *experimental design* to *test the hypothesis* that corridors increase diversity (**Figure 8.45**). With the cooperation of the U.S. Forest Service and the Savannah River Experimental Research Park in southeastern South Carolina, they were able to implement this experiment.

Haddad and Damschen studied six 50-ha (120-acre) "landscapes" established in an extensive area of pine plantations. Within each landscape, trees were cleared from a central 1-ha (2.5-acre) patch. Four *treatment patches* were located about 150 m (492 ft) from the edge of the central patch. One of these treatment patches was connected to

▲ **Figure 8.45 Connection Science**
Nick Haddad and Ellen Damschen of North Carolina State University were among the first scientists to carry out experiments confirming that corridor connections between habitat patches significantly increase biodiversity within those patches.

the central patch by a 25-m-wide corridor (82 ft). The other treatment patches were not connected to the central patch. Each of the other patches had the same total area as the connected patch and its corridor but varied in shape so that the effect of the corridor without its connection could be accounted for (**Figure 8.46**).

All cleared patches were subsequently colonized by a variety of herbaceous plant species that were not found in the adjacent pine plantation. Over the next five years, Haddad and Damschen monitored the diversity of herbs in each treatment plot. By the end of the study, herb species richness was 20% higher in the connected patches than in the unconnected patches. There was no difference in species richness among the other treatments. These results clearly demonstrated that corridors connecting habitat patches do increase species diversity.

More recently, Haddad and Damschen have documented similar trends for species of butterflies. They continue to explore the factors that influence the effectiveness of corridors. Ellen Damschen is now on the faculty at University of Wisconsin.

◀ **Figure 8.46 Experimental Design**
This is an aerial view of one of the six landscapes studied in Haddad and Damschen's experiments. The "winged" treatments are intended to simulate the effects of the corridor itself without its connection.

Source: Damschen, E.L., N.M. Haddad, J.L. Orrock, et al. 2006. Corridors increase species richness at large scales. *Science* 313: 1284–1286.

Managing Populations of Individual Species

■ Conserving individual species requires restoration and maintenance of healthy populations.

In many cases, conservation is focused on a single species or group of species. This is particularly true for species that have been designated as threatened or endangered. Conservation biologists determine whether a species is threatened or endangered by assessing its **population viability**, the probability that it will go extinct in a given number of years. The exact length of time used in a population viability assessment varies, although 100 years is widely used.

Several factors influence population viability; these factors are the focus of actions to conserve species. The foremost threat to species viability is usually diminished habitat; this includes both the availability and quality of habitat. Most recovery plans for endangered species focus on increasing the area and suitability of habitat. Sometimes introduced predators or diseases threaten population viability. In such cases, managers may work to remove or control the predator or disease.

When populations become very small, the loss of genetic diversity and the impact of inbreeding may be the greatest threat to restoring viability. In such cases, captive breeding programs are important. In these programs, care is taken to mate genetically different males and females, with the goal of enriching the genetic diversity of the

captive population. Eventually, the captive animals are reintroduced to their wild habitats. In 1987, the total population of the majestic California condor numbered only 27 individuals. A cooperative captive breeding program involving zoos and other facilities in Arizona, California, and Oregon has succeeded in building the total condor population to nearly 200 birds and the wild population to 50 birds (Figure 8.47).

Given the great number of organisms in ecological communities, it would be virtually impossible to monitor and manage the populations of every species. Instead, conservation biologists often focus their attention on **umbrella species**, species whose ecological requirements are highly correlated with the needs of many other species. Top carnivores are often good umbrella species because they require large territories and help ensure the diversity of lower trophic levels (Figure 8.48). Species such as freshwater mussels that are particularly sensitive to pollution or other environmental changes are also effective umbrella species.

▲ Figure 8.47 **Back from the Brink**
Captive breeding and reintroduction programs have doubled the wild populations of critically endangered California condors and increased their genetic diversity.

◄ Figure 8.48 **An Umbrella Species**
Much conservation in the tropical forests in Mexico and Central America has focused on the well-being of the jaguar. By doing this, conservation biologists believe they will meet the habitat needs of many other species, too.

QUESTIONS 8.6

1. Describe six general guidelines for the design and management of areas dedicated to preserving biodiversity.

2. How do conservation biologists determine if a species is threatened or endangered?

3. Why do conservation biologists consider large carnivores to be good umbrella species?

(MES) For additional review, go to
MasteringEnvironmentalScience

253

8.7 U.S. Policies for Conserving Biodiversity

BIG IDEA The inauguration of the U.S. National Park System caused national interest in the conservation of biodiversity. As a result of the Wilderness Act of 1964, over 400,000 square kilometers (100 million acres) of public land have been designated as wilderness areas. These lands are managed for biodiversity, and human intrusion is strictly managed. The earliest federal laws to protect biodiversity governed the buying and selling of species that had been legally protected by individual states. The 1973 Endangered Species Act provided much broader protection to threatened and endangered species on both public and private land. Management on private lands is playing an ever-larger role in the conservation of the nation's biodiversity.

National Parks and Wilderness Areas

■ The U.S. National Parks and Wilderness System set a high standard for managing biodiversity.

In 1864, President Abraham Lincoln signed a bill granting Yosemite and Hetch Hetchy valleys, as well as the nearby groves of giant sequoias, to the state of California to be conserved as a public park. In the words of the bill, the new park was to be "an inalienable public trust." Eight years later, the U.S. Congress established Yellowstone as the world's first national park. Although the legislation establishing Yellowstone called attention to the wildlife and beauty of the region, it emphasized that, above all else, this park was to be a "pleasuring ground for the benefit and enjoyment of the people" (Figure 8.49).

Over the next 40 years, Congress established more parks, including Sequoia, Mt. Rainier, Crater Lake, Glacier, and Grand Canyon. Unhappy with California's management practices, the federal government made Yosemite into a national park in 1896. The creation of new parks was heavily promoted by the railroads, which hoped to expand routes and recruit passengers. The railroads also managed many park concessions, such as hotels and restaurants. Park management was seen as "people management" and was carried out by the U.S. Army.

As early as 1900, John Muir and other preservationists began expressing concern that the parks' focus on people was having negative effects on the natural resources of the parks, including plant and animal diversity. The prolonged battle over the damming of the Tuolumne River in Hetch Hetchy Valley heightened these concerns.

After much political wrangling, President Woodrow Wilson signed the National Park Service Organic Act in 1916. This legislation brought all of the national parks under the management of a single agency in the Interior Department, the National Park Service. The Organic Act made it clear that the parks' natural resources and biodiversity were central to the National Park Service mission. Unlike the inscription on the entrance to Yellowstone, the new emphasis was on biodiversity, not people.

Today, the U.S. National Park System includes 58 national parks and nearly 340 other units, such as national monuments and national seashores, distributed among all 50 states. Conservation of biodiversity plays a central role in most of these preserves. In many regions, national parks provide the core for conservation planning. Parks are, of course, heavily visited by people, and accommodating human visitors remains one of the most important challenges to conserving the parks' biodiversity (Figure 8.50).

◀ **Figure 8.49 Focus on People**
Yellowstone was the nation's first national park. In the early days, park management emphasized human recreation, as indicated by the words engraved in its historic northern entrance: "For the benefit and enjoyment of the people."

In the 1950s, conservation professionals began to recognize that public lands outside the national parks, such as national forests, also supported a significant amount of biodiversity. They also noted that development for lodging and recreation diminished the biodiversity of some lands within national parks.

In 1964, with prodding from environmental groups such as the Wilderness Society, the U.S. Congress passed the Wilderness Act. This created a category of federal lands designated as wilderness, in which "the earth and its community of life are untrammeled by man." This act prohibits activities such as logging and mining in areas designated as wilderness; it also limits road construction. Initially, 36,000 square kilometers (9 million acres) of national forest and national park land were designated as wilderness. Land can be added to this system only by congressional action.

As of 2014, the U.S. Wilderness Preservation System included more than 441,000 square kilometers (109 million acres). These lands are administered by four different agencies—the National Park Service (177,600 km^2 or 68,500 mi^2), the National Forest Service (146,500 km^2 or 56,550 mi^2), the U.S. Fish and Wildlife Service (83,760 km^2 or 32,310 mi^2), and the Bureau of Land Management (35,200 km^2 or 13,590 mi^2). The Wilderness System is extremely important to the conservation of biodiversity. In fact, wilderness lands are the core habitat of over 40% of the species that are listed as threatened or endangered in the United States (Figure 8.51).

▲ Figure 8.50 **Loving Them to Death**
National parks are greatly appreciated by visitors, but high visitation also produces many stresses, including the diminished diversity of wildflowers and other herbs, the introduction of non-native species, and threats to wildlife along trails.

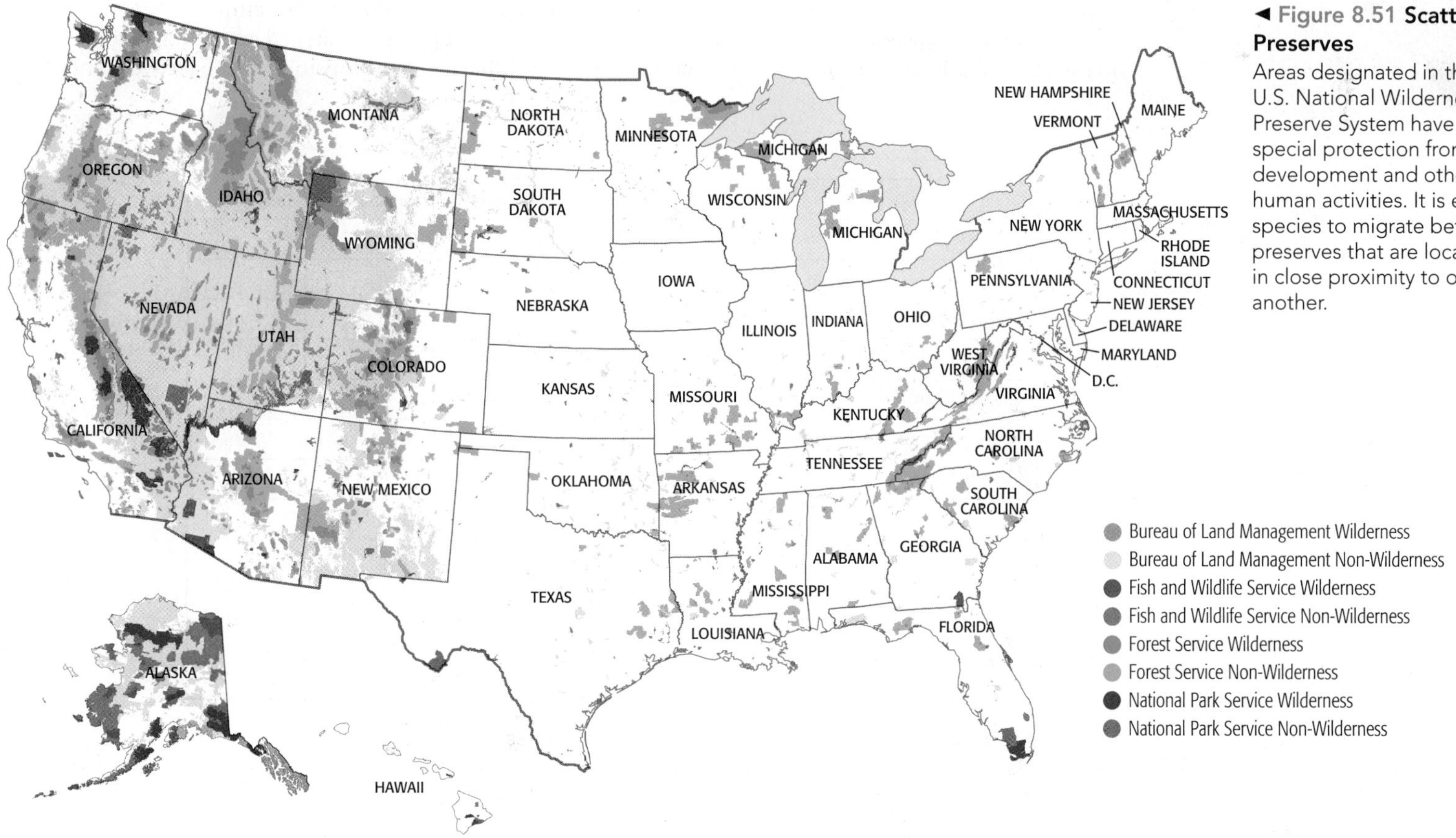

◄ Figure 8.51 **Scattered Preserves**
Areas designated in the U.S. National Wilderness Preserve System have special protection from road development and other human activities. It is easier for species to migrate between preserves that are located in close proximity to one another.

Legend:
- Bureau of Land Management Wilderness
- Bureau of Land Management Non-Wilderness
- Fish and Wildlife Service Wilderness
- Fish and Wildlife Service Non-Wilderness
- Forest Service Wilderness
- Forest Service Non-Wilderness
- National Park Service Wilderness
- National Park Service Non-Wilderness

Legislation to Protect Species

■ U.S. laws limit commerce in and injury to threatened or endangered species.

▼ Figure 8.52 A Catalyst for Action
The extinction of the passenger pigeon as a result of overhunting solidified public opinion about the need to protect species and catalyzed the U.S. Congress to pass the Lacey Act, which prohibits commerce in protected species.

In the early 19th century, the passenger pigeon was one of America's most common and most hunted birds. These birds flew in large flocks, were easily captured or shot, and fetched a high price in urban markets like New York, and Boston, and in Europe. As a result of market hunting and widespread deforestation, these once-abundant birds were extinct in the wild by 1900 (Figure 8.52). By that time, it was clear that other bird species were also threatened by overhunting and loss of habitat.

The declining numbers of birds and other wildlife led a number of states to pass laws prohibiting the hunting of threatened species. However, these laws varied enormously from state to state and were difficult to enforce. In response to the situation, the U.S. Congress passed the 1900 Lacey Act. This law, which has been reauthorized and amended several times since its passage, prohibits trade in wildlife, fish, and plants that have been taken illegally under either state or federal statutes. The U.S. Fish and Wildlife Service (USFWS) was created to enforce this law. The Lacey Act abruptly put an end to the market hunting that was responsible for the decline of so many wildlife species.

For the most part, the Lacey Act protects only those species that have commercial value and are threatened by overcollection or hunting. By the 1970s, it was clear that some sort of protection was needed for the vast majority of species that do not fall into this category. In 1972, the Marine Mammals Protection Act was passed. This law prohibits the taking of marine mammals such as seals, dolphins, and whales from U.S. waters; it also prohibits U.S. citizens from taking marine mammals in any other part of the world.

The following year, President Richard M. Nixon signed the Endangered Species Act into law. As stated in its defining legislation, the ESA was intended "to provide a means whereby the ecosystems upon which endangered species and threatened species depend may be conserved and to provide a program for the conservation of such endangered and threatened species." The ESA prohibits the taking of any threatened or endangered species, regardless of land ownership. With the ESA, the term *take* has been interpreted very broadly to include not only capture, hunting, or collection, but any conduct that might damage a species' habitat or diminish its ability to reproduce.

Under the ESA, a species is considered endangered if it is in danger of extinction throughout all or a significant part of its range. A species can be listed as threatened if it is at risk of becoming endangered (Figure 8.53). The criteria for listing (and delisting) species as either endangered or threatened are set by the U.S. Fish and Wildlife Service. The USFWS must also develop a recovery plan for each species that it lists.

There is widespread agreement that the ESA is one of the most important pieces of biodiversity legislation ever passed by any country. Certainly it has played a significant role in slowing or halting declines in populations of many threatened and endangered species. It has been responsible for the recovery and

Group	Endangered	Threatened	Total
Mammals (*polar bear, wolverine*)	68	13	81
Birds (*northern spotted owl, red-cockaded woodpecker*)	77	15	92
Reptiles (*pygmy rattlesnake, bog turtle*)	14	23	37
Amphibians (*tiger salamander, red-legged frog*)	13	10	23
Fishes (*delta smelt, chinook salmon*)	76	61	137
Invertebrates (*species of snails, insects, spiders, and shellfish*)	153	32	185
Animal Subtotal	401	154	555
Higher Plants	573	144	717
Ferns and Allies	24	2	26
Lichens	2	0	2
Plant Subtotal	599	146	745
Grand Total	1,000	300	1,300

▲ Figure 8.53 Threatened and Endangered Species
Many groups of organisms are represented by the species that are currently listed as threatened or endangered under the Endangered Species Act.

eventual delisting of other species, such as the bald eagle (Figure 8.54).

As it is currently written and administered, the ESA has many critics. Private property owners complain that it unfairly limits the use and economic value of some properties. Conservationists worry that such constraints may encourage landowners to take habitat-destroying actions, such as logging, before an ESA-listed species can be found on their property. Nearly everyone agrees that the processes by which species are eventually listed or delisted under the ESA can be bureaucratic and time-consuming.

Probably the greatest concern of conservationists is that the ESA provides no means to prevent species populations from declining to threatened or endangered status. No action to conserve or restore habitat is mandated until a species is listed as threatened or endangered. In some cases, this may be too late.

The ESA was initially scheduled for congressional evaluation and reauthorization in 1993, 20 years after its passage. Critics from all sides hoped to use the reauthorization process to amend the bill to their liking. However, each time an interest group has crafted reauthorization legislation, it has been blocked by other groups that fear their interests will be jeopardized. Without reauthorization, the 1973 ESA will remain in effect as written.

▲ Figure 8.54 **Restored!**
Threatened by insecticides and loss of habitat, the bald eagle was among the first species formally listed as endangered under the Endangered Species Act. After the implementation of a very successful restoration plan, the bird was delisted in 2008.

Conservation on Private Land

■ Private landowners can play a significant role in species conservation.

Public lands owned by state or federal governments account for about 20% of all land in the United States, although the proportion varies widely among states (76% in Nevada; less than 5% in most eastern states). The remaining land, which is owned by individuals and corporations, also contains important ecosystems and habitats. Thus, there is much interest in strategies to conserve biodiversity on privately owned lands.

Many cities, towns, and counties have zoning laws that encourage the conservation of biodiversity by limiting development or mandating greenspaces within developments. Greenspace zoning has helped many urban counties protect both habitat and water supplies (see Module 16.4). To date, Oregon is the only state to implement statewide zoning. The 1973 Oregon Land Use Act requires cities and counties to enact zoning plans using standards set by the state. Those standards include the needs of biodiversity in coastal, aquatic, and forest ecosystems.

As an alternative to zoning or mandated controls on land use, many environmental groups and government agencies rely on incentive-based programs. In some states, landowners receive tax exemptions if they protect threatened habitats or species. In other cases, private organizations or government agencies use conservation easements to purchase development rights from landowners. Under conservation easements, landowners give up certain rights to develop their land. For example,

to protect the biodiversity of a stream, a county might purchase a conservation easement on the forested land along the stream. That easement prevents the landowner from ever subdividing the land to build homes. Easements may also limit activities such as hunting and logging on the property. In any case, the landowner is directly compensated for the economic value of these activities.

In the United States there are over 1,400 separate **land trusts**, nonprofit organizations that are dedicated to conserving land for the protection of biodiversity and ecosystem services. Most land trusts focus on a few counties or a specific region within a state. A few, like the Trust for Public Lands, The Nature Conservancy, and The Conservation Fund, facilitate land conservation across the country. Land trusts use a variety of tools, including public education, conservation easements, land purchase, and active land management, to conserve biodiversity on public and private lands.

In summary, the U.S. system of national parks, national forests, and wilderness areas has been effective in conserving biodiversity. Federal laws that regulate the trade in wild species and protect threatened and endangered species and their habitat have also been important. Biodiversity on privately owned lands is receiving more and more protection through conservation easements, greenspace zoning, and the activities of land trusts.

QUESTIONS 8.7

1. How do national parks, national forests, and formally designated wilderness areas differ from one another?

2. By what means does the Lacey Act preserve the diversity of species?

3. By what means does the Endangered Species Act preserve the diversity of species?

4. What is a conservation easement? How does a conservation easement benefit private landowners and help conserve biodiversity?

(MES) For additional review, go to **MasteringEnvironmentalScience**

8.8 International Policies for Conserving Biodiversity

BIG IDEA International cooperation to conserve Earth's biodiversity has been facilitated by legally binding treaties and conventions and by economic incentives and partnerships. Treaties designed to limit illegal trade and prevent overharvesting have focused on endangered species. Economic incentives have included the promotion of ecotourism and the exchange of debt for the establishment of preserves. In developing countries, biodiversity conservation must be undertaken in ways that help alleviate poverty and support economic development.

Endangered Species Trade and Harvest

■ International treaties that limit the trade in and harvesting of endangered species have had mixed results.

The global market for wild plants and animals and their products is very large. For example, the global trade in exotic birds alone is thought to exceed $50 billion. So it is not surprising that many international conservation efforts have focused on limiting commerce in threatened or endangered species.

The Convention on International Trade in Endangered Species of Wild Fauna and Flora (CITES), which was ratified by UN member nations in 1975, is the most notable example of an effort to conserve species by limiting trade. This treaty aims to ensure that international trade in wild animals and plants does not threaten their survival, whether they are traded as live specimens, fur coats, or dried herbs. Today, CITES accords varying degrees of protection to more than 30,000 species (Figure 8.55).

CITES has slowed the overhunting and collection of many plant and animal species, but its overall effectiveness is hotly debated among conservation biologists. They point out that in some developing countries enforcement of CITES is limited by corruption and the lack of resources. In some cases, international restrictions on commerce have driven up prices in black markets. These high prices encourage even more illegal hunting.

The illegal trade in rhinoceros horn appears to be such a case. In traditional Asian medicine, rhinoceros horn is highly valued as a treatment for many ailments, including fever and impotence. Daggers with rhino horn handles are prized in many areas of the Middle East. Trade in rhinoceros horn was banned under CITES in 1975. Over the next 15 years, however, the combined populations of the world's five species of rhinoceros declined by 85%, largely due to poaching. Today the black market value of rhinoceros horn is nearly $60,000 per kilogram because of high demand and

▲ Figure 8.55 **CITES Protection**
The international demand for bird feathers and plumage has resulted in overhunting of many species worldwide. This plumage was used to adorn the hats and heads of many fashion-conscious people, including a young Marie Antoinette. Today CITES limits international trade in such items.

limited supply. Thus, incentives for continued poaching are very high (Figure 8.56).

Other international treaties have successfully limited the harvest of endangered species. For example, the International Convention for the Regulation of Whaling was signed by 42 nations in 1946. This convention established the International Whaling Commission (IWC) to oversee the whaling industry and ensure the conservation of whale populations. With strong pressure from environmental groups opposed to whaling and like-minded governments, the IWC implemented a worldwide five-year moratorium on all whaling. Despite objections from some member nations, this moratorium has been extended to the present. The actions of the IWC have helped to stabilize and even increase the populations of some whale species, although other species remain endangered due to factors such as coastal development and pollution.

Why has the ban on whaling been effective, whereas treaties to limit hunting of other species, such as rhinos, elephants, and tigers, have been less so? The most important factor is probably the capacity for treaty enforcement. Whaling is a complicated and expensive endeavor that requires large vessels and crews, so it is relatively easy to monitor.

▲ **Figure 8.56 Extinction in Store?**
Ⓐ Limits on trade in rhinoceros horn and ivory from elephant tusks have not led to diminished demand, despite having prices driven up in black markets. The perceived benefits keep demand high and have led to continued poaching. Ⓑ The black rhinoceros is particularly threatened and is now listed as critically endangered even though its population numbers have been rising since hitting a low of 2,140 in 1995.

Economic Incentives for Conservation

■ Conservationists strive for economic development that is compatible with the protection of biodiversity.

Much of the world's biodiversity is found in poor and developing countries. That biodiversity is often threatened by activities associated with economic development, such as road construction, mining, and logging. Conservation organizations and international development agencies are seeking ways to encourage economic development in these countries while maintaining or even enhancing biodiversity.

The Global Environmental Facility is an international agency that provides grants and loans for projects that promote both biodiversity and economic development. Nearly half of those projects encourage some aspect of ecotourism. Ecotourism has been a strong incentive for the creation of preserves and parks in a great many countries. It has the potential to bring economic resources to underdeveloped places and to create incentives for local people to conserve their natural resources. In some places, however, tourism has had a destructive effect on environments. In addition, the financial benefits of some ecotourism projects have not gone to the local communities or the countries that have hosted them.

The International Ecotourism Society has established criteria for sustainable ecotourism. First and foremost, ecotourism should encourage biological and cultural diversity by protecting ecosystems. It should also promote the sustainable use of biodiversity by local people and share the economic benefits with those people. Finally, ecotourism should minimize its own environmental impacts.

Many developing countries carry large burdens of debt to wealthier nations. **Debt-for-nature swaps** are international arrangements in which poor nations are forgiven a portion of their debts in exchange for the establishment of permanent parks and preserves. The debtor nation is then able to use more of its limited economic resources to meet local needs. Since their inception, debt-for-nature swaps in countries in Africa, South America, and Asia have converted nearly $500 million of debt into preserved habitat. All nations benefit from this conservation of global biodiversity.

The increasing number of preserves in developing nations is a hopeful trend. However, no wildlife preserve can protect biodiversity if the people living on its boundaries are desperately poor. When local people view the plants and animals in a preserve as the fuel and food they need to survive, the cause of biodiversity protection is lost. The World Commission on Protected Areas therefore promotes the idea that plans to protect biodiversity must incorporate the whole landscape, including its human inhabitants.

QUESTIONS 8.8

1. Describe the strengths and weaknesses of the Convention on International Trade in Endangered Species.

2. Describe three international policy initiatives aimed at encouraging both economic development and conservation of biodiversity.

(MES) For additional review, go to **MasteringEnvironmentalScience**

Integrated conservation and development projects (ICDPs) promote both the protection of wildlife and the improvement of living standards for people who live within and around areas that are managed for biodiversity. ICDPs begin with a comprehensive understanding of the local culture and economy. Investments and incentives are then structured to help the local people recognize that the protection of biodiversity is a source of prosperity. Poachers are hired as gamekeepers. Sustainable harvest of forest products is promoted. Slash-and-burn agriculture is replaced by steady jobs in ecotourism. Destructive overgrazing is replaced by buffer zones of sustainable farming. When local people begin to view biodiversity preserves as the source of their livelihood, conservation becomes a priority.

Because every region is unique, each ICDP begins with discovery. As landscape plans are designed, scientific knowledge must be supplemented with the knowledge of the indigenous people. As a project goes forward, continuous monitoring of biodiversity and ecosystem health provides new data. Stakeholders and participants continue giving feedback and receiving information. Economic development and activities to protect biodiversity are continuously adjusted as participants learn what does and doesn't work. Lessons learned are shared with other agencies and other nations in an effort to extend the management model to new sites.

In 1992, representatives from 179 countries participated in the United Nations Earth Summit in Rio de Janeiro, Brazil. The participants adopted Agenda 21, a set of principles and strategies for sustainable development in the 21st century. Agenda 21 emphasizes conservation of natural resources and biodiversity as well as the social and economic needs of specific human groups, including women, children, and indigenous peoples and their communities.

Much in keeping with Agenda 21's emphasis on both conservation and economic development, Earth Summit participants crafted the Convention on Biodiversity. This convention prescribes actions to conserve biological diversity and ensure the sustainable use of its components. It also provides guidelines to ensure that a fair portion of the economic benefits obtained from species in developing countries are returned to those countries. The Convention on Biodiversity was ratified by nearly all participants and is now international law. However, the United States has declined to ratify it, arguing that its economic guidelines are not fair.

The 2002 UN World Summit on Sustainable Development, held in Johannesburg, South Africa, marked the 10th anniversary of the Earth Summit. The agenda and products of this summit reflected developing countries' concerns that conservation efforts were not adequately considering the needs of poor people.

The 2012 UN World Summit on Sustainable Development returned to Rio de Janeiro. UN leaders at this Rio+20 summit advocated for a "green economy roadmap" that would encourage policies and practices to reduce dependence on fossil fuels, encourage efficient resource use and be socially inclusive. The primary result of the conference was a nonbinding document, "The Future We Want," which largely reaffirmed commitments made in Agenda 21. However, discussions at Rio+20 particularly reflected growing concerns about global warming. Delegates from poor and developing countries remained concerned that there was still not a sufficient commitment on the part of the UN's wealthiest member countries to protect the environment, ensure equitable supplies of food, and alleviate poverty (Figure 8.57). Environmental justice remains a persistent theme in discussions of international environmental policy.

▲ Figure 8.57 Plea for Justice
These Ethiopian farmers protested provisions in the proposed green economy roadmap in the weeks prior to the Rio+20 summit. They were particularly concerned that the needs of poor, small-scale farmers were not being considered in the roadmap.

An Integrated Conservation and Development Project

Can an integrated conservation and development project (ICDP) conserve a diverse landscape and the well-being of people in one of the poorest regions of Latin America?

The Mandidi-Tambopata Landscape, an area larger than the state of Maine, straddles the border between Bolivia and Peru on the eastern side of the Andes (**Figure 8.58**). With its wide elevational range and varied topography and climate, it supports a diverse assemblage of plant and animal species. Conservation scientists estimate that over 12,000 plant, 1,100 bird, and 300 mammal species are found in this area. It provides critical habitat for a number of endangered species, including Andean condor, jaguar, Andean bear, white-lipped peccary, military macaw, and giant otter. This landscape is also home for over 260,000 people. It is the ancestral home for eight indigenous groups dispersed from the humid tropical lowlands to remote highland areas. These are among the poorest people in one of the poorest regions of Latin America.

Due largely to its productivity and biodiversity, this region has experienced booms and catastrophic busts in exports of rubber, quinine, animal products, and timber over the past century. Timber extraction continues to employ many of the people in this region. Both Bolivia and Peru have set aside large areas as natural areas and national parks, and large tracts of undisturbed habitat remain, but these are severely threatened by gas, oil, and hydropower projects; highway construction; illegal timber extraction; mining; and deforestation for agriculture. People in this region are justifiably concerned for the future of their beautiful landscape, but they would also like the economic means to improve the quality of their lives.

In 1999, the Wildlife Conservation Society (WCS), with headquarters at the Bronx Zoo, New York, initiated an integrated conservation and development program for the Mandidi-Tambopata Landscape in collaboration with governments and local communities in both Bolivia and Peru. Conservation of biodiversity and important ecosystem services are central elements of this ICDP. WCS scientists work closely with local communities to monitor populations of threatened species and to develop comprehensive management plans for natural areas and national parks. But they have also been collaborating with indigenous communities to clarify their land rights and provide technical skills for sustainable resource use, including

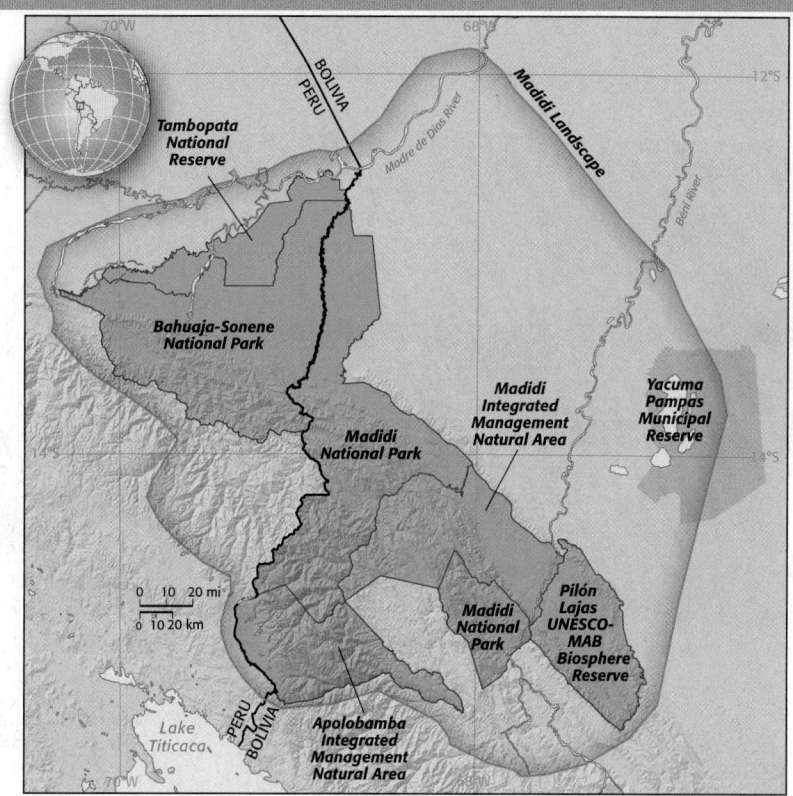

▲ **Figure 8.58 An Imperiled Landscape**
The Madidi-Tambopata Landscape comprises 110,000 km² (42,500 mi²) in southeastern Peru and northwestern Bolivia. It extends from lowland tropical forests to cloud forest and paramo in the high Andes.

hunting and timber extraction. They have also assisted in the development of integrated rangeland management practices that minimize the loss of domestic animals to wild carnivores such as jaguar and puma (**Figure 8.59**).

WCS recognizes that infrastructure projects such as highways and hydropower will be an important part of future economic development in this landscape. To be sure that ecosystem conservation receives the highest priority, it is working closely with national and regional governments to review the costs and benefits of such projects and see that they are appropriately sited.

This is very much a work in progress, but much has already been accomplished. Sustainable management plans have been developed for nearly 6 million hectares (14 million acres) of national park, indigenous territories, and municipal land. Forty-one economic initiatives have benefited nearly 2,000 families, in some cases doubling their annual income. Through the WCS biodiversity monitoring program nearly 650 new species have been added to the already long list of species resident in this landscape.

▼ **Figure 8.59 Building the Local Economy**
WCS veterinarians work with local communities in the Apolobamba region of the Madidi-Tambopata Landscape to improve the health and management of their alpaca herds.

261

Protecting a Unique Biodiversity Hotspot

Varsha Vijay is a recent graduate of Duke University with a degree in environmental sciences. While at Duke and after graduation, she worked in Ecuador with a remote Amazonian tribe called the Waorani. Varsha began an organization called *Fortificando el Intercambio* (Strengthening the Exchange) in order to share academic findings about the rain forest with the Waorani and to give them technological tools to help them preserve their territories and their culture.

How did you first get the idea for Strengthening the Exchange?

When I was a sophomore at Duke, I began working with Dr. Stuart Pimm's group of conservation biologists. My first assignment was at Ankarafantsika National Park in Madagascar. While conducting basic ecological monitoring at the field station, I became aware of the important role of the native villagers as stakeholders in the park's conservation efforts.

When an opportunity came up to work in the Amazon, I jumped at it. There I focused on conducting conservation projects to protect the Yasuní Man and Biosphere Reserve, a biodiversity hotspot that encompasses the ancestral territory of the Waorani tribe.

As I spent more time with the Waorani, I could not ignore the very real threats to their well-being. I saw the incursions on their territory by colonists following oil roads, the health effects of pollution from petroleum contaminants, and the threats to their food supply from overhunting and habitat loss.

I realized that if a conservation project was to be successful, it would need to show the Waorani how conservation and the sustainable use of resources are compatible with the tribe's own desire to preserve their culture and territory.

What are the biggest challenges you've faced with the project, and how did you solve them?

Initially, I felt extremely isolated due to the dramatic cultural differences between me and the tribe. I was raised as a vegetarian, but with them I found myself eating the traditional foods of monkey meat and saliva-fermented manioc beer.

At first I had to struggle to communicate, since I spoke no Spanish and had no familiarity with the Waorani tribal language, Wao terero. However, as I continued working with the tribe, I adapted and became fluent in Spanish. I still struggle with Wao terero, but the tribe is always quick to give me a hand.

By being open to these difficulties and sticking with it, I became closer with the Waorani than I ever thought possible.

How has your work affected other people and the environment?

I've worked to create a conservation ethic in Waorani communities by engaging them in projects that address the tribe's concerns for the sovereignty of their territory and the health and safety of their people, as well as their desire to profit from ecotourism.

Today my projects get younger members of the tribe back into the forest. They are learning and cataloguing the knowledge of medicinal plants possessed by the older members of their tribe. We also teach them the basic science of Amazonian ecosystems.

We have equipped them with GPS and other mapping technologies so they can map important places in their territories. These tools help them to conserve not only plants and animals but also important cultural landmarks.

The Waorani who have worked with me continue to talk about ideas they have for new conservation projects. I've influenced the dialogue and made conservation something that is interesting and fun!

What advice would you give to students who want to replicate your model?

You don't need an exotic location. Sometimes the best place to start is in your own community. Figure out what environmental issue you're passionate about. Then find a mentor doing the kind of work you'd like to do. If you intend to do research, the best mentor is probably a professor. If you intend to be involved in environmental activism, there are many inspiring people doing great work in the NGO [nongovernmental organization] community.

Once you've found a problem that moves you, consult with your mentor and the people affected by that issue to find a way you can use your current skills to contribute, and then build from there. Don't expect to solve everything on your own. Once you get involved, you'll realize that just being part of a solution is rewarding enough.

Summary

Biodiversity is the variety of life in all its forms, levels, and combinations. Biodiversity in landscapes is determined by variations in climate, soil, and patterns of disturbance. Diversity within communities can be measured by the numbers of species and the relative abundance of those species; the three-dimensional distribution of species and biological features is another measure of community biodiversity. Within populations, *biodiversity* is measured by genetic diversity. Biodiversity is important because it ensures the provision of ecosystem functions and services, imparts stability to ecosystem processes, and has economic value for humans. Most people believe that the protection of biodiversity is an ethical responsibility. In general, biodiversity is highest in tropical biomes and lowest in polar biomes.

Biodiversity in communities is influenced by habitat diversity, competition, predation, and patterns of disturbance. The equilibrium number of species in a particular place is ultimately determined by the rates at which species immigrate and disappear. The many threats to biodiversity include habitat loss and fragmentation, overharvesting of vulnerable species, competition and predation from non-native species, pollution, altered disturbance patterns, and global warming. Strategies for conserving biodiversity include the establishment of preserves, creating buffers and corridors to protect those preserves, and restoring healthy populations of species. A variety of domestic and international policies and laws have been established to reverse current trends that are accelerating the loss of biodiversity and to sustain the biodiversity that remains.

8.1 What Is Biodiversity?

- Biodiversity is the variety of species and their relative abundance across landscapes and within ecological communities, as well as the genetic diversity within populations.

KEY TERMS

biodiversity, landscape diversity, endemic, species richness, species evenness, structural complexity, genetic diversity, outbreeding, inbreeding

QUESTIONS

1. Describe the factors influencing landscape diversity for a park, national forest, or rangeland near your home.
2. Structural complexity of ecological communities is often correlated with species richness. Explain why this is so, using a specific community as an example.
3. Give two reasons why inbreeding within a species is a matter of concern to wildlife managers.

8.2 Why Biodiversity Matters

- Biodiversity is essential for ecosystem functions and services and for the stability of ecosystems.
- Biodiversity has great economic value, and most people agree that we have an ethical responsibility to conserve it.

KEY TERMS

sampling effect, complementarity effect, ecotourism

QUESTIONS

1. Describe two reasons why ecosystem functions, such as productivity, may increase as species richness increases.
2. Describe three ways in which biodiversity contributes to the economic well-being of communities.

8.3 Global Patterns of Biodiversity

- The species richness of Earth's biomes is highest in the tropical zone and lowest in the polar zone.
- Biodiversity hotspots are places with high numbers of endemic species that are especially threatened by human activities.

KEY TERM

biodiversity hotspots

QUESTIONS

1. Explain how the different rates of net primary production in tropical and temperate forests might be related to differences in the species richness of the forests.
2. What factors contribute to the endemism that is characteristic of biodiversity hotspots?

8.4 Differences in Biodiversity Among Communities

■ Habitat diversity, interactions among species, disturbances, and rates of species immigration and disappearance are responsible for differences in biodiversity among different ecological communities.

KEY TERM

equilibrium theory of island biogeography

QUESTIONS

1. Why do rugged mountainous areas generally support more species than relatively flat plateaus?
2. Explain why grazing and burning increase the diversity of plant species in grasslands.
3. In Minnesota, the Land of 10,000 Lakes, what factors might contribute to differences in plant and animal diversity among different lakes? Explain why.

8.5 Threats to Biodiversity

■ Habitat loss and fragmentation, overharvesting, non-native species, pollution, altered disturbance patterns, and global warming are the most important causes of species extinctions.

QUESTIONS

1. What habitats are disappearing near your home? What species might be threatened by the loss of these habitats?
2. Tundra swans were hunted nearly to extinction 100 years ago. In the summer, breeding pairs nest in isolation from one another in the Arctic; in the winter, birds congregate in large flocks on a few lakes in the southeastern United States. How did this behavior make them vulnerable to overhunting?
3. Give an example of a non-native invasive species that is influencing biodiversity in ecosystems near your home. What factors contribute to its invasiveness?
4. Explain why global warming is likely to increase the threats to biodiversity posed by habitat loss and fragmentation.

8.6 Strategies for Conserving Biodiversity

■ Biodiversity conservation depends on the establishment of preserves and protected areas, and the management of the ecosystems, disturbances, and people that surround them.
■ Conserving endangered species requires restoration of healthy, genetically diverse populations.

KEY TERMS

preserves, migration corridors, population viability, umbrella species

QUESTIONS

1. Conservation biologists argue that to preserve biodiversity inside national parks, managers must pay as much attention to lands outside the park as they do to the land within the park. Explain why this is important.
2. Name three umbrella species for ecological communities that occur near your home. What features qualify them as umbrella species?

8.7 U.S. Policies for Conserving Biodiversity

■ In the United States, biodiversity is protected by the system of national parks and wilderness areas, the Endangered Species Act, and actions by private landowners.

KEY TERM

land trusts

QUESTIONS

1. The U.S. Endangered Species Act prohibits the "taking" of a threatened or endangered species. What does this mean?
2. Describe three criticisms of the Endangered Species Act.
3. What laws or incentives exist to encourage private landowners to protect biodiversity on their lands?

8.8 International Policies for Conserving Biodiversity

■ Most international policies to conserve biodiversity are based on treaties that prohibit trade in endangered species or encourage economic incentives for conservation.

KEY TERMS

debt-for-nature swaps, integrated conservation and development projects

QUESTIONS

1. Although nearly all UN nations have signed agreements to protect their biodiversity, the actual level of protection varies considerably. What factors contribute to this variation?
2. How does the UN's Agenda 21 propose to deal with the fact that many of the world's most diverse ecosystems occur in very poor countries?

MasteringEnvironmentalScience®

Students Go to **MasteringEnvironmentalScience** for assignments, the eText, and the Study Area with practice tests, activities, and more.

Instructors Go to **MasteringEnvironmentalScience** for automatically graded tutorials and questions that you can assign to your students, plus Instructor Resources.

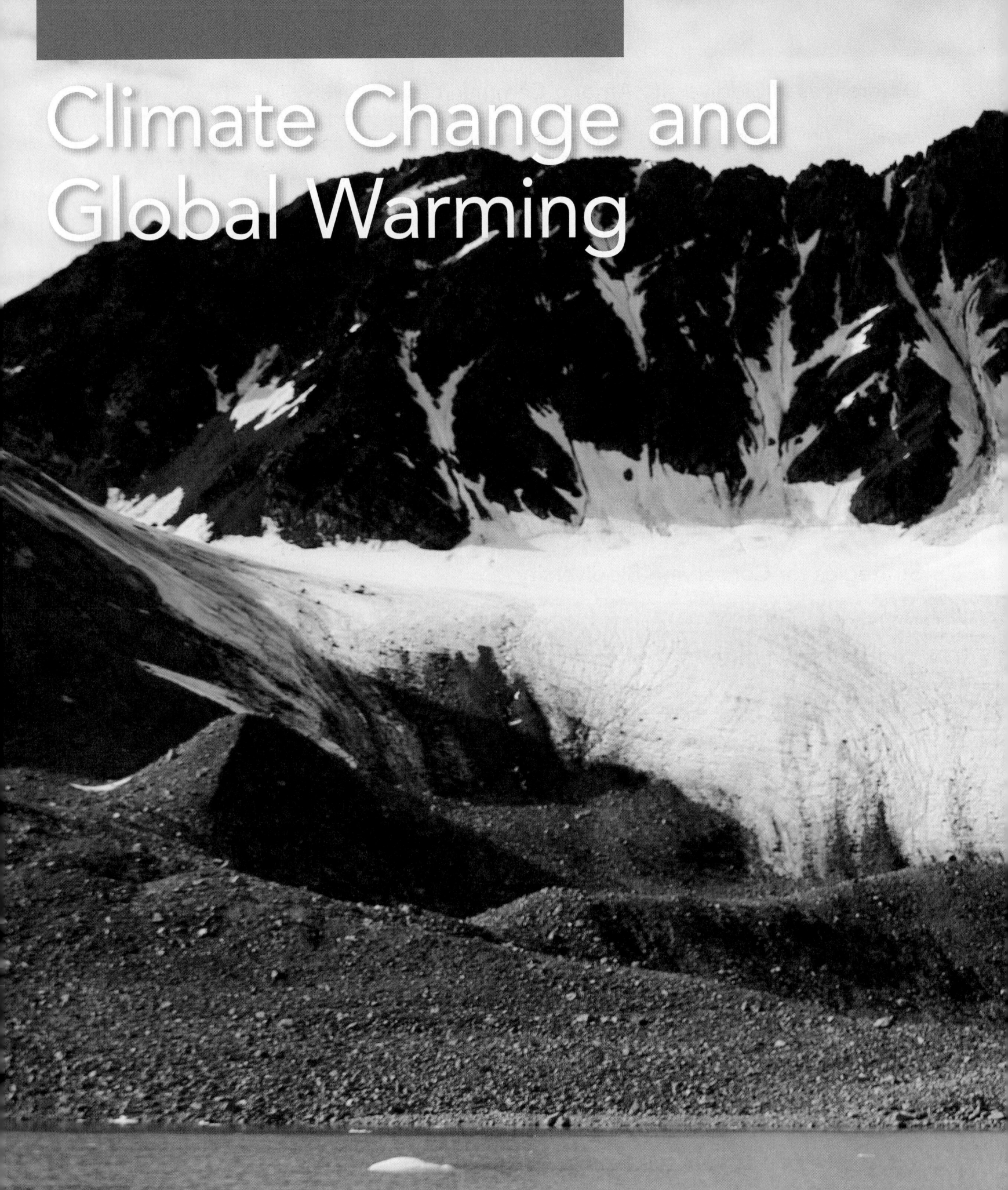

Climate Change and
Global Warming

Arctic Meltdown

Is Earth's climate warming?

You may have heard that the Inuit—the people who live in the Arctic—have dozens of words for snow (Figure 9.1). An appealing story, but one that was often dismissed as a myth. Recent research, however, suggests that certain Inuit dialects have as many as 40 words for snow and 70 words for ice. What we do know for certain is that the Inuit have an incredibly close relationship to an environment dominated by snow and ice. The Inuit depend on the snow and ice of long, cold winters for their survival. Thick ice helps them to hunt and fish and to travel from village to village (Figure 9.2). Snow is the building material for their iconic temporary shelters, igloos. The permanently frozen ground beneath the tundra provides a solid foundation for their homes.

Like all people who live in extreme environments, the Inuit must pay particularly close attention to changes in their environment. In the words of Sheila Watt-Cloutier, past chair of the Inuit Circumpolar Conference, "talk to hunters across the North and they will tell you the same story, the weather is increasingly unpredictable. The look and feel of the land is different. The sea-ice is changing. The Earth is literally melting" (Figure 9.3).

▲ Figure 9.1 **Polar People**
The Inuit live on the northernmost shores of North America and the northeastern tip of Siberia.

Earth's atmosphere is indeed warming, especially in the Arctic. Yearly average temperatures in the Arctic have increased at nearly twice the rate as those in temperate and tropical regions. Over 90% of Arctic glaciers are retreating at unprecedented rates as more ice melts or evaporates than is formed each year. And the total mass of ice floating

► Figure 9.2 **Living on Ice**
A century ago, Inuit wore sealskin parkas and depended on dogsleds for transportation over long icy winters (left). Though technologies have changed (right), the Inuit still depend on the cold winter, snow, and ice for hunting and transportation.

▲ Figure 9.3 **The Earth Is Melting**
The permafrost, permanently frozen ground, has melted beneath the foundation of this house in the northern Canadian town of Inuvik, causing it to sink. As the climate warms, we find that permafrost is clearly not all that permanent.

on the Arctic Ocean is steadily diminishing. So much ice has melted that most summers ships can navigate from the North Atlantic Ocean across northern Canada and Alaska to the Pacific Ocean. The loss of sea ice has contributed to a 30% decline in the number of North American polar bears, which hunt much of their prey from ice floes. As a result, polar bears are now listed as "threatened" under the U.S. Endangered Species Act.

The warming climate is affecting entire ecosystems. Since 1990, the growing seasons across much of the Arctic tundra have lengthened by over 30 days. The boundary between the forests of the taiga and the treeless tundra is shifting northward. Because vegetation and surface peat layers now dry out more frequently, fires are becoming more common in both ecosystems (**Figure 9.4**). Warmer temperatures and longer growing seasons have changed the vegetation of the tundra, reducing the habitat of small herbivores called lemmings. These once numerous rodents are now in decline across the Arctic. Because lemmings are the primary food of many predators, their decline is jeopardizing the survival of carnivorous birds such as the jaeger and mammals such as the Arctic fox. Whole food webs are starting to collapse.

Marine and coastal ecosystems are also affected. Rivers fed by melting glaciers are carrying more water to the sea. This fresh water has lowered the salinity of estuaries and coastal waters, changing the habitats of marine plants and animals.

Several sources of data indicate that the Arctic climate has been alternating between warm and cold periods for the past million years. Throughout those cycles, ecosystems shifted and organisms adapted. Can we expect similar shifts and adaptations in the years ahead? For some organisms, the answer may be yes. For others, the more likely answer is no. The current rate of temperature change in the Arctic is unprecedented, and there are many obstacles to species' migration. Evolutionary change for many species requires a long period of time. Also, if this rate of change continues, future temperatures will be much warmer than any experienced in the past.

As the nations of the world argue about the causes of climate change and possible responses, the Arctic is being rapidly transformed. A recent report from the Inuit Circumpolar Conference put it this way: "While many in the South characterize climate change as an environmental and/or economic issue, to us it raises questions of culture and survival." Many experts see the environmental changes in the Arctic as omens of the major changes that may soon occur worldwide.

Scientists are committed to identifying actions to slow or halt global warming as well as actions that will minimize the effects of warming that are already occurring or likely to occur.

- *How has climate changed in the distant and more recent past?*
- *How do we know Earth's climate is warming now?*
- *What are the causes of global warming?*
- *What are the consequences of global warming?*
- *How can we forecast future climate change?*
- *What actions can we take to slow global warming?*
- *What actions can we take to adjust to warming that is likely to occur regardless of our immediate actions?*
- *What national and international policies are needed to motivate and guide our actions?*

▶ Figure 9.4 **Fire on Ice**
This satellite photo shows dozens of the more than 17,000 wildfires that burned through 30 million hectares (74 million acres) of tundra and taiga across Siberia in the summer of 2012. The warmer conditions of the last several decades have caused wildfires such as these to become more common in the Arctic.

Source: http://earthobservatory.nasa.gov/IOTD/view.php?id=78305.

9.1 Long-Term Climate Patterns

BIG IDEA A long view—one that spans thousands or even millions of years—shows that Earth's climate is dynamic. Over hundreds of thousands of years, the global climate has varied in response to regular changes in Earth's orbit that influence the amount of sunlight it receives. During the past 10,000 years, Earth has grown warmer, but with notable ups and downs. In the past 1,000 years, the climate has warmed, cooled, and then warmed again. The phrase **climate change** encompasses all of these variations. Long-term changes in climate have significantly influenced human evolution and history. A study of long-term climate change provides an understanding of the factors that have determined global climate in the past and will surely shape it in the future.

The Pleistocene—The Last 2 Million Years

■ For nearly 2 million years, Earth's climate has cycled between cold and warm periods.

You might not think so from current trends, but for the past several million years, Earth's climate has been cooling. Between 3 and 5 million years ago, a huge ice sheet in Antarctica grew until it totally covered the continent. In the Northern Hemisphere, mountain glaciers grew larger and more numerous.

About 2 million years ago, the global climate began to display regular cycles in which cold temperatures alternated with warm temperatures. The cold periods are typically referred to as glacial periods, or sometimes ice ages. The periods of intervening warmth then are referred to as interglacial periods. Each of these cycles of cold and warm lasted about 100,000 years. Scientists call this time of alternating cold and warm periods the **Pleistocene epoch**.

The surface of Earth we know today started taking shape during a particularly severe glacial period that began 800,000 years ago, an ice age that brought unprecedented glaciation. Huge glaciers between 1,000 and 3,000 m (3,300 and 10,000 ft) thick spread across vast areas of North America and Eurasia. During warm periods, the glaciers retreated or disappeared entirely. Later cold periods, or ice ages, again produced glaciers. The most recent continental glaciation lasted from about 110,000 years to 12,000 years ago. At the peak of this last ice age, glaciers covered 30% of Earth's land. In North America, glaciers reached as far south as central Illinois and Long Island, New York (Figure 9.5).

These cycles of changing temperatures produced significant changes in nearly all ecosystems. Fossils found in lake and ocean sediments show major shifts in the distribution of ecosystems, both on land and in the oceans. For example, 15,000 years ago boreal forests similar to those found today in southern Canada covered large areas of Georgia and Alabama. Deciduous forests extended only as far north as the southern portions of these states.

During the ice ages, much of the water evaporating from the ocean fell as snow and became part of the glaciers. This reduced the volume of water in the ocean. During the coldest periods, sea level dropped by about 100 m (330 ft). The lower sea level exposed coastlines and created land bridges between places that today are separated by water (Figure 9.6). On the continents, water melting from glaciers produced complex river systems and expansive lakes. At the height of the last ice age, for example, Lake Agassiz covered most of what is today Manitoba, western Ontario, northern Minnesota, and eastern South Dakota.

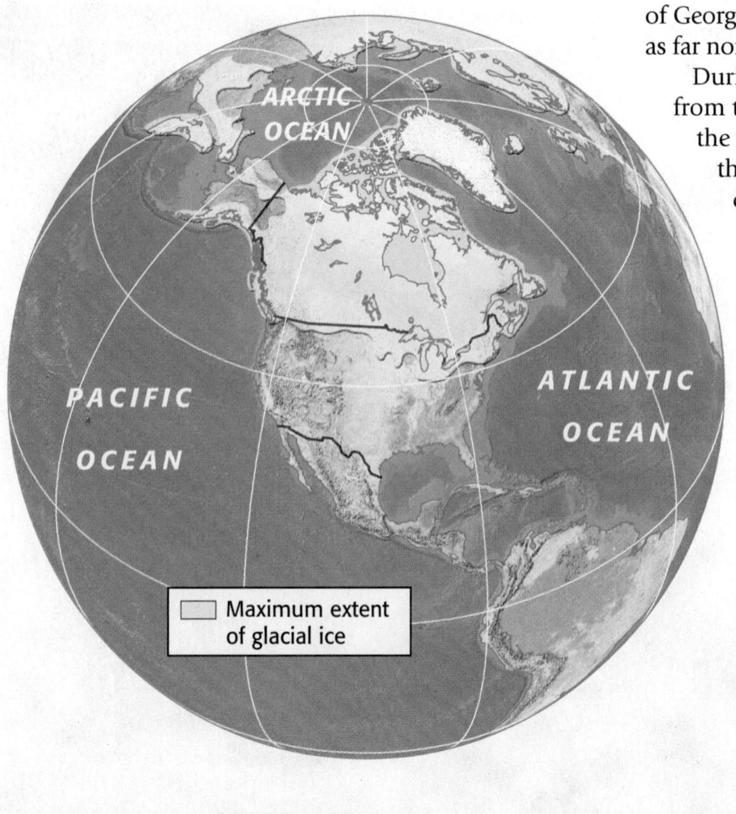

◀ Figure 9.5 **Ice Age**
At the peak of the last glacial period 15,000–19,000 years ago, massive glaciers extended across the northern portions of Eurasia and North America. Glaciers extended as far south as central Illinois and Long Island, New York.

What caused these cycles of cold glacial periods and warm interglacial periods? Changes in Earth's orbit that influence the amount and distribution of solar radiation certainly play a role. This idea was first proposed in a series of papers in the 1920s by Serbian mathematician-astronomer Milutin Milankovitch. Astronomers had for some time been aware of regular changes in the elliptical nature of Earth's orbit and in the tilt and orientation of Earth's rotational axis. These changes affect the average amount and distribution of solar radiation striking Earth's surface. Analyzing data on these various changes, Milankovitch hypothesized that major glacial and interglacial periods had varied on a 100,000-year cycle. Milankovitch's hypothesis was verified in 1976, using data from ocean floor sediments. In honor of his landmark research, scientists refer to these shifts from glacial to interglacial conditions as **Milankovitch cycles**.

By itself, orbital variation in the amount of sunlight striking Earth is not sufficient to cause major shifts in climate. During periods of reduced radiation, snow and ice begin to accumulate on the continents. The snow and ice reflect more of the sun's energy. As a consequence, the land absorbs less heat. The cooler land allows the ice to spread. Thus, the growing glaciers speed up the cooling process by reflecting more and more radiation back into space.

Other factors may also have contributed to the spread of the glaciers. Changes in ocean currents could have altered the patterns of heat transfer around the globe, leading to greater cooling toward the poles. Volcanoes that spewed gases and ash into the atmosphere could

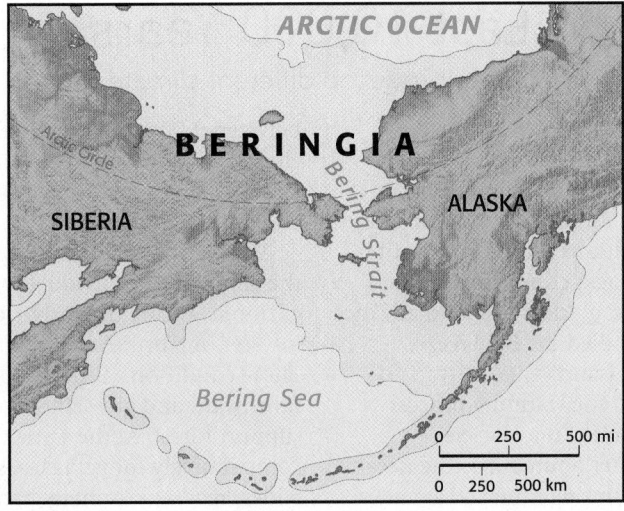

▲ Figure 9.6 **The Bering Land Bridge**
At the peak of the last ice age, sea level was much lower, and a wide isthmus of land (in beige) across what is today the Bering Strait joined northeast Asia to northwestern Alaska.

have reduced the amount of sunlight reaching Earth.

Changes in the chemical composition of Earth's atmosphere also played an important role in the cycle from glacial to interglacial conditions. The evidence is in the ice. Each year, a new layer of snow is deposited on glaciers. Variations in average atmospheric temperatures produce very predictable differences in the relative amount of different isotopes of hydrogen in snow water. Also bubbles of air are trapped in these layers as the snow is compressed into ice. The chemical analysis of both the ice and the bubbles it contains provides a year-by-year record of change in the temperature and chemistry of Earth's atmosphere (Figure 9.7).

Concentrations of carbon dioxide (CO_2) were much lower during cold glacial periods. Climate scientists theorize that colder conditions at the beginning of glacial periods encouraged upwelling that brought nutrients to the surface in large areas of the ocean. This resulted in very high net primary production (NPP) that pulled great amounts of CO_2 from the atmosphere (see Module 6.5). In interglacial times, ocean NPP was as much as 50% lower, resulting in much higher atmospheric CO_2. As you will learn in detail later in the chapter, CO_2 traps heat in the atmosphere. Decreasing CO_2 concentrations likely produced additional cooling during glacial periods.

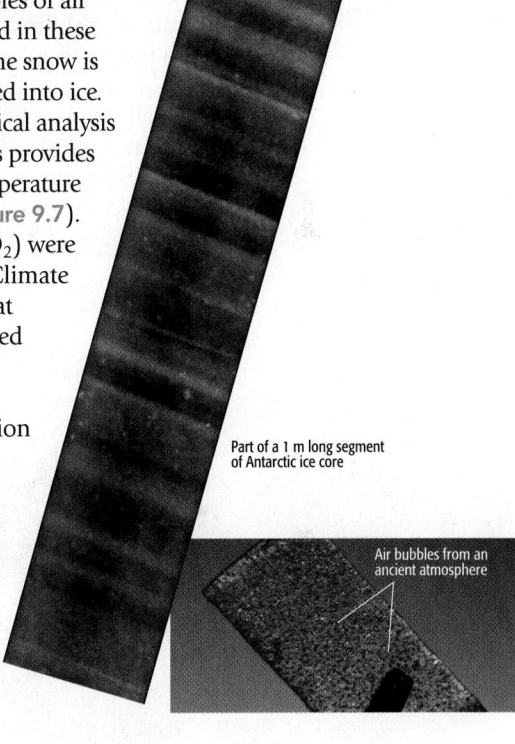

Part of a 1 m long segment of Antarctic ice core

Air bubbles from an ancient atmosphere

◄ Figure 9.7 **CO_2 and Glacial Cycles**
Milankovitch cycles are obvious in this graph of atmospheric CO_2 concentrations and temperature over the last 420,000 years, derived from cores of Antarctic ice. Concentrations were low during cold glacial periods and high during warm interglacial periods (yellow bands). Note that CO_2 remained below 300 parts per million (ppm) throughout this time.

Source: Petit, J.R., et al. 1999. Climate and atmospheric history of the past 420,000 years from the Vostok Ice Core, Antarctica. *Nature* 399: 429–436.

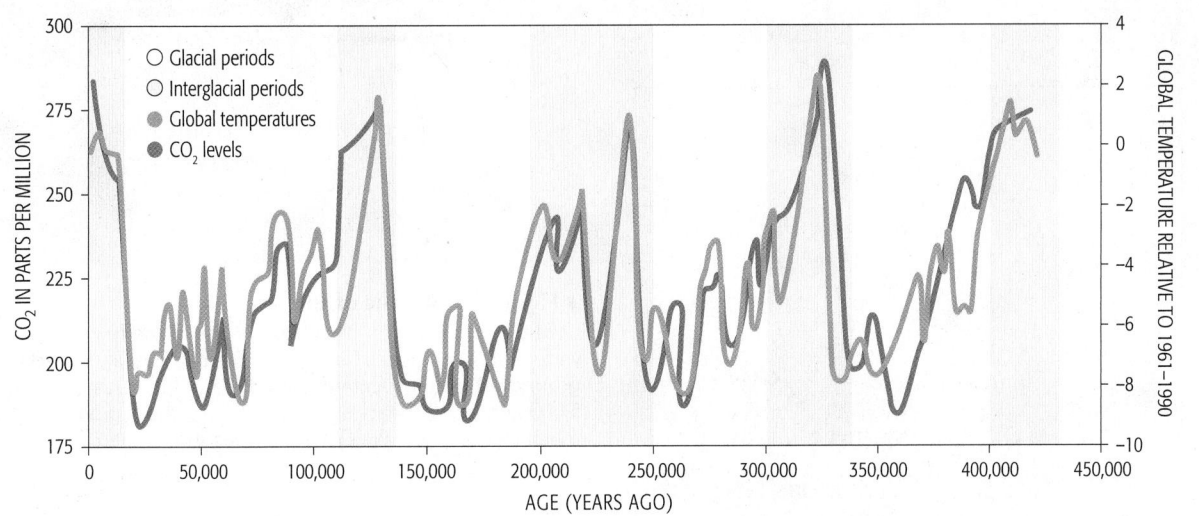

- ○ Glacial periods
- ○ Interglacial periods
- ● Global temperatures
- ● CO_2 levels

CO₂ IN PARTS PER MILLION

GLOBAL TEMPERATURE RELATIVE TO 1961–1990

AGE (YEARS AGO)

Holocene—The Last 10,000 Years

■ Periods of warming and cooling have produced different climate patterns in specific regions.

Earth's climate was still cold 15,000 years ago, but a warming trend was underway. Glaciers had begun to retreat, and the bare rock and sediments they left behind were undergoing rapid succession to forest (see Module 6.6). But climate transitions often occur in fits and starts. About 12,000 years ago, climates turned abruptly colder. In many places, glaciers began to advance again. This cold period, often called the Big Freeze, continued for more than a thousand years.

What caused the Big Freeze? The warming process itself. As the initial warming melted the glaciers, enormous volumes of fresh water poured into the ocean. Much of this water flowed out through the St. Lawrence River. The fresh water diluted the ocean water and interrupted the circulation of warm water from the tropics to the north Atlantic. This interruption produced much cooler conditions in polar and north temperate regions. It also brought an end to the Pleistocene.

The last 10,000 years of Earth's history belong to the **Holocene epoch**. During the Holocene, Earth has been in a warm interglacial period. Even so, global temperatures have varied. Climates were especially warm between 9,000 and 5,000 years ago. This period correlates closely with variations in Earth's orbit that favored solar heating. Then, average temperatures became a bit cooler 5,000 to 1,000 years ago.

The warming and cooling of the Holocene affected the climates of individual regions in different ways. For example, during the early Holocene warm period, the interior of North America was relatively warm and dry. These conditions allowed southern deserts to expand northward and grasslands to grow in regions that today support forest. At the same time, North Africa was comparatively moist; lakes, savanna, and forest were widespread across many parts of the Sahara region (**Figure 9.8**). Since then, North America has become moister, and the Sahara has become a vast desert.

The last 1,000 years have also witnessed significant climate variation. Climatologists call the years from A.D. 1000 to 1300 the **Medieval Warm Period**. During this time, average global temperatures increased by 0.2–0.5 °C (0.6–1.2 °F). However, warming was greater in some northern hemisphere locations. Evidence from tree rings suggests that Iceland was 2–3 °C (3.6–5.4 °F)

▲ Figure 9.8 **A Relic of the Past**
Lake Gaberoun in southern Libya is one of the few bodies of water in the nearly rainless Sahara desert. The lake is fed by groundwater that accumulated beneath the sands between 5,000 and 10,000 years ago. At that time, the region was much wetter, and Lake Gaberoun was surrounded by tropical savanna.

◄ **Figure 9.9 A Living Climate Record**
These are the remains of the village of Brattahlid established by the Viking leader, Erik the Red, in southern Greenland during warmer times in the late 10th century. This site was abandoned about 400 years later as the climate became much colder.

The eruptions of four very large tropical volcanoes between A.D. 1400 and 1700 are thought to have played a major role in this cooling.

We humans are creatures of the Pleistocene. The first members of the genus *Homo* appeared in Africa about 2 million years ago. Climatic changes since then have had much to do with our evolution and migration around the globe. The warm, moist climates in North Africa and the eastern Mediterranean region during the early Holocene played a key role in the development of agriculture and irrigation. Climate change over the past 1,000 years first facilitated colonization and then forced the abandonment of lands in northern Europe and Scandinavia.

Through ice ages and warm interglacial periods, climates changed as a consequence of periodic changes in Earth's orbit, volcanic eruptions, changes in Earth's snow and ice cover, and shifts in net primary production of oceans that altered atmospheric CO_2. These factors were outside the control of humans, but the climatic changes that they caused most certainly influenced human history. Today, there is no doubt that the activities of modern humans play a central role in the current global warming trends. Furthermore, Earth's atmosphere is warming at a rate faster than at any previous time in the Holocene. The Greek word for human being is *anthropos*, and because of our dominant influence on Earth's environment, many scientists suggest that we have entered an entirely new epoch, the **Anthropocene**. As we shall see, global warming is currently impacting human communities and the ecosystems upon which they depend, and those impacts are likely to grow.

QUESTIONS 9.1

1. What factors contributed to the alternating glacial and interglacial periods over the past 800,000 years?
2. What was the Big Freeze, and what was its cause?
3. Describe the general pattern of change in average global atmospheric temperature during the past 1,000 years.

(MES) For additional review, go to **MasteringEnvironmentalScience**

warmer than today. Vikings who settled Iceland and Greenland at this time found these lands hospitable for farming (**Figure 9.9**); today they are not. In western North America, prolonged drought resulted in the widespread migration of Native Americans (**Figure 9.10**). The warmer climate also led to frequent fires in western forests. Warming during this period appears to have been caused by decreased volcanic activity and an increase in solar radiation.

Between A.D. 1400 and 1500—just as the Renaissance was blossoming in Europe—a cold and wet period set in over the Northern Hemisphere. Once again, mountain glaciers in Europe and North America grew larger and advanced downslope. These colder conditions were almost certainly the reason that Viking settlements in Greenland were abandoned; growing seasons simply became too short to support agriculture. This cold period, called the **Little Ice Age**, continued well into the 1700s and the beginning of the Industrial Revolution.

▼ **Figure 9.10 Abandoned**
Dry conditions in the Medieval Warm Period forced the Anasazi peoples to abandon their farms and cliff dwellings (such as these in Mesa Verde) and migrate to moister locations in what are today New Mexico and Arizona.

9.2 Measuring Global Temperature

BIG IDEA Long-term trends in global temperature can be inferred from changes in the fossils and chemical isotopes found in sediments and glacial ice. Climate change during the past 3,000 years can be estimated from tree growth, as measured by the width of annual rings. These techniques lack the precision necessary to detect year-to-year or decade-to-decade global trends. Scientists can, however, determine differences in global average annual temperatures with great precision by analyzing data gathered from thousands of locations worldwide. These analyses reveal that Earth's temperature has warmed by 1–2 °C since the late 1800s. This warming has been more intense in some regions than in others. Changes in ocean currents and events such as volcanic eruptions complicate the pattern of global warming.

Measuring Recent Climate Change

■ Since the beginning of the Industrial Revolution, Earth's atmosphere has warmed significantly.

Using computer models, scientists estimate that the average annual air temperature near Earth's surface is about 14 °C (57 °F). Given the current interest in global warming, you may be surprised to learn that scientists do not know this number with much precision. That's because measuring global annual surface air temperature is a daunting challenge.

It is not difficult to measure air temperature at a particular place on land—all that is needed is a calibrated thermometer. Such measurements are made in well-ventilated, shaded shelters to avoid the effects of the sun's direct radiation (Figure 9.11). Determining the average temperature of a particular location over a

period of a year requires persistence and care. Individual readings must be taken in a uniform manner and at carefully determined times to avoid sampling one particular time of day or one season more than another.

Measuring air temperatures over the ocean is a bit more complicated process. In the 19th century, Ben Franklin used a mercury thermometer hung overboard to take readings as he sailed from the United States to Europe. By the mid 1900s, a system of stationary weather buoys had been deployed, a system which is still used today. In addition, weather satellites are used to provide sea surface temperatures (SSTs) and air temperatures over most of the world's oceans (Figure 9.12).

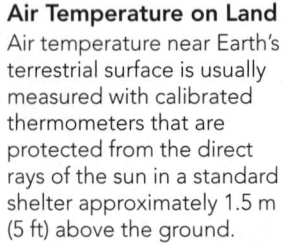

▼ Figure 9.11 **Measuring Air Temperature on Land** Air temperature near Earth's terrestrial surface is usually measured with calibrated thermometers that are protected from the direct rays of the sun in a standard shelter approximately 1.5 m (5 ft) above the ground.

▼ Figure 9.12 **Measuring Air Temperatures at Sea** Ⓐ Over 1,000 stationary buoys such as this are maintained and monitored by the United States and other countries. Sensors in the buoy's mast measure air temperature. Other sensors measure sea surface temperatures. Ⓑ Satellites also measure temperatures at the sea surface and the air just above it over vast areas.

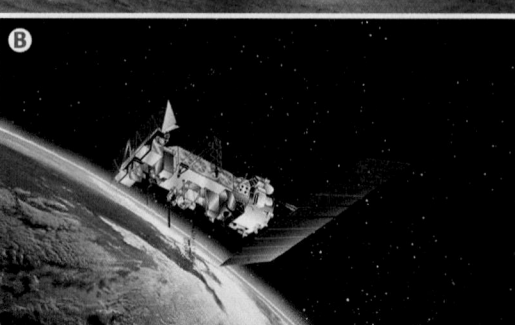

Estimating the average annual air temperature over the entire globe is a far greater challenge. For this, scientists need a large number of sampling stations at which measurements are taken in exactly the same way. And then, it is nearly impossible to determine whether such a collection of samples is biased toward warm or cold regions. Given these uncertainties, how can we possibly know if the surface air temperature of our planet is changing from year to year, century to century, or millennium to millennium?

Fortunately, it isn't necessary to know the exact value of the average global surface air temperature to determine whether it is changing. Instead, scientists study patterns of temperature variation among the many locations in the world where reliable data exist. First, scientists determine a benchmark of the average annual temperature for each location over a specific time interval, such as 1901–2000. Then they compare each year's average temperature with that benchmark. The difference between each year's average temperature and the benchmark is called a **temperature anomaly**. A positive anomaly indicates a year that is warmer than the benchmark. A negative anomaly represents a year that is cooler than the benchmark.

Temperature anomalies for each year are then averaged across the thousands of locations where data are available. This average provides a measure of global temperature change. Such an analysis reveals that although the global temperature varies considerably from one year to the next, it has been gradually warming over the past 130 years (Figure 9.13). Warming is especially notable over the past 30 years. During the period from 2001 to 2013, the average global temperature was 0.54 °C (0.97 °F) warmer than during the period from 1961 to 1973. The 10 warmest years on record have occurred since 1995. The data also suggest that temperatures will continue to rise.

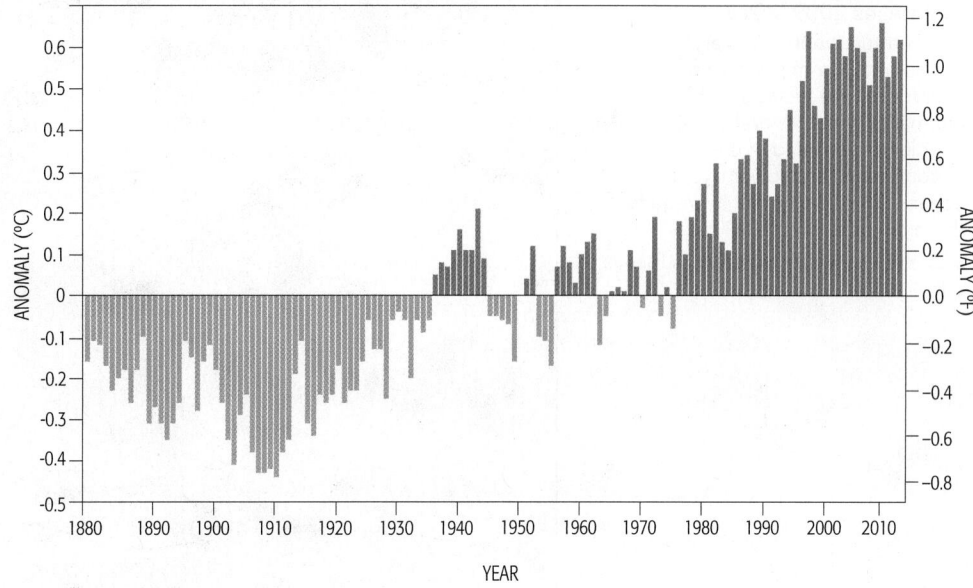

GLOBAL LAND AND OCEAN TEMPERATURE ANOMALIES, JANUARY–DECEMBER

▲ Figure 9.13 **It's Getting Warmer**
Based on data from thousands of locations, the global annual average temperature anomaly shows a steady increase in the temperature of Earth's atmosphere since 1880. Blue bars represent years when average annual temperatures were below, and red bars represent years when average annual temperatures were above the benchmark average for 1901–2000.

Source: U.S. National Oceanographic and Atmospheric Administration.

Today, we are witnessing warming considerably greater than what occurred in the Medieval Warm Period. The term **global warming** as climate scientists use it refers to a specific climate-change trend—the increase in atmospheric temperature since the late 1800s. This period of global warming coincides with the Industrial Revolution.

Causes of Natural Climate Variation

■ Temperature trends vary among regions and from year to year.

Scientists have been able to make extremely accurate measurements of the general increase in world temperatures known as global warming. But this warming trend is not uniform throughout the world. Significant differences are found from place to place and year to year.

Solar forcing. We know that 100,000-year Milankovitch cycles are caused in part by variations in Earth's orbit that affect the amount of solar radiation it receives. However, given the much shorter time frame that defines global warming, the focus for such variation has to be on the sun's radiation output rather than on variations in Earth's orbit. We also know that within a much shorter time frame, radiation output from the sun increases and decreases regularly in association with the 11-year cycle of sunspot appearance and disappearance. This solar radiation output also changes in a less regular

fashion from year to year, and over multiple decades. The phenomenon of changes to climate brought about by changes in solar radiation is referred to as *solar forcing.*

Solar forcing has been offered as an explanation to account for current warming trends. Given the important role of solar energy in Earth's climate (see Modules 3.6, 3.7), this so-called solar forcing hypothesis seems reasonable. However, taking into account the sunspot cycle, solar radiation output has actually decreased over the past 50 years, while global temperatures have increased. Perhaps more important, the variation in solar radiation striking Earth's atmosphere over the past 150 years has been calculated to be only plus or minus about 2 W/m². That is a variation of just 0.0014% of the average total solar radiation input (1361.5 W/m²). The impact of this small change on temperature of the troposphere is very small relative to observed temperature trends.

Q *Is climate change something to be concerned about, or are variations in climate normal?*

Alexcis Zare, University of Houston

Norm: A Variations in climate are normal. However, the rate at which Earth's atmosphere is warming is truly unusual, and climate scientists are concerned that warming may increase even faster if emissions of greenhouse gases continue to rise.

► **Figure 9.14 Temperature Anomalies 2009–2013**
Temperature anomalies vary considerably from place to place. During this five-year period, anomalies were generally higher in polar regions and in the centers of continents, and lower over the oceans. The benchmark average for this analysis is for 1900–2013.

Source: U.S. National Aeronautics and Space Administration.

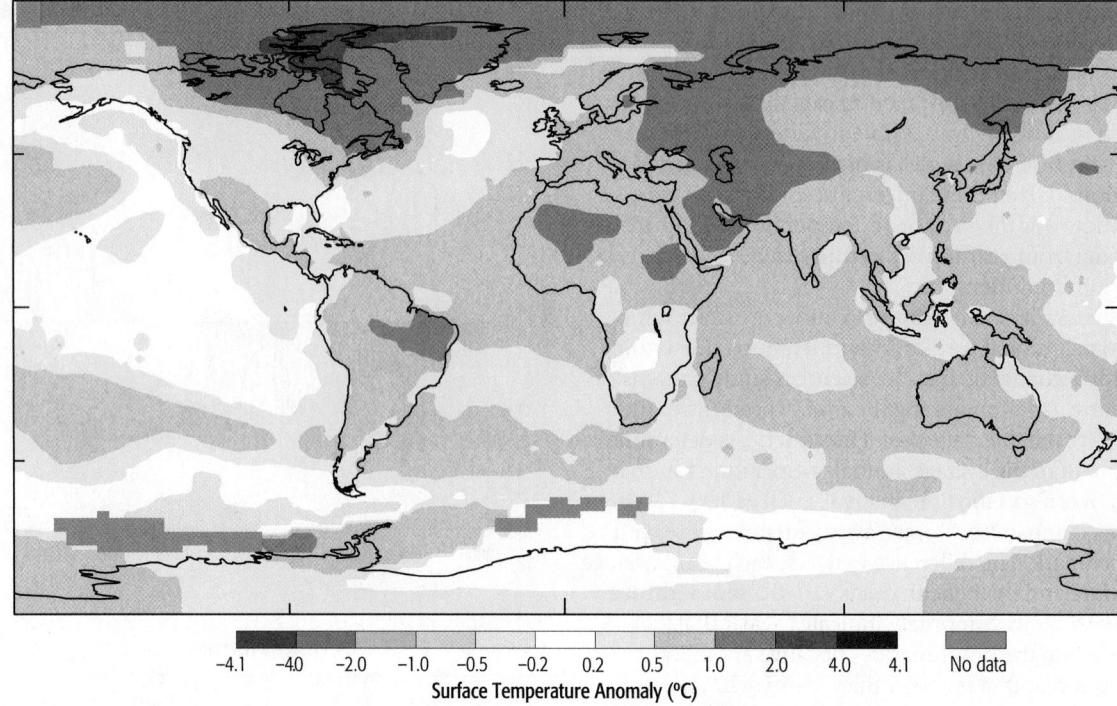

-4.1 -4.0 -2.0 -1.0 -0.5 -0.2 0.2 0.5 1.0 2.0 4.0 4.1   No data
Surface Temperature Anomaly (°C)

Regional variations on land.
Annual temperature anomalies vary considerably from location to location (Figure 9.14). In much of the Arctic, for example, temperature anomalies have exceeded 2 °C (3.6 °F). These higher-than-average temperature anomalies are related to circulation patterns in the troposphere and ocean currents, which move heat from the equator toward the poles (see Module 3.7). This movement intensifies temperature change in the polar regions. This is one of the factors contributing to the accelerated climate change in Arctic ecosystems.

Because continental regions heat up more quickly than nearby oceans, temperature anomalies are generally higher over the interior of continents, such as North America and Eurasia. In a few regions, such as over parts of the Gulf of Alaska and the ocean around Antarctica, temperature anomalies are actually negative. In these areas cooling appears to be related to local changes in ocean circulation and the upwelling of cold water.

As was the case with climate change in the Holocene, the specific effects of global warming on factors such as rainfall and length of growing season vary from location to location. In some places, rising temperatures are accompanied by increased aridity and drought. In other places, warming increases evaporation from nearby bodies of water, thereby increasing rainfall and even winter snowfall.

Regional variations in sea surface temperatures.
Weather conditions vary from year to year, often as a consequence of changes in currents that affect the temperature of water near the ocean's surface. Changes in sea surface temperatures over large regions that occur

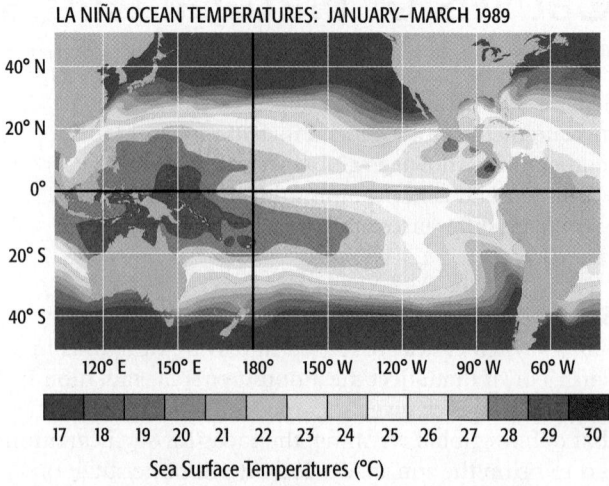

EL NIÑO OCEAN TEMPERATURES: JANUARY–MARCH 1998

LA NIÑA OCEAN TEMPERATURES: JANUARY–MARCH 1989

17 18 19 20 21 22 23 24 25 26 27 28 29 30
Sea Surface Temperatures (°C)

▲ **Figure 9.15 ENSO**
El Niño conditions are indicated by the extension of warm water across the entire equatorial Pacific Ocean. During La Niña, colder water extends along the west coast of South and Central America.

over intervals of years or decades are called oscillations. The **El Niño/Southern Oscillation (ENSO)** in the Pacific Ocean is one of the most studied examples of such variation. Over a three- to eight-year period, surface ocean temperatures cycle between relatively cold and warm conditions. ENSO affects weather conditions around the world.

During the cold part of the ENSO cycle, ocean currents bring cold water from the Antarctic Ocean to the equator along the west coast of South America. The upwelling of deep ocean water contributes to the cold temperature of the surface water. The situation changes when water warmed in the western equatorial Pacific moves eastward, overriding the colder, denser water. Because the warm part of the cycle often begins around Christmas time, South American fishermen call it El Niño (Spanish for baby boy), in reference to the Christ child. Given that name, scientists called the part of the cycle dominated by cold surface water La Niña (baby girl) (**Figure 9.15**).

El Niño's warm waters feed thunderstorms and increase rainfall across the eastern Pacific Ocean and western South America. But to the east, across the Andes and in the Amazon Basin, warm, dry conditions prevail. In North America, El Niño winters are warmer than normal in the Midwest, Northeast, and Canada, but in the Southwest and Mexico, winters are cooler and wetter. In contrast, the Southwest and Mexico often experience prolonged droughts during La Niña. Although the effects are less dramatic, temperature and rainfall patterns in Australia, Southeast Asia, and Africa are also correlated with the ENSO cycle.

Other sea surface temperature oscillations influence climate and weather in the Northern Hemisphere. The Pacific Decadal Oscillation (PDO) involves significant changes in the temperature of ocean waters from east to west across the northern Pacific. These changes occur on a 20- to 30-year cycle and strongly influence temperature and rainfall in northwestern North America. Similar oscillations in the northern Atlantic Ocean influence weather patterns across Europe and Asia.

Volcanic eruptions. Global climates can be influenced by large volcanic eruptions. Such eruptions add tons of sulfur dioxide to the lower atmosphere. Small crystals of sulfur dioxide reflect the sun's rays and cool the atmosphere. In 1815, for example, the massive eruption of Mount Tambora, a volcano in Indonesia, resulted in two years of abnormally cold temperatures in places as distant as Europe and New England (**Figure 9.16**).

Earth's climate has indeed changed through time. Over the long term, ice ages have alternated with warm interglacial periods. Climates have also undergone centuries-long fluctuations, such as the Medieval Warm Period and the Little Ice Age. Even shorter periods of fluctuation characterize the El Niño/Southern Oscillation. These fluctuations are caused by natural processes, such as shifts in Earth's orbit relative to the sun, variations in global snow and ice cover, changes in ocean currents, and volcanic eruptions.

▲ Figure 9.16 **Global Cooling**
In 1991, Mount Pinatubo on the island of Luzon, the Philippines, explosively erupted, spewing sulfur-rich ash and gases into the troposphere. This eruption was far smaller than the Mount Tambora event, but it produced much cooler average global temperatures over the next two years.

Where YOU LIVE | **Is Your Climate Changing?**
How have average annual temperature and precipitation changed over the past century where you live? Summary data for thousands of reporting stations are available at the National Oceanic and Atmospheric Administration's (NOAA) National Climatic Data Center website. To generate a graphical presentation of annual temperature and precipitation averages from 1895 to the present for your state, go to the "Climate At a Glance" page at the NCDC website. From here, enter in your state and city information, choose the time period you want to explore, and create graphs looking at the temperature or precipitation trends. In addition to line graphs, the raw data will also be provided as a table below each graph.

1. Describe the temperature trend at your location over the past century. If at all, by approximately what amount has temperature changed?
2. Is there any trend in precipitation over this time period?
3. How do these changes compare with global changes?

QUESTIONS 9.2

1. The global annual temperature anomaly for 2008 was +0.44 °C. What does this mean?

2. What causes the shift from La Niña to El Niño conditions in the ocean waters near Ecuador, South America?

3. Why does the eruption of large volcanoes often result in one to two years of global cooling?

(MES) For additional review, go to **MasteringEnvironmentalScience**

9.3 Causes of Global Warming

BIG IDEA The temperature of Earth's atmosphere is influenced by many factors. Some, such as the amount of energy radiating from the sun and Earth's orbital position relative to the sun, are outside the realm of human influence. Other factors, such as the chemical composition of the troposphere and the reflectivity of Earth's surface, are increasingly influenced by human activities. Deforestation and the burning of fossil fuels add large quantities of carbon dioxide to our atmosphere. Industrial and agricultural activities release methane and nitrous oxide. These three gases absorb a disproportionate amount of the infrared radiation emanating from Earth's surface and trap it as heat (see Module 3.7). The rapid increase in the atmospheric concentration of these three heat-trapping gases is responsible for global warming.

The Greenhouse Effect

■ Some gas molecules trap heat radiated from Earth's surface in the lower atmosphere.

Nitrogen (N_2) and oxygen (O_2) molecules make up about 99% of the volume of Earth's atmosphere. Together, water (H_2O) and carbon dioxide (CO_2) account for less than 0.5%. Yet, molecules of H_2O and CO_2 are thousands of times more efficient than molecules of N_2 and O_2 in capturing the infrared light that is constantly being radiated from Earth's surface (see Module 3.6). This captured light energy is transformed into heat. Thus, H_2O and CO_2 have an impact on the atmospheric temperature far in excess of their relative abundance. Were it not for the presence of these two molecules, the average temperature of Earth's troposphere would be 33 °C (59 °F) cooler: that would be a frigid −19 °C (−2 °F). Without them, our world would be a frozen, ice-covered rock without the

possibility of plant and animal life. Gases that efficiently capture heat are called **greenhouse gases**.

Although the analogy is not perfect, the trapping of heat in the troposphere by greenhouse gases is often referred to as the **greenhouse effect**. On a sunny winter day, the plants and soil inside a greenhouse absorb light energy. Some of this energy is radiated back as infrared radiation, warming the air inside. The glass of the greenhouse prevents this air from mixing with the cold air outside, trapping the heat inside. In a similar fashion, greenhouse gases in the troposphere are warmed, as light is absorbed and then released as infrared radiation from Earth's surface. Because the tropopause tends to prevent the mixing of gases in the troposphere with the gases in the stratosphere, the heat gets trapped (Figure 9.17). The tropopause, in this case, acts like the glass of Earth's troposphere greenhouse.

The most abundant natural greenhouse gas is water vapor. Even so, water is not considered to be adding to global warming. The concentration of water vapor in the atmosphere fluctuates widely with seasons and at different locations around the world. The effect of water vapor on the temperature of the atmosphere is complicated by water's phase transitions from solid to liquid to gas.

As air becomes warmer, its ability to hold water vapor increases. Warmer air causes more water to evaporate from bodies of water and the leaves of plants. This additional evaporation might cause runaway warming, as the increasing levels of water vapor trap even more heat. Instead, the heated air rises and cools, causing the water vapor to condense into clouds. Because they reflect incoming solar radiation back into space, clouds cool Earth's surface and atmosphere.

Carbon dioxide is the second most important natural greenhouse gas. Gas bubbles in glacial ice reveal that prior to the Industrial Revolution the tropospheric concentration of CO_2 was about 285 parts per million (0.028%). At that time, CO_2 concentrations did not vary much from place to place, although they did vary by a few parts per million with the seasons.

▼ **Figure 9.17 The Greenhouse Effect**
Because there is very little mixing of greenhouse gases (GHG) across the tropopause, heat is trapped within the troposphere.

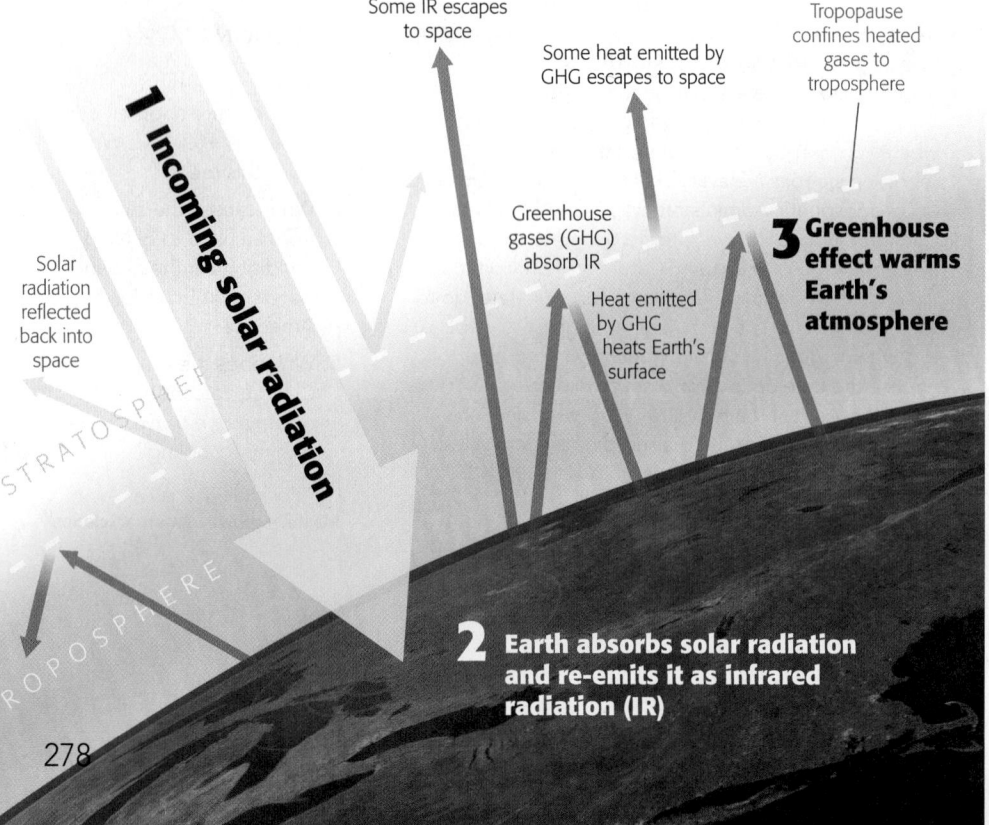

Some IR escapes to space

Some heat emitted by GHG escapes to space

Tropopause confines heated gases to troposphere

1 Incoming solar radiation

Solar radiation reflected back into space

STRATOSPHERE

TROPOSPHERE

Greenhouse gases (GHG) absorb IR

Heat emitted by GHG heats Earth's surface

3 Greenhouse effect warms Earth's atmosphere

2 Earth absorbs solar radiation and re-emits it as infrared radiation (IR)

Methane (CH_4) and nitrous oxide (N_2O) are greenhouse gases that occur naturally at comparatively low concentrations. Before the Industrial Revolution, CH_4 was found in concentrations of about 700 parts per billion (0.00007%), and nitrous oxide occurred at concentrations of 270 parts per billion (0.000027%).

Methane is released into the atmosphere naturally from the decomposition of vegetation by microbes and termites and from the digestive processes of various animals, including you. Nitrous oxide is a natural by-product of the nitrogen cycle (see Modules 12.2 and 12.6).

Human Impacts

■ Human activities have increased the abundance of several greenhouse gases in the atmosphere.

Humans first began to change the chemical composition of the atmosphere thousands of years ago when they began cutting and burning forests to expand agricultural fields. This reduced the total amount of carbon stored in trees and increased the amount of CO_2 in the atmosphere. It also increased emissions of CH_4 and N_2O. When humans domesticated cows and sheep, more CH_4 was added to the atmosphere by those animals' digestive processes, or less delicately, burps and flatulence. Later, humans cut forests and burned wood to make charcoal to fuel Iron Age furnaces. Until the Industrial Revolution, however, human populations were comparatively small. As a result, *anthropogenic*, or human-caused, greenhouse gas emissions were minuscule compared to the total volume of gases in the atmosphere.

Truly significant increases in anthropogenic greenhouse gas emissions came with the burning of fossil fuels: first coal, then petroleum and natural gas. These high-energy carbon fuels were produced and stored as plant and animal biomass in ecosystems eons ago (see Module 6.5 and Chapter 14). Their combustion releases greenhouse gases, especially CO_2. Currently, the burning of fossil fuels adds about 33 petagrams, or Pg (~26 billion tons), of CO_2 to the atmosphere, as well as smaller amounts of other greenhouse gases each year (Figure 9.18).

Deforestation contributes to the emission of greenhouse gases, as do other land uses that diminish the amount of carbon stored in Earth's ecosystems. Scientists estimate that human land use adds 3.2–4.3 Pg CO_2 to the atmosphere each year. Fires associated with deforestation and agriculture also emit CH_4 and N_2O.

In addition to naturally occurring greenhouse gases, humans have been releasing a variety of synthetic greenhouse gases into the atmosphere. These include chlorofluorocarbon (CFC) compounds, which are used as refrigerants, cleaning solvents, and fire retardants and in the manufacture of plastic foam products. Compared to other emissions, only small amounts of CFCs are added to the atmosphere each year. However, these gases absorb infrared light thousands of times more efficiently than CO_2. Because of their destructive effect on Earth's stratospheric ozone layer, the manufacture of CFCs was greatly curtailed by an international treaty, the Montreal Accord of 1987. This is covered in more detail in Chapter 10.

The impact of the emission of different anthropogenic greenhouse gases depends on two factors: (1) their capacity to absorb infrared light and retain heat and (2) the length of time they stay in the atmosphere, or their atmospheric residence time. These two factors are used to calculate a molecule's **global warming potential (GWP)**. GWP is a measure of an individual molecule's long-term impact on atmospheric temperature.

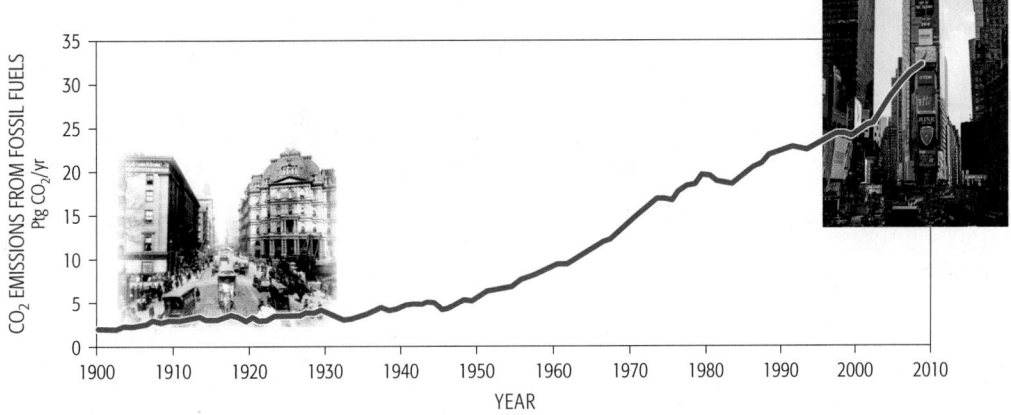

▲ Figure 9.18 **Fossil Fuels and CO_2 Emissions**
Before the beginning of the Industrial Revolution, emissions of CO_2 to the atmosphere from fossil fuels were negligible. Since 1900, CO_2 emissions from the burning of coal, oil, and natural gas have increased exponentially.

Source: Boden, T.A., G. Marland, and R.J. Andres. 2010. Global, Regional, and National Fossil-Fuel CO_2 Emissions. Carbon Dioxide Information Analysis Center, Oak Ridge National Laboratory, U.S. Department of Energy, Oak Ridge, TN, USA. doi 10.3334/CDIAC/00001_V2010.

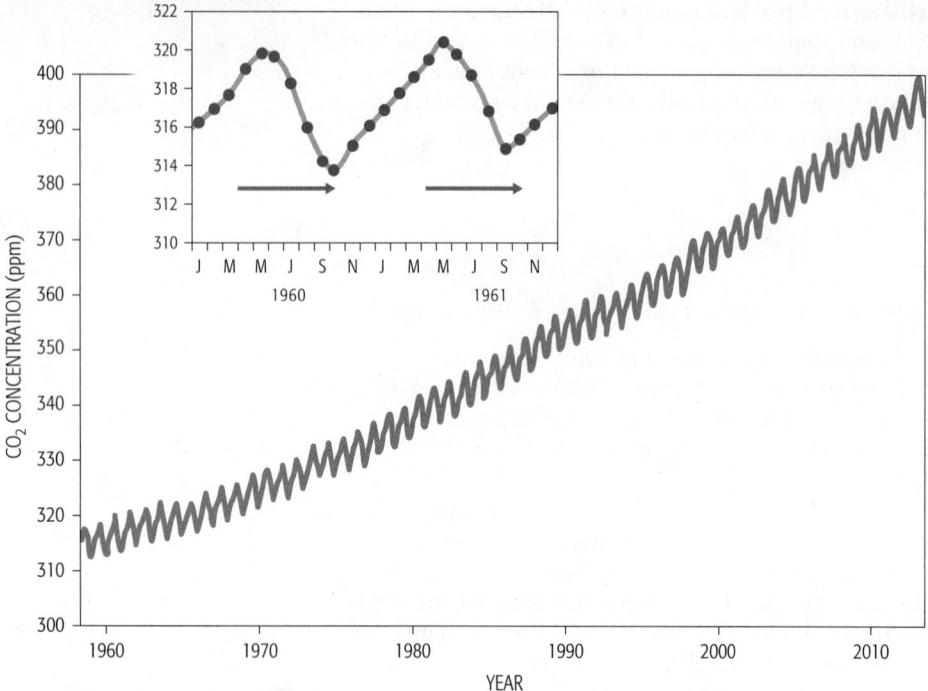

▲ **Figure 9.19 The Keeling Curve**
Measurements taken at Hawaii's Mauna Loa Observatory indicate that the level of CO_2 in Earth's atmosphere is increasing at a rate of 1–2 ppm per year, taking annual cycles in CO_2 concentration into account. The inset graph shows how the concentration of CO_2 changes during a two-year interval. The orange arrows correspond to the Northern Hemisphere growing season, when net primary production is high and more CO_2 is being withdrawn from the atmosphere.

Source: NOAA. 2014. Mauna Loa CO2 monthly mean data. http://www.esrl .noaa.gov/gmd/ccgg/trends/#mlo_data.

The GWP for different gases is calculated relative to CO_2, which is assigned a GWP of 1. An average molecule of CO_2 stays in the atmosphere for 100 years. CH_4 has a GWP of 25—it absorbs infrared photons 160 times more efficiently than CO_2, but an average CH_4 molecule resides in the atmosphere for only 14 years. With a GWP of 25, we would expect CH_4 to trap 25 times more heat than the same amount of CO_2 in a given period. N_2O has a long residence time, 120 years; its GWP is 296. CFCs absorb infrared light very efficiently and also hang around in the atmosphere for a long time; their GWPs are a whopping 1,300 to 13,000.

Prior to 1957, there were no long-term measurements of atmospheric gases. Therefore, there was no way of knowing whether anthropogenic emissions were altering our atmosphere's chemistry. That year, Charles D. Keeling of Scripps Oceanographic Institute began making systematic monthly measurements of atmospheric CO_2. Keeling chose to work at the Mauna Loa Observatory on the island of Hawaii because it was far from urban and industrial centers. The idea was to establish a baseline representing "normal" levels of atmospheric CO_2; compared to that baseline, future changes could be detected.

In 1957, the atmosphere contained about 312 parts per million (ppm) CO_2 (**Figure 9.19**). Keeling's data revealed clear seasonal cycles in the concentration of CO_2, which were related to the changes in the net productivity of ecosystems during summer and winter. After a few years of sampling, Keeling realized that CO_2 concentrations were increasing by 1–2 ppm each year. Measurements on Mauna Loa continue to the present. Since Keeling's first measurements, atmospheric CO_2 has risen by nearly 25%. In 2013, it reached 400 ppm for the first time.

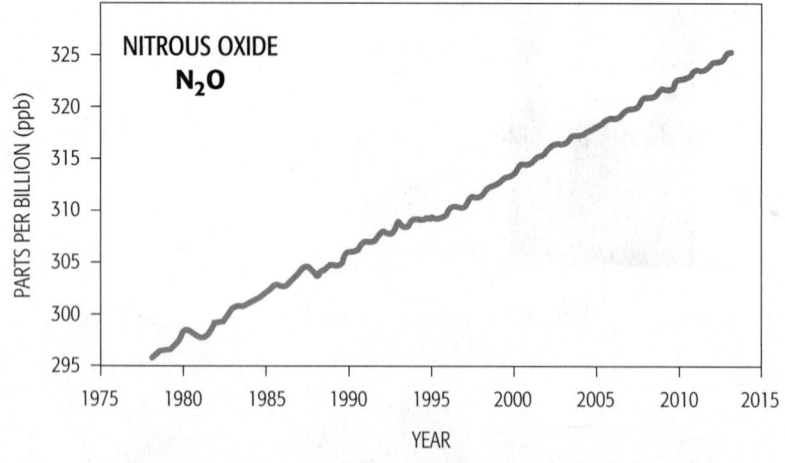

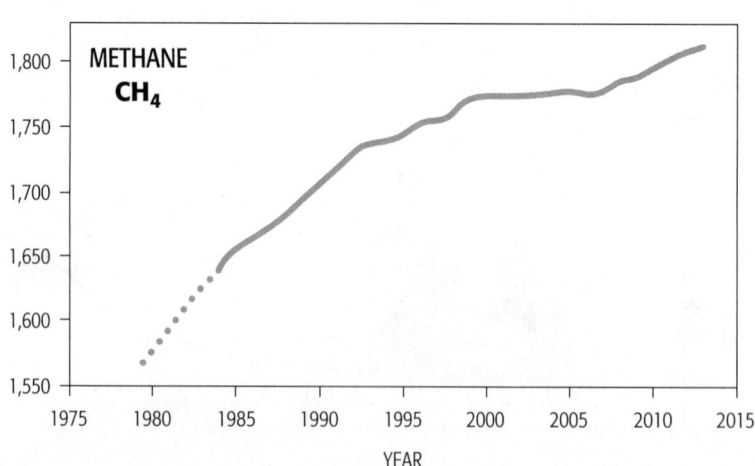

▲ **Figure 9.20 Other Greenhouse Gases**
Atmospheric concentrations of methane and nitrous oxide are increasing as a consequence of human activities such as deforestation, agriculture, and the burning of fossil fuels. Measurements of these gases did not begin until 1975.

Source: NOAA. 2013. Annual greenhouse gas index (AGGI). http://www.esrl.noaa.gov/gmd/aggi/.

More recently, atmospheric scientists monitoring other greenhouse gases have found trends that are similar to those of CO_2 (Figure 9.20). CH_4 in the atmosphere has more than doubled from its natural state to over 1,750 parts per billion (ppb). N_2O has increased to nearly 320 ppb. By comparison, CFC concentrations increased steadily into the early 1990s, but they have since stabilized as a consequence of an international agreement, the Montreal Protocol, that limits their production (see Module 10.5).

Although the issue was hotly debated in the 1980s and 1990s, today there is no doubt that global warming is a direct consequence of the increased amount of greenhouse gases in Earth's atmosphere. Nor is there any doubt that these atmospheric changes are directly linked to human activities, such as the burning of fossil fuels, deforestation, and pollution from industry and agriculture.

There are other activities associated with humans that contribute to global warming but that are not directly tied to greenhouse gas emissions. Forest fires and the burning of fuelwood and charcoal release solid particles called **black carbon** into the atmosphere. Diesel engines in many vehicles are also a source of black carbon. Black carbon particles absorb incoming solar radiation and directly warm the atmosphere. They also reduce the reflectivity of the surfaces they settle on, thereby contributing to additional warming. The black carbon deposited on snow and glacial ice is believed to be an important factor contributing to the worldwide retreat of glaciers (Figure 9.21). Taking its direct and indirect effects together, scientists estimate that black carbon emissions rank second only to emissions of CO_2 in their contribution to global warming.

▼ Figure 9.21 **Black Carbon**
The black carbon particles shown here in a patch of snow absorb solar radiation, and their increased warmth causes the snow beneath them to melt. Smoke stacks, many diesel engines, and charcoal kilns are some important sources of black carbon particles.

Sources of Greenhouse Gas Emissions

■ Heat-trapping gases are generated from a variety of human activities.

Scientists express total greenhouse gas emissions in terms of **carbon dioxide equivalents (CO₂e)**, taking into account the global warming potential of each gas. CO_2 is by far the most important, accounting for approximately 65% of warming. Increased emissions of CH_4 and N_2O are responsible for an additional 23%. Industrial pollutants such as CFCs are responsible for the remaining 12%. Warming due to black carbon is not included in these calculations.

Various human activities contribute differing amounts to total greenhouse gas emissions (Figure 9.22). Most of the electricity we use is generated by burning fossil fuels, largely coal and natural gas, so it's not surprising that electric generation produces the largest portion of emissions. Industrial processes are next in importance. These include activities such as the manufacture of concrete, which is responsible for 30–40% of all industrial emissions. Concrete production involves heating limestone and other materials to 1,400 °C (2,550 °F). This requires burning over 180 kg (400 lb) of coal for each ton of concrete. Agriculture is the source of about 12.5% of greenhouse gas emissions, primarily in the form of CH_4 and N_2O from livestock and the use of fertilizer.

Greenhouse gas emissions vary widely among countries. In 2012, the United States and China accounted for about 40% of anthropogenic emissions (Figure 9.23A). The United States has fewer than 5% of Earth's people but emits over 17.6 metric tons of CO_2 per person (Figure 9.23B). With 18% of the world's population, China emits about 6.2 metric tons of CO_2 per person each year. The nations of the European Union account for about 7.6% of greenhouse gas emissions, about 7.4 metric tons of CO_2 per person. Russia, India, Japan, and Canada also account for a significant portion of annual CO_2 emissions.

In industrialized countries, the vast majority of greenhouse gas emissions comes as CO_2 from burning fossil fuels. However in many countries, deforestation and agriculture are the predominant sources. In Brazil, for example, emissions equal 2.2 metric tons of CO_2 equivalents per year per person; 70% of this is due to CH_4 and N_2O emissions associated with deforestation. In New Zealand, greenhouse gas emissions equal over 7 metric tons of CO_2e per person. Over half of that is methane produced by that country's 40 million sheep.

Worldwide, per capita CO_2 emissions increased by about 11% between 1990 and 2011. Trends in per capita

CO_2 emissions during this time interval vary among countries (see Figure 9.23). Although total emissions in the United States increased by about 12%, per capita emissions actually declined by 7.8% during this time. In comparison, total CO_2 emissions among European Union countries shrank by about 3%, and per capita emissions shrank by 12%. Most developing countries have experienced considerable industrial and economic growth, and both total and per capita CO_2 emissions increased, by considerable margins in many cases. For example, total CO_2 emissions in China increased by 236%, while per capita emissions increased by over 180%. India and Indonesia have experienced similar increases in both total and per capita emissions. For countries such as Zimbabwe that experienced chronic social and economic problems during this time, total and per capita emissions declined.

These changes and the challenges they present are now acknowledged by nearly every nation. Governments have begun to develop policies to halt or reverse these trends; they are also finding ways to adapt to the changes that may be inevitable.

QUESTIONS 9.3

1. How does the greenhouse effect get its name?

2. Name four important greenhouse gases and the human activities that are most important in their production.

3. What factors influence the global warming potential of a particular gas?

(MES) For additional review, go to MasteringEnvironmentalScience

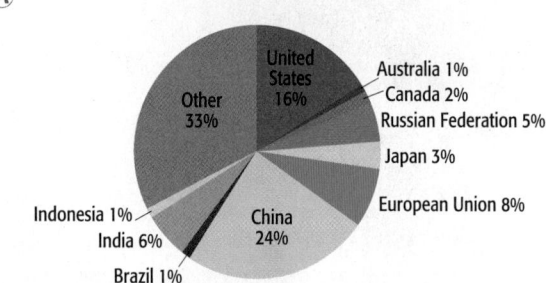

TOTAL GLOBAL CO₂e EMISSIONS IN 2012 Ⓐ

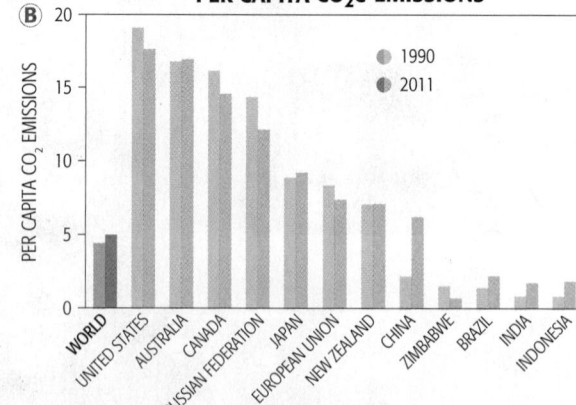

PER CAPITA CO₂e EMISSIONS Ⓑ

▲ Figure 9.23 **Annual CO₂ Emissions**
The pie chart Ⓐ indicates the percent contribution of various countries to global CO_2 emissions. Ⓑ The bar chart indicates the per capita CO_2 emissions (metric tons/person/yr) for these same countries and for the entire world in 1990 and in 2011. Emissions of other greenhouse gases are not included in these data.

Source: United Nations World Development Indicators (WDI). http://data.worldbank.org/data-catalog/world-development-indicators/wdi-2012.

► Figure 9.22 **Sources of Greenhouse Gases**
The major source of anthropogenic greenhouse gas emissions is the use of fossil fuels for industry, transportation, and power generation. The pie chart shows global averages; the specific contributions by different countries will vary depending on their economic activities.

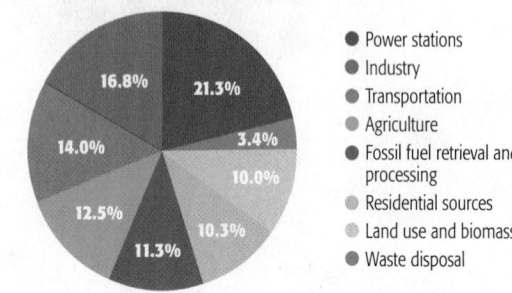

- Power stations
- Industry
- Transportation
- Agriculture
- Fossil fuel retrieval and processing
- Residential sources
- Land use and biomass
- Waste disposal

9.4 Consequences of Global Warming

BIG IDEA The changes caused by global warming vary from region to region. In some places, increasing temperatures have been accompanied by higher rainfall amounts. In others, they have brought drought. Winters have become milder and shorter in Earth's middle latitudes, and dry seasons have grown longer in some parts of the tropics. Glaciers and ice sheets are melting worldwide. Warming is causing sea levels to rise. Taken together, these changes are having a significant impact on the flora and fauna of many ecosystems.

Drier and Wetter

■ Global warming is producing wetter conditions in some places and drought in others.

The effects of rising temperatures on precipitation vary geographically. Rainfall has increased significantly in eastern North and South America, as well as in most parts of Europe and Asia. In contrast, sub-Saharan Africa, the Mediterranean region, and western North America have been drier. Since 1970, longer and more intense droughts—as measured by decreased precipitation and higher temperatures—have affected wide areas of the tropics and subtropics. At the same time, there has been a worldwide increase in the frequency of rainstorms that result in flooding, even in areas where total annual rainfall has declined (Figure 9.24). Warm air holds more moisture (see Module 3.6).

In regions in which rain is highly seasonal, such as sub-Saharan Africa, global warming appears to be changing the length of wet and dry periods. This is a matter of special concern because food production depends on the length of wet seasons. Based on current trends and climate models, growing seasons are expected to become shorter over most of sub-Saharan Africa, with the exception of lands very near the equator.

There is evidence that global warming is influencing drought cycles. For example, the El Niño/La Niña/Southern Oscillation is caused by changes in the temperature of surface waters in the equatorial Pacific Ocean. When waters off the west coast of South and Central America are cold, drought is much more common in the southwestern United States. Some climatologists think that since 1970 the length and intensity of El Niño and La Niña events have been outside the range of natural variability. Although climate models predict that such changes will occur, most scientists feel that it is not clear that they are actually underway.

Q *Is climate change the reason for increased storms and global disturbances?*
Ciara Tyce,
Georgia Southern University

Norm: **A** Climate scientists are generally careful not to attribute a particular weather event to global warming. However, increased frequency and intensity of storms and heat waves is consistent with climate model predictions.

▲ Figure 9.24 **Deluges and Droughts**
Ⓐ In 2013, torrential rains impacted crop production across much of upstate New York. Ⓑ In 2014, extreme drought in California meant farmers could not grow crops on hundreds of thousands of acres. Global warming may have contributed to both situations.

New **FRONTIERS**

Revving up Severe Weather?
Ocean temperature is an important factor in the development of tropical storms and hurricanes, and ocean temperatures have increased between 0.25 °C and 0.5 °C (0.45–0.9 °F) over the past century. Warmer sea surface temperatures appear to be associated with the observed increase in the number and strength of tropical storms in the Pacific Ocean. For example, Typhoon Haiyan, which hit the Philippines in 2013, was one of the strongest tropical storms ever recorded. But trends in the Atlantic Ocean are far less clear. The very significant damage from Hurricane Sandy in 2012 was largely a consequence of a combination of sea level rise (see the next section) and poorly managed coastal development (see Module 18.2).

Debate continues regarding the effects of global warming on past and current storm patterns, but there is consensus among scientists that continued sea surface warming will very likely increase the frequency and severity of tropical storms in the future. How much evidence do you believe we need in order to take strong action to mitigate the effects of future strong storms? How much of the risk associated in living in coastal areas should be the responsibility of property owners versus the government?

283

Melting Glaciers and Ice Sheets

■ Over 80% of Earth's glaciers are retreating.

Wilkins Ice Shelf

February 28, 2008

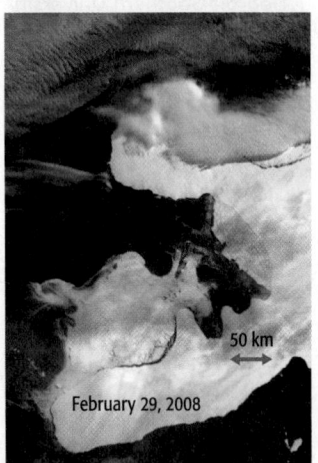

50 km

February 29, 2008

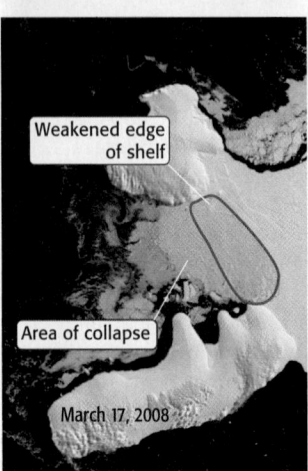

Weakened edge of shelf

Area of collapse

March 17, 2008

▲ **Figure 9.27 Antarctic Ice Shelf Collapses**

Over a 90-day period in 2008, warming caused over 2,000 km² of the Wilkins Ice Shelf to collapse into the Antarctic Ocean. In the past decade, an area twice the size of Rhode Island has been lost from this ice shelf.

Nearly 75% of Earth's fresh water is contained in various forms of ice. This includes snow cover, which varies seasonally, mountain glaciers, and the massive glaciers or ice sheets that cover most of the landmass of Greenland and Antarctica. It also includes Arctic sea ice and floating ice shelves that surround Antarctica. The total amount of this frozen water depends on the rate of snowfall and ice formation relative to the rate at which it melts. In most places today, the rate of melting exceeds the rate of ice formation. Snowpack in many mountainous regions has been shrinking, and over 80% of Earth's mountain glaciers are retreating. Average snow cover has been steadily decreasing, especially in spring and summer months in the Northern Hemisphere (Figure 9.25).

Since satellite observations began in 1978, Arctic sea ice has shrunk by about 3% per decade (Figure 9.26). As you would expect, rates of shrinkage are higher in the summer than in the dark winter, when ice rebuilds. Prior to 1985, summer minimum ice extent was nearly always greater than 8 million km². This effectively prevented ship travel across the Arctic between the Atlantic and Pacific Oceans, regardless of season. Since 2003, however, minimum sea extent has rarely exceeded 6 million km², and in early September 2012 it extended over less than 4 million km² (see Figure 9.26). During the last decade, ships have been able to easily traverse this northern passage during the late summer. Given

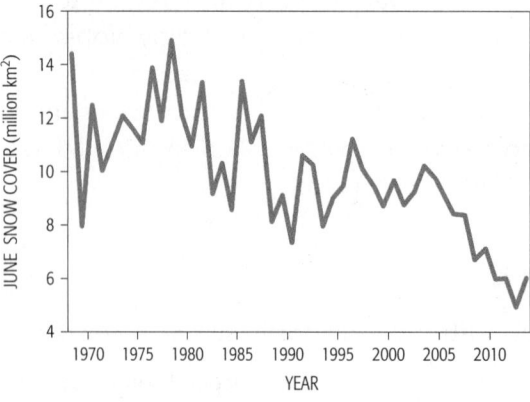

▲ **Figure 9.25 Diminishing Snow Cover**

June snow cover is indicative of snow extent in the Northern Hemisphere spring and summer. Although there is year-to-year variation, the downward trend is clear. Average snow extent for 1967–1989 was 11.34 million km², but for 1990–2013 it was only 8.57 km².

Source: The Rutgers University Snow Lab. http://climate.rutgers.edu/snowcover/table_area.php?ui_set=2.

the current rate of shrinkage, scientists are forecasting that the Arctic Ocean could be completely ice free during summer in 20 years or less.

For decades, scientists have noted that Greenland's ice sheet is shrinking. Recent data indicate that melting is far more rapid than previously thought because of the prevalence of streams and rivers beneath the ice. The glacial retreat has exposed new land along Greenland's coast. This exposure, along with longer growing seasons, has allowed farmers to plant crops that have not been grown in Greenland since the Medieval Viking period.

Greenland's small community of farmers may be benefiting from the effects of a warmer climate, but globally the loss of snow and ice is a double whammy. First, snow and ice reflect about 35% of incoming solar radiation, compared to just 10% from forests or ocean waters. A decrease in the amount of land and water covered by snow and ice accelerates global warming, since solar radiation is absorbed instead of being reflected back into space. Second, water melting from ice sheets and glaciers eventually makes its way to the ocean, where it adds to the rising sea level. In the Southern Hemisphere, the area of ice shelves extending onto the ocean from Antarctica's ice sheet has decreased dramatically (Figure 9.27). The meltwater produced has the potential to alter ocean currents, much as it did during the Big Freeze 12,000 years ago.

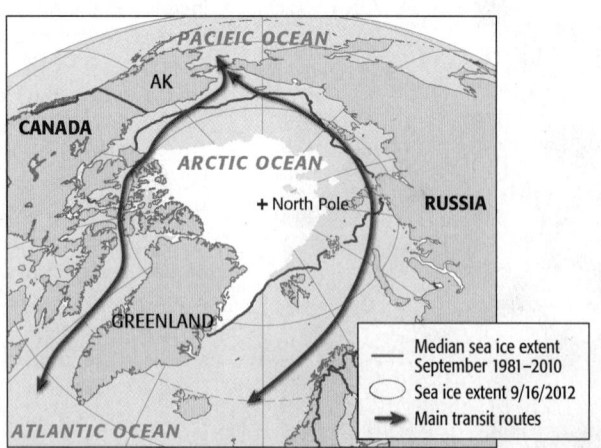

AK

CANADA

ARCTIC OCEAN

+ North Pole   RUSSIA

GREENLAND

— Median sea ice extent September 1981–2010
◯ Sea ice extent 9/16/2012
→ Main transit routes

ATLANTIC OCEAN

▲ **Figure 9.26 A Northern Passage at Last?**

Over the past several decades, the extent of summer sea ice across the Arctic Ocean has been steadily shrinking. The red line on this map indicates the average extent of summer sea ice between 1981 and 2010. The white area is the extent of sea ice in 2012; this was the lowest in recorded history. The purple arrows indicate large expanses of open water connecting the Atlantic and Pacific Oceans.

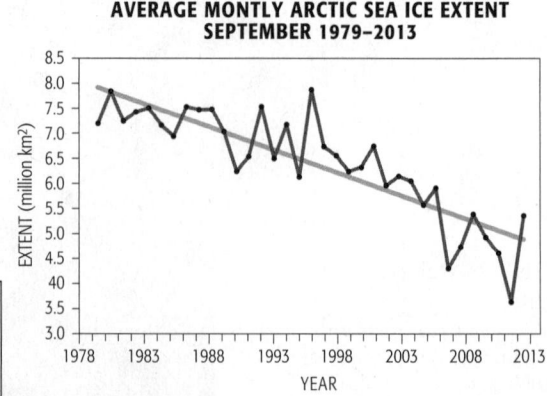

**AVERAGE MONTLY ARCTIC SEA ICE EXTENT
SEPTEMBER 1979–2013**

Rising Sea Level

■ Thermal expansion and meltwater from glaciers are increasing ocean volume.

Over the past 50 years, sea level has risen about 9 cm (3.5 in.); that is an average rate of 1.8 mm (0.07 in.) per year. More recent satellite measurements suggest that the rate may now be as high as 3 mm (0.12 in.) per year (Figure 9.28). Two factors are responsible for this change—the expansion of water as it warms and melting ice.

The ocean absorbs much of Earth's heat. Over the past century, surface ocean waters have warmed by as much as 0.5 °C (0.9 °F). This increase in temperature causes liquid water to expand slightly. In small volumes of water, the expansion is barely detectable. In the huge volume of the ocean, its effect is significant. Scientists estimate that thermal expansion accounts for as much as half of the observed rise in sea level.

The net influx of water from the continents into the oceans, as mountain glaciers and ice sheets melt, also contributes to the rising sea level. By comparison, the Arctic ice sheet floats on the ocean surface. Its melting has little effect on sea level, for the same reason that melting ice cubes do not cause your glass of iced tea to overflow.

The effects of rising sea level are already being felt in many coastal areas. Shorelines and barrier islands are eroding. Higher water levels and increasing salinity in estuaries and coastal wetlands are changing the distribution of some plant communities. Projections vary, but sea level may rise between 0.5 and 1.0 m (1.6–3.3 ft) by 2100. This rise could dramatically alter coastlines and cause many low-lying ocean islands to disappear (Figure 9.29).

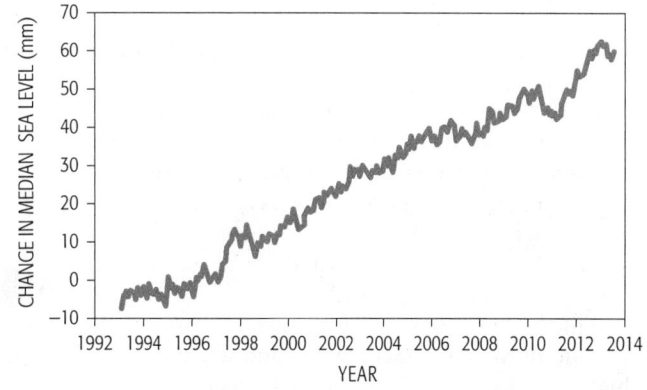

▲ Figure 9.28 **Sea Level Rise**
Precise satellite measurements indicate that sea level is currently rising at the rate of about 32 mm (1.2 in.) every decade. Scientists fear that additional warming may cause this rate to accelerate.

Source: University of Colorado Sea Level Lab. http://sealevel.colorado.edu.

▼ Figure 9.29 **Islands at Risk**
The nation of Maldives in the Indian Ocean comprises a number of islands that rise barely 1 m above sea level. The inset photo is the nation's capitol, Male. Close to 400,000 people live on these islands. Within the next century, these islands could be completely inundated by the rising ocean.

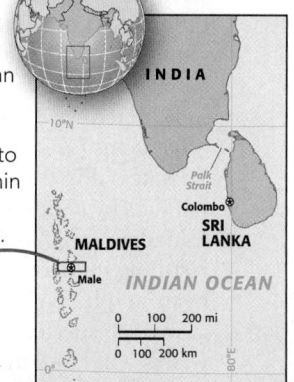

QUESTIONS 9.4

1. Describe specific climate changes associated with global warming in three different regions in the world.

2. Why does the loss of ice and snow cover accelerate global warming?

3. Describe two factors contributing to the global rise in sea level.

(MES) For additional review, go to
MasteringEnvironmentalScience

Changing Populations and Ecosystems

■ Because climate is a key component of habitats, global warming is producing significant changes in the populations of many species and in many ecosystems.

Global warming is affecting ecosystems as plants and animals respond to altered habitats. Satellite images of Canada, New England, and the Midwest since 1980 reveal that warmer spring temperatures are causing the leaves of plants to open earlier. On the ground, scientists have observed that many species of birds are nesting and laying eggs earlier than in the past. Nevertheless, there is still a risk that spring frosts will damage young leaves and kill nestling birds.

Milder temperatures are causing migratory birds to move to the Arctic earlier and stay in high latitudes later into the fall. This delayed departure exposes some species to fierce autumn storms as they migrate southward. The migration of some species no longer coincides with the availability of food, which also varies seasonally (Figure 9.30).

A synthesis of more than 100 scientific studies covering 1,400 different species found that global warming is affecting the behavior of plants and animals in both terrestrial and aquatic ecosystems (see Module 8.2). Threatened in their coastal habitats by shrinking sea ice, Canadian polar bears are hunting farther inland. This is resulting in more frequent negative interactions between bears and human communities. In rivers and streams, many fish species are migrating earlier because of warmer waters. On land, hibernating mammals are awakening earlier than in the past as they respond to temperature cues that winter is over and spring has arrived. One study

▲ Figure 9.30 **Bad Timing**
Warmer spring temperatures across Europe cause insects to mature weeks earlier than in the past. As a result, the migration of pied flycatchers from Africa no longer matches the peak abundance of insects, so the flycatchers have insufficient food for their nestlings.

shows that marmots in the mountains of Colorado are waking up three weeks earlier than they did in the 1970s. The consequences of these seasonal shifts are not clear in every case, but they certainly do pose problems when a behavior gets out of synch with the availability of food or other resources necessary for survival (Figure 9.31).

A changing climate may allow organisms that carry human diseases to expand their ranges, thus having a direct effect on human well-being. In some places, warmer and moister conditions have encouraged the spread of disease vectors, organisms that carry pathogens (see Module 18.5). For example, an increase in the range of mosquitoes has resulted in a marked increase in malaria and dengue fever in regions where these diseases have not been seen for over a century. It may also be important in the recent spread of the mosquito-borne West Nile virus in the United States.

While some plants and animals may be adjusting successfully to global warming, others are not. Many ocean species seem to be struggling. For example, the management plan to help North Sea codfish recover from overfishing is failing. Apparently, the waters off Europe's coast are too warm for the plankton necessary to support juvenile cod. Off the coast of California, populations of many seabirds are approaching collapse. It appears that warmer waters have altered populations of plankton there as well. This, in turn, has diminished the supply of small fish upon which the seabirds feed.

Ocean warming appears to be a major factor in the deterioration of many coral reefs. The warmer conditions contribute to a phenomenon called "bleaching," which is caused by a breakdown in the relationship between coral animals and their symbiotic algae. Although it appears not to be a factor in coral bleaching, many marine species are also being impacted by increasing acidity of ocean waters, a direct consequence of rising levels of CO_2 (see Module 11.5).

All forecasts suggest accelerated warming in the decades ahead. Continued loss of snow and ice, rising sea levels, and changes to patterns of rainfall seem inevitable, as do the effects of these changes on plant and animal communities. Yet, just as human actions have caused global warming, there are many human actions that can slow or even reverse these trends.

◄ Figure 9.31 **Bad Bear!**
In Yosemite National Park, black bears now emerge from hibernation earlier than in the past and before their natural food sources are available. As a consequence, they are more prone to raid picnic baskets and garbage cans.

9.5 Forecasting Global Warming

BIG IDEA Some say that the activities associated with global warming have resulted in the largest experiment humans have ever undertaken. However, unlike usual scientific experiments, this experiment has only one treatment, no replication, and no control. Given the scale and complexity of climate change, how can we forecast future trends? More important, how can we predict the effects of human actions on the nature of climate change? Climate scientists use computer simulations to forecast climate change. Depending upon the assumptions made about human actions that determine greenhouse gas emissions, these computer models forecast very different futures. By taking action to reduce greenhouse gas emissions, we can avoid many adverse changes in Earth's resources and ecosystems.

Computer Simulation of Global Warming

■ General circulation models use basic physical principles to forecast future climate change.

Climate scientists use computer programs called **general circulation models (GCMs)** to forecast climate change. GCMs use mathematical equations to simulate the physical processes that determine Earth's energy budget. These processes include the absorption and reflection of sunlight by the atmosphere, the heating of the land and ocean, infrared radiation given off by the land and ocean, and the circulation of heat from the equator to the poles (see Module 3.7). Where it is relevant, GCMs also include ocean circulation.

These models divide Earth's surface and the atmosphere above it into a grid of cells. GCM programs use specific equations to represent the transfer of gases and energy among individual cells. Starting conditions for parameters such as temperature, wind speed, and humidity must be determined for each cell prior to running the model. The effects of factors such as changes in greenhouse gas emissions can be varied from one model run to the next. Forecasts from these different runs can then be compared to evaluate the importance of the different factors.

How can we determine the accuracy of the forecasts produced by GCMs? First, it is important that the equations that drive these models be based on our best understanding of the physical processes they portray. For this, experimental observations are essential. The accuracy of forecasts can also be checked by "backcasting." Using past meteorological records, GCMs are evaluated by determining how well they predict past changes in weather and climate.

GCM forecasts are most accurate for large areas, such as entire continents or oceans, or the entire globe. As the area being considered gets smaller, accuracy diminishes. The forecasts of individual models differ somewhat for regions such as the southwestern United States or southern Africa. They agree even less with regard to forecasts for smaller areas, such as Botswana or the state of Arizona.

▼ **Figure 9.32 GCM Scenarios**

The graph below compares GCM predictions of future temperatures based on *Today's World* (yellow), *Sustainable World* (blue), and *Business as Usual* (red) scenarios. The black line represents temperature change up to the year 2000, and the shading indicates the uncertainty in the predictions. Warming is presumed to have begun in 1800, shortly after the beginning of the Industrial Revolution.

Source: IPCC. 2013. Climate change 2013: The physical basis: summary for policy makers. http://www.ipcc.ch/report/ar5/wg1/#.Uq8GqLTWuJI.

Forecasting Scenarios

■ Action to reduce greenhouse emissions will result in lower future temperatures compared to no action.

Scientists vary GCM starting conditions and equations to simulate different scenarios. These scenarios differ in their assumptions about such things as economic growth, fossil fuel consumption, and government policies that affect energy use. Different scenarios of human actions result in different greenhouse gas emissions. This triggers different climate interactions, producing different climate change forecasts. Three commonly simulated scenarios are *Today's World, Business as Usual,* and *Sustainable World* (**Figure 9.32**).

Today's World scenarios assume that anthropogenic greenhouse emissions immediately diminish so that CO_2e in the atmosphere stays at current levels. This is not a realistic scenario, but it does provide an important

point of reference. *Business as Usual* scenarios assume that individuals and governments will take no actions to reduce emissions and that annual CO_2e emissions will continue to increase. If this were to occur, by 2060 atmospheric CO_2 concentrations would be nearly 900 ppm, or more than three times their pre-industrial levels. *Sustainable World* scenarios assume that people and their governments will take significant actions to slow the increase in CO_2e emissions and return them to current rates within the next 50 years.

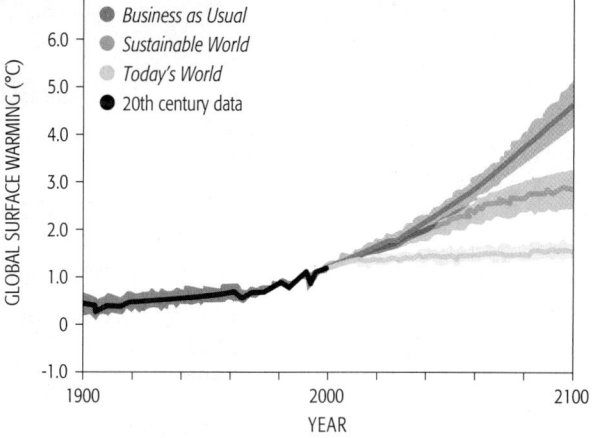

Numerous intermediate scenarios have also been modeled. Scenarios that assume little or no action to reduce greenhouse gas emissions forecast much higher future temperatures than those in which aggressive action is assumed. But even if we were to stop all emissions immediately, some warming would still occur. Climatologists call this **committed warming**, and it is about 0.1 °C per decade. *Sustainable World* scenarios predict that Earth's temperature will increase 1.8 °C (3.2 °F) by 2100. *Business as Usual* scenarios predict a temperature increase of 4.0 °C (7.2 °F) over that time.

There are, of course, uncertainties in these estimates. For example, *Sustainable World* forecasts have a 95% confidence range of 1.1–2.9 °C (2.0–5.2 °F). This means that climatologists feel there is a 95% probability (a chance of 19 in 20) that the actual future temperature will fall in this range. In *Business as Usual* forecasts, the margin of error increases considerably to 2.4–6.4 °C (4.3–11.5 °F). The high uncertainties associated with *Business as Usual*

forecasts concern climatologists because they may indicate that there will be changes in Earth's climate that are beyond our current understanding.

Changes in individual GCM cells provide a means to forecast climate change in different regions (Figure 9.33). Temperatures in 2100 are forecast to be much higher in polar regions and in the centers of continents. This geographic variation in warming is much greater in *Business as Usual* forecasts.

GCM forecasts show that rising temperatures will affect geographical patterns of rainfall (Figure 9.34). Forecasts for some regions differ, but most GCMs agree on two important points. First, rainfall will generally increase in regions at higher latitudes, such as Canada and northern Eurasia. Second, rainfall will decrease markedly in many subtropical and temperate regions that are already arid. Such places include the southwestern United States, large parts of Africa, and the Mediterranean region.

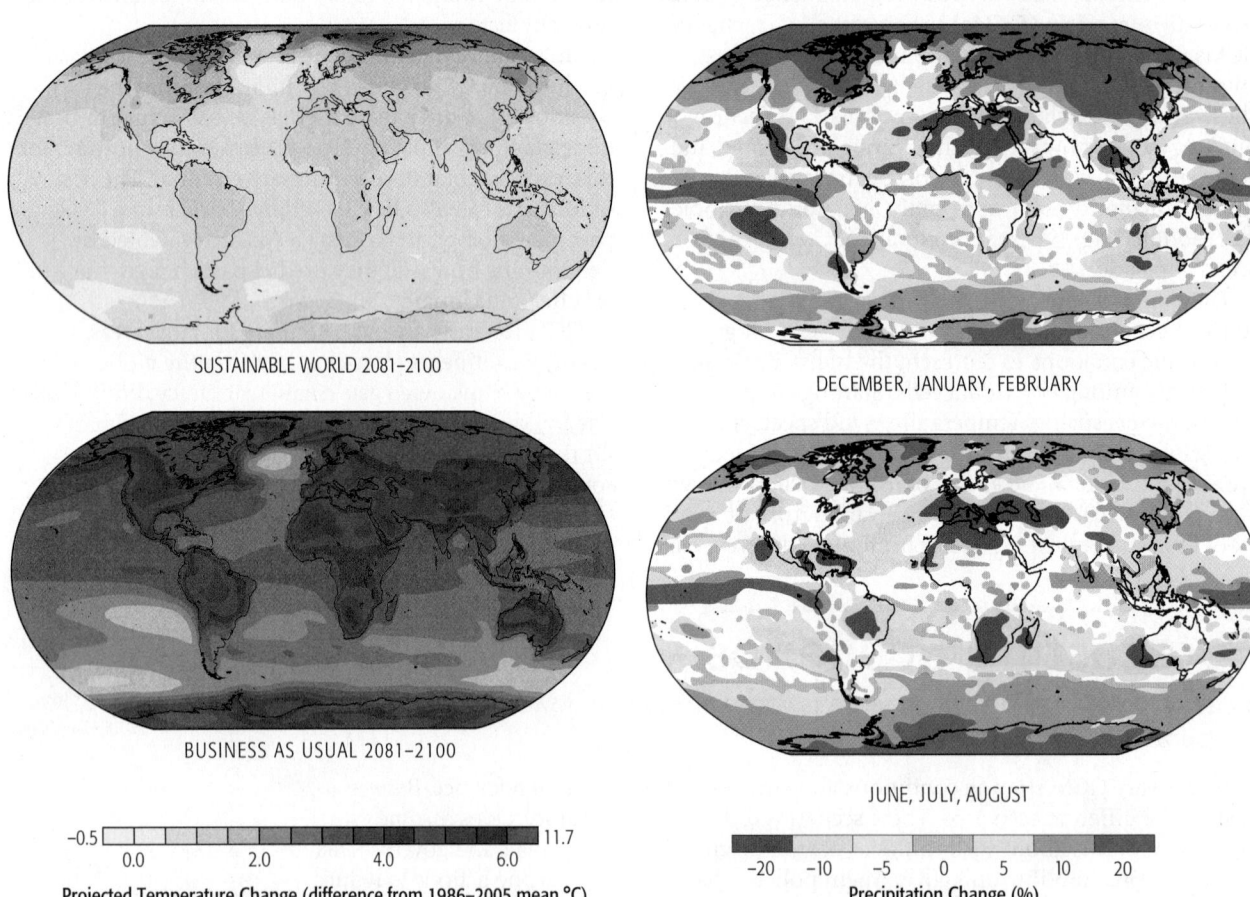

SUSTAINABLE WORLD 2081–2100

DECEMBER, JANUARY, FEBRUARY

BUSINESS AS USUAL 2081–2100

JUNE, JULY, AUGUST

-0.5					11.7
	0.0	2.0	4.0	6.0	

Projected Temperature Change (difference from 1986–2005 mean,°C)

	-20	-10	-5	0	5	10	20

Precipitation Change (%)

▲ Figure 9.33 **Regional Variation**

GCMs for the *Sustainable World* and *Business as Usual* scenarios predict that the extent of warming will vary at different locations around the world. Note that heating is greatest over continents and in polar regions, and it is much greater for the *Business as Usual* scenario at all locations.

Source: IPCC. 2014a. *Climate change 2014: impacts, adaptation, and vulnerability. Summary for policymakers.* http://ipcc-wg2.gov/AR5/images/uploads/IPCC_WG2AR5_SPM_Approved.pdf (last reviewed 7/21/2014)

▲ Figure 9.34 **Changing Rainfall**

These maps represent average predictions from several GCM models for changes in rainfall for December–February and June–August over the next 50 years. They are based on future temperature change that is intermediate between the *Sustainable World* and *Business as Usual* scenarios. The models predict that rainfall will increase in some regions and decrease in others.

Source: IPCC. 2007. *Climate change 2007: Synthesis report.* New York: Cambridge University Press.

Forecast Consequences

■ The impacts on natural resources, ecosystems, and human health are high with the *Business as Usual* scenario.

The United Nations Intergovernmental Panel on Climate Change (IPCC) compiled a list of how different levels of temperature change are likely to affect key resources and human health (**Figure 9.35**). Each of these estimated impacts is based on numerous published scientific studies. The list is daunting. Clearly, global warming will influence a great many resources and processes that have direct consequences for people. However, the impact of the temperature increase forecast by the *Sustainable*

World scenario (1.8 °C) is much less than that for *Business as Usual* (4 °C).

Like it or not, we are committed to some change. In some cases, the response to that change may be costly. However, we can take a variety of actions that will enable us and many of the ecosystems we depend on to adapt to committed warming. But taking action is also needed now to reduce greenhouse gas emissions to prevent us from moving into a very uncertain and unsustainable future.

QUESTIONS 9.5

1. Describe how scientists use general circulation models to forecast the consequences of different patterns of change in carbon emissions.

2. Differentiate between *Today's World*, *Business as Usual*, and *Sustainable World* scenarios of carbon emission.

3. What is meant by committed warming?

(MES) For additional review, go to **MasteringEnvironmentalScience**

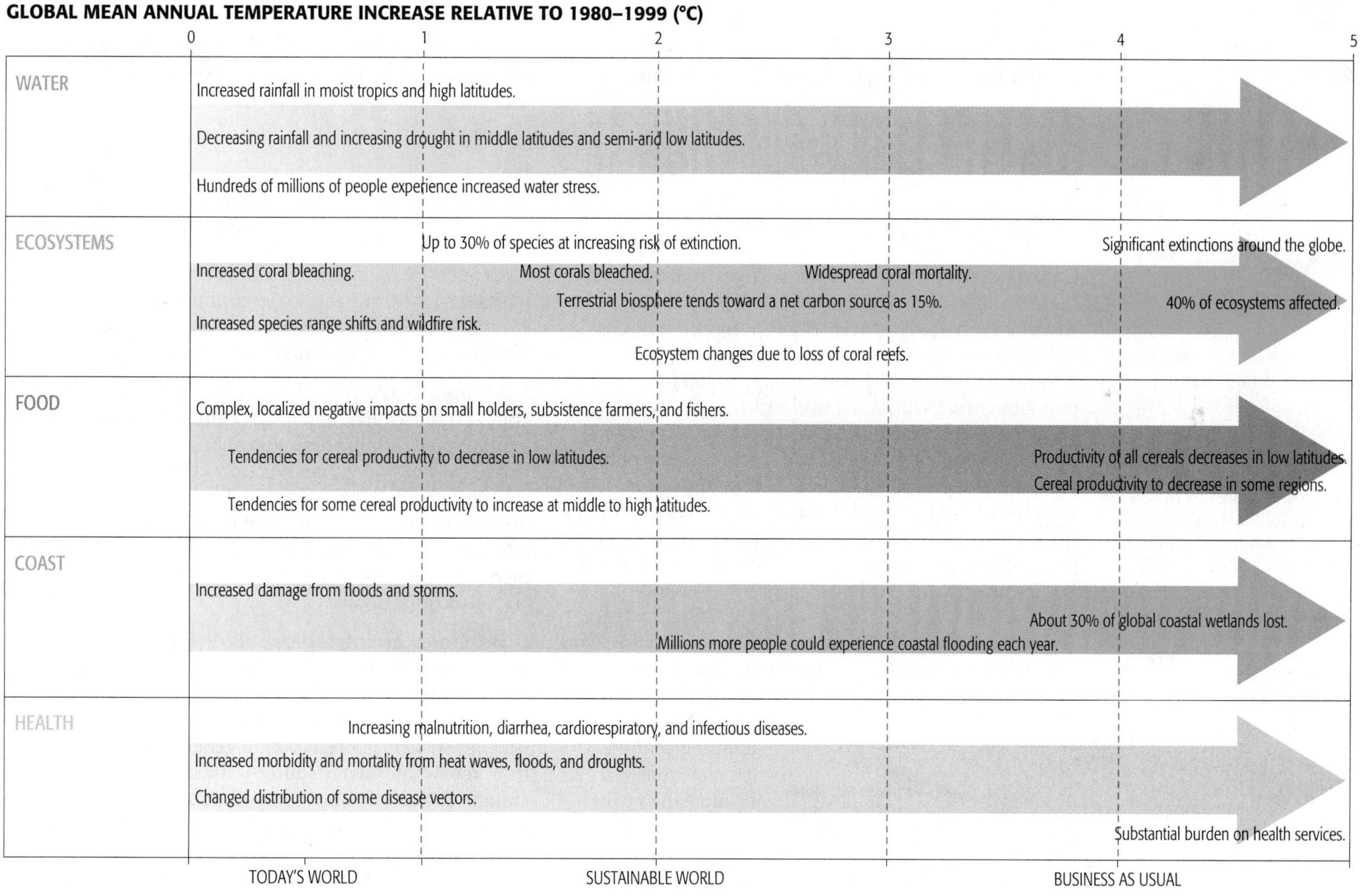

GLOBAL MEAN ANNUAL TEMPERATURE INCREASE RELATIVE TO 1980–1999 (°C)

	0	1	2	3	4	5

WATER
- Increased rainfall in moist tropics and high latitudes.
- Decreasing rainfall and increasing drought in middle latitudes and semi-arid low latitudes.
- Hundreds of millions of people experience increased water stress.

ECOSYSTEMS
- Up to 30% of species at increasing risk of extinction.
- Significant extinctions around the globe.
- Increased coral bleaching.
- Most corals bleached.
- Widespread coral mortality.
- Terrestrial biosphere tends toward a net carbon source as 15%.
- 40% of ecosystems affected.
- Increased species range shifts and wildfire risk.
- Ecosystem changes due to loss of coral reefs.

FOOD
- Complex, localized negative impacts on small holders, subsistence farmers, and fishers.
- Tendencies for cereal productivity to decrease in low latitudes.
- Productivity of all cereals decreases in low latitudes.
- Cereal productivity to decrease in some regions.
- Tendencies for some cereal productivity to increase at middle to high latitudes.

COAST
- Increased damage from floods and storms.
- About 30% of global coastal wetlands lost.
- Millions more people could experience coastal flooding each year.

HEALTH
- Increasing malnutrition, diarrhea, cardiorespiratory, and infectious diseases.
- Increased morbidity and mortality from heat waves, floods, and droughts.
- Changed distribution of some disease vectors.
- Substantial burden on health services.

TODAY'S WORLD   SUSTAINABLE WORLD   BUSINESS AS USUAL

▲ **Figure 9.35 Consequences of Global Warming**
This chart shows IPCC estimates of changes that are likely to be associated with different levels of global warming. As temperatures increase, the severity of the changes and their implications for human well-being also increase.

Source: IPCC. 2007. *Climate change 2007: Synthesis report.* New York: Cambridge University Press.

9.6 Mitigating Global Warming

BIG IDEA Emissions of anthropogenic greenhouse gases are increasing each year. Environmental scientists refer to actions that directly reduce an environmental threat such as CO_2e emissions as *mitigation*. There is general agreement that actions that reduce emissions of CO_2 are necessary to mitigate global warming. If, as in the *Business as Usual* scenario, no actions are taken, annual emissions will double by 2065. Concentrations of CO_2e in the atmosphere will triple. Mitigation to hold annual emissions at their current level, the *Sustainable World* scenario, means that concentrations of CO_2e in the atmosphere will be at about twice their pre-industrial amount in 50 years. This will result in additional warming, but within sustainable levels. Many potential actions—increased energy efficiency and conservation, changes in the use of fossil fuels, and increased use of renewable energy resources, nuclear energy, and biostorage—can help meet the challenge of moving from a *Business as Usual* world to a *Sustainable World*. By itself, no single action or technology is sufficient to meet the goal; a portfolio of actions is required (**Figure 9.36**). Beyond this 50-year time frame, it is hoped that annual emissions can actually be reduced well below current rates. Some scientists have suggested that global warming might also be mitigated using technologies to either diminish solar radiation inputs or directly remove CO_2 from the atmosphere. Many fear, however, that such climate engineering would have unintended and potentially undesirable consequences.

▲ **Figure 9.36 Field of Dreams**
This array of solar panels on the Taos campus of the University of New Mexico is one example of the kinds of actions that can significantly mitigate global warming.

Defining the Challenge

■ *Business as Usual* and *Sustainable World* scenarios imply big differences in future annual greenhouse gas emissions.

Global warming is a complex challenge. Its most direct cause is the ever-increasing amount of greenhouse gases in Earth's atmosphere. The concentration of CO_2 in Earth's atmosphere is currently about 400 ppm. Prior to the Industrial Revolution it was 285 ppm. Each year, human activities such as electricity generation, transportation, cement production, deforestation, and agriculture put about 32 Pg of CO_2 into the atmosphere. It is especially troubling that each year anthropogenic emissions of CO_2 and other greenhouse gases increase by several percent. However, there is a great difference in the greenhouse gas emissions associated with the *Business as Usual* scenario and the *Sustainable World* scenario.

Business as Usual. The *Business as Usual* scenario occurs if we take no actions to curb greenhouse gas emissions.

At the present rate of increase, in 50 years annual greenhouse gas emissions will be more than twice their current values; CO_2 emissions will be about 64 Pg per year. At this rate of increase, the concentration of greenhouse gases in Earth's atmosphere would eventually triple. Climate scientists are in complete agreement that these levels of CO_2e would have catastrophic

consequences. Average annual global temperatures would increase by 8–10 °C, and sea level would rise more than 1 m (3.3 ft). Such changes would likely be devastating for humans and the ecosystems on which they depend.

Sustainable World. If over the next 50 years we are able to stabilize greenhouse gas emissions at their current annual levels, atmospheric greenhouse gas concentrations would be a bit more than twice pre-industrial levels. The concentration of CO_2, for example, would be about 600 ppm. At this concentration, some amount of warming and climate change will be inevitable, but scientists predict that sustainable adaptation to this amount of change is quite possible.

A portfolio of strategies. Ecologist Stephen Pacala and engineer Robert Socolow argue that it is possible to take actions that will realize the *Sustainable World* scenario and, thereby, a sustainable outcome. By itself, no single strategy, such as reforestation or generating electricity from renewable energy sources, can reduce emissions enough to meet the goal. Instead, Pacala and Socolow suggest that a *portfolio*, or combination of different activities, will produce the needed reductions.

Pacala and Socolow note that the difference between current annual emissions and those projected for 2065 in *Business as Usual* is 32 Pg. They divide this total amount of necessary reduction into seven 4.6-Pg "stabilization wedges" (**Figure 9.37**). They group stabilization wedges into five categories: efficiency and conservation, fossil fuel use, renewable energy, nuclear energy, and biostorage.

▼ **Figure 9.37 Getting to a Sustainable Future**
In this graph, the difference in accumulated emissions between *Business as Usual* and *Sustainable World* scenarios is indicated by the yellow stabilization triangle, which represents an increment of 4.6 Pg CO_2/year by 2065.
Source: Pacala, S., and R. Socolow. 2004. Stabilization wedges: Solving the climate problem for the next 50 years with current technologies. *Science* 305: 968–972.

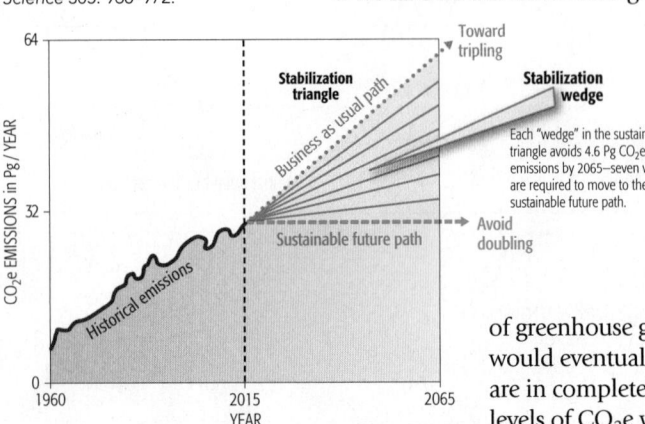

Efficiency and Conservation

■ Changes in technology and behavior can result in more efficient energy use.

By simply being more efficient in our use of energy, emissions could be reduced by as much as 16 Pg (more than 3 wedges). Pacala and Socolow propose action in four areas: transport efficiency, transport conservation, building efficiency, and efficiency in electricity production. Each of these actions would reduce the burning of fossil fuels and has the potential to form a stabilization wedge.

Transport efficiency. Today, there are about 600 million automobiles in the world. Given population growth and current trends in economic development, that number could top 2 billion by 2065. A single car traveling 10,000 miles using 30 miles per gallon of gas puts 3.6 tons of CO_2e into the atmosphere each year. If fuel efficiency were doubled to 60 miles per gallon, the emissions from each auto would be halved. New hybrid and diesel technologies, as well as lighter construction materials, have already increased mileage significantly. There are also opportunities to improve fuel efficiency in trucks and planes.

Transport conservation. In terms of the energy used per person, trains, buses, and other forms of public transportation are far more efficient than automobiles. Investment in mass transit and urban planning to make public transportation more accessible could cut the number of miles traveled by automobiles by half (Figure 9.38).

Building efficiency. Residential and commercial buildings account for a significant amount of CO_2e emissions. Buildings use energy for heating and air conditioning, heating water, lighting, and electrical appliances. Improved design and use of materials could reduce energy use and emissions considerably. A 25% improvement in efficiency in all new and existing buildings over the next 50 years would reduce the increase in emissions by a wedge.

Efficiency in electricity production. Coal-burning power plants produce about 65% of the world's electricity and account for over 25% of its CO_2e emissions. Increased efficiency in the conversion of the energy in coal into electricity could equal a wedge (see Module 14.3). Increased efficiency could come from technologies such as high efficiency turbines. More even distribution of the demand for electricity could also save energy.

▼ Figure 9.38 **More Efficient Transportation**
Together, more efficient mass transit systems and improved gas mileage could reduce emissions by as much as two stabilization wedges, or at least 9 Pg, by 2060.

Fossil Fuel Use

■ Changes in the use of different kinds of fossil fuels and capture of the CO₂ released from their burning can reduce emissions.

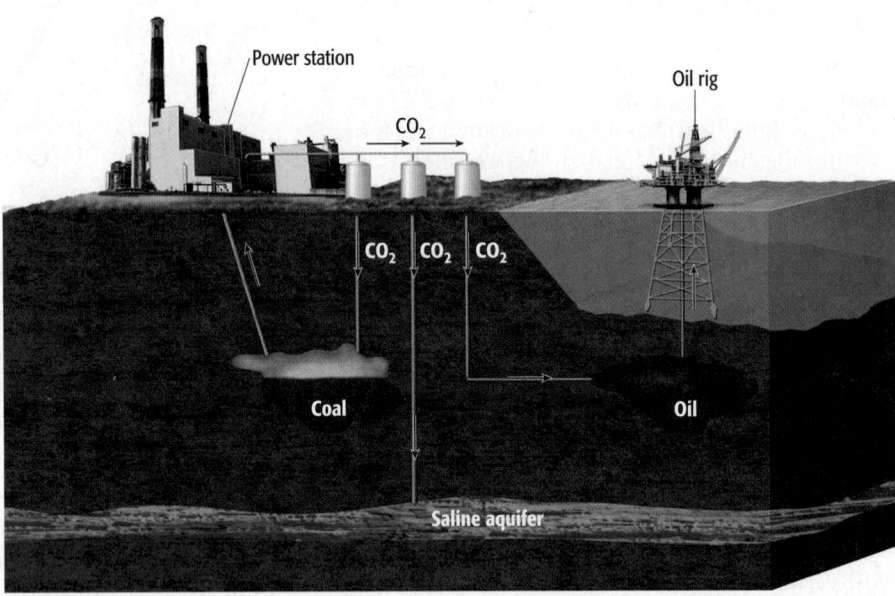

▲ Figure 9.39 **Carbon Capture**
Rather than allowing CO₂ to escape into the atmosphere, CCS systems liquefy it. The liquid CO₂ is then stored underground, either in saline deposits or in the sedimentary layers from which coal and oil have been extracted.

Changes in the kinds of fossil fuels we burn and how we manage the CO₂ that such burning emits have the potential to add several stabilization wedges. For example, power plants fired by natural gas emit about half as much CO₂e as coal-fired plants. This takes into account leakage of methane, a powerful greenhouse gas, associated with natural gas transport and use (see Modules 14.4 and 15.6). Replacing 1,400—about half—of the world's large coal-fired plants by natural gas-fired plants would equal one stabilization wedge. This would also result in a fourfold increase in electricity generation from natural gas.

Much attention is being given to new technologies called **carbon capture and storage (CCS)**. Rather than allowing CO₂ to escape to the atmosphere, CCS technologies capture it and transfer it into a form that can be permanently stored underground (Figure 9.39). Pilot projects are testing a number of storage forms, including super cold liquid CO₂ and baking soda-like salts. Many technical uncertainties about CCS remain (see Module 14.3). For example, scientists disagree about the ability of underground sediments to retain CO₂. If CCS is found to be effective, Pacala and Socolow estimate that its widespread use could equal as many as three stabilization wedges.

Renewable Energy

■ Renewable energy sources produce few CO₂e emissions.

Why burn fossil fuels at all if alternative energy sources will emit much less CO₂e? Renewable sources of energy include wind, sunlight, and biofuels.

Wind generates electricity without emitting any CO₂e. Today, wind energy produces less than 1% of energy worldwide. Increasing our capacity for wind generation by 30-fold will cut future emissions by one wedge (Figure 9.40). Currently, the use of wind energy is growing very rapidly, over 30% per year. If this rate of growth

▶ Figure 9.40 **Wind Energy**
Renewable energy sources such as these wind turbines offset the use of fossil fuels. Steady growth in the use of renewable technologies could account for one or more stabilization wedges.

continues, the goal can be reached (see Module 15.3). Wind generation at this level would, however, require a combined area larger than the state of Texas. Wind energy has also been proposed as a sustainable source of electricity to produce hydrogen that could be used to fuel future cars and trucks.

The ultimate source of most of our energy is the sun. Why not tap it directly? Using solar energy can be as simple as hanging your wet laundry out to dry. Solar energy is widely used to heat water. **Photovoltaic (PV) cells** are devices that directly convert sunlight to electrical current (see Module 15.4). At present, PV cells provide less than 0.1% of global electricity. We would need a 700-fold increase in this amount to offset the increase in emissions by a wedge.

Biofuels are derived directly from plant materials, such as corn or wood. Burning them has no net effect on CO₂e emissions, because photosynthesis in these plants withdraws CO₂ from the atmosphere. Today, the most commonly used biofuel is ethanol, or grain alcohol, which is made from corn and sugar cane (see Module 15.5). To reduce emissions by one wedge, we would need to convert one-sixth of Earth's cropland from food to biofuel production. Genetically modified crops and new technologies that allow more efficient use of crop wastes could reduce that demand.

Nuclear Energy

■ Although it emits little CO_2e, nuclear energy is a controversial alternative.

Today, nuclear energy is used to generate about 18% of the world's electricity. There are some CO_2e emissions associated with its use, but they are a very small fraction of those associated with electric generation from fossil fuels (see Module 14.5). If nuclear generation were tripled and offset an equivalent amount of fossil fuel generation, it would result in a wedge of emissions benefit. This would require building new nuclear power plants 10 times faster than they are being built today (Figure 9.41).

No energy source generates more emotion than nuclear power. That's especially true now, given the ongoing concern with the aftermath of the meltdown of reactors at the Fukushima Daiichi power plant following the tsunami in 2011. The advocates of nuclear power point out that it is a proven technology that can be deployed immediately. Its critics call attention to the high cost of building nuclear plants. They also worry about safety and the risk of the proliferation of nuclear weapons. Nuclear energy may not emit much CO_2, but it generates large amounts of very toxic radioactive waste. The puzzle of how to permanently dispose of this waste remains unsolved (see Module 14.5).

► Figure 9.41 **Nuclear Energy?**
Here the reactor core for a nuclear generating facility is under construction. Although controversial, rapid development of nuclear facilities could diminish CO_2e emissions by a 4.6 Pg wedge by 2065.

Biostorage

■ Atmospheric concentrations of CO_2 can be reduced by storing more carbon in forests and soils.

Today, deforestation is putting CO_2 into the atmosphere and diminishing the amount of carbon stored in forests (see Modules 6.5, 13.3). Unsustainable farming practices cause soil erosion and degradation, which also increase CO_2 emissions and diminish storage capacity.

Biostorage refers to actions that increase the absorption and storage of CO_2 in Earth's ecosystems. Halting deforestation within the next 50 years would provide a 4.6 Pg wedge. We could achieve the same benefit by planting trees in areas where forests are currently absent. However, this would require that new forests be planted over an area about the size of the lower 48 United States.

CO_2 emissions from soil could be greatly reduced by the use of farming practices that encourage carbon storage. Such practices include reducing soil tilling, planting cover crops to prevent erosion, and applying fertilizer more efficiently. If such practices were applied to all of the world's croplands, the reduction in emissions might be equivalent to one wedge.

Self-help books advise that, when faced with a large and complex challenge, we should divide it into smaller, more manageable chunks. Pacala and Socolow propose a strategy for doing just that. If we adopt a portfolio of different approaches, we already have the technology to hold greenhouse emissions at their current level over the next 50 years. Beyond 50 years, we must reduce greenhouse gas emissions well below their current amounts. New technologies that are in the works provide hope that that can be done.

New FRONTIERS

Climate Engineering

Climate engineering advocates believe we can mitigate global warming by using deliberate and large-scale interventions in Earth's climate system. One proposal is to simulate the cooling effects of volcanic eruptions by dispersing fine particles of sulfur or aluminum oxide in the stratosphere. Ocean fertilization advocates believe that photosynthesis and the uptake of CO_2 could be greatly increased by adding iron over large areas of the ocean. As phytoplankton and the organisms that eat them die, their remains settle to the ocean bottom. CO_2 taken up in this way would be permanently removed from the atmosphere.

Most scientists are concerned that the costs and risks of climate engineering are not well understood. Increasing atmospheric reflectivity could diminish warming, but it could disrupt Earth's climate system in other ways. Given the complexity of ocean ecosystems, unintended and undesirable consequences are likely from ocean fertilization. What ethical and political issues does climate engineering raise? Would you support its implementation?

QUESTIONS 9.6

1. Environmental scientists talk about mitigation strategies that can produce a stabilization wedge. What are they referring to?

2. Describe how several mitigation strategies might be employed that would result in a "sustainable future" outcome.

(MES) For additional review, go to **MasteringEnvironmentalScience**

Becoming Carbon Neutral

What actions can you take personally to halt global warming?

"Think global, act local." This oft-quoted phrase has become something of a cliché among environmentalists. But this advice is critical if we hope to slow and eventually halt global warming. After all, global warming is, well, global, and it is easy to overlook the fact that fundamentally, it is in large part the cumulative consequence of our individual actions. You can be part of the solution to this global challenge by striving to become carbon neutral.

The first step toward becoming carbon neutral requires knowing your "carbon footprint"—the amount of CO_2 or other greenhouse gases your actions produce. On the Internet you can find numerous carbon counters that will do the math for you. The EPA calculator found on the EPA site is a reliable calculator tailored to North American populations (search EPA Household Carbon Footprint). On average, the actions of each U.S. citizen generate about 18 tons of CO_2e each year. Generally, a college student who lives in the dorm and rides a bike is responsible for generating about 10–12 tons of carbon per year. That contrasts with the 20–25 tons generated by a person living in a three-bedroom house, who drives a car 15,000 miles and takes several short airplane trips each year. A more luxurious lifestyle generates even more carbon emissions.

What are the components of your carbon footprint? Energy use is one of the most important. It is influenced by where you live, but it also depends on your personal actions. Have you taken steps to heat or cool your dorm room efficiently? Do you use energy efficient lighting and appliances? What about those long, hot showers? Transportation is a major source of greenhouse gases. How often do you walk or ride a bike rather than driving to a place? Do you take advantage of public transportation? Do you take frequent or long airplane trips? Many people are surprised to find that air travel is the largest single contribution to their annual carbon footprint. One round-trip flight from New York to Europe generates over 2 tons of CO_2e per person. What you eat matters. The average American's diet contributes about 4 tons of CO_2e to their footprint. If you eat meat at most meals, your footprint is probably 2 tons higher; if you are a vegetarian, your footprint is about 2.5 tons lower (see Module 12.9 for a detailed discussion of the ecological impacts of diet). Finally, how much waste do you generate, and how do you dispose of it? On average, waste disposal contributes about 1.2 tons of CO_2e to our carbon footprint. Careful attention to waste reduction and recycling can cut that number by half.

It is, however, difficult for individuals or organizations to take direct actions to reduce their carbon footprint to zero. The purchase of carbon offsets provides a way to complete the job. A carbon offset is a reduction in emissions made in order to compensate for or offset an emission made elsewhere. Many nonprofit and for-profit organizations create carbon offsets by planting trees or developing sources of renewable energy. These offsets are then sold to individuals and organizations wishing to reduce their carbon footprint. Carbon offsets are not the total solution to global warming, nor are they a replacement for actions that each of us ought to take to reduce our carbon footprints. But they do provide a voluntary way for each of us to compensate for unavoidable activities that influence greenhouse gas emissions (**Figure 9.42**).

Finally, each of us can make a difference by "speaking out" on this issue. Write to legislators and decision makers. Get involved in programs aimed at diminishing your university's carbon footprint. Support organizations that are committed to a carbon neutral future.

▶ Figure 9.42 **Becoming Carbon Neutral**
Many musical groups and sports teams offset the carbon emissions their activities generate. Buy a ticket to a Rolling Stones concert or, perhaps, go to the Super Bowl, and you could also be buying offsets to CO_2e emissions.

9.7 Adapting to Global Warming

BIG IDEA Even if atmospheric greenhouse gas concentrations were to remain at current levels, additional warming is unavoidable. The associated changes in climate and environment, which will vary from region to region, include drought, increased rainfall and flooding, warming and heat waves, changing patterns of storms, and rising sea level. These changes will require adaptation in many human activities, including agriculture, water management, coastal management, industry, and public health.

Committed Warming, Inevitable Change

■ Adaptation to global warming should be an international priority.

If we were able to halt all greenhouse emissions immediately, continued warming of about 0.6 °C (1.1 °F) is likely to occur over the next 50 years. Given that emissions are likely continue to increase, we can expect warming well above this amount. This warming will change factors that directly impact human well-being.

Drier conditions and drought. Drier conditions and extended droughts are most likely to occur in regions where water is already in short supply and is often a source of political conflict. These regions include the southwestern United States, the Mediterranean region, the Middle East, and much of sub-Saharan Africa.

Agriculture in arid regions often depends on irrigation. Responses to drier conditions may include replacing crops with varieties that are drought resistant, mulching to limit evaporation from soil, and using more efficient watering technologies, such as drip irrigation (Figure 9.43). Adaptive measures that are already being implemented include more effective water pricing and technologies to convert salt water to fresh water (see Module 11.6).

Urban areas will also be affected. Many cities have already begun to meter water and restrict water use. Many are also working to minimize leakage and evaporation from water supply systems. Some cities are recharging aquifers with stormwater. In the United States, many states are now requiring cities to implement plans for sharing water among jurisdictions. Drought conditions often accentuate water pollution, and steps will need to be taken to ensure the quality of water supplies. Public education on water conservation will be critical in all drought-prone areas. These strategies are discussed in more detail in Chapter 11.

Increased rainfall and flooding. Other regions will experience increased rainfall and risk of flooding. Adaptations in agriculture may include changes in crop varieties and planting methods and the use of technologies to aerate soil and prevent erosion caused by increased runoff. Land-use planners in both rural and urban areas will need to pay closer attention to development in flood-prone areas, such as river floodplains. Managing storm runoff is already a significant challenge in many cities. Increased rainfall may require the redesign of storm sewers and drainage systems.

▼ Figure 9.43 **Adapting to Drier Weather**
Drip irrigation systems, such as the one at this vineyard, deliver water efficiently to where it is needed most—the root systems of individual plants. Such systems can help farmers in regions that are already arid adapt to future water shortages.

QUESTIONS 9.7

1. What are some examples of adaptation to global warming that are already underway in some urban areas?

2. What are some examples in agricultural areas?

(MES) For additional review, go to **MasteringEnvironmentalScience**

▼ Figure 9.44 **Adapting to Rising Sea Level**
The complex system of dikes and canals that protect the Netherlands' low-lying farmlands is being upgraded in anticipation of rising seas.

Increasing heat. In 2003 and 2007, southern Europe was hit with record heat waves—periods of two to three weeks when daily temperatures topped 38 °C (100 °F). Each heat wave was responsible for the deaths of over 10,000 people. Although there is some disagreement as to whether these heat waves can be attributed to global warming, there is no doubt that such events have become increasingly frequent. Adaptations to hotter weather may involve improved building design and heat control. More important is the need for effective programs to protect vulnerable populations, such as the poor and the elderly, who do not have access to air conditioning.

Heat waves coupled with droughts have increased the number and severity of forest fires in many regions of the world. In 2012 and 2013, record temperatures and drought were major factors contributing to extensive, severe wildfires in Australia. In the United States, increased warming in the west is likely produce a similar trend. As a consequence, the U.S. Forest Service is reviewing its forest management programs to reduce the risk of wildfire. Models to forecast likely year-to-year variations in temperature and drought in different regions are helping it deploy and manage its firefighting resources more effectively.

Patterns of storms. Climate scientists are not in agreement about the effects of global warming on the frequency and intensity of storms. Additional data and the development of improved computer models should help resolve this issue.

Nevertheless, recent storms such as Hurricanes Katrina and Sandy show that, even in wealthy countries, coastal lands and cities are extremely vulnerable to severe storms. Any increase in the frequency or intensity of storms warrants adaptive measures. Such measures might include requiring water-resistant construction, strengthening levees, and restoring wetlands that once protected coastal areas.

Rising sea level. With continued warming, some amount of rise in sea level is inevitable. Some coastal locations will suffer direct inundation, especially during storms. Many other places will be affected indirectly as rising seas change the level and salinity of groundwater. Adaptation to rising sea level has already begun in the low-lying areas of the Netherlands. There, pumping systems, canals, and dikes are being upgraded to protect agriculture and urban centers (Figure 9.44).

Regional differences. The capacity to adapt to climate change is not evenly distributed among the nations of the world. Wealthy countries have the resources to make adjustments that minimize the effects of committed warming. Many poor countries do not. Without adaptation, changes such as diminished rainfall and food supply will intensify political conflicts, threatening the security of many nations. Thus, more than the well-being of hundreds of millions of people is at risk. Adaptation to global warming must be seen as an international priority, and its costs must be borne equitably.

Adapting to Rising Seas

How effective is the conservation of natural habitat as an adaptation strategy to sea level rise and increased storm severity?

Rising sea level and increased storm activity and flooding pose ever-increasing threats to coastal communities around the world. These threats are compounded by rapid population growth and sprawling development in coastal cities. The traditional approach to protect coastal towns has been construction of sea walls and other "hardened" structures. More recently, greater emphasis has been placed on the conservation and restoration of natural habitats such as coral and oyster reefs, sea grass beds, and coastal forests and wetlands that buffer coastlines from waves and storm surges. We know these conservation strategies are effective at particular locations. Katie Arkema and her colleagues at Stanford University Natural Capital Project were interested in determining the value of such conservation practice applied on a large scale, across the entire coast of the United States (Figure 9.45).

Arkema and her team used a combination of *data synthesis* and *ecological models* to address this question. They began by calculating a hazard *index* for each square kilometer of the U.S. coastline based on the physical features that influence water movement, the types of natural coastal habitats at current sea levels, and the likelihood of coastal storms. They then calculated hazard indices based on five sea level scenarios, and for coastlines with or without natural coastal habitat.

Scenario 1 represented current conditions and scenario 2 approximated sea level change expected in a *Sustainable World* future. Scenarios 3, 4, and 5 represented sea level rise with successively greater warming, with 5 corresponding to changes expected under the *Business as Usual* trajectory for global warming. They also mapped data on human populations and property values for each square kilometer of coastline. By overlaying these maps, Arkema was able to convert hazard indices to more direct measures of imperiled human life and property damage.

As expected, the number of people and the amount of property at risk increased with increasing rates of sea level rise (Figure 9.46). The presence of natural coastal habitat diminished those risks by at least 40% in each scenario on a national scale. At least as important, Arkema and her colleagues have produced the first national map indicating where conservation and restoration of reefs, wetlands, and coastal forests have the greatest potential to protect human life and property in coastal communities (Figure 9.47).

Source: Arkema K.K. 2013. Coastal habitats shield people and property from sea-level rise and storms. *Nature: Climate Change* 3: 913–918.

▲ **Figure 9.45 Ecosystem Services**
Katie Arkema is interested in finding ways to quantify nature's benefits to people and applying that information to the management of coastal and marine ecosystems.

1. What physical features of a coastline might increase risks associated with sea level rise?

2. Was the effect of habitat protection consistent among the sea level rise scenarios? Explain your conclusion.

3. How might coastal counties use this information to plan future land use?

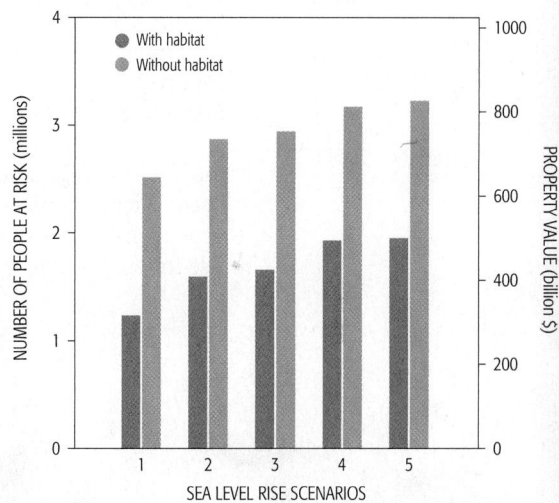

▲ **Figure 9.46 With and Without Habitat**
Bar graphs indicate the number of people and property value at risk nationally. Across all five sea level change scenarios, natural coastal habitats such as reefs, wetlands, and coastal forests substantially diminish risks to life and property in coastal communities.

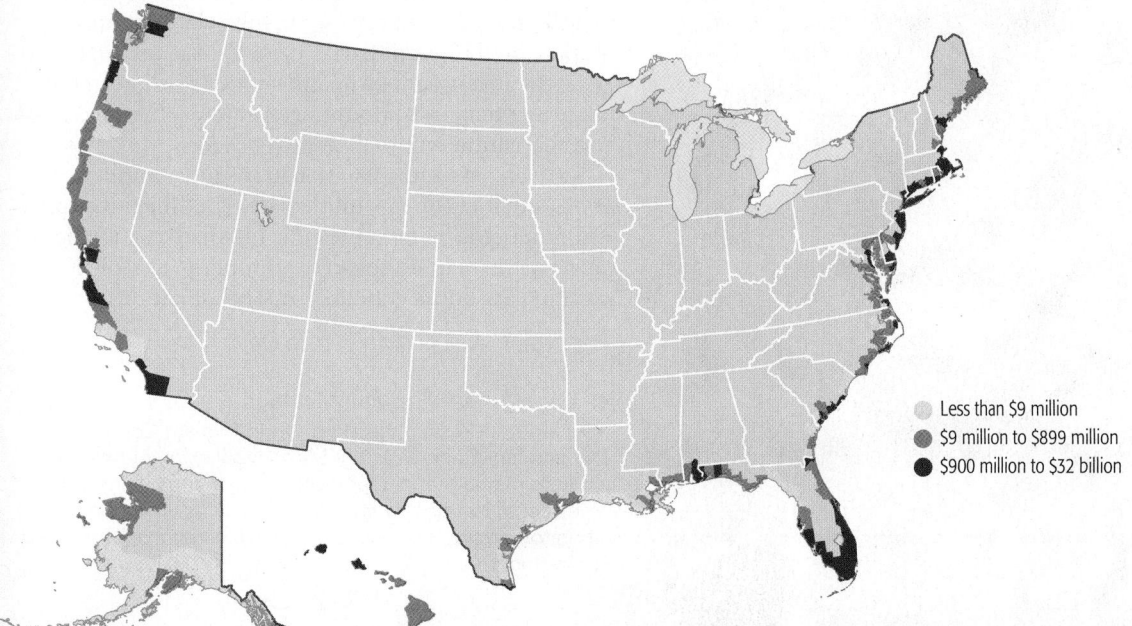

Less than $9 million
$9 million to $899 million
$900 million to $32 billion

◄ **Figure 9.47 Where It Matters Most**
The color for each coastal county indicates the total property value for which coastal habitats reduce exposure to storms and sea level rise under sea level rise scenario 4. Coastal habitats protect the greatest value and number of people in New York, Florida, and California.

9.8 Mitigation and Adaptation Policies

BIG IDEA Taking action to mitigate and adapt to global warming and climate change will be very expensive, yet the ultimate costs of not taking action will be much greater. Individual governments and international organizations such as the United Nations are formulating policies to encourage prompt action. Some of these policies depend on the enforcement of regulatory standards. Others rely on economic markets and taxes. Global initiatives such as the 1998 Kyoto Treaty set expectations for the reduction of greenhouse gas emissions. Expectations for each signatory nation were based on past emissions and economic status. Negotiations are underway on a new treaty that would require all nations to take more aggressive actions to mitigate global warming. This treaty will also establish international policies to facilitate adaptation to committed warming.

What Is the Cost?

■ The costs of action are less than the costs of *Business as Usual*.

Economists argue furiously over the correct way to estimate the costs of global warming and the process of mitigation (**Figure 9.48**). What is the correct value of ecosystem services (see Module 2.4)? How can future costs be compared to present costs? Robert Socolow estimates that the cost of moving from a *Business as Usual* to a *Sustainable World* scenario would be between 1% and 2% of global gross domestic product (GDP), or about $0.65 to $1.3 trillion per year.

The economic impact of global warming will vary greatly from region to region. An increase in global average temperature of less than 2 °C (3.6 °F) may have economic benefits for some regions. In Canada, such a rise in temperature is likely to increase grain production and lower the amount of fuel needed for heating. In contrast, increasing drought in the southwestern United States will result in greater expenditures.

If temperatures climb above 3 °C (5.4 °F), the costs will be high for all regions. Estimates of those costs vary widely; like estimates for the cost of mitigation, they depend heavily on starting assumptions. There is, however, little doubt that the temperature changes produced by *Business as Usual* scenarios would have catastrophic costs. Many estimates exceed 4% of global GDP, or more than $2.6 trillion per year.

Because the projected costs of inaction are so high, it would seem logical and wise to act now. Two issues, however, complicate this picture. First, it is very difficult to persuade people and nations to make large investments in changes when the benefits will not be realized for decades. Second, the costs of global warming will not be equitably distributed. Many of the poorest nations are likely to be hit with the highest costs. Moving to a *Sustainable World* scenario will require the world's wealthiest nations to contribute financial support in proportion to their impact on greenhouse gas emissions. Thus, the investment of wealthy nations will need to be far greater than the amount needed to pay for mitigation and adaptation within their own borders.

To overcome these issues, nations must implement policies that encourage immediate economic investment and produce long-term benefits. International treaties must ensure that the costs for responding to global warming are borne equitably among all nations.

◀ Figure 9.48 **A Forceful Message**
In his 2009 address to the Third World Climate Conference, UN Secretary General Ban Ki-Moon argued for international action to mitigate and adapt to climate change. In his words, "The cost of inaction today will be far greater than the cost of action tomorrow. Not just [for] future generations, but for this generation, too."

Policy Alternatives

■ Policies rely on a combination of regulations and market incentives.

Environmental policies aim to guide the actions of individuals and organizations in ways that influence environmental outcomes (see Module 2.5). Numerous policies are being developed and implemented to encourage the reduction of greenhouse gas emissions. Some policies set standards and regulate the actions needed to meet those standards. Others involve a combination of regulations and economic incentives.

Regulations.   Regulatory approaches generally require individuals and businesses to take steps to diminish CO_2e emissions. For example, the U.S. government regulates the mileage of automobiles with the Corporate Average Fuel Economy (CAFE) standard. CAFE requires that the average gasoline mileage of all the cars made by a manufacturer meet a certain standard. The current standard for automobiles is 27.5 miles per gallon; this will increase to 35 miles per gallon by 2020. The EPA has recently proposed regulations that could increase the CAFE standard to 54 miles per gallon by 2025. Although this regulation was originally aimed at curbing the demand for gasoline, it has a significant effect on CO_2e emissions. Other countries, such as Japan, China, and the members of the European Union, have implemented mileage standards that are nearly twice as stringent as those in the United States.

Many nations have implemented regulations aimed at diminishing dependence on fossil fuels as well as greenhouse gas emissions. The European Union, for example, requires that member countries use increasing proportions of energy from renewable sources. The 2008 Farm Bill passed by the U.S. Congress required a 400% increase in biofuel production by 2020 (see Module 15.2).

Regulatory actions sometimes depend on how emissions are classified. For example, until recently, CO_2 was not legally defined as a pollutant in the United States. Therefore, standards and regulations for its emission were outside the jurisdiction of the U.S. Environmental Protection Agency. In 2007, the U.S. Supreme Court ruled that CO_2 is indeed a pollutant, and that the EPA must set emission standards. The EPA now monitors annual CO_2 emissions from transportation, industry, and utilities, and it has established standards and rules emissions in new or modified facilities that produce greenhouse gases.

Economic incentives.   In recent years there has been much interest in using economic incentives to promote action. For example, the European Union uses a cap-and-trade policy to regulate greenhouse gas emissions from electric utilities (see Module 2.5). A regulatory cap sets a standard for the level of CO_2 emissions that each power plant is allowed to produce. Utility managers may choose to reduce their emissions to meet this standard. Alternatively, they may purchase credits from other utilities whose CO_2 emissions are below the cap. This cap-and-trade policy has created a lively market for emissions credits. In 2014, credits sold on European markets at $8–10 per ton of CO_2e emissions.

Some economists suggest that cap-and-trade approaches will not cause large enough reductions in emissions. Critics of cap-and-trade policies argue that it is difficult to verify the levels of emissions from power plants and that the process does not have adequate oversight.

Taxes are another form of economic incentive. The gasoline taxes that are already in place in most countries are a form of carbon tax. Carbon taxes would increase the existing taxes on gasoline and extend them to other forms of carbon-based energy use (Figure 9.49). Proponents of carbon taxes argue that their implementation is easily verified and that taxes can be adjusted to ensure that emission goals are met. Some opponents worry that carbon taxes would have a disproportionate impact on poor people. Other opponents are concerned about the impacts of additional taxes on economies.

▼ Figure 9.49 **Responding to Price**
Some policymakers argue that CO_2e emissions could be cut significantly by increased taxes on gasoline and other fuels, which would encourage people to choose more efficient modes of transportation, such as mass transit.

Agreeing on the Facts

■ Consensus among nations on the role of humans in global warming is critical.

One of the most important steps in policy development is reaching agreement on the nature of the problem that a policy is intended to fix. Given the highly technical nature of climate science and the uncertainties associated with climate forecasting, agreeing on policies regarding global warming is especially challenging.

To meet this challenge, the World Meteorological Association and the United Nations Environmental Program established the Intergovernmental Panel on Climate Change in 1988. The IPCC is a scientific body that synthesizes evidence on climate change from within the scientific community. The comprehensiveness of the scientific content is achieved through contributions from experts in all regions of the world and all relevant disciplines. The organization is open to participation by decision-makers and scientists from all UN member nations.

In 2014, the IPCC released a series of reports that emphatically restated its earlier conclusion that Earth's atmosphere is warming. The reports also concluded that global warming is a direct consequence of human activities that are increasing emissions of greenhouse gases. These reports stated that some amount of future warming is inevitable and that the global community will need to take steps to adapt. They also emphasized that aggressive action to mitigate emissions could diminish global warming and avert its worst impacts.

This strong statement and the consensus of IPCC members provide a solid foundation for policy development in the international community. For its work in establishing this foundation, the IPCC was awarded the 2007 Nobel Peace Prize, which it shared with former vice president Al Gore.

International Global Change Policy

■ The Kyoto Protocol is a first step to reduce greenhouse gas emissions.

The first international action designed to address global warming came at the 1992 United Nations "Earth Summit" held in Rio de Janeiro, Brazil. At that time, the United Nations Framework Convention on Climate Change (UNFCCC) was signed by 150 countries, including the United States (Figure 9.50). The goal of this agreement was to stabilize emissions at 1990 levels by the year 2000 through voluntary reductions in greenhouse gas emissions.

Within a few years, it became evident that the UNFCCC emission reduction goals would not be met.

▶ **Figure 9.50 A Work in Progress**
Ⓐ At the 1992 Earth Summit, U.S. president George H.W. Bush signed the Earth Pledge, a commitment to protect Earth's environment, and encouraged nations to take actions that would reduce the threat of global warming. Ⓑ In Kyoto, Japan, in 1997, United Nations member nations agreed on protocols aimed at reducing emissions to 1990 levels by 2010. Ⓒ The United States subsequently declined ratification of this treaty. At the 2009 UN Climate Change Conference in Copenhagen, Denmark, President Barack Obama called for actions to reduce greenhouse gas emissions by 80% by 2050.

Ⓐ 1992

Ⓑ 1997

Ⓒ 2009

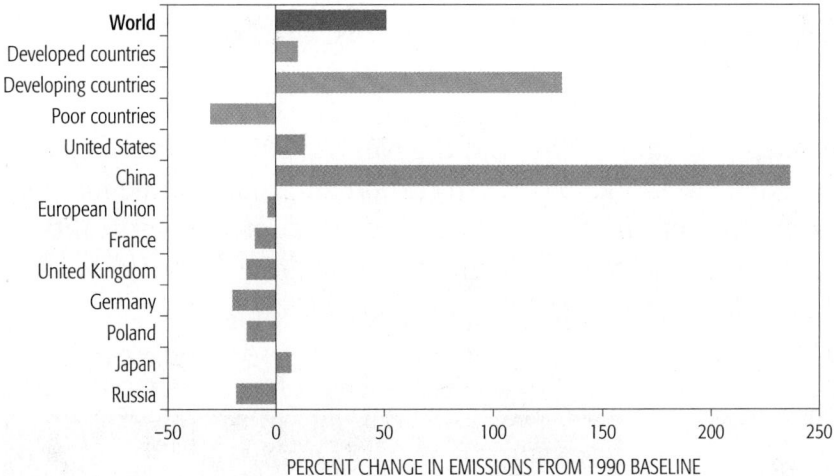

◄ **Figure 9.51 Kyoto Consequences**
The percent change in greenhouse gas emissions between 1990 and 2010 for the world and countries are shown grouped by economic status (in blue). Changes for several individual countries are also listed (in red). Percent change was greatest for developing countries, but emissions actually decreased in most of the world's poorest countries.

Data source: UN World Development Index. http://data.worldbank.org/data-catalog/world-developmentindicators/wdi-2012.

In fact, because of the booming global economy, most developed countries increased their emissions. It was clear to most UN member countries that a policy based on voluntary action would not work.

In 1997 countries met again in Kyoto, Japan, to formulate a more aggressive treaty. The Kyoto Protocol set an overall goal of reducing CO_2e emissions to 5% below their 1990 levels by 2010. For developed countries, the targets for reducing emissions were based on historic levels of emissions. For example, the target for the United States was 7% below its 1990 emissions levels. For European Union countries, the target was 8% below 1990 emission levels. No targets were set for developing countries (including China and India) because of fears that emission caps would inhibit their economic development. Instead, targets for developing countries would be added to the protocol at an unspecified later date.

In 2001 the United States indicated that it would not sign the treaty. Nevertheless, the Kyoto Protocol was ratified by the required number (130) of countries in 2004. It became international law in February 2005.

Some of the countries that signed the Kyoto Protocol have significantly diminished their greenhouse gas emissions and met their Kyoto targets. As a group, members of the European Union have experienced a 3.5% reduction compared to the 1990 baseline. While this did not quite meet the 8% target, individual countries such as France, the United Kingdom, Germany, and Poland did considerably better than target expectations. Nevertheless, greenhouse gas emissions in developed countries increased on average by 10.4%; U.S. emissions increased by 13.6%. Emissions from developing countries with rapidly growing economies grew by over 100%, led by China with a 236% increase in emissions between 1990 and 2010 (**Figure 9.51**).

Critics of the Kyoto Protocol suggest that its costs are too high relative to its benefits, which are uncertain and will not be realized for many decades. They argue that most of the reduction in emissions credited to the treaty would have been achieved without it. There is also concern that targets were not set for the world's two most

populated countries, China and India. The absence of caps may well have encouraged the relocation of greenhouse gas emitting facilities in these and other developing countries.

Advocates of the Kyoto Protocol see the treaty as a necessary first step in moving greenhouse gas emissions to sustainable levels. They argue that it is only fair that wealthier nations take the first steps. They are, after all, the nations with the longest history of greenhouse gas emissions.

Talks are now underway for the next step in what has come to be called the Kyoto Process. The sober findings and recommendations of the IPCC weigh heavily in those discussions. The United States is an active participant in those deliberations.

The original goal of the Kyoto Process was to produce a revised treaty in 2010; the goal is now to have a treaty for the UN Climate Change Conference scheduled for 2015 in Paris. Key elements of that treaty were discussed at the 2013 UN Climate Change Conference in Warsaw. First, overall emission targets are likely to be more aggressive than in the current treaty. Second, it is likely that targets for emission reduction will be set for all but the very poorest nations. Third, those targets will be set for each country based on its historic emissions as well as current emission trends.

Among the most important matters discussed at the 2013 Warsaw meeting was the need for strategies for adaptation to committed warming. All countries are likely to have some costs associated with adaptation, but the impact of those costs on the economies of the world's poorest countries are likely to be very high. Delegates agreed that treaty mechanisms were needed to provide vulnerable populations with protection against loss and damage caused by sea level rise and increased frequency of extreme weather events. To be sustainable, these mechanisms must address the triple bottom line—they must be environmentally and economically feasible, and they must be equitable. Most observers agree that costs for these mechanisms must be borne fairly by the entire community of nations. This has been a contentious matter, but hope remains that a process for sharing costs will be included in the final treaty.

QUESTIONS 9.8

1. How do the worldwide costs of global warming differ between the *Sustainable World* and the *Business as Usual* scenarios for greenhouse gas emissions?

2. Describe two policies aimed at reducing CO_2e emissions that have been enacted in the United States.

3. What is the role of the IPCC in the development of international climate policy?

4. What are the key features of the Kyoto Protocol?

(MES) For additional review, go to **MasteringEnvironmentalScience**

The University of Florida's Carbon-Neutral Football Games

Jacob Perritt-Cravey *talks about starting the Neutral Gator Initiative, a nonprofit organization that works to reduce carbon emissions while helping people in low-income communities conserve energy and lower their utility bills. Neutral Gator is committed to reducing Gainesville's contribution to global climate change by helping the University of Florida reach carbon neutrality by 2025.*

How did you first get the idea for Neutral Gator?

I was at a University of Florida football game reading the paper when I saw an article saying UF was hosting the nation's first carbon-neutral football game. UF offset that game, but they didn't have any intention of offsetting others. I thought there were so many more ways future games could be offset.

At that time I knew service work was going to be my path, and I really wanted to do something that could affect a large amount of people. I thought, "That's it! Use something people love and attach your cause to it." This was an opportunity to connect with people.

What steps did you take to get started?

First, we figured out the games' imprint and what it would take to offset a season. I didn't ask permission of the athletics department. Instead, I focused on self-funding the program. That way, when I went to the athletic department I only had to request space to set up a tent and engage fans at the games.

We then worked with a local weatherization initiative that already had a strong base of supporters. They became partners in our efforts and helped us get needed training. Then, we recruited volunteers from the University of Florida, the surrounding community, and local businesses. At UF we sent emails to service organizations, fraternities, teachers, and clubs.

At every event, we provided free training for participants, educating them on weatherization. We did projects like exchanges where people could bring their incandescent light bulbs and exchange them for compact fluorescent lights, CFLs. We partnered with a local ice cream shop that supplied a free scoop of ice cream for every five light bulbs exchanged.

We then went door to door. We modified our strategy by going strictly into low-income residences, providing more extensive energy retrofits, installing low-flow showerheads, kitchen and bathroom aerators, and so on. Most importantly, we created an education component for the residents, giving them general information on how to create utility savings through behavioral change.

What are the biggest challenges you've faced with the project, and how did you solve them?

The main challenges have been communication and implementation. We have to teach people what carbon offsets are, what a carbon footprint is, and why it's important. Then we need to get them to take impact-reducing actions. Sometimes it's having them replace their incandescent bulbs with CFLs, installing low-flow showerheads, or taking shorter showers. Those things add up. It's important to find a way for people to get involved that works for them.

How has your work affected other people and the environment?

In our first summer, our group consisted of 15 volunteers. As we continued, the amount of people grew exponentially. At our last event that summer we had an astonishing 75 people show up! All summer we did projects to educate the community and help them save money. By the end of the summer, we had offset approximately 3,000 tons of CO_2—enough to offset all the football games that season.

Since we began, the economic benefits for the low-income community have been substantial. Our weatherization efforts have retrofitted 180 low-income residences and distributed 63,000 energy-efficient light bulbs, creating a combined savings of more than $3.2 million.

What advice would you give to students who want to replicate your model on their campus?

Every community has different resources and needs. However, moving forward there are a few key steps to take:

First, know the environmental impact of what you're trying to offset. This will give you an idea of what you'll need to do to generate offsets. For example, how many trees will you need to plant? How many houses will you need to weatherize? How many light bulbs will you need to install?

Then create an education strategy. It's important to be specific about what you want people to know and do. It can be as simple as making them aware of the environmental impact of a football game, and then asking them to participate in reducing their own footprint by signing a commitment to change a personal habit or donate to your cause.

Finally, follow up with people, be consistent with your outreach, and let people know how they can learn more.

And you can always find us at: http://www.neutralgator .org for support along the way!

Summary

Earth's climate varies over very long and comparatively short timescales. Before the Industrial Revolution, these variations were a consequence of natural variation in the physical processes that determine climate, including periodic variation in Earth's orbit, carbon dioxide in the atmosphere, snow and ice cover, and ocean currents. Persistent warming since the late 1800s, however, is largely a consequence of human activities, such as burning fossil fuels and deforestation, that increase the concentration of heat-trapping gases in the troposphere. This global warming is causing significant changes in many ecosystems and threatens human well-being.

New technologies and changes in human behavior can reduce greenhouse gas emissions and slow global warming. Even so, historical emissions have committed Earth's atmosphere to some additional warming. In many places, adaptations will be required to minimize the effects of drought, flooding, and temperature extremes. Strategies to reduce greenhouse gas emissions and to adapt to inevitable change have been encouraged by changes in national and international policies, but additional actions are needed to prevent unpredictable and unsustainable outcomes.

9.1 Long-Term Climate Patterns

- Over the last several million years, Earth has experienced multiple ice ages. During the last 10,000 years, Earth's climate has generally been warming.

KEY TERMS

climate change, Pleistocene epoch, Milankovitch cycles, Holocene epoch, Medieval Warm Period, Little Ice Age, Anthropocene

QUESTIONS

1. What caused concentrations of CO_2 to decline during cold glacial periods in the Pleistocene?
2. Scientists say that changes in Earth's snow and ice cover are both a cause and consequence of changes in climate. Explain what they mean.

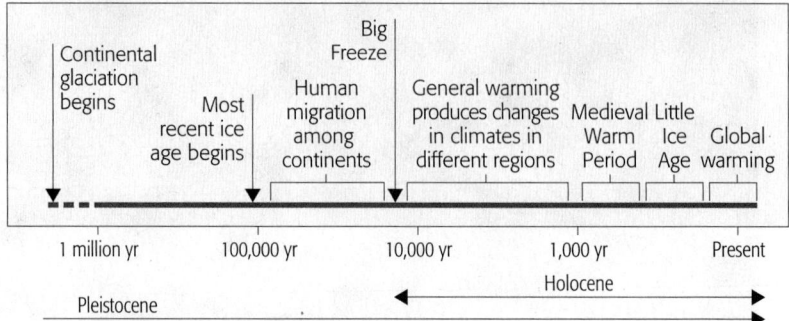

3. Over the past 100,000 years, human migrations have been significantly influenced by changes in climate. Describe two important examples of such influences.

9.2 Measuring Global Temperature

- Regional weather varies from year to year as a result of changes in ocean currents and events such as volcanic eruptions. In general, though, the average annual global temperature has increased significantly over the past 130 years.

KEY TERMS

temperature anomaly, global warming, El Niño/Southern Oscillation (ENSO)

QUESTIONS

1. Why is it difficult to measure directly the global average annual temperature?
2. Describe two factors that influence variations in annual temperature anomalies from one geographic location to another.
3. Why do El Niño conditions generally produce higher rainfall along the western slopes of the Andes Mountains?

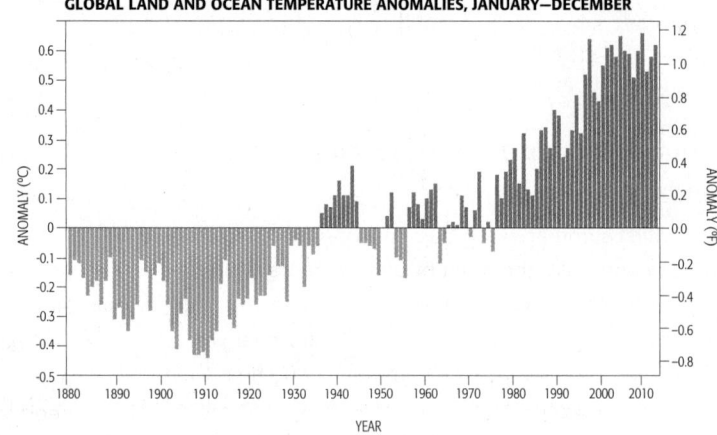

GLOBAL LAND AND OCEAN TEMPERATURE ANOMALIES, JANUARY–DECEMBER

9.3 Causes of Global Warming

- Greenhouse gases such as water (H_2O) and carbon dioxide (CO_2) efficiently absorb infrared photons radiated from Earth's surface and warm the atmosphere.
- Human activities have increased the concentration of greenhouse gases in the atmosphere.

KEY TERMS

greenhouse gases, greenhouse effect, global warming potential (GWP), black carbon, carbon dioxide equivalents (CO_2e)

QUESTIONS

1. Water molecules absorb infrared light, but water vapor is not believed to contribute to global warming. Why not?

2. In what ways does deforestation contribute to the increase in greenhouse gases in the atmosphere?

3. What is responsible for the seasonal changes in CO_2 concentrations observed in data from the Mauna Loa Observatory?

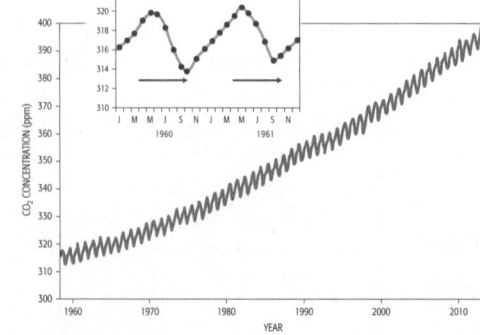

4. CO_2e emissions differ widely among countries. Discuss four factors contributing to these differences.

9.4 Consequences of Global Warming

- The effects of global warming vary among geographic locations; they include the retreat of glaciers and ocean ice packs, rise in sea level, and changes in ecosystems and populations of many species.

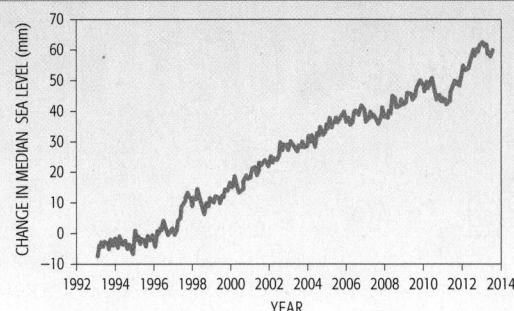

QUESTIONS

1. Global warming will likely increase rainfall in some places. Why might warming cause this change?
2. In some places, even if rainfall increases, it may actually become drier. Explain.
3. The melting of continental glaciers causes sea level to rise, but the melting of Arctic sea ice does not. Why is this so?
4. Global warming is having a significant effect on populations of many plant and animal species. Describe three ways in which this is occurring.

9.5 Forecasting Global Warming

- Three CO_2e emissions scenarios are frequently used to forecast future climate. They are (1) *Today's World*—immediate cessation of CO_2e emissions; (2) *Business as Usual*—no action taken, with CO_2 concentrations increasing beyond 900 ppm; (3) *Sustainable World*—actions are taken to slow CO_2e emissions so that atmospheric CO_2 stabilizes at about 540 ppm.

KEY TERMS

general circulation model (GCM), committed warming

QUESTIONS

1. General circulation models forecast that global warming will be most intense in polar regions and over continents. Why is this so?

2. The uncertainties for *Business as Usual* forecasts are much higher than those for *Today's World* or *Sustainable World* forecasts. Explain why.

3. Describe the potential differences between the consequences of *Business as Usual* forecasts and *Sustainable World* forecasts with regard to human health.

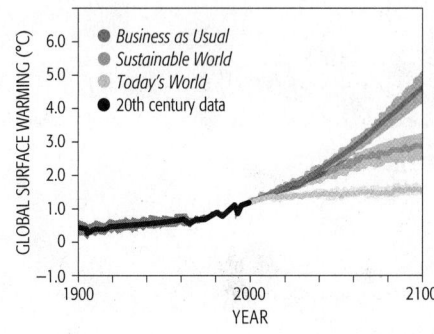

9.6 Mitigating Global Warming

- Mitigation refers to actions taken in *Sustainable World* scenarios that directly reduce the threat posed by CO_2e and associated global warming. However, no single mitigation strategy is sufficient to achieve *Sustainable World* conditions; a portfolio of actions will be required.
- The costs, benefits, and risks of proposals to mitigate global warming using technologies to reduce solar radiation or directly remove CO_2 from the atmosphere are not as yet understood.

KEY TERMS

carbon capture and storage (CCS), photovoltaic (PV) cells, biofuels, biostorage

QUESTIONS

1. The Pacala and Socolow stabilization wedge model indicates that a portfolio of strategies will be required to mitigate the increasing rate of CO_2e emissions. Explain what this means and why it is the case.
2. It will eventually be necessary to lower CO_2e emission rates below their current level. Why?
3. What are some of the strategies scientists and engineers have proposed to reduce warming without reducing emissions of greenhouse gases? Why are they controversial?

9.7 Adapting to Global Warming

- The effects of committed warming and inevitable climate change will require humans to adapt in a variety of their endeavors.

QUESTIONS

1. Describe four strategies for adaptation to global warming that might be implemented in the area where you live.
2. Explain why many world leaders view adaptation to global warming as an environmental justice issue.

9.8 Mitigation and Adaptation Policies

- Actions to adapt to and mitigate global warming have significant costs; not acting has greater costs, but they are less immediate.
- Policy alternatives include regulatory actions, such as mandated fuel economy standards and use of biofuels, and market-based strategies, such as cap and trade on CO_2e emissions.

QUESTIONS

1. Some policymakers prefer cap and trade, whereas others prefer taxes as a means of lowering CO_2e emissions. What are the pros and cons of these two approaches?
2. Because the costs of global warming are uncertain and many of the consequences are decades away, some economists argue that we should delay action at this time. What is your response to such an argument?

MasteringEnvironmentalScience®

Students Go to **MasteringEnvironmentalScience** for assignments, the eText, and the Study Area with practice tests, activities, and more.

Instructors Go to **MasteringEnvironmentalScience** for automatically graded tutorials and questions that you can assign to your students, plus Instructor Resources.

Air Quality

The Killer Smog

What causes air pollution to become deadly?

The chill in the air early on Tuesday, October 26, 1948, left no doubt among the people of Donora, Pennsylvania, that summer was over. During the night, dense cold air had drained into the Monongahela River valley. This air was held in place by a stationary layer of warm air a few hundred feet above the river. As a result, the yellowish morning smog lasted through the noon hour. This cold air remained trapped, and each day the smog became more persistent and dense. By Friday, four days later, the smog did not clear at all (Figure 10.1).

Smoky, eye-burning fog was certainly familiar to Donora's 14,000 residents. Smog began to appear on cold mornings soon after completion of the steel mill in 1900. With the establishment of Donora Zinc Works, one of the world's largest zinc-smelting facilities, the smog events became more frequent and dense (Figure 10.2). For most people, smog seemed like a small price to pay for Donora's thriving economy and jobs. But this event in 1948 was different; this town was about to experience an environmental tragedy.

On Friday evening, October 29, hundreds of local residents began to crowd area hospitals, coughing and displaying other signs of respiratory distress. Through Saturday, conditions remained unchanged, and smog and traffic congestion prevented evacuation of the town. A local physician led an ambulance by foot through the darkened streets of Donora, trying to move seriously ill patients to hospitals and the dead to a temporary morgue.

By Sunday morning, 20 people had died and thousands were ill. It was also on Sunday morning that the president of U.S. Steel ordered the shutdown of the smelter furnaces. Later that afternoon, a gentle rain began to clear the air. Even so, over the next month an additional 50 Donora citizens died of respiratory complications.

The "Donora killer smog" quickly became international news. Walter Winchell sensationalized it on his weekly radio show, and it was reported in newspapers around the world. Finger pointing began immediately. Local papers accused the zinc works of "murder" by releasing "an airborne poison" and called for the renovation or removal of the facility. The chemist for the Pennsylvania Smoke Control Bureau insisted that the smog was derived from multiple sources, including domestic and commercial furnaces, automobile emissions, and several mill operations. U.S. Steel argued that this event was caused by the unique weather conditions that trapped stagnant air in the valley and that it was not a direct consequence of their operations.

Under political pressure from industry, state and federal agencies at first rejected requests by local officials for an investigation of this episode. When the Borough of Donora and the United Steelworkers Union initiated their own study, however, the U.S. Public Health Service reversed its earlier position. That agency's study of the Donora killer smog was the first comprehensive study of the health effects of air pollution. Over the months and years that followed, a clear picture of the causes of this tragedy emerged.

▼ Figure 10.1 **Midday Darkness**
A deadly smog had enveloped Donora, Pennsylvania, when this photograph was taken at noon on October 29, 1948.

▲ Figure 10.2 **The Price of Progress**
Donora's enormous smelting facility provided economic prosperity, but it was also the source of considerable air pollution.

A variety of studies revealed that the toxic effects of the Donora smog were due to a mixture of chemicals, and nearly all of them originated from the smelter. Initial roasting of zinc ores released sulfur compounds and heavy metal dusts. As molten zinc was produced, nitrogen oxides, carbon monoxide, and fluorine were also emitted from the zinc works. However, many of the most toxic components of the smog were formed by reactions among this mix of chemicals after they dissolved in fog droplets.

Yes, the meteorological conditions that trapped air in the Monongahela Valley prevented the dilution of these toxins and greatly exacerbated their impacts on human health. But such conditions are common in river valleys and topographic basins. Investigators concluded that emissions from facilities such as Donora Zinc Works must be managed with these weather conditions in mind. The Donora killer smog aroused public concern about the health effects of air pollution and led directly to the passage of the Air Pollution Control Act of 1955 and the Clean Air Act of 1967.

Much has changed in the Monongahela Valley since 1948. Unable to compete in a growing global market and to meet state and federal air-quality regulations, the Donora Zinc Works was closed in 1957. Donora's steel mill was shuttered in 1967. Today on cold October mornings, fog still forms in the Monongahela Valley, but this fog is laced with far fewer chemicals (**Figure 10.3**).

The conditions in Donora in October 1948 were extreme, but they were by no means unique. Smog remains a common feature in cities across America and around the world. Today, national laws and international treaties regulate many air pollutants; technologies such as scrubbers and catalytic converters reduce emissions of these pollutants. Nevertheless, air pollution remains a significant threat to the well-being of humans, plants, and animals.

■ What are the sources of pollutants, and what processes influence their concentrations and dispersal in the atmosphere?

■ How do pollutants influence the chemistry of Earth's stratosphere in ways that affect human well-being?

■ What are the important pollutants in Earth's troposphere, and how do they influence the health of humans and other organisms?

■ What important pollutants influence the quality of the air we breathe indoors?

■ What national and international policies and laws have been developed to reduce air pollution?

◄ Figure 10.3 **Modern Donora**
Today, Donora's mills are gone, along with the smog that they produced. Its downtown features quaint storefronts dating from its mill town days.

10.1 Air Quality and Air Pollution

BIG IDEA **Air quality** refers to the amounts of gases and small particles in the atmosphere that influence ecosystems or human well-being. The concentrations of some atmospheric gases are relatively constant, while those of many other gases vary from place to place and through time. Variable gases and particles have the potential to be air pollutants. **Air pollution** refers to gases or particles that are present in high enough concentrations to harm humans, other organisms, or structures such as buildings or pieces of art. Some pollutants are released into the atmosphere directly from specific sources. Other pollutants are produced by reactions among chemicals that have been released into the atmosphere from more than one source. The dispersal of pollutants depends on the movement of gases in the atmosphere, the height at which pollutants are emitted into the atmosphere, and factors influencing the length of time that pollutants remain in the atmosphere.

Gases and Particles

■ The atmosphere contains a large array of gases and particles that vary in concentration over time and from place to place.

Three gases—nitrogen (N_2), oxygen (O_2), and argon (Ar)—account for over 99% of the gas molecules in Earth's atmosphere. Gases that make up less than 1% of Earth's atmosphere are called **trace gases**. The atmosphere contains hundreds of trace gases, most of which are present in concentrations expressed in parts per million (1 ppm = 0.0001%) or parts per billion (1 ppb = 0.0000001%). The concentrations of some trace gases, such as helium (He), are quite stable from place to place. The concentrations of other gases vary widely from place to place, from day to day, from season to season, or over decades and centuries. It is these variable gases that have the potential to become pollutants.

Water and carbon dioxide are variable gases. Concentrations of water vary in association with proximity to bodies of water, air temperature, and other factors influencing the rate of evaporation. Concentrations of carbon dioxide are influenced by daily and seasonal patterns of photosynthesis and respiration, as well as emissions associated with the combustion of fossil fuels (see Module 9.2).

Nitrogen oxides (NO, NO_2, and N_2O) are also variable gases. In pre-industrial times, concentrations of nitrogen oxides varied because the balance of chemical reactions differed from place to place. Since the beginning of the Industrial Revolution, the burning of fossil fuels has caused concentrations of nitrogen oxides to gradually increase. Today, their concentrations are much higher near urban and industrial centers than in rural locations (Figure 10.4).

▶ Figure 10.4 **Air Pollution**
Riding through Lanzhou, China, this cyclist wears a protective mask to keep from breathing in the heavily polluted air. With its many factories, Lanzhou is one of China's most polluted cities.

Volatile organic compounds (VOCs) are organic chemicals that can vaporize into the air. VOCs are a diverse array of chemicals with a wide variety of properties. Some VOCs are produced naturally. Methane (CH_4) is the most abundant naturally occurring VOC. Other naturally occurring VOCs give flowers and decaying organic matter their distinctive odors. Anthropogenic sources of VOCs include solvents, paints, gasoline, and the exhaust from automobiles.

In addition to gases, the atmosphere contains **particulate matter**—very small solid and liquid particles suspended in the air, abbreviated as PM. These suspended particles are called **aerosols**. Aerosols originate from both natural and human sources. The most familiar aerosols are the tiny water droplets that make up clouds and fogs. Aerosols are considered to be pollutants if they pose a health hazard when inhaled or if they affect the regional or global climate.

The size of aerosol particles is significant because it determines how long they remain in the atmosphere, the way they scatter light, and how they are deposited in the lungs of humans or other animals. Aerosol particles vary in size from 0.001 to 100 micrometers (μm = 1 millionth of a meter). Regulatory agencies like the U.S. Environmental Protection Agency (EPA) call particles smaller than 10 μm PM_{10}, those smaller than 2.5 μm $PM_{2.5}$, and so on. Because aerosols are denser than the air that surrounds them, gravity eventually

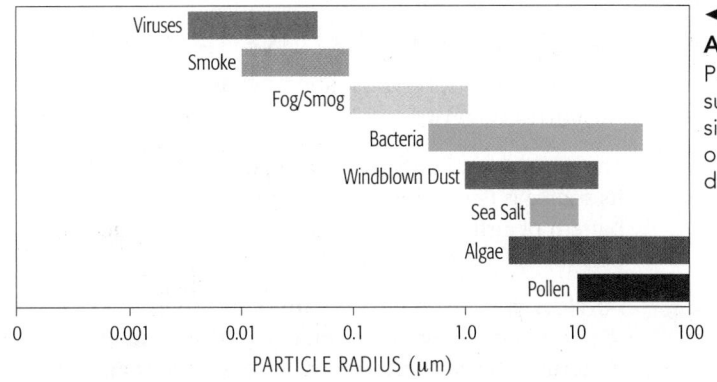

◄ Figure 10.5 **Common Aerosols**
Particles that are suspended in air vary in size. Larger particles fall out of the air faster than do smaller particles.

causes them to settle from the air onto the ground or other surfaces. The rate of such settling, or fallout, depends on the size of the particles; large particles settle much faster than smaller particles. PM_{10} and smaller aerosols are of particular concern to atmospheric scientists because they fall out of the atmosphere very slowly (Figure 10.5).

The variety of potential air pollutants is staggering. The EPA sets standards, monitors, and controls nearly 190 individual air pollutants, but this is only a fraction of the variety of chemicals added to our air each day. For example, over 400 different kinds of hydrocarbons have been identified in the exhaust of automobiles alone.

Sources of Air Pollution

■ Some pollutants are released directly into the atmosphere, but many others are generated in the atmosphere by reactions among gases and aerosols.

We usually associate air pollution with human actions, but it can also arise from natural processes. For example, volcanic eruptions emit a variety of sulfur compounds. Radon, a radioactive gas that can have serious health effects in some indoor environments, is released naturally from rocks that are rich in uranium. Aerosols such as sea salt and pollen are produced naturally, whereas dust and smoke come from a combination of natural processes and human actions.

Primary air pollutants are chemicals or particles that are directly emitted from identifiable sources. The elemental mercury released by combustion and volcanic eruptions is an example of a primary air pollutant. Other primary pollutants include the carbon monoxide and sulfur dioxide that are released when fossil fuels burn and the chlorofluorocarbons (CFCs) that are released from various industrial processes.

Secondary air pollutants are chemicals or particles that are produced in the atmosphere as a result of reactions among chemicals or aerosols. The ozone (O_3) that accumulates in city air during the daytime is a secondary pollutant. It is produced by reactions involving

NO and O_2 in the presence of sunlight (Figure 10.6). Although its specific characteristics vary from place to place, urban smog is a complex aerosol that forms from the combination of smoke, water droplets, and a mixture of chemicals. Its formation is also facilitated by sunlight. Secondary pollutants whose formation is facilitated by sunlight are called **photochemicals** or photochemical aerosols.

A few chemicals can be either primary or secondary pollutants. For example, NO is released directly into the air by automobiles. It is also produced when NO_2 and O_2 react to form O_3.

Atmospheric scientists distinguish between point and non-point sources of air pollution. **Point sources** are stationary, localized sources, such as the smokestack of a factory. Each point source has the potential to produce large amounts of pollutants. Air pollution may also be generated by numerous **non-point sources**, each of which produces relatively small amounts of chemicals or aerosols. Non-point sources of air pollution may be mobile, such as automobiles, or stationary, such as residential fireplaces and agriculture.

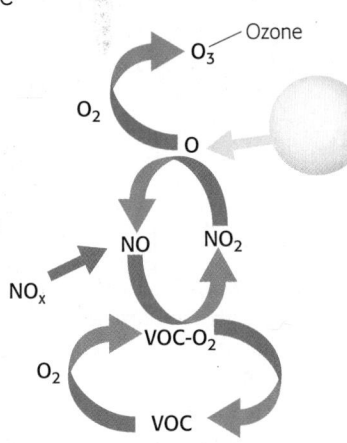

▲ Figure 10.6 **Forming a Secondary Air Pollutant**
During the daytime, primary air pollutants such as NO_2 and volatile organic compounds (VOCs) undergo a series of chemical reactions to produce ozone (O_3), a secondary pollutant that is toxic to humans.

Dispersion and Deposition of Air Pollution

■ The movement of gases and aerosols through the atmosphere is determined by air movement, emission height, and time in the atmosphere.

The distance and direction that a pollutant travels from its source is determined by three main factors—the pattern of airflow, the height of emission, and the pollutant's atmospheric lifetime.

Airflow patterns. Pollutants are dispersed in the atmosphere by diffusion, convection, and wind patterns. These processes vary at local, regional, and global scales.

At local scales, gases and aerosols diffuse along concentration gradients. The concentration of a pollutant diminishes with increasing distance from its source. Higher rates of production of a pollutant cause diffusion to occur more rapidly and result in higher concentrations of the pollutant farther away from the source.

If the air is perfectly still, a pollutant will disperse equally in all directions. But in the real world, concentration gradients are complicated by convection and turbulence. These air movements occur even at small scales, such as within a room or a portion of a city.

Convection causes pollutants to disperse more rapidly and to greater distances than would the process of diffusion alone. Convection currents result from differences in gas density, which are usually caused by differences in temperature. For example, pavement that is heated by the sun warms the air above it. This warmer, less-dense air rises, carrying air pollutants with it and dispersing them over great distances. Rising air carries pollutants above cities, where their concentrations become more dilute.

In general, air temperature decreases with increasing altitude. But sometimes weather patterns cause a layer of warm air to form above colder air; such a pattern is called a **temperature inversion** (Figure 10.7). Temperature inversions commonly occur on cold nights when air near the ground cools and is trapped beneath warm air. Temperature inversions prevent air close to the ground from mixing with the air above it. As a result, pollutants are trapped in the lower layer, causing them to accumulate below the inversion layer. This effect can be accentuated in valleys, as happened in Donora, Pennsylvania, in October of 1948.

At a regional scale, pollutants are dispersed by winds. Prevailing winds, such as the westerly winds of temperate regions and the easterly trade winds of the tropics, tend to move pollutants in one direction. For example, prevailing winds carry much of the pollution generated in the industrial cities around the southern end of Lake Michigan to the northeast.

Convection cells in the troposphere—the Hadley, Ferrel, and polar wind cells—disperse pollutants widely around the globe (see Module 3.7). These convection cells also move pollutants to high altitudes where they can migrate into the stratosphere.

Height of emission. The height at which point sources emit pollutants into the atmosphere has a large effect upon the concentration of those pollutants near the ground. Emission heights are influenced by the height of smokestacks, emission velocity, wind speed, and vertical temperature gradients (Figure 10.8). In general, the greater the emission height, the greater the distance pollutants will be carried and the more mixing and dilution will occur before they reach the ground.

Atmospheric lifetime. The lifetime of a gas molecule or a particle is the average time that it remains in the atmosphere. That lifetime depends on the chemical stability of a pollutant and factors that influence its removal from the air. The lifetimes of chemically reactive trace gases are measured in hours or days. Ammonia, which reacts quickly with oxygen in the atmosphere, has a lifetime of only an hour or two. Nitric oxide, which is more stable, survives in the atmosphere for about a day. Carbon monoxide reacts with relatively few molecules in the atmosphere; its lifetime is longer than 2 months. Lifetimes for very inert gases may exceed 1,000 years.

Aerosols are lost from the atmosphere through dry and wet deposition. In **dry deposition**, nonliquid particles are removed from the atmosphere by gravity. Because larger particles are less easily suspended in air, they tend to have a higher deposition rate and shorter lifetime than smaller particles. In **wet deposition**, trace gases and particles are captured in raindrops, snowflakes, or droplets of fog. The more soluble a pollutant is in water, the more prone it is to be lost from the atmosphere by wet deposition.

In summary, our air includes a great variety of gases and aerosols that are generated from both natural and anthropogenic sources. Their dispersion and deposition are influenced by the movement of air, the height in the atmosphere at which they are emitted, and the atmospheric lifetimes.

QUESTIONS 10.1

1. What characteristic of an aerosol is most important to its lifetime in the atmosphere?

2. Differentiate between primary and secondary air pollutants.

3. Describe three factors that influence the movement of pollutants in the atmosphere.

(MES) For additional review, go to **MasteringEnvironmentalScience**

Warm air

Cold air

▲ Figure 10.7 **Temperature Inversion**
Smoggy cold air is separated from the
warmer air above it by a thin layer of clouds.

► Figure 10.8 **Increasing Emission Plume Height**
Tall smokestacks at industrial facilities release pollutants
high into the air, thereby diminishing the concentration of
pollutants near the ground. Most of the visible emissions
from these smokestacks are steam, which is not a
pollutant.

313

10.2 Pollution in the Stratosphere

BIG IDEA The stratosphere is the layer of Earth's atmosphere extending 15–48 km (9–30 mi) above the troposphere. Although changes in the chemistry of the stratosphere do not directly affect the health of humans or ecosystems, they can influence climate and the amount of sunlight striking Earth's surface. Pollutants with particularly long atmospheric lifetimes can be dispersed through the troposphere and into the stratosphere. Some of these pollutants scatter or reflect sunlight back into space, reducing the amount of radiation reaching the troposphere and Earth's surface. Others chemically degrade stratospheric ozone. Reduced ozone results in increased amounts of ultraviolet light striking Earth's surface, with potentially serious consequences for ecosystems and human health. The Montreal Protocol on Substances That Deplete the Ozone Layer aims to halt the destruction of stratospheric ozone by these pollutants.

Aerosols and Climate

■ Sulfur compounds create aerosols in the stratosphere that reduce temperatures in the troposphere.

Many aerosols absorb or reflect light, thereby influencing climate. Some, such as dust blown from deserts and aerosols from volcanoes, are produced by natural processes. Others, such as smoke particles from burning wood and fossil fuels, are derived from human activities. In the troposphere, the lifetime of most aerosols is only a few days. In the stratosphere, aerosols may survive for several years.

Aerosols such as clouds, smoke, and dust block incoming sunlight, cooling Earth's surface. Because these particles have short atmospheric lifetimes, their effects on climate are generally brief. During the 1991 Gulf War in Kuwait, hundreds of oil wells were set afire, producing a plume of smoke that extended over much of the region (**Figure 10.9**). This dark haze produced cooler air temperatures during the months that the fires burned, but temperatures returned to normal soon after the fires were extinguished. Gaseous sulfur compounds released by volcanoes and human activities can have longer, more widespread effects on climate. Sulfur gases such as hydrogen sulfide (H_2S) and sulfur dioxide (SO_2) can be transported through the troposphere and into the stratosphere. There the gases are converted to an aerosol of tiny droplets of highly concentrated sulfuric acid (H_2SO_4). Because these droplets reflect sunlight back into space, variations in their abundance affect temperatures in the troposphere and at Earth's surface. For example, the massive eruption of Mount Tambora in Indonesia in 1815 caused two years of abnormally cold temperatures in places as distant as Europe and New England (see Module 9.2).

The combustion of fossil fuels, particularly coal, releases significant amounts of sulfur oxides; these chemicals have steadily increased the level of sulfur aerosols in the stratosphere. Atmospheric scientists believe that these aerosols have moderated the global warming caused by increasing concentrations of greenhouse gases in the troposphere. This moderating effect has led some to suggest that we might be able to reduce or even halt global warming by putting more sulfur compounds into the troposphere. However, most scientists warn that this would have devastating effects on many of Earth's ecosystems and could pose significant health risks to humans.

◄ Figure 10.9 **Aerosols Can Cool Earth's Surface**
In 1991, Kuwaiti oil wells burned for several months. The particulate emissions from these fires dispersed and produced cooler temperatures over large parts of the Middle East.

Stratospheric Ozone Destruction

■ Chemicals produced by humans in the troposphere can have destructive effects in the stratosphere.

In the troposphere, ozone (O_3) is a pollutant that has serious health effects on both humans and ecosystems. But in the stratosphere, there is a layer of ozone that is essential to the health of both humans and ecosystems. This layer of ozone filters out most of the ultraviolet radiation coming from the sun. Without the ozone layer, ultraviolet light would destroy nearly all terrestrial life (see Module 3.6).

Chemicals that contain the elements chlorine and bromine can destroy the ozone in the stratosphere. Among the most important of these chemicals are CFCs, which were once used as refrigerants and propellants in spray cans. In the troposphere where they are created, CFCs are quite stable. In the stratosphere, however, ultraviolet (UV) light causes them to release chlorine, which then catalyzes the destruction of ozone (Figure 10.10). Chlorine is constantly regenerated in this reaction. Thus, a single CFC molecule causes the destruction of thousands of ozone molecules, depleting the ozone in the stratosphere.

For reasons related to the circulation of gases in the upper atmosphere, ozone depletion is especially marked in the stratosphere over Antarctica and the Antarctic Ocean. This region is often called the Antarctic ozone hole, although it is not actually a hole but rather a zone of diminished ozone concentration (Figure 10.11).

Satellite measurements have verified the gradual global depletion of stratospheric ozone at the rate of about 4% per decade. As a consequence of depleted stratospheric ozone, levels of UV light have increased in Southern Hemisphere countries such as Australia and New Zealand. This increase is correlated with an increased frequency of skin cancers among people in these countries.

The issues related to CFCs highlight the challenges associated with many air pollutants. Chemicals that are relatively harmless in one setting can be quite destructive in another. This is particularly true of pollutants with long atmospheric lifetimes because they survive long enough to be dispersed widely. The atmospheric lifetime of a pollutant is also a major factor in the amount of time required to mitigate its effects.

QUESTIONS 10.2

1. Explain the connection between volcanic eruptions and Earth's climate in the months and years that follow them.

2. Why are CFCs stable in the troposphere but not in the stratosphere?

3. Chlorine is said to catalyze the degradation of ozone. What does this mean?

(MES) For additional review, go to **MasteringEnvironmentalScience**

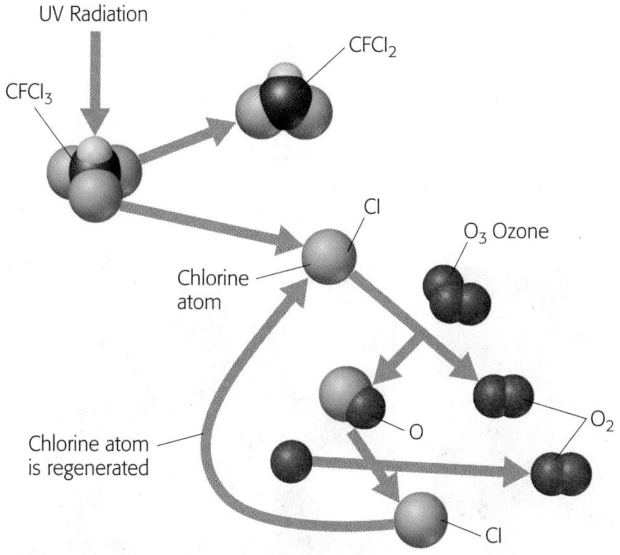

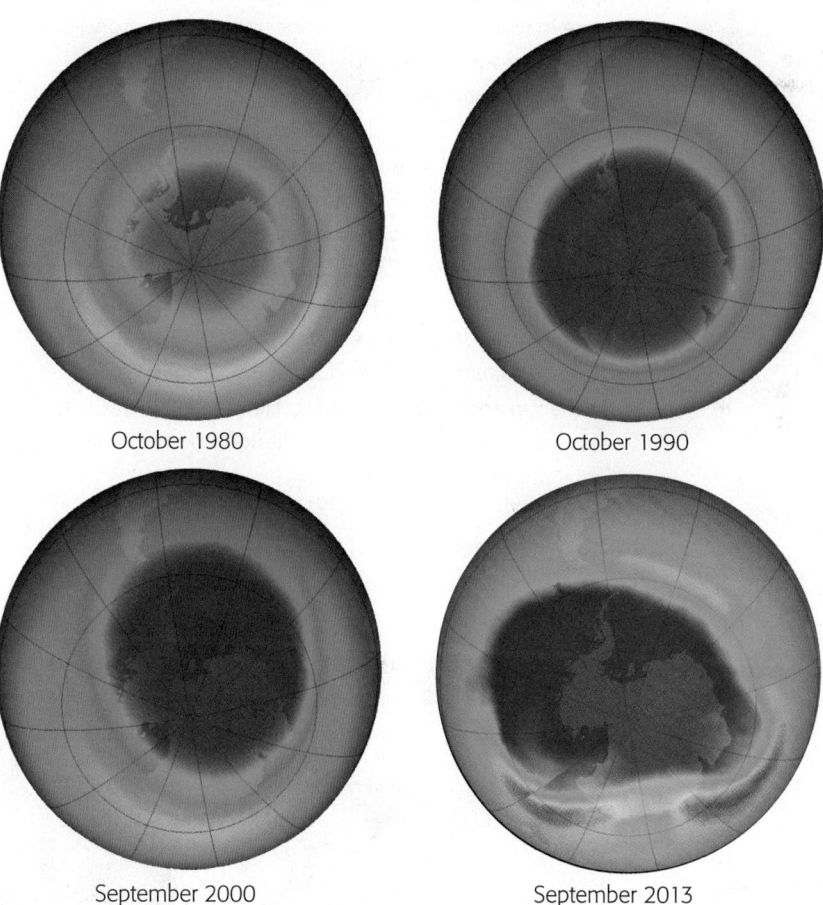

October 1980   October 1990

September 2000   September 2013

▲ Figure 10.11 **Ozone Depletion over the Antarctic**
The area of depleted stratospheric ozone known as the Antarctic ozone hole (blue-purple) is monitored each year by satellite. Although CFC emissions have declined significantly, the long residence time of these chemicals in the stratosphere means that this hole has not diminished in size.

Source: http://earthobservatory.nasa.gov/IOTD/view.php?id=82235

UV Radiation

$CFCl_2$

$CFCl_3$

Cl

O_3 Ozone

Chlorine atom

Chlorine atom is regenerated

O

O_2

Cl

▲ Figure 10.10 **Ozone Destruction**
In the stratosphere, ultraviolet light from the sun causes chlorine atoms to separate from CFCs. These atoms catalyze the conversion of O_3 to O_2. In the process, chlorine atoms are continually regenerated.

Laboratory Science Predicts Global Effects

How can throwing an old refrigerator into a landfill affect the stratosphere?

1. Why did Rowland and Molina suggest that the stability of CFCs might contribute to their impact on stratospheric ozone?

2. Explain why laboratory experiments demonstrating the destructive effects of UV light on CFCs led these scientist to the hypothesis that these chemicals might destroy ozone in the stratosphere.

Why were CFCs developed? How were their destructive effects discovered? The answers to these questions show how seemingly benign chemicals can have unintended effects. They also show how carefully *controlled experimental studies* in the laboratory can lead to understanding of chemical processes that are truly global.

In 1928, industrial chemist Thomas Midgley synthesized the first chlorofluorocarbon, called CFC-12. Within days, the makers of Frigidaire refrigerators asked Midgley to find an inexpensive, nontoxic, insoluble, and nonexplosive gas to substitute for ammonia, which was then the most widely used refrigerant. CFC-12 fit these criteria exactly.

Within three years, the chemical manufacturer DuPont had begun mass production of CFC-12 under the trade name Freon. Several similar CFC compounds were synthesized and manufactured soon after. By 1940, CFCs were in widespread use in refrigerators and air-conditioning systems around the world.

It is noteworthy that these refrigerants were only released to the atmosphere when coolants leaked or were drained. This changed in the 1940s when two other important uses for CFCs were discovered. Scientists found that these chemicals made ideal spray-can propellants. Under pressure, the CFCs flowed out of spray can nozzles with a mist of other ingredients, such as paint, insecticide, or hairspray. CFCs were also found to be ideal "blowing agents" for various kinds of foam products, such as insulation and disposable cups. Here again, it was the nontoxic and stable characteristics of CFCs that made them ideal for these applications.

By 1970, chemical companies were selling over 1.5 million metric tons of CFCs each year. Sooner or later, all of it was making its way into the atmosphere.

At that time, scientists assumed there would be no adverse environmental consequences because CFCs were nontoxic and did not react with other gases in the troposphere.

Then in 1973, chemists Frank Sherwood Rowland and Mario Molina began intensive studies of the fate of CFCs in Earth's atmosphere (**Figure 10.12**). They pointed out that because these chemicals are so stable, they have very long atmospheric lifetimes. This causes them to accumulate in the troposphere and eventually disperse into the stratosphere. Rowland and Molina demonstrated that ultraviolet light in the stratosphere would decompose CFC molecules, releasing chlorine atoms. Based on research by fellow chemist Paul Crutzen, they then *hypothesized* that this chlorine would catalyze the destruction of stratospheric ozone.

Rowland and Molina's hypothesis was hotly disputed by the chemical industry. The chairman of DuPont called it "a science fiction tale" and "utter rubbish." However, evaluations by the U.S. National Academy of Sciences supported the ozone depletion hypothesis. In fact, studies indicated that other chemicals containing chorine, bromine, and fluorine might also contribute to the destruction of stratospheric ozone. Such chemicals included carbon tetrachloride, used for dry cleaning, and methyl bromide, a widely used weed killer.

In 1985, a team of British scientists made *measurements* of ozone in the stratosphere over Antarctica. Their *data* left no doubt about the connection between CFCs and ozone depletion. These studies were the basis for an international treaty banning the use of CFCs. (That treaty is discussed in the final module of this chapter.) For their important research, Rowland, Molina, and Crutzen were awarded the 1995 Nobel Prize in Chemistry.

▶ Figure 10.12 **Ozone Layer Heroes**
Based on their laboratory research, F. Sherwood Rowland, Mario Molina, and Paul Crutzen predicted that CFCs were destroying ozone in the stratosphere. Their prediction was verified with the discovery of the Antarctic ozone hole in 1985.

F. SHERWOOD ROWLAND

MARIO MOLINA

PAUL CRUTZEN

10.3 Pollution in the Troposphere

BIG IDEA The troposphere is the 15-km (9-mi) thick layer of air nearest to Earth's surface that is constantly mixed by convection (see Module 3.6). Here, three kinds of air pollution have the greatest effect on humans and ecosystems. (1) **Acid deposition** refers to the dispersion of acid-containing gases, aerosols, and rain onto soils, plants, buildings, and bodies of water. It forms from sulfur and nitrogen oxides that are emitted by the burning of fossil fuels. (2) **Heavy metal** pollution comes primarily from anthropogenic sources, including the combustion of coal, metal smelting, and waste incineration. Mercury and lead are particularly important heavy metal air pollutants. (3) **Smog** is a complex mixture of both primary and secondary pollutants.

Acid Deposition

■ Sulfur and nitrogen oxides dissolve in water to produce sulfuric and nitric acids.

In 1872, the Scottish chemist Robert Angus Smith published a book entitled *Air and Rain: The Beginnings of a Chemical Climatology*. By taking systematic measurements across Great Britain, Smith discovered that rain falling near industrial centers was considerably more acidic than rain falling in more remote locations (see Module 3.1). Smith was the first scientist to use the phrase "acid rain." He and others asserted that acid rain was responsible for the deterioration of buildings and statues in cities and the diminished production of nearby cropland (Figure 10.13).

In the 1950s, scientists began to suspect that acid rain was killing trees in the forests of northern Europe and causing the biological diversity of lakes and streams to decline. By 1970, nearly 25,000 lakes near Sweden's cities supported only those species of plants and animals that could withstand acidic conditions. In the 1960s and 1970s, aquatic and terrestrial ecosystems in New England and southeastern Canada began to display similar symptoms (Figure 10.14). Acidic air pollution was the central focus of the United Nations Conference on Human Environment held in Stockholm in 1972. That conference catalyzed an international program to study and manage the issue.

What is the source of this acidity? Sulfur dioxide (SO_2) and nitrogen oxide (NO_x) gases are the most abundant acidifying pollutants. SO_2 reacts with oxygen and water to produce sulfuric acid; NO_x reacts with oxygen and water to produce nitric acid. These acids can adhere to dust particles or dissolve in cloud droplets. Eventually the acids are deposited on plants, soils, and waters. Because both dry deposition and precipitation deposit acids, scientists prefer to call the process acid *deposition* rather than acid *rain*.

Acid deposition has significant effects on both forest and freshwater ecosystems. Acidic rainwater leaches nutrients and minerals from vegetation and soils, diminishing the fertility of forests. Acidic rainfall and fog also damage the leaves of trees, causing defoliation. Increased acidity (lower pH) in lakes and streams increases the solubility of calcium carbonate, the major constituent of the shells of many invertebrates. Aluminum and other potentially toxic metals are usually bound to sediments but go into solution as waters become more acidic. High levels of aluminum clog the gills of fish and inhibit the development of their skeletons.

The primary sources of SO_2 are coal-fired power plants and metal smelters. NO_x is emitted from power plants and in vehicle exhaust. Power plants typically have very tall smokestacks that are designed to disperse emissions and prevent high concentrations of pollutants from accumulating nearby. These tall smokestacks release pollutants into high prevailing winds that can carry them over long distances.

▲ Figure 10.13 **Acidic Weathering**
These statues, called caryatids, have supported part of the Acropolis in Athens for over 2,000 years. But in the past century, sulfur and nitrogen emissions from nearby factories have greatly accelerated the weathering of these pillars.

◀ Figure 10.14 **Acid Rain Affects Ecosystems**
Acid deposition was a major factor in the death of these fir trees.

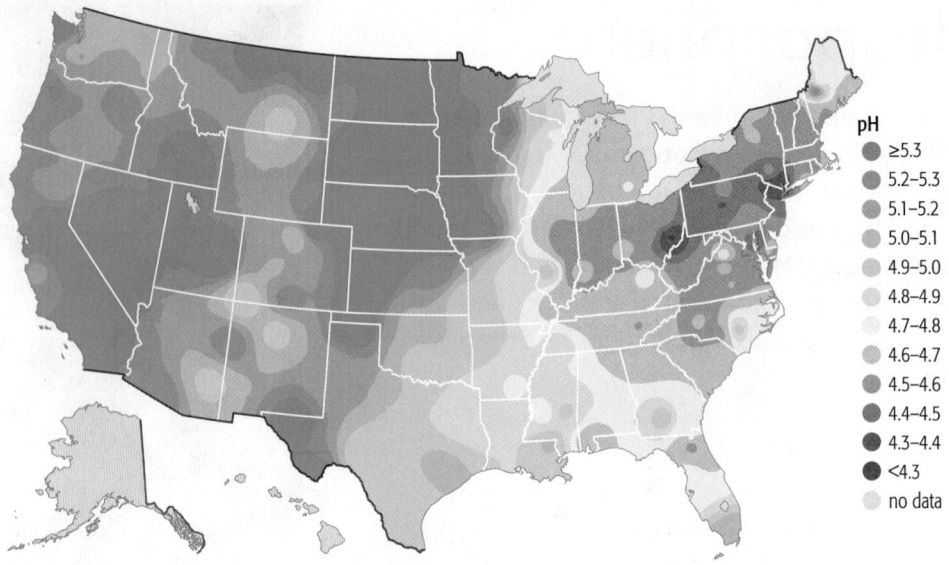

pH	
●	≥5.3
●	5.2–5.3
●	5.1–5.2
●	5.0–5.1
●	4.9–5.0
●	4.8–4.9
●	4.7–4.8
●	4.6–4.7
●	4.5–4.6
●	4.4–4.5
●	4.3–4.4
●	<4.3
●	no data

▲ Figure 10.15 **Acid Deposition in the United States**

Rainfall has a lower pH (is more acidic) near cities and industrial centers. The large amounts of SO_2 and NO_x emitted from industrial facilities and utilities in Illinois, Indiana, and Ohio are deposited as acid rain in regions to the east.

Source: National Atmospheric Deposition Program. 2014. *National Trends Network.* URL = http://nadp.sws.uiuc.edu/ntn/. Last accessed 9-14-2014.

The pattern of acid deposition in North America is greatly affected by the location of the sources and prevailing winds. SO_2 emissions are especially high in the industrial cities in northern Illinois, Indiana, and Ohio. Emissions from Ohio alone are twice the combined emissions from all of New England, New York, and New Jersey. Prevailing winds carry acidic aerosols northeast, where they rain down on New England and southeast Canada (Figure 10.15).

Between 1980 and 1990, the United States, Canada, and the countries of the European Union enacted laws and regulations to limit sulfur emissions from coal-burning facilities. With these strong incentives, utilities and industries installed **scrubbers** on smokestacks that removed over 90% of SO_2. Scrubbers are devices that contain chemicals that react with pollutants such as SO_2 and filter them from industrial exhausts.

Today, sulfur emissions in North America and Europe are less than half of the 1980 levels. Emissions of NO_x from coal-burning facilities have also declined, but at the same time emissions from vehicles have increased. As a consequence, total annual NO_x emissions in developed countries have declined only a little over the past 30 years.

In most developed countries, overall acid deposition is declining, and the pH of many lakes and streams is beginning to rise. In developing countries such as China and India, however, coal consumption and automobile use are rapidly increasing: so are SO_2 and NO_x emissions and acid deposition.

Heavy Metals

■ Human activities have greatly increased the emission of heavy metals.

Heavy metals such as mercury and lead are quite toxic to humans and other animals, but normally they are present in Earth's atmosphere in only small amounts. These elements are emitted to the atmosphere from natural sources such as volcanoes, sea salt, and dust. Analyses of ice cores from Greenland and Antarctica reveal that atmospheric concentrations of heavy metals have increased severalfold over the past 200 years. This increase is the result of human activities, including the combustion of fossil fuels, waste incineration, and industrial processes such as metal smelting. The concentrations of heavy metal pollutants diminish with increasing distance from their sources.

Because it has become so widespread and its toxicity is so well understood, mercury may be the most notorious heavy metal air pollutant. Emissions of mercury are highest around the world's industrial centers. Most emissions to the air are in the form of mercury gas. Gaseous mercury is subsequently incorporated into droplets of water and dust, which eventually fall onto Earth's lands and waters (Figure 10.16).

In terrestrial and aquatic ecosystems, mercury is transformed into soluble organic mercury. Organic mercury is then absorbed and stored in the tissues of organisms. Mercury may accumulate to toxic concentrations in animals that feed near the top of food webs, such as carnivorous fishes (see Module 6.4). Humans take in mercury when they eat these fishes. Certain occupations, such as working near mines or coal-burning utilities,

also expose humans to mercury. Prolonged exposure to even small amounts of mercury can cause damage to the nervous system, paralysis, and death.

Global emissions of mercury are estimated to be 5,000–7,000 metric tons per year. About one-third of this mercury comes from natural sources. Another third comes from human activities such as burning fossil fuels and smelting metals. Mercury is also reemitted to the environment from wildfires and ocean aerosols. This reemitted mercury is thought to be the source of the other third of annual emissions.

Over the past 30 years, anthropogenic emissions of mercury have declined. The major causes of this reduction have been the installation of smokestack scrubbers and the diminished use of mercury in industrial, commercial, and residential products and processes. In the United States in 1990, industries and medical facilities emitted more than 220 tons of mercury; today, annual emissions are below 100 tons. Nevertheless, mercury that was emitted in the past continues to be recycled through ecosystems.

Lead, like mercury, has numerous effects on the health of organisms. In humans, it can disrupt the synthesis of blood hemoglobin and cause brain and kidney damage. Even small amounts can cause neurological damage in young children.

To determine the history of lead emissions, scientists study sediments in bogs, areas of land saturated with moisture and covered with water. The levels of lead in these sediments provide a reliable record of the emissions

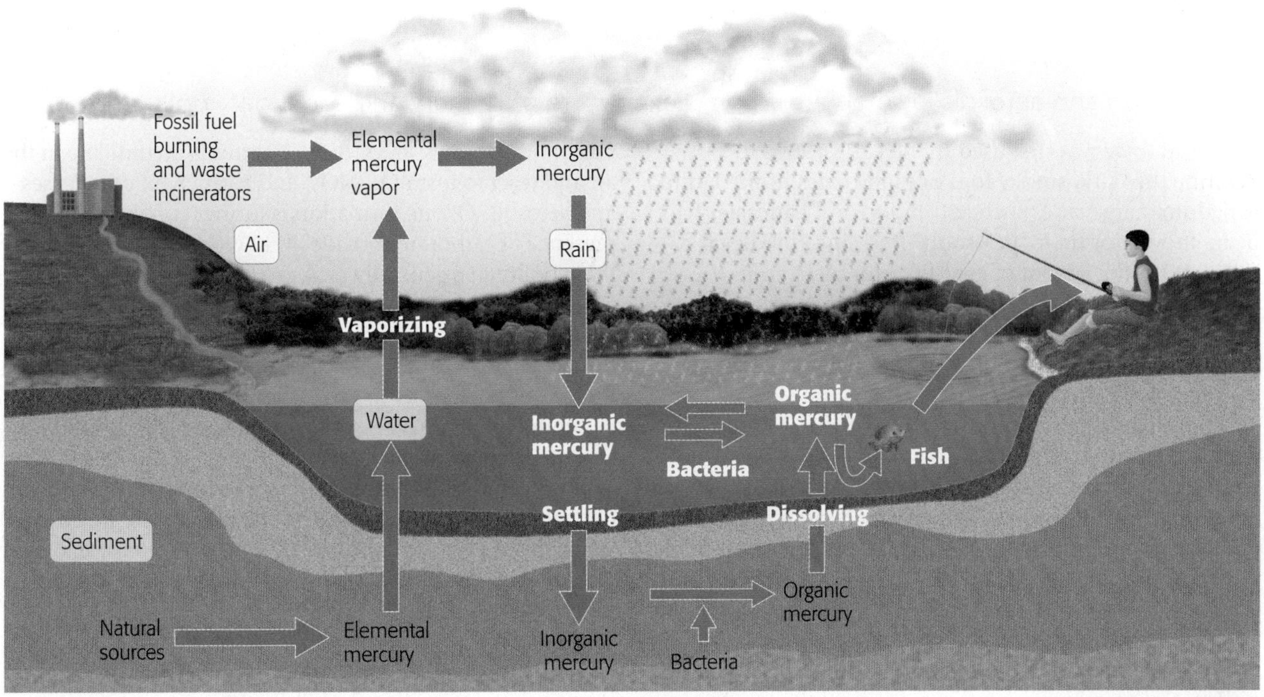

◀ Figure 10.16 **The Mercury Cycle**
Nowadays, human activities are the most important source of mercury to the environment. Because mercury is taken up by plants and animals and stored in their tissues, they play an important role in the cycling of mercury.

of this pollutant over thousands of years. The pattern of change in these emissions is a telling case study of how human technologies affect air quality. Ten thousand years ago, annual global lead emissions were 100–200 metric tons. Most of this is thought to have come from volcanism and wind erosion of desert soils.

About 8,000 years ago, the concentration of lead in bog sediments gradually began to increase. This increase was a consequence of expanding agriculture, which put more lead-laden dust into the air.

Around 5,500 years ago, humans discovered how to smelt silver from lead-containing ores. Smelting caused total emissions to increase to about 1,000 metric tons per year. Silver coins and jewelry became widespread about 3,000 years ago in the Mediterranean region and elsewhere. This led to rapid expansion of silver mining and smelting and a 10-fold increase in lead emissions. But as the Roman Empire declined, lead emissions diminished. When the Industrial Revolution began in the mid-18th century, levels of lead in sediments began to increase again.

Sediments laid down after 1920 show a dramatic increase in the amount of lead, suggesting that emissions had suddenly increased to well above 1 million metric tons per year. They had indeed. In 1921, Thomas Midgley, the chemist of CFC fame, found that adding tetraethyl lead (lead bonded to four ethylene molecules) to gasoline increased the power and efficiency of automobile engines and reduced engine knock. Gasoline containing lead was widely marketed as ethyl gasoline beginning in 1923. (Ironically, in that same year Midgley and three of his coworkers came down with serious cases of lead poisoning.) By the 1930s, 90% of all gasoline in the United States and Europe was leaded.

Some experts began to worry about the poisonous effects of lead emissions almost immediately. In 1925,

a study commissioned by the U.S. Surgeon General concluded that, although there were no grounds for prohibiting the use of ethyl gasoline, further studies were needed. None were done for more than 30 years.

When the U.S. Public Health Service finally reexamined the health effects of leaded gasoline in 1959, they concluded that there was significant reason for concern. Despite mounting evidence to the contrary, car manufacturers and gasoline producers insisted that leaded gasoline was safe.

It was not until 1976 that the U.S. EPA began regulating tetraethyl lead as a pollutant. This change in policy came not so much from the many concerns about the health effects of lead pollution but rather from the fact that lead deactivates the catalysts in automobile **catalytic converters**. These devices were required on all cars manufactured after 1975 in order to reduce emissions of carbon monoxide, hydrocarbons, and nitrogen oxides.

Since 1980, the use of leaded gasoline has been banned in most countries, and global emissions have declined by nearly 90%. In the United States, for example, lead emissions were over 200,000 metric tons in 1970; today, they are below 2,000 metric tons, most of which comes from ore smelting and battery manufacture (**Figure 10.17**). However, a number of developing countries continue to use leaded fuel, and emissions in those countries remain quite high.

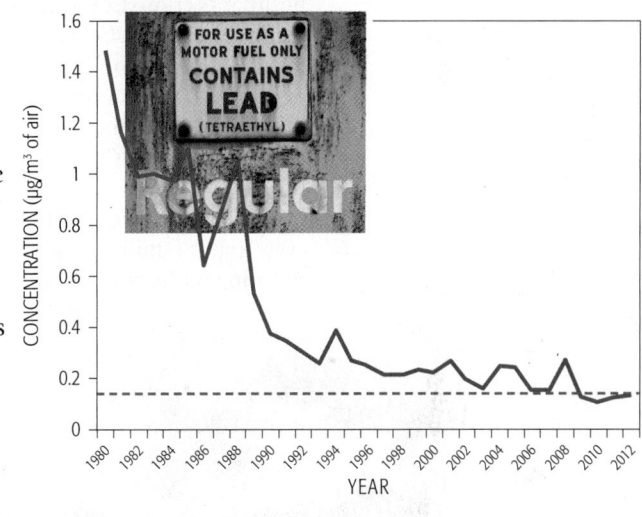

▲ Figure 10.17 **Getting the Lead Out**
The 1970 Clean Air Act required that leaded gasoline be phased out by 1980. This has resulted in a 90% reduction in concentrations of lead in the air of U.S. cities. The red dashed line represents the U.S. EPA standard of 0.15 µg/m³. Concentrations above this level are considered harmful to public health and the environment.

Source: http://www.epa.gov/air/airtrends/lead.html#pbnat

Q *Do modern cars really add significantly to air pollution?*

Sierra Seaman, University of Florida

Norm: **A** Catalytic converters on modern cars have reduced the amounts of some pollutants, but in most cities cars remain the most important source of the primary pollutants that form photochemical smog.

Smog

■ The gases and aerosols in smog are a complex mixture of primary and secondary pollutants.

The word *smog* was first used in the early 1900s to describe the dark, smoky fogs of many cities in the United Kingdom. Large cities such as London and Lancaster had been known for their "pea-soup" fogs since medieval times. With the advent of coal-fired furnaces and steam engines, episodes of foul air became particularly prolonged and intense. Smog was a major reason why people who lived in cities had significantly lower life expectancies than people who lived in rural areas.

Today, smog is a common feature of cities around the world. Atmospheric scientists differentiate between two kinds of smog, based on the mode of formation. **Industrial smog** is composed primarily of pollutants released in coal burning. These include carbon monoxide (CO), sulfur dioxide (SO_2), and particles of carbon soot. The illness and deaths in Donora, Pennsylvania, were a consequence of industrial smog.

Photochemical smog is formed when certain primary pollutants interact with sunlight. Many of these primary pollutants come from the exhaust of vehicles. Nitrogen oxide (NO) and VOCs are among the most important of these primary pollutants. These chemicals react with oxygen and water vapor to form a variety of secondary pollutants, including ozone (O_3) and nitrogen dioxide (NO_2). It is the NO_2 that gives photochemical smog its brownish color.

The formation of photochemical smog follows a daily cycle that is familiar to many city dwellers. Exhaust from morning traffic releases large quantities of NO and VOCs.

On sunny, windless days, these chemicals accumulate in the air and react to form O_3, NO_2, and a variety of other gases and aerosols. Ozone formation is further enhanced by the presence of carbon monoxide and VOCs. Concentrations of these pollutants generally peak in the late afternoon.

The lifetimes of many of the constituents of photochemical smog are only an hour or two, so their high concentrations depend on their being constantly replenished. Thus, as the sun goes down and light-dependent reactions cease, concentrations of O_3 and NO_2 and other constituents of photochemical smog decline. But if the air is still or trapped by a temperature inversion, the chemical building blocks remain. With the morning sun and the addition of more automobile emissions, the photochemical smog that forms the next day may be even more intense.

What are the smoggiest cities in the world? Based on accounts in media outlets, you might answer Los Angeles, Mexico City, or Beijing. But while these cities do have significant air pollution challenges, they are not among the very worst (Figure 10.18). Ahwaz, Iran, tops the list of cities with the worst air pollution with PM_{10} levels that are 15 times higher than those of Los Angeles. Five other Iranian cities are among the world's 10 most polluted locations. Little or no regulation of industrial and transportation emissions and mountainous topography that favors the formation of thermal inversions are among that factors that contribute to the poor air quality in these cities. The high levels of smog in cities such as Ulaanbaatar, Mongolia, and Gabarone, Botswana, are largely a consequence of the widespread use of coal, charcoal, and wood for cooking and heating. With PM_{10} levels that are more than 50% higher than Los Angeles, the city of Bakersfield, located in the southernmost part of California's San Joaquin Valley, is now the smoggiest city in the United States. With limited industrial development and situated where thermal inversions are uncommon, Santa Fe, New Mexico, ranks high among U.S. cities with the cleanest air.

Emissions of NO, CO, and VOCs have been greatly reduced in countries such as the United States where automobile exhaust systems are required to have catalytic converters. This in turn has reduced ground-level concentrations

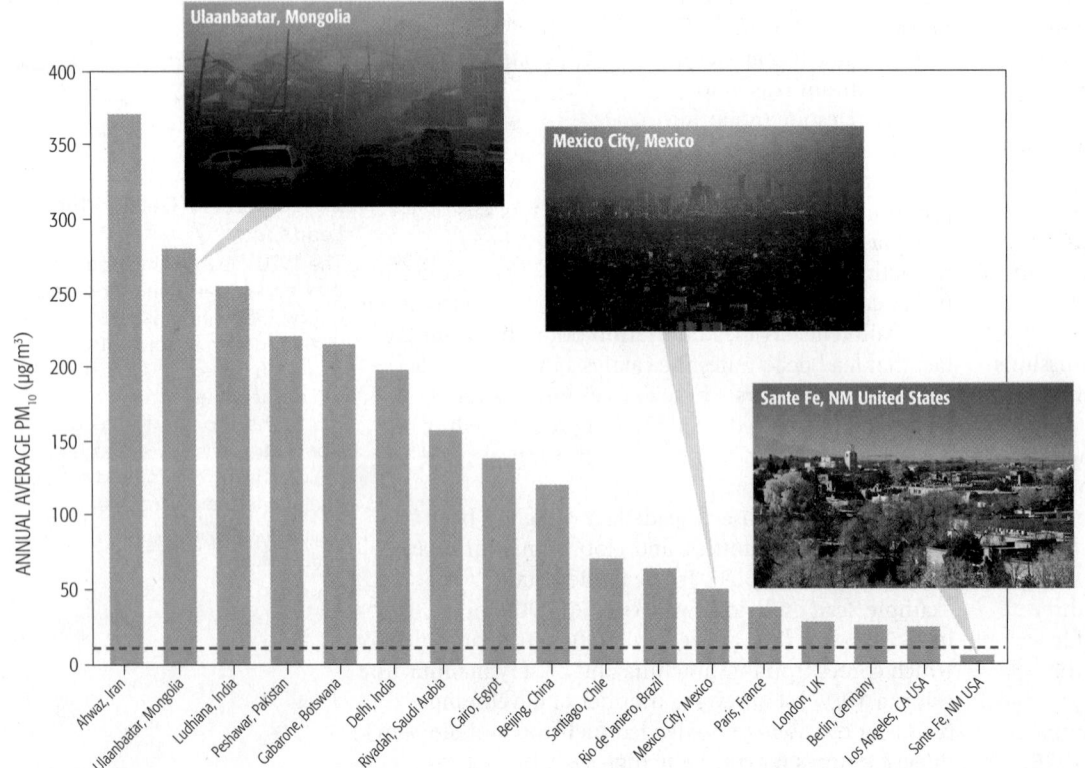

◄ Figure 10.18 **Smog Around the World**
This chart shows average annual PM_{10} exposure levels for selected cities around the world. The red dashed line indicates the U.S. EPA threshold standard for average annual PM_{10} of 12.5 μg/m³.

Source: The World Health Organization. http://apps.who.int/gho/data/view.main.AMBIENTCITY2014?lang=en

of O_3 and smog (Figure 10.19). Metals such as platinum, palladium, and rhodium embedded in the interior surface of these devices catalyze the conversion of these primary pollutants to less toxic substances. Catalytic converters on new cars remove over 90% of the primary pollutants that contribute to photochemical smog.

In the world's developed countries, improved technologies such as smokestack scrubbers have diminished the extent of industrial smog. In developing countries with rapidly growing industrial and transportation systems and many fewer air-quality regulations, both kinds of smog are increasing. Although catalytic converters have reduced emissions of NO, CO, and VOCs from automobiles, the total number of vehicles on urban highways is growing. Such rapid growth in automobile transportation is one important reason for the high smog levels in cities such as Beijing, China, and Delhi, India.

◄ Figure 10.19 **Reduced Smog**
Catalytic converters have been required on all U.S. cars since 1986. This has produced a significant decline in levels of ozone and associated smog. Ozone concentrations now hover around the EPA standard of 0.075 ppm (red dashed line). Concentrations of O_3 above this standard are deemed unhealthy.

Source: EPA. 2014. National trends in lead levels. URL = http://www.epa.gov/air/airtrends/lead.html, last accessed 10/15/2014

Air Quality Index

■ How can we know when air pollutant concentrations threaten the health of humans and the ecosystems they depend on?

In most countries, regulatory agencies such as the U.S. EPA set standards for acceptable levels of air pollutants. Such standards are based on scientific studies designed to determine the minimum concentrations and exposure durations that might impair the health of humans or other organisms. The U.S. Clean Air Act requires the EPA to set standards for the six so-called criterion pollutants listed in Figure 10.20.

In cooperation with state and local agencies, the EPA monitors levels of important pollutants at over 1,000 locations, including 300 cities. At each location, EPA calculates a measure of air health for each criterion pollutant called the **air quality index (AQI)**. The AQI for each pollutant is scaled from 0 to 500, with 100 equal to its regulatory standard. Air quality is then scaled by color code (Figure 10.21). For example, air quality at AQI less than 50 is considered good or code green, whereas AQI levels above 100 are considered unhealthy or code orange for sensitive groups such as children or the elderly.

Where YOU LIVE ---------------------------------------
What Is Your AQI?
How healthy is the air you breathe? Go to the home page of the AirNow website and enter your zip code and click "GO." You will receive your local AQI for the EPA's criterion pollutants. You can also compare the AQI for your location with others across the country using the AirCompare feature.

What is your AQI for today and how does it compare to other locations in the United States? What factors likely contribute to the quality of the air you breathe every day?

Pollutant	Threshold concentration	Duration
Ground-level Ozone (O_3)	0.075 ppm	8 hrs
Particulate Matter $PM_{2.5}$	35 $\mu g/m^3$	8 hrs
Particulate Matter PM_{10}	150 $\mu g/m^3$	24 hrs
Carbon Monoxide (CO)	9 ppm	8 hrs
Sulfur Dioxide (SO_2)	75 ppb	1 hr
Nitrogen Dioxide (NO_2)	100 ppb	1 hr
Lead (Pb)	15 $\mu g/m^3$	3 months

▲ Figure 10.20 **Criterion Pollutant Standards**
Under the U.S. Clean Air Act, the EPA sets standards for acceptable levels of seven important pollutants. These standards are defined in terms of exposure over a specific time duration.

Air Quality Index (AQI) Values	Levels of Health Concern	Colors
When the AQI is in this range:	*...air quality conditions are:*	*...as symbolized by this color:*
0 to 50	Good	Green
51 to 100	Moderate	Yellow
101 to 150	Unhealthy for Sensitive Groups	Orange
151 to 200	Unhealthy	Red
201 to 300	Very Unhealthy	Purple
301 to 500	Hazardous	Maroon

▲ Figure 10.21 **AQI and Your Health**
The EPA and other public agencies define air quality conditions over different ranges of AQI. These are color coded in order to communicate conditions clearly to the public.

QUESTIONS 10.3

1. What are the primary sources of acid deposition?

2. What are the primary sources of mercury emissions to the atmosphere?

3. Differentiate between industrial smog and photochemical smog.

(MES) For additional review, go to **MasteringEnvironmentalScience**

10.4 Indoor Air Pollution

BIG IDEA During severe episodes of smog, people are often advised to stay indoors. Although staying indoors does protect us from some forms of air pollution, indoor air generally contains a greater quantity and variety of pollutants than does outdoor air. Indoor air pollution comes from many sources, including tobacco smoke, building materials, furniture, chlorine-treated water, ovens, and even the soil and rock beneath buildings. The long-term effects of these pollutants are compounded by the fact that we spend 80–90% of our time indoors. Efforts to make buildings more energy efficient by reducing ventilation can increase the concentrations of indoor pollutants.

Combustion By-Products

■ Burning organic matter releases a large number of inorganic and organic chemicals into the air.

The complete combustion of organic carbon in the presence of oxygen produces carbon dioxide and water. In most situations, however, the process of combustion is incomplete, and many additional gases and particles are also released into the air. These **combustion by-products** include carbon monoxide, nitrogen and sulfur oxides, hundreds of VOCs, and particulate matter.

In developing countries, most people cook and heat their homes with open fires, particularly those fueled with wood or animal dung. These fires release large amounts of carbon monoxide and particulate soot. The World Health Organization estimates that the combustion by-products from open fires are responsible for respiratory and circulatory ailments that kill over 4 million people in developing countries each year (Figure 10.22). In many places, people are simply unaware of the health risks associated with open indoor fires. But even where the health effects are understood, people often cannot afford cleaner alternatives.

In developed countries, fireplaces and woodstoves are generally vented to the outside. However, improper installation, downdrafts, and leaks in flue pipes can cause indoor pollution. Pollutants can also be introduced by activities such as starting and stoking fires and removing ashes.

Gas stoves, furnaces, and space heaters used in developed countries also produce combustion by-products. Carbon monoxide and nitrogen oxide levels are generally higher in homes using gas stoves and furnaces than in those using electric stoves and heaters. However, studies of the relationship between gas cooking and respiratory ailments have been inconclusive. Unvented kerosene space heaters, which were popular in the United States in the 1970s and 1980s, also produce combustion by-products.

The U.S. EPA has developed emissions standards for all new gas- and wood-burning appliances. These standards have reduced emissions of combustion by-products to both outdoor and indoor environments.

▶ **Figure 10.22 Deaths Caused by Indoor Smoke**
Death rates (number of deaths/million/year) from indoor smoke are especially high in countries where wood is commonly used as fuel. This includes India and most African countries.

Source: World Health Organization. http://www.who.int/heli/risks/indoorair/en/webiapmap.jpg

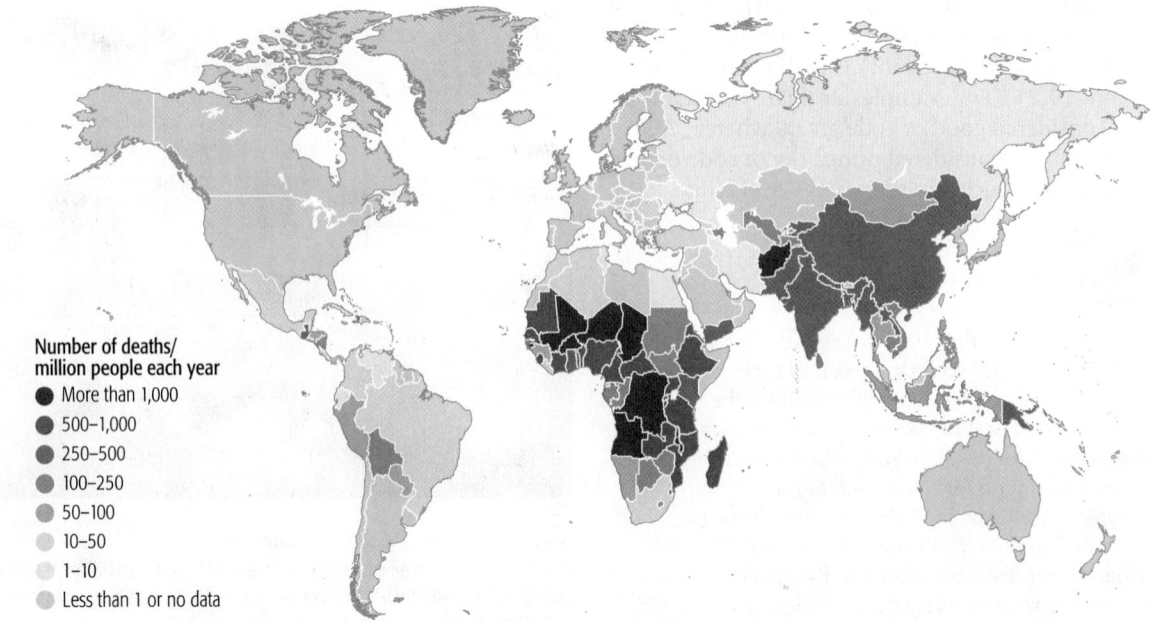

Number of deaths/ million people each year
● More than 1,000
● 500–1,000
● 250–500
● 100–250
● 50–100
○ 10–50
○ 1–10
○ Less than 1 or no data

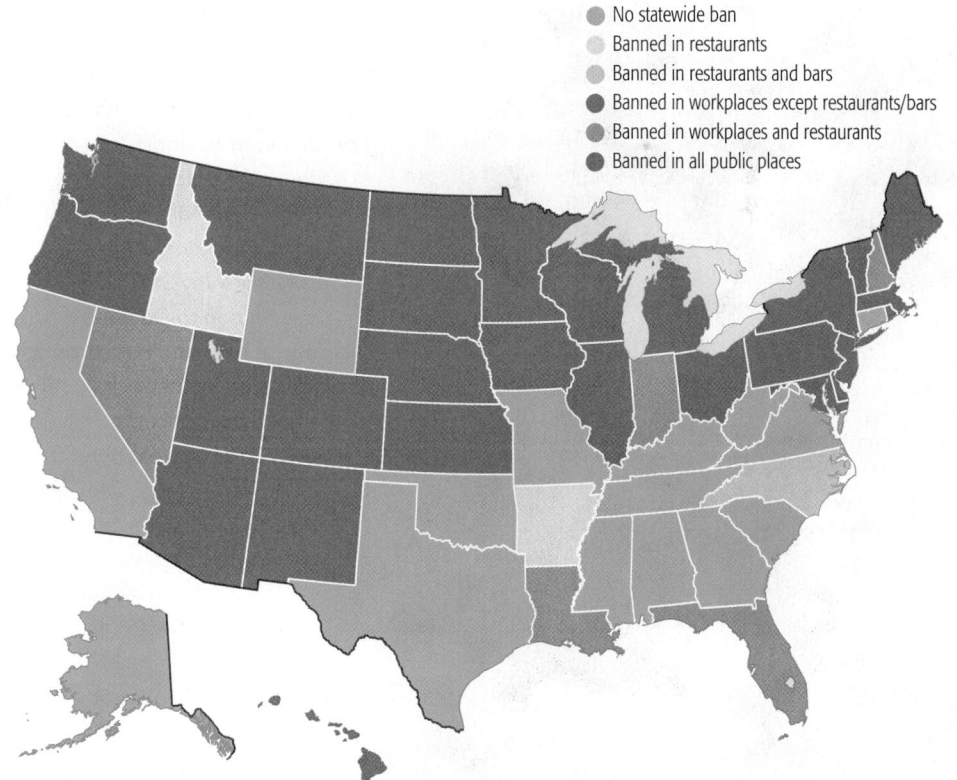

Relatively inexpensive carbon monoxide detectors are available for those who may be concerned about levels of this deadly pollutant.

Tobacco smoke is one of the most common indoor air pollutants. It is also one of the most hazardous (Figure 10.23). The serious health risks associated with smoking have been known for decades. Smoking is the primary cause of lung cancer and pulmonary emphysema and significantly increases risks of heart disease. Environmental tobacco smoke, or secondhand smoke, poses these risks to smokers and nonsmokers alike. It is also linked to chronic bronchitis, asthma, eye and ear infections, and sudden infant death syndrome. The U.S. EPA estimates that each year environmental tobacco smoke is responsible for over 40,000 premature deaths from lung cancer and heart disease among nonsmokers.

Over 250 different toxic or cancer-causing gases and particles have been identified in tobacco smoke. These include particulate matter, nicotine, carbon monoxide, nitrogen oxides, formaldehyde, hydrogen cyanide, and a host of VOCs. For example, concentrations of respirable particulate matter are three to four times higher in homes with smokers than in homes where no one smokes. Such particulate matter is implicated in many of the negative health effects of tobacco smoke.

Smoking bans are the primary means for reducing pollution from environmental tobacco smoke. Many studies have demonstrated rapid improvement in both the air quality and the health of individuals working in areas where smoking is banned. Today, many cities and states have implemented laws requiring workplaces, restaurants, and bars to be smoke free. Similar laws are pending in many other locations (Figure 10.24). Efforts to set national standards and regulations for environmental tobacco smoke have, to date, been unsuccessful.

▲ Figure 10.23 **The Most Deadly Indoor Pollutant**
Tobacco smoke contains hundreds of toxic gases and aerosols.

- No statewide ban
- Banned in restaurants
- Banned in restaurants and bars
- Banned in workplaces except restaurants/bars
- Banned in workplaces and restaurants
- Banned in all public places

▲ Figure 10.24 **State Laws Regulating Indoor Smoking**
As of 2014, 34 states have implemented either limited or complete bans on smoking in public places.

Breathing Easier in the Kitchen

How can poor countries diminish indoor pollution while encouraging economic development?

In developing countries worldwide, smoke from indoor wood-fueled stoves causes many health problems and more deaths than are caused from malaria (**Figure 10.25**). Wood smoke contains hundreds of different chemicals and aerosols, many of which cause cancer or encourage respiratory infections, tuberculosis, and other pulmonary diseases. Women and young children suffer the greatest exposure to wood smoke and so are at greatest risk. In addition to causing indoor pollution, burning wood for fuel increases the rate of deforestation in developing countries.

Using smokeless cooking fuels would greatly reduce the risks of indoor pollution. Unfortunately, kerosene and natural gas, the most reasonable alternatives, are too expensive for most families in the world's poorest countries. Charcoal is a more affordable option. Burning charcoal produces far less smoke than wood; it also releases more heat per unit weight. Charcoal can be stored indefinitely, whereas wood rots quickly in tropical regions. But charcoal is more costly than wood. And because charcoal is normally produced from wood, its use still encourages deforestation.

Enter the MIT students doing class projects in professor Amy Smith's Introduction to Development course. Amy and her students wondered whether charcoal might be produced from agricultural wastes such as banana leaves, corn cobs, and the sugarcane waste called bagasse. In many poor countries such wastes are abundant but are simply discarded. This group understood that charcoal production needed to be cheap and easily implemented with minimum technology. They came up with a simple and economical three-step process. First, the wastes are dried in the sun for several days. Next, the wastes are heated in low-oxygen kilns constructed from discarded oil drums, which are widely available in most poor countries (**Figure 10.26**). This heating produces charcoal powder. Finally, the powder is formed into charcoal briquettes using a paste made from cassava, a widely grown tropical crop. These briquettes are similar to those you might use at a backyard barbecue. They can be formed by hand, or they can be mass produced using relatively inexpensive briquetting presses.

One of the students who worked with Amy Smith on this project was business student Jules Walter, a native of Haiti. Walter teamed up with Amy Banzaert, a mechanical engineering graduate student researching this alternative charcoal, and they, along with others, have created a company called Bagazo (Spanish for bagasse) that is dedicated to the development and application of this process. With ingenuity and lots of hard work, a class project has produced a solution to three challenging problems. Most immediately, it provides families in poor countries with the means to cook their meals, reduce indoor pollution, and preserve their respiratory health. Second, it provides an alternative to the use of wood as fuel and thereby promises to diminish deforestation. Third, it does all this while significantly diminishing health costs. This is truly sustainable development!

▲ Figure 10.26 **From Agricultural Waste to Charcoal**
In the first step in charcoal manufacture, Amy Banzaert helps residents in an El Salvador village load sugarcane bagasse into an oil drum kiln. Banzaert did her master's degree research on technologies to optimize charcoal manufacture from this material.

▼ Figure 10.25 **Kitchen Pollution**
Each day, the wood fire used by this Masai family to heat water and food also exposes them to considerable indoor air pollution.

Building Materials

■ Construction materials emit many volatile organic compounds.

Many materials used in the construction of residential and commercial buildings emit pollutants that pose significant risks to human health. Adhesives, synthetic fabrics, paints, wood preservatives, and insulating materials are all sources of indoor pollutants. Formaldehyde (HCHO), VOCs, and asbestos are among the most important indoor pollutants derived from building materials. Sometimes the pollutants in a building will affect the health of so many people that the building is said to have "sick building syndrome."

Formaldehyde is a potent irritant of mucous membranes and exposure can irritate the eyes and upper respiratory tract. Prolonged exposure can cause skin rashes, headaches, fatigue, and depression. Formaldehyde pollution began to receive public attention in the early 1980s, when homebuilders began to use more urea-formaldehyde foam insulation and pressed wood products, such as strand board. Since then, improved manufacturing processes have significantly reduced formaldehyde emissions from adhesives and pressed wood products. Furthermore, emissions from these products decrease significantly with time. Thus, risks from this pollutant can be diminished by allowing building materials to age.

Significant concentrations of more than 300 different VOCs have been measured in indoor air. Their sources include adhesives, solvents, paints, room deodorizers, carpets, draperies, and furniture. Carpets alone emit nearly 100 different VOCs. Because of the diversity of these compounds, specific cause-and-effect relationships with human health are difficult to establish. However, some VOCs, such as polycyclic aromatic hydrocarbons, are known to cause cancer.

Asbestos is a fibrous mineral that is resistant to heat and fire. Until the early 1970s, it was widely used for a variety of building applications, including spray-on fireproofing and insulation, fiberboard, roofing, and flooring (Figure 10.27). In 1973, the U.S. EPA determined that asbestos fibers were a direct cause of debilitating lung disease, lung cancer, and cancer of the lining of the chest and abdominal cavity. It has since been regulated as a hazardous air pollutant.

Today, the use of asbestos is banned in all new buildings. Nevertheless, asbestos-containing materials remain in place in hundreds of thousands of older buildings. The mere presence of asbestos fibers in building materials does not pose a health risk. However, if these materials are disturbed—as during building repair, renovation, or demolition—microscopic asbestos fibers are wafted into the air. Therefore, the EPA regulates the renovation and demolition of all older buildings made with asbestos-containing materials.

▲ Figure 10.27 **Asbestos** The tiny fibers in asbestos-containing materials cause cancer in breathing passages and lungs.

Radon

■ Naturally emitted radon accumulates in closed indoor spaces and can cause lung cancer.

Radon (^{222}Rn) is a nontoxic gas produced from the radioactive decay of the element radium-226 (^{226}Ra). Radon itself undergoes radioactive decay to produce radioactive isotopes of the elements polonium (^{218}Po) and lead (^{214}Pb). Because they are electrically charged, polonium and lead bind to particulate matter, producing a radioactive aerosol. When inhaled, this aerosol can cause lung cancer.

Radium occurs in various amounts in most rocks, including granite, limestone, and basalt. As a result, it is present almost everywhere on Earth in or near soil. Outdoor concentrations of radon and its decay products are generally low and do not pose a significant health risk. However, concentrations of radon may be a thousandfold higher inside buildings with poor ventilation that are built on rock or soil that is high in ^{226}Ra (Figure 10.28). The EPA estimates that each year in the United States exposure to radon causes 15,000–16,000 lung cancer deaths.

Because of these health risks, the EPA and U.S. Surgeon General issued a public health warning in 1988 recommending that all homes be tested for radon and that steps to reduce radon levels be taken where warranted. Radon levels can be reduced by sealing basement walls and floors, installing check valves in drains, and ventilating crawl spaces.

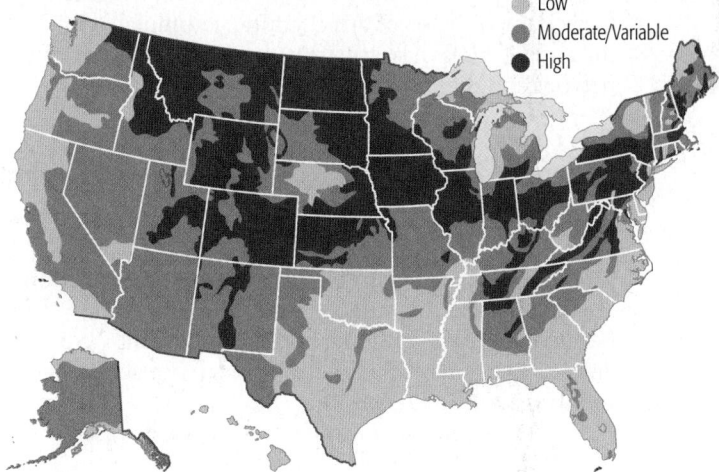

● Low
● Moderate/Variable
● High

▲ Figure 10.28 **The Geography of Radon Emissions** Radon levels are largely determined by the geological strata upon which homes are built. For example, dark brown indicates areas with large amounts of the rock basalt, which is rich in the element radon.

Source: http://energy.cr.usgs.gov/radon/usrnpot.gif

► **Figure 10.29 Indoor Pesticides**
In the past, potent pesticides were sprayed in homes with little attention to their possible effects on humans and their pets.

Pesticides

■ Chemicals used to control insects and mold also affect human health.

We use a wide range of toxic chemicals in our homes and other buildings to control unwanted insects, rodents, molds, and bacteria. If misapplied, these chemicals may affect our health. Pesticides are especially hazardous to vulnerable populations, such as small children and the elderly (**Figure 10.29**).

Pesticides used to kill termites and preserve wood are of particular concern. Between 1948 and 1988, for example, the pesticide chlordane was injected into soils and cement slabs beneath houses to control termites. Significant amounts of this chemical accumulated in the air of houses that were treated. Chlordane affects the nervous and digestive systems of people and animals, producing symptoms such as headaches, confusion, stomach cramps, and diarrhea. Since 1988, the EPA has banned all uses of this pesticide.

Prior to 1980, pentachlorophenol (PCP) was widely used to treat foundation timbers and the wood in log houses. Like DDT, PCP is a chlorinated hydrocarbon that accumulates in fatty tissues. People living in PCP-treated log houses were found to have from five to seven times as much of this pollutant in their tissues as people living in untreated homes. Such levels of exposure can produce symptoms similar to those produced by chlordane. Today, PCP-treated wood is no longer used in buildings, although it is still used to treat utility poles.

Biological Contaminants

■ Microorganisms can pollute indoor spaces.

Indoor spaces with many people and limited ventilation are more likely than outdoor air to contain viruses, bacteria, and fungi that cause illness in humans. Indoor environments may also favor the growth of bacteria and molds that cause allergies.

The 1976 outbreak of a pneumonia-like disease among some of the people attending an American Legion convention in Philadelphia is a dramatic example of building-related disease. Over several days, 221 individuals became ill and 34 died from this disease. After intensive study, the Centers for Disease Control determined that this outbreak of legionnaires' disease was caused by a previously unknown bacterium, which was subsequently named *Legionella pneumophila*. This bacterium thrives in water that condenses in the cooling towers of air-conditioning systems. It was transmitted in aerosols from the air-conditioning system that served the hotel lobby.

Since then, hundreds of outbreaks of legionnaires' and similar diseases have been reported in office buildings, college dorms, cruise ships, hospitals, and industrial facilities. Regulations requiring large air-conditioning systems to be regularly cleaned and disinfected have reduced the incidence of these diseases.

In summary, many different kinds of air pollutants are more highly concentrated in the air inside buildings than in outdoor air. Activities such as cooking and heating are often the source of indoor pollutants. Building materials themselves may emit pollutants. Limited ventilation allows pollutants to accumulate indoors.

QUESTIONS 10.4

1. What are combustion by-products?
2. Describe four types of pollution generated inside your home or dorm room that might influence your health.

(MES) For additional review, go to **MasteringEnvironmentalScience**

10.5 Air Pollution Policy and Law

BIG IDEA Prior to 1963, the regulation of air pollution in the United States was delegated to individual states. State regulations were generally weak and highly variable. Laws passed between 1963 and 1970 moved the responsibility for managing air quality to the federal government. These laws required federal agencies to set standards for acceptable emission levels and to prescribe the best technologies to meet those standards. The 1990 amendment to the Clean Air Act implemented a market-based approach to reducing emissions of SO_2, although such an approach is not appropriate for managing all pollutants. International treaties have focused on the movement of pollutants across borders and on pollutants that pose global threats, such as CFCs. Such agreements require that countries perceive that pollutants threaten ecosystems and human health. They must also ensure that the costs and benefits of pollution reduction are shared among countries fairly.

▲ Figure 10.30 **Protest for Change**
Following a deadly smog event in 1954, these Los Angeles women protested for air-quality laws and the creation of an agency to control smog.

U.S. Air Pollution Policy

■ Policies to control air pollution in the United States have evolved significantly over the past 60 years.

The public outcry over the 1948 tragedy in Donora, Pennsylvania, put pressure on politicians, but legislative action to reduce air pollution was slow in coming. As postwar industries grew and the population boomed, air quality in major cities declined rapidly. In 1952, a smog event in London resulted in the deaths of over 4,000 people. In 1953, a similar event in New York killed at least 200 people, and severe smog in Los Angeles resulted in dozens of deaths in 1954 (Figure 10.30).

Finally in 1955, the U.S. Congress passed the Air Pollution Control Act, which President Eisenhower signed. As it states, this act was intended "to provide research and technical assistance relating to air pollution control." While acknowledging that air pollution was a real danger, the act asserted that the responsibility for setting standards and regulating air quality should remain with the states. The federal government's role should be purely informational. The U.S. Surgeon General was given a small budget to conduct research and distribute information "relating to air pollution and the prevention and abatement thereof." Meanwhile, the nation's air quality continued to deteriorate.

Between 1963 and 1990, a series of new federal laws shifted the responsibility for regulating air pollution to the national government. The 1963 Clean Air Act focused on emissions of pollutants from stationary sources, such as power plants and steel mills. Subsequent legislation included regulations for emissions from mobile sources, such as trucks and automobiles.

In 1970, the Clean Air Act was amended. In fact, this "amendment" was nearly a complete rewrite of the original law. The amendment required that new and more stringent National Ambient Air Quality Standards (NAAQS) be set to protect public health. The amendment also set standards for emissions from motor vehicles.

And perhaps most important, it allowed citizens to take legal action against any person, organization, or government that violated emission standards (Figure 10.31).

What moved the federal government to act so swiftly and forcefully during this period? Certainly the rising tide of environmentalism played some role. More important, however, was the fact that a few leading cities and states—Los Angeles, California, New York City, and Pennsylvania—had begun to enact tough air pollution control regulations. Key industries such as utilities and automobile manufacturers found that it was in their interest to push for moderate and uniform federal standards that would preempt stringent or inconsistent state and local regulations.

Q *How can you tell what the air quality is for your local area?*
Matthew Poveromo, Syracuse University

Norm: **A** You can access air quality data for U.S. zip codes, states, and regions on the U.S. EPA website.

1963	Clean Air Act	Required that the secretary of Health, Education and Welfare (HEW) set firm standards for pollutant emissions from stationary sources such as power plants and steel mills.
1965	Motor Vehicle Air Pollution Control Act	Extended authority to mobile sources of pollutants such as trucks and automobiles.
1967	Air Quality Act	Expanded federal regulation of both stationary and mobile pollution sources, and it authorized the HEW secretary to seek immediate court-ordered abatement of air pollution that presented an "imminent and substantial danger."
1970	Clean Air Act Amendment	Required that National Ambient Air Quality Standards (NAAQS) be set to protect public health and welfare, and established New Source Performance Standards that strictly regulated the emissions from any new source in an area.
1990	Clean Air Act Amendment	Set new standards to reduce automobile emissions and established a cap-and-trade program for the reduction of SO_2 emissions.

▲ Figure 10.31 **Legislating Clean Air**
A wide range of federal laws mandating cleaner air in the United States were enacted over the course of 27 years.

327

▶ Figure 10.32 **U.S. SO₂ Pollution**

Concentrations of SO₂ in the United States have been declining since 1970, but the implementation of the sulfur cap-and-trade program set up by the 1990 Clean Air Act has caused the decline to accelerate. On average, concentrations are well below the EPA threshold standard of 75 ppb (red dashed line).

Source: US EPA. http://www.epa.gov/air/airtrends/sulfur.html. Last accessed 1-16-2014.

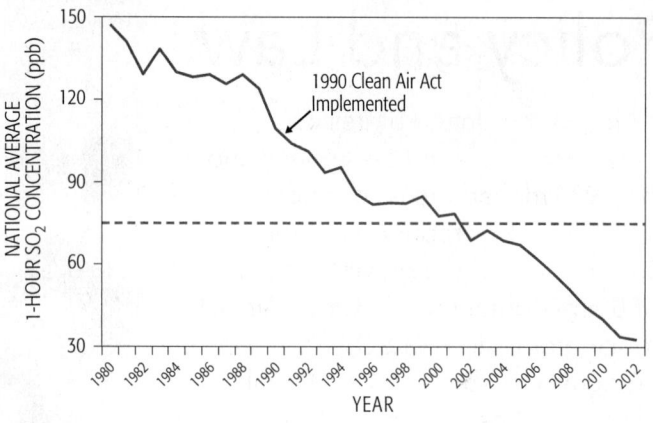

In 1990, the U.S. Congress once again significantly amended the Clean Air Act. This amendment empowered the U.S. EPA to determine the best technologies for controlling specific pollutants and to require the installation of those technologies where appropriate. The act also specified actions to reduce emissions of chlorofluorocarbons in order to prevent stratospheric ozone depletion.

One of the most notable provisions of the 1990 Clean Air Act was the establishment of a cap-and-trade program for SO₂ emissions (see Module 2.5). These provisions authorized the EPA to set caps for SO₂ emissions from facilities such as power plants. Each facility can either take action to reduce its emissions to meet its cap or buy publicly traded emission credits. Emission credits are created and sold by facilities that have decreased their emissions below their caps. Ideally, emission caps would gradually be lowered and credits for emission reductions would eventually expire.

The sulfur emissions trading program was quite controversial because it allowed individual facilities to continue to release high levels of SO₂. But advocates pointed out that the primary concern about SO₂ is

its contribution to acid deposition. Individual facilities contribute only a small amount to this large-scale problem; rather, it is a consequence of the cumulative emissions from many facilities.

Since the SO₂ cap-and-trade program began, sulfur emissions have been reduced by more than a third (Figure 10.32). This cap-and-trade program has also served as a model for the reduction of emissions of other pollutants. A cap-and-trade program is now in operation for emissions of CO₂ under the Kyoto Treaty (see Module 9.8).

Despite the success of the SO₂ cap-and-trade program, environmental groups fiercely opposed George W. Bush's 2002 Clear Skies Initiative, which proposed extending the program to include emissions of NO$_x$ and mercury. Why? Because cap-and-trade programs for NO$_x$ and mercury would be likely to cause the effects of these pollutants to be felt more severely in some communities than others. Facilities that emit NO$_x$ and mercury have a great impact on the environment in their immediate vicinity. NO$_x$ is one of the most important primary pollutants contributing to photochemical smog and high concentrations of ozone in cities, and wet and dry fallout of mercury is much higher near coal-burning facilities than farther away. Environmental groups also opposed the Clear Skies Initiative because it extended the timetables for meeting emission standards for these pollutants. In the end, the 2002 Clear Skies Initiative was not approved.

Unlike standards for outdoor air quality, standards for managing indoor air quality are set by state and local governments. The Occupational Safety and Health Administration (OSHA) and the EPA provide guidelines for standards for important indoor pollutants, such as carbon monoxide, particulate matter, and radon. About half of the states have adopted legal standards and enforcement policies based on these guidelines.

Q *Is the risk from poor air quality greater in developed or developing countries?*

Macy, Louisiana State University

Norm: A Generally, risks from poor air quality are higher in developing countries because of their lax regulations and standards. Developing nations also have large numbers of woodstoves and high rates of tobacco use.

International Air Pollution Policy

■ Agreements among countries to limit air pollution are based on a shared sense of need and an equitable distribution of costs and benefits.

Most international boundaries have little relationship to the movement of air masses or the pollutants they carry. Emissions from one country often affect the air quality of other countries downwind. In the case of pollutants such as CFCs and greenhouse gases, the combined emissions from many countries have significant global consequences.

Geneva Convention on Long-range Transboundary Air Pollution.
In the 1960s, scientists found that emissions from countries across Europe were contributing to the acidification of Scandinavian lakes. The need for action on this issue was highlighted in the 1972 United Nations Conference on the Human Environment held in Stockholm. This led the

member states of the United Nations to craft the Geneva Convention on Long-range Transboundary Air Pollution, which was ratified in 1983.

The convention laid down general principles for international cooperation in reducing air pollution. It also created an institutional framework for research and policy collaboration. This framework has been the basis for the formulation and ratification of protocols to reduce SO₂, NO$_x$, VOCs, and heavy metal pollutants.

The 1999 Protocol.
The 1999 Protocol to Abate Acidification, Eutrophication, and Ground-level Ozone set 2010 targets for emissions of four pollutants—SO₂, NO$_x$, VOCs, and ammonia. The biggest cuts were required from countries whose emissions had severe

environmental or health impacts and who could most easily afford abatement measures. The signatories to this protocol have reduced SO_2 emissions by 66%, NO_x emissions by 31%, VOC emissions by 40%, and ammonia emissions by 20% compared to 1990 levels.

The Montreal Protocol on Ozone.

One of the most successful international agreements to manage an air pollutant was the 1987 Montreal Protocol on Substances That Deplete the Ozone Layer. In it, countries agreed to a timetable for phasing out the production and use of CFCs and other ozone-destroying chemicals. Even the largest developing countries—China and India—ratified this protocol, although they were granted a longer phase-out period.

The Montreal Protocol was unusual in several respects. It incorporated procedures for continuing review and refinement as scientific knowledge changes. As a consequence, the timetable for reducing CFC emissions has been tightened several times. It also required that a strong majority of producer nations sign before it could be ratified.

The impact of the Montreal Protocol on global CFC production was immediate and significant (**Figure 10.33**). What factors made international action on the issue of ozone depletion comparatively easy? First, there was widespread agreement about the scientific evidence for the relationship between CFCs and ozone depletion. Second, ozone depletion threatens everyone—there are not winners and losers. Third, it was easy to monitor production because there were relatively few CFC-producing companies, and they were concentrated in a small number of industrialized countries. Fourth, a number of substitute chemicals that did not interact with ozone were readily available. Many of the companies that produced CFCs played a very positive role in moving the protocol forward as they positioned themselves to produce substitute chemicals.

The Montreal Protocol has served as a model for the development of other agreements, including the Kyoto Protocol to limit greenhouse gas emissions. Unfortunately, the factors that led to rapid agreement on CFCs are not in place for greenhouse gases. Compared to the scientific consensus regarding CFCs, there has been much less certainty regarding the causes and consequences of greenhouse gas emissions. Unlike the risks of CFCs, the risks of global warming are not equally distributed among nations. Unlike the production of CFCs, the production of greenhouse gases occurs worldwide, and emissions from developing countries are accelerating. And finally, although there are energy alternatives to fossil fuels, there is social and economic resistance to their adoption (see Chapter 15).

In general, the development of policies and laws to limit air pollution requires agreement that the threats to the environment or human health are significant. Policies are more likely to succeed if they ensure that the costs and benefits of reducing emissions are shared equitably among stakeholders.

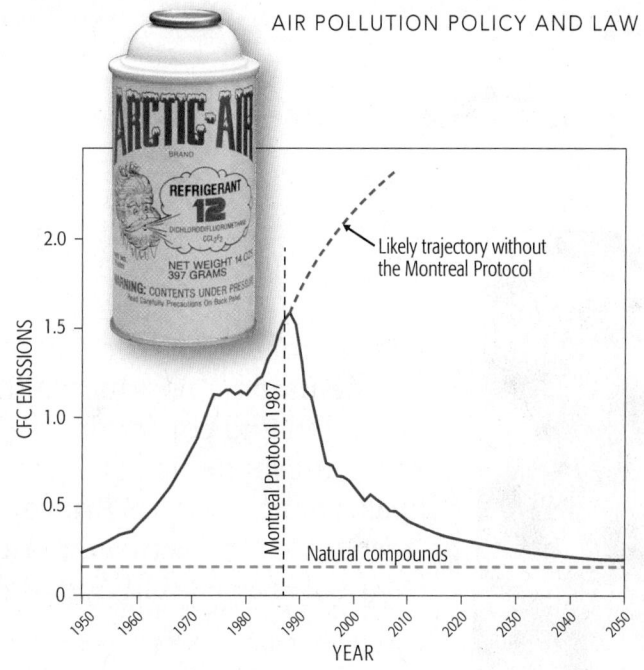

◄ **Figure 10.33 Declining CFC Emissions**
Up to the 1987 Montreal Protocol, global emissions of CFCs were increasing (indicated here in millions of metric tons of CFC-11 equivalents). Countries and industries responded quickly to the protocol's mandates, and emissions declined by 80% in just 10 years. The blue solid line indicates emissions of CFCs from natural processes.

Source: World Meteorological Organization. 2011. *Scientific Assessment of Ozone Depletion*: 2010. Global Ozone Research and Monitoring Report No. 52, 516 pp.

New FRONTIERS

Making Climate Connections

The lay public often confuses destruction of the ozone hole with global warming even though scientists go to great lengths to distinguish between these two phenomena. But recent work suggests that there may be an important connection between them.

Over the past century, Earth's atmosphere has warmed by about 0.8 °C. However, the rate of warming has varied considerably during that time. In fact, the rate has been notably slower since the early 1990s (refer to Figure 9.12). Recent studies by Francisco Estrada and his colleagues at the National Autonomous University of Mexico suggest that this recent slowdown in the rate of global warming may be a direct benefit of efforts to restore Earth's ozone layer.

Beside their destructive effects on ozone, CFCs are also very powerful greenhouse gases, 1,300–13,000 times more powerful than CO_2 (see Module 9.3). The decreased CFC emissions obtained under the Montreal Protocol have yet to diminish the size of the stratospheric ozone hole, but they have lowered tropospheric concentrations of CFCs considerably. Models developed by Estrada and his colleagues suggest that diminished tropospheric CFC levels may be responsible for the recent decline in global warming rates.

These results have sparked considerable controversy among climate scientists who note there are alternative explanations for the recent decline in the global warming rate. These include increased heat absorption of Earth's oceans and natural cycles in tropospheric temperatures that are unrelated to human activities. Nevertheless, work by Estrada and his colleagues has encouraged other scientists to look more closely at the possible benefits of the Montreal Protocol on climate change.

Disagreement and debate are essential to the scientific process. How can the public and decision makers encourage such debate without letting it stop forward progress toward solutions to important environmental challenges?

Source: Estrada, F., P. Perron, and B. Martínez-López. 2013. Statistically derived contributions of diverse human influences to twentieth-century temperature changes. *Nature Geoscience* 6: 1050–1055.

QUESTIONS 10.5

1. In general terms, how has U.S. policy toward air pollution changed since the tragedy in Donora, Pennsylvania?

2. How have the cap-and-trade provisions of the 1990 Clean Air Act Amendment operated to lower sulfur emissions?

(MES) For additional review, go to **MasteringEnvironmentalScience**

Mapping Pollutants in Little Village and Around the World

Marisol Becerra studied public policy at DePaul University in Chicago, Illinois. As a volunteer with Chicago's Little Village Environmental Justice Organization (LVEJO), Marisol led a youth group in the creation of OurMap of Environmental Justice. This interactive online map helps educate the people of Little Village about sources of pollution and contaminants in their community. For her leadership and innovative use of social media to produce social change, Marisol was awarded a 2008 Brower Youth Award, a national award that recognizes the achievements of young environmental leaders. Marisol is also a Bill and Melinda Gates Millennium Scholar.

How did you first get the idea for OurMap of Environmental Justice?

I grew up in Little Village, a community in the southwest side of Chicago. I first became aware of the Crawford power plant at age three. Back then, I thought it was a "cloud factory" because of the large clouds of smoke emitted from its smokestacks. I would always ask my parents what the building was for. They couldn't tell me because they weren't sure themselves.

During my freshman year of high school, I accompanied my mom on a "community asset and toxic tour" led by the Little Village Environmental Justice Organization (LVEJO). As part of that tour we stopped by the industrial corridor in our community and learned what each industry does and how it affects the environment.

It was then I learned the "cloud factory" was, in fact, a coal-fired power plant. I soon discovered that pollutants emitted from the plant increase the risk of several health problems, including asthma and other respiratory illnesses and premature death. When I realized the plant's emissions have an impact on the environment and public health, I was angry. I saw an injustice happening in my community, and I had to do something about it.

What steps did you take to get started?

I realized that if I wanted to make an impact, I had to get my peers involved. I decided to start Young Activists Organizing as Today's Leaders, a youth branch of LVEJO. Through this organization, I helped recruit and train local youths to give community asset and toxic tours and to lead strategy sessions for organizing the community. Members also helped to inventory and map environmental assets and toxins in Little Village.

Most of my peers interact and find information through the Internet, so I knew I could reach a larger audience that way. I decided to create a paper newsletter, which would also appear online in a blog format.

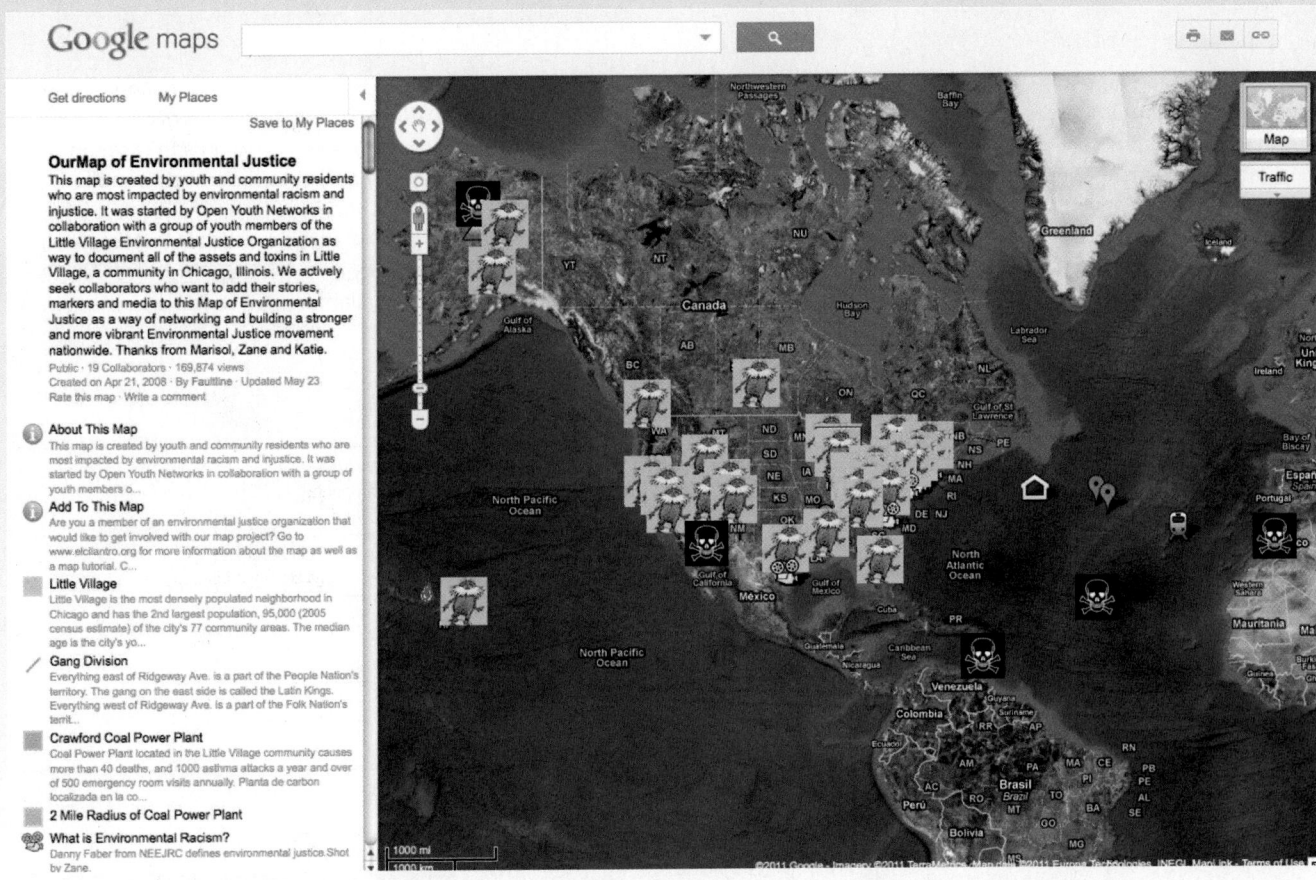

Source: http://www.airnow.gov; map data Google, INEGI.

We then began to use Google Maps, YouTube, Flickr, and websites like Myspace and Facebook to disseminate information. Using those social media tools, we made an interactive, online tour of the community asset and toxic tours we had been conducting in person. This is how OurMap of Environmental Justice was created.

What are the biggest challenges you've faced with the project, and how did you solve them?

One of the biggest challenges was bridging the gap between youths and adults. We wanted adults to take us seriously. There was a misconception in our community that youths were not interested in being involved in community issues.

We bridged the generational gap by interviewing adults and sharing their stories on our map. We broke the stereotype of "lazy youths" who just play video games or cause trouble. The more work we did, the more the community saw that we were dedicated and hardworking and wanted to collaborate with adults to extend our project's reach.

How has your work affected other people and the environment?

OurMap of Environmental Justice is used nationally and internationally. It has connected people around the world.

We now work with youth networks in the United States and other countries to educate communities about pollutants, toxins, and potentially harmful community issues, such as gang violence. The more that people know about what's going on in their community and how it's affecting others, the more likely they will be to organize and help stop the things that are adversely affecting people and the environment.

What advice would you give to students who want to replicate your model in their own communities?

Explore your community. You can do research through interviews and surveys. Identify the assets of your community, including the strengths of its members. Identify the toxic places causing harm to your community.

Then get to work by using the strengths and talents of community members. Everyone can contribute something, no matter how small. Work together to create your own Google map to shed light on community issues, to strengthen your case, and to change policy.

Summary

Air quality refers to the amounts of gases and aerosols—air pollutants—in the atmosphere that may degrade ecosystem processes and human well-being. Pollutants generated from various sources are dispersed in Earth's troposphere. Humans also generate pollutants that make their way into the stratosphere. Chlorofluorocarbons are particularly important because they cause the breakdown of stratospheric ozone, which in turn allows more UV radiation to reach Earth's surface. Indoor spaces confine air movement and provide important sources of pollution, including combustion by-products, building materials, radon, pesticides, and biological contaminants. During the past 60 years, U.S. policies regarding air pollution have evolved from programs of research and technical assistance to programs that set and enforce stringent standards for stationary and mobile sources of pollution. International treaties to manage air pollution have been most successful when the evidence for threats to human well-being is clearly established, when pollution sources are easy to identify and manage, and when cost-effective alternatives to pollutants are available.

10.1 Air Quality and Air Pollution

■ Earth's atmosphere contains a large number of trace gases that vary in concentration over time and from place to place.

KEY TERMS

air quality, air pollution, trace gases, volatile organic compounds (VOCs), particulate matter, aerosols, primary air pollutants, secondary air pollutants, photochemicals, point source, non-point source, temperature inversion, dry deposition, wet deposition

QUESTIONS

1. Why are aerosols that are PM_{10} and smaller of particular concern to public health officials?

2. What is the difference between point- and non-point-source air pollutants? How might this difference influence their dispersion in the environment?

3. What is a temperature inversion, and why does it increase the threats posed by many kinds of air pollution?

4. Differentiate between wet and dry deposition.

Pollutant types	Inorganic trace gases Volatile organic compounds (VOCs) Particulate aerosols
Pollutant sources	Primary, directly from specific sources Secondary, from reactions among other sources
Influences on pollutant dispersal	Airflow patterns Emissions altitude Lifetime in the atmosphere
Pollution in the troposphere	Acid deposition Heavy metals Smog
Pollution in the stratosphere	Sulfur compounds CFCs
Indoor pollution	Combustion by-products VOCs and asbestos from building materials Radon Pesticides Biological contaminants

10.2 Pollution in the Stratosphere

■ Sulfur-containing aerosols reflect solar radiation back to space, producing lower temperatures in the troposphere.

■ Chlorofluorocarbons are stable in the troposphere but cause the destruction of ozone in the stratosphere.

QUESTIONS

1. Explain how sulfur-rich aerosols in the stratosphere influence Earth's energy budget.

2. Describe the chain of events that connects the release of CFCs from discarded refrigerators to increases in UV radiation at Earth's surface.

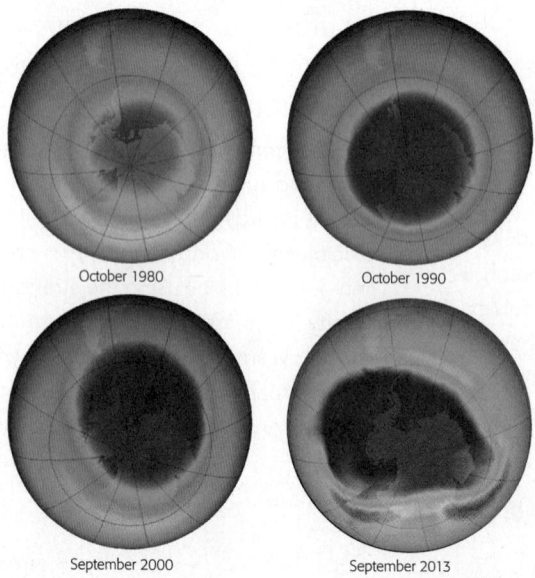

October 1980   October 1990

September 2000   September 2013

10.3 Pollution in the Troposphere

■ Acid deposition, heavy metals, and smog are the most important forms of pollution in the troposphere.

KEY TERMS

acid deposition, heavy metal, smog, scrubbers, catalytic converters, industrial smog, photochemical smog, air quality index (AQI)

QUESTIONS

1. Describe three significant impacts of acid deposition on freshwater and forest ecosystems.
2. Nowadays, acid rain contains more nitric acid than it did 25 years ago. Why is this so?
3. The concentration of lead in airborne dust increased dramatically in the 1930s, reached a peak in the early 1980s, and has been decreasing since then. Why?
4. Explain what causes the daily cycle in the amounts of photochemical smog in a large city on a windless day.

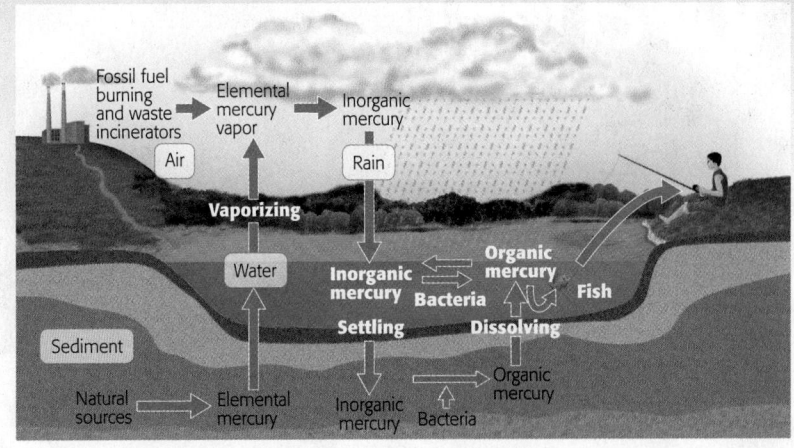

10.4 Indoor Air Pollution

■ Indoor spaces constrain air movement and are also the sources of pollution from combustion by-products, building materials, radon, pesticides, and biological contaminants.

KEY TERMS

combustion by-products, radon

QUESTIONS

1. Why does the indoor environment increase the likelihood that pollutants will have adverse health effects?
2. What is the source of radon gas, and what is the basis for its effect on human health?

10.5 Air Pollution Policy and Law

■ Policies to manage air pollution in the United States have evolved from research and technical assistance to the establishment and enforcement of air-quality standards.

■ Agreements among countries to limit air pollution depend on a shared sense of need and an equitable distribution of costs and benefits.

QUESTIONS

1. Environmental groups have been united in opposition to the use of cap-and-trade programs to manage pollutants such as mercury and lead. Why?
2. What factors have contributed to the success in implementing the 1987 Montreal Protocol on ozone?

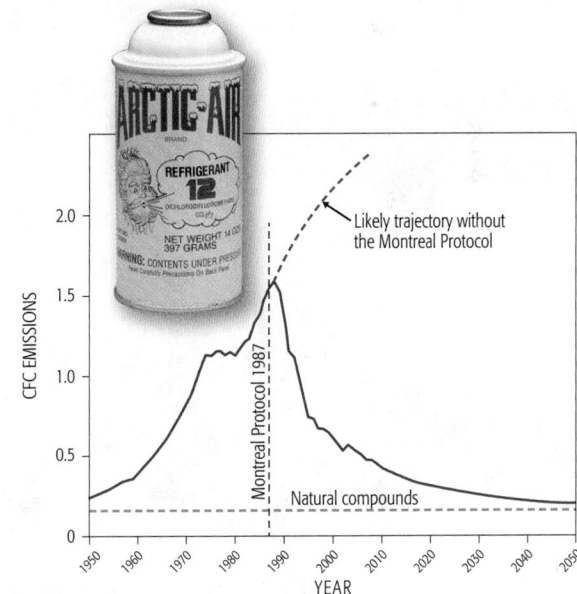

MasteringEnvironmentalScience®

Students Go to **MasteringEnvironmentalScience** for assignments, the eText, and the Study Area with practice tests, activities, and more.

Instructors Go to **MasteringEnvironmentalScience** for automatically graded tutorials and questions that you can assign to your students, plus Instructor Resources.

Water

A Disappearing Resource

How are human activities and climate change depleting an essential water resource?

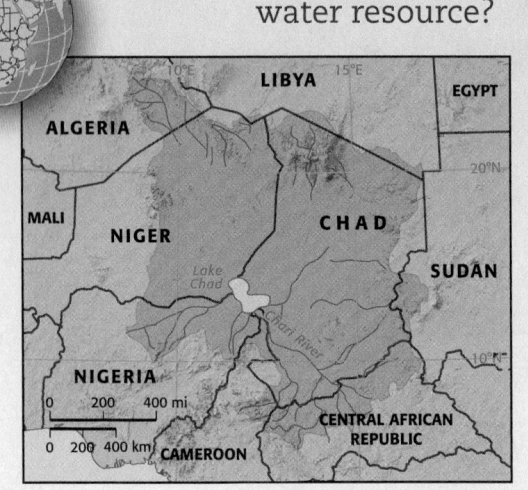

▲ **Figure 11.1 A Lake on the Edge**
Lake Chad and its drainage basin (shown in green) lie in the north central portion of the Sahel. The Sahel is a region of dry woodland and savanna that separates the Sahara in the north from moist tropical forests in the south.

The African Sahel is a 500- to 1,000-km (310–620 mi) wide ribbon of grassland, woodland, and savanna separating the very dry Sahara Desert to the north from the wet equatorial forests to the south. Near the dry northern edge of this ribbon, halfway between the Atlantic Ocean and the Red Sea, lies Lake Chad (**Figure 11.1**).

The Chari River, flowing from the much wetter savannas to the south, is the primary source of water for this shallow lake. Lake Chad has no outlet. Water is lost from it by evaporation and seepage into the deep sediments beneath its basin. Ninety-five percent of rainfall in the lake's basin comes between May and September. During these months, the influx of water from the Chari exceeds water loss, and the volume of water in the lake increases. By late April, the end of the dry season, water flow in the Chari River is meager, and the lake may shrink to half its wet season size.

Five thousand years ago, this region was much wetter than it is today. The lands surrounding the lake were densely forested. At that time, Lake Chad was equal in area to Lakes Superior, Michigan, and Huron combined. This region has since grown drier, and for the past 2,500 years its climate has fluctuated around present-day conditions. Studies of lake sediments laid down over thousands of years reveal remarkable variations in the yearly rainfall and length of rainy seasons. Wet periods were punctuated by years of drought when streams barely flowed and the lake dried up.

Throughout this time, the nomadic people living nearby adapted their activities to changes in the lake ecosystem. These people subsisted by fishing, herding cattle and goats, and tending small, temporary farms. During the dry season, they took advantage of the new, fertile land exposed by the retreating lake. During the wet season, they were able to expand their activities farther from the lake's shore. Dry periods undoubtedly placed hardships on these people, but their nomadic lifestyles allowed them to migrate to more favorable locations during droughts.

In the late 1800s, the British and French colonized the Lake Chad region, establishing four protectorates—Cameroon, Chad, Niger, and Nigeria. The Europeans established land ownership and rigid national boundaries, notions that were foreign to the indigenous nomadic people. Between 1955 and 1965, colonial rule ended, and the four protectorates surrounding the lake became independent nations.

From 1968 to 1974, severe drought prevailed across the Sahel. National boundaries and patterns of land ownership prevented the migrations that had previously allowed the nomadic communities to adjust to drought. As a consequence, they overgrazed their lands and cut down their woodlands, converting vast areas of grassland and savanna to desert throughout the Sahel. Famine was widespread.

▶ **Figure 11.2 Who Owns the Waters of Lake Chad?**
Over the past 50 years, the human population surrounding Lake Chad has increased threefold, placing increased demands on its fisheries Ⓐ and the grazing lands that surround the lake Ⓑ. Irrigated agriculture in the former lake bed Ⓒ continues to draw water from the lake, as do growing cities such as N'Djamena Ⓓ. The four countries that share the lake must negotiate transboundary issues that affect the water supply for the whole region.

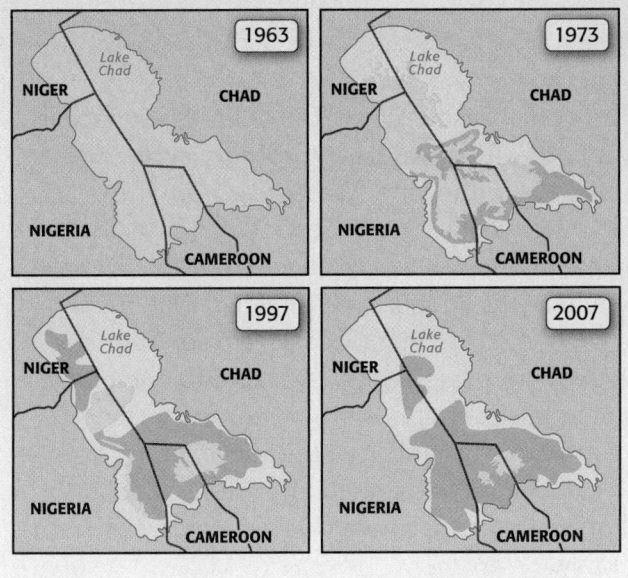

◄ Figure 11.3 **Lake Chad 1963–2007**
The area of Lake Chad has been reduced by more than 90% as a result of lower rainfall, diminished flows in the Chari River, and overutilization of lake water.

Source: Adapted from *Vital Water,* www.unep.org; sourced from satellite images provided by NASA Goddard Space Flight Center.

- Water
- Former shoreline
- Vegetation

0   25   50 mi
0   25   50 km

- *What are the natural and human processes that influence water flow?*

- *How can we restore the features of the landscape that assure the reliable provision and use of water?*

- *How can we ensure collaboration among the countries that use the water and influence its flow?*

- *How can we build resilience in hydrologic systems and the human communities that depend upon them, so both can adapt to the inevitable changes that the future will bring?*

Today, over 5 million citizens of the four countries that border Lake Chad depend directly or indirectly on the resources it provides (**Figure 11.2**). Its waters quench their thirst, irrigate their crops, and provide for basic sanitation; its fish are a major source of their dietary protein. The wetlands and woodlands surrounding the lake support a diverse array of wildlife and waterfowl that are also important sources of food for this population.

Despite some recent increases in rainfall, the average size of Lake Chad has decreased over 90% since 1963 (**Figure 11.3**). Part of this decrease can be attributed to a gradual increase in temperature and the rate of evaporation in this region, perhaps associated with global warming. Deforestation of lands in the upper reaches of the Chari River has reduced rainfall in that region, so less water now flows in the Chari. However, most of the loss of water is due to increasing numbers of people diverting water from the river and lake to meet their growing needs.

Most of the people living around Lake Chad are poor. Sixty percent of them live on the equivalent of less than $1 per day, and more than 20% of their children perish before they turn five years old. Economic development is needed to improve the well-being of these people, but that development will depend on the restoration and sustainable management of Lake Chad's natural resources.

Recently, the four countries surrounding the lake created the Lake Chad Basin Commission to restore the lake and its ecosystems. To succeed, the commission will need to understand the natural processes at work while seeking collaboration among people who so desperately rely on its waters.

11.1 Water World

BIG IDEA Our planet is called the "Blue Marble," and for good reason: Over 70% of Earth's surface is covered with water (Figure 11.4). Solar energy and gravity drive the **hydrologic cycle**, the movement of water between the oceans, the atmosphere, and the land. Water falls from the atmosphere as rain or snow. Rainwater may then evaporate back to the atmosphere, flow across Earth's surface into rivers and streams, or percolate through the soil to underlying rock. Snow may melt or accumulate as ice in the glaciers of high mountains, the Arctic, and Antarctic. Water is a critical resource that provides ecosystem goods and contributes to many ecosystem services. Despite the abundance of water on Earth, only 1/100th of 1% is fresh water available to humans. Of the rest, 97% is salt water, and the remaining fresh water is either frozen in ice caps or permafrost or locked in soil. As fresh water is distributed unevenly around the globe, arid regions with high population and agricultural demands experience a great deal of water stress, where demand for this precious resource outstrips supply.

▲ **Figure 11.4 It's Mostly Water!**
The water that covers over 70% of Earth's surface is essential to all life.

The Hydrologic Cycle and Earth's Water Budget

■ Solar energy and gravity drive the hydrologic cycle.

The hydrologic cycle describes the distribution and flux of water through Earth's ecosystems. It includes three phases of water—solid ice, liquid water, and gaseous water vapor. The energy of solar radiation causes ice to melt and liquid water to evaporate. The force of gravity moves water from the sky to the land and across the land to the sea (Figure 11.5). These simple interactions between matter and energy power this biochemical cycle so critical to life.

Precipitation (including rain, sleet, snow, and hail) forms as water vapor cools and condenses, transferring water vapor in the atmosphere to liquid water in the hydrosphere. This flux is usually measured as liquid water with a rain gauge, which shows the depth of water that falls at a particular location.

Once on the ground, liquid water has three general fates—it may evaporate back to the atmosphere, flow across Earth's surface into streams and lakes and eventually to the ocean, or percolate through soil to become groundwater. Snow and ice act as temporary water-storage reservoirs until water melts or evaporates.

Water returns to the atmosphere. Water evaporates from bodies of water, from soils, and from the leaves of plants that have absorbed water from the soil. Evaporation from leaves is called **transpiration**. Because of the large surface area of plant leaves, transpiration is a major component of evaporation in terrestrial ecosystems. **Evapotranspiration** is the sum of transpiration and evaporation and represents the entire flux of water into the atmosphere.

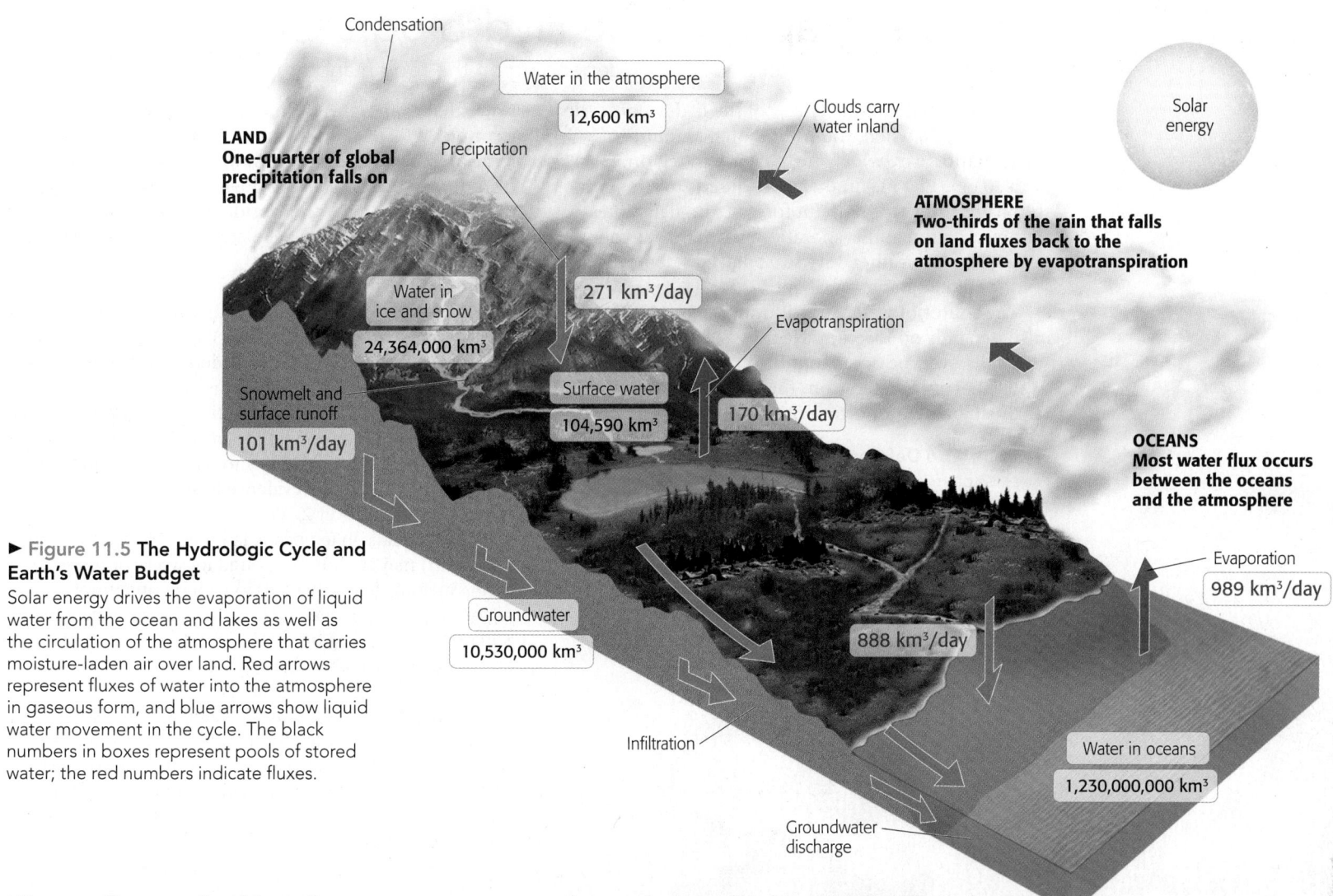

Condensation

Water in the atmosphere

12,600 km³

LAND
One-quarter of global precipitation falls on land

Precipitation

Clouds carry water inland

Solar energy

ATMOSPHERE
Two-thirds of the rain that falls on land fluxes back to the atmosphere by evapotranspiration

271 km³/day

Water in ice and snow

24,364,000 km³

Evapotranspiration

Surface water

104,590 km³

170 km³/day

Snowmelt and surface runoff

101 km³/day

OCEANS
Most water flux occurs between the oceans and the atmosphere

Evaporation

989 km³/day

Groundwater

10,530,000 km³

888 km³/day

Infiltration

Water in oceans

1,230,000,000 km³

Groundwater discharge

► Figure 11.5 **The Hydrologic Cycle and Earth's Water Budget**
Solar energy drives the evaporation of liquid water from the ocean and lakes as well as the circulation of the atmosphere that carries moisture-laden air over land. Red arrows represent fluxes of water into the atmosphere in gaseous form, and blue arrows show liquid water movement in the cycle. The black numbers in boxes represent pools of stored water; the red numbers indicate fluxes.

Water collects on Earth's surface. Rainwater that falls on land may flow across the surface as **runoff**, eventually entering lakes and streams. Although some moisture evaporates along the way, gravity pulls the water in lakes and streams inexorably toward the sea. Approximately 10% of the water that evaporates from the ocean returns to it via rivers each year.

Water moves underground. Gravity causes some water to percolate through the soil and into the rock below as **groundwater**. A layer of soil or rock that is saturated with groundwater is called an **aquifer**. The proportion of rainfall that percolates into aquifers varies widely, depending on factors such as total rainfall, vegetation cover, and soil porosity. In most locations, only 0.1–5.0% of total rainfall reaches an aquifer. Groundwater moves through rock very slowly, infiltrating the underlying rock layers, often at rates of less than a foot per year, eventually delivering water to lakes, streams, or the ocean.

Hydrologists, the scientists who study Earth's waters, compile statistics about the distribution and flow of water on our planet. They quantify the inputs and outputs of water as it moves between the atmosphere and Earth's

surface to calculate a global water budget. Components of this budget vary dramatically from desert to rain forest, so hydrologists develop regional water budgets, which provide useful information for planners.

Hydrologists have determined that the total pool of water in the atmosphere is about 12,600 km³ (3,119 mi³), or only about 0.001% of Earth's total water. The flux of water into and out of the atmosphere each year is much larger. Annually, approximately 496,000 km³ (~119,000 mi³) of liquid water falls on Earth's surface, and an equal amount evaporates to the atmosphere (see Figure 11.5). That's enough water to fill all the Great Lakes 22 times each year! At any given location and at any particular moment, the rates of precipitation and evaporation are not likely to be equal. Over the entire globe, however, these rates equal out and remain relatively constant.

About 23% of global precipitation (271 km³/day) falls on land. This is the total renewable supply of fresh water for Earth's terrestrial ecosystems. Two-thirds of this terrestrial precipitation evaporates back into the atmosphere, while the rest flows over land into streams and rivers (68 km³/day) or percolates into groundwater aquifers (33 km³/day). From this pool, an almost equal amount (101 km³/day) discharges into the ocean.

The Geography of the Hydrologic Cycle

■ Components of the water budget vary depending on the region and time of year.

How you experience the effects of the hydrologic cycle depends upon where you live. Annual amounts of precipitation vary widely from region to region, from 5 to 10 cm (2–4 in.) in the driest deserts to over 6 m (20 ft) in some tropical rain forests. And daily rates of precipitation and evaporation, part of what you experience as weather, vary seasonally, as demonstrated by the climatographs in Chapter 7.

The source of water contributing to precipitation also depends on location. Nearly 100% of the rain that falls over the ocean and near coasts comes from water that evaporated from the ocean surface. Inland, a considerable proportion of rainfall comes from

▼ **Figure 11.6 Deforestation and Rainfall**
The loss of once extensive forests has diminished the rainfall over large areas of Madagascar because less water now evaporates back to the atmosphere from this landscape.

water that evaporates from terrestrial ecosystems. For example, 25–50% of the rain that falls on the Amazon rain forest region comes from water that evaporates from those forests. This is evidenced by the fact that in many parts of the world where large areas of tropical rain forest have been cut, rainfall has declined (Figure 11.6).

The residence time of water—the average time that a water molecule resides in a particular part of the cycle or pool—varies widely. A water molecule evaporating from the surface of the ocean has a residence time of 9.5 days in the atmosphere. If it then falls as rain and is captured as runoff, it will have a residence time of 14 days in rivers and streams. Residence time for groundwater is much longer, ranging from hundreds to tens of thousands of years. In the ocean, the residence time for an average water molecule is over 2,700 years.

Snow that falls in temperate latitudes (23.5–66.5° north and south) may stay on the ground for weeks or months before melting. Snow that falls in high mountains or polar regions may accumulate as ice to form glaciers, where it may remain for a very long time. For example, ice in the lower layers of glaciers in Greenland and Antarctica fell as snow hundreds of thousands of years ago.

The majority of Earth's fresh water, 28.6 million cubic kilometers, resides at the poles in the form of ice. As human activities cause worldwide climates to warm, glaciers and ice shelves are breaking, diminishing the amount of water stored in glacial ice. As this water feeds into the world's oceans, sea levels will rise.

Watersheds

■ Scientists study a river's drainage basin to measure water flow and evaporation.

A **watershed** is the area of land from which rainfall drains into a river or lake. Watersheds are also called drainage basins. They are separated from one another by mountain ridges and other topographic divides that direct the flow of water (Figure 11.7A). The watershed is an ideal unit for understanding water flows and budgets at local and regional scales. Models of watersheds allow hydrologists to evaluate how human activities, such as irrigating crops and constructing reservoirs, affect the various components of a water budget.

How do hydrologists determine the flow of water in a watershed? To estimate the total amount of rain coming into a watershed, they use a number of widely distributed rain gauges. To estimate the amount of liquid water leaving a watershed, they monitor the flow of streams and groundwater. Evapotranspiration can be measured directly or be estimated by subtracting stream and groundwater flow from rainfall.

▲ Figure 11.7 **Watersheds**
Ⓐ Because water inputs and outputs can be measured, watersheds are ideal units for the management and study of water flow. Ⓑ The watershed of the Mississippi River includes nearly one-third of the land in the lower 48 United States. This watershed can be subdivided into six smaller watersheds that correspond to the land drained by major rivers. Each of these is divisible into hundreds of yet smaller watersheds.

Hydrologists study watersheds of many sizes. Watersheds of only a few hectares may be studied to learn how particular land uses or management practices affect the flow of water. Small watersheds drain into larger watersheds, which, in turn, drain into yet larger watersheds. Very large watersheds, such as that for the entire Mississippi River, can help us understand the cumulative effect of many activities over vast areas (Figure 11.7B).

Open watersheds are drained by rivers that eventually make their way to the sea. These include the world's major river basins, such as the Amazon, Mississippi, Nile, and Congo. **Closed watersheds** are inland basins that do not drain to the sea. Rivers in closed watersheds often end in inland lakes or seas, such as the Great Salt Lake, Lake Chad, and the Dead Sea. Closed watersheds comprise about 18% of Earth's land surface.

Because the rainfall in its watershed is measured at hundreds of locations and its flows are closely monitored, the Mississippi River watershed provides an excellent case study of a water budget. The total volume of rain falling on its watershed (in cubic kilometers per year) is calculated by multiplying the average rainfall by the area of the watershed. The volume of water discharged from the mouth of the Mississippi River is measured directly. It includes flows from surface runoff and groundwater. Pumping from aquifers must also be included in the water budget; various measurements indicate that pumping from groundwater aquifers adds an average of 2 km³ to the Mississippi each year. Evapotranspiration over such a large area is impossible to measure directly. However, because the other components of the water budget are known, it can be calculated as evapotranspiration = rain volume + groundwater pumping – river discharge.

The flow of water within a watershed is influenced by a number of factors. Impermeable surfaces, such as pavement and rooftops, shed water and increase runoff. Plant cover slows the flow of water and increases evapotranspiration. Thus, in deciduous forests, a greater proportion of rainfall makes its way into streams in the winter than in the summer.

Providing Essential Ecosystem Services

■ The hydrologic cycle provides a wide range of ecosystem services.

Total Water Use (billions of cubic meters)

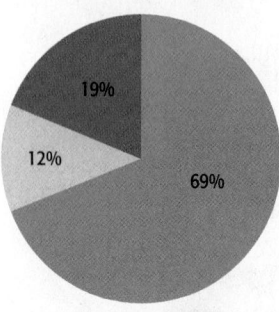

GLOBAL (3,902)

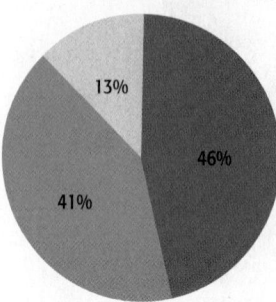

UNITED STATES (482.2)

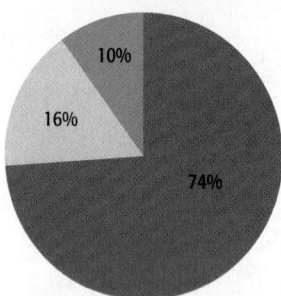

FRANCE (33.1)

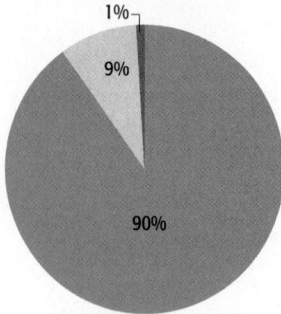

MALI (1.7)

Freshwater withdrawals by type of use

● Industry
● Agriculture
● Domestic

Water is a critical resource that provides all of Earth's ecosystems with a whole range of integrated services. The intrinsic value we humans place on waterscapes is evident in every human culture. In economic terms, it is not possible to measure the value of water, for it is essential to life and influences every sector of the U.S. economy. A recent report from the U.S. Army Engineer Research and Development Center estimated that U.S. beaches alone contribute roughly $225 billion a year to the U.S. economy through tourism and travel. More important, however, are the ecosystem services that sustain life and support our environment.

Supporting services. Water is an essential component of photosynthesis, the process that converts the sun's energy to chemical energy (food) in primary production. The hydrologic cycle also plays a significant role in the cycling of nitrogen (see Module 12.2) and carbon (see Module 6.5). Raindrops condense on small dust particles and pick up other chemicals in the atmosphere. A multitude of these compounds dissolve in water and then move through plants and soils.

The movement of water has a direct and significant impact on other biogeochemical cycles. Earth's liquid water carries with it a variety of dissolved molecules. Water flowing in streams and rivers transports these materials along with vast amounts of sediment through streams, into lakes, and on toward the ocean. These materials nourish soils and sustain a variety of food webs.

Regulatory services. Rain and snow falling through the air help to remove pollutants from the atmosphere. The oceans and other large bodies of water help to moderate air temperatures, storing then releasing heat over time. Water is also the medium in which various ions and buffers work to maintain a pH that enables and promotes cellular activity.

Provisioning services. In addition to meeting the very basic need for providing drinking water, the hydrologic cycle provisions all the seafood we eat and provides the water necessary for all agricultural activity. And it can, itself, be used as a source of energy.

Human uses and impacts. Through our various activities, we humans have appropriated a large portion of Earth's water resources for our own use. As a result, we have altered the hydrologic cycle in significant ways.

Diversions such as canals and aqueducts move water from areas where it is abundant to areas where it is scarce. Impoundments, such as human-made lakes and reservoirs, store water for a variety of uses, including irrigation, flood control, energy generation, and human consumption. Humans have also drained many wetlands and marshes to claim these lands for agriculture and urban development.

All these activities affect the hydrologic cycle. Diversions reduce the flow of water in one region and increase it in others. Impoundments slow the movement of water and greatly alter the seasonal flow of the streams that drain them. Irrigation alters water chemistry.

Scientists estimate that about one-fourth of terrestrial evapotranspiration now comes from lands used to grow crops, graze livestock, or produce wood for human use. Dams and diversions control the flow of approximately half of the runoff ($34 km^3/day$) in Earth's terrestrial ecosystems. More than half this flow is used for agriculture, industry, and municipalities. Approximately 82% of the water we use is returned to the streams from which it was taken, although the returning water often contains a variety of pollutants. The remaining 18% is lost to evapotranspiration.

Activities that use water and then return it to streams or aquifers are said to be **nonconsumptive uses**. Such uses include the generation of hydroelectric power and the disposal of wastewater in septic systems. In other activities, such as irrigation and industrial cooling, a considerable amount of water evaporates into the atmosphere. These activities are said to be **consumptive uses** because much of the water is not returned to streams or aquifers.

The proportion of water used in agriculture, industry, and municipalities varies widely from country to country, depending on the climate and degree of economic development. In dry regions, such as Mali in Africa's Sahara–Sahel region, agriculture usually accounts for a large proportion of water withdrawals. Economically developed countries, such as France and the United States, withdraw greater amounts of water for industrial purposes than do poorer countries (Figure 11.8). Note that water needs may be met indirectly through imported food and other goods. Therefore, water withdrawal patterns in a region may differ somewhat from the total water used in a region.

◄ Figure 11.8 **Freshwater Withdrawals**
Patterns of water withdrawals vary among countries, depending on the importance of industry and agriculture in their economies. Data shown do not account for indirect water use, such as water needed to grow food imported from another country, nor do they exclude direct water used to produce exported goods.
Data from: Pacific Institute.

Where Is Earth's Fresh Water?

■ Only a fraction of a percent of Earth's water is fresh water, readily accessible for human consumption.

Though 70% of the world is covered in water, only 1/100th of 1% of it is fresh water available for human use (Figure 11.9). Put another way, if all the world's water were stored in a five-gallon jug, the amount of fresh water readily accessible for human use would be less than one teaspoon. The vast majority of Earth's water is salty (97.5%), frozen (1.7%), or trapped in soil (0.001%) and is simply inaccessible without a significant input of energy for extraction or treatment. We know from the hydrologic cycle that Earth's total volume of water never changes; it simply moves from one pool to another. This has two implications: (1) the water on Earth today is the same water that was present when dinosaurs roamed the earth, and (2) without a significant input of energy for salt removal or some discovery of deep reservoirs well below Earth's surface, no new water will ever be available to replenish a shortage of clean, accessible fresh water.

Fresh water is not evenly distributed across Earth's surface (Figure 11.10). One method of measuring the availability of fresh water relative to its demand is to calculate **baseline water stress**. This represents the total annual water withdrawals (municipal, industrial, and agricultural) expressed as a percent of the total annual

available flow in a region. Higher values indicate more competition among users.

Heavily populated arid regions that use water for agriculture experience significant water stress. For example, California, the American West, southern Europe, and much of China are under extremely high baseline water stress. This inequality of water distribution has been a cause of conflict for thousands of years, including five violent disagreements over development and water rights during a three-year period alone (2010–2012) in the Middle East. Growing demand for fresh water due to the increasing human population has led some to suggest that wars over water will replace wars over oil in the future.

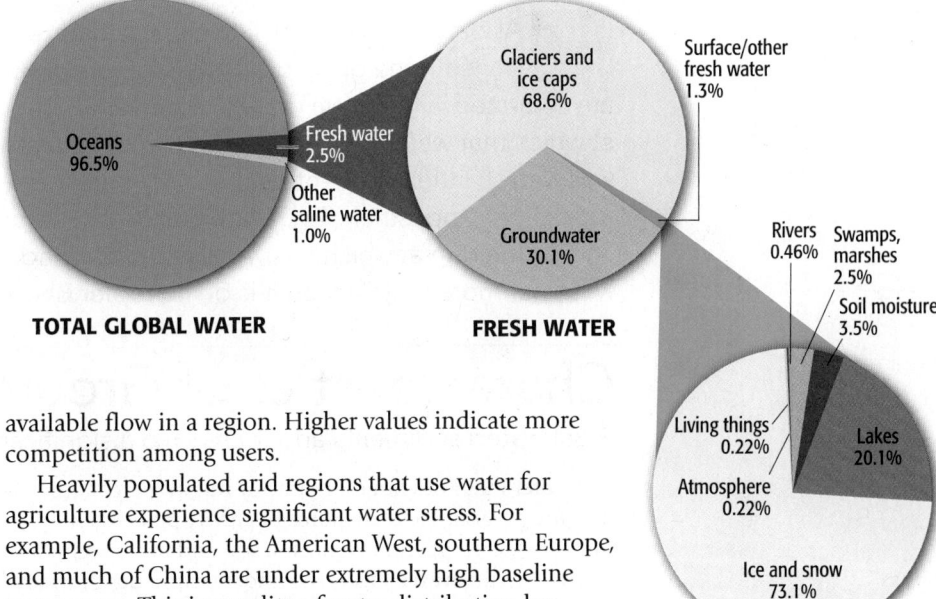

TOTAL GLOBAL WATER   **FRESH WATER**

SURFACE WATER AND OTHER FRESH WATER

▲ Figure 11.9 **Where Is Earth's Water?**
Most of Earth's water is salty or trapped in ice. Only 1/100th of 1% is fresh water available for human consumption.

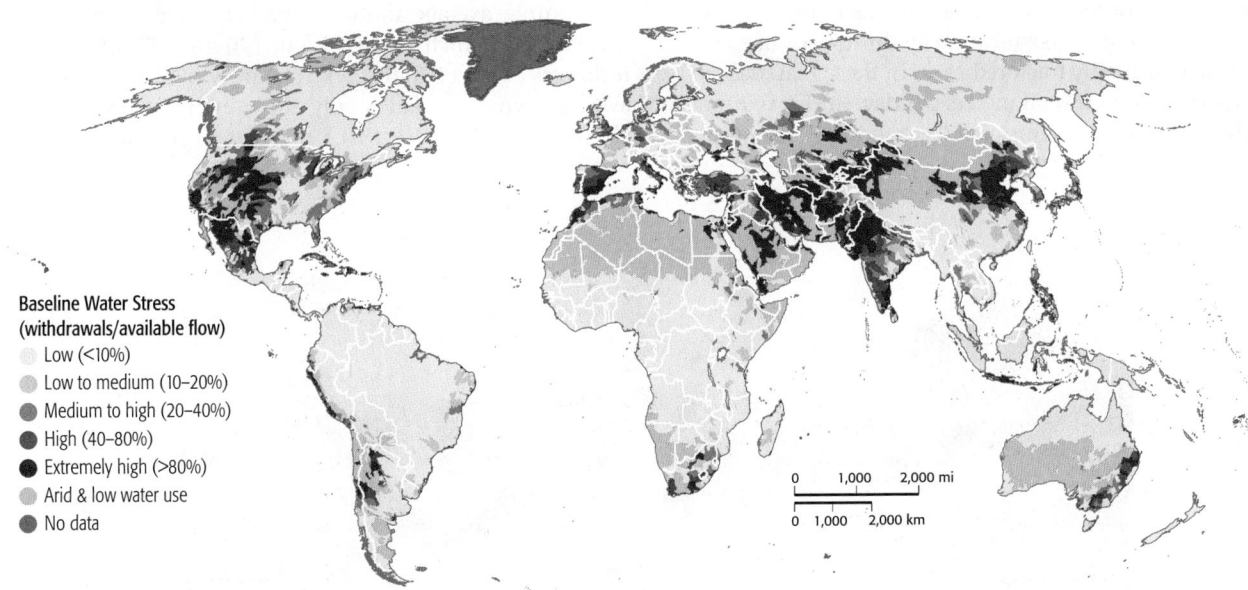

▲ Figure 11.10 **Baseline Water Stress Varies by Region**
This map identifies values for baseline water stress around the world. Data are calculated from water withdrawals (2010) divided by mean available fresh water (1950–2008). Areas with available fresh water and water withdrawal less than 0.03 and 0.012 m/m², respectively, are coded as "arid and low water use."
Source: WRI. Aqueduct Metadata Document: Aqueduct Global Maps 2.0 Francis Gassert, Matt Landis, Matt Luck, Paul Reig, and Tien Shiao

QUESTIONS 11.1

1. Describe the three possible fates of a molecule of water in a raindrop hitting Earth's surface.
2. Describe some ecosystem services associated with water.
3. Why is accessible fresh water so limited on a blue planet? What implications does this have for how we use fresh water?
4. What is the difference between consumptive and nonconsumptive use of water by humans?

(MES) For additional review, go to **MasteringEnvironmentalScience**

343

11.2 Groundwater

BIG IDEA About 0.3%, or 10 million cubic kilometers, of Earth's fresh water resides in rocks and soil beneath the ground. Most of this groundwater is in aquifers, layers of sediment and rock that are saturated with water. There is wide variation in the amount of water stored in aquifers and the rate at which that water moves through them. Aquifers in some regions formed long ago when the climate was wetter; little water enters these aquifers today. Humans use wells to extract groundwater for agricultural, industrial, and residential uses. The global rate at which humans use groundwater is about 38% of the rate at which this water is replenished. But in many locations, humans are withdrawing water far more rapidly than it is being replenished. Human activities have also polluted many aquifers.

Characteristics of Groundwater

■ Saturated sediments and rocks hold a significant portion of Earth's fresh water.

Rainwater that does not evaporate or run off percolates into the ground. There the force of gravity pulls it down through the pores and fractures in soil and rock. Eventually it arrives at the **water table**, the underground depth where rock and sediment are completely saturated with water. An aquifer is the saturated zone beneath the water table from which water can be extracted (**Figure 11.11**).

A **recharge zone** is an area where water flows directly between the soil surface and the water table. Aquifers located beneath recharge zones are said to be **unconfined aquifers**. In other aquifers, groundwater is trapped between layers of comparatively impermeable rock or sediment; such aquifers are said to be **confined aquifers**.

Once groundwater reaches an aquifer, it continues to flow downhill. Its rate of movement depends on the structure of the rocks and sediments in the aquifer. Generally, water moves most easily through coarse sediment or highly fractured rock. In these materials, water may move several meters per day. In clay sediments or unfractured crystalline rocks, groundwater may move less than a meter per year!

Groundwater flows to the surface in **discharge zones**. In places where a confined aquifer meets the soil surface, water may flow out freely in a seep or spring. Groundwater may also be discharged directly into rivers, lakes, or the ocean.

The size of aquifers varies considerably. On the major continents (excluding Antarctica), approximately 30% of the land is underlain by homogeneous aquifers that extend over thousands of square miles. However, most aquifers are smaller and are found where porous rocks and sediments are located.

There is also great variation in the recharge rate of aquifers. Recharge rates are high in aquifers near rivers and lakes and in regions with abundant rainfall. They are very low in aquifers in arid regions. In Germany, for example, average annual rainfall is about 800 mm (31.5 in.), of which about 15% or 120 mm (5 in.) recharges aquifers. In the desert country of Namibia, average annual rainfall is only 285 mm (11 in.), and only about 1%, or 3 mm (0.1 in.), percolates down to the water table.

► Figure 11.11 **Groundwater** Aquifers are regions of saturated rock or sediment from which water can be pumped. Confined aquifers occur between relatively impermeable strata. The water in confined aquifers may be under pressure causing it to rise in wells above the level of the surface of the aquifer.

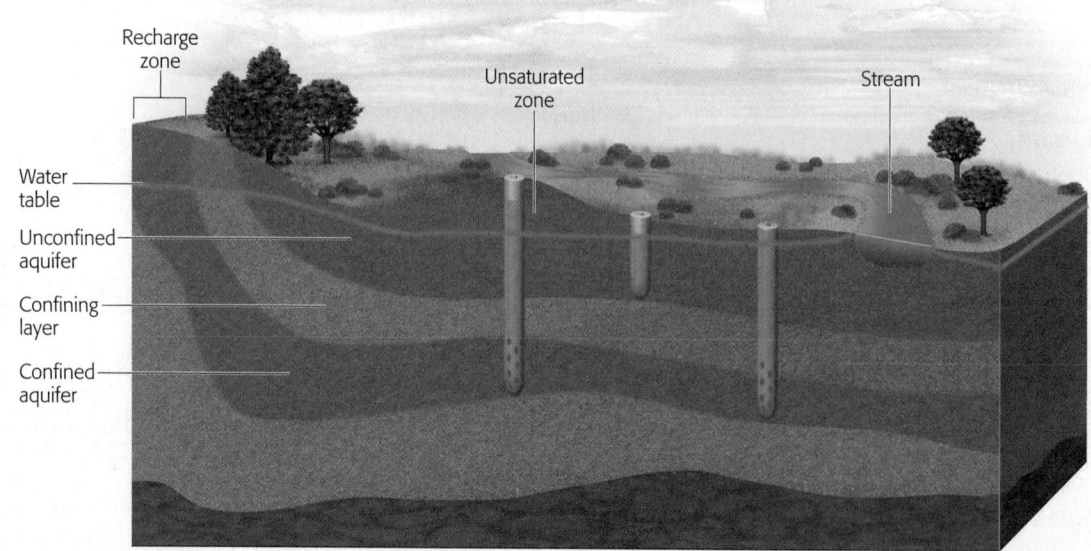

The residence time for groundwater varies from decades to thousands of years, depending on the rate at which the water flows and its depth in an aquifer. Where recharge and discharge rates are high, residence times are generally short. However, the natural recharge and discharge of many large aquifers is very slow. The water in such aquifers has accumulated over tens of thousands of years.

Some regions of the world have vast groundwater reserves that formed thousands of years ago when wetter climatic conditions prevailed. For example, the Nubian Sandstone Aquifer extends across 2 million square kilometers of the northeastern Sahara Desert, beneath the countries of Chad, Egypt, Libya, and Sudan. Water entered this aquifer during a wet period 5,000–10,000 years ago. Today, there is very little recharge to this aquifer (Figure 11.12).

Surprisingly, not all aquifers have yet been discovered. Indeed, advanced satellite technology was employed to detect a vast underground water source in Kenya in 2013. Five previously unknown aquifers, one of which is larger than the state of Rhode Island, have the potential to dramatically improve the lives of the more than 40% of Kenyan people who lack access to clean, fresh water.

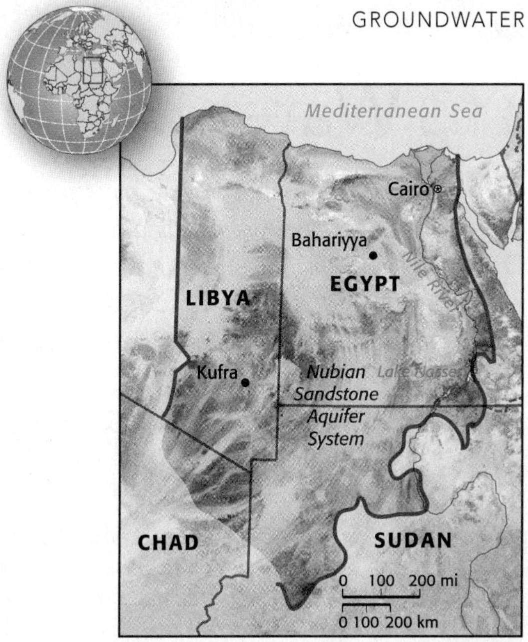

▲ Figure 11.12 **No Recharge**
Rain that fell over 5,000 years ago created the Nubian Sandstone Aquifer beneath the Sahara Desert. This aquifer is providing much-needed water in this dry region, but accelerated use of this water could deplete this nonrenewable resource.

QUESTIONS 11.2

1. What is the difference between a confined and an unconfined aquifer?

2. What factors influence the residence time of water in aquifers?

3. Is the rate at which humans are currently using groundwater sustainable? Explain why or why not.

(MES) For additional review, go to **MasteringEnvironmentalScience**

Human Uses and Impacts

■ In many places, the extraction of groundwater for agricultural, industrial, and domestic uses far exceeds the recharge rate.

Humans access groundwater by digging or drilling wells. To understand how a well works, think of sipping water through a straw. A well is like a straw, and an aquifer is like the liquid in a glass. The analogy is even better if the straw has perforations that allow liquid to come in through the sides as well as the bottom. A pump provides the force to move water up the well casing. Water from the aquifer then moves into the well through perforations in the casing, replacing the water that has been withdrawn. If the rock or sediment in the aquifer is very porous, water moving through the aquifer quickly replaces the pumped water. In such situations, water can be pumped rapidly. Where groundwater flows slowly, the rate of pumping must be slower (Figure 11.13).

Sometimes a well is drilled into a confined aquifer where the groundwater is under great pressure from the weight of the water above it. The pressure causes the water to rise above the confining layer, without the need for pumping. Such a well is called an **artesian well**.

Today, humans pump approximately 3,800 km³ (912 mi³) of water from Earth's aquifers each year. Total annual recharge to Earth's aquifers is approximately 10,000 km³ (2,400 mi³). Thus, humans currently withdraw about 38% of the water that enters the world's aquifers each year. About 68% of this water is used to irrigate crops, 21% is used in industry, and 11% goes to domestic uses. This level of groundwater use appears sustainable, since the total natural recharge rate exceeds the total rate of human withdrawal. However, human dependence on groundwater is usually highest in arid regions where recharge rates are typically very low. As a

result, aquifers in many arid regions are rapidly being depleted (see Module 11.3).

Aquifers are the primary source of drinking water in many parts of the world and so provide a critical provisioning ecosystem service. They also provide important regulatory services. As water percolates through soil and rock, microorganisms and chemicals are filtered from it, making groundwater less polluted than surface water. Nevertheless, natural processes and a variety of human activities can alter the quality of groundwater. In some regions, groundwater contains high concentrations of natural chemicals, such as arsenic, nitrates, and sulfates, which limit its use by humans. Near seacoasts, overpumping can cause seawater to move shoreward, increasing the salinity of freshwater aquifers (see Module 11.3).

Despite the filtering ability of soils and sediments, human-generated pollutants can sometimes make their way into aquifers. If storage tanks for gasoline, oil, or other chemicals leak, their contents may seep into shallow aquifers. Household chemicals and bacteria and viruses from faulty septic systems sometimes make their way into groundwater, often through nearby wells.

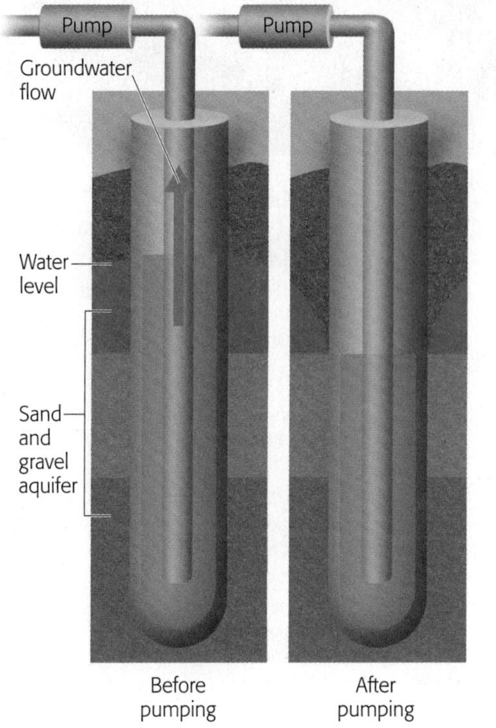

▲ Figure 11.13 **Pumping Water**
As a pump pulls water from a well, water moves into the well casing from the surrounding aquifer. This pulls the water table down in the vicinity of the well. When pumping ceases, the water table recovers. The rate of recovery depends on the permeability of aquifer rocks and sediments.

345

11.3 Water Distribution

BIG IDEA Human impacts on the hydrologic cycle are quite significant. Humans have diverted large quantities of water from rivers and streams for agricultural, industrial, and residential uses. Deforestation and extensive areas of pavement have altered the relative amounts of evaporation, runoff, and percolation to groundwater aquifers. In many places, humans are taking water from aquifers far more rapidly than the natural rates of replenishment.

Too Much Water

■ Paving and deforestation can increase the severity of flooding.

▲ Figure 11.14 **Mississippi River Floodplain**
Flood waters overflowing onto the low floodplain of the Mississippi enrich the soil with nutrients left behind in the river sediment.

Flooding is a natural phenomenon caused by a volume of rain or snowmelt greater than can be absorbed by soil or carried away by rivers. When rushing waters overflow riverbanks, they deposit rich sediments upon the floodplains and fertilize the soil, two important ecosystem services (Figure 11.14). Indeed, the annual flood cycle has supported rich agricultural societies along the Euphrates, the Nile, and the Mississippi for thousands of years. Modifications of river courses and increasing development of human structures around water bodies, however, has amplified the negative consequences of flooding, resulting in property damage and even loss of life.

Urban systems have particular challenges regulating water to avoid flooding. **Impervious surfaces**, such as pavement and roofs, are unable to absorb rainfall,

and much of the surface runoff is funneled into storm drains. In turn, stormwater systems empty directly into streams and rivers. In urban systems, 30–50% of surface runoff goes directly into water bodies as compared with only 5% in forested systems (Figure 11.15). This contributes to increased potential for flooding in heavy storms.

Paved surfaces in the United States cover an estimated 43,000 square miles, an area nearly the size of Ohio. As unpaved surfaces and forested areas decrease, flooding frequency increases. In Prince George's County, Maryland, just outside Washington, D.C., a United States Geological Survey gaging station in the Northeast Branch of the Anacostia River showed a flooding frequency of 1 to 2 times a year in the 1950s prior to urbanization. In the 1990s, following urbanization in the Washington, D.C., metro area, flooding frequency increased up to six or more times a year (Figure 11.16).

Deforestation is also associated with flooding. Vegetative cover slows surface runoff and allows water to infiltrate the soil. A single tree in the North Carolina deciduous forest can transpire up to 100 gallons in a day, so imagine the stormwater management benefits provided by many trees (see Chapter 13).

► Figure 11.15 **Landscape Runoff**
The impervious surfaces of urban landscapes generate over four times as much runoff as do forests or meadows, which have completely pervious (permeable) surfaces. In this diagram the estimates of infiltration and runoff are based on 3 inches of rainfall.

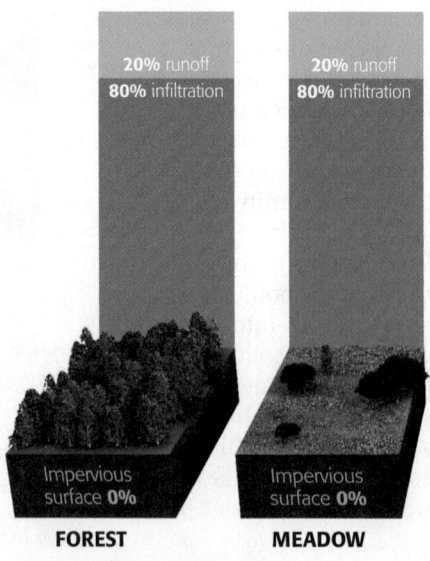

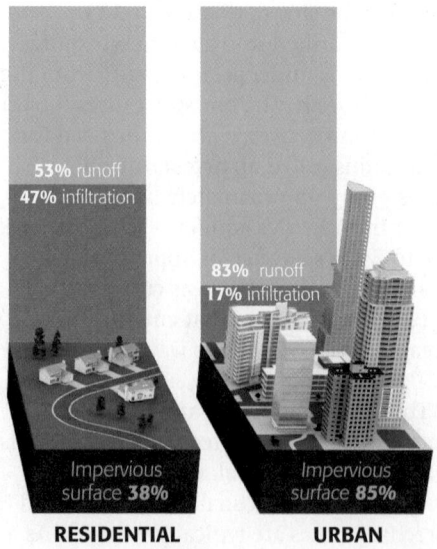

20% runoff 80% infiltration	20% runoff 80% infiltration	33% runoff 67% infiltration	53% runoff 47% infiltration	83% runoff 17% infiltration
Impervious surface 0%	Impervious surface 0%	Impervious surface 0%	Impervious surface 38%	Impervious surface 85%
FOREST	**MEADOW**	**AGRICULTURAL**	**RESIDENTIAL**	**URBAN**

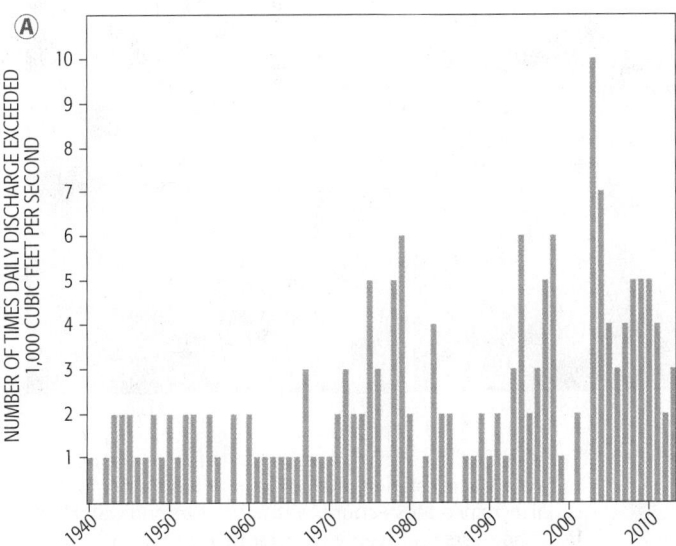

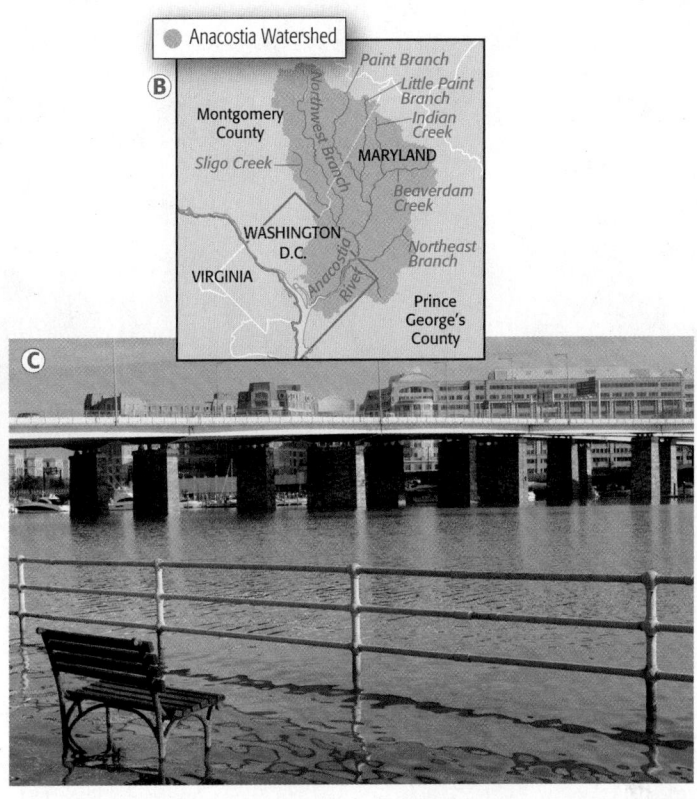

▲ Figure 11.16 **Flooding Frequency Increases with Urbanization in Washington, D.C.**
Ⓐ Flooding frequency in the North Branch of the Anacostia River triples over time as urbanization overtakes Washington, D.C. The watershed of the Anacostia Ⓑ feeds flood waters into Washington D.C. as its banks overflow Ⓒ.
Source: Data from USGS surface-water daily statistics for the nation USGS 01649500 Northeast Branch Anacostia River at Riverdale, Maryland.

To decrease the potential for flooding, many cities and towns are replacing the ecosystem services lost to them by hardscapes by increasing the area given over to vegetation and pervious surfaces. For example, Portland, Oregon, uses a pervious pavement that allows water to pass through and infiltrate the ground below. Cities like Chicago, Illinois, employ "green roofs" on many municipal buildings, and with great success (**Figure 11.17**). These living roofs can retain up to 75% of rainwater, gradually releasing it back to the atmosphere through evaporation. Green roofs also decrease the urban heat island effect and improve air quality (see Module16.5).

Too Little Water

■ Overdrawing surface and underground water can exacerbate water scarcity.

As much as 85% of the world's population lives on the driest half of the planet. Naturally occurring drought limits water supply, but water scarcity is further exacerbated by increasing human populations. To support the many competing needs for Earth's fresh water, both groundwater and surface water supplies are frequently overdrawn, with important environmental consequences.

▲ Figure 11.17 **Green Roof**
The plant cover on top of Chicago City Hall keeps summer roof temperatures 14–44 °C (25–80 °F) cooler than the roofs of surrounding buildings. Engineers calculate that on hot days Chicago temperatures would be as much as 7 °C (13 °F) cooler if all downtown buildings had green roofs.

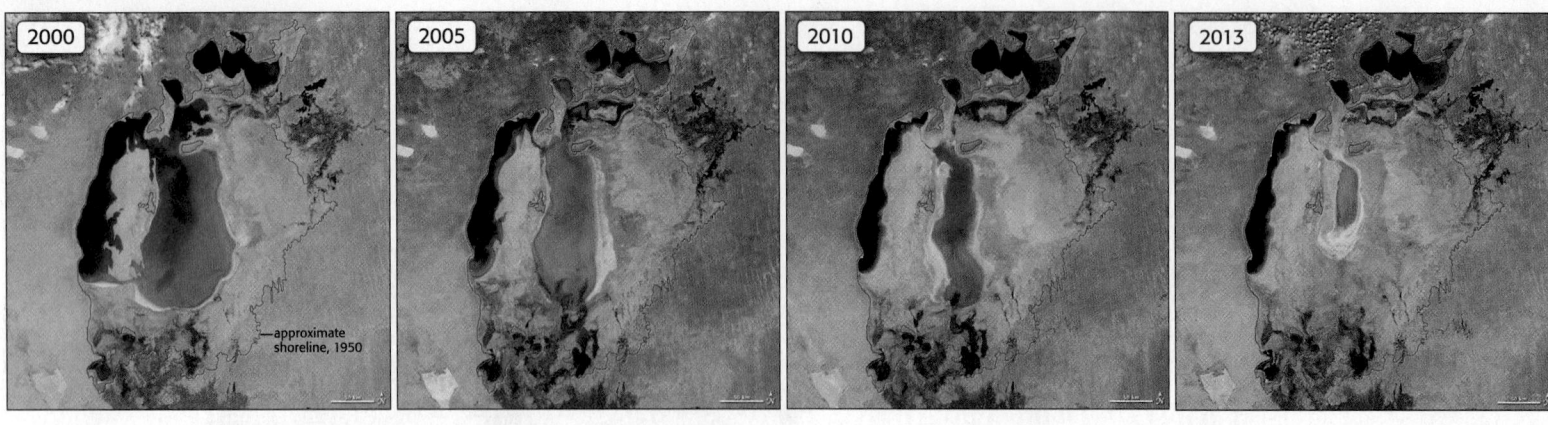

▲ Figure 11.18 **Aral Sea**

The Aral Sea is a closed basin lake that lies on the border between Kazakhstan and Uzbekistan. It has been rapidly shrinking since these countries began diverting water from the rivers that feed it in order to irrigate crops. In 1973, the water of the Aral Sea covered 68,000 km² (26,300 mi²) but today it covers less than 10% of that area.

Overdrawing surface water. For much of history, humans have diverted water from lakes and streams to fulfill their needs. When this water use is consumptive, for example with irrigation or municipal use, the flow is diminished producing consequences for ecosystems and human populations downstream. For example, prior to 1960, the Aral Sea between Kazakhstan and Uzbekistan was Earth's fourth largest lake. The Amu-Dar'ja and Syr-Dar'ja Rivers provided water to the Aral Sea at approximately the same rate that water evaporated from its surface. Since then, water from both these rivers has been diverted to irrigate more than 8 million hectares (20 million acres) of former desert. As a result, the Aral Sea has shrunk to less than half its former size (Figure 11.18).

In some cases, the overdrawing of rivers has become a contentious transboundary issue. The 1,450-mile long Colorado River and its tributaries, for example, flow through seven water-limited states and once emptied into the Gulf of California, just across the Mexican border (Figure 11.19). Only one-tenth of the river's former volume now flows into Mexico, where it is diverted into the Morelos Dam for irrigation and urban use. Agriculture consumes 78% of the Colorado's water via more than 100 dams, aqueducts, and other diversions. Municipal and industrial use take

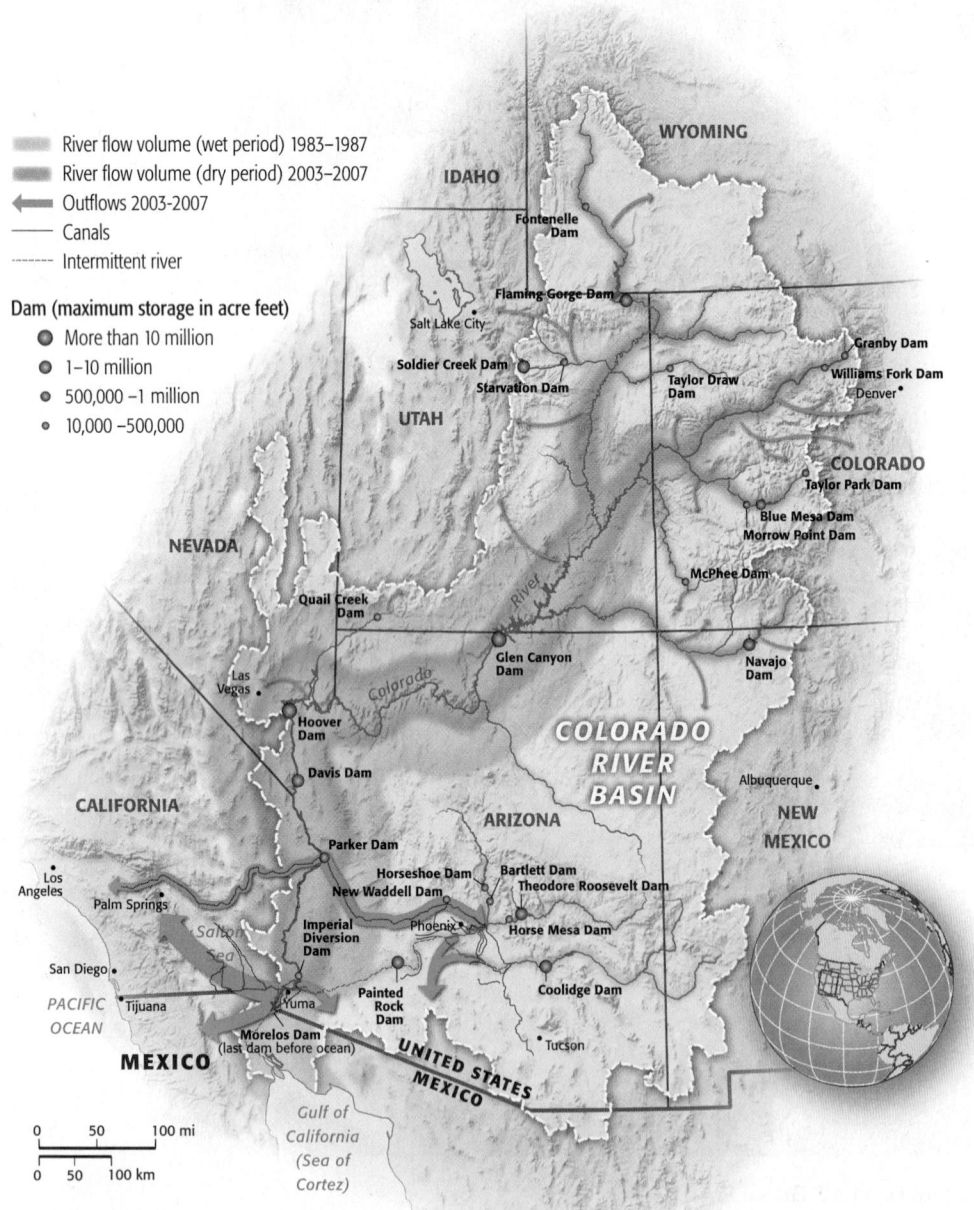

Legend:

- River flow volume (wet period) 1983–1987
- River flow volume (dry period) 2003–2007
- ← Outflows 2003-2007
- Canals
- Intermittent river

Dam (maximum storage in acre feet)
- More than 10 million
- 1–10 million
- 500,000–1 million
- 10,000–500,000

◄ Figure 11.19 **The Colorado River**

The Colorado River originates in the Rocky Mountains and flows 1,450 miles, supplying seven states and two countries with water. More than 100 dams and untold diversions remove every drop of water from the river before it reaches the sea in the Gulf of California. The Colorado River provides water for 30 million people and supplies irrigation for 4 million acres of agricultural land.

Source: Map based on National Geographic: Change the Course: Colorado River Map. Data Sources: Flow volumes: U.S. Geological Survey; Diversion flow volumes: U.S. Bureau of Reclamation; Dams, canals, pipelines, and aqueducts: USGS and National Geographic.

► Figure 11.20 **Depleting the Ogallala Aquifer**
The largest aquifer in the world, the Ogallala Aquifer underlies eight states and supplies freshwater needs for nearly 30 million people. It has been depleted by 30% already, with an additional 39% loss projected by 2060.

Source: Steward, D. R., et al. 2013. Tapping unsustainable groundwater stores for agricultural production in the High Plains Aquifer of Kansas, projections to 2110. *Proceedings of the National Academy of Sciences* 110(37): E3477–E3486. Map: from USGS digital data 1:2,000,000 Albers Equal Area projections.

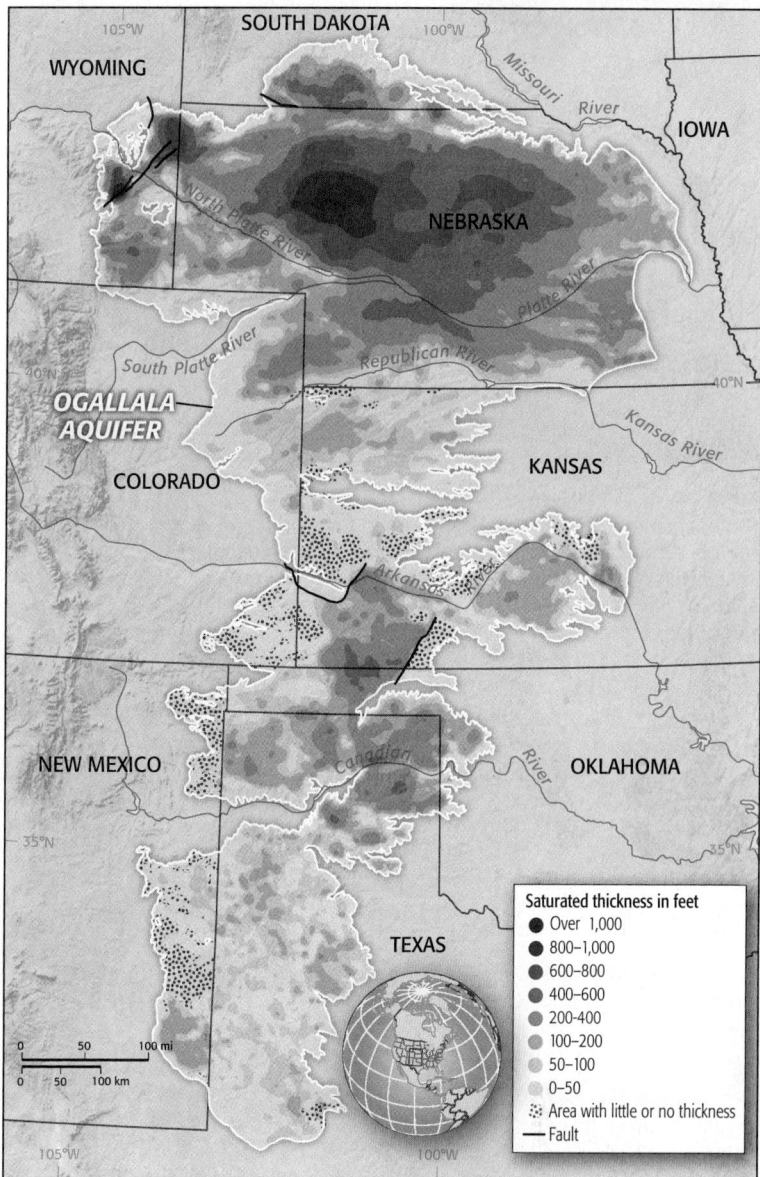

up the remaining 22% of the river water. By 1998, the Colorado River had dried up some 50 miles upstream from the sea, devastating the fishing industry as well as the rich biodiversity of the river delta leading into the Gulf of California. A 2012 agreement between the nations reduces water flow to Mexico during drought years, but allows Mexico to store some of its surplus water in the United States in good years to supplement the flow of the Colorado River into Mexico when needed. In addition, both U.S. and Mexican governments will provide 6.2 million cubic meters (5,000 acre feet) of water to the delta each year to begin to restore some of the devastated habitat. While this is a step in the right direction, this is just in a drop in the bucket relative to pre-settlement flows, which likely exceeded 6.2 billion cubic meters (5 million acre feet) per year.

Overdrawing groundwater. Currently 30% of Earth's freshwater supply is stored in underground aquifers. Recharge rates vary depending on proximity to rivers, amount of rainfall, and porosity of the overlying strata. Aquifers provide a sustainable source of fresh water only if the recharge and withdrawal rates are approximately equal. In some areas where rainfall is low and agricultural demand is high, underground water supplies are being overdrawn, leading to concern by the farmers, industries, and municipalities who depend upon it.

One aquifer under great stress is the Ogallala Aquifer, a vast store of groundwater that supplies an eight-state region in the High Plains of the United States (**Figure 11.20**). Most of the water in the Ogallala was deposited during the last ice age, over 12,000 years ago. The present-day recharge to the Ogallala comes mostly from the Rocky Mountains to the west and is very slow. Significant pumping from the Ogallala began in the 1930s, following the droughts of the Dust Bowl era. Indeed, the development of the technology to pump water from this underground source revitalized agriculture in this area. The Ogallala currently supplies 30% of the water used for irrigation in the United States. However, 30% of the water supply from what is now the world's largest aquifer has already been depleted, and 39% more is expected to be removed within 50 years based on current projections.

Subsidence and Intrusion

■ Overpumping of underground water can lead to loss of structural support and saltwater intrusion.

Aquifer depletion results in much more than just loss of stored water. Another important ecosystem service groundwater supplies is the structural support for the overlying land itself. **Subsidence** is the sinking of land above an aquifer. It can happen slowly. A pole located along the roadside in San Joaquin, California, marks 33 feet worth of subsidence that has occurred over 50 years of aquifer depletion. When subsidence occurs rapidly in a large area, it can create a **sinkhole** (**Figure 11.21**). Apart from the physical damage it causes, another cost associated with aquifer depletion is the greater energy expense necessary for pumping and drilling deeper wells.

◄ Figure 11.21 **That Sinking Feeling**
Ⓐ This pole located in San Joaquin, California, marks 33 feet of subsidence that has occurred over 50 years of aquifer depletion. Ⓑ Sinkholes such as this one are common in Florida due to its readily erodible limestone geology and heavy use of aquifer water.

349

For Venice, a city built on a lagoon, subsidence is putting this much beloved tourist destination at risk (Figure 11.22). The city is subsiding at a rate of about 2 mm a year as a result of overpumping its shallow aquifer and the compaction of soils from the weight of its buildings. Subsidence, combined with sea level rise, is projected to cause this historic treasure to sink another 80 mm into the sea over the next 20 years.

Overdrawing groundwater may compromise water quality as well. In coastal areas **saltwater intrusion**, the migration of salt water into a freshwater aquifer, may occur when groundwater is pumped faster than it can be replenished (Figure 11.23). This contaminates the freshwater supply and can wreak havoc on municipal water systems. Saltwater intrusion was detected as early as 1854 in New York City and is a problem from China to Africa and on the Atlantic and Pacific coasts of the United States.

▲ **Figure 11.22 Venice Is Sinking**
The enchanting canal city of Venice, Italy, is subsiding. The phenomenon of "acqua alta" (high water) seen here most often occurs in the winter. It affects the lowest parts of the city, such as St. Mark's Square, as water moves in and out in response to the tides.

QUESTIONS 11.3

1. Explain how flooding frequency is increased by paved surfaces.

2. Why does the Colorado River no longer flow to the sea? What are some of the consequences of this?

3. How does subsidence differ from saltwater intrusion? What process is responsible for each?

(MES) For additional review, go to **MasteringEnvironmentalScience**

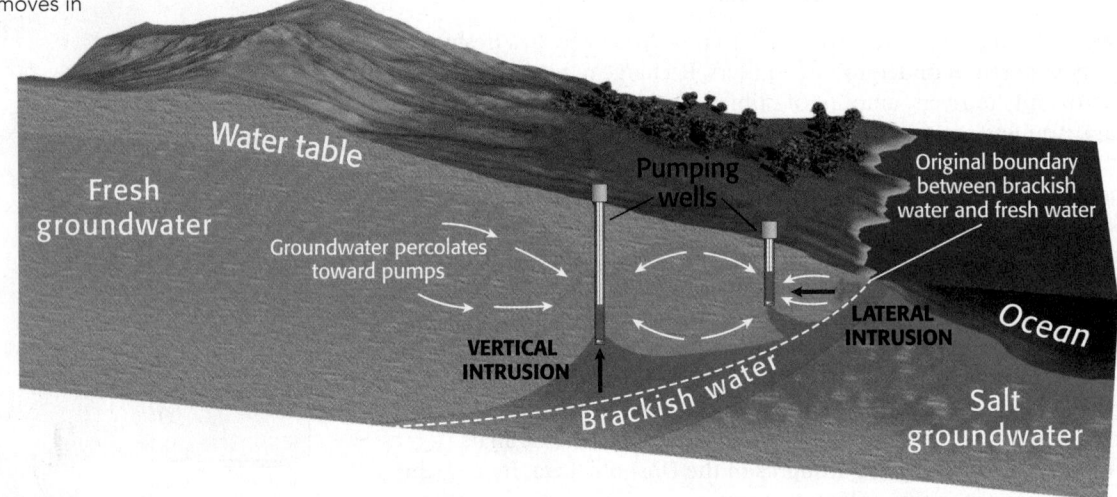

▲ **Figure 11.23 Saltwater Intrusion**
Overpumping of the aquifer in coastal regions allows salt water to migrate into the aquifer. Wells become contaminated and may be abandoned if the salt concentration rises too high.

New FRONTIERS

The Highs and Lows of Sharing Water

Artificial aquifer recharge (AR) is a proposed deterrent to saltwater intrusion and aquifer depletion. In AR, surface waters are injected directly into groundwater during periods of high availability to increase freshwater storage during periods of drought or overpumping. More than 1,200 AR wells are operational in the United States, most located in arid regions in the West. Proponents of AR liken this water storage method to a below-ground dam. However, this rapid injection method bypasses the lengthy natural recharge process that removes nearly all impurities in the water. Surface water is mixed directly with pure groundwater in AR. What negative aspects of above-ground dams does AR avoid? What problems might injection create?

Another type of water sharing involves interbasin transfers, which remove water from a donor watershed to supplement a recipient watershed. This is often done to generate hydroelectric power or to alleviate water shortages. Such large-scale projects occur around the world, but especially in the United States, Canada, Australia, China, and India. Often the economic benefits of interbasin transfers come at the expense of the donor basin and to the benefit of the recipient basin. The Great Lakes, holding 84% of the surface fresh water in the United States, have been targeted for such transfers, though the potential ecological consequences are not well understood. The technology to transfer water in this way leads to questions about who owns the water and who should decide whether or not to relocate it.

11.4 Water Quality

BIG IDEA Water quality is as important as quantity when it comes to satisfying the needs of a thirsty world. Measures of water quality include physical, chemical, and biological parameters: turbidity, dissolved oxygen, nitrogen and pathogen levels, as well as the well-being of indicator species. Pollutants come from specific and diffuse sources and cause a variety of problems as water flows from the mountains to the sea. All water bodies are subject to the negative effects of nutrient enrichment, including overgrowth of algae and subsequent oxygen depletion.

Water Pollution

■ Water pollution is defined by physical, chemical, and biological characteristics.

Even when water is readily available, the quality of the water is of critical importance to human health and to the health of ecosystems. As of 2013, 783 million people on Earth did not have access to clean water.

Water quality is measured by its chemical and biological impurities as well as physical characteristics. Standards for water quality differ with its intended use. For example, standards for a public water supply differ from those for a body of water used for recreation. Through the Clean Water Act, the Environmental Protection Agency (EPA) requires states to report all impaired waters: those that exceed total maximum daily loads (TMDLs) of listed pollutants. TMDLs represent the maximum amount of a pollutant that a body of water can support without reducing water quality below acceptable levels. Pathogens top the list as the #1 cause of impairment; they are released from sewage, septic systems, and picked up from stormwater flow (**Figure 11.24**). Metals (typically from industry) and nutrients (usually from agricultural lands and feedlots) represent the #2 and #3 causes for impaired waters.

A hydrologist measuring water quality might sample temperature, pH, conductivity (an indirect measure of salt content), and dissolved oxygen at a surface water site. Other measurements could include turbidity (clarity), nutrients, including nitrogen and phosphorous, as well as a variety of specific chemicals such as mercury and other metals. See the summary table in **Figure 11.25**.

Hundreds of chemicals are present in surface waters; however, one new area of concern for the EPA are pharmaceuticals and personal care products (PPCPs).

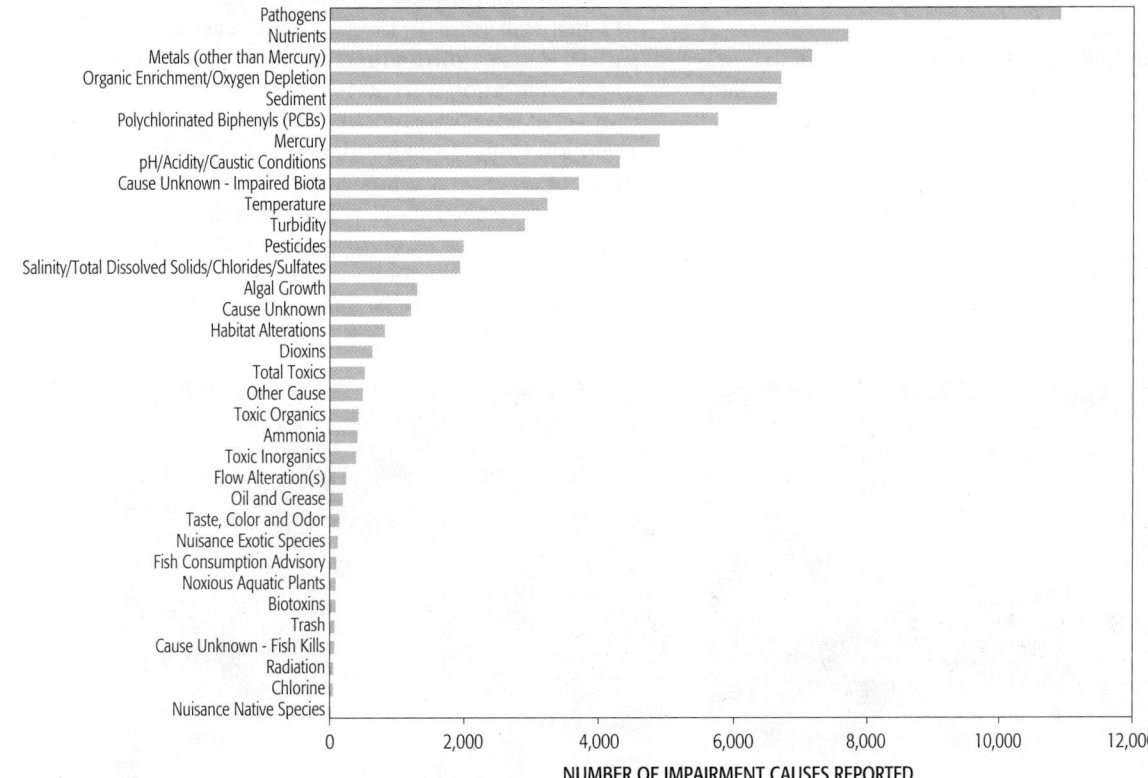

CAUSES OF IMPAIRMENT

NUMBER OF IMPAIRMENT CAUSES REPORTED

Where YOU LIVE Where does your water come from?

Your home has both a street address and a watershed address. Using EPA data, determine in which watershed your home is located.

- What is your watershed called? Where does the water that runs off your street ultimately end up? What upstream rivers lead into your watershed, and what downstream rivers take its water to the ocean?
- Using EPA data for your watershed (impaired water), characterize the water quality in a stream or river near your home.
- What are the most important factors influencing water quality in your area?

Q *Is there actually a health benefit to bottled water compared to tap water?*

Amber Apostal, Louisiana State University

Lissa: 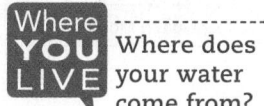 **A** Bottled water is advisable in developing countries where water quality is poor. In the United States, however, tap water must meet the EPA's stringent quality standards, so there is no health benefit to drinking bottled water. In addition, plastic bottles generate waste and consume unnecessary resources.

◄ **Figure 11.24 Pollutants Causing Impairments for U.S. Waters** The data show the number of U.S. waterways exceeding the total maximum daily load of pollutants in each of 22 categories.

Source: Data reported by states to the EPA for 2012 under Sections 305(b) and 303(d) of the Clean Water Act.

Water Quality Measurement	Definition	Source of Change	Consequence
pH	Level of acidity	Underlying geology; acid deposition, industrial processes	Some species are limited to narrow pH range
Temperature	Degree of heat or cold in the environment	Removal of riparian vegetation, thermal pollution from industry	Alters species composition
Dissolved Oxygen	Oxygen available in the water column	Reduced in slow water, high temperature, eutrophication	Alters species composition
Conductivity	Indirect measure of salt content	High evaporation or salinization from saltwater sources	Alters species composition
Turbidity	Clarity of liquid: amount of light penetration	High rainfall and runoff	Limits light penetration, increases siltation rate, clogs gills and filter feeders
Pathogens	Disease causing microorganisms, including bacteria, viruses, protozoans and other organisms, e.g. fecal coliforms (*E. coli*), *Cryptosporidium*, *Giardia*, etc.	Livestock runoff, sewage discharge, septic tanks, etc.	Spreads disease
Mercury	Heavy metal	Released from burning of coal, from coal mining	Accumulates in the food chain, may impair neurological development especially in children, hazardous for women of child-bearing age
Metals (other than mercury)	Arsenic, cadmium, iron, lead, manganese, nickel, zinc, etc.	Weathering of rock, deposition from volcanic activity; surface runoff from mining, industrial effluent, wastewater, agricultural runoff	Cancers, skin lesions, intestinal problems, various depending on metal
Nutrients			
Nitrogen	Limiting nutrient in terrestrial systems	Agriculture and livestock, aquaculture, wastewater, fossil fuel combustion, industrial processes, urban runoff	Eutrophication: algal blooms, oxygen depletion, dead zones
Phosphorous	Limiting nutrient in aquatic systems	Agriculture and livestock, aquaculture, wastewater, industrial processes, urban runoff	Eutrophication: algal blooms, oxygen depletion, dead zones

▲ Figure 11.25 **Parameters Commonly Measured in Assessments of Water Quality**

Physical, chemical, and biological parameters are included in the assessment of water quality.

This category includes human and veterinary drugs, fragrances, cosmetics, and nutritional supplements. PPCPs flow into wastewater via excretion, bathing, and trash and thus enter the hydrologic cycle. Until recently, the role of individuals directly contributing to the combined load of chemicals in the environment had been largely unrecognized. This is now an important area of research for the EPA.

In addition to the physical parameters used to assess water quality, living organisms are themselves excellent indicators of long-term water quality. Their survival depends upon the water conditions every day of their lives. With encouragement from the U.S. EPA, all 50 states have developed **bioassessment** programs, as well as criteria for ranking water quality based on those assessments. A bioassessment consists of surveys of aquatic biodiversity, including vegetation and animals such as fish, insects, and mussels that are known to be sensitive to overall water quality. These species are typically referred to as biological indicator species.

In the United States, monitoring systems are in place to measure both physical and biological parameters and to help detect changes in water quality. The U.S. EPA requires states to supply water quality information to the public, available for your region through the EPA website.

Chemicals from a variety of human activities pollute surface waters. **Point-source** pollution comes from a specific location, such as an industrial facility or municipal sewage treatment plant. **Non-point-source** pollution comes from a variety of activities occurring at different places across landscapes. Urban stormwater and fertilizer-enriched runoff from croplands are examples of non-point-source pollution (**Figure 11.26**). Pollution impacts systems all along water courses, as streams make their way from the headwaters of a watershed, through rivers, lakes, and estuaries and into the sea.

▶ Figure 11.26 **Sources of Water Pollution**

Stormwater runoff from streets and other impervious surfaces Ⓐ is an example of non-point-source pollution. Chemical-laden water released from industrial facilities Ⓑ is an example of point-source pollution.

Effects of Water Pollution on Ecosystems

■ Water pollution impacts ecosystems from the headwaters to the sea.

Despite the ability of ecosystems to respond and adapt to changes in the environment, pollution can disrupt and overwhelm ecosystem integrity. (See Chapter 7 for a description of aquatic and marine ecosystems.)

Streams.   Pollution has devastated the biodiversity of many streams. Filter-feeding animals are particularly sensitive to pollutants, especially sediments, which clog their gills. Today, over half of the 308 species of freshwater mussels native to streams in the United States are either endangered or have gone extinct.

Watersheds draining coal, copper, and other metal mines are often polluted by heavy metal contamination (Chapter 14). These mining processes expose sulfides in the rock to oxygen, which transform them into sulfuric acid. The resulting low pH releases metals from the mine tailings (rocks) that then wash into nearby streams, negatively affecting the stream ecosystem. Often the downstream biota will differ from the upstream biota because of the reduced pH and the presence of metals. Remediation often includes applying lime to neutralize the acidity.

Lakes.   Streams carry nutrients from stormwater, factories, and farms to lakes, where they accumulate in the water and sediments. **Eutrophication**, the enrichment of bodies of water with excess nutrients such as nitrogen and phosphorus, results in the rapid growth of populations of certain species of algae, with harmful consequences for the ecosystem. In the photosynthetic zone of eutrophic lakes, concentrations of oxygen are high during the day when photosynthesis peaks but very low at night due to high rates of respiration by the algae. Below the photosynthetic zone, the decomposition of large amounts of organic matter keeps the level of oxygen very low. These low levels of oxygen greatly reduce the biodiversity of eutrophic lakes.

Estuaries.   Watersheds eventually drain through estuaries to the ocean. Because estuaries occur at the base of large watersheds, pollutants of many kinds are carried into them. The excess nitrogen and phosphorus from farms eventually accumulates in estuaries, where it causes eutrophication, as it does in lakes. When the greatly increased populations of algae die, the high rates of respiration associated with their decomposition result in **dead zones**—regions of very low oxygen in which few marine animals can survive. Stormwater runoff from towns and cities adds additional pollutants that diminish ecosystem diversity and threaten human health.

Over the past few decades, communities have come to recognize the value of estuaries and the services they provide. Estuary ecosystems filter nutrients and sediments from the water and provide a buffer against floods and storm surges. As a result, local governments have begun to implement programs to restore and maintain these ecosystems. For example, the EPA's National Estuary Program works with states and communities to restore shorelines and wetlands and to reduce the inputs of pollutants. The EPA also sets standards for salinity and regulates the minimum flow of rivers into many estuaries so that enough fresh water is present to dilute the salt water from the ocean in these critical habitats.

Oceans.   The effects of pollution are particularly significant in the waters that drain into the ocean. Streams and estuaries transport nutrients from agricultural runoff and urban stormwater into coastal waters. There, those nutrients boost the primary production of algae, resulting in a "bloom," a large, often colorful accumulation of organic matter. As bacteria decompose this organic matter, they deplete the oxygen dissolved in the surrounding ocean water, creating dead zones. Such regions of ocean water contain such low levels of oxygen that few animals can survive in them (Figure 11.27). The United Nations Environment Program estimates that there are over 200 such dead zones worldwide, nearly all of them associated with the mouths of major rivers. Dead zones are a major contributor to the decline in the populations of many fish species.

Nutrient pollution has also had a devastating effect on coral reefs in many parts of the world. The addition of nutrients to these ecosystems encourages the growth of free-living algae, which may coat the corals and kill their symbiotic algae.

Even low concentrations of some pollutants, such as heavy metals, can be taken up by phytoplankton. When animals eat those phytoplankton, the pollutants are stored in the animals' tissues. At each successive trophic level, the concentration of pollutants in animal tissues increases. This process has significant consequences for top predators and humans (see Modules 6.4 and 18.3). Mercury, for example, is released into ocean waters from coal-fired power plants and the dredging of some wetlands. When mercury is taken up by phytoplankton,

▼ **Figure 11.27 Dead Zones** This false color satellite image indicates phytoplankton activity in the Gulf of Mexico near the mouths of the Mississippi and Atchafalaya rivers. The red areas correspond to large populations of algae and an extensive dead zone.

Atchafalaya River Delta

Mississippi River Delta

► Figure 11.28 **Ocean Acidification**
Ⓐ Relationship between atmospheric and oceanic CO_2 levels and ocean pH. Ⓑ Sea urchins lose the ability to fully form their spines when raised in conditions with high levels of CO_2. (Part A modified after R.A. Feely, *Bulletin of the American Meteorological Society*, July 2008.)

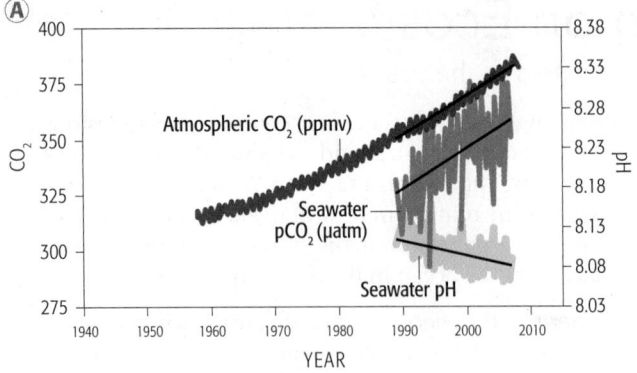

QUESTIONS 11.4

1. What pollutant is the most common cause for impairment of surface waters in the United States? Where does it come from?

2. Describe the effects of excess nitrogen and phosphorous on lakes and oceans.

3. What are the sources of mercury in surface waters? What are its consequences?

4. What is ocean acidification? What are its consequences?

(MES) For additional review, go to **MasteringEnvironmentalScience**

its concentration increases at each higher trophic level of the food web. In some places, the concentration of mercury in the tissues of carnivorous fishes, such as shark, tuna, and swordfish, is sufficiently high to pose a threat to human health.

Oil is a significant pollutant in many of the world's oceans. Although this pollution is often the consequence of oil spills, such as that from BP's deepwater drilling rig in the Gulf of Mexico in 2010, the majority of the oil in the ocean comes from non-point sources, such as urban stormwater and leakage from marinas, small boats, and pipelines.

Not all ocean pollutants arrive through the water courses that feed them. Direct atmospheric inputs can also introduce pollutants into the ocean. **Ocean acidification**, a reduction in ocean pH due to increasing CO_2 input from the atmosphere, has accelerated rapidly in the past 10 years and threatens to significantly alter ocean ecosystems.

Approximately one-quarter of the CO_2 in the atmosphere is absorbed by the oceans. As atmospheric CO_2 levels have risen, so too has the quantity that enters the oceans (Figure 11.28A). As recently as the 1980s, the oceans were considered infinite sinks of carbon: excess CO_2 from the atmosphere could be absorbed by the oceans with seemingly little effect. By the 1990s, however, a decline in ocean pH had become apparent, and ocean acidification quickly became a public policy issue.

The extent of the consequences of ocean acidification is not yet well known; however, studies now show that ocean acidification limits *calcification*, the ability of ocean organisms to add calcium carbonate to their bodies. Shelled marine organisms (e.g., snails, oysters, phytoplankton with calcareous shells, coralline algae) as well as corals, sea urchins and many other species, depend on the process of calcification for skeleton and shell formation (Figure 11.28B). Ocean acidification may result in a slowing of coral reef growth and a loss of calcareous species, the consequences of which are still to be determined. In addition, ocean acidification alters the availability of important nutrients such as nitrogen and iron, with the potential to change limiting factors and growth rates for marine organisms all the way up the food chain.

Finally, each year millions of tons of trash—anything from discarded fishing gear to plastic grocery bags—make their way into the oceans. Winds and currents have concentrated nearly 4 million tons of this refuse in an area measuring 1.2 million square kilometers (0.5 million square miles). This area, about halfway between North America and Hawaii, has come to be known as the Great Pacific Garbage Dump. Chemicals that leach out of the plastics are a significant source of pollution. More important, many marine mammals, birds, and fish die when they swallow or become entangled in this trash (Figure 11.29).

◄ Figure 11.29 **Ocean Garbage**
Each year, tens of thousands of tons of decomposition-resistant trash, such as fishing nets and plastic bags, end up in the ocean.

11.5 Water Management and Conservation

BIG IDEA Today, 700 million people live in countries experiencing water scarcity (Figure 11.30). More than three times this number lack access to clean water. The problem is most acute in some of the world's poorest countries, but many wealthy nations, including the United States, face significant water shortages. If current trends continue, by 2035 half of the people in the world will live under severe water shortage. The World Commission on Water estimates that human water use will increase by 50% by 2050. Increased water conservation is essential to the sustainable use of Earth's limited supply of fresh water. Strategies for reducing the use of water include improved techniques for irrigating crops, recycling of municipal and industrial water, and more efficient use of residential water. In some locations, it may be feasible to convert seawater to fresh water. Policies that determine the price of water also affect water conservation.

▶ **Figure 11.30 Water, Water Everywhere?** Ten percent of Earth's people do not have enough water, and nearly 40% must drink and bathe in water that does not meet even the most minimal standards of sanitation.

Regulating the Flow

■ Humans have modified streams in order to minimize the risks of both floods and water shortages.

Throughout history, much agricultural and urban development has depended on rivers and the ecosystem services they provide. River floodplains provide some of the world's most fertile agricultural lands. The great majority of the world's cities have grown up beside rivers. Rivers provide energy, transportation for commerce, and water for industrial and residential use. Nevertheless, season-to-season and year-to-year variations in river flow have presented significant challenges to both urban and agricultural development. Humans have taken a variety of actions to manage floods and periods of low flow. Many human activities have also polluted streams.

Dams and diversions. Humans construct dams to control floods, store water, and generate electricity. Dams restrict the flow of water on 180 of the 300 largest rivers in the world. In the United States alone, states regulate the activities of over 79,000 dams that are higher than 7 m (25 ft). There is no question that dams provide an important source of renewable energy. Their impoundments store water that can be diverted to irrigate dry places or used during dry seasons. But they also have significant negative impacts on stream ecosystems.

By design, dams create lakes with little or no water movement. When streams flow into these lakes, they slow down and drop the sediment they are carrying. As sediments accumulate, they reduce the capacity of the lake to store water. If streams carry high loads of sediment, lakes may fill in rather quickly. In contrast,

streams flowing out of dams carry little sediment. Therefore, these streams cannot replace the sediments lost in the natural erosion occurring on stream banks below the dams (Figure 11.31).

Dams fragment stream ecosystems, often interfering with the movement of organisms such as salmon and other migratory fish. Today, most dams are required to have fish ladders or other means of moving fish over these barriers. But because most migratory fish use the flow of the water to navigate upstream, they may lose their sense of direction in the still waters behind dams (see Module 15.5).

▼ **Figure 11.31 Dams Redistribute Sediment** The Hoover Dam, which sits at the boundary of Arizona and Nevada on the Colorado River, created Lake Mead, the largest reservoir in the United States by volume. Sediment collects behind the dam in Lake Mead, but very little flows downstream to replenish sandbars and riparian habitat.

Dams increase the total storage of water within watersheds, but they also reduce the flow downstream. Lower flows reduce the amount of water available across floodplains, thereby altering the composition of riparian forests and other floodplain ecosystems. Dams also alter seasonal patterns of water flow, which can have a significant effect on ecosystems downstream. Water diversions such as canals often move water to new locations, thereby diminishing the flow of water to the sea.

Channelization and artificial levees.

Humans have built artificial levees and channelized rivers to control the flow of water, prevent flooding, and make floodplains available for agriculture and urban development. Artificial levees elevate stream banks in order to confine water to stream channels. When a river is channelized, it is dredged and straightened (Figure 11.32). Channelization usually involves the removal of most riparian vegetation.

While channelization and the construction of artificial levees may protect the lands adjacent to a particular part of a river, they also increase the speed of the water flowing through the modified channel. This greater speed increases the potential that flooding will occur downstream. In response to this threat, additional levees or channels may be constructed on lower parts of the river, until much of the length of the river is managed. For example, a series of 5- to 15-m (16–49 ft) tall levees extends along more than 1,600 km (995 mi) of the Mississippi River, all the way to the river delta in New Orleans.

The protection offered by levees sometimes encourages development in areas that could not be otherwise inhabited. For example, the city of New Orleans is situated in a bowl located below sea level and is surrounded by the waters of two lakes as well as the Mississippi River (Figure 11.33). The failure of the levees around the city of New Orleans following Hurricane Katrina in 2005 resulted in catastrophic flooding and a national tragedy. The hurricane-induced storm surge into nearby Lake Pontchartrain overtopped and damaged the levees in many locations, flooding over 75% of the metropolitan area.

▲ Figure 11.32 **Stream Channelization**
This segment of the Emscher River, a tributary of the Rhine River in the Ruhr area of Germany, has been channelized and used as an open sewage canal since the late 1800s. Renaturalization is underway and sewage will be diverted to an underground pipe.

Wetland loss and restoration.

One of the important lessons learned from Hurricane Katrina is how critical wetlands are in mitigating the effects of flooding and storm surges upon the city of New Orleans. Over the past 200 years, the total area of wetlands in the lower 48 United States has been reduced from 90 million to 40 million hectares (220 million to 100 million acres) and a large part of that loss has been along the Gulf Coast (Figure 11.34). Much of the remaining wetland has been degraded by drainage, changes in water supply, and vegetation damage. Wetlands in many areas have been drained to control mosquitoes or to facilitate nearby development and farming. Drainage can result in significant change in vegetation cover and accelerate the decomposition of peat. It can also make wetlands vulnerable to destructive wildfires.

There are limits to the capacity of wetlands to absorb nutrients and pollutants. Over long periods, sediment, fertilizer, sewage, pesticides, and heavy metals degrade the functions of these ecosystems. Overgrazing by animals, peat harvesting, logging, and the introduction of non-native plants can also degrade wetlands.

Many countries have set policies to halt the loss of wetlands. The United States, for example, has established a "no net loss" policy. Using powers provided under the Clean Water Act, the U.S. Army Corp of Engineers reviews actions by private landowners or public agencies that may impair or destroy a wetland. When those actions are approved and permits granted, the landowner or agency must compensate for the loss by restoring or protecting additional wetlands.

◄ Figure 11.33 **Levee Failure Following Hurricane Katrina**
Once Hurricane Katrina made landfall and the levees were breached, water poured into the below-sea-level basin of New Orleans, flooding 75% of the metro area.

Mississippi River

New Orleans

Lake Pontchartrain

A

B

N

30	Floodwall along Mississippi River (23 ft)		Lake Ponchartrain
20	A	Hurricane Protection Levee and Floodwall (17.5 ft)	
10			B
0	Mississippi River		
-10	St. Louis Cathedral: 5 ft above sea level	Dillard University at sea level	University of New Orleans at sea level
-20			

ELEVATION (ft)

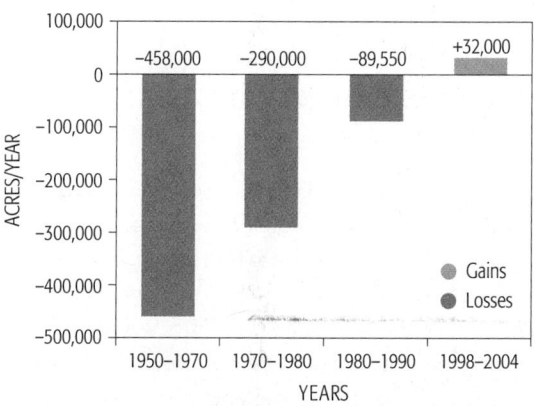

▲ Figure 11.34 U.S. Wetland Loss

Wetland loss has occurred during the past two centuries, but the rate of that loss was slowed by the passage of the 1972 Clean Water Act, which included provisions for wetland protection. In recent years, restoration efforts have been gradually increasing wetland area.

Source: Dahl, T. E. 2006. Status and trends of wetlands in the conterminous United States 1998 to 2004. Washington, D.C.: U.S. Fish and Wildlife Service.

▲ Figure 11.35 Wetland Restoration

Formerly drained and degraded, this 60-acre marsh near Seal Beach, California, has been restored and once again provides important ecosystem services.

Wetland restoration projects are underway in many locations around the world (**Figure 11.35**). Scientists are developing techniques to create new wetlands and restore the ecosystem services that have been lost. Many of these projects and techniques show great promise. However, scientists agree that it is essential to protect Earth's wetlands that remain intact.

Estuarine wetland loss.   Estuarine wetlands are subject to significant alteration of the flow of the rivers that feed into them. Dams have changed the seasonal patterns of river flow. More importantly, the diversion of water has reduced the overall flow of fresh water into many estuaries and consequently increased the salinity.

San Francisco Bay provides a striking case study of such change (**Figure 11.36**). The primary sources of water for the northern portion of this very large estuary are the Sacramento and San Joaquin rivers. Prior to 1930, water flowing into San Francisco Bay from these two rivers averaged about 34 km^3 (8.2 mi^3) per year. Between 1930 and 1970, a number of dams and canals were installed to divert water to agricultural fields and urban centers. The California Aqueduct alone transports as much as 8 km^3 (1.9 mi^3) of water to the San Joaquin Valley and the city of Los Angeles. By 1980, the annual flow into the bay was less than 13 km^3 (3.1 mi^3), and today it is less than 10 km^3 (2.4 mi^3). Over that same time, springtime salinity in the northern portion of the bay has increased by three- to fivefold. These changes have altered the composition of species in the pelagic, benthic, and wetland communities in the northern portion of the bay and in the lower stretches of the rivers that feed it.

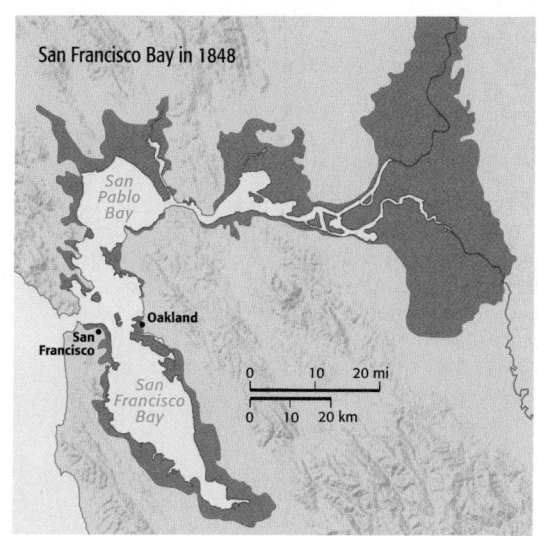

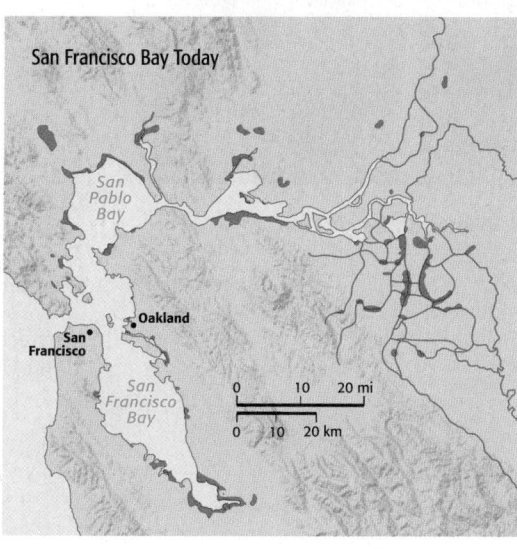

◄ Figure 11.36 Estuarine Wetland Loss

Destruction of most of the tidal wetlands in San Francisco Bay (shown in red) is a major factor contributing to the diminished water quality in this large estuary.

Source: Nichols, F.H. 2007. The San Francisco Bay and Delta—An Estuary Undergoing Change. U.S. Geological Survey. http://sfbay.wr.usgs.gov/general_factsheets/change.html.

Wetland Restoration Rates

How long does it take to restore a degraded wetland to its natural condition?

1. The investigators did not study these wetland systems continuously for 55 years. How were they able to draw conclusions about recovery rates over this time period?

2. Use the data shown to make predictions about wetland recovery. If recovery continues at the same rate plotted here, how long would you expect it take for soil organic matter to return to natural levels? Nutrient absorption capacity?

To make up for the loss and degradation of wetlands, restoration projects have been undertaken in thousands of locations. In fact, in many states it is legal to change or even destroy wetlands to make room for developments or highway construction, as long as a previously degraded wetland is restored first. Such policies presume that newly restored wetlands have the same habitats and functions and provide the same services as the natural wetlands they replaced. However, few *long-term data* are available to determine how long it actually takes for a restored wetland to develop the characteristics of undisturbed wetlands.

Cornell University doctoral student Kate Ballantine and her mentor Rebecca Schneider investigated rates of change in wetlands in central New York State. They were particularly interested in the characteristics of soil in restored wetlands, since those characteristics influence net primary production and nutrient movement (**Figure 11.37**). In their investigations, they studied restoration projects in 30 marshes; these restoration projects ranged from 5 to 55 years in age. They *sampled* soils and vegetation biomass from each restored marsh. They also sampled soils and vegetation from five undisturbed natural marshes as *controls* for comparison.

Ballantine and Schneider found that plant productivity increased rapidly in the first decade following restoration, but slowed thereafter. After 55 years, plant productivity in restored marshes was nearly equal to that in their natural

counterparts. However, key soil characteristics were much slower to recover. Even after 55 years, the levels of organic matter in the soils of restored wetlands and the capacity of those soils to absorb nutrients were less than half of those in the natural wetlands (**Figure 11.38**). Their findings are significant because these soil characteristics are highly *correlated* with long-term fertility and important ecosystem services, such as the ability of a wetland to retain nutrients and pollutants.

Ballantine and Schneider's data represent one of the longest time intervals over which wetland restoration has been studied. They *confirm* what many scientists had suspected: Wetland restoration must be viewed as a long-term process because it takes decades, or even centuries, for many important wetland functions and services to return to their pre-disturbance condition.

Reference: Ballantine, K. and R. Schneider. 2009. Fifty-five years of soil development in restored freshwater depressional wetlands. *Ecological Applications* 19: 1467–1480.

▲ Figure 11.37 **A Wetland Restoration Collaboration**
Research carried out by Kate Ballantine Ⓐ and Rebecca Schneider Ⓑ provides the best available data on the time it takes for the ecosystem functions and services in restored wetlands to recover to natural levels.

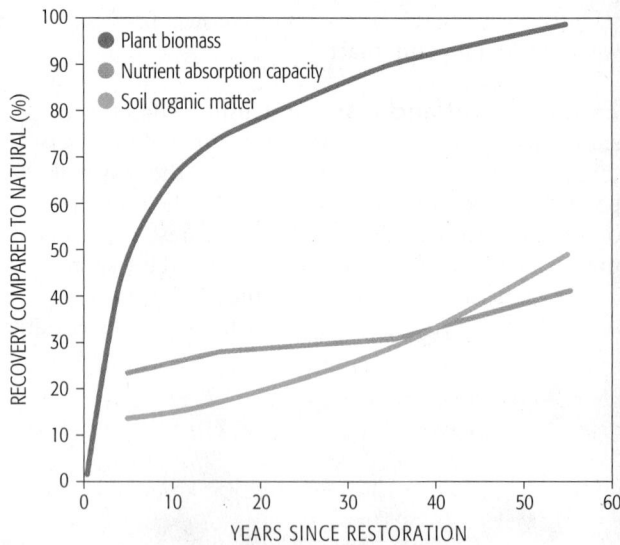

▲ Figure 11.38 **Soil and Vegetation Change**
Plant biomass, a direct measure of net primary production, recovers rapidly in restored wetlands. However, soil organic matter and nutrient absorption capacity recover much more slowly. At their current rates of recovery, these two characteristics of restored wetlands will not approach levels found in natural wetlands for a century or more.

Managing and Conserving Water Used in Agriculture

■ Improved technologies can reduce evaporation and runoff from agricultural fields.

About 70% of the water that humans use goes to irrigate crops and pastures. Much of that water is lost when it evaporates from the surfaces of reservoirs, canals, and flooded fields. The rate of that evaporation is greatest in arid regions, where irrigation is most widely used. Water that runs off into nearby streams is also lost. Yet there is much opportunity for improvement. In fact, engineers estimate that improved techniques could reduce the amount of water used to irrigate crops by more than 50%.

Irrigation efficiency is the percentage of the water applied to fields that is actually used by crop plants. Agricultural scientists estimate that average irrigation efficiency of cropland is less than 40%. In other words, over 60% of the water applied to crops simply evaporates or runs off fields without being used by plants (**Figure 11.39A**).

In most parts of the world, land is irrigated by flooding fields or channeling water down a field through shallow, parallel furrows. These methods distribute water so unevenly that farmers must apply excessive amounts of water to ensure that the entire field gets enough. Wasted water then runs off or evaporates. One method being used to reduce this waste is *surge irrigation*. In this method, an initial surge of water wets the furrows, partially sealing the surface of the soil. The next application of water flows across the field more evenly. Surge irrigation can decrease water use by 15% to 50%.

Aerial sprinkler systems are widely used in many arid regions, especially in the western United States. Much of the water distributed by these systems is lost to evaporation. New sprinkler designs can significantly diminish these losses. For example, low-energy precision application (LEPA) sprinklers have tubes extending down from the sprinkler arms that deliver water close to plants. In some situations, LEPA systems can reduce water use by more than 60%. And in addition to saving water, LEPA systems save the energy that would have been used to pump that water (see Figure 12.30, Module 12.5).

During the 1950s, scientists in Israel developed an entirely new method for delivering water to crops—drip irrigation. This process delivers water to the roots of individual plants through a network of perforated tubes or pipes installed at the soil surface. Since their conception, drip irrigation systems have become highly automated with monitors to ensure that the amount of water delivered optimizes plant growth and minimizes losses to runoff and percolation. Irrigation efficiency for drip irrigation systems can exceed 95% (**Figure 11.39B**).

▲ **Figure 11.39 A Flood vs. a Trickle**
Ⓐ Over 60% of the water applied by flood irrigation to this Arizona cornfield will simply run off or evaporate without being used by the plants. Ⓑ Drip irrigation, by comparison, feeds a minimal although sufficient amount of water directly where it is needed.

The primary deterrent to the implementation of efficient irrigation systems is cost. LEPA systems cost approximately $500 per hectare, and drip irrigation systems can cost 5 to 10 times that amount. Such costs are prohibitive in the world's poorest countries. In developed countries, inappropriate pricing of water often means that there is little incentive for such investments.

Water Reuse

■ Reuse can conserve large amounts of water and provide other benefits as well.

Q Despite my community's efforts to conserve water, droughts continue to be an issue in my city. Is there a way that we can collect more water and save it for when we need it the most?

Samantha Fullon,
Los Angeles Mission
College

Lissa: A Many cities capture stormwater runoff in small reservoirs and allow the water to percolate down to recharge aquifers.

Most industrial and residential uses of water are nonconsumptive. After the water is used, it is returned to a body of liquid water. For example, when you take a shower, the soapy water may eventually flow into a municipal sewer. From there it eventually makes its way into streams, estuaries, or groundwater. The primary challenge associated with the water coming from industries and residences is altered water quality. As Sandra Postel, Director of the Global Water Policy Project, points out, much water could be saved if we could "get the 'waste' out of wastewater." It may be difficult or impractical to restore the quality of wastewater to drinking water standards, but many uses of water do not require such high quality.

The idea of reusing municipal wastewater is not new. By 1800, wastewater from many English cities was being used to irrigate nearby farms. Such uses became limited in the early 19th century, as people came to understand how unclean water can spread disease and lead to other health problems. Modern methods of sewage treatment have put those health concerns to rest. Today, wastewater safely irrigates croplands near urban centers in many parts of the world.

Reusing wastewater provides several benefits in addition to conserving water. Because wastewater is high in nitrogen and other nutrients, using it to irrigate crops reduces the need for additional fertilizer. Furthermore, wastewater is purified as it percolates through soils. With additional treatment, wastewater can be recycled for a variety of other uses, including landscape and golf course irrigation, vehicle washing, toilet flushing, and groundwater recharge.

St. Petersburg, Florida, is one of a few cities that reuses all of its wastewater, releasing none into streams or the nearby ocean. The city has two water distribution systems. One delivers fresh water for drinking and other household uses. The other provides treated wastewater for irrigating public and residential lawns. Residents pay about 70% less for the treated wastewater (Figure 11.40).

There are important challenges regarding the reuse of municipal water. Wastewater is generated year-round, but irrigation needs vary from season to season. Therefore, many municipalities must store wastewater in ponds, which adds to processing costs. In many developing countries, sewage treatment is inadequate and may be a significant risk to public health.

Reuse of water from commercial and industrial uses presents different challenges. Because wastewater from industrial facilities often contains heavy metals or organic pollutants, it cannot be used for irrigation. These chemicals can be removed from water, but the costs are often high. Nevertheless, in response to the increasing cost of water and tighter regulations regarding wastewater disposal, industries around the world have implemented a variety of techniques to reuse wastewater and increase the efficiency of water use. While industrial production in the United States has increased nearly fivefold since 1950, the amount of water used by industries has declined by more than 50%.

Can wastewater be recycled to drinking water standards? From a technical standpoint, the answer is yes. Pilot projects in water-stressed cities such as El Paso, Texas, and Denver, Colorado, have shown it to be possible. However, public acceptance of this idea is generally low, and the costs for such processing are very high. In most cases, it is far more economical to direct wastewater to uses that require far less treatment.

▶ Figure 11.40 Reusing Water
In Collier County, Florida, nonsewage wastewater is treated and reused for yard sprinkling and similar purposes.

Desalination

■ Converting seawater to fresh water is feasible in locations where water is scarce and energy is abundant.

There is, of course, a vast amount of water—albeit salty—in the ocean. The process of removing salts and other chemicals from seawater is called **desalination**. The two main methods of desalination are **distillation**, boiling water and condensing the steam, and **reverse osmosis**, filtering water through a selective membrane. Both methods require a great deal of energy and are, therefore, quite costly. Yet where other sources of water are scarce, desalination can be cost effective.

In many countries in the Middle East, water is particularly scarce and energy, especially oil and natural gas, is particularly cheap. Thus, it is not surprising that over 75% of the world's desalination capacity has been built in that region. For example, the world's largest desalination facility is in the United Arab Emirates near the port city of Dubai. This is a dual-purpose facility: It uses natural gas to generate electricity for Dubai and then uses the heat from the generators to distill seawater. Each year this facility produces 300 million cubic meters (79 billion gallons) of fresh water (**Figure 11.41**).

In other regions, there has been interest in desalinating brackish water, which contains less salt than seawater but is still too salty to drink. Removing the salt from brackish water requires considerably less energy and expense than desalinating seawater. Reverse osmosis systems are especially well suited to this task. Hundreds of coastal communities use such facilities to produce fresh water from brackish groundwater.

Nearly 13,000 desalination plants have been built worldwide. Each year they produce 16.5 billion cubic meters (4.4 trillion gallons) of water. Although this may seem like a large amount, it represents only about 0.1% of total human freshwater use.

▼ **Figure 11.41 Desalination and Energy**
The Jabel Ali desalination plant in the United Arab Emirates generates electricity from natural gas. The heat from the natural gas turbines is used to desalinate seawater by distillation.

Getting the Price Right

■ Changes in water pricing are needed to ensure sustainable conservation.

Like the air we breathe, we generally consider access to water to be an entitlement. We rarely think about its value or its cost until we run short of it. Many of the world's water shortages are a consequence of our failure to value water appropriately.

Consider the example of water pricing in California. Agriculture uses more than 80% of the state's water resources, yet farmers pay only 5% to 20% of the costs associated with supplying that water. The remaining costs are subsidized by revenues from state and federal taxes. Some people argue that subsidies of this kind are appropriate because society at large benefits from a reliable and relatively inexpensive supply of food. But such subsidies externalize the real costs of water use. Because water is so inexpensive, farmers have little incentive to invest in technologies that conserve water, such as LEPA or drip irrigation. Worsening drought in California may begin to change that calculus, however. The U.S. Congress eliminated the federal government's supply of irrigation water to Central Valley farms in 2014 due to the extreme drought, and farmers were left with the difficult choice of what (or if) to plant given the lack of water.

Differences in pricing also influence the residential use of water. Studies show that water is used more efficiently in households with water meters than in households where water is priced at a flat monthly rate, regardless of the amount used. Many communities have increased conservation further by graduated pricing, in which the unit cost of water increases as more is used.

In summary, large amounts of water can be conserved by using water more efficiently, improving irrigation, and recycling municipal and industrial wastewater. In some locations, water supplies can be augmented by desalination. Yet many of these conservation strategies involve significant costs, so their implementation depends on how water is valued in economic markets.

QUESTIONS 11.5

1. Name and describe four types of wetlands.
2. What factors limit plant growth on wetland soils?
3. Describe three human actions that have degraded many wetlands.
4. Describe two alternatives to aerial sprinkler systems that improve irrigation efficiency.
5. What are three benefits of water reuse?
6. Describe how a natural gas–generating facility can also function to desalinate water.
7. How do differences in household water-billing practices affect consumer water use?

(MES) For additional review, go to **MasteringEnvironmentalScience**

11.6 Wastewater Treatment

BIG IDEA Wastewater includes sewage, water from sinks and other household uses, stormwater runoff, and water used by manufacturing facilities and other industries. In the past, wastewater was simply dumped into nearby waterways. Today, in most developed countries, wastewater is treated to protect the environment and prevent the spread of disease before it is returned to streams and rivers to flow back into the natural hydrologic cycle. Unfortunately, in some poor countries, wastewater still goes untreated. The most common methods of managing wastewater are municipal sewage treatment plants and septic systems. Recently, some municipalities have become interested in treating wastewater with methods that mimic the biogeochemical processes of natural ecosystems.

Municipal Wastewater Treatment

■ Wastewater treatment varies widely among developing and developed countries.

▼ Figure 11.42 Wastewater Treatment
When wastewater enters a modern treatment plant, it is pretreated to remove solids. Primary treatment removes additional solids and produces a homogeneous liquid that is high in organic compounds. In secondary treatment, microorganisms decompose those organic compounds.

Until recently, sewage and other forms of wastewater were simply dumped into nearby waterways. Over time, dilution and natural processes would eventually decompose the sewage and purify the water. In areas with concentrated human populations, however, this approach was problematic because natural processes cannot break down large amounts of waste in a short period of time. Untreated sewage and wastewater harm natural ecosystems, threaten human health, and contaminate surface water, the source of drinking water for the vast majority of Earth's people.

In most developed countries, wastewater is now treated before it is returned to the environment. In less developed countries, the treatment of wastewater is more variable. Many poor countries have virtually no wastewater treatment. In Latin America, only about 15% of the wastewater that is collected is treated in

some manner. The lack of adequate water treatment is a major cause of the high rates of waterborne illnesses, such as cholera and typhoid fever, in developing countries.

In the United States, most cities and towns manage their wastewater and sewage in **municipal sewage treatment plants (MSTPs)**. These plants use a stepwise process to remove wastes and chemicals from the water (**Figure 11.42**). When water first enters a sewage treatment plant, it is pretreated to remove large solids, such as rags, feminine hygiene products, sand, and gravel. Insoluble chemicals, such as grease and oils, may also be removed.

Pretreated wastewater then flows into large settling tanks, where it undergoes **primary treatment**. Particles in the wastewater settle to the bottom of the tank, forming a sediment called sludge. The main

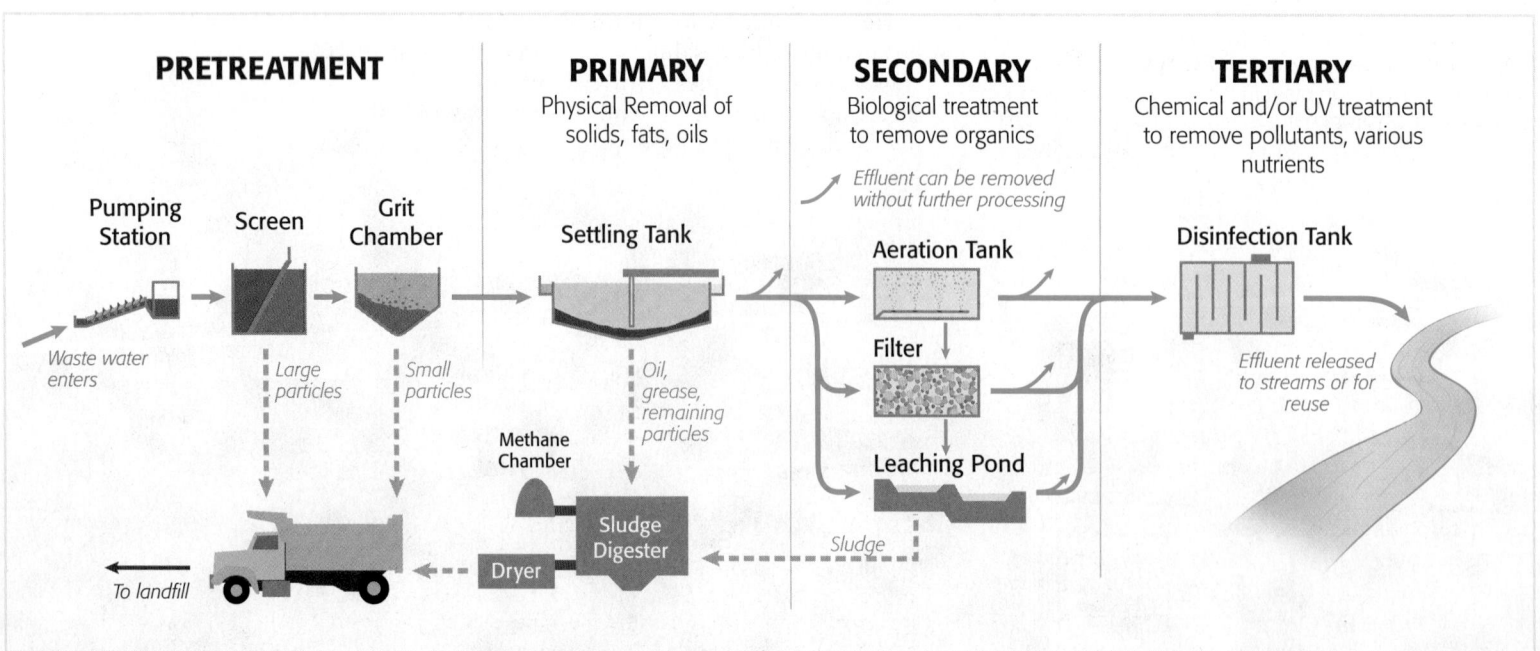

purpose of primary treatment is to produce a relatively homogeneous liquid that can be treated biologically and a sludge that can be processed separately. In many developing countries, municipalities return wastes to the environment after primary treatment.

In **secondary treatment**, bacteria and other microorganisms are used to break down the organic material dissolved in the wastewater. In most treatment plants, secondary treatment takes place in aerated tanks and basins. Some facilities use membranes or gravel filters to further separate solid and liquid wastes (Figure 11.43).

Some treatment plants also use **tertiary treatment** to remove inorganic nutrients from wastewater. In this stage of treatment, wastewater is passed through sand and charcoal filters to remove residual solids and toxins. Next, the water is stored in human-made ponds or lagoons where microorganisms remove significant amounts of dissolved nitrogen and phosphorus. Finally, the water is disinfected with chlorine, ozone, or ultraviolet radiation to reduce the number of microorganisms.

Solid wastes, or sludge, accumulate at each step in this treatment process. Most often, sludge is subjected to digestion by microorganisms, which reduce the volume of organic matter and the number of disease-causing microbes. Sludge is then dried so that it can be transported and disposed of off-site. Usually, it is dumped into landfills or spread onto open land. However, a growing number of treatment plants convert sludge into

pellets that can be used as fertilizers; these pellets are often sold to local gardeners and farmers.

In place of traditional MSTPs, some communities are beginning to use natural or constructed wetlands to purify wastewater that has had primary treatment. Wetlands are very effective at purifying water (see Module 7.7). As water slowly percolates through wetland soils, solid materials are filtered out and microbes decompose the organic matter. Nutrients, such as nitrogen and phosphorus, and contaminants, such as heavy metals, are adsorbed by soil particles or taken up by plants and stored in their tissues.

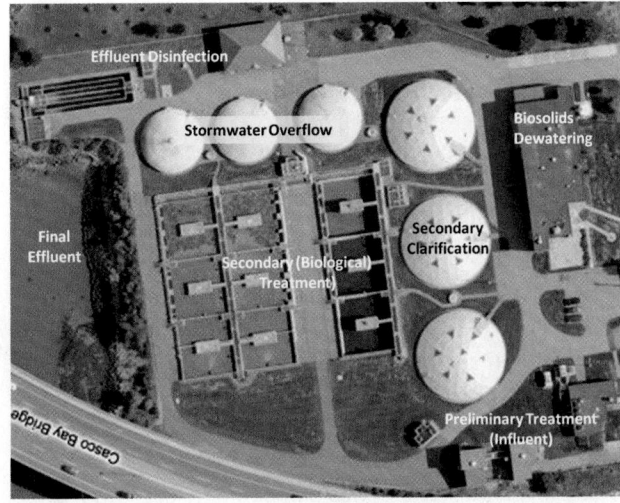

◄ Figure 11.43 **Your Neighborhood MSTP**
Communities use different components of wastewater treatment depending on their needs. This aerial view of a wastewater treatment plant in Portland, Maine shows primary and secondary treatment.

QUESTIONS 11.6

1. Describe what happens in primary, secondary, and tertiary treatment of wastewater.

2. Explain how an on-site septic system operates.

(MES) For additional review, go to **MasteringEnvironmentalScience**

On-Site Wastewater Treatment

■ If properly maintained, septic systems can isolate waste and protect water supplies.

In less densely populated areas, households often rely on **septic systems** to treat their wastewater and sewage. In these systems, sewage and household wastewater flow to an underground septic tank outside the home. Solids settle to the bottom of the tank where microorganisms begin to break down the waste. Wastewater flows to a series of perforated, underground pipes through which it is released into a **leach field**, where microorganisms in the soil finish breaking down the waste materials (Figure 11.44). Periodically, the solids that settle in the septic tank need to be pumped out and disposed of in a landfill.

Nearly 25% of the households in the United States rely on septic systems to treat their wastewater. When properly maintained, septic systems are an effective means of isolating wastes and protecting water supplies. Maintenance includes monitoring leach fields and occasional pumping of septic tanks. However, about 10% of these systems are not functioning properly. In communities where soil conditions prevent effective leaching, failure rates may exceed 70%. The U.S. EPA reports that failed septic systems are the third most common cause of groundwater contamination.

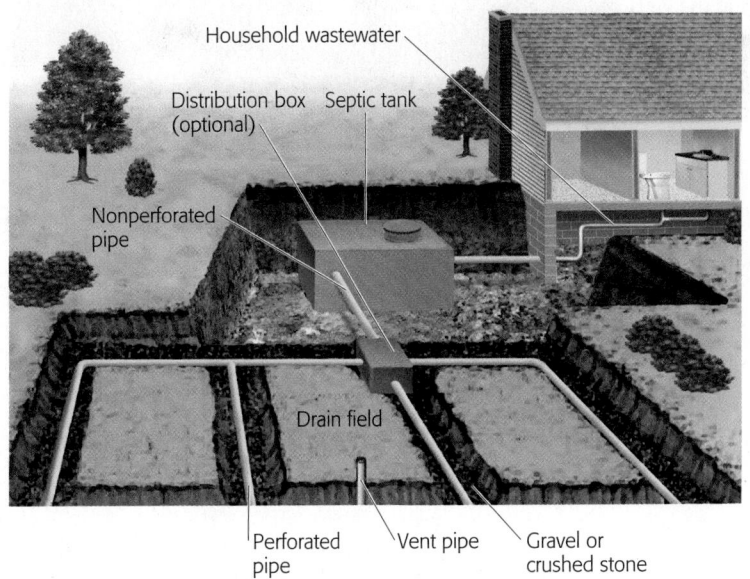

▲ Figure 11.44 **Household Septic System**
Rural wastewater is often treated in on-site septic systems. Wastewater flows into the septic tank, where solid materials are decomposed. Liquid wastes flow out of the tank and into a system of perforated pipes in the leach field, where dissolved organic chemicals are broken down by microbes.

Eco-Machines

Can simulated wetland ecosystems purify wastewater?

▲ Figure 11.45 **Wastewater Visionary**
John Todd has applied his understanding of the ecological processes in a variety of natural wetlands to the design of Eco-Machines, which use communities of living organisms to purify waste-laden water.

▶ Figure 11.46 **An Eco-Machine in Action**
In this greenhouse-size system, waste is removed from water as it moves from anaerobic ecosystems in the tanks through tanks such as these that support ecosystems that are successively more diverse and increasingly aerobic.

John Todd has been long impressed with the capacity of complex ecosystems to adapt to constant change over the eons (Figure 11.45). That capacity has been the basis for Todd's development of Eco-Machines, systems that use living organisms from diverse ecosystems to treat wastewater. His systems are solar powered by photosynthesis; they are also self-organized and adaptable to change by virtue of their diversity.

Todd's first Eco-Machine was installed in the Cape Cod town of Harwich, Massachusetts. Prior to 1990, nearly all of the town's wastewater was dumped into Flax Pond, a small lagoon in a 25-foot-deep layer of sand just above the aquifer that supplies the town's drinking water. The pond was filled with an organic "soup" that was at least 35 times more concentrated than municipal wastewater. The pond water also contained high quantities of nearly every one of the priority pollutants listed by the U.S. EPA.

To restore Flax Pond, Todd set up a series of 21 large tanks that were connected by pipes. Into these tanks he put thousands of species of plants and animals collected from a great variety of natural habitats, including pig wallows,

swamps, salt marshes, and forests. Water from the pond was circulated through the tank. Each aquarium received a different strength and composition of waste, with those nearest the pond receiving the highest concentrations. Soon, a different ecological community organized itself in each tank. This process of community development was encouraged by the installation of racks for plants and special surfaces for microbes (Figure 11.46).

Within 13 days, the water flowing out of the system was free of all EPA-listed pollutants, and concentrations of heavy metals in the water were well within the standards for safe drinking water. After inspection, the state of Massachusetts approved Todd's installation for septic waste treatment.

Since 1990, Todd has installed over 100 Eco-Machines in schools, offices, and factories in 11 countries and on five continents, from Sweden to Australia and across the United States. An Eco-Machine may be housed in a small greenhouse or deployed over several hectares, depending on the amount of effluent it must process. Each machine is divided into a number of individual tanks that correspond to various kinds of wetland habitats you might encounter in nature. Wastewater treatment begins in anaerobic tanks that support blue-green algae, bacteria, fungi, and protozoa. Successive tanks support higher plants, which are often suspended in reinforced fiberglass racks. The final processing tanks support shrubs and trees, as well as snails, clams, and fishes. As water moves through the system of tanks, the concentrations of wastes diminish and the concentration of oxygen increases.

Todd has received numerous awards for his innovation, including the Environmental Merit Award from the EPA and the Buckminster Fuller Challenge Award for visionary thinking, which seeks "to solve humanity's most pressing problems in the shortest possible time while enhancing Earth's ecological integrity." That visionary thinking is captured in Todd's own words: "The world is a vast repository of unappreciated or unknown biological strategies that have immense importance for humans if we can develop a science of integrating the stories embedded in nature in the basic systems that sustain us."

11.7 Water and You

BIG IDEA In the United States, 13% of the fresh water used is distributed through municipal systems. A significant fraction of that water is lost from leaky pipes and mains during distribution and does not reach the end user. The public has many opportunities to reduce water use by limiting waste, improving water efficiency, and engaging in conservation practices.

Municipal Water Use

■ Bathrooms and laundry rooms consume the highest proportion of water in the home.

Domestic and public water use makes up 13% of freshwater consumption in the United States. Within households, the majority of water is used in bathrooms and laundry rooms, with toilets, washing machines, and showers topping the list (Figure 11.47). Leaks also consume 14% of household water: representing wasted water that never reaches the end user.

Limiting water loss and waste is an important first step to reducing water consumption. Aging water mains and leaky pipes plague many municipal water systems, and losses of 10–30% are typical in cities in the United States and Canada. Leaks result from corrosion in aging cast iron pipes (being replaced today with less corrosive materials), poorly constructed systems, improperly adjusted valves, and mechanical damage. Globally, unaccounted-for water, or that which is lost via leaks

or unauthorized use before it reaches the customer, can be much higher. For example, Nigeria, Mexico, and Armenia record municipal water losses of more than 50%. A 2012 study by the American Water Works Association suggests that the United States will need to invest $3 trillion in water infrastructure by 2035 simply to maintain the current level of water service. Despite the initial investment, leak detection and repair programs result in long-term economic benefits by reducing water loss and delaying the development of drinking water distribution systems with greater capacity.

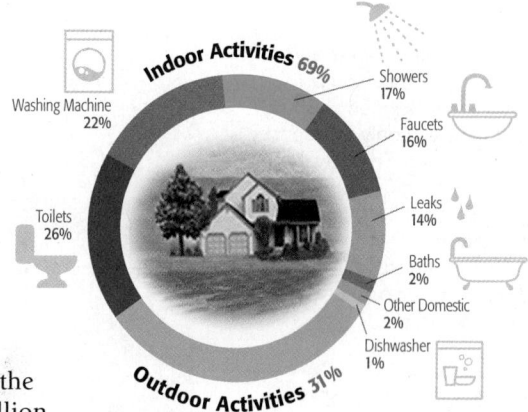

▲ Figure 11.47 **How Do We Use Our Water?**
Americans use most of their household water in the bathroom, with more than 61% supplying toilets, showers, faucets, and baths. Fixing leaks and updating water fixtures with low-flow technology can significantly reduce water use in the home.

Water Efficiency and Conservation

■ Technology has greatly improved water efficiency in home fixtures and appliances, but behavioral change is usually required for water conservation.

Increasing water efficiency is the next step in reducing water consumption and it often employs technological advances. In 1992, water efficiency standards were established by the U.S. EPA that cut water use of toilets, showerheads, water faucets, and water-using appliances to less than one-third of pre-1992 levels. Established in 2006, the U.S. EPA's WaterSense® program has further improved water efficiency without mandating new standards. WaterSense is a federal partnership with companies and communities that provides consumers with water-efficient options (Figure 11.48). Appliances and fixtures that meet the WaterSense Label criteria are 20% more water efficient than other products in that category. From 2006 to 2012, the WaterSense program saved 487 billion gallons of water: enough to supply all the homes in Colorado and Arizona for a year. None of these increases in efficiency sacrifice performance, and consumers notice only the water savings. The EPA reports that if all old, inefficient toilets in the United States were replaced with WaterSense labeled models, we could save 520 billion gallons of water per year, or the amount of water that flows over Niagara Falls in 12 days.

Water conservation is the final step in a water saving program. Conservation refers to using less water rather than simply using water more efficiently. Conservation methods in the home often include behavioral change, for example, taking shorter showers, turning off the water when brushing teeth, reusing vegetable rinse water to flush the toilet or water houseplants, and washing only full laundry loads (Figure 11.49). Rain barrels and cisterns also conserve water. They capture rainwater runoff and store it for later use, often on the lawn and in the garden.

Water-conserving **graywater** systems are gaining acceptance, particularly in the arid parts of the United States, despite some regulatory issues. Graywater systems separate "wash water" coming from sinks, washing machines, dishwashers, showers, and other light-use sources and filter it for uses that do not require drinking-water standards. Graywater is then piped back into the home and used primarily for flushing toilets and irrigating the lawn and garden. Simplified graywater systems collect bathroom sink water for use in toilet

▲ Figure 11.48 **EPA's WaterSense Program**
Look for the WaterSense® label to ensure high efficiency water fixtures and products. A WaterSense low-flow showerhead can save the average U.S. family up to 2,900 gallons of water a year.

► Figure 11.49 **Saving Water at Home**
For improved water efficiency, install low flow fixtures in bathrooms, choose water-saving appliances in the kitchen and laundry, mulch flower beds and irrigate landscaping with soaker hoses or at the coolest time of day. To conserve water, take shorter showers and turn off the tap while you brush. Plug the sink when washing dishes, wash only full laundry loads, and use rinse water for sprinkling houseplants or flushing toilets. Install a rain barrel for your outdoor irrigation needs. Wash your car on the lawn.

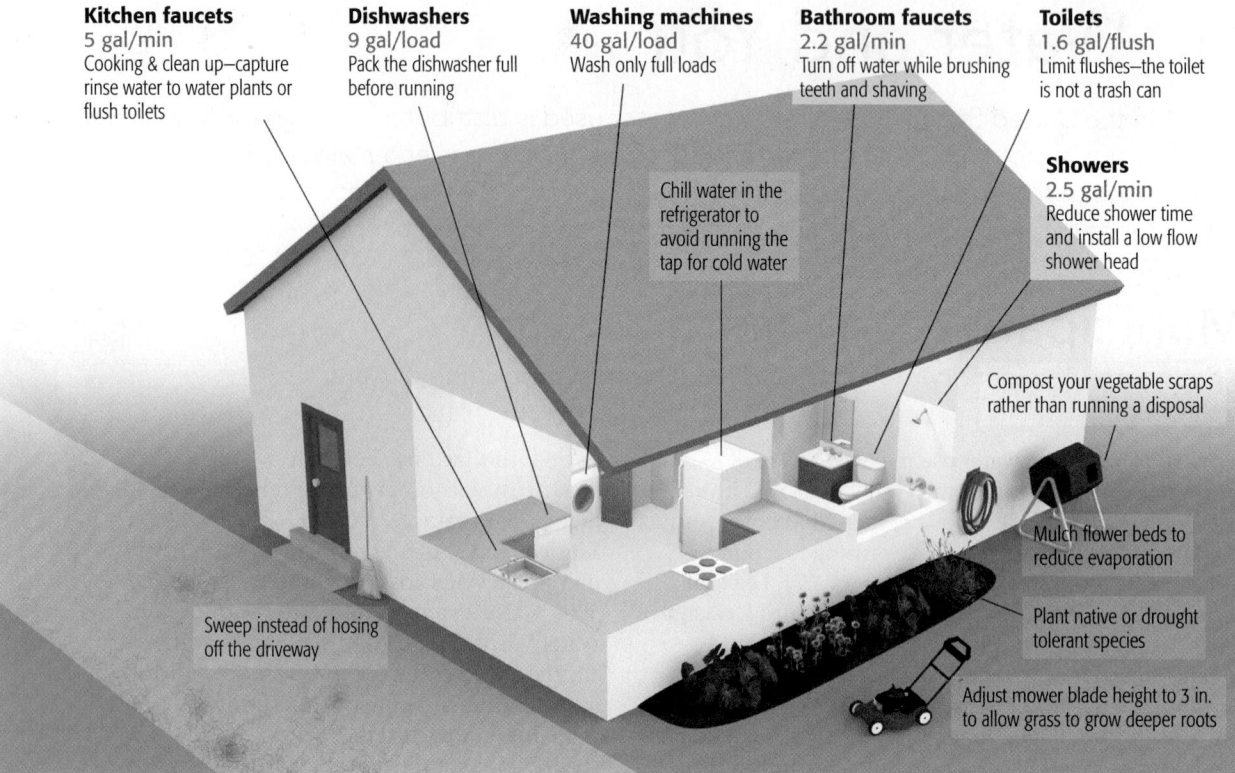

Kitchen faucets
5 gal/min
Cooking & clean up—capture rinse water to water plants or flush toilets

Dishwashers
9 gal/load
Pack the dishwasher full before running

Washing machines
40 gal/load
Wash only full loads

Bathroom faucets
2.2 gal/min
Turn off water while brushing teeth and shaving

Toilets
1.6 gal/flush
Limit flushes—the toilet is not a trash can

Showers
2.5 gal/min
Reduce shower time and install a low flow shower head

Chill water in the refrigerator to avoid running the tap for cold water

Compost your vegetable scraps rather than running a disposal

Mulch flower beds to reduce evaporation

Plant native or drought tolerant species

Sweep instead of hosing off the driveway

Adjust mower blade height to 3 in. to allow grass to grow deeper roots

QUESTIONS 11.7

1. Differentiate between avoiding water loss/waste, water efficiency, and water conservation. Give examples of each.

2. Describe the EPA WaterSense program and discuss its benefits.

3. What is graywater? How is it collected and used?

4. Describe three methods of reducing drinking water use in residential and commercial landscaping.

(MES) For additional review, go to **MasteringEnvironmentalScience**

flushing. The reuse of graywater for these purposes reduces drinking water use, at the same time relieving stress on septic fields and waste treatment systems. So called "blackwater," from flushing toilets, is routed directly to the sewer system and never mixes with the graywater. Rain chains with cisterns and cooling tower condensate also provide sources of graywater for irrigation around commercial buildings.

Composting toilets can eliminate blackwater production altogether. Much like a backyard composting system, composting toilets employ aerobic bacteria (those that require oxygen) and fungi to break down wastes and to convert them to nutrient-rich fertilizer. They are vented to outdoors (as are typical toilets) and produce no odor. Composting toilets eliminate the need for plumbing and are most practical in scarce water settings and those that utilize overworked septic systems.

Waterless urinals limit water use, but *do* utilize existing plumbing systems. They drain into the sewer pipes by gravity rather than flushing with water. A sealant in the bowl prevents sewer gases and their odors from escaping the drain into the restroom. Waterless urinals are becoming more common in public places and on university campuses and result in significant water savings where they are installed.

The need for irrigation in home gardens can be reduced by selecting plants appropriate to the climate and using mulch in the landscaping. Mulch usually consists of an organic material, such as shredded bark or pine needles, that helps to retain soil moisture and suppress weeds.

Over time, the mulch decays and enriches the soil. Irrigating landscaping during the early morning or late evening hours reduces evaporative losses, and soaker hoses or other options that deliver water straight to the plants rather than into the air help to reduce water waste as well.

Where YOU LIVE ------------------------------------
How much water do you use? The average American uses 80–100 gallons (302–378 liters) of water a day. The World Health Organization estimates that only 8.7 liters are needed for survival and 20–50 liters per person per day are needed to meet basic human needs. How much water do you use in a day? Take a guess. Now record your every direct use of water during the next 24 hours, using the information in **Figure 11.49**. Calculate the water you use in the bathroom, the laundry, and the kitchen. Don't forget the water your use outside to wash the car, water the lawn, and so on. Now, add up your total 24-hour water use. How far off were you from your guesstimate? If you wanted to decrease your water consumption by 20%, what efficiency measures might you implement? Keep in mind that fixtures and appliances with the WaterSense label use at least 20% less water than their standard counterparts. What conservation measures would you be willing to adopt?

Improving Urban Water Use

How can we reduce the amount of water that cities waste?

Modern cities use a great deal of water, but much of it is wasted by inefficiency and negligence. For example, between 6% and 15% of the water used in U.S. cities is lost through leaks in distribution pipes and faulty plumbing. Depending on climate, between 30% and 60% of municipal water is used to irrigate lawns and landscaping.

In the past, cities usually implemented water conservation measures only in response to droughts or seasonal water shortages. But as water has become scarcer and costlier, cities around the world have begun to implement permanent measures to ensure that water is used more efficiently. Boston and nearby cities provide a case study of the kinds of strategies that can be used to conserve water.

The Massachusetts Water Resources Authority (MWRA) provides water to 2.5 million people and more than 5,500 businesses in the greater Boston metropolitan area. Nearly all of that water comes from the Quabbin and Wachusett reservoirs, which are located west of the city (**Figure 11.50**). In a 1986 study, MWRA determined that the rate at which water was being withdrawn from these reservoirs put the region at risk of a water shortage. In response to this situation, it immediately established an integrated water conservation program.

Distribution pipes belonging to the MWRA and local communities were checked for leaks and then repaired. Water meters were upgraded to help track and analyze community water use. Over 370,000 homes were retrofitted with low-flow plumbing devices, such as water-efficient showerheads. The MWRA even championed a change in the state plumbing code to require that water-saving, 1.6-gallon-per-flush toilets be installed in all new residential and commercial construction. Perhaps most important, the MWRA launched an extensive public information campaign

▲ **Figure 11.50 Boston's Water Supply**
The Quabbin Reservoir located in central Massachusetts provides two-thirds of the water used by the greater Boston metropolitan area. The volume of water in the reservoir varies from year to year, depending on the amount of rainfall in its watershed.

and a variety of educational programs for the public schools. To pay for these programs, as well as the cleanup of Boston Harbor, the MWRA boosted the price of water. These higher rates have made Boston citizens much more aware of their water use.

By 1989, MWRA water use was below the safe reservoir yields. By 2011, water use had declined by more than 40% from 1986 levels (**Figure 11.51**). Furthermore, MWRA calculated that the conservation program cost only a fraction of what it would have had to spend to increase the capacity of the reservoirs or buy additional water from other watersheds in order to maintain the 1986 level of water use.

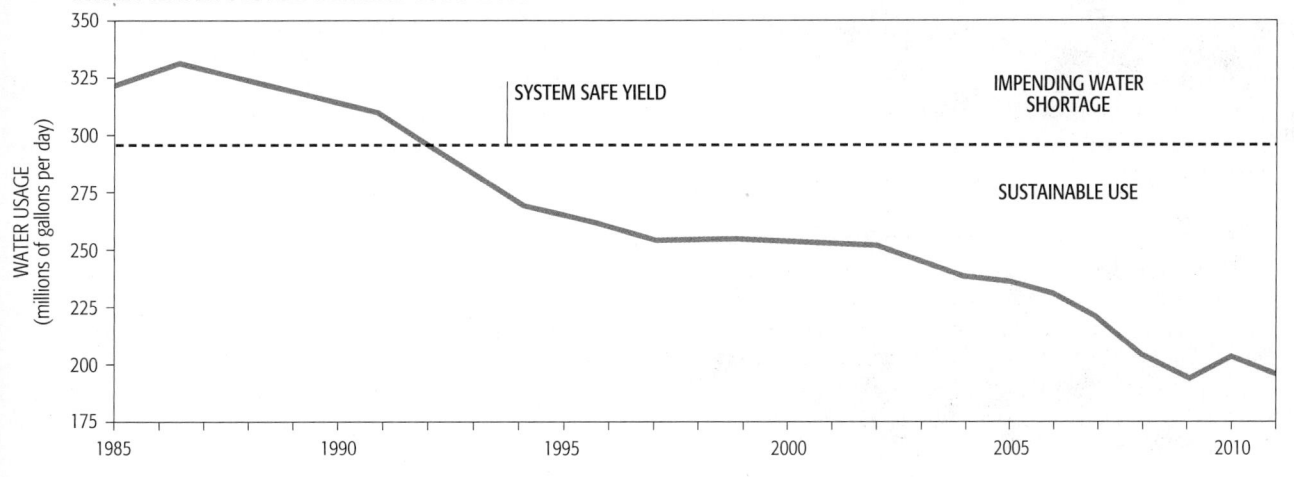

MWRA WATER SYSTEM DEMAND 1986–2011

(Line graph. Y-axis: WATER USAGE (millions of gallons per day), ranging from 175 to 350. X-axis: YEAR, ranging from 1985 to 2010. A dashed horizontal line at approximately 295 is labeled SYSTEM SAFE YIELD. Region above the dashed line labeled IMPENDING WATER SHORTAGE; region below labeled SUSTAINABLE USE.)

◄ **Figure 11.51 Sustainable Water Use**
The safe yield of water from the Quabbin and Wachusett reservoirs is determined by the average annual rainfall in their basins. In the mid-1980s, water use in Boston exceeded the safe yield. A combination of strategies implemented by the MWRA has lowered water consumption well below the safe yield.

11.8 Water Conservation Policy and Law

BIG IDEA A variety of national and international policies and laws influence water use and water quality. In the United States, the laws governing access to water differ from region to region. Conflicts over water ownership and allocation are greatest in the West, where water is scarcest. Water quality in the United States is regulated by the Clean Water Act and the Clear Drinking Water Act. These laws provide a variety of measures for restoring and maintaining water standards. More than 60% of the water in Earth's streams and lakes is shared by two or more countries; many individual treaties regulate the use and quality of these waters. The member nations of the United Nations have passed a convention on the use of ocean waters, but have been unable to agree on laws to regulate the non-navigational uses of fresh water. Global warming is likely to diminish supplies of fresh water in many regions and increase international conflict over water use. These predictions reinforce the need for international agreements governing water use.

Water Use in the United States

■ Water allocation policies strongly affect water use.

In the United States, each state sets its own water rights policy. In the East, access to water is determined by **riparian water rights**, a system derived from English common law. Under this system, all landowners with property adjacent to a body of water have a right to make reasonable use of it. Reasonable use is defined by each state, but usually includes domestic water use, swimming, boating, and fishing. Rights can be sold or transferred only to adjacent landowners, and water cannot be transferred out of a watershed.

In most western states, where water is generally scarce, water use is determined by **prior appropriation water rights**. These rights are not connected to land ownership and can be bought and sold like other property.

Under this system, the first person to use a quantity of water from a water source for a beneficial use has the right to continue to use that quantity of water for that purpose indefinitely. Rights to the remaining water are allocated to subsequent users, so long as they don't impinge on the rights of previous users. When water runs short, users with the earliest appropriation get their full allocation; later users may get no water. Conflicts regarding water use within and across state lines have been common, and they have increased in number as demand for water in the West has grown.

Prior appropriation water rights were created to accommodate the needs of farmers, miners, and frontier towns and cities. As a result, most states define beneficial use in terms of agricultural, industrial, or household purposes. When these laws were adopted, there was little awareness of environmental needs. Ecological purposes, such as maintaining floodplain ecosystems or conserving wildlife, are not classified as beneficial uses in most western states. For this reason, some argue that this approach to water allocation is outdated.

The 1973 Endangered Species Act (ESA) has often been used to protect the flow of water in rivers and streams that provide habitat for threatened or endangered species. For example, Friant Dam and associated diversions take water out of California's San Joaquin River for agricultural uses. The rights to that water have been assigned to farmers in the San Joaquin Valley. Since the completion of this project in 1984, long segments of the river have been completely dry (Figure 11.52). Recently, the U.S. Fish and Wildlife Service and the courts have required that water flows be restored to the San Joaquin River in order to recover stocks of endangered Chinook salmon. The courts also required the federal government to compensate farmers for water rights lost in association with this restoration.

▼ Figure 11.52 **Reviving a Dead River**

Diversions for agriculture have left this segment of California's San Joaquin River dry for over 60 years. Rights to this water have been allocated to area farms. To meet the conditions of the Endangered Species Act, the state must restore water to the river.

Water Quality in the United States

■ The Clean Water Act has evolved to emphasize integrated watershed strategies.

In the late 19th century, streams near cities were often so badly polluted that they were described as "open sewers." The 1899 Refuse Act was the first attempt to manage water quality at the national level. This law stipulated that wastes could not be dumped into navigable waters without a permit from the Army Corps of Engineers. The Oil Pollution Control Act of 1924 regulated the dumping of oil from boats but not from stationary sources, such as industrial facilities or oil wells. These laws may have diminished water pollution somewhat. But as the population continued to grow, pollution from rapidly growing cities and industries and the increasing use of agricultural fertilizers and pesticides continued to foul U.S. streams and lakes.

In 1948, the U.S. Congress passed the Water Pollution Control Act, the first comprehensive national law to manage water quality. This law provided grants to local governments to clean up water and authorized the federal government to sue polluters in court. It did not, however, set standards for water quality, nor did it assign the responsibility for enforcing the law to a specific agency.

The degradation of U.S. water supplies continued through the 1960s. A number of events, most notably the fire on Ohio's Cuyahoga River, garnered widespread public attention to this issue (Figure 11.53). This attention moved Congress to pass the Clean Water Act (CWA) of 1972 to ensure "the protection and propagation of fish, shellfish, and wildlife and recreation in and on the water." This act requires the EPA to regulate the discharge of pollutants into waterways. It also regulates activities that influence water quality, such as dredging streams, lakes, and wetlands. Indeed, the CWA now provides the basis for the protection of wetlands across the country. The Safe Drinking Water Act of 1974 required the EPA to set standards for allowable quantities of chemicals in water; it also authorized local governments to monitor and maintain these standards.

Since its original passage, the CWA has been modified to accommodate new knowledge and changing priorities. Programs have been implemented to engage landowners in efforts to minimize the runoff of pollution from their lands. Regulations have also been established to diminish pollution from urban storm-sewer systems and construction sites. Perhaps most important, instead of focusing on the regulation of individual sources of pollution, CWA programs now take an integrated approach to managing watersheds. These programs address a full array of issues and emphasize the involvement of communities and important stakeholders in the development and implementation of strategies to restore and maintain clean water.

International Water Law

■ A comprehensive international convention on freshwater resource use has yet to be ratified.

Rivers are among the most common boundaries between countries. Over 250 large watersheds are shared among two or more countries, and more than 400 international treaties regulate the allocation and use of those waters.

In 1997, the UN General Assembly adopted the Convention on the Non-navigational Uses of International Watercourses. This convention proposes three general principles for the use of international surface waters and groundwater. First, allocation among countries should be equitable and reasonable. Second, every country is obligated to prevent significant harm to those waters. Third, countries must notify and consult with one another regarding planned usage. However, this convention has yet to be ratified by enough countries to make it law.

The UN Convention on Law of the Sea was drafted in 1982. It requires countries bordering oceans to prevent the pollution of coastal waters and to cooperate in the management of living resources, such as migratory fish. This treaty was originally ratified by 60 countries and became international law in 1994. The United States signed the treaty in 1994 and recognizes it as general international law. But as of 2014, the U.S. Senate has not ratified the treaty because of concerns about how the convention would be governed.

Global warming is likely to cause even more challenges for allocating the use of water among nations. A 2013 report from the UN Intergovernmental Panel on Climate Change predicts that water availability will decline by 10–30% in many of the world's already water-stressed regions. Such shortages are likely to lead to international conflicts. These trends show the need for a comprehensive international water convention.

Q How can I determine the cleanliness of the water in my area?

Hayley Baum, University of Central Florida

Lissa: **A** The U.S. EPA provides water quality data for your watershed and for drinking water in your community on its website.

QUESTIONS 11.8

1. How do riparian water rights differ from prior appropriation water rights?

2. Describe three important changes to the original U.S. Clean Water Act passed in 1972.

3. What are the three primary principles for water management across international borders named in the 1997 UN Convention on the Non-navigational Uses of International Watercourses?

(MES) For additional review, go to **MasteringEnvironmentalScience**

▼ Figure 11.53 **A Flaming River**
In 1969, flammable pollutants on the Cuyahoga River caught fire and burned for days. This event caught widespread public attention and catalyzed support for the passage of the Clean Water Act in 1972.

Water Conservation Competition

Martin Fugueroa is majoring in biology with an emphasis on Human Biology and a minor in sustainability. During his sophomore year at University of California, Merced, Martin Figueroa created the Water Conservation Competition—a month-long water battle between the 14 dorms on campus, which challenges students to reduce their water use. Students can view their daily usage throughout the competition on an online, real-time dashboard that Martin developed with a local technology company. Martin continues to host this competition, which conserves over a million of gallons of water each year.

How did you first get the idea for the UC Merced Water Battle?

UC Merced is located in one of the driest climates in California—the Central Valley, a region also famous for its rich agriculture. When I arrived on campus my freshman year, I noticed a drastic difference in the landscape—parts of the campus had amazing grass and green lands, while other sections were arid and dry. This difference is all due to water resources.

On my drive home to Los Angeles, I observed the large tubes required to pump water to the drier regions of California; just above these were billboards calling for more water resources in the Central Valley. This sparked my desire to learn about where our water originates and how it is distributed.

I then enrolled in a course on sustainability and current environmental issues. This class inspired me to take action on my campus and influence administrators to implement new sustainability standards. My water conservation efforts started out as a campaign, which developed into a competition in the hopes of increasing student participation and awareness. I wanted students to think about water and how our usage impacts the future of our planet.

What steps did you take to create the competition?

The first step was getting approval and support from campus administrators, water stakeholders, and sustainability and housing departments, who would then give me access to facilities and meters needed to track the water usage. I then contacted Aquacue, a water technology company, to help create the water battle dashboard where students can view their water usage in real-time, as well as notify us about leaks. This dashboard greatly increased student participation and allowed students to visually understand their impact.

I also assembled a committee of administrators, stakeholders, and two student groups, Green Campus and Engineers for a Sustainable World. With the help of several classes and professors, we created marketing materials including flyers, short films, commercials, and QR codes, which link students to the water dashboard. Social media provided a great forum for students to encourage their dorm-mates to reduce water consumption, while also fueling competition among the dorms. We kicked off the competition with several tabling events where students could ride our amazing bike blender, drink a free smoothie, and learn about the ways students can conserve water.

What are the biggest challenges you've faced with the project, and how did you solve them?

The biggest challenge for competitions is always gaining support for the cause and increasing student participation. Students are not always willing to change their daily habits (like taking shorter showers). We tackled this problem with incentives: We gave students free smoothies at every tabling event, we hosted a free pizza party for the winning dorm (this is really important for college students—what college student doesn't like free food?!), and we also acquired a $1,000 grant to be directed to a nonprofit chosen by the winning hall.

We also faced challenges with our technology, as we were the first university to implement the real-time water dashboard. By partnering with Aquacue and our tech committee, any glitches were fixed right away.

How has your work affected other people and the environment?

The Water Conservation Competition was a huge success for both the students and the community of Merced. In the first year of the water battle, we reduced our water consumption by 14%, saving 89,000 gallons of water. We also saved 1.4 million gallons of water from 16 water leaks detected by the water usage technology and reported by students. Nine dorms and 565 students participated.

The greatest result was that students continued conserving water even after the competition ended—students reduced water usage by an additional 6% reduction in the month following the battle.

We launched the competition again the next year—11 buildings and over 700 students participated, resulting in a 9% reduction in water usage. In its third year, we organized the biggest campus-wide water conservation competition in the world (to our knowledge), involving all of our student residence halls (14 buildings) and more than 2,000 students.

What advice would you give to students who want to replicate your model on their campus?

Every campus can replicate this water battle and make it even bigger and better. The first criterion is having the passion and commitment for conservation work.

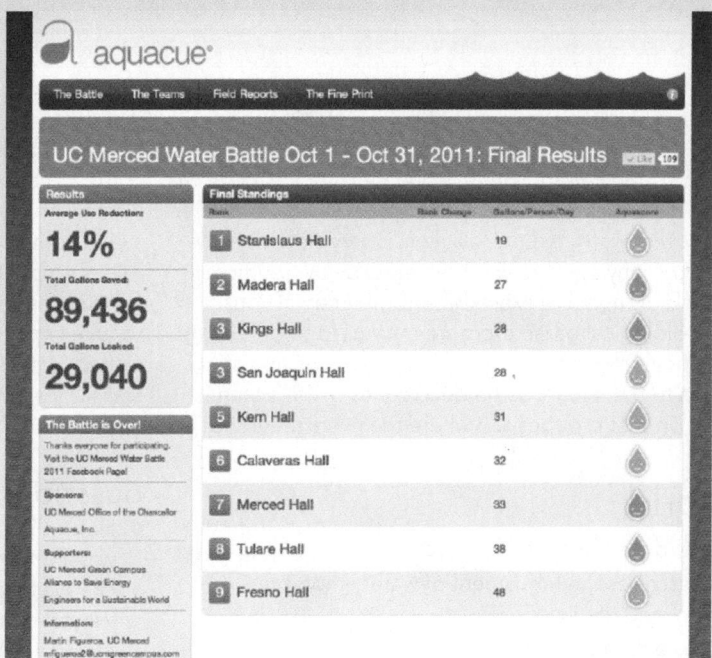

Source: Badger Meter.

Implementing a competition and gaining support will be challenging, but witnessing the impact of a project like this on a college campus is very rewarding, so stick with it!

In the end, it's not about the numbers or the amount of money saved; it is about inspiring students to get involved. Many of us are not well informed about critical environmental issues, but we are in charge of changing our future—we must help our local communities become more sustainable.

What is your major and has it helped you in your leadership role?

With my Biology major, I have learned about human impacts on our environment. I've analyzed how our habits are harmful to both the environment and human health. My education at UC Merced has shown me that creating a sustainable environment does not just require topnotch engineering, but must also involve medicine, psychology, writing, history, sociology, and many other fields of work. All disciplines need to work together to accomplish and promote a more sustainable society.

Summary

Energy from the sun and the force of gravity cause water to move between the oceans, atmosphere, and land. The majority of Earth's water is either salty or frozen and only 1/100th of 1% of Earth's water is fresh water easily accessible for human consumption. Watersheds define the area of land from which water drains into a stream, lake, or ocean. Most of Earth's groundwater is found in aquifers. The flow of water through aquifers is very slow, as is the rate of recharge to aquifers. Many aquifers are threatened by overpumping and pollution. Flooding and drought are exacerbated by human actions.

Physical, chemical, and biological parameters are measured to determine water quality. Pollutants come from point and non-point sources and affect water quality from streams to the ocean. Large dead zones in the ocean result from nutrient pollution that is deposited where the rivers meet the ocean. Wastewater includes sewage, stormwater runoff, and water used in industry and manufacturing. In the United States, most wastewater is treated in a stepwise process that first removes sediment and then removes dissolved organic matter and inorganic chemicals. In many rural areas, wastewater is treated in on-site septic systems. Humans have affected the majority of stream ecosystems by building dams, diverting water, and releasing pollutants.

An increasing number of people are experiencing water shortages that require implementation of water conservation strategies and policies. In the United States, the EPA has increased water efficiency standards of fixtures and appliances, and implemented the WaterSense® program to further increase water efficiency. Behavioral changes can also conserve precious fresh water.

11.1 Water World

- Solar energy and gravity drive the hydrologic cycle.
- Earth's water budget describes the movement of water among the oceans, the atmosphere, and the land.
- Accessible fresh water is extremely limited.
- Humans use over half of the water flowing in Earth's streams.

KEY TERMS

hydrologic cycle, transpiration, evapotranspiration, runoff, groundwater, aquifer, hydrologist, watershed, open watershed, closed watershed, nonconsumptive use, consumptive use, baseline water stress

QUESTIONS

1. How would you expect the water budget of a rainforest to change following deforestation? What fluxes would be reduced? Increased? What would be the long-term consequences?

2. About 10% of the water that evaporates from the ocean returns to it in rivers. How does the other 90% get back to the ocean?

Freshwater withdrawals by type of use
- Industry
- Agriculture
- Domestic

GLOBAL (3,902)

19%
12%
69%

3. How would you expect human water use to differ between an industrial country, such as Japan, and a developing country, such as Bangladesh?

11.2 Groundwater

- The groundwater that resides in soil and rock beneath Earth's surface is replenished very slowly.

KEY TERMS

water table, recharge zone, unconfined aquifer, confined aquifer, discharge zone, artesian well

QUESTIONS

1. Explain why sometimes water will gush to the surface of a newly drilled well without pumping.

2. Why is groundwater considered to be a nonrenewable resource in many regions?

3. Why does groundwater pumping near coasts often result in the intrusion of seawater?

11.3 Water Distribution

- Accessible fresh water is extremely limited. Human actions can influence both flooding and drought.

KEY TERMS

impervious surfaces, subsidence, sinkhole, saltwater intrusion

QUESTIONS

1. What features of a land area and its human residents might predict susceptibility to water stress?

2. An expanding city seeks to avoid flooding as it paves roads and constructs buildings. What three steps might they take to manage their storm water in an environmentally friendly way?

3. A politician argues that the Ogallala aquifer is a resource to be used, and that limiting water withdrawals unnecessarily reduces the economic value of farmland in her county. Provide several opposition arguments that detail the consequences of overdrawing groundwater.

11.4 Water Quality

- Hydrologists measure physical, chemical, and biological parameters to determine the quality of water.
- Pollution from point and non-point sources impacts water quality from the headwaters to the oceans.

KEY TERMS

bioassessment, point source, non-point source, eutrophication, dead zone, ocean acidification

QUESTIONS

1. Describe four important measures of water quality and what they indicate about a body of water. As a hydrologist, what parameters might you select to measure water quality in a stream flowing past a textile plant in a rural area?

2. Compare point and non-point source pollution and give examples of each for the city of Cleveland, Ohio, located on the banks of Lake Erie.

3. Why does adding nutrients to aquatic systems encourage the growth of algae? Explain how increased growth of organisms that produce oxygen can result in oxygen-poor dead zones that support no aerobic life.

375

Blowing in the Wind

What caused the collapse of agricultural systems in the Dust Bowl?

Today, if you fly low over the land described as the Llano Estacado, you will see evidence of how humans have shaped the land for food production. This rural area that encompasses northeastern New Mexico and the western Texas Panhandle is covered with large ranches and irrigated farms. In 1541, when the Spanish explorer Francisco Coronado first looked upon the same land, he saw only a vast windy prairie. It was he who gave it the name *Llano* (pronounced "yano") *Estacado*, the "staked plain," because he and his group had to put down stakes to navigate its flat, featureless terrain (Figure 12.1).

The short-grass prairie of Coronado's day had a dense cover of long-lived perennial grasses that used the limited rainfall of the region very efficiently. Organic litter and the tightly knit root systems of the grasses formed a dense *sod* that covered the mineral soil (see the photo in Figure 12.1). The prairie sod effectively protected this landscape from erosion by both wind and water. Despite periodic droughts, this ecosystem was sufficiently productive to support large herds of bison on a rainfall of little more than 50 cm (20 in.) a year, often less.

For much of the 19th century, the Llano Estacado remained undisturbed. Judging the lands to be of little value, the U.S. government promised much of it to the

Comanche and Apache Indians in the 1867 Medicine Lodge Treaty. However, under pressure from growing numbers of Texas cattlemen, the Native Americans were soon driven off the plateau, and the bison were exterminated. In 1875, the area was opened to cattle grazing.

From Prairie to Farmland. In 1882, the State of Texas sold 1.2 million hectares (3 million acres) of the Llano to a Chicago real estate company, which lured would-be farmers to the region with the slogan, "Riches in the soil, prosperity in the air, progress everywhere—an Empire in the making!" Even then, farmers knew that in this hot region 20 in. of rainfall per year was the bare minimum required for growing crops without irrigation. But promoters countered with yet another slogan: "Rain follows the plow," suggesting that plowing the land would actually increase rainfall.

The people who began to settle this land were called sodbusters. Their metal plows cut through the dense sod, exposing the mineral soil below. Where once the prairie grasses had grown, the settlers planted wheat.

Soon the U.S. government opened up additional land in 160-acre parcels that were free to settlers. The Enlarged Homestead Act of 1909 opened up even more land for settlement. Not surprisingly, the most productive lands were settled first. Each successive wave of immigrants was forced to farm increasingly marginal land.

Between 1910 and 1920, the area of farmland in the Llano Estacado doubled; between 1925 and 1930, it tripled. By 1930, over 90% of the short-grass prairie was under the plow. Farmers used windmills to provide water for their homes and livestock, but such wells could not irrigate 160 acres of land. Nevertheless, rainfall up to this time was generally sufficient to get by—though often only barely.

From Farmland to Wasteland. By any measure, farming practices in the Llano Estacado were unsustainable. Plowing greatly accelerated the decomposition of the organic matter in the soil. This, in turn, reduced the ability of the soil to hold essential nutrients and water. Fields were burned in the fall, which increased the loss of nutrients. The broken sod no longer protected the soil from erosion by wind and water. Farmers could not afford to plant cover crops over the winter, exposing the soils to further erosion.

Just as the nation was sinking into the Great Depression, the region was overcome by drought. Beginning in 1930 and continuing through 1936, crops across the Llano and nearby regions failed for lack of rain. Lacking plant cover and sod, the now fragile soils of these "agricultural ecosystems" were exposed to dry winds that moved vast amounts of dust across the continent (Figure 12.2).

▼ **Figure 12.1 The High Plains**
The Llano Estacado rises from 900 m (3,000 ft) at its southeastern tip in Texas to over 1,500 m (5,000 ft) in New Mexico. Prior to human settlement, a diverse short-grass prairie community of herbs and low shrubs stabilized the soil.

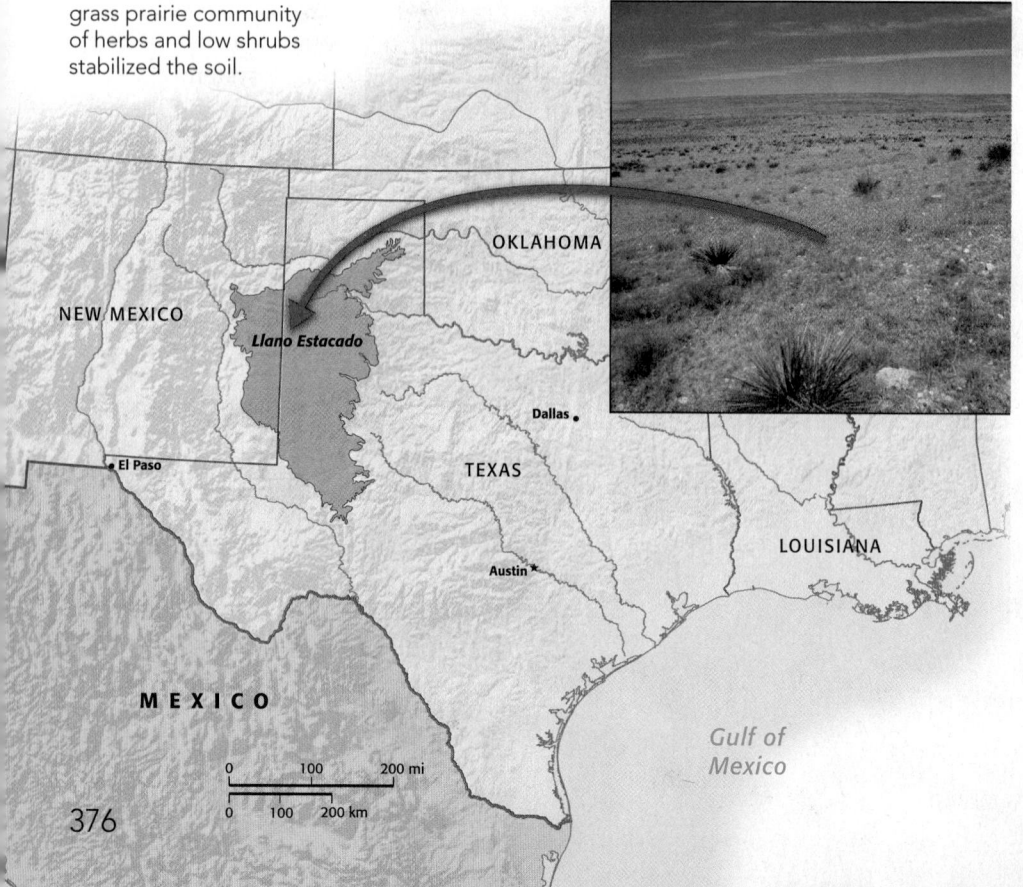

In response to the crisis, President Theodore Roosevelt initiated the Prairie States Forestry Program in 1934 in an effort to save these valuable agricultural lands. Over the next 10 years, the U.S. Forest Service planted millions of trees in a "shelterbelt" that stretched across the Great Plains to Canada, establishing a windbreak to protect the farms from further erosion. But for this generation of farmers, the damage was done. Millions of the Llano's acres were stripped of their precious topsoil and with it their ability to grow crops. With their land barren and their homes in foreclosure, hundreds of thousands of people were forced to move away.

And Now?   Today, cattle graze in many parts of the Llano Estacado that once supported row crops. In other areas, improved farming methods allow crops to grow, but this production depends on irrigation and the application of large amounts of fertilizer (**Figure 12.3**).

The story of the Llano Estacado and the dust storms that swept through so much of the southern Great Plains in the 1930s is not unique to that time or place. Today dust-bowl conditions exist in northern China and south of the Sahara in Africa, brought about by overgrazing and drought. In Nigeria, for example, nearly 1 million acres are lost to desert each year, resulting in significant conflicts between farmers and herders. In the United States, extreme drought

▲ Figure 12.3 **The Llano Estacado Today**
Today, irrigation and the application of fertilizers allow crops to grow in some parts of the Llano Estacado.

conditions in 2012 rivaled those of the dust-bowl years and affected nearly 80% of the country's agricultural land. Overall, U.S. crop yields were at a 20-year low in much of the breadbasket of the Midwest. Studies showed that soils with better water-holding capacity, including fields that were managed by conservation tillage, experienced fewer losses than conventional fields.

The exponential growth of human populations over the past century is directly related to the increased availability of food. Rapid growth in the amount of cultivated land and its increased productivity have provided that food. However, as the collapse of agricultural ecosystems in the Llano Estacado demonstrates, agricultural ecosystems are far more fragile than their natural counterparts.

- What are the factors that determine the productivity of farmland?
- What are the ecological impacts of agriculture?
- What factors put agricultural systems at risk of failure, such as the failure witnessed in the Dust Bowl era?
- What steps can be taken to ensure that agriculture and our food supply are sustainable?

▼ Figure 12.2 **The Dust Bowl**
Without the sod and the prairie plants it supported, Llano soils were easily converted to dust and blown away. During the dry years between 1930 and 1936, massive dust storms created midday darkness over the plains.

377

12.1 Origins and History of Agriculture

BIG IDEA Modern humans, *Homo sapiens*, have been around for about 250,000 years. During the vast majority of that time, members of our species lived as hunter-gatherers. Only in the past 10,000 years did humans begin to practice agriculture. **Agriculture** is the system of land management used to grow domesticated plants and animals for food, fiber, or energy. Agriculture originated independently in several regions as a consequence of changing climates, diminishing wild resources, and human cultural and technological progress. Since the Middle Ages, agricultural production has increased as a consequence of refined technologies, continued plant and animal breeding, and changes in land management. Increased mechanization of the Industrial Revolution accelerated production, and today the use of fertilizers, herbicides, and pesticides have brought about a Green Revolution in agricultural production.

Why Did Agriculture Begin?

■ Climate change, cultural progress, and population growth played important roles in the origins of agriculture.

▼ **Figure 12.4 Hunter-gatherers**
Today, the bushmen of southeastern Africa make their living by hunting for game and gathering plants from the Kalahari Desert. This !Kung tightens his bowstring with his hands and teeth.

By 40,000 years ago, communities of hunter-gatherers had well-developed language skills, technology, and culture, including stone tools, bone needles and fish hooks, jewelry, art, and music. There is ample evidence that most of the food eaten by hunter-gatherers came from wild plants. In most of these cultures, the diet was quite diverse and nutritionally balanced (**Figure 12.4**). Across North America, for example, Native American hunter-gatherers ate over 3,000 different plant species. Of course, hunter-gatherers had to have an exceptional knowledge of the plant resources upon which they depended.

Then about 10,000 years ago, humans began to domesticate plants and animals. Anthropologists have long wondered why agriculture arose at that time. There is little doubt that climate changes associated with the close of the last ice age were an important factor. Climates were getting warmer, and dry seasons were growing considerably longer. Scientists argue that these conditions favored the evolution of plants with adaptations that made them ideal targets for domestication. These included annual grasses and herbs that grew quickly and produced abundant seeds in wet periods, surviving dry periods as seeds. The climate also favored perennial plants with underground bulbs or tubers that

allowed them to survive dry periods. Such seeds, bulbs, and tubers were excellent sources of food.

Why did agriculture develop at all? With the advantages of agriculture that we perceive today, it may seem odd even to ask such a question. Anthropologists once assumed that the domestication of plants and animals opened up new and better opportunities for human well-being. Yet the evidence shows just the opposite: The change from hunting and gathering to farming did *not* improve the quality of human life. Skeletal remains reveal that hunter-gatherers were generally taller, better nourished, and longer-lived than people in farming communities. It is true that a hectare of farmland provides humans with a greater quantity of food than a hectare from which humans get food by hunting and gathering. But the food from the farmland is of lower nutritional value. In addition, agriculture requires more human labor—it takes more energy to tend a farm than it does to hunt food in the wild.

Most anthropologists now agree that human communities turned to agriculture because their growing populations were placing greater demands on ecosystem resources, leading to shortages of wild food. Agriculture was a better alternative because it allowed people to produce more food for their larger communities.

These theories about when and why humans adopted agriculture are, of course, not mutually exclusive. Climate change was a necessary precondition for agricultural development, while population growth and increasingly scarce wild resources provided incentives for this important cultural shift.

How Did Agriculture Begin?

■ Humans and their domesticated plants and animals have coevolved.

Although there is evidence that hunter-gatherers may have purposefully planted the seeds of some wild plants earlier, the first evidence of planned cultivation of crops dates from 10,000 years ago in the Fertile Crescent region of Mesopotamia. This region, which today comprises Iraq, Turkey, Jordan, Syria, and Lebanon, was then considerably less arid. The first crops to be domesticated included wheat, barley, peas, lentils, and chickpeas. These plants were easy to grow, and their edible seeds were easy to store. Their domestication resulted in few genetic changes; their wild relatives can be easily recognized. It is likely that the chickpeas you eat today are almost identical to those consumed 10,000 years ago.

Agriculture developed independently in many different regions of the world. Rice cultivation began in China about 9,500 years ago. Soon after that, millet and various beans, including soybeans and mung beans, were domesticated in China. Approximately 7,000 years ago, people in the Sahel region of Africa domesticated rice and a sugar-rich grass called sorghum. Agriculture using other crops also developed in such places as West Africa, India, and New Guinea (Figure 12.5).

Agricultural societies developed in three regions of North and South America between 5,000 and 6,000 years ago. People living in the Andes Mountains in parts of modern-day Peru, Bolivia, and Ecuador domesticated

▼ Figure 12.5 **Origins of Agriculture**
Between 5,000 and 10,000 years ago, groups of people who lived on each of the continents began the process of domesticating crop plants.

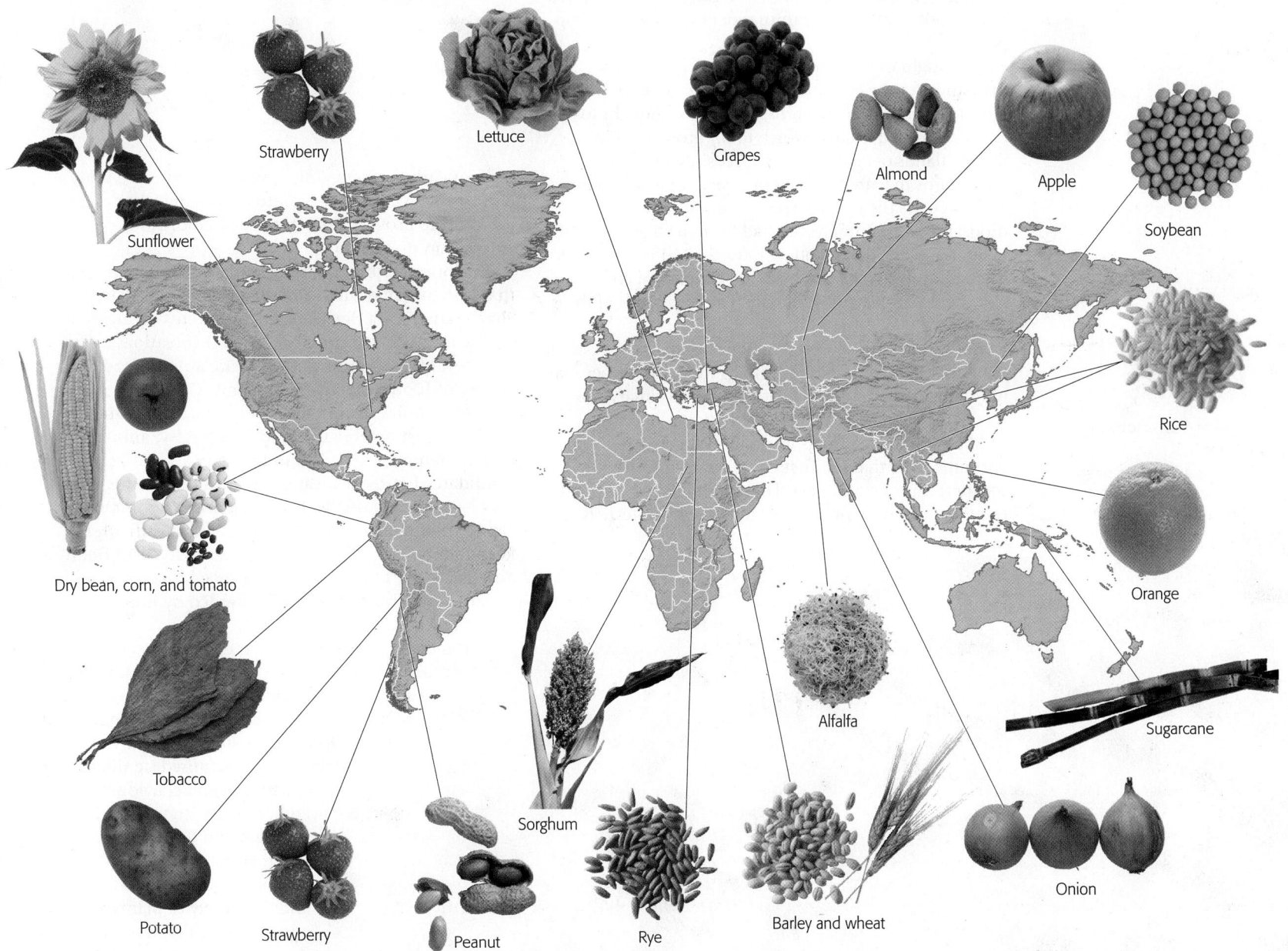

Sunflower

Strawberry

Lettuce

Grapes

Almond

Apple

Soybean

Dry bean, corn, and tomato

Rice

Tobacco

Orange

Alfalfa

Sugarcane

Sorghum

Potato

Strawberry

Peanut

Rye

Barley and wheat

Onion

379

potatoes. People in central Mexico began the cultivation of corn. Native Americans in regions that are now part of the United States grew squash and sunflowers.

The general process of plant domestication was probably similar in these various regions. At first by trial and error and later by conscious effort, humans directed the evolution of domestic plants. They did this by selecting and planting seeds or tubers from individual plants that possessed traits that improved the growth of the plant or the quality of the food. For example, limited seed dispersal was a key step in the domestication of grains, such as wheat and corn. It was easier for early farmers to harvest seeds from stalks that failed to shatter; as a result, seeds from those plants were used to plant the next year's crops. Generation after generation, this artificial selection led to crop plants that tended to have a greater percentage of their seeds remain on the stalk, until they could be harvested. Similarly, by planting and harvesting plants at uniform times, early farmers selected for uniform patterns of germination and growth in crop plants.

The origin of domestic corn, or maize, provides a case study for the evolution of domestic grain plants. The ancestor of modern cultivated corn is thought to be a grass called *teosinte*, which today grows wild in the highlands of central Mexico. Teosinte produces large grains that are arranged on a stalk that fragments and disperses the seeds when they ripen. Over several thousand years, humans selected for traits that eventually led to the corn you eat today (**Figure 12.6**). At the time of first European contact, over 300 varieties of corn were being cultivated by the indigenous peoples of the Americas.

About 12,000 years ago, humans began to domesticate animals. The first domestic animals were dogs. Dogs may have provided a source of food, but their primary role was to increase the effectiveness of human hunters.

Sheep, goats, and pigs were domesticated at about the same time that the first crops were cultivated in southern Europe and Asia. These animals undoubtedly provided essential protein to the carbohydrate-rich

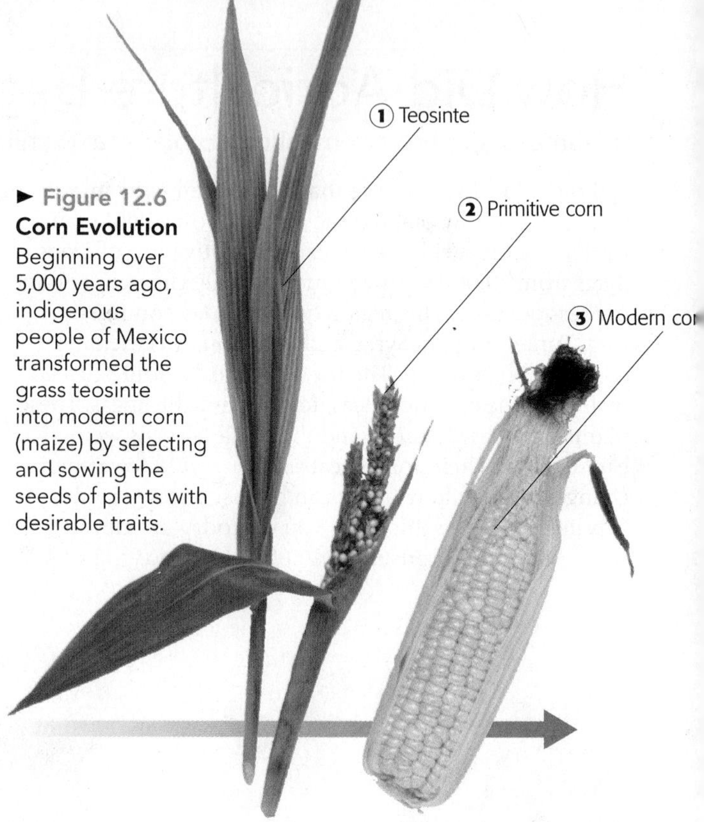

① Teosinte

② Primitive corn

③ Modern cor[n]

▶ **Figure 12.6 Corn Evolution** Beginning over 5,000 years ago, indigenous people of Mexico transformed the grass teosinte into modern corn (maize) by selecting and sowing the seeds of plants with desirable traits.

diets of early farmers. Cattle were domesticated about 8,000 years ago; in addition to producing meat and milk, cattle provided much-needed labor for the cultivation of crops.

Evolutionary biologists note that the wild animals that humans chose to domesticate generally shared five characteristics—a flexible diet, a fast rate of growth, an ability to breed in captivity, a calm disposition, and sociality. Most domestic animals are able to eat a wide variety of foods, mostly from plants. Compared to humans, domestic animals mature quickly. Domestic animals must be able to breed in captivity; animals that are reluctant to breed in captivity are unlikely candidates for domestication. Domestic animals must also be relatively docile. Large, aggressive animals cannot be managed in captivity; it is also difficult to keep highly territorial animals in captivity. And finally, animals that live in herds with a social hierarchy, such as cattle and sheep, can also be herded by humans (**Figure 12.7**).

You could ask, who was domesticated by whom? By nurturing and protecting them, humans provided a significant competitive advantage to the plants and animals on the evolutionary road to domestication. But plant cultivation and animal care required significant changes in human behavior and even in human physiology. Thus, the origin and development of agriculture has been a coevolutionary process for us and the variety of plants and animals on which we now depend (see Module 6.2).

▼ **Figure 12.7 Herding Instinct** The ancestors of domestic sheep were naturally social animals that moved in herds.

And Then What? Agricultural History

■ Technological and cultural innovations have increased agricultural production and reduced the need for human labor.

The earliest agriculture depended on intensive human labor, using stone, and later, ceramic, bronze, and iron tools (**Figure 12.8**). The invention of the plow coincided with the domestication of cattle—cattle provided the power needed to pull plows, reducing the amount of human energy invested in producing crops.

Through time, innovations such as wheeled plows, water pumping devices, dams, and reservoirs allowed for the exploitation of more land. Many new crops, including cotton, almonds, figs, and sugarcane, were introduced.

The period from A.D. 1400 to 1700 brought an influx of American crops to Europe and Asia, including corn, potatoes, tobacco, and tomatoes. The highly efficient moldboard plow was imported from China, greatly reducing the amount of labor required to grow crops. Farmers also began to use a system of crop rotation, alternating between grain and legume crops. Crop rotation sustained productivity because legumes fix nitrogen from the air, restoring nitrogen to soils that have been depleted by harvest.

The Industrial Revolution brought about improvements to plows and reapers that facilitated cultivation and harvest while diminishing the human labor required for these activities. Beginning in the late 1900s, tractors powered by fossil fuels greatly increased both the speed and scale of farming operations.

Throughout the industrial period, the rapidly increasing understanding of the science of plants, animals, and soils resulted in numerous agricultural innovations. **Agronomy** is the science that applies knowledge from fields such as genetics, physiology, chemistry, and ecology, to agriculture. For example, a growing understanding of genetics resulted in the development of varieties of crops and domestic animals that were better suited to particular environments, more resistant to pests and disease, and more productive.

Over the past 50–60 years, the **Green Revolution** has increased global agricultural productivity many times over. Thanks to the development of modern fertilizers, herbicides, and pesticides following World War II, large-scale nutrient supplementation, weed and pest control have increased production. The Green Revolution came with a cost to the environment, however. Increased fossil fuel use as well as runoff of nitrogen and chemicals into aquatic systems have altered nutrient cycles and depleted natural resources. Organic methods began to be implemented in the 1970s to counter some of the negative effects of the Green Revolution, and are gradually gaining support today.

Several trends are apparent in the history of agriculture. Domestic plants and animals have coevolved with human cultures and technologies. The increased food production provided by agriculture has allowed the growth of human populations and increasingly complex human societies. During the Industrial Revolution and the Green Revolution, the use of energy from fossil fuels brought even greater productivity to agricultural systems, but not without environmental costs.

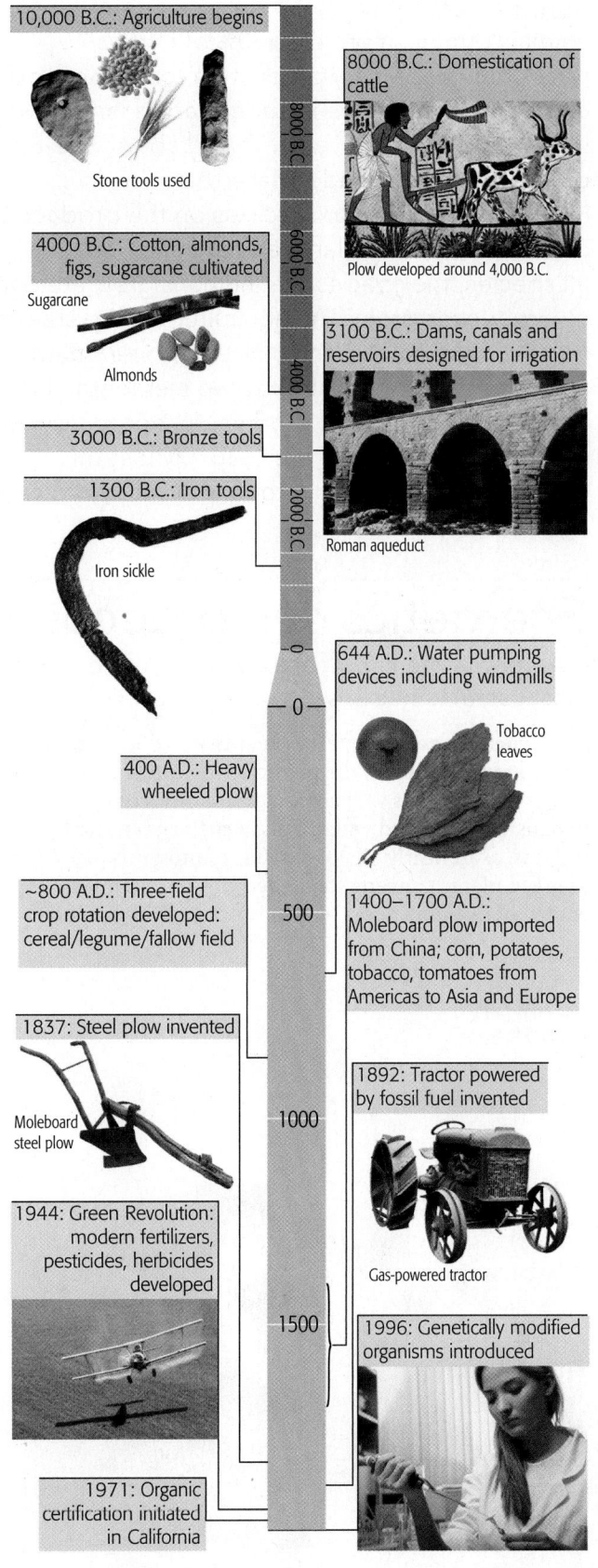

10,000 B.C.: Agriculture begins

Stone tools used

8000 B.C.: Domestication of cattle

Plow developed around 4,000 B.C.

4000 B.C.: Cotton, almonds, figs, sugarcane cultivated

Sugarcane

Almonds

3100 B.C.: Dams, canals and reservoirs designed for irrigation

3000 B.C.: Bronze tools

1300 B.C.: Iron tools

Iron sickle

Roman aqueduct

644 A.D.: Water pumping devices including windmills

Tobacco leaves

400 A.D.: Heavy wheeled plow

~800 A.D.: Three-field crop rotation developed: cereal/legume/fallow field

1400–1700 A.D.: Moleboard plow imported from China; corn, potatoes, tobacco, tomatoes from Americas to Asia and Europe

1837: Steel plow invented

Moleboard steel plow

1892: Tractor powered by fossil fuel invented

Gas-powered tractor

1944: Green Revolution: modern fertilizers, pesticides, herbicides developed

1996: Genetically modified organisms introduced

1971: Organic certification initiated in California

◄ **Figure 12.8 Timeline of Agricultural History**
Agriculture began around 10,000 B.C. and developed slowly to incorporate innovations that decreased human labor and increased land productivity. Rapid advances came during the Industrial Revolution, with implementation of fossil fuel engines. The scientific understanding of crop breeding and pest resistance furthered production in the late 19th century. The Green Revolution employed modern fertilizers, herbicides, and pesticides that dramatically improved production in the 1950s forward. By the 1970s, organic methods were being promoted in California and have spread to a small proportion of farms across the United States. Genetically modified crops were introduced in 1996 and by 2000 were cultivated around the world.

QUESTIONS 12.1

1. What factors are likely to have played a role in the origins of agriculture?

2. What kinds of adaptations would have made a plant an ideal target for domestication?

3. What features made certain animals likely candidates for domestication?

4. What have been the major trends in agriculture through the years? How is modern day agriculture different from that of the earliest agriculture? How do you think environmental costs of agriculture have changed through the years?

(MES) For additional review, go to **MasteringEnvironmentalScience**

12.2 Agroecosystems

BIG IDEA Today, we humans make up less than half of 1% of the total biomass of Earth's animals but use more than 30% of Earth's terrestrial net primary production to obtain food and fiber. In fact, if we include grazing lands and orchards, humans have converted about one-third of Earth's ice-free land surface to agriculture (**Figure 12.9**). Agriculture focuses on the production of a few species of plants and animals. Yet, like all species, the species that humans grow are parts of larger ecosystems. An agricultural ecosystem depends on energy flow and nutrient cycling, following the same principles that govern natural ecosystems. Of particular importance to agriculture are the cycling of two elemental nutrients essential to life: nitrogen and phosphorous. To support agricultural production, agroecosystems often require significant inputs of water and nutrients because they are routinely lost through the cycle of planting to harvest. The well-being of human communities throughout the world depends on the sustainable management of agroecosystems.

▲ Figure 12.9 **A Managed Ecosystem**
An agricultural ecosystem needs to be managed and maintained in order to be productive.

Energetics of Agroecosystems

■ Agroecosystems funnel energy into the production of plants and animals that are most useful to humans.

An **agroecosystem** is an ecosystem that includes crops and domestic animals, the physical environments in which they grow, as well as the communities of other organisms associated with them. In such agricultural systems, as in nearly all ecosystems, photosynthesis in green plants converts solar energy to the chemical energy of carbohydrates (see Module 3.3). This primary production is at the base of relatively simple food chains in which humans feed directly on plants or on herbivorous animals, such as cattle (**Figure 12.10**; see Module 6.4).

In either case, the main objective in managing agroecosystems is to funnel a major portion of the ecosystem's net primary production into food or fiber for human use. Simplification is a key strategy to meet this objective in conventional agroecosystems. Farmers and ranchers work to optimize the growth of a single species and to minimize the growth of other species that may eat it or compete with it. This simplicity, however, comes at the expense of the biodiversity and resilience found in natural ecosystems.

While agroecosystems can be highly productive, humans cannot use all of the primary production of a crop or all of the biomass of a herd of domestic animals. The fraction of total production that can be used by people is the **harvest index**. For example, the net primary production of a cornfield can be as high as Earth's most productive natural biomes. But more than half of that primary production is located in corn stalks and roots, which have limited use to humans. The harvest index for the part of the corn plants that can be eaten—the kernels—is between 30% and 50%. The harvest indexes for other grain crops are similar.

When humans eat meat, such as beef or chicken, they are in effect consuming net primary production that has been transformed by another organism. As with any such transformation, energy is lost in the process. The amount

▼ Figure 12.10
Agroecosystem Food Chain
Because they eat plant materials directly, vegetarians function as primary consumers in agricultural ecosystems. Meat-eaters function as secondary consumers. Depending on the efficiency of the primary consumer, 60–90% of the energy from the primary producer is lost through the livestock step before it reaches the meat-eating human consumer. Therefore, eating meat requires more cropland per person because of the energy loss through the additional step in the food chain.

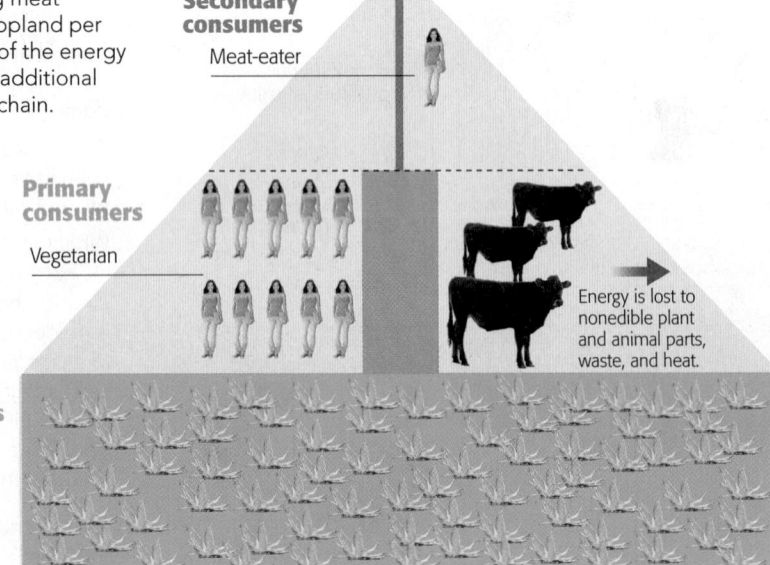

Secondary consumers
Meat-eater

Primary consumers
Vegetarian

Energy is lost to nonedible plant and animal parts, waste, and heat.

Primary producers

of energy lost depends on the domestic animal's **trophic-level efficiency**, the fraction of energy the animal consumes that is actually stored as biomass.

For most wild animals, trophic-level efficiency is less than 10%. Thus, 1 unit of energy in the meat of a deer or wild bison required more than 10 units of primary production (see Module 6.4). Trophic-level efficiency for domestic animals is generally greater than 10%. This is partly due to selective breeding. In addition, domestic animals are often given food that is more easily digested than the food that is available to animals in the wild. Another reason for the higher efficiency of domestic animals is that they are usually confined, so they expend less energy searching for food than their wild counterparts. For enclosed chickens, trophic-level efficiency may exceed 40%. In other words, 40% of the

energy in the corn eaten by a chicken is transformed into the tissues of the chicken.

Trophic-level efficiency for corn-fed cattle in enclosed feedlots may be as high as 20–25%. A corn-fed steer in an enclosed feedlot requires energy equal to the annual net primary production of about 0.56 ha (1.4 acre) of cornfield. Yet, in the context of modern agriculture, this energy calculation is incomplete. For example, it does not include the energy required to fuel the machines used in irrigation, cultivation, harvest, production, and distribution. These energy investments vary widely with growing conditions and cropping practices but can be as high as 30–70% of net primary production. The energy expended by the farmers who grow the corn or care for the cattle could also be added to these calculations of energy expenditures.

Cycling of Nitrogen and Phosphorous

■ Agroecosystems significantly influence nitrogen and phosphorous cycles; both elements act as limiting factors in agroecosystems.

Like their natural counterparts, agroecosystems depend on the cycling of nutrients. Of particular interest are two elements that affect the productivity of crop and pasture lands: nitrogen and phosphorus. These two elements are critical components of organic molecules needed to support and sustain life and both are present in fertilizers used to increase productivity. The effects of the movement of these nutrients extend well beyond agroecosystems.

The nitrogen cycle. Nitrogen accounts for 78% of the molecules in Earth's atmosphere but less than 0.003% of its crust. It enters the biosphere through the process of **nitrogen fixation** (**Figure 12.11**), a process that converts nitrogen gas (N_2) to ammonia (NH_3). A small amount of nitrogen fixation is caused by lightning; however, most is carried out by nitrogen-fixing bacteria.

Once fixed, nitrogen undergoes a number of transformations that allow it to cycle through the biosphere and thus agroecosystems. Ammonia dissolves in water to form ammonium (NH_4^+). Specialized bacteria in the soil then convert NH_4^+ to nitrite (NO_2^-) and nitrate (NO_3^-), a process known as **nitrification**. Plants take up ammonium, nitrite, and nitrate through their roots and use them to manufacture nitrogen-containing organic molecules, such as amino acids and nucleic acids. Animals must obtain their nitrogen by eating plants or other animals. When decomposers break down the remains and wastes of plants and animals, ammonia (NH_3) is recycled back into the soil.

Some nitrogen-fixing microbes, such as many cyanobacteria, live independently in soil and water. Others coexist in symbiotic relationships with plants, including **legumes**—plants that belong to the pea family (see Module 6.3). Farmers take advantage of this relationship by introducing legumes such as peanuts, alfalfa, and peas into the crop rotation and using the microbes to add nitrogen back into the soil.

Nitrogen in soil and water is returned to the atmosphere by **denitrification**. In this process,

▼ Figure 12.11 **The Nitrogen Cycle**
Nitrogen pools (Tg = 10^{12} g) are indicated in black and fluxes (Tg/yr) are indicated in red. Vast amounts of nitrogen occur in the atmosphere as N_2 gas. Specialized microorganisms play especially important roles in the nitrogen cycle, including nitrogen fixation, nitrification, and denitrification. Human impacts on the nitrogen cycle are enclosed with a blue dashed line. Note the significant amount of nitrogen flux associated with food production.

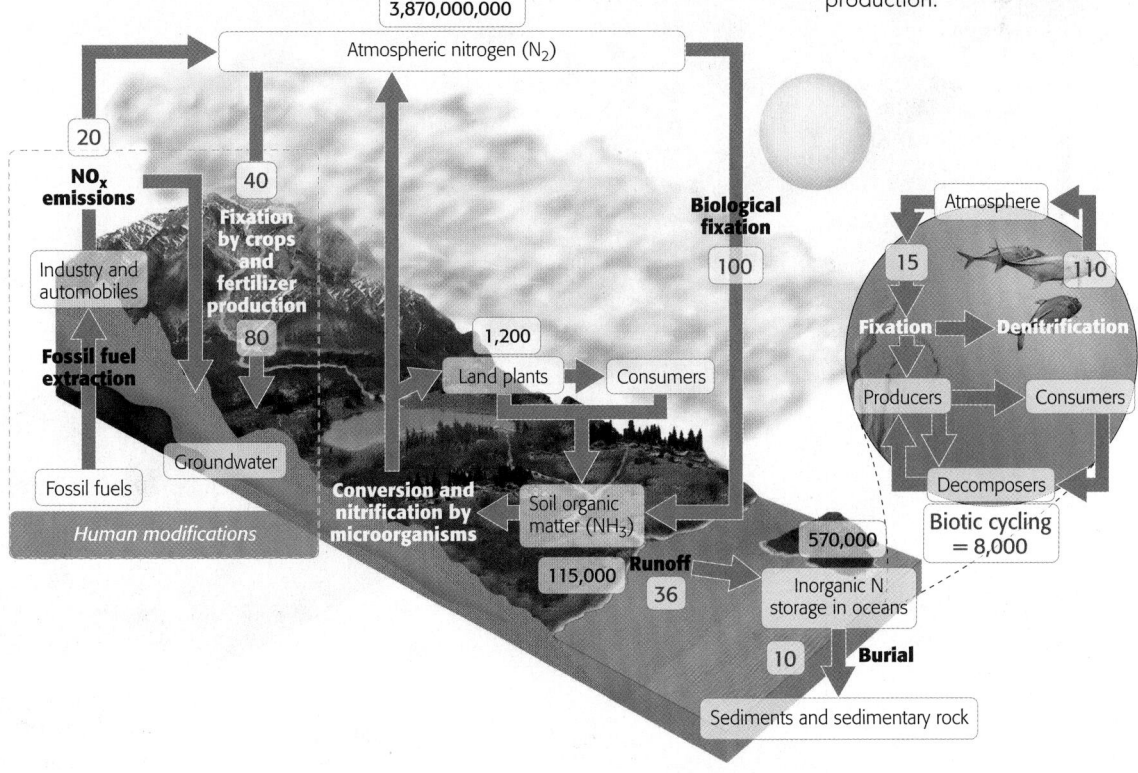

specialized bacteria convert NO_3^- to N_2 gas. The total nitrogen in Earth's organisms, soils, and waters is determined by the balance between nitrogen fixation and denitrification.

The phosphorus cycle.
Phosphorus accounts for about 0.13% of the mass of Earth's crust, making it the 11th most abundant element in the lithosphere. The vast majority of this phosphorus occurs in rocks as calcium phosphate, which is also known by its mineral name, apatite. Apatite is especially common in sedimentary rocks that form in some marine environments. As these rocks weather, the calcium phosphate in them dissolves, releasing phosphate ions (HPO_4^{2-}) into the soil. This very slow weathering process is the ultimate source of all the phosphorus that cycles through the biosphere (Figure 12.12).

Phosphorus makes up 0.5–1.0% of the tissues of organisms; this phosphorus is located in phosphate-containing organic compounds, including DNA, RNA, ATP, and the phospholipids in cell membranes. As organic matter decomposes in soil, inorganic phosphate is released and dissolves in soil water. Plants absorb this phosphate and use it to manufacture organic phosphate compounds. All other terrestrial organisms obtain organic phosphorus from the things they eat. Waste and dead organisms replenish the stores of organic phosphate in the soil, completing the terrestrial portion of the phosphorus cycle.

Each year, streams and rivers carry a significant amount (21 Tg) of phosphorus to the sea. Yet only about 10% of this phosphorus is available to aquatic organisms as dissolved phosphate. The remainder is tightly bound to soil particles and other sediments that sink to the ocean bottom; these sediments are destined to become sedimentary rocks.

Human impacts.
The availability of nitrogen and phosphorus are limiting factors in the productivity of agroecosystems. Human activities have increased the availability of both these nutrients to agroecosystems and hence have altered the fluxes and pools of these nutrient cycles.

For nitrogen, the increased cultivation of legumes and the production of chemical fertilizers from atmospheric nitrogen have nearly doubled the global rate of nitrogen fixation (Figure 12.13). In the early part of the 20th century, chemists discovered the **Haber–Bosch process**, a nonbiological method of nitrogen fixation. Initially developed for the manufacture of explosives, the Haber–Bosch process was quickly adapted to the production

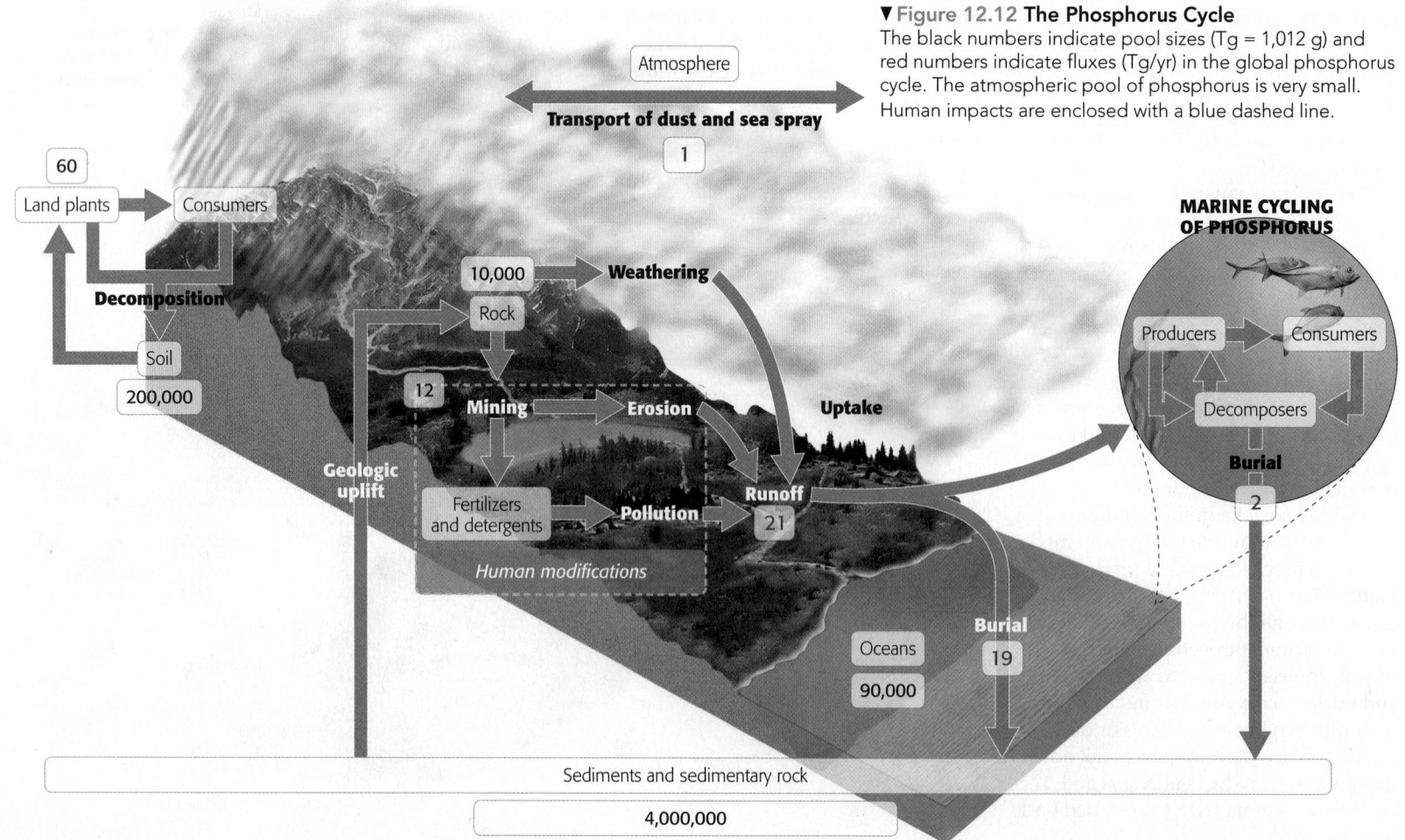

▼ Figure 12.12 **The Phosphorus Cycle**
The black numbers indicate pool sizes (Tg = 1,012 g) and red numbers indicate fluxes (Tg/yr) in the global phosphorus cycle. The atmospheric pool of phosphorus is very small. Human impacts are enclosed with a blue dashed line.

TERAGRAMS OF NITROGEN PER YEAR

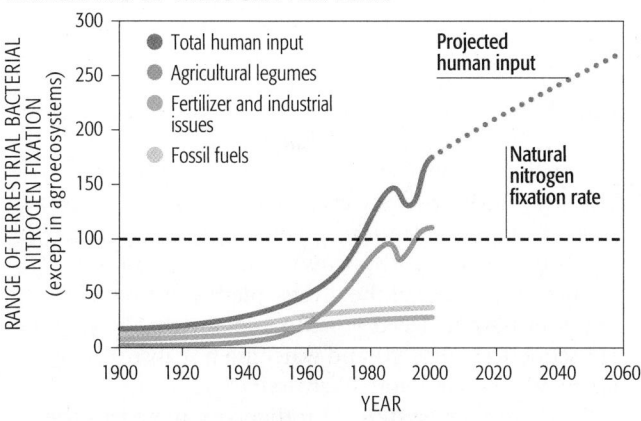

Range of terrestrial bacterial nitrogen fixation (except in agroecosystems)

- Total human input
- Agricultural legumes
- Fertilizer and industrial issues
- Fossil fuels

Projected human input

Natural nitrogen fixation rate

YEAR

▲ **Figure 12.13 Nitrogen in Fertilizer**
Worldwide, nitrogen fertilizer application has increased 15-fold since 1950. Altogether, human input of nitrogen into Earth's ecosystems exceeds natural rates of nitrogen fixation.

Source: United Nations. 2007. Millennium Ecosystem Assessment—latest available data.

of chemical fertilizers. Since 1950, the Haber–Bosch manufacture of fertilizer has increased the global rate of nitrogen fixation by about 40% compared to pre-agricultural rates, and that percentage continues to rise.

Phosphorus is also a critical component of fertilizer. Currently, about 12 Tg of phosphorus in the form of phosphate is mined each year, either from guano deposits or sedimentary rocks high in phosphates (**Figure 12.14**). Guano, the excrement of sea birds such as such as gulls, terns, boobies, and puffins, is very rich in phosphate. Over the centuries, large deposits of guano have accumulated at nesting sites (**Figure 12.15**).

The use of nitrogen and phosphate has significantly contributed to the increased agricultural production that has supported the exponential increase in our population. It has also increased the amounts of nitrogen and phosphate pollutants in streams, lakes, and coastal ecosystems. Excess nitrogen and phosphorous in agricultural runoff results in an increase in the growth of algae and can reduce the biodiversity of many stream, lake, and coastal ecosystems. In fact, there are over 200 dead zones worldwide, forming where rivers empty into the ocean, a result of these agricultural nutrient inputs (see Module 11.4).

▲ **Figure 12.15 Phosphate from Guano**
Phosphate from deep-water upwelling increases net primary production and supports a diverse ecosystem. The excrement of predatory sea birds such as these puffins accumulates as guano.

► **Figure 12.14 Phosphate Mining**
Today, phosphate is mined from marine sedimentary rocks that were formed millions of years ago from sediments produced in areas of ocean upwelling.

QUESTIONS 12.2

1. Describe four factors that contribute to high rates of nutrient loss from agroecosystems.

2. What process ultimately limits the amount of nitrogen available to organisms in an ecosystem?

3. What is the primary source of phosphorus to organisms in ecosystems?

(MES) For additional review, go to **MasteringEnvironmentalScience**

Dynamic Homeostasis
■ Agroecosystem simplicity causes vulnerability.

There can be no question that the dominance of our species on this planet is largely due to our development and management of agroecosystems. That said, we must be mindful that their productivity is finite. The energy they can provide us depends on how well we manage them. Dynamic homeostasis—the ability of an ecosystem to maintain stable values of key processes, such as net primary production and nutrient cycles—is dependent on how well we manage these systems. For example, the rate at which nutrients are lost or flow out of agroecosystems is often considerably greater than the rate at which they are naturally replenished. To maintain high levels of primary production, farmers must compensate for this imbalance between nutrient inflow and outflow by applying fertilizers.

Nutrient losses from agroecosystems are typically high for four reasons: harvest, continual disturbance, irrigation, and low biodiversity.

Harvest. Each year, nutrients that are stored in plant tissues are harvested and removed from the ecosystem. The rate at which these nutrients are harvested is generally far greater than the rate at which they are restored. Thus, with each successive year, nutrient stocks decline, resulting in diminished primary production.

Continual disturbances. Many systems of growing crops require that soils be plowed, sometimes more than once each year. This continual disturbance makes soils more vulnerable to erosion and nutrient loss. The movement of air into soil improves the growth of some plants, but it also causes some nutrients to escape into the atmosphere more rapidly.

Irrigation. In many regions, crops receive supplemental water in the form of irrigation. Water that runs off agricultural fields or seeps into groundwater often carries large amounts of water-soluble nutrients that have been leached from the soil (Figure 12.16). This depletes the soil of nutrients. It can also pollute streams and lakes that are fed by the runoff. Furthermore, fresh water is limited, and irrigation competes with other important uses of fresh water.

Low biodiversity. Agroecosystems are composed of only a few species, those that maximize the production of the things that humans need or want. This low biodiversity affects key ecosystem functions, such as nutrient retention and efficiency of water use. For example, the root systems of most crops are far simpler than those of the diverse plant communities that they have replaced (see Modules 6.1 and 8.1). As a consequence, nutrients and water are not absorbed as efficiently and are more likely to be lost.

In natural ecosystems, biodiversity increases the capacity for dynamic homeostasis (see Module 1.3). Since the biodiversity of agroecosystems is very low, it is not surprising that their capacity for dynamic homeostasis is often significantly diminished. In other words, compared to most natural ecosystems, agroecosystems are quite fragile. This weakness is obvious when agroecosystems suffer from periods of low rainfall or harsh weather. It is also evident in their vulnerability to invasion by competing weeds or voracious herbivores. We are learning how to better mimic natural processes in agriculture, thereby requiring fewer human inputs and reducing damage to surrounding water courses. We must keep in mind that the key feature to the success of agroecosystems, their simplicity, also makes them fragile and vulnerable to collapse.

▲ Figure 12.16 **Water Erosion**
This spectacular landscape, sculpted by water erosion, was a poorly managed agricultural field in the 1800s. Today "Providence Canyon," a Georgia state park, serves as a powerful reminder of the importance of careful management of agricultural land.

12.3 The Growth of Crop Plants

BIG IDEA Like all green plants, agricultural crops depend on photosynthesis for growth and reproduction. The process of photosynthesis requires light, water, and carbon dioxide. Plants also require various nutrients that they obtain from soil. The ecological needs and tolerances of the various crop species determine when and where particular crops can be grown. The growth of crops is also greatly influenced by the community of other organisms living in the agroecosystem. Interactions with some of these organisms are beneficial to crop production, while interactions with others are detrimental.

Plant Growth and Reproduction

■ Plants depend on light, water, and essential nutrients.

The energy needed for plants to grow and reproduce is provided by photosynthesis, which is carried out in the green tissues of leaves and some stems (see Module 4.1). In photosynthesis, plants use the energy from light to convert carbon dioxide and water into high-energy carbohydrates. Carbon dioxide from the atmosphere enters leaves through tiny pores in the leaf surface called stomata. As stomata open to take in carbon dioxide, they also allow water vapor and oxygen, a by-product of photosynthesis, to escape from the moist interior of the leaf.

Continued photosynthesis requires replacement of this water. As water evaporates from the leaf, a diffusion gradient is established that provides a mechanism that pulls water from the soil into the roots and then the stems and leaves of the plant. This process is called **transpiration** (Figure 12.17).

The plant's root system brings fine roots into direct contact with soil particles and soil water. The cells of these fine roots have extensions called root hairs that greatly increase the surface area through which water and nutrients can be absorbed. As the water moves through the plant, it brings with it a steady supply of dissolved nutrients picked up from the soil. These nutrients plus the carbohydrates produced through photosynthesis enable a plant to grow and process the materials needed for it to survive.

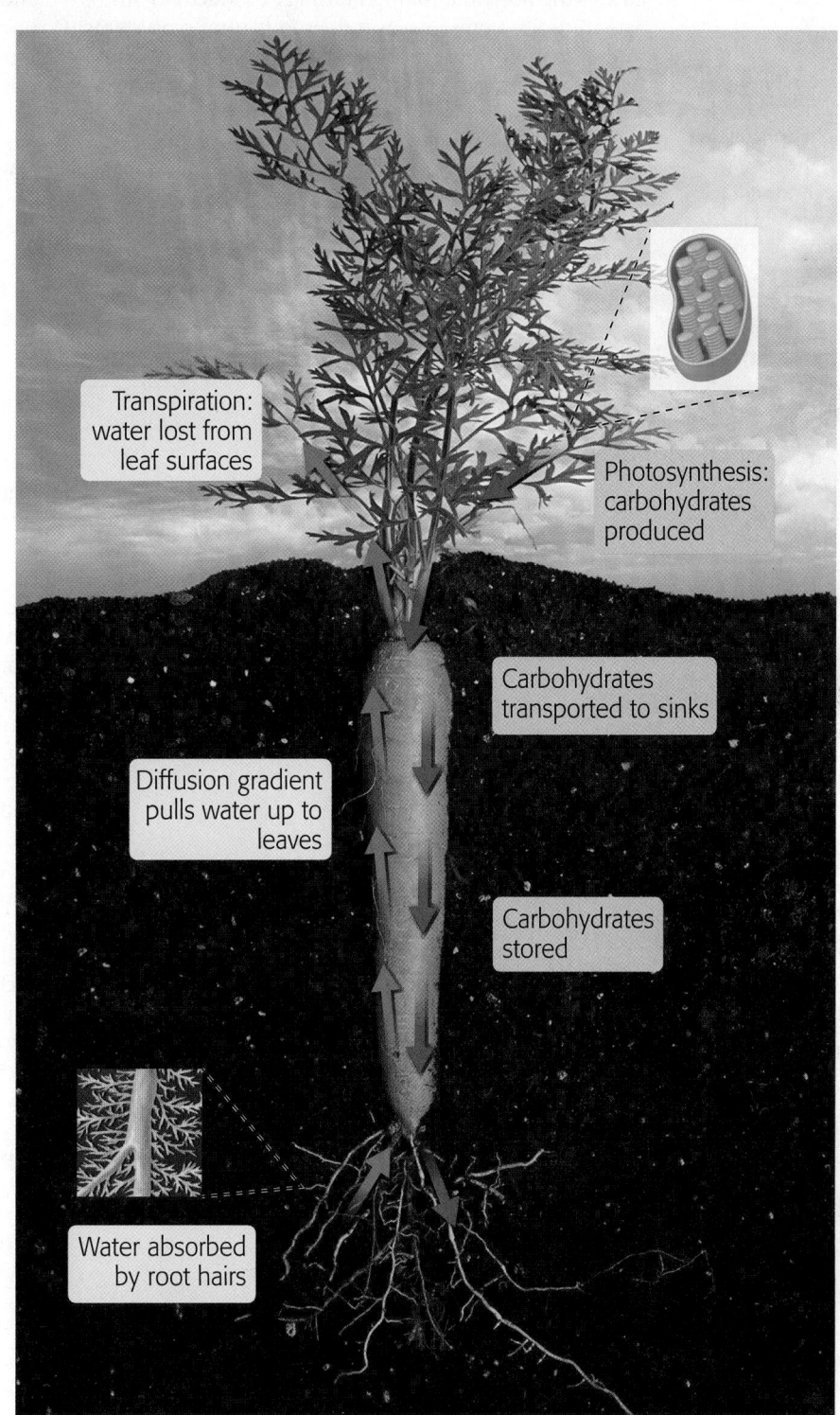

Labels on figure:
- Transpiration: water lost from leaf surfaces
- Photosynthesis: carbohydrates produced
- Carbohydrates transported to sinks
- Diffusion gradient pulls water up to leaves
- Carbohydrates stored
- Water absorbed by root hairs

▶ Figure 12.17 **How Plants Capture Energy and Acquire Resources**
Vascular plants, including most crop plants, have leaves with specialized structures that transform energy from the sun into energy-rich carbohydrates. Specialized vascular tissues stretch from roots to leaves and transport water, carbohydrates, and various nutrients the plant needs to survive.

Where YOU LIVE **What's growing in your neighborhood?**

Research the average precipitation and temperature for your area and use this information to locate your biome on Figure 12.18. What crops have the appropriate tolerance range to grow where you live? What do farmers in your area cultivate? Do the crops match with those in the tolerance range? If not, how might agriculturists have extended the range of those crops beyond what is natural? What might be the consequences of these actions?

What Grows Where and Why?

■ The climate and soil of a region determine which crops can be grown.

Most agriculture takes place in the tropical and temperate climate zones. The seasonal patterns of temperature and rainfall within a region determine when crops can be grown. In tropical countries such as Brazil, seasonal changes in temperature and rainfall are small, and crops can be grown year-round. The long, warm growing season creates the potential for high primary productivity in agricultural systems, just as it does in natural ecosystems in the tropics. In subtropical climates such as the Sahel of Africa, the growth of crops must be timed to coincide with wet seasons and avoid dry seasons. In temperate latitudes, the growing season is usually limited to the months when freezing temperatures are unlikely.

Like plants in natural biomes, each kind of crop plant has a specific set of tolerances for climatic factors, including temperature, rainfall, and length of day. For this reason, the climate of a region is a major factor in determining where each crop may be grown (Figure 12.18). Spring wheat, for example, grows best in cool, moist conditions. It can tolerate relatively short growing seasons, but the soil must remain moist for complete grain development. In the southeastern and midwestern United States, wheat is grown in early spring and harvested before the onset of hot weather. In the northern United States and southern Canada, wheat is grown in the summer. Wheat also thrives in much of northern Europe and Russia at this time.

Corn requires warmer growing conditions and a longer growing season than does wheat. Its photosynthetic chemistry works most efficiently at temperatures above 30 °C (86 °F). Thus, in North America, corn is grown farther south and planted later in the growing season than wheat.

Rainfall is a critical factor in determining the kinds of crops that are best suited to a region, though increasingly irrigation is used to extend the range of crop tolerance. Some kinds of plants use water more efficiently than others, allowing them to thrive in dry conditions. Desert-adapted food plants, such as the prickly pear cactus grown for its edible stems (cactus pads) and fruits, can be cultivated without irrigation in regions that are far too dry for most crops.

Plant growth can be limited by too much water as well as by too little water. Plant roots need oxygen to grow, and soils saturated by water contain little oxygen. As a result, most crops cannot grow in waterlogged soils. On the other hand, cranberries are cultivated in standing water that would suffocate the roots of most other crop species.

The characteristics of soil limit the kinds of crops that can be grown in a region. Soil moisture and chemistry, including pH, also differ from place to place, depending on the prevailing weather, topography, past vegetation, and character of the underlying bedrock. For example, acid soils tend to form in regions where granite is the underlying material; blueberries thrive in acidic soils and are a good choice in those sites. Limestone tends to produce more alkaline soils; asparagus is a crop plant that requires alkaline soil.

Plants vary greatly in their need for soil nutrients. Crops such as corn and wheat require large amounts of nitrogen, which is often supplied by fertilizers. Because of their symbiotic relationship with nitrogen-fixing bacteria, legumes such as soybeans and peanuts can be cultivated on nitrogen-poor soils without the heavy application of nitrogen fertilizer.

Agricultural systems and the human cultures they support ultimately depend on the local climate and soils. While we can supplement and manipulate some factors within agricultural systems, agroecosystems must live within the same climatic limitations as natural ones. The traditional foods of a region, such as the rye bread of northern Europe, the corn of Central America, and the rice of Southeast Asia, come from plants with adaptations that make them well suited for cultivation in that part of the world.

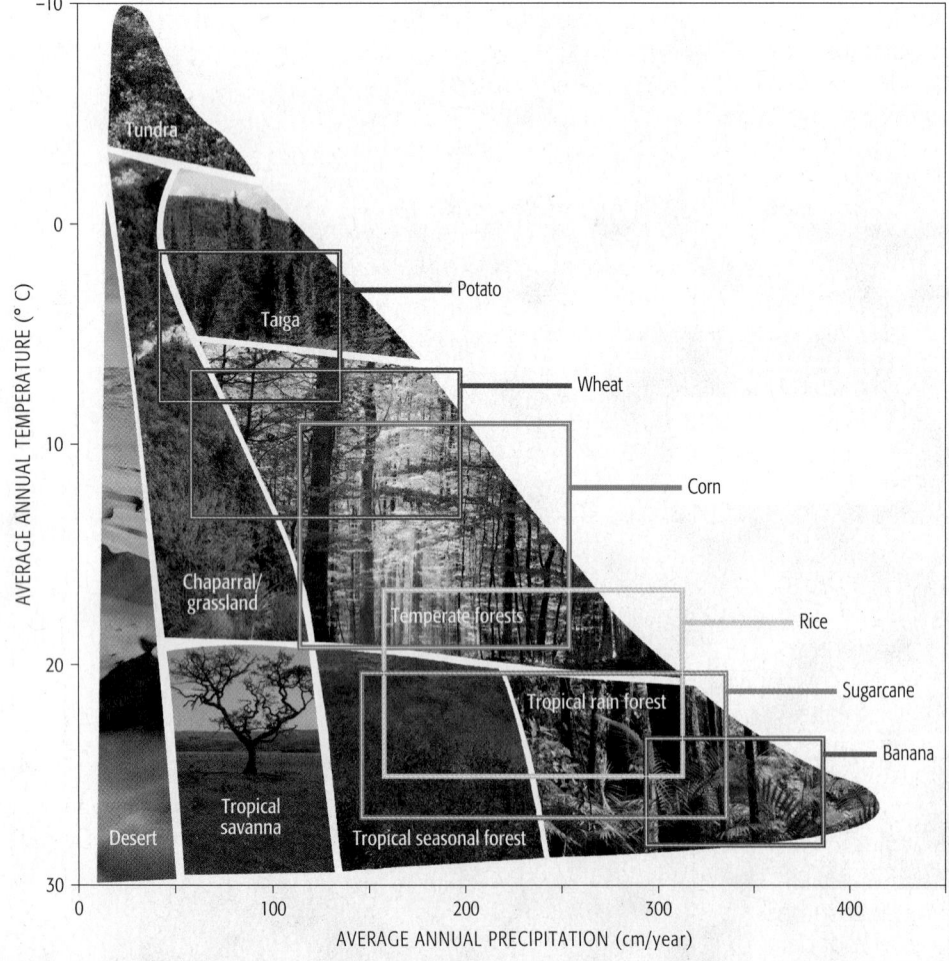

▲ **Figure 12.18 Climate Limits the Growth of Crops**
Each kind of crop has a particular range of tolerance for rainfall and temperature. The colored boxes overlaying this biome diagram indicate the biome—and therefore the range of rainfall and temperature—in which a particular crop plant may be cultivated.

The Role of Other Organisms

■ The growth of crop plants is influenced by the surrounding ecological community.

In most agroecosystems, attention is focused on the production of a single crop species. However, the growth of those species also depends on a community of other organisms, including pollinating insects, bacteria, and fungi that live in the soil.

Before seeds and fruits can form, the flowers of crop plants must be pollinated. Insects pollinate vegetable, fruit, and nut crops, which make up about a third of the human diet (see Module 6.3). Honeybees are among the most important insect pollinators, but the loss of most of their natural habitat has resulted in very sparse populations of wild honeybees. In many cases, domesticated honeybees raised by beekeepers provide this essential ecosystem service for crops requiring pollination (Figure 12.19). Today, honeybee populations are threatened by the inappropriate use of pesticides, bacterial diseases, and infestations of parasitic mites. Because honeybees are so important to our food supply, these threats must be taken very seriously.

A diverse community of soil organisms recycles the nutrients in agroecosystems. Earthworms are among the most visible of these soil organisms. Because they feed on bits of dead leaves and other organic materials in the soil, earthworms are very effective at converting agricultural crop residues back into inorganic soil nutrients. Their vertical burrows allow air to enter the soil and improve soil texture, vastly benefiting plant roots. Because earthworms constantly bring soil from deep in the root zone back to the surface of the ground, they play an important role in soil development (Figure 12.20).

Mites, springtails, beetles, and a myriad of beneficial bacteria also contribute to the breakdown of crop residues and the subsequent release of nutrients to the soil. Fungi are particularly effective in this recycling function. They release enzymes that break down the cellulose in plant cell walls, a biochemical task few other organisms can do. Without fungi in the soil, most of the carbon, nitrogen, and other elements essential for life would remain in dead leaves and stems, and new growth would come to a standstill.

Scientific understanding of the community of soil organisms and the beneficial work they accomplish is limited, and this is a challenge for agricultural science. We do know that the excessive application of herbicides and fungicides to agricultural fields can kill vital soil organisms and interfere with nutrient cycling. The result is a greater need for artificial cultivation and fertilizers and a lower net yield.

Organisms involved in mutualistic associations with plant roots are especially important to agricultural productivity. Specialized fungi form associations with plant roots called mycorrhizae. The fine threads of these fungi grow through the soil and around the hairlike extensions of the roots of their plant partners. The mycorrhizal fungi transport inorganic nutrients such as phosphates and potassium into plant roots, while the plants provide the fungi with carbohydrates derived from photosynthesis. In nutrient-poor soils, many crop plants fail to thrive without symbiotic fungi.

Of course, not all of the organisms in agroecosystems are beneficial to the growth of crops. A variety of non-crop plants also grow in agricultural fields, where they compete with crops for nutrients, water, and light. Farmers call these competitors weeds. Many animals that grow in fields feed upon crops, thereby lowering production. Herbivorous insects may consume a significant portion of crop production. Roundworms called nematodes may parasitize roots. Crop plants are also susceptible to a variety of viral, bacterial, and fungal diseases. Much of agroecosystem management is aimed at minimizing the impacts of these competitors, herbivores, and parasites.

▲ Figure 12.20 **Soil Fauna**
Earthworms and a diverse array of smaller soil organisms speed up the recycling of nutrients as they feed on organic material in the soil.

◄ Figure 12.19 **Domestic Honeybees**
Each spring, hives of honeybees are brought to orchards to ensure that the trees are pollinated and produce fruit. In addition to honey, domestic honeybees provide pollination services to farmers worth over $14 billion each year.

QUESTIONS 12.3

1. What do plants need to grow and reproduce? How do they collect and transport each of these items?

2. Name three crop plants that are typically grown in tropical climates and three that are typically grown in temperate climates.

(MES) For additional review, go to **MasteringEnvironmentalScience**

12.4 Managing Soil Resources

BIG IDEA The uppermost layer of Earth's crust is **soil**, a mixture of organic matter and mineral particles. Soils are formed by the constant weathering of the rocks below them and the continual addition of organic matter from the plants that grow on top of them. As a consequence, soils have abundant mineral-rich, rocklike particles at the bottom and nutrient-rich, organic materials toward the surface. The roots of plants draw nutrients and water from soil. The ability of a soil to support plant growth is determined by the total amount of nutrients it contains, as well as by factors such as acidity and aeration that determine how well plants can use those nutrients. In ecosystems such as forests and prairies, natural processes maintain soil structure and fertility. In agroecosystems, these processes are compromised, so humans must take steps to sustain soil productivity.

Soil Origins and Structure

■ Organic matter and weathered rock form soil.

Soil is composed of mineral particles and organic matter, as well as water, air, and a community of living organisms. The mineral particles are classified according to their size (Figure 12.21). **Sand** particles are 0.05 to 2 mm in diameter. Although sand particles have been weathered, the chemical structure of these fragments still resembles that of their parent bedrock. **Silt** particles are smaller, with diameters ranging from 0.002 to 0.5 mm. These particles have experienced more weathering than sand. The smallest particles are **clay**; their diameters are less than 0.002 mm. Clay particles are highly weathered and have unique physical and chemical properties. Clays are composed of platelike particles of silicon-rich minerals. Electrical charges in these minerals allow clay particles to bind to one another and to form chemical bonds with many soil nutrients. Soil **texture** refers to the relative amounts of sand, silt, and clay in soil.

Soils form as a result of two processes: the weathering of the bedrock below them and the deposition and decomposition of leaf litter and other organic materials from the soil surface. As a consequence, natural soils have a distinctive vertical structure, or **soil profile**. At the surface, soil profiles are dominated by organic matter. Beneath that is a mixture of organic matter and weathered mineral particles. At the bottom, soil profiles contain mainly less-weathered, rocklike particles. Soil scientists divide soil profiles into layers called horizons (Figure 12.22).

The **O horizon** at the very top is not really soil, but consists of leaves or crop residue and the organic products of their decomposition called **humus**. Below this, the **A horizon** is made up of a mixture of organic matter and mineral particles. The upper, humus-rich portion of the A horizon is topsoil. Hundreds of species of organisms live in the O and A horizons, including bacteria, fungi, and animals such as worms and insects. The activities of these organisms contribute to the decomposition of humus, breaking down complex organic molecules and releasing inorganic nutrients,

such as ammonia and phosphate. Plant roots are most dense in the A horizon because essential nutrients are most abundant here.

Below the A horizon is the **B horizon**, also called subsoil. This layer is often especially rich in clay particles that form from minerals dissolved from the weathering of sand and silt. Minerals such as iron and aluminum leach out of the A horizon and accumulate with the clays in the B horizon. These minerals often give the B horizon a red or yellow hue. Sometimes a soil profile has a light-colored layer of silt and sand between the A and B horizons, where this leaching occurs. This is referred to as the effluvial or E horizon.

The lowest layer of soil is the **C horizon**, or weathering zone. Here bedrock is broken up by the action of soil water and sometimes by the growth of plant roots. The C horizon differs substantially from place to place, depending on the type of bedrock.

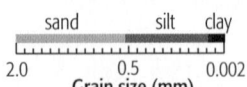

▲ **Figure 12.21 Soil Texture** Depending on their size, the mineral particles in soil are classified as Ⓐ sand, Ⓑ silt, or Ⓒ clay.

► **Figure 12.22 Soil Horizons** Weathering and the decomposition of organic matter create distinct soil layers. The O horizon is primarily organic matter and fine roots. The A horizon contains organic matter and highly weathered mineral soil. The B horizon contains abundant clay and far less organic matter. The C horizon contains particles of rock that are chemically similar to the parent rock below.

Soil Fertility

■ The ability of soils to support plants depends upon the essential nutrients they contain, as well as their pH, degree of aeration, and overall structure.

Soil fertility refers to the variety of soil characteristics that support plant growth. These characteristics include the availability of essential nutrients, the pH (or acidity), the amount of aeration, and the overall structure of the soil.

The availability of nutrients in soil is influenced by several factors. The total amount of a nutrient in soil sets an upper limit on its availability. Because organic matter contains significant amounts of nutrients such as nitrogen and phosphorus, fertility is often correlated with the amount of organic matter in soils. Clay tends to increase soil fertility, because clay particles chemically bind to many nutrients and prevent them from being leached away by water flowing through the soil.

Fertility is also determined by factors such as moisture content, aeration, and pH. These factors influence the decomposition of organic matter and the chemical transformation of nutrients. Many of the nutrients in soil are not in a form that plants can use. For example, over 99% of soil nitrogen exists in complex organic molecules, such as proteins, that plant roots are not able to access. Before plants can take up this nitrogen, it must be converted to either ammonium (NH_4) or nitrate (NO_3).

The acidity, or pH, of soil is a measure of the concentration of hydrogen ions (H^+) relative to other cations (positively charged ions) in the soil (see Module 3.1). Soils with pH values less than 7 are acidic, and those with pH values greater than 7 are alkaline. The decomposition of organic matter is the largest source of H^+ to the soil. Continual decomposition of organic matter tends to increase soil acidity, or lower soil pH. Weathering of soil particles releases large amounts of other cations, such as potassium (K^+), calcium (Ca^{++}), magnesium (Mg^{++}), and iron (Fe^{+++}). These cations tend to raise soil pH, making it more alkaline. Other factors, such as the amount of aluminum and sulfur, also influence soil pH. In some wetlands where organic matter abounds and other ions are in short supply, soil pH may be as low as 3. In contrast, in many deserts where soils have only small amounts of organic matter and large amounts of other ions, pH may approach 9 (Figure 12.23).

Soil pH influences a variety of processes that determine soil fertility, including the availability of nutrients. Most nutrients are available to plants within the range of 6.5–7.5 pH. Soil pH also has a direct effect on the growth of plant roots. The roots of some species grow best in acidic soils, whereas others do best in alkaline soils. Soil pH also affects the growth of many bacteria and fungi, thereby influencing the decomposition of organic matter and the availability of essential nutrients. Finally, because it affects how clay particles bind to one another, soil pH can directly influence the physical structure of the soil itself.

Soil aeration refers to the ability of atmospheric gases, particularly oxygen and carbon dioxide, to diffuse into and out of the soil. As a general rule, as soil aeration increases, so does the growth of roots and soil microbes. Poor aeration can affect the availability of some essential nutrients. For example, when soils lack oxygen, certain bacteria convert the nutrient nitrate to nitrogen gas (N_2). This nitrogen diffuses to the atmosphere and cannot be used by plants.

Soil aeration is directly related to a structural property of soil that farmers call tilth. **Tilth** refers to the physical arrangement of soil particles that facilitates aeration, seedling emergence, and the growth of roots. Tilth is affected by soil texture, particularly by the extent to which clay particles are bound to one another. While soil texture cannot be managed, soil tilth or structure can be, depending on the type and timing of various tillage operations. Most often tillage results in compaction and loss of soil pores. Tilth can generally be improved by increased amounts of organic humus. This is one of the main reasons that gardeners often add organic amendments such as peat moss or cow manure to soil.

Conventional farmers often supplement their soils with synthetic fertilizer to increase nitrogen (N), phosphorous (P), and potassium (K), as well as other essential nutrients to improve crop yield. Organic farmers rely on natural sources of nutrients such as animal and crop wastes, cover crops, and crop rotation to maintain soil fertility.

▲ Figure 12.23 **Desert Soil** High rates of evaporation cause salts to accumulate in desert soils. As a result, desert soils have a higher pH than soils that develop in moister conditions.

AGRICULTURE AND THE ECOLOGY OF FOOD

► **Figure 12.24 Techniques for Preventing Soil Erosion**
Ⓐ In contour farming, planting the crop rows follow the contours of hills causing water to flow across, minimizing erosion. Intercropping is used as well, planting rows of perennial alfalfa (green) alongside annual corn (gold) to further limit erosion. Ⓑ Terracing allows for farming of very steep terrain. Ⓒ Shelterbelts or windbreaks slow the wind, reducing soil erosion and protecting fragile crops. Ⓓ No-till techniques allow this cotton crop to grow in a field that has not been plowed, maintaining maximum topsoil cover. Ⓔ Hungarian grazing rye is sown as a cover crop to improve soil fertility and prevent erosion during the winter months.

Ⓐ Contour farming/intercropping

Ⓒ Shelterbelts/windbreaks

Ⓑ Terracing

Ⓓ No till

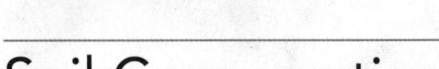
Ⓔ Cover crops

Soil Conservation

■ Agroecosystems lack many of the processes that sustain soil fertility.

Soil structure and fertility are sustained in natural ecosystems in a variety of ways. Prairies and forests build topsoil through the constant accumulation of organic matter and protection of the soil already in place. Organic matter is constantly being added to soils, replenishing the stock of essential nutrients such as nitrogen and phosphorus. Plant roots and chemicals from organic decomposition ensure steady weathering of soil minerals, thereby ensuring continual soil development. The leaves and stems of plants intercept rainfall and their root systems hold soil in place, minimizing erosion.

All of these processes are altered when natural ecosystems are replaced by agroecosystems. The sustained productivity of agroecosystems, therefore, depends on sustainable management of soils. Because it is the zone of highest fertility and root activity, topsoil is especially vital to agricultural production. However, most agricultural practices diminish the organic inputs that build topsoil and accelerate the loss of organic matter and erosion of topsoil. When the majority of crops are harvested, most of the aboveground organic matter is removed from fields. Plowing and tilling increase humus decomposition, further depleting stores of nutrient-rich organic matter. These activities also expose topsoil to rain and wind, accelerating erosion.

Water erosion is equally problematic in unprotected agricultural fields. When raindrops fall on exposed soil, they dislodge soil particles, which are carried along by the surface flow of rainwater. If running water cuts a small gully, more water moves through that channel, cutting the gully deeper and deeper (see Figure 12.16). If agricultural practices fail to prevent erosion, topsoil built

up over several centuries can be lost in a few heavy rains. Not only does this diminish fertility of agricultural fields, but it also contributes large amounts of nutrients and sediment to the waters of streams and rivers.

The National Resource Conservation Service recommends that farmers follow four basic principles for maintaining healthy soils: (1) keep the soil covered as much as possible, (2) disturb the soil as little as possible, (3) keep plants growing throughout the year to feed the soil, and (4) diversify as much as possible using crop rotation and cover crops. Different conservation practices are used to implement these principles depending on conditions. For hilly terrain, **contour farming**, or plowing along the contour of the land, rather than up and down hills, can prevent the formation of gullies (Figure 12.24A). On steep mountain slopes, **terracing**, or cutting a series of wide steps into the slope, helps to retain water and limit runoff in sites that would otherwise be very difficult to cultivate (Figure 12.24B). For windy sites with limited tree cover, **shelterbelts** or **windbreaks** slow the wind and help avoid the blowing of topsoil (Figure 12.24C). Under any conditions, **no-till** techniques allow for planting with special drills in an unplowed field, preventing the soil erosion that occurs with plowing and enabling direct drill application of fertilizer (Figure 12.24D). **Intercropping**, or alternating bands of different crops in the same field, binds soil particles in place between the primary crop rows (as seen in Figure 12.24A). And **cover crops**, such as winter wheat or rye, are planted to hold soils that would otherwise lie exposed to the weather between seasons, reducing the risk of erosion (Figure 12.24E). They are not intended for harvest.

QUESTIONS 12.4

1. What are the characteristics of the different layers in a soil profile?

2. What four factors influence soil fertility?

3. Why are the soils in agroecosystems prone to erosion and loss of fertility?

(MES) For additional review, go to **MasteringEnvironmentalScience**

Developing Perennial Wheat

Can perennial wheat provide a more sustainable replacement for annual wheat?

Scientists are touting perennial crops as the biggest breakthrough since the origin of agriculture 10,000 years ago. Perennials live for many years, while annuals, the source of many of our food grains (corn, wheat, and rice) and 70% of the calories in the human diet, live for only a single growing season. Named "one of the five crop researchers who could change the world" by the journal *Nature* in 2008, Jerry Glover is hard at work testing the hypotheses that perennial crops are (1) environmentally sustainable and (2) economically viable (**Figure 12.25**).

Annuals die when they are harvested, so each year they must be replanted from seed. Why is that a problem? First and foremost because before planting annual crops, fields are usually cleared and plowed unless no-till techniques are being used. These activities expose the soil, increasing runoff and soil erosion; year by year, these processes diminish soil fertility. Second, because plant leaves, stems, and roots are constantly being removed, the permanent store of carbon on annual croplands is minimal. Third, adverse weather conditions, such as drought or late frost, pose a greater risk to annual plants than to perennial plants. Finally, the need to clear, plow, fertilize, and plant new crops each year requires the use of fossil fuels and makes the cost of growing annuals greater than the cost of growing perennials.

Unlike annual crops, perennial crops maintain important ecosystem services throughout the year. For example, they intercept, retain, and efficiently use rainwater, diminishing the need for irrigation. Perennials have longer growing seasons and deeper roots and thus maintain more carbon on the land. These *observations* led to Glover's *hypothesis* that perennial crops are more environmentally sustainable than annuals.

Today, annual crops are grown on over 70% of Earth's croplands. As Glover sees it, soils on the world's best croplands are at low risk of being degraded by the cultivation of annual crops, but these lands make up only about 20% of farmable land. The remaining land is at high risk of degradation by the production of annual crops. Much of this lower-quality cropland is capable of producing perennial crops.

▲ Figure 12.25 **Agricultural Revolutionary** Jerry Glover is an agroecologist at the Land Institute, an independent agricultural research center near Salina, Kansas. He is holding a sample of perennial wheat.

Glover and his colleagues at the Land Institute have focused much of their attention on a perennial called intermediate wheatgrass (*Thinopyrum intermedium*), which they hope can be used as a replacement for annual wheat. The roots of this grass live for many years; each year its roots send up new shoots that produce seeds. Intermediate wheatgrass has many of the characteristics required of crop plants, such as having stiff, erect stems, synchronous seed maturation, and large, easily harvested nutritional seeds.

To *test his hypothesis* that perennial wheat is economically viable, Glover has developed several breeding populations of intermediate wheatgrass. In three generations of careful selective breeding in which the plants that produced the largest seeds were cross-pollinated, researchers found that seed size more than doubled (**Figure 12.26**). These results suggest that with intensive selection as well as further genetic evaluation, this perennial species may be capable of increasing its yield to marketable levels. The seeds of these plants have a nutritional value that is similar to or better than that of the seeds of annual wheat, although the seeds of wheatgrass cannot be used alone to make leavened bread. Glover is confident that continued selective breeding will improve yield, and the dream of perennial wheat will be realized.

1. Scientists have hypothesized that perennial wheat can be bred to the point where its yield is economically viable. What evidence do they have to support this hypothesis?

2. What are the characteristics of perennials that suggest that they would be a good choice for sustainable agriculture? How might Glover test his hypothesis that perennial crops are more sustainable than annual crops?

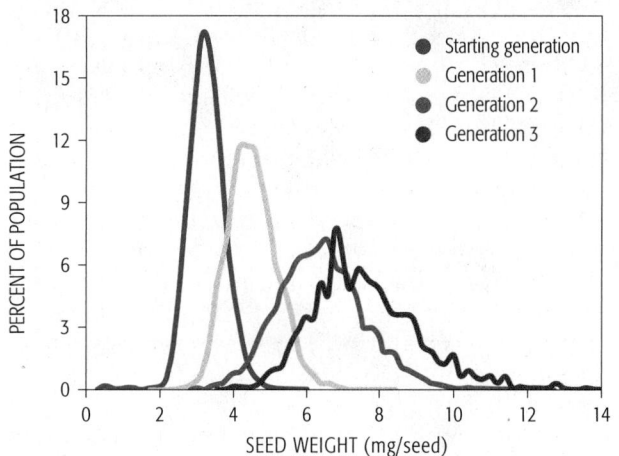

◄ Figure 12.26 **Effects of Perennial Wheat Breeding Program** After three generations of careful selective breeding, perennial wheat seed weight more than doubled, suggesting that this perennial grain crop may have the potential to produce much higher yields.

Source: L. DeHaan. Spring 2013. To select a needle in a haystack. *The Land Report* 105: 7.

12.5 Water and Agriculture

BIG IDEA Water is essential to the growth of all plants. In fact, modern agroecosystems account for 70% of all human water use. Water availability in a region is determined by rainfall and factors that influence runoff and percolation into soil. The amount of water that soil can hold against the force of gravity depends on soil texture and organic matter content. Irrigation has greatly increased the productivity of many existing agroecosystems and allowed the expansion of agriculture in dry regions. At the same time, irrigation has increased demands on water supplies, diminished water quality, and increased the salinity of soils in many places. Increasingly, agronomists and farmers are growing crops and using technologies that make use of water in more sustainable ways.

Water in Soil

■ The surfaces of soil particles hold water against the force of gravity.

Water comes into an agricultural field as rainfall or irrigation. As in natural ecosystems, the water entering an agroecosystem has three fates. It may simply run off the soil surface into nearby streams; it may percolate through the soil to groundwater aquifers; or it may evaporate to the atmosphere from soil or plants (see Module 11.1). Interactions between soil and water play a key role in determining the ultimate path of this water.

How much of the water that falls on a field becomes runoff and how much percolates into the soil? The answer depends on several factors, including the topography of the land, the intensity of the rainfall, and the porosity and permeability of the soil. Runoff is generally greater on steep slopes than on gradual slopes. The more porous the soil, the more space is available to hold water. The more permeable the soil, the easier it is for water to percolate through it.

In general, sandy soils are more permeable than clay soils, which are more porous. Organic matter usually increases porosity. Soils that have been compacted by heavy equipment or other activities are less porous and have increased runoff.

Water that percolates into the soil is drawn downward by the force of gravity. Water flowing through the soil is called **gravitational water**. However, the soil is able to hold some water against the force of gravity.

Soil holds water because of hydrogen bonding between soil particles and water molecules (see Module 3.1). Electrical charges on the surface of soil particles bind very tightly to water molecules, forming a microscopic film of water around the soil particles. The water that is bound to soil particles is called **hygroscopic water** and is unavailable for plant uptake. However, water molecules that are bound to soil particles also form hydrogen bonds with other water molecules, and those molecules form bonds with yet other water molecules. This water, held in the tiny spaces (micropores) in the soil by water-to-water hydrogen bonds, is available to plants and is called **capillary water** (Figure 12.27). A sponge holds water against the force of gravity by exactly the same principle. Water molecules form strong bonds with the surfaces of the sponge, and more water is retained due to water–water hydrogen bonding.

The amount of energy required to remove the water in soil depends on the proximity of water molecules to soil particles. Within the hygroscopic film, the molecules closest to the surface of soil particles are held most tightly. Those farther away are held less tightly. At some distance from a soil particle, the electrostatic forces holding water are weaker than the force of gravity, and water percolates downward.

Field capacity is the amount of water that a given volume or weight of soil can hold against the force of gravity. The capacity of soil to hold water is determined primarily by the total surface area provided by soil particles. The tiny particles of clay have much greater surface area relative to their volume than the coarser particles of silt and sand. Thus, fine-textured soils have higher water-holding

▼ **Figure 12.27 Soil Water** Field capacity is the amount of water that a soil can retain against the pull of gravity. When soil water is below the wilting point, plants are no longer able to take up water fast enough to keep pace with transpiration.

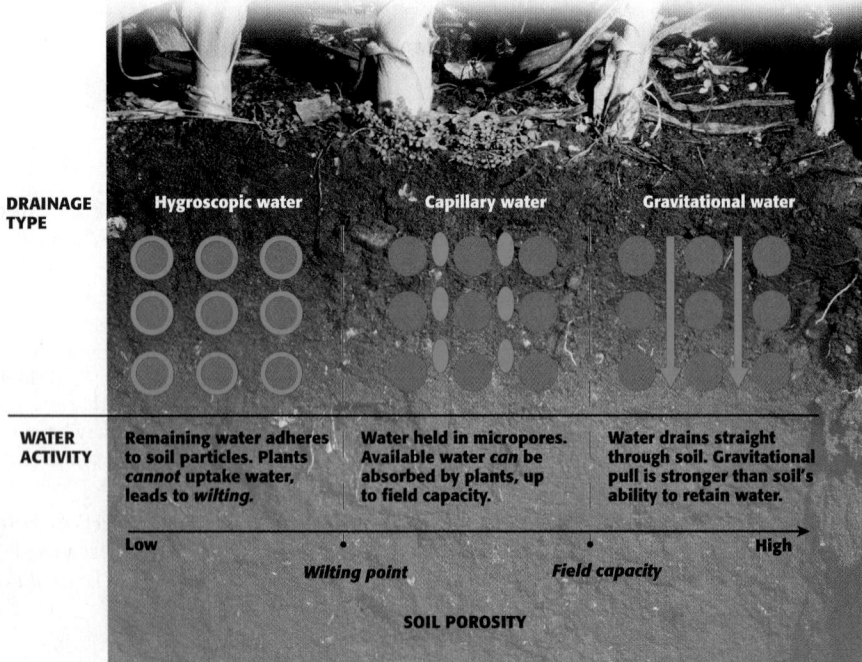

DRAINAGE TYPE	Hygroscopic water	Capillary water	Gravitational water
WATER ACTIVITY	Remaining water adheres to soil particles. Plants *cannot* uptake water, leads to *wilting*.	Water held in micropores. Available water *can* be absorbed by plants, up to field capacity.	Water drains straight through soil. Gravitational pull is stronger than soil's ability to retain water.

Low ————— Wilting point ————— Field capacity ————— High

SOIL POROSITY

capacities than coarse soils. Particles of organic matter have a great deal of surface area, so water-holding capacity generally increases with increasing organic matter content. This is another reason why gardeners and farmers amend soil with organic manure and humus.

To obtain water from soil, plants must overcome the forces that hold water to soil particles. When soils are near their water-holding capacity, plant roots can easily take up water. However, as soil water is depleted, the soil holds the remaining water ever more tightly. At some point, plant roots cannot take up enough water to keep pace with the rate of transpiration from their leaves. When this occurs, the plant wilts. The **Soil wilting point** is the point at which there is no longer enough water in the soil for plants to replace the water that is being lost to transpiration (see Figure 12.27).

If you forget to water your houseplants, the soil surrounding their roots gets drier and drier. Eventually their roots cannot take up water fast enough to replace the water evaporating from their leaves, and they wilt. The soil in their pots is then at its wilting point. If you water your plants promptly, they can recover; if you do not, they will dry out completely and die.

As water moves through soil, it is altered chemically. Chemicals from soil particles or produced by the decomposition of organic matter may dissolve in water. Other chemicals dissolved in the water may adhere to soil surfaces. Thus, the chemical properties of the water that runs off agricultural fields or percolates into groundwater are different from those of the rainwater or irrigation water coming into the fields.

Irrigation

■ Water diversion and pumping has allowed humans to grow crops where natural supplies of water are limited.

The **potential evapotranspiration (PET)** in a region is an estimate of the average amount of water that would evaporate from a hypothetical agricultural field over the course of a year. Like rainfall, PET is measured in centimeters or inches. In many dry regions of the world, PET is much higher than total rainfall. In these regions, the availability of water is a major factor limiting the productivity of agroecosystems. Even where annual rainfall exceeds PET, water deficits may occur during seasons when there is little rainfall, high temperatures, and low relative humidity, which increases the rate of evaporation.

Irrigation has increased crop production in many regions where water deficits are common. Irrigation has a long history, dating back over 5,000 years to farms in Mesopotamia and Egypt. As pipelines and pumps became more sophisticated, humans were able to transport water from sources that were farther and farther away. The Romans developed remarkable aqueducts and complex canal systems to transport agricultural water; many of these remain in use to this day.

Irrigation has played a significant part in the Green Revolution of the past six decades. It has improved the productivity of existing croplands and allowed the cultivation of many lands that were previously thought to be unsuitable for crops. Recently, humans have gone to great lengths to transport water from places where it is abundant to places where it is not. The Central Arizona Project, for example, carries water from the Colorado River up and down hills for 540 km (336 mi) to irrigate agricultural fields in central Arizona (Figure 12.28). Indeed, crop irrigation now accounts for about 70% of all human water use and is steadily increasing (Figure 12.29). In the United States, water pumps for crop irrigation account for nearly 20% of the energy used in agriculture.

▶ Figure 12.28 **Irrigating the Desert**
The Central Arizona Project is an 80-ft-wide concrete-lined canal that supplies water to central Arizona farms and to the rapidly growing desert cities of Phoenix and Tucson.

IRRIGATED ACREAGE IN UNITED STATES

(line graph: y-axis "IRRIGATED LAND (millions of acres)" from 30 to 60; x-axis "YEAR" from 1970 to 2005)

◀ Figure 12.29 **Increasing Irrigation**
The distinctive circular pattern of center pivot irrigation is visible from an airplane in this southwest Utah landscape. The area of irrigated land in the United States has increased by 32% from 1969 to 2007 (inset).

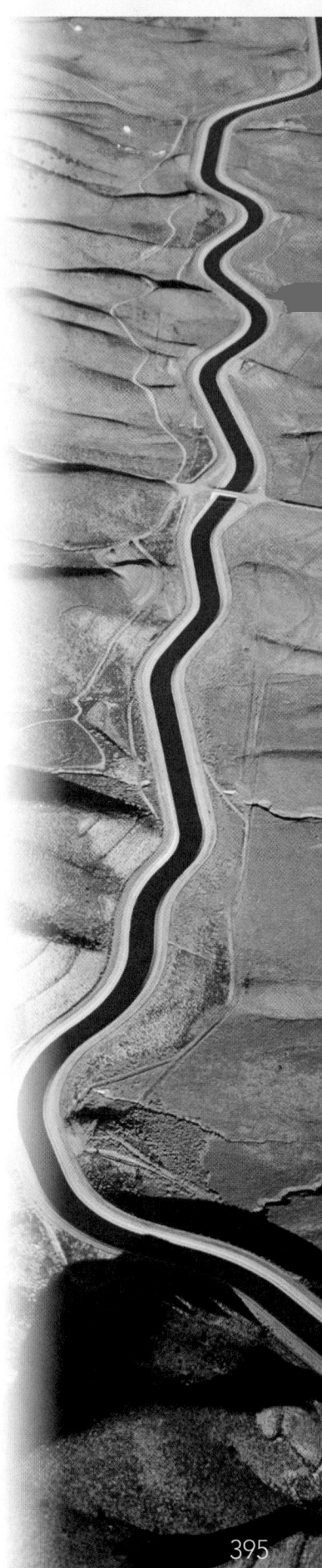

395

▲ Figure 12.30 Aerial Irrigation
Although aerial spraying distributes water evenly, it is inefficient because much of the water evaporates before reaching the plants Ⓐ. The LEPA system reduces the spray distance to the ground and therefore conserves irrigation water Ⓑ.

Where surface water is unavailable or is consumed by other demands, crops may be irrigated by pumping groundwater. Such irrigation has been especially important in the High Plains region of the American West. In many places, the rate at which aquifers are recharged is considerably less than the rate at which water is being pumped for irrigation (see Module 11.3).

Throughout history, the most common form of irrigation has been surface irrigation, in which gravity moves water through canals, across fields, and between furrows. The energy costs for this form of irrigation are comparatively low. However, surface irrigation often distributes water across fields unevenly and is not effective on complicated topography.

Today, overhead spraying is widely used to put water onto fields (Figure 12.30). Overhead spraying distributes water across fields more evenly than surface irrigation but requires more energy to pump water. Water sprayed from overhead evaporates more rapidly than water flowing along the soil surface. As a result, overhead spraying often decreases the efficiency of water use. Low-pressure spray sprinklers were introduced in the 1970s followed by low energy precision application (LEPA) systems in the early 1980s. These water conservation devices are mounted on drop-down tubes that decrease the sprinkler distance to the soil surface and use water more efficiently (see Figure 12.30).

In addition to its impact on the water supply, irrigation presents other environmental challenges. Most conventional methods of irrigation apply more water to fields than can be held by the soils or used by plants. The excess irrigation water drains into streams and rivers or percolates into groundwater, carrying with it the chemical residues from fertilizers and pesticides.

Where evaporation rates are high, the salts contained in irrigation water are concentrated and crystallize on the soil surface. This process is called **soil salinization**. Over many growing seasons, irrigated lands accumulate more and more salts. Eventually, salts brought in with irrigation water build up to the point that they interfere with plant growth. This is a long-standing problem that is difficult to solve. Arid farmlands rarely have access to sufficient irrigation water to wash these excess salts out of the root zone.

Conserving Water in Agroecosystems

■ New irrigation technologies can provide significant water savings.

As energy costs and competing demands for limited supplies of water increase, the need to conserve water resources in agroecosystems has become clear. To increase water use efficiency, decrease runoff, and minimize salinization, many farmers are turning to drip irrigation. In **drip irrigation** systems, pipes with small openings at the base of each plant feed water to the root zone, avoiding the dry air above the field (Figure 12.31). This approach not only reduces the amount of water wasted by evaporation but also reduces the rate of salinization, because less evaporation means that less salt is left behind in the soil.

An even more efficient approach to farming in arid regions is to plant crops that demand less water. Plant breeding programs have developed new genetic strains of many common crop plants that are able to grow on reduced amounts of water.

Many agronomists suggest that entirely different plant species that are adapted to arid lands should be substituted for water-demanding species, such as corn and soybeans. For example, jojoba (pronounced ho-HO-ba) is a large shrub that thrives in the Sonoran desert of Arizona, southern California, and northwestern Mexico. Jojoba produces edible seeds from which fuel-quality oil can be extracted. Thus, jojoba produces products that are similar to corn and soy but requires virtually no irrigation.

QUESTIONS 12.5

1. What soil factors determine the amount of water that is available to crop plants?

2. Why does irrigation cause soil salinization in arid regions?

(MES) For additional review, go to **MasteringEnvironmentalScience**

► Figure 12.31 Efficient Water Use
Drip irrigation systems deliver water directly to the roots of crop plants through a network of tubes. These systems use water 5–10 times more efficiently than conventional irrigation systems.

12.6 Livestock in Agroecosystems

BIG IDEA Livestock represent a significant portion of agricultural operations, both in industrialized and developing nations. Humans use nearly 20% of Earth's land as pasture to support domestic animals. In addition, nearly 30% of cropland—another 5% of Earth's land—is used to produce hay and grain to feed livestock. Trophic-level efficiency, the ability of domestic animals to convert their food into food that humans can consume, varies among animal species. It also varies within species, depending on the source of food and level of activity. The environmental impacts of raising large numbers of livestock include air and water pollution, significant emissions of greenhouse gases, and the spread of disease.

Trophic-Level Efficiency

■ The amount of energy that livestock provide to humans depends on the species, the quality of their food, and the activity level of the animals.

With the exception of dogs and cats, most domestic animals are herbivores. Much of the diet of herbivores is composed of cellulose, a polymer of glucose that is not easy for most animals to digest (see Module 3.2). As a consequence, the trophic-level efficiency of herbivores in a food chain is generally lower than that of carnivores.

The trophic-level efficiency of domestic species found in agroecosystems varies considerably, depending on how efficiently their digestive systems manage cellulose. Among the livestock typically used in food and fiber production are cattle, sheep, goats, and pigs from the mammalian order Artiodactyla. These herbivores have complex, multichambered stomachs that harbor symbiotic microbes that assist in the decomposition of cellulose. By comparison, horses and their kin—from the mammalian order Perissodactyla—have much simpler digestive systems that are less efficient at using cellulose (**Figure 12.32**). For this reason, an area of pasture can support about twice as many cattle as horses. Because of the inefficiency of their digestive systems, horses are usually fed a grain-rich, low-cellulose diet.

The source of food used for livestock also influences trophic-level efficiency. Cattle raised on wild rangeland typically have trophic-level efficiencies well below 10%. Cattle raised on pastures that have been planted with more digestible grasses and herbs have higher trophic-level efficiencies. The highest efficiencies are found in cattle that are fed grains, such as corn and oats. Chickens do not have especially efficient digestive systems, but they have high trophic-level efficiencies because they feed almost entirely on grain.

Finally, the trophic-level efficiency of domestic animals is strongly influenced by their degree of physical activity. Like humans who exercise often, domestic animals that can move about freely burn much of the energy in their food in respiration. Many nutrition experts consider the leaner meat from such animals to be a healthier food choice than the fattier meat from animals that are raised in confined conditions.

Conversion efficiencies are somewhat more relevant to human consumers than trophic-level efficiencies because they measure grams of *edible* meat produced per gram of feed input. Conversion efficiencies are significantly lower than trophic-level efficiencies, particularly for beef (**Figure 12.33**).

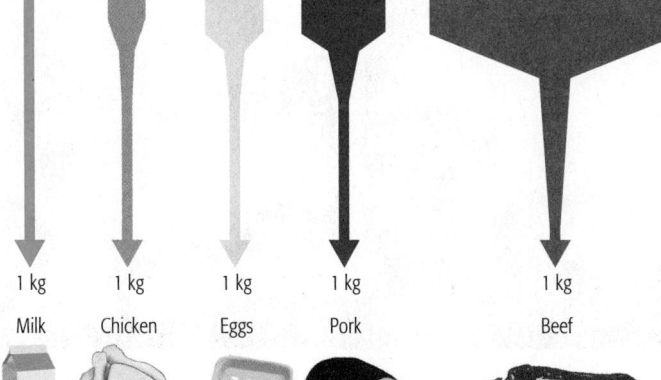

FEED INPUT

1.1 kg	2.8 kg	4.5 kg	7.3 kg	20.0 kg

1 kg	1 kg	1 kg	1 kg	1 kg
Milk	Chicken	Eggs	Pork	Beef

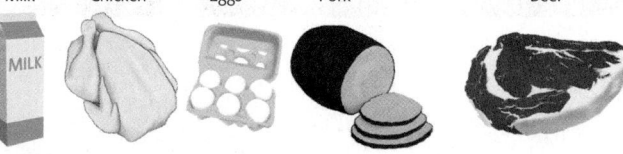

PRODUCE OUTPUT (edible weight)

▲ Figure 12.32 **Comparing Digestive Systems**
A cow softens high-cellulose material in the first of the four chambers of its stomach and then re-chews this cud, enabling it to fully digest cellulose. A horse, by comparison, has a relatively simple digestive tract that is not very efficient at digesting high-cellulose plant material.

◄ Figure 12.33 **Comparing Conversion Efficiencies**
Domestic animals differ in their conversion efficiencies. In this diagram the breadth of the arrows depict the relative amount of feed needed to produce a single kilogram of a particular food. Cows require the greatest feed input per kilogram of edible weight, pork is intermediate, and chickens are most efficient.

Environmental Impacts

■ Managing animal waste is a significant challenge, but it can be an opportunity too.

Since the 1960s, industrial agriculture has shifted toward higher efficiency mass production of domesticated animals in confined animal feeding operations (CAFOs). This technique presents a number of challenges to the environment and human health. These include issues related to waste production, changed patterns of land use, and the transmission of diseases.

Waste production. The wastes produced by CAFOs can have significant environmental impacts. According to the Environmental Protection Agency, the amount of waste generated by 700 dairy cows is equivalent to the septic waste produced by a city of 10,000 people. Hog lots and poultry operations also produce large quantities of waste. In the United States, most states have regulations regarding the management of waste for large numbers of animals raised in confined conditions. In many cases, however, such management is far less stringent than the management of human waste.

When exposed to the weather, animal manures release inorganic nutrients, such as nitrates, ammonia, and phosphates, that can pollute surface waters and nearby streams. The decay of livestock waste also releases a variety of foul-smelling organic gases. Typically, wastes from cattle feedlots and the confined feeding facilities for swine and poultry are stored in open lagoons (Figure 12.34A). Such lagoons may leak or fail completely, particularly during rainy periods. Wastes are allowed to decompose over a period of months, after which they are often distributed as fertilizer onto cropland or pasture. Once again, excess nutrients often find their way into nearby streams.

Methane gas is a by-product of the digestion of nearly all herbivores. Methane is a powerful greenhouse gas that is 23 times as effective as CO_2 at trapping heat in the atmosphere (see Module 9.3). The Intergovernmental Panel on Climate Change (IPCC) estimates that methane emissions from domestic livestock account for more than 5% of current global warming.

Methane, which can be burned like natural gas, is also a potential source of energy (see Module 14.4). Realizing this, managers of many domestic animal facilities are beginning to install waste treatment systems that capture methane and use it to generate electricity (Figure 12.34B). Such facilities have the added benefits of providing more secure waste confinement and releasing fewer odors than conventional lagoons.

Land use. Livestock operations often cause major changes in land use. In many developing nations, forests are being cleared to create pasture for sheep, goats, and cattle. Such deforestation contributes to climate change, reduces watershed protection, and threatens biodiversity. As human populations that rely on herd animals grow, there is a need for larger herds. As a result, more and more animals are concentrated in a fixed area of land, resulting in overgrazing. Under crowded conditions, sheep, goats, and cattle degrade pastures by stripping vegetation from the ground and exposing soils to erosion.

▼ Figure 12.34 **Energy from Waste**
In many livestock operations, animal waste is simply pumped into lagoons Ⓐ, where it decomposes. As an alternative to lagoons, waste can be processed in containers that trap methane gas, which can then be burned to generate electricity Ⓑ.

Transmission of diseases. As humans and their domestic animals have coevolved, so have the disease organisms that infect them. Livestock such as sheep, cattle, and pigs are alternative hosts for parasites that can also afflict humans, such as flukes and tapeworms. Forms of the bacterium *Escherichia coli* that can cause deadly infections in people grow in the intestines of cattle, goats, and pigs with no ill effects. Viruses that are responsible for various forms of flu can originate from populations of swine or domestic birds (Figure 12.35).

Confined animals easily spread disease from one individual to another. To reduce outbreaks of these diseases and to maximize animal growth rates, antibiotics are often added to animal feed. Because low levels of these drugs are administered routinely, the only strains of bacteria that can thrive in the poultry house or feedlot are those that are resistant to antibiotics. In the absence of competition with drug-susceptible strains, the resistant bacteria spread throughout the animal population. Thus the animals become a source of antibiotic-resistant bacteria. These bacteria may spread to human food supplies when the animals are slaughtered; they may also spread to surface waters when animal manure washes into rivers and streams. Most of these bacteria do not cause human disease, but some, such as *E. coli*, can cause

severe problems. Diseases caused by antibiotic-resistant bacteria are a special concern because traditional antibiotic therapies may not work against them.

The impact of human diets. Some argue that a vegetarian diet would eliminate environmental consequences of livestock. Others argue that meat and animal products such as milk, cheese, and eggs have long been a part of most human diets and are important sources of protein and other nutritional elements. In regions that are too arid, hilly, or cold to support row crops, domestic animals are vital to human subsistence. If managed properly, livestock can facilitate nutrient recycling from crop residues back to fields, and grazing animals can be an environmentally sensible means to control weeds. However, high-density animal facilities, where thousands of animals are raised in confined spaces, present significant challenges to sustainable management.

In general, as societies develop economically, the proportion of their population's total diet that comes from meat and animal products increases. Thus, the economic development and human population growth expected over the next several decades suggest that the management of domestic livestock will continue to be important.

▲ Figure 12.35 **Domestic Animals and Human Health** Humans share many diseases with their domestic animals. For example, some strains of flu originate in poultry that are raised and sold in unsanitary conditions.

QUESTIONS 12.6

1. How does a domestic animal's digestive system influence its trophic-level efficiency?

2. Describe three important environmental challenges associated with livestock production.

(MES) For additional review, go to **MasteringEnvironmentalScience**

12.7 Managing Genetic Resources

BIG IDEA With industrial development, humans have become dependent on fewer and fewer species of crops and domestic animals. While selective breeding, hybridization, and cloning have produced highly productive strains of domestic plants and animals, they have also reduced the genetic diversity within some species. This genetic homogeneity increases the risk that disease or environmental change could jeopardize human food supplies. Older varieties of crops and livestock have greater genetic diversity; that diversity may be important for future adaptation. Genetic engineering holds promise for improving the rate of production, nutritional value, and disease resistance of some crops. However, care must be taken to avoid adverse environmental and health impacts that could arise from these genetic techniques.

Genetic Diversity and the Stability of Agroecosystems

■ Deliberate efforts to enhance crop diversity are creating a more stable food supply.

Plantings of a single crop species are called **monocultures**. The practice of planting multiple crop species in the same field is called **interplanting** or **polyculture**. Just as in natural ecosystems, increasing the diversity of crop plants in an agroecosystem can increase the ecosystem services that it provides. Differences in root systems of interplanted crops can result in more efficient uptake of water and nutrients and increased productivity. Furthermore, the spread of insect pests and fungal diseases is retarded in polycultures compared to monocultures.

The diets of people in industrialized societies include foods from only a few of the thousands of plants that have been used for food throughout human history. In the United States, most of the items on the shelves of grocery stores are from only six commonly cultivated plants: corn, wheat, rice, potatoes, soybeans, and barley.

In addition, modern plant breeding has reduced the genetic diversity within the few species that are grown. Today, agronomists use hybridization, asexual propagation, and cloning to produce crop plants that are genetically similar, or even identical, to one another. The goal of this standardization is to maximize production and ensure homogeneous growth. The high production that comes from this homogeneity has contributed to the greatly increased food production of the Green Revolution. However, this homogeneity also can lead to serious problems when diseases or other unanticipated threats arise.

The story of the Irish potato famine illustrates the danger of relying on a single source of food. In the early 1800s, potatoes became the primary food source for nearly all of the peasant farmers of Ireland. In 1845, a rapidly spreading fungal blight attacked the Irish potato crop, creating a terrible famine across the country. One million people starved to death. Two million people fled from Ireland, many of them immigrating to the Americas during the famine years between 1845 and 1855.

Reliance on a few crop species is unnecessary because farmers around the world continue to cultivate a rich variety of food plants, and research into new crop species with great agricultural potential is ongoing. Grain amaranth (also called quinoa) and the winged bean of Southeast Asia are two examples of rediscovered crop plants with exceptional promise (Figure 12.36).

Diversity within varieties and genetic strains of crops is also important. In the 1970s, for example, corn plantings in the American Midwest were so genetically homogeneous that a fungus called the southern corn leaf blight raced through field after field of identically susceptible corn stalks. Millions of bushels of grain were lost. Although other corn hybrids were resistant to the blight, too many square miles of farmland had been planted with a single, susceptible strain. A more biologically diverse planting strategy that incorporated many crop varieties would have afforded greater protection against the blight.

Like any population of organisms, an agricultural crop depends on genetic diversity to adapt to changing environmental conditions. The strategy of planting crops that are genetically diverse helps slow the spread of routine diseases and also helps prevent total crop failure within a region. Diseases tend to attack particular genetic strains of a particular plant species. When susceptible strains are separated by other types of plants, diseases cannot spread as easily from field to field. In agriculture, as in the rest of nature, biological diversity promotes ecosystem stability.

◄ Figure 12.36 **Alternative Crops**
Grain amaranth, a staple food of the ancient Aztecs, is high in carbohydrates and protein. Agronomists are interested in expanding the use of amaranth plants because they are resistant to the diseases and pests associated with other crops.

Fungal disease in bananas is another example that illustrates the ongoing consequences of low genetic diversity in crops. Most of the bananas that are grown for export around the world belong to a single variety, the Cavendish. This variety is very productive, and its bunches simultaneously ripen to uniform yellow color. Because Cavendish bananas were bred to be seedless, they cannot be sexually propagated. New plants are started from cuttings, so every plant is genetically identical to all others. Because Cavendish bananas have no genetic diversity and because so many acres are planted in this one variety, the banana crop is vulnerable to the same kind of disaster that befell the U.S. corn crop in 1970.

A deadly fungus known as race 4 Panama disease has now begun to spread through Cavendish banana plantations around the world. Affected plants wilt and die. Spores of the fungus spread in the wind and remain in the soil, where they can attack future plantings. This new strain of Panama disease threatens Cavendish banana plantations worldwide and could mean the loss of the banana most Europeans and Americans have come to love. In places where people depend on bananas as a staple food, such as Uganda, this loss would cause economic hardship and hunger.

Fortunately, there are a number of other varieties of bananas, many of which are resistant to the Panama fungus (Figure 12.37). Banana growers often cultivate these other varieties for local consumption but not for export. If the Cavendish can no longer be grown, other varieties are available to take its place. Rather than replacing the Cavendish with a single variety, it would be wise for growers and consumers around the world to rely on a greater diversity of bananas.

Until the last century, there was great genetic diversity among most cultivated plants. These "heirloom" varieties may not be ideally suited for today's agricultural economy, but they retain genetic variations that may be needed in the future. To preserve the option of incorporating genetic diversity in future crops, the rich variety of these older vegetables, grains, root crops, and orchard trees must be protected. To preserve heirloom crops, seed banks have been established in the United States and in other countries. Seed banks store the seeds from heirloom crops in cold, dry conditions that can keep them viable for many years (Figure 12.38).

Heirloom breeds of animals, such as Icelandic sheep and bourbon red turkeys, are also valuable to the preservation of agricultural diversity (Figure 12.39). Hobby breeders and small-scale farms provide an international service by maintaining breeds of domestic birds and animals that would otherwise be allowed to go extinct. Heirloom breeds may provide much-needed genetic diversity if disease or environmental change threatens the limited variety of breeds used in large farming operations today.

▲ Figure 12.38 **Storing Genetic Variation for the Future**

A technician at the Kew Botanical Garden's Millennium Seed Bank in the United Kingdom checks seeds stored at −4 °F. Seeds stored in cold, dry conditions remain viable for many years. In the future, these seeds may provide the genetic resources needed to improve crop plants.

▲ Figure 12.37 **Banana Varieties**
Though only a few banana varieties are grown for international export, small farms in the tropics grow many different varieties. This genetic diversity ensures that we shall continue to enjoy bananas in the future.

▲ Figure 12.39 **The Cost of Genetic Homogeneity**
Nearly all domestic turkeys are of the same white-feathered variety Ⓐ, which has been selected for its high production of breast meat and rapid growth in crowded conditions. These turkeys must be artificially inseminated, hatched in incubators, and raised in protected houses. Bourbon red Ⓑ is an heirloom breed that retains the ability to fly, to find its own food, to mate, and to rear offspring.

Genetically Modified Organisms

■ Genetic engineering of domestic plants and animals can improve yields but also is controversial.

Since the dawn of agriculture, humans have used artificial selection and crossbreeding to create new genetic combinations in plants and animals. Over time, this has created the great majority of domesticated plants and animals grown today. However, this form of breeding is limited to selecting among the traits that already exist within a species or within closely related species that can be crossbred.

With the development of modern genetic technology, it has become possible for geneticists to develop new varieties by taking a gene from the cells of one species and inserting it into the cells of another. The new varieties have a mixture of traits not achievable with more traditional methods of genetic improvement. An organism whose DNA has been altered in this way is a **genetically modified organism**, or **GMO**.

In principle, GMO technology is similar to traditional methods of crop breeding and animal husbandry. The difference is that molecular biology now allows DNA from any organism, even human DNA, to be inserted into the genetic code of a crop plant or farm animal. The speed and precision of genetic modification using the new molecular techniques give agricultural scientists unprecedented power to change the species we use for food. The adoption of this technology has been rapid: By 2012, the acreage of GMO crops around the world had increased 100-fold to 170.3 million hectares since their introduction in 1996 and now compose 90–93% of all cotton, corn, and soy grown in the United States (Figure 12.40).

One of the earliest applications of GMO technology was a genetically engineered variety of rice. Rice is a staple crop that is especially important in Southeast Asia. In the poorest communities of that region, rice may provide up to 80% of dietary calories. Although rice is nutritious, its grains do not contain beta-carotene, a yellow pigment that the human body converts to vitamin A. Because vitamin

▲ Figure 12.41 **A Golden Opportunity**
Unlike the grains of normal rice Ⓐ, the grains of golden rice Ⓑ contain beta-carotene, which colors them yellow. The plants of golden rice have been genetically modified to produce seeds that contain beta-carotene, an essential nutrient for humans.

A is essential for healthy skin and vision, insufficient dietary beta-carotene can lead to skin disorders, blindness, and other health problems. This is the deficiency the geneticists wanted to address.

Beta-carotene is made in the leaves of rice plants but is not normally stored in the edible part of the seed. In 1999, scientists inserted into the genetic material of an existing variety of rice two genes that control the production of beta-carotene, thereby creating a genetically modified plant called golden rice. Golden rice grows as any other variety of rice and provides as many calories, but it offers the additional nutritional value of beta-carotene. Beta-carotene lends its yellow color to the rice grain, giving the variety its name (Figure 12.41).

GMO technology also led to the development of several crop plants, including corn, potato, soybeans, and cotton, with the capacity to produce an insect-killing compound that is normally found only in bacteria. The bacterial source of the gene is *Bacillus thuringiensis*, or Bt, so the plant is called Bt corn. *B. thuringiensis* causes disease in insects, attacking the lining of their gut with a protein toxin. This toxin is not soluble in the human stomach, so it does not harm people. Organic gardeners have long used spores of *B. thuringiensis* to prevent insect damage in their gardens. Geneticists have been successful in inserting the gene for this toxin into the DNA of corn and several other crop plants. These genetically modified plants produce the toxin in all of their tissues. Insects feeding on Bt crops are poisoned, so there is no need to apply additional pesticides.

Another effective use of GMO technology confers resistance to the herbicide glyphosate (know better by the trade name Roundup®). "Roundup Ready" cotton, corn, soy, sugar beets, alfalfa, canola, and others all have a gene that allows these crops to survive the application of Roundup until a certain stage of life. Farmers can then use glyphosate to eliminate competing weeds that are not resistant to the effects of the herbicide.

▼ Figure 12.40 **Genetically Modified Crops Crowd the Fields**
GMO corn, cotton, and soybeans have crowded out all other varieties in the United States. More than 90% of these plants are now genetically modified, mostly for resistance to pests (BT) and herbicides (HT).
Source: Data from USDA, Economic Research Service and National Agricultural Statistics Service.

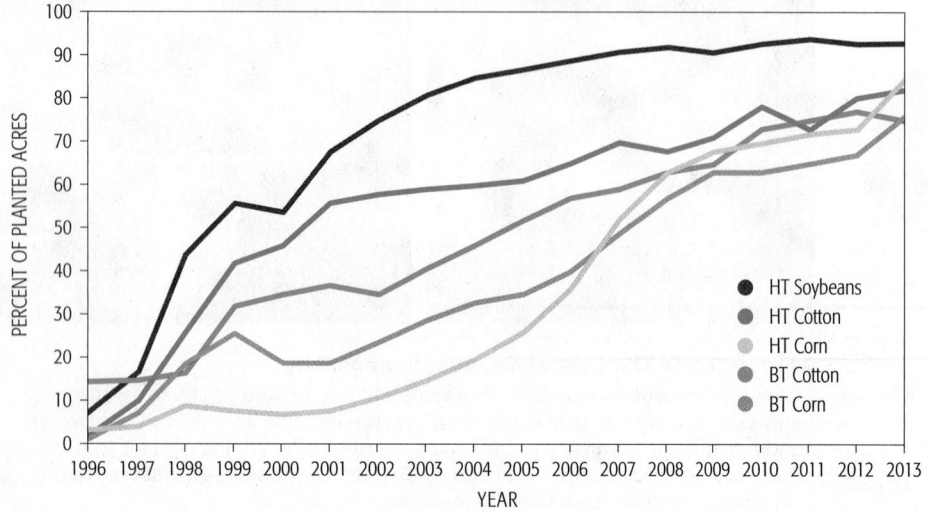

ADOPTION OF GENETICALLY ENGINEERED CROPS IN THE UNITED STATES, 1996–2013

- HT Soybeans
- HT Cotton
- HT Corn
- BT Cotton
- BT Corn

PERCENT OF PLANTED ACRES

YEAR

Genetic tools have also been used in efforts to genetically modify farm animals. Scientists at the Baylor College of Medicine in Houston found that genes for human growth hormone introduced into the chromosomes of pigs cause the animals to grow larger and mature more rapidly. As they mature, these genetically modified swine consume 25% less food and generate less manure than traditional breeds; thus there are potential environmental benefits to growing these genetically modified pigs.

Some people embrace GMO technology for its potential to produce more food at lower costs and to reduce environmental impacts. Others wonder about the nature of the foods they are eating and the long-term effects on the diversity of life. There are also persistent questions about the safety of meats modified by human hormones, of plant foods with modified biochemistry, and the environmental side effects of genes moved from one species to another. American consumer groups are demanding labeling of GMO foods (Figure 12.42). In response to consumer concerns, the supermarket chain Whole Foods Market became the first major food retailer in the United States to require labeling of all genetically modified foods sold in its stores. Announced in 2013, they gave suppliers five years in which to implement the labeling. Worldwide, many of the 28 member states of the European Union have banned or greatly restricted imports of American crops produced with GMO technology.

Some negative side effects of GMO have been demonstrated. Pollen from Bt corn can limit the growth of butterfly larvae, although direct effects on butterfly populations have not been demonstrated. A growing number of weeds have developed resistance to Roundup, and the initial reduction in the volume of herbicides used to control competing weeds has been reversed. Additionally, some concerns surround possible adverse health effects and allergenic impacts of GMO foods. For example, testing of a soybean engineered to produce the essential nutrient methionine was shown to create an allergic reaction in blood samples of people known to be allergic to nuts. The gene inserted into the soy DNA had been transferred from a Brazil nut, which inadvertently had brought with it the nut's allergenic qualities.

Most of the objections to GMO foods, however, are based on more philosophical concerns. For example, some people question the right of humans to change the fundamental nature of other species. Others question the ethics of eating animals that have been artificially endowed with human genetic material.

Like any powerful technology, genetic modification of plants and animals conveys both great promise for improvements and serious concerns for health and safety. Although some fears may be overblown, public concerns in the United States and abroad have generated careful examination of the potential good and the potential problems that could arise from biotechnology in agriculture.

▲ Figure 12.42 **A Right to Know**
Protestors outside the headquarters of the FDA's Center for Food Safety and Applied Nutrition in College Park, Maryland, made their feelings clear with this sign. They joined the call for the labeling of all GMO foods.

New FRONTIERS

Patenting Life

We are entering a new frontier with the dawn of genetically modified crops. These living organisms have been developed through years of intensive research and financial investment. They also disperse their newly engineered genes into the world through pollen, carried by wind and beyond human control. To whom do the genes belong? Currently, the agricultural company Monsanto owns the patents to the majority of GM crops in the United States, including all of their genetic material.

Farmers growing non-GM crops near GM crops face the threat of cross contamination by GM pollen. If pollinated in this way, the seeds from the non-GM crop will include Monsanto's genetic material. Monsanto has filed lawsuits against farmers for patent infringement when they found seeds with Monsanto's genetic material in them. On the other hand, organic farmers have filed lawsuits against Monsanto for contaminating their non-GM crops with GM pollen.

How can Monsanto protect its investment in the development of GM crops when their patented product escapes from those who purchased it?

How can organic farmers protect their non-GM crops from contamination by GM pollen from neighboring fields?

What guidelines might be developed to reduce conflict between these parties and protect both of their interests?

And will the U.S. Supreme Court's recent judgment against a company claiming a patent for a human gene (*Association for Molecular Pathology et al. v. United States Patent and Trademark Office et al.*) change the legal landscape for agricultural research?

QUESTIONS 12.7

1. Why is genetic diversity within crop species important to sustainable agriculture?

2. How does GMO technology differ from traditional methods of plant and animal breeding?

(MES) For additional review, go to **MasteringEnvironmentalScience**

12.8 Managing Competitors and Pests

BIG IDEA Weeds, pests, and plant diseases have plagued farmers since the beginning of agriculture. Scientific studies of eight of the world's most commonly cultivated crops indicate that 42% of all agricultural productivity is consumed by pests and diseases. This is true in spite of all the pesticides and pest control practices used to protect crops. Early pesticides killed a wide variety of organisms and often led to the evolution of pesticide-resistant insect pests. Today, agronomists try to use pesticides that break down quickly and affect a narrow range of species. They also time the application of pesticides carefully in order to have the greatest effect. To avoid the undesirable effects of pesticides, many farmers now use biological controls, such as organisms that parasitize or prey on pests.

Chemical Pest Control

■ Pesticides have been both a salvation and a threat to humanity.

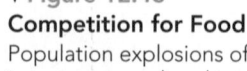

▼ **Figure 12.43**
Competition for Food
Population explosions of insect pests such as locusts Ⓐ have destroyed crop plants since the beginning of agriculture. Weevils Ⓑ attack stored grain. The blackened grains on these wheat spikes Ⓒ are infected with a fungus called ergot; this fungus produces an alkaloid chemical that can be quite toxic to humans.

A variety of organisms in an agroecosystem compete with humans for the energy in crop plants. Specialized insects attack every part of the growing plants. Wireworms attack roots, cutworms eat stems, caterpillars and locusts eat leaves, beetles munch on flowers, and fly larvae bore into fruits. As a result of these pests, total production is reduced and fruit and vegetables look unappetizing. In some cases, insects can cause entire crops to fail. Outbreaks of pests, such as the locust plagues described in the Hebrew Bible, have long been associated with famine in agricultural societies (Figure 12.43).

Even after crops are harvested and stored, pests continue to eat away at the human food supply. Losses of stored wheat, rice, and legumes are of particular concern in tropical countries, where grain storage facilities are often inadequate to protect crops. Moist tropical climates allow pests to flourish all year long. Insects, especially grain beetles, are a constant problem. Rats are even more destructive, since they can chew through plastic and paper containers to get at stored foods. If moisture is not adequately controlled, molds can grow. Molds reduce the palatability of crops and may also produce toxins that threaten humans with disease.

Since ancient times, farmers have used various measures to control insect pests. In the mid-20th century, the developments in chemistry brought a variety of powerful pesticides to agriculture. Foremost among these developments was dichloro-diphenyl-trichloroethane, or DDT, an artificially synthesized, oil-soluble toxin.

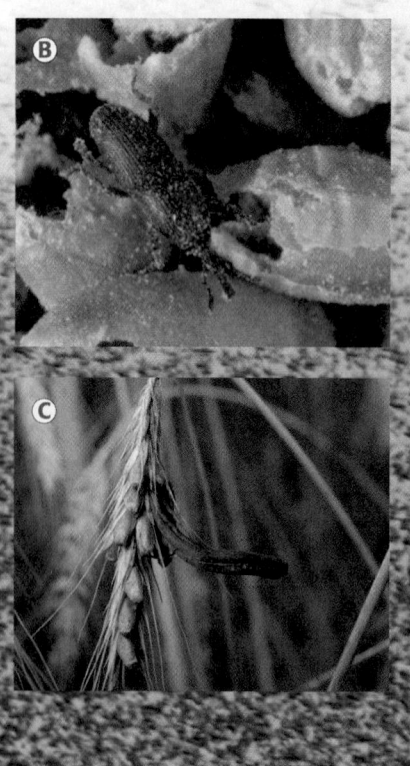

DDT belongs to a class of chemicals called chlorinated hydrocarbons, so named because chlorine is attached to a chemical framework of carbon and hydrogen atoms.

DDT is a **broad-spectrum pesticide**, meaning it kills a wide variety of insects. In the 1950s, it was discovered to be much more toxic to insects than to humans, so scientists concluded it could be used to control pests with little harm to human populations. It was, therefore, enthusiastically adopted for many uses. Often it was applied by aerial spraying, in which low-flying airplanes released the pesticide over wide areas. Public health agencies sprayed DDT on marshes and around swimming pools and parks to control the spread of mosquito-borne diseases, such as malaria. Farmers sprayed DDT on fields to control crop pests, and foresters sprayed woodlands to control destructive caterpillars. DDT was readily available to homeowners to kill cockroaches, flies, and other household pests.

However, DDT is a long-lived compound that is not easily broken down in soil or water. As a result, it continues to kill insects for a long time after its application. Since it easily attaches to body fats and does not break down in the body, DDT is stored within the body. All the DDT consumed over the lifetime of an animal remains in the animal's fatty tissues. Thus, as one organism eats another, it consumes tissues laden with DDT. The higher an organism is in the food chain, the greater its exposure to DDT. The process by which the concentration of a pesticide increases through a food chain is called **biomagnification** (Figure 12.44).

The negative consequences of the wholesale application of DDT were not noticed until wildlife began to succumb to its long-term effects. Birds, particularly those such as eagles and hawks that are at the top of the food chain, were the first indicators that DDT was poisoning the entire food chain. In some cases, DDT killed the birds outright. In other cases, high levels of DDT in the bodies of adult birds made their eggshells so thin that they could not reproduce.

Author and naturalist Rachel Carson became alarmed by the loss of birds and began investigating other signs of ecological trouble. Her best-selling book, *Silent Spring*, alerted the public to the dangers of the indiscriminate use of toxic chemicals in agriculture. She warned that DDT might exterminate birds from the natural world, resulting in a "silent spring" devoid of their songs. As a result of Carson's work, the United States banned the use of DDT in 1972.

Carson also pointed out that broad-spectrum pesticides can be counterproductive in the long run. As a result of natural selection, pests develop adaptations that allow them to live in their chemically altered environment. For example, when DDT is first sprayed on a wetland for mosquito control, most of the mosquitoes die. The only mosquitoes that survive are those that happen to carry genes conferring some tolerance to the pesticide. These resistant mosquitoes reproduce, producing more resistant mosquitoes. Over time, the mosquito population becomes increasingly resistant to DDT. Eventually, the pesticide becomes useless against its intended target.

DDT concentration: increase of 10 million times

DDT in fish-eating birds 25 ppm

DDT in large fish 2 ppm

DDT in small fish 0.5 ppm

DDT in zooplankton 0.04 ppm

DDT in water 0.000003 ppm

▲ Figure 12.44 **Biomagnification**
When DDT is sprayed on agricultural fields and marshes, it eventually flows into streams, lakes, and oceans. Because DDT is stored in animal tissues and not metabolized, it becomes increasingly concentrated in the tissues of animals at higher trophic levels.

▲ Figure 12.45 **A Little Help from Our Friends**
A praying mantis protects a crop by consuming insects such as this grasshopper.

Broad-spectrum pesticides disturb agroecosystems by killing helpful insects as well as harmful ones. Before DDT was widely used, farmers had little appreciation for the level of pest control exercised by naturally occurring insect predators. Helpful insects, such as robber flies, assassin bugs, and praying mantises, protect crops by consuming large numbers of insect pests (Figure 12.45). After DDT is sprayed on a field, pest species generally breed much faster than predatory insects, so their populations rebound more quickly. Released from the control of their natural enemies, pest populations commonly reach higher levels after spraying than before.

Today, agronomists recognize the long-term effects of DDT and similar pesticides. As a result, they have adopted three strategies to maximize the effects of pesticides while minimizing their environmental impacts. First, agronomists try to use short-lived pesticides that break down quickly in water or soil. Such pesticides do not remain in the environment to kill other organisms after their intended targets have been killed. Second, they try to use pesticides that target particular species or groups of species. Pesticides with greater specificity kill the pest but not other organisms, so they cause fewer damaging side effects. Third, they study pest life cycles carefully to determine the time and place that pests are most susceptible to control measures. If the pest populations are too low to cause significant damage, no chemical control is used.

Biological Pest Control

■ Biologically based methods can control pests without collateral damage to ecosystems.

A detailed understanding of agroecosystem food chains can often lead to biological controls that avoid chemical toxins altogether. Most pests have predators and parasites, and farmers can use these helpful species to consume pests. The use of predators and parasites to manage pests is called **biological pest control**.

For example, braconid wasps are widely used by farmers and gardeners to control populations of leaf-eating caterpillars (Figure 12.46). These tiny parasitic wasps lay their eggs beneath the skin of a suitable host caterpillar. When the wasp larvae hatch, they feed on the live tissues of their host. Caterpillars that are parasitized by braconid wasps fail to mature. Thus the tiny wasp prevents the population growth of these garden pests without harming other species in the environment.

Another mechanism of biological control involves manipulating the chemical communications of pest insects. Many insects use volatile chemical signals, called **pheromones**, to send a message from one individual to another. Pheromones may be used to communicate the location of mates or sources of food. A common garden pest called the Japanese beetle uses pheromones to find both mates and food sources. Traps baited with artificially synthesized Japanese beetle pheromones can lure hundreds of beetles. Beetles attracted to the scent bump into the top of the trap fall through a funnel into a container.

One of the greatest success stories in the realm of pest management involved the release of millions of pest animals. The screwworm fly was once a virulent pest on livestock across much of the southern United States. Beginning in the 1960s, the U.S. Department of Agriculture (USDA) began a program of releasing massive numbers of male flies that had been sterilized by radiation. When wild female flies mated with these sterile males, their eggs were not fertilized and produced no offspring (Figure 12.47). Populations rapidly dropped, and by 1966 the USDA declared the country free of this pest. Since then, this approach has been successfully employed for a variety of other fly pests.

▲ Figure 12.46 **Biological Pest Control**
These braconid wasp juveniles (white) are parasitizing a hornworm caterpillar, a major crop pest that feeds on the leaves of many crop plants, including tomatoes, eggplants, peppers, and tobacco.

▲ Figure 12.47 **Controlling Screwworm Flies**
Screwworms lay their eggs on livestock; when the eggs hatch, their larvae invade the tissues of their host. Populations of this pest have been controlled by releasing large numbers of sterile male flies.

Agroecosystem Management of Pests

■ Control of the crop plant's environment further reduces the need for pesticides.

Good agricultural practice can make the crop environment less suitable for pest species, reducing the need for chemical pest control. Such practices include careful crop inspection and quarantines, crop rotation, maintenance of non-crop areas to support natural predators and pollinators, and integrated pest management.

Quarantines prevent the introduction of exotic pests in the first place. The laws of agricultural states such as Florida, Arizona, and California require fruits and live plants to be carefully inspected before they are brought across their borders. These laws are intended to prevent the importation of agricultural pests, such as the Mediterranean fruit fly, from sub-Saharan Africa, and the citrus canker, believed to have originated in Southeast Asia. If even one of these pests is detected inside the state boundaries, agricultural officials act quickly to isolate it and destroy all fruit or vegetation that might harbor it before the pest is able to spread.

Crop rotation—the practice of replacing one crop species with another on an annual planting cycle—helps to control pest populations because pests tend to specialize in particular crop plants. European corn borers, for example, are insects that feed only on corn plants. The adult corn borer is a moth that lays eggs on corn leaves. Young larvae feed on corn leaves and stems, and older larvae spend the winter in crop debris. If corn is grown in the same field year after year, corn borer populations build up over time. Farmers can easily disrupt this buildup by planting soybeans every other year. When corn borer larvae emerge in the spring, they will encounter young soybeans, a crop they cannot use as food. In the following year, the pests that grew on bean plants will encounter nothing but corn. Thus, crop rotation helps to control all types of specialized insect pests. For insects like the corn borer that require crop residues for shelter during the winter, burning crop wastes or plowing them under can remove residual pests before the crop is planted again, though burning will result in a loss of critical organic matter.

Weeds can be controlled by mechanically disturbing the soil between crop rows or by mulching. Mulching involves covering the ground with plant material or plastic sheeting to prevent weeds from germinating near crop plants. Organic mulches, such as wheat straw and pine needles, have the added benefit of shading the soil and reducing the loss of moisture during hot weather. Organic mulches also degrade over time, releasing nutrients back into the soil.

Farmers can also adopt a landscape approach that provides natural predators and pollinators with uncultivated areas that offer protection and support. This means maintaining non-crop areas planted with trees, shrubs, and grasses that can serve as a habitat for beneficial wildlife. Beneficial predators need a place to reproduce, overwinter, and feed when no crop pests are present.

Chemical, biological, and cultural means of pest control are not mutually exclusive strategies. **Integrated pest management (IPM)** is the effective combination of all three kinds of pest control, carefully designed to manage pest populations while also minimizing environmental damage (Figure 12.48). It is not the goal of IPM to completely eradicate pests. Rather, IPM aims to keep pest populations small enough to minimize crop loss while also controlling the cost of their management. This approach requires extensive knowledge of the pest species, including its life cycle, population levels, and interactions with other organisms in the agricultural ecosystem. IPM methodologies usually must be tailored to the natural enemies, plants, soils, topography, and climate of a particular region. Extensive local experimentation is essential. IPM is a growing movement for the environmentally sound management of row crops, orchards, and even lawns and golf courses.

QUESTIONS 12.8

1. Why is the concentration of DDT magnified in organisms at higher trophic levels?

2. Describe three examples of biological pest control.

(MES) For additional review, go to **MasteringEnvironmentalScience**

Integrated Pest Management

Pest resistant crop species and genotypes

Careful monitoring of pest populations

Cultivation practices that discourage pests

Natural control of pest populations

Prudent use of pesticides

▲ Figure 12.48 **Agroecosystem-based Management**
IPM combines several strategies to minimize impacts of weeds and pests on crop species. This not only reduces negative environmental effects of pesticides, it also saves farmers money.

12.9 The Ecology of Eating

BIG IDEA As omnivores, we humans can choose where to eat on the food chain. From well-established principles of ecosystem ecology, we know that our ecological footprint increases as we shift from an herbivorous to a carnivorous lifestyle (see Module 5.4). Our ecological footprint is also influenced by the place where our food is grown and the amount of energy and materials used to transport, process, and market our food. The choices that individual consumers make about what to eat and how it is produced affect the environment and directly influence human well-being.

The Food Footprint

■ What we eat determines a significant portion of our ecological footprint.

Historically, humans chose diets that reflected the biological resources that were available where they lived. In southern Asia, where the land and climate are suitable for agriculture and farms support dense human populations, many cultures have a largely vegetarian diet. At the other extreme, in Arctic environments where cultivation of crops is impossible, indigenous cultures depend primarily on animal foods. In Lapland, for example, people obtain almost all their food from reindeer, fish, and other animals (Figure 12.49).

In general, as countries and regions develop economically and become more urban, meat and fish make up a larger proportion of their total diet. In most developed countries, food accounts for about 25% of the total ecological footprint. That proportion is greater in countries where people consume lots of meat.

Placing ourselves toward the top of the food chain by consuming more meat and less plant foods poses two potential problems. First, it takes more land to produce a unit of energy in meat than to produce an equivalent

unit of energy in vegetables or fruits. The amount of land required to support one person who gets half of his or her dietary calories by eating meat from domestic animals could instead be used to support two or three vegetarians. The second problem is that eating meat may expose humans to toxins, such as mercury, that undergo biological magnification in the food chain. Poisoning from mercury in tuna, for example, results from humans choosing to eat large carnivores from the sea, thereby placing themselves at the end of a long food chain. (The majority of the mercury is a by-product of coal-burning, see Module 14.3).

The ecological impact of our diet is also influenced by the energy and materials required to transport, process, and market our food. In places where people consume food that comes mostly from local or regional sources, the contribution of food transportation to total ecological footprint is comparatively small, just 1–2%. However, in places where a large amount of food is acquired from distant regions, the cost of transporting food may exceed 10% of the footprint.

▶ **Figure 12.49 Contrasting Diets**

Thai farmers grow a diverse array of fruits and vegetables. Along with rice, these plants make up the majority of food purchased in this floating market in Thailand Ⓐ. In Lapland, where there are few options for agriculture, reindeer herders rely on animals for most of their food and clothing Ⓑ.

A century ago, even in industrialized countries, diets varied with the seasons. Peas, asparagus, strawberries, and many leafy vegetables were available in the spring and early summer. Tomatoes, corn, and squash were summer and autumnal fare. Winter meals typically included lots of cabbage and potatoes. Nowadays, go to the fresh vegetable section of a supermarket in Seattle, Chicago, or Boston, and you are likely to find all of these items, regardless of season.

We enjoy this dietary diversity at a price. The fresh tomatoes we eat in midwinter are either grown in the Southern Hemisphere or in a heated greenhouse (Figure 12.50). In either case, considerable energy from fossil fuels is needed to produce and transport them. Refrigeration and chemicals that retard fruit and vegetable ripening add additional costs in energy and materials. If fresh items are unavailable, canned or frozen forms certainly are available. Such processing can add an additional 20–30% to the energy cost of processed foods compared to the alternative of consuming them fresh, in season, from local sources.

Organic foods.   You can also alter your dietary ecological footprint by considering how what you eat is grown or raised. **Organic foods** are produced to meet strict standards that limit the use of fertilizers and pesticides. In the United States, these standards are set by the USDA, and plant and animal products must be independently certified against these standards to bear the "organic" label (Figure 12.51). Nutritional experts disagree as to whether or not organic foods are healthier than nonorganic foods. Yet there is no doubt that a diet of mostly organic vegetables and meat has a much smaller ecological footprint than a conventional diet. A recent study in Britain revealed that eating a diet of 87% organic vegetables and meats reduced the "food ecological footprint" by 23% compared to an average diet containing about 1% organic foods.

Hidden water.   You may also want to consider the basic resources that go into food production (Figure 12.52). More water goes into making food than just that involved in its washing and preparation. The water footprint encompasses all water inputs in production, including those for crop growth, consumption, and wastewater. Beef is especially water intensive, as nearly 8% of global human water use goes toward irrigation for cattle feed crops alone. The production of a pound of meat protein usually requires at least 100 times as much water as the production of a pound of vegetable protein, and processed foods such as chocolate require much more water. Animal production systems influence the water footprint as well, but the outcome may be complex. For example, grass-fed beef, which are not "finished" in CAFOs, eat primarily grasses rather than water-intensive corn (maize) and therefore utilize significantly less water for their feed. However, industrial beef production takes less than two years from birth to slaughter, and this growth efficiency balances out the more efficient water use of longer-lived grass-fed beef cattle.

► Figure 12.50 **Tasty but Costly**
Tomatoes purchased during cold winter months are likely to have been grown far away or in a hothouse. In either case, considerably more energy is required for their production than is needed to produce tomatoes that are purchased when they are in season locally.

U.S. ORGANIC FOOD SALES

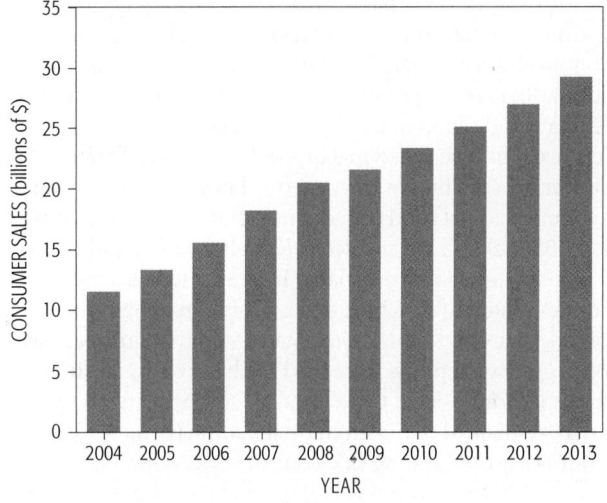

◄ Figure 12.51 **It's Organic**
American consumption of organic foods has increased by a factor of 15 since 1997. To be labeled as an organic food, products must meet strict standards.

GALLONS OF WATER USED IN PRODUCTION

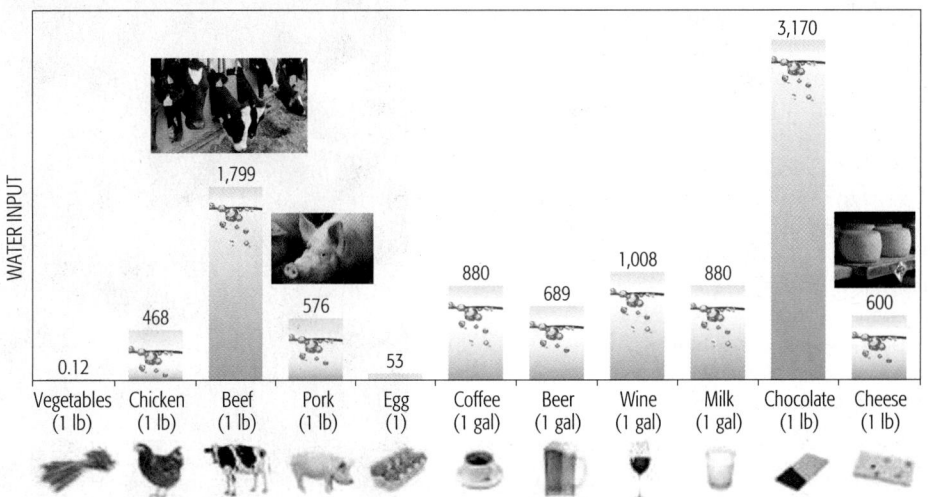

▲ Figure 12.52 **The Hidden Water in Your Food**
Vegetable production requires a fraction of the water needed to grow and process meat. Production of beverages such as coffee and wine are also high in water demand.

Food safety. The methods by which foods are grown and processed have significant consequences for public health. Often, contamination scares have been traced back to toxic strains of *E. coli* bacteria on produce. In a serious spinach contamination case in 2006, in which 4 people died and over 100 were hospitalized, the virulent strain of *E. coli* was traced to cattle feedlots in Salinas, California, located near the fields where the spinach was grown. The bacteria causes no illness in cattle; however, the widespread use of antibiotics in cattle at feedlots has favored the development of strains of *E. coli* that are resistant to antibiotics. These bacteria are passed in the cattle's waste, so the water draining from large cattle feedlots often contains high concentrations of bacteria. Thus, the resistant strain of bacteria was especially common in the water used to irrigate the spinach on the farm. The problem was further exacerbated by placing spinach in closed plastic bags, which provided ideal conditions for bacterial growth.

International shipping of processed food and food commodities has raised concerns as well. In 2007, a substantial amount of processed wheat imported from China was discovered to be contaminated with a toxic fertilizer additive called melamine. In the United States, the melamine contamination caused sickness in animals that ate imported pet foods but did not affect people. However, in 2008 melamine was discovered in baby food and other processed foods that originated in China. Its source was contaminated milk. China exports more than $2 billion of food per year to the United States, and that number has been growing rapidly. Since the FDA has personnel sufficient to inspect only 1% of these imports, any lapse in the food safety systems of our international trading partners raises health concerns among U.S. consumers as well.

Ecological Eating

■ Consumer choices can diminish environmental impacts and encourage sustainable agriculture.

The challenges of sustainable agriculture can be met by improved education and more environmentally conscious public policies. But the most powerful force for change lies in the choices made by consumers. The agricultural economy follows preferences of grocery shoppers and restaurant patrons, so each person exercises choices in the way food is grown, processed, and delivered. People in wealthy countries have the luxury of a great variety of food choices, which in turn influences agricultural practices. With these choices come a greater responsibility to support a sustainable future. Here are some simple approaches for you to consider when making choices about foods.

Reduce food waste. As simple as it may sound, the first step in reducing your food footprint is to cut down on food waste. In the United States, a full 40% of the food produced never gets eaten. This is 10 times as much food waste as produced in Southeast Asia and 50% more than the United States in 1970. Of that food wastage, 11% is thrown out at grocery stores, 33% at restaurants, and a whopping 44% in the home. Consumer choices can significantly alter these numbers.

University eateries with buffet-style dining have discovered that going "trayless" reduces waste because patrons do not take more than they can carry (Figure 12.53). This practice reduces food waste by 20–30%; it also limits the amount of water, chemical detergents, and energy needed to wash the trays.

Where YOU LIVE How can you reduce the food footprint of your campus?

First determine where your food comes from. Is it grown locally and/or with sustainable agriculture techniques? What happens to food waste on campus? Is the food composted or sent to a landfill? Do your dining facilities recycle food packaging? Do they use washable china, flatware, and cups or disposable or compostable takeout containers? Does your campus employ any waste reduction initiatives, such as refills in reusable cups for a reduced price? Now think about what you can do to help.

▲ Figure 12.53 **Trayless Dining**
The simple practice of not offering food trays conserves water and cuts down on food waste in campus dining facilities.

Eat more plant-based foods. By placing yourself toward the bottom of the agricultural food pyramid, you reduce the size of your ecological footprint. Less agricultural land is required to support a vegetarian diet than one based largely on meat and dairy products. The fewer animal foods that are eaten, the fewer tons of grain must be harvested, dried, and transported to feedlots, and the fewer the tons of manure that threaten water quality. Fewer cattle mean that fewer tons of heat-trapping methane are released into the atmosphere. The choice between a plant-based diet and an animal-based diet does not need to be an either-or choice. Shifting even a little of your diet toward plant-based foods makes a difference.

Eat local foods. Find out where farmers are producing foods close to your home. Most cities have farmers markets that specialize in local, in-season produce. You may be surprised at the difference between locally grown fruits and the supermarket varieties, which are sometimes bred more for their capacity for shipping and storage than for their taste and which travel many more miles to get to your plate (**Figure 12.54**). Try participating in community-supported agriculture, or a CSA: You buy a share in the produce of local farmer, which is delivered each week.

Eat fresh foods. Fresh foods, particularly those grown locally, involve less processing and throwaway packaging. They also generally require less energy to move from the farm to the dinner table.

Enjoy foods in season. Learn what is in season in your area, and enjoy each fruit and vegetable in its own time. Out-of-season foods, such as fresh strawberries in the middle of winter, are a luxury purchased at the high cost of the fuels consumed to refrigerate and ship them long distances—sometimes halfway around the world.

Choose sustainably grown products. Choosing products of sustainable agriculture is not usually a significant sacrifice for the consumer but may require some research. Coffee, for example, can be grown in the shade of native trees that continue to provide wildlife habitat, or it can be grown in monoculture stands that replace local trees. Certified organic produce and pasture-raised meats are generally considered more environmentally sound choices as well. Find out which producers are making real efforts at environmental protection. Be wary of green-sounding labels that do not represent real differences in agricultural practice. Although there are strict standards for organic foods, there are no standards for foods labeled "natural."

Ask questions. When dining out, ask about the sources of food. Many restaurants support local growers and select sustainably grown foods. Some items on the menu may be more sustainably produced than others. Environmental publications and websites provide a wealth of information about foods that support or undermine sustainability in agriculture. At the meat counter or vegetable cooler of your grocery, ask where the foods come from. The more you seek information, the more responsive the growers and sellers will be to your desire to support the safety and sustainable production of food.

Q *Are organic foods and liquids really better for us?*
Stanley Uzoka, Kennesaw State University

Lissa: **A** The health benefits of organic foods are debated, but there is no question that their production has a much lower impact on the environment.

▼ **Figure 12.54 Eat Local**
This chart shows the food miles traveled by produce shipped to a grocery store in Statesboro, Georgia, as compared to travel to the local farmers market. The average America meal travels 1,500 miles from farm to plate.

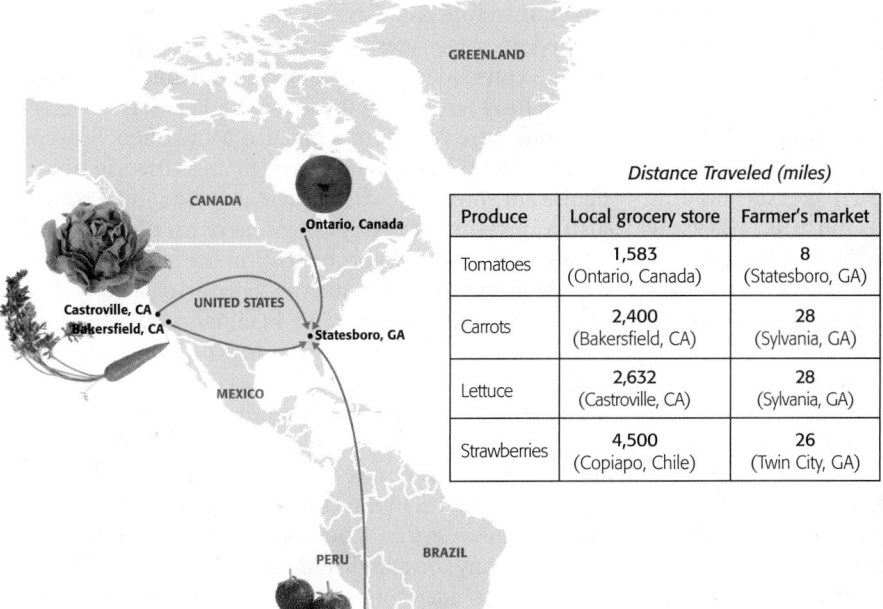

Distance Traveled (miles)

Produce	Local grocery store	Farmer's market
Tomatoes	1,583 (Ontario, Canada)	8 (Statesboro, GA)
Carrots	2,400 (Bakersfield, CA)	28 (Sylvania, GA)
Lettuce	2,632 (Castroville, CA)	28 (Sylvania, GA)
Strawberries	4,500 (Copiapo, Chile)	26 (Twin City, GA)

QUESTIONS 12.9

1. Describe how changes in our eating habits have increased our ecological footprints.

2. What are the advantages of eating locally grown food?

(MES) For additional review, go to **MasteringEnvironmentalScience**

12.10 Food for the Future

BIG IDEA There is no doubt that the increased agricultural production associated with the Industrial Revolution and Green Revolution has played a significant role in the growth of human populations over the past century. Nevertheless, a significant portion of Earth's people remains undernourished. Can we provide adequate nutrition to all of the world's people? Current trends suggest that our numbers will increase by another 2 to 3 billion over the next 40 years. Can we ensure adequate food to match this growth in human numbers? In order to answer these questions in the affirmative, we must sustainably manage our agricultural resources, minimize waste, and ensure that the worldwide distribution of food is equitable. Sustainable agriculture includes careful stewardship of soil and water resources, maintenance of the genetic diversity of crops and domestic animals, the fostering of stability in agroecosystems, and minimization of negative environmental impacts. Most of all, sustainable agriculture requires long-term vision. To guarantee adequate food for all people, governments must develop stable, honest institutions, manage economic markets for agricultural economies, and improve systems of food distribution and storage.

Q *Are we going to be able to keep up with demand for food as the population increases, cities grow, and the amount of available farmland decreases?*

Rebecca Sullivan, University of West Florida

Lissa: **A** To meet future nutritional needs we must ensure equitable food distribution by fighting poverty, corruption, and the potential negative effects of global markets

Sustainable Agriculture

■ Feeding a growing human population on a finite supply of arable land without causing irreparable damage to our environment is a major challenge.

Fossil fuels, synthetic pesticides, chemical fertilizers, and genetic modification of crops help boost crop yields. Each of these innovations has contributed to increases in the world food supply. Yet each also has the potential to cause problems that reduce the productivity of agricultural lands and damage the environment upon which they depend. Sustainable agriculture aims to provide future generations with high levels of food production with minimal environmental costs.

Agricultural production ultimately depends on the quality of soil and the availability of water resources. The sustainability of each of these resources is very dependent on the management of the other. Most erosion and loss of soil nutrients is associated with the movement of water across farmlands. This process is also the source of virtually all water pollution associated with agriculture. The demand for irrigation water can be greatly diminished by growing drought-resistant crops. Strategies such as drip irrigation can decrease water use by as much as 90% while reducing erosion and nutrient loss by similar amounts. Cover crops and no-till cultivation and harvest techniques further protect soil resources. Although there are short-term costs involved in the implementation of some of these technologies and strategies, in the long term they are both economical and ecologically sustainable.

As discussed several times in this chapter, the simplicity of agroecosystems is a major factor in their productivity. Yet that simplicity makes them especially vulnerable to disturbance. To ensure resistance to and resilience from competitors, pests, and climate change, sustainable agriculture seeks ways to maintain the genetic diversity of domestic species. In many situations, the overall productivity and stability of agroecosystems can be increased by planting polycultures of multiple species.

Greater diversity can also diminish the demand for water, fertilizer, and pesticides.

Sustainable management of agroecosystems pays attention to landscape-level factors that influence production and environmental impacts. Polyculture farming and crop rotation increase production and reduce the loss of soil nutrients. Adequate buffers of native prairie or forest along streams and rivers reduce the impacts of agriculture on water resources.

Agronomists have understood each of these elements of sustainable agriculture for several decades. One might, therefore, ask why they have not been universally applied. In some cases, as in some developing countries, this is a consequence of ignorance or misunderstanding among local farmers. In those cases, education alone may transform agricultural practice. More often, however, the problem is that management changes involve significant short-term costs. These may be capital costs, such as investment in drip irrigation or no-till cultivation technologies, or they may be costs in the production and harvest of specific crops, such as those associated with polyculture and intercropping. In either case, management that is focused solely on short-term economic gain or that ignores off-site environmental impacts will not recognize the benefits of sustainable agricultural practices.

Government policies and laws can provide incentives and regulations to ensure the sustainable management of agroecosystems. In the United States, the Farm Bill that is reauthorized by Congress about every five years provides many incentives and regulations. Farm Bill programs provide economic incentives for soil and water conservation, as well as funding for the maintenance of genetic resources. The 1985 Farm Bill, for example, established the Conservation Reserve Program (CRP),

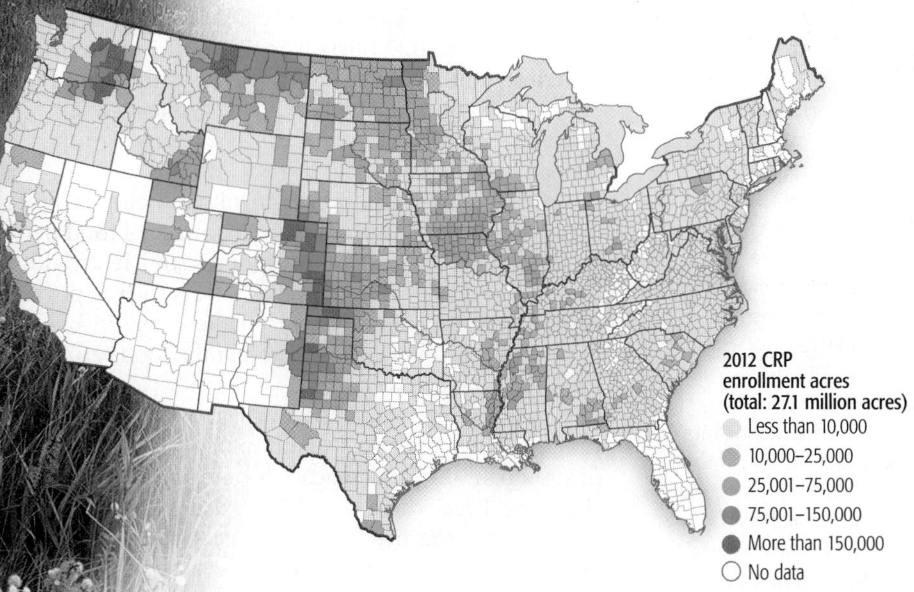

2012 CRP
enrollment acres
(total: 27.1 million acres)

○ Less than 10,000

● 10,000–25,000

● 25,001–75,000

● 75,001–150,000

● More than 150,000

○ No data

▲ Figure 12.55 **Conservation Reserve Land**
This diverse community of prairie plants (backdrop) has stabilized soils and diminished erosion on former cropland that is now protected under the Conservation Reserve Program. This map displays the amount of land enrolled in the program. Note that much of the conserved land is in the basins of the Mississippi River and its many tributaries.

which pays farmers to retire eroded croplands and lands near streams from production for a minimum of 10 years (Figure 12.55). Natural succession on these lands restores soil resources and protects water quality. As of 2013, about 110,000 km² (about 27.1 million acres) of land were enrolled in CRP. When the price of commodities is high, however, farmers are driven to cultivate more land, as it is more profitable than removing acreage from production in CRP. Indeed, high corn prices and other factors have decreased CRP enrollment by 7 million acres since 2008.

Farm Bill critics point out that this legislation often creates incentives for actions that are counter to sustainable agricultural management. For example,

crop subsidies may encourage the cultivation of crops in marginal areas. Such subsidies often influence the supplies and prices of agricultural commodities in ways that create disincentives for sustainable cultivation both here and in other countries. Other critics see the five-year time horizon as a significant challenge to the long-term vision required for sustainable management. Programs such as crop subsidies and the CRP are constantly revisited, and policies lack the permanence implied by sustainable management. This has caused some to argue that the major provisions of the Farm Bill should be formulated over a 50-year rather than 5-year timeline.

▼ Figure 12.56 **Malnutrition in a World of Plenty?**
These children in Yama, Niger, are starving because of a regional shortage of food. Nevertheless, the global supply of food is sufficient to feed everyone adequately.

Feeding a Hungry World

■ Food distribution is a major obstacle to feeding all of Earth's people.

Food security is defined by the United Nations' Food and Agriculture Organization (FAO) as physical, social, and economic access to sufficient, safe, and nutritious food to meet dietary needs and food preferences needed for an active and healthy life. With a growing human population and 870 million hungry people in the world (>10% of the population), food security has become a major concern. In sub-Saharan Africa, undernutrition is a chronic problem. Uncertain rainfall, political unrest, and growing populations have seriously threatened food security in the region. Children are especially vulnerable to shortages of protein, which is needed for normal development (Figure 12.56).

Food security is not a just a problem for developing countries. Here at home, roughly 15% of Americans are food insecure, lacking access and means to supply enough food for themselves and their families. And at the other end of the food availability spectrum, overnutrition is an increasing problem in

413

Where
YOU
LIVE

What can you eat for $4 a day?

Poverty brings with it a complex set of nutritional issues. It is a cruel irony that because of risk factors associated with poverty and food insecurity, the low-income sector is especially vulnerable to obesity.

If you were on a food budget of $4 per day, what could you purchase to maximize your caloric intake? List all food items and their caloric values and sum them for a daily total. Now on that same budget, what could you purchase to consume as many nutritious calories as possible? List all food items and their caloric values. How do these calorie totals compare?

the United States, where nearly 20% of children and adolescents are overweight. The Centers for Disease Control reports that obesity occurs in one out of three American adults. Since obesity increases the risk of developing coronary artery disease, stroke, type 2 diabetes, and some kinds of cancers, overnutrition is also a serious health issue.

Careful analysis of current agricultural trends indicates that there currently is not a global shortage of food per se: In total, global agriculture produces sufficient yields to feed the world's people. However, factors such as poverty, corrupt governments, economic markets, and problems in shipping and storing food prevent the food supply from reaching everyone.

Sustaining agricultural productivity in the world's poor countries remains a significant challenge. Ignorance of sustainable farming practice is only part of that challenge. Even where knowledge exists, people often lack the resources needed to improve local agriculture. The U.S. Agency for International Development (AID), Heifer International, and many other international relief organizations are working to help people in famine-prone countries increase their food security through locally controlled agricultural projects.

Where economic hurdles prevent impoverished people from starting a farm, small loans from international development agencies have proven especially effective. Microcredit programs supply start-up costs for orchards, farms, and livestock operations in the poorest nations (**Figure 12.57**). A loan as small as $100 can help an African farmer purchase seeds, oxen, and a plow to begin farming. As these small loans are paid off, the capital can be loaned out again to help others.

Corrupt and unstable governments can interfere with the equitable distribution of food. This was a problem as far back as Roman times and remains a serious problem in several countries today. Prior to 2000, Zimbabwe was a major producer of food, particularly corn, for much of southern Africa. In that year, the government of President Robert Mugabe implemented policies that were supposedly aimed at reallocating the ownership of farmland in a more equitable fashion. Instead, corruption in that process resulted in land ownership shifting to Mugabe's political cronies. Between 2000 and 2004, Zimbabwe's corn production declined by nearly 75%.

▲ Figure 12.57 **A Little Help Can Go a Long Way**
A microloan helped this Cambodian woman start her watermelon shop in Phnom Penh.

As of 2013, attempts by relief agencies to distribute donated food to this country's many starving people have been thwarted by corrupt government officials.

Food donation by wealthy nations is the most direct effort to address world hunger. The U.S. government spends over $19 billion per year in international aid. Though these gifts make up a small part of the gross domestic income of industrialized economies, and though they have been criticized for promoting various political agendas, there is no doubt that government-funded assistance has saved the lives of many hungry people.

People around the world have become directly or indirectly dependent on global markets for their food. For this reason, policy decisions in food-producing countries that affect the supply and price of food commodities can affect the availability of food worldwide.

For example, U.S. government policies have encouraged the increased use of corn to produce ethanol as a supplement to gasoline to fuel automobiles (see Module 15.3). By 2012, nearly half of U.S. corn production was used to make ethanol (**Figure 12.58**). During the 10 years following the increase in ethanol production (2003–2012), the price of corn increased by nearly 300%. Corn is a staple food that is also used to

▼ Figure 12.58 **Growing Corn for Fuel**
Since 2003, the proportion of the corn crop in the United States used for ethanol production has increased fourfold.

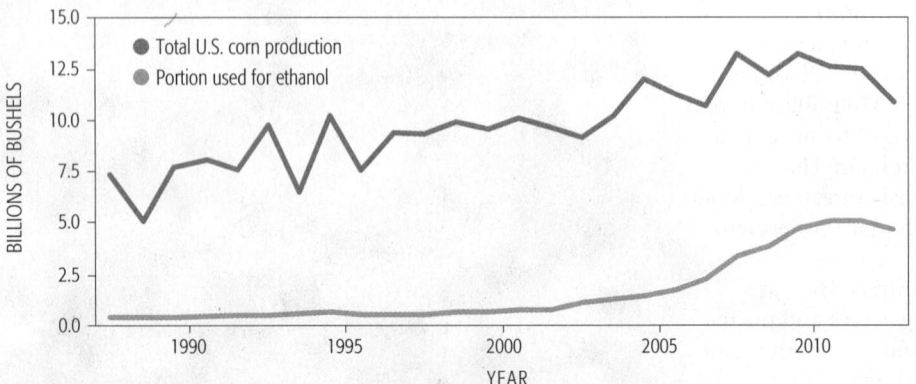

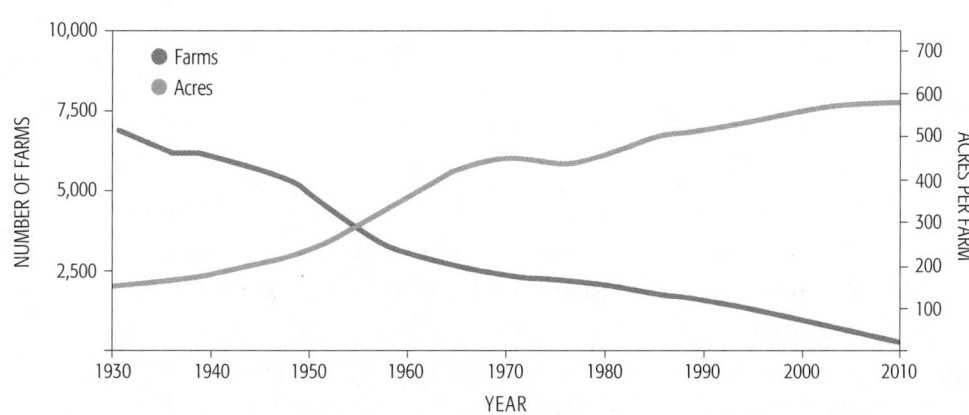

◄ **Figure 12.59 Fewer, Bigger Farms**
Today there are 60% fewer farms than in 1930, but the average farm is now three times larger.

Source: http://nationalatlas.gov/agriculture.html

QUESTIONS 12.10

1. Discuss three strategies that farmers can use to make their activities more sustainable.

2. Describe four factors that may limit the supply of food to people in poor countries.

(MES) For additional review, go to **MasteringEnvironmentalScience**

feed livestock and is a basic ingredient in many processed foods. Because of the wide use of corn and because the United States produces 40% of the world's corn, the U.S. policies regarding corn affected the price and availability of a wide variety of foods around the world.

Global markets and competition have a significant effect on agriculture within individual countries. In Mexico, for example, 2 million small-scale farming operations, most smaller than 5 hectares (12 acres), have historically produced most of the corn needed for domestic consumption. Tortillas and other traditional foods made from corn comprise one-third of the national diet. This traditionally stable agricultural system is changing rapidly as a result of the international trade stimulated by recent trade agreements. Small, hand-cultivated corn crops cannot compete with grain imported from the subsidized and heavily mechanized farming operations in the United States. More and more of Mexico's farmers are abandoning their traditional way of life and seeking work in cities or across the border in the United States. The traditional Mexican farm economy and the connection between local crops and local foods have been destabilized by unprecedented global competition.

Similar losses of small farms are occurring in the United States. Economies of scale, Farm Bill subsidies, and increasing energy costs have favored large farms instead of small family operations. As a result, the average size of an American farm has increased, while the numbers of farmers has declined (Figure 12.59). As more and more family farmers sell out to larger-scale operations, fewer and fewer farms control more and more of agricultural markets.

Finally, food distribution and storage are often major obstacles to feeding the world's hungry. Grain is a prime example. While grain supplies are scarce in the hungry nations of the world, grain-producing nations sometimes produce such large surpluses that the food spoils for lack of storage and shipping capacity (Figure 12.60).

The effects of global warming on world food supply are at best uncertain. Warmer temperatures and longer growing seasons may increase crop production in temperate climates. However, most climate models indicate that conditions for agriculture will worsen in tropical regions. It is these regions where food shortages are most acute.

In summary, sustainable management of Earth's agroecosystems is essential if we are to provide adequate nutrition to all people in the future. Some experts worry that increases in agricultural production may not be able to keep pace with projected growth in human populations. Climate change adds to this uncertainty. Sustainable management requires a long-term vision and a willingness to accept short-term costs that will reap long-term benefits. Providing food for all will also require economic and social policies that ensure that food is distributed equitably.

▼ **Figure 12.60 Wasted Food**
Grain produced in a bumper-crop year is piled on the ground due to lack of storage capacity.

415

STOGROW: A Student-Run Campus Farm at St. Olaf College

Hillary King was a Studio Art major who served as a STOGROW farmer for two years. STOGROW (St. Olaf Garden Research and Organic Works) is a student-initiated, student-run campus farm in operation since 2005. The goals of the farm are to practice sustainable farming methods; provide fresh, local vegetables, fruits, herbs, and flowers to the St. Olaf community; foster agricultural awareness; and provide education about sustainable food production. STOGROW is one of a growing number of student-run farms at colleges across the country, but it is unique in that it sells its produce exclusively to St. Olaf's food service provider.

How did you first get involved in STOGROW?

My freshman year at St. Olaf, I became aware of STOGROW and attended a few volunteering opportunities and the annual harvest event. Though my garden experience was quite limited, I was intrigued by the idea of farming and stayed involved and informed in the happenings on the farm until I was hired as a farmer the spring of my junior year. The farm was founded in 2005 by two students, Dayna Burtness and Dan Borek, who exercised incredible drive to research, plan, and persist in making a student-run, on-campus farm a reality. They forcefully pursued this project, convincing St. Olaf that the farm would provide education opportunities, reduction of carbon in the campus food system, and future potential for growth and success.

What steps did you take to get started?

Upon entering an already instituted organization, it was important to examine the records of previous years to anticipate challenges, plan improvements, and identify resources. Over the duration of my two years as a STOGROW farmer, our goals have been centered on boosting production to diminish recent debt, increasing awareness and accessibility of the farm, and creating successful ways to engage and involve students and volunteers.

Looking to the future, we hope to adjust the structure of farm leadership with the introduction of a non-student farm manager position to improve stability and sustainability, and to encourage further growth. The college has plans to move the farm to a location closer to the central campus, allowing for the potential expansion of greenhouse and field space and research opportunities that will further integrate the faculty and student body.

What are the biggest challenges you've faced with the project, and how did you solve them?

Weather is an unavoidable challenge for any farmer. When dealing with an unpredictable force, we have learned to be flexible with our planning, experimental with our methods, and resourceful as we seek out advice from farmers and gardeners in the community. This year, for example, a cool spring and late start led us to forgo spring crops altogether and focus on larger quantities of crops that could be harvested in late summer or early fall.

The largest challenges we currently face are the sustainability of the farm model and the presence of the farm within the campus community. We are working to keep the learned knowledge, experience, and projects of the farmers alive at STOGROW, even as the farmers and student body change each year. As a student organization that is comprised solely of a few student-farmers who work from spring to fall, it is essential that we have an on-campus presence with a continuing voice from year to year in which the student body can be involved, even during the winter months. Offering something concrete for students to interact with will create a support system for the farm, generate new ideas, and allow the farm to expand its reach by forming projects led by non-STOGROW-employed students.

How has your work affected other people and the environment?

The farm provides an amazing learning experience for farmers and volunteers, many of whom, like myself, have never previously farmed. Involvement in STOGROW is unique because it provides real, physical impact and immediate results; the completion of day-to-day projects, the gradual growth of plants, and the appearance of the farm's fresh veggies in the cafeteria.

Environmentally speaking, we are the closest food provider for Bon Appetit, our campus food service, thereby reducing its carbon footprint. Bon Appetit's efforts to make the student body aware of the benefits of local food are evident in notations placed around the cafeteria, often labeling produce from Northfield farms. Bon Appetit purchases 100% of what we grow, allowing us the opportunity to focus on a few main crops while giving a little leeway to experiment with a few new things. Harvest days on the farm are truly the most rewarding, and the appearance of our produce in prepared dishes or the salad line feels like a triumphant event! St. Olaf students can be proud of Bon Appetit's constant efforts to support local farmers, and STOGROW has proved itself to be an inspiring program for both the farmers and the student community.

Personally, the farm has been inspiring and rewarding not only as a visible vehicle for change but also a creative outlet. Working with my hands and surrounding myself with the life cycles of plants and animals has greatly impacted my world views and been an aesthetic and conceptual source for much of my art work.

What advice would you give to students who want to replicate your model on their campus?

Keep it simple. It is important to focus on a few things (plant varieties, campus involvement activities, etc.) that the farm does really well on a consistent basis before turning to new projects or expansion. Having a firm base and introducing new branches (honeybees, livestock, student-led projects) one at a time will ensure their stability and success. However, dreams and aspirations are essential too! They provide the work with passion, recruit new interest, and promote growth and learning.

What are your plans for the future now that you have graduated? How did this experience influence your future?

My plan is to keep on making things, whatever they may be. When I was hired at STOGROW, I was surprised to find that, when working on the farm, I felt a satisfaction of my creative impulses similar to when I was working on pieces in the art studio. Art and farming allow me to physically work toward something that can be shared with others, doing important work in communities. I am currently working with a contemporary art exhibit at Minneapolis's Walker Art Center that utilizes farming, foraging, processing and craft, and fine arts to comment on domesticity and how we choose to "make ourselves at home." Creativity is expressed through endless activities: cooking, gardening, drawing, movement, or writing, and I hope to lead a life in which the activities of "homemaking" and "fine arts" are not compartmentalized. I may never have my own farm, but growing food has become an addiction, and I am excited to contribute to the local food system of any community I find myself in!

Hillary King, *Rooted: Revelation of the Farmer*, Oil on wood panel, Senior Art Show, 2013

Summary

Agriculture originated independently in several regions of the world during the past 10,000 years as a consequence of changing climates, diminishing wild resources, and changing human cultures and technologies. People chose for domestication plants and animals that possessed useful features; over many generations, humans enhanced those features through the process of artificial selection. During the Industrial Revolution, new technologies increased the production of existing farmlands and allowed agriculture to expand onto new land. In the past century, fertilizers and pesticides have increased both the productivity of land and the amount of land in production. Today, humans use over 30% of Earth's terrestrial net primary production to obtain food and fiber and in so doing affect the flux of nitrogen and phosphorous by using fertilizers to increase productivity.

Agroecosystems support many fewer species than natural ecosystems and are prone to nutrient loss because of harvest removals, continual soil disturbance, irrigation, and low biodiversity. Crop plants depend on light, water, and nutrients to produce carbohydrates, proteins, and oils that are stored in various plant parts. The range of tolerance for climatic and soil conditions determines where and when particular crops are grown. Although agroecosystems are usually focused on the growth of a single species, those species depend on other species, including pollinating insects and soil insects, bacteria, and fungi. Characteristics of soil that influence plant growth include the depth and structure of soil horizons and soil fertility. Soil characteristics such as texture and organic matter also influence the availability of water. Irrigation has allowed farming in regions where evapotranspiration far exceeds rainfall.

A large amount of agricultural production from crops and pasture land supports domestic animals that provide meat and other products to humans. Their environmental impacts include increased demand for land, disposal of waste products, and spread of disease. Artificial selection, cloning, and genetic engineering have diminished the genetic diversity of many crop plants and domestic animals, putting some agroecosystems at greater risk of failure due to disease and climate change. Pesticides and herbicides can improve crop production, but they also present significant environmental challenges. Integrated pest management is an alternative approach designed to control weeds and pests and minimize environmental impacts.

What we chose to eat, where our food is grown, and how our food is processed have an important effect on our ecological footprint. Although the world's food supply appears to be adequate to feed our current population, chronic hunger and poor nutrition are problems for many people. Climate change adds more uncertainty. Sustainable agriculture and equitable food distribution are important to meet future needs.

12.1 Origins and History of Agriculture

- Agriculture originated independently in several places as a consequence of climate change, cultural progress, and human population growth.
- Humans, crop plants, and domestic animals have coevolved.

KEY TERMS

agriculture, agronomy, Green Revolution

QUESTIONS

1. It appears that early agriculture may not have improved human well-being compared to hunting and gathering. What was the likely driver for adoption of this style of life?
2. Describe how artificial selection likely led to the evolution of crop plants and domestic animals.
3. Explain how the Industrial Revolution and the Green Revolution significantly increased agricultural productivity.

12.2 Agroecosystems

- Agricultural ecosystems depend on energy flow and nutrient cycling following the same principles as natural ecosystems.

KEY TERMS

agroecosystem, harvest index, trophic-level efficiency, nitrogen fixation, nitrification, legumes, denitrification, Haber–Bosch process

QUESTIONS

1. Why is trophic-level efficiency generally higher for domestic animals than for their wild counterparts?
2. Why are agroecosystems generally more fragile than natural ecosystems?
3. How has conventional agriculture altered N and P cycles? What are the consequences beyond agroecosystems?

12.3 The Growth of Crop Plants

- Plant growth and reproduction depend on adequate light, water, and nutrients.
- Differences in range of tolerance for environmental factors determine when and where different crop plants grow best.

KEY TERM

transpiration

QUESTIONS

1. What are the roles of light, water, and essential nutrients in crop plant growth?
2. Name four ways that other organisms influence the growth of crop plants.

12.4 Managing Soil Resources

- Soil structure, organic matter, and nutrient content influence the growth of crop plants.
- Unlike natural ecosystems, agroecosystems lack processes to sustain soil fertility.
- Soil conservation practices include limiting soil exposure and disturbance as well as diversifying crops.

KEY TERMS

soil, sand, silt, clay, texture, soil profile, O horizon, humus, A horizon, B horizon, C horizon, soil fertility, tilth, contour farming, terracing, shelterbelts, windbreaks, no-till, intercropping, cover crops

QUESTIONS

1. Why is soil nutrient content highest in the O and upper A horizons of the soil profile?
2. Describe three ways that soil pH influences plant growth.
3. What actions can a farmer take to minimize soil erosion and fertility loss?

12.5 Water and Agriculture

- Soil texture and organic matter determine the amount of water that a soil can hold against the force of gravity.
- Irrigation has allowed expansion of agriculture in arid regions, but it creates important environmental challenges.

KEY TERMS

gravitational water, hygroscopic water, capillary water, field capacity, soil wilting point, potential evapotranspiration (PET), soil salinization, drip irrigation

QUESTIONS

1. Why do sandy soils generally have a lower field capacity than soils rich in clays and organic matter?
2. Why does irrigation tend to diminish soil fertility?
3. Describe three ways in which modern farmers are using water more efficiently.

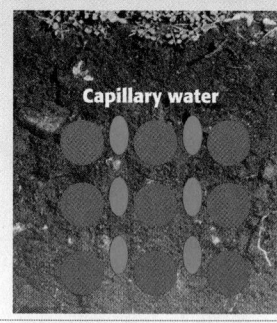

Capillary water

12.6 Livestock in Agroecosystems

- Domesticated animals in agroecosystems present several environmental challenges, including land for food and range, waste disposal, and disease.

QUESTIONS

1. Why is the trophic-level efficiency of free-range livestock considerably lower than similar animals reared in feedlots?
2. The use of antibiotics to control disease in domestic animals has increased significantly over the past 20 years. Why is this a matter of concern among many public health experts?

12.7 Managing Genetic Resources

- Selective breeding, cloning, and genetic engineering have produced very productive strains of domestic plants and animals, but they have also diminished genetic diversity.

KEY TERMS

monocultures, interplanting, polyculture, genetically modified organism (GMO)

QUESTIONS

1. What are the advantages of genetically uniform agricultural species?
2. Why is such uniformity risky in the long term?
3. What are the potential advantages and disadvantages of GMO technologies?

12.8 Managing Competitors and Pests

- Pesticides are used to control populations of weeds, pests, and disease organisms that compete for the food in agroecosystems.

KEY TERMS

broad-spectrum pesticide, biomagnification, biological pest control, pheromone, integrated pest management (IPM)

QUESTIONS

1. Describe three changes in pesticide development and use that are designed to minimize negative environmental impacts.
2. Rather than simply relying on pesticides, many agronomists recommend the use of integrated pest management (IPM). What does this involve?

12.9 The Ecology of Eating

- The ecological impacts of our eating habits are determined by whether our diet emphasizes plants or meat, how our food is grown, and where our food comes from.

KEY TERM

organic foods

WATER INPUT (gal)

1,799

468

0.12

Vegetables (1 lb) Chicken (1 lb) Beef (1 lb)

QUESTIONS

1. What factors have increased the incidence of outbreaks of strains of bacteria such as *E. coli* and *Salmonella* due to consuming tainted meats and produce?
2. Describe six ways that you can diminish the ecological impacts associated with your food.

12.10 Food for the Future

- Chronic hunger affects more than 10% of Earth's people even though global food production is sufficient to feed everyone.

KEY TERM

food security

QUESTIONS

1. If you had the opportunity to craft a farm bill for congressional consideration, what four features would you include to improve the sustainability of farming practices?
2. Explain why increased use of biofuels in automobiles has resulted in food shortages in many countries.

MasteringEnvironmentalScience®

Students Go to **MasteringEnvironmentalScience** for assignments, the eText, and the Study Area with practice tests, activities, and more.

Instructors Go to **MasteringEnvironmentalScience** for automatically graded tutorials and questions that you can assign to your students, plus Instructor Resources.

Forest Resources

13

421

The Tragedy of Forest Loss in Haiti

Why does a country need its forests?

On September 16, 2004, tropical storm Jeanne settled over the Dominican Republic and Haiti. In the next 48 hours, Jeanne dumped over 50 cm (20 in.) of rain on the northern portions of both countries. That's more than a third of the rain that typically falls during an entire year. In the Dominican Republic there was flooding and damage to crops, but the effects were short lived. In a matter of days, life was back to normal. However, just across the border in Haiti, Jeanne created a national disaster. There was extensive flooding, and massive landslides buried entire villages (Figure 13.1). Over 3,000 lives were lost in the city of Gonaïves alone.

Why did the same storm produce such different consequences for two countries on the same island? It was not because Haiti received more or heavier rain than the Dominican Republic—if anything, less rain fell on Haiti. The difference, in one word, was forests. The Dominican Republic has forest, and Haiti has virtually none (Figure 13.2).

▲ Figure 13.1 **Jeanne Hammers Hispaniola**
Tropical storm Jeanne brought heavy rains to the island of Hispaniola. This produced flash flooding in Haiti's deforested watersheds that resulted in the loss of thousands of lives and extensive property damage.

▲ Figure 13.2 **The Haiti–Dominican Republic Border**
The countries of Haiti and the Dominican Republic are located next to each other on the island of Hispaniola. The river channel in this photo forms the boundary between deforested Haiti on the left and the Dominican Republic on the right. The Dominican Republic still has over 60% of its forests.

Haiti was not always a barren land. When Christopher Columbus visited its shores in 1492, 75% of Haiti's land was covered by lush tropical rain forest, and the rest of the land supported seasonal dry tropical forest. Even a century ago, despite land clearing and agricultural development, 60% of the country was covered by forest. At that time, Haiti's coastal waters were known for their azure clarity, diverse coral reefs, and productive fisheries. Like Cuba and Jamaica, Haiti was a premier tourist destination in the Caribbean, and its economy was thriving.

In the first half of the 20th century, commercial logging and the expansion of rubber plantations increased the pace of deforestation. From 1957 to 1986, Haiti was under the brutal dictatorships of François (Papa Doc) and Jean-Claude (Baby Doc) Duvalier. During their regimes, trees were cut for revenue and the process of deforestation accelerated greatly. Loss of tourism dollars and corruption during the Duvalier dictatorships inaugurated a spiral of poverty that further amplified deforestation.

By 1990, most of Haiti's forests had been cut. As the country's rapidly growing population has become increasingly dependent on subsistence farming, the forests have been almost completely cleared from the mountainous terrain (Figure 13.3). The meager forest that remains has been logged, mostly to produce charcoal, the primary source of fuel for 90% of the population. Today, less than 3% of Haiti's landscape supports forest, compared to 60% in the Dominican Republic, a country that did not suffer the same unfortunate political rule.

▲ Figure 13.3 **An Impoverished Land**
Deforestation and subsequent erosion have left Haitian soils with little organic matter or nutrients. As a result, the steep, poorly terraced slopes support only meager crops.

- What ecological and economic values do forests provide?

- How do individual trees, forest stands, and forested landscapes grow and mature?

- What factors are responsible for deforestation and forest degradation?

- Can forests be managed sustainably? If so, how?

- What are the policies and laws needed to support sustainable forest management?

Haiti's deforestation has had catastrophic consequences. Without forests to retain and filter water, even moderate rains produce floods, and groundwater and stream waters are laden with sediment and pollution. In most places, Haiti's once-clear coastal waters are now murky and filled with sediment. The sediment and pollution have degraded estuarine and coastal ecosystems; coral reefs are nearly gone, and the fish have nearly vanished.

It is no coincidence that Haiti is, by a considerable distance, the poorest country in the Western Hemisphere. Over 60% of its national income comes as aid from foreign countries such as the United States, and 65% of its people survive on less than $1 a day.

There have been many efforts to restore forests to Haiti, but the task has proved daunting. Agriculture and erosion have stripped soils of their capacity to hold water and nutrients, and as a result, many native trees have difficulty growing. Goats, pigs, and chickens, which are the primary sources of protein for Haiti's rural poor, feed on the sprouts of most tree seedlings, making it even more difficult for new trees to become established (Figure 13.4). When trees do manage to survive a few years, they are usually harvested to provide fuelwood and charcoal. As if all this were not enough, Haiti's governmental institutions are corrupt and lack the capacity to mount and enforce countrywide programs of reforestation and conservation.

Extreme poverty is itself one of the root causes of deforestation, as well as one of the most powerful barriers to forest restoration and sustainable management. A forest manager working on a reforestation project in Haiti's southern mountains recalls meeting a group of men who were cutting the young trees from a demonstration plot.

The forester pointed out that if the trees were left to grow for a few more years, they would provide much greater economic value as structural timber and would also restore soil fertility. To this, the men replied, "But we must boil our water and cook our food today," describing the immediate services the young trees would supply.

A long-term perspective is essential to sustainable forest management. Yet, obsession with immediate needs is one of the unfortunate consequences of poverty. Herein lies a great paradox: Much of Haiti's poverty and human suffering is the result of the loss of its forests, but in order to restore those forests, poverty must be alleviated.

Community-based programs that couple forest restoration with economic development and the well-being of families are beginning to show promise. The 2010 earthquake that killed over 225,000 people in Haiti attracted considerable international attention to the plight of the Haitian people. Today, there is widespread agreement that long-term plans for Haiti's recovery must include the restoration of its forests.

Haiti's history demonstrates the importance of forests to humans. In addition to providing economically valuable resources, forests provide ecosystem services that are essential to human well-being. Forest restoration and conservation require scientific understanding of climate, soils, and tree growth. As Haiti's story shows, these actions also depend on an understanding of human behavior, economic incentives, and institutional structures.

▼ Figure 13.4 **A Tree's Enemy**
Rural Haitians depend on domestic animals, especially goats, for nutritional protein. However, these animals contribute to deforestation by eating most tree seedlings.

13.1 The Values of Forests

BIG IDEA Biologists estimate that forests contain over 75% of the world's biodiversity. This biodiversity is essential to many of the ecosystem services that forests provide, such as flood prevention, maintenance of air and water quality, carbon storage, recreation, and aesthetic beauty. Forests provide humans with a diverse array of wood and paper products that add over $500 billion to the world economy each year. In addition, forests are an important source of food, clothing, and medicine.

Ecosystem Services

■ Forests prevent floods and provide clean water, recreation, and inspiration.

The world's forests provide a remarkable array of ecosystem services. They purify water and regulate its flow, withdraw and store carbon from the atmosphere, conserve biological diversity, and are a source of beauty and inspiration to humans. Most of these services are not bought and sold, so they are not assigned a monetary value. However, when forests are disturbed or destroyed, the loss of these services can have significant economic consequences.

Forests play a key role in the hydrologic cycle. The roots of trees absorb large quantities of water, which eventually evaporates from their abundant leaves. In a tropical forest, transpiration from a single large tree can release over 11,000 liters (2,905 gallons) of water each day. Forests are constantly recycling moisture back into the atmosphere, where it can then recondense as rain. Thus, when forests are lost over a large area, rainfall in that area often declines (see Module 11.1).

When rain falls on a forested landscape, the trees, shrubs, and complex mixture of plants and leaf litter on the forest floor slow the flow of water across the land. Thus, even on steep slopes, forests diminish the direct runoff of rainfall and reduce the risk of flooding (Figure 13.5). Before entering streams, most water flows through the litter of the forest floor and then through soils. The litter and soil remove sediment and dissolved chemicals from the water. A study of hundreds of streams in locations across the United States revealed that streams flowing through forests contained less than 15% of the nitrogen and phosphorus found in streams flowing through farmland or cities.

Forests store large amounts of carbon. In the process of photosynthesis, forest plants remove carbon dioxide from the atmosphere to make organic compounds, which are stored in the tissues of their roots, stems, and leaves. Worldwide, forests store approximately 400 petagrams (1 Pg = 10^{15} g) of carbon; each year they take up and release approximately 60 Pg of carbon.

▶ Figure 13.5 **Forests and Water**
As water flows through a forested watershed, trees moderate its flow and remove dissolved minerals from it.

Deforestation reduces total global forest carbon storage by about 1 Pg each year. That carbon is eventually added to the atmosphere as CO_2. This CO_2 accounts for about 15–20% of the anthropogenic carbon emissions that are contributing to global climate change (see Module 9.3).

Forests are the primary habitat for 60–80% of all terrestrial species—microbes, plants, and animals. (Scientists are uncertain of the exact proportion because so many species have yet to be identified.) In addition to its intrinsic value, this biodiversity provides humans with an array of foodstuffs, clothing, and building materials, as well as hundreds of medicinal chemicals (see Module 8.2).

Few would disagree that forests are among the most beautiful places on Earth, and many of us are quite willing to pay for this beauty. Ecotourism in forested regions adds over $200 billion to the global economy each year. In many developing countries, it is a major portion of the gross domestic product (GDP) (see Module 8.2). In fact, ecotourism is one of the few ways in which we humans pay directly for the ecosystem services that forests provide.

Wood Products

■ The wood extracted from Earth's forests is primarily used for fuel, paper, and building materials.

Each year, approximately 3.4 billion cubic meters of wood are harvested from the world's forests. To give a sense of scale, a log that is 2 feet in diameter and 16 feet long yields about 1 cubic meter of wood; an oak tree with a trunk that is 2 feet in diameter might yield 2 to 3 cubic meters of wood. The volume of the world's annual harvest of wood is equivalent to about 1.5 billion of those oak trees.

Nearly 53% of the world's wood consumption is used for fuel, about 31% is used to produce various building materials and solid wood products, and about 16% is used to make paper. However, these proportions vary widely among different regions of the world (Figure 13.6).

▼ Figure 13.6 **Wood Use**
In North America and Europe, most wood is harvested for lumber used in construction and for fiber used to make paper. In Asia and Africa, most wood is harvested to meet local needs for cooking and heating.

Source: Food and Agriculture Organization of the United Nations (FAO), 2010. Global Forest Resources Assessment 2010: Main Report. FAO Forestry Paper Rome: FAO, 378 pp.

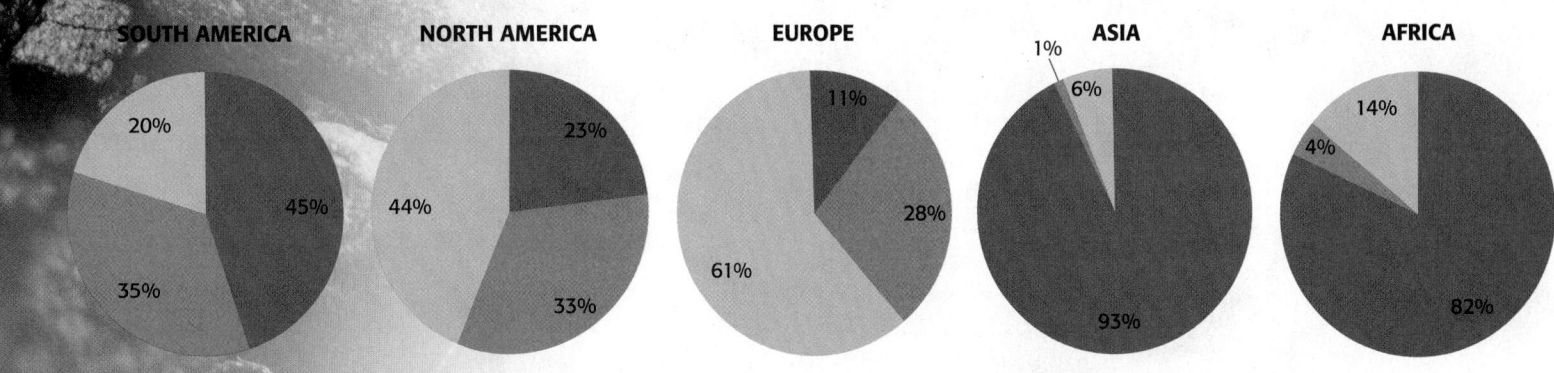

| SOUTH AMERICA | NORTH AMERICA | EUROPE | ASIA | AFRICA |

SOUTH AMERICA: 20%, 45%, 35%
NORTH AMERICA: 23%, 44%, 33%
EUROPE: 11%, 28%, 61%
ASIA: 1%, 6%, 93%
AFRICA: 14%, 4%, 82%

● Fuelwood   ● Fiber   Lumber

Fuel from wood. More than half of all the wood that is harvested is burned for fuel or made into charcoal. Yet, wood provides less than 7% of the total energy used by humans. In Africa, nearly 90% of wood is consumed for fuel, representing over 40% of the continent's total annual energy use.

In many developing countries, wood is harvested, dried, and burned to heat homes, food, and water. More often, wood is converted to charcoal, which is then burned for fuel (Figure 13.7). Charcoal has many advantages over wood. Weight for weight, it produces nearly twice as much heat as wood. Once it is lit, charcoal burns much more evenly than wood. Because charcoal is impervious to termites, it can be stored for longer periods. Charcoal fire releases less smoke than wood fire, so it is far less irritating to use, particularly indoors (see Module 10.4). Unfortunately, burning charcoal does release pollutants, such as carbon monoxide.

Rapidly growing interest in biofuels is encouraging the development of technologies that convert wood cellulose into methanol or ethanol (see Module 15.3). When fully developed, these technologies are likely to create a much greater demand for wood from the world's forests.

Building materials and solid wood products.

Each year just over 1 billion cubic meters of wood are harvested for use as building materials and solid wood products, such as furniture, containers, and musical instruments. About 68% of this wood comes from temperate evergreen and deciduous forests—43% from North America, 21% from Europe, and 4% from Asia. The majority of wood from these regions is used for lumber, plywood, and strand board used in construction.

Tropical forests account for only 32% of solid wood production, most of which comes from Southeast Asia and South and Central America. Yet tropical forests provide some of the most valuable woods, such as ebony, teak, mahogany, and jacaranda. Most often these woods are harvested from old-growth forests (Figure 13.8).

▶ Figure 13.7 **Charcoal Production**
Charcoal is produced by burning wood in special kilns that provide only limited oxygen.

▼ Figure 13.8 **High-value Wood**
Trucks loaded with teak logs that have been cut illegally from old-growth tropical forests are a common sight on Myanmar's (Burma's) roads. Each log is worth hundreds of dollars.

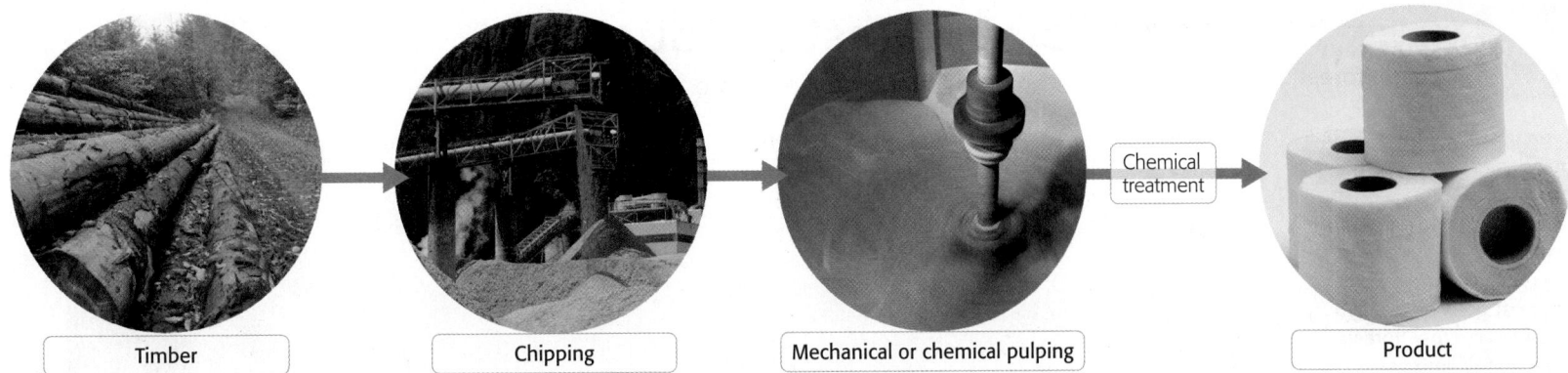

| Timber | Chipping | Mechanical or chemical pulping | Chemical treatment → | Product |

▲ Figure 13.9 **Producing Paper**
Wood fiber is the primary ingredient in paper manufacture. Each of the steps involved in producing paper has significant impacts on our environment.

Paper production.   Each year, approximately 540 million cubic meters of wood are converted into pulp, which is then used to produce over 388 million tons of paper. That is equal to about 50 kg (110 lb) of paper for every person on Earth, although per capita use of paper varies widely among regions. Each year an average American uses over 320 kg (700 lb) of paper, whereas an average African uses less than 1 kg (2.2 lb).

In general, only smaller logs (less than 30 cm, or 12 in., diameter) are used to make paper. A significant amount of this wood comes from plantations of fast-growing trees, such as pine and eucalyptus, that are grown for this purpose.

To produce paper, wood is cut into small chips, which are then chemically digested into pulp (Figure 13.9). Wood is made up of cellulose fibers that are bound together by a chemical called lignin. Chemical digestion separates the cellulose fibers from one another. After digestion, the fibers are washed and concentrated into a slurry. The slurry is then dried and pressed to produce large rolls of paper. To make white paper, the fibers must be chemically bleached before they are added to the slurry.

In addition to using great quantities of wood fiber, the production of paper has a significant impact on air and water quality. The production of a ton of paper requires 7,000–10,000 gallons of water, and the caustic chemicals used in the process are a source of both water and air pollution (Figure 13.10).

Because paper can be decomposed to its original fibers and then reconstituted, it is relatively easy to recycle. A ton of paper containing 100% recycled fiber saves about 20 plantation trees (8 in. diameter) and about 5,000 gallons of water. The production of recycled paper uses fewer caustic chemicals than that of regular paper and so causes far less air and water pollution (see Module 17.3).

▶ Figure 13.10 **Paper Production and Water Pollution**
Wastewater from a paper mill pours directly into Dongting Lake in China's Hunan Province. Such pollution has severely damaged the lake's ecosystems.

Q Are we making a dent in the number of trees being cut down each year with paper recycling?
Rebecca Sullivan, University of West Florida

Norm: A   Currently, Americans recycled about two-thirds of the paper they used—that's 210 kg (460 lb) per person. That saves four mature trees and over 1,300 gallons of water for each person.

Where YOU LIVE **What is your national forest worth?**

Even in states where grasslands or deserts predominate, forests are ecologically and economically important. You can determine the most important kinds of forest growing near your hometown, type "forests of [your state]" into your web browser. This will give you access to the website for your state's division of forestry.

Describe the forests growing closest to your hometown. What are the dominant tree species? What ecological conditions (e.g., soils, proximity to water) are typical of these forests? What economic values do these forests provide? And what are the most important threats to the sustainable management of these forests?

Non-Wood Forest Products

■ Hundreds of non-wood commodities come from forests.

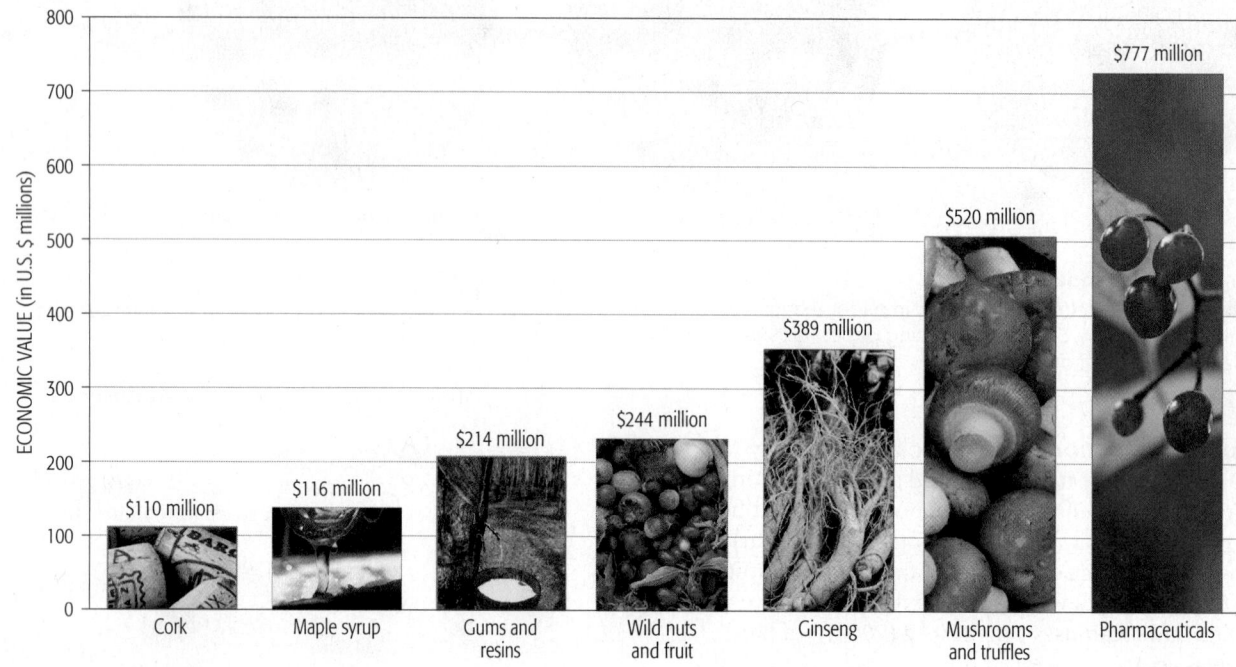

▲ Figure 13.11 **More Than Just Wood**
These are a few examples of economically important non-wood forest products and their approximate economic value before they are converted to finished products.

QUESTIONS 13.1

1. List and describe four important ecosystem services provided by forests.

2. What are the most important wood products derived from African forests? How do they compare to the wood products derived from North American forests?

3. Name three non-wood forest products that you use routinely.

(MES) For additional review, go to **MasteringEnvironmentalScience**

A remarkable variety of natural products can be harvested from forests without cutting trees. Wild nuts and fruits, mushrooms, and understory plants provide a continuous source of supply for people living in forested regions. Eighty percent of people in the world's poorest countries depend on non-wood forest products directly for health and nutrition (**Figure 13.11**).

Non-wood products are also important sources of income. The United Nations Food and Agriculture Organization (FAO) lists over 150 non-timber forest products that total over $17 billion in international trade each year. Such products include honey, gum Arabic, rattan and bamboo, cork, forest nuts, mushrooms, essential oils, and raw materials to make pharmaceuticals.

Natural rubber is an especially important forest product. Nearly 6.0 million metric tons (6.5 U.S. tons) of rubber, with an estimated value of $14.7 billion, is extracted from tropical forest trees each year (**Figure 13.12**). Today, most natural rubber comes from Asia and Africa, where extensive tracts of tropical forest have been replaced by plantations of rubber trees.

In summary, forests are critical to human well-being in a variety of ways. They provide many essential ecosystem services, including carbon storage and the provision of clean water. They provide wood for fuel, paper, and construction, as well as a wide array of non-wood products.

▲ Figure 13.12 **Natural Rubber Extraction**
Latex tapped from rubber trees (*Hevea brasiliensis*) provides nearly half of the world's rubber. Although some latex is harvested from trees in natural forest ecosystems, most of it comes from plantations.

13.2 Forest Growth

BIG IDEA Much of a forest's value is derived from its patterns of growth. Forest growth can be considered at three levels—individual trees, communities of trees called forest stands, and large forested landscapes. Individual trees begin as seeds; as they grow taller, they are supported by growth in the diameter of their stems. Forest stands develop over time, undergoing predictable changes in species composition and structural complexity. Across landscapes, natural and human-caused disturbances produce a mosaic of forest patches undergoing different stages of development. Such mosaics influence forest biodiversity as well as the goods and services provided by the forest.

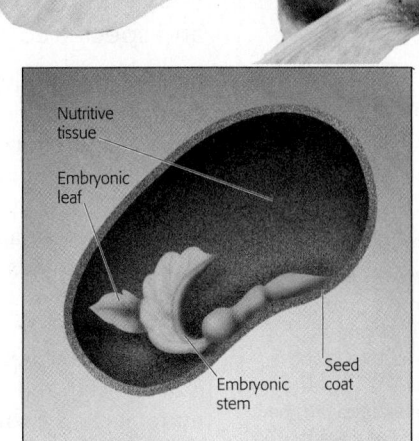

▲ **Figure 13.13 Trees Begin Here**
The seed of a tree includes the embryonic root, shoot, and leaves surrounded by nutritive tissue and a protective seed coat.

The Life History of a Tree

■ Wood supports the vertical growth of trees.

The life of an individual tree begins as a **seed**, which is an embryo embedded in nutritive tissue and surrounded by a protective coat (**Figure 13.13**). Trees are classified into two major groups, based primarily on the structures in which their seeds are produced. The seeds of **conifers** grow in cones. The seeds of **angiosperms** grow in flowers and are eventually surrounded by fruits that grow from the flower (**Figure 13.14**). Each species of tree has its own particular type of cone or flower.

◀ **Figure 13.14 Cones or Flowers**
In conifers, such as ponderosa pines, seeds develop in cones and are released when the cones open Ⓐ. In angiosperm trees, such as tulip poplar, seeds are produced in flowers Ⓑ.

When seeds are mature, they may be dispersed by wind, water, or animals. If a seed lands in soil that has the appropriate temperature and moisture conditions, it will germinate, extending a shoot and leaves above ground and roots into the soil. This upward and downward growth is called **primary growth**. Primary grow occurs because of rapid cell division and growth at the very tips of the shoots and roots (see Module 4.2).

When a tree first germinates, the carbohydrates it needs to grow come from the nutritive tissues in the seed. As the tree continues to grow, its leaves begin to produce carbohydrates by photosynthesis. These carbohydrates are transported down the stem to the roots in a thin layer of specialized tissue called **phloem**. Phloem encircles the stem just beneath the protective bark. The tree's roots absorb water and mineral nutrients from soil and transport them up the stem in a specialized tissue called **xylem**. Most of the cells of xylem are hollow with very thick walls comprising the chemicals cellulose and lignin. The woody interior portion of the stem is made up of xylem (**Figure 13.15**).

Each year, tree stems produce additional layers of xylem, or wood, which causes them to increase in diameter. This increase in girth is called **secondary growth**. In temperate and boreal climates with distinct growing seasons, the xylem cell walls that grow in spring are thinner than those that grow in summer, so the wood produced in spring is less dense than that produced in summer. This difference in growth results in distinct annual rings. In tropical climates where trees grow year-round, xylem cell walls have uniform thickness throughout the year, so annual rings do not form.

▲ **Figure 13.15 Tree Stem Cross Section**
Most of the interior of a tree stem or bole is composed of xylem. Phloem, the tissue that transports food and nutrients up and down the tree, forms a thin layer just beneath the bark.

The Life History of a Forest Stand

■ Communities of trees develop through four stages that differ in structure and species composition.

Trees play an especially important role in successional change in forest ecosystems (see Module 6.6). After a disturbance such as fire or logging, a forest goes through four distinct stages of succession (Figure 13.16). In the **establishment stage**, tree seedlings germinate on the recently cleared ground. Tens of thousands of tree seedlings may germinate on just one acre. With abundant light and soil resources, seedlings quickly grow into saplings. The species of trees that grow best under these conditions are likely to grow quickly and to compete effectively for water and nutrients in the soil.

Toward the end of the establishment stage, the leafy branches of the fastest growing trees begin to close in on one another, creating dense shade beneath the canopy. Competition for soil resources becomes more intense. Because of the intense competition for light and water, trees grow more slowly; trees that are not able to access enough light or water die. This inaugurates the next stage of forest development, the **thinning stage**.

During the thinning stage, the largest trees continue to grow and the total biomass in the forest continues to increase, but the total number of trees declines from several thousand per acre to a few hundred. Intense competition for light and soil resources severely limits the establishment of new trees. Because most of the trees became established at about the same time, thinning stage forests are said to be even-aged.

Toward the end of the thinning stage, trees are much larger and much more widely spaced. As these large trees die, they leave large openings, or gaps, in the forest canopy. At this point, total biomass may actually decline somewhat.

This is the beginning of the **transition stage** of forest development; this stage is also called the gap phase.

In the transition stage, increased light and reduced competition for soil resources allow tree seedlings to become established in the gaps. A gap receives less light than an open field, so shade-tolerant trees typically grow best there. The total biomass of the forest remains relatively constant as new trees replace those that die, but the diversity of trees usually increases. Because trees die at different times, the ages of the new trees vary greatly. Therefore, transition forests are said to be uneven-aged.

If transition forests are not cut or visited by a major disturbance, such as a fire or hurricane, they accumulate ever-increasing amounts of decaying logs and standing snags. The **old-growth stage** of forest development is characterized by very large, old trees and abundant standing and fallen woody debris. This is the most mature stage of forest development. The complex structure of the old-growth forest creates a variety of unique habitats. Many kinds of animals feed on and nest in the woody debris. The rich soils and favorable moisture and temperature conditions in the understory of such forests support a very diverse assemblage of herbs.

The time span over which these four stages occur varies considerably, depending on the conditions influencing tree growth and longevity. Where tree growth is rapid, as in tropical climates, the change from establishment to thinning stages may occur in less than 5 years, and transition-stage forests may be less than 100 years old. In most temperate evergreen and deciduous forests, the succession takes longer. In these cooler climates, it usually takes several hundred years for old-growth forests to develop.

▼ Figure 13.16 **Forest Stand Development**
Ⓐ Establishment stage. After a disturbance, the seeds of numerous trees sprout and begin to grow. Ⓑ Thinning stage. The fallen logs are the remains of smaller trees that were unable to compete for light and soil resources. Ⓒ Transition stage. As trees begin to die, gaps in the canopy allow light onto the forest floor, allowing new trees to become established. Ⓓ Old-growth stage. Living trees and abundant logs and understory litter create high structural complexity and promote biodiversity.

The Life History of a Forested Landscape

■ Landscapes are a mosaic of patches representing different stages of forest stand development.

Forested landscapes may extend over thousands of square kilometers. Such forests are made up of a mosaic of individual forest stands in various stages of development as a result of past disturbances. These forest patches vary in size because natural and human-caused disturbances occur at different spatial scales. An individual tree fall may produce a patch that is 20 m wide, whereas a large wildfire may produce a patch that extends over many kilometers.

Because some plant and animal species are typically found in establishment- or thinning-stage forests, and others occur primarily in transition or old-growth stands, you might predict that a mixture of patch types would promote biodiversity. However, transition and old-growth forest patches generally support many more species with very specific habitat requirements than do patches of less mature forest. The biodiversity of transition and old-growth forest patches tends to diminish if they are small and disconnected from one another (see Module 8.4).

Across large landscapes, net primary production and many other ecosystem processes are determined by the arrangement and relative abundance of patches in different stages. Patches in the establishment stage store relatively little carbon and have lower net ecosystem production than do later stages. Patches in this stage may be subject to erosion and nutrient loss. If many patches in the establishment stage are located near streams, large amounts of sediment and nutrients may be lost to stream waters.

In stands that are in the thinning stage, trees are rapidly taking up nutrients, and the loss of nutrients to streams is small. Net ecosystem production is high, and the amount of stored carbon is rapidly increasing. By the time patches reach the transition and old-growth stages, net ecosystem production slows. Even so, these stands store large amounts of carbon. They also take up large amounts of water, thereby limiting sedimentation and flooding in streams.

Over time, the forest landscape changes as disturbances create new patches and individual patches undergo succession from one stage to the next. Ecologists define the **minimum dynamic area** of a forested landscape as the area necessary to maintain all different patch types and populations of their associated species, given the typical patterns of disturbance. Minimum dynamic area is an important consideration in forest management and conservation.

If disturbances occur at irregular intervals and vary widely in size, the relative abundance of different patch stages will change considerably over time. The minimum dynamic area for such landscapes will be quite large. This appears to be the case in some regions where large fires naturally occur at intervals of hundreds of years. For example, researchers William Romme and Donald Despain have used tree ring data to determine how the lodgepole pine forest of the Yellowstone Plateau changed between 1700 and 1985. Their data show a steady progression of change in the dominance of different stages between major fire events (Figure 13.17). To conserve the biodiversity and ecosystem processes of such an area, forest parks and preserves must be quite large.

In contrast, where disturbances occur at regular intervals and create similar-sized patches, the relative abundance of patch stages is more constant through time. As a result, the minimum dynamic area is smaller. Where forests are managed for timber, harvests are usually carried out to maintain this kind of landscape. This ensures a steady supply of trees ready for harvest and constantly replenishes the landscape. However, trees are usually harvested during the middle or late thinning stage. Transition and old-growth patches are rare or missing, as are their unique biodiversity and the ecosystem services that they provide.

In summary, forests can be considered at scales ranging from the growth of individual trees, to stands initiated by a single disturbance, to landscapes with many stands in different stages of development following disturbances. Sustainable forest management involves decisions and actions at each of these scales.

QUESTIONS 13.2

1. What is the difference between primary and secondary tree growth?

2. Describe the four stages of forest stand development following a disturbance such as fire or logging.

3. Explain the minimum dynamic area concept.

(MES) For additional review, go to **MasteringEnvironmentalScience**

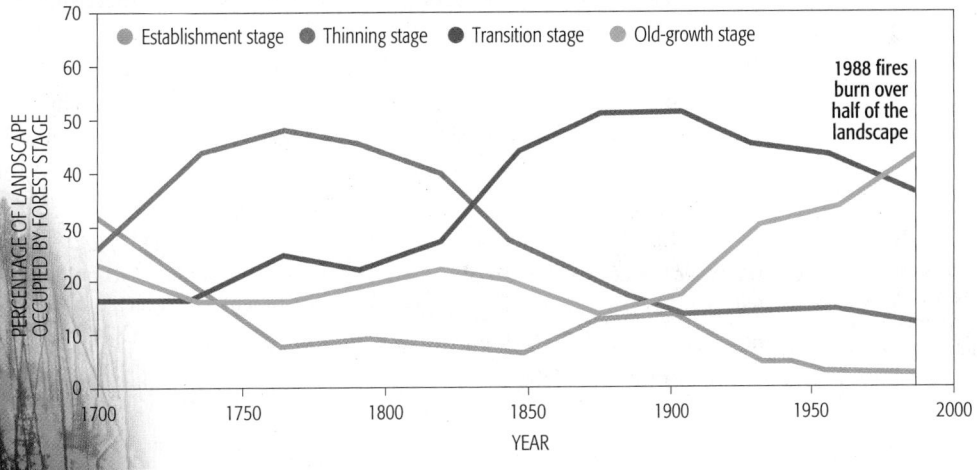

◄ Figure 13.17 **Forest Landscape Change**
Between 1700 and 1740, fires burned over half of the forest landscape of the Yellowstone Plateau. In 1740, the landscape was predominately covered by establishment and thinning stage forest. By 1850, transition forest patches became the dominant forest patch type. Just prior to the extensive 1988 Yellowstone fires, old-growth forest patches had become the most abundant forest stage on this landscape.

Source: Romme, W.H. and D.G. Despain. 1989. Historical perspective on the Yellowstone fires of 1988. *BioScience* 39: 695–699.

EXTENSIVE FIRES 1700–1740

INFREQUENT SMALL FIRES

● Establishment stage ● Thinning stage ● Transition stage ● Old-growth stage

1988 fires burn over half of the landscape

PERCENTAGE OF LANDSCAPE OCCUPIED BY FOREST STAGE

YEAR

CO₂ and the Growth of Forest Stands

Can rising carbon dioxide increase net primary production in forest ecosystems?

1. Why might experiments be more difficult to perform under natural forest conditions than in a greenhouse?

2. Some scientists argue that increased net primary production with elevated CO₂ might actually slow global warming. Explain their reasoning.

At its current rate of increase, the concentration of CO₂ in our atmosphere will reach 550 parts per million (ppm) by the year 2050. That will be twice the concentration of CO₂ prior to the Industrial Revolution. Emissions of CO₂ from burning fossil fuels and clearing land are the most important factors influencing global warming, yet atmospheric CO₂ is also used in the process of photosynthesis. Is it possible that higher concentrations of CO₂ will increase the net primary production (NPP) of the world's forests?

In highly controlled greenhouse conditions, ecologists have shown that a variety of forest plants have significantly higher rates of photosynthesis at CO₂ concentrations of 550 ppm than at current concentrations. Would such high concentrations of CO₂ increase NPP in natural forest ecosystems where other factors, such as variations in water and soil nutrients, are also important? Ram Oren and his colleagues at Duke University sought to answer this question with an elegant *experiment* in a natural forest ecosystem.

Oren and his team carried out their research at Duke University's Forest–Atmosphere Carbon Transfer and Storage (FACTS) facility, which is a natural forest area in which it is possible to control the level of CO₂ in the atmosphere (**Figure 13.18**). FACTS consists of a set of eight *experimental plots* inside a forest dominated by loblolly pine. Around each plot is a ring of large pipes that can pump CO₂ into the atmosphere; each ring has a diameter of 30 m (98.4 ft). Inside each ring, sensors measure CO₂ concentrations at numerous locations, from the soil surface to the top of the forest canopy. A system of computers adjusts the pumping rates to maintain the desired concentrations of CO₂.

In four of these rings, Oren and his team set the concentration of CO₂ at 550 ppm. The four other rings received no additional CO₂ and tracked current concentrations (360–380 ppm). Then

▲ **Figure 13.18 Enriching the Air**
Ram Oren and his team used the FACTS facility at Duke Forest to determine how increased levels of CO₂ affect forest productivity. Ring-shaped arrays of vertical pipes deliver and maintain CO₂ at concentrations that are nearly twice the normal level. Control arrays disperse air with normal levels of CO₂.

every year from 1994 to 2010, the team monitored the growth of trees, shrubs, and herbs in each of the eight rings.

Oren *hypothesized* that increased CO₂ would have its greatest effect on NPP when water and soil nutrients, such as nitrogen, were abundant. During dry periods and in places where nitrogen was in short supply, he *predicted* that elevated CO₂ concentrations would have little effect. To test this prediction, the levels of soil water and nitrogen were monitored in each ring. Two of the 550-ppm rings were divided in half; each year nitrogen fertilizer was applied to one half and the other half received none.

Over the course of the study, elevated CO₂ concentrations increased NPP by 22–30%. As predicted, this effect was highest in the places with the highest levels of soil nitrogen (**Figure 13.19**). The effect of elevated CO₂ on NPP also depended upon the availability of water: The effect was high in wet years and very low in dry years.

These results suggest that increasing concentrations of CO₂ will enhance the ability of forests to take up and store carbon. However, this effect will not be uniform. Rather, it will depend on the availability of other resources, such as water and soil nitrogen.

Source: McCarthy, H.R., R. Oren, K.H. Johnsen, et al. 2010. Reassessment of plant carbon dynamics at the Duke free-air CO₂ enrichment site: interactions of atmospheric CO₂ with nitrogen and water availability over stand development. *New Phytologist* 185: 514–528.

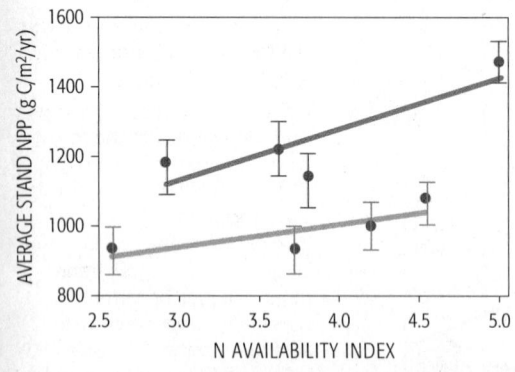

▲ **Figure 13.19 Elevated CO₂ Boosts NPP**
At each level of nitrogen availability, net primary production was higher under elevated CO₂ (red dots) compared to normal CO₂ conditions (blue dots). The vertical lines through each dot indicate the range of variation in treatment results over the course of the study.

13.3 Deforestation

BIG IDEA The United Nations Food and Agriculture Organization defines **deforestation** as "the removal of forest cover to an extent that allows alternative land uses." More simply, deforestation is the process in which humans replace a forest with something else, such as agriculture, pasture, or pavement. Each year 0.2% of the world's forests disappear, but within particular regions and forest types the rate of deforestation is often much higher. Reasons for deforestation include the expansion of agriculture, unsustainable harvest, road construction, and urban development. These activities are driven by poverty, misguided government policies and subsidies, and corruption.

▲ Figure 13.20 **The Loss of Ireland's Forest**
The Nephin Beg Mountains in Ireland were once covered by dense forest. Between 1600 and 1800, this forest was systematically cut for fuel and shipbuilding.

Historical Change

■ The pace of deforestation is quickening.

Scientists estimate that 10,000 years ago, forests covered 65.8 million square kilometers (25.4 million square miles), or nearly 50% of Earth's land. By 1750, the beginning of the Industrial Revolution, about 16 million square kilometers of forest had been permanently cleared for agriculture and urban centers (Figure 13.20). Today, the remaining forests cover less than 40 million square kilometers (15.5 million square miles), or 30% of Earth's land.

In southern and central Europe, deforestation began more than 2,000 years ago as human populations spread across the continent. Initially, land was mostly cleared for agriculture. Between A.D. 1100 and 1500, the rate of forest loss accelerated greatly in association with the warmer climate and rapid population growth. Cities were often built in the midst of forests, which were quickly cut for building materials and fuelwood.

Between 1500 and 1700, shipbuilding created an enormous demand for forest products (Figure 13.21). Much of the remaining forests were cut to provide wood, resins, and turpentine for the rapidly expanding navies and commercial shipping industry. Forest cutting expanded even more between 1750 and 1800, as industrial development and steam power increased the demand for wood and charcoal. By 1800, coal had replaced wood and charcoal as the industrial fuel, but the process of deforestation continued as agriculture expanded and the demand for wood for paper and construction increased.

Today, worldwide rates of deforestation remain high. The Food and Agriculture Organization of the United Nations estimates that about 130,000 km² (50,100 mi²) were deforested each year between 2000 and 2010. Taking into account the amount of land that was naturally or artificially replanted in trees, this resulted in an annual net loss of 52,000 km² (20,100 mi²), an area about twice the size of Massachusetts.

The rate of net forest loss between 2000 and 2010 was actually somewhat less than the average net loss between 1990 and 1999, which was 83,000 km² (32,000 mi²) per year. This decrease was partly due to a nearly 20% decline in the rate of deforestation in countries like Brazil, Peru, and Colombia. It was also due to increased efforts to reforest lands in several countries. For example, between 2005 and 2010, China reported planting trees on 2,000 km² (772 mi²) of land.

▲ Figure 13.21 **Built of Oak**
The development of Europe's navies contributed to widespread deforestation. Six thousand mature oak trees were required to build a single fighting ship such as these.

Q *What percentage of the world's forests have we lost in the past 10 years?*
Sierra Seaman, University of Florida

Norm: **A** Between 2000 and 2010, Earth lost about 2% of its remaining forests. The rate of loss for tropical and old-growth forests was considerably greater.

Causes of Deforestation

■ Conversion to agriculture and overharvest are accelerated by poverty, road and urban development, and political corruption.

Deforestation occurs as a direct consequence of agricultural expansion, wood harvest, road building, and urban development. There is much controversy regarding the relative importance of each of these causes. This is partly due to limited data. It is also due to the fact that these causes often operate simultaneously or sequentially on the same landscape.

Agricultural expansion.   In tropical regions, the conversion of forested land to cropland and pasture accounts for more than half of all deforestation. Much of this conversion is attributed to the practice of **shifting agriculture**. In shifting agriculture, forests are cut and the slash—the downed logs and branches—is burned. Burning releases nutrients to the newly cleared soil, making it ideal for agriculture. Crop production is initially high in such clearings. But after several years, productivity declines and fields are abandoned. Given enough time, the bare fields undergo natural succession and forest and soil productivity are eventually restored (see Module 6.5; Figure 13.22).

Shifting agriculture has been practiced around the world for thousands of years. Where farmed plots are small and the periods between forest cutting are long enough for soil productivity to be restored, this form of agriculture may be sustainable. However, the ever-growing numbers of subsistence farmers and changes in their cutting and cultivation practices are having unsustainable consequences. Today 300–500 million people practice some form of shifting agriculture. This has encouraged cutting on marginal lands as well as ever-shorter periods between forest cutting. Under these conditions, fragile tropical soils become infertile and are easily eroded, and reforestation becomes very difficult.

Expansion of global markets for livestock and cash crops, such as sugar, soybeans, and corn, is also responsible for much deforestation. Approximately 60% of the deforestation in Brazil is associated with the establishment of cattle ranches that provide beef to meet growing global demand, driven in part by the fast-food industry. In recent years, much of this pasture has been converted to high-value crops, such as soybeans and sugarcane, which are used for ethanol. This conversion has, in turn, led to even more forest cutting to create more pasture.

Wood harvest.   Much deforestation is a direct consequence of cutting trees for fuelwood. Over 90% of fuelwood is consumed in underdeveloped countries, where wood is the most important source of energy for heating and cooking. For example, most of the wood harvested in Africa is used locally for fuel. Population growth has increased this demand and has been responsible for the conversion of millions of acres of arid woodland to desert.

Global demand for wood and wood products has also accelerated deforestation in many places. Commercial logging for high-quality woods, such as teak, mahogany, and ebony, is a major factor in the loss of tropical forests in Indonesia and Central America. Global demand for paper has encouraged the clearing of large tracts of old-growth forest and their replacement with plantations of fast-growing trees, such as eucalyptus and pine.

▼ **Figure 13.22 Shifting Agriculture**
This mosaic of landscape patches in Indonesia is the result of shifting agriculture. The lighter color patches are being actively farmed. Other patches have been abandoned and are undergoing succession back to forest.

Roads and urban development.

Although it accounts for a comparatively small portion of total global deforestation, the development of urban infrastructure, including roads and housing, contributes to significant forest loss in many areas. This is particularly true in many developed countries. In the United States, approximately 1,800 km^2 (695 mi^2) is deforested for transportation and urban development each year.

To encourage economic development, many countries have undertaken road and railway expansion projects that have a significant effect on deforestation. Such projects are often done with financial support from developed countries. Such roads provide access to uncut forest and facilitate more forest cutting (Figure 13.23).

Economic and political causes.

Poverty is often cited as the most important factor contributing to deforestation worldwide. This is, however, an oversimplification. It is true that deforestation rates are highest among the world's poorest countries and that poverty causes people to migrate to forest frontiers where they engage in subsistence farming. But many other political and economic factors also drive deforestation and forest degradation. For example, agricultural subsidies and timber concessions aimed at encouraging economic growth often encourage deforestation in developing countries.

Many developing countries have passed laws to regulate forest cutting and limit deforestation. However, deforestation continues because of governmental corruption and because governments have very limited capacities to enforce their laws. Illegal logging accounts for over half of the timber production in Brazil, Russia, Indonesia, and many African countries. This logging accelerates deforestation and deprives the citizens of these countries of much needed revenue. Illegal logging is estimated to cost the world's developing countries $15–$20 billion each year.

Industrial countries in North America and Europe lead the world in rates of reforestation. At the same time, their patterns of consumption are driving deforestation in many developing countries. Although this is partly due to the demand for wood and wood products, it is also driven by the demand for other commodities. The high consumption of beef associated with the expansion of the fast-food industry, for example, is directly correlated with deforestation to create pasture for cattle in the Amazon. Likewise, the demand for gasoline in developed countries creates incentives to replace forests with agricultural crops that can be used to produce gasoline substitutes, such as ethanol.

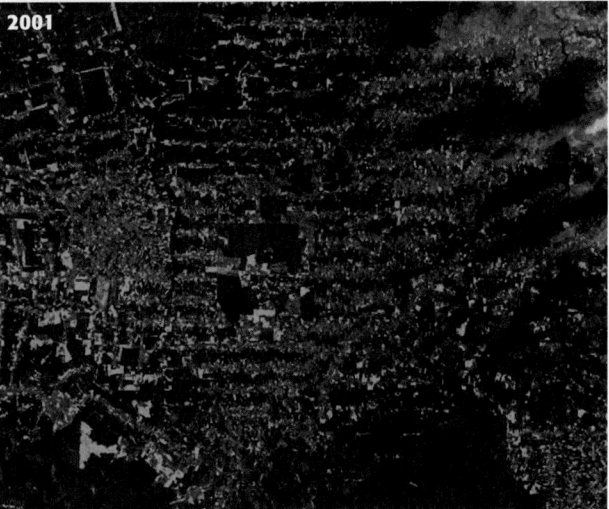

◄ Figure 13.23 **Roads Facilitate Deforestation** This sequence of satellite photographs shows the rapid deforestation associated with roads (light green lines) in a portion of the Amazon Basin in the Brazilian state of Rondonia between 1975 and 2001. The images are approximately 100 km (62 mi) wide.

QUESTIONS 13.3

1. What factors have contributed to the global loss of forests over the past 10,000 years?

2. Describe the factors that can cause shifting agriculture to lead to extensive and permanent deforestation.

3. Why has road development accelerated deforestation in many countries?

(MES) For additional review, go to **MasteringEnvironmentalScience**

How Can Deforestation Be Halted?

■ Slowing worldwide deforestation will require multiple strategies.

Just as deforestation has multiple causes, controlling deforestation will require multiple strategies. These include improved agricultural practices, management of population growth in forested regions, reduced demand for virgin wood products, reforestation and sustainable forest management, and sustainable economic development.

Intensive agricultural methods, including the use of high-yield crops and the heavy application of fertilizers, herbicides, and pesticides, can produce more food on less land, thereby decreasing the need to convert forests to farmland. However, these methods depend on chemical and energy inputs that often have unsustainable consequences.

Many governments and development organizations are working to limit migration and population growth in frontier forest areas. To be successful, such efforts must be coupled with programs that alleviate poverty and provide economic opportunity. Ecotourism has provided

important economic incentives for forest conservation (see Module 8.8).

In the past, governments and development banks, such as the World Bank, have had inconsistent policies, often funding projects to establish forest preserves through one department while funding road construction projects through other departments. Recently, there has been an increasing awareness of the need to limit road construction through regions where forest preservation is a high priority.

Perhaps most important, many organizations and governments in developed countries are implementing programs aimed at limiting the consumption that encourages deforestation. For example, the McDonalds Corporation has committed to using recycled paper products and to procuring its beef only from suppliers that can certify that their activities have not caused deforestation.

New FRONTIERS

Spying on Forests

Countries such as Brazil, Peru, and Indonesia have passed laws and made commitments to slow the rates of deforestation on their lands. To verify that those commitments are being met and laws are being obeyed, ecologists have borrowed techniques from Central Intelligence Agency. Satellite systems currently provide images of Earth's entire land cover several times each year at spatial resolutions of 30 m (100 ft). Matthew Hansen and his colleagues at the University of Maryland used these data to map forest cover worldwide each year between 2000 and 2012. With a high-speed computer and geographic information systems (GIS) software, they were then able to determine whether forest was lost or gained at every location on Earth from one year to the next. Their estimates of global annual rates of forest loss and gain are similar to those reported by the FAO. More important, they provided an independent assessment of each country's forest cover trends.

There was good news and bad news in this analysis. The good news is that Brazil has cut its deforestation rate by half, from over 40,000 km²/yr to less than 20,000 km²/yr, since 2000. Unfortunately, Brazil's performance has been more than offset by a two- to threefold increase in deforestation in other tropical countries such as Indonesia, Malaysia, Paraguay, and Zambia (**Figure 13.24**).

▲ Figure 13.24 **Going, Going …**

This image maps forest loss and gain over part of Sumatra, Indonesia, between 2000 and 2012. "Mixed" indicates areas where was forest lost, but subsequently regrown. Note that some of the most extensive loss was in the Tesso Nilo National Park.

Source: Hansen, M.C., et al. 2013 High-resolution global maps of 21st century forest cover change. *Science* 342: 850–583.

As we will see later in this chapter, country-by-country estimates of forest cover change are beginning to play a very important role in international negotiations on carbon emissions and biodiversity loss. So, these monitoring tools will become ever more important. What do you think about spying on the world's forests to save them?

Restoring Forests and Community Well-Being in Haiti

How can very poor people in rural communities be engaged in the restoration of their forests?

In places such as the Mayan Forest of Central America and the Atlantic Forest of Brazil, halting deforestation and conserving the biodiversity and ecosystem services of the remaining forests is the central priority. In Haiti, opportunities for forest conservation are all but gone, and there are numerous obstacles to forest restoration. Perhaps most important, the extreme poverty of Haiti's people provides few incentives for forest stewardship.

Recent community-based initiatives show some promise of overcoming these obstacles. Near the watersheds of Cormier and Fond de Boudin in Haiti's southern mountains, the Comprehensive Development Project (CODEP) is working with nearly 650 farmers to reforest entire watersheds. The first steps in this reforestation are to stop further erosion and to reclaim soils (Figure 13.25). This is done by digging contour canals. Grasses are planted on the downhill side to prevent erosion, and fast-growing trees like eucalyptus are planted above. The canals retain water and collect falling leaves. In two or three years, soils have been restored from the natural compost in these canals so that gardens, fruit trees, and coffee can be grown

▼ Figure 13.25 **First Steps**
On barren slopes, CODEP volunteers have dug canals along the contours of barren slopes. Grass has been planted on the downslope side of canals, and eucalyptus trees will be planted upslope. Canals will slow water flow and capture tree litter to restore soil organic matter.

successfully (Figure 13.26). CODEP works with individual communities to provide training in techniques to limit erosion, restore soil fertility, and plant trees.

Just as important, CODEP programs focus on economic development and the well-being of families. Participating communities volunteer a day of work each week. In return, they are guaranteed to receive much-needed pay for working an additional two days. As they work, farmers earn credits that can be used to purchase water cisterns, hurricane-proof homes, tools, and fertilizer. They are also eligible for microeconomic loans. CODEP helps support several schools that teach environmental reclamation and have plant nurseries on site.

CODEP also works with communities to build and maintain fishponds. Tilapia grown in such ponds provide a source of dietary protein, thus diminishing the demand for grazing animals. When the fish are sold at market, they provide families with additional income.

As of 2013, CODEP partners have 10 million trees in the ground, of which approximately 20% are fruit trees. They have also constructed 30 working fish ponds, over 200 water cisterns, and 20 homes. In response to the 2010 earthquake, they have built an additional 39 homes in partnership with the Building Goodness Foundation of Charlottesville, Virginia. CODEP has been working in the area for 20 years and it is just now ready to say that the development effort is working—because the people are healthier, they have improved nutrition, and young folks are staying in the area rather than moving to large cities like Port-au-Prince. The community-based approach, CODEP people believe, is the key to restoring both forest ecosystems and the economy.

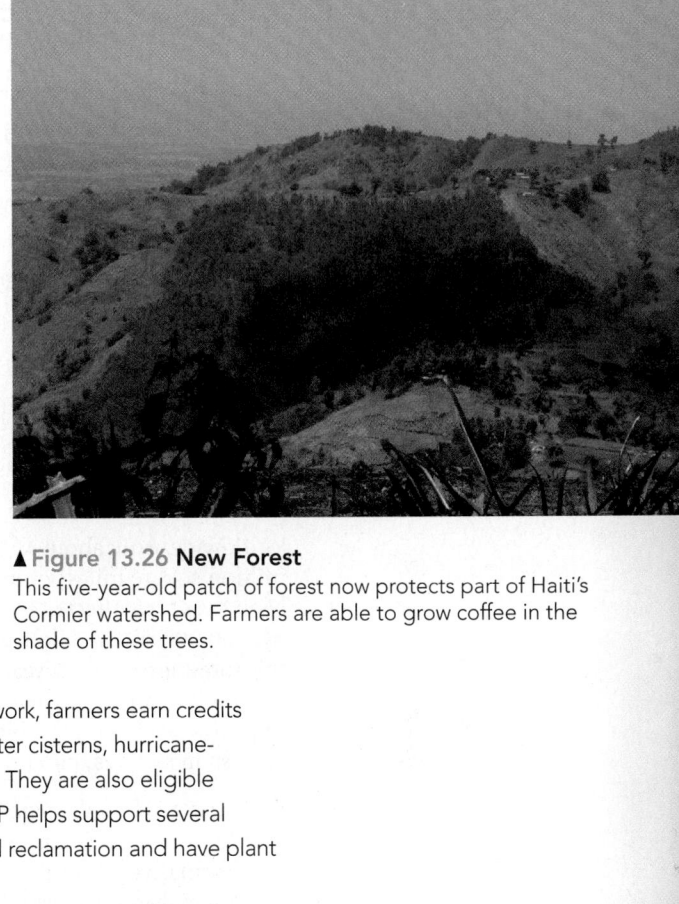

▲ Figure 13.26 **New Forest**
This five-year-old patch of forest now protects part of Haiti's Cormier watershed. Farmers are able to grow coffee in the shade of these trees.

13.4 Forest Degradation

BIG IDEA As the total area of Earth's forests is decreasing, the overall condition of the remaining forests is also deteriorating. **Forest degradation** occurs when changes in forest structure and composition diminish biodiversity or key ecosystem services. Logging and other human activities are simplifying once complex and diverse native forests. Agriculture, roads, and urban development are dividing forest tracts into ever-smaller pieces and preventing the migration of forest species. In many forests, natural disturbance processes have been altered, thereby reducing biodiversity. The introduction of non-native species threatens forest biodiversity and the ecosystem services it supports.

Forest Health in Peril

■ Cutting, fragmentation, altered disturbance patterns, and non-native species have degraded forests worldwide.

Worldwide, only about 19 million square kilometers (7.3 million square miles) of **primary forest**—forest that has been largely unaffected by humans—remains. Currently, over 60,000 square kilometers (23,000 square miles) of this forest are cut each year. Although some of this land is naturally or artificially reforested, the **secondary forests** that grow back have less biodiversity and provide fewer ecosystem services than primary forests.

Cutting and high grading. The temperate deciduous forests of eastern North America provide an informative example of the large-scale conversion of primary forest to secondary forest. Between 1650 and 1850, nearly 65% of the forests in this region were cut and converted to farmland. Forests that were not cleared were degraded by heavy livestock grazing and **highgrading**, the removal of particular species of trees for fuel and construction.

Following the Civil War, many of the farms in this region were abandoned. Old fields were quickly invaded, first by herbs and shrubs, and then by forest trees (see Module 6.6). Today, forest cover in the eastern United States is about 70% of what it was in 1600. However, less than 0.1% of that cover is primary forest, and most of that primary forest is in small, widely scattered patches. Much of the forest may have been restored, but it is not

nearly as diverse as it once was. Recent studies show that the diversity of herbs and birds in primary deciduous forests is 20–50% higher than in secondary forests.

Fragmentation. Throughout the world, forest landscapes are becoming increasingly fragmented. Forest patches are smaller and more widely separated. Roads, agriculture, and urban development make it more challenging for forest organisms to travel between patches. As predicted by the equilibrium theory of island biogeography, this fragmentation has diminished the biodiversity of animal groups such as migratory songbirds (see Module 8.5).

Altered patterns of disturbance. The natural frequency and severity of fires vary widely among different forest types. Nearly everywhere, humans have modified fire regimes, often in ways that have degraded forests.

In the mid-19th century, the ponderosa pine forests extending from eastern Washington and Oregon to Arizona and New Mexico were typically quite open with a grassy understory. This pattern of growth allowed light fires to burn across the surface of the ground every two to four years. Between 1880 and 1900, heavy livestock grazing in much of this region reduced the grass cover

▶ **Figure 13.27 Fire Suppression Creates Risk** This sequence of photos was taken in the same ponderosa pine stand over during the last century. Prior to 1900, frequent fires in the understory kept the stand open Ⓐ. By 1949, in the absence of fire, young trees had invaded the understory Ⓑ. Today, if a fire ignites in this stand, it will likely be very severe Ⓒ.

Ⓐ 1900

Ⓑ 1949

Ⓒ 1999

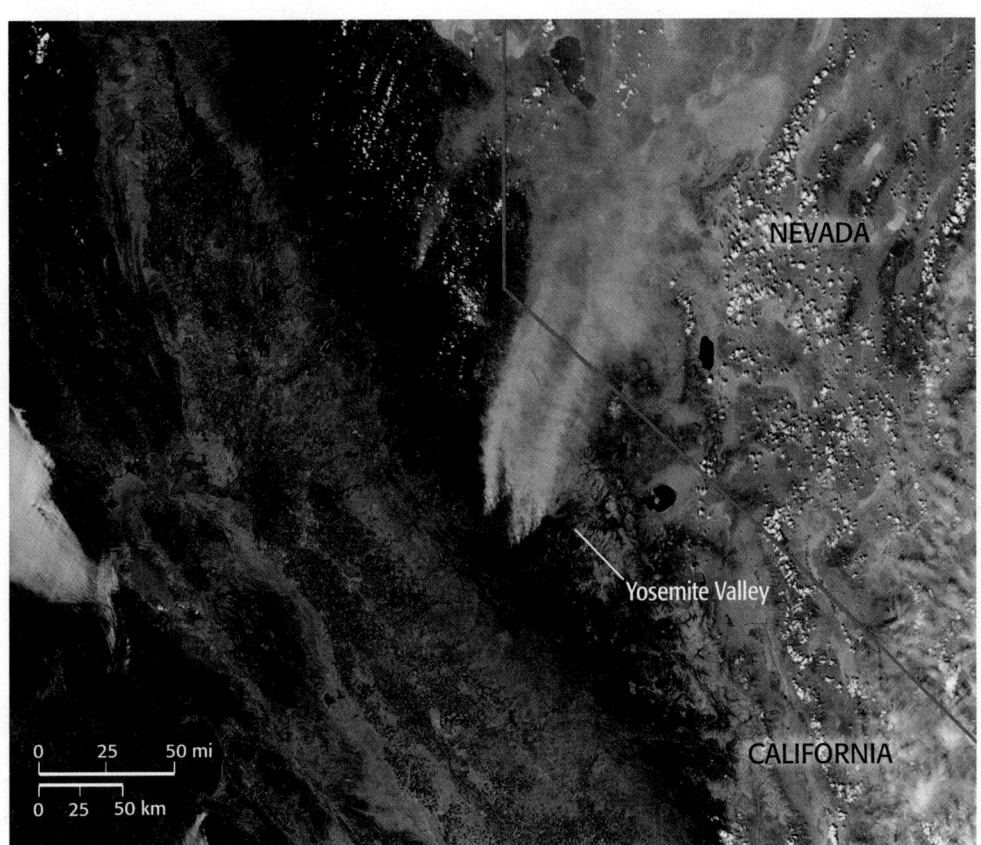

◄ **Figure 13.28 Megafire**
The 2013 Rim Fire burned 104,000 ha (257,000 acres) in the Sierra Nevada of central California, including a large area in the northern part of Yosemite National Park. As you can see in the satellite image, smoke from the blaze stretches into Nevada. The size and severity of this fire, which was unprecedented in the Sierra Nevada, was attributed to a combination of severe drought and great quantities of brush and other flammable materials that had accumulated during years of fire suppression.

QUESTIONS 13.4

1. How do primary and secondary forests differ in terms of biodiversity and the history of past human disturbances?

2. Describe two threats to the health of forests in the United States.

(MES) For additional review, go to **MasteringEnvironmentalScience**

so that there was insufficient fuel to carry fires. Without regular fires, tree seedlings and woody shrubs invaded the forest floor (Figure 13.27). In the decades that followed, fire suppression favored the continued growth of the woody understory. Because of this accumulation of potential fuel, when fires occur in these forests nowadays, they are often quite large and severe (Figure 13.28).

Recently, there has been a dramatic increase in the frequency of fires in tropical rain forests, such as those in the Amazon and Central Africa. Usually these fires are set by humans; many are set to burn slash left after logging, but the fires escape into nearby forests. These fires may not totally destroy the forest overstory, but they do kill many understory species and diminish overall forest diversity.

Non-native species. The introduction of non-native species has degraded forests in many regions. In 1748, a gardener introduced an Asian species called tree of heaven to Pennsylvania. Tree of heaven thrived in the climates of the eastern United States and spread rapidly. It is particularly well adapted to colonizing and outcompeting native trees in forest openings and disturbed locations.

Non-native fungal diseases and insect pests have had an especially significant impact on the health of many forests. The fungus that causes chestnut blight has all but eliminated the American chestnut from eastern forests, and the aphidlike hemlock adelgid seems to have doomed the eastern hemlock to the same fate (see Module 8.5).

The gypsy moth is native to Europe and Asia, where its larvae feed on the spring leaves of deciduous trees, especially oaks. It was introduced to the United States in 1868 by a Massachusetts scientist interested in crossing it with silkworms. Lacking its natural predators, the moth has quickly spread through eastern forests. Gypsy moth larvae can consume nearly all of the leaves of canopy trees (Figure 13.29). This weakens trees and makes them more vulnerable to drought and fungal disease.

Halting deforestation—keeping forests as forests—is a central conservation goal. However, that objective alone is insufficient to preserve the biodiversity and ecosystem services provided by Earth's forests. Sustainable forest conservation must also include the protection of primary forests, connection of forest patches, restoration and maintenance of natural disturbance cycles, and management of non-native species.

▶ **Figure 13.29 Gypsy Moth Caterpillar**
Caterpillars of gypsy moths feed on the leaves of many trees, often completely defoliating them. Over several years, such defoliation can result in the death of trees.

Q Can harvesting trees for heating fuel be a sustainble process?

Glenda Sullenger, St. Charles Community College

Norm: A Yes it can, so long as harvest rates do not exceed the rate of forest recovery and care is taken to manage pollution from wood smoke.

13.5 Defining Sustainable Forest Management

Humans manage forests for different purposes and with differing impacts. At one end of the spectrum, forest preserves are managed with minimal human intervention. At the other end of the spectrum, plantation forests are intensively managed to produce wood and fiber, somewhat like an agricultural crop. Between these two extremes, mixed-use forests are managed at varying levels of intensity and for a variety of objectives, including wood fiber, recreation, water protection, and biodiversity. Forest management across this spectrum entails four categories of action: allocation, harvest, rationing, and investment. Each of these categories includes decisions and actions that influence forest sustainability. The United Nations has established seven criteria to guide nations in the sustainable management of their forests.

Allocation

■ Which forests should be dedicated to which management goals?

Allocation is the management of the location and relative amounts of different forest uses. Managers allocate forestlands according to their specific objectives. For example, landowners who are interested in using forests for commercial wood production, but are also interested in maintaining water quality, might locate forest preserves near streams and plantations away from streams. To sustain biodiversity, they might create corridors of mixed-use forests to allow wildlife to migrate between the preserves (Figure 13.30).

Environmental features such as topography, economic factors, and legal restrictions influence the allocation of forest use. Approximately 17% of the world's forests are designated as preserves, although the extent of protection of such forests varies widely from country to country.

Currently, 8% of the world's forests are allocated to intensively managed plantations; such forests produce approximately 30% of commercial wood fiber.

Environmentalists disagree about the role that plantations should play in global conservation. Because plantations provide fewer ecosystem services than other forms of forest management, some argue that it is undesirable to allocate increased amounts of forests to plantations. Others note that plantation forests provide a disproportionate amount of the world's commercial wood fiber. If the allocation to plantations were increased to 12%, such forests could possibly meet over 50% of the global demand for wood fiber. Such an increase could relieve the pressure for legal and illegal logging in other forests.

▼ Figure 13.30 **Forest Corridors**

The corridors of forest across this landscape provide migration routes for a variety of forest species, thereby helping to maintain biodiversity.

Harvest

■ How should goods and services be taken from forests?

Harvest is the extraction of goods or services from a forest. Often there are several ways to harvest goods. When deciding what methods of extraction to use, managers must consider what is to be taken from the forest and also what is to be left behind.

Clear-cutting, the complete removal of forest canopy trees, is certainly the most controversial harvest practice. In the western United States, extensive clear-cutting of old-growth forests growing on steep slopes has produced significant erosion and flooding; it has also had devastating effects on wildlife habitat. In other types of forests, however, clear-cutting on relatively flat terrain may be a sustainable practice (Figure 13.31). In such situations, forest managers often clear-cut patches of 10–20 ha (25–50 acres) to simulate patches of natural disturbance. Because clear-cutting produces trees of uniform age, it is often called **even-aged management**.

To sustain biodiversity and ecosystem services, forest managers are increasingly turning to less intensive methods of harvest. In various kinds of **uneven-aged management**, trees, dead snags, and woody debris are retained to produce a more heterogeneous age structure. The simplest forms of uneven-aged management are seed cuts, in which select trees are left behind to provide seeds to accelerate the establishment stage of forest recovery. Other forms of uneven-aged management preserve habitat for particular species and maintain tree diversity while allowing the removal of valuable wood or non-wood forest products.

Rationing

■ What amount of forest goods and services should be dedicated to human use?

Rationing is the process of determining the amounts of forest goods or services to be used by humans. Examples of rationing include decisions about how much wood to harvest each year and how many visitors to allow in a forest reserve.

In intensively managed forests, rationing of wood harvests is determined by **rotation time**, the average interval between successive cuts. Rotation time depends on the intended use of the wood. For example, paper is most efficiently produced from small-diameter trees; therefore, rotation times for paper production are quite short, varying from 7 to 25 years. In contrast, the production of high-quality wood for furniture requires much larger trees and may require rotation times exceeding a century (Figure 13.32).

▲ **Figure 13.31 Unsustainable Harvest?**
Clear-cutting forests on steep terrain and fragile soils can produce severe erosion and very slow rates of reforestation. (A) The loss of forests from such landscapes increases runoff. Selective clear-cutting, however, may be appropriate on gentle slopes to regenerate forests in which the dominant trees like these pines grow best in full sunlight (B).

▼ **Figure 13.32 Short and Long Rotation Times**
Plantations of eucalyptus trees, which are grown for wood fiber, grow so rapidly that their rotation times may be as short as seven years (A). In contrast, oak forests that are grown for high-quality timber may have harvest times that are longer than 100 years (B).

Increasingly, forest managers are striving to select rotation times that are similar to the return times that forest communities require to recover from natural disturbances, such as fires. Research has shown that such rotation times are more likely to support overall forest biodiversity.

Sustainable forest management must address human use of all forest goods and services. Thus, managers set bag limits on wildlife and the collection of understory herbs. Restrictions on intensive recreational use of public forests and preserves are an important type of forest rationing.

Investment

■ What actions should be taken to restore and maintain forest health and productivity?

Investment is the wide array of management strategies used to restore, maintain, or protect a forest's capacity to provide goods and ecosystem services. The application of fertilizer to improve plantation growth and the construction of roads for logging access are examples of investments directed toward commercial wood extraction (Figure 13.33). Understory thinning and managed burns to reduce the risk of catastrophic wildfires are examples of investments aimed at forest protection. The installation of nesting cavities in longleaf pine trees for endangered red-cockaded woodpeckers is an example of a management investment aimed at conserving biodiversity.

Sustainable forest management begins with a clear articulation of management goals and then ensures that those goals can be achieved for generation after generation. Such goals set priorities for the forest goods and services that managers wish to sustain. Most important, sustainable forest management is not defined by specific choices, such as clear-cutting versus selective harvest. Rather, it encompasses a broad array of choices about when, how much, and by what methods products are harvested, and what to invest to restore and maintain forests.

▼ Figure 13.33 **Investing in the Future**
Replanting trees following harvest is an example of investment management.

Criteria for Sustainable Forest Management

■ Sustainable forest management is concerned at least as much with what it leaves behind as with what it extracts.

Sustainable forest management maintains the biodiversity, productivity, capacity for regeneration, and health of forests and forest landscapes for current and future generations. In 1994, delegates from United Nations member nations meeting in Montreal, Canada, used this definition to identify standards for sustainable forest management. They proposed seven criteria for setting sustainable management goals. For each criterion, they identified several indicators of success. These have come to be known as the Montreal Criteria and Indicators. The seven criteria are listed below, followed by a description of some of the important indicators.

1. Biological diversity.

Sustainable forest management requires biological diversity from the level of individual trees to entire forested landscapes. Managers should maintain genetic diversity within individual tree species. Individual stands should be managed to maintain a diverse number of native species and to minimize the impact of invasive non-native species. At the landscape level, indicators include the abundance of different forest types and stages of succession, the extent of forest fragmentation, and the protection of rare, unique, or threatened forest types.

2. Productive capacity

Individual forests and forested landscapes should maintain their capacity to produce a full array of forest goods and services. Indicators of productive capacity include the total area of forested land, the relative amounts of lands managed for different purposes, and the rate at which timber and non-wood products are removed from forests compared to the rate at which they are regenerated.

3. Ecosystem health and vitality

Factors that present risks to individual species and forest landscapes must be managed. Indicators of ecosystem health include the amount of forest that is abnormally affected by factors such as insects, disease, fire, land clearance, and animal grazing, and the amount of forest that is affected by air pollutants, such as ozone or nitrogen oxides.

4. Soil and water resources

Soil and water resources, as well as forests' protective and productive functions, must be conserved. Key indicators of unsustainable management include the area of forestland with significant soil erosion, the area of forestland with significantly diminished soil fertility, and the number of polluted streams and lakes in forested regions. An indicator of sustainable management would be the area of forestland managed for protective functions, such as water management and avalanche control (Figure 13.34).

5. Global carbon cycles

Forests should be managed to maintain their ability to take up and store carbon. This criterion recognizes the important role of forests in determining CO_2 fluxes to and from the atmosphere. The most important criterion is the total amount and stability of biomass on forested landscapes. Other criteria include the patterns of carbon uptake and loss among different forest types and successional stages and the storage of carbon in forest products, such as paper and wood.

6. Socioeconomic benefits

Forests should maintain and enhance long-term socioeconomic benefits. One set of indicators is the amounts of wood and non-wood forest products produced, consumed, and recycled. Another set of indicators is the amount of forestland conserved for recreation and tourism and the rate of their visitation and use. Other indicators include direct and indirect employment in forest management or the production of forest products, the amount of land used for subsistence purposes, and the amount of forestland managed to protect cultural, social, and spiritual needs. In this regard, an especially important indicator is how well forest-dependent communities, including indigenous peoples, adapt to changing economic conditions.

7. Legal, institutional, and economic framework

The overall policies and laws of countries should facilitate the conservation and sustainable management of forests. Legal indicators include laws that clarify property ownership and the rights of indigenous people; laws should also set standards and regulate management practices. The ability to enforce those laws and regulations is an important measure of success, as is the public's involvement in and awareness of forest management issues. Other important indicators include the capacity to measure and monitor changes in forest status and the capacity to conduct and apply research to improve forest management and the delivery of forest goods and services.

▶ Figure 13.34 **Sustainable Forestry**
This photo shows forested buffers along streams in an area that is being logged. Such buffers prevent erosion and protect water quality.

QUESTIONS 13.5

1. Describe four categories of forest management activities.

2. For each forest management category, describe a management action that would likely have unsustainable consequences.

3. What are the seven Montreal criteria for sustainable forestry?

(MES) For additional review, go to **MasteringEnvironmentalScience**

The University of Winnipeg's Campaign Against Logging in Parks

Robin Bryan lives close to the boreal forest of the East Shore Wilderness Area in Manitoba, Canada, the world's largest single land storehouse of carbon and most abundant source of fresh water. While a student at the University of Winnipeg, Robin led a campaign against logging in Manitoba's provincial parks. In 2008, he was rewarded for his efforts when Manitoba banned logging in four of the five parks that had logging operations.

How did you first get the idea for a campaign against logging in Manitoba's provincial parks?

I grew up in Manitoba, a province containing the largest tracts of pristine, unbroken forests on the planet. Here a network of clean rivers and lakes interweave among lush evergreen foliage and rock ridges. These vast forests are also home to a host of hoofed mammals, migratory birds, amphibians, and fish.

In my early teens, I learned that commercial logging and mining operations were degrading these amazing natural wonders. Our provincial government was not taking meaningful actions to protect the forests, and their lack of sustainable ecological planning was placing our ecological resources in danger. As I began learning about this situation, I knew I had to change it.

I became involved with the Wilderness Committee, where the mentors I met sparked my interest in becoming a conservation activist. My first major activity was to participate in the ongoing campaign to stop clear-cut logging in Manitoba's provincial parks.

What steps did you take to get started?

First, I carried out vigorous research to gain knowledge and perspective on the issues of sustainability, logging practices, and policies. This knowledge helped build my confidence as an activist. I continued by learning about volunteer management, fundraising, organization structure, persuasive writing, and public education.

Because the success of my work rested upon my ability to speak clearly, I began practicing speaking to people one-on-one and to small groups. Eventually I overcame my fear of public speaking and was able to feel confident when speaking with elected officials, organizing rallies, delivering classroom presentations, raising funds, and organizing volunteers to write and send letters to the government.

What are the biggest challenges you've faced with the project, and how did you solve them?

As a person who is passionate about environmental protection, it is hard to face the fact that many people in our communities have limited education or experience with environmental issues, causing them to misunderstand how important these issues are. Because of public apathy, things are often slow to change, especially when it comes to changing major legislation.

During long periods of hard work with little visible impact, burnout can become a problem for many activists. When facing burnout, it was important for me to set boundaries between my activism and my personal life. I needed to recognize the limits of what I could contribute to the cause while meeting my own needs and maintaining my health. It was also important to surround myself with peers who could share my inspirations and frustrations. This was crucial to unwinding and feeling connected to my community. In the end, I learned to push through the rough patches and stick to the campaign until the job was done.

How has your work affected other people and the environment?

Over the course of four years, I worked with the Wilderness Committee and the community to end the practice of logging within provincial park boundaries. In 2008, legislation banned logging in all but one park in Manitoba, resulting in the protection of one million acres of land. I truly feel that ending logging in Manitoba's parks was a victory for all Manitobans. It was a testament to the power of proactive citizen involvement in a democracy to build change for our ailing environment.

What advice would you give to students who want to replicate your model on their campus?

Start by gaining knowledge. Build confidence in your abilities as an activist by practicing your communication skills. You must overcome any fears you have of speaking out in public about contentious subjects such as environmental protection.

To see a major campaign succeed, be prepared to work hard for years until your vision is realized. Be prepared to face apathy and criticism, but try to have compassion for those who oppose what you stand for.

Getting involved with an organization as a volunteer or an entry-level employee is a great way to access resources and begin developing your own vision. Books like *The Young Activist's Guide to Building a Green Movement and Changing the World*, recently produced by Sharon Smith of the Brower Youth Awards and Earth Island Institute, can also provide insight.

The most important thing is to find your passion, aim high, and don't give up until you've done what you set out to do.

Summary

Forests contain a major share of Earth's biodiversity. They also provide essential ecosystem services on which we all depend. They are the source of wood products such as fuelwood, lumber, and fiber for paper manufacture; they are also the source of numerous non-wood products, including pharmaceuticals, mushrooms, wild nuts and fruits, and rubber. Forests grow as individual trees, germinate from seeds, and then increase in height and girth. Forest stands develop through a series of successional stages, beginning with the establishment of young trees and culminating in uneven-aged, old-growth forest. Forested landscapes comprise patches representing these different developmental stages. Over the past 10,000 years, the total area of Earth's forests has been reduced by 40%, with much of the deforestation occurring over the past 300 years. Most deforestation is due to clearing for agriculture and harvest for fuelwood and commercial wood products. In many countries, poverty, road construction, and corruption encourage deforestation.

The health of the world's remaining forests is declining due to unsustainable management, fragmentation, altered patterns of disturbance, and invasion of non-native diseases and pests. Sustainable forest management includes decisions about which forests should be used in which ways, how much product should be harvested, how harvesting should be done, and what investments should be made to ensure future forest diversity and productivity. As human demands for forest products and ecosystem services increase, sustaining Earth's forests will require clear standards for the management of forests and the marketing of forest products. The world's forests can play a key role in mitigating and adapting to global warming.

13.1 The Values of Forests

- Forests provide essential ecosystem services, including flood protection, clean water, recreation, and inspiration.
- Humans depend on forests for both wood and non-wood products.

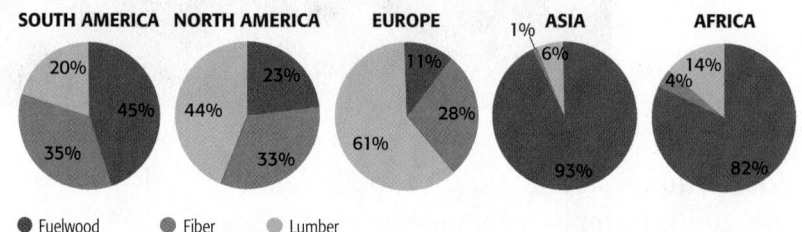

SOUTH AMERICA 20% 45% 35%
NORTH AMERICA 23% 44% 33%
EUROPE 11% 28% 61%
ASIA 1% 6% 93%
AFRICA 14% 4% 82%

● Fuelwood   ● Fiber   ● Lumber

QUESTIONS

1. Explain how diminished biodiversity influences three important ecosystem services provided by forests.
2. How is demand for various wood products likely to change as economic development proceeds in poor regions, such as sub-Saharan Africa?

13.2 Forest Growth

- Trees germinate from seeds and grow in both girth and height.
- Over time, populations of trees in forest stands undergo successional development, passing through the establishment, thinning, transition, and old-growth stages.
- Forested landscapes are composed of patches representing different forest stand developmental stages.

KEY TERMS

seed, conifer, angiosperm, primary growth, phloem, xylem, secondary growth, establishment stage, thinning stage, transition stage, old-growth stage, minimum dynamic area

QUESTIONS

1. If the bark and phloem are stripped away from the trunk of a tree, it will quickly die. Why?
2. How does the likelihood of a tree seed germinating and becoming established change through the various stages of forest stand development?
3. What role might the minimum dynamic area concept play in determining ideal boundaries for conservation preserves?

Bark
Xylem or wood
Phloem
Annual rings

13.3 Deforestation

- Each year, more than 130,000 km² are deforested, primarily as a consequence of agricultural expansion and overharvest.
- Deforestation is facilitated by poverty, road construction, and governmental corruption.

KEY TERMS

deforestation, shifting agriculture

QUESTIONS

1. The rate of net forest loss has diminished in the past decade compared to the previous decade. Why has this occurred?
2. In what ways does poverty contribute to deforestation?

13.4 Forest Degradation

- Cutting, fragmentation, altered disturbance patterns, and non-native diseases and pests have degraded forests worldwide.

KEY TERMS

forest degradation, primary forest, secondary forest, high-grading

QUESTIONS

1. How has the suppression of fire degraded the health of forests in parts of the western United States?
2. Why are non-native tree diseases and insects often so much more virulent than their native counterparts?

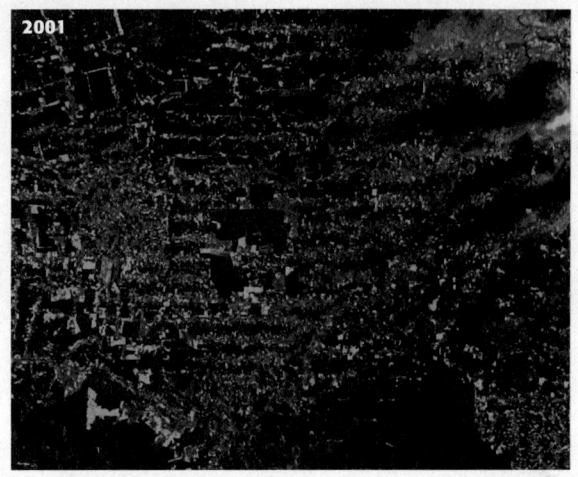

13.5 Defining Sustainable Forest Management

- Sustainable forest management requires the determination of which forests should be dedicated to which goals, how goods and services should be harvested, what amounts of goods and services should be harvested, and what investments should be made to ensure future forest health.

KEY TERMS

allocation, harvest, clear-cutting, even-aged management, uneven-aged management, rationing, rotation time, investment

QUESTIONS

1. Investment is a particularly important part of sustainable forest management. Why?
2. For forests near you, give an example of actions associated with each forest management category.

MasteringEnvironmentalScience®

Students Go to **MasteringEnvironmentalScience** for assignments, the eText, and the Study Area with practice tests, activities, and more.

Instructors Go to **MasteringEnvironmentalScience** for automatically graded tutorials and questions that you can assign to your students, plus Instructor Resources.

Nonrenewable Energy and Electricity

14

449

The History of an Oil Field

Just when is a nonrenewable resource exhausted?

How much oil is available on Earth? In theory, the answer to that question is determined by the absolute amount of oil in Earth's geologic strata. In reality, the answer is far more complicated. Oil deposits exist in different forms and in different places. Some of these deposits are easily tapped, and others can be exploited only with great difficulty. Thus, how much oil is actually available is determined by the technologies that are available to extract it and the cost of applying those technologies relative to the price of the resource. Many of those costs are directly related to the health of the environment. These principles are demonstrated in the history of the human use of a single oil field—the Kern River Oil Field near Bakersfield, California (Figure 14.1).

The Yokut Indians were the first humans to benefit from the Kern River Oil Field. As far back as 8,000 years ago, they

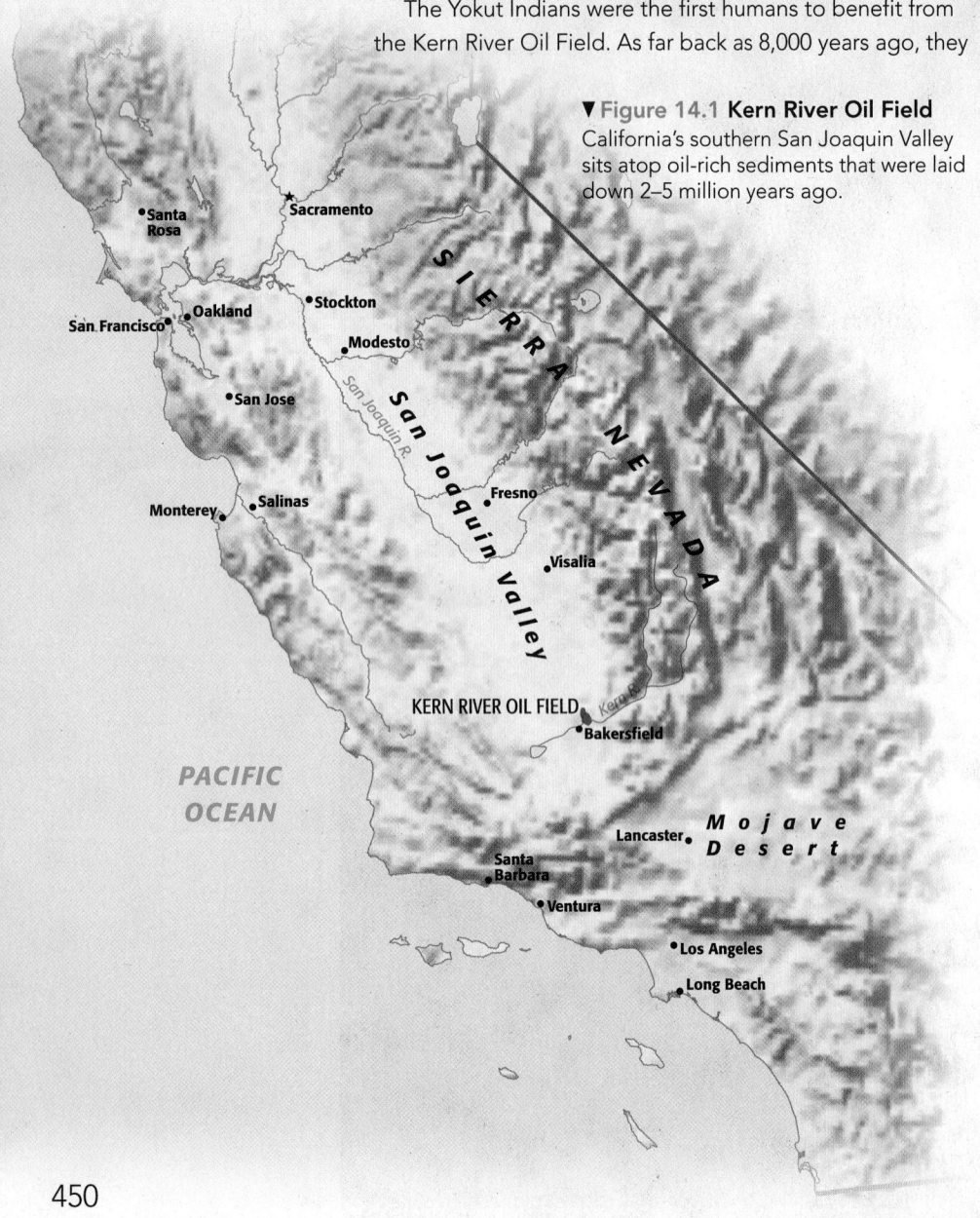

▼ Figure 14.1 **Kern River Oil Field**
California's southern San Joaquin Valley sits atop oil-rich sediments that were laid down 2–5 million years ago.

mined the tar that oozed from the ground and used it as an adhesive and for waterproofing and decoration. The Yokut traded the oily substance far and wide and thus might be considered to be one of the first oil-exporting nations. Of course, they had no idea of the vast amounts of oil that had accumulated hundreds and thousands of feet beneath their villages. Even if they had, they lacked the technologies to extract and use it.

When California was granted statehood in 1849, people of European descent were still collecting thick oil and tar from seeps and shallow pits in this area. But the demand and, therefore, the price for this goo were modest and brought little income to the local farming economy. No one saw uses for this resource beyond those devised by the Yokuts.

In the mid-1800s, lamps and small heaters were fueled with whale oil. With growing human populations, the demand for whale oil was increasing. This demand caused the price of whale oil to rise, just as whale populations around the world were rapidly declining.

The first oil well was drilled in Pennsylvania in 1859. Within a few years, technology was developed for distilling crude oil to make a fuel called *kerosene*, which could be used as an alternative to whale oil. The demand for kerosene grew quickly, as did its price. This demand spurred interest in locating and exploiting other sources of oil.

In 1899, this interest motivated James and Jonathan Elwood to dig by hand a 45-ft-deep well on the banks of the Kern River. They celebrated as thick crude oil squirted up to the surface. News spread quickly, and soon others began digging and drilling nearby with similar success. Within a couple years, the small farming community of Bakersfield was crowded with men seeking their fortune in the Kern River Oil Field, one of the largest oil reservoirs ever discovered in the United States.

By 1906, oil production in the Kern River Oil Field had risen to a peak of 16 million barrels a year. But the oil wells built at that time could access only the oil that flowed easily through rock pores and fractures. During the next 50 years, oil production declined, slowly at first and then more rapidly (Figure 14.2). By 1955, most of the 9,183 wells on the oil field were producing considerably less than 1 barrel of oil each day. Early on, little energy was required to harvest large amounts of energy-rich oil. After a half-century of constant drilling, however, energy and economic costs exceeded the benefits of production. Most of the wells were shut down.

Those decades of oil exploitation had a long-lasting impact on the environment. The large amounts of sediment- and chemical-laden water that were pumped to the surface

▲ Figure 14.2 **Kern River Oil Field 1910**
By 1910, over 9,000 wells had been drilled, but production had already peaked and was beginning to decline.

with the oil were simply dumped into small streams, which eventually flowed into the Kern River. Spilled oil was seen as a wasted resource but not as a pollutant. Consequently, the costs associated with this pollution were never captured in the price of the oil from this field. To this day, large areas of land in the oil field remain contaminated with oil residues.

In the 1960s, oil companies began experimenting with new technologies to coax oil from unproductive wells. At the Kern River Oil Field, companies began using large boilers and complex pumping systems to inject steam into wells. Technology to the rescue! By 1968, the field was producing 68,000 barrels of oil each day or nearly 25 million barrels each year. With these new technologies, the cost of producing each barrel of oil was several times higher than those produced by conventional operations. Oil-fired steam generators required 25–40% of the crude oil pumped from the field. But the rapid growth of the U.S. population and its affection for automobiles increased the demand for oil and kept the price high. The high price of oil more than compensated for the costs of the new technologies.

In compliance with clean-water regulations, wastewater was managed in treatment facilities and no longer dumped into streams. Thus, the cost of this pollution was captured in the cost of oil production. Indeed, clean water from these treatment facilities was sold to nearby farmers for irrigation. On the other hand, these new technologies spewed large amounts of pollutants into the air. That air pollution was viewed as a cost of doing business; its environmental costs were ignored.

Oil prices collapsed in the 1980s. At the same time, air-quality regulations were becoming stiffer. Operations at the Kern River Oil Field were again tenuous. Yet once again, technological innovation provided a fix. Oil companies built facilities to generate electricity that were fueled by natural gas, a by-product of oil production. Natural gas burns cleaner than oil. This electricity was a source of revenue. The electric facilities also supplied steam that was used to increase production from the wells. In 2000, the Kern River Oil Field produced nearly 40 million barrels of oil. However, this level of production could not be sustained. Since then, production has fallen to less than 30 million barrels each year (**Figure 14.3**).

Since 1899, over 2 billion barrels of oil have been extracted from the Kern River Oil Field. Scientists estimate that this field could yield another 475 million barrels. But actually producing that much oil will depend on continuing improvements in technology and high oil prices.

Oil and other fossil energy resources are being consumed by humans at rates that are thousands of times faster than the rates at which they are being extracted. These are considered finite resources, so at some point their supply will diminish, no matter how resourceful our engineering and technical expertise in extracting them. By comparison, other energy sources such as sunlight, wind, and water are replenished as we use them. Our use of each of these resources has an impact on our environment.

- *How do we characterize the various sources of energy we depend on?*

- *Where do we obtain the finite energy resources we use today?*

- *How does the amount of these resources compare to the amount we use?*

- *How do technology and economic factors influence the availability of these energy resources?*

- *What are the environmental consequences of the production and use of different energy resources?*

▼ Figure 14.3 **Not Dead Yet!**
New technologies and increasing demand and prices for oil have provided the incentives to maintain production in the Kern River Oil Field.

14.1 Energy Production

BIG IDEA Day in and day out, the average American family uses an amount of energy equivalent to the work that 3,000 laborers and 400 draft horses would have provided on a large farm 200 years ago. Most of that energy comes from resources that exist in limited amounts. The energy from those resources usually undergoes one to several transformations before it can be used. The kinds of energy resources that people use vary considerably from region to region and within different economic sectors. In the future, patterns of energy use will depend heavily on the relative supply of different energy sources. They will also depend on the environmental impacts associated with the extraction, conversion, delivery, and consumption of these energy resources.

Energy Sources

■ For more than 100 years, most of our energy needs have been met by nonrenewable fossil fuels.

Scientists classify energy by its source. **Primary energy** is the energy contained in natural resources, such as coal, oil, sunlight, wind, and uranium. Primary energy can be changed into other forms of energy through the process of **energy conversion**. These other forms of energy are classified as **secondary energy**. Thus, the primary energy of oil may be converted into secondary forms of energy, such as electricity or the kinetic energy of an automobile. Further energy conversions are also possible. For example, electricity may be used to charge a battery to power a cell phone. **End use** refers to the final application of energy, such as running an appliance or driving a car.

As far as energy conversion goes, there is no escape from the second law of thermodynamics—no energy conversion is 100% efficient (see Module 3.3). In each conversion step, some energy is lost as heat. **Energy conversion efficiency** is the percentage of primary source energy that is captured in a secondary form of energy. For example, when coal is burned to generate electricity, 70% of the primary energy in the coal may be transformed into unused heat. Therefore, the efficiency of the conversion from chemical to electrical energy is only 30%.

Energy end-use efficiency is the product of the efficiencies of all the energy conversions from the primary source to the end use. For example, if 100 units of coal energy are converted to electricity, which is then used to power an incandescent lightbulb that produces 1.2 units of light energy, the end-use efficiency is 1.2%. (Figure 14.4).

▶ Figure 14.4 **End-Use Efficiency**
The electricity to power an incandescent lightbulb comes from the potential chemical energy of coal that is burned in a coal-fired power plant. In these energy transformations, 100 units of chemical energy produce 1.2 units of visible light energy. End-use efficiency for this process is, therefore, 1.2%.

70 units lost as heat as the chemical energy in coal is transformed to electrical current

6 units lost as heat due to resistance in transmission lines

Primary energy source

Chemical energy 100 units

Electric current 30 units

Generation

Transmission

The **reserves** of a primary energy source represent the total amount that can be exploited. Energy experts define **production** as the amount of an energy source extracted from reserves during a particular time. For example, oil production is usually expressed in barrels or metric tons per year. **Consumption** refers to the amount of a primary energy source that is actually used during a particular time. Globally, rates of production roughly equal rates of consumption in a particular year. However, production and consumption rates often differ considerably within individual regions or countries. For example, in 2013, 3.25 billion barrels of oil were produced in the United States. Compare this to the 6.77 billion barrels that were consumed. To keep pace with its consumption, the United States clearly depends on production from other places such as the Middle East and South America.

Energy experts further classify forms of primary energy as nonrenewable or renewable. **Nonrenewable energy** is derived from sources that exist in limited quantities or from sources that are replenished at rates well below the rate of consumption. When nonrenewable energy sources are consumed, there is less available for future use. Forms of nonrenewable energy include fossil fuels—coal, petroleum, and natural gas—as well as nuclear energy. Fossil fuels were formed from the remains of organisms that lived eons ago; the geological processes that formed those fuels took millions of years. These processes continue today, but their rates are miniscule compared to the rates of fossil fuel consumption. Nuclear energy relies on uranium, an element that exists in limited amounts. In contrast, **renewable energy** is derived from sources that are not depleted when they are used, such as sunlight and wind, or that can be replenished in a short period of time, such as fuelwood.

▲ Figure 14.5 **Fossil Fuels Dominate**
That fossil fuels provide nearly 80% of the energy we consume is evident in this view of the oil storage and refining facility at Long Beach, California.

Today, the human energy system is dominated by fossil fuels, despite growing concerns over their supply and impact on the environment. Fossil fuels currently meet 79% of global energy needs (**Figure 14.5**). Nuclear energy provides about 5% of primary energy use, and hydropower contributes 6%. Biomass energy, primarily fuelwood used in poor countries, accounts for about 10% of primary energy use. Other forms of renewable energy, such as wind and solar, contribute less than 1% (**Figure 14.6**).

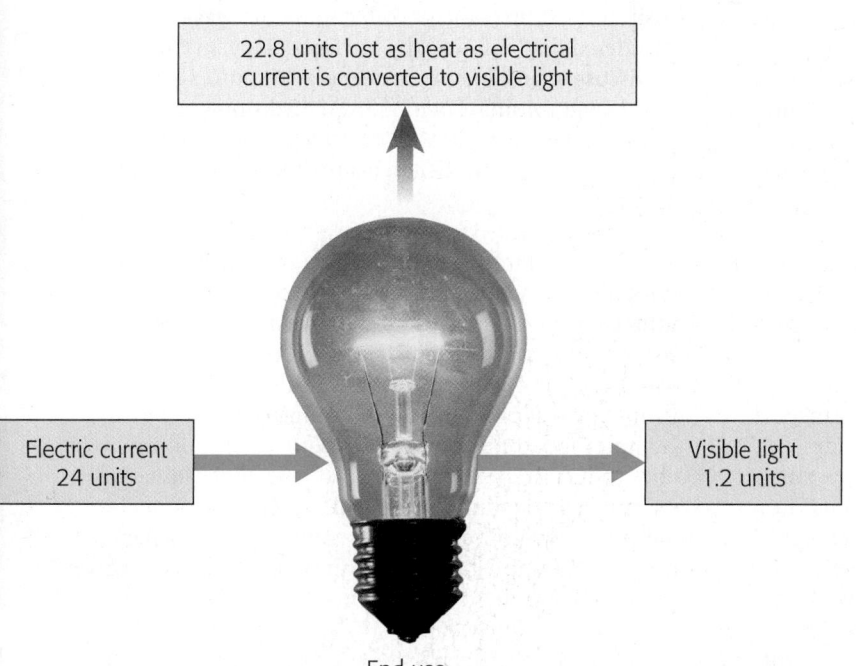

22.8 units lost as heat as electrical current is converted to visible light

Electric current 24 units

Visible light 1.2 units

End use

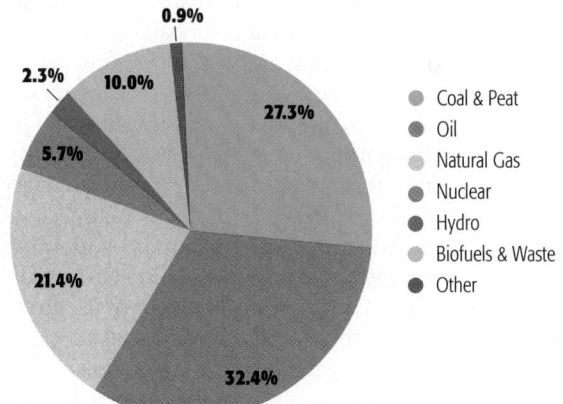

0.9%
2.3%   10.0%
27.3%
5.7%
21.4%
32.4%

- Coal & Peat
- Oil
- Natural Gas
- Nuclear
- Hydro
- Biofuels & Waste
- Other

▲ Figure 14.6 **Primary Energy Sources**
Fossil fuels—coal, oil, and natural gas—account for nearly 80% of the primary energy resources consumed by humans.

► Figure 14.7 Increasing EROI

This train is moving coal from the Powder River region of Wyoming to coal-fired power plants over 1,000 miles away. Transportation adds to the energy costs of coal production, and it is also an additional source of greenhouse gas emissions.

The Economics of Energy Resources

■ The amount of energy required to produce an amount of a primary energy resource is one measure of sustainability.

There are growing concerns over the long-term availability of some of the fossil fuels that dominate our energy economy. The **proved reserves** of a nonrenewable energy source are the quantities of an energy resource that could be recovered from known deposits *using current technology at current prices*. The **reserves-to-production ratio (R/P)** is the proved reserves figure for a given fuel divided by a particular year's level of production or use. The reserves-to-production ratio provides an estimate of how many years a fuel will last if the level of production remains constant and no additional reserves are discovered. For example, proved reserves of oil are currently estimated at 1.67 trillion barrels, and current rates of production are about 31.5 billion barrels a year. Therefore, the reserves-to-production ratio for oil is about 53. In other words, at current rates of production, we will exhaust the current reserves of oil in 53 years. The current global reserves-to-production ratio for natural gas is about 56 and for coal it is close to 110.

In reality, we do not really know when we will "run out" of oil, natural gas, or coal. Estimates of proved reserves are constantly changing as new deposits are discovered and new technologies are developed to extract more fuel from deposits already in production. In fact, it is unlikely that we will ever totally exhaust the supply of these primary energy sources. Before the last barrel of oil is pumped or the last lump of coal dug out of the ground, it will cease to be economical to exploit that source of energy.

Energy must be invested to produce any form of primary energy. Energy is required to extract, transport, and refine fossil fuels, to build and maintain dams, or to manufacture solar panels. The **energy return on investment (EROI)** is the useful energy provided from an energy resource divided by the amount of energy it took to produce it. If EROI for a primary energy source is less than 1.0, then more energy must be invested than can be produced. We will use EROI as one measure of the sustainability of different primary energy sources, but a caution is in order. Estimates of EROI for particular forms of primary energy sometimes vary widely based on what are or are not included as energy inputs. For example, the energy required to mitigate pollution from an energy source is generally not included in EROI estimates.

EROI close to 1 or even below 1 may be economically acceptable if the economic value of the energy commodity is greater than the cost of the resources used to produce it. But low EROI is often an indication that important environmental costs such as air pollution or emission of greenhouse gases are being externalized (Figure 14.7).

Because it represents such a large fraction of our total primary energy use, we begin our energy discussion with a secondary energy source, electricity. Each of the following modules in this chapter then focuses on a nonrenewable primary energy source. Carbon-based, fossil fuel resources—coal, oil, and natural gas—are considered first, followed by nuclear energy. In Chapter 15 we discuss renewable primary energy resources— solar energy, biomass, wind power, hydropower, and geothermal energy—as well as strategies to conserve energy and diminish the environmental impacts of our energy use.

Changes in reserves, production technologies, and environmental impacts are providing strong incentives to reorder the relative proportions of renewable and nonrenewable energy sources that we use. Discovering, extracting, transporting, refining, and consuming nonrenewable energy sources create a range of environmental problems. The burning of fossil fuels is the single most significant contributor to air pollution and production of the greenhouse gases that cause global warming (see Modules 9.3 and 10.3). The use of nuclear energy relies on mining activities that damage the land, and it requires the long-term storage of highly radioactive spent fuel. The environmental impacts associated with the use of nonrenewable energy resources present external costs to society, costs that are not easily captured in the economic markets for these resources.

QUESTIONS 14.1

1. Differentiate between energy conversion efficiency and energy end-use efficiency.

2. Explain how energy experts estimate the number of years that a nonrenewable resource is likely to be available.

3. Why are low values of EROI a matter of concern?

(MES) For additional review, go to **MasteringEnvironmentalScience**

14.2 Electric Power—Generation, Distribution, and Use

BIG IDEA Of all the forms of energy we use in our lives, none is more ubiquitous than electricity. Electrical energy is the energy of flowing charged particles. About 40% of global primary energy supply is consumed to generate electricity. The most important primary energy sources are coal, natural gas, nuclear, solar energy, wind, and flowing water. Electricity may also be generated by chemical reactions in batteries and fuel cells. The relative contribution of each of these sources to the total amount of generated electricity is influenced by patterns of availability, economic costs, and environmental impacts. Most electric power is distributed to end users through complex transmission networks. The environmental impacts of electric power are largely associated with the primary energy sources used to generate it, its transmission, and its distribution. The chemicals in batteries are also important pollutants.

Generating Electricity

■ Most of the electricity we use is generated by the kinetic energy that comes from primary energy sources.

An electric current is the flow of electrically charged particles, usually electrons flowing through a wire. Current is measured in **amperes**, or amps. The current is the amount of charge that passes through the wire in a certain amount of time. The charged particles in an electric current flow between terminals that have opposite electrical charges. A flowing current has the capacity to do work. The amount of work depends on the difference in electrical potential energy between the two terminals. Electrical potential is measured in **volts**. The greater the difference in electrical potential between the terminals, the greater the voltage of the current.

The flow of electricity through a wire is similar to the flow of water in a river. The total amount of water flowing in the river is the river's current. The difference in elevation between the river's source and its outlet to the sea determines the potential energy of the river.

The rate at which energy is converted from one form to another is power. Electric power is the rate at which electricity is generated or the rate at which electrical energy is transformed into another form, such as heat or the kinetic energy of an appliance. Electric power is measured in **kilowatt hours** (kWh) and is calculated by multiplying electrical potential (volts) by current (amperes). Power (kWh) = voltage (volts) × current (amperes).

Most electricity is generated in power plants from one of several primary energy sources. For example, primary energy sources, such as coal, wood pellets, or nuclear fission, are used to heat water and produce pressurized steam. This steam is then used to turn a turbine that drives an **electric generator**, or dynamo (Figure 14.8). Alternatively, moving water or wind or ignited natural gas may directly drive turbines. Spinning magnets in the generator produce an **electromagnetic field**, or area of electromagnetic force. The electromagnetic field provides the voltage that causes an electric current to flow. Electricity is also generated from hydrogen using fuel cells and from solar energy using photovoltaic cells.

▼ Figure 14.8 **Steam Turbine** Steam turbines are attached to electrical generators. Turbines are turned by pressurized steam that can be generated from a number of primary energy sources.

Batteries and Fuel Cells

■ Chemical reactions can be used to generate electricity.

For many applications we obtain electrical energy from batteries. Batteries convert stored chemical energy into electrical energy. The alkaline battery in your flashlight, for example, has zinc powder at the negative terminal and magnesium oxide at the positive terminal, all soaked in an alkaline solution (Figure 14.9). A car battery takes advantage of the electrical potential between plates of lead and lead oxide that are immersed in sulfuric acid.

As a battery is used, the strength of its electrical potential diminishes. In disposable batteries, this potential cannot be reestablished. In other kinds of batteries, such as those in a car or cell phone, the electric potential can be restored by using the electric current from a charger.

Fuel cells convert the chemical energy of a fuel into electricity through a reaction with oxygen. Fuel cells differ from batteries in that they require continuous input of fuel and oxygen to sustain the chemical reaction that generates electricity. Hydrogen is the most common fuel, but natural gas and other hydrocarbons can also be used.

Fuel cells consist of two electrodes separated by an electrolyte, a solution containing positively and negatively charged ions (Figure 14.10). Different types of fuel cells use different kinds of electrolytes, but they function in the same

basic way. Hydrogen is supplied to the negative electrode (anode), and oxygen or air goes to the positive electrode (cathode). A catalyst on the porous anode promotes the dissociation of H_2 molecules into protons (H^+) and electrons (e^-). The protons migrate through the electrolyte to the cathode. Here they react with O_2 and electrons provided to the cathode by the external circuit to produce water. The electrical energy provided by electrons running through the external circuit can be used to do useful work such as power an electric motor or light a building.

There has been a great deal of interest in the development of hydrogen fuel cell automobiles. However, widespread use of cars will ultimately depend on the development of efficient systems for the generation, distribution, and storage of hydrogen gas (see Module 15.8). Nevertheless, fuel cells do currently have important applications. They are very reliable compared to other electrical generation systems because they have no moving parts and do not depend on combustion and storage of a fossil fuel. They have been used for decades as both a primary and backup source of electricity in commercial, industrial, and a few residential settings. They are particularly useful in remote situations such as spacecraft, submarines, weather stations, and research stations.

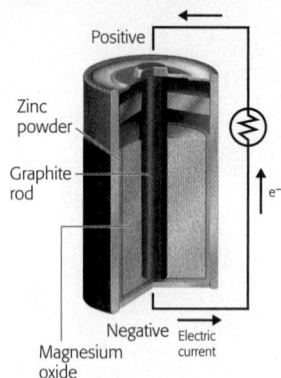

▲ Figure 14.9 **Stored Electrical Energy**
Alkaline batteries use zinc powder and magnesium oxide to generate electrical potential.

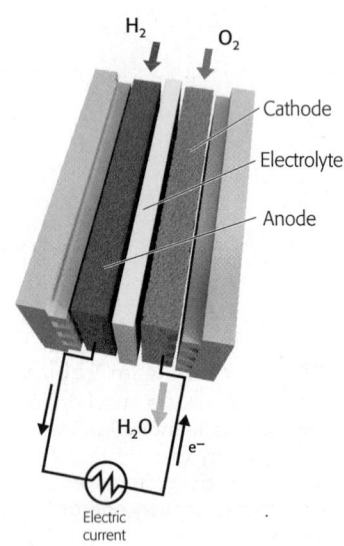

▲ Figure 14.10 **Hydrogen Fuel Cell**
The dissociation of H_2 into protons (H^+) and electrons (e^-) is used to establish an electrical current that provides power to do work.

Transmission of Electricity

■ Electricity is transmitted from centralized locations via a grid network.

Most power plant generators operate at low voltage. Power plant **transformers** reduce the amperage and increase the voltage so that electricity can move efficiently over long distances through transmission lines. The voltage in major transmission lines is high, 230,000–760,000 volts. At distribution points, transformers convert high-voltage electricity back to lower voltage electricity. The local overhead transmission lines that deliver electricity to your home typically carry 20,000–30,000 volts. Transformers near your home reduce that voltage for end use. In the United States, the electric circuits in a home typically operate at 110–120 volts.

Multiple power plants, perhaps drawing on a variety of primary energy sources, supply electricity to a network of transmission lines and transformers called an **electric power grid** (Figure 14.11). Electric power grids can be quite large. The ever-present power lines that surround us are part of an electric power grid that may extend over tens of thousands of square miles.

All of the components of an electric power grid create some resistance to the flow of electricity, much as friction in the walls of a pipe retards the flow of water through it. This resistance converts some of the electrical energy into heat, which is then lost to the system. The total energy loss associated with the transmission of energy from a power plant to its end use is about 7–10%.

In natural ecosystems, the flow of energy is regulated by feedbacks among the various ecosystem components, such as predators and their prey. The flow of energy in electrical power grids is also regulated by feedbacks, although these feedbacks are often incomplete and not nearly as efficient as their natural counterparts. For example, the flow of electricity through a grid is driven by consumer demand. The more electricity used by consumers, the more electricity the power plants on the grid must generate. If total demand exceeds production, the system can fail, producing brownouts or blackouts.

▶ Figure 14.11 **The Grid**
Electrical grids are composed of the transformers and transmission lines that distribute electricity from power plants to homes and businesses. Grids generally extend over several states and include multiple generating facilities.

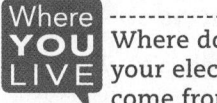

Base load is the minimum amount of electricity that a utility must provide to meet consumer needs. The demand for electricity never drops below this level. **Peak load** is the maximum amount of electricity demanded by consumers. Peak load is often several times greater than base load, and the variety of power plants on the grid must be able to adjust quickly to this difference in demand. For example, in the middle of winter the demand for electricity is usually highest on cold nights, when the use of heating and lighting increases consumption. The problem of matching supply with demand is compounded by the fact that peak output from some renewable energy sources, such as wind, solar, and tides, may not coincide with periods of peak consumer use. Hence, these energy sources require some way to store energy they collect.

In general, the cost of generating electricity increases as demand increases from base load to peak load. Nevertheless, most residential and commercial consumers pay a single price for their electricity based on the total energy used over a month as measured by an electric meter. These consumers have neither the knowledge nor the incentive to adjust their energy use to these varying energy costs. In contrast, many industrial consumers negotiate variable rates with utilities. They are able to meter their electrical use throughout the day and maximize use during periods when rates are low.

Most existing electric power grids were designed to distribute electricity from large, centralized generating facilities that derive their power from nonrenewable sources of energy, such as coal, natural gas, and nuclear energy. Facilities that use renewable energy sources, such as wind, solar, or wave energy, usually generate less power and are often widely separated rather than being centralized in one location. Redesigning electrical power grids to make efficient use of these widely separated power sources is an important challenge to the transition from nonrenewable to renewable sources of energy.

Where YOU LIVE — Where does your electricity come from?

The mix of different primary energy resources used to supply electricity varies a lot from place to place. The U.S. Energy Information Administration (EIA) provides this information for each state by year. Search the electricity data. What proportions of nonrenewable and renewable energy sources currently provide electricity to your state? How does this compare with the energy mix used in 1990?

SEEING SOLUTIONS

A Smart Grid

Can better communication between utilities and consumers lead to more efficient use of electrical energy?

Do you know how much electricity you used today? You probably don't, and neither does the utility that provides your electricity! There is widespread agreement that the existing system of demand-driven generation and distribution of electricity is wasteful and inefficient and that the root of this problem is poor communication. Consumers communicate to the utility that provides their electricity once a month, when their meter is read. Utilities communicate back to consumers once each month through electric bills. Moment to moment, individual consumers have little sense of how much energy they are using or at what cost. Similarly, utilities have no way of knowing the moment-to-moment use of energy by individual consumers. If consumers knew when they were using the most energy or what activities required the most energy, they could change their habits.

Poor communication is particularly problematic during periods of peak demand. It is during these times that utilities must bring the most expensive and often environmentally problematic power sources onto the grid. Better communication among all parts of the grid could not only diminish peak demand but allow utilities to adjust the mix of power sources to minimize costs and environmental impacts.

Smart grid refers to new technology designed to provide moment-to-moment communication between electric utilities and consumers. Like your "smart" phone, smart grid technology is smart because it is computerized, and it is able to gather and transmit data from each component of the network. Components include a variety of power sources, grid substations, and transformers; they also can include your home electric meter (**Figure 14.12**). With appropriate infrastructure and software, utilities can adjust the price consumers pay for electricity every 5–10 minutes according to demand.

For this real-time pricing to work, customers must be able to see the price changes. To meet that need, homes can be outfitted with meters and remote devices that are in constant wireless communication with the grid. These devices signal when power on the grid is abundant and cheap and when power is more expensive. Consumers can even have the option of receiving this information on their cell phones via text messaging. In the future, homeowners may even be able to use batteries, such as those in electric cars, to store bargain electricity purchased during slack hours for use at peak times.

Smart grid technology can also help decentralize the grid and facilitate greater use of renewable sources of electric power. This will have the added advantage of shortening the distance between power sources and consumers, increasing transmission efficiency. Elements of smart grid systems such as smart meters are already in place in cities like Houston, Texas, and Boulder, Colorado.

▲ Figure 14.12 **A Meter That Communicates**
An electric meter such as this allows utilities to adjust usage of the homeowner's heating and air-conditioning systems and appliances to lower costs and improve efficiency. The homeowner provides guidelines for these adjustments through the utility website.

QUESTIONS 14.2

1. What do volts and amperes measure with regard to the flow of electrical energy?

2. How does a fuel cell differ from a battery?

3. Describe two problems with the existing electric power grid that reduce the efficiency of electricity use.

4. Describe two ways that the current system of distributing electricity impacts the environment.

(MES) For additional review, go to **MasteringEnvironmentalScience**

▼ Figure 14.13 **PCB Contamination**
Wastewater from a General Electric Corporation facility that manufactures electrical transformers has contaminated the sediments over large stretches of the Hudson River with PCBs. Pollutants are no longer being released, but dredging activities such as this to remove polluted sediment will continue through 2016.

Environmental Impacts

■ Transmission of electricity has uncertain health effects.

The most significant environmental impacts of using electric power are those associated with the primary energy sources used to generate it. Over 50% of global electricity is generated in conventional coal-fired plants, and the impacts of such coal generation on the environment are greater than any other primary energy source. Greenhouse gas emissions per kilowatt hour of electricity for coal are more than twice those for natural gas and over 10 times greater than most renewable energy sources. Coal production and consumption also have significant negative effects on air and water quality and the sustainability of terrestrial and aquatic ecosystems.

As for the devices we use to store electricity, all batteries must eventually be disposed of, whether they are rechargeable or not. Each year, three billion batteries are discarded in the United States alone, and the majority of them end up in landfills. They represent less than 1% of total municipal waste volume, but because they contain a variety of heavy metals, they contribute a much larger share to total landfill pollution. For example, the nickel–cadmium batteries commonly used in electronic games and cell phones account for 75% of the cadmium found in landfills. Most states and counties now have programs to collect and recycle most batteries.

By comparison, hydrogen fuel cells are touted as environmentally clean because water is their only end product. But there are significant environmental impacts associated with the generation of H_2. Hydrogen can be generated by electrolysis of water. However, with current technology, nearly 3 kWh of electricity input are needed to produce the H_2 to generate 1 kWh of fuel cell output (EROI = 0.33). Currently, H_2 is obtained most efficiently from natural gas, but EROI is still less than 1. Natural gas can be burned directly in a gas-fired turbine to generate electricity more efficiently than fuel cells. Either way, greenhouse gases are released to the atmosphere.

The transmission and distribution of electricity also have important environmental consequences. The electric currents flowing through transmission lines create electromagnetic fields around the lines. There has been considerable debate over the possible effects of those electromagnetic fields on human health. Some scientific evidence suggests that children who live near power lines may have a higher frequency of leukemia. Other alleged health effects have, however, been hard to substantiate. In any case, no cause-and-effect relationship has been established between electromagnetic fields and specific ailments.

One of the most important environmental impacts of electric power transmission is associated with the transformers used to modify line voltage. The electrical wiring within a transformer is immersed in liquid coolants. Until recently, these coolants often contained chemicals called polychlorinated biphenyls (PCBs). Although effective as coolants, PCBs have a wide range of toxic effects on wildlife and humans. Today the use of PCBs is banned, but in the past, large quantities of these chemicals were released into the environment. Because they are remarkably resistant to decomposition, PCBs remain a threat to ecosystem health in many places (Figure 14.13).

In summary, nearly half of all energy end uses take advantage of electricity. Current systems of generating and distributing electricity rely on large power plants and complicated electric power grids. This results in the inefficient use of electric power and constrains efforts to limit environmental impacts.

14.3 Coal

BIG IDEA Humans have been using coal longer than any other fossil fuel. Archeological evidence shows that coal was being used in England as far back as the Bronze Age, 5,000 years ago. Energy from coal inaugurated the Industrial Revolution in the mid-1700s, and today it meets 27% of global energy needs. Coal is the most abundant of all fossil fuels, and supplies appear to be sufficient to meet global demands for at least a century. Today, most coal production is used to fire power plants that generate electricity. The use of coal is associated with a variety of environmental problems. Coal mining alters the landscape and pollutes water; the combustion of coal emits air pollution and greenhouse gases. Technological advances are already reducing some of these impacts and have the potential to diminish others.

Lignite
3.0–4.5 kWh/kg

Sources and Production

■ Coal was formed from plants that lived in swamps hundreds of millions of years ago.

Most of Earth's coal began to form between 300 and 400 million years ago, when large swampy forests covered much of the land. Over millions of years, large volumes of plant material fell into these swamps, where it was covered by sediment and water before it could decay. At first, this organic material was broken down anaerobically into a wet, partially decomposed mixture material called **peat**. Over time, thick layers of sediment accumulated on top of the peat. The higher temperature and pressure caused by the weight of these sediments forced much of the water out of the peat and packed the carbon compounds from the original plants closer together, eventually forming coal. Today, the layers of sediment that contain coal are called **coal seams**.

The carbon and energy content of coal deposits vary, depending on their age and the extent to which they were subjected to underground heat and pressure. **Lignite** is a type of coal that is generally associated with younger deposits that have not been subjected to as much heat and pressure as other types of coal. As a result, lignite is comparatively soft and contains considerable moisture. Lignite contains 25–35% carbon; when it is burned, it yields 3.0–4.5 kWh of energy per kilogram (3.0–4.5 kWh/kg). **Sub-bituminous coal** contains 35–45% carbon; its energy content is, therefore, somewhat higher than lignite.

Bituminous coal is much denser than either sub-bituminous coal or lignite because it was formed under higher heat and pressure. Bituminous coal has a carbon content of 45–86% and yields 5–8 kWh/kg when it is burned. **Anthracite coal**, which is dark and shiny, has the highest energy content of all types of coal. It has a carbon content of 86–97% and yields more than 8 kWh/kg when burned. Anthracite is much less common than bituminous coal and is usually found only at great depths (**Figure 14.14**).

Bituminous coal and anthracite coal are often called "hard coal." Because hard coal has higher energy content and contains fewer polluting chemicals, it is more widely used as a primary energy source than either lignite or sub-bituminous coal.

The EROI for coal varies widely between 40 and 80, depending on the accessibility of coal seams, coal type,

and energy required for transportation. EROI is highest for anthracite from shallow mines that is burned within a short distance. However, the vast majority of reserves is lower quality coal that requires much higher mining and transportation energy. Worldwide, the proved reserves of coal equal over 860 billion metric tons. About 47% of this is anthracite and bituminous coal; the remainder is sub-bituminous coal and lignite. In 2013, approximately 7.8 billion metric tons of coal were mined and used, giving a reserve-to-production ratio of 110. Even with coal production increasing at a rate of 1.5–2.0% each year, most experts agree there is enough coal to last for at least another century.

The geological conditions under which coal formed were geographically widespread, so coal deposits are widely distributed. Nevertheless, 67% of the world's recoverable coal reserves are located in just four countries—the United States, Russia, China, and India (**Figure 14.15**).

Bituminous
5–8 kWh/kg

Anthracite
8 or more kWh/kg

▲ Figure 14.14 **Types of Coal**
Energy content per unit weight is highest in anthracite coal, less in bituminous, and lowest in lignite.

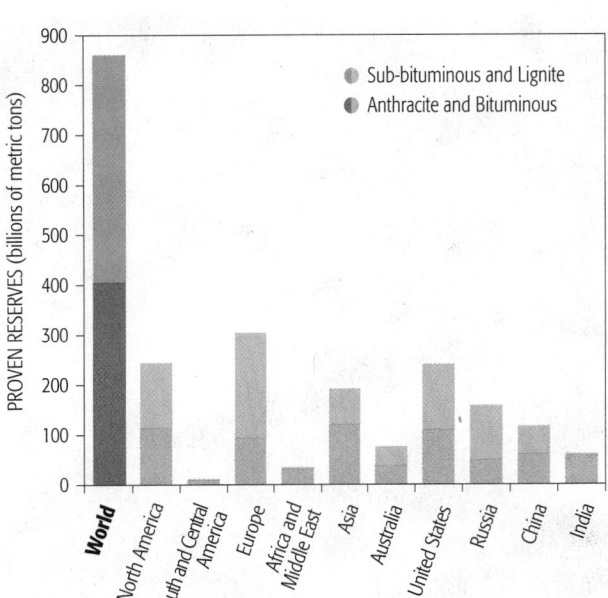

◄ Figure 14.15 **World Coal Reserves**
Coal is relatively abundant in North America, Europe, and Asia, although much of it is low-quality sub-bituminous and lignite.

Data from: British Petroleum, 2013. Statistical review of world energy June 2013. http://www.bp.com/content/dam/bp/pdf/statistical-review/statistical_review_of_world_energy_2013.pdf. Last accessed 1-30-14.

Q In what year will the supply of coal be depleted?
Ciara Tyce, Georgia Southern University

Norm: A At current rates of use, coal supplies could last more than 100 years. But can Earth's ecosystems tolerate the effects of burning that much coal?

Coal is removed from sediments by underground and surface mining. Underground mining is used to extract coal from sediment layers that are deeper than 200 ft. The deepest coal mines are well over 1,000 ft below the surface. Mine operators dig multiple shafts into the ground to reach the coal seam (Figure 14.16).

Surface mining, also called strip mining, is used to extract coal from deposits that are less than 200 ft underground. Giant earth-moving machines remove topsoil and layers of rock, known as overburden, to expose coal seams. Mountaintop removal is a particularly controversial form of surface mining (Figure 14.17). In this form of mining, the entire top portion of a mountain is removed to expose a coal seam, and the overburden is dumped into surrounding valleys. This mining technique has been widely used in the Appalachian Mountains of West Virginia and Kentucky.

Once coal is removed from the ground, it usually undergoes some form of cleaning to remove dirt, rock, sulfur, and other impurities. It may then be transported to coal-burning power plants and other end users by train, barge, or truck. Alternatively, it may be mixed with water and sent by pipeline.

Globally, 67% of coal production is used to generate electricity, whereas 30% goes to industrial end uses, mostly steel production. In the United States, more than 90% of coal production is used to generate electricity. This is a considerable change from 50 years ago, when more than half of U.S. coal production was used to heat homes and for industrial purposes.

The chemical energy in coal is transformed into electricity in coal-fired power plants. There coal is burned to produce heat, which is captured in a boiler and used to produce pressurized steam. This steam turns a turbine, which then powers an electrical generator (Figure 14.18). After passing through the turbine, the steam condenses into liquid water; its remaining heat is dissipated to the air in cooling towers or to the water in a river or lake.

▲ Figure 14.16 **A Modern Coal Mine**
Underground coal mines employ an array of shafts and tunnels. Some of these are used to transport miners and equipment into and out of the mine, and others are used to ventilate the mine and haul the coal out.

Exhaust from burning coal is passed through a **scrubber**, a device designed to remove particulates and some pollutant gases. The exhaust then rises through a smoke stack and is released into the atmosphere. The solid material left behind is called **fly ash**.

In the oldest coal-fired power plants, energy conversion efficiency may be less than 25%. This means that only a quarter of the chemical energy in coal is converted into electrical energy. In contrast, most modern coal-fired facilities use more efficient combustion and heat-capturing technologies, which allow them to operate at efficiencies of more than 40%.

► Figure 14.17 **Moving Mountains**
In mountaintop removal coal mining, the upper portion of a mountain is removed to expose the coal seam. The overburden is then deposited in surrounding stream valleys.

Smoke stack

Steam released from cooling tower

"Scrubbers" remove NO_x and SO_x

Hot pressurized steam

Boiler

Generator turbines

Electric power

Cooling water

Coal supply

Firing system

Coal hopper

▲ **Figure 14.18 Coal-Fired Power Plant**
In a modern coal-fired facility, coal is burned at high temperature to produce pressurized steam, which turns turbines attached to a generator. Scrubbers remove pollutants such as SO_x and NO_x from the exhaust, which is then released from the smoke stack. Steam, not smoke, is being emitted from the conical cooling towers. The aerial photo shows a modern lignite-fired power plant in Grevenbroich, Westphalia, Germany.

Coal and the Environment

■ Coal mining and coal combustion present significant environmental challenges.

From its extraction to its various end uses, coal presents more environmental challenges than any other energy source. Underground coal mining is a very hazardous occupation due to the potential for cave-ins, flooding, dust, and gas explosions. In 2013, mining accidents in the United States killed 20 coal miners, and accidents in China killed more than 1,300 miners. Even without accidents, thousands of underground coal miners suffer from life-threatening respiratory diseases caused by inhaling coal dust.

Mining activities also have a direct effect on the environment. **Mine tailings**, the rock and debris from mining operations, often contain high concentrations of sulfide. When exposed to oxygen, these sulfides are transformed to sulfuric acid, which runs off into nearby streams, where it harms fish and other aquatic organisms (Figure 14.19).

Surface mining destroys the terrestrial ecosystems above the coal seams. Once the coal is removed, however, the overburden may be returned to the pit. The overburden can then be covered by a layer of topsoil and the area revegetated to prevent erosion (Figure 14.20). The coal-mining industry in the United States has received especially harsh criticism for the environmental impact of mountaintop removal. In West Virginia and Kentucky, overburden from mountaintop removal has permanently buried more than 700 miles of mountain streams, thereby affecting wildlife, flooding nearby communities, and degrading water quality far downstream (see Figure 14.17).

▲ **Figure 14.19 Acid Mine Drainage**
Sulfides leached from coal mine tailings pollute a nearby stream.

▲ **Figure 14.20 Mine Restoration**
Topsoil has been replaced and grasses seeded on this former open pit mine in West Virginia so that it can be used for grazing. However, it is not nearly as fertile as the farmland in the distance.

QUESTIONS 14.3

1. Compare the four types of coal with regard to their carbon content and energy yield when burned.

2. What are the environmental impacts of the most commonly used methods of coal mining?

3. Describe three environmental challenges caused by the combustion of coal.

(MES) For additional review, go to **MasteringEnvironmentalScience**

The exhaust and fly ash from coal fires contain a number of toxic chemicals that are harmful to many organisms, including humans. Fine particulate soot from coal fires can cause respiratory distress. In addition, coal usually contains mercury, which accumulated in the wetlands where it was formed. When coal is burned, this mercury is released into the atmosphere. Concentrations of mercury in streams and lakes near coal-fired power plants are often several times higher than normal. Eventually, this mercury accumulates in the tissues of animals, including humans, where it can cause a number of serious health problems, including neurological disorders (see Module 10.3). Fly ash also contains high concentrations of mercury, which makes it complicated to dispose of. This ash is often stored in open-air pits or waste ponds. Without adequate safeguards, toxic fly ash can spill into local waterways.

Coal contains significant quantities of organic nitrogen and sulfur, which are converted to nitrogen oxides (NO_x) and sulfur oxides (SO_x) when coal is burned. In rainwater, nitrogen oxides are converted to nitric acid, and sulfur oxides are converted to sulfuric acid (see Module 10.3).

New technologies can reduce the amounts of pollutants produced by coal-powered plants. Improved methods of cleaning coal before it is burned reduce the emissions of many pollutants. In many modern plants, sophisticated scrubbers remove significant amounts of NO_x, SO_x, and mercury from the exhaust. However, the efficiency of such scrubbers varies from pollutant to pollutant. For example, scrubbers can reduce SO_x emissions by over 90% but typically reduce mercury emissions by only 30–40%.

The burning of coal is a significant contributor to global warming. For every unit of energy production, the combustion of coal releases 80% more carbon dioxide (CO_2) than is released by the combustion of natural gas and 30% more than is released by the combustion of oil (see Module 9.3). A typical 1,000-megawatt coal-fired power plant produces 6 million metric tons of CO_2 a year. This is equivalent to the annual CO_2 emissions from 2 million cars.

By 2030, the worldwide consumption of coal is expected to double, largely to meet growing demands for electricity in developing countries. China alone is currently commissioning two to three new large coal-fired power plants each month. The coal-fired power plants that are slated for construction between now and 2030 are expected to emit as much CO_2 as has all the coal consumed in the world since the beginning of the Industrial Revolution to the present. Such projections have caused great interest in the development of alternative energy sources to offset coal consumption.

New FRONTIERS

Cleaning Up Coal

Wouldn't coal-fired energy without emissions of greenhouse gases and toxic pollutants be wonderful? "Clean coal" is a suite of technologies that some argue will do just that.

Here is how the process might work (**Figure 14.21**). Rather than burning it directly to generate steam, coal is heated to very high temperatures (800–1,800 °C) in a closed vessel to produce raw "syngas," a gas composed of lots of chemicals, but rich in hydrogen (H_2), carbon monoxide (CO), and CO_2. The CO reacts with water to produce more H_2 and CO_2. Raw syngas is then processed through a sequence of filters and reactors to remove nitrogen, sulfur, and mercury. These elements might then be recovered for other beneficial uses. The CO_2 is then captured, pressurized, and transported for sequestration in belowground geological formations. The remaining H_2-rich syngas can then be burned in a gas turbine to drive a generator and produce electricity.

Some of these clean coal technologies are better developed than others. For example, over 150 coal gasification facilities are currently in operation in North America and Europe. Pollutant emissions are somewhat lower than conventional coal-fired facilities. However, the technologies for their complete removal and reuse are very much in the planning and development stage. The same is true for the carbon capture and sequestration (CCS) step in this process (see Module 9.6). Significant concerns remain regarding the energy required to capture and transport CO_2, as well as the ability of geologic formations to sequester it permanently. Critics argue that the overall EROI for clean coal electricity may be considerably less than 1.

"Clean coal is an essential part of our energy future." So states a brochure from one of America's largest producers of coal fired electricity. Environmental groups reply that clean coal is an "insulting oxymoron." Environmental impacts from mining and fly ash remain, and they argue that resources dedicated to this approach would be better spent on other pollution and global warming mitigation strategies.

▼ Figure 14.21 **Clean Coal System**
A simplified diagram of a coal-integrated gasification combined cycle (IGCC) system with carbon capture and storage. Although coal gasification facilities such as the Indiana Wabash River IGCC plant are now in operation, several other components of this system are as yet undeveloped.

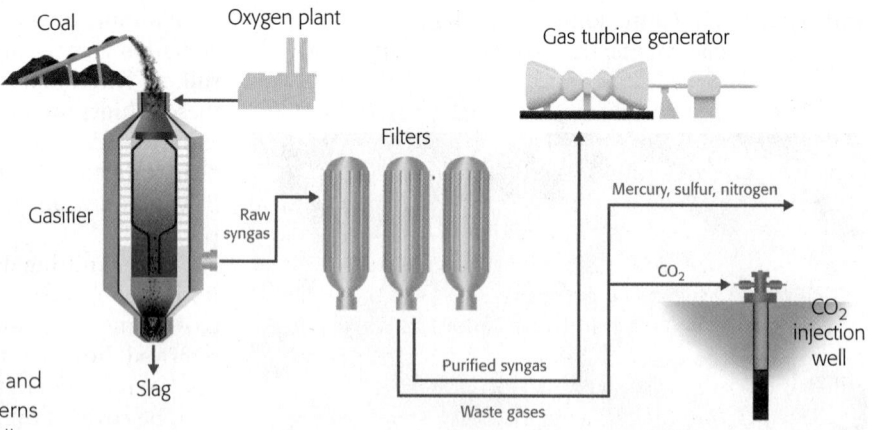

The Intergovernmental Panel on Climate Change (IPCC) takes an intermediate stance. Given its abundance and low cost, coal is likely to be a major part of the global energy portfolio for the foreseeable future, the IPCC suggests that clean coal could play an important role in global warming mitigation. But it cautions that much more research is needed. What roles should you and the public have in determining how energy is generated?

14.4 Oil and Natural Gas

BIG IDEA The first commercial oil well was drilled in Titusville, Pennsylvania, in 1859. Within 100 years, oil had surpassed coal as the leading source of energy in the world economy. Today, oil accounts for nearly 35% of global energy use. However, oil and natural gas reserves were created under very specific conditions, so they are unevenly distributed around the world. Experts disagree about the exact amounts left in these reserves but are in complete agreement that supplies are dwindling as demand is growing. The majority of this growth is associated with economic development and increased demand from the transportation sector. There are growing concerns over the environmental impacts of oil and natural gas use, as well as with the effects of its supply on international security and conflict. Despite these concerns, it appears that our reliance on these fossil fuels will remain high for some time.

Sources

■ Oil and natural gas formed under specific geologic conditions that were common in only a few places.

Like coal, oil is a fossil fuel. Crude oil, or petroleum, is formed from the remains of microorganisms that lived in shallow seas and coastal swamps hundreds of millions of years ago. These areas were characterized by abundant algae and zooplankton and very high rates of net ecosystem production (see Module 6.5). As these microorganisms died, they sank to the bottom, forming thick layers of organic matter. Over the course of tens of thousands of years, this organic matter was buried under layer after layer of sediment, where it was cut off from oxygen and subjected to increasing amounts of heat and pressure. These conditions transformed the organic matter into **kerogen**, a waxy substance that is a precursor to oil and natural gas. Underground at depths of 2–3 miles and at temperatures above 200 °F, kerogen became crude oil.

In deep deposits, where kerogen was subjected to very high temperatures and pressures, it was converted into natural gas rather than oil. Natural gas is a mixture of hydrocarbon gases composed mainly of methane (CH_4), with smaller amounts of other gases, such as propane. Natural gas is colorless and odorless. However, gas distributors add small amounts of sulfur-containing chemicals so that gas leaks can be easily detected by the smell. The newly formed crude oil and gas were less dense than the water in the ground, so it tended to migrate upward through porous sediments. In some places, oil reached the surface, where it seeped out into tar pits. In other places, oil became trapped beneath a layer of impermeable rock, forming **oil reservoirs** (Figure 14.22). Natural gas is often trapped in these reservoirs. These reservoirs are not underground lakes of oil and gas. Rather, reservoirs are composed of permeable rocks with numerous pores and fractures that hold small drops of oil, similar to the way in which a sponge holds water.

Reservoirs such as these are said to be **conventional**, and they have historically been the most important sources of oil and gas. However, crude oil and natural gas are found in other, **unconventional** geologic settings. For example, oil and gas can accumulate in shale, a kind of sedimentary rock formed from muddy deposits rich in organic matter. Crude oil can also be produced from tar in sandy deposits. Extensive tar sand deposits are found in Canada, Kazakhstan, and Russia.

For an oil reservoir to form, three geological conditions must be met. First, there must have been a shallow sea with large quantities of aquatic organisms that died and sank to the bottom, were covered in sediments, and were eventually converted to oil by subterranean heat and pressure. Second, the oil had to migrate upward into a porous rock reservoir. Third, the reservoir rock must have been covered by a layer of impermeable rock, or a cap rock, that prevented the oil from seeping to the surface.

▼ **Figure 14.22 Oil Reservoirs** Oil and gas reservoirs formed in porous rocks where there was enough organic matter to produce oil and an impermeable rock cap prevented its upward movement. Natural gas, which is lighter than oil, was trapped above the oil.

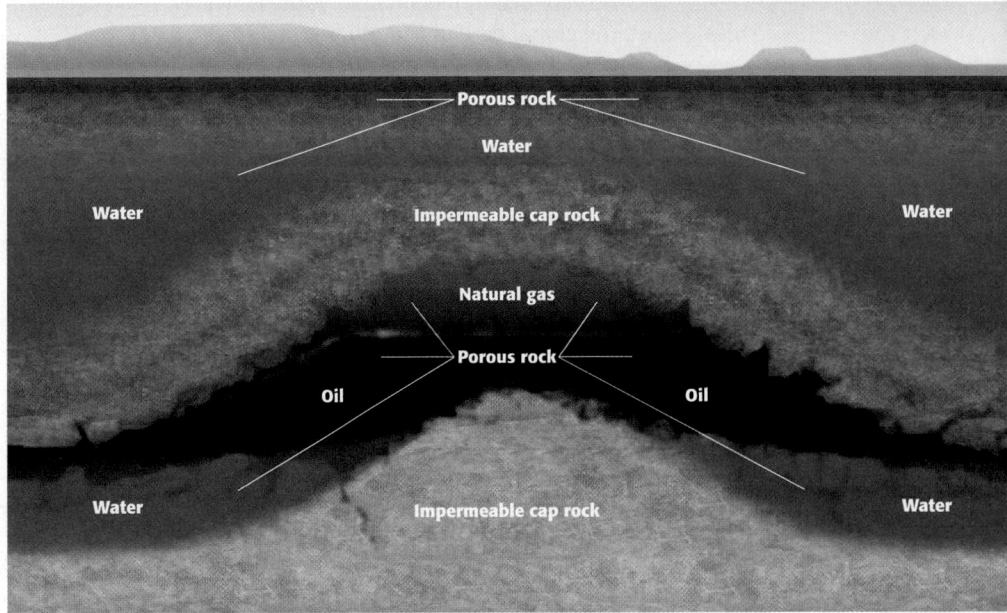

Porous rock

Water

Water

Impermeable cap rock

Water

Natural gas

Porous rock

Oil

Oil

Water

Impermeable cap rock

Water

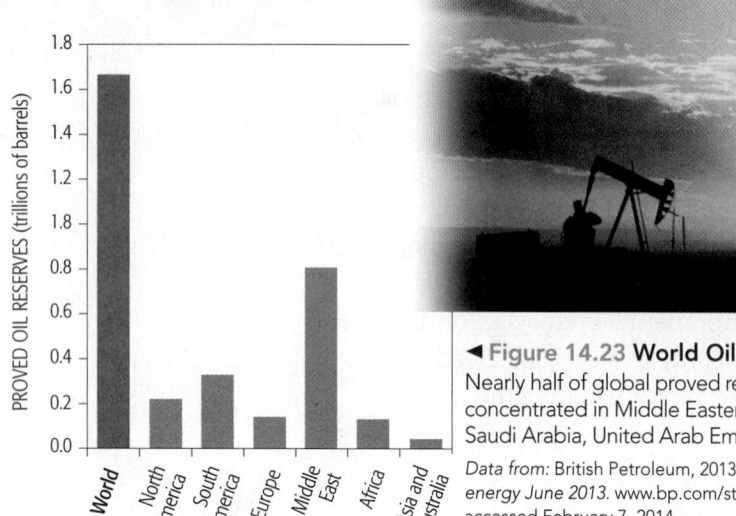

◀ **Figure 14.23 World Oil Reserves**
Nearly half of global proved reserves of oil are concentrated in Middle Eastern countries, including Saudi Arabia, United Arab Emirates, Kuwait, and Iraq.

Data from: British Petroleum, 2013. *Statistical review of world energy June 2013.* www.bp.com/statistical review, p. 6. Last accessed February 7, 2014.

Q *When are we going to run out of gas?*

Alexcis Zare, University of Houston

Norm: (A) We cannot set an exact time when gasoline will no longer be available, but there is no doubt that the supply is rapidly diminishing while the demand is increasing.

The geological conditions necessary for oil formation are not evenly distributed on Earth. As a consequence, oil reservoirs are concentrated in certain geographical regions. Today, close to half of known oil reservoirs are located in just five Middle Eastern countries—Saudi Arabia, Iran, Iraq, Kuwait, and the United Arab Emirates. Venezuela has about 18% of proved reserves, and tar sands in Canada account for about 10% of proved reserves. The United States possesses about 2% of Earth's known oil reserves (**Figure 14.23**).

Like all nonrenewable resources, the world's supply of oil is finite. Over the past century, global oil consumption has totaled about 1.2 trillion barrels of oil. Proved oil reserves that can be recovered using available technologies at current world oil prices are estimated to equal about 1.67 trillion barrels. Worldwide, about 31.5 billion barrels of oil are produced each year. At this rate of production, the reserve-to-production ratio is 52.9, which suggests that oil reserves will be depleted in less than 53 years. However, the global demand for oil is increasing so rapidly that global consumption could exceed 100 billion barrels per year in less than 20 years. At that rate of consumption, the lifetime of known reserves would be much shorter.

Peak oil is a term used to describe the point of maximum production in an oil field; beyond that point, production from the field

will only decline. The peak oil concept can also be applied to oil production at the national scale. Oil production in the United States peaked around 1970. Whether or not global oil production is near its peak depends on how much oil remains in the ground and whether it can be accessed; these are both areas of uncertainty. Some analysts predict that continued exploration and the development of new technologies will cause proved reserves to increase to as much as 2 or even 3 trillion barrels. For example, new technologies and higher prices have facilitated production of unconventional oil reservoirs such as oil shale in North Dakota, tar sands in Canada, and extra heavy crude oil in Venezuela. As a consequence, R/P has increased by 70% since 2009.

In 2013, global proved reserves of natural gas were estimated to be 187.3 trillion m^3. Current world production of natural gas is about 3.36 trillion m^3/yr. At this rate of consumption there is about a 56-year supply (R/P = 187.3/3.36 = 55.7). However, if the rate of natural gas use continues to accelerate as it has over the past decade, production will double by 2030. As with oil production, many experts speculate that we may pass the peak of natural gas production within a decade or two.

Nearly two-thirds of known natural gas reserves are located in the Middle East, Russia, and Turkmenistan (**Figure 14.24**). The location of these reserves presents an important challenge because they are quite distant from the regions with the greatest demand for natural gas. Countries in North America currently consume close to 30% of the annual natural gas production but possess only about 6% of the global reserves. In contrast, the countries of the Middle East consume about 12% of total production but possess over 40% of the reserves.

With the exploitation of unconventional shale reservoirs over the past 15 years, proved reserves of natural gas in the United States have nearly doubled to about 8.5 trillion m^3. However, R/P is only 12.5 at the current production rate of 0.68 trillion m^3/yr.

▶ **Figure 14.24 Natural Gas Reserves**
The world's largest natural gas reserves generally occur where oil is also abundant. Note that almost 90% of Europe and Eurasia's natural gas reserves are found in only two countries, Russia and Turkmenistan.

Data from: British Petroleum, 2013. *Statistical review of world energy June.* http://www.bp.com/en/global/cor-porate/about-bp/energy-economics/statistical-review-of-world-energy-2013.html. *Last accessed February 30, 2014.*

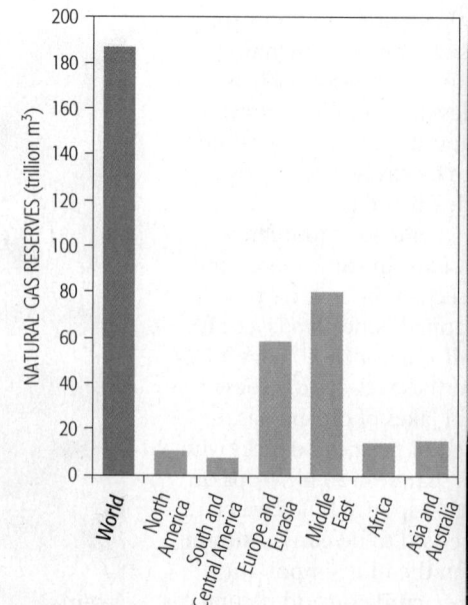

Oil Production

■ Crude oil is pumped from underground reservoirs and sent to oil refineries to be converted into gasoline and other petroleum products.

When an oil company locates an oil reservoir, it begins drilling oil wells. At first, there is usually enough built-up pressure in the reservoir to push crude oil up to the surface. This is called **primary oil recovery**, and it can usually extract about 20% of the oil in a reservoir. After that, oil companies use secondary recovery methods to extract oil, such as injecting water into a reservoir to increase pressure. **Secondary oil recovery** methods can extract another 10–20% of the oil reservoir. Historically, oil wells were capped once primary and secondary methods had extracted as much oil as they could. Even though 60–70% of a reservoir's original oil remained, the cost of retrieving it was too high.

Nowadays, most of the world's easily accessible oil reservoirs have been tapped. As a result, oil companies are often resorting to tertiary recovery methods to extract another 10–20% of oil from depleted reservoirs. **Tertiary oil recovery** methods stimulate the flow of additional oil by injecting CO_2, steam, or hot water into reservoirs.

The cost and amount of energy required to extract oil depends upon the method of extraction. For primary oil recovery, EROI is about 25, meaning that 25 units of energy are obtained for each unit of energy used to extract it. For secondary oil recovery, EROI is 8–12, and for tertiary oil recovery, EROI may be less than 6.

Some rocks, such as oil shale, are not porous enough to allow sufficient amounts of oil to flow into wells. Over the past 20 years, engineers have developed a technology called **hydraulic fracturing**, or fracking, to produce oil and natural gas from these deposits. Fracking involves the use of high-pressure liquids to fracture the rocks. Sand mixed with these liquids holds fractures open, allowing oil and gas to flow (Figure 14.25). Fracking is being used to produce oil from the Bakken oil shale in North Dakota. The EROI for this oil is 5–6.

High oil prices have facilitated oil production from the tar sands in Alberta, Canada. These deposits are very near the surface and are accessed by a process akin to shallow open pit mining (Figure 14.26). The boreal forest and its peat soils are cleared and the sand is removed and processed. Approximately 2 tons of tar sand are required to produce a barrel (42 gallons) of oil.

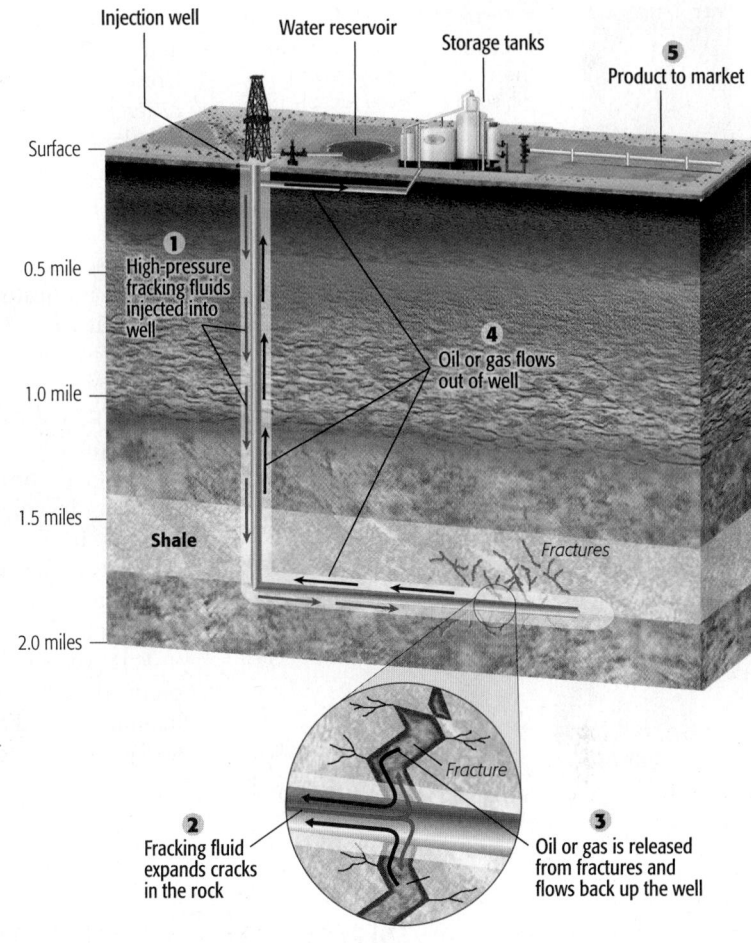

▲ Figure 14.25 **Fracking**
Hydraulic fracturing or fracking is typically used to extract oil or natural gas from shale deposits several thousand feet beneath the surface. Well shafts are drilled horizontally into the shale deposits and injected with a mixture of liquids and sand.

◄ Figure 14.26 **Mining Oil from Tar Sands**
In Alberta, Canada, large expanses of boreal forest are being cleared and the underlying soils excavated in order to produce oil from tar sands.

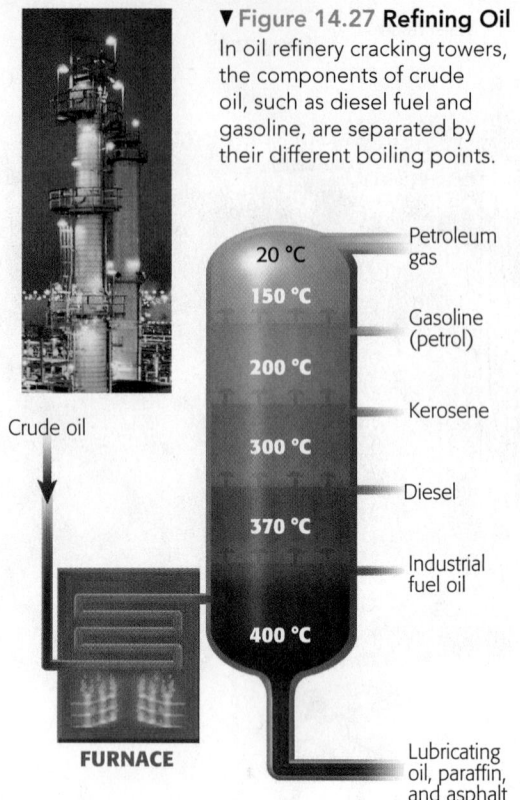

▼ Figure 14.27 **Refining Oil**
In oil refinery cracking towers, the components of crude oil, such as diesel fuel and gasoline, are separated by their different boiling points.

Crude oil

20 °C — Petroleum gas
150 °C
— Gasoline (petrol)
200 °C
— Kerosene
300 °C
— Diesel
370 °C
— Industrial fuel oil
400 °C
— Lubricating oil, paraffin, and asphalt

FURNACE

After crude oil is brought to the surface, it is sent through pipelines or loaded onto oil tankers and shipped to refineries, where it is converted into a variety of useful petroleum products. Crude oil is made up of hundreds of different types of hydrocarbon molecules, each with a different boiling point. To separate crude oil into different products, oil refineries heat it in large distillation columns, also called *cracking columns* (Figure 14.27). Lighter hydrocarbon molecules, which have relatively low boiling points, evaporate and rise as vapor to the upper portions of the distillation column, where they cool and condense into a liquid. These lighter hydrocarbons are used for jet fuel and gasoline. Heavier hydrocarbon molecules, which have higher boiling points, condense near the bottom of the distillation column. These heavier hydrocarbons are used for fuel oil, lubricating oil, and asphalt.

About 15–16% of oil production is used for nonenergy products, such as solvents, fertilizers, pesticides, and plastics. The remainder is converted into products that are burned as sources of energy.

About 70% of every barrel of crude oil sent to a refinery is converted into gasoline or diesel fuel to power cars and trucks. The chemical energy in these fuels is converted to kinetic energy by **internal combustion engines**. In such engines, fuel is explosively ignited in a closed cylinder, driving a piston that turns a crankshaft. In gasoline engines, an electric spark ignites the fuel. In diesel engines, high pressure ignites the fuel.

Aviation fuel accounts for 7% of all oil production. Most of this fuel is used to power commercial and military jet airplanes. Jet engines are a form of internal combustion engine in which ignited fuel is used to create the high-pressure stream of air that provides thrust.

Forty years ago, residential and commercial heating consumed nearly 15% of total oil production. Since then, there has been a significant increase in the amount of electricity and natural gas used for heating. Today, the kerosene and oil used to heat homes and offices account for less than 7% of oil production.

Natural Gas Production

■ Natural gas is pumped from underground reservoirs, purified, and sent to consumers by pipeline.

In the past, it was not uncommon for oil companies to simply burn the natural gas that escaped in the process of drilling for oil. Today, oil companies recognize the value of this resource and capture it in pipelines. If a pipeline is not available, the gas is reinjected into the oil deposit. In Alaska, for example, oil well operators reinject the natural gas pumped from oil wells with the expectation that it will be reextracted when a natural gas pipeline connects the oil field to consumers. In addition, the reinjected gas helps to increase pressure in the oil reservoir, thereby improving the production of the oil field.

Fracking is now being used in many places to produce natural gas from deposits of shale. Since 2002, for example, gas-rich shale deposits at many locations have increased proved gas reserves in the United States by over 40%; U.S. natural gas production has jumped 25% over this same period (Figure 14.28).

After natural gas is pumped from underground reservoirs, it is sent by pipeline to gas refineries, where impurities, such as water, sulfur, and other gases, are removed. The purified natural gas is then delivered to consumers through another set of pipelines.

Transporting natural gas across the ocean presents a significant challenge, since it is not possible to send

it through pipelines. Instead, it is often converted into **liquefied natural gas (LNG)**. When natural gas is chilled to −160 °C (−260 °F), it changes into a liquid that takes up only 1/600th of the space that it occupies in its gaseous form. In this form, it becomes economical to use ocean tankers to carry natural gas to distant consumers. LNG production does, however, add to the EROI costs of using natural gas.

Globally, the largest consumers of natural gas are the industrial and electric power sectors. Industry consumes 44% of total natural gas production. Much of this is used in high-temperature boilers and furnaces, but some is used as the raw material, or feedstock, for the production of plastics, fertilizers, synthetic fiber, and other products. About 25% of natural gas production is used as a direct source of energy in households.

Currently, 31% of global natural gas production is being used to generate electricity, and that fraction is growing. Although their construction and operating costs are high compared to coal-fired power plants, natural gas–fired power plants release many fewer pollutants and no fly ash.

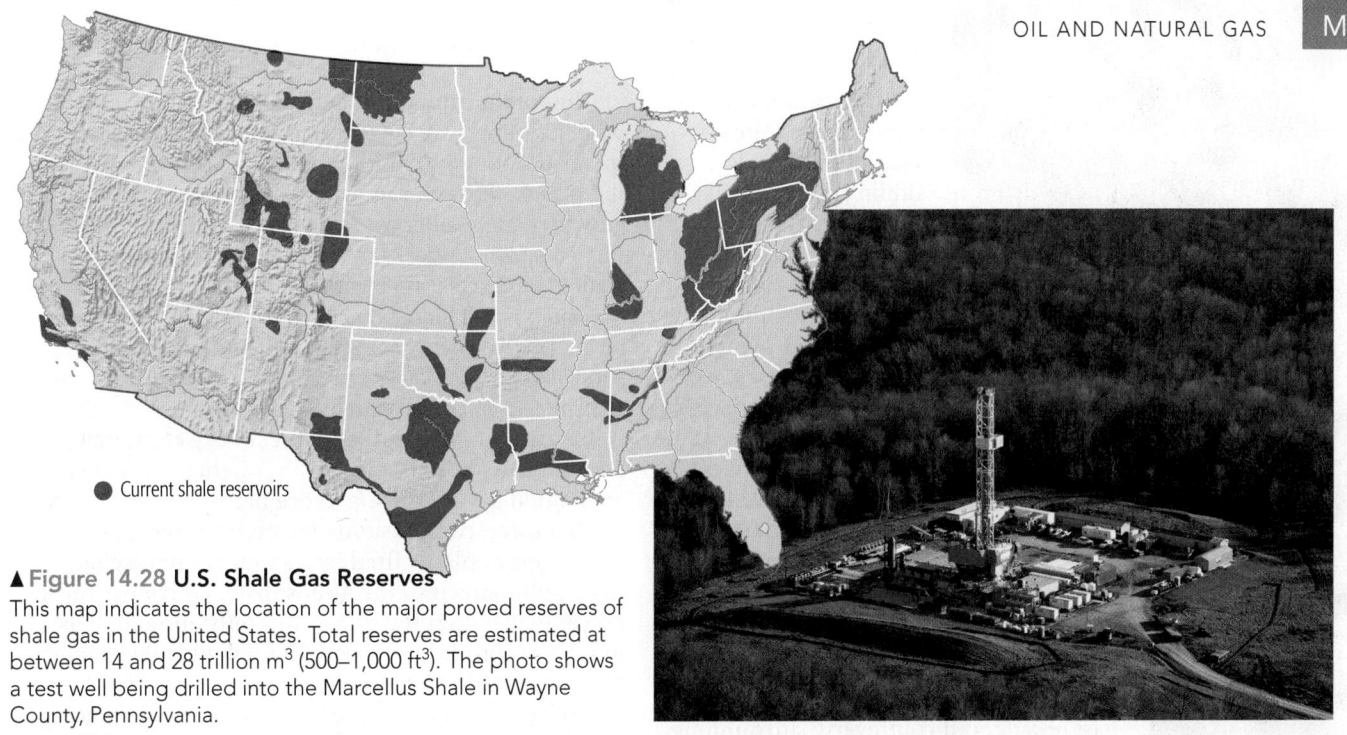

▲ **Figure 14.28 U.S. Shale Gas Reserves**
This map indicates the location of the major proved reserves of
shale gas in the United States. Total reserves are estimated at
between 14 and 28 trillion m³ (500–1,000 ft³). The photo shows
a test well being drilled into the Marcellus Shale in Wayne
County, Pennsylvania.

● Current shale reservoirs

Oil, Natural Gas, and the Environment

■ Transportation of oil and natural gas and combustion of the products derived
from them have significant environmental consequences.

Environmental degradation. Locating and
drilling for oil and natural gas present risks to habitat
and wildlife in sensitive environments. In 2008, a large
area of Amazon rain forest bridging the countries of
Colombia, Ecuador, Peru, and Brazil was approved
for oil exploration. Conservation scientists fear that
roads, pipelines, and associated activities will threaten
biodiversity across this area.

Since 1977, there has been ongoing debate about
drilling for oil in the Arctic National Wildlife Refuge
(ANWR) on Alaska's northern coastal plain. A U.S.
Department of Energy Report suggests that the sediments
beneath ANWR's tundra might yield as much as
876,000 barrels a day by 2025 if drilling were to begin

immediately. Environmentalists argue that this represents
a small portion of U.S. oil demand and that the threats to
the refuge's fragile ecosystems from a possible oil spill are
too great relative to this potential benefit (Figure 14.29).
This controversy is likely to continue in the years ahead.

Many oil and natural gas reserves are located in coastal
waters. Drilling for these reserves puts ocean ecosystems
at risk. In 2001, the world's largest floating oil-drilling rig
suffered an explosion and sank in waters about 150 km
(93 mi) from the coast of Brazil. Although little crude oil
escaped into the environment, more than 1 million liters
(265,000 gallons) of diesel fuel used to power the rig
were released into surrounding waters, killing thousands
of fish and seabirds.

◄ Figure 14.29 **The ANWR
Controversy**
Significant reserves of oil
sit beneath this portion of
the U.S. Fish and Wildlife
Service Arctic National
Wildlife Refuge on Alaska's
northern coastal plain.
Environmentalists fear
that drilling for oil and
the construction of an oil
pipeline would have serious
consequences for this fragile
ecosystem.

467

▲ Figure 14.30 Infamous Oil Spill—Lasting Impacts
The explosion of the Deepwater Horizon drilling platform opened an undersea well that spilled 5 million barrels of oil into the Gulf of Mexico. Remnants of this spill still regularly wash onto the shores of the Gulf Coast states.

QUESTIONS 14.4

1. Describe the process by which oil reservoirs form.

2. How does the formation of natural gas differ from that of oil?

3. Differentiate among primary, secondary, and tertiary methods of oil recovery.

4. What is hydraulic fracturing, and why has it become an attractive technology for oil and natural gas production?

5. Describe two important environmental consequences of the extraction and combustion of oil and natural gas.

(MES) For additional review, go to **MasteringEnvironmentalScience**

In April 2010, an explosion and fire on BP's Deepwater Horizon drilling platform killed 11 workers and began spewing oil into the waters of the Gulf of Mexico. In the months that followed, nearly 5 million barrels (210 million gallons) of crude oil were released, severely affecting the fisheries of the Gulf, as well as its beaches and coastal wetlands (Figure 14.30). This was the largest marine oil spill in the history of the petroleum industry.

Pollution. Fracking to produce oil and natural gas has increased fivefold over the past decade, and controversy surrounding its use has also increased significantly. Many who live near drilling sites complain about noise and air pollution associated with drilling equipment. There is evidence that the fluids used in fracking have contaminated aquifers in some locations. Each well requires the use of 3.5–26 million liters (1–7 million gallons) of water; of this, about one-third is returned to the surface. This water contains large amounts of fracking chemicals, and there are concerns about its safe management. Added to these concerns, for proprietary reasons, natural gas companies are unwilling to disclose the actual chemical composition of fracking fluids. The natural gas industry argues that the risks of fracking are minimal and are more than balanced by the environmental benefits of natural gas. Research is underway to evaluate these risks more thoroughly.

Transport and spills. Major spills associated with oil transportation have done serious damage to wildlife and important marine fisheries. The 1988 grounding of the oil supertanker *Exxon Valdez* spilled over 40 million liters (10.6 million gallons) of crude oil into Prince William Sound, Alaska. The impacts on coastal and marine ecosystems were immediate and tragic; plant and animal communities still have not recovered fully. Recent derailments and explosions have raised concerns about the risks associated with train transportation as well.

Emissions. The emissions from gasoline and diesel-powered automobiles and trucks are a major cause of air pollution. The exhaust of automobiles contains toxic pollutants, such as carbon monoxide, hydrocarbons, and nitrogen oxides (NO_x). Most diesel engines also release hydrocarbon particulates into the atmosphere. Along heavily used transportation corridors, NO_x interacting with oxygen produces high levels of ozone (O_3) in the lower atmosphere.

The combustion of oil-derived fuels accounts for 31% of all human-caused greenhouse gas emissions. Currently, there are about 600 million automobiles and trucks in the world. With economic development and population growth, that number could double by 2030. However, higher gasoline prices and more stringent regulatory standards are creating incentives for the production of more fuel-efficient vehicles and vehicles powered by alternative energy sources. If these trends continue, some experts predict that greenhouse gas emissions from transportation could actually decrease.

Because it is a versatile and relatively clean fuel, and because it is abundant in North America, the demand for natural gas is growing. Since natural gas is mostly methane and contains only traces of nitrogen and sulfur, power plants fired by natural gas release few toxic pollutants into the atmosphere. Of course, those power plants do release CO_2, but only about half as much as is produced by coal-fired plants that generate the same amount of electricity. Thus, nearly all plans to reduce greenhouse gas emissions include replacing coal-fired power plants with power plants fueled by natural gas.

Despite these desirable attributes, the leakage of methane is a persistent concern with production and transport of natural gas. Scientists estimate that the extraction and processing of natural gas account for about 15% of the methane that human activities add to the atmosphere each year. Because the global warming potential (GWP) of methane is 25 times higher than that of carbon dioxide, these methane emissions have the potential to cancel out the favorable effects of natural gas use on CO_2 emissions (see Module 9.3). Work is underway to reduce leakage at each stage of natural gas extraction and processing.

Political turmoil. More than 60% of world oil and natural gas reserves are located in countries with political conditions that could constrain oil exploration and production, thereby threatening the global oil supply and international security. Such political conditions could also lead to actions that would have significant implications for the environment. For example, as Iraq retreated from Kuwait in 1991, it set afire hundreds of oil wells. These fires devastated the surrounding ecosystems and created enormous amounts of air pollution. Today, many drilling, transportation, and refining facilities are vulnerable to terrorist attacks. A single oil terminal in Saudi Arabia, for example, loads almost 10% of the world's oil production onto oil tankers every day. A successful attack on such a facility would have serious economic consequences and devastate many Persian Gulf ecosystems.

There is much discussion but little consensus about future trends in oil and natural gas supply and demand. Nevertheless, the words of a former Saudi Arabian oil minister ring true: "The stone age did not end for lack of stone, and the oil age will likely end long before the world runs out of oil."

Monitoring Pollution from Oil Spills

How can tissues from seagulls be used to measure oil pollution?

In November 2002, the oil tanker *Prestige* sank in heavy seas, releasing 60,000 metric tons of crude oil into the marine environment along Spain's northwest coast. There was no doubt whatsoever that this oil spill had serious immediate impacts on coastal ecosystems. Although much of the visible evidence of the spill was gone by the summer of 2004, scientists suspected that residual pollutants from the oil spill were continuing to affect these ecosystems.

Alberto Velando and his colleagues at the University of Vigo, in Vigo, Spain, *hypothesized* that chemical analysis of the red blood cells of seagulls might provide a means of monitoring the levels and distribution of oil spill pollutants in these ecosystems (**Figure 14.31**). They focused their attention on the concentrations of specific polycyclic aromatic hydrocarbons (PAHs) that are released as crude oil decomposes. These chemicals accumulate in animal tissue and have well-known toxic effects. Because gulls feed near the top of marine food webs, their tissues are likely to accumulate larger amounts of PAHs than animals at lower trophic levels (**Figure 14.32**).

Velando chose to examine PAH concentrations in red blood cells because these cells are produced constantly and have a life span of less than four weeks. Thus, PAH in a gull's red blood cells indicates that it has had relatively recent exposure to the pollutants. In addition, blood samples can be obtained without harming the gulls.

Gulls feed within 40 km (25 mi) of their nesting sites; therefore, any differences in their blood PAH levels reflect the environment in that geographic range. In 2004, Velando's team found that PAH concentrations in red blood cells of gulls nesting near oiled areas were more than 120% higher than those from gulls nesting outside the oil spill zone. Over the next two breeding seasons, PAH levels dropped by nearly a third.

To *verify* that differences in PAH concentrations were a direct consequence of oil pollution, separate field experiments were conducted outside the range of the oil spill. One set of gulls was given food that contained small amounts of oil, and a control group of gulls was given food that did not contain oil supplements. Then blood samples were taken from each group of gulls. PAH levels were indeed higher in the *treatment group* than in the *control group*. As expected, PAHs generally decreased within 4–6 weeks

▲ **Figure 14.31 Team Leader**
Alberto Velando is a professor at Spain's Vigo University, where he studies the effects of pollution from oil spills on seabird populations.

of exposure, reflecting the death and replenishment of red blood cells.

The research by Velando's group provides *evidence* for the distribution and duration of oil pollution following this particular oil spill. More importantly, it suggests that nondestructive sampling of blood from seabirds such as gulls can provide an effective means of monitoring oil pollution.

Source: Perez, C., A. Velando, I. Munilla, M. Lopez-Alonso, and D. Oro. 2008. Monitoring polycyclic aromatic hydrocarbon pollution in the marine environment after the Prestige oil spill by means of seabird blood analysis. *Environment Science & Technology* 42: 707–713.

1. Why did Velando select blood cells rather than other tissues for chemical analysis?

2. Velando and his colleagues used both laboratory experiments and field samples in their study. Why were both important?

▲ **Figure 14.32 Biomonitors**
Levels of PAHs in seagull blood indicate whether birds have been feeding in areas polluted by spilled oil. This graph shows that blood PAH levels in gulls from colonies located near polluted sites were twice those found in gulls from colonies near unpolluted sites.

14.5 Nuclear Power

BIG IDEA The awesome power of nuclear energy was first demonstrated when the United States dropped atomic bombs on Japan near the end of World War II. Only a few years later, scientists were developing peaceful uses for nuclear energy, including electric power generation. The first commercial nuclear power plants were commissioned in the 1950s. In nuclear power plants, fission of an isotope of the element uranium is used to heat and pressurize steam, which then turns generator turbines. Although the activities associated with the use of nuclear power emit few greenhouse gases, mining activities and waste management have real and potential environmental impacts. Concerns about the safety of nuclear power plants have been heightened by accidents at reactors in Japan and the former Soviet Union. Nevertheless, the increasing costs of other forms of nonrenewable energy and concerns about global warming are increasing interest in the expanded use of nuclear power.

Sources and Production

■ Nuclear power takes advantage of heat generated by the fission of atoms.

Nuclear energy is energy contained in matter itself (see Module 3.3). In nuclear fission, the nucleus of an atom splits to form two or more smaller nuclei. In the process, a small amount of the nucleus's mass is converted to electromagnetic and kinetic energy. Nuclear fission occurs naturally when radioactive isotopes of particular elements decay. Indeed, it is this process of decay that heats Earth's interior.

Electrical power is generated by taking advantage of the energy released by the decay of an isotope of the element **uranium**, ^{235}U. When a ^{235}U nucleus is split, high-energy neutrons are emitted, as well as other forms of energy. If these neutrons are sufficiently abundant, they can split the nuclei of other ^{235}U atoms. The newly split atoms emit yet more neutrons, which then split more nuclei, creating a chain reaction (see Figure 3.22). If this reaction is uncontrolled, the result is explosive, as in the atomic bomb. In nuclear power plants, the chain reaction is controlled, and the released energy is used to generate electricity.

Uranium ore is found in many locations. Each year, mining operations produce about 58,000 metric tons of uranium oxide. Most of this is produced in Australia, Kazakhstan, Russia, South Africa, and Canada (Figure 14.33). Proved economically accessible reserves of uranium ore are thought to contain about 5.3 million metric tons of uranium. This suggests that reserves are sufficient for another 91 years (R/P = 5,300,000/58,000 = 91.3). However, experts note that at higher energy prices or with improved technologies, uranium may be economically extracted from lower grade deposits or even from seawater. Thus, uranium reserves may be as high as 15 million metric tons.

Today nuclear reactors supply about 11.5% of global electricity. This electricity comes from over 430 nuclear reactors, located in 31 countries. The United States has the highest number of nuclear reactors: 100. These reactors supply about 19% of its electricity. In contrast, France has 58 nuclear reactors, but they provide almost 75% of that nation's electricity.

The cost of operating a nuclear power plant is generally much lower than the cost of operating a power plant fueled by coal or natural gas. However, the start-up costs for a nuclear power plant are much higher—a typical nuclear power plant costs $8–9 billion. In most countries, including the United States, these costs are met with the help of various government subsidies.

Uranium is obtained from open pit mines. Uranium ore is rarely concentrated in large amounts, so mining operations must excavate large amounts of material to recover comparatively small amounts of ore

▼ **Figure 14.33 Proved Uranium Reserves**
Like other primary energy sources, uranium ore is found in large quantities in a relatively few locations around the world.

Data from: Uranium 2013: Resources, Production and Demand. A Joint Report by the OECD Nuclear Energy Agency and the International Atomic Energy Agency.

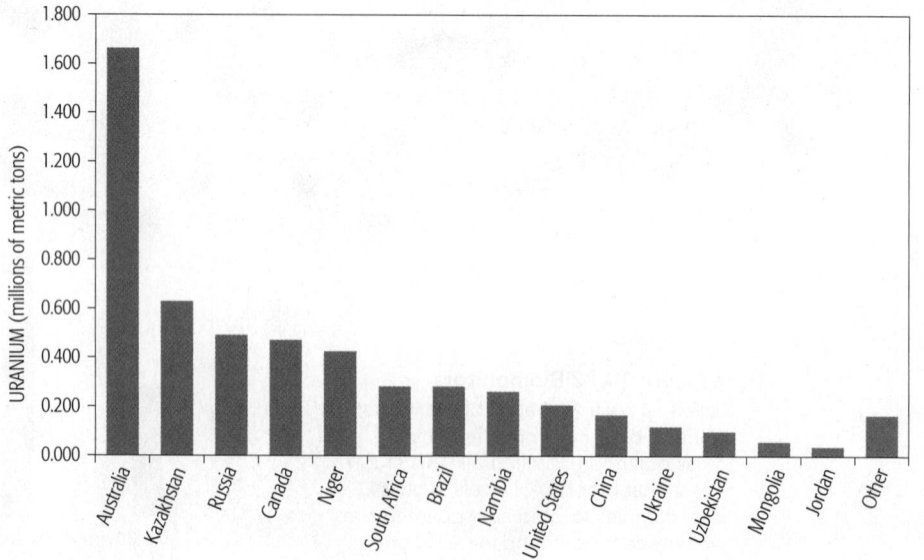

▲ Figure 14.34
Uranium Excavation
Uranium open pit mine near Arlit, Niger, has environmental impacts on nearby ecosystems and human communities.

(**Figure 14.34**). Once mined, uranium ore is crushed into a fine powder and washed with a strong acid or alkali. Next, the resulting liquid is concentrated into a solid material called *yellowcake*, which contains about 75% uranium oxide. Yellowcake is then transported to plants where it will be processed to produce uranium fuel.

At the processing plant, the uranium in the yellowcake is converted to gaseous uranium hexafluoride. Only 0.7% of the uranium in yellowcake is composed of ^{235}U atoms. The remainder is ^{238}U atoms, which do not undergo nuclear fission. Atoms of ^{235}U weigh slightly less than atoms of ^{238}U, and this weight difference is exploited to increase the amount of ^{235}U relative to ^{238}U. When the uranium hexafluoride is spun in cylindrical centrifuges, the heavier isotope moves to the cylinder walls, and ^{235}U is concentrated in the center. In this way, the relative concentration of ^{235}U is increased to about 4%.

This material is then formed into fuel pellets that are about the size of jelly beans. Fuel pellets are packed into corrosion-resistant metal tubes to make fuel rods. Next, 100–300 fuel rods are joined together to make a fuel assembly (**Figure 14.35**). Inside a nuclear power plant, up to 3,000 fuel assemblies are placed in a **reactor core,** where the fission reaction will take place.

In the reactor core, the fuel assemblies produce high-energy neutrons that set the nuclear fission chain reaction in motion. The energy released through the fission process heats pressurized water or liquid sodium surrounding the assemblies. Control rods in the fuel assemblies regulate the rate of the chain reaction and the heat generated from the fuel assemblies. Control rods contain elements such as silver or cadmium that intercept fast neutrons. That hot liquid is then used to convert water into steam. In the final step, this steam is used to turn turbines to generate electricity (**Figure 14.36**).

Note that the steam that drives the turbines does not come into direct contact with the radioactive fuel assemblies. Nuclear power plants operate much like power plants that use fossil fuels, except the water is boiled with heat from a nuclear reaction instead of heat from burning fossil fuels.

▼ Figure 14.36 Electricity from Atoms
Heat generated from the controlled fission of ^{235}U is used to heat either water or sodium to very high temperature and pressure. This superheated liquid is then used to generate steam that powers turbines, which in turn drive electric generators.

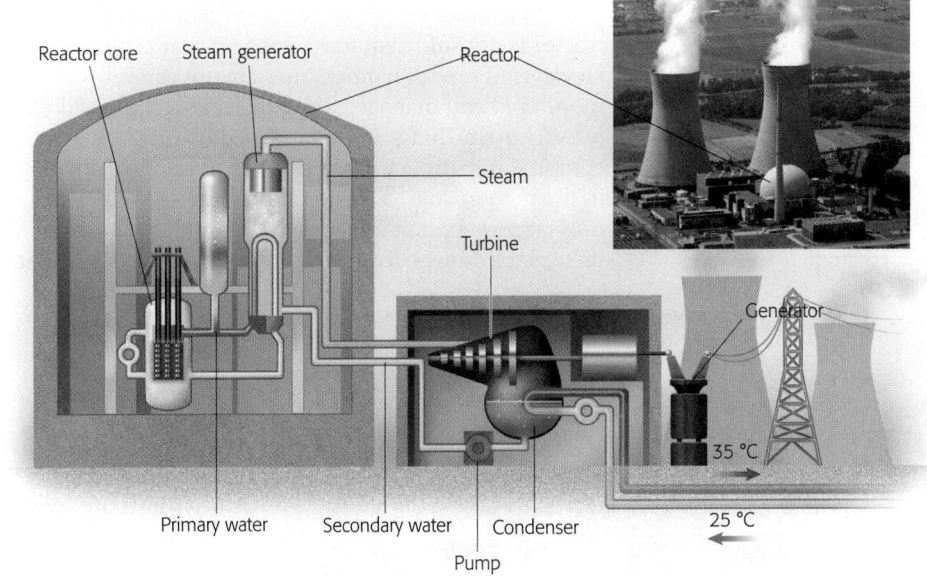

Reactor core   Steam generator   Reactor
Steam
Turbine
Generator
35 °C
25 °C
Primary water   Secondary water   Condenser
Pump

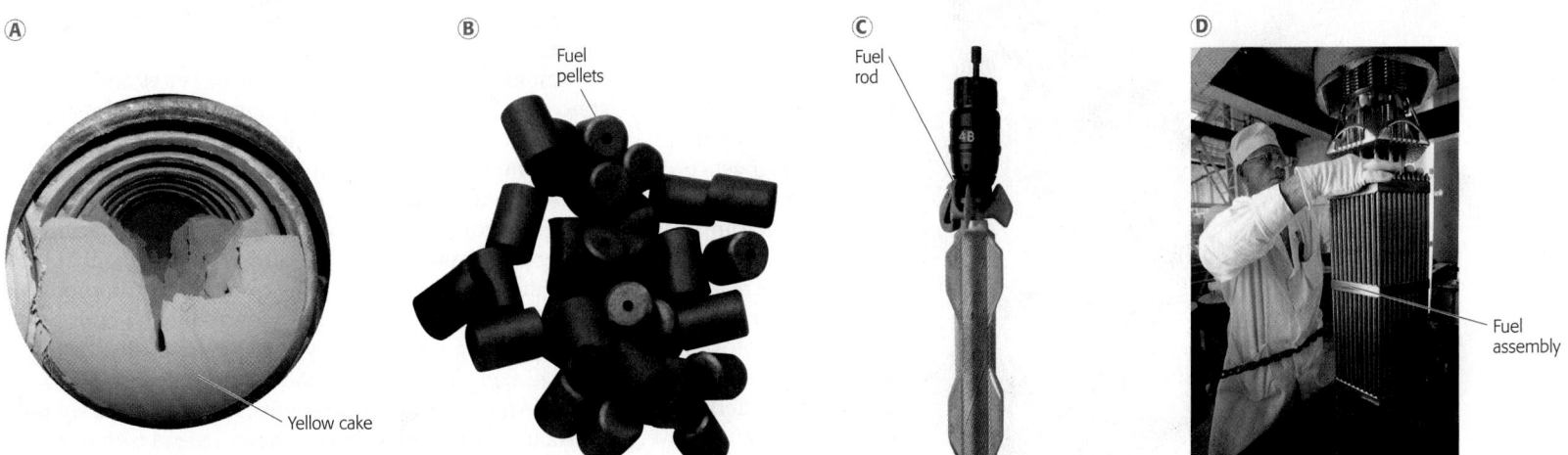

Ⓐ
Ⓑ Fuel pellets
Ⓒ Fuel rod
Ⓓ

Yellow cake

Fuel assembly

▲ Figure 14.35 Producing Nuclear Fuel
Uranium ore is formed into yellowcake Ⓐ, which is converted into uranium hexafluoride and spun in a centrifuge. The uranium enriched in ^{235}U is then formed into fuel pellets Ⓑ that are inserted into fuel rods Ⓒ. Fuel rods are joined together into a fuel assembly Ⓓ.

Nuclear Power and the Environment

■ The environmental benefits and costs of nuclear power are hotly disputed.

Q *Given the damage to the plants in Japan, is nuclear energy worth the risks?*
Patty Adams, Metro State College of Denver

Norm: A Nuclear energy is appealing because it generates few greenhouse gas emissions. However, public resistance, the high cost of building new plants, and the difficulties associated with disposing of radioactive waste are serious impediments to its use. The disaster at Japan's Fukushima nuclear facility raised additional concerns that safety risks have been underestimated.

Even in the best mining sites, uranium is not very plentiful. To extract the uranium, large quantities of rock must be removed and processed. This process disturbs large areas of land and leaves behind extensive piles of rock and soil, or mine tailings. As with other kinds of mines, water flowing through these tailings is often acidified and has severe negative impacts on nearby streams and lakes.

Excess heat is one of the important by-products of nuclear power plants. This heat is most often dispersed in the form of steam that exits through large cooling towers. In fact, those cooling towers have come to symbolize nuclear power generation itself. In some cases, heat is dissipated in nearby river, lake, or ocean water. Elevated temperatures in waters near nuclear plants can alter the quality of the aquatic or marine habitats and diminish biodiversity.

Even without controls, the rate of fission in a nuclear reactor is not sufficient to trigger a nuclear explosion. However, if reactor temperatures are not carefully monitored and managed, the core can overheat and literally melt. That is what happened in 1979 when equipment malfunctions and operator errors resulted in the meltdown of a reactor at the Three-Mile Island nuclear power plant in Pennsylvania. In this incident, the reactor dome—the heavily reinforced container that

holds the reactor—remained intact and prevented any significant leakage of radioactive materials. Had the dome failed, a large area of southeastern Pennsylvania and parts of New Jersey could have been contaminated.

In April 1986, a far more serious failure occurred in a reactor at the Chernobyl nuclear power plant in the former Soviet Union (now Ukraine). Operator negligence resulted in a reactor meltdown, accompanied by a gas explosion that destroyed the reactor dome (Figure 14.37). Additional explosions sent a plume of highly radioactive material into the atmosphere and over parts of western Russia, Belarus, eastern Europe,

and Scandinavia. The amount of radioactive fallout released in this disaster was more than 30 times greater than that released by the atomic bombs in Hiroshima and Nagasaki, Japan. A large area of Ukraine, Belarus, and Russia was contaminated, causing the evacuation and relocation of over 330,000 people. This event was the direct cause of at least 50 deaths. The World Health Organization estimates that radioactive fallout from Chernobyl has caused at least 9,000 extra cases of cancer among the 7 million people exposed to the highest levels of radiation.

Three Mile Island and Chernobyl alerted government agencies and people around the world to the potential risks associated with the use of nuclear power. In the United States, public opinion regarding the use of nuclear power turned strongly negative. Electric utilities shelved all plans to build additional nuclear power plants. Some European countries, such as Italy and Spain, decided to decommission existing nuclear facilities and forego construction of any new facilities.

A recent catastrophe in Japan has raised new doubts about reactor safety. On March 11, 2011, a major earthquake struck the northeastern island of Honshu. Although the nearby Fukushima Daiichi nuclear power plant had been designed to withstand such events, it was flooded by the subsequent tsunami. This caused explosions and fires, damaged at least two reactor containment vessels, and resulted in the release of large amounts of radioactivity (Figure 14.38). Nearly 19,000 people died in the earthquake and tsunami, but no deaths have yet been directly attributed to the power plant catastrophe. Although the radiation released was less than 30% of that from Chernobyl, the World Health Organization predicts elevated cancer rates for people living within 50 km (30 mi). High levels of radiation forced the evacuation of over 400,000 people in communities within 20 km (12 mi). As of 2014, over 160,000 evacuees are still living in temporary housing and are unlikely to be able to return to their homes in the foreseeable future.

Despite these events, soaring energy prices and concerns about greenhouse gas emissions and global warming have caused governments, power utilities, and even some environmental groups to reconsider their opposition to nuclear power. While acknowledging the significant consequences of accidents at nuclear facilities, these groups note that the specific circumstances that led to the accidents at Three Mile Island and Chernobyl have been remedied by improved technologies and management protocols. They also argue that the Fukushima Daiichi catastrophe occurred in a unique geologic setting; in most locations, nuclear plants are not at risk for the same level of earthquakes and tsunamis.

Others argue that the benefits of nuclear power do not outweigh its potential and real costs. They note

▼ Figure 14.37 **1986 Meltdown**
As a consequence of operator error, Reactor number 3 at the Chernobyl nuclear power plant overheated and exploded, spreading radioactive fallout across much of northern Europe.

▲ Figure 14.38 **Fukushima Daiichi**
These are the remains of two of the reactor buildings at the Fukushima nuclear plant. Three years after the earthquake and tsunami, millions of gallons of radioactive water continue to pour into the ocean. High amounts of radioactive cesium have been detected in fish caught within 100 km of the facility.

that there are still many uncertainties about the safety of nuclear power plants. Several independent reviews revealed that substandard construction and human error contributed to the Fukushima Daiichi catastrophe. Many are concerned about the threat of terrorist attacks.

Nuclear power does not rely on the combustion of fossil fuels and, therefore, does not directly emit CO_2 or other atmospheric pollutants. Small amounts of emissions are associated with uranium mining, transportation, and processing. Scientists calculate that annual greenhouse gas emissions could be reduced by 1 billion metric tons if there were a threefold increase in the number of nuclear power plants built to replace coal-fired plants. By itself, this is not sufficient to avoid significant future global warming, but it might be included in a portfolio of actions to achieve this goal.

One of the most daunting challenges associated with nuclear energy is the safe disposal of **spent nuclear fuel**. Through time, the ^{235}U in fuel rods is depleted and the rods' heating capacity diminishes. Although they can no longer be used in reactors, these fuel rods accumulate an array of other radioactive elements and will remain radioactive for over a million years. This spent fuel must be carefully stored and monitored. Currently, it is stored in storage pools or stainless steel casks at the nuclear reactor sites (Figure 14.39).

The total volume of high-level radioactive waste can be reduced through **fuel recycling**, in which the fuel is reprocessed and the elements that can undergo fission chain reactions are reconcentrated. This process has been implemented by some countries, such as France. However, as waste is recycled, concentrations of another element, **plutonium**, increase. Like uranium, plutonium can undergo fission and generate heat. Unlike ^{235}U, however, it is relatively easy to use plutonium to make nuclear bombs.

Because of the threat that plutonium production poses to global security, the United States and most other European countries have agreed not to recycle spent fuel.

Many experts believe that geologic repositories are the best long-term solution to the spent fuel challenge. A **geologic nuclear repository** is a layer of rock in which nuclear waste and all of its toxic and radioactive elements are permanently sealed from the biosphere. By international agreement, a repository must protect humans and the ecosystems they depend on from exposure to radiation for hundreds of thousands of years.

Sweden and Finland have established small experimental geologic nuclear repositories in granite, 300 m beneath their coastal waters. But no other nuclear repositories are in operation. This is partly due to political and social opposition to their location. The United States has been working to develop a geologic nuclear repository at Yucca Mountain, Nevada, a site about 90 miles north of Las Vegas. Here, spent fuel in durable stainless steel casks would be placed in long tunnels dug through dry rocks about 300 m (1,000 ft) below the surface. Since 2011, however, funding for the Yucca Mountain repository has been frozen by Congress, in large part because of opposition to the project by the state of Nevada. Repository development at this site or any or any other U.S. site appears unlikely for the foreseeable future.

It is fair to say that no primary energy source generates more controversy and debate than nuclear power, especially in the United States. Because of high construction costs and widespread public resistance, no new nuclear facilities have been built in the United States since 1978. In recent years, however, many electric utilities have begun planning for new reactor facilities in order to reconcile rising demands for energy with the need to mitigate global warming. As a result, the debate over nuclear power has intensified. Meanwhile, 30 new nuclear reactors are under construction around the world, mostly in Asia, and additional growth appears very likely.

▶ Figure 14.39 **Storing Spent Fuel**
In wet storage, spent fuel assemblies are immersed in large pools of water. In dry storage, fuel assemblies are encased in thick-walled metal casks.

QUESTIONS 14.5

1. Describe the chain reaction that provides energy for nuclear power plants.
2. How is the heat from the fission of ^{235}U captured to produce electricity in a nuclear power plant?
3. Describe two ways of storing high-level radioactive waste.

(MES) For additional review, go to **MasteringEnvironmentalScience**

Summary

Primary energy sources are natural resources such as sunlight, oil, and coal that have not been transformed by humans. These sources may be converted to secondary energy sources such as electricity or the kinetic energy of machines. Energy conversion efficiency is the percentage of energy in a primary source that is captured in a secondary energy form. Energy end-use efficiency is the amount of energy available to the final application compared to the amount of energy from the primary source that was required to produce it. Nonrenewable primary energy resources are used more rapidly than their natural rates of production. Such primary energy sources include coal, oil, natural gas, and nuclear energy. Proved reserves of nonrenewable resources are the amounts that can be recovered in the future from known sources using existing technologies at current prices. The ratio of proved reserves to extraction or production rate provides an estimate of the length of time that reserves of a nonrenewable resource will last, assuming current technologies and prices.

Electricity is our most important secondary energy source and powers a wide variety of end uses. It can be generated by spinning magnets in an electric generator or by chemical reactions in a battery. Electric power is distributed through a system of transmission lines and transformers called the electric power grid. Environmental impacts of electrical transmission include energy lost in the power grid and chemical pollutants from transformers and batteries.

Coal, the most abundant of all fossil fuels, occurs in underground reserves that vary in energy yield per unit weight. Environmental impacts of coal use include the direct effects of mining, as well as air and water pollution from coal-fired power plants. Crude oil is formed from the remains of marine microorganisms that were slowly transformed by heat and pressure in geologic sediments. Oil is pumped from underground reservoirs and refined to produce fuels such as gasoline, kerosene, and diesel oil. Environmental impacts include spills associated with oil drilling and transportation as well as pollution associated with the combustion of oil-derived fuels. Natural gas is primarily composed of methane, and it is usually found in association with crude oil reserves. The combustion of natural gas produces fewer pollutants than that of coal or oil, but the methane released when natural gas is extracted and transported contributes to global warming. Nuclear power uses the heat released by the fission of uranium to produce steam that is used to drive turbines and generate electricity. Environmental impacts of using nuclear power include the direct effects of uranium mining and the potential release of highly radioactive elements and toxic chemicals into the air and water.

14.1 Energy Production

- Over the past century, our demand for energy has increased dramatically. That demand has largely been met through the use of nonrenewable fossil fuels, such as coal, oil, and natural gas.

- Primary energy is the energy contained in natural resources. Secondary energy sources, such as electricity, are produced by transforming primary energy sources.

- The future availability of nonrenewable energy sources depends on the size of proved reserves, the rate of extraction, and the prices consumers are willing to pay.

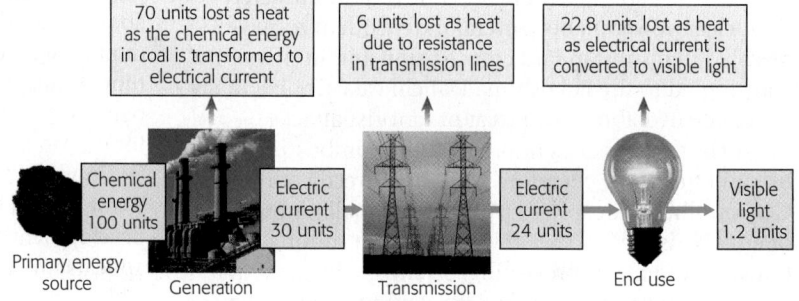

70 units lost as heat as the chemical energy in coal is transformed to electrical current

6 units lost as heat due to resistance in transmission lines

22.8 units lost as heat as electrical current is converted to visible light

Primary energy source — Chemical energy 100 units — Generation — Electric current 30 units — Transmission — Electric current 24 units — End use — Visible light 1.2 units

KEY TERMS

primary energy, energy conversion, secondary energy, end use, energy conversion efficiency, energy end-use efficiency, reserves, production, consumption, nonrenewable energy, renewable energy, proved reserves, reserves-to-production ratio (R/P), energy return on investment (EROI)

QUESTIONS

1. The end-use efficiency of most cars is less than 10%. What does this mean, and what factors are likely to be contributing to this low number?

2. What factors influence the relative sizes of the end-use energy sectors in different countries?

3. The current reserves-to-production ratio (R/P) for crude oil is about 53 years. What factors are likely to influence the actual future availability of oil?

4. How does energy return on investment (EROI) generally change as reserves of a nonrenewable energy source are depleted? Why?

14.2 Electric Power—Generation, Distribution, and Use

- Primary energy sources are used to power electric generators and produce electric currents.

- Batteries and fuel cells use chemical energy to generate electricity.

- Electricity is distributed on power grids, and its flow is regulated by feedbacks between consumers and power generators.

- The direct effects of the electromagnetic fields generated by the power grid on human health are uncertain.

QUESTIONS

1. The EROI for hydrogen fuel cells is typically less than 1. Why is this so?

2. What is the role of the transformers in power grids?

3. What do base load and peak load refer to, and how are they related to the cost of electricity?

4. Design experts complain that electricity is wasted because of poor feedback between consumers and producers of electricity. Why?

KEY TERMS

amperes, volts, kilowatt hours, electric generator, electromagnetic field, fuel cells, transformers, electric power grid, base load, peak load

14.3 Coal

- Underground mining and surface coal mining have direct impacts on the environment. The combustion of coal produces a variety of air and water pollutants.
- Coal burning is a major source of greenhouse gas emissions.

KEY TERMS

peat, coal seams, lignite, sub-bituminous coal, bituminous coal, anthracite coal, scrubber, fly ash, mine tailings

QUESTIONS

1. What is mountaintop-removal mining and why is it a major concern to environmentalists?
2. What is carbon capture and storage (CCS), and how might it be used to reduce greenhouse emissions from coal burning?
3. Why are some energy experts concerned about the use of CCS?

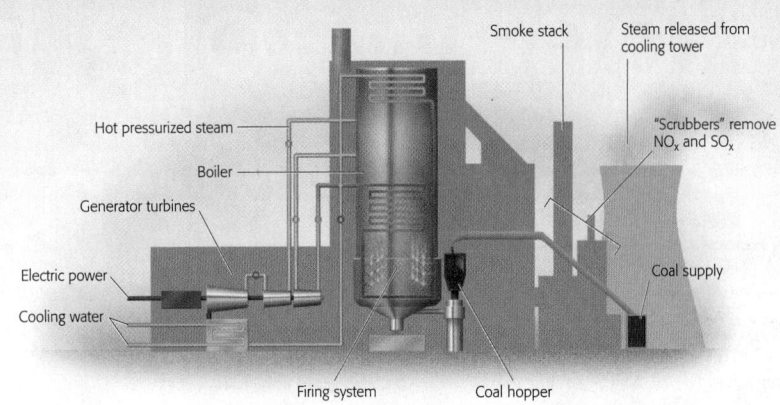

14.4 Oil and Natural Gas

- Oil is produced from the decomposition of microbes in ocean sediments; oil is trapped in reservoirs beneath impermeable cap rocks.
- Natural gas is composed of methane and is formed in nearly the same way as oil and is usually associated with crude oil reservoirs.
- Major environmental impacts caused by the use of oil include spills associated with oil pumping and transportation and air and water pollution from fuel combustion.
- Natural gas combustion releases relatively few pollutants, but methane is a powerful greenhouse gas.

KEY TERMS

kerogen, oil reservoir, conventional reservoir, unconventional reservoir, peak oil, primary oil recovery, secondary oil recovery, tertiary oil recovery, hydraulic fracturing, internal combustion engine, liquefied natural gas (LNG)

QUESTIONS

1. What evidence supports the conclusion that we are very close to peak global oil production?
2. What factors might extend the availability of oil beyond the time spans estimated by the ratio of production to reserves?
3. Describe three reasons why continuing to rely heavily on oil-derived fuels presents serious challenges for future sustainability.

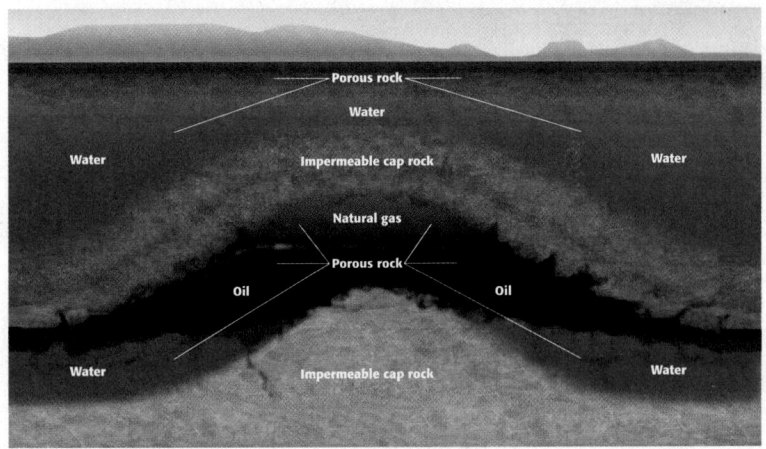

4. Why is the conversion of natural gas to electricity considerably more efficient than the conversion of coal to electricity?
5. Describe two ways in which the use of natural gas adds to global warming.

14.5 Nuclear Power

- The controlled fission of ^{235}U in the core of a nuclear reactor heats water or sodium, which is then used to produce steam to drive turbines and generate electricity.
- Many decision makers are concerned about the potential for accidents and the release of highly radioactive materials from nuclear power plants.
- The normal operation of nuclear power plants produces significant amounts of spent nuclear fuel that remains radioactive for hundreds of thousands of years.

KEY TERMS

uranium, reactor core, spent nuclear fuel, fuel recycling, plutonium, geologic nuclear repository

QUESTIONS

1. What are the steps required to transform uranium ore to the fuel rods that power a nuclear reactor?
2. Describe two kinds of high-level radioactive waste storage.
3. How does a high-level nuclear waste storage facility differ from a nuclear waste repository?
4. What are the primary environmental advantages of nuclear power compared to fossil fuels?
5. What are the primary environmental concerns associated with nuclear power?

MasteringEnvironmentalScience®

Students Go to **MasteringEnvironmentalScience** for assignments, the eText, and the Study Area with practice tests, activities, and more.

Instructors Go to **MasteringEnvironmentalScience** for automatically graded tutorials and questions that you can assign to your students, plus Instructor Resources.

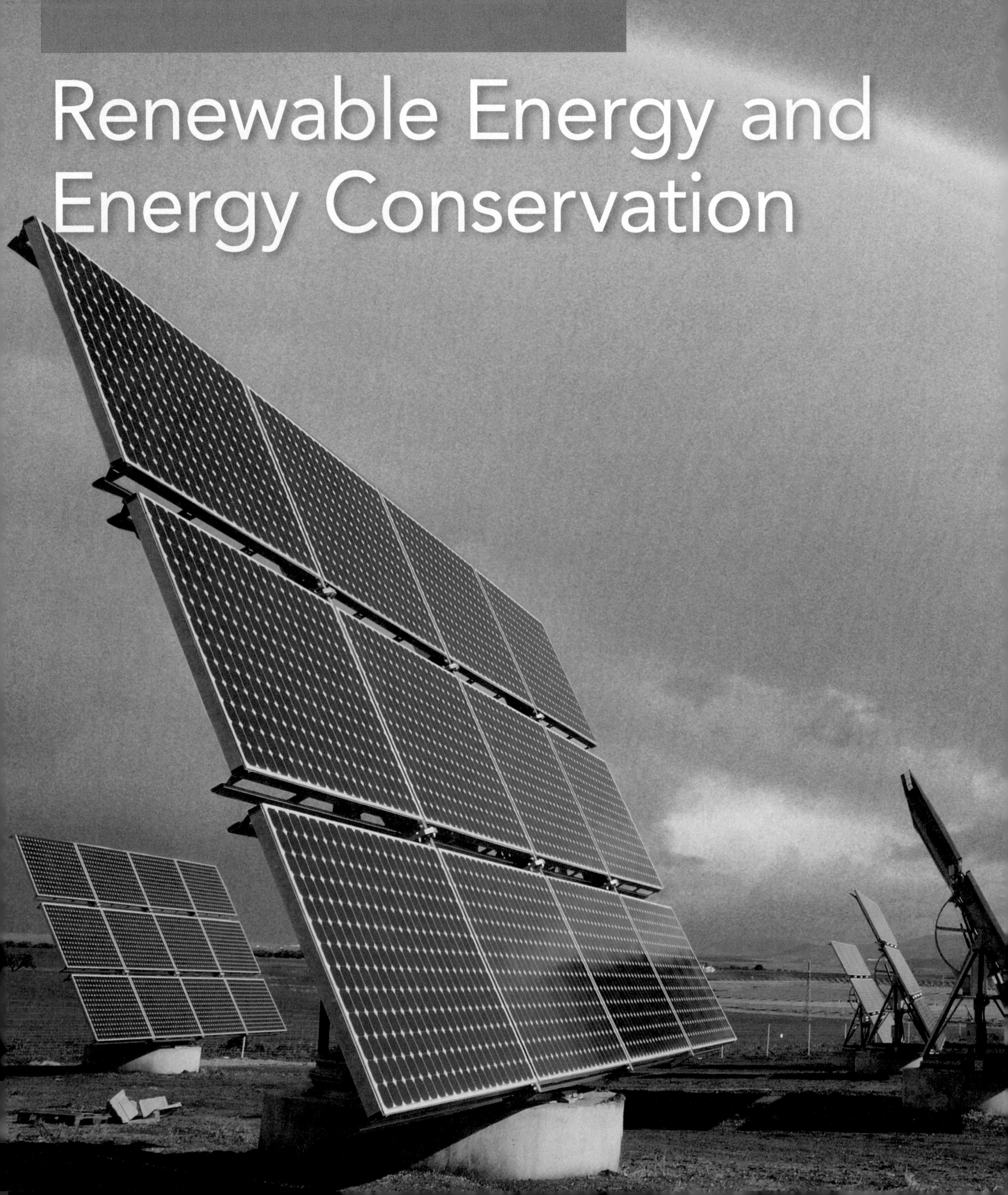

Renewable Energy and Energy Conservation

15

Human Energy Consumption Through the Ages

Can we reverse historical trends in energy use?

We humans, and our most direct ancestors, have been around for about a million years, and over that time the amount of energy that each of us uses has increased by over 300-fold. Indeed, the trajectory of human population growth over the past million years has very much been shaped by changes in the kinds and amounts of energy we consume (Figure 15.1, see also Module 5.2).

(A) The numbers of our earliest ancestors were limited just like any animal in a terrestrial food web, that is, by the energy in the food they could capture and eat (Figure 15.1A). And, by modern standards, that amount was meager indeed. Early hominids gathered plant material or killed animals equivalent to about 1.5 gigajoules (GJ) per person each year. Based on this, we estimate that these early hominids survived on about 4 megajoules (MJ) or about 1,000 food Calories each day (1 food Calorie = 1,000 calories). The U.S. Department of Agriculture (USDA) recommends that a moderately active adult eat 2,000–2,200 Calories each day, so we can assume that obesity was not a problem for these early hominids.

(B) Modern humans, folks who looked pretty much like us, appeared about 100,000 years ago. Improved tools and increased intelligence, among other things, allowed these people to access more food energy—about 2.7 GJ/person/year (Figure 15.1B). These people also had access to an additional source of energy, fire. Based on the habits of hunter-gatherers today, we estimate they consumed an additional 3 GJ of energy from burning biomass. Making use of fuel to cook food enabled them to store food and use it more efficiently. This fuel also provided the warmth needed to survive in colder regions, and populations were able to disperse worldwide.

(C) The domestication of plants and animals marked another significant change in human energy consumption (see Module 12.1). Around 5,000 B.C.E. (7,000 years ago), citizens of ancient Mesopotamia were probably consuming about 3.5 GJ of food/person/year (Figure 15.1C). This probably translated into an individual daily diet of over 2,000 Calories. Fuel use for cooking and heating appears to have increased to over 8 GJ/person/year. Added to this was about 5 GJ of energy from domestic animals and flowing water used in agriculture. Total energy consumption was over 20 GJ/person/year or the energy equivalent of 3.5 barrels of oil.

(D) By the beginning of the Renaissance (500 years ago), average human energy consumption had increased by another 65% to 33 GJ/person/year (Figure 15.1D). Energy consumed as food remained relatively constant, but social diversification, the development of cities, and technologies like the smelting of metals more than doubled the energy consumed by agriculture and industry. People at this time also began to travel more, adding another 5 GJ to the per capita total.

▼ Figure 15.1 **Human Population Growth and Energy Use**

Human populations increased through three different periods over the last million years (see Module 5.1). Changes in both the amount and uses of energy over that time span are depicted in the bar graph, correlated to the beginning and end points of each period. Refer to the letters in both graphs as you read through this essay.

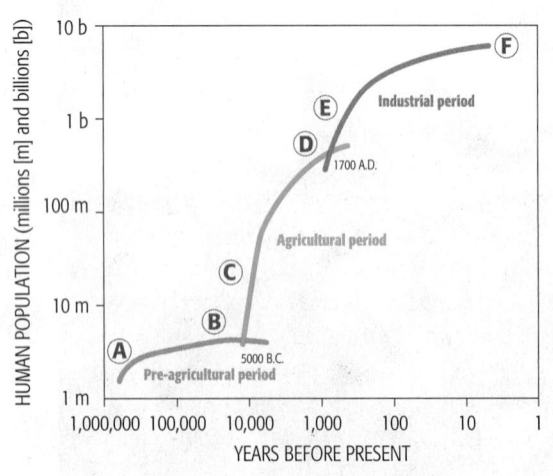

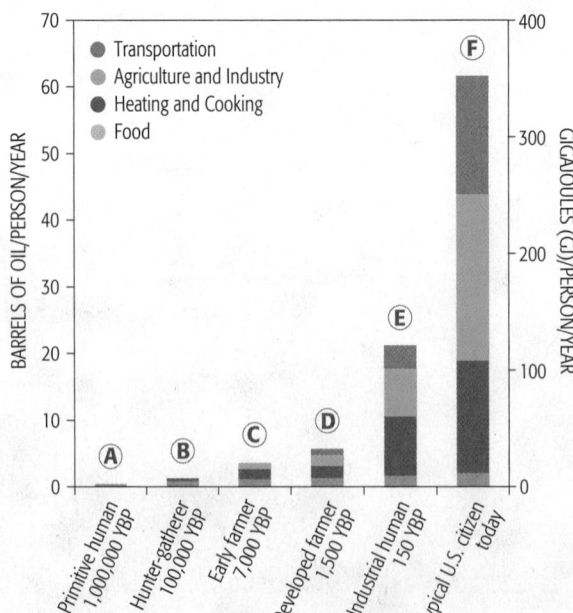

E During the Industrial Revolution, the amounts and patterns of human energy consumption were transformed once again, at least for those living in industrialized countries. Fossil energy—coal and petroleum—became the most important source of energy for industry and agriculture. In 1850, the average American consumed over 120 GJ/year, the energy equivalent of 21 barrels of oil (Figure 15.1E). Domestic energy use increased nearly fivefold from pre-industrial times, as people began to cook and heat their homes with fossil fuels. These same fuels now powered agriculture and industry. Coal-fired trains and steamboats moved people and cargo over great distances.

▼ Figure 15.2 **Electrified**
The photo shows a typical evening in New York's Times Square. Over 40% of primary energy consumed in the United States is used to generate electricity.

Now, the average American consumed over 20 GJ (3.5 barrels of oil) each year for transportation.

F The pace of change continues to accelerate. Today, the average American consumes almost 350 GJ of energy each year, and fossil fuels are the primary energy source for over 80% of that amount (Figure 15.1F). A large portion of that consumption involves electricity produced from other primary energy sources (Figure 15.2). We use it to illuminate and air condition our homes and workspaces, as well as to produce and power the great variety of gadgets and contraptions that now grace our lives. We expend as much energy to cool our homes and commercial spaces in the summer as we do to heat them in the winter. Over the past 150 years, energy consumption for agriculture and industry has increased by 300%. And do we travel! Moving around town or around the world, we each consume over 90 GJ/year, nearly 30% of our total energy use. The primary source for almost all of that travel is oil.

The energy we consume as food averages to almost 7 GJ/person/year. That is the daily equivalent of 17 MJ or about 4,000 Calories. Very few of us actually eat that much but, at over 2,700 Calories (11.5 MJ), the majority of us eats in excess of what the USDA recommends. What happens to the remaining 1,300 Calories (5.5 MJ)? It is "lost" as spoilage, plate waste, and in cooking.

Those of us who live in the world's wealthiest countries enjoy a very high level of well-being compared to our ancestors, and it is very tempting to attribute that directly to our very high level of energy consumption. But many of the greatest threats to our well-being—air and water pollution, biodiversity loss, and global warming—are also a direct consequence of our very high energy use. Paradoxically, we are consuming ever larger amounts of the energy available to us to cope with the negative impacts of the very large amount of energy we consume. Furthermore, we are increasingly aware that supplies of some of the nonrenewable primary energy sources that we most depend on are dwindling.

The majority of Earth's citizens consume far less energy than we Americans do, but they rightfully aspire to our level of well-being. Must average global consumption of nonrenewable energy resources rise to our level to meet those aspirations? Energy experts and environmental scientists agree that that is not a sustainable path. Not only are supplies of fossil fuels insufficient for the long term, but ever-increasing reliance on them poses significant threats to the well-being of humans and the ecosystems upon which they depend. Alternatively, can we meet global aspirations by relying more on renewable primary energy sources and reducing energy consumption by increasing the efficiency of the energy use? This is a compelling question.

- *What are current patterns and trends in the use of renewable primary energy resources?*

- *What are the barriers to increased reliance on renewable energy resources?*

- *What technologies exist or are in development to access renewable primary energy sources such as solar energy, biomass, wind energy, water movement across the land and in the sea, and Earth's heat?*

- *What challenges are associated with the use of these technologies?*

- *Can we maintain a high level of human well-being while reducing end-use energy consumption through conservation?*

- *What economic and policy changes are needed to encourage sustainable energy consumption?*

15.1 Renewable Energy Overview

BIG IDEA Renewable sources of energy could transform the way we heat our homes, power our industries, and fuel our cars. Our current reliance on fossil fuels, such as oil, coal, and natural gas, to meet 80% of our energy needs has significant environmental consequences. The limited supply of many of these nonrenewable resources is the source of much international conflict. Greater reliance on renewable energy sources can diminish these environmental impacts, as well as conflicts over energy supplies. But although renewable energy sources are abundant on a global scale, their availability in particular locations is often limited. To accelerate use of renewable energy sources, we must develop new and sometimes unfamiliar technologies, overcome economies of scale, minimize the externalized costs associated with nonrenewable energy sources, and increase consumer understanding of the benefits of renewable energy use.

The Transition to Renewable Energy

■ Can we reduce demand for limited energy sources and diminish the environmental impacts of human energy use?

Renewable primary energy resources are based on energy flows that are continuously replenished, and the supplies of renewable energy resources are potentially vast. One hour of sunlight striking Earth's surface contains more energy than is consumed by all humans in an entire year. The winds that blow across Earth's surface could theoretically meet current global energy demand 15 times over (Figure 15.3).

The primary challenge of using many renewable energy resources is that they are diffuse and intermittent. For example, although the overall quantity of solar energy is vast, that energy is spread out across Earth's surface. The solar energy available in a particular location is far less concentrated than the energy in fossil fuel reserves. In addition, sunlight is only available during the day, and then only when the weather is clear.

Solar 5,900 terawatts

Wind 25 terawatts

◀ Figure 15.3 **Energy Abundance**
Compared to total human energy use, renewable sources of energy such as wind and sunlight are abundant. However, these resources are not concentrated in forms that are easy to access. Note that the relative size of these circles is not to scale.

Global consumption
1 terawatt
Biomass, geothermal energy, wave tidal energy, and hydropower.

In order to make use of most renewable energy resources, we must capture and convert their diffuse energy flows into more concentrated forms. Technological research and development are increasing the accessibility of many renewable resources. Limited supplies and higher prices of nonrenewable resources are providing additional incentives for rapid growth in the use of renewable energy.

At present, nonrenewable resources, such as fossil fuels and nuclear energy, account for about 80% of global primary energy production. Wood and charcoal burning, mostly in developing countries, contribute another 9.3%. Only about 9.7% of global primary energy production comes from renewable resources, such as hydropower, solar power, wind, biofuels, and geothermal energy (Figure 15.4).

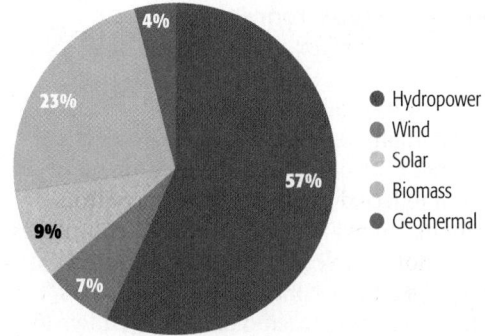

◄ **Figure 15.4 Renewable Energy Consumption**
Hydroelectric power provides more than half of the renewable energy consumed in the world. This graph does not include the fuelwood used in developing countries.

Data from: Renewables 2013: *Global Status Report: 2013 Update.* http://www.ren21.net/REN21Activities/GlobalStatusReport.aspx. Last accessed 2-28-2014.

Legend: Hydropower, Wind, Solar, Biomass, Geothermal
Pie values: 57%, 4%, 23%, 9%, 7%

Challenges

■ There are four general categories of challenges to the widespread use of renewable energy.

As the benefits of energy conservation and an increased reliance on renewable energy become increasingly apparent, many wonder why they are not being used to a greater extent. Energy experts believe that a more widespread adoption of renewable energy resources is being hindered by a variety of political, economic, technical, and social challenges. These include unfamiliar technologies, economies of scale, externalized costs, and limited consumer knowledge and understanding.

New technologies.   Many technologies designed to use renewable energy sources, such as concentrating solar power stations, are still in the early stages of development and are just now being deployed on large scales. The novelty of these technologies affects their economic viability in several ways. Because we do not have much experience with these new technologies, there are many uncertainties and risks associated with investing in them. This makes investment capital for renewable energy

technologies relatively scarce. As a result, companies developing these technologies are forced to pay higher interest rates for investment funds, driving up the cost of development.

Economies of scale.   The scale of production for many renewable energy technologies remains small, which keeps their cost of production high. The limited use of solar panels is a good example of this chicken-and-egg problem. Until recently, solar panels were produced on a small scale, which kept the costs per unit of production high. Because the panels were expensive to produce, their price to the consumer was high, so demand for this technology remained low. Low demand, in turn, maintained small-scale production, keeping production costs high, and so on.

As a consequence of increased understanding of the environmental benefits and higher costs for electricity delivered on the grid, demand for solar panels has increased significantly in recent years. As a result, their production costs have fallen, which in turn has lowered their price and increased the sales of the panels (Figure 15.5). As the price continues to fall, the use of solar energy can become more competitive with the use of fossil fuels.

Q *Considering all that we know (good and bad) about each form of clean energy production or energy-saving systems, which do you think is the best for the environment and most efficient in its production?*

Brandon MacDonald, Citrus Community College

Norm: (A) All renewable energy sources have advantages and disadvantages. Sustainable energy use will depend on an appropriate mixture of them all.

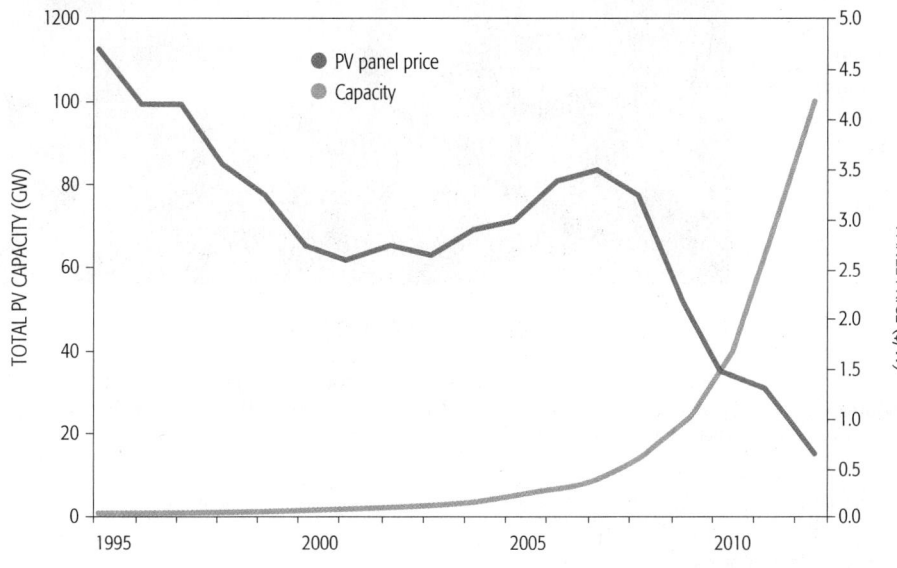

Chart legend: PV panel price, Capacity
Left axis: TOTAL PV CAPACITY (GW) — 0, 20, 40, 60, 80, 100, 1200
Right axis: PANEL PRICE ($/W) — 0.0, 0.5, 1.0, 1.5, 2.0, 2.5, 3.0, 3.5, 4.0, 4.5, 5.0
X axis: 1995, 2000, 2005, 2010

◄ Figure 15.5 **Economy of Scale**
Since 1995, demand for solar photovoltaic (PV) panels has increased and so has PV panel production, indicated here as PV capacity. As a consequence, the price of panels has come down by almost 80%. The bump in price between 2005 and 2008 was due to a shortage of silicon needed for PV cell manufacture.

1. What distinguishes a renewable from a nonrenewable energy resource?
2. Describe four general challenges to the widespread use of renewable energy resources.

(MES) For additional review, go to **MasteringEnvironmentalScience**

Externalized costs favor nonrenewable energy.   The price we pay for a gallon of gas or a kilowatt-hour of electricity does not capture all of the costs associated with its production and use. Instead, many costs are externalized (see Module 2.3). For example, government subsidies for the development of fossil fuels are not reflected in the price of energy from these sources. In the United States, oil and coal companies receive billions of dollars a year in direct and indirect government subsidies. These subsidies lower the market price of these fuels, reducing incentives for consumers to conserve energy and use renewable energy resources.

Many of the environmental impacts of fossil fuels are externalized. For example, the combustion of coal in electric power plants and of gasoline in cars results in air pollution that causes asthma, emphysema, and other respiratory ailments. The costs associated with these health problems are not reflected in the price that consumers pay for electricity or gasoline. The presence of external costs makes the use of fossil fuel seem to be cheaper than it actually is and makes renewable alternatives appear less competitive.

Limited consumer understanding of the benefits of renewable energy.   Consumers often have limited knowledge about the source of their electricity. They also have little influence over the choices made by the power company from which they get their electricity. When buying automobiles and household appliances, many consumers do not understand how to interpret the energy usage figures on the labels for these products.

In some cases, decisions about energy efficiency and the installation of renewable energy devices are made by people who are not responsible for future energy costs. For example, a contractor building a house will not have to pay the energy bills for that house; therefore, the contractor has little incentive to install a highly efficient but expensive refrigerator or washing machine. The family that buys the house does have to pay the energy bills, but they may not have been consulted about which appliances to purchase or how much insulation to put into the house to save energy.

In the next several modules, we explore individual renewable energy sources and the technologies used to exploit them. We begin with that most renewable of resources, solar radiation. Biomass and wind energy are very direct products of solar energy, and they are discussed next. We then discuss the energy available in the movement of water across the land and in ocean tides and waves. Geothermal energy derived from Earth's heat completes our discussion of renewable energy resources and technologies.

New FRONTIERS

Glowing Green

Thousands of cities and towns around the world have developed policies and plans to increase the proportion of renewables in their total energy portfolio. Indeed, cities like Sydney, Australia, and Yamanashi, Japan, have set goals to achieve 100% renewable energy for heating, cooling, and electrical power over the next several decades (Figure 15.6). They propose to employ several strategies to meet these goals. New regulations in these cities set standards for energy efficiency and the use of renewable energy sources. Sydney, for example, has implemented 15 specific regulations for sharing of excess renewable energy among public buildings. Regulations are accompanied by financial incentives to businesses and homeowners for the installation of renewable energy technologies such as solar hot water heaters and ground source heat pumps (see Module 15.7). By working with utilities to decentralize electrical grids, businesses and individuals can actually provide renewable wind and solar energy back to the community. To meet 100% renewable targets, community governments must be more involved in private sector energy generation, distribution, and use. What challenges do you see to this increased involvement? Would you be willing to champion a transition to solar energy in your community?

▲ Figure 15.6 **Going Green, Glowing Green**
For an hour on the evening of March 23, 2013, Sydney's iconic opera house glowed green while the lights throughout the rest of the city were shut off as a symbol of its commitment to renewable energy and to reducing global warming.

15.2 Solar Energy

BIG IDEA Solar energy is the radiant energy of the sun, an energy source with enormous potential. The sunlight that strikes Earth in one hour contains more energy than all of the coal, oil, natural gas, and other sources of energy consumed by human societies in an entire year. A variety of methods are already being used to convert solar energy into light and heat for homes and commercial buildings, as well as into electricity. Because sunlight is free and nonpolluting, it has many economic and environmental advantages over fossil fuels. However, solar energy is diffuse and unevenly distributed around the globe; its availability also varies with the seasons and weather. The main challenge facing the use of solar energy is the development of economical ways to capture, convert, and store it. Solar energy technologies also need to overcome a number of other challenges for this energy source to make a more significant contribution to meeting our energy needs.

Sources and Production

■ Solar energy is now widely used to heat and light homes and other buildings.

For centuries, people have been designing and building houses and other structures that take advantage of sunlight to provide light and warmth to interior spaces. **Passive solar technologies** use the energy of sunlight without relying on electrical or mechanical devices, such as pumps or fans. Passive solar approaches include orienting buildings so they receive maximum sunlight and using building materials that absorb sunlight to keep interior spaces warm. Modern energy-efficient buildings often feature passive solar approaches, and these are discussed in more detail in Chapter 16.

Active solar technologies use mechanical devices to heat water and buildings or electrical devices to generate electricity. Rooftop solar panels are among the most common active solar devices (**Figure 15.7**). These panels use pipes filled with water, air, or antifreeze to collect the heat of sunlight and then transfer that heat to an interior surface, where it is used for heating water or space heating. Other active solar devices include

lens collectors, reflective light pipes, and hybrid solar lighting; these devices use mirrors and optical fibers to transmit sunlight into interior spaces that are far from windows.

Sunlight can also be used to generate electricity, or **solar power**. There are two main methods of generating solar power. The first method relies on **photovoltaic (PV) technology**, also referred to as solar cells or PV cells. PV cells convert light energy into electricity through the photovoltaic effect, in which light energy causes certain materials to emit electrons, thereby generating an electric current. PV cells typically contain pairs of silicon plates embedded with different metals (**Figure 15.8**). When sunlight strikes a PV cell, it is absorbed by the first silicon plate, causing the plate to emit electrons. As those electrons move toward the second silicon plate, they create an electric current.

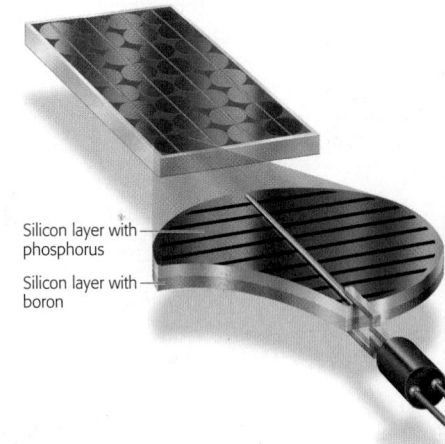

Silicon layer with phosphorus

Silicon layer with boron

◀ **Figure 15.7 Active Solar**
Energy captured by solar panels can be stored in batteries for use around the clock. However, in most states utilities compensate homeowners for solar electricity that they put back onto the grid.

▲ **Figure 15.8 Photovoltaic (PV) Cell**
Light energy is captured to produce an electrical current in the silicon wafers of a PV cell.

Solar energy is now widely used for heating and lighting applications throughout the world. In the United States, several million homes and commercial buildings use solar water heating equipment, which can typically supply 40–80% of the hot water used in the building. Solar water heaters are especially cost-effective for businesses such as hotels, fitness clubs, and restaurants, which use large amounts of hot water. Approximately 250,000 homes and buildings in the United States use solar energy for space heating.

Today, solar power technologies such as PV cells and CSP systems supply only a fraction of a percent of the world's electricity. However, interest in the development of these technologies is growing rapidly. For the past 10 years, the global production of PV cells has increased by more than 25% a year. Global investment in solar power has grown from about $12 billion in 2004 to over $150 billion in 2013.

Because solar PV cells can be set up as stand-alone power supply systems that are independent of the electric power grid, they are becoming increasingly important as sources of electricity in remote areas. In the United States, solar PV technology is used to power water-pumping stations and cell-phone towers that are located far from other sources of electric power. In remote areas of developing countries, solar PV cells are providing power to people who never before had access to electricity (Figure 15.10).

PV cells and concentrating solar power systems are attracting the interest of large-scale electric power companies. The largest CSP system currently in use is a 350-megawatt facility in southern California that generates enough electricity for more than 250,000 homes. In some areas with high solar potential, plans are underway to develop **solar photovoltaic farms**, in which thousands of PV panels are arranged to generate electricity on a large scale. A photovoltaic farm recently built in Portugal uses 2,500 very large solar panels to provide electricity to 30,000 homes. Another facility under construction in New Mexico will eventually generate enough electricity for 240,000 homes.

▲ Figure 15.9
Concentrating the Sun's Energy
Concentrating solar power systems focus sunlight onto water pipes, producing steam that is used to turn generator turbines. Rows of parabolic trough solar collectors stretch in the California desert heating synthetic oil to produce steam for electrical power.

Individual PV cells about 5 cm (2 in.) across are grouped together in modules that are a few feet across. A typical module can generate approximately 75 watts of energy, enough to power a small radio. Modules are frequently combined to form PV panels, and PV panels are grouped together to form PV arrays.

Modules and panels of PV cells are commonly used to power small devices. For example, PV cells are used in calculators and wristwatches, as well as road signs and emergency call-boxes along the sides of highways in remote locations. There is even a solar PV backpack that can generate electricity to recharge batteries for flashlights and other devices while you are hiking.

The second method of generating electricity from sunlight relies on **concentrating solar power (CSP) systems**. CSP systems are typically large-scale projects that use mirrors to concentrate the sun's rays on a tower or a series of pipes that hold water or another fluid (Figure 15.9). The concentrated sunlight heats the fluid to boiling, producing steam. The steam is then used to spin a turbine and generate electricity, as in a conventional power plant. The primary advantage of CSP over PV systems is that heated liquids can be stored for several hours, allowing electricity to be generated at night.

▶ Figure 15.10 **Rural Solar Power**
In the high Andes of Peru, a family is able to power lights and small appliances with a solar panel.

Advantages and Disadvantages

■ Solar energy is widely available and its use has relatively few environmental costs.

Solar energy has numerous advantages. It is free and infinitely renewable, and its use does not directly produce local air pollutants or greenhouse gases. Solar PV panels can be installed almost anywhere—on the roofs or walls of homes and businesses, in vast solar photovoltaic farms in the desert, or in remote locations where there is no connection to the electric power grid.

The production of solar energy systems is generating significant new employment in research and development, manufacturing, and installation. The U.S. solar energy industry already employs more than 140,000 people, and this number is growing rapidly.

Despite these advantages, solar energy still faces a number of challenges that limit its contribution to our energy portfolio. One challenge is the high up-front cost of most solar energy systems. In 2014, for example, the cost of a PV solar system that could provide power to a medium-sized house in the United States was $10,000–$20,000. Once installed, such a system might save a family several hundred dollars each month relative to the cost of conventional power. But at that rate of savings, it would take a family many years to repay the cost of installing the PV solar system.

Cost is also a challenge for large-scale uses of solar power. Electricity derived from large-scale CSP and PV systems currently costs about $0.12–$0.30 per kilowatt-hour. This is significantly more than electricity from wind ($0.08/kWh), natural gas ($0.06/kWh), or coal ($0.10/kWh). In recent years, however, the cost of producing electricity with PV cells and CSP systems has been declining rapidly. In addition, solar power systems are most productive during daylight hours that coincide with important times of peak energy demand. The market price of electricity during these hours is very close to the cost of CSP and PV power.

As with wind power, the large-scale production of solar power has some geographical limitations. Although solar energy can be used almost anywhere (Germany is one of the world's leading producers of solar power), it is most cost-effective in areas with high solar radiation,

such as the southwestern United States and other desert regions (Figure 15.11).

The manufacture of PV panels is energy and materials intensive; at present, much of this energy comes from fossil fuels. Assuming a 20-year lifetime, EROI for a solar PV panel is about 6; it produces six times more energy than was used in its manufacture. Critics note that the production of a solar PV panel requires the use of a number of hazardous materials and heavy metals. Here again, the environmental impacts of these materials are considerably less than the impacts associated with the extraction of fossil fuels or uranium.

The use of solar energy is growing rapidly. Since 2000, the production of PV cells has increased at a rate of 40% each year. The cost of producing electricity with PV panels and CSP systems is expected to continue to decline with advances in production methods, standardization of installation practices, and increased use of less expensive production materials. Some energy experts predict that by the year 2020, CSP and PV electricity will be available at $0.05/kWh and will make up 20% of the U.S. energy portfolio.

QUESTIONS 15.2

1. Differentiate between passive and active methods of using solar energy.

2. Describe how a photovoltaic (PV) cell generates an electric current.

3. What are the most important challenges to the widespread use of solar power?

(MES) For additional review, go to **MasteringEnvironmentalScience**

▼ Figure 15.11 **U.S. Solar Resources**

Low rainfall and limited cloud cover make the southwestern United States an ideal region for solar power use.

Source: U.S. Department of Energy, National Renewable Energy Laboratory. http://www.nrel.gov/csp/troughnet/images/map_normal_radiation.gif. Last accessed February 14, 2014.

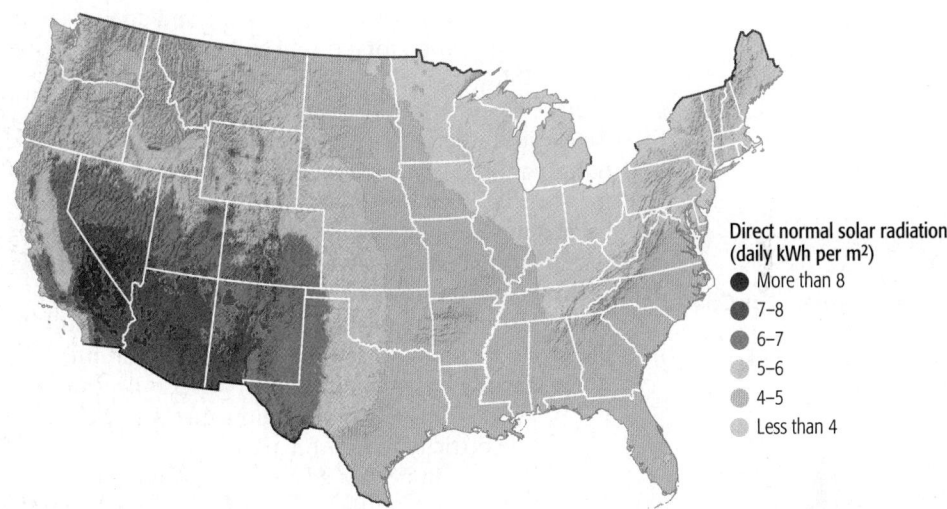

Direct normal solar radiation (daily kWh per m²)
- More than 8
- 7–8
- 6–7
- 5–6
- 4–5
- Less than 4

New **FRONTIERS**

Running on Sunshine

On October 6, 2013, 40 cars left the starting line in the northern Australian city of Darwin, on a 3,021-km (1,877 mi) race—the 14th biennial World Solar Challenge—to Adelaide on Australia's south coast. Each car was powered by sunlight and nothing else. Like the Tour de France bicycle race, the World Solar Challenge is divided into stages that test different aspects of vehicle design and race strategy. Cars must be able to endure a wide range of driving conditions including mountainous terrain and the blazing sun of Australia's Outback.

Each car was limited to no more than 6 square meters of solar panel, which have to be fixed to the car chassis. Each entry takes advantage of the latest technologies in solar panels, high-efficiency motors, batteries, superlight materials, and aerodynamic design. But even the most efficient solar panels cannot power a car continuously over long stretches. Batteries are essential to store solar energy, and their design and weight are critical. Among the starters in 2013, 16 cars completed the race. The Nuon Solar Team from the Netherlands won the race in just 33 hours and 3 minutes at an average speed of 90.7 km/hr (56.4 mi/hr).

Competitions such as this aim to encourage innovation in solar technologies, including efficient, lightweight solar panels and batteries. They also foster creative approaches to vehicle design and construction, which can help improve the energy efficiency. Compared to other vehicle types, what advantages would solar cars provide? Would you be willing to purchase a solar car, even if it meant some adjustment to the way you think about cars?

Q Why don't we use more biofuels produced from sources like recycled French fry grease?
Said "Sadie" Byboth, Louisiana State University

Norm: A Many communities recycle cooking and deep-fry oil to produce biodiesel. However, the supply of recycled oil is limited relative to total demand.

15.3 Biomass Energy

BIG IDEA Long before humans learned to harness the energy in wind, water, or fossil fuels, they were using energy from plants to cook food, keep warm, and illuminate the darkness. **Biomass energy**, also called **bioenergy**, is the energy contained in firewood and other plant matter. Bioenergy is derived from solar energy, which plants use to carry out photosynthesis and grow. Today, humans use the energy of biomass in a variety of ways. In many developing countries, the direct combustion of firewood and charcoal is an important source of energy. Newer uses of biomass include the conversion of crops, such as corn and sugarcane, to liquid fuels that can power vehicles. Biomass can also be burned in power plants to produce electricity. Biomass energy is renewable because once plants are harvested, they can grow again. When biomass is burned, the carbon dioxide that is released into the atmosphere can be taken up again by new plant growth; thus biomass has the potential to be a carbon-neutral form of energy. However, the potential environmental advantages of the increased use of bioenergy depend on what plant material is used and how those plants are grown, harvested, and used for energy.

Sources and Production

■ Fuelwood, charcoal, and ethanol are the most commonly used forms of biomass energy.

The biomass energy of wood, charcoal, and dried animal dung is widely used for cooking and heating. Worldwide, these traditional fuels make up nearly 10% of primary energy consumption. They are especially important in developing countries, where they constitute 30–50% of primary energy consumption. In sub-Saharan Africa, for example, over 80% of the wood that is harvested is burned for fuel (see Module 13.1).

Agricultural and forestry wastes are also forms of biomass. Wastes from wood processing can be burned to produce heat or generate electricity. Many sawmills burn wood scraps and sawdust to produce heat for drying lumber in wood kilns. Most paper mills use wood scraps to fire generators that provide electricity to run the machinery.

In developed countries wood in the form of wood pellets is being used in place of coal, heating oil, and natural gas used to generate electricity and supply heat (**Figure 15.12**). Wood pellets can be produced from waste wood and sawdust or from the debris left over after timber harvest. However, the majority of wood pellet production is from the harvest of whole trees.

Wood and agricultural waste can also be converted into a combustible gas, or syngas, by a gasification process that breaks the biomass down under high temperatures and in the absence of oxygen. The syngas is then burned to generate electricity.

These sources of bioenergy make up an additional 2% of global primary energy consumption. In addition to wood and agricultural wastes, there are liquid **biofuels** derived from biomass that can be used in diesel and internal combustion engines. Ethanol is by far the most widely used biofuel. Currently, it is derived from two main crops, sugarcane and corn. Both of these crops are rich in carbohydrates. Those carbohydrates are broken down to ethanol through a process of fermentation that is very similar to the process used to brew beer (see Module 4.1). Ethanol can be used to power automobiles.

Until 2006, Brazil was the world leader in ethanol production and use. Ethanol provides about 50% of that country's automobile fuel. Some Brazilian cars run on pure ethanol; most others burn a blend of 75% gasoline and 25% ethanol (75/25). Brazil's ethanol comes almost exclusively from sugarcane (**Figure 15.13**). The production of biofuel from sugarcane also produces a by-product called bagasse. Bagasse is burned to provide heat and electric power for the ethanol production process.

Recently, the United States has become the top ethanol producer in the world. Spurred in part by government subsidies and rising gasoline prices, ethanol production in the United States increased ninefold between 2000 and 2013. In 2014, nearly 52 billion liters (14 billion gallons) were expected to be produced. Ethanol production in the United States is based almost entirely on corn. Most of the ethanol used in the United States is blended with

▼ Figure 15.12
A Replacement for Coal?
Pelletized wood can be used as fuel in stoves and coal-fired power plants.

gasoline in a 90/10 mix known as E10. Many major metropolitan areas mandate the use of E10 as a way to improve air quality. Another product known as E85, 85% ethanol and 15% gasoline, is available at a few hundred gasoline stations in the upper Midwest. E85 is only suitable for use in later-model cars designed to run on this blend.

Biodiesel is a biofuel derived from plant oils and animal fats that can be burned in diesel engines. Biodiesel is usually produced from vegetable crops, such as soybeans or canola, but it can also be produced from palm oil, coconut oil, and even waste cooking fats. In the production of biodiesel, the chemical glycerin is separated from the plant oils; this glycerin is sold as a by-product for soap production. The remaining oil is then combined with ethanol or methanol in a chemical reaction that produces biodiesel. Most modern diesel engines can run on fuel blends that include up to 20% biodiesel. With relatively minor adjustments, those engines can burn 100% biodiesel. Today, biodiesel production is considerably less than the ethanol produced in most countries, but the demand for biodiesel to fuel trucks, buses, boats, and diesel automobiles is growing rapidly.

▲ **Figure 15.13 Pure Ethanol**
At this Brazilian service station, ethanol (Alcool) is much cheaper than gasoline. In Brazil, ethanol is produced from sugarcane and is burned directly in many automobiles.

Advantages and Disadvantages

■ Using biomass energy influences patterns of land use and can affect the price and availability of food.

Bioenergy offers a number of obvious environmental advantages over other forms of energy. The use of bioenergy is, in theory, carbon neutral. Although the combustion of biomass and biofuels releases carbon to the atmosphere, an equal amount is removed from the atmosphere by plants during photosynthesis. The use of some forms of bioenergy may help reduce acid rain and smog because their combustion emits much lower amounts of NO_x and SO_x than does the combustion of gasoline and diesel. However, biodiesel derived from soy oil actually emits more NO_x than conventional diesel fuel. Wood and agricultural waste combustion can produce more air pollution than coal.

Biomass is abundant in many parts of the world. Supplies of this resource are more evenly distributed than supplies of fossil fuels, such as oil and coal, or other renewable resources, such as hydropower, solar, and wind. In the United States alone, more than 40 million tons of crop residues that could be burned to generate electricity or converted to biofuels go unused every year. Biofuels such as ethanol and biodiesel can be produced domestically and therefore help to lower dependence on imported oil. On the other hand, there are growing concerns over a number of possible negative impacts associated with bioenergy.

Theoretically, the use of bioenergy is carbon neutral, but in reality the use of many biofuels is not. While it is true that the amount of carbon dioxide released in the combustion of biomass is roughly equal to the amount of carbon that plants take up during photosynthesis to produce that biomass, this comparison ignores greenhouse gas emissions at other points in the biofuel production cycle. Significant quantities of fossil fuels are consumed in growing, shipping, and processing wood pellets and crops for biofuel. For example, wood pellets are widely used by countries in the European Union in order to meet requirements to reduce use of fossil fuels. However, most of these pellets come from whole trees harvested and processed in the United States and Canada. Drying and transportation of these pellets can produce greenhouse gas emissions equal to or greater than those released by the combustion of an equal amount of coal.

Considerable amounts of fossil fuels are consumed to plant, grow, and process corn for ethanol. This includes the fuel used by farm machinery, in the production of fertilizers and pesticides, and in pumping water for irrigation (**Figure 15.14**). Furthermore, only the kernels—a small portion of total corn production—are actually suitable for ethanol production. Some critics assert that when the fossil fuels used to produce corn are included in calculations, burning a gallon of ethanol instead of a gallon of gasoline actually *increases* greenhouse gas emissions by 50%. Based on different assumptions about corn yields and the amount of energy used in production, others argue that burning corn ethanol *decreases* greenhouse gas emissions by 10–20%.

▼ **Figure 15.14 Energy Return on Investment**
The production of ethanol from corn uses considerable amounts of nonrenewable energy, including the gasoline used to run machinery and the fossil fuels used to manufacture fertilizer.

► Figure 15.15 **Biofuels Fuel Deforestation**
In Southeast Asia, large areas of tropical forest Ⓐ have been cleared for plantations of oil palm Ⓑ, an important source of biodiesel. Although biodiesel offsets the use of fossil fuels, an oil palm plantation such as this stores only 20% of the carbon that was stored in the tropical forest it replaced.

Sugarcane production in tropical countries such as Brazil requires far less machinery, fertilizer, and irrigation than does corn production. Furthermore, the sugar content of cane is much higher than that of corn. Most experts agree that burning ethanol derived from sugarcane results in 87–96% fewer greenhouse gas emissions than burning an equivalent amount of gasoline.

Some question whether the production of biofuels in the United States actually reduces our dependence on fossil fuels. From a net energy perspective, does producing biofuels use more energy than is available in the finished product? One recent study estimates that the production of corn ethanol requires 29% more energy than is contained in the ethanol that is produced. However, a number of other studies suggest that corn ethanol yields 30–40% more energy than its production consumes. Even with a net energy gain of 30–40%, it would take all of the corn grown in the United States to produce just 15% of the fuel used by the American automobile fleet. That same level of oil savings could be achieved through a 4-mile-per-gallon increase in the fuel efficiency of the average car.

The increased use of corn, soy, and palm oil as biofuels could increase the demand for irrigation water, pesticides, and fertilizers on farms, thereby causing higher rates of soil erosion and chemical runoff into surrounding waterways. There is also concern that the rapid increase in the global demand for biofuels is increasing the rate of deforestation. Large areas of tropical rain forest in Indonesia and Malaysia have recently been cut and burned for the expansion of palm oil plantations; some of that palm oil is used to produce biodiesel for global markets (Figure 15.15). This trend has implications for both biodiversity and global warming.

The use of corn and soybeans for biofuel has a significant impact on the world economy, increasing the cost of food and even causing food shortages. In 2000, about 5% of U.S. corn went into ethanol production. The Energy Production Act of 2005 required that ethanol be mixed with gasoline sold in the United States. By 2013, nearly 40% of the U.S. corn harvest was being used to produce ethanol. The increased demand for corn drove up the price of corn and had ripple effects throughout the food industry. Corn is used to produce everything from animal feed for chicken and beef production to high-fructose corn syrup for soda and processed foods. Because the United States accounts for over 40% of global corn production, this has affected global food prices and supply (Figure 15.16).

Recent legislation has mandated even greater use of biofuels in the United States, and the number of ethanol refineries in the United States is growing rapidly. Should these trends continue, as much as 50% of the U.S. corn crop could go toward ethanol production by 2016.

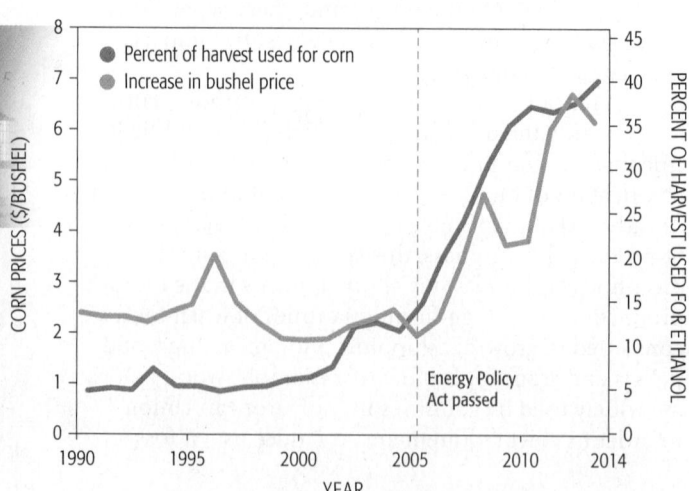

► Figure 15.16 **Biofuels vs. Food**
With passage of the Energy Policy Act, the percentage of the U.S. corn harvest (red line) dedicated to ethanol production rapidly increased. This was accompanied by a significant increase in the price of corn (blue line). Because corn is used in a variety of foods and for livestock feed, this has directly contributed to increased food prices.

Future Development

■ Bioenergy produced from sources other than food crops may have fewer environmental impacts.

Much public and commercial interest in bioenergy has been focused on the production of ethanol from starches and sugars in various crops. But the vast majority of the world's plant biomass is composed of another carbohydrate—cellulose. Thus, there is great interest in the production of **cellulosic ethanol**, ethanol that is produced from agricultural residues, wood, grasses, or other cellulose-rich feedstocks. Some energy experts estimate that the cellulose in wastes from agriculture and forestry operations could, in theory, supply enough ethanol to replace 40–50% of U.S. gasoline consumption.

The challenge of producing cellulosic ethanol is finding a way to digest cellulose on a large scale. Like starch, cellulose is a polymer of the sugar glucose. Most organisms can digest starch, but very few organisms produce enzymes capable of breaking the bonds between the glucose molecules in cellulose. The process is further complicated by the presence of other organic molecules, such as lignin, that interfere with cellulose digestion.

A number of promising pilot projects are underway. For example, a plant in Ottawa, Canada, uses wheat, oat, and barley straw to produce close to 1 million gallons of cellulosic ethanol a year. The first step is to separate the lignin in the plant matter from the cellulose. This lignin is burned to produce heat that is used in the fermentation process and electricity that is used in the production facility. The sugars are then fermented to produce cellulosic ethanol. This process is much more efficient than the production of corn ethanol, in which facilities typically burn fossil fuels for the fermentation process and obtain their electric power from power plants fired by fossil fuels. Proponents suggest that the combustion of cellulosic ethanol produced by such a process would reduce greenhouse emissions by 80–90% compared to corn ethanol.

Other sources of cellulosic ethanol are also being considered. Switchgrass (*Panicum virgatum*) is native to much of temperate North America. It requires little fertilizer or irrigation and grows well on land that is marginal for other crops (Figure 15.17). Because it is perennial, switchgrass does not require regular plowing and planting. Switchgrass can be harvested annually and processed in the same manner as the straw from other grasses.

At present, cellulosic ethanol is more costly to produce than corn ethanol for three reasons. First, unlike corn ethanol, cellulosic ethanol is not yet produced on a large scale, so production costs tend to be higher. Second, in the United States, the production of corn ethanol receives a subsidy of $0.55 per gallon; such a subsidy is not yet available for cellulosic ethanol. Third, the enzymes used to produce cellulosic ethanol are expensive, although their costs are expected to decline as their production increases.

▲ Figure 15.17 **Corn Ethanol Alternative**
Perennial switchgrass is a potential source of biomass for ethanol production. This native grass grows well on marginal agricultural land. A single year's growth is shown in this photograph.

The current production of corn ethanol is helping to spur investment in technology and infrastructure (such as pipelines) that could be shifted to the production of cellulosic ethanol when it becomes more commercially viable. There are, however, concerns that growing biofuel crops like switchgrass will diminish the land area available for food production in a world where 10% of people are chronically hungry (see Module 12.10).

QUESTIONS 15.3

1. Describe four different forms of bioenergy.

2. What are the environmental benefits and challenges associated with the use of corn ethanol as a fuel?

3. What are the environmental benefits and challenges of biodiesel produced from soy and palm oil?

(MES) For additional review, go to **MasteringEnvironmentalScience**

Omega

Can marine algae produce biofuels while cleansing municipal wastewater?

Throughout his career, Jonathan Trent of the NASA Ames Research Center has been interested in finding ways to use fresh water and marine microorganisms to solve problems. He has discovered a wide range of biochemicals that are produced by these microbes and then found applications for these biochemicals, ranging from health care to data computation.

To solve some of our most vexing environmental challenges, Trent has turned to marine algae. Trent noted that many algae produce and store large amounts of fatty lipids. Those lipids are ideal for making biodiesel. He also recognized that the production of these lipids is normally limited by the availability of nutrients, such as nitrogen and phosphorus. In addition, he observed that algae are difficult to harvest from the open ocean.

Trent hypothesized that if marine algae could be confined in a closed space and provided with essential nutrients and plenty of sunlight, they could be a superb source of biofuel. To do this, he devised a system for growing algae called OMEGA—offshore membrane enclosures for growing algae. In this system, algae are grown in large bags composed of a plastic membrane that is permeable to water, oxygen, and carbon dioxide, but not to nutrients (**Figure 15.18**). As a source of nutrients, the bags are filled with wastewater from municipal sewage. The algae growing in the OMEGA bags use sunlight to carry out photosynthesis, producing sugars that they then turn into lipids. As by-products, the algae produce oxygen and clean water, which are released through the membrane into the ocean.

Trent's research on this challenge has taken several forms. It took three years of investigation to identify the right plastic polymer for the construction of OMEGA bags. Work is still continuing on a plastic that is biodegradable and can withstand pounding waves and cold temperatures. Field trials are currently underway in California's ocean waters to test bag performance and identify appropriate strains of algae and the ideal conditions for lipid production (**Figure 15.19**).

Nevertheless, the early results are very encouraging. An acre of ocean covered with OMEGA bags could produce 1,000–4,000 gallons of biodiesel per year. In comparison, an acre of oil palms produces 600 gallons of biodiesel; an acre of soybeans produces just 60 gallons. Over 2,000 gallons

▲ **Figure 15.18 Jonathan Trent**
Biochemical engineer Jonathan Trent in a prototype OMEGA bag.

of fresh water are required to produce 1 gallon of ethanol from corn. In contrast, the OMEGA system cleanses wastewater. Furthermore, OMEGA will not compete with agriculture for water or land or encourage additional deforestation.

Much research remains before OMEGA bags are ready to be deployed commercially. With recent grants from Google, NASA, and the California Energy Commission, Trent expects that OMEGA biodiesel will be available at the filling station in just a few years.

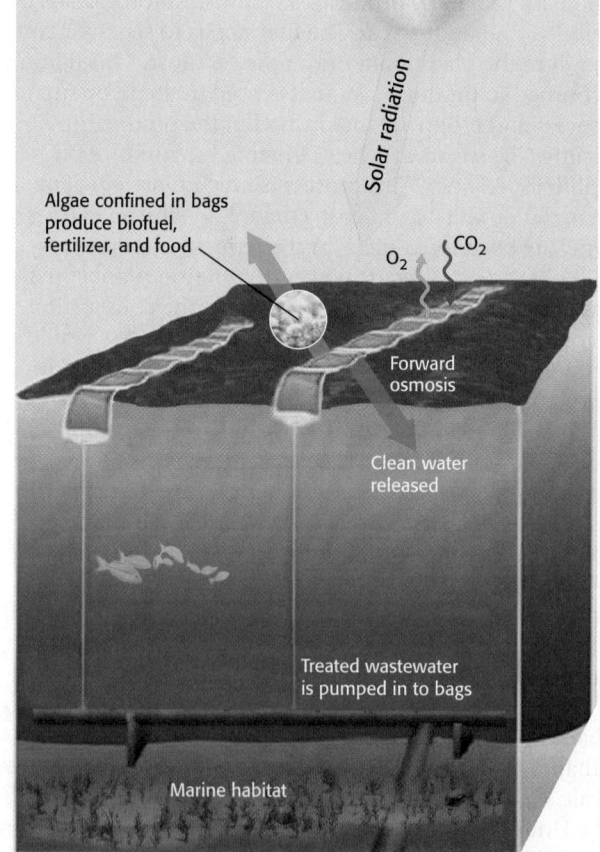

▲ **Figure 15.19 Experimental System**
These OMEGA bags are about 3 m (10 ft) across and are deployed in clusters representing different experimental treatments. The bags float because the wastewater in the bags is less dense than the surrounding seawater.

15.4 Wind Power

BIG IDEA **Wind power** is one of the earliest forms of energy harnessed by human societies. For thousands of years, humans have used the energy of wind to help move ships across water. For centuries, they have used windmills to power devices that pump water, mill grain, and saw wood (Figure 15.20). Today, wind power is being used to generate electricity and is becoming an increasingly important source of electric power around the world. Some countries already get as much as 30% of their electricity from this energy source. Using wind power to generate electricity offers a number of environmental advantages over the use of fossil fuels. Yet, as with all large-scale energy technologies, the use of wind power does have environmental impacts, especially on birds and other wildlife. In addition, wind power is naturally intermittent and somewhat less reliable than other sources of electricity. The degree to which wind power can be used to meet our electricity needs will depend in part on how well these challenges are addressed.

▲ Figure 15.20 **Wind Power** Across much of the western United States, windmills such as this still pump water for livestock and irrigation.

Sources and Production

■ The kinetic energy of moving air can be converted into electricity.

The energy that drives the wind comes from the sun. Differences in latitude, geography, and topography cause sunlight to heat Earth's surface unevenly. This uneven heating produces areas of high and low pressure. As air masses move from regions of high pressure to regions of low pressure, they create winds (see Module 3.7). Thus wind power is actually a form of solar energy.

The kinetic energy of moving air can be transformed into electricity through the use of devices called **wind turbines**. A wind turbine is a mechanical assembly that is mounted on a tower to take advantage of higher wind speeds farther off the ground. The wind's kinetic energy is captured by the turbine's rotor, which consists of two or three propeller-like blades mounted on a shaft. The shaft is connected to a gearbox, generator, and equipment to monitor the turbine (Figure 15.21). A rotor blade works like an airplane wing. When wind blows across the blade, a combination of lift and drag causes the rotor to spin like a propeller. The turning shaft then drives the generator, producing electricity.

The large-scale wind turbines used by electric power companies may have rotor blades that are up to 100 m (328 ft) long. Each large turbine can generate enough electricity to power 1,400 homes. These turbines are

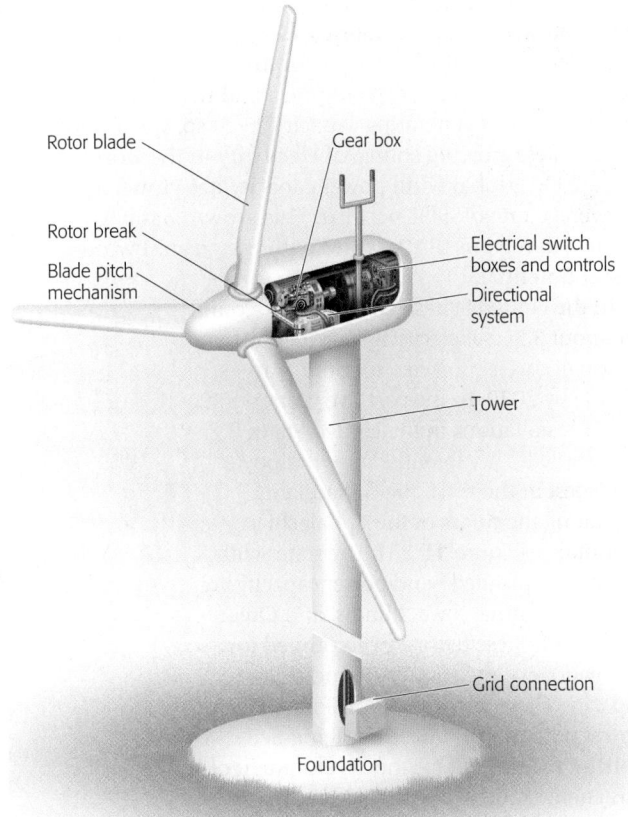

Rotor blade

Gear box

Rotor break

Blade pitch mechanism

Electrical switch boxes and controls

Directional system

Tower

Grid connection

Foundation

► Figure 15.21 **Wind to Electricity**
Through a system of shafts and gears, the propellers of a wind turbine turn a generator to produce electricity.

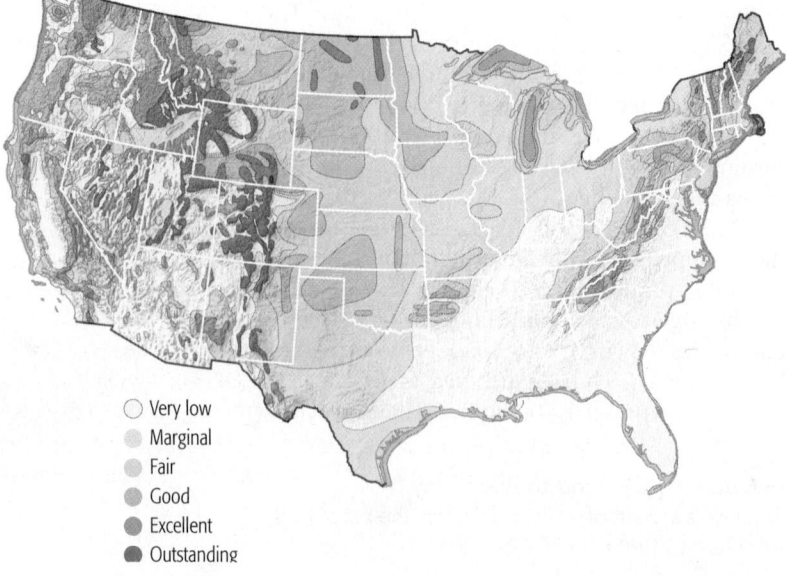

▲ **Figure 15.22 Wind Farm**
This wind farm in the California desert contains more than 4,000 separate wind turbine generators and provides enough electricity to power the city of Palm Springs and surrounding communities.

usually built close together in areas with the strongest and most regular winds. Such clusters of wind turbines are called **wind farms** (Figure 15.22).

Today, wind power accounts for just over 1% of total global primary energy consumption and nearly 3% of global electrical generating capacity. Even so, wind power is the fastest growing source of electricity in the world. Since 2000, global wind power capacity has grown at an average rate of 24% per year. This growth has been greatest in regions of the world with the greatest wind power potential.

In the United States, wind power accounts for about 3.5% of electricity production, although this is projected to more than double by 2020. In the past five years, wind power installations in the United States have tripled. The potential for wind power is greatest in the West, the Great Plains, and along the ridges of the Appalachian Mountains (Figure 15.23). The states with the largest installed wind power capacity are Texas, California, Iowa, Illinois, and Oregon. As of 2014, these five states accounted for nearly 60% of the country's wind power capacity. These states have considerable wind power potential and have implemented policies that encourage the development of this energy source.

Outside of the United States, Germany, Spain, India, Denmark, and China have the most installed wind power capacity.

Of these countries, Denmark relies on wind power to the greatest extent, using it to produce more than 20% of its electricity. Danes plan to rely on wind for more than half of their electric power by 2030. Germany currently gets more than 10% of its electricity from wind power.

▼ **Figure 15.23 U.S. Wind Resources**
In the United States, wind energy is most plentiful in the western mountains and central plains.

Source: U.S. Department of Energy, National Renewable Energy Laboratory. http://windeis.anl.gov/guide/maps/map2.html. Last accessed 10-10-2011.

○ Very low
Marginal
Fair
Good
Excellent
Outstanding

Advantages and Disadvantages

■ Wind power is economical but intermittent.

The most obvious advantage of wind power is that it requires no fuel to generate electricity. With wind power, there is no need for mining or drilling; there are no pipelines, no radioactive wastes, no need for water in cooling towers, and no water pollution. Energy is of course required for the manufacture of turbines and associated equipment, but EROI = 20–22 over 20 years. Because wind turbines do not burn fuel, they do not release air pollutants or greenhouse gases. By one estimate, a single 1-megawatt wind turbine operating for one year in average wind conditions offsets 1,800 metric tons of CO_2, 9.4 metric tons of SO_2, and 5.6 metric tons of NO_2; these amounts are equivalent to the emissions associated with the generation of 1 megawatt of electricity from a coal-fired plant.

Wind energy has the advantage of being a domestic energy source. Because the electricity produced by wind farms is spread among dozens, or even hundreds, of individual turbines, wind power is not subject to the kinds of supply disruptions that can occur when the bulk of a region's electricity is produced by a few large power plants. Wind energy facilities are also less susceptible to disruption by natural disasters and terrorist attacks than are other energy sources.

As with all forms of energy production, wind power has some definite disadvantages. Because of the location of some wind farms, fairly large numbers of birds and bats have been killed in collisions with turbine towers and rotor blades. For example, approximately 7,000 wind turbines are located in California's Altamont Pass, east of San Francisco. Studies undertaken in the 1990s found that each year as many as 40 golden eagles were killed in collisions with those wind turbines; smaller numbers of red-tailed hawks and American kestrels were also killed. In the eastern United States, wind farms are often located along ridges to take advantage of higher winds. Because these ridges also serve as natural corridors for migrating birds, the location of these turbines sets up a potential conflict.

Some wind power advocates acknowledge the risk of bird and bat mortality associated with wind farms but note that this impact is very small compared to the millions of birds that are killed every year by domestic cats, by collisions with cars, and by flying into tall buildings and cell phone towers.

Others critics object to wind power projects on aesthetic grounds, arguing that wind turbine towers and their associated infrastructure diminish scenic views. Since some of the best locations for wind power development are along mountain ridges or just offshore, the issue of aesthetics can mobilize public opposition. Thus, wind power companies need to give careful consideration to the siting of towers and turbines long before they begin to develop their facilities.

In general, wind-generated electricity is useful for meeting base-load consumer demands. Yet from a practical standpoint, wind power can be less reliable than other forms of electric generation because winds can be intermittent. In considering a potential site for wind power development, companies must consider wind consistency as well as wind speed. Nevertheless, newer wind turbines are designed to generate electricity in even modest breezes; these turbines are programmed to "find the wind" and turn into it. As a result, most new turbine designs are able to generate electricity 65–80% of the time. There is also growing interest in placing wind turbines offshore to take advantage of higher wind speeds and more reliable winds in these areas.

The remote location of potential wind farms is a challenge to the development of wind power. Many of the areas that have the strongest and most reliable winds, such as the Great Plains, are quite distant from the urban centers where the demand for electricity is greatest. Electricity produced from wind farms in such locations would have to be transmitted over long distances, resulting in a significant loss of energy along the way.

Q
Is wind power really the best form of energy?
Ashley Baxter, Louisiana State University

Norm: (A) Although wind energy is cost-effective and has comparatively few negative environmental impacts, wind power is intermittent, and the windiest areas are often distant from areas with the greatest demand for energy.

Future Development

■ Wind power has become cost-competitive with other sources of electricity.

In the last 25 years, the cost of generating electricity with wind power has declined by more than 90%. In some locations with favorable wind conditions, wind farms can produce electricity for as little as $0.03 per kilowatt-hour (kWh). This cost is competitive with coal-fired power plants and less expensive than natural gas–fired and nuclear power plants. The cost of wind power is expected to continue to decline as a result of improvements in turbine design and increasing understanding of where to place turbines to maximize power production.

Because wind turbines are powered by a free fuel, wind power costs do not fluctuate with changing fuel prices. In contrast, the cost of producing electricity in

power plants fueled by natural gas has fluctuated wildly in recent years in response to changes in the price of this fuel.

Although the use of wind power is growing rapidly, at present it represents only a small fraction of our national and global energy portfolio. In the future, the development of wind power is expected to continue to grow rapidly. Challenges to the development of wind power include concerns about the aesthetics of wind turbines and the intermittent nature of winds. Nevertheless, some experts predict that wind power could be supplying as much as 10% of the world's electricity by 2020.

QUESTIONS 15.4

1. How is the kinetic energy of wind converted into electrical energy?

2. Describe two important advantages of wind-generated electricity.

3. What are the most significant environmental problems associated with wind-generated electricity?

(MES) For additional review, go to **MasteringEnvironmentalScience**

15.5 Hydropower

BIG IDEA **Hydropower** is the energy of water moving under the force of gravity. Historically, hydropower was used to turn waterwheels that powered machinery, such as the equipment in textile mills and sawmills (**Figure 15.24**). Today, hydropower is used primarily to produce **hydroelectric power**, electricity generated by using the kinetic energy of moving water to turn a turbine. Hydropower currently accounts for 57% of all renewable energy use. The increased use of hydropower is uncertain due to the costs of building new hydroelectric facilities and growing concerns over their potential environmental and social impacts.

► **Figure 15.24 Hydropower** For centuries, humans have harnessed the kinetic energy of falling water to power machines that grind grains and saw wood.

Sources and Production

■ Hydropower facilities provide a significant amount of the world's renewable energy.

Water is constantly moving through the hydrologic cycle—evaporating from oceans and lakes, forming clouds, falling to the surface as rain or snow, and then flowing downhill through streams and rivers toward the ocean. The kinetic energy of flowing water, or hydropower, can be converted into mechanical and electrical energy. Hydropower is renewable because the liquid water in lakes and streams is constantly being replenished in the hydrologic cycle, which is itself powered by energy from the sun (see Module 11.1).

The primary use of hydropower today is to generate electricity, or hydroelectric power. There are two main ways to do this. The most common way is to impound, or store, water behind a dam. When water is released from the dam, it flows downhill through a turbine, spinning the turbine and generating electricity (**Figure 15.25**). Such a facility is known as an **impoundment hydroelectric plant**.

► Figure 15.25 **Impoundment Hydroelectric Plant** Water stored behind the wall of a hydropower dam accumulates potential energy. The force of gravity carries water through a tunnel or pipe known as a penstock, where it spins turbine blades attached to an electrical generator.

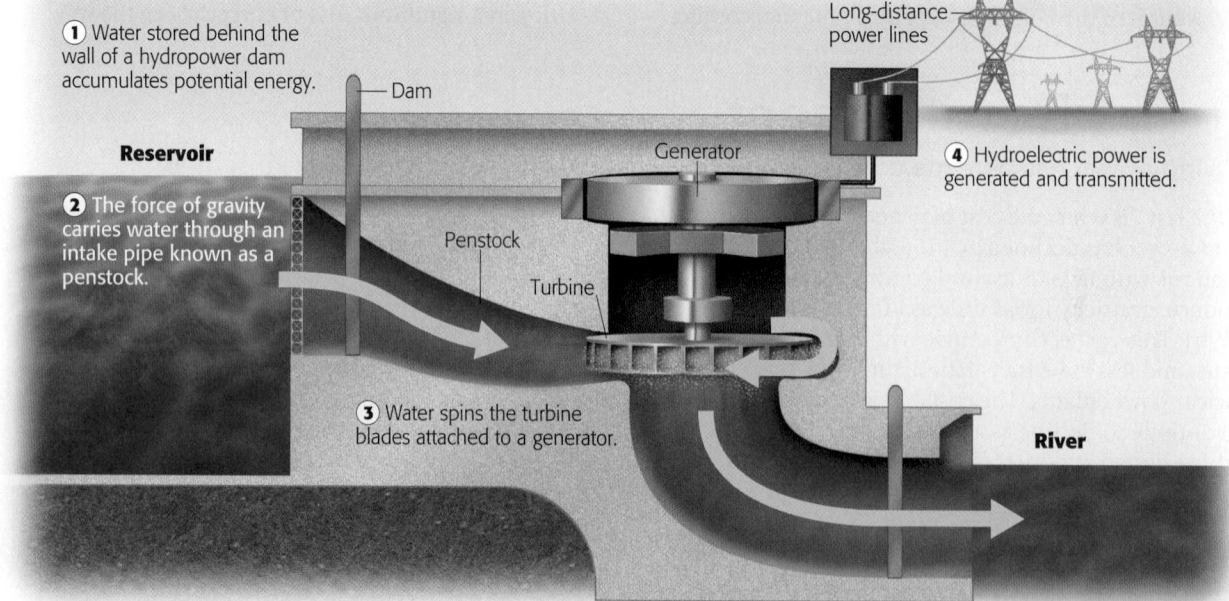

① Water stored behind the wall of a hydropower dam accumulates potential energy.

Long-distance power lines

Dam

Reservoir

Generator

④ Hydroelectric power is generated and transmitted.

② The force of gravity carries water through an intake pipe known as a penstock.

Penstock

Turbine

③ Water spins the turbine blades attached to a generator.

River

Water from a river is diverted into pipes that flow to a power station. There the force of the water spins a turbine attached to an electrical generator.

A second type of hydropower facility is a **run-of-river hydroelectric plant**. In such plants, a portion of a river's water is diverted into a series of pipes that lead to a powerhouse. There the force of the flowing water spins a turbine, generating electricity (Figure 15.26). Run-of-river hydroelectric facilities are usually less costly to build and operate than are impoundment facilities. They are also generally less disruptive to aquatic ecosystems. However, because they are dependent on the sustained flow of river water, they are not practical on rivers that experience wide variations in flow or in regions that are prone to occasional drought.

Hydropower is currently used to meet about 16% of the global demand for electricity and accounts for a little over 2% of the global primary energy supply. Hydropower resources are unevenly distributed around the world, with some countries relying on this energy source to a much greater extent than others. As of 2010, China was by far the world's largest producer of hydroelectric power, followed by Brazil, Canada, and the United States (Figure 15.27).

A somewhat different picture emerges when you consider the importance of hydropower in meeting a country's electrical needs. Norway, for example, obtains 99% of its electricity from hydropower, although it produces just one-third as much hydroelectric power as Canada. The South American country of Paraguay uses hydropower to generate 100% of its electricity, and Brazil

generates almost 84% of its electricity in this fashion. The United States gets less than 7% of its electricity from hydropower, with the greatest production in the Pacific Coast states of Washington, Oregon, and California.

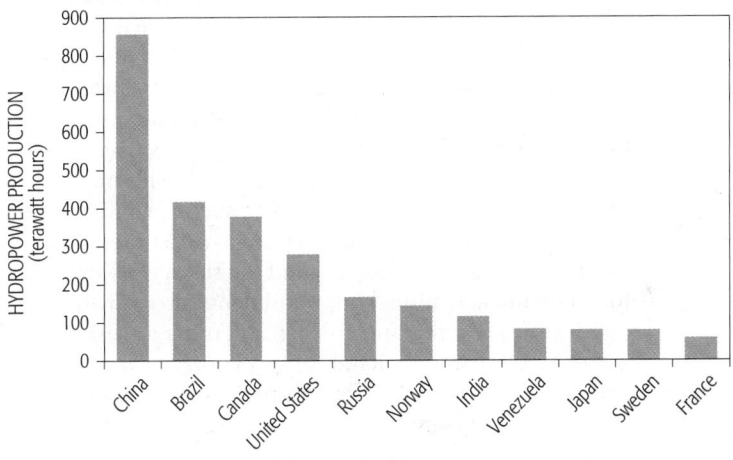

▲ Figure 15.27 **Hydropower Production**
These 11 countries account for about 75% of the world's hydropower production. With the completion of the Three Gorges Dam facility, China is now the world's number one producer and consumer of hydroelectric power.

Data from: British Petroleum, 2013. *Statistical review of world energy June 2013.* http://www.bp.com/content/dam/bp/pdf/statistical-review/statistical_review_of_world_energy_2013.pdf. Last accessed January 30, 14.

Advantages and Disadvantages

■ Hydropower emits few pollutants or greenhouse gases but has negative impacts on river and floodplain ecosystems.

The primary advantage of hydropower is that it does not rely on fossil fuel to generate electricity. Since hydropower derives its energy from the hydrologic cycle, it is both renewable and clean. Unlike the electric power produced by burning coal, the production of hydroelectric power does not directly result in the emission of local air pollutants such as sulfur dioxide or greenhouse gases such as carbon dioxide.

2000

2006

▲ Figure 15.28 **World's Largest Dam**
Ⓐ The satellite image taken in 2000 shows the Yangtze River flowing (left to right) past the construction site for the Three Gorges Dam.
Ⓑ The 2006 image shows the completed dam. Thousands of people lived in the fertile floodplain of the area that is now flooded by the impoundment.

It is relatively inexpensive to run hydropower plants because their fuel is essentially free. Once the dam and energy infrastructure are in place, hydroelectric facilities can generate electric power for very low costs. In addition, hydropower is a domestic energy resource.

Hydropower provides flexibility in meeting fluctuating power demands. Managers of impoundment facilities can rapidly increase or decrease the production of electric power by increasing or decreasing the amount of water released to the turbines. This allows power grid managers to match power generation to power demands. This flexibility makes hydropower a good match with more intermittent sources of electric power, such as wind energy.

Hydropower impoundments provide additional benefits, such as flood control and the storage of water for cities and agriculture. Impoundments also provide opportunities for boating, fishing, and other recreational activities. On the other hand, some people feel that hydropower impoundments diminish the beauty of river ecosystems and regret the loss of wildlife and recreation associated with free-flowing rivers.

Despite its many advantages, hydropower is frequently criticized for its effects on the environment and human communities. The construction of dams and creation of reservoirs often submerge forests, farmlands, or even whole communities. For example, China's recently completed Three Gorges Dam created a reservoir that flooded lands more than 600 km (370 mi) upstream from the dam; the resulting reservoir displaced more than 1.3 million people (Figure 15.28).

Hydroelectric power facilities often have serious impacts on surrounding ecosystems. Dams for hydroelectric impoundments can interfere with patterns of fish migration. For example, salmon migrating to spawning grounds on the Columbia River in the Pacific Northwest must use special fish ladders, a series of ascending pools that allow them to swim around the major dams (Figure 15.29). These salmon depend on river currents to orient themselves for their upstream journey. Even when they are able to get past a dam, the absence of currents in reservoirs presents a significant barrier to their migration.

▼ Figure 15.29 **Breaking Down the Barrier**
This fish ladder allows salmon to migrate over the Bonneville Dam in Cascade Locks, Oregon.

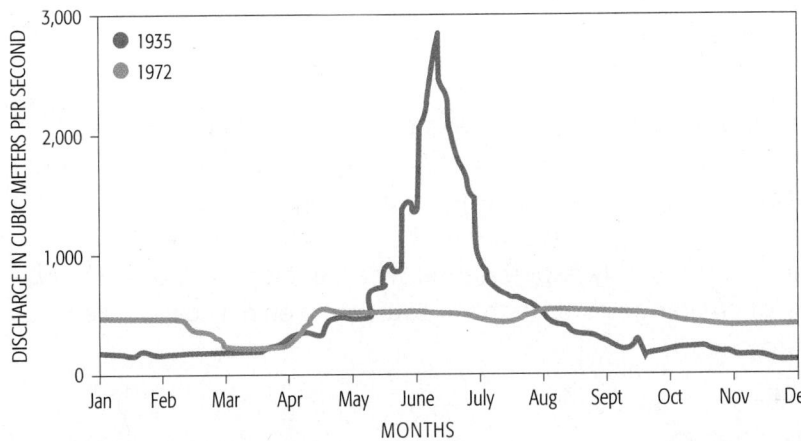

Prior to completion and filling of the Glen Canyon Dam in 1964, the Colorado River experienced high flows during the spring and early summer. After the construction of the dam, water flows become relatively constant throughout the year.

Data from: Andrews, E.D. and L.A. Pizzi. 2000. Origin of the Colorado River experimental flood in Grand Canyon. *Hydrological Sciences* 45: 607–627.

Hydropower facilities usually diminish downstream flows and alter the natural seasonal variation in those flows (see Module 11.5). In temperate regions, as in most of North America, the amount of water flowing in rivers is normally highest in late winter and spring as a consequence of snowmelt and rains (Figure 15.30). As free-flowing rivers spread across their floodplains, they deposit nutrient-rich silt that nourishes the trees. These seasonal floods occur when the trees in the floodplain are dormant and able to tolerate water-saturated soils. In contrast, peak flows from hydropower facilities occur primarily in summer, when the demand for electric power is highest. Floodplain trees are much less tolerant of flooding during this, their growing season.

Dams and impoundments alter the downstream movement of sediment. When a river flows into a reservoir, its velocity and turbulence diminish, as does its ability to suspend sediments. As a result, the sediments it was carrying settle to the bottom of the reservoir. Over time, sediments start to fill the reservoir, and the reservoir's capacity to store water decreases. Where the sediment load of inflowing rivers is high, the rate of sediment accumulation may be rapid and the life span of the impoundment for hydropower may be limited.

When water flows out of an impoundment, it carries little sediment downstream. This increases the rate of erosion below the dam and greatly alters the normal processes of sandbar formation and floodplain succession. In addition, the lack of nutrient-rich sediments reduces the fertility of floodplain forests downstream.

Hydropower facilities also influence water quality. Water exiting a dam is usually colder and contains less dissolved oxygen than that in a free-flowing river. These changes can negatively affect aquatic organisms downstream.

Although the activities associated with hydropower emit little carbon dioxide, there is concern that they do emit large amounts of other greenhouse gases. When dams are built, their reservoirs often submerge large amounts of vegetation. This vegetation subsequently decomposes under anaerobic conditions, releasing methane to the atmosphere. One hydropower facility in Brazil flooded 2,500 square kilometers (965 square miles) of Amazonian rain forest. The methane released from the decaying vegetation in this impoundment was nearly equal to the global warming potential (GWP) of emissions from a coal-fired power facility producing a similar amount of electricity (see Module 9.4).

QUESTIONS 15.5

1. What is the difference between impoundment and run-of-river hydroelectric facilities?

2. Describe three advantages of using hydropower to generate electricity compared to that of other primary energy sources.

3. Describe four environmental challenges created by the use of hydroelectric power.

(MES) For additional review, go to **MasteringEnvironmentalScience**

Future Development

■ Growth in hydropower production is likely to slow in the near future.

About one-fifth of the world's potential hydropower is already being used. The greatest potential for the expansion of hydropower is in Asia, where only about one-tenth of the hydropower resources have been exploited. However, many regions of Asia are also densely populated, so hydropower projects that are technically feasible may not be economically or socially acceptable. Although hydropower facilities are inexpensive to operate, they do require large up-front investments of capital, which many governments cannot afford. In addition, hydropower projects in Asia and elsewhere have often resulted in social conflicts over land use. In some cases, there have been violent clashes between police and residents being displaced by the flooding of reservoirs.

In developed countries such as the United States, environmental concerns and regulations are limiting the growth of hydropower. For example, many migratory fishes,

such as salmon and steelhead, are classified as endangered under the U.S. Endangered Species Act. The presence of endangered fishes limits the development of hydropower in many parts of California, Oregon, and Washington.

One approach to increasing the contribution of hydropower with fewer environmental and social drawbacks is to add hydroelectric-generating facilities to thousands of already existing dams. Of the 80,000 dams in the United States, only 2,400, or just 3%, are designed to generate electric power. Most of the dams without the capacity to generate electricity are small. To take advantage of the potential hydropower of these dams, it will be necessary to build hydro-facilities that are much smaller than most existing hydroelectric plants. Widespread use of small hydro-facilities would necessitate significant changes in our electricity distribution system (see Module 14.2).

15.6 Ocean Energy

BIG IDEA Oceans, which cover more than 70% of Earth's surface, have the potential of providing several forms of renewable energy. These include the kinetic energy of tides and waves and the thermal energy of the ocean's surface waters. Several systems have been proposed for capturing and converting the energy from these sources to electricity. Some of these systems have already been commercially proven, while others are still highly experimental. The development of each of these systems must overcome a variety of challenges before these sources of energy can make a larger contribution to meeting our energy needs.

Sources and Production

■ Ocean energy systems take advantage of tides, waves, and differences in ocean temperature.

Systems for generating electricity from the kinetic and thermal energy of the oceans can be grouped into three categories—tidal power, wave power, and ocean thermal energy conversion. At present, the only one of these that has been developed on a commercial scale is tidal power.

Tidal power.   The gravitational pull of the moon and sun causes the daily cycle of tides. In most coastal areas, there are high and low tides twice a day; the movement of water in this daily rise and fall represents a vast amount of kinetic energy. **Tidal power** systems seek to convert that kinetic energy into electricity.

The most common method of harnessing tidal energy is to erect a **tidal barrage**, or dam, across the narrow outlet of a tidal basin. As the tide rises and falls, water passes into and out of the bay or estuary through the sluice gates of the barrage. Turbines in the sluice gates drive dynamos to generate electricity. A tidal barrage system is only feasible in locations where the difference between high and low tide is at least 4–6 m (15–20 ft). There are only about 40 such locations on the planet. The world's largest tidal barrage system, a 240-megawatt system, has been in operation at La Rance, France, since the 1960s (**Figure 15.31**). A smaller 20-megawatt tidal barrage has been in operation in Canada's Bay of Fundy since 1960.

A **tidal fence** is an array of underwater turnstiles or revolving doors set across a narrow channel. Vertical hydrofoil blades are mounted on a rotor shaft, which is linked to a generator. As water moves through the channel, the blades are pushed, spinning the rotor and generating electricity. Because seawater is much denser than air, a tidal current of 8 knots (roughly 9 miles per hour) provides energy equivalent to wind speeds of more than 200 miles per hour. A large-scale tidal fence project is currently planned for the San Bernardino Strait in the Philippines. The initial installation will include almost 300 tidal fence turbines.

Tidal turbines look and operate much like wind turbines, except they operate underwater and are spun by the movement of tides and currents. Rows of tidal turbines can be placed anywhere that underwater currents flow at speeds of about 4–6 miles per hour. Recently, six tidal turbines were installed under New York's East River. Electricity from the experimental turbines was used to provide power to a supermarket and a parking garage. The first commercial-scale tidal turbine system has recently been commissioned in Strangford Lough, Northern Ireland (**Figure 15.32**).

▼ Figure 15.31 **Tidal Barrage**
Each day, the tide rises and falls 13.5 m (40 ft) at the barrage in La Rance, France. The barrage turbines spin and generate electricity as water moves in and out of the bay.

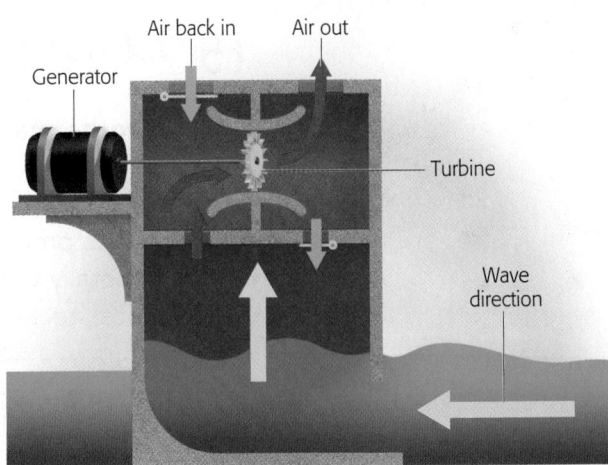

▲ Figure 15.33 **Onshore Wave Energy**
Waves entering the lower chamber of this system drive a stream of air past a generator turbine.

▲ Figure 15.32 **Tidal Turbine**
This tidal energy turbine was installed at the European Marine Energy Center test site in the Orkney Islands of northern Scotland in 2012. It can generate 1.2 megawatts of renewable energy, enough to service about 1,000 homes.

stretches and relaxes the hoses, producing enough pressure on the water in the hoses to turn a turbine and generate electricity.

Onshore wave energy systems are built along the shoreline to harness the energy of breaking waves. One system, known as an oscillating water column, uses a partially submerged concrete or steel structure that is open to the sea below the waterline (Figure 15.33). The structure encloses a column of air above a column of water. As waves enter from below, they cause the water column to rise and fall. The motion of the water forces the air past a turbine, which turns and drives an electric generator, producing electricity.

Ocean thermal energy conversion. Earth's oceans absorb large amounts of solar energy, so water near the surface is much warmer than water at greater depths. **Ocean thermal energy conversion (OTEC) systems** exploit the sharp differences in temperature between surface water and deeper water to generate electricity. OTEC systems have the most potential in areas where the difference in temperature between surface and deep ocean water is about 20 °C (36 °F). Such conditions are found mainly in tropical coastal areas.

A closed-cycle OTEC system uses warm surface water to vaporize a fluid with a low boiling point, such as ammonia. The expanding vapor turns a low-pressure turbine to generate electricity. Cold water from the deep ocean is used to condense the ammonia vapors back to a liquid, which is then recycled in the system.

An open-cycle OTEC system places warm surface water in a low-pressure container, causing it to boil. The steam is used to turn a low-pressure turbine and generate electricity. The steam is then brought into contact with cold water from the deep, which causes it to condense into liquid water. Both closed-cycle and open-cycle OTEC systems are still highly experimental, although small-scale systems have been successfully demonstrated.

Wave power. Winds blowing over the water's surface form waves, which can travel great distances across the ocean. **Wave power** exploits the kinetic energy of waves to generate electricity. Scientists and engineers are examining the potential of wave energy at both offshore and onshore locations.

Offshore wave power systems are located in deep water. One offshore system uses the bobbing motion of waves to power a pump that generates electricity. Another system relies on a series of hoses attached to floats that bob up and down with the waves. This bobbing motion

Advantages and Disadvantages

■ Tidal and wave energy systems can affect coastal and estuarine ecosystems.

Like other renewable sources of energy, ocean energy is attractive because it is virtually pollution free and its greenhouse gas emissions are very low. The primary costs for using tides, waves, or differences in ocean temperature are those associated with the generating devices themselves.

Tidal barrage systems are most effective where the tidal amplitude is large. However, such locations are also typically associated with highly productive marine estuary ecosystems. A dam or tidal barrage across the mouth of such an estuary can seriously limit the movement of many organisms.

Systems dedicated to capturing tidal and wave energy are limited geographically. None of these systems is as yet cost-competitive with other sources of electric power.

QUESTIONS 15.6

1. Describe three mechanisms used to convert tidal energy into electricity.

2. What are the advantages and disadvantages currently associated with the use of energy from ocean tides and waves?

(MES) For additional review, go to **MasteringEnvironmentalScience**

15.7 Geothermal Energy

BIG IDEA **Geothermal energy** refers to heat energy that is generated and stored in the Earth itself. Unlike hydropower, wind power, solar energy, and bioenergy, geothermal is a renewable source of energy that does not originate from the sun. Geothermal energy can be used as a source of heat and to produce electricity. As with other renewable energy resources, the use of geothermal energy generally has fewer environmental impacts than the use of fossil fuels. There are, however, geographical, technical, and economic limitations to the widespread use of geothermal energy.

Sources and Supplies

■ Geothermal energy is used in three different ways.

Geothermal energy comes from the hot interior of the Earth, which is heated by the radioactive decay of elements (see Module 3.5). In certain places, hot magma is so close to the surface that it heats groundwater, sometimes to its boiling point. Usually, this hot water is in geothermal reservoirs that are located a mile or more beneath the surface. In a few places, heated water and steam comes to the surface naturally, as in the hot springs and geysers in Yellowstone National Park, Iceland, and New Zealand.

The simplest application of geothermal energy is to use the hot water beneath Earth's surface as a direct source of heat and hot water for industrial, commercial, or residential purposes. To access this water, wells are drilled into a geothermal reservoir and the hot water is pumped to the surface. That water often contains lots of corrosive chemicals, so it cannot be used directly. Instead, it is used to heat liquid in a heat exchanger, and that heated liquid is piped to where it is needed. The cooler water is usually reinjected underground to be heated again. This method is now being used to heat homes, large buildings, and greenhouses; to maintain water temperature in fish farms; and to carry out industrial processes, such as pasteurizing milk (Figure 15.34).

The kinetic energy of the hot water and steam from geothermal reservoirs can also be used to turn a turbine and generate electricity. Geothermal electricity is generated in power plants with three different designs—dry steam, flash steam, and binary cycle.

Dry-steam power plants generate electricity by tapping underground sources of steam (Figure 15.35). The Geysers Power Plant in northern California is the largest geothermal electricity plant in the world. It uses the dry-steam method to meet the electrical power needs of over 1 million homes.

In flash-steam power plants, wells are drilled into geothermal reservoirs that contain high-pressure hot water with temperatures exceeding 180 °C (360 °F). As the hot water from these reservoirs flows to the surface, the pressure decreases and some of the water converts to steam. The steam is used to turn a turbine and generate electricity. The leftover water and condensed steam are usually reinjected back into the reservoir to maintain pressure.

Binary-cycle power plants use geothermal reservoir water at lower temperatures, from 105–180 °C (225–360 °F), to boil another fluid. The other fluid is then used to generate steam and turn a turbine.

Overall, the production of geothermal electricity meets slightly less than 1% of global electricity demand. Two countries, the United States and the Philippines, produce roughly half of all geothermal power in the world, with geothermal providing nearly one-fourth of the Philippines's electricity. Although Iceland produces less than either of these countries, geothermal power accounts for over 30% of its total energy production.

Ground-source heat pumps (GSHPs) are yet another way to exploit Earth's heat for energy. GSHPs take advantage of the fact that the upper 10 ft of Earth's crust maintains a nearly constant temperature of 10–15 °C (50–60 °F), regardless of the time of year. In temperate regions, this ground temperature is warmer than outside air in winter and cooler than outside air in summer. GSHPs use this temperature difference to warm buildings in the winter and cool them in the summer. In the United States, there are currently over 1 million GSHP systems installed in homes, businesses, and institutions such as universities and hospitals.

▲ Figure 15.34 **In Hot Water** Alligators thrive in geothermally heated waters on this Colorado farm where temperatures routinely drop below freezing.

▼ Figure 15.35 **Geothermal Electricity** In a dry-steam power plant, underground hot water and steam are used to generate electricity.

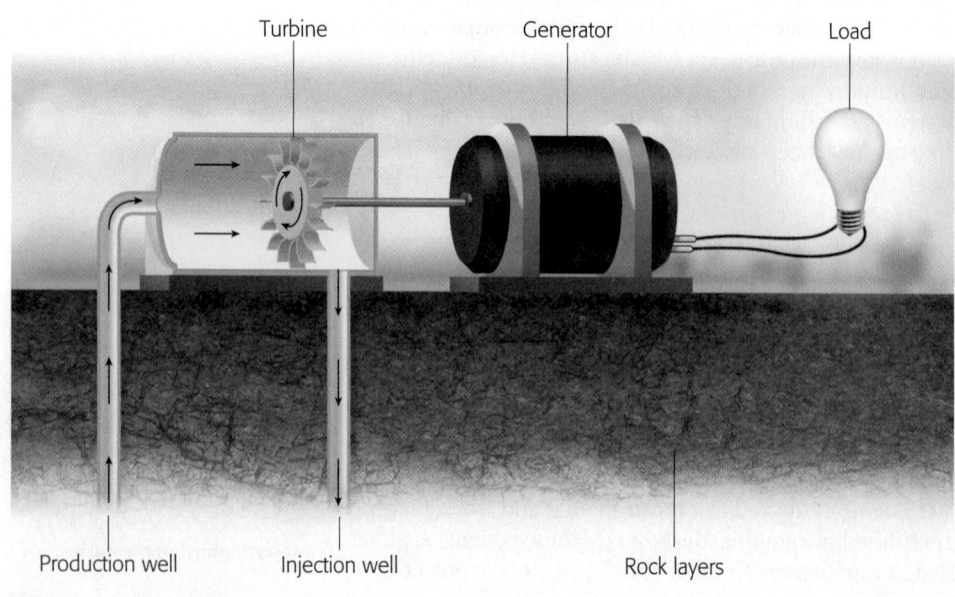

Turbine   Generator   Load

Production well   Injection well   Rock layers

Advantages and Disadvantages

■ Favorable locations for geothermal energy use are geographically restricted.

Geothermal energy is highly reliable and relatively inexpensive. Whereas wind power can be intermittent and solar energy can only be exploited during daytime, geothermal energy is constantly available. Electricity generated by geothermal power plants typically costs only $0.03–$0.04 per kilowatt-hour, making its cost competitive with coal-fired power plants and less expensive than most other sources of electric power.

Geothermal energy is also nearly pollution free. Geothermal wells can release small amounts of carbon dioxide, methane, hydrogen sulfide, and other gases, but these are only a small fraction of the pollutants generated through the equivalent use of fossil fuels, such as coal.

The cost of installing a GSHP system may be as much as twice the cost of installing conventional air-conditioning. However, GSHP systems consume 30–70% less energy for equivalent heating and cooling. The 1 million GSHP units in the United States offset over 1.6 million metric tons of greenhouse gas emissions each year. This is equivalent to taking 1 million cars off the road.

However, there are challenges to the use of geothermal energy. The development of large-scale geothermal power projects can disrupt local ecosystems and may be incompatible with wildlife in protected areas. One of the few places in the United States where underground steam resources can be tapped to produce dry-steam geothermal power is in the Yellowstone National Park region of Wyoming. This resource is not being exploited for power production because of the impact that its development might have on the iconic geysers and the wildlife in the area.

Geothermal energy is renewable only if the rate at which it is used does not exceed its rate of replenishment. Although Earth's heat is largely unlimited, the amount of hot water in geothermal reservoirs *is* limited. Some geothermal power projects have had to shut down temporarily because the geothermal resources of hot water and steam were pumped from the ground faster than they could be replenished.

Although geothermal power plants are inexpensive to operate once completed, they have high up-front costs and can be fairly risky investments. Only one out of every five exploratory wells is successful in locating adequate geothermal energy reserves for commercial development.

Finally, geothermal energy resources are far more geographically limited than are other renewable energy resources (Figure 15.36). In the United States, only four states—California, Nevada, Hawaii, and Utah—produce geothermal electric power. California alone accounts for almost 90% of this production. An exception to this limitation is GSHP systems, which can be used almost anywhere.

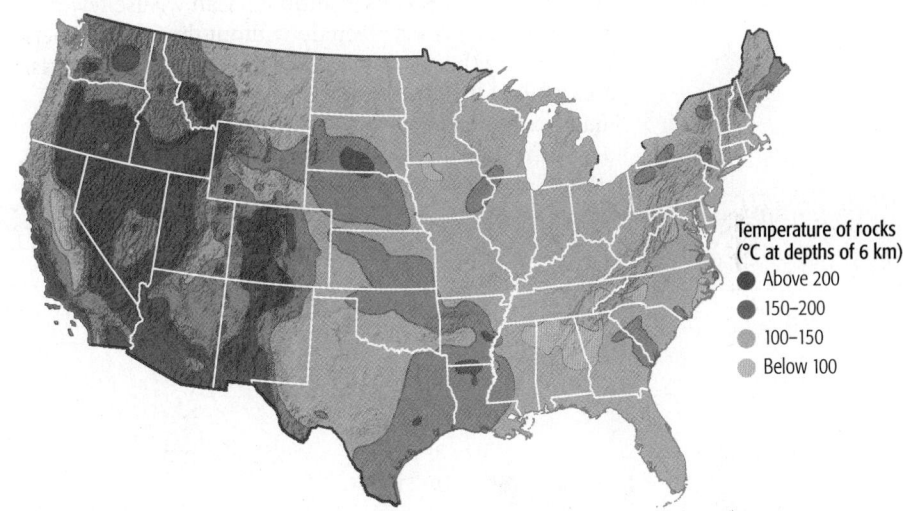

▶ Figure 15.36 **U.S. Geothermal Resources**
Ground-source heat pumps may be used across the United States, but geothermal reservoirs sufficient for geothermal power plants exist mainly in the western states.

Temperature of rocks (°C at depths of 6 km)
● Above 200
● 150–200
● 100–150
● Below 100

Source: U.S. Department of Energy, National Renewable Energy Laboratory. http://www1.eere.energy.gov/tribalenergy/guide/images/photo_geomap.jpg. Last accessed February 14, 2014.

QUESTIONS 15.7
1. Describe the three most common applications of geothermal energy.
2. What are the environmental benefits and challenges associated with geothermal energy use?

(MES) For additional review, go to **MasteringEnvironmentalScience**

Future Geothermal Energy Use

■ The use of GSHPs is growing rapidly.

In the next few decades, the use of geothermal energy to generate electricity is expected to become increasingly important in regions that are endowed with easily exploitable geothermal reservoirs. To ensure that geothermal resources are used sustainably, power plant operators are making an increased effort to reinject used water and condensed steam into reservoirs. In one case, operators even resorted to injecting municipal wastewater

into a geothermal reservoir to sustain the resource. If managed properly, geothermal energy can be expected to continue to make a relatively small but locally important contribution to meeting our energy needs.

Increasing energy costs and tax rebates are providing incentives for wider use of GSHPs. New installations in the United States are currently increasing at the rate of 10–20% each year.

Q *Hybrid vehicles and solar panels are both great energy alternatives. Unfortunately, being a college student, I cannot afford either. On top of that, my work requires me to drive a truck. Is there anything I can do that would make a big difference in decreasing my negative impact on the environment without spending so much money and switching cars?*

Daniel Cowell, Citrus Community College

Norm: (A) You will save money and energy by walking, biking, or using public transportation when those alternatives are feasible. When driving, reduce your speed and take care to accelerate and decelerate smoothly.

15.8 Energy Conservation and Efficiency

BIG IDEA Between now and the year 2030, global energy demand is projected to increase by more than 75%. The rate of that growth will be much greater in the world's developing countries. Most proposals for meeting this demand focus on increasing the supplies of energy resources. Thus, there is a rush to find and exploit new sources of fossil fuels and to increase the accessibility of various forms of renewable energy. Another approach to meeting this challenge is to reduce demand by conserving energy. Energy can be conserved by lifestyle changes, by reducing energy waste, and by increasing energy use efficiency—using primary energy resources to accomplish tasks more efficiently. Most experts acknowledge that energy conservation can significantly alleviate energy demand. This would have favorable consequences for the world's economy and its environment.

Defining Energy Conservation

■ Energy conservation can be thought of as an energy resource in the same way as oil or hydropower.

By managing the demand for energy, we can reduce the amount of energy we use or eliminate the need to increase the supply of primary energy sources. The first question in energy conservation is, "Can we use fewer goods and services, or even do without them altogether?" For all of us, the answer to this question is probably yes. A careful audit would reveal that each of us consumes many energy-demanding goods and services out of habit or even negligence. Does this room really need to be this warm or cold? Did I turn off the lights? Wouldn't it be just as easy to take the bus as it is to drive a car?

As a general rule, the fewer commodities and services we use, the more energy we save. Yet energy conservation is not about following a particular formula or lifestyle. Rather, it begins with a conscious examination of what we consume and an understanding of the trade-offs between perceived and real benefits of that consumption versus its energy and environmental costs. Each of us must make our own decisions with regard to what and how much to forgo.

Consumers in developed countries spend thousands of dollars a year on energy to fuel cars, to heat homes, and to power appliances and electric devices. Generally, it is not the energy itself that consumers want but rather the services it provides. Consuming gasoline allows us to travel by car; consuming electricity allows us to light our homes, keep food cold, and power televisions and computers; consuming natural gas allows us to take hot showers and keep our homes warm.

Energy efficiency is the process of using less energy to obtain the same goods or services. Rather than reducing our use of a commodity or forgoing the use of a service, we decrease the amount of energy required to produce it. Consuming less energy to get the same services does not reduce individual satisfaction in the way that consuming less food or less entertainment would.

There are opportunities to increase the efficiency of energy use at each step in the energy supply chain. Such opportunities are distributed among a myriad of commodities and services, each of which may contribute relatively small energy savings. However, the combined effect of these potential energy savings is quite large.

One measure of overall energy use efficiency is the **economic energy intensity (EEI)** of a country or group of countries. This is the amount of energy used

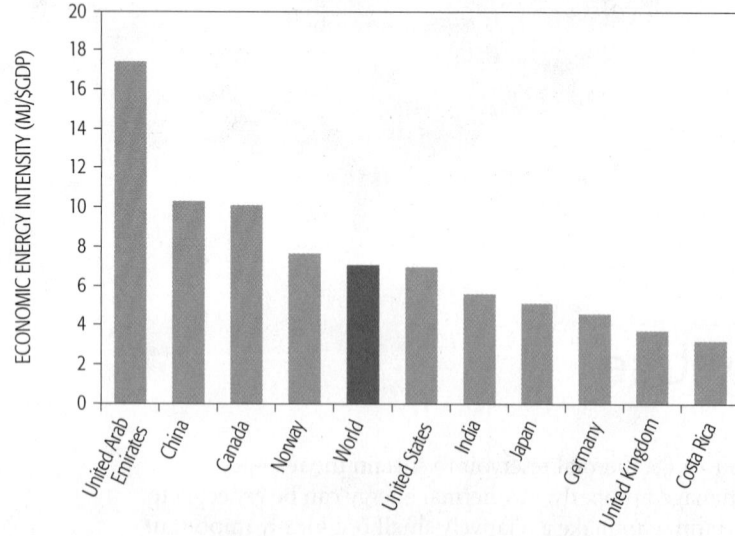

▲ Figure 15.37 **Economic Energy Intensity**
The United Arab Emirates uses over four times as much energy per unit of economic output compared to Germany or the United Kingdom, where efficient energy use has been a national priority.

Data source: U.S. Energy Information Administration: International Energy Statistics.

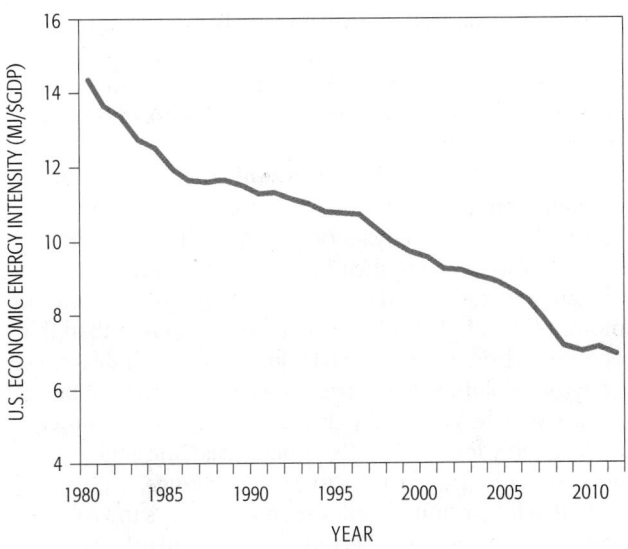

◀ **Figure 15.38 Increased U.S. Efficiency**
The U.S. economic energy intensity has steadily declined over the past three decades.

Data source: U.S. Energy Information Administration: International Energy Statistics

per dollar of economic output (MJ/$ GDP). In 2012, the United Arab Emirates had the highest EEI of any country in the world (Figure 15.37). Energy intensity is generally higher for countries in cold climates, like Canada, than those in more temperate climates, like the United States. Conservation measures can reduce the economic energy intensity of a country. For example, improved energy efficiency in the residential, commercial, industrial, and transportation sectors has cut the energy intensity of the U.S. economy by over 50% since 1980 (Figure 15.38).

More Efficient Lighting and Appliances

■ Opportunities for energy efficiency exist in virtually all sectors.

About 25% of the energy consumed by the residential sector goes to lighting. Most of that lighting comes from **incandescent lights**, in which an electric current causes a thin metal filament to glow. Only about 10% of the energy consumed by these fixtures is used to produce visible light. The remaining 90% is radiated as heat, as is painfully familiar to anyone who has accidentally touched a lit incandescent bulb.

Compact fluorescent lights (CFLs) emit three to five times more light per kilowatt-hour than do incandescent fixtures. Although CFLs are more expensive than incandescent lights, they last 8 to 15 times longer. **Light-emitting diodes (LEDs)**, which emit light as electric current passes through silicone-rich materials, are even more efficient and long-lasting than CFLs. The cost of LEDs is currently about twice that of CFLs, but that difference is expected to diminish considerably in the next few years.

Electric motors, which are found in a wide variety of machines and appliances, consume more than 60% of all generated electrical energy. In conventional electric

motors, about 80% of that electricity is converted to kinetic energy; the remaining 20% is lost as heat. Electric generators, in which the kinetic energy of a turbine is converted to electricity, also have an efficiency of about 80%. A new generation of motors and generators has a very different wiring configuration that allows them to operate at efficiencies of greater than 90%. This technology is already widely used to maximize the efficiency of generators in wind turbines.

Some appliances waste energy even when they are not being used. When many modern appliances are turned "off," they are not actually off but instead are on "standby." They continue to use energy to run timers, clocks, computer memory, and remote on/off switches. This hidden use of energy is often called "vampire power." The typical American home has about 20 appliances that, turned off, add $2–$300 to the electric bill each year. Efficient clocks and switches can diminish vampire power use; however, the best strategy is to unplug devices such as computers and coffeemakers when they are not in use.

Automobile Efficiency

■ Automobiles with traditional internal combustion engines are inefficient in their use of energy.

No machine has caused greater awareness of energy efficiency than the automobile. "What's the mileage?" is one of the first questions asked by most car buyers these days. Yet today's cars are highly inefficient. In an average automobile with an internal combustion engine, only

about 10% of the chemical energy in a tank of gasoline is used to drive the wheels. The rest of the energy is lost to idling, dissipated as heat, or used for air conditioning and other accessories.

QUESTIONS 15.8

1. Does energy conservation always require personal sacrifices? Explain.

2. What is economic energy intensity, and why does it vary among countries?

3. Why are compact fluorescent light fixtures more energy-efficient than conventional incandescent lightbulbs?

4. How does a gasoline–electric hybrid car differ from a conventional car? From an electric car?

(MES) For additional review, go to **MasteringEnvironmentalScience**

There is general agreement that the automobiles we produce today can be made far more efficient. In the years after the 1973–1974 oil embargo, there were dramatic improvements in automobile fuel efficiency. By the 1990s, however, fuel efficiency gains had leveled off. In the years between 2005 and 2012, truck and SUV fuel efficiency even declined (Figure 15.39). While automobile design and engineering have improved since the 1980s, much of the focus has been on increasing power and comfort, not on fuel efficiency. Recently, however, the high cost of oil and concerns about its environmental impact have renewed public interest in automobile efficiency.

One of the most direct ways to improve fuel efficiency is to make vehicles lighter. Physicist Amory Lovins advocates the use of ultralight carbon composite materials that are also ultrastrong. These composites can reduce the weight of an automobile by over 50% without sacrificing safety. Automobiles with lighter bodies could use smaller engines to achieve the same performance, greatly increasing fuel efficiency. Lovins responds to concerns about the safety of such vehicles by pointing out that increased weight does not always translate into increased safety—otherwise bicycle helmets would be made of steel rather than lightweight foam and carbon fiber material.

Much greater fuel efficiency is being realized by significant changes in the way cars are powered. Over the past decade, **gasoline–electric hybrid automobiles**, such as the Toyota Prius, have become widely available. Hybrid vehicles are equipped with both an internal combustion engine and an electric motor powered by batteries. On the highway, a hybrid's primary source of power is its gasoline-powered engine. When only low power is needed, as in slow city driving, the hybrid relies on the batteries to power its electric motor. The batteries are constantly being recharged by the kinetic energy created when the car brakes, a process known as regenerative braking.

Recently, **plug-in hybrid automobiles** have become available commercially, including a new model of the Toyota Prius. These vehicles operate much like gasoline–electric hybrids, except their batteries can be recharged by plugging the car into an electric outlet. This gives the electric motor of plug-in hybrids greater range and power than that of other hybrid vehicles, thereby increasing gas mileage. The Chevrolet Volt functions as either an electric or gasoline–electric vehicle. When fully charged, it can operate entirely off its battery for 40–50 miles. Then a gasoline engine powers an electric generator to extend its range.

Fully electric automobiles such as the Nissan Leaf and Tesla have recently become widely available. These cars run entirely off their batteries, which are charged with electricity from the grid. New battery technologies have improved the range to as much as 300 miles on a charge. These vehicles produce no tailpipe emissions of pollutants or greenhouse gases. However, whether these vehicles actually reduce pollution or greenhouse gas emissions depends on the source of electricity. If it comes from coal-fired plants, then pollution and emissions are only being shifted from one place to another. Also, it takes more energy and potentially scarce mineral resources to produce electric cars compared to conventional vehicles

Although they do not really represent a new automobile design, high-efficiency diesel vehicles operate with far greater fuel efficiency than did earlier diesel engines. Earlier diesel engines were already more fuel efficient than conventional internal combustion engines. High-efficiency diesel vehicles combine that advantage with improvements that make them cleaner, quieter, and more powerful than earlier models.

Can we power automobiles without using fossil fuels? Advocates of vehicles that rely on hydrogen-powered fuel cells believe that we can. Rather than burning gasoline in an internal combustion engine, hydrogen vehicles use onboard fuel cells to convert the chemical energy in hydrogen fuel to electric power (see Module 14.2). Prototype vehicles are already in operation, and some manufacturers have begun very limited commercial production. The greatest challenges to the widespread development of this technology are obtaining and distributing hydrogen. The production of hydrogen from organic materials or water requires energy that must be generated from some other primary source. Furthermore, the infrastructure for distributing and selling hydrogen remains limited and may be very costly to develop on a large scale.

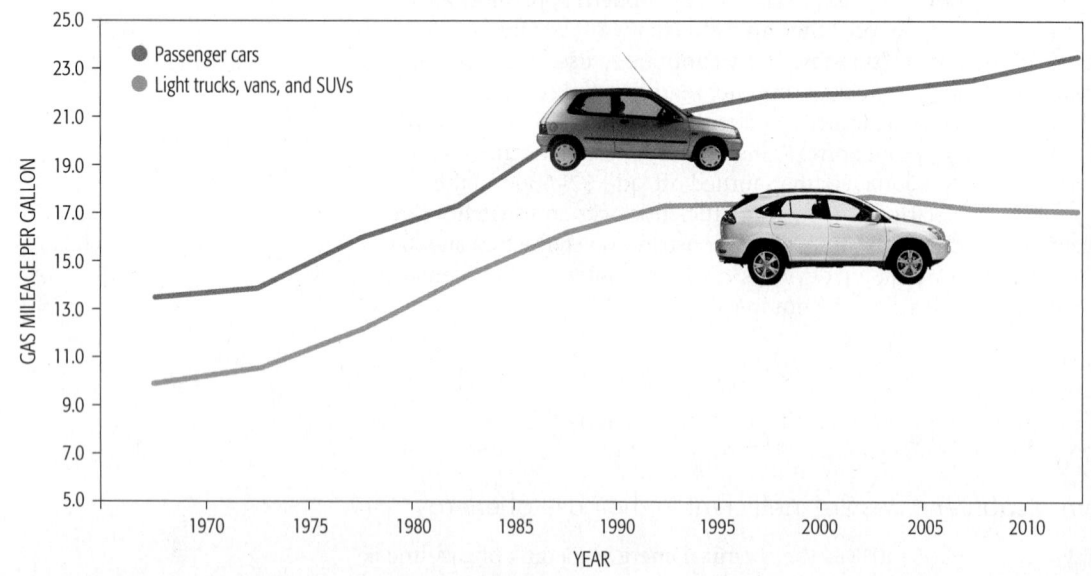

▲ Figure 15.39 **U.S. Fuel Efficiency**
Gas mileage for U.S. passenger vehicles (red) has steadily improved. Mileage for trucks and SUVs increased until about 1995. Since then, it has remained flat, or even decreased.

Source: U.S. Bureau of Transportation Statistics. 2012. Transportation Statistics Annual Report. http://www.rita.dot.gov/bts/sites/rita.dot.gov.bts/files/tsar_2012.pdf.pdf. Last accessed February 25, 2014.

Industrial Ecology

How do the environmental impacts associated with electric vehicles compare with those of conventional vehicles?

Electric vehicles (EVs) powered from renewable electricity sources could reduce greenhouse gas and tailpipe pollutant emissions compared to gas-powered vehicles. But are these potential benefits offset by impacts associated with the manufacture and eventual disposal of electric cars? Anders Hammer Strømman and his colleagues at the Norwegian University of Science and Technology addressed this question using a life-cycle assessment (LCA) comparison of typical electric, gas, and diesel-powered vehicles (Figure 15.40). LCA involves the careful analysis of environmental impacts and the materials and energy consumed at each stage in the life of a vehicle, from "cradle to grave."

For their *analysis*, Strømman and his coworkers compared generic vehicles based on the energy characteristics of the Nissan Leaf EV, and gas and high-efficiency diesel versions of the Mercedes A-series. These vehicles are similar in size and performance characteristics. They then divided the life of each car into three *phases*: production, use, and end of life. They used *computer simulation models* of each stage of car production to assemble *core data* on material use and energy consumption for all car components, including the drivetrain, batteries, engines, and electrical systems. They assumed a standard use lifetime of 150,000 km (93,200 mi). Their use *phase simulations* assumed that the EV was charged overnight using electricity generated from the average European Union mix of primary sources (26% coal,

▲ Figure 15.40 **Taking the Long View**
Industrial ecologist Anders Hammer Strømman analyzed energy and natural resource use over the entire lifetime of automobiles and found EVs come up short in terms of relative efficiency.

26% natural gas, 28% nuclear, and 18% renewable). But they also compared this to electricity generated entirely from coal or from natural gas.

Over the LCA lifetime, they found EVs reduced greenhouse gas emissions by 20–24% compared to gas and 10–14% compared to diesel-powered cars (Figure 15.41). These benefits were significantly reduced for EVs powered by natural gas electricity. EVs powered entirely by coal electricity actually increased greenhouse gas emissions by 17–27% compared to gas and diesel vehicles. Life-cycle greenhouse gas emissions for the production phase were over twice those for other vehicle types.

When they examined some other environmental impacts, Strømman's team found that EVs compared poorly to conventional vehicles. EVs require three to four times the amount of rare metals such as lithium and manganese and the production and potential release of three to four times the amount of toxic pollutants compared to conventional vehicles. From these analyses, Strømman concludes that improving EV environmental performance will require reducing production supply chain impacts and promoting the generation of electricity from renewable sources.

Source: Hawkins, T.R., B. Singh, G. Majeau-Bettz, and A.H. Strømman. 2012. Comparative environmental life cycle assessment of conventional and electric vehicles. *Journal of Industrial Ecology* 17: 53–64.

1. Why would EV lifetime greenhouse gas intensity in the United States differ from region to region?

2. How might these results change if Strømman and his team had doubled the standard use lifetime to 300,000 km (186,400 mi)?

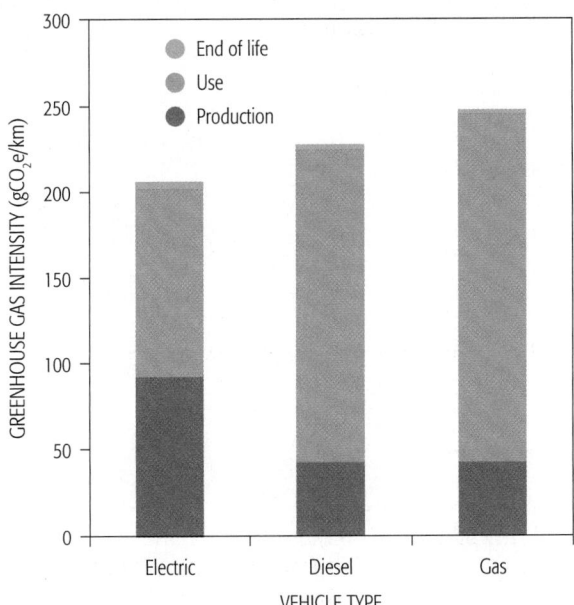

◄ Figure 15.41 **Lifetime Greenhouse Gas Intensity**
Greenhouse gas emissions per kilometer averaged over the full lifetime of electric, high-efficiency diesel- and gasoline-powered vehicles. Note that the production phase accounts for a much greater proportion of lifetime emissions for electric vehicles compared to other vehicle types.

Q *Even with the utilization of renewable energy sources, is it really possible to maintain our current lifestyle without further degrading the environment?*

Darrik Carlson, Citrus Community College

Norm: A *Sustainability will require a greater use of renewable energy sources, as well as more efficient use of all energy resources. Lifestyles may need to change, but the quality of our lives need not.*

Where YOU LIVE **What are your state's renewable portfolio standards?**

The U.S. Energy Information Agency (EIA) maintains a database of renewable portfolio standards for each state. What mix of energy sources does your state mandate? How does this compare to actions you have taken to reduce your own dependence on nonrenewable energy sources?

15.9 Sustainable Energy: Economics and Policy

BIG IDEA Renewable energy resources and energy conservation strategies would appear to offer many environmental, economic, and social benefits. So why haven't we pursued these resources and strategies to a greater extent? Part of the challenge to the wider use of renewable energy and energy conservation is simply being able to assess their advantages. Additional challenges include the lack of familiarity with new technologies, economies of scale, externalized costs, and limited consumer knowledge and understanding. Meeting these challenges will depend in large part on political decisions and the economic circumstances surrounding different energy technologies. Government policies can encourage conservation and boost the demand for and production of renewable energy resources. Such policies can help to build a diversified and decentralized energy economy that addresses the environmental and economic issues of our current energy system.

Measuring Economic Benefits

■ Conservation metrics compare the costs and benefits of energy conservation and renewable energy use.

Energy conservation is often described as a "win–win" approach: It saves money and also has many environmental advantages. There are many ways to improve energy efficiency, such as replacing inefficient lightbulbs, installing sensors for lights and heating, upgrading to more efficient appliances, and installing more insulation or energy-efficient windows. Yet all of these actions cost money. How can we assess the benefits of each action relative to its cost?

Energy economists use a measure known as the **cost of saved energy (CSE)** to compare the costs and benefits of energy alternatives. CSE is calculated by dividing the cost of an energy investment by the energy savings resulting from that investment. For example, a company is considering spending $2,000 to install CFL lighting in its office. That upgrade would save 10,000 kWh of

electricity per year. In this case, the CSE would be $2,000 divided by 10,000, which equals $0.02, or $0.02 per kilowatt-hour. If the company is already paying $0.08 per kilowatt-hour for its electricity, then it could save $800 a year by going ahead with the lighting upgrade. This investment would pay for itself in less than three years ($3 \times $800 = $2,400$).

By themselves, actions such as upgrading lighting, adding insulation, or installing solar panels may not be cost-effective. But taken together, a combination, or portfolio, of actions might be cost-effective. Energy experts use **conservation supply curves** to evaluate such portfolios (Figure 15.42). These graphs compare the amount of energy saved by each action to the cost of that energy. Whether or not to include an action in a portfolio can be judged in relation to the price of energy.

▶ Figure 15.42 **Conservation Savings**
This is a hypothetical conservation supply curve for energy-saving features that might be used in building a new home. Each step represents savings from a conservation feature. Measures 1–4 are cost-effective because the cost of conserved energy is less than the price of energy (dashed line). If energy prices were to rise, then measures 6 and 7 might become cost-effective.

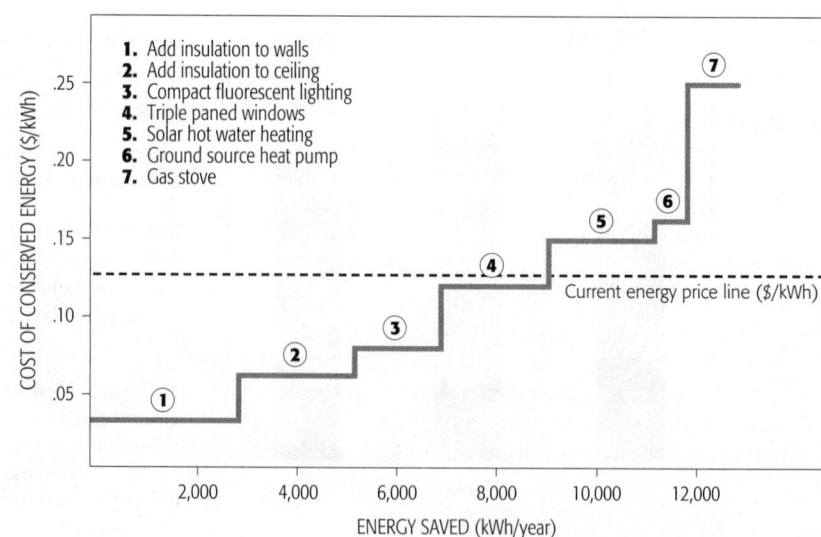

1. Add insulation to walls
2. Add insulation to ceiling
3. Compact fluorescent lighting
4. Triple paned windows
5. Solar hot water heating
6. Ground source heat pump
7. Gas stove

Policy Options

■ A variety of policy options can be used to speed the transition to greater energy conservation and use of renewable energy.

Advocates of renewable energy sources and increased energy conservation frequently argue for the elimination of existing government subsidies to oil, coal, and natural gas as a critical first step in the energy transition. If even a small portion of the hundreds of billions of dollars in annual government subsidies were redirected to the development of renewable energy resources, then some of the investment and production barriers limiting the expanded use of those resources could be overcome.

Some energy economists favor imposing taxes on fossil fuels so that their price reflects more of the external costs the use of those fuels imposes on society. Currently, federal and state gasoline taxes are used to fund highway construction and maintenance, thereby partially capturing those costs in the price of automobile fuel. The same could be done to cover costs associated with the mitigation of pollution or global warming.

One way that governments can encourage the development and commercialization of renewable energy resources is through **renewable portfolio standards (RPS)**. An RPS is a government mandate that a certain percentage of energy use come from renewable energy resources. RPS mandates help build market demand for renewable technologies, increasing their scale of production and driving down costs. As of 2013, there were 37 states in the United States with an RPS mandate in place. Governments can also mandate energy-efficiency standards for automobiles, appliances, and even buildings. For example, both the United States and the European Union set gasoline-mileage standards that must be met by automobile manufacturers.

To help consumers make informed decisions, utilities could be required to disclose the energy sources they use to generate electricity and the amount of emissions released in the process. Consumers could then have the option of selecting an electric power provider that is using a greater amount of renewable energy to generate electricity.

To improve consumer awareness of energy efficiency, greater effort could be put into consumer education and labeling. The U.S. Environmental Protection Agency's Energy Star® program is an example of a very successful education program. The Energy Star program certifies the energy efficiency of virtually all home appliances, from hot water heaters, to refrigerators, to air-conditioning systems (Figure 15.43). Manufacturers are encouraged to build appliances that incorporate innovations such as improved insulation, solid-state technologies, and more efficient motors and compressors. The Energy Star program provides consumers with easily understood measures of energy efficiency to guide their purchases. In 2013 alone, energy savings from this program exceeded $18 billion and avoided greenhouse gas emissions equivalent to 33 million cars.

The combination of policy, economic, technical, and social changes described above can accelerate the use of renewable energy resources and energy efficiency. However, because this shift will take decades, it is important to begin the process now.

By itself, no single renewable energy resource is going to replace our dependence on fossil fuels; there is no single technical "silver bullet" to meet our energy needs. Instead, we need what one energy expert has called a "web of innovations"—improved energy efficiency and advances in many energy technologies, including the development of new ways to store and use the energy obtained from renewable resources. It is this web of innovations that holds the promise of ushering in the transition needed to address the great environmental challenges of our current energy system.

QUESTIONS 15.9

1. Explain how cost of saved energy (CSE) is calculated.

2. Describe three policies that might be employed to encourage greater energy conservation.

(MES) For additional review, go to **MasteringEnvironmentalScience**

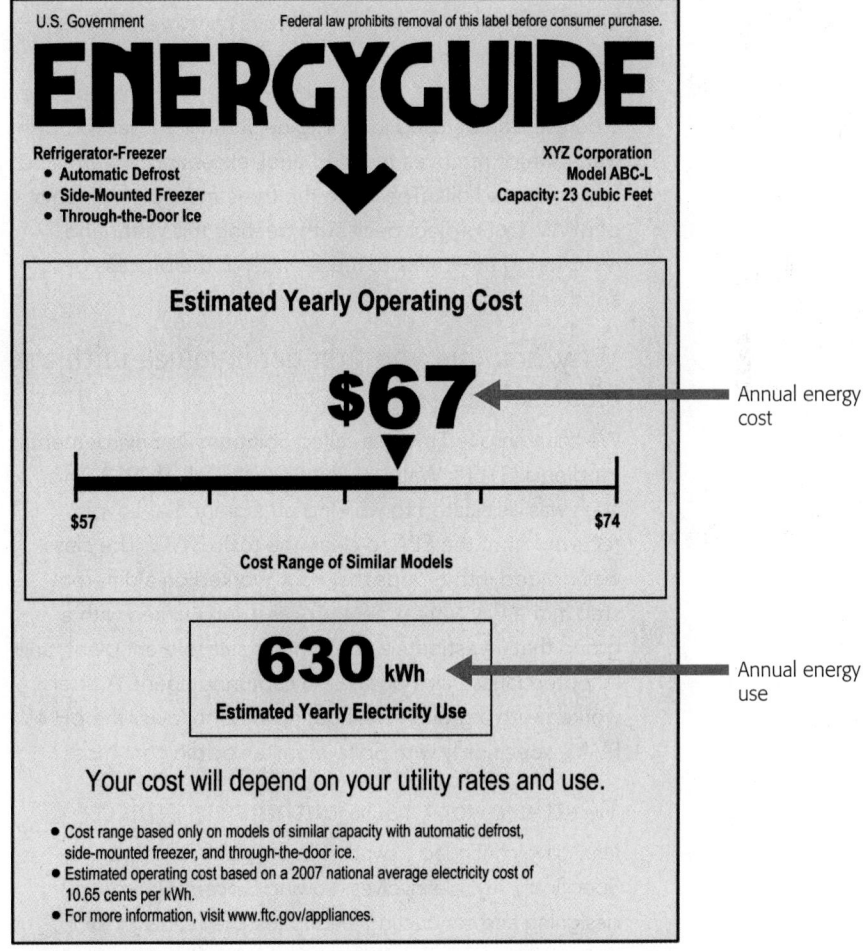

▲ Figure 15.43 **Incentives for Efficiency**
Appliances that meet EPA Energy Star standards clearly display average annual energy use and likely annual cost to the consumer.

Biodiesel Project at Loyola University Chicago

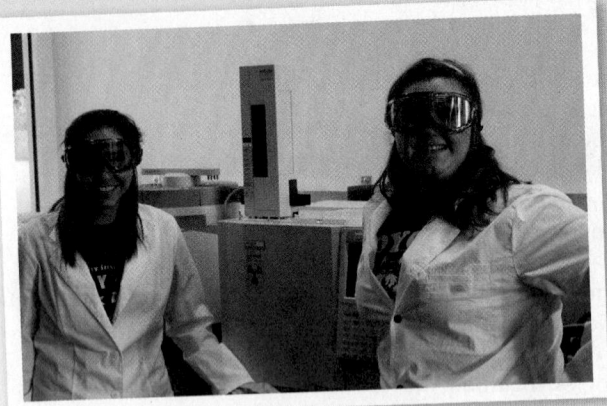

Jen Kelso is an Environmental Science and Sociology major and **Amber White** is a Biology and Environmental Science major at Loyola University in Chicago. They have both worked with Loyla's biodiesel initiative since 2012.

Can you give us an overview of the biodiesel project at Loyola?

Loyola's biodiesel project began in 2007 as a student-led initiative on campus. We use waste vegetable oil from our dining halls and add methanol and potassium hydroxide to create biodiesel. Biodiesel is accepted as a sustainable replacement for petroleum diesel. Our biodiesel is used to run our campus shuttles, which take students between our two campuses. Biodiesel is often seen as an effective carbon emissions reducing agent, but during the process a toxic waste product is produced: Biodiesel Wash Water (BWW). Before biodiesel can be used in an engine, it must be washed with water, which removes the methanol, excess potassium, and any unreacted oils. These are the three main components of BWW. Our project deals with treating this water and recapturing the toxins to reuse either in the process or another form.

Why/how did you first get involved with the biodiesel project?

We both enrolled in class called Solutions To Environmental Problems (STEP): Water at Loyola in the fall of 2012. This class was dedicated to working on a grant that Loyola received from the EPA to clean the toxic BWW. The class was divided into groups that each worked on a different step in the BWW treatment process: Jen worked with a group that investigated *Salicornia*, a salt-tolerant plant, and its potential use as a potassium capturing agent. Amber worked with a group investigating ways to lower the pH of BWW, specifically with *Sphagnum*, an acidic peat moss.

What are your roles within this project?

Jen: I have continued to work with the plant *Salicornia*. Specifically, my role involves growing *Salicornia* plants and designing and conducting trials to test their capacity for potassium uptake. The *Salicornia* project is the last step of our

BWW treatment process, so I also work on figuring out what is the best way to integrate the entire process together smoothly.

Amber: I have worked with finding sustainable ways to lower the pH of BWW. This is important because the lower the pH, the more unreacted fatty acids precipitate out of the water. This has included *Sphagnum* (mentioned before), used coffee grounds, and carbon dioxide. We have recently been focusing on carbon dioxide and the potential to use generator exhaust as a replacement to lab-grade carbon dioxide.

What are the biggest challenges you've faced with the project, and how did you solve them?

By far, the biggest challenge has been not having consistent access to BWW. In summer of 2013, we moved our biodiesel production program to a new, larger location. However, the new system has yet to be connected and put into use (as of May 2014). This has been a significant problem because our project specifically works with cleaning BWW, but if we're not producing biodiesel, were definitely not making BWW. We have worked around this obstacle by running growth trials with *Salicornia* using a synthetic BWW we made in

the lab. We have also constructed an exhaust-capturing system and been running generator exhaust through regular water. We can use this to hypothesize about our results of running exhaust through BWW so when we get BWW we can hit the ground running.

How has your work affected other students, your surrounding community and the environment?

Our work will directly affect Loyola students because as long as we need to do research on the project, there will be a STEP: Water class running, which allows more students to get involved on the project, exposing them to water pollution issues. The project also directly affects the Loyola student population because the biodiesel is used in our shuttles, so reducing our waste lowers Loyola's carbon footprint. The greater Chicago community is also affected because our waste is currently disposed of through the municipal water treatment system. If we can treat our own water, we can remove this stress from the Chicago system. Finally, this impacts the greater environment because there is currently no outlined treatment process in the biodiesel industry for BWW. The industry standard for treatment is neutralizing the water and dumping it down the drain. Our process could be implemented in both small- and large-scale biodiesel production plants to make the production process more sustainable and prevent toxic chemicals from being released into the environment.

What advice would you give to students who want to replicate your model on campus?

We would recommend starting a conversation among you and your friends about a cause you're invested in, and then seeking out support from a faculty or staff member. Once you have a project designed, getting assistance from the administration should follow. The biggest part is having dedicated faculty/staff to help you in your project. They'll be there longer than the students and can help keep the project going after the initial group of students leave. They also know what the university/college research scene looks like and how to get involved. They will be your most valuable resources.

What do you hope for the biodiesel project in the coming months? Years?

We would like to see the project be completed by spring of 2016. For the next few months, we want to work on connecting the treatment steps into an integrated treatment process. We would also like to run more trials with our BWW so we know exactly how our system will respond to our specific BWW. Eventually in the process we would like to create a plan for a system that can be used in both small- and large-scale biodiesel production processes.

Summary

Renewable energy refers to energy sources that are replenished continuously as they are used. These sources include the energy contained in falling water, wind, sunlight, biomass, Earth's heat, and ocean tides. Renewable energy sources are abundant on a global scale, but they are often diffuse and intermittent at particular locations. The environmental impacts associated with the use of renewable energy sources are generally less than those associated with the use of nonrenewable energy sources. Nonetheless, they do affect the environment. Use of renewable energy sources is growing rapidly with the development and scaling up of new technologies.

The environmental impacts associated with the use of renewable energy sources are generally less than those associated with the use of nonrenewable energy sources. Use of renewable energy sources is growing rapidly with the development and scaling up of new technologies. The growing costs of nonrenewable energy sources, such as oil and coal, as well as concerns about their environmental impacts, are contributing to the increased use of renewable energy. Diminishing supplies and increasing costs of some primary energy sources and environmental impacts associated with their use are encouraging the development of strategies and technologies to conserve energy and diminish per capita energy demand. Increased energy conservation will include the development of technologies for more efficient use of electrical energy and more fuel-efficient transportation, as well as policies to encourage the adoption of these technologies.

15.1 Renewable Energy Overview

- There is great potential for growth in the use of renewable energy if we can meet the challenges of technology development, economies of scale, externalized costs, and limited consumer knowledge.

QUESTIONS

1. Many renewable energy sources are intermittent. What does this mean, and why does it pose a challenge to the use of these sources for electrical power generation?
2. The economic costs of environmental impacts of energy use are often externalized. Why does this favor the use of nonrenewable energy sources compared to renewable energy sources?

3. How might improved consumer knowledge result in greater use of renewable energy sources?

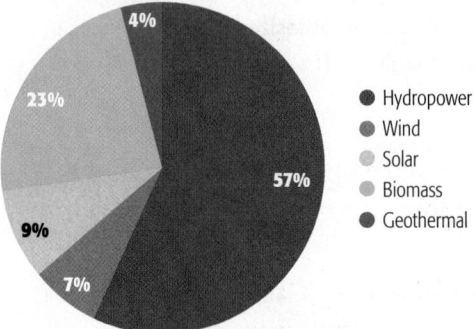

- Hydropower
- Wind
- Solar
- Biomass
- Geothermal

15.2 Solar Energy

- Solar energy has relatively few environmental impacts, and its use will likely grow as its costs decrease.

KEY TERMS

passive solar technology, active solar technology, solar power, photovoltaic (PV) technology, concentrating solar power (CSP) systems, solar photovoltaic farms

QUESTIONS

1. Describe how a building might be designed to passively collect solar energy.
2. PV cells and CSP systems are both used to generate electricity. How do they differ?

Silicon layer with phosphorus

Silicon layer with boron

15.3 Biomass Energy

- Fuelwood, charcoal, and ethanol are the most widely used forms of biomass energy.

KEY TERMS

biomass energy, bioenergy, biofuels, biodiesel, cellulosic ethanol

QUESTIONS

1. In theory, biofuel combustion releases CO_2 but does not contribute to global warming. Why is this so?
2. How might the production of biodiesel and ethanol contribute to global warming?
3. What are the challenges and advantages of producing ethanol from a perennial plant, such as switchgrass, compared to corn?

15.4 Wind Power

- Wind power use is rapidly growing, but wind energy is intermittent and concerns remain regarding its effects on wildlife.

KEY TERMS

wind power, wind turbines, wind farms

QUESTIONS

1. In what regions of the United States has wind energy been exploited to the greatest extent? Why?
2. What are two challenges to the distribution of electricity that are presented by wind power?

15.5 Hydropower

■ The use of hydropower produces few greenhouse gas emissions and air pollution, but it has significant impacts on river and floodplain ecosystems.

KEY TERMS

hydropower, hydroelectric power, impoundment hydroelectric plant, run-of-river hydroelectric plant

QUESTIONS

1. What region of the United States produces the greatest amount of hydroelectricity? Why?
2. Hydroelectric dams increase sediment accumulation upstream and diminish it downstream. Explain why this is true and the problems that it presents.

3. What factors are likely to limit the growth in the number of hydroelectric facilities in the future?

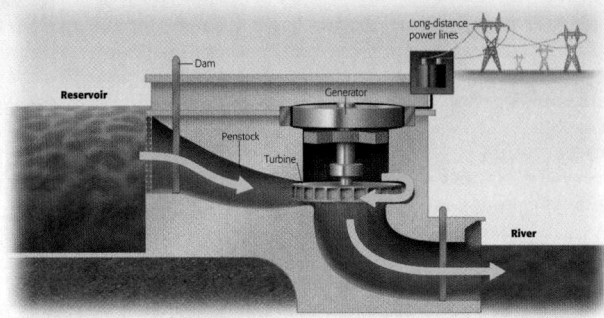

15.6 Ocean Energy

■ The kinetic energy of tides and ocean waves, as well as differences in ocean temperature, can be tapped to produce electricity.

KEY TERMS

tidal power, tidal barrage, tidal fence, tidal turbine, wave power, ocean thermal energy conversion (OTEC) systems

QUESTIONS

1. Explain how tidal power systems capture the energy of horizontal water movement whereas wave power systems capture the energy of vertical water movement.
2. Differentiate between open-cycle and closed-cycle OTEC systems.

15.7 Geothermal Energy

■ The energy of Earth's heat can be used directly to heat and cool buildings and to generate electricity.

KEY TERMS

geothermal energy, ground-source heat pumps (GSHPs)

QUESTIONS

1. How do flash-steam geothermal power plants differ from dry-steam power plants?
2. What factors limit the widespread use of geothermal energy?

15.8 Energy Conservation and Efficiency

■ Energy conservation requires an understanding of the trade-offs between perceived and real benefits and the costs of consumption.

KEY TERMS

economic energy intensity (EEI), incandescent lights, compact fluorescent lights (CFLs), light-emitting diodes (LEDs), gasoline-electric hybrid automobile, plug-in hybrid automobile

QUESTIONS

1. Economic energy intensity is much higher in some states than in others. Describe two factors that might cause such differences.
2. What is vampire power? Give examples of this kind of energy use in your home and classroom.
3. Is it true that electric automobiles produce no greenhouse emissions? Explain.

15.9 Sustainable Energy: Economics and Policy

■ Conservation metrics and a portfolio of policies can speed the transition to increased conservation and renewable energy use.

KEY TERMS

cost of saved energy (CSE), conservation supply curve, renewable portfolio standard (RPS)

QUESTIONS

1. Explain how to use cost of saved energy (CSE) to determine the time required to repay the cost of investments in conservation or renewable energy.
2. How would you use a conservation supply curve to determine the best combination of conservation strategies relative to cost?

MasteringEnvironmentalScience®

Students Go to **MasteringEnvironmentalScience** for assignments, the eText, and the Study Area with practice tests, activities, and more.

Instructors Go to **MasteringEnvironmentalScience** for automatically graded tutorials and questions that you can assign to your students, plus Instructor Resources.

Urban Ecosystems

Malmö: Sweden's Green City

How can a deteriorating city become a model for sustainability?

In Malmö, Sweden, the Turning Torso building towers above all other buildings (Figure 16.1). This new 54-story apartment building, powered completely by locally produced renewable energy, is intended to mimic the human body, with a "spine" that twists 90° from top to bottom. In many ways, this building also symbolizes the remarkable history of change in this, Sweden's third-largest city.

In A.D. 1275, Malmö was established as a fortress city to monitor and control the movement of ships through the narrow Øresund strait that connects the Atlantic Ocean with the Baltic Sea (Figure 16.2). By 1450, Malmö's strategic location had allowed it to become a major center of fishing and commerce in northern Europe. During the 19th century, Malmö became Europe's largest shipbuilding center. This industry anchored Malmö's economy and supported the development of other industries, including food processing and the production of leather goods and textiles. Throughout these seven centuries, Malmö experienced continuous growth in its population and economy.

In the late 20th century, Malmö fell on hard times. An economic recession in the 1980s forced the closure of Malmö's shipyards, leaving behind a coastline of abandoned docklands. The economic impact of their closure cascaded through the city's economy. By 1990, the city's population had declined by 15%. Buildings and public spaces were beginning to deteriorate. Indicators of community decline, such as unemployment, crime, and suicide rates, were the highest in all of Sweden. It hit rock bottom in 1995, and the city had to begin to re-envision itself for a different future.

Malmö decided to renew itself as a sustainable city, spurred on by strong leadership from its City Council and a variety of citizen groups. In 1997, the City Council adopted its Local Agenda 21, modeled after the United Nations Global Agenda 21. Global Agenda 21 is a set of strategies that countries can use to diminish their negative impacts on the environment while increasing the well-being of their people. Similarly, Malmö's goal was to improve its local environment and, thereby, the health of its population. The city was committed to becoming a model for sustainable development in both its day-to-day work and its long-term planning.

Public education and engagement are key components of Local Agenda 21. Immediately after its adoption, the city implemented an outreach and public participation program

◄ Figure 16.1 **A Symbol of Sustainability**
The Turning Torso building, designed by architect Santiago Calatrava, dominates the skyline of Malmö's Western Harbor district. It was built to replace a large shipbuilding crane that had long been a prominent feature of the city's skyline, a symbol of its identity. The crane fell into disuse and was removed as Malmö's fortunes changed in the late 20th century. This building's environmentally sustainable design is emblematic of a new era of planning and renewal throughout the city.

▼ Figure 16.2 **A Strategic Location**
Malmö's location on the Øresund strait contributed to its development as an important European center for fishing and commerce. Today, the 16-km long Øresund Tunnel-Bridge connects Copenhagen to Malmö. Its path is visible in this satellite photograph.

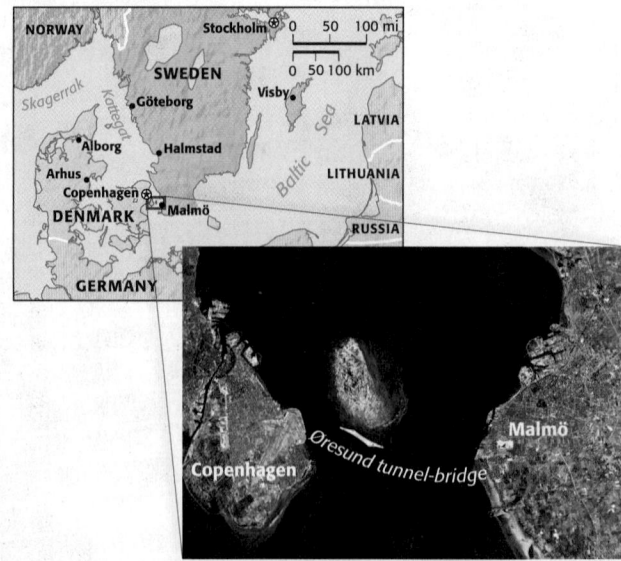

to assist in the development of its environmental strategy. Exhibitions showcasing elements of urban sustainable development were displayed at local libraries, and a monthly Agenda 21 newsletter highlighting new projects was distributed throughout the city. The city published a children's book focusing on environmental issues and human health and distributed it to every Malmö child between the ages of six and eight. With widespread public support and regional, national, and international funding, Malmö has inaugurated programs to improve energy use, reduce waste, and develop open space.

The spirit of Turning Torso is especially evident in Malmö's partnership with the private utility eON Sweden, which has the goal of obtaining 100% of the city's electrical energy from renewable sources by 2030. Today, solar collectors provide hot water to most of the city's buildings. Innovative heat pump systems extract heat from seawater and groundwater. A 2-megawatt wind turbine and photovoltaic panels on a single large building provide virtually all of the electricity to over 5,000 homes and businesses in the city's Western Harbor area. All of the apartments in the Turning Torso have units to convert organic waste into biogas, which is used for cooking and to fuel vehicles. With these sources of renewable energy, the expansion of its biofuel-powered public transportation system, and a growing commitment to cycling (one out of every four trips is by bicycle in the city), Malmö aims to be climate neutral with its CO_2 emissions by 2020.

Outside, Malmö is using its existing system of city parks to anchor a larger system of green spaces. Sixteen green corridors provide streams, ponds, and other wildlife habitats, as well as a system of bike trails (Figure 16.3). These green spaces as well as 10,000 m^2 of green roofs in the Augustenborg district have reduced floodwater levels by 20% and increased the city's biological diversity by 50%. The green spaces also provide opportunities for recreation and human-powered transportation.

All of these efforts have been coupled with programs to improve the health and well-being of Malmö's citizens. These include educational programs dealing with nutrition, smoking, and alcohol use, as well as a goal to provide free 100% organic meals for all school children by 2020. With support from the Swedish government, the city has begun a program to offer affordable housing to

singles, small and large families, and the elderly in compact developments with easy access to public spaces and services.

Malmö's Local Agenda 21 is also a commitment to influence sustainable development in other communities. In 1998, the city founded Malmö University, which has a focus on the environment, conservation of natural resources, ethnicity, and gender. In addition, Malmö has joined other cities in the Sustainable Energy Communities and Urban Areas in Europe (SECURE) project, which provides opportunities for sharing ideas on energy conservation.

Yet challenges remain. In parts of the city, crime rates are still high. A rapid influx of immigrants presents challenges on many fronts. The city is struggling to meet its Agenda 21 goals in the suburban communities that are rapidly growing outward from the city center. Nevertheless, Malmö's health indicators are improving, and its unemployment rate has dropped from 16% in 1996 to less than 8% in 2011. Since then it has remained below 8% despite a worldwide economic recession.

- *What factors are responsible for decaying urban infrastructure and rapid suburban growth, and how can cities deal with these problems?*

- *How can the principles of dynamic homeostasis associated with ecosystems be applied to cites and the environments that surround them?*

- *Can cities around the world become sustainable from the standpoint of the environment, economy, and community?*

▼ Figure 16.3 **The City of Parks**
Ⓐ Malmö's Green Plan envisions 16 green corridors connecting the city's many parks. Ⓑ As part of the extensive system of bike trails, this bicycle counter records the number of bikes that pass by each day.

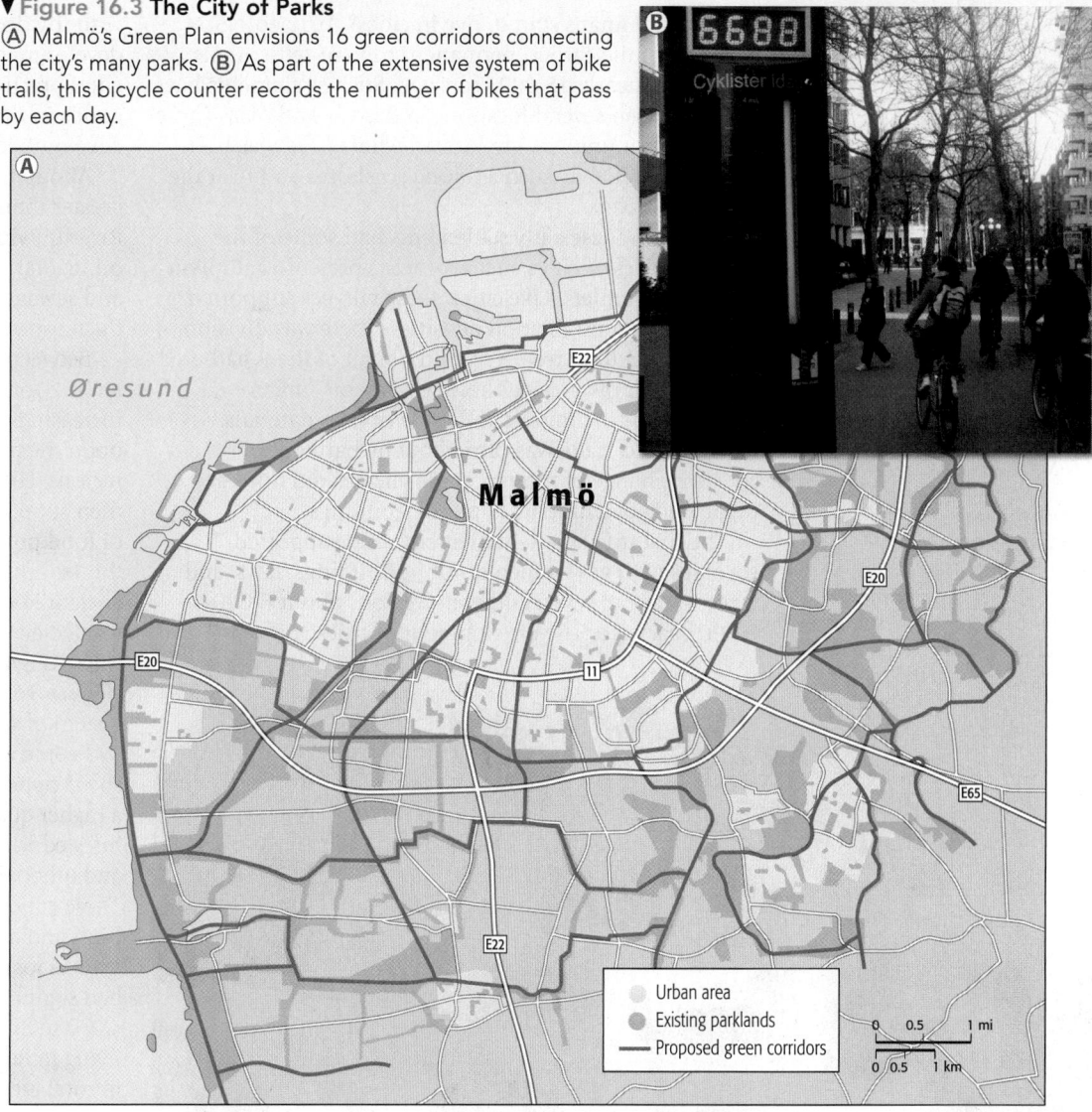

16.1 Urbanization

BIG IDEA The earliest permanent human settlements were located in areas with abundant supplies of food and other resources. Some settlements eventually became cities—centers of trade and industry. **Urbanization**, the concentration of human populations in densely populated cities, depended on the growth and distribution of the food supply, as well as the development of technologies to deliver water and remove wastes. Thus, the process of urbanization hastened the development of technology, market economies, and social and governmental structures. Cities facilitated the growth of human populations, yet the poor sanitation and diseases associated with them often diminished the well-being of city dwellers. The Industrial Revolution along with immigration further hastened the growth of cities, as improved technologies for food production allowed a dwindling number of people to produce enough food to support the rapidly growing urban populations. The growth of cities has, however, significantly altered Earth's ecosystems and led to the unsustainable use of many resources.

Ancient Cities

■ Permanent settlements require abundant and reliable resources.

When did humans start to live in cities? Archaeological evidence of the earliest permanent human settlements dates from nearly 15,000 years ago. Such settlements have been found in several locations in Europe and Asia. These simple communities were located where food was both ample and reliable, such as along seashores and near the mouths of rivers.

Although these early settlements had some of the characteristics of cities, archaeologists prefer to call them villages or hamlets. Like cities, these villages supported relatively high population densities. There was division of labor among their residents, and not all of them gathered or grew their own food. These settlements functioned as centers for the import and trade of raw materials, although this trade was carried out by barter rather than by the exchange of money. These villages did not have organized governments or public buildings.

The first true cities—settlements that supported societies with governments, public buildings, laws, and taxes—arose in several different regions about 6,000–8,000 years ago. The development of cities coincided

with the domestication of plants and animals and the development of agriculture. Residents of these early cities practiced specialized trades and bought and sold goods in monetary economies. That trade encouraged the development of writing and systems for keeping records.

About 4,500 years ago, large cities with populations greater than 40,000 sprang up in the valley of the Indus River in what is today Pakistan. These cities were laid out on regular grids. They also had sophisticated sanitation and sewage systems, including well-planned drainage and flush toilets.

Between 4,000 and 2,000 years ago, the number and size of cities grew, and their organization became increasingly complex. These ancient cities depended upon the flow of energy and materials from surrounding regions. Urban populations were ultimately limited by their supplies of resources, such as water and the amount of food produced in surrounding agricultural lands. In the 1st century, for example, the million plus residents of ancient Rome obtained their food from farms that were located across southern Europe and northern Africa. Water was brought into the city by a system of 11 large aqueducts (Figure 16.4), and waste was carried out by a complicated system of sewers known as the Cloaca Maxima.

People were probably drawn to cities by economic or social opportunities, but most city dwellers did not enjoy a higher quality of life than their rural kin. Cities were infested with insects, rats, and other vermin. Despite sewers and other amenities, water supplies were badly polluted. These conditions facilitated the spread of disease. Studies of birth and death dates recorded on thousands of tombstones from across the Roman Empire reveal that people in cities had significantly reduced life spans. During the period from A.D. 0 to 300, for example, the life expectancy for people living in Rome was about 27 years, and that of people living in rural areas was more than 40 years.

▼ **Figure 16.4 Water for Ancient Rome**
Two thousand years ago, a complex system of canals and aqueducts carried water from distances of more than 50 km (31 mi) to supply the millions of people living in Rome and other urban centers.

Development of Modern Cities

■ Urban development was both necessary for and dependent on the growth of industry and technology.

As empires waxed and waned over the succeeding centuries, cities continued to be centers of commerce and governance. During medieval times, many cities and towns were politically independent from the surrounding rural lands. Because cities provided freedom and opportunity from the oppressive rule of feudal lords, they attracted many residents. Even so, life for the citizens of most cities was difficult, even brutish. Death rates from the multiple plagues that swept Europe between the 13th and 17th centuries were 5 to 10 times higher in cities and towns than in the rural countryside.

Beginning around 1750, the Industrial Revolution increased the rate of urbanization, as more and more people migrated into cities. An ever-increasing proportion of the population was employed in urban industry, either directly or indirectly. At the same time, technological advances in agriculture made it possible for fewer people to produce increasing amounts of food. This increased supply of food fed the growing numbers of city dwellers.

In 1740, nearly 70% of British laborers were farmers. By 1840, only 25% of British workers were employed on farms; the rest worked in cities.

The rapid urbanization of the early Industrial Revolution did not necessarily bring improved well-being to city dwellers. Up to 100 years ago, large industrial cities in Europe and North America were characterized by polluted air and water, extensive poverty and slums, and high rates of a great many diseases.

Around the year 1900, the health penalty for urban life began to diminish in the developed world. In many regions, cities built water and sewage treatment facilities, established effective methods of waste management, and improved health care facilities. About the same time, the use of fossil fuels, fertilizers, herbicides, and pesticides produced the Green Revolution, allowing fewer and fewer people to produce the food needed by the rapidly increasing number of urban dwellers (see Module 12.1).

Current Trends

■ Urbanization has led to substantial change in Earth's ecosystems.

Since World War II, urbanization has proceeded at a rapid rate. In 1950, 29% of the world population lived in urban areas. Today, 53% are urbanites, and that number is projected to reach nearly 70% by 2050. This change has been directly coupled to the global demographic transition in which birth, death, and population growth rates have declined (see Module 5.2).

The growth of urban populations has been especially notable among the world's wealthiest nations. Currently, 79% of the people in these countries live in cities. In the United States, 83% of the people live in urban areas. The population of U.S. cities is growing by 1.24% each year, while the population of rural areas is shrinking by 0.63% each year.

Over the next 30 years, the vast majority of Earth's human population growth will occur in its poorest countries. Urban populations in the world's least developed countries are growing by nearly 4% each year. At this rate, city populations in these countries will double in 17 years. Because they have limited economic resources, many cities in these countries are currently unable to provide clean water, manage wastes, or meet the basic health needs of their citizens (Figure 16.5). As in ancient times, urban citizens in most poor countries suffer diminished well-being, as indicated by high rates of infectious diseases and infant mortality and diminished life expectancy.

▼ Figure 16.5 **The Urban Poor**
Port Au Prince, Haiti, has more than 2 million residents, many of whom suffer from infectious diseases because they lack sanitary facilities and have no access to clean water.

▼ **Figure 16.6 Megacities**
Megacities, metropolitan areas with over 10 million people, are widely distributed and growing in number.

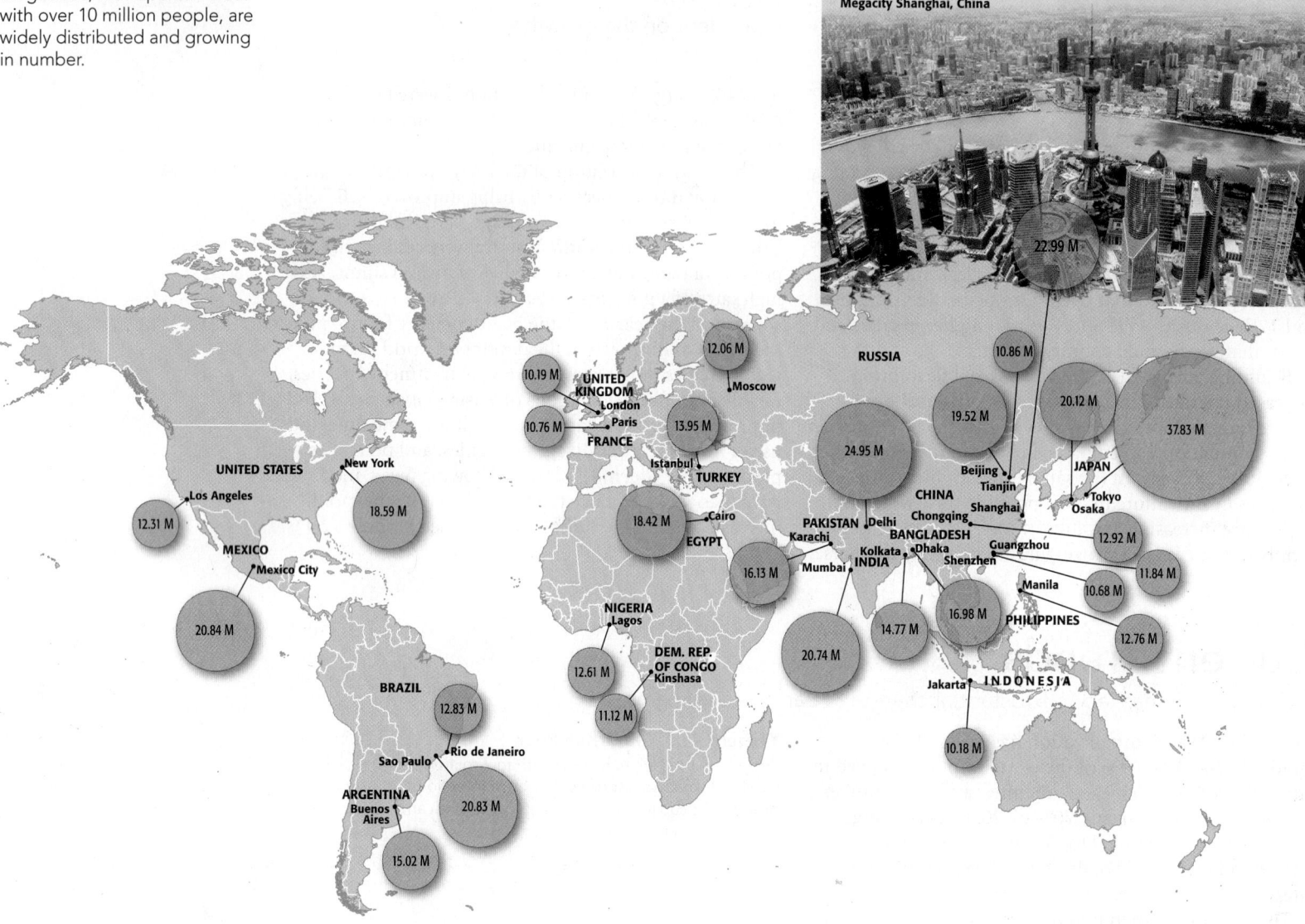

Megacity Shanghai, China

QUESTIONS 16.1

1. Name three features that early villages and hamlets share with modern cities.

2. What is the evidence that the well-being of people living in ancient Rome was less than that of people living in rural areas?

3. How do urban population growth rates differ between wealthy and poor countries?

(MES) For additional review, go to
MasteringEnvironmentalScience

The redistribution of Earth's people into urban centers has affected nearly every aspect of human life. It has redefined the nature of human interactions and communities, powered economic growth, and altered modes of recreation and education. It has required increased growth and complexity of local governance, a change that many sociologists argue has been a major factor in global trends toward increased democracy.

The effects of urbanization are especially evident in **megacities**, urban areas with more than 10 million inhabitants (Figure 16.6). Although less than 10% of Earth's urban population dwells in such cities, megacities have a disproportionate importance in terms of global commerce, governance, and environmental impacts.

Urbanization has caused significant restructuring of Earth's ecosystems. In order to support the needs of large concentrations of people, humans have redirected the flow of water all over the world. For example, over 90%

of the water consumed by the people in the Los Angeles metropolitan area comes from more than 200 miles away. Most of the energy that powers cities comes from nonrenewable fossil fuels. Most of the food for the growing number of city dwellers is supplied by a dwindling number of farmers but requires increasing investments of energy from fossil fuels. Some urban waste is altering the chemistry of our water and air. Other types of urban waste accumulate in huge garbage heaps and landfills.

Clearly, many of these urbanization trends are not sustainable. Nevertheless, it is impossible to imagine Earth supporting our current numbers, much less our projected population growth, without cities. To resolve this paradox we must pursue strategies that lower the demand for water, energy, and other natural resources and reduce the streams of waste. With appropriate investments, urbanization in the world's poorest countries can actually improve human well-being.

16.2 Urban Ecosystems

BIG IDEA An **urban ecosystem** is a region of Earth that supports the commerce, industry, and community interactions of a large number and high density of humans. Urban ecosystems include the physical environment and all the organisms living in cities and the less densely populated areas that surround them. It may seem paradoxical to refer to a city as an ecosystem. Yet, like natural ecosystems, cities are dependent on the interactions among living and nonliving components and on the sustained flows of matter and energy (see Module 1.3). Human business and communities require a constant inflow of energy and materials and produce a steady stream of wastes. Urban ecosystems are sustained by the inflow and outflow of ideas and money and by the movement of people along transportation corridors. The physical environment and biodiversity of urban ecosystems change significantly from city centers to surrounding rural areas. The complex structure and artificial surfaces of city centers produce climates that are very different from those in rural locations. Urban development also alters the flow of water and diminishes water quality.

Defining Urban Ecosystems

■ Urban ecosystems change gradually from city centers to rural landscapes.

On many maps of the United States, Los Angeles appears as a large "smudge" on the southwestern coast of California. In satellite photographs, it is easily identified by its sprawling developments and freeways. Yet if you drive across this urban landscape, it is hard to define the actual boundaries of Los Angeles (Figure 16.7).

Pulling out of your parking place in front of the Los Angeles City Hall, you know you are in a city. Skyscrapers tower above you as you maneuver through the heavy traffic on South Main Street toward Interstate 10, known locally as the San Bernardino Freeway. Heading east on Interstate 10, tall buildings give way to a mixture of residential and commercial neighborhoods. Less than a mile from downtown, the freeway crosses a wide, dry, concrete-lined canal—the Los Angeles River.

About 4 miles farther on, a small sign at the edge of the freeway reads "Los Angeles City Limits." Aside from that sign, nothing in the landscape indicates a boundary. You journey east for miles through a mix of housing subdivisions, shopping malls, and commercial districts, such as West Covina, Pomona, Ontario, and Fontana, although no clear lines delineate these communities. As you proceed east, buildings and homes seem to have been built more recently, and patches of shrubs and desert become more frequent.

◄ **Figure 16.7 Los Angeles, The City of Angels**
As can be seen in this aerial photo, Los Angeles sits at the center of an extensive area of development, stretching from the coast to the mountains. The metropolitan area that surrounds the city center is the 20th largest in the world and the second largest in the United States. It encompasses 12,562 km² (4,850 mi²) and is inhabited by some 13.4 million people.

519

▶ **Figure 16.8 Into the Burbs**
The metropolitan area of Los Angeles is tied together by over 20,770 miles of public roads and highways. Downtown L.A. is dominated by high-rise buildings and traffic-clogged streets with few natural spaces Ⓐ. The iconic City Hall can be seen in the lower right of the photo. Ⓑ Drive 15 miles out and you find suburbs dominated by low-stature construction and more green space. Ⓒ Continue 122 miles east of the city center, and you'll come to the exurban community of Palm Desert, which marks the outer fringe of this sprawling metropolitan area.

After an hour of driving, the traffic thins and you pass more open spaces. Finally, about 15 miles east of the city of Palm Springs, the strip malls, new housing developments, and golf courses give way to the vast, wild expanses of the Mojave Desert (Figure 16.8).

Scientists who study urban ecosystems would use more specific terminology to describe your drive. The U.S. Census Bureau defines an area as **urban** if its average population density is at least 500 people per square mile. Certainly, the entire corridor from Los Angeles city center to Palm Springs qualifies as urban. A city is an area that is defined by legal and political boundaries and has a formal government and taxes. Thus, the city of Los Angeles is bounded by very precise city limits. Within those limits, the decisions and policies of the Los Angeles mayor and city council hold sway over its more than 3.8 million residents.

Extending out in all directions from the city center is the Los Angeles metropolitan area. A **metropolitan area** is a large central city and the adjacent communities that have a high degree of social and economic integration with that central city. The transportation system and movement of commuters between the suburbs and the city is a very visible example of such integration. **Suburbs** are the parts of the metropolitan area located outside

of the central city but often lacking clear boundaries between where urban ends and suburban begins. Beyond the suburbs are **exurbs**, areas where patches of commercial and residential development are interspersed in a matrix of rural land (Figure 16.9).

All of these areas are components of an urban ecosystem. However, the processes that support urban ecosystems often extend beyond the boundaries defined by population density. According to the U.S. Census Bureau, the metropolitan area of Los Angeles ends near Palm Springs, but more than 90% of the water and electricity consumed within that area come from places that are more than 100 miles from that boundary. This urban ecosystem is clearly larger than its metropolitan area.

Urban ecosystems also include the complex array of interactions represented in the structure of economies, governments, and society—the human community. The city limits established by urban governments define the boundaries of certain political and legal jurisdictions but often have little to do with patterns of transportation, energy use, or citizens' sense of community. Rather than trying to draw firm boundaries to the urban ecosystem, urban ecologists prefer to describe gradients of change from the city proper into its rural surroundings. Thus urban centers grade into less densely populated suburbs and then into sparsely populated exurbs. In the centers of large cities such as Hong Kong, human population density may exceed 10,000 people per square kilometer (25,900 per square mile), while in rural areas there may be 1–10 people per square kilometer.

For those who associate ecosystems with the natural world, the phrase "urban ecosystem" may seem contradictory. Built by humans and supporting only the most meager biodiversity, cities seem to exemplify the unnatural. Indeed, cities exist for totally anthropocentric purposes—to support the commerce, industry, education, and social interactions necessary to sustain the well-being of their human residents. Nevertheless, thinking of a city as an ecosystem can help to clarify its inner workings (Figure 16.10).

Compared to natural ecosystems, urban ecosystems are especially open to inflows and outflows of matter and energy. Indeed, urban ecosystems depend on inputs of resources from local, regional, and global locations, whereas urban outputs have consequences ranging from local to global scales. Every day, significant amounts of materials needed for the urban ecosystem flow in and out. Many different kinds of finished products are imported to support the needs of city residents. Raw materials are brought in and used to manufacture finished products that are then consumed or exported for sale. Some amount—often a large amount—of this material eventually ends up as waste. Unlike natural ecosystems, most of the waste materials from cities are deposited outside the political boundaries of the city.

Another characteristic of natural ecosystems is the biodiversity represented in food webs—the complex array of interactions that affect the flow of matter and energy through ecosystems. Interactions among different organisms also play important roles in urban ecosystems, although urban ecosystems typically have far less biodiversity than natural ecosystems.

At this point in time, fully 90% of Earth's surface has been impacted by humans in some way, so few truly "natural" ecosystems remain. In 2003, ecologist Michael Rosenzweig conceptualized *reconciliation ecology*, a new branch of ecology focused on increasing biodiversity in human-dominated systems. He suggests that by supporting greater diversity in developed systems, both human enterprise and biodiversity will succeed. Urban ecology encompasses that portion of reconciliation ecology that addresses the most heavily human-dominated systems, and strategies to increase biodiversity in the following pages are examples of win–win solutions.

METROPOLITAN AREA

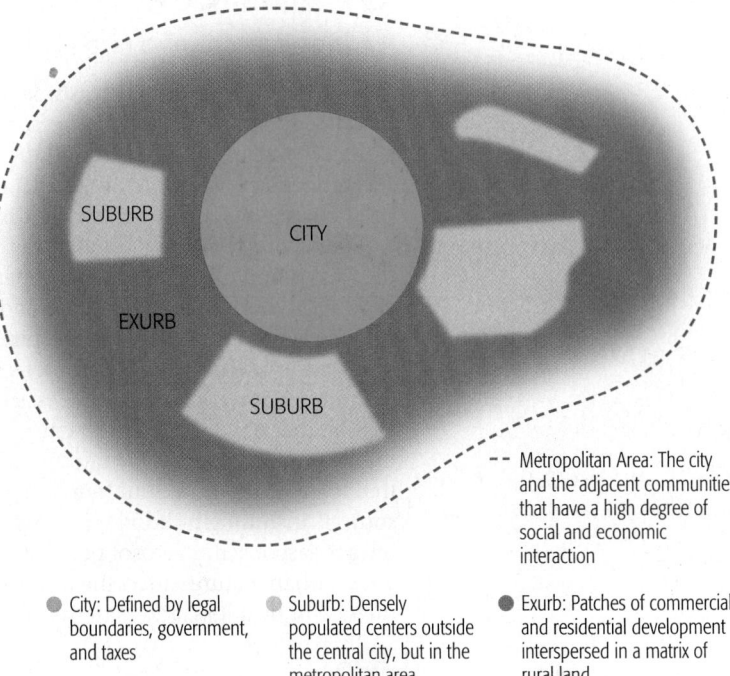

-- Metropolitan Area: The city and the adjacent communitie that have a high degree of social and economic interaction

● City: Defined by legal boundaries, government, and taxes

● Suburb: Densely populated centers outside the central city, but in the metropolitan area

● Exurb: Patches of commercial and residential development interspersed in a matrix of rural land

▲ Figure 16.9 **Urban Terminology**
Geographers define metropolitan areas, cities, suburbs, and exurbs in specific ways. Note that a suburb might also be defined as a city if it has legally defined boundaries and a government.

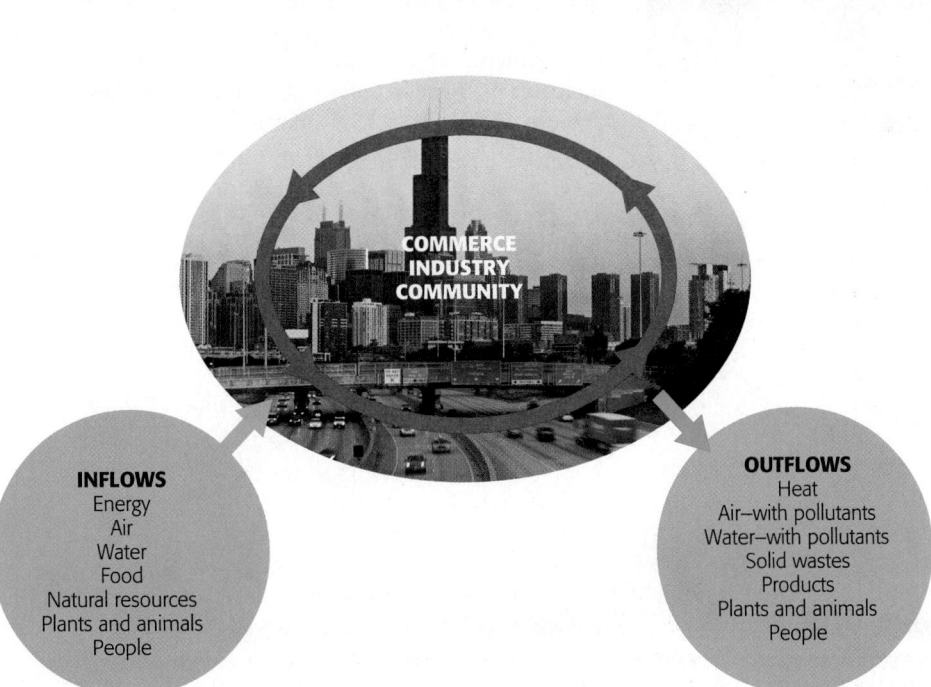

◄ Figure 16.10 **The Urban Ecosystem**
Ecosystem components and processes are as applicable to cities as they are to natural areas. The ecosystem services that sustain commerce, industry, and community well-being in urban ecosystems depend on inflows and outflows of matter and energy, as well as the movements of plants, animals, and people into and out of the city.

COMMERCE INDUSTRY COMMUNITY

INFLOWS
Energy
Air
Water
Food
Natural resources
Plants and animals
People

OUTFLOWS
Heat
Air—with pollutants
Water—with pollutants
Solid wastes
Products
Plants and animals
People

Urban Climate

■ Cities capture and retain heat.

The transition from rural landscape to city center is often correlated with significant changes in climate. Cities are generally warmer and less humid than their surroundings. The tendency for cities to be warmer than surrounding rural areas is called the **urban heat island** effect.

Four characteristics of urban areas contribute to the heat island effect. First, cities generally have less vegetation than rural and natural landscapes and therefore provide less shade and evapotranspiration. Second, in many parts of cities, naturally porous soils are replaced with impervious or sealed surfaces, such as concrete or asphalt that retain heat. Third, cities have complex three-dimensional structures of buildings and "street canyons" that tend to trap heat. And fourth, a great many urban sources, including buildings, factories, and automobiles, release gaseous and aerosol pollutants and waste heat. These urban features affect the absorption of solar energy, the rate of heat loss, and patterns of air movement.

The urban heat island effect is generally larger at night than during the day and larger in the winter than in the summer. During the day, the surfaces of buildings and pavement absorb solar radiation. These surfaces become hotter than soils and vegetation, which are cooled by evapotranspiration. During the night, these artificial surfaces lose heat by radiation. However, the loss of heat is slowed because buildings block the way to the cold night sky (Figure 16.11). The radiated heat simply cannot escape the buildings in a large city. Tall buildings also block winds, further inhibiting cooling. In addition, waste heat from air-conditioning, automobiles, and industrial operations contribute to the warmth in cities.

Heat island effects vary across urban ecosystems, depending on building density, vegetation cover, and other urban features. City parks and urban greenways decrease urban temperatures (Figure 16.12), as do white roofs that reflect the sun's rays. Even with this variation, urban heat island effects extend for some distance from urban centers. A study of 70 urban areas in the eastern United States and Canada revealed that the effects of urban heating extended up to 10 km (6 mi) from cities. Within this range, the growing season for vegetation—the interval between the last spring and first autumn frost—was about 15 days longer than in more rural areas.

The heat island effect has significant consequences for human communities. In warmer climates, buildings and homes in cities require more air conditioning and refrigeration than in rural locations. For example, the heat island effect is estimated to add more than $100 million to energy costs in the city of Los Angeles. Furthermore, death rates due to heat-related illness are much higher in large urban areas than in rural settings. A simple change in the color of urban roof tops can help alleviate the problem: White paint reflects the sun's rays and increases energy efficiency of buildings by up to 25% (Figure 16.13). The U.S. Department of Energy launched a cool roof initiative for federal buildings in 2010.

► Figure 16.11 **Urban Heat**
In this thermal satellite image of Atlanta, Georgia, buildings and streets appear yellow to red, indicating that they are warmer than the open lawns, parks, and ponds, which appear green to blue. Heat absorption by buildings and streets and shading by tall buildings produce complex variations in climate within city centers.

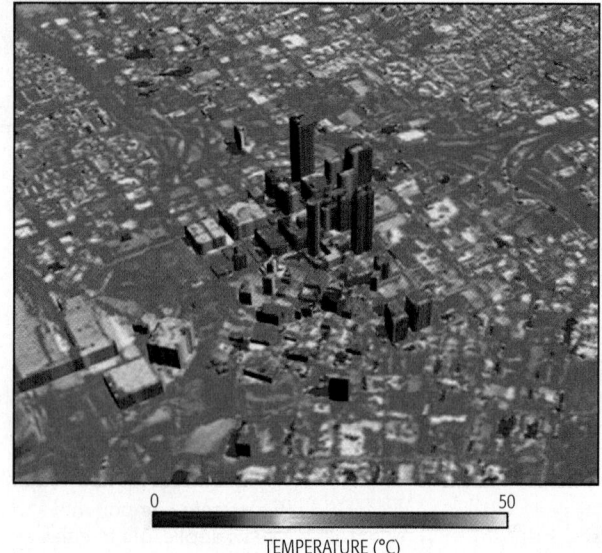

TEMPERATURE (°C)

► Figure 16.12 **Urban–Rural Temperature Gradient**

On average, temperatures are highest in city centers and cooler in suburban and rural areas.

Adapted from: Figure 2, p. 157 of Lemmen, D.S. and F.J. Warren (eds.). 2004. Climate Change Impacts and Adaptation: A Canadian Perspective. Climate Change Impacts and Adaptation Program, Natural Resources Canada, Ottawa, ON, 174 pp. http://www.nrcan.gc.ca/files/earthsciences/pdf/perspective/pdf/report_e.pdf . Last accessed July 18, 2014.

URBAN HEAT ISLAND PROFILE

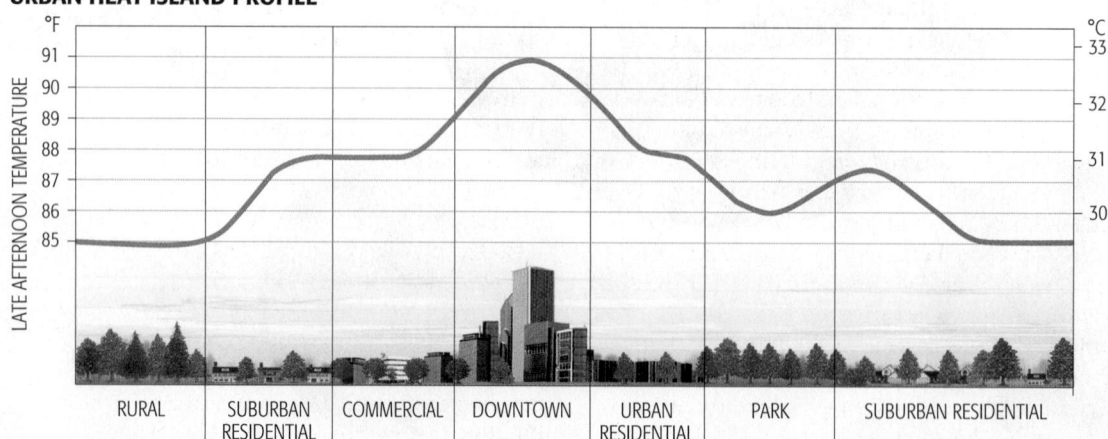

◀ **Figure 16.13 A Cool Roof**
In New York City, the White Roof Project is coating roofs with white paint to reflect the sun's rays and reduce energy costs and the heat island effect.

QUESTIONS 16.2

1. What is the difference between a metropolitan area and a city?
2. What factors cause the climate of urban areas to differ from that of rural areas?
3. How does the relative amount of rainwater runoff in cities differ from that in nearby rural areas? What is responsible for this difference?

(MES) For additional review, go to **MasteringEnvironmentalScience**

Cities also affect the amount of rain falling on the landscapes that surround them. In areas that are located downwind from large cities (within 60 km, or 40 mi), the total average rainfall is generally 10–30% higher than in areas that are upwind. This is partly due to the heat island effect, which causes air to rise as it passes over a metropolitan area. It is also due to urban aerosols that serve as nuclei for the formation of raindrops. Note that the causes of this urban rain shadow effect differ from those of rain shadows that occur east of mountain ranges.

Air movement in urban centers is limited. This facilitates the formation of stagnant air masses and temperature inversions that hold polluted air over cities. The negative health effects of this air pollution are exacerbated by the higher temperatures of cities (see Module 10.1).

Urban Hydrology

■ Impervious surfaces limit stormwater infiltration and increase runoff.

A dominant trend along the rural-to-urban gradient is **impervious surface cover (ISC)**, the percentage of land covered with sealed surfaces, such as pavement and rooftops. These impervious surfaces affect the hydrologic cycle in several ways.

High ISC diminishes the infiltration of stormwater, thereby increasing surface runoff (see Module 11.1). Overall, the higher the ISC, the more precipitation that turns into surface runoff. This greater surface runoff increases the likelihood of flash floods. During non-storm periods, high ISC reduces the recharge of aquifers and the flow in streams. This reduces groundwater storage and the amount of water available in streams and lakes between storms. In addition, high ISC diminishes water quality. Between rainstorms, impervious surfaces collect a wide variety of chemicals and particles, which then pollute stormwater.

Impervious surfaces alter the pattern of water flow in urban streams. Because infiltration is limited, a greater proportion of rainfall runs off, so peak flows are much higher. Increased ISC also shortens flood lag time, the time delay between the peak of rainfall and the peak of flow in streams. Thus, urban streams are more prone to flooding.

In addition to ISC, construction and other forms of urban development alter the flow of urban streams. Over periods of years, the height and width of stream channels change in response to changes in the amount of sediment and patterns of flooding. During the early stages of urban development, streams carry heavy loads of sediment from construction sites; these sediments tend to fill stream channels. Once a city is built up, the source of new sediment diminishes and the frequency and intensity of flooding increases. This increases erosion, causing stream channels to become deeper and wider.

To facilitate development and manage peak water flows, many urban streams have been straightened or channelized. In many cases, their channels have been sealed with concrete to halt erosion. These changes further slow the infiltration of water, diminish flood lag times, and increase peak flows.

Urban development generally decreases water quality. Urban runoff and streams usually carry increased concentrations of chemicals. Illegal waste discharges, leaking sewer systems, and failing septic systems are also important sources of pollution (see Module 11.4). The specific effects on water quality depend on whether the development is residential, commercial, or industrial. The presence and management of wastewater treatment plants and the extent of stormwater drainage affect the degree to which water quality can be protected

Some urban communities have implemented natural solutions to their hydrological problems. The city of Indianapolis, Indiana, for example, installed 25,400 ft^2 of stormwater planters along its 8-mile Indianapolis Cultural Trail (Figure 16.14). These troughlike *bioswales* are vegetated landscape features designed to slow stormwater runoff, remove silt, increase pervious surface area for stormwater filtration, remove pollutants, and beautify the city streets.

▼ **Figure 16.14 Bioswales**
Vegetated planters along city streets in Indianapolis help to soak up rainwater, reduce flooding, and filter pollutants from runoff.

Q *Is the population in cities just getting too dense?*

Jaymee Castillo, University of California, Los Angeles

Lissa: **A** Actually, in many parts of the United States, it is the diminishing density of cities and growing suburban sprawl that present the greatest environmental challenges.

16.3 Urban Land Use

BIG IDEA Around the world, urban populations have grown more rapidly than rural populations. However, the amount of land occupied by urban regions has grown far more rapidly than urban populations, particularly in developed countries such as the United States. This means that each urban citizen is using an ever-increasing amount of land. This widespread growth in the amount of land included in metropolitan areas is called **urban sprawl**. Sprawl is a consequence of people moving out of city centers and is influenced by the nature of transportation networks and commercial and residential development in suburbs. Government policies, economic factors, cultural issues, and poor planning often encourage urban sprawl. Sprawl wastes energy, increases air and water pollution, and results in the loss of farmland and natural ecosystems. It also has negative economic effects on human communities and is correlated with diminished human health and well-being.

Urban Population Distribution

■ Suburban land use is increasing because of population growth and changing patterns of development.

In the 1970 U.S. census, the population of Las Vegas, Nevada, was recorded as 358,000 people. This number did not include the 100,000 tourists who were then visiting the city each day. At that time, the urban area of Las Vegas encompassed about 38 km^2 (15 mi^2). By 2010, Las Vegas's population had grown fivefold, to over 1.9 million people, with an additional population of 500,000 tourists each day. And during those 40 years, the area of the city had grown even faster than its population, ballooning to 290 km^2 (112 mi^2), nearly 7.5 times its 1970 size (Figure 16.15).

Over the past 40 years, the population in Las Vegas has grown more rapidly than that of any other large American city. However, the disproportionate growth in its land area is, in fact, typical of most metropolitan areas in the United States and around the world.

Between 1950 and 2010, the amount of land in the United States that the Census Bureau classified as urban grew from 60,700 km^2 (23,430 mi^2) to 282,912 km^2 (109,233 mi^2). That is an increase of 466%. Some of this change was the result of the growing population in urban areas, which increased by 295%. But if urban population growth had been the sole factor determining this change, total urban land area would have increased to only 179,100 km^2 (69,150 mi^2). The remaining urban sprawl was caused by a significant increase in per capita land consumption.

▼ **Figure 16.15 Sprawling Las Vegas**
This sequence of satellite photos shows the urban development (green) of Las Vegas in 1973, 1991, and 2006. Over this time, the population of the city grew from 358,000 to more than 1,700,000.

Per capita land consumption is the average amount of land used by each urban citizen. It includes the land occupied by homes as well as the land used for other urban purposes, including roadways, public buildings, factories, commercial offices, and public parks. In 1950, per capita urban land consumption was 0.071 ha (hectares; 0.18 acres) per person. By 2010, it had increased to 0.11 ha (0.27 acres) per person, an incremental change of 0.04 ha per person.

Through the 2000 census, metropolitan areas were divided into a central city, the largest incorporated unit in the area, and the suburbs outside of the central city. From 1950 to 2000 central cities have increased in size, but the number of people living within them has shrunk by almost 8% (Figure 16.16). The decline in the population density of central cities has been even more striking. In 1950, population density in U.S. central cities averaged about 2,900 people per square kilometer (7,500 per square mile). In 2000, the population density of central cities was only 1,160 people per square kilometer (3,000 per square mile). During this same period, the number of people living in suburbs increased by 43%, and population density increased slightly, from 66 people per square kilometer (175 per square mile) to 77 people per square kilometer (200 per square mile). A new urban classification system in the 2010 census no longer allows for the delineation between suburbs and central cities. However, the trend toward sprawl away from the central city has continued as the percentage of the U.S. population living in metropolitan areas continues to increase.

These data reflect two persistent trends. First, people have been moving out of U.S. urban centers. In many cases, this decline in population has been coupled with **urban decay**, a decline in economic and community activity and deterioration of infrastructure (Figure 16.17). Second, urban populations have shifted to the suburbs, pushing commercial and residential development ever farther from central cities.

A comprehensive 2014 study (Smart Growth America, Measuring Sprawl 2014) includes several additional measures of urban sprawl beyond population density in metropolitan areas. Additional factors include the ratio of jobs to population size, the mix of job types within a 1-mile radius, as well as measures of walkability and street accessibility. These factors were combined with measurements of population and development density to calculate a Sprawl Index Score. By these measures, New York City and San Francisco, California, are the most compact, large metro areas with the least sprawl in the United States. Atlanta, Georgia, and Nashville, Tennessee, have the greatest sprawl for large metro areas in the United States.

PERCENT OF U.S. POPULATION LIVING IN METROPOLITAN AREAS: 1910 – 2010*

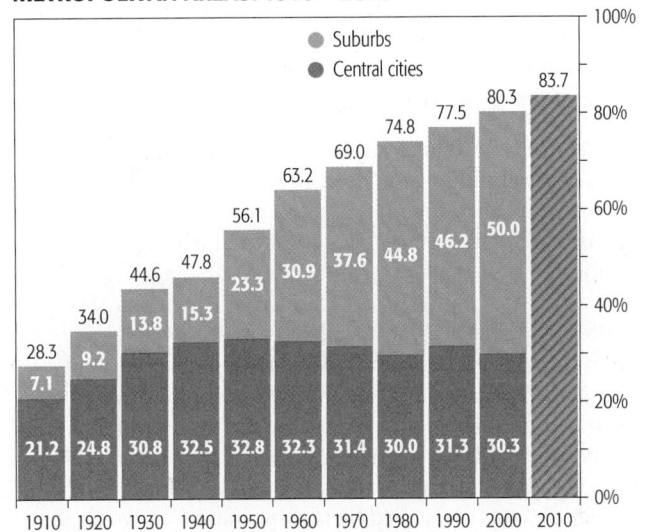

*Data from the 2010 census does not delineate between central cities and their suburbs.

◄ Figure 16.16 **Urban–Suburban Shift**
The percentage of the total U.S. population living in metropolitan areas has grown steadily over the past century. Within these areas, growth has been greatest in the suburbs.

Source: Hobbs, F. and N. Stoops. 2002. U.S. Census Bureau, Census 2000 Special Reports, Series CENSR-4, Demographic Trends in the 20th Century, U.S. Government Printing Office, Washington, DC. http://www.census.gov/prod/2002pubs/censr-4.pdf. Last accessed 10-1-2011.

▼ Figure 16.17 **Urban Decay**
This abandoned apartment building in Detroit is emblematic of the diminished economic activity and the deterioration of buildings and infrastructure in many city centers.

Causes of Urban Sprawl

■ Government policies, economic growth, transportation, cultural issues, and poor planning drive urban sprawl.

About half of urban sprawl can be attributed to the increased population in urban areas. The remainder is the result of an overall increase in per capita land consumption in urban areas combined with decreased population density in central cities.

Most suburban land development is significantly different from the high-rise development typical of central cities. In the suburbs, residential development generally focuses on single-family dwellings, often situated on relatively large lots. Suburban commercial development is usually spread out in office parks and strip malls with large parking lots. All of these suburban developments are served by transportation networks that take up even more land. This shift in land use has been driven by government policies, transportation, economic factors, social issues, and poor planning.

Many government policies have encouraged dependence on automobiles and population decentralization. Federal and state funding have encouraged the development of a complex system of highways and secondary roads with little regulation, allowing sprawl to leapfrog farther and farther away from central cities (Figure 16.18).

Commitment to the automobile as the primary mode of transportation is a major factor contributing to U.S. urban sprawl. Limited public transportation alternatives reinforce this commitment, which in turn discourages investment in public transportation. It has been further reinforced by comparatively low gas prices; owing to low gas taxes, U.S. gas prices are about half those in most European countries.

Government policies have facilitated the "American Dream" of single-family home ownership. For example, various government programs have facilitated home financing, and the U.S. tax code allows homeowners to deduct mortgage interest and property taxes. Much of this growth has been spurred by the rapid growth in per capita wealth. In the United States, per capita GDP has more than doubled over the past 30 years. Much of that economic development has been driven by the expansion of the automobile industry and home and commercial construction.

Many of the economic costs associated with sprawl are externalized from the direct costs of commercial and residential development. These external costs often include those associated with road construction and maintenance and the mitigation of environmental impacts, such as increased air and water pollution (Figure 16.19). Because property taxes are a large source of their revenue, and many of the costs of sprawl are externalized, urban governments often have strong incentives to encourage sprawl.

▲ Figure 16.18 **Transportation Encourages Sprawl**
In California, the construction of Interstate 15 (seen at the top of this photo) has encouraged the development of many suburban communities, such as Corona, California. From Corona, people may commute for more than an hour each way to work near the center of the Los Angeles metropolitan area.

◀ **Figure 16.19 Sprawling Consequences**
Sprawling developments, such as in this community in the suburbs of Altanta, encroach on rural and natural landscapes Ⓐ. In addition to the direct effects of increased air and water pollution, sprawl and its associated dependency on the automobile discourage physical activity, as seen in suburban Pennsylvania Ⓑ.

A variety of social issues have also encouraged flight from central cities and suburban sprawl. These include push factors, such as high crime rates and environmental deterioration associated with urban decay, as well as pull factors, such as the dream of owning a home, better school systems, and the beauty of natural landscapes in some suburban areas. The influx of minorities into American cities has also been associated with urban flight. For example, in 2000 in Atlanta, net black migration into the eight central counties of Atlanta ranged from 0.1 to 14.2%. During that same period, whites migrated *out* of these counties at a net rate of 0.2 to 5.3%.

Finally, much sprawl is a direct consequence of poor planning. Political and legal concerns about the rights of private property owners and developers are often powerful disincentives for effective planning. To improve the planning process, decision makers should envision a city as an integrated ecosystem in which decisions about specific locations have consequences for the functioning and well-being of the entire system and its human communities. Such a vision is discussed in the final module of this chapter.

Consequences of Sprawl

■ Sprawl degrades human environments, reduces biodiversity, and diminishes human health and well-being.

Urban sprawl affects the human and natural environment in numerous ways. Most important among these impacts are the inefficient use of energy and increased pollution. Increased automobile traffic increases urban air pollution, including photochemical smog and high concentrations of ozone (see Module 10.3). It is also an important source of the greenhouse gases that cause global warming. The great expanses of impervious surfaces associated with many sprawling developments increase stormwater runoff and diminish water quality.

Roadways and suburban developments result in the loss and fragmentation of rural and natural landscapes, leading to the loss of biodiversity. In many regions of the United States, the loss of forest cover is directly attributable to urban sprawl. Often the open areas and green spaces within suburban landscapes are poorly maintained and too small and disconnected to support diverse communities. For example, the loss and fragmentation of open lands has been an important factor in the decline of songbird diversity in many suburban and exurban areas.

Many studies have shown that sprawl has a negative impact on human health. In addition to the direct effects of increased air and water pollution, sprawl and its associated dependency on the automobile discourage physical activity. Obesity and high blood pressure are more prevalent in sprawling communities than in communities where walking and bicycling are encouraged. Several studies suggest that sprawl also takes a psychological toll on human communities. Crowded highways increase commuter "road rage." For those who cannot drive, access to nature and to public services and facilities is often very limited.

Economic studies show that the per capita costs of building and maintaining neighborhood facilities, such as schools, public buildings, and emergency services, increase as per capita land use increases and population density decreases. These costs are reflected in the rapid growth in house sales and property taxes in many sprawling metropolitan areas.

As human populations grow, some sprawl is inevitable. However, continued growth in per capita land use is not inevitable. Today there is widespread agreement that the growth of urban areas can and should be more sustainable, from the standpoint of both the environment and human communities. Careful planning aimed at reducing the many negative consequences of urban sprawl is a critical step in this direction. Changes in urban transportation can help diminish negative environmental and community impacts. These issues are discussed in the following modules.

QUESTIONS 16.3

1. What is the connection between per capita land consumption and urban sprawl?

2. Describe five factors that have contributed to increased rates of urban sprawl.

3. What are four negative consequences of urban sprawl for ecosystem and human well-being?

(MES) For additional review, go to **MasteringEnvironmentalScience**

16.4 Urban Planning

BIG IDEA Until the last century, the growth of most towns and cities involved only minimal planning, and that planning was done by government officials. Today, much urban growth is guided by professional urban planners. The profession of **urban planning** unites disciplines such as engineering, architecture, ecology, and sociology to design built environments and transportation systems for urban communities. In most cities, urban planners work with city governments, developers, and citizen groups. Urban planners use zoning and growth boundaries to plan for growth and limit sprawl. To be environmentally, economically, and socially sustainable, urban planning must pursue a diverse portfolio of strategies.

Urban Plans and Planning

■ Dedicated planning programs are relatively new to many cities.

Most towns and cities, even ancient ones, reflect some amount of planning. Most cities are organized around one of three basic plans—a grid pattern, a radial pattern, or a combination of the two. Cities that developed near rivers are often organized in a grid pattern, in which streets and alleys were laid out at right angles and buildings were grouped into rectangular blocks. Cities built on hilltops or around a central market are typically arranged in a radial pattern. Major boulevards radiate outward from the city center like spokes of a wheel, connected by a complicated array of cross streets. Many large cities built during the Industrial Revolution have a combination of both (Figure 16.20).

Historically, defense and aesthetics have also been important themes in the organization of cities. Fortress cities were common throughout medieval Europe, and their protective walls imposed constraints on both city design and growth. During Renaissance and Enlightenment times, parks, town squares, and other public areas became more common features of many cities. This led to the development of park or garden cities, in which aesthetics was a fundamental organizing principle (Figure 16.21).

During the last century, urban planning became an organized profession and area of academic scholarship. Today, nearly all cities with more than a few thousand residents have urban-planning departments. These agencies are staffed by professionals with interdisciplinary training. Much of the work of these planners is focused on the linked challenges of urban decay and sprawl.

In the beginning, formal urban planning in most cities was a top-down process. Urban planners created plans that were then approved and enforced by city governments. In recent decades, urban planning in most communities has become more democratic. In addition to providing technical input, urban planners engage citizens groups, community organizers, and developers in many important decisions.

In the past, most urban planning focused on transportation and the design of public spaces. Today, urban planning seeks to influence development on private lands by setting rules for the location of residential and commercial development and by using strategies to encourage more sustainable development.

▼ Figure 16.20 **Urban Organization**
San Diego, California Ⓐ, provides an example of an urban grid, whereas Paris, France Ⓑ, has a radial organization. Barcelona, Spain Ⓒ, combines grid and radial patterns of organization.

528

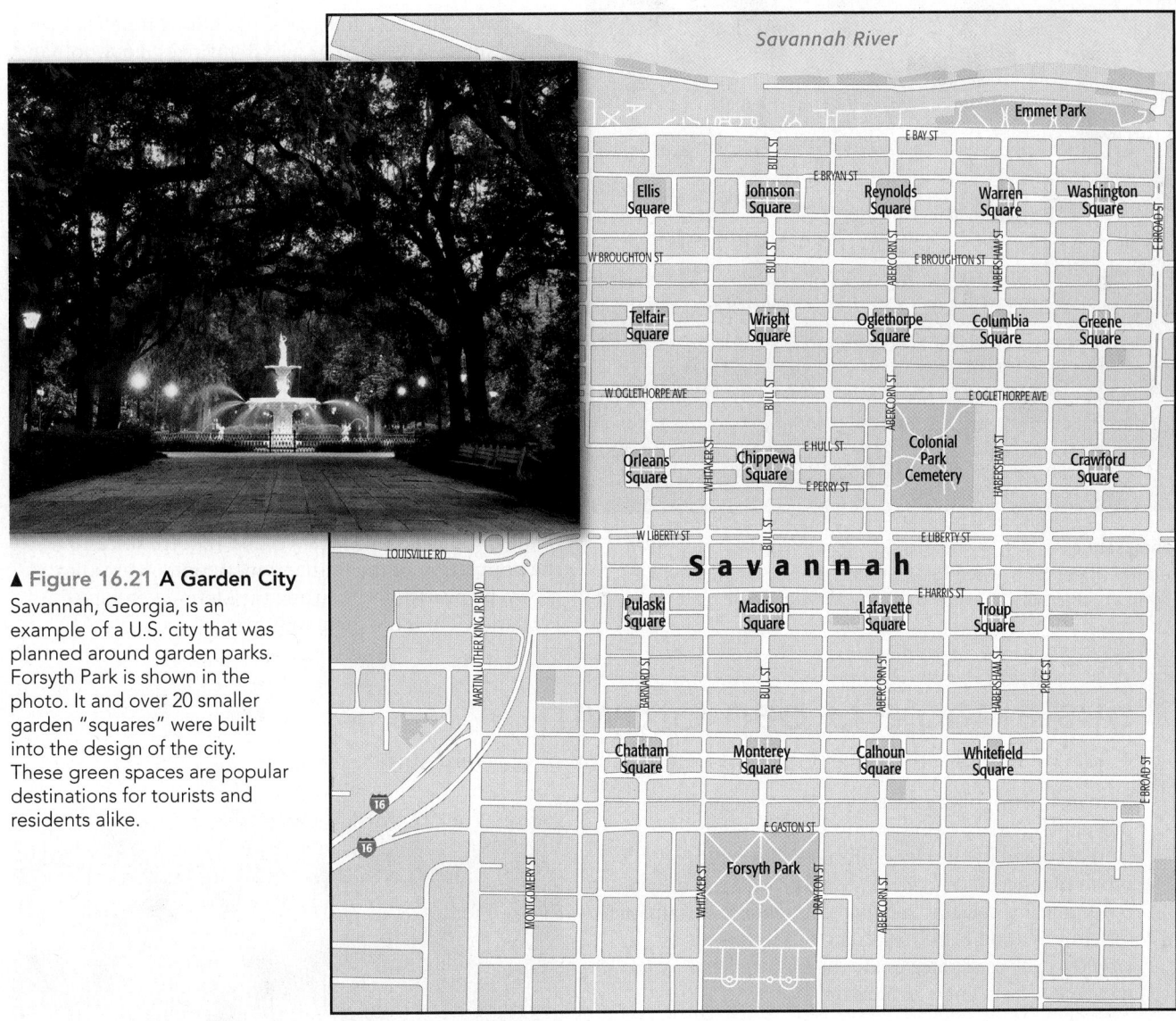

▲ Figure 16.21 A Garden City
Savannah, Georgia, is an example of a U.S. city that was planned around garden parks. Forsyth Park is shown in the photo. It and over 20 smaller garden "squares" were built into the design of the city. These green spaces are popular destinations for tourists and residents alike.

Bounding Growth

■ Zoning and urban growth boundaries provide ways to manage planned growth and development.

One of the most important tools for urban planning is **zoning**, the designation of specific areas for different categories of land use. Urban zones are usually based on categories of land use, such as residential, commercial, industrial, or open space. Zones may also regulate other characteristics, such as building height, the size of residential lots, and residential density. In the United States, zoning is usually regulated by counties or municipalities, although sometimes states set rules or limits on urban zoning. In France and Germany, zoning is regulated by strict national codes.

Zoning serves several purposes. Planners use it to segregate types of land use that are thought to be incompatible. For example, zoning typically separates residential neighborhoods from high-density commercial developments. Zoning may also be used to provide open space or to conserve the historical character of communities. For example, many countries have *greenbelt* policies, which specify land use to include wild or undeveloped areas surrounding urban centers. Notable examples include the European Greenbelt, a 12,500-km ribbon of continuous natural habitat located along the path of the former Iron Curtain.

Some landowners believe that zoning restricts their right to use their land as they see fit. However, planners can make the case that having such constraints provides home and business owners the advantage of knowing what sort of development can occur near their property.

An **urban growth boundary (UGB)** is a form of regional zoning that delineates areas suitable for high-density urban development from those suitable for lower-density development. Large residential and commercial developments are permitted only within the boundaries of the UGB. Essentially, UGBs determine where there will be cities and where there will be rural landscapes. In nearly all cases, UGBs are administered by a regional planning agency established by a county or

▲ Figure 16.22 **An Urban Renaissance**
Portland, Oregon, has revitalized its city center by limiting suburban sprawl through the use of urban growth boundaries and by investing in public transportation and development.

state government with the goal of limiting urban sprawl. Often, the area affected by a UGB involves metropolitan areas with multiple jurisdictions.

Many urbanized areas across the United States have designated urban growth boundaries, but Oregon is the only state that has mandated UGBs for all of its metropolitan areas. The UGB for the 25 cities and 3 counties comprising the Portland, Oregon, metropolitan area, for example, is administered by the Metropolitan Service District, or Metro. Many credit the Metro and UGB with the revitalization of downtown Portland, including increased urban employment and investment in urban infrastructure and mass transit (Figure 16.22).

Few doubt that UGBs can contain sprawl and many of the costs associated with it. However, critics point out that UGBs can cause a shortage of housing units within city limits, which often results in home prices escalating beyond the reach of poorer families. Faced with rapid population growth, regional planning agencies are frequently forced to relocate UGBs. Indeed, since 1979 the population of the Portland metropolitan area has grown by nearly 50%, causing the Metro to enlarge the Portland UGB more than 40 times.

Q How can we encourage people to move to urban centers without resulting in sprawl?

David Oborn, *University of Tennessee, Chattanooga*

Lissa: A Urban growth can occur without sprawl by planning compact communities that offer a variety of transportation choices and focus on human well-being.

Sustainable Urban Growth

■ Sustainable planning depends on a combination of different strategies.

In many metropolitan areas, population growth is inevitable. With it, some growth in the amount of land occupied by the city is probably desirable. Yet urban planners recognize that such growth can only be economically, socially, and environmentally sustainable if a combination of 10 "smart growth" strategies is employed. These strategies are the basic tenets of a social movement called **new urbanism**.

Promote compact communities. Urban planning should encourage human settlement patterns that lower per capita land consumption. The goal is not wall-to-wall high-rise development. Rather, planners should encourage compact developments embedded in natural landscapes with abundant green spaces. In some ways, this is a return to the modes of city development embodied in many European cities and older U.S. communities (Figure 16.23).

Mix land uses. In the 1960s and 1970s, much planning and zoning focused on razing old neighborhoods and separating residential areas from commercial functions. Among other things, this reinforced dependence on automobiles. New urbanists argue that diversity is important in urban neighborhoods and urban environments. While incompatible land uses should not be combined, an intermix of residential, commercial, and public land uses promotes a sense of community and minimizes transportation needs (Figure 16.24).

▲ Figure 16.23 **Compact Communities**
This urban development in Toronto, Canada, has used architecture and land planning to create appealing urban spaces.

Create a range of housing opportunities and choices. Traditional zoning and urban growth boundaries often have the effect of increasing property values so that housing is out of reach for low-income and even middle-income families. Today, many cities promote the urban village strategy, in which relatively dense, multifamily housing and single-family housing are located near shops, cafes, and public transit stops. Tax incentives and subsidized mortgage financing are often used to ensure that housing remains accessible to lower-income families.

Foster communities that provide a sense of place. Much criticism of suburban sprawl is directed at its environmental, cultural, and social sameness. Strip malls and suburban housing projects all seem to look alike. Sustainable cities thrive on community involvement and promote the appreciation of unique environments and the social context of cities.

Conserve open space, farmland, natural beauty, and critical environmental areas. The intensive development of rural lands results in the loss of biodiversity and many ecosystem services. Yet all too often, zoning regulations and even UGBs leave the impression that most rural land will eventually be open to development. One strategy for halting rural land loss is for communities to purchase open spaces. A less costly alternative is to buy the development rights to open spaces but allow farming and forest management activities to continue on the land.

Strengthen existing communities first. Sustainable development gives priority to the efficient use of land that is already committed to urban development. This is the impetus for the renewal of inner-city spaces. Urban infill is a planning strategy that focuses on the development and restoration of undeveloped or abandoned building lots. By strengthening existing communities, cities can grow and evolve through many incremental changes.

Provide a variety of transportation choices. Urban sprawl encourages dependence on automobiles. Yet dependence on automobiles increases the need for more roads, thereby encouraging more sprawl. To break this cycle, transit-oriented development has become a guiding theme for many urban planners. Strategies for balanced transportation systems are discussed later in this chapter.

Create walkable neighborhoods. Pedestrian-oriented neighborhoods promote human well-being and a sense of community. They are also necessary to the success of public transportation systems. Cities such as New York have converted abandoned rail lines to pedestrian corridors.

Make development decisions predictable, fair, and cost-effective. Many cities are notorious for decision making that occurs in secret in backrooms and involves only a few powerful individuals. Such processes are likely to be driven by special interests and usually do not serve the broader interests of the environment and the community. In sustainable development, decision making should be transparent,

or open to the public, and should take account of environmental, social, and cultural needs.

Encourage community collaboration in development decisions. To be sustainable, decisions regarding urban planning and development require public support and participation (Figure 16.25). Involving and empowering the community not only draws on local knowledge and resources but is also critical to resolving the inevitable conflicts over appropriate goals and the means of achieving them.

These strategies represent the ideal framework for smart growth; however, they are not without unintended consequences. For example, critics of New Urbanism argue that it eliminates consumer choice, increases the potential for crime with a grid street system rather than traditional suburban cul-de-sac design, and increases sprawl by conserving open space.

▼ Figure 16.25 **Community Involvement**
Open hearings such as this before the Boise, Idaho, City Council provide an opportunity for public input into development decisions. That involvement builds a true sense of community among citizens.

▲ Figure 16.24 **A New Urban Community**
Southern Village, near Chapel Hill, North Carolina, was developed based on new urbanism principles. Shops, churches, and other meeting places populate the town square and are in easy walking distance from surrounding homes and apartments.

QUESTIONS 16.4

1. Describe the three basic plans of urban organization.

2. What are urban growth boundaries, and how are they generally administered?

3. What are the basic tenets of the new urbanism movement?

(MES) For additional review, go to **MasteringEnvironmentalScience**

Atlanta's Beltline: Abandoned Railway to Transformative Park Network

How can a blighted railroad corridor dotted with abandoned industrial sites transform into a solution for many common urban problems?

The city of Atlanta, Georgia, holds the auspicious title for the U.S. city with the greatest growth in urban area from 2000 to 2010. Long ago, the city burst through its original boundaries and spread in every direction to occupy a metro area now the size of New Jersey. Atlanta originated as a railroad settlement in the 1830s and was eventually circled by 22 miles of railroad tracks that brought goods to and from the industrial sites located along the outskirts of the city. As the city grew and its focus shifted to a less industrial economy, the railway waned in importance as the industrial sites it served were abandoned. Like many urban systems, Atlanta struggles with sprawl, inadequate public transportation, limited green space, and neighborhoods fragmented by major physical barriers.

Atlanta's award-winning BeltLine began as a thesis project developed by Ryan Gravel in 1999 as he graduated with a master's degree in Architecture and City Planning from the Georgia Institute of Technology. Ryan's vision was to transform the blighted 22-mile railway corridor ringing Atlanta's inner city into a continuous multiuse trail system. The BeltLine would reconnect 45 neighborhoods, revitalize and expand 40 parks, and provide much needed public transportation via a streetcar system. As an added benefit, Ryan anticipated economic redevelopment of the central city surrounding the BeltLine (**Figure 16.26**).

Though Ryan's initial vision received accolades, it would take many years, significant and persistent political will, millions of dollars, and a mobilized community to transform it into reality. Several years after graduating, Ryan joined an architecture firm, where he discussed his thesis project with colleagues. They put together some concept maps and a letter to send to the mayor, the governor, regional planners, and anyone else who might be able to help. City Councilwoman Cathy Woolard, chair of the transportation committee, gave the project her full support. Together, Gravel and Woolard held meetings in neighborhoods across the city, and Friends of the BeltLine was born.

Over the next six years, the plan gained the support of the mayor's office and funding through public–private partnerships.

The first trail opened in 2008. As of January 2014, the BeltLine had four developed trail segments running through 11 miles of new green space. Ultimately, the BeltLine will include 22 miles of pedestrian friendly rail transit, 33 miles of multiuse trails, 1,300 acres of parks, 5,600 units of affordable housing, 1,100 acres of remediated brownfields (industrial wasteland space), public art, and historic preservation (**Figure 16.27**). The project is expected to be completed over the next two decades.

The economic consequences of the BeltLine are already evident. Property values surrounding the BeltLine were up as much as 30% by 2005, before any part of the project was even complete. In addition, the BeltLine's management group estimates almost $1 billion has been invested in new development surrounding the BeltLine since 2005. Atlanta's BeltLine is hailed by some as "the country's best smart growth project"—an engine for new economic development and revitalization of what was once an urban blight. The success of Atlanta's BeltLine shows yet again that what benefits the environment often also benefits the economy.

▼ Figure 16.26 **A Man with a Plan**
Ryan Gravel, the architect whose thesis inspired Atlanta's BeltLine, stands in the foreground of what had once been an abandoned railway line. Eleven miles of trails surrounded by new green space have been created since the project broke ground in 2006, with much more on the way.

Ryan Gravel

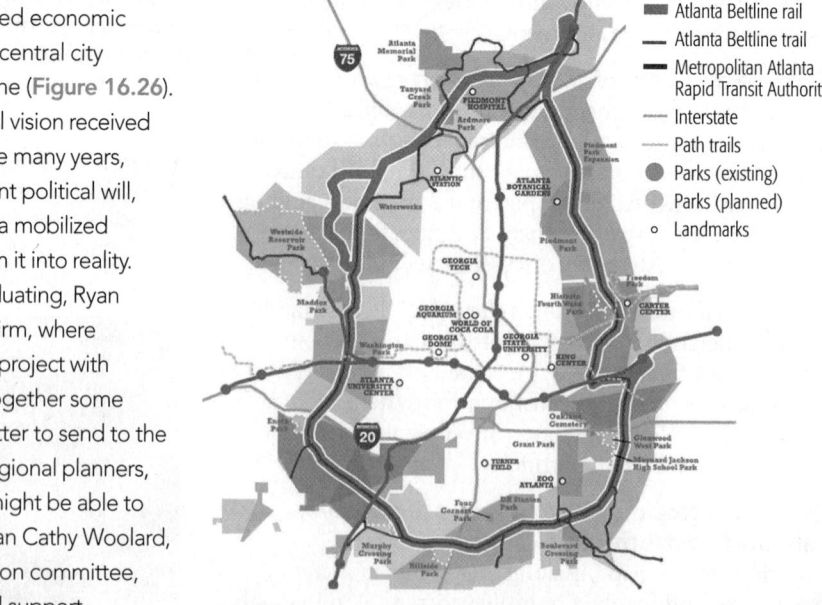

Legend:
- Atlanta Beltline rail
- Atlanta Beltline trail
- Metropolitan Atlanta Rapid Transit Authority
- Interstate
- Path trails
- Parks (existing)
- Parks (planned)
- Landmarks

▲ Figure 16.27 **Mapping a Vision**
Atlanta's proposed BeltLine encompasses 1,300 acres of parks, 33 miles of trails, and 22 miles of public transportation. The project is expected to be completed in 20 years.

16.5 The Built Environment: Sustainable Building

BIG IDEA The "built" environment, in the broadest sense, encompasses all the infrastructure, buildings, and landscapes that we humans have engineered to support all our activities. In this module, we are particularly interested in the design of the structures in which we live and work and the choices made when building or restoring them. The building sector, which includes both residential and commercial buildings, is responsible for 41% of the energy used in the United States as well as significant consumption of raw materials and production of waste. Green building approaches can help reduce the ecological footprint of buildings through consideration of the site, building orientation, and choice of materials. Many different rating systems have been developed to evaluate the sustainability of commercial and residential buildings as the interest in green building practices has grown rapidly in recent years.

Green Building

■ Site selection, building orientation, and choice of materials can significantly improve the performance and sustainability of buildings.

Buildings are significant consumers of natural resources and producers of waste. Indeed, according to the U.S. Department of Energy's *Buildings Energy Data Book* (2011), 41% of all energy used in the United States in 2010 was consumed by the building sector. Homes accounted for 54% of that daily energy use while commercial and industrial buildings made up the other 46%. Likewise, 40% of all raw materials used in the United States go into the construction of new buildings and 35 million tons of construction waste is generated annually in the United States, representing a quarter of our municipal solid waste stream.

Because of its great demand for natural resources, the building sector represents an important target area for resource reduction efforts. This opportunity has not gone unrecognized; the green building market has grown 850% from 2005 to 2012, recently hitting the $85 billion mark.

Sustainable building promises key environmental benefits, including ecosystem and biodiversity protection, improved air and water quality, reduction of solid waste, and conservation of natural resources. A study of 22 green federal buildings showed significant reductions in energy use and operational costs, as compared to national averages (Figure 16.28). Commercial business leaders, too, are beginning to recognize the great economic benefits of sustainable building. A 2013 report by the World Green Building Council showed that while the up-front costs of going green could be as much as 12.5% higher than a standard building, over the lifetime of a structure, green buildings not only save money, they also increase in value and improve employee productivity and satisfaction.

Several key principles guide sustainable building: site selection, building orientation, and choice of materials.

Benefits of Building Green

25%	**19%**	**27%**	**36%**
Less energy use	Lower aggregate costs	Higher occupant satisfaction	Fewer CO_2 emissions

▲ Figure 16.28 **Building Green**
The Regional Headquarters of the Environmental Protection Agency in downtown Denver, Colorado, is housed in a nine-story high-performance green building. It features a green roof planted with grasses, perennials, and groundcovers, a portion of which is shown here. This green roof is expected to minimize heat island effects, absorb carbon dioxide, and contribute to an estimated 26.7% reduction in stormwater runoff.

▶ Figure 16.29 **Here Comes the Sun**
In green building, the proper orientation of a building can maximize its use of the sun's energy in winter and minimize its effects in summer. Landscaping can also help to regulate exposure to sun and wind and thus support energy efficiency.

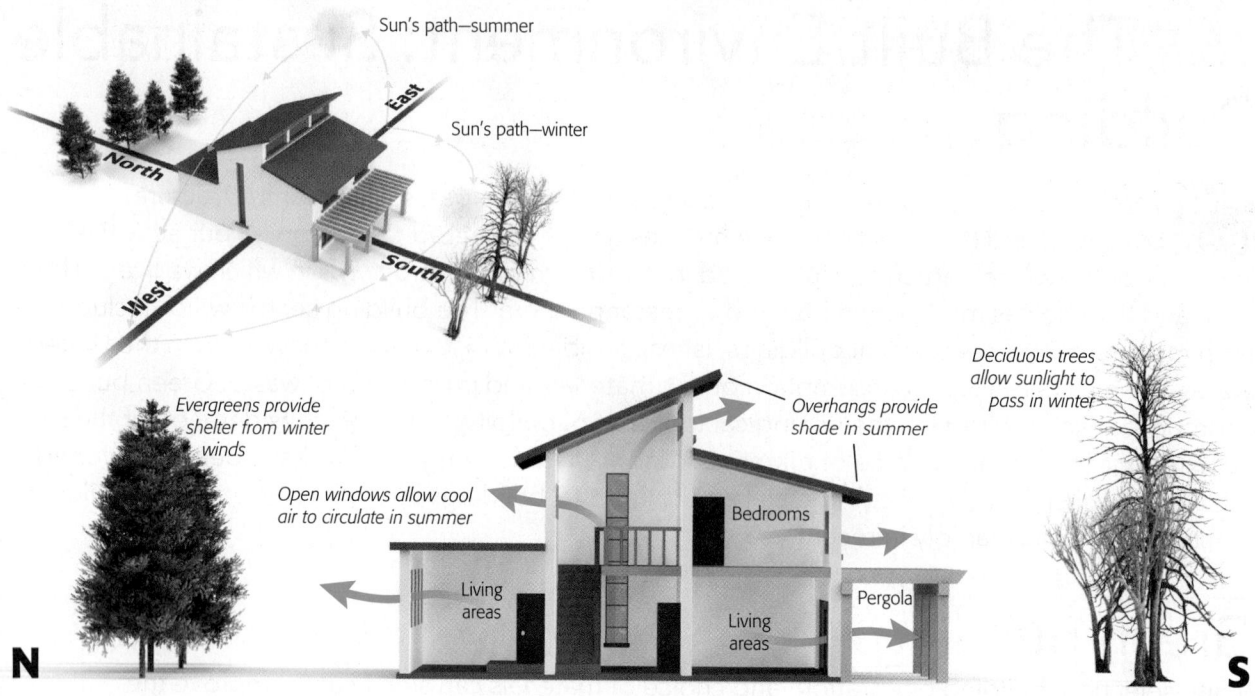

Sun's path—summer

Sun's path—winter

North

East

West

South

Evergreens provide shelter from winter winds

Deciduous trees allow sunlight to pass in winter

Overhangs provide shade in summer

Open windows allow cool air to circulate in summer

Bedrooms

Living areas

Living areas

Pergola

N

S

Site selection.

For new construction, site selection is important. Many developers prefer to build on undeveloped land because it offers fewer constraints. Such building is often referred to as greenfield development. At the opposite end of the spectrum, a more sustainable approach is infill development. Infill involves building on a previously developed but now vacant site that is located in an already built-up area. Infill has many advantages. Locating new buildings close to existing infrastructure (streets, parking, public transportation, water, phone lines, etc.) increases property values and reduces sprawl. As of 2009, greenfield projects still outnumbered infill projects in all but four metro regions in the United States. However, construction of housing in previously developed areas is actually increasing. Residential infill construction was higher during 2005–2009 than during the previous five years for three out of every four metro regions in the country.

Building orientation.

The orientation and associated landscaping of a building can provide significant energy savings over time. In the northern hemisphere, an east–west orientation of a building along its long axis takes best advantage of solar radiation when it is needed (Figure 16.29). In the summer, when the sun angle is high, long overhangs above south-facing windows can shade the building and decrease the need for cooling. In the winter, when the sun angle is lower, sunlight can penetrate the windows, warming the building and reducing heating costs. Some large buildings accomplish this effect with automatic retractable overhangs or awnings for each window. Because west- and east-facing walls receive direct sunlight, keeping windows to a minimum in this

orientation will better regulate temperature and reduce glare in the building.

Landscaping can also help with energy efficiency. In the northern hemisphere, planting deciduous trees on the south and west sides of a building provides summertime shading, while enabling the winter sun to penetrate the windows once the leaves have dropped. Evergreens such as pines and spruces can be used to protect the building from winter winds to the north.

Choice of materials.

Several considerations, including impact on health and the environment, influence the selection of materials in sustainable building design. One such consideration is a material's **embodied energy**, a term used to evaluate the sum total of energy used in the extraction, harvesting, manufacturing, and delivery of construction materials. Some materials, such as aluminum, plastics, and stainless steel, have high energy requirements, due primarily to energy inputs during the extraction process (Figure 16.30). By comparison, clay bricks and tiles as well as timber and concrete from local sources have significantly less embodied energy, depending on their origin. They generally have less extensive energy requirements for extraction, harvesting, and production.

The decision to choose more sustainable building materials is often complicated by other factors. For example, in the southern United States, steel siding on a commercial building has 3.5 times the embodied energy of wood siding. However, even though wood represents the choice with the lowest embodied energy, wood siding is labor intensive to install, requires regular maintenance including painting, and is more expensive than steel. Ultimately, developers must balance aesthetics, functionality, and cost—including the cost to the environment—when making a choice.

Some incentives have been put in place to increase the sustainability of both new and old buildings in the United States. The federal government established mandatory high performance standards for all federal buildings in 2006 that require a 30% energy reduction in new buildings, 20% less water use indoors and 50% less outdoors, strict indoor air quality standards, and reduced environmental cost of building materials. For the general public, President Obama rolled out the Better Buildings Challenge in 2011 to incentivize sustainable building design in commercial and industrial buildings. The challenge requires adopters to reduce their energy portfolios by 20% over 10 years and to share their strategies with the market. By 2013 more than 110 organizations had signed on and were making marked progress. If all U.S. commercial and industrial buildings were to take this challenge, the savings would total more than $80 billion per year after 10 years.

Rating Systems for Sustainable Building

■ A variety of rating systems evaluate the sustainability of buildings around the world.

Starting in the 1990s, both trade and governmental agencies became interested in establishing criteria for defining and evaluating sustainability. Different organizations developed rating systems that quantified the performance of green buildings, both in the United States and abroad. Most focus on the impact of the building on the environment, while some also assess the impact of the interior environment on the building's inhabitants. They include

- BREEAM® (Building Research Establishment's Environmental Assessment Method), established in the United Kingdom in 1990 and now used worldwide
- Energy Star®, established by the U.S. Environmental Protection Agency in 1992, includes a building rating system as well as a rating system for energy efficiency in appliances
- LEED® (Leadership in Energy and Environmental Design), established in the United States in 1998
- Green Globes® System, established in Canada in 2000 and now also used in the United States

One of the most popular green building rating systems, LEED was developed and is maintained by the nonprofit U.S. Green Building Council. LEED evaluates the design, construction, and operation of green buildings with third-party verification. Buildings receive points for features such as water conservation, daylighting (using the sun instead of electric lighting), recycling construction waste, use of locally sourced materials, and much more (Figure 16.31). The U.S. Green Building

Material		Primary energy requirement worldwide (GJ/tonne*)
Very high energy	Aluminum	200–500
	Plastics	50–100
	Copper	100+
	Stainless steel	100+
High energy	Steel	30–60
	Lead, zinc	25+
	Glass	12–25
	Cement	5–8
	Plasterboard	8–10
Medium energy	Lime	3–5
	Clay bricks and tiles	2–7
	Gypsum plaster	1–4
	Concrete: In situ / Blocks / Precast	0.8–1.5 / 0.8–3.5 / 1.5–8
	Sand-lime bricks	0.8–1.2
	Timber	0.1–5
Low energy	Sand, aggregate	<0.5
	Flyash, volcanic ash	<0.5
	Soil	<0.5

*tonne = metric ton = 1,000 kg

▲ Figure 16.30 **Embodied Energy**
This chart compares the embodied energy of some common building materials based on the input of energy required for extraction or harvesting, production, and delivery. While the numbers are highly generalized from the world market, they do provide a basis for comparison.

Data from: Thomas, Randall ed. Max Fordham LLP. 2006. *Environmental Design: An Introduction for Architects and Engineers*, 3rd edition. p. 72.

Main credit categories		Description
	Sustainable sites	Encourages strategies that minimize the impact on ecosystems and water resources
	Water efficiency	Promotes smarter use of water, inside and out, to reduce potable water consumption
	Energy & atmosphere	Promotes better building energy performance through innovative strategies
	Materials & resources	Encourages use of sustainable building materials and reduction of waste
	Indoor environmental quality	Promotes better indoor air quality and access to daylight and views

▲ Figure 16.31 **LEED-ing the Way**
Buildings are assessed in five credit categories for points in LEED certification.

Council also accredits professionals who can help to administer the LEED certification process. Buildings can receive basic, silver, gold, or platinum ratings, depending on how many points are earned in the various categories. For example, the EPA's office in Denver (see Figure 16.28) received a Gold certification. LEED currently applies to new construction as well as retrofits, including commercial and industrial buildings, schools, health care facilities, homes, and even neighborhoods.

QUESTIONS 16.5

1. List several benefits of green building.

2. Why is it environmentally beneficial to place deciduous trees in front of south-facing windows in the northern hemisphere?

3. Of what value are green building rating systems?

(MES) For additional review, go to **MasteringEnvironmentalScience**

Greening the Empire State Building

Can a prewar trophy building become an example of sustainability?

When you think about New York City's 102-story Empire State Building (ESB), the world's best-known office building and a symbol of American economic prosperity, you probably do not think "green." Though the Empire State Building lost its title as the tallest building in the world in the 1970s, it now holds another honor as the tallest LEED-certified building in the United States (Figure 16.32). Completed in 1931, the ESB offered developers significant challenges in retrofitting it for sustainability. It had an aging 80+ year old infrastructure built with little regard to energy efficiency and operated as a multi-tenant facility in which the building owner had little influence over the energy consumption habits of its tenants.

In 2008, the building's owner and his team set out to optimize the performance of the ESB, while reducing greenhouse gas emissions and providing economic benefits. An additional goal was for the ESB to serve as a model that others could replicate. The team wanted to show that it was economically beneficial for building owners and tenants to pursue sustainability goals in existing buildings.

To accomplish their lofty goals, the team focused on solutions with the highest potential for return on investment. That included window refurbishment, improved insulation, a chiller overhaul, daylighting, and tenant energy management.

The first step was to refurbish the building's 6,514 windows on site and increase their insulative value by up to 400%. The windows were pulled apart, cleaned, and fitted with a spacer and heat-reflective film layer. They were then filled with a mix of insulating gases before being sealed and reinstalled. This allowed for the reuse of 96% of all the glass in the building and saved a mountain of waste from the landfill. This also meant little disruption for tenants, as 50–75 windows were processed each day and reinstalled following refurbishment.

The next target was insulation. The radiator in front of each of the 6,514 windows was temporarily removed to place an inexpensive layer of insulation between it and the

▲ Figure 16.32 **A Tall Order**
At 443.2 m (1,454 ft), the Empire State Building towers over much of Manhattan and today holds the title of the tallest LEED-certified building in the United States.

exterior wall, reflecting 24% of heat back into the building. This also helped to reduce the summer cooling load. With significantly reduced energy demand the team no longer needed to enlarge and replace the chillers but overhauled them instead to save over $17 million.

Tenants were also encouraged to increase the energy savings of the building. Skanska, the construction firm that occupies the 32nd floor of the ESB, earned LEED Platinum for its efforts. They structured their interior to incorporate daylighting in 90% of their office space and have seen a 57% reduction in energy use.

So did the retrofit accomplish its goals? YES! The building will save more than 38% in energy costs ($4.4 million) each year following the renovation. The owner is on track for a rapid three-year return on investment: All energy savings beyond this point will become true cost savings. And the retrofit design has proven to be easily transferrable to other commercial buildings. Already this model has been replicated in nearly 100 properties across the country, with significant energy savings and CO_2 emissions reductions projected.

16.6 Urban Transportation

BIG IDEA Over the past 50 years, suburban sprawl has been a major factor in the exponential growth in automobile use. In a vicious cycle, sprawl creates automobile dependency and road construction, which then encourage additional sprawl. Automobile dependency has high economic, environmental, and social costs. Public transportation, including bus and train systems, is more cost-effective and has many fewer environmental impacts than travel by automobile. Successful public transportation systems depend on strategies to encourage nonmotorized travel, such as walking and bicycling. Economic incentives for balanced transportation systems are greatly enhanced when the environmental, economic, and social costs of automobile dependency are clearly understood.

Automobile Dependency

■ Sprawl encourages automobile use, which in turn encourages sprawl.

Around the world, urban dwellers have become increasingly dependent on automobiles. Urban sprawl has been a driving factor in this trend. But this trend is also the result of the unwillingness of city and national governments to invest in alternative forms of transportation. Most cities have responded to the growing dependence on automobiles by expanding road and highway systems. Investment in roads and highways has, in turn, displaced other modes of transportation, including public transit, bicycling, and walking. Furthermore, the proliferation of roads facilitates additional sprawl. Travel times increase, and people become even more dependent on automobiles.

As might be expected, cities in developing countries have considerably lower automobile use than those in developed countries. In such cities, per capita consumption of gasoline in private autos is less than 100 L (26 gal) each year. In contrast, in most American cities per capita consumption of gasoline in private autos is 1,000–3,000 L (260–800 gal) per year. Yet a country's level of wealth is not the only factor determining the degree of automobile dependency in cities. Differences in suburbanization and the availability of transportation alternatives are also important.

Among large cities in the world's wealthiest countries, automobile use generally increases as per capita land consumption increases (**Figure 16.33**). However, there are significant differences among cities with similar per capita land consumption. For example, individuals in U.S. cities consume nearly twice as much gas in their cars as the citizens of Australian cities with similar levels of per capita land consumption. This is because Australians

have greater access to public transportation and drive smaller, more fuel-efficient cars.

Automobiles offer great mobility and flexibility but not without environmental, economic, and social costs. Increased dependence on automobiles diminishes the sustainability of urban ecosystems. Among the many negative consequences of automobile dependency are increased congestion and accident rates and diminished air quality. Automobile exhaust is the primary cause of photochemical smog and other air pollutants in cities. Increased automobile use also drives up greenhouse gas emissions. The spiraling relationship between suburban sprawl and automobile use contributes to the loss of forests and farmland. Expanding roadways increase the proportion of impermeable surfaces and the management challenges of stormwater runoff.

Studies show that the economic consequences of automobile dependency include high infrastructure costs and lower worker productivity due to increased commute times. Social impacts include diminished neighborhood and community interactions, diminished public safety, and increased road rage and other health problems.

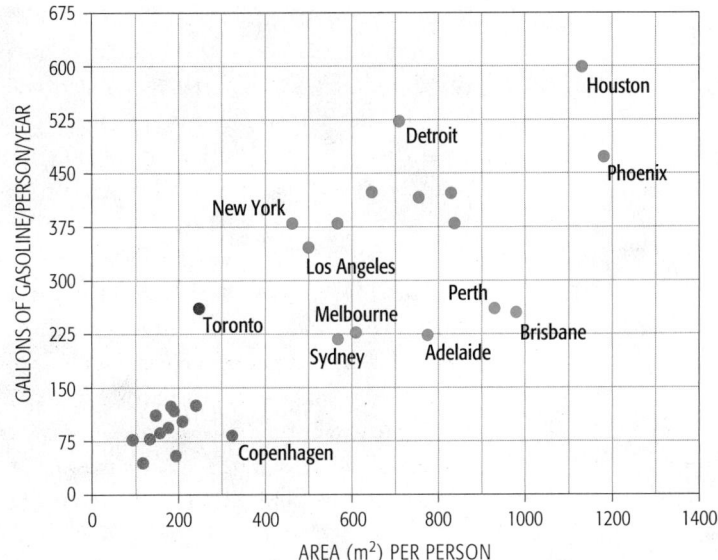

▶ **Figure 16.33 Suburban Automobile Dependency**
The amount of gasoline used by each person generally increases with per capita land consumption. However, at any level of per capita land consumption, gasoline consumption in U.S. cities (green dots) is higher than that in Australian cities (blue dots). This is a consequence of easier access to public transportation in Australia.

Data source: Newman, P.W.G., and J.R. Kenworthy, 1989. Gasoline consumption and cities. *Journal of the American Planning Association* 55: 24–37.

Balancing Transportation Options

■ Pedestrian-oriented cities encourage public transportation.

▲ Figure 16.34 **Light Rail**
With an annual ridership of 1.3 million, the 12.8-mile (20.6-km) MetroRail light-rail system in Houston, Texas, is one factor accounting for diminished automobile traffic congestion in its city center.

▼ Figure 16.35 **Bus Rapid Transit**
As with BRT systems in other cities, buses in Bogotá, Colombia, have their own lane and boarding stations, which are physically separated from the rest of the highway. Note the traffic backup to the left. Bogotá has set daily driving restrictions based on the last digits of automobile license plates to help limit the congestion.

In U.S. cities, fewer than 5% of trips are made by walking or biking. Only 12% of trips take advantage of public transportation. In European cities, 20% of trips rely on walking and biking, and 35% rely on public transit.

To diminish the vicious cycle of automobile dependency and sprawl, many U.S. cities are taking actions to provide a more balanced array of transportation options. Different modes of public transit vary in their energy use and operating cost. However, on a per passenger basis, they are all more efficient than automobiles.

Public buses have long been a transit option in many cities, although poorly maintained vehicles and limited routes have often kept ridership low. Where bus fleets have been modernized and routes configured to maximize efficiency, bus use has increased dramatically, with a corresponding decrease in automobile use. In large cities, for example, planners estimate that each bus replaces 250–300 cars each day.

Many larger urban centers depend on rail transportation systems. Extensive subway and aboveground rail systems, such as Washington D.C.'s Metro, New York's subways, Atlanta's MARTA, and San Francisco's BART, carry 25–35% of commuters in these cities. Light rail systems using smaller, electrically powered trains are being developed in many metropolitan areas (Figure 16.34).

Bus rapid transit (BRT) combines many of the advantages of bus and rail transportation. In BRT systems, buses move on dedicated roadways that allow them to bypass road traffic. BRT roadways feature sophisticated boarding stations, prepaid fares, and frequent service (Figure 16.35). BRT systems can be developed at much lower cost than light- or heavy-rail systems.

One of the biggest economic impacts of automobiles is the space they require for parking and roads. Although the actual number of parking spaces in the United States is not known, it is estimated that automobile-dependent cities end up providing three to five parking spaces for each car. Building codes typically mandate a certain number of parking spaces per building area in an attempt to meet the expectation of automobile drivers for abundant, free parking. Automobiles on a multilane freeway can transport about 2,500 people per hour. In contrast, a bus lane can transport 5,000–8,000 people, light rail can carry 10,000–20,000 people, and a heavy-rail subway can transport up to 50,000 people each hour.

Urban planners have found that for public transit to be successful, nonmotorized transportation, such as walking and bicycling, must be facilitated. Heavy traffic and multilane streets present major obstacles to pedestrian and cycle activity. This discourages movement within communities and also limits access to centers of public transportation.

Transit-oriented development emphasizes the role of streets for more than automobile traffic. Many cities have transformed former roadways into pedestrian malls. This has encouraged walking and cycling and revitalized downtown business districts (Figure 16.36). In an increasing number of American cities, dedicated walking trails and bike lanes encourage nonmotorized transportation.

To transform cities into pedestrian-oriented places, attention must be given to aesthetics and safety. Many cities have undertaken downtown beautification programs that include trees, gardens, and public art. Some cities require that a small portion of the budget of public buildings be dedicated to landscaping and public art. To improve public safety, cities can provide public call boxes, efficient evening lighting, and pedestrian police.

▶ Figure 16.36 **Pedestrian Friendly**
This pedestrian mall in downtown Boulder, Colorado, has become an attraction for residents and visitors and has moved automobile traffic out of the city center.

Urban Transportation Economics

■ Sustainable balanced transportation requires a full assessment of the costs of automobile dependency.

One of the reasons for our continued dependence on automobiles is that neither car owners nor city officials have to pay many of the costs associated with auto use. Low U.S. gas prices and free parking make automobiles appear to be the cheapest form of transportation. Drivers do not factor the cost of roadways into their travel expenses. When cities construct new roads and freeways, the costs are often subsidized by national governments. The costs of pollution on the health of humans and ecosystems are also externalized.

Development of balanced transportation systems can be greatly facilitated by strategies to internalize the real costs of auto dependency. Gasoline taxes and road tolls are mechanisms to do this. In its Central City Plan, Portland, Oregon, placed a cap on the number of downtown parking spaces and increased parking fees. In addition, several large parking lots were torn out and replaced by pedestrian parks.

Congestion pricing can be useful in limiting automobile traffic and encouraging the use of public transportation. Oslo, Norway, for example, raises about $75 million each year by charging drivers for the privilege of driving in the city. London has developed the world's largest market-based system to limit automobile use. For the privilege of driving in certain downtown areas between 7 A.M. and 6:00 P.M., private vehicles must pay £10 (~$16.50) each day. Closed-circuit television monitors record license plate numbers, and fines for nonpayment can exceed £130 (~$215) (Figure 16.37).

The success of green transportation systems ultimately depends on changes in planning goals. As former New Jersey governor Christine Todd Whitman once commented, "If you design communities for automobiles, you get more automobiles. If you design them for people, you get walkable, livable communities."

▼ Figure 16.37 **Congestion Market**
During peak traffic hours, motorists driving in London's Central Zone are charged a daily fee.

Where
YOU
LIVE   How do you get around on campus?

How does the number of bikes on your campus compare to the number of automobiles? Is public transportation available to you? List the pros and cons of public transportation. Do you feel the benefits are worth the cost? Explain.

QUESTIONS 16.6

1. Describe two factors that influence automobile dependency worldwide.

2. Compare the proportion of automobile use relative to public transportation in U.S. and European cities.

3. How do different modes of transportation in urban areas compare with regard to the number of people transported past a place per hour?

(MES) For additional review, go to **MasteringEnvironmentalScience**

16.7 Urban Biodiversity

BIG IDEA Urban ecosystems provide habitats for a diverse array of native and non-native plants and animals. Most of these species are generalists and are able to tolerate a wide range of environmental conditions and to disperse easily from one location to another. Food webs in urban ecosystems may be complex and are often subsidized by garbage and other inputs from humans. Because they moderate urban climates and provide habitat for other species, trees play a key role in cities. Urban gardens and green roofs provide food for humans, diminish stormwater runoff, and improve water and air quality. City parks, greenways, and waterways are especially important in the maintenance of urban biodiversity and ecosystem services. Such habitats are most effective when they are designed and managed as integrated networks.

▲ **Figure 16.38 An Urban Tree**
Tree of heaven produces tiny windblown seeds that can germinate and thrive in a variety of urban habitats.

Urban Wildlife

■ Urban species are widely dispersed and tolerant of a wide range of environments.

In Betty Smith's acclaimed 1943 novel, *A Tree Grows in Brooklyn*, a single tree—a specimen of tree of heaven (*Ailanthus altissima*)—is a central metaphor for a young city woman's coming of age. Tree of heaven is also representative of the vast majority of plant and animal species that thrive in urban environments (Figure 16.38). The tree is well known for its ability to tolerate a wide variety of environments, including sidewalk cracks, abandoned industrial sites, and disturbed roadsides. It is an invasive, non-native tree that produces large numbers of small, wind-dispersed seeds. It is also a source of food and habitat for squirrels, a variety of birds, and a host of insects that share its broad range of tolerance and ability to disperse.

Ecologists classify most of the species that thrive in urban centers as habitat generalists. These organisms are able to take advantage of habitats that have no counterpart in nature, such as mowed lawns, vacant fields, garbage dumps, and even cracks in sidewalks. Tree of heaven, dandelion, and other "weedy" species are habitat generalists, as are robins, rats, and pigeons. Certain falcons and hawks are interesting exceptions. Although these birds of prey are highly specialized, they are able to feed on the plentiful birds and rodents that

live in cities; sometimes these birds nest in skyscrapers, finding the canyon-like structure of many downtown areas similar to rocky cliffs in their native habitat (Figure 16.39).

The diversity of creatures in urban ecosystems decreases along the gradient from rural areas to the city center. In exurban and suburban communities, flora and fauna from surrounding farms and wild lands may be common and have a significant impact on ecosystems. Some of these impacts, such as the grazing of ornamental plants by deer or the raiding of garbage containers by raccoons and opossums, might be classified as benign nuisances. Others, such as the worry of attacks on humans and pets by mountain lions and bears or the spread of infectious diseases such as plague and rabies, are much more serious.

Moving from highly urbanized to suburban and rural settings, the number of plant and animal species generally increases (Figure 16.40). Even so, urban food

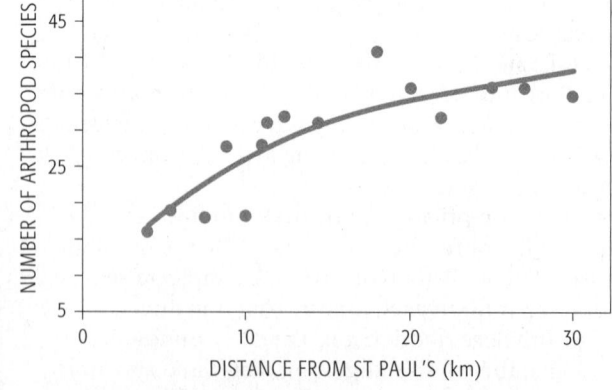

▲ **Figure 16.40 Urban–Rural Gradient Biodiversity**
In a study of 15 London gardens, the diversity of insects was found to increase with distance from the city center, in this case designated as St Paul's cathedral.

Source: Davis, B.N.K. 1979. The ground arthropods of London Gardens. *London Naturalist* 58: 15–24. London Naturalist is published by the London Natural History Society http://lnhs.org.uk/.

► **Figure 16.39 An Urban Predator**
Rodents and pigeons provide abundant food for urban peregrine falcons such as this one.

webs can be quite complex. Plants, including ornamental flowers and shrubs, street trees, and weeds, provide some of the energy for such food webs. Human inputs, in the form of pet food, birdseed, and all sorts of garbage, are often a major source of energy for urban food webs.

The mowed lawn is a distinctive urban habitat that is useful for recreation and urban aesthetics. Because lawns are permeable surfaces, they facilitate water infiltration. However, in their fervor to have the greenest lawn in the neighborhood, homeowners often use large quantities of fertilizer and pesticides that contribute to stormwater pollution. Furthermore, lawn maintenance can require large quantities of water in dry climates. Xeriscaping, or landscaping with drought tolerant plants, provides an attractive solution to the multifaceted problem of the green lawn in arid climates (Figure 16.41).

▲ Figure 16.41 **Xeriscaping**
Many communities in arid regions are encouraging homeowners to replace water-demanding lawns with drought-adapted landscaping. Xeriscaping is standard practice in this residential southern Arizona community.

Green Infrastructure

■ Trees, gardens, and green roofs provide a great many services in urban ecosystems.

In 1792, the citizens of Philadelphia took the revolutionary step of petitioning their city government to plant trees along city streets. Prior to that time, planting trees near buildings was prohibited for fear that they would increase the risk of catastrophic fires. The Philadelphians argued that the risk of fire was more than offset by the "healthful effects" of trees—summer shade and beauty. Street trees were not widely used in either European or American cities until the 1800s.

Today, the benefits of urban trees are widely recognized. In summer, deciduous trees provide shade that mitigates urban heat island effects. In winter, they allow solar warming. Evergreen trees are used as windbreaks in many cities. Both deciduous and evergreen trees provide nesting sites and habitat for a diverse array of urban birds. Perhaps most important, trees are greatly valued by city dwellers for their beauty (Figure 16.42).

Most cities have programs for planting and managing trees that are overseen by professionals trained in urban

◀ Figure 16.42 **An Urban Forest**
Trees such as these along Barcelona's famous La Ramba beautify cities; they also provide shade and diminish the urban heat island effect.

URBAN BIODIVERSITY   MODULE 16.7

541

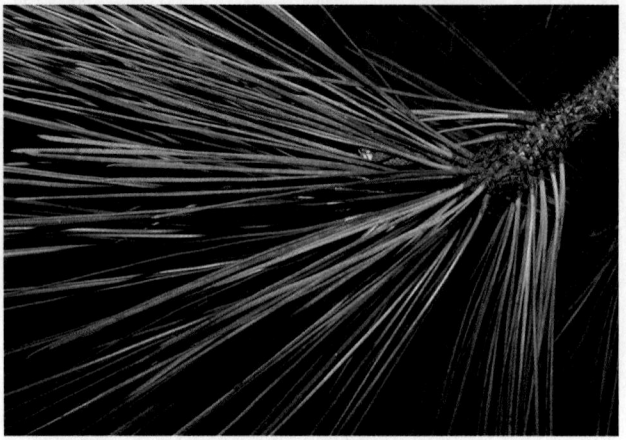

▲ Figure 16.43 **Air Pollution and Urban Trees**
Ozone pollution has damaged the green needles of this urban white pine, causing the tips to turn brown.

▲ Figure 16.44 **A Community Resource**
Battery Park City's Liberty Garden in New York City is planted next to West Street, an eight-lane highway, and is used by over 60 gardeners.

forestry. Urban foresters must understand which tree species are most likely to thrive at a particular location within a city. Some tree species require open, sunny environments, and others can thrive in shady urban canyons.

Urban environments present special challenges for tree growth. Urban soils are often compacted and poorly drained and are likely to be surrounded by impermeable surfaces. Without special soil preparation and drainage, tree roots may experience extremes of wet and dry conditions. Pollution and disease also affect the growth of urban trees. For example, moderate to high levels of ozone and photochemical smog can suppress the growth of many trees (Figure 16.43). Because urban trees are often grown in single species groups, they are particularly vulnerable to insect and fungal pathogens.

Urban gardens have become an increasingly prominent feature in many urban landscapes. In many inner cities, trash-strewn vacant lots have been transformed into productive vegetable gardens. Urban flower gardens can beautify abandoned industrial sites,

and urban vegetable gardens can provide fresh, healthful food for local neighborhoods. Such gardens have been the focus for recreation and educational activities in many communities (Figure 16.44).

Green walls represent an interesting variation on the urban garden theme. A nonprofit group, the Urban Farming Food Chain, uses recycled materials to manufacture modular wall panels that provide a growth medium and automatic irrigation for vegetables. "Edible walls" can provide fresh produce to communities that lack open land.

In the 1960s, German architects began to experiment with green roofs. These are roofs constructed of multiple layers to support the growth of plants while protecting the interior of the building from the elements. Green roofs greatly diminish stormwater runoff. Properly designed, they can retain up to 75% of rainwater, gradually releasing it back to the atmosphere through evaporation. Green roofs also decrease the urban heat island effect. See if you can identify any green roofs in the New York City landscape shown at the beginning of this chapter.

City Parks, Greenways, and Waterways

■ In integrated networks of urban green spaces, parks, and preserves are connected by greenways and waterways.

The 1859 dedication of Central Park in New York City was a landmark event that ultimately affected the planning of hundreds of other cities. Central Park was designed by Frederick Law Olmstead, a landscape architect and pioneer in city planning who was committed to blending the natural environment of a place with the aesthetic qualities of the city. To give city dwellers access to nature, Olmstead included wooded landscapes, open "pastures," and ponds in the design of Central Park. His design also integrated transportation corridors, public spaces, and formal

gardens (Figure 16.45). Olmstead was well aware of other important benefits of city parks, including water management and escape from city heat.

Although Olmstead is best known for his design of city parks, he also played a central role in the development of the profession of city planning. He recognized the importance of streams and wetlands in managing stormwater flows and maintaining water quality. In the late 19th century, he planned the Back Bay Fens project for the city of Boston. This project converted a sewage-filled swamp into a network of artificial streams and

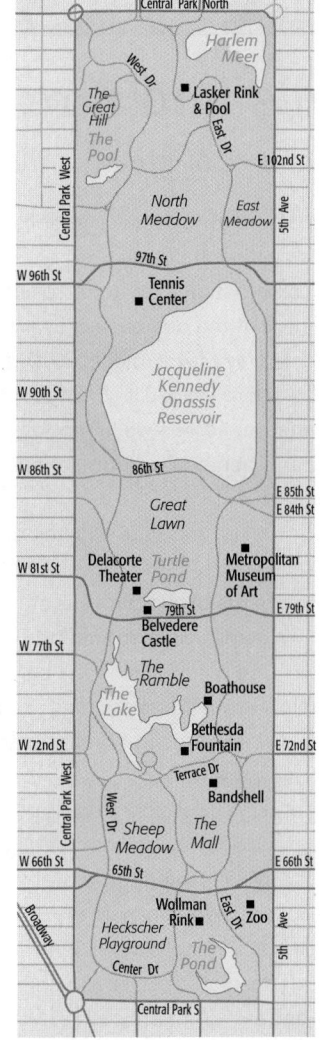

▲ Figure 16.45 **Central Park, New York**
Frederick Law Olmstead's design for New York City's signature park set a standard for the development of city parks around the world.

QUESTIONS 16.7

1. Ecologists describe most urban plant and animal species as "generalists." What do they mean, and why do such species thrive in urban ecosystems?

2. Describe three ways in which urban trees enhance urban ecosystems.

(MES) For additional review, go to **MasteringEnvironmentalScience**

restored marshes that were able to hold floodwater. The project also reduced the flow of pollution into the Charles River. The Fens was carefully planted so that it would serve as a recreational park.

It is ironic that many of the streams and rivers that once nurtured the growth of cities have since been nearly destroyed by the urban and industrial development associated with those cities. In many places, urban streams have been channelized, and riparian vegetation has been cut back or eliminated entirely. In recent years, a number of cities have undertaken major projects to restore streams and rivers. These projects aim to reduce pollution and flooding and are also central components of comprehensive urban renewal (Figure 16.46).

Biodiversity, ecosystem services, and community benefits can be greatly enhanced when urban green spaces are viewed as hubs and links in an integrated network. Hubs include parks and preserves that are large enough to harbor sustainable populations of flora and fauna and to sustain key ecosystem processes. Hubs are also destinations for human recreation and educational activities, such as gardens and museums. Greenways and stream corridors are the links that provide pathways for wildlife and people to travel between the hubs. Greenways also provide people with opportunities for recreational activities, such as biking and hiking. Historic sites might also be included in such a network.

◄ Figure 16.46 **A Restored Urban River**
In the early 1920s, the City of San Antonio, Texas, considered paving over its namesake river. Instead, it invested resources to restore the San Antonio River and develop the River Walk that is now lined with shops and restaurants. Flows are maintained in this dry city using treated municipal wastewater.

Green Walls and Bird Abundance

Can the establishment of green walls increase bird abundance in urban systems?

1. As an urban planner, what recommendations might you make to increase the size of bird populations in an urban area, based on this research?

2. What experimental studies might you propose to follow up on the results of this research?

Great Britain and other parts of Europe have recently experienced a significant decline in populations of many common birds. A combination of factors are likely responsible for the decline, but the decrease in available forage in urban ecosystems may play a role. In 2010, a research team from Staffordshire University in England decided to investigate whether "green walls" could compensate for this loss and benefit bird populations in urban ecosystems (**Figure 16.47**).

Caroline Chiquet, John Dover, and Paul Mitchell of Staffordshire University set out to test the hypotheses that (1) walls with substantial amounts of climbing vegetation (green walls) and their 10-m semicircle surroundings would attract more birds and be more species rich than bare walls and their comparable surroundings and that (2) green walls with evergreen plants would attract more birds than deciduous green walls during the winter.

To test these hypotheses, 27 outdoor green walls of varying size were selected. To be included in the study, a green wall had to have a minimum of 3 m² of uninterrupted vegetation, with the subject walls being located a minimum of 200 m apart. Twenty-seven similar but bare walls were identified as controls throughout the same region. Species and abundance of birds visiting each wall and the 10-m semicircle surrounding it were recorded in the summer of 2010 and the winter of 2011. Birds visiting the walls and surrounding areas were observed, identified, and counted during 20-minute periods, both in the morning and in the evening.

So did the presence of green walls affect bird abundance in an area? The data showed that birds were six times more prevalent on green walls than bare walls in the morning hours averaged across seasons (**Figure 16.48**). Within-season analysis showed that more birds were associated with walls with evergreen foliage in the winter than with either bare or deciduous foliage, though birds did not differ in abundance among wall types during the summer. These research findings show a *correlation* between bird abundance and green walls rather than *causation*, so we cannot conclude that the green walls are the reason for greater bird abundance. It is likely, however, that bird abundance is associated with higher resource availability on green walls. Food, nesting sites, and cover were associated with green walls and, therefore, they were of greater utility to the birds.

This research is *observational* rather than *experimental*, because the system was not manipulated

▲ Figure 16.47 **Strictly for the Birds**
Caroline Chiquet samples the insect population on one of the green walls in Stoke-on-Trent that was part of her bird study.

in any way. In an observational study it is critically important to have significant replication of the units of study (in this case, green walls and controls) given the preexisting differences between them. This simple observational study suggests that installing new green walls within the urban setting may result in increased bird abundance and, therefore, help to mitigate the negative effects of the loss of tree cover on bird density in urban ecosystems.

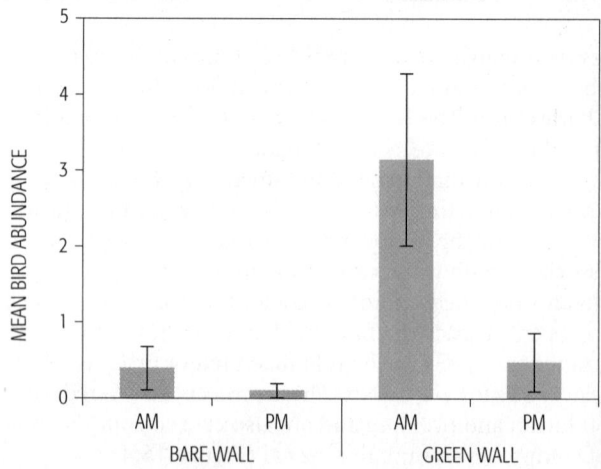

▲ Figure 16.48 **The Value of Green Walls**
Data show that bird abundance was more than six times higher on or near green walls than on bare walls during the morning. Columns indicate means with ± standard error bars, which show variation among measurements.

16.8 The City as a Sustainability Strategy

BIG IDEA Over the next 30 years, an additional 2 billion people will be added to our population. Most of that growth will occur in cities. Cities concentrate human impacts in ways that are not sustainable. Cities rely on inputs of energy, natural resources, and water, and the per capita demands for these inputs are increasing. Cities are the source of nearly 80% of human greenhouse gas emissions and 90% of Earth's garbage. They are also home to a disproportionate share of human poverty and misery. However, by mimicking important features of natural ecosystems, cities can play a central role in the development of a sustainable future. These features include the efficient capture and use of energy and matter, biodiversity, self-regulation, adaptation, resiliency, and beauty.

Efficient Capture and Use of Energy and Matter

■ A city's metabolism should be circular rather than linear.

Writer and urban philosopher Herbert Girardet notes that unlike natural ecosystems, which have a circular metabolism, most urban ecosystems have a linear metabolism. By this, Girardet means that cities require very large inputs of energy and materials; cities then use or recycle these resources inefficiently, resulting in large outputs of waste (Figure 16.49).

The circular metabolism of a forest ecosystem is illustrated by the cycling of nitrogen (see Module 12.2). Nitrogen enters the forest in a raindrop, is absorbed by tree roots and is incorporated into the protein structures of a leaf. In the fall, the leaf drops to the ground and decays, releasing the nitrogen into the soil to be taken back up again by a plant in the spring. This circular process can go on indefinitely and the residence time of nitrogen in this system is a century or more. The annual influx of nitrogen is meager in comparison to the amount of nitrogen needed to sustain the forest's annual productivity. Most of the nitrogen the forest comes from recycling.

In contrast, the residence times for most materials coming into urban ecosystems are measured in weeks or months. Natural resources are often manufactured into finished goods that are used only once before they are discarded. Thus, the annual per capita demand for resources is particularly high among urbanites. This rate of resource use is environmentally unsustainable. This inefficiency and waste also inflicts economic and social costs.

Stockholm, Sweden, is one of several cities that is "circularizing its metabolism" by engaging in zero waste planning. In its model ecological district, Hammarby Sjöstad,

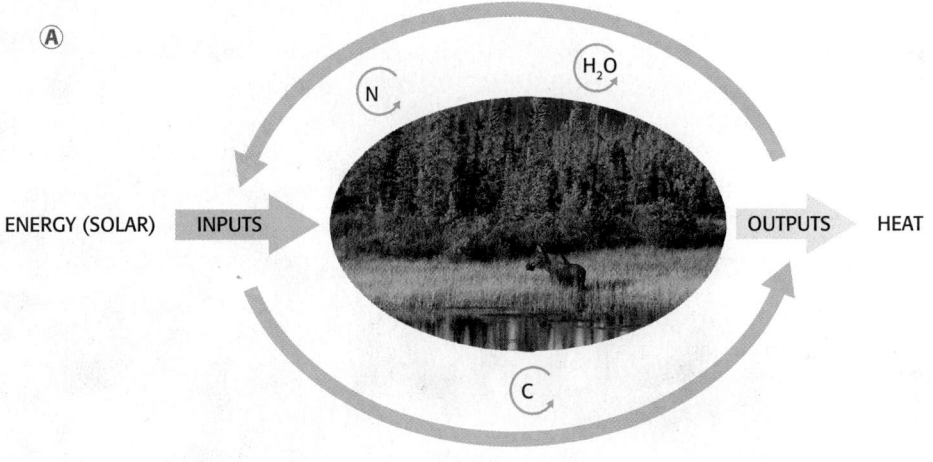

Ⓐ

ENERGY (SOLAR) → INPUTS → OUTPUTS → HEAT

CIRCULAR METABOLISM

Ⓑ

FOOD

ENERGY (COAL, OIL, NUCLEAR) → INPUTS → OUTPUTS

GOODS

ORGANIC WASTES

EMISSIONS

INORGANIC WASTES

LINEAR METABOLISM

▲ **Figure 16.49 Circular vs. Linear Metabolisms**
Compared to natural ecosystems, the flow of materials and energy in urban ecosystems is linear. Ⓐ In a natural ecosystem, most material outputs are recycled back into the system with very little need for new external inputs. Renewable energy is captured and used efficiently. Ⓑ In an urban ecosystem, most material outputs flow out of the system causing a demand for continuous external inputs. Intensive energy demands are typically met with large inputs of nonrenewable energy sources.

► Figure 16.50 **Circularizing a City's Metabolism**
The waterfront at Hammarby Sjöstad resembles that of other European cities. But all waste is collected in special bins Ⓐ and subsequently sorted Ⓑ to maximize recycling. Along streets, stormwater is diverted through gardens to moderate flows and remove pollutants Ⓒ. Methane captured from wastewater is used as a biofuel to power city buses Ⓓ. All these practices mimic ecosystem functions for the benefit of the city and efficiently convert energy to power the system.

residents put their solid wastes into an underground collection system that separates recyclable materials from organic and other wastes. The recyclables are sold to factories, which use them to manufacture new products. Combustible wastes are burned, and their heat is used to generate electricity and hot water. The hot water is used for residential heating. Stormwater from streets is collected in a special purification system, and green roofs purify water flowing from buildings. Wastewater, including human waste, is kept separate from stormwater and treated so that methane can be captured and used as a source of energy. All of Hammarby Sjöstad's buses are powered by this biogas (Figure 16.50).

Ecosystem Characteristics

■ Biodiversity, self-regulation and adaptation, resiliency, and beauty characterize sustainable cities.

Diversity. Ecosystem functions and services, such as the cycling of nutrients and the purification of water, depend on a diverse community of microbes, plants, and animals (see Module 1.3). Yet historically, urban planning has given little attention to biodiversity. The resulting lack of biodiversity in urban centers has economic costs, since humans must pay to restore lost ecosystem services. It also has social costs. In his nonfiction book *Last Child in the Woods*, Richard Louv argues convincingly that separation from nature harms the physical and emotional health of both children and adults.

Parks, greenways, and streamside buffers can help restore biological diversity to urban ecosystems. They diminish urban heat island effects, mitigate flooding, and purify stormwater. They also provide opportunities for recreation and education.

In addition to its restoration of ecosystem function, biodiversity can be an indicator of economic value. A 2013 study found that with an increase of one less common bird species in Lubbock, Texas, the sale price of an urban home gained $32,000! (Farmer et al. 2013).

Many urban planners envision even greater integration of natural ecosystems in urban settings. Urban Biosphere Reserves that have been proposed for several European cities have the goal of connecting rural landscapes with networks of urban green spaces. The Green–Blue Meander in the heavily populated Ranstad region of the Netherlands is one such project. Here, waterways connect fragmented areas of forest and pastureland to urban greenways and parks. Legacies of past industrialization, such as abandoned lands, are being restored. Planners are optimistic that this development will enhance the quality of life in urban centers and serve as a boundary for sprawl.

Self-regulation and adaptation.
Natural ecosystems have a remarkable capacity for self-regulation and adaptation. For example, forests moderate weather extremes and ensure regular flows in the streams that drain them. In urban ecosystems, responses to environmental change and patterns of growth are determined largely by human interactions, which lack many of the homeostatic mechanisms of natural ecosystems. For example, many of the everyday costs of sprawl are not included in the economics and commerce associated with development. Therefore, incentives for sustainable urban planning and development are often overlooked. Political boundaries, such as city limits, often result in processes of decision-making that are fragmented across urban landscapes.

Sustainable self-regulation of urban ecosystems requires a connection between planning and process. The scales over which decisions are made must be in agreement with the scales over which processes occur. The flows of water and waste materials, for example, often extend across city and county boundaries. In response, cities and counties in many urbanized areas have organized metropolitan or regional councils that have authority to coordinate planning and regulate the actions of their members.

When environmental costs are included, urban planning and development decisions are far more likely to be sustainable. For example, congestion pricing is a very successful strategy for decreasing automobile use and encouraging use of public transit (see Module 16.5).

Whereas natural systems have built-in feedback loops that lead to adaptation, urban ecosystems require intentional knowledge-seeking and learning to develop adaptive strategies. Cities and agencies within urban ecosystems must systematically monitor key processes and measures of sustainability. They must also foster the means to learn from this information and adapt. Citizens, too, must be well informed. Thus, educational programs such as those implemented by Malmö as described in the opening essay are critical elements in sustainable urban development.

Resiliency.
Through the process of succession, natural ecosystems have a remarkable capacity for renewal and restoration (see Module 6.5). Disturbances such as fires and hurricanes have occurred in Earth's ecosystems for millions of years. Species in natural ecosystems possess myriad adaptations that make individual populations, communities, and ecosystem processes resistant to disturbance and allow ecosystems to recover rapidly when disturbances occur. The inevitability of disturbance and the capacity for recovery are, quite literally, embedded in the DNA of the organisms living within the ecosystem.

Disturbances are inevitable in urban ecosystems, too, a fact that must be acknowledged in urban planning. Many communities have developed plans for responding to the effects of earthquakes, fires, and floods. Some coastal communities in particular have even begun to plan for climate change. Yet as demonstrated by the recent hurricanes along the Gulf Coast (Katrina, 2005) and Northeast Coast (Sandy, 2012) and the massive earthquake and tsunami of 2011 in Japan, such planning is often inadequate. Cities often fail to plan for economic and social disturbances, such as the collapse of major industries or major shifts in the demand for residential housing.

Diversity—the variety of species and complexity of processes—has been found to be an important factor contributing to the ability of ecosystems to resist and recover from disturbance. Urban diversity—variety in the economy, the culture, and political viewpoints—has a similar effect on the resilience of cities.

Beauty.
We are drawn to nature by its beauty. Indeed, that beauty is often a central motive for our preservation of natural places. Beautiful urban environments also motivate preservation and sustainable behavior.

For example, the recent transformation of an abandoned railway line into a park on the west side of Manhattan has greatly increased the beauty of that community. The new High Line Park is a garden walkway that offers a view of the Hudson River and a retreat from noisy city streets. Foot traffic is heavy, and residents and visitors often compare the new High Line Park to Central Park.

In an essay called "The Land Ethic," Aldo Leopold wrote, "A thing is right when it tends to preserve the integrity, stability, and beauty of the biotic community." Cities are human-centered ecosystems. They exist to support the complex environmental, economic, and social needs of human communities. Nevertheless, if we were to craft an urban ethic, its central creed might well be that *a city is sustainable if it tends to imitate the integrity, stability, and beauty of a natural community.*

Where YOU LIVE How does your campus compare to an urban ecosystem?

Calculate the weekday population density by estimating the number of students, faculty, and staff and dividing by the campus area. Does it qualify as an urban area? What are the inflows and outflows to the system? Does your campus employ any methods to reduce the heat island effect (i.e., white or green roofs, others)? Investigate how your campus handles stormwater runoff around large parking lots and structures. Look for bioswales and pervious pavers.

QUESTIONS 16.8

1. What is meant by the assertion that a city's "metabolism" should be circular rather than linear?

2. Describe three ways in which the governance of urban ecosystems can become more sustainable.

MES For additional review, go to **MasteringEnvironmentalScience**

547

Stockton College's Greening Urban Spaces

Jessica Franzini is a graduate of Stockton College in Galloway, New Jersey. While an undergraduate, she led New Jersey Water Watch's first vacant lot transformation project, working with municipal officials and volunteers to transform a vacant lot in Galloway City, New Jersey, into a green community space. After that project, Jessica went on to serve as codirector of Stockton College's Stockton Save. Since graduating, she has served as program director of the New Jersey Tree Foundation's Urban Airshed Reforestation Project (UARP). In that role, she has planted hundreds of trees in Camden, New Jersey, while uniting communities, improving the environment, and teaching community members about urban forestry.

How did you first get the idea to green an urban space?

Early in college, I took a class called Global Issues. When we learned how small the U.S. population is in relation to how much we contribute to environmental degradation, I realized that it's not just the size of a country's population that's important. What really matters are the lifestyle choices that the people in a country make—individually and collectively. In that course we learned about sustainable cities and areas in Europe where public transit is used more frequently, people bike instead of drive, and food is grown locally. I realized that was the kind of place I wanted to live in, and that there was a lot I could do to make a difference, especially in an urban space.

What steps did you take to get started?

I began by getting involved with a student club on campus. Through that club, I volunteered to coordinate a project. I decided to look in Galloway, New Jersey, near Stockton College, where I found some abandoned land covered in trash. My goal was to transform that desolate area into a green community space. I started meeting with city officials and other municipal workers, who went out to show me which areas were owned by the city. The municipality was able to help with all the things that seemed so overwhelming to me as a student.

Having to organize volunteers really intimidated me. But as I started attending weekly meetings, I saw that everyone loved my idea, so I became more confident. We got a lot of support by setting up tables at one of the busiest areas on campus. We then rolled out a large canvas and provided paint for people to illustrate something they wanted to change about the environment. Everyone who came over to the table got information on all the projects we were doing, including my vacant lot project in Galloway. Surprisingly, people really responded to this!

I then organized volunteers from Stockton and Galloway to recycle and dispose of the trash, build park benches, and plant a dogwood tree and the other plants that had been donated to us. If you have a positive attitude and invite others to share part of themselves right away, you will find people who want to support you.

What are the biggest challenges you've faced with the project, and how did you solve them?

I have to admit, I didn't know what I was doing when I started this project. My project was the first of its kind, and I didn't know how to take care of trees, and neither did anybody else. We weren't quite prepared.

We got lots of volunteers, beautiful plants, and had an awesome event. However, we soon realized you can't just plant a tree and expect it to grow. You have to take care of it. We hadn't thought of a maintenance plan at all. Because of this experience, we learned and were able to do further greening projects more effectively.

How has your work affected other people and the environment?

With the New Jersey Tree Foundation, I've worked with community members to plant hundreds of trees in some of the most dangerous communities in the country. I've seen the trees and the people come alive. Being able to help people who want to improve their community and take pride in it makes me feel like I make a difference every day.

What advice would you give to students who want to replicate your model on their campus?

First, see what's already out there that could be a good resource. Connect to a group that is already active, or create your own group with specific goals in mind.

When you know what materials or support you will need, you can seek donations or small grants to make the work

happen! I recommend reaching out to local businesses, organizations, and corporations as partners. And remember, volunteers go a long way! A small group of people can work together to motivate more students as volunteers to move projects forward in a fun and exciting way.

Regular meetings are helpful. Decide a project point-person, and develop a committee with those who want to help. Part of being a good leader is learning to delegate and share tasks with others. Students are empowered when they can truly contribute to the planning and implementation of a worthwhile project.

Most importantly, continue to build experience. Without my college experience in organizing the community to transform a vacant lot, I would never be able to do the work I do now. Learning about plants in class just isn't enough. You have to get out in the world and do something!

Summary

Over the past 10,000 years, the development of industry and technology has encouraged the growth of human populations in urban centers. The growth of cities has, in turn, stimulated the development of new industries and technologies. Urban ecosystems support the commerce, industry, and community interactions of a large number and high density of humans. By 2050, 70% of the world's population is expected to live in cities. Cities have precise political boundaries, but the climate, hydrology, and flow of materials in metropolitan areas change along a gradient from densely populated city centers to less densely populated suburbs, exurbs, and rural areas.

The built environment uses many resources; therefore buildings represent a great opportunity for sustainability improvements. Green building design focuses on site, building orientation, and choice of building materials to reduce the ecological footprint of buildings. Rating systems such as LEED serve as a method of evaluating success and allow for comparisons among buildings.

Over the past 50 years, government policies, economic growth, dependence on automobiles, and poor planning have produced urban sprawl in many metropolitan areas. Increased urban land consumption has resulted in the loss of ecosystem services and diminished human well-being. Urban planners can use zoning, urban growth boundaries, and the development of parks and greenways to reduce urban sprawl and many of its negative impacts. Sustainable urban planning must provide alternatives to the use of automobiles, such as pedestrian corridors, subways, light rail, and bus rapid transit. Economic incentives such as higher parking fees and congestion pricing can diminish the use of automobiles in city centers.

Urban ecosystems provide habitats for a diverse array of organisms; however, urban food webs are generally simple. Cities that support biodiversity and use energy and matter more efficiently are more resilient, adaptable, and beautiful. Such cities will be essential to the well-being of future human populations.

16.1 Urbanization

- The first cities with formal governance and division of human labor appeared in association with the domestication of plants and animals and the development of agriculture.

- Urbanization is an especially prominent trend in the world's poorest countries; in these countries the well-being of people living in cities is less than that of people in rural areas.

KEY TERMS

urbanization, megacities

QUESTIONS

1. The people who lived in ancient cities were often less healthy than people who lived in the rural countryside. What factors contributed to this so-called urban health penalty?

2. In what ways has urbanization required the restructuring or reorganization of many of Earth's ecosystems?

16.2 Urban Ecosystems

- Urban ecosystems are dependent on interactions between organisms and their nonliving environment and on the flows of matter and energy; they are also sustained by the flows of ideas and money and by the movement of people.

KEY TERMS

urban ecosystem, urban, metropolitan area, suburb, exurb, urban heat island, impervious surface cover (ISC)

QUESTIONS

1. In the United States, rainfall is often higher to the east of large metropolitan areas. What causes this pattern of rainfall?

2. Why must urban ecosystems be especially open to flows of energy and materials?

3. How would you expect the water quality of urban streams to differ from that of streams in nearby forested areas?

16.3 Urban Land Use

- Many U.S. metropolitan areas have experienced urban sprawl; human populations have grown most in suburban and exurban areas, resulting in diminished population density in urban centers and increased per capita land consumption.

- Government policies, economic growth, highway development, and poor planning have encouraged urban sprawl.

- Urban sprawl has diminished biodiversity and a variety of ecosystem services that influence human well-being.

KEY TERMS

urban sprawl, per capita land consumption, urban decay

QUESTIONS

1. Urban sprawl has produced very different trends in urban centers than in suburbs. Explain.

2. Many experts feel that urban sprawl has been encouraged by a variety of economic externalities. Explain the basis for this viewpoint.

3. Describe the impacts of urban sprawl on three ecosystem services that influence human well-being.

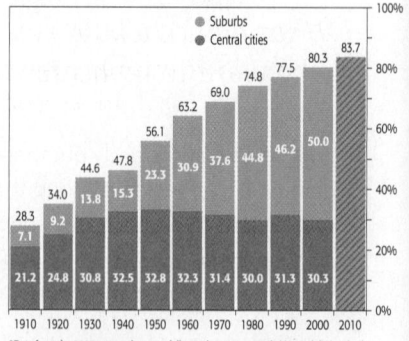

PERCENT OF U.S. POPULATION LIVING IN METROPOLITAN AREAS: 1910 – 2010*

Data from the 2010 census does not delineate between central cities and their suburbs.

16.4 Urban Planning

- Urban planning involves strategies to manage development in order to maintain key ecosystem services and human well-being while facilitating economic growth.
- The smart growth, or new urbanism movement, has 10 main strategies for strengthening human communities and sustaining ecosystem services.

KEY TERMS

urban planning, zoning, urban growth boundary (UGB), new urbanism

QUESTIONS

1. How would you describe the organization of your hometown or the town nearest to your home? What factors might have contributed to that organization?
2. Some criticize urban zoning as an infringement on the rights of private landowners. What public benefits does zoning provide that outweigh such criticism?
3. In what ways do urban planners in your hometown engage the public?

16.5 The Built Environment: Sustainable Building

- Choice of site, building orientation, and selection of building materials can significantly improve building performance and reduce impacts on the environment.

KEY TERM

embodied energy

QUESTIONS

1. What are the advantages of infill housing development relative to new development in greenfield areas? What are the disadvantages?
2. Find a description of a LEED building in your state online. What features of the building make it more sustainable than the average building?
3. If you were to design a sustainable home for an urban site in the southern hemisphere, how would it differ from one designed for the northern hemisphere? Explain. Refer to Module 3.6 for seasonal sun angles.

16.6 Urban Transportation

- Many urban areas are turning to public transportation to diminish automobile dependency, revitalize central cities, and slow sprawl.

QUESTIONS

1. What factors contribute to the higher dependence on automobiles in most U.S. cities compared to most European cities?
2. Why is the use of public transportation often dependent on facilitation of nonmotorized transportation, such as walking or bicycling?
3. What role do externalized costs play in urban automobile dependency?

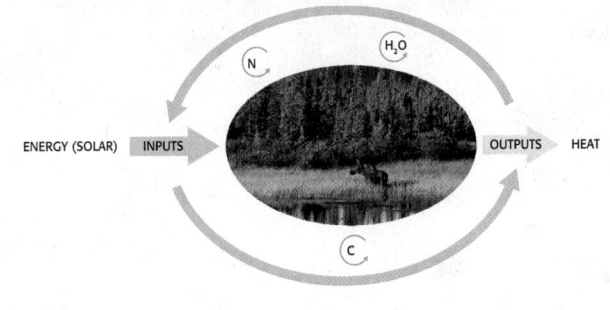

16.7 Urban Biodiversity

- Biodiversity generally increases along the gradient from urban centers through suburbs to rural landscapes.
- Green infrastructure—trees, gardens, and green roofs—provides important ecosystem services, including improved water quality and flows, climate moderation, and aesthetics.

QUESTIONS

1. Describe three strategies to increase the biological diversity of cities.
2. In what ways do green roofs improve urban environments?
3. Describe the benefits that are provided by a park or greenway in your hometown.

16.8 The City as a Sustainability Strategy

- Urbanization has created many challenges to ecosystem and human well-being, but thoughtfully planned cities will be essential to accommodate future human populations in a sustainable fashion.

QUESTIONS

1. Describe three strategies to increase the residence time for materials coming into urban ecosystems.
2. Why must beauty be an important part of planning for urban sustainability?

CIRCULAR METABOLISM

MasteringEnvironmentalScience®

Students Go to **MasteringEnvironmentalScience** for assignments, the eText, and the Study Area with practice tests, activities, and more.

Instructors Go to **MasteringEnvironmentalScience** for automatically graded tutorials and questions that you can assign to your students, plus Instructor Resources.

Waste Management

17

553

What a Waste!

What are the environmental costs of throwing a soda can in a landfill?

How many aluminum cans did you open today? Have you ever wondered where they came from or what was their fate once you disposed of them?

Aluminum cans were first used by soda manufacturers and brewers in 1965. Compared to tin and steel cans, these new cans had a number of advantages: Aluminum resists corrosion and has an extended shelf life. The sides and tops of aluminum cans are made of a single piece of metal, which simplifies their manufacture and packaging. Their structure allows businesses to imprint labels, logos, and other information on designs that surround the entire can. No wonder the aluminum can has become one of the most ubiquitous and durable products of our age!

Despite their advantages, aluminum cans have also become a prominent emblem of one of our most widespread environmental challenges: the disposal of waste. Each year, the typical American uses and disposes of an average of 340 aluminum cans. This is equal to about 5 kg (11 lb) of aluminum per person. In 2014, that added up to 101 billion cans, or 1.54 million metric tons of aluminum.

The production of aluminum cans has significant impacts on Earth's environment, as can be seen by analyzing their life cycle (Figure 17.1). The first step in making a can involves the mining of aluminum-rich bauxite ore. Bauxite forms from highly weathered soils in moist tropical areas. Because it is found near the surface, bauxite is generally strip-mined in a process that permanently removes forests from large tracts of land: for every ton of bauxite, 5–20 tons of earth must be moved. In addition, bauxite mining often results in water pollution and acid mine drainage.

In the second step, bauxite ore is treated with steam and caustic soda to yield aluminum oxide, or alumina. This refining process requires thousands of gallons of water for each ton of bauxite. It also produces considerable waste in the form of a toxic slurry called "red mud." One metric ton (2,200 lb) of bauxite yields about 500 kg (1,100 lb) of alumina.

▼ Figure 17.1 **Life Cycle of an Aluminum Can**
Aluminum is concentrated in tropical soils as bauxite. Bauxite mining requires the removal of forests and causes a range of other environmental impacts Ⓐ. Bauxite is then refined to alumina in facilities such as this one in Queensland, Australia Ⓑ.

In the third step, alumina is smelted to remove metallic aluminum, which is then formed into bars, or ingots. The smelting process requires very strong electrical currents and uses large quantities of electrical energy—by itself, aluminum production uses about 1% of the electrical energy generated in the United States. For this reason, smelting facilities are usually located close to large power plants.

In the fourth step, ingots of aluminum are heated and rolled into aluminum sheets. These sheets are then formed and cut to make cans. Once the cans are complete, they are labeled, filled with beverages, shipped to stores, and sold to the public.

In the final step of an aluminum can's life, it is either recycled or discarded. Although aluminum cans are 100% recyclable, every year more than 640,000 metric tons of these cans end up in landfills, incinerators, and along roadsides. The high volume of these discarded cans obviously increases the demand for waste management facilities and the cost of maintaining those facilities.

Each step in the production of aluminum cans uses energy and produces waste. Yet when aluminum cans are produced from recycled material, steps one through three—mining, refining, and smelting—are skipped. If all aluminum cans were recycled, bauxite mining and its associated environmental impacts would be greatly reduced. Skipping the refining and smelting stages would save 10 billion kilowatt hours of electricity each year, an amount equivalent to the electricity used by 600,000 U.S. homes. Recycling would also avoid the emission of between 25 and 50 million metric tons of greenhouse gases, or 0.3–0.6% of the total human-caused emissions per year. However, it must be noted that the process of recycling aluminum does require energy and has some environmental impacts.

If throwing away something as simple as a soda can creates these environmental impacts, imagine the effects of discarding complex consumer items, such as automobiles and kitchen appliances. In addition to generating solid waste, Americans generate large amounts of hazardous waste, electronic waste, and industrial waste. Each of these forms of waste requires a different approach to management. Furthermore, some wastes contain toxic chemicals that are a potential threat to human health.

- *What kinds of waste do communities generate and in what amounts?*
- *How are various forms of waste managed?*
- *How can we reduce the environmental impacts of waste production and management?*
- *What strategies can we employ to reduce the volume of waste we produce?*
- *By focusing on the entire life cycle of products, can we reduce waste amounts and their environmental impacts?*
- *What policies and laws are in place or are needed to ensure sustainable waste management?*

▼ **Figure 17.1 Life Cycle** (*cont.*)

In smelting "pots," the oxygen in alumina reacts with carbon electrodes that carry a very high electrical current, producing carbon dioxide and pure aluminum metal Ⓒ. Each ton of aluminum requires a great deal of electricity and nearly a half ton of carbon electrodes. The ingots of aluminum are rolled into sheets, then formed into cans. At the end of its life cycle, an aluminum can is either discarded as waste or recycled Ⓓ.

17.1 Solid Waste

BIG IDEA **Municipal solid waste (MSW)** consists of materials that are discarded from homes, small businesses, and institutions such as hospitals and universities. This is the stuff that most of us call trash or garbage. In the past, MSW was dumped into open pits, rivers, or the ocean or burned in open piles. In developed countries such as the United States, these methods of disposal have been replaced by modern landfills and incineration facilities, which have fewer negative environmental impacts. In less developed countries, however, open dumping and burning of MSW still persist. Because MSW contains large amounts of organic material, it can be incinerated or managed in special landfills to produce electricity. **Industrial solid waste (ISW)** is produced in industrial processes before a finished product reaches the consumer. The amounts of ISW that are generated are heavily influenced by patterns of consumption and management.

Municipal Solid Waste

■ The quantity and composition of solid waste differ from region to region and have changed over time.

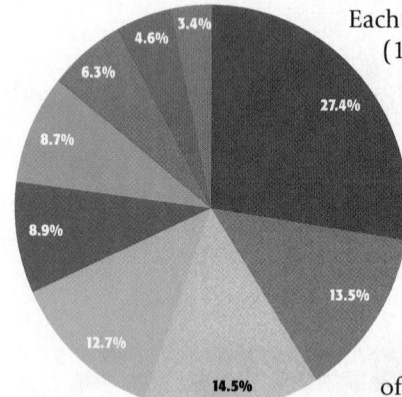

- ● Paper and paperboard
- ● Yard trimmings
- ● Food waste
- ● Plastics
- ● Metals
- ● Rubber, leather, and textiles
- ● Wood
- ● Glass
- ● Other

▲ **Figure 17.2 MSW Stream** Paper is the largest single component of municipal waste in the United States. Nonorganic wastes such as metals and glass make up less than 15% of the waste stream.

Source: U.S. EPA. 2013. *Municipal Solid Waste (MSW) in the United States: Facts and Figures.* http://www.epa.gov/osw/nonhaz/municipal/. Last accessed April 2, 2014.

Each year, an average American generates 725 kg (1,600 lb) of municipal solid waste. About 27% of this waste is paper products, including newspapers, office papers, unwanted mail, and books (Figure 17.2). Other important components of this waste stream are yard wastes, food scraps, and containers, packaging, and other objects made of plastic, metal, rubber, and cloth.

The United States ranks high among the world's developed countries in per capita waste production. For example, although the composition of their MSW is similar, the average French person throws away only 520 kg (1,145 lb) of trash per year, about 28% less than the average American.

In less developed countries, the consumption of packaged goods is not as pervasive, and reuse is more common. Consequently, waste streams are smaller and are dominated by food wastes. For example, the average person in Bangalore, India, generates about 146 kg (320 lb) of MSW each year. This is less than a quarter of the amount generated by an average American. Approximately 75% of the Indian MSW is food scraps and other organic wastes that decompose rapidly. Paper accounts for less than 2% of their waste. In Mexico City, the average person produces 248 kg (547 lb) of MSW each year; about 60% of this is food waste and 11% is paper.

The composition and amount of the waste we generate have changed over time. One hundred years ago, trash collectors in New York City picked up about 530 kg (1,170 lb) of waste per resident each year. Nearly three-fourths of that waste was coal ash from home furnaces. The rest was mainly food waste and paper. By 1960, most homes were heated with fuel oil, natural gas, or

electricity, and the per capita production of MSW in America had fallen to about 443 kg (963 lb) per year. Over the next 30 years, the increased use of packaging and disposable containers in the United States contributed to a 40% increase in the per capita waste stream. Since 1990, local, state, and federal agencies, as well as many environmental groups, have mounted campaigns to reduce the annual flow of waste. As a consequence, per capita MSW production peaked in 2000. Since then, it has declined (Figure 17.3).

GROWN TOGETHER

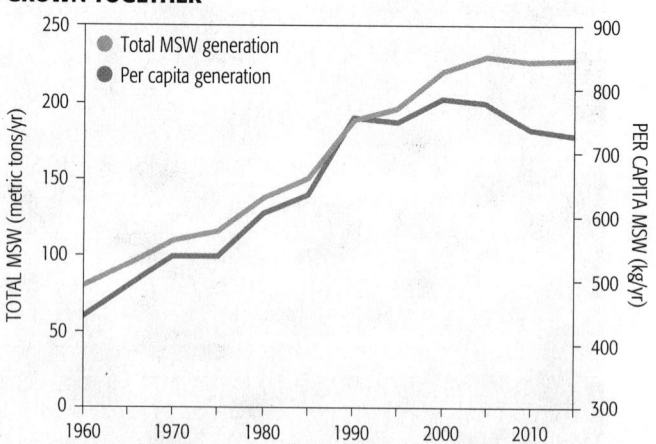

▲ **Figure 17.3 Growth of Waste** The total MSW stream has actually decreased slightly since 2005, and per capita MSW in the United States has decreased by almost 10% since 2000. This indicates that campaigns to encourage reuse and recycling are having a positive effect.

Source: U.S. EPA. 2012. *Municipal Solid Waste (MSW) in the United States: Facts and Figures.* http://www.epa.gov/osw/nonhaz/municipal/. Last accessed April 2, 2014.

◄ **Figure 17.4 Open Dumps**
In many parts of the world, municipal solid waste still ends up in open dumps such as this one in Manila, the Philippines. The very poor scour such dumps for any reusable or recyclable items.

A century ago, MSW was simply dumped into open pits or rivers; often it was burned in open piles. The decaying garbage attracted animals such as rats and produced foul odors that could be smelled for miles around. Furthermore, these dumps were a source of hundreds of polluting chemicals. Large cities, such as New York City, loaded their waste onto barges and dumped it into the ocean a few miles from shore.

As urban populations grew, so did the amount of waste, odors, and associated pollution. In developed countries, public opposition to open dumps and ocean dumping has resulted in better methods of managing MSW. These include sanitary landfills, waste-to-energy incineration, composting, and programs to reduce waste at its source. In many developing countries, however, dumping into pits and rivers and open burning persist to this day (Figure 17.4).

Sanitary Landfills

■ Modern landfills have many advantages over traditional open dumps.

In 1935, the city of Fresno, California, opened the first modern landfill in the United States. Managers at that landfill pioneered the use of trenches to deposit waste, compacting to compress waste into smaller chunks, and daily burial to keep waste underground and combat rodent and odor problems. Because it set the standard for all subsequent landfill construction, the Fresno Municipal Sanitary Landfill was declared a National Historic Landmark in 2001.

Modern **sanitary landfills** are designed to prevent three important problems associated with waste dumps—the accumulation of methane gas, the contamination of groundwater, and noxious odors (Figure 17.5).

Over time, the waste in the landfills decomposes under generally anaerobic conditions, releasing large amounts of methane or natural gas. Because this gas is flammable and explosive, it can be dangerous if it is not properly vented, as is often the case in older landfills. Today, as landfills are covered, they are fitted with pipes to vent the gases. At most retired landfills, gas wells capture this methane, which can then be used for energy.

Water percolating through the accumulated waste picks up a wide variety of chemicals, many of which are quite toxic. This water is called **leachate**. To protect groundwater from contamination by leachate, landfills are lined with plastic and impervious clay. Leachate is pumped from pipes in the landfill liner and stored in tanks or ponds. It can then be managed like other forms of wastewater (see Module 11.6).

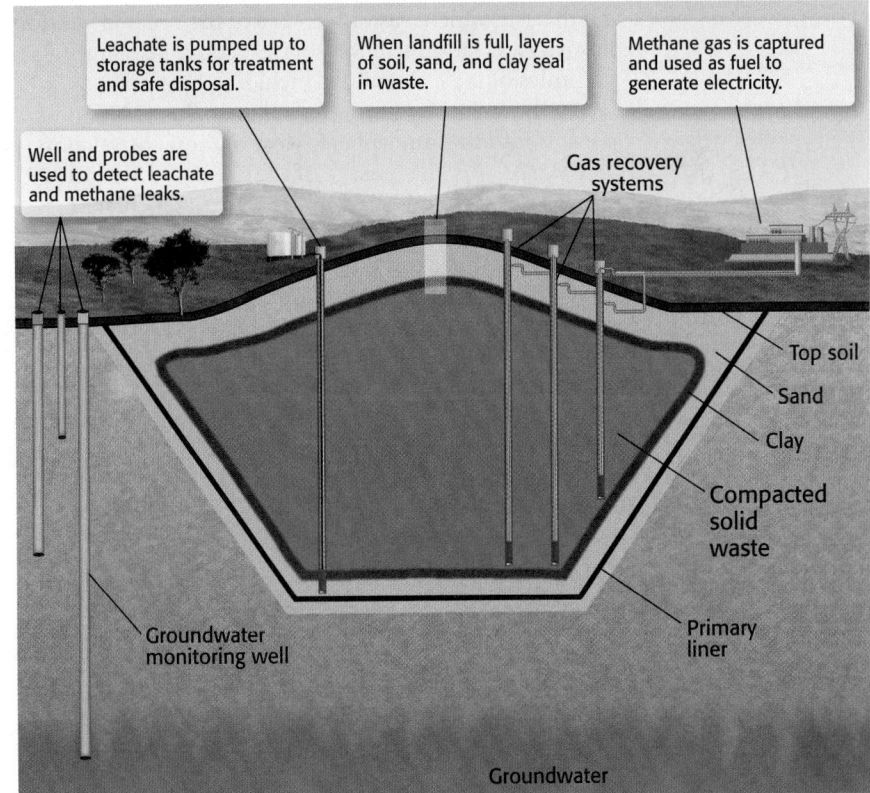

Leachate is pumped up to storage tanks for treatment and safe disposal.

When landfill is full, layers of soil, sand, and clay seal in waste.

Methane gas is captured and used as fuel to generate electricity.

Well and probes are used to detect leachate and methane leaks.

Gas recovery systems

Top soil

Sand

Clay

Compacted solid waste

Groundwater monitoring well

Primary liner

Groundwater

▲ Figure 17.5 **Sanitary Landfill**
A cutaway view of a modern sanitary landfill designed to remove hazardous methane gas and prevent the leakage of toxic leachates.

To reduce odors, layers of waste are covered with soil, coal fly ash, or similar materials. The layers of waste are then allowed to settle and decompose. When sanitary landfills are full, they are capped by a thick layer of clay. When retired landfills are deemed safe, they can be converted into public parks, open spaces, or even building sites (Figure 17.6).

Sanitary landfills are a definite improvement over open dumping, but they do pose environmental risks and challenges. If landfill liners are punctured or the systems for collecting leachate fail, groundwater may be contaminated. The methane leaking from landfills is a greenhouse gas with a warming potential that is more than 1,300 times greater than that of carbon dioxide. Even where there is abundant open space for landfills, public opposition to their location is often fierce. Thus, some cities must transport their garbage hundreds of miles for disposal. Finally, the loss of useful materials (such as paper, plastics, and metals) in landfills is not sustainable over the long run.

▲ **Figure 17.6 Landfill Restoration**
This public park near Virginia Beach, Virginia, was once a sanitary landfill. Appropriately enough, the park is called Mt. Trashmore.

Waste to Energy

■ About 80% of MSW is organic matter, a potential source of energy.

Communities have long burned municipal solid waste to reduce its volume and weight. A century ago, this was done in open dumps. By the mid-20th century, waste was being burned in hundreds of incinerators located across the United States. Incinerators consumed organic matter, and the ash, which was only 5–6% of the original volume of waste, was landfilled. But this ash contained large quantities of toxic chemicals, including heavy metals. Furthermore, incinerators released toxic flue gases and aerosols into the atmosphere. Most of these incinerators were closed after the passage of the Clean Air Act in 1970.

In the 1980s, a new generation of waste incinerators called **waste-to-energy facilities (WTEs)** was developed. These are designed not only to reduce waste volume but also to produce energy. Furthermore, they are designed with complex filtration and scrubbing devices to reduce emissions of particulate aerosols and other gaseous pollutants.

WTE facilities use the heat produced by burning waste to generate electricity. MSW is fed into furnaces that operate at temperatures near 850 °C (1,560 °F). At these temperatures, toxic organic gases are broken down entirely to carbon dioxide and water. The superhot flue gas from these furnaces is used to heat water and produce pressurized steam to over 400 °C (752 °F). The steam is then piped to turbine generators to produce electricity (Figure 17.7).

In the United States, over 10% of MSW is currently processed in 87 WTE facilities. These facilities have a total generating capacity of about 2,500 megawatts, or about 0.3% of total U.S. generating capacity. Waste incineration and WTE facilities are much more common in Japan and some European countries where land is scarce. Denmark and Sweden have been leaders in the development of

combined heat and power facilities. Those facilities pipe the excess heat from electric power plants to district heating systems, which use it to heat homes and businesses. Currently, Denmark obtains about 5% of its electricity and nearly 14% of domestic heat from the incineration of waste.

In the United States, WTE facilities must meet the EPA's Maximum Achievable Control Technology regulations. Nevertheless, concerns remain about the emissions of hazardous chemicals, and waste ash disposal remains a challenge. Public opposition to incinerator sites is often higher than for landfills.

Although methane from conventional landfills can be burned to generate electricity, concentrations are generally low and not economical. **Bioreactor landfills** are specifically designed and managed to enhance methane production so that it can be used more efficiently and economically to generate electricity. In some cases, air is pumped through a system of pipes into the landfill. This enhances decomposition and methane production and reduces leachate toxicity and production of toxic gases. Methane production can also be enhanced by recirculating leachate through the landfill. Some bioreactor landfills use both of these methods.

With regard to greenhouse gas emissions, how do WTE facilities and bioreactor landfills compare to other sources of electrical energy? A recent study by researchers at the EPA National Risk Management Research Laboratory revealed that emissions from WTE facilities ranged from 110 to 420 g CO_2e/MJ compared to 700 g CO_2e/MJ from intensively managed bioreactor landfills. This is considerably lower than emissions from coal (>3,600 g CO_2e/MJ) and natural gas.

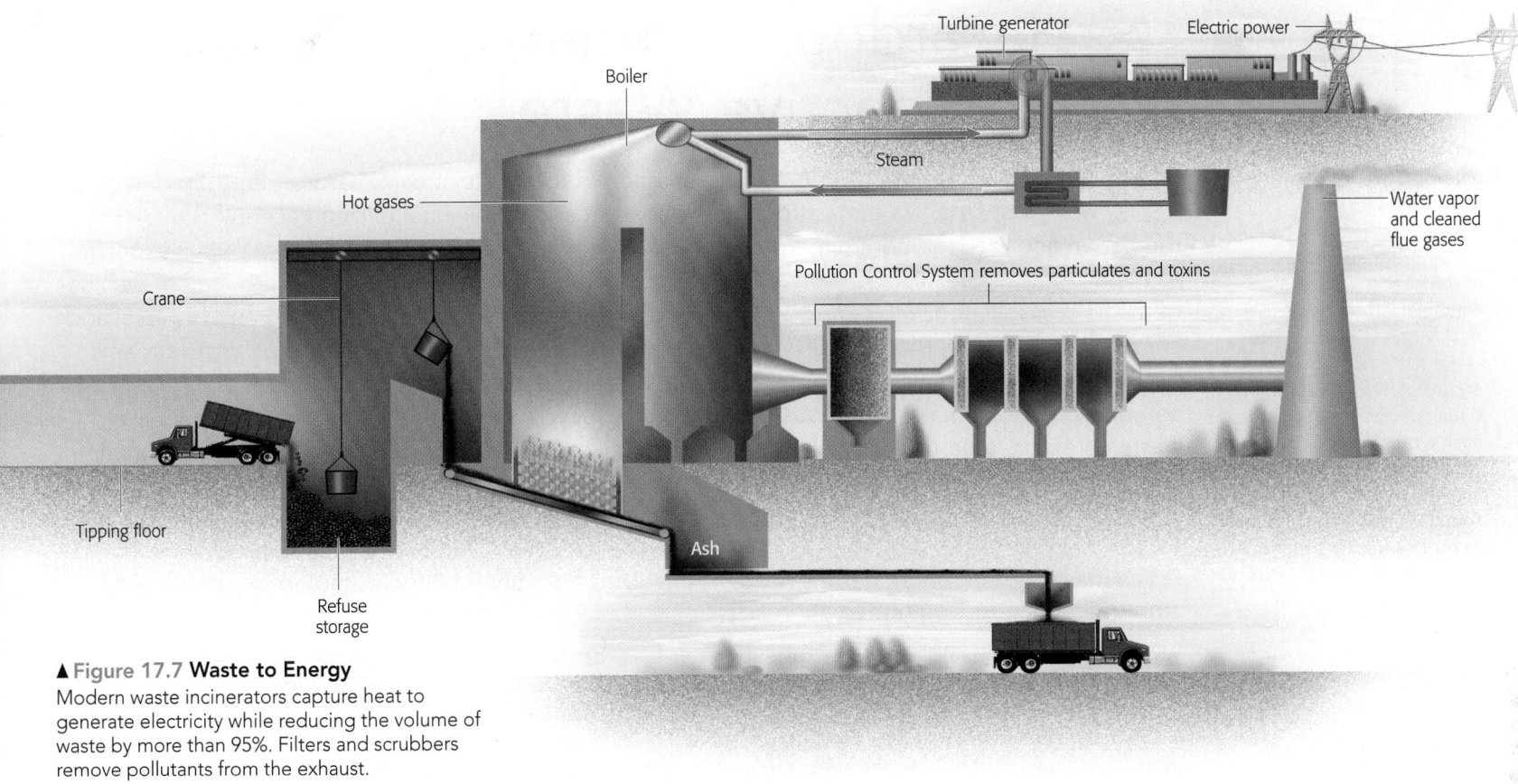

▲ Figure 17.7 Waste to Energy
Modern waste incinerators capture heat to
generate electricity while reducing the volume of
waste by more than 95%. Filters and scrubbers
remove pollutants from the exhaust.

Industrial Solid Waste

■ Waste streams from manufacturing facilities vary widely.

There is a wide variety in the kind and amount of waste
generated in making the many thousands of products
used today. ISW includes all of the waste generated at
each stage in the manufacture of products. For example,
the ISW generated in making an aluminum can includes
the wastes associated with mining bauxite, refining
bauxite into alumina, smelting alumina into aluminum
ingots, shaping aluminum ingots into rolls of aluminum
sheeting, and forming and cutting sheeting into cans.

In contrast to municipal waste, industrial waste is usually
collected and managed by the private sector. This makes data
on the specific nature and fate of ISW difficult to acquire.
However, the U.S. EPA estimates that U.S. industrial facilities
generate and dispose of about 7.6 billion tons of waste each
year. Large industrial facilities typically discard their waste in
specially designed ISW facilities. Yet some industrial waste,
particularly that generated by smaller enterprises, makes its
way to municipal landfills. Most municipal facilities have a
special area dedicated to wastes from building construction
and demolition.

Many industries have recognized that simply
treating ISW as something to be discarded has adverse
consequences for the environment as well as for
their economic bottom line. Thus, they are seeking
opportunities to reduce, reuse, and recycle at every stage
of production (see Module 17.4).

New **FRONTIERS**

Building a Truly Sustainable Landfill
In addition to the potential energy stored in its
organic matter, many of the materials in MSW are
potentially valuable. For example, why couldn't
we reclaim "waste" metals rather than extracting
them from environmentally destructive mines?
Waste engineers have proposed the development
of "Biocells," bioreactor landfills that would be
operated in two phases, first for energy and then
for the recovery of other valuable materials. The
goal during the first bioreactor phase is to maximize
methane production and extraction during a
contracted period of 10–20 years. This would be
done by more intensive aeration and leachate
recycling than in typical bioreactor landfills. Biocells
would be designed so that organic compost and
recyclable metals could be mined in the second
phase, once the bioreactor methane was depleted.

So far, fully operational biocells have not been
developed. However, a pilot-scale biocell is now in
operation near Calgary, Alberta. In the short term,
biocells will likely be more expensive to operate
than conventional landfills. Do you think the expense
would be justified? Why are biocells likely to be more
economical than landfills over their entire lifetime?

QUESTIONS 17.1

1. How do municipal
   solid waste streams in
   developed countries
   differ from those in less-
   developed countries? What
   causes these differences?

2. How does a sanitary landfill
   differ from a simple open
   garbage dump?

3. How does industrial solid
   waste differ from municipal
   solid waste?

(MES) For additional review, go to
MasteringEnvironmentalScience

Where **YOU** Are you living
LIVE with hazardous
waste?

The U.S. EPA maintains a list
of the kinds of household
hazardous wastes. It also
provides information
on waste characteristics
and appropriate means
of disposal. Identify five
containers of materials that
qualify as hazardous waste
in the cupboards, basement,
and/or garage of your home.
What specific threat does
each type of waste pose?
These wastes are typically
managed by individual
counties. Each county lists
its procedures for hazardous
waste disposal on its website.
How will you dispose of your
five waste containers where
you live?

17.2 Hazardous, Biomedical, Electronic, and Radioactive Waste

BIG IDEA Although they make up only a small portion of the waste stream, hazardous and radioactive wastes pose grave threats to the environment and public health. Until recently, the regulation and management of hazardous waste sometimes allowed for significant public exposure to these materials, resulting in concerns about public health and calls for tighter regulation. Because of the risks associated with hazardous materials, they require specialized handling and disposal and are costly to manage. Biomedical wastes must be processed separately from other waste streams because they can spread infections or disease. Discarded electronics like cell phones and computers contain many hazardous but potentially recyclable materials. Disposal of radioactive wastes presents a particular challenge because they can remain radioactive for long periods of time.

Hazardous Waste

■ The management of dangerous waste has evolved from unregulated dumping to permanent storage, reduction, and remediation.

Wastes that are flammable, corrosive, explosive, or toxic are classified as **hazardous wastes**. The U.S. EPA identifies three categories of hazardous wastes. *Source-specific wastes* are wastes that are unique to particular industries. Sludge and wastewater from petroleum refiners and pesticide manufacturers are examples of source-specific wastes. Over 70% of the hazardous wastes produced in the United States come from the chemical and petroleum industries. *Nonspecific-source wastes* are produced by common manufacturing and industrial processes. Solvents used for cleaning or degreasing, including those used by commercial

drycleaners, are examples of nonspecific-source wastes. *Discarded commercial chemical wastes* include commercial chemical products, such as unused paints, pesticides, and pharmaceuticals.

Although less hazardous than the waste produced by industry, household hazardous wastes also represent a significant threat to the environment and public health (Figure 17.8). Household hazardous wastes include unused paint, old batteries, solvents, cleaners, and other chemicals. On average, each person in the United States generates about 5 kg (11 lb) of household hazardous waste each year. Compared to the stream of municipal

▶ **Figure 17.8 Household Hazardous Waste**
Each year, U.S. households dispose of more than 1.5 million tons of hazardous waste. This waste comes from a diverse array of products.

solid waste, this may seem like a small amount. However, household wastes tend to accumulate in basements and garages. The EPA estimates that the average home has accumulated more than 45 kg (100 lb) of hazardous waste. As much as 90% of this hazardous material is eventually poured down drains or into storm sewers or discarded in municipal dumps or landfills.

Approaches to managing hazardous waste have changed as our understanding of the dangers associated with them has grown. In the past, most hazardous wastes were simply dumped into the ocean, lakes, and rivers or open pits and allowed to degrade. This misguided approach to waste management is neatly described by this bit of doggerel: "The solution to pollution is dilution." In the 1970s, several highly publicized cases brought the health risks associated with such dumping to public attention. For example, in 1976, the Love Canal housing development in Niagara Falls, New York, was discovered to have been sited on a former dump site with soils contaminated by a variety of toxic chemicals. Public outcry over these cases led to stricter regulations for the management of hazardous wastes, as well as efforts to clean up previously contaminated sites. Nevertheless, thousands of previous dumping sites continue to contaminate soils and groundwater.

Current methods for disposing of hazardous wastes have greatly reduced the risks to the environment and public health. Many hazardous wastes are stored in specially designed landfills called **permanent retrieval storage sites**. Wastes in these sites can be monitored to ensure they do not leak into the surrounding environment; they can also be recovered for reuse or for more permanent disposal. Chemical processing and high-temperature incineration are used to convert some hazardous waste to less toxic ash or gases.

In **bioremediation**, living organisms—microorganisms, plants, or fungi—break down hazardous wastes into less harmful components. The most common use of bioremediation is to clean up spills of hazardous chemicals.

▲ Figure 17.9 **Bioremediation**
These workers are spraying nutrients to encourage the growth of oil-devouring microbes on the shores of Prince William Sound. This area was polluted with oil spilled from the *Exxon Valdez* tanker in 1989.

For example, many kinds of microorganisms can break down the various hydrocarbons in petroleum products. Mixtures of such organisms are now routinely used to help clean up oil spills (Figure 17.9). In other situations, microorganisms growing in reactor vessels are used to break down particular hazardous chemicals.

Another way to dispose of liquid hazardous wastes is to inject them into deep wells in porous rocks, such as sandstone or limestone. More than 375,000 waste-injection wells have been formally listed with the EPA. These wells are tightly regulated and monitored to prevent the contamination of groundwater. Nevertheless, there is concern that considerable amounts of hazardous liquids are being injected into unregulated wells.

Biomedical Waste

■ Some medical wastes can spread infections and disease.

Hospitals and medical facilities generate large quantities of waste, including solid waste such as packaging, hazardous chemical waste, and radioactive waste. They also generate **biomedical waste**, specifically defined as waste that can spread infections or disease, or waste that may decompose and produce toxins and noxious odors. Such waste includes human and animal body parts and tissue, blood, used bandages and dressings, discarded gloves, and "sharps," such as needles, scalpels, and lancets that can penetrate skin.

Because of the infectious nature of biomedical waste, it must be handled and disposed of separately from other waste streams. It is generally collected in leak-proof containers that are strong enough to resist breakage, even with rough handling. These containers must prominently display the internationally recognized biohazard symbol (Figure 17.10).

Biomedical waste requires special treatment. Microbiological cultures and other materials are often sterilized under high temperature and pressure in an autoclave. Most biomedical waste is ultimately disposed of in special incinerators that destroy pathogens and sharps.

In recent years, there have been concerns about the disposal of another medical waste, drugs and pharmaceuticals. This issue is explored in Chapter 18.

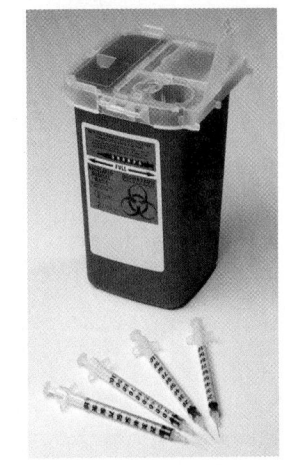

▶ Figure 17.10 **Collecting Biomedical Waste**
A familiar sight in doctor's offices, sharps containers such as this are designed to collect needles and other potentially infectious items that might penetrate skin.

Electronic Waste

■ E-waste regulation has only recently begun in some places.

One of the most prominent trends of the past 50 years is our ever-increasing dependence on electronic devices, such as computers and cell phones. When these devices eventually fail, they are discarded. Changes in technology, such as the transition from analog to digital television, have made millions of electronic devices obsolete; these devices must also be discarded or put into storage. Old televisions, computers, cell phones, and other electronic devices disposed of by businesses and households make up electronic waste, or **e-waste** (Figure 17.11).

E-waste is classified as hazardous because it contains large amounts of heavy metals, including lead, cadmium, and mercury. Although e-waste represents only 2% of the trash in landfills, the EPA estimates that it is the source of more than 70% of the heavy metals in them. Some devices that were manufactured before 1977 also contain significant amounts of polychlorinated biphenols (PCBs), which were used to insulate electronic components such as capacitors and transformers. PCBs have been shown to cause cancer in many animals and in humans.

In 2012, the stream of e-waste in the United States exceeded 2.5 million metric tons (about 2.9 million U.S. tons). More than 80% of discarded e-waste ends up in landfills. A significant amount of e-waste is simply "stored," although most stored devices are eventually discarded. About 75% of all of the personal computers ever sold are now gathering dust in storage.

Despite the fact that discarded electronic devices contain many valuable components and chemicals,

▲ Figure 17.11 **E-waste**
The electronic age has generated new forms of hazardous waste. Circuit boards, cell phones, and other components contain heavy metals.

recycling rates for e-waste are less than 25%. Nevertheless, e-waste processing is a rapidly growing commercial sector, and the rate of recycling is increasing. Shiploads of e-waste are sent to Asia and Africa, where products are disassembled and useful materials are salvaged (Figure 17.12). While shipping these hazardous wastes overseas benefits the U.S. environment, it often presents significant health risks to workers in underregulated facilities elsewhere.

► Figure 17.12 **Recycling E-waste**
Workers in Asia reuse and recycle parts from electronic devices that have been discarded in developed nations. These activities expose the workers to many toxic substances.

Radioactive Waste

■ Radioactive wastes require long-term monitoring to avoid human health risks.

Radioactive waste is material that is contaminated by isotopes of elements that emit destructive forms of radiation. Because most radioactive waste decays very slowly, the management of this waste poses unique challenges.

Low-level radioactive waste contains comparatively low amounts of radioisotopes and represents the greatest volume of radioactive waste. Much of this waste comes from the use of radiation in hospitals and research laboratories; typically, it consists of contaminated clothing, filters, wiping rags, medicinal swabs, tubes, syringes, and the carcasses and tissues of laboratory animals. The residues from nuclear reactor wastewater are also an important form of low-level waste. Radioactivity in such wastes ranges from just above natural background levels to very high levels, as in the case of parts from inside nuclear reactors. Much low-level waste is stored at the site where it is generated. When enough material accumulates, it is shipped to low-level radioactive disposal sites (**Figure 17.13**). In the United States, such sites are licensed and regulated by the Nuclear Regulatory Commission.

Nuclear power plants and weapon labs generate **high-level radioactive waste**, which is far more dangerous than low-level waste and requires long-term storage and management. The majority of high-level waste is in the form of used fuel rods from nuclear power plants (see Module 14.5). At present, all high-level radioactive waste is stored on-site at the world's nuclear power facilities. In the United States, such waste is stored at over 100 locations. When spent fuel is first removed from reactor cores, it is placed in wet storage pools. At most facilities, it is subsequently transferred to stainless steel casks for dry storage (**Figure 17.14**).

Wet and dry storage facilities must be licensed and are constantly monitored. Nevertheless, there are concerns that such storage may be hazardous to the health of people working at the sites. Such facilities may also be vulnerable to terrorist actions. Although spent fuel is not explosive, it does contain large amounts of radioactive and toxic heavy metals. It is feared that a terrorist attack could disperse these metals widely in the environment.

The safe management of high-level nuclear waste is one of the most daunting challenges associated with the use of nuclear power. Because spent fuel rods still contain large amounts of radioactive material, some experts argue that they should be recycled into new fuel rods. However,

▲ Figure 17.13 **Storing Low-Level Nuclear Waste**
Articles that are contaminated with low levels of radioactive chemicals are encased in concrete casks or containers, which are then buried in clay-lined excavations like the one shown here.

reconstituted fuel rods are enriched with plutonium, an element that can be used to manufacture nuclear bombs. For security reasons, therefore, most nations have agreed that such recycling should not be pursued.

Several nations are pursuing the development of permanent geologic repositories for nuclear waste. In these repositories, used fuel rods encased in special metal casks would be placed in deep caves in geologic strata that are isolated from the biosphere. It is hoped that the metal casks and the surrounding rock will prevent the radioactive materials from leaking into the environment. In the United States, one such repository was proposed for Yucca Mountain, Nevada. However, licensing of this repository has been halted because of uncertainties about the eventual fate of the nuclear waste and of opposition from nearby communities and other interest groups.

In summary, industry, medical facilities, electrical utilities, and electronic devices of all kinds generate wastes that pose a direct threat to human health. While current management focuses on the containment and storage of these hazardous wastes, increasing emphasis is being placed on reducing the production of such wastes and reducing their toxicity in the environment.

QUESTIONS 17.2

1. What is a permanent retrieval storage site? How does it differ from an open dump?

2. Why must biomedical waste be handled separately from other waste streams?

3. Although it comprises only 2% of the total MSW waste stream, e-waste is a major challenge. Why?

4. What is the difference between low-level and high-level radioactive waste? How does this difference affect the methods of managing these wastes?

(MES) For additional review, go to **MasteringEnvironmentalScience**

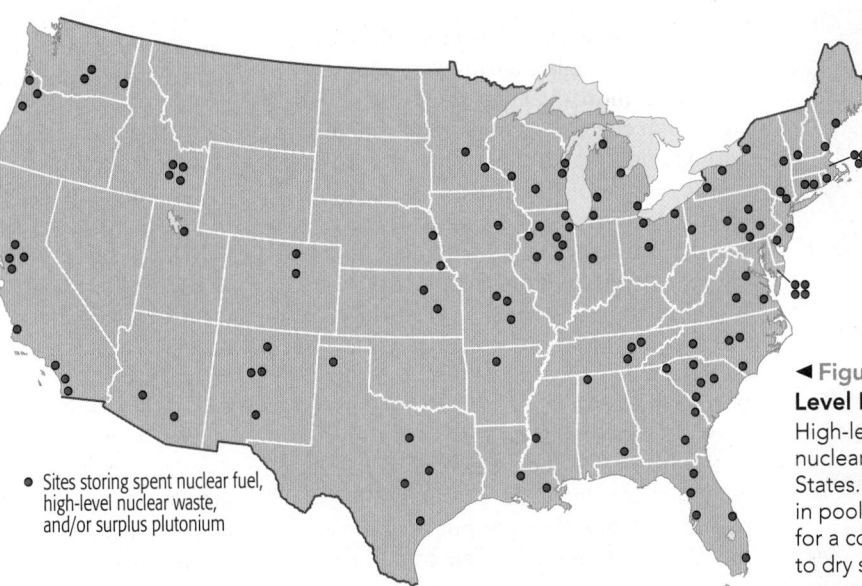

◀ Figure 17.14 **Storage Sites for High-Level Nuclear Waste**
High-level nuclear waste is stored at nuclear power facilities across the United States. At first, spent nuclear fuel is stored in pools of water. After the fuel has cooled for a couple of years, it may be transferred to dry storage casks.

• Sites storing spent nuclear fuel, high-level nuclear waste, and/or surplus plutonium

17.3 Waste Reduction, Reuse, and Recycling

BIG IDEA We can significantly reduce the total volume of waste and its associated environmental impacts by (1) reducing the amount of stuff we use, (2) reusing items whenever possible, and (3) recycling items at the end of their useful life. Reduce, reuse, and recycle programs are steadily decreasing waste streams, but much more can be done.

Q *How much recycling gets done in the United States every year compared to what is consumed?*

Shayna Jefferson, University of Florida

Norm: **A** The United States recycles almost 40% of its municipal waste. Japan and many countries in Europe recycle considerably more than half of their waste.

Reducing the Waste Stream

■ Reduce, reuse, and recycling efforts can greatly diminish the original source of municipal solid waste.

Over the past 50 years, the total amount of waste generated by each person in the United States has nearly doubled. Much of this increase is a direct consequence of our growing affluence. We acquire more things, and those things are eventually discarded. Many of the products we buy, such as disposable bottles, diapers, cameras, and razors, are not designed to last beyond a single use. We view many items as obsolete long before their useful life is up. Pop songs, editorials, and blogs refer to ours as the "disposable society."

A significant portion of the recent increase in MSW is related to changes in how the things we acquire are packaged. Each year, the United States produces nearly 25 billion kilograms (56 billion pounds) of plastic for containers and packaging. That is enough plastic to make 2 trillion plastic water bottles; if arranged end-to-end, those bottles would reach from Earth to Mars and back. Certainly, some of this plastic serves a useful, perhaps even essential, function. But much of it is simply cosmetic. Currently, less than 10% of plastic containers are recycled; the rest ends up in landfills or incinerators.

Reducing waste. Given the immense volume of packaging waste, even small changes in individual containers can have a significant effect on the waste stream. Over the past two decades, the net weight of a 2-liter plastic soda bottle has been reduced from 68 to 51 grams (2.4–1.8 ounces) by using less plastic. This has eliminated 110 million kilograms (243 million pounds) of plastic from the waste stream each year. Consumers can diminish their individual waste streams by buying in bulk with less packaging or by purchasing items in reusable containers. These practices save money, too.

In response to requests from consumers and environmental groups, major retailers are beginning to put pressure on manufacturers and suppliers to reduce packaging waste. In 2008, the world's largest retail chain, Walmart, announced policies to eliminate landfill waste by 2025. It works with suppliers to create less packaging overall and to increase package-recycling rates. Virtually all of its cardboard and paper waste is now recycled. By focusing on efficient packing and giving discounts to customers with reusable bags, Walmart has reduced plastic shopping bag waste from its stores by 38%.

Reusing waste. The U.S. EPA estimates that 2–5% of the material sent to landfills is actually reusable. This includes items as varied as shopping bags, clothing, furniture, and building materials. For hundreds of years, poor people in both developing and developed countries have made a meager living harvesting items from trash heaps (**Figure 17.15**). Even some rather well-to-do people save money and diminish their demands on the environment by retrieving usable items and material from dumps. In the United States, more than 6,000 reuse centers run by organizations such as Goodwill, the Salvation Army, and Habitat for Humanity provide an outlet for the resale of these materials. This reuse reduces the waste stream by increasing the residence time of materials in communities.

Recycling waste. Recycling programs have become common in many developed countries. In 2013, the rate of recycling in the United States reached 34.5%, more than double the rate in 1990 (**Figure 17.16**). E-waste recycling rates have risen from less than 10% in 2000 to almost 25% in 2013. In Japan and many European countries, recycling rates exceed 50%. Recycling diverts material from landfills and incinerators, allowing it to be remanufactured into other items. This reduces the demand for resources and saves energy. For example, making a can from recycled aluminum requires only 5% of the energy needed to make a can from bauxite ore. Recycling items such as batteries and electronics also diminishes the flow of toxic chemicals into the environment.

▼ Figure 17.15 **Dumpster Diving**
Considerable amounts of economically valuable material that enter the MSW stream are retrieved for reuse or recycling. Such trash picking is estimated to add nearly $1 billion to the U.S. economy.

MSW RECYCLING RATES, 1960–2012

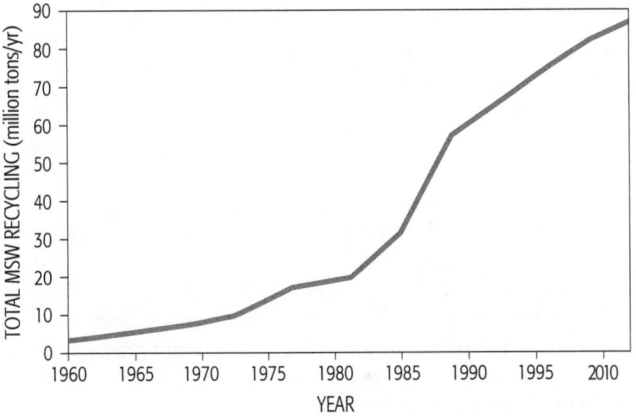

◀ Figure 17.16 **Recycling Rates**
Since 1985, recycling rates have increased significantly. This increase is a result of a growth in the technology and infrastructure to support recycling. In addition, many communities have mounted programs to encourage recycling.

Source: U.S. EPA. 2013. Municipal Solid Waste (MSW) in the United States: Facts and Figures. http://www.epa.gov/osw/nonhaz/municipal/. Last accessed April 2, 2014.

In **closed-loop recycling**, the material in a product is reused to produce a similar product (Figure 17.17). Metals such as aluminum can be recycled in this manner. In **open-loop recycling**, a material is recycled into a different product that may or may not be discarded after it is used. Recycled glass and plastics, for example, are often not suitable for their original use but may be manufactured into a variety of other products (Figure 17.18). Other factors may also limit recycling. For example, most paper can only be recycled a few times before its fibers are degraded beyond further use.

Composing

■ Compost production is a unique kind of open-loop recycling.

On average, nearly 80% of MSW is composed of decomposable organic matter. Indeed, paper, yard trimmings, and food waste alone constitute almost 60% of MSW in the United States. With relatively simple treatment, most of this material can be converted to **compost**, decayed organic matter that can be used to improve the fertility and water-holding capacity of soil, thereby improving plant growth.

Small-scale or household composting can significantly reduce the volume of yard waste and food scraps entering the waste stream. Such composting can be carried out even in small yards. Organic wastes such as yard trimmings and most kitchen wastes are collected in piles or composting bins, where they are allowed to "cure," or decompose into stable humus. This compost can then be tilled into garden soils to improve aeration and fertility or used directly as a medium for plant growth.

Red worms or garden worms are especially effective composting agents. In **vermicomposting**, worms are placed with organic waste in specially designed bins. Worms eat nearly 50% of their weight in organic matter each day, and they convert it to humus that is especially nutrient rich and valued by gardeners.

◀ Figure 17.17 **Closed-Loop Recycling**
The aluminum in cans can be recycled to produce new cans.

◀ Figure 17.18 **Open-Loop Recycling**
The wood-like boards in this park bench are a composite material made from recycled plastics.

▲ Figure 17.19 **Composting**
Many municipal waste facilities are now able to convert mixed paper, food, and yard waste into compost, which can be sold commercially.

Municipal governments and institutions often carry out large-scale composting, sending food and yard waste to facilities designed to speed up the decomposition process (Figure 17.19). Most often, waste is arranged in long "windrows" that are turned frequently to improve aeration. This procedure can accommodate a wide variety of wastes, and it diverts foul-smelling, methane-producing organic matter from landfills and cuts the cost of waste disposal. It produces compost that can be used for landscaping or sold for additional revenue.

Large-scale composting facilities must manage potential environmental impacts carefully. Leachates from compost piles and windrows can contaminate surface water and groundwater. Facilities must also take steps to minimize odors and populations of pest animals.

Where YOU LIVE

What you can do to reduce your waste footprint?

The things you buy and how you dispose of them can add or subtract several acres from your total ecological footprint (see Module 5.4). Before purchasing any item, thoughtfully answer the following six questions.

1. *Do I really need this item?* Most of us possess lots of stuff that we now realize we did not really need. But this question applies to some very ordinary acquisitions. Do you need to buy water in a plastic bottle? Do you need to buy food in wrapped single servings? Do I really need to print that email?
2. *Was this item sustainably produced?* Buy refurbished and recycled materials, and support manufacturers and retailers who are committed to best waste management practices.
3. *Can I reuse this item?* Look for durability and avoid things that are likely to become obsolete.
4. *Can I recycle this item?* Look for items made from easily recyclable materials.
5. *Will I recycle this item?* Go out of your way to ensure all waste that can be recycled is. Support recycling programs in your community.
6. *Can this item be safely discarded without risk to the environment or human health?* Avoid items that are likely to pose an environmental or health risk at their end of life. Take advantage of community programs to collect batteries, e-waste, and household hazardous wastes.

QUESTIONS 17.3

1. What is the difference between closed-loop and open-loop recycling?
2. How would you respond to the assertion that recycling programs are too costly?
3. Describe three potential environmental impacts of composting that must be carefully managed.

(MES) For additional review, go to **MasteringEnvironmentalScience**

Q *Are landfills obsolete?*
Macy, Louisiana State University

Norm: A Unfortunately not. But municipal waste streams are diminishing, and the demand for new landfills is declining.

Challenges to Municipal Recycling

■ The costs of recycling programs are offset by environmental and social benefits.

The economic feasibility of municipal recycling and composting programs depends on several factors, including markets for recycled material, the cost of waste disposal, and government policies toward resource extractive industries, such as mining and forestry. Consumer demand for recycled products is important: Recycling programs cannot succeed if consumers do not purchase recycled products.

Where they have been implemented, deposits on recyclable products such as bottles and cans have significantly increased recycling rates. Convenience also influences the rate of recycling. Communities with curbside collection and disposal facilities that require a minimum of sorting have higher recycling rates than those that do not.

Critics of recycling often argue that the costs associated with collecting and transporting recyclable materials may be nearly as great as the savings in production. However, such calculations usually do not account for most of the environmental and social costs associated with landfills and incinerators.

Over the past two decades, increased public awareness and aggressive community programs have resulted in significant increases in the amount of waste being recycled. Recycling rates in some communities exceed 60%. As a consequence, the amount of material being discarded in landfills has remained flat, and per capita waste generation is actually decreasing (Figure 17.20).

▶ Figure 17.20 **Landfills on the Decline**
Increased reuse, recycling, and composting have diminished the amount of MSW going to landfills, allowing more than 6,000 facilities to close. Since 2000, however, the rate of recycling has begun to level off, and the number of landfills has remained nearly constant.
Source: U.S. EPA. 2013 Municipal Solid Waste (MSW) in the United States: Facts and Figures. http://www.epa.gov/osw/nonhaz/municipal/. Last accessed April 2, 2014.

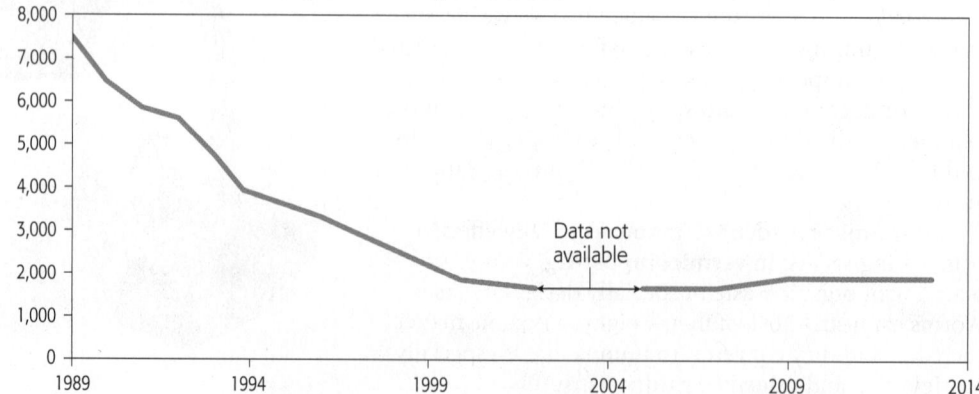

NUMBER OF LANDFILLS IN THE UNITED STATES, 1989–2013

Data not available

To Recycle or Not to Recycle

Why do we choose to recycle some things and not others?

Rummaging through trash and recycling bins at Boston University's School of Management, Remi Trudel noticed an interesting pattern (**Figure 17.21**). Whole sheets of paper were recycled and paper fragments were thrown in the trash, and this was true regardless of the kind of paper. Trudel wondered whether his colleagues and students in the school were choosing to recycle or trash paper and other recyclable materials based on their perceptions of the usefulness of those materials. With Jennifer Argo at University of Alberta's School of Business, he conducted an experiment designed to address three questions. First, do people indeed decide whether to recycle used materials based on the physical condition after their use? Second, does the amount of those materials influence their decision to recycle or not? Third, are such decisions based on perceptions of the usefulness of the materials?

The *experimental design* was as follows. Undergraduate students served as *experimental subjects*. Each of 150 students was given a pair of scissors and an 8.5 × 11 in. sheet of paper, and assigned to one of three groups. The paper contained a description of the scissors and identifying marks matching students with their particular piece of paper. Group 1 students were asked to evaluate the scissors without cutting the paper. Group 2 students were asked to evaluate the scissors by cutting the paper along dotted lines leaving eight small pieces of paper. Group 3 students were also asked to cut the description paper and a second sheet of paper with similar dotted lines; this group ended up with 16 small pieces of paper. Each student was then asked to complete an *evaluation survey* regarding their scissors. Surveys were collected by a research assistant who then asked students to dispose of all their paper on their way out. Two identical bins, a trash bin and a recycling bin, were situated next to the exit. The bins had flip lids so that students could not see inside.

After students exited the laboratory, they were approached by a second research assistant and asked to fill out a short *follow-up survey*. They were asked to think back to the paper they discarded on their way out and assess on a 1–7 numerical scale the extent to which "the paper was like garbage" (1 = not much at all, 7 = very much), "the paper contributes to the amount of waste generated in society," "how much effort you would have to exert to

▲ **Figure 17.21 Marketing for Recycling**
Remi Trudel's research is focused on consumer judgment and decision making around social issues such as whether or not to recycle.

recycle the piece of paper," and "the impact disposing the paper in the garbage would have on the environment."

Students in the group 1 (full sheet of paper) were nearly twice as likely to recycle as students in the other two groups. The amount of paper (group 2 vs. group 3) had no significant influence on whether students recycled (**Figure 17.22**). The most important difference among groups with regard to the second survey was in their perception of the usefulness of the paper. Even though the paper quality was the same in each case, groups 2 and 3 viewed their cut-up paper being more like garbage than did group 1 (**Figure 17.23**).

Trudel and Argo's studies indicate that understanding of the potential usefulness of waste materials, regardless of their form, has a very significant influence on recycling behavior.

Source: Trudel, R. and J.J. Argo. 2013. The effect of product size and form distortion on consumer recycling behavior. *Journal of Consumer Research* 40. DOI: 10.1086/671475.

1. What scientific methods were employed in this study?

2. Why were students asked whether they viewed the waste paper as garbage only after they had left the laboratory?

3. Trudel and Argo believe that the more students understand about recycling processes, the more likely they will be to recycle. How might you test this hypothesis?

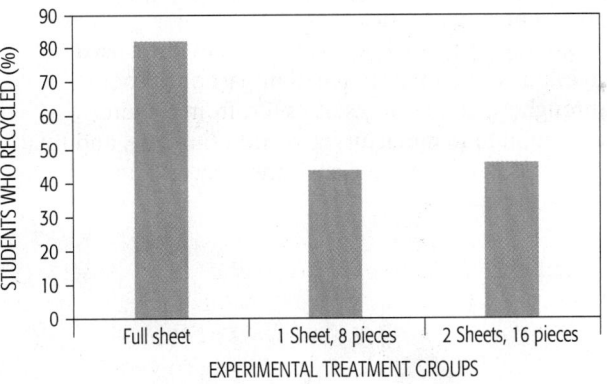

◄ **Figure 17.22 Does Physical Condition Matter?** The treatment group that discarded whole sheets of paper were much more likely to recycle than groups that discarded cut-up pieces of paper.

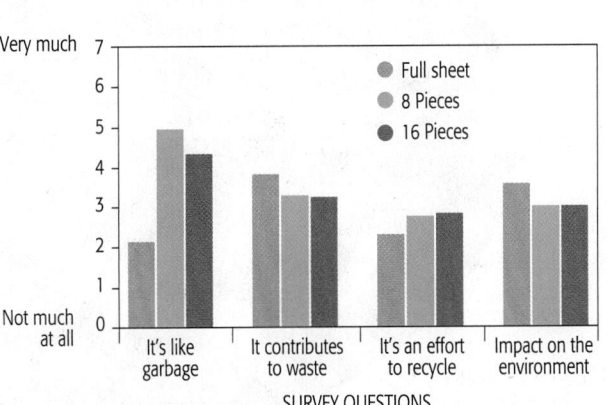

◄ **Figure 17.23 What Do You Believe About Waste?** Among the post-experiment survey questions, the groups differed most with regard to the extent that they saw their waste paper as garbage.

17.4 Managing Product Life Cycles

BIG IDEA Much of waste management focuses on how to properly dispose of waste once it is produced, but a more compelling approach is to avoid generating waste in the first place. Generating less waste reduces the pressure on landfills and other waste management facilities and saves the energy and resources that are lost when items are discarded. It also minimizes the environmental impacts that occur throughout the life of products. A variety of approaches can be used to reduce the stream of waste and to produce products that can be recycled.

Life-Cycle Assessment

■ Product manufacture, use, and disposal systems should be less linear and more circular.

Energy and matter move through natural ecosystems in a circular fashion. Energy and matter stored at one trophic level are consumed and stored by organisms at the next level. When organisms decompose, the energy and nutrients stored in their tissues are recycled back into the ecosystem's food web. In most ecosystems, the ultimate wastes are simply carbon dioxide, water, and heat.

In contrast, most human systems of manufacturing products and managing waste are largely linear. Raw materials, often from nonrenewable sources, are used to create products. These products are then packaged and marketed in ways that demand additional raw materials. Products are used until they are either obsolete or worn out and are then discarded in landfills or burned in incinerators. This approach to manufacturing is energy intensive and has significant environmental impacts.

Life-cycle assessment (LCA) is a method used to evaluate the environmental impact of a product throughout all the stages of its life, from resource extraction to manufacturing, distribution, use, and final disposal. This allows the environmental and resource impacts of a product to be examined in the context of the entire system.

A complete LCA includes three interrelated components: an inventory analysis, an impact analysis, and an improvement analysis. An *inventory analysis* quantifies the inputs of energy and raw materials and the outputs of solid waste, air and water pollution, and other environmental releases associated with each stage of a product's life cycle. An *impact analysis* evaluates the various impacts—environmental, economic, health, social, and cultural—associated with the inputs and outputs identified in the inventory analysis of a product. An *improvement analysis* identifies opportunities to reduce the environmental impact of a product by changing aspects of its life cycle, such as its design, manufacturing methods, consumer recycling, and waste management.

Life-cycle analysis begins with the creation of a flow diagram representing the steps in the system used to manufacture, use, and dispose of a particular product. Each step is then analyzed in finer detail as a subsystem with its own set of energy and material inputs and outputs (Figure 17.24).

►Figure 17.24 **Life-Cycle Analysis**
The full consequences of the manufacture, use, and disposal of a product can be understood by a close examination of the inputs and outputs associated with each stage of its life, from "cradle to grave."

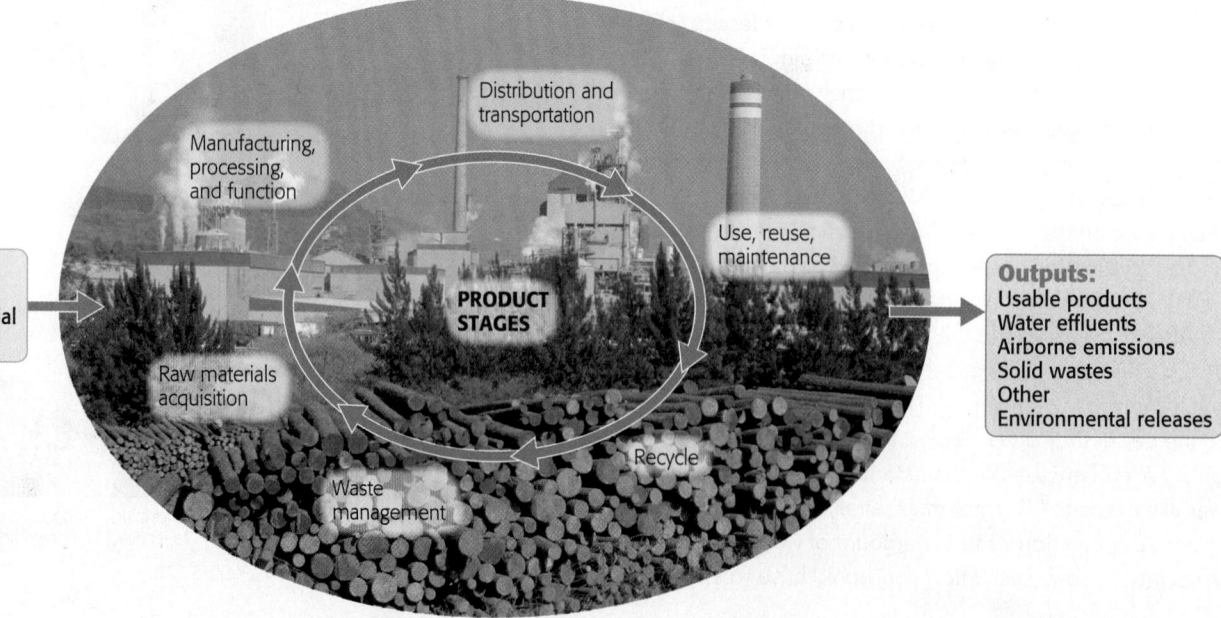

Manufacturing, processing, and function

Distribution and transportation

Use, reuse, maintenance

PRODUCT STAGES

Raw materials acquisition

Waste management

Recycle

Inputs:
Water
Raw material
Energy

Outputs:
Usable products
Water effluents
Airborne emissions
Solid wastes
Other
Environmental releases

An analysis of the life cycle of a typical cell phone illustrates the complexity of product manufacture, use, and disposal (Figure 17.25). Cell phones are composed of a diverse array of raw materials that must be extracted and processed. Mining of elements such as copper, gold, mercury, and cadmium has important direct environmental impacts and is also associated with air and water pollution. Cell phone manufacture requires a wide variety of raw materials, including large amounts of water. Wastes produced in manufacturing and distribution include packing materials and a variety of by-products associated with its manufacture. Sony Corporation calculates the average lifetime of a cell phone to be 3.5 years. During that time, the phone consumes electrical energy. When a phone fails or, more likely, becomes obsolete, it is usually discarded; fewer than 10% of cell phones are recycled. Disposal consumes more energy and materials.

The environmental costs associated with a cell phone also include outputs of air and water pollution. From cradle to grave, a single cell phone is responsible for the production of 664 kg (1,465 lb) of air pollutants, such as ozone, NO_x, and SO_2, and 19.8 kg (43.6 lb) of water pollutants. Greenhouse gas emissions associated with the cell phone's manufacture, use, and disposal are equal to about 20.65 kg (47 lb) of CO_2. The overall consequences of these life-cycle impacts are magnified by the sheer number of cell phones—over 10 billion—in various stages of life. Cell phones alone account for emissions of over 5 million metric tons of CO_2e each year.

▼ Figure 17.25 **A Complex Life**

Greenhouse gas emissions is one measure of the environmental impacts associated with manufacturing, using, and disposing of a typical cell phone over its life cycle. A life-cycle assessment by Sony Corporation indicated that greenhouse gas emissions over a 3.5-year life cycle of a phone are about 21 kg CO_2e.

Source: Sony Ericsson Communications. 2009. Sustainability Report. http://blogs.sonymobile.com/wp-content/uploads/2012/09/SESustainabilityReport2008.pdf. Last accessed April 2, 2014.

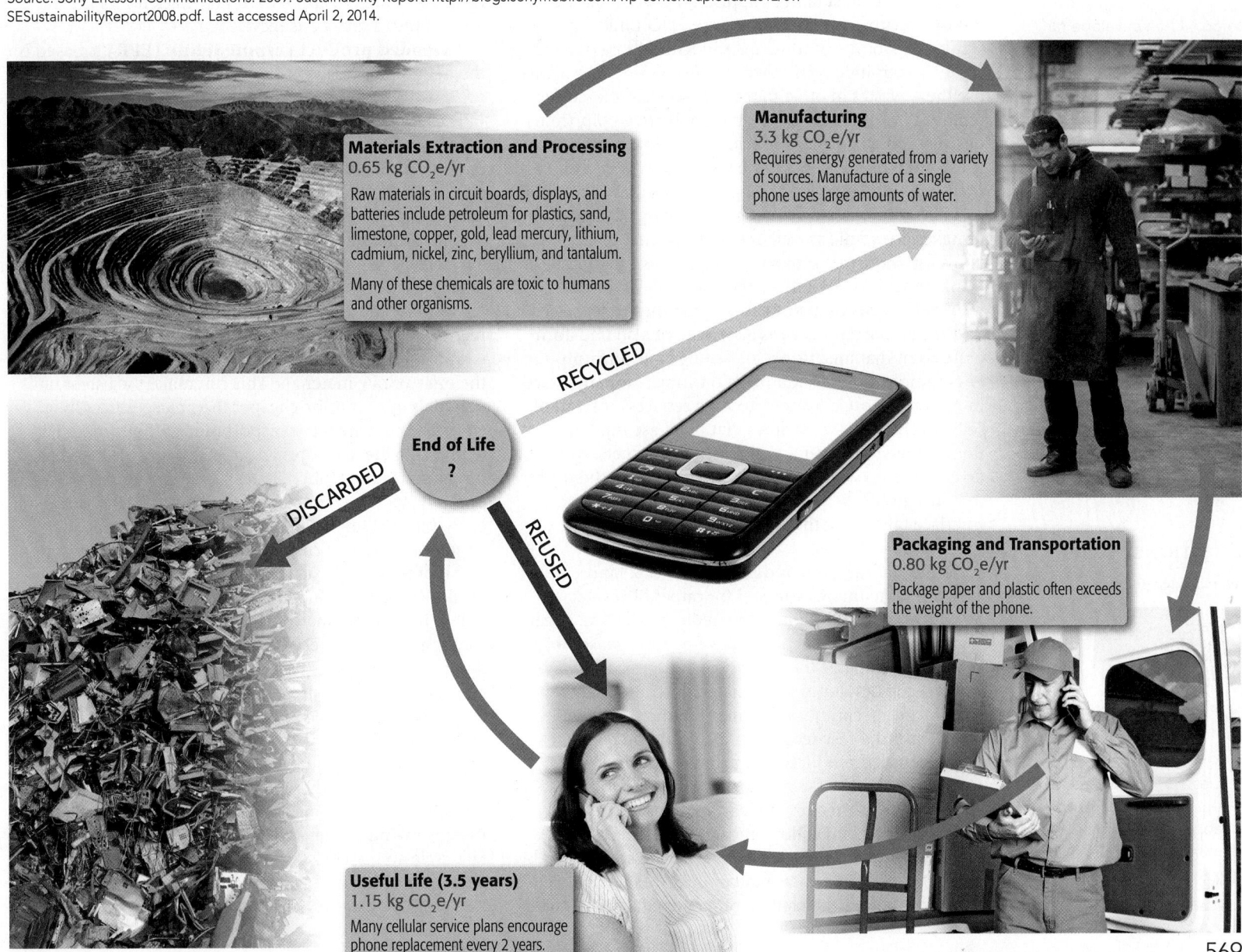

Materials Extraction and Processing
0.65 kg CO_2e/yr

Raw materials in circuit boards, displays, and batteries include petroleum for plastics, sand, limestone, copper, gold, lead mercury, lithium, cadmium, nickel, zinc, beryllium, and tantalum.

Many of these chemicals are toxic to humans and other organisms.

Manufacturing
3.3 kg CO_2e/yr

Requires energy generated from a variety of sources. Manufacture of a single phone uses large amounts of water.

RECYCLED

End of Life ?

DISCARDED

REUSED

Packaging and Transportation
0.80 kg CO_2e/yr

Package paper and plastic often exceeds the weight of the phone.

Useful Life (3.5 years)
1.15 kg CO_2e/yr

Many cellular service plans encourage phone replacement every 2 years.

Reimagining Product Life Cycles
■ Waste is just that!

▲ **Figure 17.26 Designed for Disassembly**
A team of Stanford University engineering students designed a laptop computer that can be disassembled in two minutes, without any tools. The electronic parts (purple) can be separated to go into an envelope for transport to an e-cycling program. The remaining plastic and metal parts can be placed in ordinary recycling bins.

In recent years, a variety of alternative approaches to product manufacture and disposal have gained recognition. Even small gains in efficiency at each life-cycle stage can have significant effects on environmental impacts and total energy and resource use. Greater efficiency may come from reduced resource use, reuse, or recycling.

Many manufacturers have embraced a concept called **industrial ecology**, in which industrial systems are redesigned to mimic the cycling of materials in natural ecosystems. Industrial ecology combines the disciplines of engineering, chemistry, biology, and economics to find ways to make the generation and disposal of industrial wastes more efficient and environmentally sustainable. Wherever possible, the industrial wastes generated by one process are used in another process or by another industry.

One of the most famous applications of industrial ecology occurs in an industrial zone in Kalundborg, Denmark. Dozens of companies in Kalundborg transfer some of their industrial wastes to other companies, which use those wastes in other processes. For example, sulfur removed from natural gas at an oil refinery is shipped to a nearby producer of sulfuric acid. Ash from a power plant is used to make cement. Waste heat from a power plant is used to heat greenhouses that grow vegetables year round.

Interest in industrial ecology is growing because it makes economic as well as environmental sense. However, industrial ecology requires cross-functional thinking and collaboration; these activities are often restricted when companies are geographically separated.

Thinking of a product as a service can also transform patterns of manufacturing, marketing, and consumption. For example, Interface Carpet, a flooring company based in Atlanta, Georgia, leases "flooring services" instead of selling carpeting. Companies that purchase Interface's flooring services never actually own any carpets, so they do not have to worry about how to dispose of them after their useful life. When it is time to replace carpeting, Interface takes back their old carpets and remanufactures them into new carpets. Over 1.5 million tons of conventional carpeting is dumped into U.S. landfills each year. The Interface model for carpet life-cycle management could significantly reduce this waste stream.

Many companies are using **design for disassembly (DfD)** to circularize the life cycle of their products. DfD involves the development of materials, components, construction techniques, and management systems that permit the easy recovery of parts and materials from discarded products. Dfd principles include fewer parts, making parts from only one material, simple and accessible fasteners, and easy-to-remove batteries and electronics. DfD allows most of the material in a discarded product to be reused in a new product, thereby reducing the need for raw materials. DfD is most effective when manufacturers institute programs that encourage consumers to return products at the end of their useful life.

Electronics companies such as Sony and Panasonic are now producing cell phones that are designed for easy disassembly and recycling. Work is underway to produce DfD computers and television sets (Figure 17.26). DfD principles are also being incorporated into the design and architecture of many new buildings.

Germany requires that automobiles be manufactured so that all parts can be easily recycled. This has encouraged German manufacturers, such as BMW, Mercedes-Benz, and Volkswagen, to use DfD principles in the design of their automobiles. BMW, for example, has designed its cars so that potentially polluting fluids, such as oil, brake fluid, and coolants, can be easily removed; the design of their electronics allows the easy recovery of recyclable parts. At specialized take-back facilities, over 95% of the car's mass is recycled.

Extended product responsibility (EPR) is based on the belief that those who make a product should assume responsibility for that product throughout its life cycle. EPR regulations sometimes take the form of electronic take-back laws, which require manufacturers to take back electronic devices at the end of their useful life at no cost to the consumer. The state of Maine recently enacted take-back legislation requiring manufacturers to pay for the return and reuse or disposal of computers and televisions.

At present, several economic and policy barriers are limiting the shift to a more cyclical approach to manufacturing. For example, many of the costs associated with mining and generating energy are not captured in the price of raw materials. This can make the use of raw materials appear to be cheaper than reuse or recycling. Laws that regulate the transport of hazardous materials complicate the development of industry-wide take-back programs. In the United States, these economic and policy barriers are being addressed on a state-by-state basis, but this can lead to confusion for manufacturers operating at the national and international level.

All of these approaches promote new modes of thinking about product manufacture, use, and disposal. Rather than designing consumer products with a once-through, linear model of resource use, products are designed with reuse in mind from the very start. A more cyclical approach to product life cycles has multiple environmental and economic advantages over current approaches. Fewer raw materials need to be extracted to manufacture products, reducing the environmental impacts of mining and land disturbance. Energy-intensive refining and processing are reduced or even eliminated, thereby saving energy and money and reducing pollution. Life-cycle approaches also tend to emphasize durability in consumer products, which reduces the total production of all forms of waste.

QUESTIONS 17.4

1. Describe the three components of a complete life-cycle assessment for a product.

2. Wastes can be reduced by marketing a product as a service. What is meant by this statement?

3. Describe two barriers to the adoption of cyclical approaches to product manufacture.

(MES) For additional review, go to **MasteringEnvironmentalScience**

17.5 Waste Management Policy and Law

BIG IDEA Federal laws to regulate disposal and management of most waste streams have been enacted in the past 50 years. Although the U.S. EPA sets standards for its disposal, MSW management decisions are made at the city and county level. MSW recycling rates have increased most when all the costs of landfills are understood. Hazardous wastes and radioactive wastes are regulated by the federal government, however, policies and regulations for biomedical waste and e-waste management, where they exist, are generated by state and local governments.

Municipal Solid Waste

■ MSW policies and management decisions are mostly set by city and county governments.

The U.S. federal government first became involved in MSW policy with the passage of the 1899 Rivers and Harbors Act. Using the federal government's constitutional authority over navigable streams, this act prohibited the dumping of waste materials into rivers without a permit. Nevertheless, until 1965 policy related to the disposal of waste was mostly a local affair. Landowners and city and county governments typically disposed of wastes in the most expeditious way, usually in open dumps.

Federal oversight of MSW management increased considerably with the passage of the Solid Waste Disposal Act in 1965. This legislation provided funding for the development of state and local waste management programs. More important, it required states to set standards for the regulation of MSW facilities. The 1970 Resource Recovery Act transferred the oversight of MSW to the U.S. EPA. The following year, EPA administrator William Ruckelshaus initiated his Mission 5,000 Program, aimed at closing 5,000 open dumps within a year and converting all open dumps to sanitary landfills by 1980. Mission accomplished!

Federal and state regulations remain an important element in waste management. In 2009, a bill to mandate glass and aluminum recycling was introduced to the House of Representatives but was not brought to a vote. State-level programs to encourage recycling range from simple education programs, to prohibitions on placing recyclable aluminum or plastic in landfills, to mandatory deposits on bottles and cans.

Nevertheless, most policies and decisions affecting the reduction, reuse, and recycling of MSW remain with local governments. Limited space for landfills and public objections to their expansion have created powerful incentives for reducing waste and recycling. The use of strategies such as placing collection centers in convenient locations and providing curbside collection of glass, aluminum, and paper have helped many cities and counties increase their recycling rates to well above 50% (Figure 17.27).

Changing attitudes about the economics of these programs have been a major factor in their success. In the past, local governments often considered composting and recycling programs to be an economic drain if they were not able to generate revenues to cover their costs. However, when governments calculate all of the costs involved in building and maintaining landfills, they discover that programs to reduce, reuse, and recycle municipal solid waste are a true bargain.

▼ Figure 17.27 **Encouraging Recycling**
Convenient and prominent disposal systems are the foundation of successful recycling programs.

Hazardous, Biomedical, Electronic, and Radioactive Waste

■ Federal laws regulate disposal of hazardous and radioactive waste but not biomedical or e-waste.

QUESTIONS 17.5

1. What are the most important provisions of the Solid Waste Disposal Act and the Resources Recovery Act?

2. Describe two factors that often influence the commitment of city and county governments to recycling programs.

3. Describe the two federal laws that regulate the management of hazardous waste.

(MES) For additional review, go to **MasteringEnvironmentalScience**

In the United States, the management and disposal of hazardous wastes are governed by two federal laws. The Resource Conservation and Recovery Act (RCRA) of 1976 set national goals for protecting human health and the environment from hazardous wastes; it also sets goals for conserving natural resources by reducing and recycling wastes. This law required the EPA to set standards for the disposal of hazardous wastes.

The Comprehensive Environmental Response, Compensation, and Liability Act (CERCLA) of 1980 established the "polluter pays" principle. Under this principle, companies are responsible for the cost of cleaning up sites that have been polluted by their activities. CERCLA also created a tax on industries that produce hazardous wastes; revenues from this tax go into the Superfund. The Superfund pays for cleaning up hazardous wastes in abandoned sites when the company that was responsible for the pollution cannot be identified (Figure 17.28). There are estimated to be hundreds of thousands of such sites in the United States. Between 1980 and 1985, the Superfund grew to $1.9 billion. When CERCLA was reauthorized in 1986, Congress increased the Superfund to $9.3 billion, to be funded by the federal government.

As many businesses, governments, and environmental leaders now recognize, the best approach to managing hazardous waste is to not generate it in the first place. In late 2006, the European Union adopted the Registration, Evaluation, and Authorization of Chemicals (REACH) policy. REACH requires chemical manufacturers and other industries to test for toxic and hazardous substances and to avoid using them.

At present, the United States does not have national regulations regarding e-waste. However, several states have taken action to diminish its stream. For example, in 2003 California passed its Electronic Waste Recycling Act. This act requires manufacturers to reduce the amounts of certain hazardous substances in the electronic devices they produce. It also requires consumers to pay recycling fees when they purchase new devices. The recycling fees go to a special fund to compensate for the costs of e-waste recovery and recycling. Finally, the act mandates that state agencies purchase their electronic devices from manufacturers that have reduced the amount of hazardous chemicals in their products.

The European Union and some states in the United States have recently adopted take-back legislation for some electronic devices. **Take-back regulations** require electronic manufacturers to take back their products at the end of their useful life at no charge to the consumer. The manufacturer is then responsible for the proper disposal or reuse of the materials in those products.

As for biomedical waste,, the Congress passed the Medical Waste Tracking Act in 1988, which set standards for the management and disposal of biomedical waste. This act expired in 1991, and such waste is now regulated by individual states. Regulation standards vary widely from state to state.

►Figure 17.28 **Superfund Site**
Vermont's Elizabeth Copper Mine is now closed, but the pollution from it remains. The company that opened the mine no longer exists, so the cleanup of this site is being funded by the EPA Superfund.

Is voluntary participation in e-waste recycling partnerships good for business?

In July of 2011, a presidential taskforce released its National Strategy for Electronics Stewardship laying out plans to enhance the management of electronics from production to disposal. Under this strategy, the U.S. Environmental Protection Agency committed to encouraging voluntary partnerships with electronics manufacturers and retailers to make responsible electronics recycling easily accessible to consumers. A year later, the EPA moved forward on that commitment with the launch of its Sustainable Materials Management (SMM) Electronics Challenge, an initiative intended to encourage responsible reuse and recycling practices throughout the electronics industry (Figure 17.29). Electronics manufacturers and retailers that participate in the challenge must commit to develop responsible e-waste reuse and recycling programs through the use of independently certified recyclers and to post publicly electronics collection and recycling data. The EPA is encouraging participation through an awards and recognition program.

Participating refurbishers and recyclers must first undergo a rigorous certification process overseen by an approved, independent certification body. Certification standards include environmental accountability, worker and public health, and material accountability throughout restoration, recovery, and recycling processes. E-waste processing in unregulated facilities in the United States or abroad is strictly forbidden. Industry-specific issues of data security are also included in these standards.

A typical certified recycler may accept a wide variety of e-waste. When received, all items are carefully inventoried and evaluated. If devices are serviceable, they are refurbished and resold; if not, they are stripped down to individual components. Some parts like DVDs and disk drives might be resold; other parts like motherboards and printed circuits are sent to refiners for recovery of precious metals (Figure 17.30). The company also recovers other metal components such as aluminum, steel, and copper. The recycler must carefully audit each of these steps. In addition, it is held to high standards of environmentally

▲ Figure 17.29 **Announcing the Challenge** On September 20, 2012, EPA Administrator Lisa P. Jackson launched the SMM Electronics Challenge to "ensure that we are doing all we can to repurpose or safely dispose of the cell phones, computers and other devices we use every day—all while helping to build a robust market for electronics recycling in the United States."

sustainable management including sustainable energy and water use and recycling of nonelectronic waste such as cardboard and plastic.

This program is still in its early stages, but the number of participating manufacturers and retailers is steadily growing. They already include such giants as Best Buy, Staples, Dell, LG Electronics, Panasonic, and Sony. Participants commit to increase the amount of used electronics they themselves collect and to send 100% of these materials to certified refurbishers and recyclers. In return, the challenge provides a very visible and quantifiable way for manufacturers and retailers to showcase their commitment to safe and environmentally sustainable electronics reuse and recycling. The EPA named the first three Electronics Challenge Champions in late 2014. Dell Inc. was recognized for its commitment to the development of waste-free packaging, Sprint received the award for its commitment to cell phone recycling, and Best Buy was praised for its convenient collection and recycling services.

◀ Figure 17.30 **Valuable Stuff** Every year, Americans discard electronics with circuit boards containing more than $60 million in gold and silver. Recycled properly, refiners can recover these and other precious metals for reuse.

Trash 2 Treasure/ Post-Landfill Action Network

Alex Freid majored in Political Science and Philosophy at the University of New Hampshire, where he and a few friends started *UNH Trash 2 Treasure* during his sophomore year. As the first student-led, self-sustaining program of its kind in the country, it became the pilot program for PLAN: Post-Landfill Action Network. PLAN is a national non-profit organization that provides student organizations on campuses across the country access to a variety of tools, resources, and advice on how to launch student-led waste reduction programs.

How and why did you first get involved in waste reduction efforts on your campus?

I have been involved in a number of waste reduction efforts over the years. In high school, I led efforts to increase recycling rates and to get the cafeteria to switch to re-usable silverware. My passion comes from the knowledge that we are operating in completely unsustainable ways: we live on a finite planet and we are mining resources at ever increasing rates and then throwing those resources away, rendering them completely useless. My job is to find workable solutions and to help build the demand and infrastructure around them. So at the end of my freshman year, when I saw dumpsters overflowing with usable materials: food, clothing, furniture, electronics, school supplies and more, I set out to build a solution around that. My goal was to design a program that would be fun and easy for people to participate in and that would be self-sustaining so that it could last beyond my own personal involvement.

Where did you first get the idea for Trash 2 Treasure?

Through my research, I found that students at UNH threw away 25 tons of trash in the average month, but 105 tons in May. I also found that fewer than 100 of the 2,100 colleges and universities across the country had solutions to this problem. Most of these programs gave items to students or local non-profits, and a few had well-designed yard sale programs. However, those programs were often funded by the university, and managed by a Sustainability Office or Residential Life.

At the time, our campus, like many others, was facing large budget cuts. So I set out to design a model that could be adopted on any campus. With a little bit of start-up funding and a committed group of students, we could build a program that needed little to no direct support from UNH.

What steps did you take to get started?

We had to do a lot of negotiating the first year because we had to prove to the administration that we could pull this off. We weren't given all of the drop-off locations we hoped to have, we could not hold the sale in the location we wanted, we were almost denied our permit for the sale by the Chief of Police, we struggled to find summer storage locations. The list goes on. That's not to say we didn't have support at UNH, many people were rooting for us, but we really had to push to make it happen. Once we were able to prove the benefits of the program to the campus and the community, we received an outpouring of support all across the board. This program has now become part of UNH's identity as a sustainable institution.

It was a lot of work to start this program because we literally built it from scratch. That's why we created PLAN, we want to help students with all aspects of this program – from designing a budget, to negotiating with administrators, to managing volunteers, to marketing materials, to organizing and holding the yard sale, and so on.

What are the biggest challenges you've faced with the project, and how did you solve them?

I'd say the biggest challenges we are currently facing are building this organization to scale. We are still in the development phase, and we need a lot of up-front funding to help students launch these programs in their first year. Beyond funding, each campus comes with its own unique challenges. Our goal in building PLAN is to help students through these hurdles as much as possible and to lend a level of credibility to the work that students are doing on their campuses. We want for this to grow big enough that every campus knows about and wants to be a member of PLAN.

How has your work affected other people and the environment?

In the last three years, UNH Trash 2 Treasure has diverted over 110 tons of usable materials from the landfill, properly recycled over 2,000 electronics, donated over 5 tons of food and clothing to local shelters, generated over $54,000 in revenue to be used for future sustainable programs on campus, saved the University over $10,000 in disposal fees and saved parents and students over $216,000 in back-to-school shopping expenses.

What advice would you give to students who want to replicate your model on campus?

Please contact us at www.postlandfill.org. We'd love to work with you! But here are some quick tips for getting started:

Build your network: Organizing is all about building and maintaining relationships. Take the time to get to know people. You can't be a good organizer if you don't have a solid relationship with your fellow activists and with the people who make decisions on your behalf.

Don't take no for an answer: Anyone can say "no" to change but it takes thought, time, consideration, and energy to get to "yes". Often times it's the path of getting there that's more important than the answer itself.

What do you hope for PLAN in the coming years?

I believe that waste is a fundamentally solvable issue. We are smarter than this—and we have the tools, the technology and the resources to do so. What we currently lack is the logistics, the infrastructure, and the demand in our society to implement these solutions on a larger scale. My vision for PLAN is hundreds of campus communities and thousands of student thinkers and leaders working together, sharing advice and ideas, academic research and best practices to build zero-waste campuses as microcosms of the way our society should function. We can design programs to salvage what's reusable and recycle what's not. We can build relationships and networks with innovative companies and technologies to recycle unusual items like vinyl banners or electronics. And we can re-design the system so as to shift our values and our mind-sets. I envision PLAN facilitating the student-led, zero-waste, campus movement.

Summary

Human activities generate large amounts of waste including municipal and industrial solid waste, and hazardous waste. Municipal solid waste consists of wastepaper, food wastes, yard trimmings, plastics, and the like. The composition of the MSW waste streams varies widely among countries, reflecting different patterns of resource consumption and waste management practices. In many developing countries, MSW ends up in open dumps or in rivers. In the United States, the majority of MSW is now placed in sanitary landfills that are lined with clay to minimize leakage of pollutants to aquifers. In some places, MSW may be incinerated in waste-to-energy facilities.

Wastes that are flammable, corrosive, explosive, or toxic are designated as hazardous. These wastes include a variety of household materials such as paints, insecticides, and solvents, as well as toxic wastes from industry. In recent years, discarded electronic devices (e-waste) have added significantly to the hazardous waste stream. In the past, hazardous wastes were simply dumped into open pits or rivers. Today, laws in the United States and other countries regulate the generation and disposal of hazardous waste, as well as the cleanup of polluted sites. In the United States, the disposal of such waste is monitored and regulated by the EPA. Hazardous wastes may be placed in permanent retrieval storage sites or disposed of in licensed incineration facilities. Radioactive wastes are generated by hospitals, laboratories, and nuclear reactors; these wastes require special treatment and long-term monitoring.

Over the past four decades, the U.S. EPA and environmental groups have worked to reduce the MSW stream by the reduction of packaging; the reuse of items such as shopping bags, clothing, furniture, and building materials; and the recycling of materials such as paper, plastic, glass, and metals. In many cities and counties, organic wastes such as food and yard trimmings are separated and converted to compost for gardens and agricultural uses. As of 2014, the total MSW stream had begun to decline and recycling rates were nearing 40%.

Much waste management remains focused on the disposal of waste and the isolation of hazardous materials from the environment. In contrast, life-cycle assessment examines the full life cycle of products from materials acquisition to disposal. This approach also identifies ways to minimize waste at each step in the process.

17.1 Solid Waste

- The United States ranks high among the world's countries in per capita generation of MSW, but the total and per capita waste streams appear to be decreasing.
- Most MSW is now treated in sanitary landfills that are designed to prevent pollutants from leaking into groundwater.

KEY TERMS

municipal solid waste (MSW), industrial solid waste (ISW), sanitary landfill, leachate, waste-to-energy facility (WTE), bioreactor landfill

QUESTIONS

1. Waste managers have been pleased to see total amounts of MSW and per capita rates of MSW generation decrease in recent years. What factors have contributed to these happy trends?
2. How can a landfill be a source of energy?

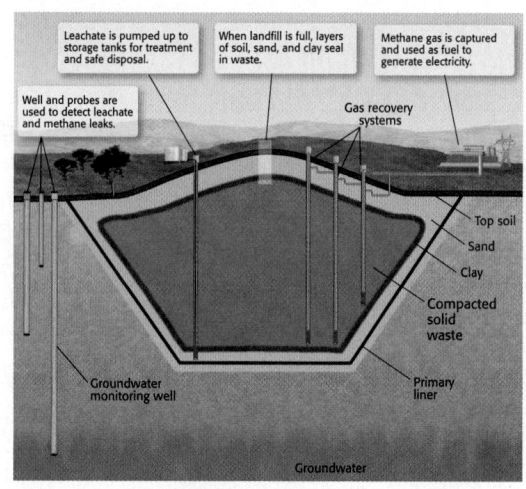

17.2 Hazardous, Biomedical, Electronic, and Radioactive Waste

- Although management of hazardous waste has improved in the United States, thousands of historically polluted sites remain to be cleaned up.
- Because of its potential to spread infection and disease, biomedical wastes must be managed separately from other waste streams.
- E-waste—discarded electronic devices—is a growing part of the global waste stream. Take-back regulations and recycling laws are encouraging the reuse and recycling of e-waste.
- Radioactive wastes are contaminated with isotopes of elements that emit dangerous radiation. These wastes are generated by hospitals, research and weapons laboratories, and nuclear power plants.

KEY TERMS

hazardous waste, permanent retrieval storage site, bioremediation, biomedical waste, e-waste, low-level radioactive waste, high-level radioactive waste

QUESTIONS

1. Describe three approaches for the disposal or cleanup of hazardous waste.
2. Describe two strong economic incentives for recycling e-waste.
3. What is the difference between a high-level nuclear waste storage facility and a permanent waste repository?

17.3 Waste Reduction, Reuse, and Recycling

- Programs to reduce packaging, reuse products, and recycle the materials in products are further reducing the amount of MSW.
- Composting is an efficient way to dispose of organic waste and produce a valuable product.

KEY TERMS

closed-loop recycling, open-loop recycling, compost, vermicomposting

QUESTIONS

1. Most cities and counties post their waste management data on their government websites. In the city or county where you live, what proportions of MSW end up in landfills versus being recycled or composted?
2. What programs does your city or county have in place to encourage waste reduction, reuse, and recycling?

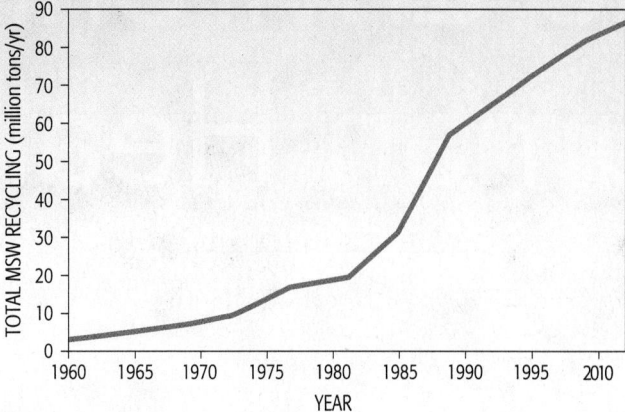

MSW RECYCLING RATES, 1960–2012

17.4 Managing Product Life Cycles

- New approaches to waste management focus on ways to avoid generating waste by assessing the full life cycles of products.
- The generation of waste can be diminished by treating products as services, by designing products for disassembly and recycling, and by extended product responsibility.

KEY TERMS

life-cycle assessment (LCA), industrial ecology, design for disassembly (DfD), extended product responsibility (EPR)

QUESTIONS

1. Industrial ecology strives to make manufacturing processes cyclical rather than linear. What is meant by this statement?
2. Describe three approaches that industrial ecologists are using to diminish manufacturing waste.

17.5 Waste Management Policy and Law

- The U.S. EPA sets standards for MSW disposal, but actual management is overseen by city and county governments.
- Federal laws place responsibility for the management of hazardous waste on the public and private individuals and organizations that produce it.
- Biomedical and e-waste are regulated at the state and local level.

KEY TERM

take-back regulations

QUESTIONS

1. How has U.S. legislation regulating the generation, disposal, and cleanup of hazardous waste changed over the past 50 years?
2. How is financial responsibility for hazardous waste disposal assigned under the Comprehensive Environmental Response, Compensation and Liability Act (CERCLA)?
3. What advantages and disadvantages do you see to federal versus state and local regulation and management of waste streams?

MasteringEnvironmentalScience®

Students Go to **MasteringEnvironmentalScience** for assignments, the eText, and the Study Area with practice tests, activities, and more.

Instructors Go to **MasteringEnvironmentalScience** for automatically graded tutorials and questions that you can assign to your students, plus Instructor Resources.

The Environment and Human Health

579

DDT vs. Malaria

How can we diminish the risk from one health hazard without increasing the risk of another?

Malaria has afflicted the people living near the headwaters of Africa's Kafue River for more than 100,000 years (Figure 18.1). Today, this region is in the nation of Zambia. It is here that the deadliest form of the disease, falciparum malaria, is thought to have evolved; historically, it was more prevalent here than anywhere else on Earth. This disease, noted for intense fevers that recur at two-day intervals, can infect anyone but is especially deadly in small children. Although actual numbers are unknown, it appears that prior to 1900, malaria killed about 20% of the children in the region who were younger than five years old. Since that time, changes in human activities and the environment have had a significant impact on the prevalence of malaria.

In 1926, Europeans discovered copper ore near the Kafue River, and operations at the Roan Antelope Copper Mine began soon afterward. The town of Luanshya was established in 1930 to accommodate British engineers and overseers, as well as several thousand local people who worked the mines. Company records show that well over 50% of Luanshya's 5,000-plus citizens—white and black—contracted malaria in each of the next two years (Figure 18.2).

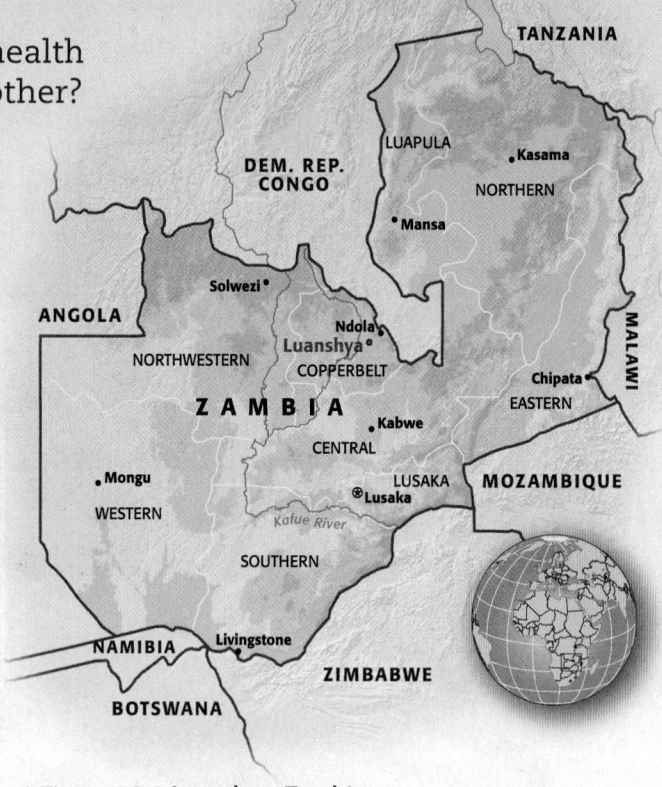

▲ Figure 18.1 **Luanshya, Zambia**
Luanshya is located in the Copper Belt Province in north-central Zambia. The swamps and seasonally wet savannas of this region provide ideal habitats for mosquito larvae.

▼ Figure 18.2 **Malaria Incidence Rate**
The incidence of malaria among residents of Luanshya began to diminish in 1931 with the implementation of measures to control mosquito populations. Several factors, including the global ban on DDT use, have contributed to the resurgence of malaria since 1970.

Source: Utzinger, J., Tozan, Y., and Singer, B.H. 2001. Efficacy and cost effectiveness of environmental management for malaria control. Tropical Medicine and International Health 6: 677–687.

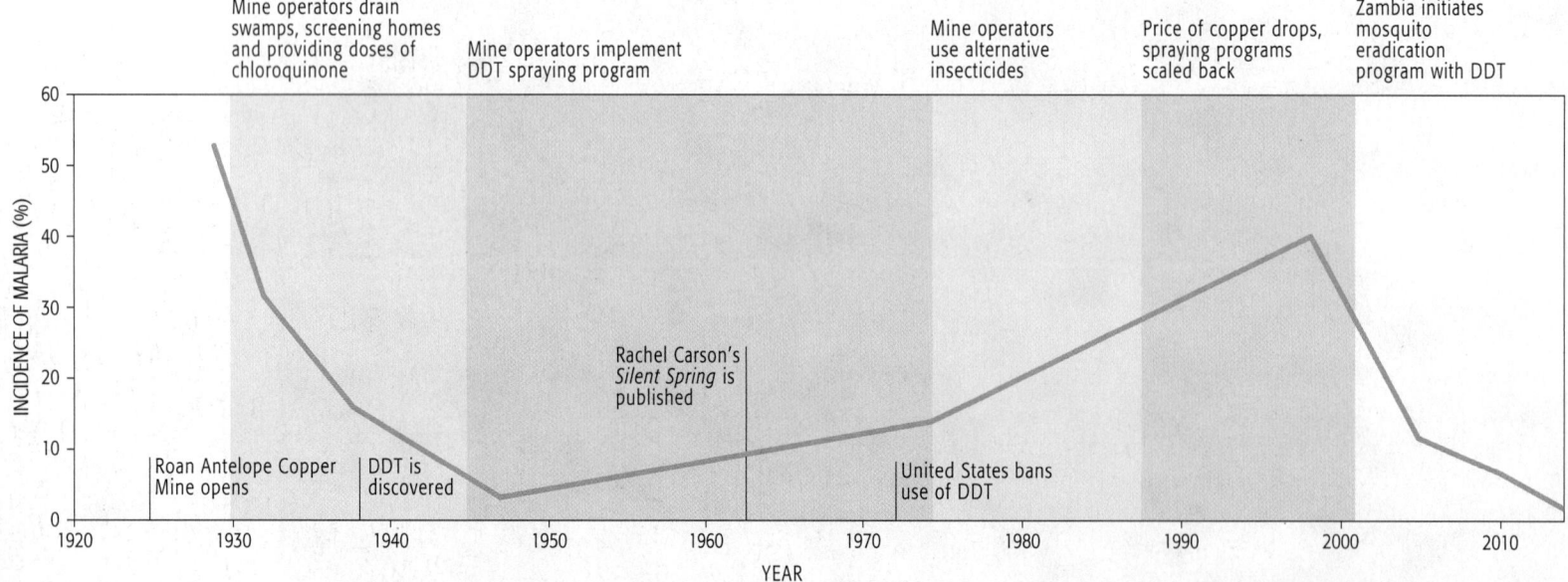

It was clear to mine owners that malaria was by far the greatest obstacle to the extraction of the region's copper, and they immediately began an aggressive program to stop transmission of this disease. Mining officials knew that the parasite that causes malaria is transmitted by mosquitoes of the genus *Anopheles* (Figure 18.3). The officials recruited hundreds of men to drain the swamps and marshy "dambos" where mosquito eggs are laid and their larvae mature. To prevent the entry of adult mosquitoes, the company put screens on the windows and doors of houses. Workers and their families were given regular doses of chloroquinone, a drug that provided considerable resistance to the disease. With these measures in place, by 1939, only 13% of Luanshya's residents were contracting malaria each year; mortality from the disease had declined by 90%.

In 1946, Roan Antelope Mining began spraying the recently developed insecticide DDT (dichlorodiphenyltrichloroethane). By 1948, the incidence of malaria had dropped to less than 2%. That year no mortalities from the disease were recorded in Luanshya.

In 1947, the World Health Organization's Expert Committee on Malaria reported that with DDT there was "at last a method of controlling malaria in many areas at costs within the economic means of the people." So it seemed. Throughout the 1950s and 1960s, malaria rates in central Africa remained low. Public health officials were optimistic that the disease might soon be eradicated.

Yet even then, evidence of the serious effects of DDT was accumulating. In 1962, Rachel Carson's book *Silent Spring* called attention to the fact that this chemical accumulates in the tissues of many animals, causing serious environmental effects. She also noted that DDT causes cancer and developmental abnormalities in humans. In 1972, the United States banned all uses of DDT. Other countries soon followed suit. The World Health Organization (WHO) withdrew its support for the use of DDT to control malaria in 1974. Alternative insecticides were available, but they were significantly more expensive. Initially, mining companies used their considerable wealth to maintain spraying programs using these alternatives. But in the late 1970s, the price of copper began to plummet, and spraying programs were scaled back. The incidence of malaria in the Copper Belt region began to rise once again.

By 1976, when Zambia began its first thorough national surveys of infection rates, the incidence of malaria in the Luanshya region was back up to 12% per year. In the following years, the incidence increased at an alarming rate. By 2000, the incidence of malaria in the Copper Belt region was 37.5%. More than 20% of children were dying before their fifth birthday, and 40% of those deaths were due to malaria.

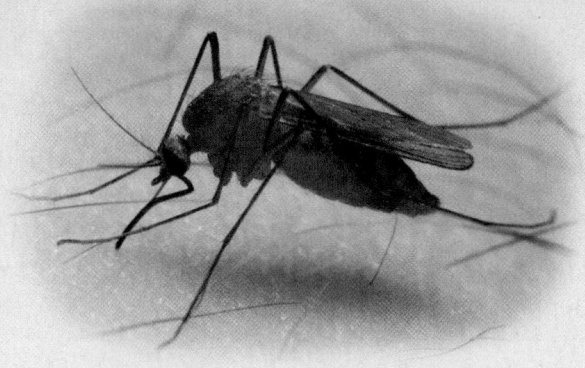

▲ Figure 18.3 *Anopheles* Mosquito
This insect is the primary vector for the parasite that causes falciparum malaria. Larvae develop and mature in standing bodies of water. Females take multiple blood meals. In doing so, they transfer the parasite from infected to noninfected individuals.

In 2000, the Zambian government began to implement measures to control this epidemic. These included the distribution of insecticide-treated mosquito bed nets and the conservative use of DDT. Spraying of DDT was confined to the interior walls and ceilings of homes, where mosquitoes land at night (Figure 18.4). These measures reduced the incidence of malaria in Luanshya and nearby villages by more than 75%. Across Zambia, childhood mortality due to malaria declined by 33%. Recognizing these benefits, WHO reversed its policies and in 2006 began active support of the use of DDT to limit malaria.

Many public health officials worry that the benefits of using DDT do not offset the risks that this chemical poses to the environment and human health. For example, prenatal exposure to small amounts of DDT increases the likelihood of premature births; such births account for 17% of infant mortality in Zambia.

Scientists point out that the prohibition of DDT was probably not the sole cause of the recent malaria epidemic in places like Luanshya. Over the last 80 years, malaria parasites have become increasingly resistant to chloroquinone. Since 1970, the mean annual temperature in this region has increased by 2 °C; in many places, this increase in temperature appears to have improved breeding conditions for *Anopheles*. Since 1990, human populations in sub-Saharan Africa have doubled, and the higher population density facilitates the transmission of disease. In addition, 20–30% of the population is infected with HIV/AIDS, which diminishes resistance to malaria infections. Meanwhile, the debate over the use of DDT to control malaria is likely to continue. It will be informed by our changing understanding of all the environmental factors that affect the spread of the disease.

- *How can we best evaluate and measure the health of human populations?*

- *How can we assess the risks of hazards that may influence such health measures?*

- *What are the characteristics of important physical, chemical, and biological hazards to human health?*

- *How are changes in our environment influencing the risks of these hazards?*

▼ Figure 18.4 Protecting Against Mosquitoes
Programs to spray DDT on the interior walls and ceilings of homes have been implemented in many African countries. Female mosquitoes are active primarily at night and have a habit of lighting on walls before seeking out a blood meal.

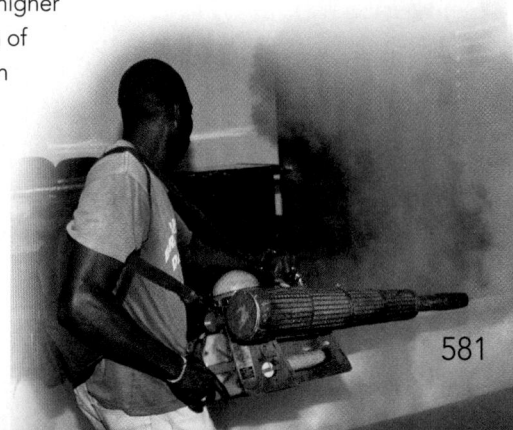

18.1 Introduction to Public Health

BIG IDEA The overall health of human populations is most often assessed with statistical measures such as life expectancy and infant mortality. Leading causes of death in populations depend on factors such as age structure and social and economic conditions. Environmental factors affect the **risk**, or likelihood, of particular hazards impairing health or causing death. However, human perceptions of the magnitude of risk associated with a particular hazard are often significantly different from the actual risk.

Measuring Public Health

■ Likelihood of death and causes of death are important measures of the health of human populations.

Where YOU LIVE How healthy is your state's population?

By measures such as death rate and infant mortality rate, the health of the U.S. population ranks among the top 10% of countries in the world. In 2013, the death rate was 0.799% or 7.99/1,000 individuals and the infant mortality rate was 0.617% or 6.17/1,000 live births. However, there is considerable variation in these measures among states. To obtain the latest official data for your state, query "death rate" or "infant mortality rate," followed by the name of your state. How does your state compare to the national average with regard to each of these measures? What factors might contribute to these differences?

How does a doctor assess the status of your health? Perhaps by checking your body temperature, pulse rate, and blood pressure. Each of these measures provides information about your health because your doctor understands how they are connected to your basic bodily functions. Your body temperature is determined by your metabolic rate and the presence or absence of infections; your pulse and blood pressure are direct measures of your heart's function.

In a similar fashion, population-level statistics can be used to assess the health of human communities. Life expectancy at birth—the average number of years that a newborn child is likely to live—is probably the most widely used measure of the overall health of human populations (see Module 5.2). Since 1950, worldwide life expectancy at birth has increased from 46 years to more than 70 years.

Mortality rates also measure the health of a community. Neonatal mortality rate is the percentage of newborns dying before 28 days of age, and infant mortality rate is the percentage of newborns that die during their first year of life. Although they focus on the health of infants, public health officials consider these statistics to be particularly sensitive indicators of factors that influence the overall health of a population, such

as water quality, sanitation, nutrition, and health care. In 1950, the infant mortality rate of the world's human population was 15.3%; currently, it is 3.8%.

An elevated body temperature indicates illness but tells a doctor relatively little about what is ailing a patient. In a similar way, mortality statistics alone tell little about what factors are causing rates to increase or decrease. Public health officials can infer much more about the health of a population from statistics on the leading causes of death (**Figure 18.5**). Worldwide, cardiovascular diseases, including heart attacks, strokes, and congestive heart failure, are among the most important causes of human deaths. Infectious diseases are a close second, causing 25% of all deaths. Among the most important infectious diseases are diarrheal diseases, human immunodeficiency virus/acquired immunodeficiency syndrome (HIV/AIDS), tuberculosis, malaria, and respiratory infections, such as influenza and pneumonia.

Leading causes of death are greatly influenced by the age of the population. In the United States, for example, accidents are the leading cause of death for people between 1 and 44 years of age. For people between 45 and 64, the leading cause of death is cancer. For people 65 and older, it is cardiovascular disease.

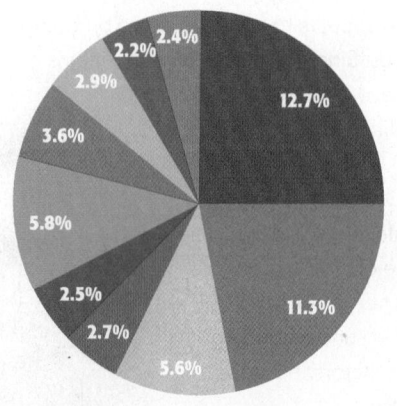

Noninfectious Disease
- Heart attack
- Stroke
- Chronic lung disease
- Lung and tracheal cancers
- Diabetes

Infectious Disease
- Pneumonia
- Diarrheal diseases
- HIV/AIDS

Other
- Premature birth
- Road injury

◀ Figure 18.5 **Worldwide Causes of Death**
These 10 leading causes of death account for about 51% of deaths worldwide. Percentages relative to total number of deaths are indicated in each slice. Infectious diseases account for about 25% of all deaths, but pneumonia, diarrheal disease, and HIV/AIDS are the only infectious diseases among the leading 10. Similarly, cardiovascular diseases claim about 30% of all deaths, but only heart attacks and strokes are among the leading 10.

Source: World Health Organization. 2013. The top 10 causes of death. Fact Sheet no. 310. http://www.who.int/mediacentre/factsheets/fs310/en/index2.html#. Last accessed April 15, 2014.

50 WEALTHIEST COUNTRIES

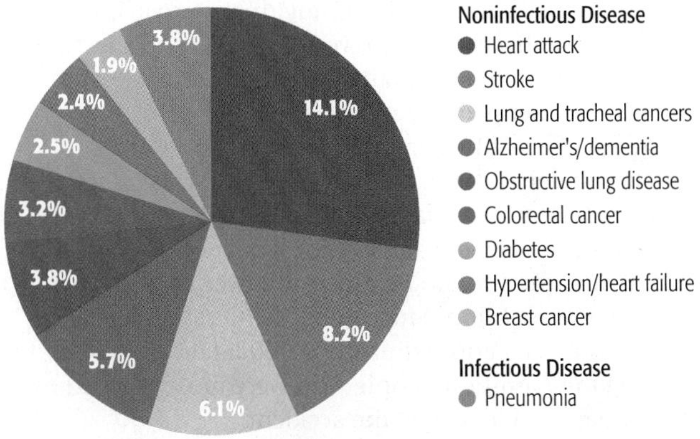

3.8%
1.9%
2.4%
2.5%
3.2%
3.8%
5.7%
6.1%
8.2%
14.1%

Noninfectious Disease
- Heart attack
- Stroke
- Lung and tracheal cancers
- Alzheimer's/dementia
- Obstructive lung disease
- Colorectal cancer
- Diabetes
- Hypertension/heart failure
- Breast cancer

Infectious Disease
- Pneumonia

50 POOREST COUNTRIES

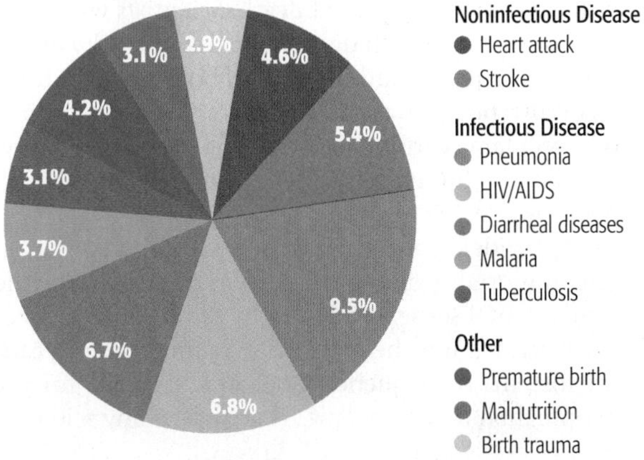

3.1% 2.9%
4.2%
3.1%
3.7%
6.7%
6.8%
9.5%
5.4%
4.6%

Noninfectious Disease
- Heart attack
- Stroke

Infectious Disease
- Pneumonia
- HIV/AIDS
- Diarrheal diseases
- Malaria
- Tuberculosis

Other
- Premature birth
- Malnutrition
- Birth trauma

◀ **Figure 18.6 Wealth and Causes of Death**
Differences in environmental and social conditions and the age structure of populations cause significant differences in the leading causes of death in wealthy and poor countries. The leading 10 represent about 50% of all deaths in both sets of countries.

Source: World Health Organization. 2013. The top 10 causes of death. Fact Sheet no. 310. http://www.who.int/mediacentre/factsheets/fs310/en/index2.html#. Last accessed April 15, 2014.

Changes in leading causes of death are directly related to changes in the environment. In 1900, for example, diarrheal diseases were responsible for more than 8% of all deaths in the United States. Today they account for less than 0.1%. Diarrheal diseases are caused by several different intestinal parasites, and their decreasing prevalence over the last century is a direct consequence of improved management of drinking water and waste.

The leading causes of death in poor countries differ significantly from those in wealthy countries (Figure 18.6). In the world's poorest countries, poor sanitation and nutrition result in a greater prevalence of infectious diseases. Malnutrition is among the 10 leading causes of death. In these countries, about 40% of deaths occur among children less than 15 years old; only about 20% of deaths occur among those older than 70. In wealthy countries, over 70% of deaths are among people older than 70, and cardiovascular diseases, cancers, and Alzheimer's disease are more prevalent. Pneumonia is the only infectious disease to appear among the 10 leading causes of death for both.

[Q] *How much smoking do you have to do throughout your life to be considered "at-risk"?*

Shayna Jefferson, University of Florida

Norm: [A] Smoking creates a variety of risks to your health. There is not a clear threshold for an amount of smoking or length of smoking below which there is no risk.

Hazards and Risk

■ Risks posed by environmental hazards may be immediate or delayed.

Most causes of death are influenced by hazards in the environment, although the specific relationships depend on the kind of hazard. For example, the prevalence of diarrheal diseases in developing countries is directly tied to inadequate sanitation for water and food. Respiratory diseases are exacerbated by poor indoor and outdoor air quality. The incidence of malaria is strongly influenced by climate.

The hazards that influence human health can be grouped into four categories—physical, chemical, biological, and cultural (Figure 18.7). Physical hazards include geological events such as earthquakes, weather events such as blizzards and floods, and the effects of radiation, such as that from ultraviolet light. Chemical hazards include pollutants in water, air, and food. Biological hazards are the result of a diverse array of organisms that cause or transmit infectious disease. Cultural hazards are a consequence of where people live, their occupations, and their lifestyle choices. Urban lifestyles, for example, increase the risk of being injured in automobile collisions. Lifestyle choices also affect an individual's exposure to hazards. Choosing to use cigarettes, for example, significantly increases exposure to the chemical toxins in tobacco smoke.

▲ **Figure 18.7 Environmental Hazards**
Air pollution is a chemical hazard Ⓐ. Tornados such as this one in rural Kansas are physical hazards Ⓑ. Measles and other infectious diseases are biological hazards Ⓒ. Cultural hazards refer to lifestyle choices such as cigarette smoking Ⓓ.

QUESTIONS 18.1

1. How do the leading causes of death in developed countries differ from those in developing countries?

2. Using an example, explain what is meant by latent risk.

3. Describe five factors that can cause the perception of the risk associated with a hazard to differ from its actual risk.

(MES) For additional review, go to **MasteringEnvironmentalScience**

Risk is the mathematical probability of a harmful outcome, such as injury or death. Scientists who study environmental health discuss the risk posed by an environmental hazard in terms of both the likelihood of its occurrence and its consequences. Thus, the probability of a very large earthquake occurring near an urban center may be low, but its consequences in terms of injuries or lives lost may be high.

One widely used measure of environmental risk is the rate of mortality due to a particular hazard. In the most fundamental sense, death has the same consequence for everyone: It is the end of life. Yet in terms of years of life lost, the consequences of death are greater for young people than for old people. Therefore, many scientists

and ethicists argue that mortality risk should be measured in terms of lost years of life. Using this approach, a hazard that affects children would be viewed as having a higher risk than an equally likely hazard that affects primarily older people.

Some effects of a hazard are immediate, while others are delayed; the delayed effects are called **latent consequences**. For example, more than 50 people died almost immediately as a consequence of the 1986 accident at the Chernobyl Nuclear Power Plant in the former Soviet Union. Since then, WHO estimates that there have been approximately 9,000 latent cancer deaths among the 7 million people who were most exposed to the radiation released by the accident.

Risk Perception and Reality

■ Human perceptions of risk often differ from reality.

Our world is filled with both natural and human-caused hazards. Risk measurement and assessment are powerful tools for making decisions about how to prioritize our responses to these hazards. Nevertheless, it is sometimes difficult to quantify risk. In addition, individual and societal responses to hazards are often determined by perceptions of risk that do not correspond to reality. Some of the key factors that influence perceptions of risk are described below.

The myth of zero risk.   The U.S. EPA sets standards for acceptable levels of cancer-causing chemicals in the air and drinking water. Yet survey after survey shows that most people believe there is no safe level of exposure to a hazard such as a cancer-causing agent. Instead, they believe that the risk should be reduced to zero. In practice, it is virtually impossible to reduce the risk of most natural and human-caused hazards to zero. Furthermore, the incremental costs of reducing the risk of a hazard generally increase substantially as the level of risk from that hazard is decreased.

Public awareness.   Publicity has a significant effect on perceptions of the degree of risk posed by hazards. For example, Alar is a synthetic chemical that was widely used by farmers between 1963 and 1989 to encourage ripening in several fruit crops, including apples (Figure 18.8). Studies between 1984 and 1989 found that Alar elevated the risk for cancer in laboratory animals. In 1989, a segment on the CBS television show *60 Minutes* portrayed Alar as a significant and imminent threat to public health. Public reaction was swift, and sales of apples plummeted. The EPA banned the use of Alar very soon after the broadcast.

In retrospect, public response to this hazard was disproportionate to the actual risk. The laboratory tests that prompted the Alar scare used very large doses of the chemical. Even consuming large quantities of juice from Alar-treated apples would cause a lifetime cancer risk of only about 5 per million. This means that an individual who consumes large amounts of Alar-laden apples

has 0.0005% greater chance of getting cancer than an individual who does not. This level of risk is far lower than that posed by many other commonly used chemicals.

Risk–risk trade-offs.   The benefits of reducing the risk posed by one hazard are often considered in isolation from other, related risks. For example, the leaders of many African countries feel that the concerns of some environmental groups about the risks of DDT neglect the risk posed by malaria if DDT cannot be used. Certainly, the benefits of actions to reduce the risk of a hazard always incur costs or other risks. These are called **risk–risk trade-offs**. However, such costs are often poorly understood. In addition, the people receiving the benefit of reducing one risk may not be the ones who must pay the costs associated with elevating another risk.

Control.   People generally perceive a risk to be lower or more acceptable if they feel they have some level of control over that risk. For example, each year in the United States, about 43,000 people die in auto collisions, a rate equal to about 1.4 deaths per 100 million miles traveled. The risk associated with commercial air travel based on miles traveled is considerably less than one-tenth of this amount. Even so, many people perceive the risk of automobile travel to be lower than that of air travel; this is partly because they believe that their driving skills and habits give them more control over that risk.

Risk and time.   People tend to view latent risks as being less than immediate risks, even if they are equal or even greater. For example, it is now widely known that 90% of all lung cancers are the result of smoking. However, a large number of people in their teens and twenties ignore this risk because it is not likely to be realized until later in life.

In summary, our perception of the risk associated with a particular hazard is often quite different from the actual risk. This significantly influences how we as individuals and societies set priorities for and respond to environmental hazards.

▼ **Figure 18.8 Overreaction to Alar?**
Publicity about the possible health effects of apples sprayed with Alar led to its being banned by the EPA. Subsequent studies revealed that the risks to humans from Alar were minimal.

18.2 Physical Hazards in the Environment

BIG IDEA Physical hazards are threats to human health that derive from physical processes in the environment. These include geologic hazards such as earthquakes and volcanic eruptions; weather hazards such as hurricanes, floods, and extreme temperatures; and wildfires. Although physical hazards are not among the 10 leading causes of death, individual events may claim hundreds or even thousands of lives in a very brief period of time. Such events are often difficult to predict or prevent. We often call these events "natural disasters," but their impacts on human life and property are greatly influenced by how we design and build our homes and cities and the ways in which we modify the ecosystems that surround us.

Geological Hazards

■ Earthquakes and tsunamis arise from natural causes, but their impacts are often increased by human actions.

Earthquakes are caused by movements in Earth's crust that are usually associated with geological faults (see Module 3.4). Around the world, crustal movements are occurring constantly, but the energy they release varies widely. This variation in energy is responsible for the great range in the magnitude, or intensity, of earthquakes.

The energy released by earthquakes is transmitted through Earth's crust and interior as waves that geologists detect and measure with seismometers. Earthquake energy is usually reported as a number on the **Richter magnitude scale**, in which each successive number indicates a 32-fold increase in energy. **Earthquake intensity** is a measure of the severity of shaking at a given location, and it is determined by its effects on people, structures, and the natural environment. Earthquakes of similar magnitude may have different intensities depending on local geological conditions and the quality of building construction. Low-magnitude earthquakes are far more common than those of higher magnitudes. Thousands of earthquakes of magnitude 3 or less occur on Earth every day, but

most go unnoticed. Earth experiences two to three magnitude 5 earthquakes each day; these earthquakes release sufficient energy to damage poorly constructed buildings near the source, or epicenter, of the earthquake (**Figure 18.9**).

▼ **Figure 18.9 Earthquake Magnitude and Intensity** The Richter magnitude scale expresses the energy released during an earthquake. The earthquake intensity scale indicates its effects on people and buildings. The 2011 earthquake centered off the east coast of Japan had a magnitude of 9.0 and an intensity of X.

Richter magnitudes	Earthquake intensity	Earthquake effects	Estimated frequency of occurrence
1.0–3.0	I	Generally not felt except under very favorable conditions	Thousands each day
3.0–3.9	II–III	Felt by most people indoors; vibrations similar to a passing truck	100,000 each year
4.0–4.9	IV–V	Felt by nearly everyone; dishes and windows disturbed or broken; unstable objects overturned	13,000 each year
5.0–5.9	VI–VII	Frightening movement felt by all; heavy furniture moved; considerable damage to poorly built structures	1,300 each year
6.0–6.9	VII–IX	Considerable damage to many buildings; great damage to poorly constructed buildings; buildings may be shifted off foundations	100–130 each year
7.0 and higher	VIII–XII	Widespread damage, even to single story wood buildings; at extreme bridges destroyed and rails bent	10–15 each year

► Figure 18.10 **The Geography of Earthquake Hazards**

Severe earthquakes are most likely near the boundaries of Earth's major tectonic plates. This includes the west coast of the Americas and much of Asia and the Middle East.

Source: Global Seismic Hazard Assessment Program. http://www .seismo.ethz.ch/static/gshap/. Last accessed April 16, 2014.

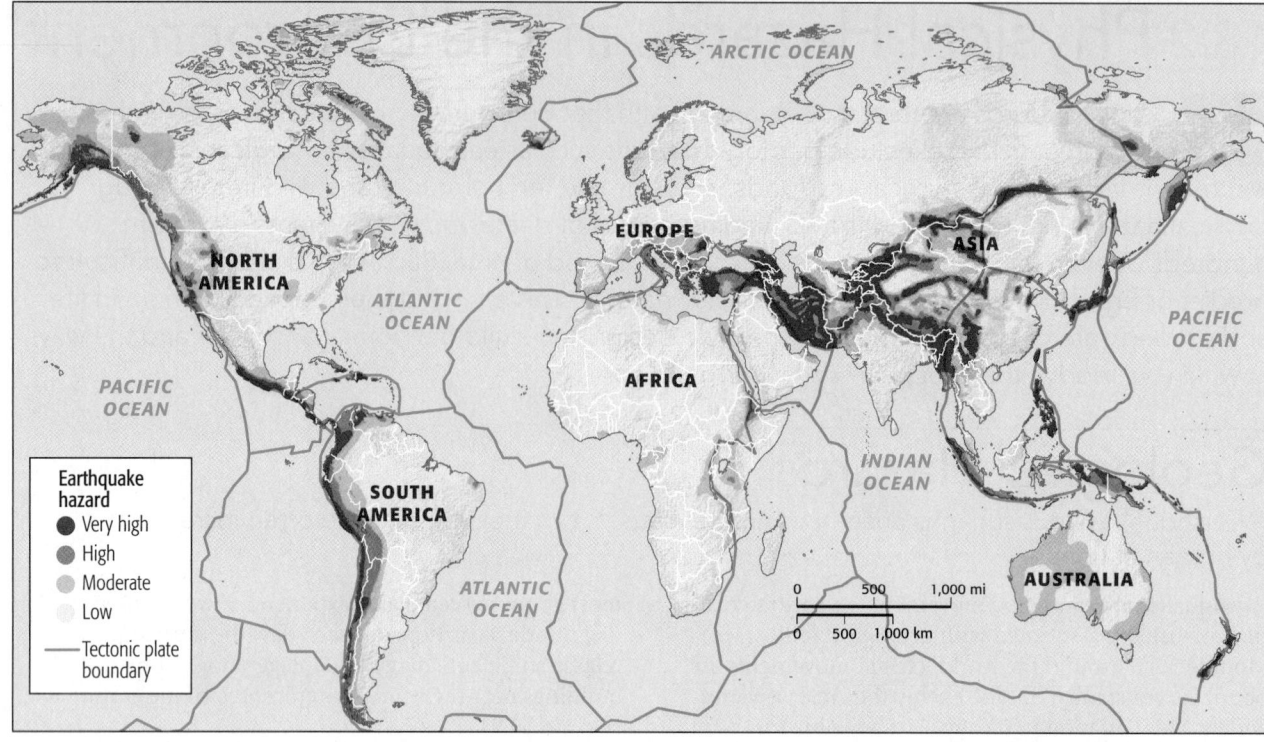

Over 90% of earthquakes of magnitude 6 or more occur in association with the boundaries of Earth's major tectonic plates (Figure 18.10). Earthquakes above magnitude 6 often cause considerable damage and loss of life. The effects of these earthquakes depend on their proximity to towns and cities and on the quality of human developments and building construction. Thus the risk of catastrophic disasters is increased by the rapid growth of very large cities in the vicinity of major faults, such as Mexico City, Los Angeles, and Tokyo.

The combined effects of location, poor construction, and population density are exemplified by comparing two recent earthquakes. In January 2010, a magnitude 7.0 quake struck Haiti near its densely populated capital, Port au Prince. This quake killed more than 225,000 people and left about 2 million homeless. Property damage was estimated to exceed $100 billion. Standards for the construction of buildings are notably absent in Haiti, and few structures were designed to withstand even moderate earthquakes. Three months later, the adjacent communities of Mexicali, Mexico, and Calexico, California, experienced a magnitude 7.2 earthquake. Although this quake was considerably stronger than the quake in Haiti, there were only two deaths, primarily because Mexico and the United States enforce strict building standards in earthquake-prone areas (Figure 18.11).

Efforts to reduce the risks associated with earthquakes have focused on the design of earthquake-resistant buildings and infrastructure. Modern materials, such as steel-reinforced concrete, are able to withstand the sideways forces of earthquakes. Building codes in many earthquake-prone areas now require that homes and apartments be designed to withstand earthquakes up to magnitude 7.8. Such standards help to protect both property and human lives.

► Figure 18.11 **A Tale of Two Cities**

In 2010, a magnitude 7.0 earthquake in Haiti devastated its capitol city, Port au Prince Ⓐ. That year a magnitude 7.2 earthquake near Mexicali, Mexico, resulted in property damage and loss of human life that were modest in comparison Ⓑ. Building codes are far stricter in Mexico and the United States than in Haiti.

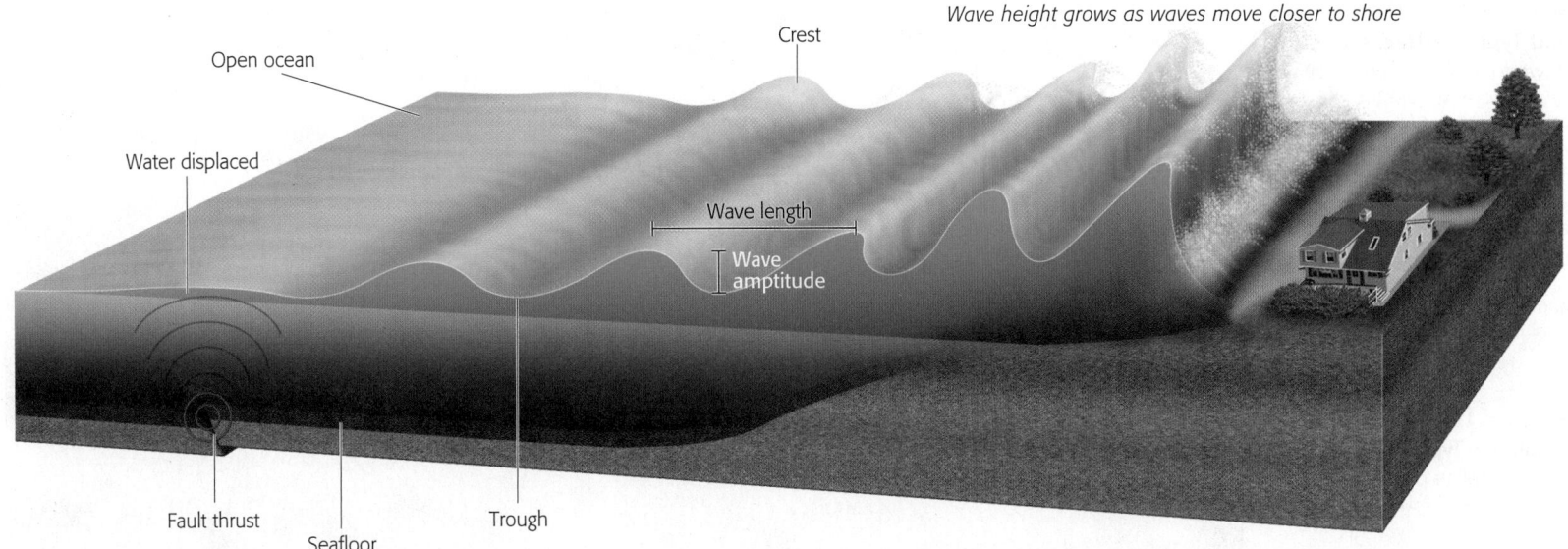

Wave height grows as waves move closer to shore

Open ocean

Crest

Water displaced

Wave length

Wave amplitude

Fault thrust

Seafloor

Trough

▲ Figure 18.12 **Forming a Tsunami**

Tsunami waves may be hardly noticed as they move across the open ocean, but their height is magnified as they move into the shallow water near coasts.

Earthquakes and the movement of sediment on the ocean floor can trigger tsunamis. Tsunamis are often called "tidal waves," although they are unrelated to tides. Tsunami waves may be only 1–3 m high as they move through deep ocean areas. As they move into shallow water, however, their height is magnified, and massive amounts of water may move onto shore (Figure 18.12).

On March 11, 2011, a magnitude 9.0 earthquake occurred in the deep ocean about 70 km (43 mi) off the coast of Honshu, Japan's largest island. Within minutes, a powerful tsunami, with waves as high as 40 m (132 ft), washed across the heavily populated region around the industrial city of Sendai (Figure 18.13). Over 400,000 buildings were destroyed and 18,500 people lost their lives in this event. This tsunami was responsible for the destruction of three reactors at the nearby Fukushima Daiichi nuclear power plant. The implications of this event for nuclear power generation are discussed in Chapter 14 (Module 14.5).

Tsunamis are random events, but their consequences are significantly influenced by human modifications of coastal ecosystems. In 2004, a large earthquake west of the Indonesian island of Sumatra caused a tsunami that struck the coasts of many countries surrounding the Indian Ocean, destroying coastal cities and towns and killing more than 150,000 people. The destruction and loss of life were greatest where coastal mangrove swamps had been eliminated by human development. These swamps can buffer the impacts of tsunamis and storm surges.

Volcanic eruptions also create hazards to human health. Lava flows may envelop homes or villages, and explosive eruptions may directly threaten human life. For example, the violent 1980 eruption of Mount St. Helens in the state of Washington killed 57 people. Plumes of ash from volcanoes can pose a threat to human communities. In 2010, an ash plume from a volcano in Iceland drifted over northern Europe. Because of the risk that the ash would cause jet engines to fail, airports across the region were forced to close for nearly a week.

Other hazards associated with Earth's crust are landslides and mudslides, caused by a variety of factors that undermine the stability of soils and rock on steep slopes. Worldwide, such events are responsible for several thousand deaths each year. On slopes where materials are on the verge of movement, landslides can be initiated by heavy rains, snowmelt, stream erosion, and earthquakes. Human activities such as heavy logging can also increase landslide risk. Above normal rainfall and stream erosion are thought to have been significant factors in a March 2014 mudslide near Oso, Washington, that claimed 43 lives. Logging in the slide area appears to have increased slope vulnerability as well.

▲ Figure 18.13 **The 2011 Tsunami**
This photo shows some of the destruction caused to the Japanese coastal city of Sendai after the tsunami struck. The economic cost of this event has exceeded $235 billion, making it the costliest natural disaster ever.

▶ **Figure 18.14 Hurricane and Typhoon Tracks**
Powerful tropical storms are much more frequent in the Northern Hemisphere than in the Southern Hemisphere. They pose the greatest risk to eastern Asia, the Caribbean, and eastern North America.

Source: National Oceanic and Atmospheric Administration. http://csc.noaa.gov/hurricanes/.

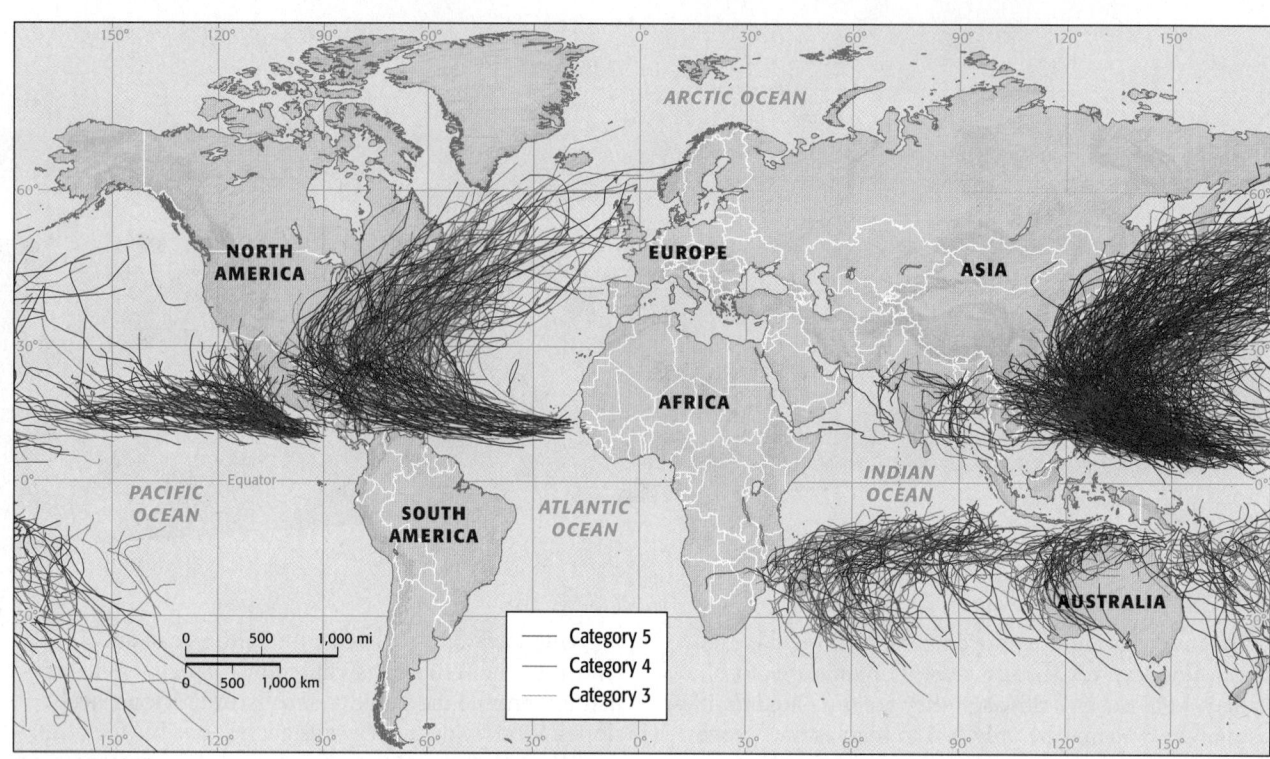

▼ **Figure 18.15 The 2008 Myanmar Typhoon**
These satellite photographs compare the Irrawaddy River and its delta in Myanmar before Ⓐ and after Ⓑ a powerful storm in 2008. Extensive clearing of land in the river floodplain (beige in photograph A) magnified the extent of flooding and the loss of human life.

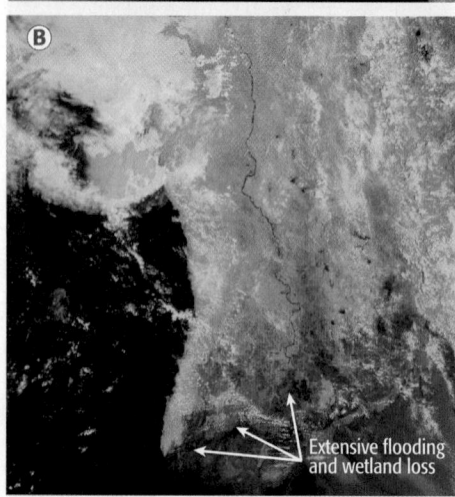

Weather Hazards

■ The overall risks from extreme weather events are low, but the impact of individual events can be very high.

Tropical storms, tornados, floods, blizzards, droughts, and heat waves are among the most important weather hazards to human well-being. Between 2001 and 2006, extreme weather events around the world claimed the lives of about 22,000 people per year, or about 0.03% of all deaths. This may seem low, but individual events such as major hurricanes may claim the lives of thousands of people at once.

Since 1990, tropical storms and their associated floods have accounted for more than 75% of weather-related deaths. Intense tropical storms in the Atlantic and eastern Pacific oceans are called hurricanes; similar storms in the western Pacific and Indian oceans are called typhoons or cyclones. These storms originate over warm ocean water north and south of the equator. Prevailing winds carry them westward, so they are most likely to make landfall along the east coasts of continents (Figure 18.14). Tropical storms bring heavy rains and high water levels called **storm surges** that can cause devastating floods in low-lying regions and on islands.

Death tolls from tropical storms and floods are typically highest in poor countries that have undergone significant deforestation and lack the resources to protect their people during and after storms. For example, in May 2008, a severe typhoon struck the country of Myanmar in Southeast Asia. At least 150,000 people died (six times the global annual average number of weather-related deaths) in the low-lying delta of the Irrawaddy River. This high death toll was partly due to deforestation of hillsides and coastal swamps, which resulted in landslides and accentuated flooding (Figure 18.15). It was also due to the failure of the Myanmar government to provide aid to the people in hundreds of affected villages and towns. The lack of clean water facilitated the spread of disease, causing many deaths soon after the storm.

Hurricane Katrina, which struck the U.S. Gulf Coast in 2005, demonstrated that such events can cause high death tolls in developed countries, too. Katrina was an especially intense storm when it made landfall along the delta of the Mississippi River near New Orleans. The storm surge overwhelmed the levees and canals that normally prevent river water from flooding the parts of New Orleans that lie at or below sea level. Over a period of days, 80% of the city became flooded, and more than 1,800 people lost their lives. For comparison, Hurricane Sandy was responsible for relatively few deaths, 117 deaths, in 2012.

Several factors contributed to the high death toll and widespread loss of property associated with Hurricane Katrina. Investigations revealed numerous flaws in the design and construction of the levees. Destruction of coastal wetlands and floodplain forests likely contributed to the remarkably high storm surge. Aid to the stranded population from federal, state, and local agencies was slow and disorganized. Indeed, this disaster resulted in the resignation of the director of the Federal Emergency Management Agency within days of the storm.

One of the important lessons from Hurricane Katrina and similar disasters is that the risks they pose to human health and well-being are not spread evenly across all segments of the population. For example, the risk of death or serious injury was much higher among New Orleans's poor and black people than among the wealthier, white population. The unevenness has continued in the years since this disaster; reconstruction and restoration have been much slower in the city's poorer areas than in wealthier areas (Figure 18.16).

Unlike tropical storms, extreme temperature events, such as blizzards and heat waves, generally account for a very small proportion of weather-related deaths. The 2003 heat wave in Europe was a major exception to this generalization. During August of that year, temperatures across southern Europe were routinely close to 40 °C (104 °F). Many people lacked air-conditioning and were unaccustomed to such heat. It is estimated that 35,000 people, most of whom were elderly, died as a consequence (Figure 18.17). Most of these deaths occurred in large cities.

Extreme weather events such as intense hurricanes and heat waves have become more frequent in the past several decades. Scientists warn that this may be a consequence of global warming (see Module 9.5). This issue is discussed in more detail later in this chapter (in Module 18.5).

Despite increases in the frequency and severity of extreme weather events, both the worldwide number of deaths and mortality rates from weather hazards have decreased steadily over the past century. Between 1920 and 1950, about 430,000 people died each year as a consequence of extreme weather events. During this period, the mortality rate from these events was about 200 deaths per million people each year. This compares with about 5 deaths per million people from such events each year since 1990.

Most of this decrease is due to diminished mortality associated with droughts, hurricanes, and floods. For example, during the period 1920–1929, drought was responsible for an average of 472,000 deaths worldwide each year. Since 1990, deaths directly attributable to drought have been fewer than 1,000 each year. (These statistics do not include deaths related to crop failures and starvation in drought-stricken areas.) These changes reflect improved infrastructure and water supply in many regions, improved forecasting of events such as hurricanes, as well as increased and improved relief efforts on the part of many organizations and governments.

▲ Figure 18.16 **Effects of Hurricane Katrina**
Wetland loss and poor levee construction magnified the impact of this event on the city of New Orleans. The effects of flooding from Hurricane Katrina were much more severe in poor parts of the city than in wealthier areas.

▲ Figure 18.17 **The 2003 Paris Heat Wave**
These Parisians are cooling down under jets of water at Paris Plage on the banks of the Seine River, protecting themselves against the scorching heat that spread across Europe during the summer of 2003.

QUESTIONS 18.2

1. How is earthquake magnitude expressed?

2. How have human activities influenced the impact of tsunamis on many coastal communities?

3. Describe three kinds of weather events that can pose a risk to human well-being.

4. What factors have contributed to the increase in the loss of life and property associated with wildfires over the past two decades?

(MES) For additional review, go to **MasteringEnvironmentalScience**

Fire in the Environment

■ Patterns of human development and land management are increasing the risks posed by fire.

Although global data are lacking, it is well known that fires of all kinds cause a great many human deaths each year. In the United States in 2012, for example, fires were responsible for 2,855 deaths. The statistics of deaths caused by fire include a disproportionate number of children and elderly people, who often have difficulty escaping burning structures.

Since 1990, deaths associated with wildfires in shrublands and forests have constituted an increasingly larger share of the total number of deaths caused by fire. This is partly due to the increased frequency of wildfires in many regions resulting from past management practices and climate change (see Module 13.4). It is also due to the expansion of human developments into wildland ecosystems that are naturally fire-prone.

Recent catastrophic wildfires in southern California exemplify this trend. In October 2007, a series of wildfires burned through the shrubby woodlands near San Diego. In total, they burned over 2,000 km² (770 mi²), destroyed 1,500 homes, and caused the evacuation of more than 1 million people (Figure 18.18). Nine people died as a direct result of these fires, and nearly 100 others were injured. These fires were ignited from several different sources and were fanned by hot, dry winds. However, their impact on life and property was greatly increased by suburban sprawl and housing developments built well into the highly flammable chaparral. Flammable landscaping and building materials, such as cedar roofing, carried fires from suburban areas into more urban areas that had historically been safe from fire. Events such as this are causing city and regional planners to pay more attention to standards for building and landscaping in what has come to be known as the wildland–urban interface.

Because geological, weather, and fire hazards are often a consequence of environmental processes that are outside human control, they are frequently called "natural disasters" or "acts of God." Yet in many places the destruction and mortality associated with these events have been reduced by improved stewardship of natural ecosystems and wise development practices. Enforcement of strict building codes and thoughtful development that anticipates risks can significantly diminish the human misery associated with these physical hazards.

▼ Figure 18.18 **Wildfire and the Wildland–Urban Interface**
In southern California, the development of homes in fire-prone areas has increased the risk of wildfire to human life and property. Such development has also provided fuel to carry fires into more urban areas.

18.3 Chemical Hazards in the Environment

BIG IDEA **Toxicology** is the scientific discipline that studies chemical poisons, or toxins, and their effects on human health. Toxins are classified according to their effects on human health; these effects include cancers, mutations, allergies, and neurological and developmental disorders. Risks to individual humans are determined by the dose and duration of exposure; they are also influenced by an individual's age, gender, genetic vulnerability, overall health, and socioeconomic status. Toxicological risks are influenced by the persistence, transport, and dispersal of chemicals in the environment. Testing is an important step in managing the risks from environmental toxins. However, testing is costly, and some forms of testing are controversial.

What Is a Toxin?

■ The toxicity of chemicals depends on dosage and duration of exposure.

The 16th-century Swiss physician and alchemist Paracelsus, who pioneered the use of chemicals and minerals in medicine, wrote the following: "Poison is in everything, and no thing is without poison. The dosage makes it either a poison or a remedy." Although this is a bit of an overstatement, Paracelsus was on to something important. When delivered in large amounts, the great majority of chemicals have adverse effects on human health. At low dosages, many chemicals have no effect on human health; some even have beneficial effects.

Consider some common examples. At doses below 1,000 mg, aspirin relieves pain and diminishes inflammation; larger doses of aspirin have adverse consequences, including internal bleeding and kidney failure. We depend on the oxygen in air to breathe. Normally 21% of air is composed of oxygen; air that is more than 50% oxygen is toxic.

Because dose is so important, scientists carry out dose–response studies to establish the toxicity of different chemicals. **Dose–response curves** show the relationship between increasing dosages of a chemical and some measure of health. One measure of health used to define toxicity is the percentage of mortality in an exposed population. More often, studies measure the proportion of individuals in a population who display specific symptoms or illnesses, such as respiratory difficulties or cancer. Because body size often affects dose responses, dosage is usually expressed in terms of the amount of the chemical per unit body weight, such as milligrams per kilogram (mg/kg).

Populations generally show little or no effect from low doses of chemicals. **Toxicity threshold** refers to the dose above which there is a measurable decline in the health measure. Scientists often characterize the toxicity of a chemical in terms of the dose at which 50% of an exposed population is killed or displays a specific symptom or illness. In the case of mortality, this is called the **median lethal dose or LD$_{50}$** (Figure 18.19).

In some cases, exposure to a toxin produces immediate effects. In other cases, the consequences are delayed. Toxins

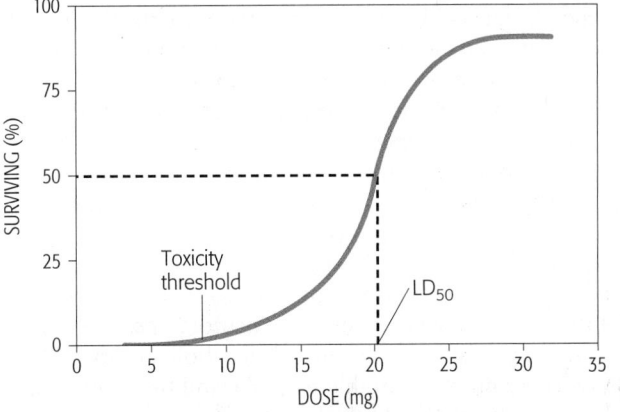

◀ Figure 18.19 **Toxicity Threshold and LD$_{50}$** Toxicologists estimate the toxicity of a chemical by exposing organisms to increasing doses of it. The toxicity threshold is the dose that produces a measurable impact on survival. As the dose of a toxin is increased, the number of individuals affected by it increases. The median lethal dose, or LD$_{50}$, is the amount of a chemical required to kill half of the individuals in a population.

with delayed effects are said to pose latent risks. For example, latent cancer rates refer to the proportion of individuals in a population who contract cancer during some time period following exposure to a substance; latent death rates measure the number of individuals who die during that period.

Dose responses are also influenced by the duration of exposure to a chemical. Contact that occurs only once or for only a short time is called **acute exposure**. In humans, exposures of fewer than 14 days are considered to be acute. Contact with a chemical for periods of weeks, months, or years is called **chronic exposure**. Generally, the toxicity threshold for a chemical is lower for chronic than for acute exposures.

The effects of excessive alcohol consumption illustrate the difference between acute and chronic exposure. Drinking a large amount of alcohol in a very short period of time can elevate blood alcohol levels, causing poisoning and death within a few hours. The LD$_{50}$ for blood alcohol content is about 0.45%. (In most states, individuals with a blood alcohol content at or above 0.08% are considered intoxicated.) In contrast, chronic consumption of large amounts of alcohol at levels below the LD$_{50}$ can eventually cause liver damage and death.

Human Vulnerability to Toxins

■ Genetics and environment cause differences in human vulnerability to toxins.

The U.S. Centers for Disease Control and Prevention estimates that more than 90% of all lung cancer is caused by cigarette smoking. Yet only 10–15% of individuals who smoke more than two packs of cigarettes each day develop lung cancer. Why are some people more vulnerable to certain toxins than others? The effects of toxic substances on different individuals vary widely, depending upon genetic makeup, age, health, and socioeconomic status (Figure 18.20).

Each person possesses a unique combination of genes that affects his or her vulnerability to toxins. As a consequence, the same exposure to potential toxins may affect individuals in different ways. For example, there is growing evidence that genetic differences among individuals play an important role in their susceptibility to chemicals that can alter the structure of DNA and thereby cause cancer.

Age influences vulnerability to toxins in four main ways. First, age-related behaviors influence the likelihood and route of toxin uptake. For example, crawling babies and toddlers spend much of their time close to floors; they are also likely to put foreign objects in their mouths. As a consequence, they are more likely to ingest lead-laden chips of paint than older children and adults. Second, developmental status influences the effects of toxins on individuals. The human nervous system undergoes enormous changes during the first 10 years of life, making children in this age group particularly vulnerable to the effects of many neurotoxins, including lead. Third, patterns and rates of metabolism vary with age. These influence both the uptake and breakdown of toxins in the human body. Fourth, age influences the

length of time that a person is exposed to a toxin and the amount that accumulates in the body. The longer a toxin is circulating in the body, the more time it has to cause a negative effect.

Gender also plays a role in vulnerability to toxins. For example, methyl mercury and other forms of organic mercury are more toxic to men than women. In some cases, gender-based differences in behavior or occupation influence vulnerability. For example, cigarette smoking was historically more common among men than women, accounting for much higher incidence of lung cancer among men. These gender-based differences in smoking rates have diminished in recent years, as have gender-related differences in the incidence of smoking-related illness.

The overall health of individuals contributes significantly to their vulnerability to toxins. For example, people with lung diseases, such as emphysema, are at much greater risk of being harmed by exposure to moderate levels of photochemical smog and ozone than people with healthy lungs.

Socioeconomic status affects vulnerability to toxins in two ways. First, because poor people may have limited access to health care, they are more likely than their better-off peers to suffer from health problems that increase the impacts of toxins. Second, poor people are more likely to live in environments that increase their risk of exposure to toxins. For example, they are more likely to live in older homes with lead paint and plumbing. Poor people are also more likely to live close to industrial and waste facilities than is the population at large.

▶ Figure 18.20 **Population Response to Toxin Exposure**
An individual's sensitivity to a toxin is influenced by many factors. Within a population, individual responses to a particular toxin dose form a bell-shaped curve.

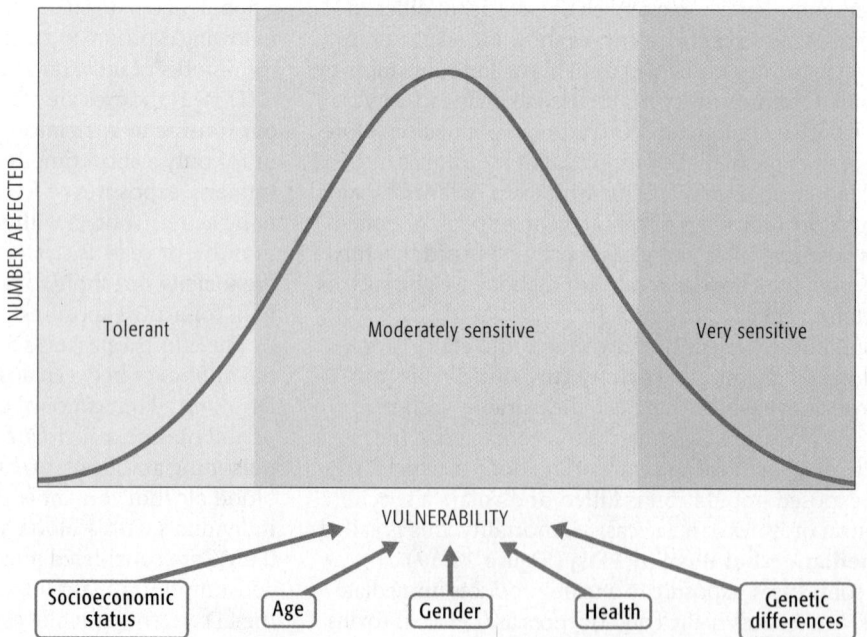

Toxin Transport and Fate

■ Persistence, mobility, and uptake by organisms in the environment influence toxin dose and duration.

The impact of particular substances on human health depends to a large degree on factors that influence their movement through the environment. These factors include their persistence, volatility, and solubility in water and their uptake and fate in other organisms.

The persistence of a toxin in the environment can be described in two different ways: residence time and half-life. Residence time describes the average length of time that a molecule of a substance spends in a particular part of the environment, such as the atmosphere or the ocean (see Module 3.5). Carbon monoxide, for example, has a residence time of about 60 days in the atmosphere, and mercury has a residence time of about one year.

Many substances undergo reactions that produce other substances with different levels of toxicity. This is particularly true of organic compounds that can be chemically altered by sunlight or metabolized by organisms. Toxicologists describe the persistence of a particular form of such a substance in terms of half-life. Half-life is the time that it takes for half of the molecules of a particular substance to be transformed into another form. For example, DDT breaks down to produce other organic compounds that vary in toxicity. Estimates of the global half-life of DDT released into the environment vary from 30 to 100 years. Because of high microbial activity, its half-life in soils is shorter, between 30 days and 2 years. Its half-life in aquatic ecosystems is much longer.

Transport in the environment is affected by a substance's volatility, or tendency to evaporate into the atmosphere, and its solubility, or tendency to dissolve in water. Volatile substances with long half-lives are likely to be dispersed widely by air circulating in the troposphere. This process has contributed to the global distribution of mercury. Chemicals that are soluble in water are more likely to pose problems in streams, rivers, and oceans.

The transport and persistence of toxins in the environment is significantly affected by their uptake and ultimate fate within individual organisms and food webs. **Bioaccumulation** refers to the process by which chemicals are stored in the tissues of living organisms. Fat-soluble chemicals are likely to bioaccumulate, as are chemicals that are not easily broken down or metabolized. Organisms that feed on other organisms whose tissues store a toxin are likely to store even higher concentrations of that toxin. **Biomagnification** is the process by which the concentration of a toxin increases at each higher level of a food web (see Module 12.8).

Toxicologists refer to the total amount of a chemical present in the tissues of an organism as **body burden**. The body burden for chemicals that bioaccumulate, such as DDT and mercury, tends to be higher than that of chemicals that are continuously excreted; the body burden of chemicals that bioaccumulate also tends to increase with age. However, chronic exposure to chemicals that are easily excreted, such as arsenic, can produce a high body burden for these chemicals.

Each of us is exposed to thousands of chemicals, some of which become part of our total body burden. Nevertheless, little is known about the bioaccumulation or typical body burdens for most of these chemicals. Toxicologists suspect that high body burdens of some chemicals influence the toxicity of other chemicals, but there have been few thorough studies of such interactions.

Toxicologists designate particularly long-lived and widely dispersed organic chemicals as **persistent organic pollutants (POPs)**. Twelve substances called the "dirty dozen" are recognized as especially problematic because they are widely dispersed and their impacts on human health are well established (Figure 18.21).

▼ Figure 18.21 **The Dirty Dozen**
These 12 persistent organic pollutants (POPs) are among the most troublesome chemical hazards to human health. Ten of these chemicals are produced to kill either fungi or insects.

Persistent organic pollutants	Type of chemical
Aldrin	Insecticide
Chlordane	Insecticide
DDT	Insecticide
Dieldrin	Insecticide
Dioxins	Industrial by-product
Endrin	Insecticide
Furans	Industrial by-product
Heptachlor	Insecticide
Hexachlorobenzene	Fungicide, industrial by-product
Mirex	Insecticide, fire retardant
PCBs	Industrial chemical
Toxaphene	Insecticide

Kinds of Toxins

■ Toxins are classified according to their effects on specific organs and tissues.

O₂ (Oxygen)

CO (Carbon monoxide)

CN (Cyanide)

▲ **Figure 18.22 Asphyxiants** Because their chemical structures are quite similar to oxygen, carbon monoxide, and cyanide bind tightly to blood hemoglobin, thereby preventing the uptake of oxygen. In sufficient doses, carbon monoxide and cyanide cause asphyxiation.

Environmental toxins include a wide variety of inorganic and organic chemicals. Toxins may be classified by their typical route of entry into the human body. Air pollutants are typically absorbed through the respiratory tract. These pollutants may accumulate in the lungs or be carried to other parts of the body by the circulatory system. Toxins in both air and water may be absorbed through the skin. Toxins in food and water may be absorbed through the digestive tract.

Toxicologists generally classify toxic chemicals according to the way they affect the body, or their mode of toxicity. **Corrosive toxins** react with and directly destroy tissue. Their effect is usually immediate, so their toxicity is acute. Corrosive toxins include caustic chemicals, such as hydrochloric and sulfuric acids, household bleach, lye, and ammonia. They also include chemicals like chlorine, which reacts with tissues to form hydrochloric acid.

Asphyxiants are chemicals that deprive tissues of oxygen and cause suffocation. Carbon monoxide and cyanide are powerful asphyxiants that block the hemoglobin in blood from absorbing oxygen (Figure 18.22). Other kinds of asphyxiants interfere with the uptake of oxygen in the lungs or affect the transport and use of oxygen in other parts of the body.

Carcinogens are chemicals that cause or promote cancer. Cancer is a disease in which the growth of affected body cells is no longer controlled, leading to the formation of tumors. Carcinogens increase the risk of getting cancer by altering cellular DNA or metabolism in ways that encourage uncontrolled, malignant cell division. The environment contains thousands of potentially carcinogenic substances. Dozens of carcinogens have been identified from industrial and tobacco smoke alone. Dioxin and polychlorinated biphenyls (PCBs) are widespread and well-studied carcinogenic pollutants.

Substances that produce mutations and birth defects are called **teratogens**. Scientists estimate that about 10% of all birth defects are caused by prenatal infections or exposure to teratogenic substances. Teratogens, like carcinogens, often cause changes in cellular DNA. Thus, many carcinogens are also teratogens.

A great number of pollutants known as **allergens** stimulate a range of responses from the body's immune system. These responses can vary in intensity from a running nose and sneezing, to a skin rash and respiratory distress, to allergic shock and suffocation. Because the immune system is especially sensitive to foreign proteins, other organisms and their products are often powerful allergens (Figure 18.23).

Neurotoxins affect the development or functioning of the nervous system. Important neurotoxins include heavy metals such as mercury, lead, and manganese and pesticides such as DDT and various organophosphates.

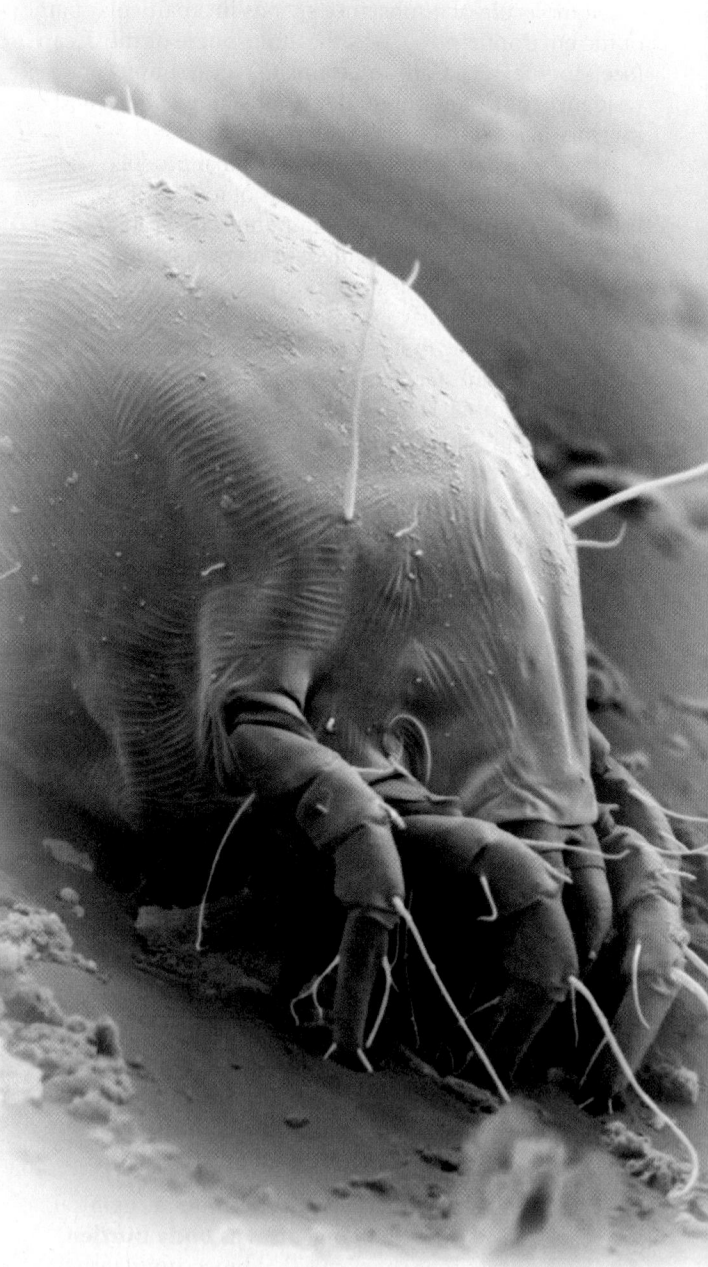

▲ **Figure 18.23 A Common Allergen**
Microscopic dust mites, which are common in carpets and upholstery, cause allergic reactions in many people.

▶ Figure 18.24 **Declining Sperm Count**

In a synthesis of 101 studies since 1934, researchers have detected a significant decline in human sperm production. They speculate that this may be due to increasing amounts of endocrine-disrupting toxins in the environment. But they also note significant scatter in the data and suggest that many factors are likely to contribute to variation in sperm counts.

Source: Swan, S.H., E.P. Elkin, and L. Fenster. 2000. The question of declining sperm density revisited: An analysis of 101 studies published 1934–1996. Environmental Health Perspectives 108: 961–966.

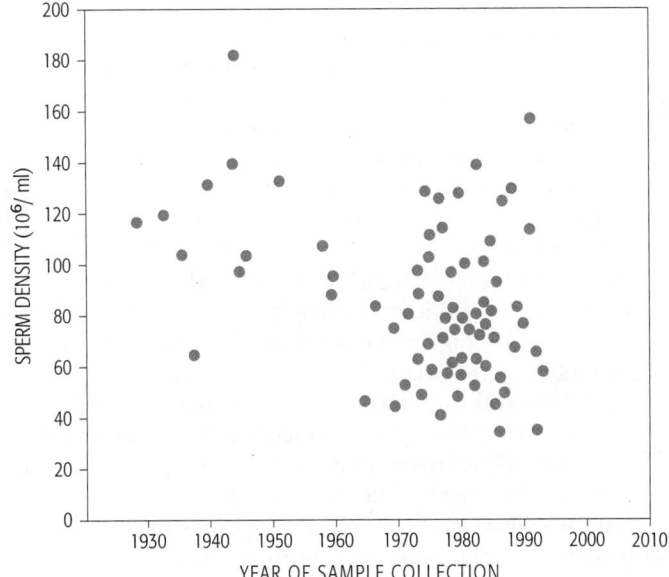

In recent years, toxicologists have discovered a number of substances called **endocrine disrupters** that interfere with the hormones that control growth and development, egg and sperm production, and reproductive function. One of the most important hormones influencing development and the production of egg and sperm is estrogen. Nearly every day, most humans are exposed to hundreds of chemicals that have been shown in laboratory studies to influence the production of estrogen. Whether these chemicals affect human health at the typical level of environmental exposure is hotly debated. However, there is growing evidence that chronic exposure to even low amounts of these chemicals is influencing the frequency of abnormal patterns of development in human populations (Figure 18.24).

The controversy surrounding the chemical bisphenol A (BPA) illustrates the challenges associated with proving that a chemical actually influences endocrine function in humans. BPA is found in plastic water bottles, baby bottles, and food containers, as well as in the linings of food and infant formula cans. Numerous animal studies have demonstrated that low concentrations of BPA have the potential to affect the endocrine system. Such effects include increased reproductive problems, decreased sperm counts, accelerated sexual development, obesity, and elevated rates of some cancers. The U.S. Food and Drug Administration (FDA) and the chemical industry have argued that these studies are not relevant to human exposure and toxicity and that BPA is safe at the concentrations present in these plastic containers. In 2008, however,

Canada announced plans to phase out the use of this chemical in products used by infants. Manufacturers, including Nalgene and Playtex, and retailers such as Wal-Mart have agreed to remove BPA from their products. In late 2008, the FDA's Science Advisory Board concluded that its earlier assessment was flawed and recommended that the safety of this chemical be reevaluated, especially with regard to infants.

Many, if not most, medicines are potentially toxic. Tests reveal that drinking water in many cities contains a variety of pharmaceutical chemicals, including antibiotics, antidepressants, artificial hormones, and anti-inflammatory drugs. For example, a 2007 study of Philadelphia's drinking water identified 56 different medicinal chemicals. These drugs enter the water supply through the disposal of unused medicine and in human and animal wastes. To be sure, the concentrations of these drugs are very low, far below medicinal doses. Nevertheless, some of these drugs have the potential to bioaccumulate and biomagnify in food webs. Toxicologists are also concerned about their possible effects on fetuses and children.

New **FRONTIERS**

Proving a Chemical Disrupts Endocrine Function

In 2014, 12 FDA scientists published a study in which rat fetuses and newborns were exposed to varying levels of BPA. The highest doses were 100,000 times higher than typical human exposures and produced lower body weights, abnormal reproductive development, and altered hormone levels. None of these effects were observed in rats exposed to lower doses. Critics quickly argued that this study was flawed because it was too short term and did not include measures of brain development. This controversy illustrates that, although it is often very difficult to confirm the toxicity of a chemical, it is equally hard to demonstrate that a chemical is risk free (Figure 18.25). Given what is now known, would you buy a plastic water bottle that contained BPA?

▲ Figure 18.25 **Precaution for Infants**

Although BPA was toxic only at very high does, the FDA set rules banning its use in the packaging of infant formula. Why is special attention being given to exposure in infants?

Toxin Testing and Regulation

■ Toxicity testing presents both ethical and practical challenges.

Ideally, whenever a new substance is proposed for commercial use, its toxicity and potential modes of transport in the environment would be fully evaluated. In practice, this is done for only a relative few chemicals. Historically, scientists have often determined that a chemical is toxic only after it is in wide use and its toxic effects have become apparent.

Scientists estimate that commercial substances contain between 150,000 and 200,000 different chemicals. Over 2,000 new chemicals are synthesized and introduced each

year. Some of these, such as nanomaterials, represent entirely new classes of compounds.

The U.S. EPA is charged with monitoring over 75,000 industrial chemicals, but at present it has evaluated and set standards for fewer than 10% of them. Chemicals are generally selected for testing based on their similarity to chemicals that are known to be toxic. Most chemicals are assumed to be nontoxic until proven otherwise.

Many government and environmental leaders have argued that the release of chemicals into the environment

QUESTIONS 18.3

1. Give an example of the adage "dosage makes a chemical either a remedy or a toxin."

2. Differentiate between acute and chronic exposure to a toxin.

3. What is the difference between bioaccumulation and biomagnification?

4. List and describe seven classes of toxins.

(MES) For additional review, go to **MasteringEnvironmentalScience**

▼ Figure 18.26 **How Much Testing Is Enough?**
The chart compares two approaches to managing the potential hazards of chemical products with uncertain risks.

should be guided by this precautionary principle: If a chemical has the potential for severely harming the public or the environment, the responsibility for proving that it is safe should fall to those who advocate its use (see Module 2.5) (Figure 18.26). Those who manufacture and market chemicals and the products that contain them argue that most of these substances are benign; testing each chemical to establish this fact would considerably increase its price.

The thorough testing of chemicals for their toxicity to humans presents significant challenges. There is unanimous agreement that testing the effects of potential toxins on human subjects is unethical. Thus, nearly all chemical testing is done with animals or microorganisms. Ideally, such testing would be done on animals that are a similar to humans as possible. However, such testing is very expensive, and, for some, it raises serious ethical questions.

Many scientists worry that the effects of a chemical on a bacterium, fish, or rabbit may not mirror its effects on the human body. To counter this concern, toxicologists focus their testing on **model organisms** in which particular processes or metabolic functions have been demonstrated to be similar to those of humans. For example, the basic process of DNA replication in the bacterium *Escherichia coli*, a widely used model organism, has been shown to be similar to that of nearly all other organisms, including humans. Thus, a chemical that has a detrimental effect on the manufacture of DNA in *E. coli* might be suspected of being a carcinogen in humans. Similarly, the early development of fish embryos is similar to that of human embryos. Chemicals that affect development in zebra fish, another important model organism, might have similar effects in humans.

Toxicologists are quick to acknowledge that the absence of a toxic response in a model organism does not guarantee that a chemical is nontoxic to humans. In addition, dose–response thresholds and LD_{50} of toxins in model organisms often differ from those in humans. Consequently, there is no simple formula for using the responses of a model organism to set standards for human exposure to a particular chemical.

The effects of chronic exposure to low levels of a chemical are particularly difficult to evaluate. Toxicologists typically administer relatively large doses of a chemical over a brief time and then use their response data to extrapolate to much lower, longer-term exposures. For example, if an acute dose of 100 mg/kg of a substance causes cancer in 50% of a population of zebra fish, a toxicologist might predict that 1 mg/kg might cause cancer in 0.5% of the human population. The toxicologist might further predict that chronic exposure to 1 mg/kg of the substance could increase the rate of cancer.

The standards established for levels of toxins in the environment should take into account how a chemical moves in the environment. A general understanding of properties such as volatility, water and fat solubility, and storage in animal tissues can provide the basis for predicting whether a chemical is likely to bioaccumulate and how it will be dispersed in ecosystems. These properties are, however, very difficult to measure.

A great many substances are potential toxins to animals and humans. Their actual toxicity depends on their dispersion, the dose received, and the vulnerability of individuals. Adequate toxicity testing for the great many substances in our environment is a daunting challenge.

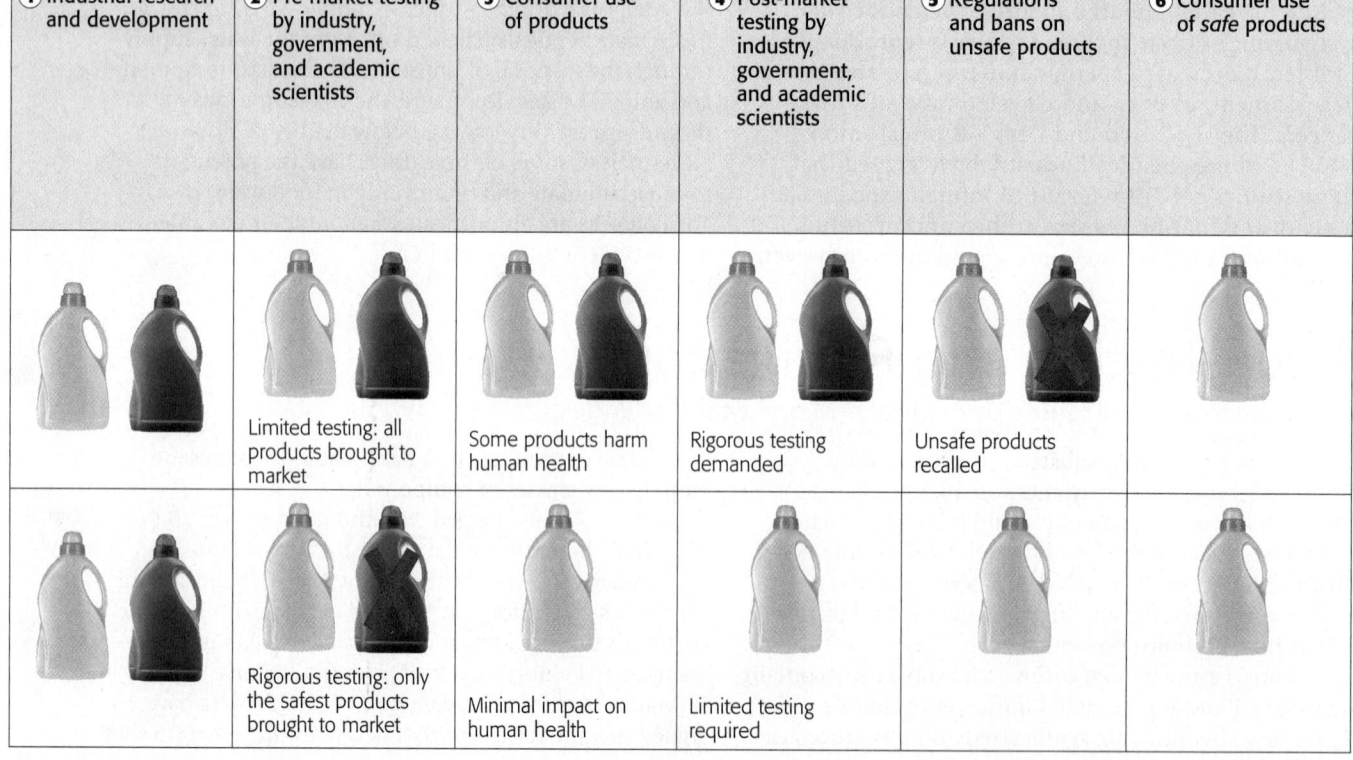

	① Industrial research and development	② Pre-market testing by industry, government, and academic scientists	③ Consumer use of products	④ Post-market testing by industry, government, and academic scientists	⑤ Regulations and bans on unsafe products	⑥ Consumer use of *safe* products
Sequence of events						
"Innocent until proven guilty" approach		Limited testing: all products brought to market	Some products harm human health	Rigorous testing demanded	Unsafe products recalled	
Precautionary principle approach		Rigorous testing: only the safest products brought to market	Minimal impact on human health	Limited testing required		

Nature's Chemicals vs. Synthetic Chemicals

Are the chemicals that naturally occur in plants and animals necessarily less toxic than chemicals synthesized by humans?

▲ Figure 18.27 **A Source of Challenging Ideas**
Over his long career in toxicology, Bruce Ames's research and synthesis has catalyzed work by hundreds of students and colleagues.

In her book *Silent Spring*, Rachel Carson wrote, "for the first time in the history of the world, every human being is now subjected to contact with dangerous chemicals, from the moment of conception until death." She was expressing a widely shared view that the toxic chemicals that influence our health are mostly those that have been artificially synthesized by humans. Dr. Bruce N. Ames, professor of biochemistry and molecular biology at the University of California, Berkeley, thinks otherwise, and his thoughts carry great weight with toxicologists (Figure 18.27).

Ames has published more than 500 scientific papers and received numerous awards, including the Tyler Prize for Environmental Achievement and the National Medal of Science. He is particularly well known for developing the Ames test, an assay that uses bacteria to measure the cancer-causing potential of chemicals. Ames's *challenge* to conventional wisdom was published in a paper that demonstrated the importance of exhaustive *review* and *synthesis* of existing literature in science. Since its *publication*, this paper has had a significant impact on the field of toxicology.

Ames first pointed out that thousands of naturally occurring toxins have been present throughout the evolution of vertebrates. Indeed, we humans possess an array of defenses against this multitude of naturally occurring chemicals. For example, in our organs that are regularly exposed to toxins, such as the mouth, stomach, intestines, and lungs, layers of surface cells are continuously being shed. Our cells produce an array of chemicals, such as antioxidants, that can neutralize the effects of many toxins. Our cells can also repair strands of DNA that have been

altered by chemicals. Perhaps most important, we possess keen senses of smell and taste that warn us not to consume natural toxins. Despite these adaptations, many natural chemicals remain quite toxic to us. For example, molds produce a variety of known carcinogens. Chemicals in comfrey tea and some herbal medicines are mutagens and can cause liver damage (Figure 18.28).

Natural chemicals often have the same mechanisms of transport and toxicity as synthetic toxins. For example, natural chemicals can bioaccumulate in the tissues of animals and biomagnify in food webs in much the same way as synthetic chemicals, such as DDT. Cabbage and broccoli contain chemicals that, at high concentrations, can stimulate rapid cell division in a fashion very similar to that of dioxin. Fortunately, the quantities of vegetables that we typically eat result in doses far below the toxic threshold of these chemicals. Ames points out that, at average levels of consumption in the United States, the mutagenic and carcinogenic effects of the alcohol in beer and wine far exceed the risks posed by exposure to dioxin at the dose limits set by the EPA.

Ames's arguments and assertions were not intended as a defense of synthetic chemicals or their haphazard release into the environment. Rather, he warned that we should not assume that naturally occurring chemicals are benign. Like many of their synthetic counterparts, most of these chemicals have not been tested for toxicity. Equally important, he pointed out that *studies* of naturally occurring chemicals have much to teach us about the transport, uptake, and toxic mechanisms of all chemicals.

Source: Ames, B.N., M. Profet, and L.W. Gold. 1990. Nature's chemicals and synthetic chemicals: Comparative toxicology. *Proceedings of the National Academy of Sciences* 87: 7782–7786.

1. Are there reasons why we might generally be more concerned about synthetic chemicals than natural chemicals?

2. Given that we synthesize many naturally occurring substances, what is it that distinguishes a synthetic from a natural chemical?

◄ Figure 18.28 **Natural Toxins**
Tea made from ephedra, commonly called Mormon tea, is used by many people to treat colds and hay fever. However, it can cause heart attacks and strokes in some people.

18.4 Biological Hazards in the Environment

BIG IDEA Worldwide, infectious diseases cause more than 26% of human deaths. In the world's 50 poorest countries, such diseases cause more than 70% of deaths. Environmental factors exert a strong influence on the incidence, spread, and impact of infectious diseases in human populations. Climate, water availability, and water quality affect disease-causing organisms, as well as the animals that transmit diseases. The role of environmental factors varies greatly, depending on the manner by which a disease is transmitted. Infectious diseases can be managed by improving sanitation and water quality, by controlling organisms that cause and transmit disease, and by taking action to diminish vulnerability in human populations.

Infectious Disease and the Environment

■ Environmental factors influence populations of the organisms that cause disease, as well as those that spread disease.

Infectious diseases are caused by **pathogens**—viruses, bacteria, protozoa, fungi, or multicellular parasites that invade host organisms and impair their functioning. The extent to which a pathogen impairs the growth and reproduction of its host is called its **virulence**.

Virulence is heavily influenced by the nature of the interaction between a pathogen and its host. For example, a pathogen that destroys cells in its host's body is likely to be more virulent than a pathogen that is able to reproduce without destroying cells. A pathogen that has only recently come into contact with its host is likely to be more virulent than a pathogen that has a long history of coevolution with its host.

Virulence is also influenced by factors that influence the immune system and overall health of the host. Organisms that would typically cause no disease or only mild disease can be quite virulent in individuals whose immune systems have been impaired by malnutrition or another disease. The tuberculosis bacterium, for example, is much more virulent in individuals who are malnourished or have HIV/AIDS.

Pathogens are transmitted from person to person by a variety of mechanisms. Viruses and bacteria that cause respiratory diseases, such as the common cold and tuberculosis, are transmitted in the tiny droplets of saliva dispersed by sneezing and coughing. Diseases of the digestive tract, such as amoebic dysentery and cholera, are transmitted in contaminated food and water. Sexually transmitted diseases, such as syphilis and HIV/AIDS, are

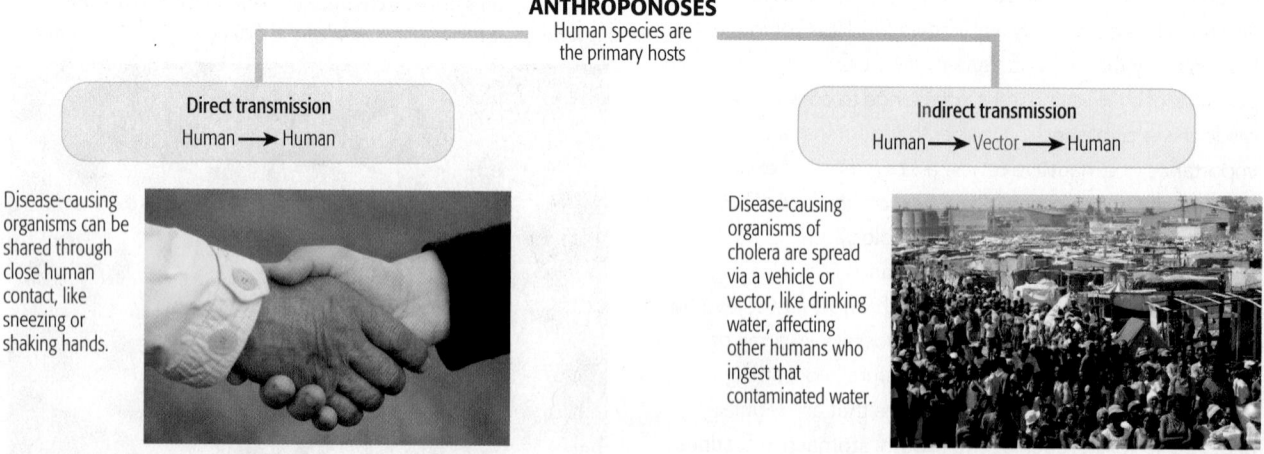

ANTHROPONOSES
Human species are the primary hosts

Direct transmission
Human ⟶ Human

Disease-causing organisms can be shared through close human contact, like sneezing or shaking hands.

Indirect transmission
Human ⟶ Vector ⟶ Human

Disease-causing organisms of cholera are spread via a vehicle or vector, like drinking water, affecting other humans who ingest that contaminated water.

▲ Figure 18.29 **Anthroponoses**
Diseases that are specific to humans are called *anthroponoses*. They may be transmitted directly or indirectly by an intermediate animal called a *vector* or in contaminated water or food.

transmitted in bodily fluids, most often during sexual intercourse.

Many pathogens are transmitted by intermediary organisms called vectors (see Module 6.2). *Passive vectors* transmit diseases by simple contact. For example, a housefly alights on animal feces, picks up pathogenic bacteria on its appendages, and then deposits the bacteria on food that humans eat. The pathogen does not infect the passive vector or enter its body. *Active vectors* transmit pathogens that have entered their bodies; in some cases, the pathogens even complete part of their life cycle inside the vector. Active vectors often deliver pathogens to humans through a bite. Malaria, sleeping sickness, and viral encephalitis are blood-borne diseases that are transmitted by active vectors, such as mosquitoes, biting flies, fleas, and ticks.

The spread of an infectious disease is influenced by its virulence, its ease of transmission, and the mobility of infected individuals. For example, the highly virulent Ebola virus is easily transmitted from infected and symptomatic individuals to healthy individuals by contact with body fluids. However, once the disease is contracted, it incapacitates and kills its victims quickly. Since infected individuals are usually too ill to travel far from the source of their infection, outbreaks of Ebola tend to be confined to small geographic areas. In fact, healthcare workers or family caring for the gravely ill

are the ones most likely to be infected. In contrast, the influenza virus is widely dispersed by the sneezing and coughing of individuals who may be quite mobile in the early stages of the disease. As a consequence, strains of influenza can spread around the globe rapidly.

An **anthroponosis** is an infectious disease that occurs exclusively in humans. The pathogens that cause anthroponoses, such as falciparum malaria and yellow fever, have coevolved with humans. Humans are the sole host of these pathogens (**Figure 18.29**).

Many other infectious diseases are caused by pathogens that have hosts in addition to humans. For example, Lyme disease occurs in humans, deer, and small rodents. Various strains of flu are shared between humans and domestic animals, such as pigs and poultry. A disease that is shared by humans and other animals is called a **zoonosis** (**Figure 18.30**). The evolution and transmission of zoonoses are influenced by their multiple hosts.

Environmental factors such as climate, soil, and the availability of water affect the growth and reproduction of pathogens, as well as their vectors and host animals. As a consequence, the environment plays a significant role in the prevalence and spread of infectious disease. The following sections explore the role of the environment in specific kinds of respiratory, diarrheal, and blood-borne diseases.

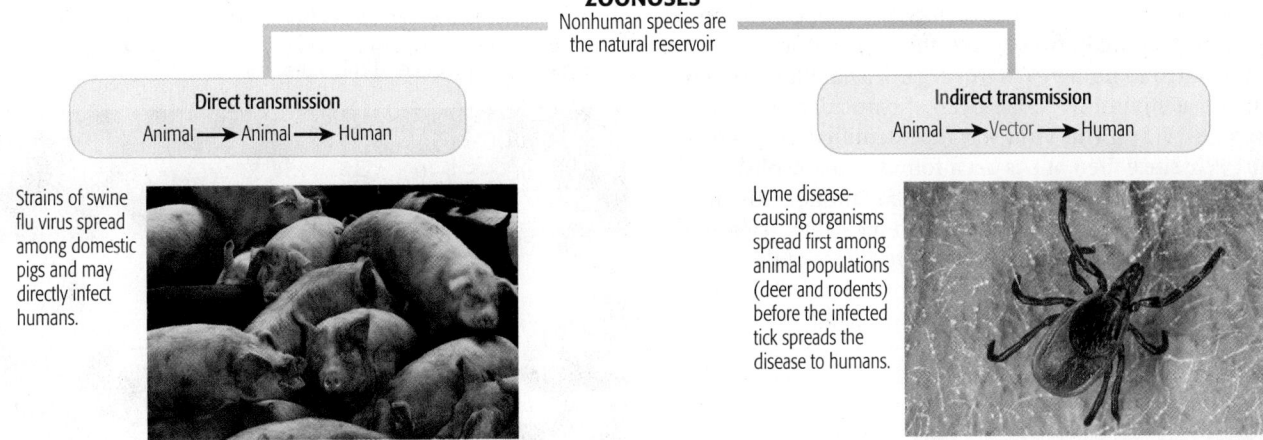

ZOONOSES
Nonhuman species are the natural reservoir

Direct transmission
Animal ⟶ Animal ⟶ Human

Indirect transmission
Animal ⟶ Vector ⟶ Human

Strains of swine flu virus spread among domestic pigs and may directly infect humans.

Lyme disease-causing organisms spread first among animal populations (deer and rodents) before the infected tick spreads the disease to humans.

▲ **Figure 18.30 Zoonoses**
Diseases shared between humans and other animals are called *zoonoses*. These may be transmitted directly or indirectly through animal vectors.

Respiratory Disease

■ Influenza provides an example of a disease that is spread through the air.

Fever, chills, muscle pain, and general discomfort—we are all familiar with the symptoms of influenza, or the flu. Every year this common disease is responsible for several hundred thousand human deaths. In some years, the global death toll has been in the millions.

Flu is caused by several strains of the influenza virus and is most often transmitted from human to human in respiratory aerosols, the tiny droplets given off by coughing and sneezing. Human viral strains are often shared with birds, pigs, and some other mammals. Viral particles are typically transmitted from other animals to humans by direct contact with tissues or feces.

Influenza is most prevalent in human populations in the wintertime. This is probably because people tend to congregate indoors when it is cold, and their proximity encourages the transmission of the virus. Colder temperatures also allow the viral particles to survive on exposed surfaces, such as doorknobs. In addition, extensive holiday travel facilitates the spread of flu.

Each strain of influenza has a large range of genetic variability and therefore is able to undergo rapid evolution from year to year. It is this genetic variability that accounts for the great variability in the virulence of the flu. In 1918, a particularly virulent strain known as the Spanish flu suddenly appeared. Over the next 18 months, the Spanish flu killed more than 50 million people.

New genetic strains of influenza often appear first in other host animals, from which they may subsequently be transmitted to humans. In this regard, avian flu, formally known as strain H5N1, has been of particular concern (Figure 18.31). This virus was first identified in waterfowl in 1997; since then, it has been found in many bird species. This virus is easily transmitted among birds, and certain genetic forms of it cause high mortality. Scientists have noted a steady increase in the virulence of this virus, and many are concerned that it may accelerate the decline of populations of bird species that are already endangered.

A few hundred humans have contracted avian flu, apparently from handling infected waterfowl or poultry. The symptoms of this flu have been severe, and the mortality rate among infected individuals has been about 60%. Indeed, its effects on humans are considerably more severe than those of the virus responsible for Spanish flu. Fortunately, the virus that causes avian flu cannot be efficiently transmitted from human to human. Nevertheless, WHO and the U.S. Centers for Disease Control and Prevention have invested billions of dollars in developing vaccines and programs to limit the spread of this disease.

The situation with avian flu is quite different from the recent global outbreak of swine flu (virus H1N1). This viral strain is common in pigs but is rarely transmitted directly to humans. In 2009, however, a genetic strain of H1N1 that was easily transmitted from person to person appeared in the United States and Mexico. Early indications were that this virus might be quite virulent, and WHO declared a state of emergency. The swine flu did spread widely, but hospitalization and mortality rates were about the same as those of other seasonal flu viruses.

▼ Figure 18.31 **Avian Flu**
The micrograph shows the virus (gold structures) that causes the avian flu in kidney cells (green). The lab worker at right is testing for the presence of the virus.

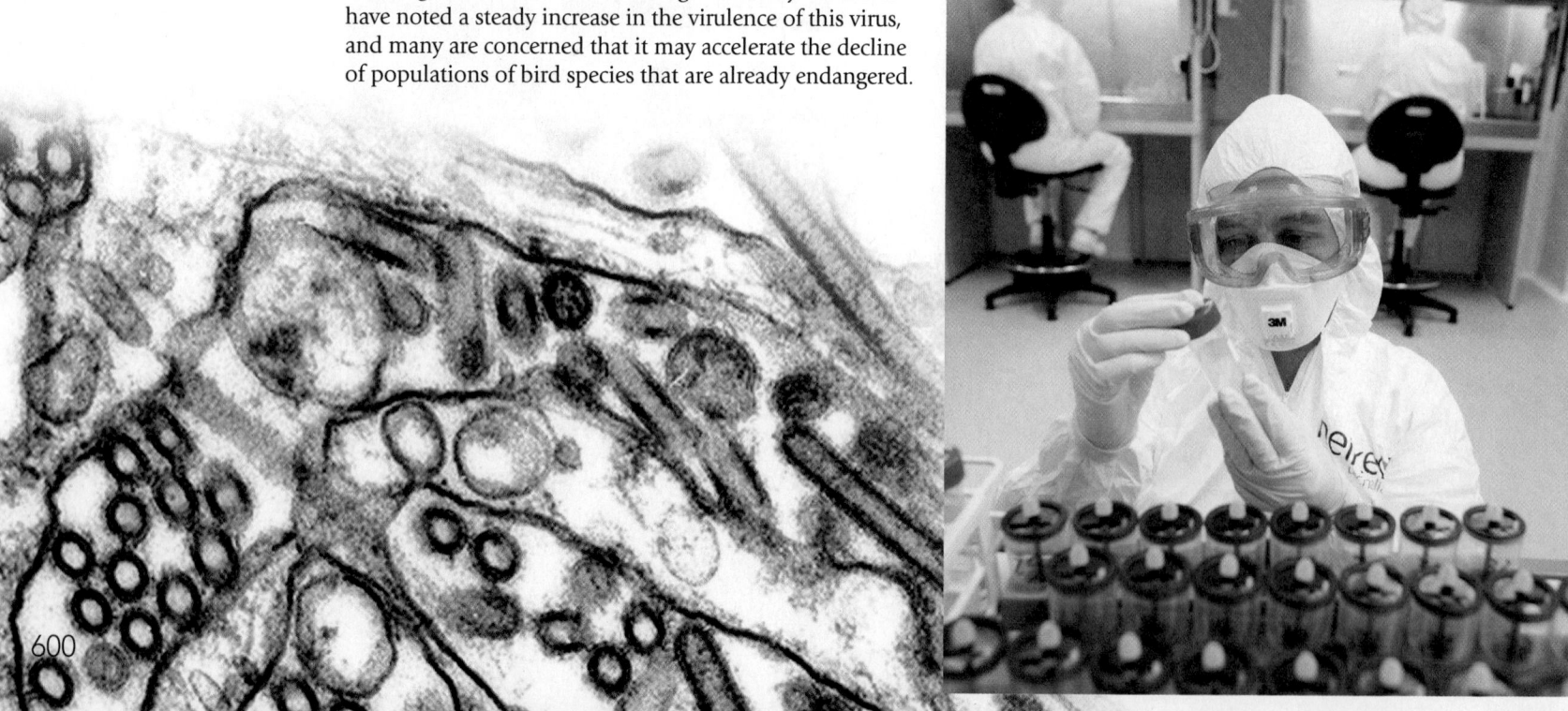

Diarrheal Diseases

■ Poor sanitation and polluted water are responsible for most diseases that infect the stomach and intestines.

Worldwide, about 80% of all infectious diseases are caused by waterborne diarrheal pathogens. These pathogens include a variety of viruses, bacteria, and protozoans that infect and inflame the lining of the stomach and small intestine. In many developing countries, untreated sewage is simply discharged into surface waters. In India, for example, only about 10% of towns and cities have partial sewage-treatment facilities; only 1% of people live in towns with full wastewater-treatment facilities. Polluted water and poor sanitation contribute to more than 2 billion human infections of diarrhea each year. In developing countries, these infections are the leading cause of death among infants and children younger than five. The immediate cause of death among most victims of diarrheal diseases is dehydration.

Changes in temperature and rainfall cause the prevalence of some diarrheal diseases to vary from year to year. For example, the incidence of cholera in Bangladesh is closely correlated with changes in the temperature of its coastal waters. These temperatures, in turn, vary as a consequence of the distant El Niño Southern Oscillation (see Module 9.2). When temperatures in Bangladeshi estuaries are warm, populations of the cholera bacterium (*Vibrio cholerae*) in plankton, shellfish, and fish are also high, and the prevalence of cholera increases (Figure 18.32). A similar pattern of cholera prevalence has been observed in cities on the west coast of South America.

Outbreaks of diarrheal disease are not confined to developing countries. *Cryptosporidium* is a waterborne pathogen that occurs worldwide. This tiny protozoan escapes most filtration systems and is quite resistant to the chlorination that is widely used to kill microbes in drinking water. Serious outbreaks of this organism

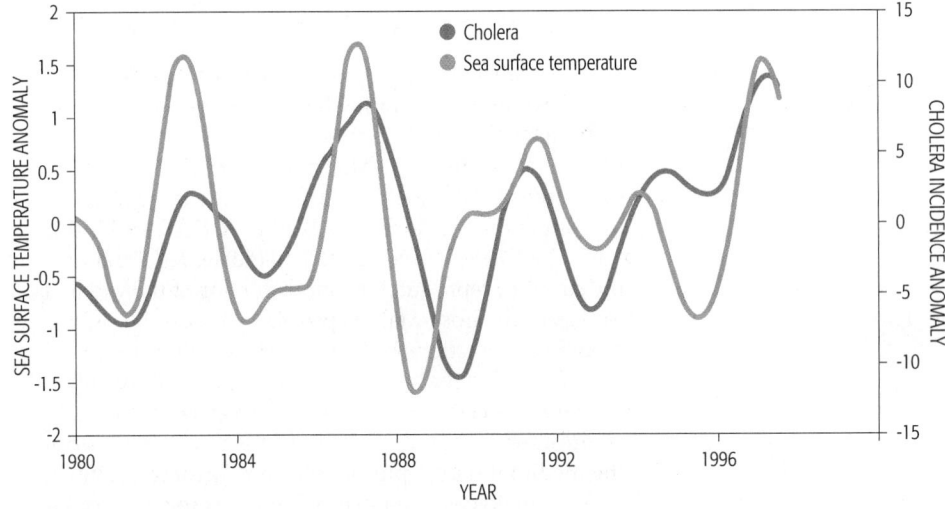

▲ Figure 18.32 **Cholera and Climate**
The incidence of cholera in Bangladesh, corrected here for natural seasonal variation, rises and falls in association with changes in the surface temperature of the sea. The increase in cholera associated with warmer water is likely due to flooding during El Niño events.

Source: Pascual, M., X., Rodó, S.P. Ellner, et al. 2000. Cholera dynamics and El Niño-Southern Oscillation. *Science* 289: 1766–1769.

have occurred in a number of European and American cities over the past four decades. In 1993, for example, *Cryptosporidium* in the drinking water of Milwaukee, Wisconsin, resulted in 403,000 cases of diarrhea and 4,400 hospitalizations.

As recently as 1980, the annual global mortality due to diarrheal diseases exceeded 5 million. Since then, improved sanitation and the development of oral rehydration therapies have reduced this mortality to about 1 million deaths each year (Figure 18.33).

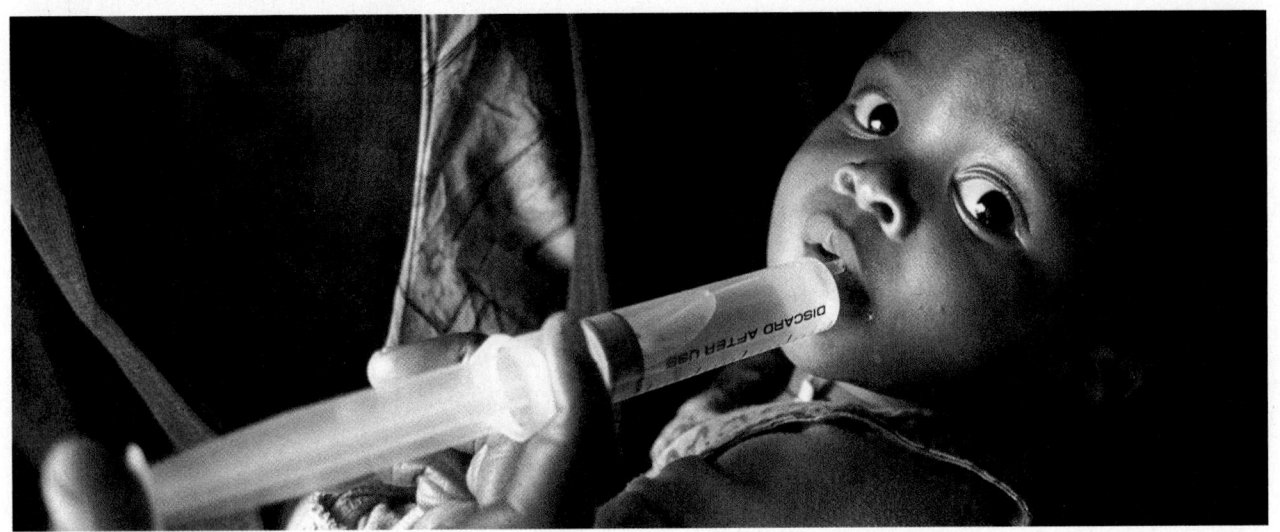

◄ Figure 18.33 **Rehydration Therapy**
Diarrheal diseases cause dehydration, which can be particularly fatal in infants. Oral treatment with relatively simple rehydration fluids containing sugar and salts now saves thousands of lives.

Blood-Borne Diseases

■ Malaria is an example of a blood-borne disease that is spread by an insect vector.

Humans are afflicted with a number of diseases that are transmitted from individual to individual in blood. Although a few blood-borne diseases, such as HIV/AIDS, are acquired by sexual contact (or in the case of drug users, by sharing needles), most are transmitted by animal vectors, such as mosquitoes, ticks, and biting flies. The spread of vector-mediated diseases depends on the survival, reproduction, and behavior of the vector as well as of the host. Malaria provides an example of the complex interactions between a vector and host.

Long before the cause of malaria was understood, humans associated the disease with aspects of the environment. Indeed, the name "malaria" derives from the medieval Italian phrase *mala aria*, meaning "bad air." People recognized that malaria was more prevalent among populations living near standing water, so they presumed that it was caused by malevolent vapors arising from swamps.

In 1880, a French physician, Charles Laveran, discovered that malaria is caused by parasitic protozoa in the genus *Plasmodium*. In 1898, scientists proved that these parasites are transmitted by mosquitoes in the genus *Anopheles*.

The life cycle of *Plasmodium* is closely tied to the ecology of its vector, the female *Anopheles* mosquito. The cycle of malarial infection begins when a female mosquito bites a person infected with malaria and takes in blood containing *Plasmodium* (Figure 18.34). (Only the female mosquitoes feed on blood, which they need in order to reproduce.) Over about a week's time, the parasites in the mosquito undergo a complex process of differentiation and reproduction in which large numbers of parasites, in the form of sporozoites, migrate to the mosquito's salivary glands. When the infected mosquito bites again, its saliva transmits the parasites to a new human victim. Each parasite invades a red blood cell, where it grows and divides. Every 48 or 72 hours, the parasites burst out of their red blood cells and invade new blood cells. This synchronous activity produces the regular 48- or 72-hour fevers that are characteristic of the particular form of malaria.

▼ Figure 18.34 **Life Cycle of the Malaria Parasite**
Malaria is caused by the protozoan parasite *Plasmodium*, which has a complex life cycle involving the female *Anopheles* mosquito and humans.

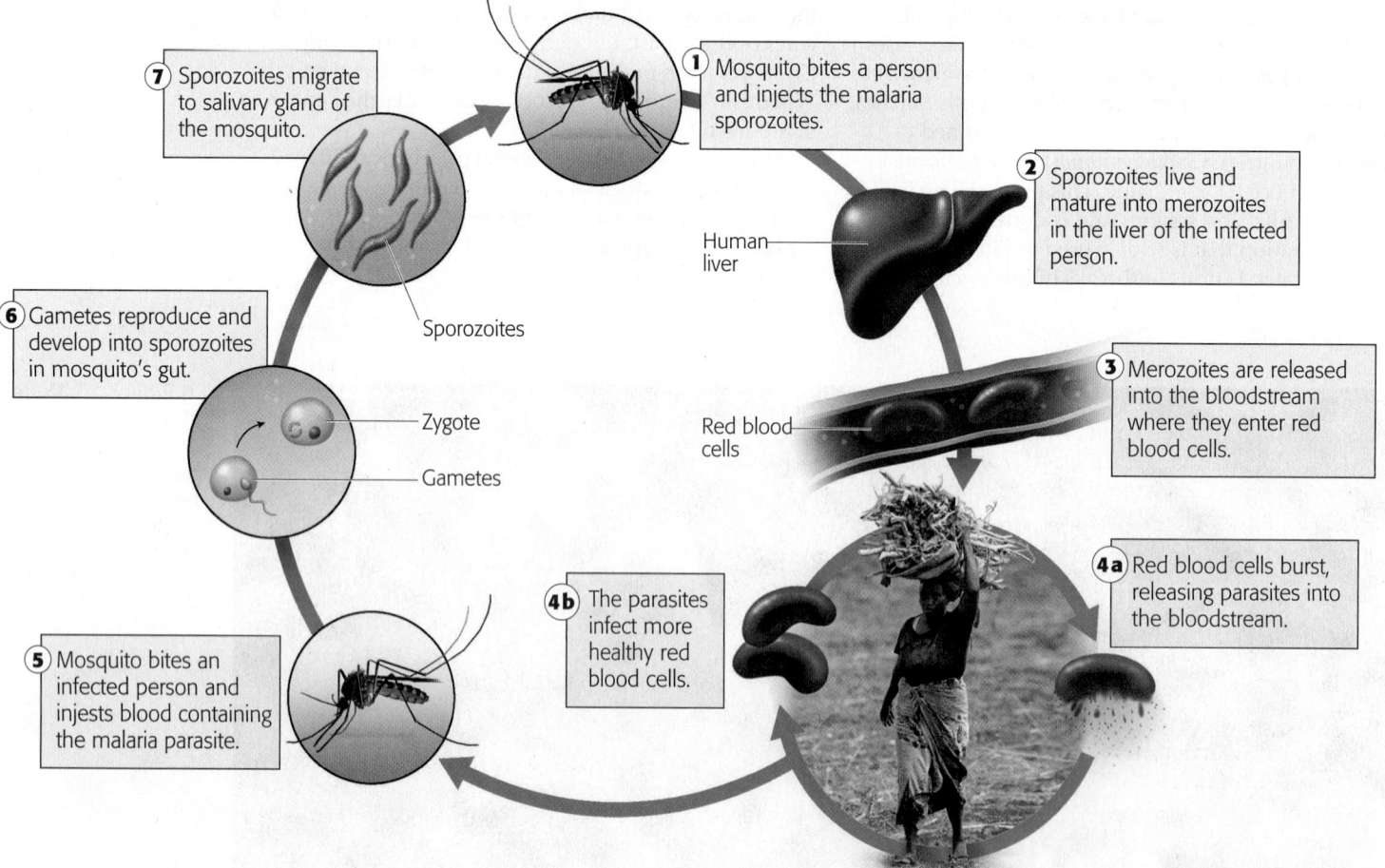

7 Sporozoites migrate to salivary gland of the mosquito.

1 Mosquito bites a person and injects the malaria sporozoites.

2 Sporozoites live and mature into merozoites in the liver of the infected person.

Human liver

Sporozoites

6 Gametes reproduce and develop into sporozoites in mosquito's gut.

Zygote

Gametes

Red blood cells

3 Merozoites are released into the bloodstream where they enter red blood cells.

4a Red blood cells burst, releasing parasites into the bloodstream.

4b The parasites infect more healthy red blood cells.

5 Mosquito bites an infected person and injests blood containing the malaria parasite.

The geographic distribution of malaria is closely tied to the habitat requirements of the *Anopheles*. Like all mosquitoes, *Anopheles* requires pools of standing water for reproduction and early larval development. It does not, however, tolerate temperatures close to or below freezing during any part of the year.

As recently as 150 years ago, malaria was common in warm temperate regions, such as those in the southeastern United States and Mediterranean Europe. At that time, human vulnerability to this disease could only be controlled by regular doses of quinine, a chemical extracted from the bark of cinchona trees native to South America (Figure 18.35).

When it was discovered that *Anopheles* mosquitoes were the sole vector of malaria, disease control began to focus on eliminating mosquito breeding habitat. Wetlands and swamps in many regions of the world were dredged and drained for this reason. In the 1940s, broadcast spraying of DDT was found to be quite effective in limiting *Anopheles* mosquitoes and malaria. Its widespread use was a major factor in the subsequent worldwide decline of this disease.

As discussed in the introductory essay on Zambia, the prevalence of malaria in some regions has increased dramatically over the past several decades. The global ban on DDT has probably contributed to this increase, but other changes in the environment have also been important. Through time, the *Plasmodium* parasite has become increasingly resistant to quinine and related chemicals, such as hydroquinone, while the mosquito vectors have become increasingly resistant to DDT and other insecticides. Changing climate and the growth and

◀ Figure 18.35 **A Natural Defense Against Malaria** The indigenous people of the lowland forests of the Andes discovered that the bark of the cinchona tree provided resistance to malaria. Scientists subsequently learned that it is the chemical quinine that gives cinchona bark this medicinal property.

urbanization of human populations have likely facilitated the spread of this disease in some places.

Because the environment influences the spread and virulence of infectious diseases among humans, management of the environment is important in the control of such diseases. Limiting human contact and mobility are important ways to control respiratory disease. Improved sanitation and water quality have diminished the prevalence of diarrheal diseases. Alteration of vector habitats and direct control of vector populations are important strategies in the control of vector-borne diseases such as malaria.

Evolutionary Change

■ Coevolution between pathogens and their human hosts tends to diminish disease severity.

Pathogens exert very strong natural selection on their host populations. This leads to the evolution of genetically based traits that increase the host's resistance to the diseases caused by the pathogens. As the host becomes more resistant, natural selection exerts pressure on pathogen populations that favors the evolution of traits that increase its rate of transmission in host populations. This tit-for-tat selection is called coevolution (see Module 6.2). Such coevolution is evident in a number of human diseases.

Among the four different kinds of malaria that occur in humans, falciparum malaria is by far the most deadly. This kind of malaria is most prevalent in Africa, where it kills more than 1 million people each year. It is no coincidence that where falciparum malaria is prevalent, so is a human genetic disorder called sickle-cell disease. In this disease, red blood cells contain an abnormal form of hemoglobin that clumps together, causing the normally round cells to take on a sickle shape (Figure 18.36). The sickle-shaped cells tend to get

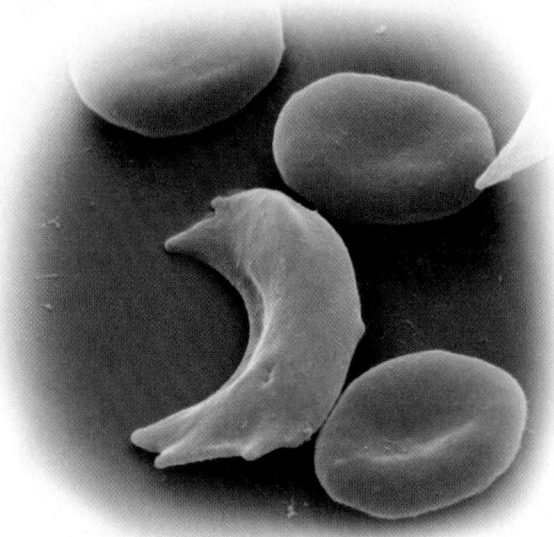

▲ Figure 18.36 **Sickle-Cell Disease** In individuals with severe sickle-cell disease, some red blood cells are misshapen and prone to causing blockages in blood vessels. Blood cells in individuals with a mild form of the condition are normal but resistant to invasion by the malaria parasite.

603

caught in blood vessels, causing blockages that may be fatal. Individuals with a mild form of sickle-cell disease have normal red blood cells and are not at risk for such blockages. However, when both parents have mild forms of the disease, their children have a 25% chance of inheriting the extreme form of sickle-cell disease.

You might expect that there would be very strong natural selection against a genetic trait such as sickle-cell disease. Yet where falciparum malaria is common, as much as 20% of the population has the mild form of this disease (Figure 18.37). Why is this trait so prevalent there? Malaria parasites cannot live in the red blood cells of people with mild forms of sickle-cell disease, making these people resistant to malaria. Thus natural selection maintains a high frequency of individuals who are resistant to malaria, even though their children have a significant risk of inheriting the extreme form of sickle-cell disease.

Recently, the evolution of the falciparum malaria parasite has been affected by the various medical treatments used to prevent the disease. For the past 50 years, doctors have prescribed a mixture of two drugs—chloroquine and sulfadoxine-pyrimethamine—to people living or traveling in areas where the falciparum parasite is prevalent. Over time, natural selection has favored the evolution of falciparum parasites that are resistant to these drugs. As a result, these drugs are no longer effective in much of Africa.

Over long spans of time, coevolution between a pathogen and its host tends to diminish the negative impacts of the disease caused by the pathogen. However, when hosts are exposed to a new pathogen, the resulting disease is often extreme or even fatal. Population growth and environmental changes are now exposing humans to a number of pathogens with whom they have not coevolved. Human immunodeficiency virus (HIV), for example, is thought to have arisen from a similar virus in chimpanzees, called simian immunodeficiency virus (SIV), which was transferred to humans sometime between 1880 and 1940. (That transfer was likely a consequence of hunting and eating chimps, although debate remains on this issue.) SIV is far less deadly in chimps than HIV is in humans.

Ebola hemorrhagic fever, or simply Ebola, is another disease that has recently appeared in humans. More 50–90% of the individuals infected with this disease die within a few weeks. This disease is also quite deadly in various species of monkeys. Its natural host appears to be several species of fruit-eating bats that live in tropical forests in central Africa. The sudden appearance of this disease in humans seems to be related to the movement of human populations into forested regions where these bats are common. The factors affecting its recent appearance and spread are discussed more in Module 18.5.

In summary, human health and well-being are affected by infectious respiratory, diarrheal, and blood-borne diseases. The prevalence of these diseases is, in turn, influenced by variations in the environment. Diseases caused by parasites that have coevolved with humans tend to be less severe than those caused by parasites that have only recently appeared in human populations.

QUESTIONS 18.4

1. Why might zoonotic diseases be more difficult to eradicate than anthroponotic diseases?

2. What factors contribute to the higher prevalence of flu during the wintertime?

3. Why are mosquitoes becoming increasingly resistant to insecticides such as DDT in many regions?

(MES) For additional review, go to **MasteringEnvironmentalScience**

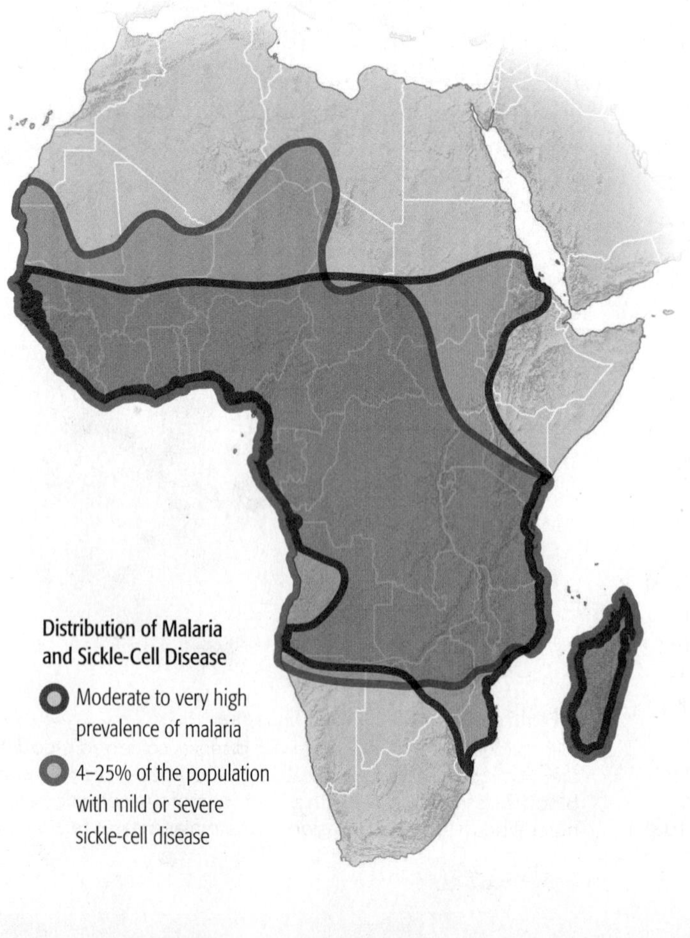

◀ Figure 18.37 **Distribution of Malaria and Sickle-Cell Disease**

Falciparum malaria is most prevalent in equatorial east Africa, where more than 20% of the population has either the mild or severe form of sickle-cell disease.

Distribution of Malaria and Sickle-Cell Disease

◉ Moderate to very high prevalence of malaria

◉ 4–25% of the population with mild or severe sickle-cell disease

18.5 Environmental Change and Human Health

BIG IDEA The intimate and complex connections between our environment and our health are compelling reminders of the need to sustain key ecosystem processes and services. But Earth's environment is rapidly changing in ways that are affecting these ecosystem processes. As human populations increase, the risks associated with many physical, chemical, and biological hazards are increasing, particularly in urban centers. In many parts of the world, the negative impacts of air and water pollution are increasing. Changes in landscapes and waterways are altering the habitats of zoonoses and disease vectors, thereby increasing the prevalence of some infectious diseases. Climate change is also increasing the prevalence of some infectious diseases, as well as the risks of some weather hazards. Sustainable development and management will require us to understand how changes in the global environment affect human health.

Human Population Size

■ Population growth and urbanization increases our vulnerability to many hazards.

Over the past century, the world's human population has grown from about 1.3 billion to about 7.3 billion. Much of that population is now concentrated in urban centers. In 1950, fewer than 25% of Earth's people lived in cities or towns. Today, more than 50% of the world's population is urban, and that proportion is likely to increase to more than 60% in the next 20 years.

Changes in human populations are altering the environment in ways that affect public health. The risks posed by nearly all physical hazards are magnified by population growth, especially the growth of urban centers. Much urban development, especially in poor countries, is susceptible to geological hazards such as earthquakes and landslides. City dwellers are also at greater risk of harm from extreme weather. Because cities tend to trap heat, temperatures in cities are much higher than those in rural areas (see Module 16.2). For example, more than 90% of the deaths caused by the European heat wave of 2003 occurred in cities of over 10,000 people.

The 2014 outbreak of Ebola in West Africa is a powerful example of the effects of population growth and urbanization on the spread of infectious disease. Ebola is a zoonosis caused by a virus that infects a variety of African tropical forest mammals (**Figure 18.38**). Fruit bats, which can carry the virus but display no disease symptoms, are thought to be the natural reservoir for the Ebola virus. Transmission from bats to other mammals likely occurs when the bats drop partially eaten fruits that are then picked up by other animals such as monkeys, gorillas, and chimpanzees. Humans can become infected by eating tainted fruit or the meat of infected animals.

The appearance of Ebola in humans is thought to have occurred as population growth encouraged migration into previously uninhabited territories. The first known outbreak of Ebola among humans occurred in 1976, in the rural village of Nzara in South Sudan. Then, the virus infected 284 and killed 151 people. Later that year, Ebola appeared again in a remote village in the Democratic Republic of Congo. Between 1976 and 2013, similar outbreaks occurred in a number of rural communities in tropical Africa. None of these involved more than 500 individuals, and the total number of known infections prior to December 2013 numbered 1,716. Outbreaks were limited by traditional quarantine methods and the fact that they occurred in isolated situations.

Between December 28, 2013 and November 1, 2014 (as this book was going to press), over 13,500 people had been infected and 4,960 had died from the disease. What changed in 2014 to catalyze this tragic epidemic? The first case in this epidemic is thought to have been a 2-year-old boy in a small village in southeastern Guinea, near the borders of Sierra Leone and Liberia. Mourners from the child's funeral are thought to have carried the disease back to urban areas in each of these countries. This appears to have been the first time this disease was transported to humans living in a dense urban setting. These are three of the world's poorest countries, and each has suffered through civil war in the past few years. Crowded conditions, poor sanitation, and limited access to health care made control of the disease especially difficult.

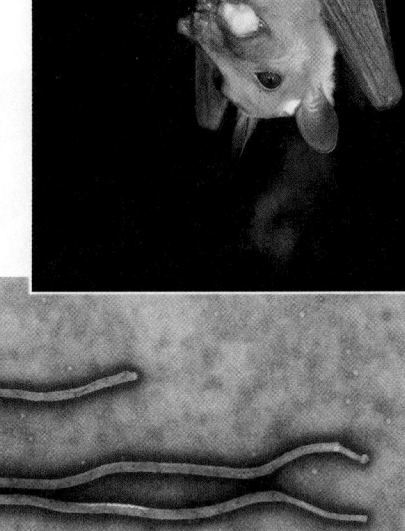

▲ Figure 18.38 **A Deadly Passenger**
Ⓐ African fruit bats are one vector known to carry the Ebola virus. Ⓑ The virus is a rather simple one, composed of only 13 genes. It is transmitted from animal to animal in body fluids such as saliva and blood. Humans infected with the virus display symptoms such as fever, sore throat, and headaches within 21 days of infection. It causes death from dehydration and liver and kidney failure in 50–90% of those who are infected.

Air and Water Pollution

■ Changes in air and water quality are increasing the incidence of respiratory and waterborne infectious diseases.

Population growth and increasing urbanization have reduced the air quality in many regions. The United Nations Environmental Program estimates that over 2 million people die each year from respiratory diseases directly related to urban air pollution, such as asthma, emphysema, and lower respiratory infections. More than 60% of these deaths occur in the urban centers of rapidly developing countries, such as China and India. Because developing countries often lack the resources to enforce air-quality standards or to mitigate air pollution, the proportion of total deaths caused by air pollution is likely to increase as urban growth accelerates.

Respiratory diseases caused by air pollution are also a problem in developed countries, as is clear from the increasing prevalence of asthma among children and adults in the United States during the past 20 years (Figure 18.39). During this period, air quality has improved in some major urban centers, such as Los Angeles, but has become considerably worse in areas undergoing rapid urban development, such as the southeastern United States and California's Central Valley. The movement of people from rural to urban areas has probably contributed to the increasing prevalence of asthma.

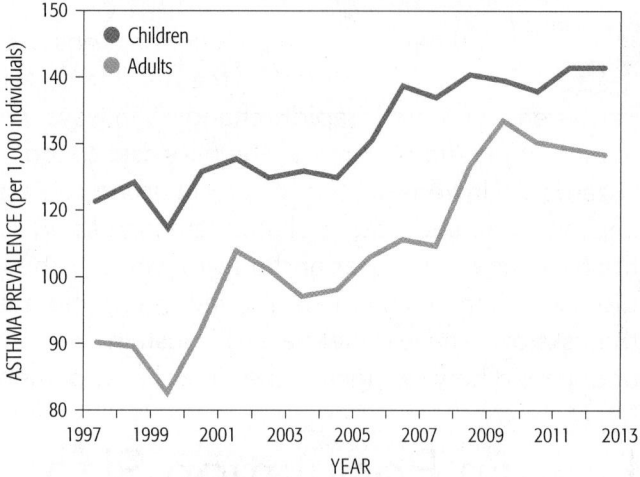

▲ Figure 18.39 **Asthma on the Rise**
The increasing prevalence of asthma in U.S. children and adults is a consequence of the increasing urbanization of the population and the diminishing air quality in many regions.

Source: U.S. Centers for Disease Control. http://www.cdc.gov/asthma/nhis/default.htm. Last accessed April 16, 2014.

Increasing populations also have an impact on water quality. Although technologies for the management of industrial and municipal wastewater have improved considerably, about a third of Earth's people still do not have access to clean water. This fraction is likely to increase because of the rapidly growing populations in poor countries that have a limited ability to adopt these technologies.

The development of water diversions and reservoirs alters ecosystems and often brings increased risks to human health. Next to malaria, schistosomiasis is the second-most important infectious disease in Africa. Schistosomiasis is caused by a parasitic fluke worm that also lives in freshwater snails (Figure 18.40). Over the past three decades, irrigation projects and the construction of dams have increased the habitat available to these snails, as well as human contact with water containing the parasitic worms. As a result, the incidence of schistosomiasis in Africa has increased significantly. For example, in the 25 years following the completion of the Aswan High Dam on the Nile River in Egypt, the rate of schistosomiasis infection in the surrounding population jumped from 5 to 77%.

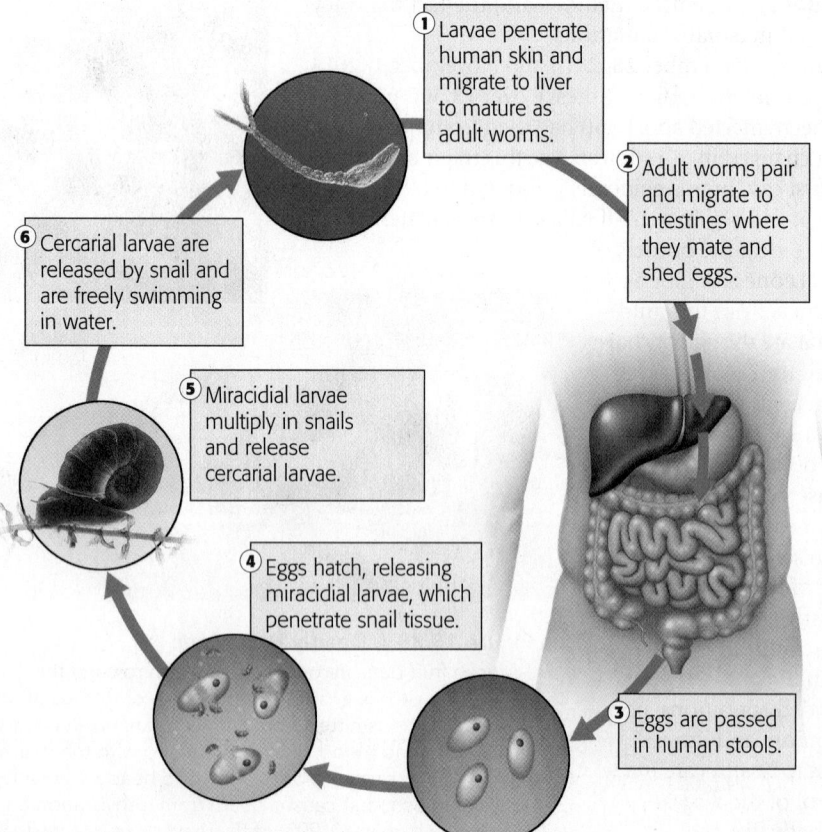

① Larvae penetrate human skin and migrate to liver to mature as adult worms.

② Adult worms pair and migrate to intestines where they mate and shed eggs.

③ Eggs are passed in human stools.

④ Eggs hatch, releasing miracidial larvae, which penetrate snail tissue.

⑤ Miracidial larvae multiply in snails and release cercarial larvae.

⑥ Cercarial larvae are released by snail and are freely swimming in water.

◀ Figure 18.40 **Life Cycle of the Schistosomiasis Parasite**
Schistosomiasis is caused by a parasitic worm. Freshwater snails are an intermediate host for this parasite. Poor sanitation and abundant habitat for snails maintain the prevalence of this disease worldwide. (The terms *miracidial* and *cercarial* refer to different stages of the fluke worm's life cycle.)

An Integrated Approach to Disease Management

Can the prevalence of schistosomiasis in sub-Saharan Africa be reduced?

North African countries, such as Egypt, are on track to eradicate schistosomiasis, but the same cannot be said of the poorer nations of sub-Saharan Africa. WHO estimates that the disease affects more than 100 million people in sub-Saharan Africa; in many communities, more than half the people are infected.

Programs to control schistosomiasis are often thwarted by poverty and poor governance. Given the complexity of the challenges presented by the disease, it is unlikely that a single organization will be able to alter the current situation. However, collaborations among nongovernmental organizations, pharmaceutical companies, and public agencies have begun to find solutions. The Bill and Melinda Gates Foundation has played a key role in catalyzing these collaborations.

In regions where the threat of schistosomiasis has been significantly reduced, collaborative groups have found it essential to use an integrated approach that involves three aspects of the problem: ecology, epidemiology, and education.

Ecology. In much of sub-Saharan Africa, access to water is a challenge. Unfortunately, many of the projects that have increased the availability of water have also created favorable habitat for snails, thereby increasing the prevalence of schistosomiasis. For example, reservoirs and irrigation canals often support large populations of the intermediate host snails.

This ecological challenge is a focus of research at the International Water Management Institute (IWMI) in Colombo, Sri Lanka. IWMI studies show that snail populations can be reduced by removing vegetation from canals and reservoirs and by applying molluscicides, chemicals that limit the growth and reproduction of snails. Where possible, irrigation canals should be lined with cement. Research by IWMI indicates that continued maintenance is essential for the success of ecological management (**Figure 18.41**).

Epidemiology. To improve the health outcomes for individuals and slow the spread of the parasite to other people, infected individuals should be identified and treated early. A single treatment with the drug praziquantel (PZQ) is often sufficient to eliminate the infection. However, WHO estimates that in 2008 fewer than 6% of infected Africans received treatment. To significantly reduce the spread of the disease, at least 30% of infected individuals must be treated. This level of treatment would require more than 200 million doses of PZQ per year.

The U.S. State Department's Agency for International Development and the U.K.'s Department for International Development have committed to providing funding for more than 180 million doses of PZQ each year from 2010 to 2014. Merck Serono, the manufacturer of PZQ, has agreed to donate an additional 20 million doses each year. The Bill and Melinda Gates Foundation is providing significant funding to the Schistosomiasis Control Initiative (SCI) in London and the Carter Center in Atlanta, Georgia, for the distribution of PZQ.

Education. Public education is essential to the successful control of schistosomiasis (**Figure 18.42**). The Gates Foundation has provided funding to SCI and the Carter Center for educational programs aimed at improving public understanding of the importance of sanitation and treatment programs in limiting the spread of this disease.

Although schistosomiasis remains a serious threat to hundreds of millions of people in sub-Saharan Africa, an integrated approach holds the promise of change. How will we know if this integrated approach is successful? In 2008, the Gates Foundation provided funding to establish the Schistosomiasis Center for Operational Research and Evaluation (SCORE) at the University of Georgia. SCORE provides funding to a network of researchers to evaluate the effectiveness of ecological, epidemiological, and educational programs.

▲ Figure 18.42 **Education Is Essential** Students in a Nigerian primary school hold samples of their own urine. Blood in the urine is the first sign of schistosomiasis. Left untreated, this disease will eventually damage intestines, liver, bladder, and lungs.

◄ Figure 18.41 **Long-Term Management Is Critical** This canal in Sudan has been lined to prevent the growth of vegetation that might provide habitat for the snails that carry schistosomiasis.

Landscape Change

■ Altered landscapes have provided opportunities for some zoonotic diseases to expand.

Human activities have significantly altered many landscapes and waterways. These alterations have affected the habitats of many animals that are disease vectors or that share diseases with humans, thereby causing the frequency of some diseases to increase.

According to the Centers for Disease Control, one of the most important emerging diseases in the United States is Lyme disease. This disease was not discovered until the mid-1970s. In the 1980s, most cases were confined to New England. Since then, its prevalence has increased exponentially across much of the eastern United States.

Lyme disease often begins with a high fever, headache, and diagnostic skin rash. Left untreated, it can cause nervous disorders, heart problems, and arthritis. It is caused by bacteria in the genus *Borrelia*, which are transmitted to humans by the bite of infected deer ticks. It is a zoonosis that is shared among humans and forest mammals such as deer mice, chipmunks, and white-tailed deer.

Recent studies indicate that *Borrelia* and deer ticks were probably abundant in New England forests in precolonial times. By the middle of the 19th century, forest clearing had nearly removed the deer ticks and the disease they transmit. Since then, reforestation has allowed the populations of forest mammals to recover. Although lagging behind them, populations of deer ticks and the prevalence of *Borrelia* have gradually increased as well (Figure 18.43).

The recent, rapid increase in the incidence of Lyme disease is attributable to equally recent and rapid changes in the forested landscapes and distribution of people on those landscapes. Fragmentation of forests for suburban

development has created habitats that are especially favorable for deer mice and white-tailed deer. Just as important, this development has brought large numbers of people into direct contact with these animals and their ticks (Figure 18.44).

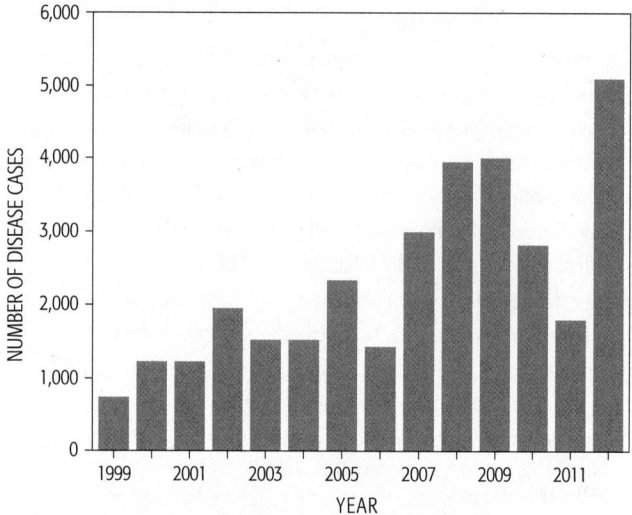

▲ Figure 18.44 **A Disease on the Increase**
This graph shows the number of confirmed cases of Lyme disease reported in the state of Massachusetts. The rapid increase corresponds to the rapid growth in the numbers of deer and to suburban development that has brought more people into forested habitats.

Source: U.S. Centers for Disease Control. http://www.cdc.gov/lyme/stats/chartstables/reportedcases_statelocality.html. Last accessed April 16, 2014.

▼ Figure 18.43 **The Geography of Lyme Disease**

The deer tick is the vector for *Borrelia*, the pathogen that causes Lyme disease. The risk of Lyme disease is greatest in the suburban areas of southern New England, where fragmented forests and development provide ideal habitat for all of the animals that are part of this zoonosis. Those animals include deer mice, white-tailed deer, and humans.

Source: U.S. Centers for Disease Control. http://www.cdc.gov/mmwr/preview/mmwrhtml/figures/R807A2F1.GIF. Last accessed April 16, 2014.

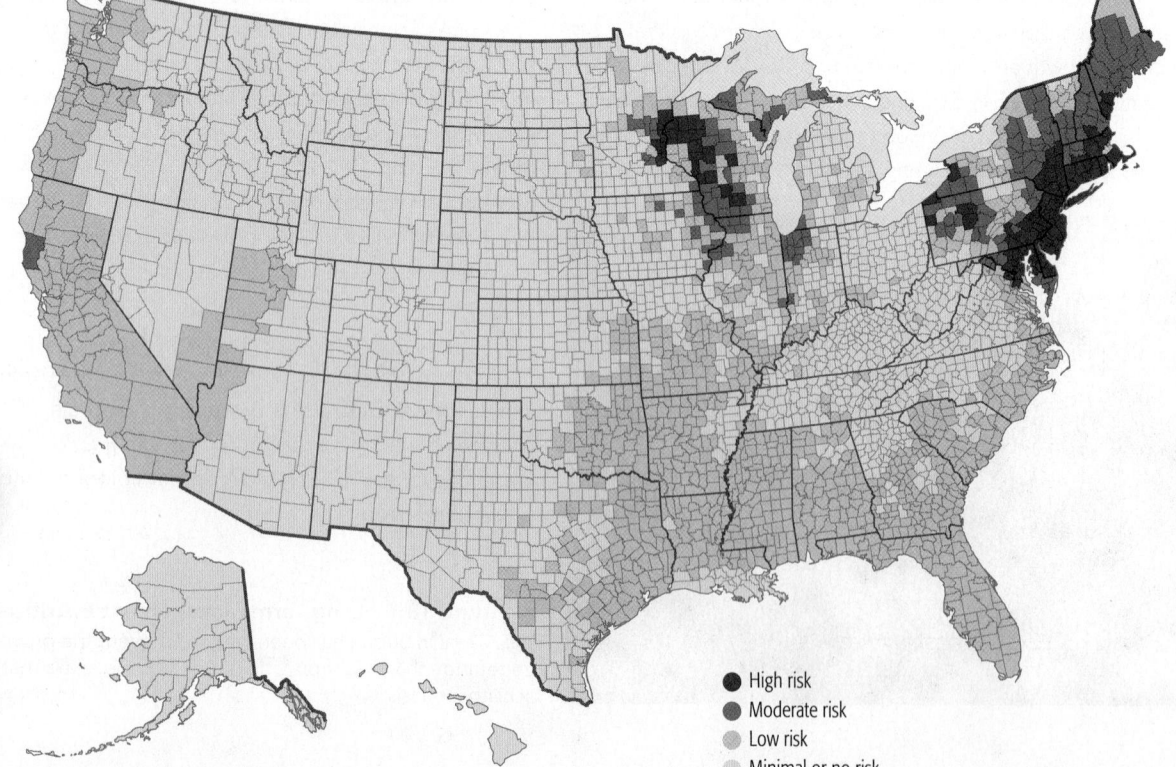

● High risk
● Moderate risk
● Low risk
● Minimal or no risk

Climate Change

■ Global warming is altering habitats for disease organisms and their vectors.

The effects of global warming differ considerably among the various regions of the world (see Module 9.5). These changes are already influencing factors that impact human health. In some areas, heat waves have become more frequent. There is strong evidence to support the hypothesis that warming will make severe hurricanes and typhoons more likely. Global warming is also affecting the habitat of some disease-causing organisms and their vectors, thereby increasing the risk of these diseases.

Public health scientists and officials are especially concerned with ongoing climate change at the transition between Earth's tropical and temperate zones. In general, this transition appears to be moving to higher latitudes. The frequency and severity of cold winter temperatures are decreasing, and in many places humidity and moisture levels are increasing. These changes have facilitated the expansion of populations of the mosquitoes that transmit diseases such as malaria and dengue fever.

The effect of global warming on the worldwide distribution of dengue fever is especially well documented. The virus that causes this debilitating disease is transmitted by *Aedes aegypti*, a mosquito that cannot tolerate freezing temperatures. The number of cases of dengue fever reported in North and South America has increased from 66,000 in 1980 to 552,000 in 2006. This increase is especially significant in countries in the transition region between the tropical and temperate zones. In Mexico, for example, cases of dengue fever have increased by 600% since 2001 (Figure 18.45).

In 2012, 357 cases of dengue fever in the United States were reported to the Centers for Disease Control and Prevention. Most of these cases were in individuals who had traveled to tropical countries, but an increasing number of cases in south Texas and Florida are appearing in individuals who have not traveled. If global warming continues at its current rate, the Intergovernmental Panel on Climate Change estimates that in 50 years an additional 3.4 billion people will be at risk for dengue fever.

In summary, human-caused environmental changes, including diminished air and water quality, modified landscapes, and global warming, are increasing the risks that we face from many physical, chemical, and biological hazards. People in the world's very poor and developing countries have been particularly hard-hit by these changes. Current trends threaten the well-being of future generations, regardless of their economic status. Reversing these trends will require an understanding of how our actions affect biotic interactions and ecosystem processes and a commitment to diminishing those impacts.

Q *Are we prepared to deal with new epidemics as global climate change causes vector-borne diseases to become more widespread?*

Chris Mills, North Carolina State University

Norm: A The U.S. Centers for Disease Control and Prevention is paying special attention to the potential impact of global warming on vector-borne diseases. Immunization and programs that manage the habitats of disease vectors, such as mosquitoes, will be important parts of climate-change adaptation.

QUESTIONS 18.5

1. How is urbanization changing the risks associated with hazards such as earthquakes, air pollution, and infectious disease?

2. The connection between global warming and the increased prevalence of infectious diseases appears to be especially strong in zoonotic diseases and diseases spread by insect vectors. Why might this be the case?

MES For additional review, go to **MasteringEnvironmentalScience**

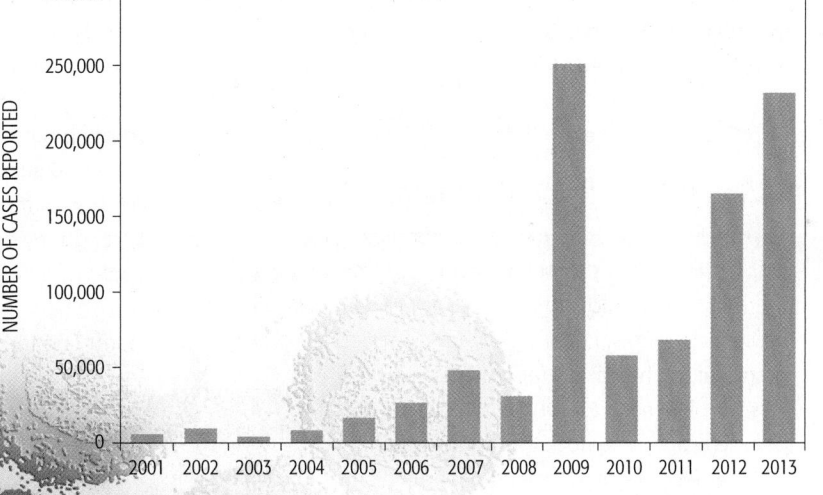

▲ Figure 18.45 **Climate Change and Prevalence of Dengue Fever**

Over the past several decades, warmer and wetter conditions in Mexico have provided favorable habitat for the mosquito that transmits dengue fever. This has increased the number of cases of dengue fever and contributed to a significant increase in its geographic range across the country. Very wet conditions and floods in 2009 provided ideal conditions for mosquitoes and the spread of dengue. The photomicrograph shows the virus that causes dengue fever.

Source: Pan American Health Organization. http://www.paho.org/hq/index.php?option=com_content&view=article&id=264&Itemid=40170&lang=en. Last accessed April 16, 2014.

609

Taking Public Health to Rural Haiti

Will Perez is a medical student with a degree in public health from Brown University. After completing his undergraduate degree, Will deferred medical school for a year to live in rural Haiti. There he developed public health programs and trained youths to become community health workers who could bring health care to regions of Haiti where there are no doctors. Will's training programs are now in 12 schools in rural Haiti and reach 20,000 Haitians.

How did you first get the idea to begin public-health training programs in Haiti?

While in school, I became interested in changing medical policy—the way medicine is practiced, the way it is distributed, and how it is defined. After reading Tracy Kidder's *Mountains Beyond Mountains* and a *New York Times* article with a photo of a Haitian woman baking biscuits of clay and water to feed her children, I committed myself to becoming a doctor for the poor.

I wanted to improve the health of rural Haitians and relieve the death toll imposed on them by the constraints of poverty. Ninety percent of Haiti's children suffer from waterborne diseases and intestinal parasites. Half of the children in Haiti are unvaccinated, and just 40% of the population has access to basic health care.

What steps did you take to get started?

I began by researching my cause, contacting experts in the field to learn about Haiti from every possible angle. I found mentors who would support me and guide me in making difficult decisions. I raised funds and applied for every grant, fellowship, and award that I could. I also took every opportunity to educate people around me about what I was doing.

I learned Haitian Creole, the language spoken by rural Haitians. When I moved to Haiti, I lived in the village where I worked and spoke only Creole. By living in the village and speaking to the people in their own language, I learned about their true needs. No one knows the needs of the people better than the people themselves. Once I figured this out, a seemingly impossible mission became a somewhat more manageable task.

What are the biggest challenges you've faced with the project, and how did you solve them?

I learned my first lesson in foreign aid soon after I arrived in a small village in Haiti, prepared to execute a tuberculosis prevention program I had developed. The program was a complete flop because it didn't serve the most pressing needs of the village—daily battles with scabies, ringworm, fungal infections, and parasites.

Undeterred, I went door to door to figure out what issue I should tackle. The answer? Bedbugs. I set my sights on this less-than-glamorous foe and managed to eradicate bedbugs and ringworm from the village during my first year. This earned me the respect of the villagers, but led to new challenges.

Although the program was hugely successful, I had empty pockets. I could either pack up and leave or raise money to support my program. I decided to raise money. I started writing a blog, telling my story of living in Haiti. Through my blog, I raised over $200,000 to support my new goal of training community health workers.

I trained 16 orphaned youths and formed a public health team. The team went on to provide treatment to hundreds

of children and adults who suffered from tuberculosis. Malaria rates in the village went down by 60%. Using simple preventative measures, the team also reduced the incidence of waterborne pathogens, bedbugs, and parasitic infections.

How has your work affected other people and the environment?

Through education, I am equipping the Haitian people with the tools they need to save their lives. These efforts are important because there are no doctors—the people are their own medical community.

Aside from engaging the community in public health issues and providing sustainable methods of treatment for hundreds of people, the youth I trained have become leaders in their communities and are now looked upon with incredible respect. Their work is inspiring to their peers, as can be seen in the long list of young adults waiting for their chance to join the public health group.

My training program is the only one like it in the southwestern region of Haiti. Recognizing that prevention is by far the most cost-effective and surest way of eliminating the many diseases in the country, other health organizations throughout Haiti have begun to adopt my public-health training program.

What advice would you give to students who want to replicate your model on their campus?

Identifying what resources are available to you is the single most important thing you can do to make your ideas become a reality. Begin by making contacts. Talk to people who have been working on similar issues and are willing to mentor you or to collaborate with you. Professors, deans, and student groups are great places to start. Deans love to hear what students are doing, and they can steer you in the direction of fellowships, grants, and financial aid to support your work.

It takes courage to identify a need and then work to fulfill it. It won't happen overnight, and it may feel as though you keep running into walls. Don't be deterred. Take note of the obstacles and then target each one individually. You'll find that a seemingly impossible task can become achievable by simply taking it one step at a time.

Summary

Environmental changes, whether natural or human-caused, have significant consequences for human health. The health of human populations is often measured by statistics about the likelihood of death and leading causes of death; infant mortality is a particularly sensitive measure of the overall health of populations. Noninfectious conditions, such as heart disease and cancer, account for more than 75% of the deaths in developed countries. Infectious diseases, such as malaria and diarrhea, are responsible for over half of the deaths in developing countries. Health risks are measured by the likelihood of a hazard occurring and by the consequences of that hazard. Public perceptions of the risks associated with a hazard are influenced by expectations that risks can be totally eliminated, by publicity, by neglect of risk–risk trade-offs, by perceived levels of control over the hazard, and by time delays in the consequences of the hazard.

Physical hazards in the environment include geological events, such as earthquakes and tsunamis, weather hazards, such as hurricanes and heat waves, and wildfires. Chemical hazards refer to toxins, chemicals that diminish human health. The toxicity of chemicals depends on dosage and duration of exposure. These factors are influenced by the manner in which toxins are transported in the environment and the way in which toxins are absorbed and stored in the body. Genetic and environmental factors also influence human vulnerability to toxins. Toxins are classified according to their effects on specific organs and tissues.

Biological hazards include infectious diseases caused by pathogenic organisms. Some pathogens infect only humans. Other pathogens are shared between humans and other animals, some of which play an active role in transmitting disease. Respiratory diseases are transmitted through the air. Diarrheal diseases are associated with poor water quality. Blood-borne diseases are generally transmitted by animal vectors, such as mosquitoes. Over time, coevolution between pathogens and their human hosts tends to diminish the severity of diseases. Environmental changes, such as increasing human populations and urbanization, diminished air and water quality, and global warming, are significantly affecting the risks posed by many environmental hazards

18.1 Introduction to Public Health

- The health of human populations is generally assessed by statistics related to the likelihood of death and the leading causes of death.
- The risk of an environmental hazard is determined by the probability of its occurrence and its consequences.

KEY TERMS

risk, latent consequences, risk–risk trade-offs

QUESTIONS

1. Why do public health experts consider infant mortality to be an especially important measure of the health of a human population?
2. Over the past 60 years in the United States, the proportion of people dying of cardiovascular disease and cancer has increased compared to infectious disease. What trends are likely responsible for this change?
3. Two different environmental hazards may be equally likely but pose very different risks. How can this be?
4. What are risk–risk trade-offs?

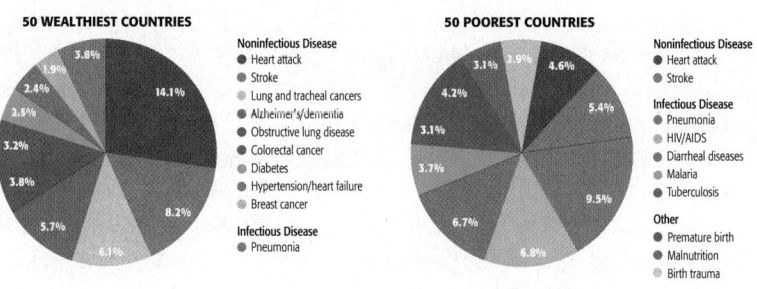

18.2 Physical Hazards in the Environment

- Although physical hazards such as earthquakes, hurricanes, and wildfires are often called natural disasters, they are heavily influenced by the activities of humans.

KEY TERMS

Richter magnitude scale, earthquake intensity, storm surge

QUESTIONS

1. Describe three factors that influence the risks posed by earthquakes in different locations.
2. The consequences of natural hazards such as hurricanes are often different for poor people than for the wealthy. Do you consider this to be an environmental justice issue? Explain your answer.
3. What factors have contributed to the steady decrease in the proportion of deaths caused by weather hazards?

18.3 Chemical Hazards in the Environment

- The toxicity of a chemical depends on the dose and duration of exposure; the vulnerability of individuals to particular toxins is influenced by genetic makeup and environmental factors.
- Toxins may influence one or a combination of organs and tissues.

KEY TERMS

toxicology, dose–response curve, toxicity threshold, median lethal dose (LD_{50}), acute exposure, chronic exposure, bioaccumulation, biomagnification, body burden, persistent organic pollutants (POPs), corrosive toxins, asphyxiants, carcinogens, teratogens, allergens, neurotoxins, endocrine disrupters, model organisms

QUESTIONS

1. Some individuals can tolerate levels of a toxin that are above its LD_{50}. What factors might explain such high tolerance?
2. What factors influence the body burden for chemical toxins like DDT and arsenic?

3. Discuss the pros and cons for using model organisms to determine the toxicity of particular chemicals to humans.

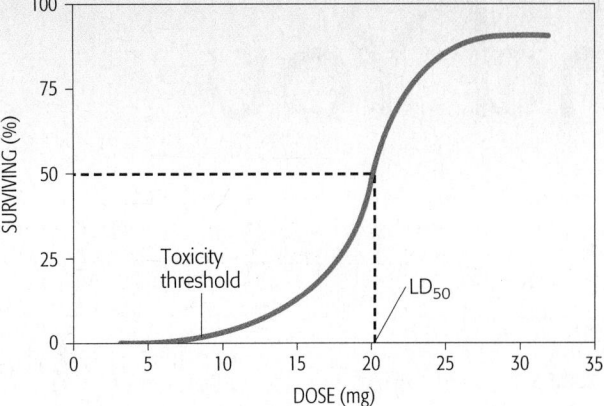

18.4 Biological Hazards in the Environment

- Environmental factors influence populations of organisms that cause diseases, as well as the populations of organisms that transmit and contract diseases.
- The relative importance of an environmental factor to the prevalence of an infectious disease depends on the disease's mode of transmission and the organs or tissues that it affects.

KEY TERMS

pathogen, virulence, anthroponosis, zoonosis

QUESTIONS

1. Why do some pathogens tend to be far more virulent among older individuals than younger individuals?
2. Why is it unlikely that we will ever be able to eradicate the common cold and the flu?

3. Explain why scientists view the prevalence of sickle-cell disease in some human populations and the occurrence of falciparum malaria to be an example of coevolution.

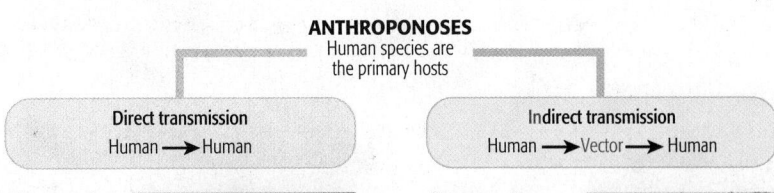

Disease-causing organisms can be shared through close human contact, like sneezing or shaking hands.

Disease-causing organisms of cholera are spread via a vehicle or vector, like drinking water, affecting other humans who ingest that contaminated water.

18.5 Environmental Change and Human Health

- The risks posed by physical, chemical, and biological hazards are influenced by changes in the size and density of our populations, the quality of the air we breathe and the water we drink, and global climate.

QUESTIONS

1. Describe two factors that may be responsible for the increased prevalence of asthma among children and adults in the United States.
2. What factors contributed to the rapid spread of Ebola in Sierra Leone, Liberia, and Guinea in 2014?
3. Explain the connection between suburban development and the increased prevalence of Lyme disease in the eastern United States.

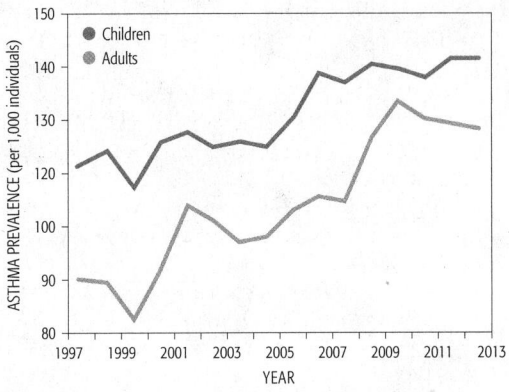

MasteringEnvironmentalScience®

Students Go to **MasteringEnvironmentalScience** for assignments, the eText, and the Study Area with practice tests, activities, and more.

Instructors Go to **MasteringEnvironmentalScience** for automatically graded tutorials and questions that you can assign to your students, plus Instructor Resources.

The Environment
and You

19

19.1 Hope for the Environment

BIG IDEA If you don't watch where you're going, you'll end up where you're headed. Over the past century, Earth's ecosystems and the human communities that depend on them have experienced enormous change (**Figure 19.1**). Significant changes are also certain to occur in the future: Earth's atmosphere will become warmer, already scarce resources will become scarcer, biodiversity will decrease, and there will be more people. Although each of these changes is certain, the extent and consequences of them are much less certain. If some human actions and policies continue, the magnitude and consequences of these changes could be great. Other actions and policies can diminish their magnitude and minimize their consequences; it is these actions and policies that define our path to a sustainable future.

▲ Figure 19.1 **A World of Change**
One of your authors, Norm, grew up near the city of Fresno in the California San Joaquin Valley. His childhood home was surrounded by vineyards and orchards Ⓐ that have long since been replaced by residential development Ⓑ.

Warmer Climates

■ Yes, some global warming is inevitable, but we can act to keep it within sustainable and adaptable limits.

Since 1900, greenhouse gas emissions associated with human activities such as burning fossil fuels and deforestation have warmed Earth's atmosphere by nearly 1 °C. What's more, climate scientists agree that the concentration of greenhouse gases such as carbon dioxide and methane will continue to increase. As a consequence, Earth's atmosphere will become even warmer in the years ahead. Just how much warmer is uncertain. It is also uncertain how the overall trend of global warming will affect the temperatures and rainfall in specific regions.

If deforestation and the burning of fossil fuels continue to increase at their current pace, greenhouse gas emissions will double by 2050. Over the next century, this could produce average global warming in excess of 8 °C (14 °F) and cause the sea level to rise by 1 m (3.3 ft) or more. Climate change in polar regions and over continents could be even more extreme. The consequences for Earth's ecosystems could be catastrophic, including altered hydrologic cycles and widespread loss of species. The consequences for human well-being could also be devastating, including extended periods of drought, diminished food production and

famine, coastal flooding, and an increased prevalence of many infectious diseases.

We can diminish the magnitude of warming.
By itself, no single action or policy is adequate to mitigate global warming. But a combination of actions can slow and eventually reverse this trend. This combination should include increased energy efficiency and conservation, diminished reliance on fossil fuels, and increased use of renewable sources of energy, such as wind, solar, and biomass (**Figure 19.2**).

We can diminish the consequences of the warming that does occur through adaptation. Conservation of biodiversity in natural and managed ecosystems can increase their resilience to climate change and maintain the important ecosystem services they provide (**Figure 19.3**). In coastal regions, conservation of wetlands and shoreline ecosystems can minimize the effects of increased storm activity and flooding. In arid regions, water-efficient irrigation techniques and shifts to drought-tolerant crops can sustain the supply of water and food.

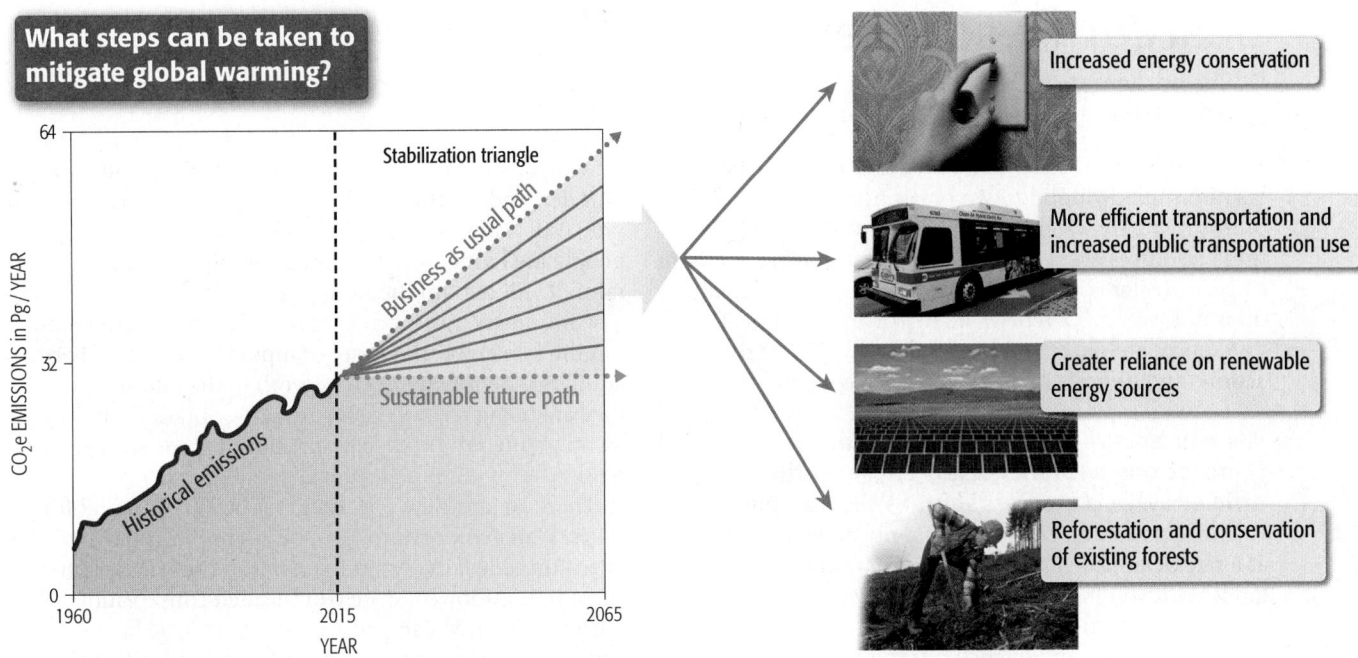

▲ Figure 19.2 **Reducing Greenhouse Emissions**
Greenhouse gas emissions can be reduced to sustainable levels using multiple strategies, such as increasing energy efficiency; using renewable energy sources, including solar, wind, biofuels, and hydropower; and by conserving Earth's forests.

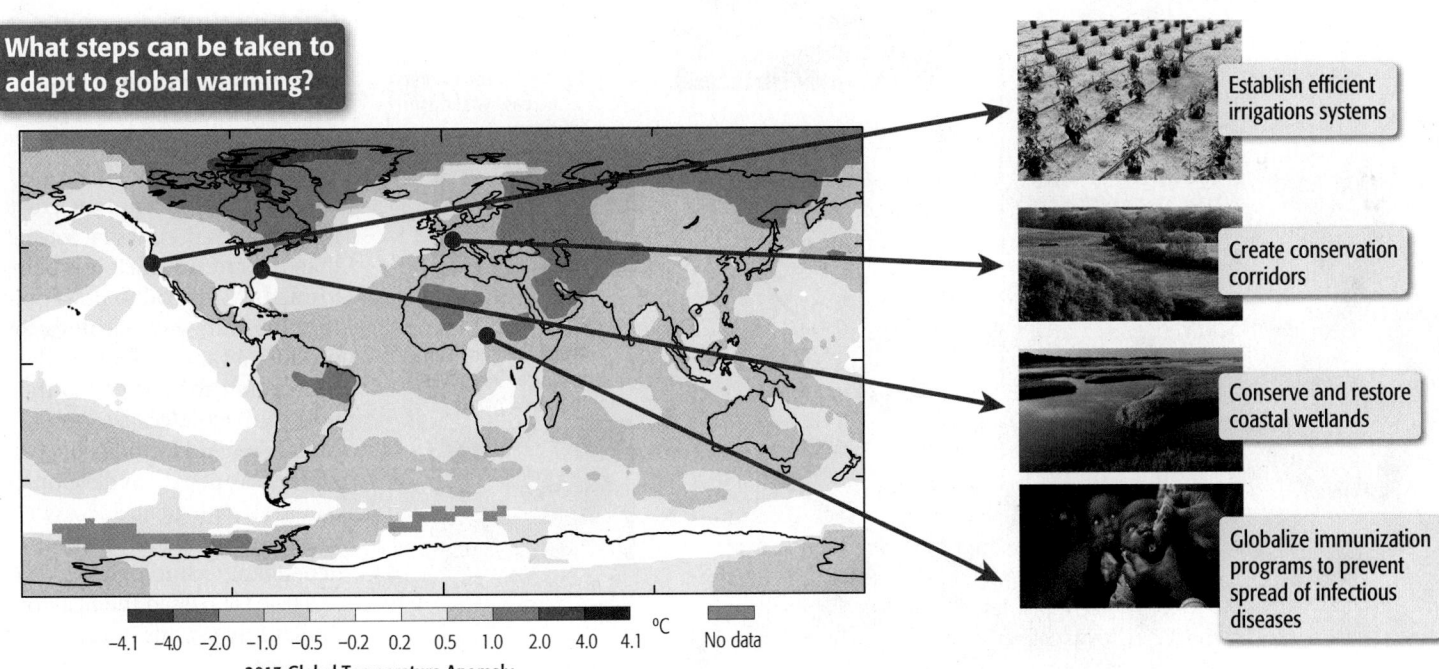

▲ Figure 19.3 **Sustainability Requires Adaptation**
Earth's atmosphere has already warmed by nearly 1 °C, and additional warming is very likely. Within the range of moderate warming (1–3 °C), adaptive actions can produce sustainable outcomes for ecosystems and human well-being.

Scarcer Resources

■ Yes, some natural resources will become scarcer, but we can act to accelerate the transitions to more efficient use and sustainable alternatives.

Today, supplies of many renewable and nonrenewable resources are dwindling. Two of the best-known examples of this trend are water and oil. We depend on both of these resources every day, but they differ in two important respects. Water is absolutely essential to our life functions; oil is not. Water is a renewable resource; oil is not. These differences are important in determining sustainable actions and policies for each of these resources.

▼ Figure 19.4 Sustaining a Renewable Resource
The supply of water on land is limited by the rate at which rain falls. A variety of strategies are available to increase the efficiency with which we use this water and to ensure its quality.

We can ensure ample water for future generations with increased conservation, efficiency, and stewardship. Water moving in Earth's streams and rivers is constantly being replenished at a rate determined by the amount of rain falling on land. We divert more than half of the water in streams to generate electricity, manufacture goods, irrigate crops, wash our clothes, and quench our thirst. The water flowing downstream from these uses is often polluted. Our demand for this precious resource is increasing each year, especially in arid regions where more and more people are living. But the annual supply of water in rainfall will remain the same.

Groundwater can meet some of the growing demand, but the rate at which we are pumping water out of Earth's aquifers is generally much faster than the rate at which they are being recharged. We can also obtain fresh water by removing salt from seawater, but the process is very costly and requires a great deal of energy.

There are a variety of ways to reduce the demand for water. Water conservation can be as simple as paying attention to leaking water fixtures and the length of time spent in the shower. More complicated conservation measures include using wastewater to irrigate lawns and replacing lawns with drought-tolerant landscaping. Agricultural irrigation systems and manufacturing processes can use water more efficiently.

Water stewardship requires that we stop using Earth's surface waters and ground waters as waste disposal systems. The real costs of conservation and stewardship must be included in the price we pay for this precious resource (Figure 19.4).

Conservation and new technologies can greatly diminish demand for oil. Even today, oil is slowly forming in certain parts of Earth's crust. But the rate at which we are extracting oil from the crust is thousands of times faster than the rate at which it is being formed. We use oil to power our cars, lubricate our machines, and produce a variety of materials from paints to plastics. For the next couple of decades, per capita demand for oil is likely to increase as people in developing countries gain greater access to automobiles and other oil-dependent technologies.

Have we passed the peak of availability for this nonrenewable resource? The answer to this question is hotly debated among energy experts and in newspaper op-eds. If we define *availability* to mean oil that can be extracted by conventional technologies and at relatively low prices, there is no debate: We passed the peak of availability more than a

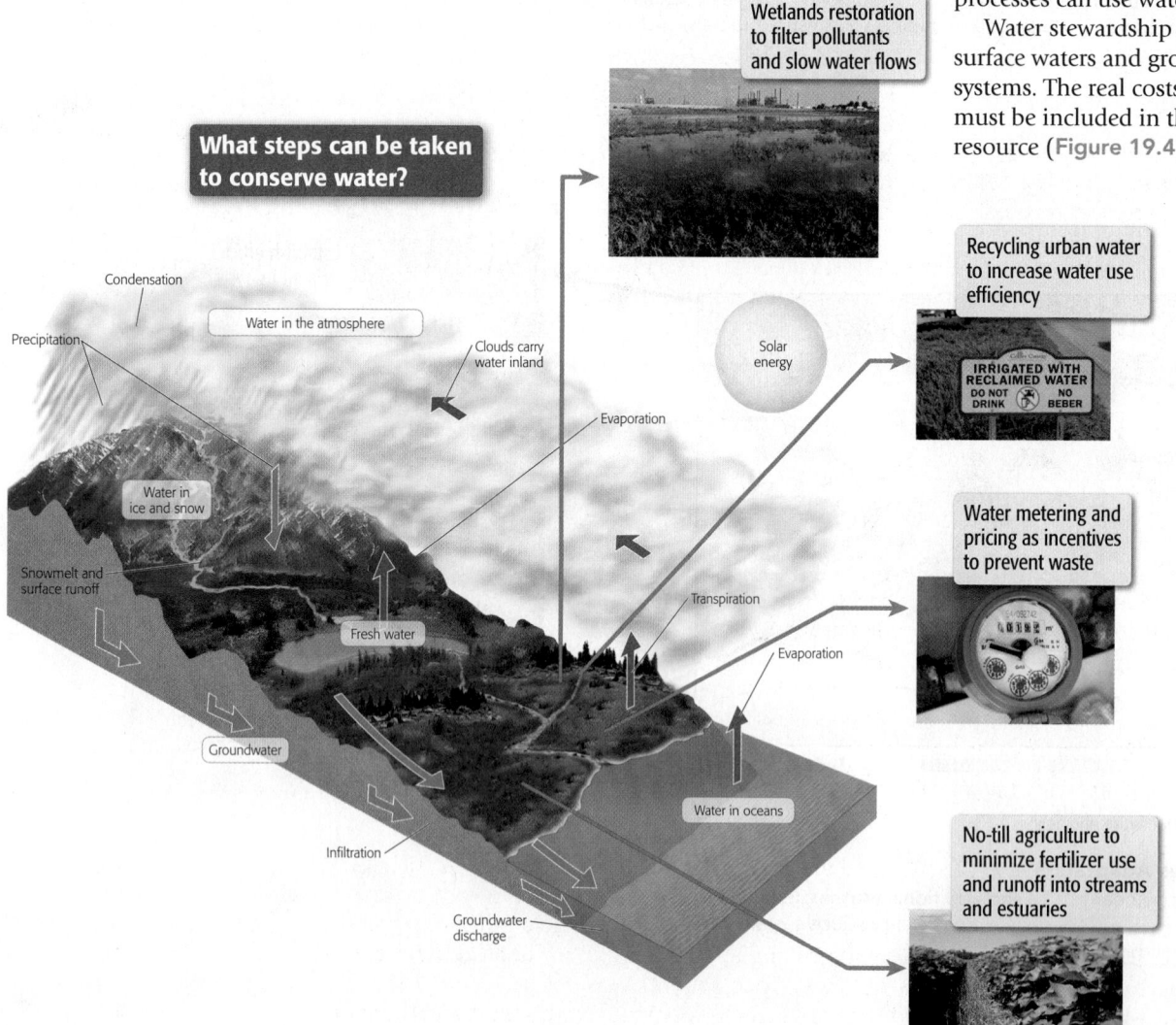

What steps can be taken to conserve water?

Wetlands restoration to filter pollutants and slow water flows

Recycling urban water to increase water use efficiency

IRRIGATED WITH RECLAIMED WATER DO NOT DRINK   NO BEBER

Water metering and pricing as incentives to prevent waste

No-till agriculture to minimize fertilizer use and runoff into streams and estuaries

Condensation
Water in the atmosphere
Precipitation
Clouds carry water inland
Solar energy
Evaporation
Water in ice and snow
Snowmelt and surface runoff
Transpiration
Fresh water
Evaporation
Groundwater
Water in oceans
Infiltration
Groundwater discharge

decade ago. The rate of discovery of new reserves that we can tap using available technologies has declined by over 85% since 1970.

New and emerging technologies allow us to extract oil from tar sands, oil shale, and deep ocean sediments. Of course, more energy and money are required to obtain this oil. Oil shale and tar sand operations require very large amounts of water and the excavation of large areas of land surface. But as worldwide demand for oil has increased and supplies have dwindled, the economic value of each barrel of oil has increased. This, in turn, has compensated for the higher costs of obtaining oil with new technologies.

Whatever technologies we use to obtain oil, high rates of use will continue to pollute the air we breathe and to increase greenhouse gas emissions. As we learned from the 2010 collapse of the BP Deepwater Horizon oil rig and devastating oil spill in the Gulf of Mexico, these new technologies have significant environmental risks and costs. It may take decades for the ecosystems of the Gulf to recover from that disaster.

To reduce our individual use of oil, we can take public transportation, drive fuel-efficient vehicles, and just simply walk and bicycle more. As alternatives to gasoline, we can use electric cars or those powered by hydrogen fuel cells and tap into renewable energy sources. Also, changes in patterns of urban development can significantly decrease automobile use (Figure 19.5).

What sustainable actions can we take in light of shrinking oil supplies?

Support fully electric vehicles and recharging stations

Increase availability of convenient public transportation

Increase use of biofuels derived from sustainable sources

Provide walkways and pathways for pedestrians and bicyclists

Given the high costs of oil to our environment and to our well-being, why do we continue to extract and burn it at such high rates? The answer, of course, is that those costs are almost never included in the price we pay for oil. When environmental and human health costs are factored into the cost of resources, it becomes clear that conservation, stewardship, and sources of renewable energy are truly bargains. Regulatory and market-based policies can ensure that the prices of energy resources include their costs to the environment and our well-being; such policies can also provide incentives for sustainable stewardship and conservation.

▲ Figure 19.5 **Sustainable Response to a Shrinking Resource**
In the future, petroleum extraction is likely to diminish and gasoline prices are likely to increase. Nevertheless, human well-being can be sustained by a variety of actions that will have the added benefit of decreasing greenhouse gas emissions.

Less Biodiversity

■ Yes, we are likely to lose more of Earth's biodiversity and the ecosystem services it provides, but we can act to halt such loss.

The extinction rates of plant and animal species have been increasing rapidly during the past 100 years, and those rates are likely to remain higher than natural rates of extinction for the foreseeable future. Habitat fragmentation and loss continue, especially in tropical forests and oceans where species diversity is particularly high. Non-native invasive species, air and water pollution, overharvesting, and global warming will very likely continue to contribute to the loss of biodiversity in many areas.

Most of us lament the loss of species for ethical reasons, but their loss is also likely to have serious consequences for our own well-being. For example, we cannot take the diverse array of plant and animals that feed and medicate us for granted; Earth's biodiversity remains a reservoir of future foods and medicines.

The loss of biodiversity in ecosystems such as wetlands and forests diminishes their ability to purify water and to moderate the flow of water during droughts and floods. Deforestation increases greenhouse gas emissions and global warming. The extinction of top carnivores and other vulnerable species can diminish diversity across entire food webs. Such changes not only decrease the stability of ecosystems but also affect the availability of many resources that we humans depend on. In addition to threatening endemic species, deforestation diminishes the flow of water and reduces the quality of water that we drink. The collapse of fish stocks threatens a primary source of food for hundreds of millions of people.

The challenges to sustaining Earth's biodiversity are magnified by other environmental challenges, such as global warming, unsustainable use of resources, and pollution. Climate change threatens entire ecosystems as well as populations of individual species. Water diversion and pollution are major factors in the loss of aquatic and marine species. Oil extraction and use is a significant factor in both climate change and water pollution.

We can reverse the trends in global biodiversity loss.

Although we have much to learn, we already possess the knowledge and resources to halt biodiversity loss. We are protecting endangered species and ecosystems through an ever-increasing number of parks and protected areas. We are increasing the effectiveness of those reserves by making them as large as possible, by connecting them with corridors, and by buffering them with areas of low-impact use. Importantly, we are learning how to accommodate the needs of people, especially poor people, while meeting conservation goals (Figure 19.6).

We can also provide economic incentives for conserving biodiversity, using tools such as the regulation of trade in endangered species, payments for ecosystem services, and ecotourism. But, in the end, we will protect what we value, and we will value what we understand. Therefore, education will be an important key to our success.

▶ **Figure 19.6 Stemming the Loss of Species**
A variety of actions can slow the rate or even halt the loss of forest habitats, which is a major cause of species loss worldwide.

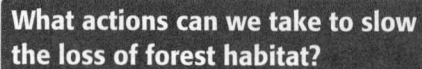

What actions can we take to slow the loss of forest habitat?

Reforestation to restore lost and degraded habitats

Buy wood products that are certified to have come from sustainably managed forests

Enforce international treaties to prevent illegal logging

Encourage sustainable harvest of non-timber products such as rubber

More People and Bigger Footprints

■ Yes, there will be more of us, but we can act to limit our growth in numbers and to diminish our ecological footprint.

Since 1900, the number of us has increased sixfold to about 7.3 billion. That number is certain to increase in the decades ahead. However, birth and death rates for the global human population have been declining for over 40 years, and the difference between these rates, the global population growth rate, is steadily getting smaller. Nevertheless, by 2050, more than 9 billion of us will populate this planet. How many more is less certain, as are the environmental consequences of our increased numbers.

Can Earth sustain so many of us? Some people are convinced that population growth is the "mother of all environmental challenges." But they are missing half of the story. It is at least as alarming that our ecological footprint is growing. Our ecological footprint is the area of Earth's surface needed to provide the resources that each of us uses and to absorb the waste and pollution that each of us creates. By some calculations, the sum of the individual ecological footprints of all the people currently living in the world exceeds Earth's land area. Whether or not that is exactly true is a matter of debate, but nearly everyone agrees that the current growth of our population and ecological footprint is not sustainable.

Our situation is complicated significantly by the great differences in the rate of population growth and the size of ecological footprints between rich and poor regions. Population growth rates are high and ecological footprints are relatively small among the people of the world's poorest countries. In many ways, poverty encourages high population growth rates, and the demands of large populations tend to encourage poverty. Because it diminishes poverty and environmental impacts and improves human well-being, reducing the rate of population growth is especially important in poor countries. Economic development, improved human well-being, and education—especially of women—are key factors in lowering population growth rates in these places. However, economic development is usually associated with rapid growth in ecological footprints.

By comparison, population growth rates are low and ecological footprints are large among the people of the world's wealthiest nations. In fact, ecological footprints of the average American or European are 5 to 10 times larger than those of people in poor countries. And it is especially worrisome that those footprints continue to grow. Greenhouse gas emissions, resource extraction, biodiversity loss—each of these challenges increases with the size of our population and our ecological footprint. Limiting the growth of ecological footprints is important for both the poor and the rich. Some amount of footprint growth is probably needed to improve the well-being of people in poor countries. To diminish their disproportionate impact on Earth's ecosystems and to ensure opportunity for improved well-being of the poor, the consumption of resources by people in many of the world's wealthiest nations must diminish.

The growth in our numbers and in the size of our ecological footprints can be diminished while improving or maintaining our well-being. Policies aimed at controlling population growth are often controversial, especially in developing countries. However, noncontroversial strategies, such as increased educational opportunities for women and improved sanitation and health care, are proving to be very effective in lowering birth and population growth rates.

We can all monitor our own activities to reduce our individual demands and impacts on Earth's ecosystems. We can replace high-impact technologies with lower-impact technologies and use renewable resources in place of nonrenewable resources. Through increased conservation and efficiency— reduce, reuse, and recycle—we can shrink our demands for energy and materials. There are many ways to do these things without diminishing our well-being (**Figure 19.7**).

Climate change, diminishing resources, biodiversity loss, and increased human numbers and impacts are among the most important environmental challenges, but they are certainly not the only ones. It is possible to reverse current trends for all of these challenges. Such change will require that our individual actions embody our understanding of and commitment to sustainability. It will also require environmental leadership that encourages sustainable behavior in others.

QUESTIONS 19.1

1. What do you think will be the most challenging part of reducing the rate of global warming? Of adapting to global warming?

2. How is our use of oil and water resources similar?

3. What do you think are the most important steps that can be taken to conserve water and oil?

4. In many developing nations, the rate of population growth is declining. What factors are contributing to this trend?

(MES) For additional review, go to **MasteringEnvironmentalScience**

◄ Figure 19.7 **Sustainable Human Impact**
Humans can diminish the rate of growth of their populations, diminish the impacts of increasing influence, and develop environmentally sustainable technologies.

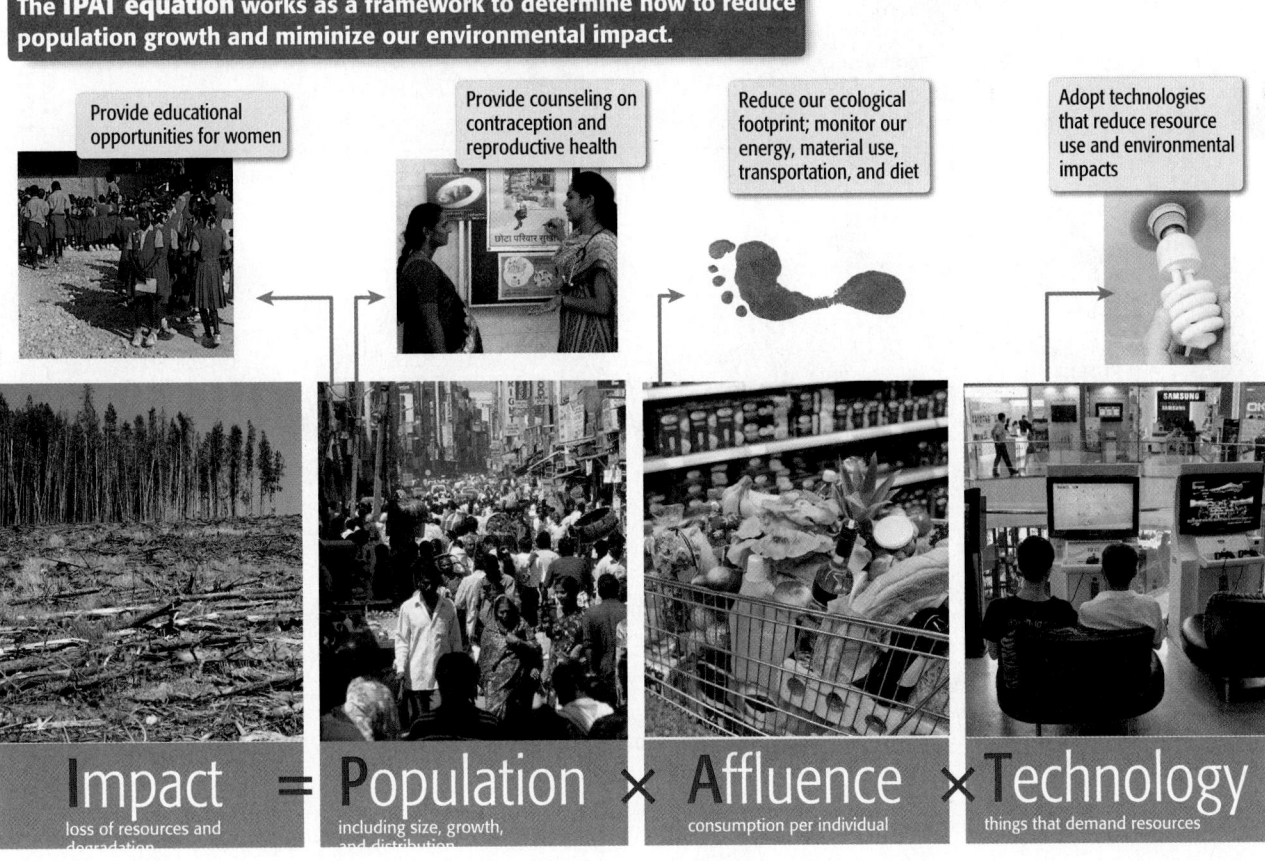

The **IPAT equation** works as a framework to determine how to reduce population growth and miminize our environmental impact.

Provide educational opportunities for women

Provide counseling on contraception and reproductive health

Reduce our ecological footprint; monitor our energy, material use, transportation, and diet

Adopt technologies that reduce resource use and environmental impacts

Impact = Population × Affluence ×Technology

loss of resources and degradation

including size, growth, and distribution

consumption per individual

things that demand resources

19.2 And You?

BIG IDEA In 2014, human activities put over 30 billion metric tons of greenhouse gases (CO₂e) into Earth's atmosphere, added nearly 20 million tons of nitrogen to Earth's rivers and streams, and were responsible for the extinction of as many as 20,000 species of microorganisms, plants, and animals. These numbers are so big that it is easy to fall into the trap of believing that individual actions can have little effect on them. But these numbers are, quite simply, the sum of each of our individual actions. Changes in these numbers—for better or worse—will ultimately be determined by the actions taken by each one of us.

Continue to Learn and Improve Your Understanding

■ The knowledge you have acquired provides the basis for life-long learning.

▼ Figure 19.8 Systems Thinking
Trying to conserve an individual species without attention to its place in the biological community is an example of piece thinking. Successful conservation depends on systems thinking that pays close attention to how individual species interact with other important ecosystem elements.

Congratulations! You have taken a college-level course in environmental science. That alone means you already know more about the environment and its importance to human well-being than 99% of Earth's people. Undoubtedly, you have your own views about which environmental challenges are most compelling and which actions are most appropriate to meet those challenges. Nevertheless, you have the necessary knowledge and vocabulary to read critically news accounts of environmental issues and to follow and weigh in on debates about environmental

policy. So, do it. But also know that your knowledge brings with it additional responsibilities.

Learn more. More than anything else, this textbook and your course have prepared you to continue learning. You've only scratched the surface with regard to environmental topics such as human population growth, global warming, and energy. Although you can learn about current issues from news media and popular websites, you should read primary sources to understand

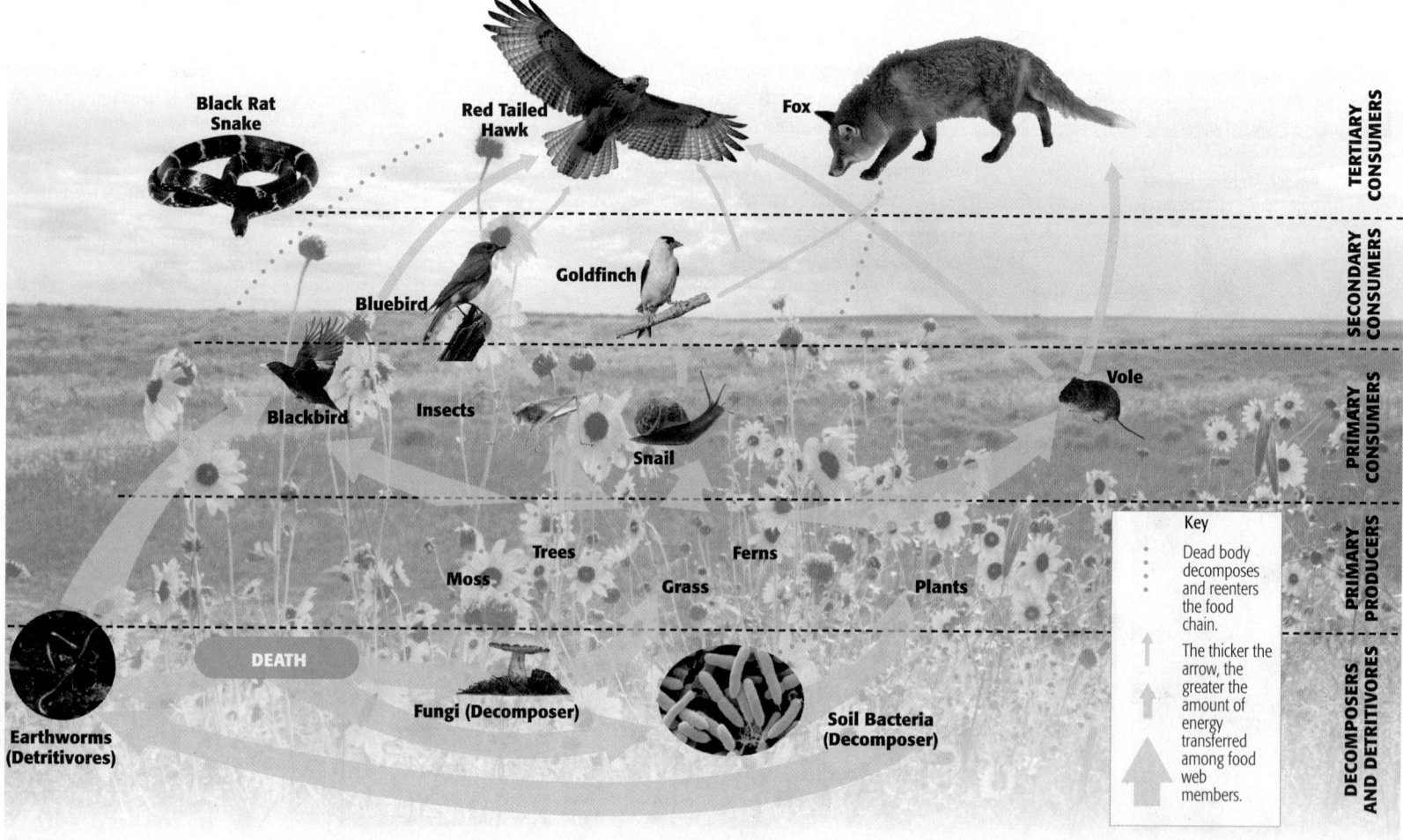

environmental topics in depth. Major science journals, such as *Science* and *Nature*, publish reports of original research on environmental topics each week. In addition to primary research articles, these journals provide research news and policy perspectives on environmental issues. A variety of other journals are important sources of information on more specific topics.

Translate your knowledge into understanding.
Environmental science and public policy are rife with numbers, vocabulary, and acronyms. It's easy to get bogged down in these details and miss the important connections and broader perspectives. How do your eating habits influence global warming? How can reducing rural poverty slow the rate of species extinction? Which energy policies are likely to have the greatest immediate effect on greenhouse gas emissions? To find answers to these questions, you need knowledge of the basic processes in ecosystems and human social systems, but you also need to understand how these processes are connected to one another. Knowledge often comes as separate pieces of information, but understanding

requires that the pieces be placed in the context of integrated systems (Figure 19.8).

When uncertain, be a skeptic. But don't lose sight of what you know for certain.
All knowledge and understanding is open to scrutiny and revision—this is the most fundamental truism of science. It is also a peculiar property of science that as we learn more, we often expose new areas of uncertainty. Much is yet to be learned about all of the topics we have covered, and uncertainty abounds on many important issues.

It is easy to be overwhelmed by these uncertainties and lose sight of the things that are certain. For example, uncertainties remain regarding the exact extent of global warming in the next few decades, as well as the local and regional consequences of this warming. There is much left to learn about the contributions of different greenhouse gases to warming and about the factors that influence climate at different locations on Earth's surface. Even so, you can be quite certain that Earth's atmosphere is warming and that human activities are directly responsible for this change (Figure 19.9).

◄ Figure 19.9 **Science as a Way of Knowing**
The scientific method provides a framework for clarifying what we know and for identifying areas where our knowledge is incomplete.

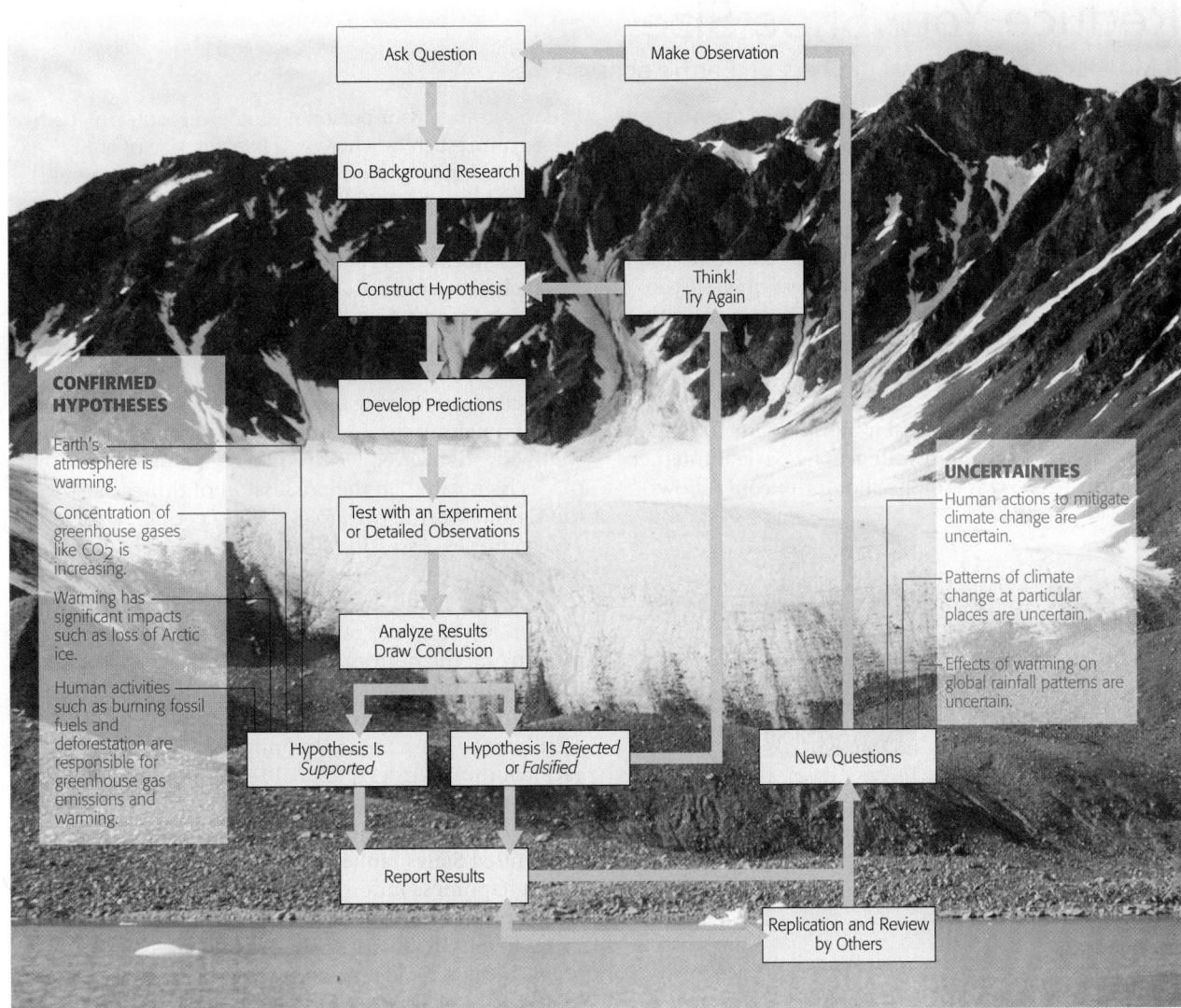

CONFIRMED HYPOTHESES

Earth's atmosphere is warming.

Concentration of greenhouse gases like CO_2 is increasing.

Warming has significant impacts such as loss of Arctic ice.

Human activities such as burning fossil fuels and deforestation are responsible for greenhouse gas emissions and warming.

UNCERTAINTIES

Human actions to mitigate climate change are uncertain.

Patterns of climate change at particular places are uncertain.

Effects of warming on global rainfall patterns are uncertain.

Ask Question — Make Observation — Do Background Research — Construct Hypothesis — Think! Try Again — Develop Predictions — Test with an Experiment or Detailed Observations — Analyze Results Draw Conclusion — Hypothesis Is *Supported* — Hypothesis Is *Rejected* or *Falsified* — New Questions — Report Results — Replication and Review by Others

► **Figure 19.10 What Is Your Shoe Size?**

Make a list of actions that influence your impact on the environment, and consider how you might alter them to diminish your ecological footprint.

Shoe size	Big foot 15 EEE	Intermediate 10 D	Small 7 B	Petite 5 A
Water use	Routinely leave water running; long showers; water used wastefully.	Some conservation concern; some water saving devices installed.	Water conservation a priority; effective use of water-saving devices; gray water recycling.	Completely sustainable water use; rainwater capture and gray water recycling.
Energy use	No use of energy saving lighting or appliances; heating and cooling used excessively; poor insulation.	Some energy saving lighting and appliances; limited attention to energy use; heating and cooling used generously.	Energy saving lighting and appliances widely used and on only when required; heating and cooling carefully monitored.	All lighting and appliances are energy efficient; effective use of natural light and renewable energy sources; heating and cooling used sparingly.
Waste disposal	All waste simply thrown out.	Some products are recycled; large amount of waste ends up in the landfill.	Most products are recyclable; some materials are composted.	Attention is given to reuse; nearly all products are recyclable; organic wastes are composted.
Transportation	Drive a large vehicle nearly everywhere and pay little attention to fuel efficiency.	Drive most places in a vehicle with moderate size and average fuel efficiency.	You drive only 25% of the time in a fuel-efficient vehicle.	You drive as little as possible and take advantage of public transportation whenever possible.
Food	Eat large amounts of meat, frozen and processed foods; pay no attention to where or how food is produced.	Several meatless meals each week with more fresh foods; try to buy local produce in season.	Limited consumption of meat and frozen or processed foods; try to buy sustainably and locally grown products.	Vegetarian or very limited meat consumption; eat locally produced fresh foods in appropriate seasons; choose only sustainably grown products.

▼ **Figure 19.11 Money and Happiness**

The average Gallup World Poll happiness index from 2010 to 2013 is compared here to average individual income for 160 countries. This index was correlated with income among countries with income levels below $20,000 per year. There was no correlation among countries with average incomes above that level. Freedom of choice was among several other factors that influenced differences in the happiness index.

Source: Helliwell, J.F., R. Layard, and J.D. Sachs. 2013. The World Happiness Report 2013. Sustainable Development Solutions Network. New York, NY.

Reduce Your Shoe Size

■ Monitor and reduce your impact on Earth's ecosystems.

Don't just stand there, do something! Given the number of challenges before us, this is apt advice. You can start by making some adjustments to your own activities. Reduce, reuse, and recycle are obvious steps, and you should make them happen. Actions that are less obvious might include paying closer attention to energy and water use in your dorm or choosing to eat meat less often. If you examine your activities systematically, you can prioritize changes to ensure the maximum benefit. You can also monitor the consequences of your actions to be sure you are meeting your goals.

Calculating your ecological footprint is an effective way to evaluate and adjust your actions. Several Internet calculators (e.g., www.ecologicalfootprint.com) allow

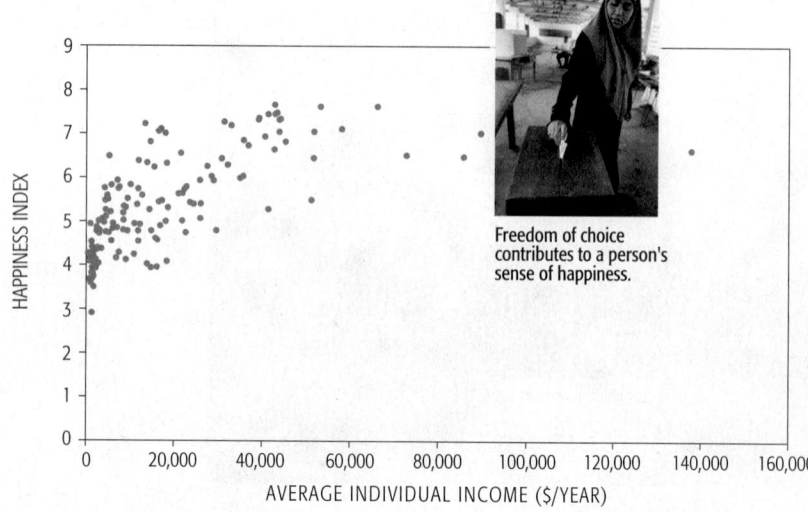

Freedom of choice contributes to a person's sense of happiness.

you to estimate your personal ecological footprint, using information such as where you live, the size of your home, the kinds of food you eat, and how far you travel and by what means (**Figure 19.10**). The details of these calculators vary, but they all provide a means of ranking the consequences of changes in your various actions.

Much like getting on the scales now and then when you go on a diet, the very act of measuring and monitoring your environmental impacts is important. Without such measures, it is difficult to assess whether your actions are having any effect.

Ultimately, each of us must set criteria for our own well-being. However, we live in a consumer society that places high value on the possession of things. Each year, the Gallup World Poll surveys 1,000 people in each of 160 countries. Assessing life satisfaction or happiness is a prominent part of this poll. Each respondent is asked to imagine a ladder with steps numbered from 0 to 10. The top of the ladder represents the best possible life for you and the bottom represents the worst possible life for you, they ask on which step of the ladder you feel you stand at this time. From these data, Gallup calculates a well-being or happiness index value for each country. Averaged for 2010–2013, the sub-Saharan African country of Togo had the lowest index value at 2.93. Denmark had the highest index value at 7.69. The United States ranked 17th among these countries with a happiness index of 7.08. It is certainly true that happiness measured in this way increases with increasing individual wealth, but only to a certain point (**Figure 19.11**). Above average individual income levels of about $20,000 per year, there is no correlation between happiness and wealth. Happiness index values in a number of countries with average incomes well below this level were equal to or greater than those in some of the wealthiest countries. Studies by researchers at the United Nations Sustainable Development Solutions Network reveal that the Gallup happiness index is also strongly correlated with people's perceptions of freedom of choice, future opportunity, personal safety, general job satisfaction, community support, and healthy life expectancy. Perceptions of inequality, beliefs that wealth and opportunity are not fairly distributed within a country, were negatively correlated with happiness measured in this way.

Recent studies show that our sense of well-being is influenced by less tangible things, such as the richness of relationships with other people and the quality of the environment. Does this mean you must live a Spartan lifestyle? No, but you should be aware that the lifestyle you choose has consequences for your environment and the well-being of others.

QUESTIONS 19.2

1. What kinds of organizations on your campus are carrying out activities to address environmental issues?

2. What environmental issue seems most important to you? Why?

3. What steps can you take now to help address that issue?

(MES) For additional review, go to **MasteringEnvironmentalScience**

Give What You Can

■ Invest your efforts in environmental programs that matter to you.

One of the best ways to effect change is by supporting one or more of the thousands of environmental programs and organizations around the world. The amount of time or money that you have to give is not an important issue. To do nothing because you can only do a little is a tragic mistake. Yet it is nearly as tragic to give without a sense of purpose or direction. Whether you are giving 10 hours of precious time or donating 10 dollars, you will find that your sense of accomplishment and your impact on the things you care most about will be magnified by informed priority setting. The Agents of Change features throughout this book provide wonderful examples and guidance on how to do this.

Environmental organizations vary widely in their goals and strategies. They range from global to local in their efforts; and from a focus on a single-issue to a wide range of issues. Organizations also vary in their approaches to meeting environmental challenges. Some are committed to education and advocacy; others, to legislative, regulatory, or judicial action. Organizations may act through grassroots activism or by facilitating market-based solutions. Your priorities should be based on the changes you would like to see in the world and on what you believe to be the most effective ways to make those changes happen. As you work on volunteer activities, your priorities may change as you discover interests and skills you never knew you had.

See your donations of time and resources as an investment in the future. Be interested in the return on your investment. Monitor the performance of organizations relative to your interests and priorities, and use this as a guide for your future commitments.

Think and Act for the Future

■ "Whatever you do may seem insignificant to you, but it is most important that you do it." (Mohandas Gandhi)

Even small actions can add up to a significant, positive impact (Figure 19.12). Your housemate, your family, and your friends are all influenced by the example you set, and through this influence you may multiply the magnitude of your actions many times over. Think and act for the long term. Sustainable living is, most importantly, about your ability to meet your own needs without compromising the ability of others to meet their needs now or in the future.

Living sustainably is not an endpoint or destination; it is truly a journey. Your personal wants and needs will likely change, along with changes in your knowledge, values, and understanding. The social and technological world around you will likely change as well. You may feel that the changes you can make in your life today are limited, but you certainly can begin the journey now.

GEORGIA SOUTHERN UNIVERSITY

▼ Figure 19.12 **Small Actions Make a Big Impact**
At Georgia Southern University, where one of your authors, Lissa, is based, nearly 5,000 students in Environmental Biology have each engaged in 3 hours of environmental service since 2006. They assist with school garden programs, tailgate recycling, and relamping with compact fluorescent lightbulbs. They plant trees, work at a local farmers' market, conduct litter cleanups, and much more. These individual contributions have resulted in a collective effort of 15,000 hours of service to the environment.

Q *How does one become involved in the sustainability movement?*

Matthew Poveromo, Syracuse University

Norm: **A** In the words of Mohandas Gandhi, "be the change you wish to see in the world."

Lissa: **A** Start today with a simple positive action. Get involved with environmental service in your community and share your strengths. Partnerships are critical to success because they allow you to pool people and resources toward the same goal.

19.3 Be an Agent of Change

BIG IDEA Leadership is nothing more or less than action by someone that changes the behavior of others. A high school student convinces local merchants to participate in a program to recycle small batteries. A young mother organizes a workshop aimed at reducing her community's greenhouse gas emissions. A college student develops a curriculum on water for third graders at a nearby elementary school. These are all examples of environmental leadership. Effective environmental leaders share five important qualities: (1) They are motivated by clearly articulated values that define their goals and determine the appropriate means of achieving them. (2) They value diversity. (3) They are willing to act and focus on outcomes. (4) They are humble and willing to learn. (5) They are committed to making a difference and confident that they can. You can be an environmental leader; you can be an agent of change.

Articulate a Vision Based on Your Values

■ Effective leaders have a clear sense of the change they wish to see in the world.

Imagine that you are a student trustee at your college or university. The board of trustees is considering a resolution to commit your school to reducing its net emissions of greenhouse gases to zero by the year 2025. This is consistent with your school's stated commitment to manage its campus in an environmentally sustainable fashion. To accomplish this, your school will need to spend several million dollars to upgrade its heating and cooling system and convert its coal-burning hot-water system to natural gas. Critics of this resolution point out that those funds could be used to hire new faculty or provide more student scholarships. Because you represent the student body and because you are knowledgeable about environmental issues, board members listen for your voice on this matter. Where do you stand?

By itself, a decision to act to meet environmental challenges can seem straightforward. If you are convinced that global warming is a significant threat and that universities should be models of sustainable management, then your answer must be yes. But you also live in a world of competing values and needs. In addition to its commitment to sustainability, your university is committed to building a strong and diverse faculty and student body.

Leadership requires decisions and action in the context of competing values and demands. It is tempting for leaders to yield to the loudest or most recent voice, or to set priorities based primarily on cost or ease of accomplishment. But decision making based only on these sorts of criteria is the antithesis of effective leadership. Posting reminders on light switches about using less energy is a visible signal of your university's interest in sustainability, but significant reductions in energy use and greenhouse gas emissions will require more costly investments in infrastructure (Figure 19.13). Effective environmental leaders have a clear vision of the goals they wish to achieve, and their decisions and actions are focused on that vision. Their decisions about appropriate goals and actions are guided by ethical principles or values.

So, how do you decide? If sustainability is an important value and part of your vision for your school's future, then action is essential. In most situations, however, the

▲ Figure 19.13 **Multiple Strategies**
These are some of the many actions your university could take to reduce its greenhouse gas emissions. The ones that make the biggest impact are often also the most costly or the ones that require the greatest change in behavior.

alternatives are rarely so stark as either sustainability or support for faculty and students. If sustainability and greenhouse gas emissions were the only values of concern, you could argue that your college or university should simply close its doors and send everyone home. Of course, such a decision would be absurd, but it does make it clear that nearly all of our activities—education, home ownership, transportation, business, and so on—have environmental costs. Short of giving up these activities altogether, we must decide which environmental costs to accept and to what extent they are acceptable.

Yes, a commitment to reducing greenhouse gas emissions will have real costs—it would be a rather hollow commitment if it did not. But it also can have real benefits, such as long-term savings in the cost of energy, opportunities for research and teaching, and increasing the reputation of your school as a green institution. The exact time frame for reaching zero emissions and the allocation of costs in your school's budget might be matters for discussion and will depend upon your school's needs in other areas.

Cultivate Diversity

■ Human diversity is a reservoir of ideas.

Ecosystems and social systems are incredibly diverse and complex. Yet all too often, human management of these systems has centered on simplification. For example, diverse grasslands have been converted to agricultural ecosystems dominated by a single species, and complex, meandering rivers have been confined to simple, straight channels. Sustainable management seeks to maintain diversity and complexity and to maintain connections within and among ecosystems.

Effective environmental leadership also values diversity. While leaders may occasionally wish that everyone's vision and values were identical to theirs, they know that this will hardly ever be the case. They understand that civil argument and dispute often clarify choices and the values that underpin them. They find they can learn from critical and dissenting voices (Figure 19.14).

To craft an appropriate plan for reducing greenhouse emissions while also meeting other priorities, your school's board of trustees will need to listen to the voices of diverse stakeholders. Some people may feel that recycling programs are more important than energy-efficiency upgrades. Others may prefer to spend money on resources and additional faculty to enhance environmental course offerings.

Because environmental decisions include many trade-offs, they often cause strong emotions and even conflict among people who otherwise share similar goals. That is why conflict resolution and negotiation skills are important attributes of a leader. But good leaders also recognize that resolving conflict is not the most important goal in environmental decision making. Decisions that are focused on minimizing conflict or trying to make everyone happy run the serious risk of producing less-than-ideal solutions or even daunting problems.

▲ Figure 19.14 **Many Voices**
Effective leaders encourage and listen to diverse points of view.

Focus on Outcomes

■ The environment is changing constantly, and action is inescapable.

A decision *not to act* to meet an environmental challenge is never neutral. For example, some members of your school's board of trustees may argue that action on greenhouse emissions should be deferred until uncertainties regarding their connections to global warming are resolved. But this decision not to act is itself a very powerful action. It means that the school's greenhouse emissions will continue to increase and assumes that the resources and support necessary for action will be available in the future.

Leaders emphasize their ultimate goals—the changes they wish to see in the world—not the means of reaching them. It is important that your school take actions such as converting the steam plant from coal to natural gas, but the real measure of success is the outcome—a reduction in greenhouse gas emissions.

Effective environmental leaders define their desired outcomes in objective and quantifiable terms that allow them to monitor the consequences of their actions. Just as important, they want others to understand and agree on the definitions of desired outcomes, as well as the approaches to monitoring them.

Be Humble and Adaptable

■ Uncertainty, new knowledge, and environmental change require adaptive leadership.

You may hear some people say, "You cannot manage what you don't understand." This statement seems logical, but it is not really true. Although our ability to manage a problem improves as our knowledge increases, some amount of uncertainty is inevitable. Indeed, what understanding we do have is constantly subject to change as we learn more.

Effective environmental leadership requires the humility to understand that your actions, or even your goals, may be incorrect. It is committed to making decisions and taking actions that can be modified as conditions change or new knowledge becomes available. To do this, leaders must be willing to learn from their successes and failures and to encourage such learning among those they lead (Figure 19.15).

Adaptive leadership is likely to be important for actions taken by your school's board of trustees. You might, for example, encourage the board to identify the primary areas of uncertainty about how their actions will affect greenhouse gas emissions and, ultimately, global warming. The board could then determine which of these uncertainties are most important or those areas where new knowledge might indicate the need for changes in their actions. Finally, you might recommend a regular review process to ensure that new knowledge is incorporated into future actions.

Adaptive leadership can be a precarious process. You must acknowledge your uncertainty and limited knowledge and, at the same time, maintain the confidence of others. Ultimately, your success as a leader depends on your ability to earn the trust of followers and stakeholders.

◄ Figure 19.15 **Measures of Success**
Leaders are willing to take action, but also constantly monitor the outcomes of those actions to ensure success. Author Norm's university implemented its Sustainable Duke program in 2007 with the goal of reducing its net greenhouse gas emissions to zero by 2025. Its CO_2e emissions are closely monitored and published on the Sustainable Duke website.

Be Confident, Committed, and Hopeful

■ "There is no greater mistake than to do nothing because you can only do a little." (Edmund Burke)

Effective environmental leaders believe that their actions, large or small, matter. More than that, they are able to convey that same spirit to those they lead. It also requires a willingness to risk making someone unhappy—chances are good that you will (Figure 19.16).

Avoid both hopeless pessimism and naïve optimism. Extreme pessimism discourages action and may even encourage destructive actions: "If we're doomed, why not?" Naïve optimism understates the very formidable challenges that we face and is likely to underestimate the magnitude of actions needed to confront those challenges. A hopeful sense of the possible is the most appropriate approach. Through your individual actions and leadership, that hope and those possibilities become self-fulfilling prophecies.

QUESTIONS 19.3

1. What five qualities are essential for environmental leaders?

2. Why is it important to consider diversity when seeking solutions to environmental issues?

MES For additional review, go to **MasteringEnvironmentalScience**

▼ Figure 19.16 **Agents of Change**
The authors are grateful to each of you with the vision and commitment to make a difference in the world.

Earth Rebirth

Andrew Sartain majored in Interdisciplinary Perspectives on the Environment at the University of Oklahoma where he started a non-profit organization called Earth Rebirth.

How did you first get involved in sustainability efforts?

I was born with a love and admiration for the world and have been obsessed with nature since the age of five. I wanted to show people that the environment plays a role in our lives and can actually make our lives better. I learned that real opportunity for change comes from creating platforms of change for others to use. Animals were always my driving passion, not people. But it soon became obvious that in order to make a true difference, I had to use the will and ingenuity of people. Like many others, I have a dream of sustainability—but I had to learn that a thought only meets its potential when you act on it.

What is your mission and vision for Earth Rebirth?

Earth Rebirth's mission is to bridge the gap between economic and environmental problems. By focusing on the three environmental factors that people recognize as important to them—food, energy, and water—Earth Rebirth seeks to enhance resource availability and social equality through education and action.

My vision for Earth Rebirth is to create an international brand around an ideology of opportunity within environmental economics. While other environmental agencies create awareness of environmental problems, Earth Rebirth educates people about environmental solutions and how to apply them. ER teaches people that environmentalism is not about doing less, cutting back on materials, or ending a way of life. Environmentalism is about living more efficiently, optimizing what we have, and working in unison with our surroundings.

What steps did you take to start this nonprofit company?

The first year of Earth Rebirth was very much just me working alone with one friend. We built a simple website with information, but I wanted more. I set out to incorporate Earth Rebirth as a nonprofit company in 2011. I reserved a room, held meetings—often with one or two random people showing up weekly. I organized basic events, helped with tornado clean up, spoke to several classes, and ER gradually grew. Over several years, we have built a strong community reputation and garnered significant interest from OU faculty and students. Earth Rebirth gained approval to offer internships, and in recent semesters, I have been in charge of four teams of interns receiving class credit at OU for their work with Earth Rebirth.

Tell us about some of your Earth Rebirth projects.

Earth Rebirth has three main programs—*Garden Your Own Growth*, *Homemade Sustainability*, and *Taking H2Ownership*. Garden Your Own Growth works with children, schools, and after-school programs to show families and kids the benefits of taking your diet into your own hands. GYOG is piloting a new garden manual for distribution to parents and teachers across the country.

Homemade Sustainability teaches individuals and businesses how to increase energy efficiency and limit waste through audits, offers energy-saving products to homeowners, and organizes workshops to improve practices.

Taking H2Ownership educates people on water use, pollution control, and restoration. ER released a web-based global database called *Pin a Problem* in which any problem can be reported and mapped. Users vote to determine the problem of highest priority. We organize water clean ups and research pollution using feedback from this database.

What is the biggest challenge you've faced and how did you overcome it?

People are the greatest challenge for any start up. Earth Rebirth has many ideas both for-profit and as a part of our nonprofit sector, but none of these ideas can meet their potential without the right people. I started with a nonprofit organization, to get the right people involved for the right ideas—not for the guarantee of financial return.

So much of the first three years of Earth Rebirth was a game of patience. Business partners who didn't understand the vision, public meetings with no attendees to talk to, weekly meetings with no returning volunteers, volunteers who disappeared when needed—people will always be the biggest challenge of any global movement. Patience will eventually provide opportunity. And the right people will find you.

What advice would you give to students who want to start their own sustainability organization?

If you are passionate about making a difference, do not give up. Jane Goodall said, "The greatest danger to our future is apathy." Every experience is a learning experience. At times you will feel that you aren't moving in the right direction. Just keep pumping forward. The only direction that isn't right is NO direction.

Don't overthink the complexity of your challenge. Evaluate your end goals and reverse engineer step-by-step plans. Reserve a room where you meet every week, even if it is just you as a leader sitting alone for an hour once a week dreaming, occasionally getting the chance to talk to one or two patrons. Hold events—bake sales, clean ups, awareness parties—whatever. Talk about your cause to everyone. The more you paint your picture to others, the more vivid its colors become.

What do you hope for Earth Rebirth in the coming months? Years?

We would like to replicate what we have built in Oklahoma all over the country so that we can join forces with others who share Earth Rebirth's way of thinking and act in unison. ER hopes to expand to other universities by finding the unique dreamers on campuses, both professors and students, who want to put their ideas into action. Earth Rebirth hopes to establish branches at universities using internships and faculty sponsors. We hope to partner with every student and professor reading these words. Find us! If there is anything the world needs, it is for you to dream big, and to act bigger.

http://earthrebirthnow.org/

Summary

Greenhouse emissions are increasing and our climate is warming, renewable and nonrenewable resources are dwindling, and the loss of biodiversity is accelerating. These changes in Earth's environment are occurring not only because our population is growing but also because the average ecological footprint of humans is growing. Fortunately, we possess the knowledge and know-how to slow and even reverse these trends. To accomplish this, each of us will need to understand and make a personal commitment to sustainability. You can do this by continuing to learn about environmental issues, by translating your knowledge into understanding, by reducing your own ecological footprint, by committing your time and resources to the environmental issues that matter most to you, and by thinking and acting for the future. You can be an agent for change by clearly articulating your values and vision for the future, by cultivating diversity, by focusing on outcomes consistent with your values and vision, and by being humble, adaptable, confident, committed, and hopeful.

19.1 Hope for the Environment

- In the future, there are likely to be increasing numbers of humans, and Earth's environment is likely to be affected by human-caused changes, including global warming, scarce resources, diminished biodiversity, and pollution.
- We possess the knowledge and technologies to slow and reverse these trends.

QUESTIONS

1. Environmental scientists like to say, "Everything is connected to everything." Describe three ways in which climate change is connected to each of the following: the availability of oil, the availability of water, and biodiversity.

2. Even though the rate of human population growth is decreasing, human impacts on our planet are increasing. Why?

3. Describe three ways in which the loss of biodiversity might diminish your well-being.

4. Sustainability implies equitable treatment and justice among humans. How can we ensure that efforts to reduce ecological footprints in different countries are equitable and just?

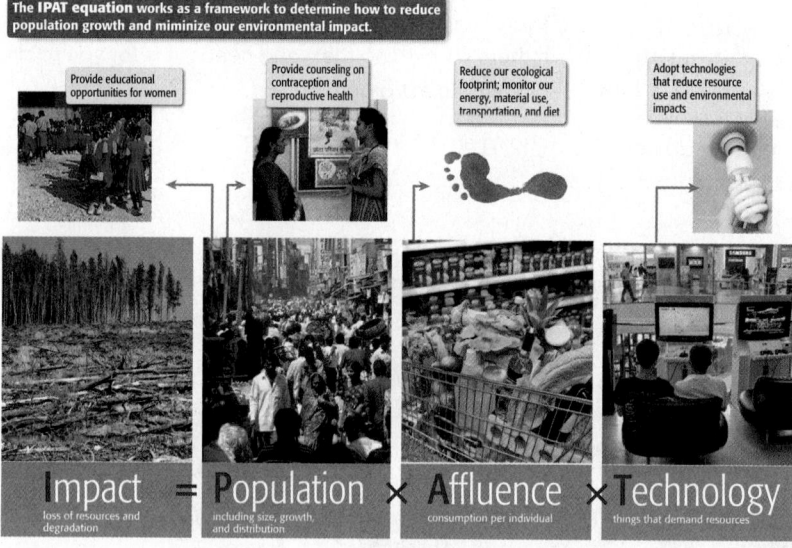

The **IPAT equation** works as a framework to determine how to reduce population growth and minimize our environmental impact.

Provide educational opportunities for women

Provide counseling on contraception and reproductive health

Reduce our ecological footprint; monitor our energy, material use, transportation, and diet

Adopt technologies that reduce resource use and environmental impacts

Impact = Population × Affluence × Technology

loss of resources and degradation

including size, growth, and distribution

consumption per individual

things that demand resources

19.2 And You?

- Your own actions matter.
- You can be the change you wish to see in the world by continuing to learn, reducing your ecological footprint, giving your time and resources, and thinking and acting for the future.

QUESTIONS

1. What environmental challenge matters most to you? Explain how systems thinking is important to meeting that challenge.

2. Describe three changes in your own activities that you could make today that would diminish your ecological footprint.

3. What features or characteristics matter most to you in selecting an environmental project or organization in which to invest your time or resources?

4. How will you go about monitoring the success of that project or organization?

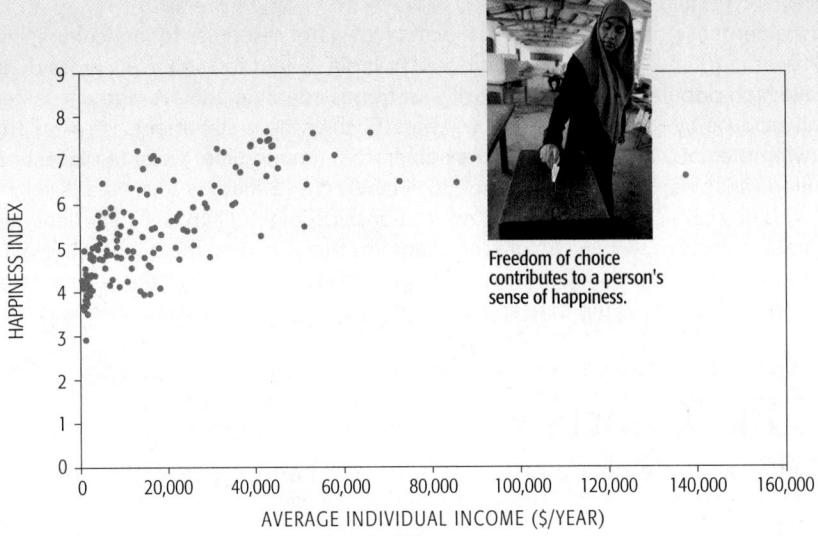

Freedom of choice contributes to a person's sense of happiness.

19.3 Be an Agent of Change

- You can be an agent of change: Leadership is action by you that changes the behavior of others.
- Leadership is based on values and vision, commitment to diversity, focus on outcomes, humility, adaptation, confidence, commitment, and hope.

QUESTIONS

1. What core values are most important to your goals for environmental sustainability and your choice of actions to achieve those goals?

2. What things could you do to build a diverse organization or program?

3. How would you work in a project or organization to be sure that it was adaptable to changes in knowledge and circumstances?

4. Environmental philosopher David Orr describes hope as "a verb with its sleeves rolled up." How will you translate your hopes for the future into actions today?

MasteringEnvironmentalScience®

Students Go to **MasteringEnvironmentalScience** for assignments, the eText, and the Study Area with practice tests, activities, and more.

Instructors Go to **MasteringEnvironmentalScience** for automatically graded tutorials and questions that you can assign to your students, plus Instructor Resources.

Graphs

Consider these observations: The allocation of water resources for agriculture, industry, and residential use is very different for the United States compared to India. Countries with a low gross domestic product (GDP) tend to have high population growth rates. If past trends continue, the average annual temperature of Earth's atmosphere will increase by 3 °C over the next 50 years. Each of these statements is based on quantitative measurements of environmental features. Because they differ from time to time, place to place, or group to group, these features are called variables. The statements about variables are verified by scientists using a variety of statistical methods.

It is hard, however, to visualize the relationships implied in these statements simply by looking at long lists of numbers. Instead, we rely on graphs, diagrams that compare the quantitative data for environmental features such as water use, economic activity, population growth rates, or average annual temperature. The kind of graph depends in part on the nature of the data and, most importantly, on the question we wish to answer with those data.

1. Pie Charts

How is a whole group or variable divided among categories?

Pie charts are used when we wish to represent the proportions of different categories within a whole variable. For example, how is water consumption in a country apportioned among different categories of use (Figure A1)? In this example, the relative amounts of water used by industry, agriculture, and households are quite different among these three countries, and each country differs significantly from the global amounts. The absolute amount of water used by each country, listed in parentheses, is of secondary interest and is not part of the graph.

2. Bar Charts

How does the amount of a variable differ among different groups, places, or times?

In bar charts, the amounts of a variable for each of several groups are indicated by the length of bars. Bars may be oriented horizontally or vertically. Figure A2 shows that total reserves of coal and reserves of different types of coal differ among continents. Figure A3 indicates the infant mortality rate for each of the world's 10 poorest countries. This graph shows clearly that infant mortality in the Central African Republic is over twice that of Madagascar. Bar charts can also be used to compare multiple data sets and group types. For example, Figure A4 compares the number of women with the number of men in different 5-year age classes for Ethiopia and the United States. The proportion of the population that is represented by middle- and old-age individuals in Ethiopia is much less than in the United States. Furthermore, gender differences are small in Ethiopia, whereas in the United States women generally outnumber men.

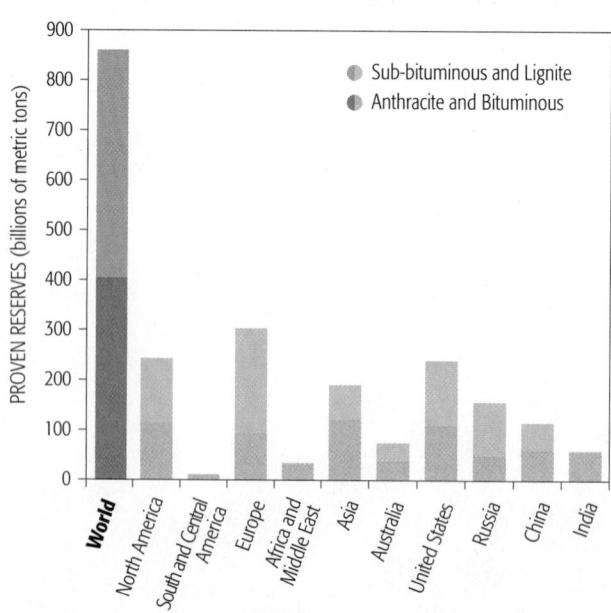

▲ Figure A2

▼ Figure A1

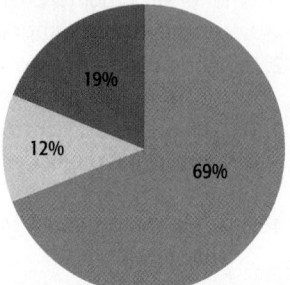

Total Water Use (billions of cubic meters)

GLOBAL (3,902)

UNITED STATES (482.2)

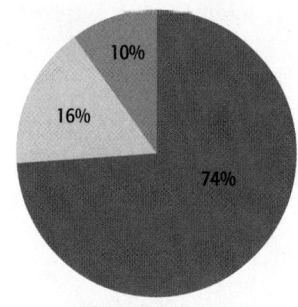

FRANCE (33.1)

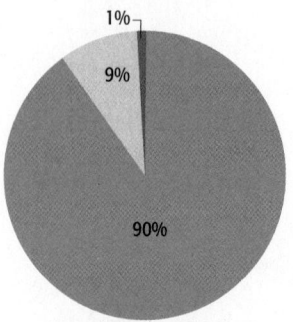

MALI (1.7)

Freshwater withdrawals by type of use

● Industry
● Agriculture
● Domestic

WORLD'S POOREST NATIONS

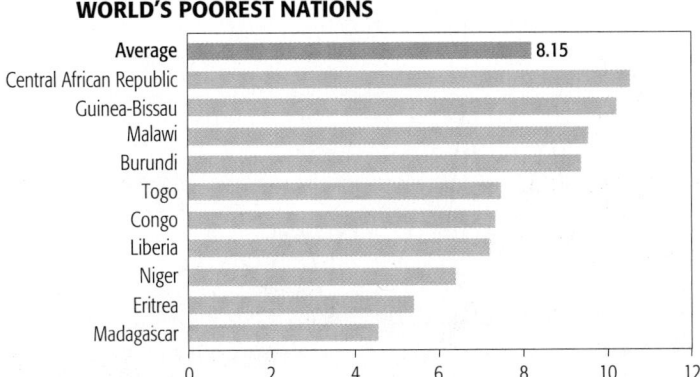

% INFANT MORTALITY

▲ Figure A3

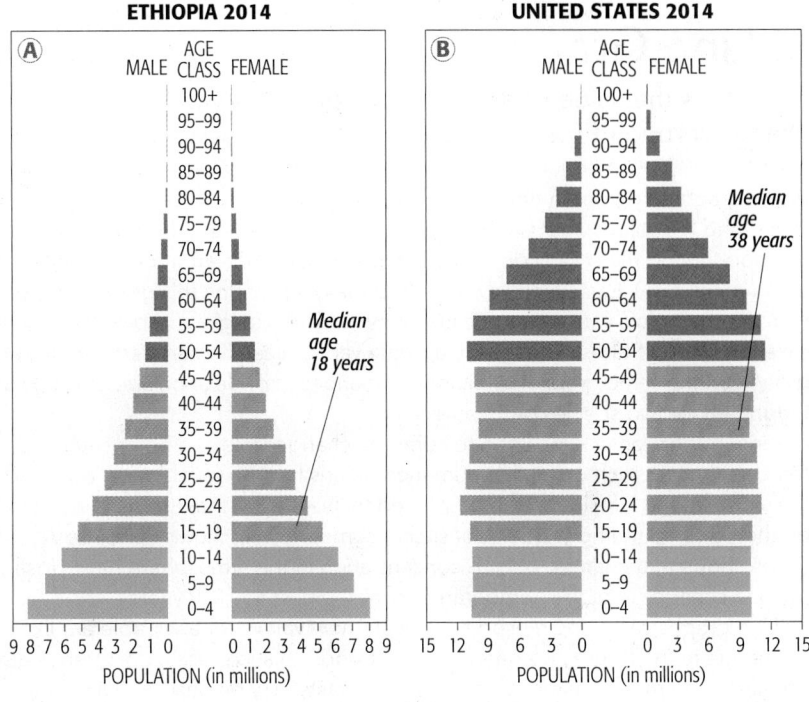

▲ Figure A4

3. Scatter Plots

How do the relationships between variables vary among different groups or populations of things?

The relationship of one variable to another for different groups can be indicated as dots on a graph where values for one variable are indicated on the horizontal or *x* axis and values for the other variable are indicated on the vertical or *y* axis. In **Figure A5**, the percentage of girls among secondary school students is compared to birth rates for each of 78 countries, each represented by a dot. While there is considerable scatter among the dots, there is a clear trend toward lower birth rates as the percentage of girls in secondary school increases. We can say, therefore, that birth rate is negatively correlated with girls' education for these 78 countries. The trendline is a best fit line that shows the direction of the relationship between the variables. **Figure A6** compares ratings of happiness by people in different countries with average individual wealth. Here, there is also considerable scatter, but, where income exceeds $20,000/year, there is no clear trend.

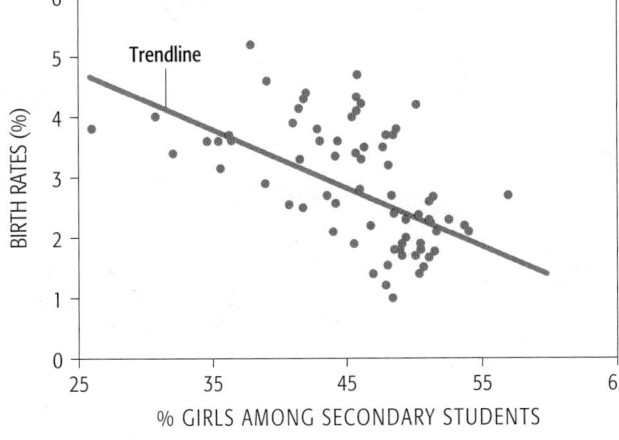

▲ Figure A5

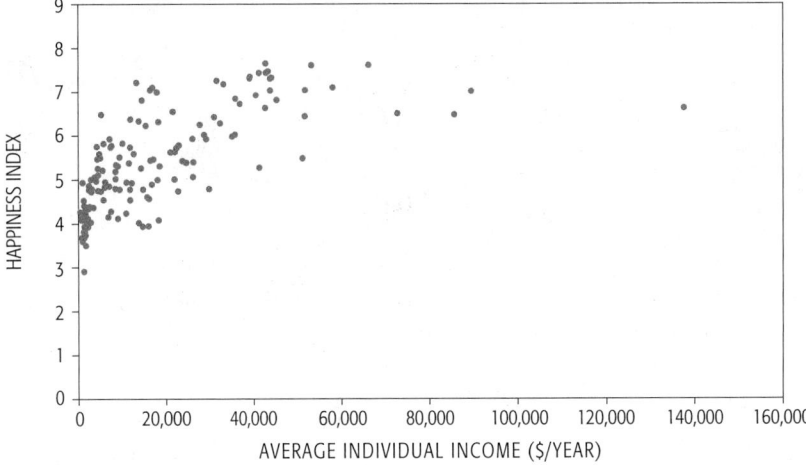

▲ Figure A6

4. Line Graphs

How does the value of a variable for one or several groups change through time?

In line graphs, changes in time or, in some cases, age are indicated along the x axis, and the values of a variable are indicated on the y axis. **Figure A7**, for example, shows the steady decrease in mean ozone concentrations in the United States since 1980. Each yearly average is actually a point on this graph, and we connect the points with a line because they refer to a single group or place, in this case the United States. Changes in a single variable can be compared for several groups on the same graph. For example, changes in population size are shown in **Figure A8** for two species of *Paramecium*.

Line graphs can also be used to compare changes in two different variables through time. In such cases, the numerical values for one variable are indicated on the y axis on the left side of the graph and the values for the other are indicated on the y axis on the right side. Just such a comparison is shown in **Figure A9**. This graph shows that changes in CO_2 concentration in the atmosphere have closely tracked changes in global temperature over the past 450,000 years.

Line graphs can also be used to indicate how values for a variable are likely to change in the future compared to past trends. In such cases, past trends are usually based on actual measurements. Forecasts may be based on a simple extension of such trend lines or on more complicated predictive models. In **Figure A10**, past changes in the temperature of Earth's atmosphere are indicated relative to that in 1900. The black line is based on real measurements up to the year 2000 from thousands of stations. The colored lines represent changes that are forecast from models based on different assumptions about future human actions. *Business as Usual* (red) implies no change in behavior, *Sustainable World* implies significant action to reduce greenhouse gas emissions, and *Today's World* implies that all greenhouse emissions cease immediately. The shaded area around each line indicates variation in measurements in the case of past trends and uncertainty about future trends in the case of the forecasts.

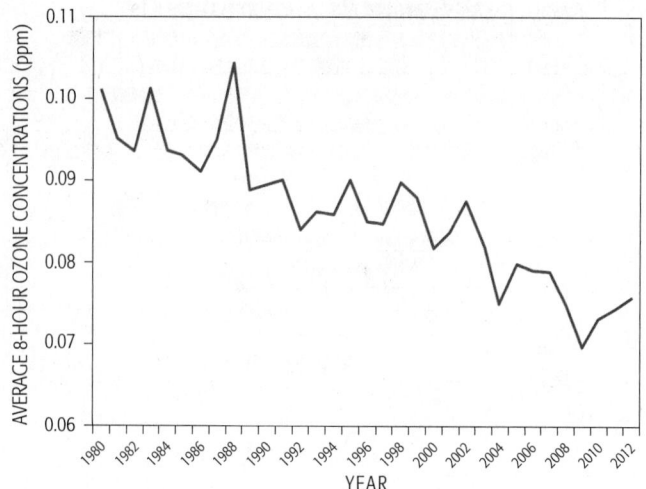

▲ Figure A7

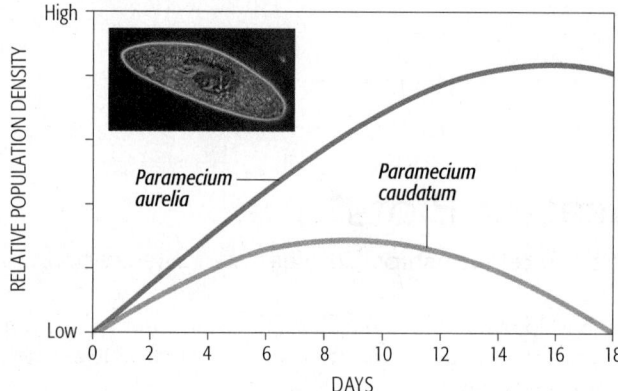

▲ Figure A8

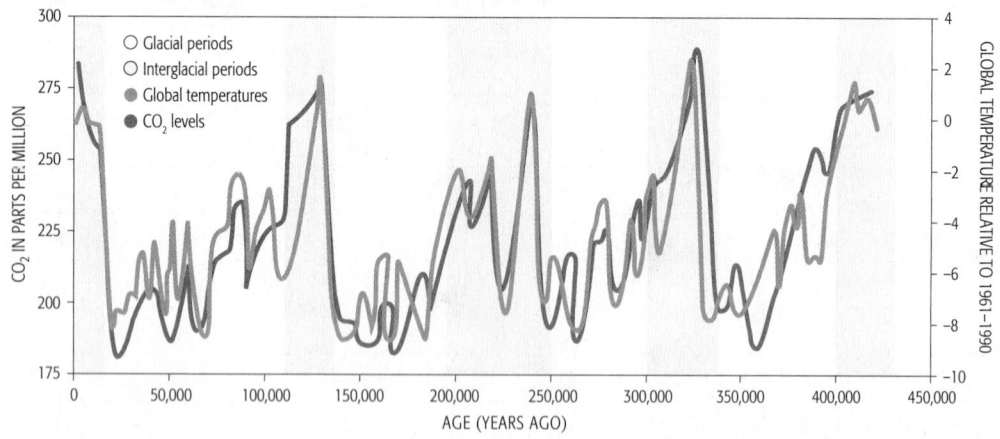

▲ Figure A9

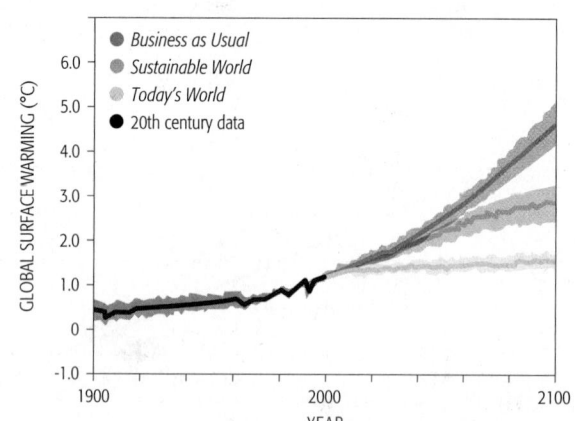

▲ Figure A10

Metric System

Measurement	Unit and Abbreviation	Metric Equivalent	Metric to English Conversion Factor	English to Metric Conversion Factor
Length	1 kilometer (km)	= 1,000 (10^3) meters	1 km = 0.62 mile	1 mile = 1.61 km
	1 meter (m)	= 100 (10^2) centimeters	1 m = 1.09 yards	1 yard = 0.914 m
		= 1,000 millimeters	1 m = 3.28 feet	1 foot = 0.305 m
			1 m = 39.37 inches	
	1 centimeter (cm)	= 0.01 (10^{-2}) meter	1 cm = 0.394 inch	1 foot = 30.5 cm
				1 inch = 2.54 cm
	1 millimeter (mm)	= 0.001 (10^{-3}) meter	1 mm = 0.039 inch	
Area	1 square meter (m^2)	= 10,000 square centimeters	1 m^2 = 1.1960 square yards	1 square yard = 0.8361 m^2
			1 m^2 = 10.764 square feet	1 square foot = 0.0929 m^2
	1 square centimeter (cm^2)	= 100 square millimeters	1 cm^2 = 0.155 square inch	1 square inch = 6.4516 cm^2
Mass	1 metric ton (t)	= 1,000 kilograms	1 t = 1.103 ton	1 ton = 0.907 t
	1 kilogram (kg)	= 1,000 grams	1 kg = 2.205 pounds	1 pound = 0.4536 kg
	1 gram (g)	= 1,000 milligrams	1 g = 0.0353 ounce	1 ounce = 28.35 g
	1 milligram (mg)	= 0.001 gram		
Volume (solids)	1 cubic meter (m^3)	= 1,000,000 cubic centimeters	1 m^3 = 1.3080 cubic yards	1 cubic yard = 0.7646 m^3
			1 m^3 = 35.315 cubic feet	1 cubic foot = 0.0283 m^3
	1 cubic centimeter (cm^3 or cc)	= 0.000001 cubic meter	1 cm^3 = 0.0610 cubic inch	1 cubic inch = 16.387 cm^3
		= 1 milliliter		
	1 cubic millimeter (mm^3)	= 0.000000001 cubic meter		
Volume (liquids and gases)	1 kiloliter (kl or kL)	= 1,000 liters	1 kL = 264.17 gallons	
	1 liter (l or L)	= 1,000 milliliters	1 L = 0.264 gallons	1 gallon = 3.785 L
			1 L = 1.057 quarts	1 quart = 0.946 L
	1 milliliter (ml or mL)	= 0.001 liter	1 mL = 0.034 fluid ounce	1 quart = 946 mL
		= 1 cubic centimeter	1 mL = approx. 1/4 teaspoon	1 pint = 473 mL
				1 fluid ounce = 29.57 mL
				1 teaspoon = approx. 5 mL
Temperature	Degrees Celsius (°C)		°C = 5/9(°F − 32)	
Energy and Power	1 kilowatt-hour	= 34,113 BTU = 860,421 calories		
	1 watt	= 3,413 BTU/hr = 14.34 calorie/min		
	1 calorie	= the amount of heat necessary to raise the temperature of 1 gram (1 cm^3) of water 1 degree Celsius		
	1 joule	= 9.481×10^{-4} BTU = 0.239 cal = 2.778×10^{-7} kilowatt-hour		
Pressure	1 pound per square inch (psi)	= 6894.757 pascal (Pa) = 0.068045961 atmosphere (atm) = 51.71493 millimeters of mercury (mm hg = Torr) = 68.94757 millibars (mbar) = 68.94757 hectopascal (hPa) = 6.894757 kilopascal (kPa) = 0.06894757 bar		
	1 atmosphere (atm)	= 101.325 kilopascal (kPa)		

Sources and Further Reading

Chapter 1

Bateman, I.J., A.R. Harwood, G.M. Mace, et al. 2013. Bringing ecosystem services into economic decision making: Land use in the United Kingdom. *Science* 341:45–50.

Chapin, F.S., III, P.A. Matson, and P.M. Vitousek. 2014. *Principles of Terrestrial Ecosystem Ecology* (2nd ed.). Springer-Verlag, New York.

Chapin, F.S., III, M.S. Torn, and M. Tateno. 1996. Principles of ecosystem sustainability. *The American Midland Naturalist* 148:1016–1037.

Daily, G.C. 1997. *Nature's Services: Societal dependence on natural ecosystems.* Island Press, Washington, DC.

Daily, G.C., S. Alexander, P.R. Ehrlich, et al. 1997. *Ecosystem Services: Benefits Supplied to Human Societies by Natural Ecosystems.* Issues in Ecology, Number 2. Ecological Society of America, Washington, DC.

Daily, G.C., S. Polasky, J. Goldstein, et al. 2009. Ecosystem services in decision making: Time to deliver. *Frontiers in Ecology and the Environment* 7:21–28.

Diamond, J. 2005. *Collapse: How Societies Choose to Fail or Succeed.* Viking, New York.

Golley, F.B. 1993. *A History of the Ecosystem Concept in Ecology: More Than the Sum of the Parts.* Yale University Press, New Haven, CT.

Haanaes, K., M. Reeves, I. von StrengVelken, et al. 2012. *Sustainability Nears a Tipping Point, a report on the findings of the 2011 Sustainability & Innovation Global Executive Study and Research Project.* MIT Sloan Management Review. Reprint #: 53380. http://sloanreview.mit.edu/reports/sustainability-strategy/introduction, last accessed 9-18-2014.

Holdren, J.P. 2008. Science and technology for sustainable well-being. *Science* 319:424–434.

Kates, R.W. and T.M. Parris. 2003. Long-term trends and a sustainability transition. *Proceedings of the National Academy of Sciences* 100:8062–8067.

Kates, R.W., W.C. Clark, R. Corell, et al. 2003. Sustainability science. *Science* 292:641–642.

Kuhn, T.S. 1996. *The Structure of Scientific Revolutions* (3rd ed.). The University of Chicago Press, Chicago, IL.

Millennium Ecosystem Assessment. 2005. *Ecosystems and Human Well-Being.* Island Press, Washington, DC. URL = http://www.maweb.org/documents/document.356.aspx.pdf, last accessed 9-10-2014.

National Research Council, Board on Sustainable Development. 1999. *Our Common Journey: A Transition toward Sustainability.* National Academy Press, Washington, DC.

Ostrom, E. 2009. A general framework for analyzing sustainability of social-ecological systems. *Science* 325:419–422.

Pimm, S.L. 2001. *The World According to Pimm.* McGraw Hill, New York, NY.

Chapter 2

Andrews, R.N.L. 2006. *Managing the Environment, Managing Ourselves: A History of American Environmental Policy.* Yale University Press, New Haven, CT.

Attfield, Robin. 2003. *Environmental Ethics: An Overview for the Twenty-First Century.* Polity Press, Cambridge, UK.

Bowler, Peter J. 1993. *The Norton History of the Environmental Sciences.* W.W. Norton, New York, NY.

Callicott, J.B. 1994. *Earth's Insights: A Survey of Ecological Ethics from the Mediterranean Basin to the Australian Outback.* University of California Press, Berkeley, CA.

Carson, R. 1962. *Silent Spring.* Houghton Mifflin, Boston, MA.

Costanza, R., R. d'Arge, R. de Groot, et al. 1997. The value of the world's ecosystem services and natural capital. *Nature* 387:253–260.

Costanza, R., R. de Groot, P. Sutton, et al. 2014. Changes in the global value of ecosystem services. *Global Environmental Change* 26:152–158.

Daily, G.C., S. Polasky, J. Goldstein, et al. 2009. Ecosystem services in decision making: Time to deliver. *Frontiers in Ecology and the Environment* 7:21–28.

Daly, H. 2005. Economics in a full world. *Scientific American* 293:100–107.

DesJardins, J.R. 1997. *Environmental Ethics: An Introduction to Environmental Philosophy* (2nd ed.). Wadsworth Publishing Co., Belmont, CA.

Ehrlich, P.R. 1968. *The Population Bomb.* Ballantine Books, New York, NY.

Freeman, A.M. 1993. *The Measurement of Environmental and Resources Values: Theory and Methods.* Resources for the Future, Washington, DC.

Hardin, G. 1968. The tragedy of the commons. *Science* 162:1243–1248.

Hawkin, P. 1993. *The Ecology of Commerce: A Declaration of Sustainability.* HarperCollins, New York, NY.

Hawkin, P., A. Lovins, and H. Lovins. 1997. *Natural Capitalism: Creating the Next Industrial Revolution.* Little, Brown and Company, New York, NY.

IPCC. 2007. *Climate Change 2007: Synthesis Report. Contribution of Working Groups I, II and III to the Fourth Assessment Report of the Intergovernmental Panel on Climate Change.* IPCC, Geneva, Switzerland.

Slaper, T.F. 2011.The Triple Bottom Line: What Is It and How Does It Work? *Indiana Business Review* 86:4–8.

Speth, J.G. 2004. *Red Sky at Morning: America and the Crisis of the Global Environment.* Yale University Press, New Haven, CT.

United Nations. 1987. *Report of the World Commission on Environment and Development: Our Common Future.* URL = http://www.un-documents.net/wced-ocf.htm, last accessed 9-14-2014.

Vitousek, P.M., H.A. Mooney, J. Lubchenco, et al. 1997. Human domination of Earth's ecosystems. *Science* 277:494–499.

Kubiszewski, I., R. Costanza, C. Franco, et al. 2013. Beyond GDP: Measuring and achieving genuine progress. *Ecological Economics* 93:57–68.

Leopold, A. 1949. *A Sand County Almanac.* Oxford University Press, Oxford, UK.

Light, A. and H. Rolston, III. 2003. *Environmental Ethics: An Anthology.* Blackwell Publishing, Oxford, UK.

Ostrom, E. 2009. A general framework for analyzing sustainability of social-ecological systems. *Science* 325:419–422.

Simpson, J.W. 2005. *Dam! Water, Power, and Preservation in Hetch Hetchy and Yosemite National Park.* Pantheon Books, New York.

Wilson, E.O. 1999. *The Diversity of Life.* W.W. Norton, New York, NY.

The World Watch Institute. 2008. *State of the World: Innovations for a Sustainable Economy.* W.W. Norton, New York, NY.

Chapter 3

Bonan, G. 2008. *Ecological Climatology: Concepts and Applications* (2nd ed.). Cambridge University Press, Cambridge, UK.

Craig, J.R., D.J. Vaughn, and B.J. Skinner. 2010. *Earth Resources and the Environment* (4th ed.). Benjamin Cummings, San Francisco, CA.

Hidore, J.J., J.E. Oliver, M. Snow, et al. 2010. *Climatology: An Atmospheric Science* (3rd ed.). Prentice Hall, Upper Saddle River, NJ.

Keller, E.A. 2011. *Introduction to Environmental Geology* (5th ed.). Prentice Hall, Upper Saddle River, NJ.

Lewis, J.S. and R.G. Prinn. 1984. *Planets and Their Atmospheres: Origin and Evolution.* Academic Press, Orlando, FL.

Lovelock, J.E. 1979. *Gaia: A New Look at Life on Earth.* Oxford University Press, Oxford, UK.

Montgomery, C. 2010. *Environmental Geology* (9th ed.). McGraw-Hill, New York, NY.

Schlesinger, W.H. and E.S. Bernhardt. 2013. *Biogeochemistry: An Analysis of Global Change* (3rd ed.). Academic Press, Orlando, FL.

Wade, L.G. 2012. *Organic Chemistry* (8th ed.). Prentice Hall, Upper Saddle River, NJ.

Chapter 4

Darwin, C. 1859. *The Origin of Species by Means of Natural Selection.* John Murray, London.

Deevey, E.S., Jr. 1947. Life tables for natural populations of animals. *The Quarterly Review of Biology* 22:283–314.

Denholm, I., G.J. Devine, and M.S. Williamson. 2002. Insecticide resistance on the move. *Science* 297:2222–2223.

Endler, J.A. 1986. *Natural Selection in the Wild.* Monographs in Population Biology 21. Princeton University Press, Princeton NJ.

Erickson, G.M., P.J. Currie, B.D. Inouye, et al. 2006. Tyrannosaur life tables: an example of nonavian dinosaur population biology. *Science* 313:213–217.

Grant, B.R. and P.R. Grant. 1989. Natural selection in a population of Darwin's finches. *The American Naturalist* 133:377–393.

Hua, X. and J.J. Wiens. 2010. Latitudinal variation in speciation mechanisms in frogs. *Evolution* 64:429–443.

Hutchinson, G.E. 1959. Homage to Santa Rosalia or why are there so many kinds of animals? *The American Naturalist* 93:145–159.

Hutchinson, G.E. 1978. *Introduction to Population Ecology.* Yale University Press, New Haven, CT.

Reece, J.B., L.A. Urry, M.L. Cain, et al. 2013. *Campbell Biology* (10th ed.). Benjamin Cummings, San Francisco, CA.

Rundle, H.D. and P. Nosil. 2005. Ecological speciation. *Ecology Letters* 8: 336–352.

Schilthuizen, M. 2001. *Frogs, Flies, and Dandelions, Speciation—The evolution of new species.* Oxford University Press, New York, NY.

Schluter, D. 2001. Ecology and the origin of species. *Trends in Ecology and Evolution* 16:372–380.

Smith, T.M. and R.L. Smith. 2009. *Elements of Ecology* (7th ed.). Benjamin Cummings, San Francisco, CA.

Thompson, H. 2012. The chestnut resurrection. *Nature* 490:22–23.

Van Dover, C.L. 2000. *The Ecology of Deep-Sea Thermal Vents.* Princeton University Press, Princeton, NJ.

Wilson, E.O. 1999. *The Diversity of Life.* W.W. Norton, New York, NY.

Chapter 5

Balter, M. 2006. The baby deficit. *Science* 312:1894–1897.

Bongaarts, J. and S. Sinding. 2011. Population policy in transition in the developing world. *Science* 333:574–576.

The Central Intelligence Agency. 2014. *The World Fact Book.* URL = https://www.cia.gov/library/publications/the-world-factbook/, last accessed 9-14-2014.

Chertow, M.R. 2001. The IPAT equation and its variants: Changing views of technology and environmental impact. *Journal of Industrial Ecology* 4:13–29.

Cohen, J.E. 1995. *How Many People Can the World Support?* W.W. Norton, New York, NY.

Cohen, J.E. 2003. Human population grows up. *Scientific American* 293:48–55.

Deevey, E.S. Jr. 1961. The human population. *Scientific American* 203:195–204.

Ehrlich, P.R. 1968. *The Population Bomb.* Ballantine Books, New York, NY.

Ehrlich, P.R. and A.H. Ehrlich. 2009. The population bomb revisited. *The Electronic Journal of Sustainable Development* 1:63–71.

Holdren, J.P. and P.R. Ehrlich. 1974. Human population and the global environment. *American Scientist* 62:282–292.

Lee, R. 2011.The outlook for population growth. *Science* 333:569–573.

Malthus, T.R. 1798. *An Essay on the Principle of Population.* J. Johnson, London. Reprinted at URL = http://www.esp.org/books/malthus/population/malthus.pdf, last accessed 9-14-2014.

Meadows, D.H. and D. Meadows. 1972. *The Limits of Growth: A Report for the Club of Rome's Project on the Predicament of Mankind.* Universe Books, New York, NY.

Meadows, D.H., J. Randers, and D. Meadows. 2004. *Limits to Growth: The 30-year Update.* Chelsea Green Publishing Co., White River Junction, VT.

Monfreda, C., M. Wackernagel, and D.Deumling. 2004. Establishing national natural capital accounts based on detailed ecological footprint and biological capacity accounts. *Land Use Policy* 21:231–246.

Parker, J. 2011. The 9 billion-people question. *The Economist*, Special Report, February 26, 2011. URL = http://www.economist.com/node/18200618, last accessed 9-14-2014.

Pearl, R. and S. Gould. 1936. Human population growth. *Human Biology* 8:399–419.

Petersen, W. 1960. The Demographic Transition in the Netherlands. *American Sociological Review* 25:334–347.

Poston, D.L., Jr., E. Conde, and B. DeSalvo. 2011. China's unbalanced sex ratio at birth, millions of excess bachelors and societal implications. *Vulnerable Children and Youth Studies: An International Interdisciplinary Journal for Research, Policy and Care* 6:314–320.

Raftery, A.E., L. Nan, H. Ševčíková, et al. 2012. Bayesian probabilistic population projections for all countries. *Proceedings of the National Academy of Sciences* 109:13915–13921.

Roberts, L. 2011. 9 Billion? *Science* 333:540–543.

Shah, A. 1998. *Ecology and the Crisis of Overpopulation: Future Prospects for Global Sustainability.* Edward Elgar Publishing, Ltd, Cheltenham, UK.

The State of World Population. 2013. *Motherhood in Childhood: Facing the Challenge of Adolescent Pregnancy.* Information and External Relations Division of UNFPA, the United Nations Population Fund. URL= http://www.unfpa.org/swp, last accessed 9-18-14.

Thompson, W. 1929. Population. *The American Journal of Sociology* 34:959–975.

United Nations. 2012. *World Population Prospects, the 2012 Revision.* URL = http://esa.un.org/wpp/index.htm, last accessed 9-18-2014.

United Nations Population Fund News. 2013. *Dispatch. First school, then marriage and babies.* URL = http://www.unfpa.org/public/home/news/pid/15477, last accessed 9-18-2014.

U.S. Census Bureau. 2014. URL = http://www.census.gov/, last accessed 9-14-2014.

Wackernagel, M. and W. Rees. 1996. *Our Ecological Footprint: Reducing Human Impact on the Earth.* New Society Publishers, Gabriola Island, British Columbia, Canada.

Wackernagel, M., N.B. Schulz, D. Deumling, et al. 2002. Tracking the ecological overshoot of the human economy. *Proceedings of the National Academy of Sciences* 99:266–271.

Chapter 6

Anthony, R.G., J.A. Estes, M.A. Ricca, et al. 2008. Bald eagles and sea otters in the Aleutian Archipelago: indirect effects of trophic cascades. *Ecology* 89:5725–5735.

Carey, A.B., W. Colgan III, J.M. Trappe, et al. 2002. Effects of forest management on truffle abundance and squirrel diets. *Northwest Science* 76:148–157.

Carey, A.B., S.P. Horton, and B.L. Biswell. 1992. Northern spotted owls: Influence of prey base and landscape character. *Ecological Monographs* 62:233–250.

Christensen, N.L., Jr. 2014. An historical perspective on forest succession and its relevance to ecosystem restoration and conservation in North America. *Forest Ecology and Management* 330:312–322.

Cooper, W.S. 1923. The recent ecological history of Glacier Bay, Alaska: II. The present vegetation cycle. *Ecology* 4:223–246.

Cramer, V.A., R.J. Hobbs, and R.J. Standish. 2008. What's new about old fields? Land abandonment and ecosystem assembly. *Trends in Ecology and Evolution* 23:105–112.

Elton, C. and M. Nicholson. 1942. The ten-year cycle in numbers of lynx in Canada. *Journal of Animal Ecology* 11:215–244.

Estes, J.A., J. Terborgh, J.S. Brashares, et al. 2011. Trophic downgrading of planet Earth. *Science* 333:301–306.

Fortin, D., H.L. Beyer, M.S. Boyce, et al. 2005. Wolves influence elk movements: Behavior shapes a trophic cascade in Yellowstone National Park. *Ecology* 85:1320–1330.

Gause, G.F. 1932. Experimental studies on the struggle for existence. *Journal of Experimental Biology* 9:389–402.

Johnston, D.W. and E.P. Odum. 1956. Breeding bird populations in relation to plant succession on the piedmont of Georgia. *Ecology* 37:50–62.

Korstian, C.F. and T.S. Coile. 1938. *Plant Competition in Forest Stands.* Duke University School of Forestry Bulletin 3, Durham, NC.

MacArthur, R.H. 1958. Ecology of some warblers of northeastern coniferous forests. *Ecology* 39:599–619.

Maser, C., A.W. Claridge, and J.M. Trappe. 2007. *Trees, Truffles and Beasts: How Forests Function.* Rutgers University Press, Brunswick, NJ.

Oosting, H.J. 1942. An ecological analysis of the plant communities of piedmont, North Carolina. *American Midland Naturalist* 28:1–126.

Pickett, S.T.A. and P.S. White. 1985. *The Ecology of Natural Disturbance and Patch Dynamics.* Academic Press, Orlando, FL.

Pimm, S.L. 1979. The structure of food webs. *Theoretical Population Ecology* 16:144–158.

Reisewitz, S.E., J.E. Estes, and C.A. Simenstad. 2006. Indirect food web interactions: sea otters and kelp forest fishes in the Aleutian Archipelago. *Oecologia* 146:623–631.

Ripple, W.J. and E.J. Larsen. 2000. Historic aspen recruitment, elk and wolves in northern Yellowstone National Park, USA. *Biological Conservation* 95:361–370.

Romme, W.H. and D.G. Despain. 1989. Historical perspective on the Yellowstone fires of 1988. *BioScience* 39:695–699.

Scheffer, M. and S.R. Carpenter. 2003. Catastrophic regime shifts in ecosystems: Linking theory to observation. *Trends in Ecology and Evolution* 18:648–656.

Smith, T.M. and R.L. Smith. 2009. *Elements of Ecology* (7th ed.). Benjamin Cummings, San Francisco, CA.

Stiling, P. 2002. *Ecology: Theories and Applications*. Prentice-Hall, Upper Saddle River, NJ.

Terborgh, J. and J. Estes. 2010. *Trophic Cascades: Predators, Prey and the Changing Dynamics of Nature*. Island Press, Washington, DC.

Turner, M.G. 2005. Landscape ecology: What is the state of the science. *Annual Review of Ecology, Evolution and Systematics* 36:319–344.

Vucetich, J.A. and R.O. Peterson. 2013. *Ecological Studies of Wolves on Isle Royale*. International Wolf Center, Michigan Technological University, Houghton, MI.

Wilson, E.O. 1999. *The Diversity of Life*. W.W. Norton, New York, NY.

Chapter 7

Archbold, O.W. 1995. *Ecology of World Vegetation*. Chapman-Hall, New York, NY.

Barbour, M.G. and W.D. Billings. 2000. *North American Terrestrial Vegetation*. Cambridge University Press, Cambridge, UK.

Breckle, S.-W. and G. Lawlor. 2002. *Walter's Vegetation of the Earth*. Springer-Verlag, New York, NY.

Chabot, B.F. and H.A. Mooney. 1985. *The Physiological Ecology of North American Plant Communities*. Chapman Hall, New York, NY.

Chapin, F.S., III, M.W. Oswood, K. Van Cleve, et al. 2006. *Alaska's Changing Boreal Forest*. Oxford University Press, New York.

Christensen, N.L. 2005. Fire in the parks: A case study for change management. *The George Wright Forum* 22:12–31.

Gibson, D.J. 2008. *Grasses and Grassland Ecology*. Oxford University Press, Oxford, UK.

Hari, P. and L. Kulmala. 2008. *Boreal Forest and Climate Change*. Springer-Verlag, Berlin.

Kellman, M.C. 1980. *Plant Geography*. Methuen & Co., London, UK.

Knight, D.H. 1994. *Mountains and Plains: The Ecology of Wyoming Landscapes*. Yale University Press, New Haven, CT.

Montagnini, F. and C.F. Jordan. 2005. *Tropical Forest Ecology: The Basis for Conservation and Management*. Springer-Verlag, Berlin.

Quinn, R.D., S.C. Keeley, and M.D. Wallace. 2006. *Introduction to California Chaparral*. University of California Press, Berkeley, CA.

Tscharntke, T., C. Leuschner, M. Zeller, et al. 2007. *Stability of Tropical Rainforest Margins: Linking Ecological, Economic and Social Constraints of Land Use and Conservation*. Springer-Verlag, Berlin.

Von Humboldt, A. and A. Bonpland. 1807. *Essay on the Geography of Plants*. Reprinted by University of Chicago Press, Chicago, IL (Translation by S. Romanowski and Introduction by S.T. Jackson).

Whitaker, R.H. 1975. *Communities and Ecosystems*. MacMillan, New York, NY.

Chapter 8

Balmford, A., A. Bruner, P. Cooper, et al. 2002. Economic reasons for conserving wild nature. *Science* 297:950–953.

Baskin, Y. 1997. *The Work of Nature: How Biodiversity Sustains Us*. Island Press, Washington, DC.

Butman, C.A., J.T. Carlton, G.W. Boehlert, et al. 1995. *Understanding Marine Biodiversity*. National Academy Press, Washington, DC.

Chapin, S.F., III, E.-D. Schulze, and H.A. Mooney. 1992. Biodiversity and ecosystem processes. *Trends in Ecology and Evolution* 7:107–108.

Christensen, N.L., Jr. 2014. An historical perspective on forest succession and its relevance to ecosystem restoration and conservation in North America. *Forest Ecology and Management* 330:312–322.

Daily, G.C. 1997. *Nature's Services: Societal Dependence on Natural Ecosystems*. Island Press, Washington, DC.

Damschen, E.I., N.M. Haddad, J.L. Orrock, et al. 2006. Corridors increase plant species richness at large scales. *Science* 313:1284–1286.

Ehrlich, P.R. and A.H. Ehrlich. 1992. *Extinction: The Causes and Consequences of the Disappearance Species*. Random House, New York.

Estes, J.A., J. Terborgh, J.S. Brashares, et al. 2011. Trophic downgrading of planet Earth. *Science* 333:301–306.

Groom, M.J., G.K. Meffe, and C.R. Carroll. 2006. *Principles of Conservation Biology* (3rd ed.). Sinauer Associates, Inc., Sunderland, MA.

Hoekstra, J.M., J.L. Molnar, M. Jennings, et al. 2010. *The Atlas of Global Conservation*. University of California Press, Berkeley, CA.

Hunter, M.L., Jr., and J. Gibbs. 2007. *Fundamentals of Conservation Biology*. Blackwell Publishing, Malden, MA.

Kinzig, A.P., S.W. Pacala, and D. Tilman. 2001. *The Functional Consequences of Biodiversity: Empirical Progress and Theoretical Extensions*. Monographs in Population Biology 33. Princeton University Press, Princeton, NJ.

Lovejoy, T.E. and L. Hannah. 2006. *Climate Change and Biodiversity*. Yale University Press, New Haven, CT.

MacArthur, R.H. and E.O. Wilson. 1967. *The Theory of Island Biogeography*. Monographs in Population Biology 1. Princeton University Press, Princeton, NJ.

Millennium Ecosystem Assessment. 2005. *Ecosystems and Human Well-Being*. Island Press, Washington, DC. URL = http://www.maweb.org/documents/document.356.aspx.pdf, last accessed 10-1-2011.

Naeem, S., F.S. Chapin III, R. Costanza, et al. 1999. *Biodiversity and Ecosystem Functioning: Maintaining Natural Life Support Processes*. Issues in Ecology, Number 4. Ecological Society of America, Washington, DC.

Naeem, S. and S. Li. 1997. Biodiversity enhances ecosystem reliability. *Nature* 390:507–509.

Norse, E.A., L.B. Crowder, and M.E. Soule. 2005. *Marine Conservation Biology: The Science of Maintaining the Sea's Biodiversity*. Island Press, Washington, DC.

O'Riordan, T. and S. Stoll-Kleemann. 2002. *Biodiversity, Sustainability and Human Communities: Protecting Beyond the Protected*. Cambridge University Press, Cambridge, UK.

Pereira, H.M., P.W. Leadley, V. Proença, et al. 2010. Scenarios for global biodiversity in the 21st century. *Science* 330:1496–1501.

Pimm, S.L., C.N. Jenkins, R. Abell, et al. 2014. The biodiversity of species and their rates of extinction, distribution and protection. *Science* 344:987–997.

Pimm, S.L., G.J. Russel, J.L. Gittleman, et al. 1995. The future of biodiversity. *Science* 269:347–350.

Ripple, W.J., J.A. Estes, R.L. Beschta, et al. 2014. Status and Ecological Effects of the World's Largest Carnivores. *Science* 343:151–161.

Sellars, R.W. 1997. *Preserving Nature in the National Parks: A History*. Yale University Press, New Haven, CT.

Tilman, D. and J.A. Downing. 1994. Biodiversity and stability in grasslands. *Nature* 367:363–365.

Wilcove, D.S., D. Rothstein, J. Dubow, et al. 1998. Quantifying threats to imperiled species in the United States. *BioScience* 48:607–615.

Wilson, E.O. 1999. *The Diversity of Life*. W.W. Norton, New York, NY.

Wilson, E.O. 2002. *The Future of Life*. Alfred A. Knopf, New York, NY.

Chapter 9

Arkema, K.K., G. Guannel, G. Verutes, et al. 2013. Coastal habitats shield people and property from sea-level rise and storms. *Nature: Climate Change* 3:913–918.

Behringer, W. 2010. *A Cultural History of Climate*. Polity, Malden, MA.

Blunden, J. and D.S. Arndt. (eds.). 2014. State of the Climate in 2013. *Bulletin of the American Meteorological Society* 95:1–257.

Bond, T.C., S.J. Doherty, D.W. Fahey, et al. 2013. Bounding the role of black carbon in the climate system: A scientific assessment. *Journal of Geophysical Research: Atmospheres* 118: 5380–5552.

Boxall, A.B.A., A. Hardy, S. Beulke, et al. 2009. Impacts of climate change on indirect human exposure to pathogens and chemicals from agriculture. *Environmental Health Perspectives* 117:508–515.

Broecker, W.S. and G.M. Henderson. 1998. The sequence of events surrounding Termination II and their implications for the cause of glacial-interglacial CO_2 changes. *Paleoceanography* 13:352–364.

Brown, L.R. 2011. *World on the Edge: How to Prevent Environmental and Economic Collapse*. W.W. Norton & Co., New York, NY.

Burke, M.B., D.B. Lobell, and L. Guarino. 2009. Shifts in African crop climates by 2050, and the implications for crop improvement and genetic resources conservation. *Global Environmental Change* 19:317–325.

Caldeira, K. and M.E. Wickett. 2003. Anthropogenic carbon and ocean pH. *Nature* 425:465.

Clausen, E. 2001. *Climate Change: Science, Strategies and Solutions.* The Pew Center for Global Climate Change, Arlington, VA.

Epstein, P.R. 2002. Climate change and infectious disease: Stormy weather ahead? *Epidemiology* 13:373–375.

Fagan, B. 2000. *The Little Ice Age: How Climate Made History (1300–1850).* Basic Books, New York, NY.

Flannery, T. 2007. *The Weather Makers: How Man Is Changing the Climate and What It Means for Life on Earth.* Text Publishing Company, Melbourne, Australia.

Gore, A. 2006. *An Inconvenient Truth: The Planetary Emergency of Global Warming and What We Can Do About It.* Rodale, Emmaus PA.

Hidore, J.J., J.E. Oliver, M. Snow, et al. 2010. *Climatology: An Atmospheric Science* (3rd ed.). Prentice Hall, Upper Saddle River, NJ.

Holt, T.O., N.G. Glasser, D.J. Quincey, et al. 2013. Speedup and fracturing of George VI Ice Shelf, Antarctic Peninsula. *The Cryosphere* 7:797–816.

IPCC. 2007. *Climate change 2007: Synthesis report.* New York: Cambridge University Press.

IPCC. 2013. *Climate Change 2013: The Physical Science Basis.* Intergovernmental Panel on Climate Change. URL = http://www.ipcc.ch/report/ar5/wg1/, last accessed 9-15-2014.

IPCC. 2014a. *Climate Change 2014: Impacts, Adaptation and Vulnerability.* Intergovernmental Panel on Climate Change. URL = http://ipcc-wg2.gov/AR5/, last accessed 9-15-2014.

IPCC. 2014b. *Climate Change 2014: Mitigation of Climate Change.* URL = http://www.ipcc.ch/report/ar5/wg3/, last accessed 9-15-2014.

Kopp, R.J. and W.A. Pizer. 2007. *Assessing U.S. Climate Policy Options.* Resources for the Future, Washington, DC. URL = http://www.rff.org/cpfreport, last accessed 10-1-2001.

Krupp, F. and M. Horn. 2009. *Earth: The Sequel.* W.W. Norton, New York, NY.

Loarie, S.R., P.B. Duffy, H. Hamilton, et al. 2009. The velocity of climate change. *Nature* 462:1052–1057.

Lomborg, B. 2001. *The Skeptical Environmentalist. Measuring the Real State of the World.* Cambridge University Press, Cambridge, UK.

Lomborg, B. 2010. *Smart Solutions to Climate Change: Comparing Costs and Benefits.* Cambridge University Press, Cambridge, UK.

Mudelsee, M. 2001.The phase relations among atmospheric CO_2 content, temperature and global ice volume over the past 420 ka. *Quaternary Science Reviews* 20:583–589.

National Geographic Society. 2008. Changing climate. *National Geographic Special Report,* 22 June 2008.

Nelson, E.J., P. Kareiva, M. Ruckelshaus, et al. 2013. Climate change's impact on key ecosystem services and the human well-being they support in the U.S. *Frontiers in Ecology and the Environment* 11:483–493.

Orr, J.C., V.J. Fabry, O. Aumont, et al. 2005. Anthropogenic ocean acidification over the twenty-first century and its impact on calcifying organisms. *Nature* 437:681–686.

Pacala, S. and R. Socolow. 2004. Stabilization Wedges: Solving the Climate Problem for the Next 50 Years with Current Technologies. *Science,* 13:968–972.

Pan, Y., R.A. Birdsey, J. Fang, et al. 2011. A large and persistent carbon sink in the world's forests. *Science* 333:988–993.

Petit, J.R., J. Jouzel, D. Raynaud, et al. 1999. Climate and atmospheric history of the past 420,000 years from the Vostok ice core, Antarctica. *Nature* 399:429–436.

Post, E., M.C. Forchhammer, M.S. Bret-Harte, et al. 2009. Ecological dynamics across the arctic associated with recent climate change. *Science* 325:1355–1358.

Rignot, E. and P. Kanagaratnam. 2006. Changes in the velocity structure of the Greenland Ice Sheet. *Science* 311:986–990.

Schlesinger, W.H. and E.S. Bernhardt. 2013. *Biogeochemistry: An Analysis of Global Change* (3rd ed.). Academic Press, Orlando, FL.

Schneider, S.H. and T.L. Root. (eds.). 2002. *Wildlife Responses to Climate Change: North American Case Studies.* Island Press, Washington, DC.

Sherwood, S. and Q. Fu. 2014. A drier future? *Science* 343:737–739.

Soverow, J.E., G.A. Wellenius, D.N. Fisman, et al. 2009. Infectious disease in a warming world: How weather influenced West Nile virus in the United States (2001–2005). *Environmental Health Perspectives* 117:1049–1052.

Speth, J.G. 2004. *Red Sky at Morning: America and the Crisis of the Global Environment.* Yale University Press, New Haven, CT.

Stein, B.A., A. Staudt, M.S. Cross, et al. 2013. Preparing for and managing change: Climate adaptation for biodiversity and ecosystems. *Frontiers in Ecology and the Environment* 11:502–510.

Swart, R., R. Biesbroek, S. Binnerup, et al. 2009. *Europe Adapts to Climate Change: Comparing National Adaptation Strategies.* Partnership for European Environmental Research (PEER), Helsinki, Finland.

United Nations. 2014. *Kyoto Protocol.* URL = http://unfccc.int/kyoto_protocol/items/2830.php, last accessed 9-14-2014.

United Nations. 1992. *U.N. Framework Convention on Climate Change.* URL = http://unfccc.int/resource/docs/convkp/conveng.pdf, last accessed 10-1-2001.

U.S. Global Change Research Program, T.R. Kar, J.M. Melillo, et al. (eds.). 2009. *Global Climate Change Impacts in the United States.* U.S. Global Change Research Program. Cambridge University Press, Cambridge, UK.

Vecchi, G.A. and G. Villarini. 2014. Next season's hurricanes. *Science* 343:618–619.

Vitousek, P.M., H.A. Mooney, J. Lubchenco, et al. 1997. Human domination of Earth's ecosystems. *Science* 277:494–499.

The World Watch Institute. 2009. *State of the World: Into a Warming World.* W.W. Norton, New York, NY.

Chapter 10

Akimoto, H. 2003. Global air quality and pollution. *Science* 302:1716–1719.

Brooks, B.O. and W.F. Davis. 1992. *Understanding Indoor Air Quality.* CRC Press, Boca Raton, FL.

EPA. 2014. National Summary of Sulfur Dioxide Emissions. URL = http://www.epa.gov/cgi-bin/broker?_service=data&_debug=0&_program=dataprog.national_1.sas&polchoice=SO2#so2nat, last accessed 9-14-2014.

EPA. 2009. NOx Budget Trading Program. URL = http://www.epa.gov/airmarkt/progsregs/nox/docs/NBPbasicinfo.pdf, last accessed 9-14-2014.

Godish, T. 2004. *Air Quality* (4th ed.). Lewish Publishers, Boca Raton, FL.

Jacobson, M.Z. 2002. *Atmospheric Pollution: History, Science, and Regulation.* Cambridge University Press, Cambridge, UK.

Metcalfe, S. and D. Derwent. 2005. *Atmospheric Pollution and Environmental Change.* Hodder Arnold, London, UK.

Molina, M.J. and F.S. Rowland. 1974. Stratospheric sink for chlorofluoromethanes: Chlorine atom catalyzed destruction of ozone. *Nature* 249:810–812.

National Atmospheric Deposition Program. 2014. *National Trends Network.* URL = http://nadp.sws.uiuc.edu/ntn/, last accessed 9-14-2014.

Parson, E.A. 2003. *Protecting the Ozone Layer: Science and Strategy.* Oxford University Press, Oxford, UK.

Schuster, P.F., D.P. Krabbenhoft, D.L. Naftz, et al. 2002. Atmospheric mercury deposition during the last 270 years: A glacial ice core record of natural and anthropogenic sources. *Environmental Science and Technology* 36:2303–2310.

Shotyk, W., D. Weiss, P.G. Appleby, et al. 1998. History of atmospheric lead deposition since 12,370 14C yr BP from a peat bog, Jura Mountains, Switzerland. *Science* 281:1635–1640.

Snyder, L.P. 1994. "The death-dealing smog over Donora, Pennsylvania": Industrial air pollution, public health policy, and the politics of expertise, 1948–1949. *Environmental History Review* 18:117–139.

Sommerville, R.C.J. 1996. *The Forgiving Air: Understanding Environmental Change.* University of California Press, Berkeley, CA.

World Health Organization. 2014. *Indoor Air and Household Energy.* URL = http://www.who.int/heli/risks/indoorair/indoorair/en/index.html, last accessed 9-14-2014.

World Meteorological Organization. 2011. *Scientific Assessment of Ozone Depletion: 2010.* Global Ozone Research and Monitoring Project Report 52. World Meteorological Organization, Geneva Switzerland.

Chapter 11

AWWA. 2012. *Buried No Longer: Confronting America's Water Infrastructure Challenge.* American Water Works Association, Denver, CO. URL = http://www.awwa.org/Portals/0/files/legreg/documents/BuriedNoLonger.pdf, last accessed 10-14-2014.

Ballantine, K. and R. Schneider. 2009. Fifty-five years of soil development in restored freshwater depressional wetlands. *Ecological Applications* 19:1467–1480.

Barlow, Paul M. 2003. *Ground Water in Freshwater-Saltwater Environments of the Atlantic Coast.* Circular 1262. U.S. Department of the Interior U.S. Geological Survey.

Bernhardt, E.S. and M.A. Palmer. 2007. Restoring streams in an urbanizing world. *Freshwater Biology* 52:738–751.

Brown, L.R. 2011. *World on the Edge: How to Prevent Environmental and Economic Collapse.* W.W. Norton & Co., New York, NY.

Caldeira, K. and M.E. Wickett. 2003. Anthropogenic carbon and ocean pH. *Nature* 425:465.

Dahl, T.E. 2006. *Status and trends of wetlands in the conterminous United States 1998 to 2004.* U.S. Department of the Interior; Fish and Wildlife Service, Washington, DC. URL = http://www.fws.gov/wetlands/_documents/gSandT/NationalReports/StatusTrendsWetlandsConterminousUS1998to2004.pdf, last accessed 9-14-2014.

Ensign, S.H. and M.W. Doyle. 2006. Nutrient spiraling in streams and river networks. *Journal of Geophysical Research* 111, G04009, doi:10.1029/2005JG000114.

EPA. 2000. *EPA Guidelines for Management of Onsite/Decentralized Wastewater Systems.* U.S. Environmental Protection Fact Sheet EPA-832-F-00-012. URL = http://water.epa.gov/aboutow/owm/upload/2004_07_07_septics_septic_guidelines_factsheet.pdf, last accessed 10-14-2014.

EPA. 2002. *Biological Assessments and Criteria: Crucial Components of Water Quality Programs.* URL = http://water.epa.gov/scitech/swguidance/standards/criteria/aqlife/biocriteria/index.cfm, last accessed 10-1-2011.

EPA. 2004. *Primer for Municipal Wastewater Treatment Systems.* EPA 832-R-04-001. Office of Wastewater Management, Washington DC.

EPA. 2009. *Water on Tap: What You Need to Know.* EPA 816-K-09-001. Office of Water, Washington, DC.

EPA. 2013. *Water: Total Maximum Daily Loads (303d).* URL = http://water.epa.gov/lawsregs/lawsguidance/cwa/tmdl/index.cfm, last accessed 9-20-2014.

EPA. 2014. *National Summary of Impaired Waters and TMDL Information.* URL = http://iaspub.epa.gov/waters10/attains_nation_cy.control?p_report_type=T, last accessed 9-20-2014.

FAO. 2010. *State of World Fisheries and Aquaculture 2010.* Food and Agriculture Organization of the United Nations, Rome, Italy. URL = http://www.fao.org/docrep/013/i1820e/i1820e.pdf, last accessed 9-14-2014.

FAO. 2014. *Aquastat. FAO's Global Water Information System.* Food and Agriculture Organization of the United Nations, Rome, Italy. URL= http://www.fao.org/nr/water/aquastat/main/index.stm, last accessed 9-20-2014.

Frazier, L. 2005. Paving paradise: The peril of impervious surfaces. *Environmental Health Perspectives* 113:A456–A462.

Gleick, P.H. 2011. *The World's Water Volume 7: The Biennial Report on Freshwater Resources.* Island Press, Washington, DC.

International Union of Geological Sciences. 2005. *Groundwater: Reservoir for a Thirsty Planet?* URL = http://yearofplanetearth.org/content/downloads/Groundwater.pdf, last accessed 9-14-2014.

Jackson, R.B., S.R. Carpenter, C.N. Dahm, et al. 2001.*Water in a Changing World.* Issues in Ecology, Number 9. Ecological Society of America, Washington, DC.

Konrad, C.P. *Effects of Urban Development on Floods.* U.S. Geological Survey. Fact Sheet 076-03. URL = http://pubs.usgs.gov/fs/fs0760, last accessed 9-20-2014.

Maddigen, R.D., W.S. Chern, and C.G. Rizy. 1982. The irrigation demand for electricity. *American Journal of Agricultural Economics* 64:673–680.

McLeod, K. and H. Leslie. 2009. *Ecosystem-Based Management for the Oceans.* Island Press, Washington, DC.

Millennium Ecosystem Assessment. 2005. *Ecosystems and Human Well-Being.* Island Press, Washington, DC. URL = http://www.maweb.org/documents/document.356.aspx.pdf, last accessed 9-14-2014.

Mitsch, W.J., J.G. Gosselink, C.J. Anderson, et al. 2009. *Wetland Ecosystems.* John Wiley and Sons, Hoboken, NJ.

Nichols, F.H., J.E. Cloern, S.N. Luoma, et al. 1986. The modification of an estuary. *Science* 231: 567–573.

OECD. 2008. *Environmental Outlook to 2030.* Organization for Economic Cooperation and Development, Paris, France. URL = http://www.oecd.org/document/20/0,3746,en_2649_37465_39676628_1_1_1_37465,00.html, last accessed 9-14-2014.

Orr, J.C., V.J. Fabry, O. Aumont, et al. 2005. Anthropogenic ocean acidification over the twenty-first century and its impact on calcifying organisms. *Nature* 437:681–686.

Outwater, A. 1996. *Water: A Natural History.* BasicBooks, New York, NY.

Pacific Institute. 2014. *Water Conflict Chronology List.* URL = http://www.worldwater.org/conflict/list/, last accessed 9-20-2014.

Postel, S. 1997. *Last Oasis: Facing Water Scarcity.* W.W. Norton, New York, NY.

Postel, S. 1999. *Pillar of Sand.* W.W. Norton & Co., New York, NY.

Postel, S.L., G.C. Daily, and P.R. Ehrlich. 1996. Human appropriation of renewable fresh water. *Science* 271:785–788.

Reisner, M. 1993. *Cadillac Desert: The American West and its Disappearing Water* (Revised Ed.). Penguin Books, New York, NY.

Steward, D.R., P.J. Brussa, X. Yangb, et al. 2013. Tapping unsustainable groundwater stores for agricultural production in the High Plains Aquifer of Kansas, projections to 2110. *PNAS* 110:E3477–E3486.

UNEP. 2008. *Africa: Atlas of our Changing Environment.* United Nations Environment Programme, Nairobi, Kenya.

U.N. Food and Agriculture Organization. 2014. *AQUASTAT: FAO's global water database.* URL = http://www.fao.org/nr/water/aquastat/main/index.stm, last accessed 9-14-2014.

USGS. 1996. *Nutrients in the Nation's Waters: Identifying Problems and Progress.* A National Water–Quality Assessment of Nutrients. URL = http://pubs.usgs.gov/fs/fs218-96/, last accessed 9-20-2014.

USGS. 2001. *A Primer on Water Quality.* URL = http://pubs.usgs.gov/fs/fs-027-01/, last accessed 9-20-2014.

USGS. 2004. *Climatic Fluctuations, Drought, and Flow in the Colorado River Basin.* Fact Sheet 2004-3062. URL = http://pubs.usgs.gov/fs/2004/3062/pdf/fs2004-3062_version2.pdf, last accessed 10-14-2014.

USGS. 2005. Trends in the Water Budget of the Mississippi Basin. Fact Sheet 2005-3020. URL = http://pubs.usgs.gov/fs/2005/3020/pdf/FS2005_3020.pdf, last accessed 10-14-2014.

Van Bers, C., D. Petry, and C. Pahl-Wostl. (eds.). 2007. *Global Assessments: Bridging Scales and Linking to Policy.* Report on the joint TIAS-GWSP workshop held at the University of Maryland University College, Adelphi, USA GWSP Issues in Global Water System Research, No.2. GWSP IPO, Bonn.

Wu, J., J. Huang, X. Han, et al. 2004. The Three Gorges Dam: An ecological perspective. *Frontiers in Ecology and Environment* 2:241–248.

Chapter 12

Brown, L.R. 2011. *World on the Edge: How to Prevent Environmental and Economic Collapse.* W.W. Norton & Co., New York, NY.

Collins, A. and R. Fairchild. 2007. Sustainable food consumption at a sub-national level: An ecological footprint, nutritional and economic analysis. *Journal of Environmental Policy and Planning* 9:5–30.

Cox, T.S., J.D. Glover, D.L. Van Tassel, et al. 2006. Prospects for developing perennial grain crops. *BioScience* 56:649–659.

Egan, T. 2006. *The Worst Hard Time: The Untold Story of Those Who Survived the Great American Dust Bowl.* Houghton Mifflin Co., New York, NY.

Ellis, E.C. and N. Ramankutty. 2008. Putting people in the map: Anthropogenic biomes of the world. *Frontiers in Ecology and Environment* 6:439–447.

FAO. 2006. *Livestock's Long Shadow: Environmental Issues and Options.* Food and Agriculture Organization of the United Nations, Rome, Italy. URL = http://www.fao.org/docrep/010/a0701e/a0701e00.HTM, last accessed 9-14-2014.

FAO. 2009. *The State of Food Insecurity in the World.* Food and Agriculture Organization of the United Nations, Rome, Italy. URL = http://www.fao.org/publications/sofi/en/, last accessed 9-14-2014.

FAO. 2010. *State of World Fisheries and Aquaculture 2010.* Food and Agriculture Organization of the United Nations, Rome, Italy. URL = http://www.fao.org/docrep/013/i1820e/i1820e.pdf, last accessed 9-14-2014.

Glover, J.D., C.M. Cox, and J.P. Reganold. 2007. Future farming: A return to roots? *Scientific American* 297:66–73.

Glover, J.D., J.P. Reganold, L.W. Bell, et al. 2010. Increased food and ecosystem security via perennial grains. *Science* 328:1638–1639.

Imhoff, M.L., L. Bounoua, T. Ricketts, et al. 2004. Global patterns in human consumption of net primary production. *Nature* 429:870–873.

International Service for the Acquisition of Agri-biotech Applications. 2012. *ISAA Brief 44-2012: Executive Summary.* Global status of commercialized biotech/GM crops: 2012. URL =http://www.isaaa.org/resources/publications/briefs/44/executivesummary/, last accessed 9-20-2014.

Losey, J.E., L.S. Rayor, and M.E. Carter. 1999. Transgenic pollen harms monarch larvae. *Nature* 399:214.

Mann, C.C. 2002. Transgene data deemed unconvincing. *Science* 296:236–237.

Mazoyer, M. and L. Roudart. 2006. *A History of World Agriculture: From the Neolithic Age to the Current Crisis.* Monthly Review Press, New York.

Norris, R.F., E.P. Caswell-Chen, and M. Kogan. 2003. *Concepts in Integrated Pest Management.* Prentice Hall, Upper Saddle River, NJ.

OTA. 2011. *The Organic Trade Association's 2011 Organic Industry Survey.* Organic Trade Association, Greenfield, MA.

Paoletti, M.G. and D. Pimentel. 1996. Genetic engineering in agriculture and the environment: Assessing risks and benefits. *BioScience* 46: 665–673.

Pollan, M. 2006. *The Omnivore's Dilemma: A Natural History of Four Meals.* The Penguin Press, New York.

Pollan, M. 2008. *In Defense of Food: An Eater's Manifesto.* The Penguin Press, New York.

Riehl, S., Z. Mohsen, and N.J. Conard. 2013. Emergence of agriculture in the foothills of the Zagros Mountains of Iran. *Science* 341:65–67.

Shiva, V. 2000. *Stolen Harvest: The Hijacking of the Global food Supply.* South End Press, Cambridge, MA.

Snow, A.A., D.A. Andow, P. Gepts, et al. 2005. Gentically engineered organisms and the environment: Current status and recommendations. *Ecological Applications* 15:377–404.

U.S. Census Bureau. 2011. *The 2011 Statistical Abstract: Agriculture.* U.S. Census Bureau, Washington, DC. URL = http://www.census.gov/compendia/statab/cats/agriculture.html, last accessed 9-14-2014.

USDA – Economic Research Service. 2014. *Adoption of genetically engineered crops in the U.S.* URL =http://www.ers.usda.gov/data-products/adoption-of-genetically-engineered-crops-in-the-us.aspx#.UgfL0FNTTA4, last accessed 9-20-2014.

USDA National Agriculture Statistics Service. 2013. *Total Corn Production and Corn Used for Fuel Ethanol Production.* URL = http://www.afdc.energy.gov/data/10339, last accessed 9-20-2014.

Wallander, S., M. Aillery, D. Hellerstein, et al. 2013. *The Role of Conservation Programs in Drought Risk Adaptation.* Economic Research Report Number 148. Economic Research Service, U.S. Department of Agriculture, Washington, DC.

Wolfenbarger, L.L.R. 2000. The ecological risks and benefits of genetically engineered plants. *Science* 290:2088.

The World Watch Institute. 2011. *State of the World: Innovations that Nourish the Planet.* W.W. Norton, New York, NY.

Chapter 13

Bormann, F.H. and G.E. Likens. 1979. *Pattern and Process in a Forested Ecosystem.* Springer-Verlag, New York, NY.

Cashore, B., G. Auld, and D. Newsom. 2004. *Governing Through Markets: Forest Certification and the Emergence of Non-State Authority.* Yale University Press, New Haven, CT.

Chamberlain, J., R. Bush, and A.L. Hammett. 1998. Non-timber forest products: The *other* forest products. *Forest Products Journal* 48:10–19.

Christensen, N.L. 2009. Future forest, future fires. *Yellowstone Science* 17:40–44.

Christensen, N.L., Jr. 2014. An historical perspective on forest succession and its relevance to ecosystem restoration and conservation in North America. *Forest Ecology and Management* 330:312–322.

Covington, W.W. and M.M. Moore. 1994. Southwestern ponderosa pine forest structure: Changes since Euro-American settlement. *Journal of Forestry* 92:39–47.

Diamond, J. 2006. *Collapse: How Societies Choose to Fail or Succeed.* Penguin Books, New York, NY.

FAO. 2010. *Global Forest Resources Assessment 2010.* Food and Agriculture Organization of the United Nations, Rome, Italy. URL = http://www.fao.org/forestry/fra/fra2010/en/, last accessed 9-14-2014.

Farmer, P. 1994. *The Uses of Haiti.* Common Courage Press, Monroe, ME.

Frelich, L.E. 2002. *Forest Dynamics and Disturbance Regimes: Studies from Temperate Evergreen-Deciduous Forests.* Cambridge University Press, Cambridge, UK.

Hansen, M.C., P.V. Potapov, R. Moore, et al. 2013. High-resolution global maps of 21st-century forest cover change. *Science* 342:850–853.

Harris, L.D. 1984. *The Fragmented Forest: Island Biogeography Theory and the Preservation of Biotic Diversity.* University of Chicago Press, Chicago, IL.

Hunter, M.L., Jr., and F. Schmiegelow. 2010. *Wildlife, Forests and Forestry: Principles of Managing Forests for Biological Diversity* (2nd ed.). Prentice Hall, Upper Saddle River, NJ.

Keeley, J.E., G. Aplet, N.L. Christensen, Jr., et al. 2009. Ecological Foundations of Fire Management. *U.S. Forest Service General Technical Report.* PNW.GTR-779.

Kohm, K.A. and J.F. Franklin. 1997. *Creating a Forestry for the 21st Century: The Science of Ecosystem Management.* Island Press, Washington, DC.

Laarman, J.G. and R.A. Sedjo. 1992. *Global Forests: Issues for Six Billion People.* McGraw Hill, New York, NY.

McCarthy, H.R., R. Oren, K.H. Johnsen, et al. 2010. Re-assessment of plant carbon dynamics at the Duke free-air CO_2 enrichment site: Interactions of atmospheric $[CO_2]$ with nitrogen and water availability over stand development. *New Phytologist* 185:514–528.

Medley, K.E., C.M. Pobocik, and B.W. Okey. 2003. Historical changes in forest cover and land ownership in a Midwestern U.S. landscape. *Annals of the Association of American Geographers* 93:104–120.

Pan, Y., R.A. Birdsey, and J. Fang. 2011. A large and persistent carbon sink in the world's forests. *Science* 333: 988–993.

Perry, D.A., R. Oren, and S.C. Hart. 2008. Forest Ecosystems (2nd ed.). Johns Hopkins Press, Baltimore, MD.

Romme, W.H. and D.G. Despain. 1989. Historical perspective on the Yellowstone fires of 1988. *BioScience* 39: 695–699.

Sayer, J., J. McNeely, S. Maginnis, et al. 2007. *Local Rights and Tenure for Forests: Opportunity or Threat for Conservation.* Rights and Resources Initiative, Washington, DC.

Smith, W.H. 1990. The health of North American forests: Stress and risk assessment. *Journal of Forestry* 88:32–35.

U.S.D.A. Forest Service. 2002. *U.S. Forest Facts and Historical Trends.* U.S.D.A. Forest Service, Washington, DC. URL = http://fia.fs.fed.us/library/briefings-summaries-overviews/docs/2002_ForestStats_%20FS801.pdf, last accessed 9-14-2014.

U.S.D.A. Forest Service. 2008. *Forest Resources of the United States.* U.S.D.A. Forest Service, Washington, DC. URL = http://nrs.fs.fed.us/pubs/7334, last accessed 9-14-2014.

Westerling, A.L., H.G. Hidalgo, D.R. Crayan, et al. 2006. Warming and earlier spring increase western wildfire activity. *Science* 313:940–943.

Westoby, J. 1989. *Introduction to World Forestry.* Basil Blackwell Ltd., Cambridge, UK.

Williams, M. 1989. *Americans and Their Forests: A Historical Geography.* Cambridge University Press, Cambridge, UK.

Chapter 14

Aucott, M.L. and J.M. Melillo. 2013. A preliminary energy return on investment analysis of natural gas from the Marcellus shale. *Journal of Industrial Ecology*, doi:10.1111/jiec.12040.

BP. 2014. *BP Statistical Review of World Energy, June 2014.* BP, London, UK. URL = http://www.bp.com/content/dam/bp/pdf/Energy-economics/statistical-review-2014/BP-statistical-review-of-world-energy-2014-full-report.pdf, last accessed 9-14-2014.

Brandt, A.R., G.A. Heath, E.A. Kort, et al. 2014. Methane leaks from North American natural gas systems. *Science* 343:733–735.

Christie, B. 1999. *Kern River field at 100: The city that oil built.* The Bakersfield Californian, April 27, 1999.

Cleveland, C.J. (Topic Editor). 2009. Smart grid. In: *Encyclopedia of Earth. Environmental Information Coalition.* National Council for Science and the Environment, Washington, DC. URL = http://www.eoearth.org/article/Smart_grid, last accessed 9-14-2014.

Deffeyes, K.S. 2001. *Hubbert's Peak: The Impending World Oil Shortage.* Princeton University Press, Princeton, NJ.

EIA. 2011. *Annual Energy Review 2010.* U.S. Department of Energy, Energy Information Administration, Washington, DC. URL = http://205.254.135.24/totalenergy/data/annual/index.cfm, last accessed 9-14-2014.

Hall, C.A.S. and J.W. Day, Jr. 2009. Revisiting the limits of growth after peak oil. *American Scientist* 97:230–237.

IAEA. 2009. *World Distribution of Uranium Deposits (UDEPO) with Uranium Deposit Classification.* International Atomic Energy Agency, Vienna, Austria. URL = http://www-pub.iaea.org/MTCD/publications/PDF/TE_1629_web.pdf, last accessed 9-14-2014.

IEA. 2008. *Energy Efficiency Indicators for Public Electricity Production from Fossil Fuels.* International Energy Agency, Paris, France. URL = http://www.iea.org/papers/2008/En_Efficiency_Indicators.pdf, last accessed 9-14-2014.

IEA. 2011. *World Energy Outlook 2011.* International Energy Agency, Paris, France. URL = http://www.iea.org/ebc/feb2011/birol.pdf, last accessed 9-14-2014.

IEA. 2013. *Key World Energy Statistics.* International Energy Agency, Paris, France. URL = http://www.iea.org/publications/freepublications/publication/KeyWorld2013.pdf, last accessed 9-14-2014.

Inman, M. 2013. The true cost of fossil fuels. *Scientific American* April 2013:59–61.

IPCC. 2005. Carbon Capture and Storage: Summary for Policymakers. *Intergovernmental Panel on Climate Change.* URL = http://www.ipcc.ch/pdf/special-reports/srccs/srccs_wholereport.pdf, last accessed 9-14-2014.

Kerr, R.A. 2010. Natural gas from shale bursts onto the scene. *Science* 328:1624–1626.

Krupp, F. and M. Horn. 2009. *Earth: The Sequel.* W.W. Norton, New York, NY.

Longwell, H.J. 2002. The future of the oil and gas industry: Past approaches, new challenges. *World Energy* 5:100–104.

Lovins, A.B. 2004. Energy efficiency, taxonomic overview. *Encyclopedia of Energy* 2:382–401.

Nature. 2006. Special report: Chernobyl and the future. *Nature* 440:982–989.

Pacala, S. and R. Socolow. 2004. Stabilization wedges: Solving the climate problem for the next 50 years with current technologies. *Science* 13:968–972.

Perez, C., A. Velando, I. Munilla, et al. 2008. Monitoring polycyclic aromatic hydrocarbon pollution in the marine environment after the Prestige Oil Spill by means of seabird blood analysis. *Environmental Science and Technology* 42:707–713.

Ristinen, R.A. and J.J. Kraushaar. 2005. *Energy and the Environment* (2nd ed.). John Wiley and Sons, Hoboken, NJ.

Scientific Committee on emerging and Newly Identified Health Risks. 2007. *Possible Effects of Electromagnetic Fields (EMF) on Human Health.* European Commission, Health and Consumer Protection, Brussels, Belgium. URL = http://ec.europa.eu/health/ph_risk/committees/04_scenihr/docs/scenihr_o_007.pdf, last accessed 10-14-2014.

State of California. 2008. *2007 Preliminary Report of California Oil and Gas Production Statistics.* State of California Department of Conservation, Sacramento, CA.

U.S. EPA. 2007. *Energy Trends in Selected Manufacturing Sectors: Opportunities and Challenges for Environmentally Preferable Energy Outcomes.* U.S. Environmental Protection Agency, Washington, DC. URL = http://www.epa.gov/sectors/pdf/energy/report.pdf, last accessed 9-14-2014.

World Nuclear Association. 2011. *Comparison of Lifecycle Greenhouse Gas Emissions of Various Electricity Generation Sources.* World Nuclear Association, London, UK.

Chapter 15

AWEA. 2011. *Annual Wind Industry Report.* American Wind Energy Association, Washington, DC. URL = http://www.awea.org/learnabout/publications/reports/AWEA-US-Wind-Industry-Market-Reports.cfm, last accessed 9-14-2014.

BioFuels Journal. 2011. *Annual and monthly U.S. ethanol production.* URL = http://www.biofuelsjournal.com/info/bf_articles.html?type=ec&ID=25474, last accessed 9-14-2014.

BP. 2014. *BP Statistical Review of World Energy, June 2014.* BP, London, UK. URL = http://www.bp.com/content/dam/bp/pdf/Energy-economics/statistical-review-2014/BP-statistical-review-of-world-energy-2014-full-report.pdf, last accessed 9-14-2014.

Bradford, T. 2006. *Solar Revolution: The Economic Transformation of the Global Energy Industry.* MIT Press, Cambridge, MA.

Congressional Budget Office. 2009. *The Impact of Ethanol Use on Food Prices and Greenhouse-Gas Emissions.* Congressional Budget Office, Washington, DC. URL = http://www.cbo.gov/sites/default/files/cbofiles/ftpdocs/100xx/doc10057/04-08-ethanol.pdf, last accessed 9-14-2014.

Center for the New Energy Economy. 2013. *State Renewable Portfolio Standards Hold Steady or Expand in 2013 Session.* URL = http://www.aeltracker.org/graphics/uploads/2013-State-By-State-RPS-Analysis.pdf, last accessed 9-14-2014.

Chow, J. 2003. Energy resources and global development. *Science* 302:1528–1531.

EIA. 2011. *Annual Energy Review 2010.* U.S. Department of Energy, Energy Information Administration, Washington, DC. URL = http://205.254.135.24/totalenergy/data/annual/index.cfm, last accessed 9-14-2014.

Flavin, C. 2008. *Low-Carbon Energy: A Roadmap.* Worldwatch Report 178. Worldwatch Institute, Washington, DC.

Frankfurt School of Finance and Management. 2014. *Global Trends in Renewable Energy Investment 2014.* Frankfurt School-UNEP Centre/BNEF. URL = http://www.unep.org/pdf/Green_energy_2013-Key_findings.pdf, last accessed 9-14-2014.

Hawkins, T.R., B. Singh, G. Majeau-Bettez, et al. 2012. Comparative environmental life cycle assessment of conventional and electric vehicles. *Journal of Industrial Ecology* 17:53–64.

IEA. 2013a. *Technology Roadmap: Wind Energy.* International Energy Agency, Paris, France. URL = http://www.iea.org/publications/freepublications/publication/Wind_2013_Roadmap.pdf, last accessed 9-14-2014.

IEA. 2013b. *World Energy Outlook 2013.* International Energy Agency, Paris, France. URL = http://www.iea.org/Textbase/npsum/WEO2013SUM.pdf, last accessed 9-14-2014.

Kreith, F. 2014. *Principles of Sustainable Energy Systems.* CRC Press. Boca Raton, FL.

Krupp, F. and M. Horn. 2009. *Earth: The Sequel.* W.W. Norton, New York, NY.

Pacala, S. and R. Socolow. 2004. Stabilization wedges: Solving the climate problem for the next 50 years with current technologies. *Science* 13:968–972.

Pearce, F. 2006. Fuels gold: Are biofuels really the greenhouse-busting answer to our energy woes? *New Scientist* 23 September 2006:36–41.

Perlack, R.D. 2005. *Biomass as Feedstock for a Bioenergy and Bioproducts Industry: The Technical Feasibility of a Billion Ton Annual Supply.* U.S. Department of Energy, Office of Scientific and Technical Information, Washington, DC. URL = http://www.osti.gov/energycitations/product.biblio.jsp?osti_id=885984, last accessed 9-14-2014.

REN21. 2011. *Renewables Global Status Report.* Renewable Energy Policy Network for the 21st Century, Milan, Italy. URL = http://www.ren21.net/REN21Activities/Publications/GlobalStatusReport/tabid/5434/Default.aspx, last accessed 9-14-2014.

Ristinen, R.A. and J.J. Kraushaar. 2005. *Energy and the Environment* (2nd ed.). John Wiley and Sons, Hoboken, NJ.

Stoft, S. 1995. *The Economics of Conserved-Energy "Supply" Curves.* University of California Energy Institute Working Paper PWP-028, University of California Energy Institute, Berkeley, CA.

U.S. DOE. 2010. *2009 Renewable Energy Data Book.* U.S. Department of Energy, Washington, DC. URL = http://www1.eere.energy.gov/maps_data/pdfs/eere_databook.pdf, last accessed 9-14-2014.

Worldwatch Institute. 2007. *Biofuels for Transport: Global Potential and Implications for Sustainable Agriculture and Energy in the 21st century.* Worldwatch Institute, Washington, DC.

Chapter 16

Abbott, C. 2001. *Greater Portland: Urban Life and Landscape in the Pacific North West.* University of Pennsylvania Press, Philadelphia, PA.

Akbari, H., S. Menon, and A. Rosenfeld. 2009. Global cooling: increasing world-wide urban albedos to offset CO2. *Climatic Change* 95:275–286.

Andrews, C.J. 2008. Greenhouse gas emissions along the rural-urban gradient. *Journal of Environmental Planning and Management* 51:1–20.

Beatley, T. and K. Manning. 1997. *The Ecology of Place: Planning for Environment, Economy and Community.* Island Press, Washington, DC.

Benedict, M.A. and E.T. McMahon. 2006. *Green Infrastructure: Linking Landscapes and Communities.* Island Press, Washington, DC.

Berkowitz, A.R., C.H. Nilon, and K.S. Hollweg. 2003. *Understanding Urban Ecosystems: A New Frontier for Science and Education.* Springer-Verlag, New York, NY.

Chiquet, C., J.W. Dover, and P. Mitchell. 2013. Birds and the urban environment: the value of green walls. *Urban Ecosystems* 16:453–462.

Diamond, H.L. and P.F. Noonan. 1996. *Land Use in America.* Island Press, Washington, DC.

Duerksen, C. and C. Snyder. 2005. *Nature-Friendly Communities: Habitat Protection and Land Use Planning.* Island Press, Washington, DC.

Faeth, S.H., P.S. Warren, E. Shochat, et al. 2005. Trophic dynamics in urban communities. *BioScience* 55:399–407.

Farmer, M.C., M.C. Wallace, and M. Shiroya. 2013. Bird diversity indicates ecological value in urbanhome prices. *Urban Ecosystems* 16:131–144.

Grimm, N.B. and C.L. Redman. 2004. Approaches to the study of urban ecosystems: The case of central Arizona—Phoenix. *Urban Ecosystems* 7:199–213.

Grimm, N.B., S.H. Faeth, N.E. Golubiewski, et al. 2008. Global change and the ecology of cities. *Science* 319:756–760.

Grimm, N.B., D. Foster, P. Groffman, et al. 2008. The changing landscape: ecosystem responses to urbanization and pollution across climatic and societal gradients. *Frontiers in Ecology and the Environment* 6:264–272.

Hobbs, F. and N. Stoops. 2002. *Demographic Trends in the 20th Century.* Special Report CENSR-4, U.S. Census Bureau, Washington, DC.

Hopkins, R. 2008. *The Transition Handbook: From Oil Dependency to Local Resilience.* Chelsea Green Publishing, White River Junction, VT.

Kalnay, E. and M. Cai. 2003. Impact of urbanization and land-use change on climate. *Nature* 423: 528–531.

Kareiva, P., S. Watts, R. McDonald, et al. 2007. Domesticated nature: Shaping landscapes and ecosystems for human welfare. *Science* 316: 1866–1869.

Kaye, J.P., P.M. Groffman, N.B. Grimm, et al. 2006. A distinct urban biogeochemistry? *Trends in Ecology and Evolution* 21:192–199.

Mackun, P. and S. Wilson. 2011. *Population Distribution and Change 2000–2010.* Census Brief C2010BR-01, U.S. Census Bureau, Washington, DC.

Marzluff, J.M., E. Shulenberger, W. Endlicher, et al. 2008. *Urban Ecology: An International Perspective on the Interaction Between Humans and Nature.* Springer-Verlag, New York, NY.

McDonnell, M.J. and S.T.A. Pickett. 1990. Ecosystem structure and function along urban-rural gradients: An unexploited opportunity for ecology. *Ecology* 71:1232–1237.

McDonnell, M.J. and S.T.A. Pickett. 1993. *Humans as Components of Ecosystems: The Ecology of Subtle Human Effects and Populated Areas.* Springer-Verlag, New York, NY.

Newman, P. and I. Jennings. 2008. *Cities as Sustainable Ecosystems: Principles and Practices.* Island Press, Washington, DC.

Pickett, S.T.A., M.L. Cadenasso, J.M. Grove, et al. 2001. Urban ecological systems: Linking terrestrial ecological, physical and socioeconomic components of metropolitan areas. *Annual Review of Ecology and Systematics* 32:127–157.

Pickett, S.T.A., M.L. Cadenassob, J.M. Grovec, et al. 2011. Urban ecological systems: Scientific foundations and a decade of progress. *Journal of Environmental Management* 92:331–362.

Price, J. 2008. Remaking American environmentalism: On the banks of the L.A. River. *Environmental History* 13:536–555.

Saieg, P. 2013. Energy efficiency in the built environment. In: *State of the World 2013. Is Sustainability Still Possible?* The Worldwatch Institute. Island Press, Washington, DC.

United Nations, 2014. *World Urbanization Prospects: The 2014 Revision.* URL = http://esa.un.org/unpd/wup/index.htm, last accessed 9-14-2014.

U.S. Department of Energy. 2011. Energy efficiency and renewable energy. *Buildings Energy Data Book.* URL = http://buildingsdatabook.eren.doe.gov, last accessed 9-20-2014.

Wear, D.N., M.G. Turner, and R.J. Naiman. 1998. Land cover along an urban-rural gradient: implications for water quality. *Ecological Applications* 8:619–630.

The World Watch Institute. 2007. *State of the World: Our Urban Future.* W.W. Norton, New York, NY.

Chapter 17

Arsova, L. 2008. The state of garbage in America. *BioCycle* 49:22.

Blackman, W.C., Jr. 1993. *Basic Hazardous Waste Management.* Lewis Publishers, Boca Raton, FL.

Graedel, T.E., and B.R. Allenby. 2002. *Industrial Ecology* (2nd ed.). Prentice Hall, Upper Saddle River, NJ.

Kaplan, P.O., J. Decarolis, and S. Thorneloe. 2009. Is it better to burn or bury waste for clean electricity generation? *Environmental Science and Technology* 43:1711–1717.

Lilienfeld, R. and W. Rathje. 1998. *Use Less Stuff: Environmental Solutions for Who we Really Are.* Ballantine, New York, NY.

McDonough, W. and M. Braungart. 2002. *Cradle to Cradle: Remaking the Way We Make Things.* North Point Press, New York, NY.

NSWMA. 2004. *Modern Landfills: A Far Cry from the Past.* National Solid Wastes Management Association, Washington, DC.

Rathje, W. and C. Murphy. 2001. *Rubbish! The Archeology of Garbage.* University of Arizona Press, Tucson, AZ.

Sidique, S.F., F. Lupi, and S.V. Joshi. 2010. The effects of behavior and attitudes on drop-off recycling activities. *Resources, Conservation and Recycling* 54:163–170.

Trudel, R. and J.J. Argo. 2013. The effect of product size and form distortion on consumer recycling behavior. *Journal of Consumer Research*, doi:10.1086/671475.

U.S. Environmental Protection Agency. 2011a. *Electronics Waste Management in the United States Through 2009.* URL = http://www.epa.gov/waste/conserve/materials/ecycling/docs/fullbaselinereport2011.pdf, last accessed 9-4-2014.

U.S. Environmental Protection Agency. 2011b. *The National Biennial RCRA Hazardous Waste Report.* U.S. Environmental Protection Agency, Washington, DC. URL = http://www.epa.gov/osw/inforesources/data/br11/national11.pdf, last accessed 9-14-2014.

U.S. Environmental Protection Agency. 2012. *Municipal Solid Waste Generation, Recycling, and Disposal in the United States: Facts and Figures for 2012.* U.S. Environmental Protection. URL = http://www.epa.gov/osw/nonhaz/municipal/pubs/2012_msw_fs.pdf, last accessed 9-14-2014.

Saling, J.H. and A.W. Fentiman. 2002. *Radioactive Waste Management* (2nd ed.). Taylor and Francis, New York, NY.

Wagner, T.P. 2012. Examining the concept of convenient collection: An application to extended producer responsibility and product stewardship frameworks. *Waste Management* 33:499–507.

Walmart. 2013. *2013 Global Responsibility Report.* URL = http://corporate.walmart.com/global-responsibility/environment-sustainability/global-responsibility-report, last accessed 9-16-2014.

Weitz, K.A., S.A. Thorneloe, S.R. Nishtala, et al. 2002.The impact of municipal solid waste management on greenhouse gas emissions in the United States. *Journal of the Air and Waste Management Association* 52:1000–1011.

Chapter 18

Adams, P.F., W.K. Kirzinger, and M.E. Martinez. 2013. Summary health statistics for the U.S. population: National Health Interview Survey, 2012. *National Center for Health Statistics. Vital Health Statistics* 10(259):1–95.

Akinbami, L.J., J.E. Moorman, P.L. Garbe, et al. 2009. Status of childhood asthma in the United States, 1980–2007. *Pediatrics* 123:S131–145.

Ames, B.N., M. Profet, and L.W. Gold. 1990. Nature's chemicals and synthetic chemicals: Comparative toxicology. *Proceedings of the National Academy of Sciences* 87:7782–7786.

Boelee, E. and H. Madsen. 2006. *Irrigation and schistosomiasis in Africa: Ecological aspects.*

IWMI Research Report 99. International Water Management Institute, Colombo, Sri Lanka.

Boxall, A.B.A., A. Hardy, S. Beulke, et al. 2009. Impacts of climate change on indirect human exposure to pathogens and chemicals from agriculture. *Environmental Health Perspectives* 117:508–515.

Carson, R. 1962. *Silent Spring.* Houghton Mifflin, Boston.

Epstein, P.R. 2002. Climate change and infectious disease: Stormy weather ahead? *Epidemiology* 13:373–375.

Golanky, I.M. 2007. *Death and Death Rates due to Extreme Weather Events.* International Policy Network, London, UK.

Graham, J.D. and J.B. Wiener. 1995. *Risk vs. Risk: Tradeoffs in Protecting Health and the Environment.* Harvard University Press, Cambridge, MA.

Hamed, M.A. 2010. Strategic control of schistosome intermediate host. *Asian Journal of Epidemiology* 3:123–140.

Heron, M. 2013. Deaths: Leading causes for 2010. *National Vital Statistics Reports* 62 (6):1-97. URL = http://www.cdc.gov/nchs/data/nvsr/nvsr62/nvsr62_06.pdf, last accessed 10-14-2014.

IPCC. 2007. *Climate Change 2007: Synthesis Report.* Intergovernmental Panel on Climate Change. URL = http://www.ipcc.ch/publications_and_data/publications_ipcc_fourth_assessment_report_synthesis_report.htm, last accessed 9-14-2014.

Millennium Ecosystem Assessment. 2005. *Ecosystems and Human Well-Being.* Island Press, Washington, DC. URL = http://www.maweb.org/documents/document.356.aspx.pdf, last accessed 9-14-2014.

Moeller, D. 2004. *Environmental Health* (3rd ed.). Harvard University Press, Cambridge, MA.

NIH. 2008. *National Toxicology Program Monograph on the Potential Human Reproductive and Developmental Effects of Bisphenol A.* National Institutes of Health Publication 08-5994, National Institutes of Health, Bethesda, MD.

Patz, J.A., H.K. Gibbs, J.A. Foley, et al. 2007. Climate change and global health: Quantifying a growing ethical crisis. *EcoHealth* 4:397–405.

Pimentel, D., M. Tort, L. D'Anna, et al. 1998. Ecology of increasing disease. *BioScience* 48:817–826.

Roberts, D.R., S. Manguin, and J. Mouchet. 2000. DDT house spraying and re-emerging malaria. *Lancet* 356:330–332.

Rubin, B.S., M.K. Murray, D.A. Damassa, et al. 2001. Prenatal exposure to low doses of Bisphenol A affects body weight, patterns of estrous cyclicity and plasma LH levels. *Environmental Health Perspectives* 7:675–680.

Soverow, J.E., G.A. Wellenius, D.N. Fisman, et al. 2009. Infectious disease in a warming world: How weather influenced West Nile virus in the United States (2001–2005). *Environmental Health Perspectives* 117:1049–1052.

Stern, P.C. and H.V. Fineberg. 1996. *Understanding Risk: Informing Decisions in a Democratic Society.* National Academy Press, Washington, DC.

Stockholm Convention. 2014. Stockholm Convention on Persistent Organic Pollutants. 2014. URL = http://chm.pops.int/default.aspx, last accessed 9-14-2014.

Sunstein, C.R. 2002. *Risk and Reason: Safety, Law and the Environment.* Cambridge University Press, Cambridge, UK.

Swan, S.H., E.P. Elkin, and L. Fenster. 2000. The question of declining sperm density revisited: An analysis of 101 studies published 1934-1996. *Environmental Health Perspectives* 108: 961–966.

United Nations. 2009. *World Urbanization Prospects, the 2009 Revision.* URL = http://esa.un.org/unpd/wup/index.htm, last accessed 9-14-2014.

United Nations. 2010. *World Population Prospects, the 2010 Revision.* URL = http://esa.un.org/wpp/Other-Information/faq.htm, last accessed 9-14-2014.

Utzinger, J., Y. Tozan, and B.H. Singer. 2001. Efficacy and cost effectiveness of environmental management for malaria control. *Tropical Medicine and International Health* 6:677–687.

Utzinger, J., Y. Tozan, F. Doumani, et al. 2002. The economic payoffs of integrated malaria control in the Zambian copperbelt between 1930 and 1950. *Tropical Medicine and International Health* 7:657–677.

Wadia, P.R., N.J. Cabaton, M.D. Borrero, et al. 2013. Low-dose BPA exposure alters the mesenchymal and epithelial transcriptomes of the mouse fetal mammary gland. *PLoS ONE* 8(5):e63902, doi:10.1371.

WHO. 2009a. *Dengue and Dengue Haemorrhagic Fever.* Fact Sheet No. 117. World Health Organization, Geneva Switzerland. URL = http://www.who.int/mediacentre/factsheets/fs117/en/. Last accessed 9-14-2014.

WHO. 2009b. *Global Health Risks: Mortality and Burden of Disease Attributable to Selected Major Risks.* World Health Organization, Geneva, Switzerland.

WHO. 2013. *World Health Statistics 2013.* World Health Organization, Geneva, Switzerland. URL = http://www.who.int/whosis/whostat/EN_WHS2013_Full.pdf, last accessed 9-14-2014.

World Bank. 2005. *Report of the Scientific Working Group Meeting on Schistosomiasis.* World Bank, Washington, DC. URL = http://apps.who.int/tdr/publications/tdr-research-publications/swg-report-schistosomiasis/pdf/swg_schisto.pdf, last accessed 9-14-2014.

Chapter 19

Brown, L.R. 2011. *World on the Edge: How to Prevent Environmental and Economic Collapse.* W.W. Norton & Co., New York, NY.

National Wildlife Federation. 2014. Campus Ecology. http://www.nwf.org/Global-Warming/Campus-Solutions.aspx, last accessed 9-14-2014.

Carlson, S. 2006. In search of the sustainable campus: With eyes on the future, universities try to clean up their acts. *Chronicle of Higher Education* 53:A10.

Chapin F.S., III, M.S. Torn, and M. Tateno. 1996. Principles of ecosystem sustainability. *The American Midland Naturalist* 148:1016–1037.

Dasgupta, P. 2007. The Idea of Sustainable Development. *Sustainability Science* 2:5–11.

Easterlin, R.A., L.A. McVey, M. Switek, et al. 2010. The happiness-income paradox revisited. *Proceedings of the National Academy of Sciences* 107:22463–22468.

Gordon, J.C. and J.K. Berry. 2006. *Environmental Leadership Equals Essential Leadership: Redefining Who Leads and How.* Yale University Press, New Haven.

Helliwell, J.F., R. Layard, and J.D. Sachs. 2013. *World Happiness Report 2013.* UN Sustainable Development Solutions Network, New York.

Kahneman, D., A.B. Krueger, D. Schkade, et al. 2006. Would you be happier if you were richer? A focusing illusion. *Science* 312:1908–1910.

Kates, R.W. and T.M. Parris. 2003. Long-term trends and a sustainability transition. *Proceedings of the National Academy of Sciences* 100:8062–8067.

Millennium Ecosystem Assessment. 2005. *Ecosystems and Human Well-Being.* Island Press, Washington, DC. URL = http://www.maweb.org/documents/document.356.aspx.pdf, last accessed 9-14-2014.

National Research Council, Board on Sustainable Development. 1999. *Our Common Journey: A Transition toward Sustainability.* National Academy Press, Washington, DC.

Ostrom, E. 2009. A general framework for analyzing sustainability of social-ecological systems. *Science* 325:419–422.

Speth, J.G. 2004. *Red Sky at Morning: America and the Crisis of the Global Environment.* Yale University Press, New Haven, CT.

United Nations. 1987. *Report of the World Commission on Environment and Development: Our Common Future.* URL = http://www.un-documents.net/wced-ocf.htm, last accessed 9-14-2014.

United Nations. 2010. *World Population Prospects, the 2010 Revision.* URL = http://esa.un.org/wpp/Other-Information/faq.htm, last accessed 9-14-2014.

University Leaders for a Sustainable Future. 1990. The Talloires Declaration. URL = http://www.ulsf.org/pdf/TD.pdf, last accessed 10-14-2014.

A

A horizon
A layer of soil containing a mixture of organic matter and mineral particles.

abiotic
Elements in the ecosystem that are nonliving.

acid deposition
The dispersion of acid-containing gases, aerosols, and rain onto soils, plants, buildings, and bodies of water.

active solar technology
The use of mechanical devices to heat water and buildings or to power electrical devices to generate electricity.

acute exposure
Contact with a toxin or pathogen that occurs only once or for only a short time.

adaptation
Inherited structures, functions, or behaviors that help organisms survive and reproduce.

adaptive radiation
The process by which several species evolve from a single ancestor to occupy new ecological niches.

aerosols
Very small solid and liquid particles suspended in the air.

age-specific birth rate
The number of children born in a year per 1,000 women within defined age groups.

age-specific fertility rate
The number of offspring produced by females during a particular age or range of ages.

age-structure pyramid
A graphical illustration of how populations are apportioned according to age and gender.

agriculture
The system of land management used to grow domesticated plants and animals for food, fiber, or energy.

agroecosystem
Crops and domestic animals, the physical environments in which they grow, and the communities of other organisms associated with them.

agronomy
The science that applies knowledge from other fields of study, such as genetics, physiology, chemistry, and ecology, to agriculture.

air pollution
Gases or particles that are present in high enough concentrations to harm humans, other organisms, or structures, such as buildings or pieces of art.

air quality
The amount of gases and small particles in the atmosphere that influences ecosystems or human well-being.

air quality index (AQI)
A measure of air health used by the U.S. EPA that is based on the abundance of seven pollutants.

allergens
Substances that stimulate a range of responses from the body's immune system.

allocation
In forest management, decisions regarding the location and amounts of different forest uses.

amperes
A measure of electric current; the flow of electrically charged particles, usually electrons flowing through a wire.

anaerobic respiration
The process in which carbohydrates are partially broken down in the absence of oxygen to produce carbon dioxide and smaller carbohydrate molecules.

angiosperm
Plants whose seeds are borne in flowers.

animism
The belief that living and nonliving objects possess a spirit or a soul.

anthracite coal
With the highest energy content, this coal has a carbon content of 86–97%.

anthropocene
A term used by some scientists to designate the current geologic epoch in which human impacts on Earth's environment are pervasive.

anthropocentric ethics
The assignation of intrinsic value only to humans; it defines right actions in terms of outcomes for human beings.

anthroponosis
An infectious disease that occurs exclusively in humans.

aphotic zone
A division of the pelagic zone in lake and ocean waters that receive little or no sunlight.

aquatic
Related to freshwater environments.

aquifer
A layer of soil or rock that is saturated with groundwater.

arithmetic growth
Where the number of new individuals added at each generation is constant.

artesian well
A well drilled into a confined aquifer where the groundwater is under sufficient pressure to cause water to rise above the confining layer.

asexual reproduction
Simple cell division that produces offspring genetically identical to their single parent.

asphyxiants
Chemicals that deprive tissues of oxygen and cause suffocation.

atmosphere
The layer of gases above a planet's surface.

atmospheric pressure
The force caused by the pull of gravity on a column of air.

atom
The most basic subunit of an element; supplies an element's physical and chemical properties.

B

B horizon
The subsoil below the A horizon that is especially rich in clay particles that form from minerals dissolved from the weathering of sand and silt.

baby boom
The period immediately following World War II in which the birth rate increased sharply.

baseline water stress
A method of measuring the availability of fresh water relative to its demand. Calculated as total annual water withdrawals (municipal, industrial, and agricultural) expressed as a percent of the total annual available flow in a region. Higher values indicate more competition among users.

base load
The minimum amount of electricity that a utility must continuously provide to meet consumer needs.

benthic zone
The water and sediment on the lake or ocean bottom.

bias
Refers to estimates of a parameter that are skewed in some particular direction.

bioaccumulation
Refers to the process by which chemicals are stored in the tissues of living organisms.

bioassessment
Surveys of aquatic biodiversity that are known to be sensitive to overall water quality.

biocapacity
A measure of the area and quality of land available to supply a population with resources.

biocentric ethics
A system of values based on the belief that the value of other living things is equal to the value of humans. It extends intrinsic value to *individual* organisms beyond human beings; organisms do not need to benefit humans in order to have value.

biodiesel
A biofuel derived from plant oils and animal fats that can be burned in diesel engines.

biodiversity
The variety of life in all its forms and combinations and at all levels of organization.

biodiversity hotspots
Places with high biodiversity and a large number of endemic species where human threats to habitats are high.

bioenergy
See biomass energy.

biofuels
Fuels derived directly from plant materials, such as corn or wood.

biogeochemical cycle
The flow of matter through an ecosystem.

biological pest control
The use of predators and parasites to manage pests.

biomagnification
The process by which the concentration of a toxin increases at each higher level of a food web.

biomass energy
The energy contained in living and dead organic matter.

biome
Communities of similar organisms found in a particular climatic zone.

biomedical waste
Hazardous waste that can spread infections or disease, or waste that may decompose and produce toxins and noxious odors.

bioreactor landfill
Landfills that are managed to generate energy and maximize the recovery of valuable materials.

bioremediation
A process in which living organisms—microorganisms, plants, or fungi—are used to break down waste into less harmful components.

biosphere
All of the organisms on Earth and the nonliving environment with which they interact.

biostorage
The absorption and storage of CO_2 in plants and animals in Earth's ecosystems.

biota
Elements in the ecosystem that are living organisms.

birth rate
The number of births in the population per unit of time expressed as a percentage of the population size.

bituminous coal
Denser than either sub-bituminous coal or lignite, it has a carbon content of 45–86%.

black carbon
Solid carbon particles released during forest fires and the burning of fuelwood and charcoal that absorb light and warm the atmosphere; believed to be an important factor contributing to the worldwide loss of snow and ice.

body burden
The total amount of a chemical present in the tissues of an organism.

bog
Wetlands with peat deposits that support a variety of evergreen trees and shrubs; primary source of water is rainfall.

boreal forest
Coniferous forest in biomes where the growing season is less than 4 months long and winters are long, dry, and bitterly cold.

breeding barrier
An obstacle—geographic, temporal, behavioral, or structural—that prevents one species from breeding with another.

broad-spectrum pesticide
A pesticide that kills a wide variety of organisms.

C

C horizon
The lowest layer of soil or weathering zone.

calorie
The energy required to raise the temperature of 1 g of water by 1 °C (1 cal = 4.18 J).

cap and trade
A market-based policy in which a regulatory limit, or cap, is set on an action; the rights to exceed that limit are traded in markets.

capillary water
Water that is held in the soil by water-to-water hydrogen bonds.

capital
The total amount (mass) of an element or molecule in a pool. *See* pool.

carbohydrate
Organic molecules made of carbon, hydrogen, and oxygen.

carbon capture and storage (CCS)
Technologies that capture CO_2 produced by burning fossil fuels before it escapes to the atmosphere and then transfers it into a form that can be permanently stored underground.

carbon dioxide equivalents (CO_2e)
An expression of total greenhouse gas emissions, taking into account the global warming potential of each gas relative to CO_2.

carcinogens
Chemicals that cause or promote cancer.

carrying capacity
The population size at which a population uses resources equal to the rate at which those resources are supplied, the birth rate equals the death rate, and the population growth rate becomes zero.

case law
The various decisions made by the individual courts collectively that establish precedents and influence future court decisions.

catalyst
Substance that promotes chemical reactions without being consumed in the reaction.

catalytic converters
Part of an automobile exhaust system that causes conversion of harmful gases to less harmful products. Required on all cars manufactured after 1975 to reduce emissions of carbon monoxide, hydrocarbons, and nitrogen oxides.

cellular respiration
The process by which energy in carbohydrate molecules is retrieved and used to carry out cell functions and facilitate growth.

cellulose
Polysaccharides composed of hundreds of glucose molecules that are bonded in a way that prevents digestion by most animals.

cellulosic ethanol
Ethanol that is produced from agricultural residues, wood, grasses, or other cellulose-rich feedstocks.

channel
The waterway through which a stream normally flows.

chaparral
Biome of evergreen shrublands and low woodlands; occurs where winters are mild and moist and summers are hot and dry.

chemical energy
The potential energy associated with the formation or breakage of bonds between atoms.

chemosynthesis
A process by which energy is derived from inorganic chemical reactions.

chronic exposure
Contact with a toxin or pathogen over periods of months or even years.

clay
Smallest particles of minerals (diameters less than 0.002 mm) that are highly weathered, with unique physical and chemical properties.

clear-cutting
The complete removal of forest canopy trees; the most controversial harvest practice.

climate
The atmospheric conditions such as temperature, humidity, and rainfall that exist over large regions and relatively long periods of time.

climate change
Refers to long-term variations in climatic variables such as temperature and precipitation.

climatograph
A graph that shows the pattern of seasonal changes in temperature and precipitation for a particular location.

climax community
Assemblages of plants and animals that are able to perpetuate themselves generation after generation.

closed basin lake
Lakes with no outlet stream.

closed watershed
Inland basins that do not drain to the sea.

closed-loop recycling
The material in a product is reused to produce a similar product; for example, aluminum cans are recycled to produce more aluminum cans.

coal seams
Layers of sediment that contain coal.

coevolution
Evolutionary change in which interactions between species select for adaptations in each species; for instance, plants evolve chemical defenses to thwart herbivores and herbivores become tolerant or even take advantage of plant chemical defenses.

combustion by-products
Due to the incomplete combustion of organic carbon, many gases and particles are released into the air. These by-products include carbon monoxide, nitrogen and sulfur oxides, hundreds of volatile organic compounds (VOCs), and particulate matter.

commensalism
Refers to associations that benefit only one species, leaving the other species unaffected.

committed warming
Warming that would occur at a rate of about 0.1 °C per decade even if we were to stop all greenhouse gas emissions immediately.

commodity
A good or service bought and sold in an economic market.

compact fluorescent lights (CFLs)
Lights in which an electric current causes a gas to glow; CFLs emit three to five times more light per kilowatt-hour than do incandescent fixtures.

competitive exclusion principle
A postulate that two species that directly compete for essential resources cannot coexist; one species will eventually displace the other.

complementarity effect
Species within a community utilize resources at different times or from different locations; groups of species are thus able to use resources more efficiently than would any single species.

compost
Decayed organic matter that can be used to improve the fertility and water-holding capacity of soil, thereby improving plant growth.

compound
Molecules that are made of more than one element.

concentrating solar power (CSP) systems
A method of generating electricity from sunlight that relies on mirrors to concentrate the sun's rays on

a tower or a series of pipes that hold water or another fluid. The concentrated sunlight heats the fluid to boiling, producing steam. The steam is then used to spin a turbine and generate electricity.

conduction
The direct transfer of heat by means of the collisions of molecules.

confined aquifer
Groundwater trapped between layers of comparatively impermeable rock or sediment.

conifer
A classification of trees and shrubs whose seeds are borne in cones.

consequence-based ethics
A system of beliefs that emphasizes the importance of *outcomes*. Right and wrong are defined in terms of pleasure or pain, benefit or harm, and satisfaction or dissatisfaction.

conservation supply curve
Graph used to compare the amount of energy saved by each action to the cost of that energy.

conservationist
The opinion that public resources should be used and managed in a sustainable fashion to provide the greatest benefit to the greatest number of people.

consumer
In reference to food webs, a species that feeds on other live organisms.

consumption
In reference to a primary energy source, the amount that is actually used during a particular time.

consumptive use
Water use by humans in which the water is not returned to streams or aquifers.

continental climate
The climate conditions in which daily and seasonal temperatures vary widely at locations within the interior of continents, away from large bodies of water.

contingent valuation
The estimates of the value of an ecosystem service gained by surveying people's willingness to pay for that service.

contour farming
A soil-conserving agricultural technique used in hilly terrain in which plowing occurs along the contour of the land, rather than up and down hills.

control group
Refers to those subjects in an experiment that are not subjected to a treatment.

convection
Warm regions in a gas or liquid become less dense and rise, causing the gas or liquid to circulate.

conventional international laws
Laws established by formal, legally binding conventions or treaties among countries.

conventional reservoir
Quantities of oil and natural gas found in porous sediments and contained by a nonporous cap rock.

convergent boundary
The point at which tectonic plates collide.

convergent evolution
The process by which natural selection favors similar features among otherwise unrelated species.

core
The center of Earth; it is composed of a mixture of nickel and iron, with smaller amounts of other heavy elements.

Coriolis effect
In reference to the atmosphere, the apparent change in wind direction due to Earth's rotation. Air moving over the ground in a straight path toward the equator appears to be deflected from east to west. Air moving toward the poles appears to be deflected west to east.

correlations
Quantitative relationships between different variables.

corrosive toxins
Chemicals that react with and directly destroy tissue; the effect is usually immediate.

cost of saved energy (CSE)
A measure that compares the costs and benefits of energy alternatives by dividing the cost of an energy investment by the energy savings resulting from that investment.

covalent bond
A strong bond formed when atoms share electrons.

cover crops
Crops such as winter wheat or rye that are planted to hold soils that would otherwise lie exposed to the weather between seasons, reducing the risk of erosion.

crust
A thin veneer of solid and relatively light rocks above Earth's mantle.

cultural services
The spiritual and recreational benefits that ecosystems provide.

customary international laws
Accepted norms of behavior or rules that countries follow as a matter of a long-standing precedent.

cycling time
The average time that it takes an element or molecule to make its way through an entire biogeochemical cycle.

cytoplasm
A plasma membrane that regulates materials flowing in and out of the cell.

D

dead zone
Regions of ocean water near the mouths of rivers with very low oxygen in which few marine animals can survive; generally a consequence of eutrophication.

death rate
The number of individuals dying per unit of time expressed as a percentage of the population size.

debt-for-nature swaps
International arrangements in which poor nations are forgiven a portion of their debts in exchange for the establishment of permanent parks and preserves.

decomposer
An organism that feeds on nonliving organic matter.

deep ecology movement
The view that all elements of the environment have equal intrinsic value and deserve moral respect in their own right.

deforestation
The process by which humans cut a forest and replace it with a non-forest use, such as agriculture, pasture, or pavement.

demographer
One who studies factors influencing the size, density, and distribution of human populations.

demographic transition model
Describes the four stages of change in birth rates, death rates, and growth rates of human populations associated with economic development.

denitrification
Nitrogen in soil and water is returned to the atmosphere; specialized bacteria convert NO_3^- to N_2 gas. The total nitrogen in Earth's organisms, soils, and waters is determined by the balance between nitrogen fixation and denitrification.

deoxyribonucleic acid (DNA)
The hereditary material that is passed from generation to generation in all organisms.

desalination
The process of removing salts and other chemicals from seawater.

desert biome
Land with a sustained and significant moisture deficit.

design for disassembly (DfD)
The development of materials, components, construction techniques, and management systems that permit the easy recovery of parts and materials from discarded products.

detritus
Organic matter such as dead leaves and nutrient-rich soil that wash in from adjacent terrestrial ecosystems.

dew point
The warmest temperature at which the relative humidity of a mass of air is 100%.

directional selection
Natural selection that favors the survival of individual organisms at one extreme end of the range of variation within a population.

discharge zone
Groundwater flowing to the surface from places where a confined aquifer meets the soil surface.

discount rate
The rate at which economic value declines with time.

disruptive selection
Situations in which individuals with genetic traits that fall at the extreme ends of the range are more likely to survive than individuals with average traits.

distillation
A method of desalination by boiling water and condensing the steam.

divergent boundary
Separate tectonic plates that are moving apart; boundaries coincide with ocean ridges.

dose–response curve
The relationship between increasing dosages of a chemical and some measure of health.

doubling time
The length of time required for a population to double in size.

drip irrigation
A system that provides water to plants through pipes with small openings at the base of each plant that feed water to the root zone.

dry deposition
Nonliquid particles are removed from the atmosphere by gravity.

duty-based ethics
A set of beliefs that the rightness or wrongness of actions should be determined by a set of rules or laws.

dynamic homeostasis
The process by which systems adjust to changes in ways that minimize how much features or processes vary from their normal values.

E

earthquake intensity
A measure of the severity of shaking at a given location; it is determined by its effects on people, structures, and the natural environment.

ecocentric ethics
A system of values based on the belief that *collections* of organisms or critical features in the environment have intrinsic value.

ecofeminism
The view that exploitation and abuse of both the environment and women derive from male domination.

ecological community
Species that interact within a specific area.

ecological disturbance
Events such as fires, hurricanes, or the logging of forests that may result in the loss of many or all of the species in an ecological community.

ecological footprint
The land area required to produce the ecosystem goods and services for entire human populations or for the average individual in such populations.

ecological legacy
Environmental features, such as soil and woody debris, that are left behind following a disturbance.

ecological niche
The various activities that define that organism's role in an ecosystem.

ecological valuation
Determination of the ecological value of the potential cost of the loss or degradation of an ecosystem service.

ecology
The branch of environmental science that focuses on the abundance and distribution of organisms in relation to their environment.

economic energy intensity (EEI)
A measure of overall energy use efficiency by a country or group of countries that considers the amount of energy used per dollar of economic output.

economic system
The institutions and interactions in a society that influence the production, distribution, and consumption of goods and services.

economic value
A commodity's value determined by the price that consumers are willing to pay for it.

ecosystem
All of the organisms and their physical and chemical environment within a specific area.

ecosystem functions
The flow of matter and energy and the processes influencing the distribution and abundance of organisms.

ecosystem integrity
The web of interactions that regulate ecosystem functions and ecosystem services.

ecosystem services
The multitude of resources and processes that ecosystems supply to humans.

ecotourism
A form of tourism in which visitors come to see the natural beauty and biodiversity of an area in a sustainable fashion that supports local economies.

El Niño/Southern Oscillation (ENSO)
Weather conditions that vary as a consequence of changes in temperature of surface waters in the tropical regions of the Pacific Ocean.

electric generator
A system of spinning magnets that generates an electrical field.

electric power grid
A network of transmission lines and transformers that distributes electricity from generating facilities to end users.

electromagnetic field
An area of electromagnetic force created by spinning magnets in a generator; used to provide the voltage that causes an electric current to flow.

electromagnetic radiation
The energy of light.

electromagnetic spectrum
The full range of electromagnetic radiation wavelengths; the energy of photons decreases from short wavelength gamma rays to long wavelength radio waves.

electron
Negatively charged particles surrounding the nucleus of an atom.

element
A chemical that cannot be broken down or separated into other chemicals.

embodied energy
The sum total of energy used in the extraction, harvesting, manufacturing, and delivery of construction materials.

emigration rate
The number of organisms moving out of an area calculated as a percentage of population size.

end use
The final application or use of a primary energy source.

endemic
Species that are found only in specific environments or in restricted localities; they are the consequence of evolution that has produced very specialized ecological niches.

endocrine disrupters
Substances that interfere with the hormones that control growth and development, egg and sperm production, and reproductive function.

energy
The capacity to do work.

energy budget
A system of accounting that measures all the energy entering and leaving an ecosystem or the entire Earth.

energy conversion efficiency
The percentage of energy in one form that is captured when it is converted to another form.

energy conversion
Process of changing primary energy into other forms of energy.

energy end-use efficiency
The product of the efficiencies of all the energy conversions from the primary source to the end use.

energy flow
The transfer and transformation of high-energy organic molecules within ecological communities.

energy return on investment (EROI)
Measure of the useful energy produced from an oil well, natural gas field, or coal mine divided by the amount of energy it took to obtain it.

entropy
Describes the disorder or disorganization in a system.

environment
All of the physical, chemical, and biological factors and processes that influence the growth and survival of organisms.

environmental ethics
A system of beliefs based on the moral relationship of humans to the environment and its nonhuman contents. Systems of environmental ethics vary regarding who or what has value.

environmental justice
The principle that, in the management of natural resources and the environment, people should be treated fairly regardless of race, gender, or economic status.

environmental policy
The decisions and actions that influence environmental conditions or processes.

environmental science
The study of all aspects of the environment, including physical, chemical, and biological factors.

environmental resistance
Factors that slow population growth like limited resources or the accumulation of toxins.

enzyme
Protein that serves as a catalyst.

equilibrium
The condition when competing influences in a system are balanced. For example, when inputs of an element equal outputs from a pool, the capital of the element in that pool remains constant, and the pool is said to be in equilibrium.

equilibrium theory of island biogeography
States that the diversity of species on an island is largely determined by the relationship between two factors: the rate at which new species migrate to the island and the rate at which species disappear from the island or become extinct.

establishment stage
The stage of forest development in which tree seedlings germinate on grounds recently cleared.

estuary
Region of transition from freshwater to marine ecosystems resulting in a mix of fresh water and saltwater.

eukaryote
Organisms whose cells contain an organized nucleus and membrane-bound organelles. They range from single-celled algae and protozoa to plants, fungi, and animals that are made of millions of cells.

eutrophication
High concentrations of rapidly growing algae in streams and lakes caused by elevated amounts of nutrients like nitrogen. Abundant organic matter from the algae encourages high rates of respiration by decomposing organisms.

evapotranspiration
The process in which water evaporates to the atmosphere from soils, plant leaves, and water bodies.

even-aged management
Use of clear-cutting to produce trees of uniform age.

e-waste
Electronic devices disposed of by businesses and households.

experiment
Scientific investigation in which groups of subjects are subjected to a specific treatment and compared to a control group of untreated subjects.

experimentation
Scientific investigation using experiments with treatments and controls.

exploitation competition
In the competition for resources, the ability to take up or use resources more efficiently.

exponential growth
The number of new individuals added to a population in each generation as a multiple of the number present in the previous generation.

extended product responsibility (EPR)
The belief that those who make a product should assume responsibility for that product throughout its life cycle.

externality
The costs and benefits associated with the production of a commodity that affect people other than buyers and sellers.

exurb
Areas where patches of commercial and residential development are interspersed in a matrix of rural land.

F

facilitation
Species alter the environment around them in ways that make it more habitable for other species. It is especially important in the early stages of succession.

fen
Wetlands that are fed primarily by groundwater.

Ferrel cell
Each of two atmospheric convection cells located between latitudes of 30 and 60 degrees.

fertility rate
The number of offspring produced by an average female in a population.

fertility transition
The third stage of the demographic transition in which continued economic development produces social and cultural changes that lead to lower birth rates.

field capacity
The amount of water that a given volume or weight of soil can hold against the force of gravity.

first law of thermodynamics
Energy can be transformed from one form to another, but it can be neither created nor destroyed; also called the law of energy conservation.

fitness
A measure of an organism's reproductive success.

floodplain
Land along rivers that experiences periodic flooding.

flux
The rate at which matter moves from one pool to another. *See* pool.

fly ash
Solid material left when coal is burned.

food chain
Feeding relationships among organisms are often depicted as simplified food webs. In a food chain, one species or trophic level is eaten by another, which is eaten by yet another, and so on.

food security
Physical, social, and economic access to sufficient, safe, and nutritious food to meet dietary needs and food preferences needed for an active and healthy life (as defined by the United Nations' Food and Agriculture Organization).

food web
The feeding interactions among species in a community.

forest degradation
Changes in forest structure and composition diminish biodiversity or key ecosystem services.

free market economy
An economy in which the government does not influence the marketplace with subsidies, taxation, or regulation.

fuel cells
Devices that convert chemical energy of a fuel into electricity through a reaction with oxygen. Unlike batteries, they require continuous input of fuel.

fuel recycling
Process in which nuclear fuel is reprocessed and the elements that can undergo fission chain reactions are reconcentrated.

fundamental niche
The complete range of environmental conditions, such as requirements for temperature, food, and water, over which a species might possibly exist.

G

gamete
Reproductive cell with half the genetic complement of the parent that fuses with another to form a zygote in sexual reproduction.

gasoline-electric hybrid automobile
Vehicle equipped with both an internal combustion engine and an electric motor powered by batteries.

gene
Segment of DNA that carries the code used to synthesize, or build, specific proteins.

general circulation model (GCM)
Computer models that use mathematical equations to simulate the physical processes that determine Earth's climate and energy budget.

generation time
The average difference in age between mothers and their offspring.

genetic diversity
The genetic variation among individuals within a population.

genetic drift
A change in the frequency of an inherited trait in a population that is brought about by chance events.

genetically modified organism (GMO)
An organism whose DNA has been altered by the insertion of a gene from a different species into the organism's cells.

genome
An organism's complete set of DNA.

genuine progress indicator (GPI)
An alternative measure of national wealth using GDP plus or minus the economic value of enhancements or degradations to the environment and to human communities.

genus
The taxonomic term used for a group of closely related organisms. Each genus comprises a number of species. *See* species.

geologic nuclear repository
A layer of rock in which nuclear waste and all of its toxic and radioactive elements are permanently sealed from the biosphere.

geothermal energy
Heat energy that is generated and stored in Earth itself.

global warming
Refers specifically to the increase in atmospheric temperature since the late 1800s and the Industrial Revolution.

global warming potential (GWP)
A measure of an individual molecule's long-term impact on atmospheric temperature relative to a molecule of CO_2.

gradient
Slope of the land that determines the speed of a stream as it flows downhill.

gravitational water
Water that flows through the soil as it percolates into the soil, drawn downward by the force of gravity.

graywater
The water from washers, sinks, and baths.

Green Revolution
The period since about 1950 during which global agricultural productivity increased many times over.

greenhouse effect
The absorption of radiation by greenhouse gases and trapping of that heat in the troposphere.

greenhouse gases
Gases that efficiently capture heat in the troposphere: water vapor and carbon dioxide, methane, nitrous oxide and ozone.

gross domestic product (GDP)
Determination of the total value of goods and services produced by the citizens of a country.

gross primary production (GPP)
The total amount of CO_2 that photosynthetic organisms convert to organic carbon each year.

ground-source heat pumps (GSHPs)
Method of using Earth's heat for energy by taking advantage of the fact that the upper 10 ft of Earth's crust maintains a nearly constant temperature of 10–15 °C (50–60 °F), regardless of the time of year.

groundwater
Liquid water that has percolated through the soil and into underlying rocks.

gyre
A pattern of circulation in surface ocean waters that is clockwise in the Northern Hemisphere and counterclockwise in the Southern Hemisphere.

H

Haber–Bosch process
A nonbiological method of nitrogen fixation adapted to the production of chemical fertilizers.

habitat
The complex environment in which an organism is found and upon which it depends for survival.

Hadley cell
Each of two atmospheric convection cells on either side of the equator.

half-life
The length of time that it takes for half of a collection of atoms to undergo a chemical transformation. For example, the length of time required for half of the molecules of a radioisotope to decay.

harvest
The extraction of goods or services from a forest or cropland.

harvest index
The fraction of total production that can be used by people.

hazardous waste
Waste that is flammable, corrosive, explosive, or toxic.

heat
Energy represented in the random motion of molecules.

heavy metal
High molecular weight metals such as mercury and lead; these are often toxic.

hedonic valuation
Economic value that is determined by the difference in the market price of real estate that is affected by different environmental alternatives.

herbivore
Species that eat plants.

high-grading
In forest harvest, the removal of a particular species of tree for fuel and construction.

high-level radioactive waste
Waste generated by nuclear power plants and weapons labs that requires long-term storage and management.

Holocene epoch
A name for the last 10,000 years of the Pleistocene epoch when Earth was in a warm interglacial period.

human well-being
A multifaceted concept that includes life's basic necessities, such as food and shelter, as well as good health, social stability, and personal freedom.

humus
Leaves or crop residue and the organic products of their decomposition.

hydraulic fracturing
Process to extract oil using high-pressure liquids to fracture the rocks; also called hydraulic fracking.

hydrocarbon
Organic molecules composed entirely of carbon and hydrogen atoms.

hydroelectric power
Electricity generated by using the kinetic energy of moving water to turn a turbine.

hydrogen bond
The bond formed by the attraction between the oxygen atom in one water molecule and a hydrogen atom in another water molecule.

hydrologic cycle
The distribution and flux of water through Earth's biogeochemical system.

hydrologist
Scientist who studies the movement of Earth's water.

hydrophyte
Species of plants that have adaptations that allow them to grow in anaerobic wetland soils.

hydropower
The energy of water moving under the force of gravity.

hygroscopic water
Water that is bound to soil particles and cannot be easily used by plants.

hyporheic zone
The region of saturated sediment beneath a stream and its adjacent riparian zone.

hypothesis
A proposition that can be tested by careful observation or experimentation.

I

igneous rock
Rocks that form as magma solidifies.

immigration rate
The number of individuals entering an area per unit of time as a percentage of population size.

impervious surface cover (ISC)
The percentage of land covered with sealed surfaces, such as pavement and rooftops.

impervious surfaces
Sealed surfaces, such as pavement and roofs, that are unable to absorb rainfall.

impoundment hydroelectric plant
Electricity generated by impounded water, such as water behind a dam, flowing downhill through a turbine.

inbreeding
Mating between closely related individuals.

incandescent lights
Lights in which an electric current causes a thin metal filament to glow.

incentive
Policies that are implemented by encouraging action through the offering of something appealing to take that action.

industrial ecology
Industrial waste systems are redesigned to mimic the cycling of materials in natural ecosystems.

industrial smog
Formed primarily of pollutants released in coal burning. It includes carbon monoxide (CO), sulfur dioxide (SO_2), and particles of carbon soot.

industrial solid waste (ISW)
Waste produced in industrial processes before a finished product reaches the consumer.

infant mortality rate
The percentage of infants within a population who die before age 1.

inorganic compound
Compounds that are not made up of carbon and hydrogen.

instrumental value
The view that a thing is valued as a means to some other end.

integrated conservation and development projects
Programs that promote both the protection of wildlife and the improvement of living standards

for people who live within and around areas that are managed for biodiversity.

integrated pest management (IPM)
The effective combination of all three kinds of pest control—chemical, biological, and cultural—carefully designed to minimize environmental damage.

intercropping
Alternating bands of different crops in the same field to bind soil particles in place between the primary crop rows.

interference competition
In the competition for resources, the fending off of would-be competitors with aggressive or territorial behavior.

intermittent stream
Streams in which water flows only during certain seasons or following major rain events.

internal combustion engine
An engine in which fuel is explosively ignited in a closed cylinder, driving a piston that turns a crankshaft.

interplanting
The practice of planting multiple crop species in the same field.

interspecific competition
Interactions among different species as they compete for shared resources.

intertidal zone
The area along seacoasts that is submerged at high tides and exposed during low tides.

intertropical convergence zone
The band of rising air at the equator.

intraspecific competition
Where members of the same species pursue limited resources.

intrinsic value
The view that a person, organism, or object is valued as an end unto itself.

investment
In forest management, the array of strategies used to restore, maintain, or protect a forest's capacity to provide goods and ecosystem services.

ion
Electrically charged atoms or molecules for which the atom donating the electron has a net positive charge and the atom receiving the electron has a net negative charge.

ionic bond
Chemical bonds that form between positively charged ions and negatively charged ions.

IPAT equation
An equation that summarizes the relationship between human impact and population size, affluence, and technology: $I = P \times A \times T$, where I represents environmental impact, P represents population, A represents affluence, and T represents technology.

irrigation efficiency
The percentage of the water applied to fields that is actually used by crop plants.

isotope
Atoms of an element having different numbers of neutrons.

J

joule (J)
The energy required to support a 1-kg mass against the force of gravity. Named in honor of a 19th-century scientist, James Prescott Joule.

judicial international laws
Standards for the actions of countries based on the decisions of international courts and tribunals.

K

kerogen
A waxy organic substance that is a precursor to oil and natural gas.

keystone species
Species that play a particularly important role regarding the abundance of other species and that are considered essential to the health of the ecosystem.

kilowatt hours
Electric power measurement that is calculated by multiplying electrical potential (volts) by current (amperes): power (kWh) = voltage (volts) × current (amperes).

kinetic energy
The energy of motion.

L

land trusts
Nonprofit organizations that are dedicated to conserving land for the protection of biodiversity and ecosystem services.

landscape diversity
The differences in the variety and abundance of species from place to place.

latent consequences
Effects of a hazard that are delayed.

latent heat transfer
Molecules with the highest kinetic energy evaporate, leaving behind molecules with lower kinetic energy and temperature.

leach field
An area of ground where wastewater released from perforated pipes percolates through the soil, allowing microorganisms to break down the waste materials. *See* septic system.

leachate
Water containing chemicals removed as it percolates through soil or a landfill sediment.

legumes
Plants in the pea family many of which have nitrogen-fixing nodules.

lentic
Ecosystems associated with lakes and ponds.

levee
Embankment along stream channels in wide floodplains.

life expectancy
The average age to which a baby born at a given time will live; provides a summary of differences in survivorship.

life-cycle assessment (LCA)
Methodology used to evaluate the environmental impact of a product throughout all the stages of its life from extraction as a natural resource to disposal.

light-emitting diodes (LEDs)
Lights in which an electric current passes through silicone-rich materials; more efficient and long-lasting than CFLs and incandescent lights.

lignite
A type of coal that is generally associated with younger deposits that have not been subjected to as much heat and pressure as other types of coal; contains 25–35% carbon.

lipid
Organic molecules made of long chains of carbon and hydrogen atoms and a shorter region with one to several oxygen molecules; usually fats and oils.

liquefied natural gas (LNG)
The liquid form of natural gas created when natural gas is chilled to $-160\,°C$ ($-260\,°F$).

lithosphere
An area of interaction between the crust and the upper reaches of the mantle; the zone of geologic activity that has shaped Earth's surface and continues to do so.

Little Ice Age
The time between A.D. 1400 and about 1750 when a cold and wet period set in over the Northern Hemisphere.

littoral zone
The shallow, sloped area of a lake or the ocean that is relatively close to shore.

logistic growth curve
Changes in population size that follow a predictable and repeatable S-shaped trajectory.

lotic
Ecosystems dominated by flowing fresh water.

low-level radioactive waste
Waste that contains comparatively low amounts of radioisotopes.

M

macromolecule
A large complex molecule composed of small organic molecules linked to one another.

macronutrient
Elements that organisms require in comparatively large amounts, such as carbon, oxygen, hydrogen, nitrogen, phosphorus, and sulfur.

magma
A liquid-like state of rocks in the upper mantle.

mantle
The layer of less dense rock above Earth's core. The mantle, which is 2,900 km (1800 mi) thick, occupies about 70% of Earth's total volume.

marginal value
The difference in people's willingness to pay for one action compared to an alternative.

marine
Related to ocean environments.

maritime climate
Regions near oceans or large lakes that experience little variation in temperature.

market-based policy
The use of economic markets to guide environmental actions.

market economy
The production and consumption of goods and services in markets guided by prices based on a system of currency.

marsh
Areas, either tidal or nontidal, that are periodically or continuously flooded.

mass
The property responsible for the gravitational attraction of all matter to all other matter.

mass-balance accounting
A method that allows scientists to account for changes in the abundance of an element within the ecosystem.

median lethal dose (LD$_{50}$)
Toxicity of a chemical in terms of the dose at which 50% of an exposed population is killed or displays a specific symptom or illness.

Medieval Warm Period
The years from A.D. 1000 to 1300 when the average global temperatures may have increased by 0.5 °C, with greater warming in northern latitudes.

megacities
Urban areas with more than 10 million inhabitants.

mesosphere
The layer of the atmosphere where the air temperature can reach lows of $-173\,°C$ ($-140\,°F$) at about 90 km (55 mi).

metamorphic rock
Rocks that form when great heat and pressure transform the physical and chemical properties of sedimentary or igneous rocks.

metropolitan area
A large central city and the adjacent communities that have a high degree of social and economic integration.

micronutrient
Elements that organisms require in very small amounts, such as molybdenum, manganese, and boron.

migration corridors
Connections among habitat fragments that facilitate migration of species; e.g., strips of forest connecting larger forest patches.

Milankovitch cycles
Hundred thousand year long cycles in Earth's climate caused by changes in its orbit relative to the Sun; named for Milutin Milankovitch.

mine tailings
Rock and debris from mining operations; often contain high concentrations of pollutants.

minimum dynamic area
The area necessary to maintain all different individual forest stand types and populations of their associated species, given the typical patterns of disturbance.

model organisms
Organisms in which particular processes or metabolic functions have been demonstrated to be similar to those of humans.

molecule
Two or more atoms held together by chemical bonds.

monocultures
Plantings of a single crop species.

mortality rate
Another term used for the death rate.

mortality transition
The second stage of the demographic transition in which improved economic conditions relieve food shortages, produce better living conditions and health care, and expand access to education. The death rate decreases while the birth rate stays high or even increases.

multinational development banks
Institutions that provide financial and technical assistance to countries for economic, social, and environmental development.

municipal sewage treatment plant (MSTP)
Facilities run by a city or town that use a stepwise process to remove wastes and chemicals from the water.

municipal solid waste (MSW)
Materials that are discarded from homes, small businesses, and institutions such as hospitals and universities; what most of us call trash or garbage.

mutation
Chance changes in the structure of an organism's DNA that produces new forms of inherited features.

mutualism
A symbiotic relationship between two species in which both benefit.

N

natural capital
All of Earth's resources that are necessary to produce the ecosystem services on which we depend.

natural selection
The process by which individuals in a population that are most fit survive and leave more offspring, causing their adaptations to become more common. Individuals that are less fit leave fewer offspring, so their features become less common or cease to exist.

negative feedback
Directional change in a process that alters the system and reverses the direction of that change.

neoclassical economic theory
Determination of the price and production of goods and services through the interaction of supply and demand.

neo-Malthusian movement
Activities related to the belief that as the human population increased, it would soon outpace the supply of food.

neritic province
Regions of the ocean extending from the low tide mark to the edge of continental shelves.

net ecosystem production (NEP)
The amount of organic carbon left each year after subtracting the respiration of nonphotosynthetic organisms from the net primary production.

net migration rate
A calculation of the difference between immigration and emigration per 1,000 individuals in the population.

net primary production (NPP)
Gross primary production minus the organic carbon respired by green plants; the amount of organic carbon available to all the nonphotosynthetic organisms, or consumers, in an ecosystem.

neurotoxins
Substances that affect the development or functioning of the nervous system.

neutron
The electrically neutral particles in the nucleus of atoms.

new urbanism
A social movement that accepts that growth can only be economically, socially, and environmentally sustainable if a combination of smart growth strategies is employed.

niche differentiation
The dividing up of the resources that make up a fundamental niche by potential competitors so that they are able to coexist.

nitrification
A process in which specialized bacteria in the soil convert NH^+ to nitrite (NO_2^-) and nitrate (NO_3^-).

nitrogen fixation
A process in which atmospheric N_2 is converted to NH_3; most nitrogen fixation is carried out by nitrogen-fixing bacteria, which convert N_2 to NH_3 and then into nitrogen-containing organic molecules, such as amino acids and nucleic acids.

nonconsumptive use
Activities that use water and then return it to streams or aquifers.

non-point source
Pollution that comes from a variety of activities occurring at different places, such as urban runoff or fertilizer-enriched runoff.

nonrenewable
The amount a resource in an ecosystem declines with virtually any level of use.

nonrenewable energy
The use of an energy source at rates faster than the rate of replenishment.

no-till
An agricultural technique in which an unplowed field is planted using special drills to prevent soil erosion.

nuclear energy
The potential energy contained in the structure of matter itself.

nuclear fission
Occurs when the nucleus of an atom is split, producing two or more smaller nuclei and a great deal of electromagnetic and kinetic energy. The sum of the masses of the resulting products is slightly less than the mass of the original atom. The difference in the mass of the system before and after fission is equal to the electromagnetic and kinetic energy that is released.

nuclear fusion
Occurs when two atoms collide with so much energy that their nuclei fuse, forming an atom of a new element. The fusion of relatively small atoms, such as hydrogen, carbon, and oxygen, releases energy as very small amounts of mass are converted to energy.

nucleic acid
Polymers of chemical subunits called nucleotides.

O

O horizon
The top layer of natural soil consisting of humus.

observation
Act of viewing or noting a fact or occurrence for a scientific purpose; the information or record obtained by such an act.

ocean acidification
The decrease in ocean water pH associated with rising levels of CO_2 in the atmosphere.

ocean current
Movement of ocean waters driven by Earth's rotation, winds, and differences in water temperature.

ocean thermal energy conversion (OTEC) systems
Systems that generate electricity by exploiting differences in the temperature of surface and deep ocean waters.

oceanic province
All ocean waters beyond the continental shelves.

oil reservoir
Oil trapped in rocks between layers of impermeable rock.

old-field succession
The reforestation of abandoned farmland.

old-growth stage
The stage of forest development characterized by very large, old trees and abundant standing and fallen woody debris.

open basin lake
Lakes drained by a well-defined stream or river.

open-loop recycling
A material is reused in a different product that is discarded after it is used; for example, plastic soda bottles may be recycled into other plastic products.

open watershed
The area of land drained by rivers that eventually make their way to the sea.

opportunity cost
The cost of buying a thing is equal to the economic value of the best alternative use of that money.

organelle
A membrane-enclosed part of a cell in a eukaryote that performs a specialized function.

organic foods
Foods produced to meet strict standards that limit the use of fertilizers and pesticides.

organic molecule
Molecules composed of carbon atoms that are covalently bonded to hydrogen and other atoms; the primary structural and functional molecules of organisms.

outbreeding
Mating between individuals that are not closely related.

ozone layer
The zone in the stratosphere that contains comparatively large amounts of oxygen in the form of ozone (O_3).

P

paramo
High mountain ecosystems in the tropics that are dominated by a tundra-like collection of grasses; temperature at night is frosty, with warm daytime temperatures.

parasite
Organism that lives in or on other plants or animals without killing them directly.

particulate matter
With regard to air pollution, very small solid and liquid particles suspended in the air; abbreviated as PM.

passive solar technology
The use of the energy of sunlight without relying on electrical or mechanical devices, such as by orienting buildings so they receive maximum sunlight and using building materials that absorb sunlight to keep interior spaces warm.

pathogen
Viruses, bacteria, protozoa, fungi, or multicellular parasites that invade host organisms and impair their functioning.

peak load
The maximum amount of electricity demanded by consumers during a day.

peak oil
A term used to describe the point of maximum production in an oil field.

peat
Organic material broken down anaerobically into a wet, partially decomposed mixture.

pelagic zone
Waters between the surface of the ocean or lakes and bottom or benthic zone waters.

per capita land consumption
The average amount of land used by each urban citizen; the size of a metropolitan area divided by its population.

perennial stream
A stream in which water flows year-round.

permafrost
Permanently frozen soil at depths below 30–100 cm (11–39 in.)

in boreal forest and tundra ecosystems.

permanent retrieval storage site
Specially designed landfills in which wastes are monitored to ensure they do not leak into the surrounding environment.

persistent organic pollutants (POPs)
Particularly long-lived and widely dispersed organic chemicals.

pH scale
A quantitative representation of the relative amounts of hydrogen and hydroxyl ions in a substance.

pheromone
Volatile chemical signals used to send a message from one individual organism to another.

phloem
A thin layer of specialized plant tissue that transports carbohydrates down the stem to the roots.

photic zone
The upper portion of the pelagic zone that receives sufficient sunlight to support photosynthesis.

photochemical smog
Formed when certain primary pollutants interact with sunlight. Nitrogen oxide (NO) and volatile organic compounds (VOCs) are among the most important of these primary pollutants.

photochemicals
Secondary pollutants whose formation is facilitated by sunlight.

photosynthesis
Formation of carbohydrates from carbon dioxide and a source of hydrogen that occurs in the chlorophyll-containing tissues of plants exposed to sunlight.

photovoltaic (PV) cells
Devices that directly convert sunlight to electrical current.

photovoltaic (PV) technology
A method of generating solar power that converts light energy into electricity through the photovoltaic effect, in which light energy causes certain materials to emit electrons, thereby generating an electric current.

phylogenetic tree
A diagram that depicts scientific understanding of the evolutionary ancestry of organisms.

phytoplankton
The various algae suspended in the water in the pelagic zone; uses light energy to carry out photosynthesis.

pioneer species
The earliest colonists of an ecosystem. They are widely dispersed, they are able to grow under very harsh and resource-poor conditions, and as they grow, they alter their environment in ways that allow other plants to become established.

planetoids
Planet-like objects composed of ice and rock.

planned economy
An economy in which the government regulates the prices of goods and services and the level of production.

Pleistocene epoch
A time of alternating cold and warm periods starting about 2 million years ago; each cycle of cold and warm lasted about 100,000 years.

plug-in hybrid automobile
Vehicle operating much like gasoline electric hybrids, except the battery can be recharged by plugging the car into an electric outlet.

plutonium
An element that can undergo fission and generate heat; it is a by-product in uranium-fueled nuclear reactors.

point source
Pollution coming from a specific location, such as an industrial facility or municipal sewage treatment plant.

polar cell
Convection cells in the troposphere between latitudes of 60° and 90°.

polar zone
The climate zone that falls above 60° latitude north and south; its climate is determined by the polar wind cells.

policy cycle
The stepwise process by which governments and communities develop and revise policies and laws.

polyculture
See intercropping.

polymer
Simple molecules linked together to form much larger chains or networks.

polysaccharide
Polymers of simple sugar molecules.

pool
In reference to nutrient cycling, a part of an ecosystem in which matter may reside, such as the atmosphere or soil.

population growth rate
The multiple by which an exponentially growing population increases.

population viability
The probability that a species will not go extinct in a given number of years.

positive feedback
Directional change in a process that alters the ecosystem to reinforce the change.

potential energy
Energy stored in a system and available to do work in the future.

potential evapotranspiration (PET)
An estimate of the average amount of water that would evaporate from a hypothetical agricultural field over the course of a year.

precautionary principle
When there is reasonable evidence that an action or policy may place human health or the environment at risk, precautionary measures should be taken, even in the context of uncertainty.

precision
Refers to the likelihood that estimates derived by sampling a subset of a population are representative of the actual values for the entire population.

predators
Species that hunt and kill their prey.

predictions
In science, if, then statements that forecast the outcome of a test of a hypothesis.

preservationist
The opinion that parks and public lands should *preserve* wild nature in its pristine state for its own value.

preserves
Areas that are protected by various means to maintain biodiversity.

prey switching
As predation depletes a species' numbers, predators switch their attention to other, more abundant species, a behavior which provides predators with a steady supply of food and ensures that none of the prey species will be totally eliminated.

primary air pollutants
Chemicals or particles that are directly emitted from identifiable sources.

primary consumer
Herbivores; organisms that feed mostly on plants.

primary energy
Contained in natural resources, such as coal, oil, sunlight, wind, and uranium.

primary forest
Forest that has been largely unaffected by humans; worldwide

only about 19 million square kilometers (7.3 million square miles) remain.

primary growth
The upward and downward growth due to rapid cell division and growth at the very tips of plant shoots and roots.

primary oil recovery
The extraction of crude oil pushed to the surface by built-up pressure in the reservoir.

primary producer
Those organisms that transform energy from sunlight or certain inorganic chemicals into high-energy carbohydrates.

primary succession
After a disturbance has removed virtually all ecological legacies, plants, animals, and microscopic organisms slowly colonize the disturbed area. Because it occurs slowly, often over centuries or even millennia, primary succession is difficult to study directly.

primary treatment
In reference to wastewater treatment, a process that produces a relatively homogeneous liquid that can be treated biologically and a sludge that can be processed separately.

prior appropriation water rights
Water law in which the person who first uses a quantity of water from a water source for a beneficial use has the right to continue to use that quantity of water for that purpose indefinitely. Rights to the remaining water are similarly allocated to subsequent users. When water runs short, users with the earliest appropriation get their full allocation; later users may get no water.

production
Refers to the amount of a primary energy source extracted during a given period of time

profit
The difference between the cost to produce a commodity and its price in the marketplace.

prokaryote
Organisms, including bacteria and their relatives, whose cells lack an organized nucleus and membrane-bound organelles.

protein
Polymers made of nitrogen containing organic molecules called amino acids.

proton
Positively charged particles in the central nucleus of an atom.

proved reserves
The quantities of a primary energy source that could be recovered from known deposits using current technology at current prices.

provisioning services
Ecosystem services that supply humans with resources, such as food, water, and air.

pull factors
Conditions that encourage people to immigrate into a country.

push factors
Conditions that force people to emigrate from a country.

R

radiation
The release of electromagnetic energy.

radioactive
Isotopes that have an unstable nuclei; they spontaneously "decay," emitting various combinations of high-energy protons, neutrons, electrons, and radiation.

radon (^{222}Rn)
A gaseous element produced from the radioactive decay of the element radium-226 (^{226}Ra).

rain shadow
The dry region on the leeward side of mountains; descending air on leeward slopes produces less rain and snow than on the windward side.

range of tolerance
The span between the minimum and maximum amounts of an environmental feature such as temperature or the amount of a chemical that an organism can tolerate; the range of conditions over which an organism can maintain population growth rates.

rationing
In forest management, a process of determining the rate at which forest goods or services can be used by humans.

reactor core
Where fuel assemblies for a nuclear power plant are placed and where the fission reaction takes place.

realized niche
The range of conditions over which a species is able to compete effectively against other species.

recharge zone
An area where water flows through the soil surface and percolates to the water table.

regulating services
The ways that ecosystems control important conditions and processes, such as climate, the flow of water, and the absorption of pollutants.

regulations
A set of specific rules that establish standards for performance, programs to ensure compliance, and protocols for enforcement.

regulatory mandate
Policies or laws that set legal standards for actions.

relative humidity
Measure of the extent to which air is saturated with water vapor, expressed as a percentage; vapor pressure/saturation vapor pressure × 100.

renewable energy
Energy derived from sources that are not depleted when they are used, such as sunlight and wind, or that can be replenished in a short period of time, such as fuelwood.

renewable portfolio standard (RPS)
A government mandate that a certain percentage of energy use come from renewable energy resources, the objective of which is to help build market demand for renewable technologies.

reproductive isolation
Members of one species do not interbreed with members of other species, even those that are closely related.

reserves
The total amount of a resource that can be exploited.

reserves-to-production ratio (R/P)
Proven reserves figure for a given primary energy source divided by a particular year's level of production or use.

residence time
The average time that an atom of an element or molecule of a compound spends in a pool; easily calculated as the size of the equilibrium pool divided by the flux through the pool.

return time
The average time between disturbances at any given place.

reverse osmosis
A method of desalination by filtering water through a selective membrane.

ribonucleic acid (RNA)
Various nucleic acids that contain ribose and uracil as structural elements and that play a central role in the synthesis of proteins using information encoded in DNA.

Richter magnitude scale
The numerical scale used to report the energy released in an earthquake in which each successive number indicates a 32-fold increase in energy.

riparian water rights
A system for assigning water possession derived from English common law whereby all landowners with property adjacent to a body of water have a right to make reasonable use of it.

riparian zone
The region of transition between a stream and the terrestrial ecosystems that surround it.

risk
Likelihood of particular hazards impairing health or causing death.

risk–risk trade-offs
The benefits of actions to reduce the risk of a hazard incur other costs or risks.

rock cycle
The process in which elements within Earth's crust and mantle are slowly converted from one type of rock to another.

rotation time
The average interval between successive cuts in a forest.

runoff
The action of rainwater on land flowing across the surface until it eventually enters lakes and streams.

run-of-river hydroelectric plant
A hydropower facility that uses the force of river water diverted into a series of pipes to spin a turbine to generate electricity.

S

saltwater intrusion
The migration of salt water into a freshwater aquifer, which may occur when groundwater is pumped faster than it can be replenished.

sampling effect
The increased likelihood as species diversity increases that a species has a big effect on a process in an ecosystem.

sand
Mineral particles 0.05–2 mm in diameter; the chemical structure of these particles still resembles their parent bedrock.

sanitary landfill
A system of municipal solid waste disposal designed to prevent the

contamination of groundwater and to reduce the odors and aesthetic problems associated with open dumps.

saturation vapor pressure
A measure of the maximum amount of water that air can hold at a given temperature.

science
A process that poses and answers questions objectively in order to increase knowledge and lessen uncertainty by using careful observation, experimentation, and synthesis to produce results that can be duplicated by others.

sclerophyllous
Vegetation with thick, hard leaves resistant to water loss and wilting.

scrubbers
Devices containing chemicals that react with pollutants such as SO_2 and filter them from industrial exhausts.

second law of thermodynamics
Every energy transformation increases disorder, or entropy. In any energy transformation, energy always tends to go from more usable or higher quality forms, to less usable or lower quality forms, such as heat.

secondary air pollutants
Chemicals or particles that are produced in the atmosphere as a result of reactions among other chemicals or aerosols.

secondary energy
Forms of energy such as electricity or the kinetic energy of an automobile that are derived from energy contained in natural resources such as flowing water or oil.

secondary forest
Forests created by natural succession or reforestation following human disturbance.

secondary growth
Cell growth and division that increases the diameter of plant stems and roots.

secondary oil recovery
Injection of water to extract another 10–20% of the oil reservoir after the primary recovery stage.

secondary succession
The process of change following disturbances that leave behind legacies such as soil, woody debris, or plant seeds.

secondary treatment
In reference to wastewater treatment, a process using bacteria and other microorganisms to break down the organic material dissolved in the wastewater.

sedimentary rock
Rock that forms when sediments, such as sand, silt, and the remains of dead organisms, become "glued together" under pressure.

seed
A plant embryo embedded in nutritive tissue and surrounded by a protective coat.

septic system
System to treat household sewage and wastewater; solids settle to the bottom of the tank and wastewater flows to a series of perforated underground pipes.

sexual reproduction
Mating between two parents that produces offspring that are genetically unique.

shelterbelts
Strips of trees planted as windbreaks along the edge of an agricultural field to slow the wind and reduce soil erosion.

shifting agriculture
The process in which forests are cut and the slash is burned to release nutrients to the newly cleared soil to create soil for agriculture. When crop production declines, the fields are abandoned and the process is repeated in other forested areas.

silt
Mineral particles that have experienced more weathering than sand and that have diameters ranging from 0.002 to 0.5 mm.

sinkhole
Rapid sinking of land above an aquifer as it is depleted, resulting in a hole into the earth.

smog
Fog or haze in the atmosphere caused by the combination of many different pollutants.

soil
The uppermost layer of Earth's crust that supports plant growth; a mixture of organic matter and mineral particles.

soil fertility
The variety of soil characteristics that support plant growth.

soil profile
The distinctive vertical structure of soil layers.

soil salinization
The process that occurs when salts contained in irrigation water concentrate and crystallize on the soil surface due to evaporation.

soil wilting point
The point at which there is no longer enough water in the soil for plants to replace the water that is being lost to transpiration.

solar photovoltaic farms
In areas with high solar potential, thousands of PV panels are arranged to generate electricity on a large scale.

solar power
The use of sunlight to generate electricity.

solubility
The ability of a chemical to dissolve in a particular liquid.

sovereignty
The concept that a country may behave as it pleases within its borders as long as it does not violate international laws to which it has agreed.

species
A group of similar organisms that can potentially mate and produce fertile offspring and are reproductively isolated from other groups.

species evenness
A measure of the relative abundance of the different species in a community.

species richness
The total number of species in each community; the simplest measure of biodiversity.

spent nuclear fuel
Nuclear fuel rods that are depleted in ^{235}U; but still contain large amounts of hazardous and radioactive chemicals.

stability transition
The fourth stage of the demographic transition that is characterized by low birth, death, and population growth rates.

stabilizing selection
Selection that favors traits that fall somewhere in the middle of the range of variation within a population.

starch
Polysaccharides composed of hundreds of glucose molecules. The bonds between these glucose molecules are easily broken in animal digestion.

storm surge
Storm-caused increase in water levels that can produce devastating floods in low-lying regions and on islands.

stratosphere
The layer of the atmosphere that extends from 15 to about 48 km (9–30 mi) above Earth's surface.

structural complexity
The three-dimensional distribution of species and biological features within an ecosystem.

sub-bituminous coal
Coal containing 35–45% carbon; its energy content is somewhat higher than lignite.

subsidence
The sinking of land above an aquifer as the aquifer is depleted.

subsistence economy
A basic economic system in which a society meets its needs from its environment without accumulating wealth.

suburb
Part of the metropolitan area located outside of the central city.

succession
The process of post-disturbance change in an ecological community.

succulent
Plant with thick, fleshy stems or leaves that store large quantities of water.

sugar
Carbohydrates with the general chemical formula $(CH_2O)_n$, where n is between 3 and 7. A basic source of energy in most organisms and an ingredient in the synthesis of numerous other organic molecules.

supporting services
The basic ecosystem processes, such as nutrient cycles and soil formation, that are needed to maintain other services.

survivorship
The probability of surviving to a particular age.

sustainability
Meeting the needs of the present in an equitable and fair fashion without compromising the ability of future generations to meet their own needs.

sustainability science
The study of the interactions between ecosystems and human social systems, with a special focus on long-term changes in environment and human populations and society.

swamp
A wetland dominated by shrubs or trees and fed by flowing water.

symbiosis
Intimate interdependencies between species.

synthesis
The development of concepts and theories based on connections among numerous facts and sources.

T

taiga
Russian word meaning "little sticks" and used synonymously with boreal forest.

take-back regulations
Requirements that electronic manufacturers take back their products at the end of their useful life, at no charge to the consumer.

taxonomy
Classification of organisms through description, identification, and naming.

tectonic plate
A piece or plate of Earth's crust that floats on top of the mantle. The positions of the plates are not fixed; rather, they slowly move in relation to each other.

temperate deciduous forest
A biome in regions with moderate summers and cold winters that is dominated by broad-leaved trees that lose their leaves in the autumn and grow a new set in the spring. Precipitation is relatively even throughout the year.

temperate evergreen forest
In this biome with less precipitation and warmer temperatures than the temperate deciduous forest, trees keep their leaves throughout the year.

temperate grassland
Grasses interspersed with a diverse array of herbs in a biome too dry to support forest, woodland, or shrubland, but wet enough to prevent the land from becoming desert.

temperate rain forest
Lush and diverse evergreen forest in a region of heavy winter rains, summer fogs, and ample year-round moisture and mild temperatures.

temperate zone
The climate zone that falls between 25° and 60° latitude north and south, corresponding to the area of prevailing westerly winds in the Ferrel wind cells.

temperature
A measure of the average kinetic energy of a collection of molecules.

temperature anomaly
The difference between each year's average temperature and the average annual temperature over 20–30 benchmark years. A positive anomaly indicates a year that is warmer than the benchmark average, and a negative anomaly represents a year that is cooler than the benchmark average.

temperature inversion
A pattern of warmer air formed above colder air. Temperature inversions prevent air close to the ground from mixing with the air above it.

teratogens
Substances that produce mutations and birth defects.

terracing
An agricultural technique used on steep mountain slopes in which a series of wide steps are cut into the slope. This helps to retain water and limit runoff in sites that would otherwise be very difficult to cultivate.

tertiary oil recovery
Methods used to stimulate the flow of additional oil by injecting CO_2, steam, or hot water into reservoirs.

tertiary treatment
Referring to wastewater treatment, a process to remove inorganic nutrients from wastewater by passing them through sand and charcoal filters to remove residual solids and toxins.

texture
The relative amount of sand, silt, and clay in soil.

theory
Principles devised to explain a group of facts or phenomena.

thermohaline circulation
The three-dimensional circulation of ocean water around the world due to the differences in temperature and salinity, causing an exchange of surface water and water from the depths.

thermosphere
The layer of the atmosphere that extends about 480 km (300 mi) into space.

thinning stage
A stage of forest development during which intense competition for light and water results in the death of smaller trees. The largest trees continue to grow and the total biomass in the forest continues to increase.

tidal barrage
A dam across the narrow outlet of a tidal basin that uses the ebb and flow of tidal waters to power turbines and dynamos to generate electricity.

tidal fence
An array of underwater turnstiles or revolving doors set across a narrow channel. As water moves through the channel, its energy is used to generate electricity.

tidal power
The kinetic energy of ocean tides used to generate electricity.

tidal turbines
Turbines operating underwater that are spun by the movement of tides and currents.

tilth
The physical arrangement of soil particles that facilitates aeration, seedling emergence, and the growth of roots.

total fertility rate
The sum of age-specific fertility rates across all ages; the *potential* number of offspring that an average female in a population can produce if she survives to old age.

toxicity threshold
The dose above which there is a measurable decline in a health measure.

toxicology
The scientific discipline that studies chemical poisons or toxins and their effects on human health.

trace gases
Gases contained in the atmosphere, most of which are present in concentrations expressed in parts per million (1 ppm = 0.0001%) or parts per billion (1 ppb = 0.0000001%).

tragedy of the commons
The overexploitation of common resources that occurs when individual interests prevail over the common good.

transcription
DNA code for a particular gene rewritten as a segment of RNA.

transform fault boundary
The zone along which tectonic plates slide past one another. This sliding may generate earthquakes.

transformers
Devices used to adjust amperage and voltage so that electricity can move efficiently over long distances through transmission lines and be delivered to end users.

transition stage
The stage of forest stand development in which tree seedlings become established in the gaps created by the death of large trees; also called the gap phase.

translation
The process of forming a protein molecule using the information contained in RNA, which serves as the template for the assembly of a specific protein.

transpiration
The action of water evaporating from leaves; a major component of evaporation in terrestrial ecosystems.

travel-cost valuation
A valuation method based on the amount of money that people are willing to pay for transportation and lodging to visit a particular place.

treatment group
Refers to those subjects in an experiment that are subjected to a treatment.

tree line
Region of transition from boreal forest to tundra-like vegetation on a mountain or in polar regions.

triple bottom line (TBL)
An accounting framework in which sustainability is assessed by environmental, social, and economic outcomes.

trophic cascade
The influence of carnivores across multiple trophic levels; for example, loss of carnivores may allow populations of herbivores to increase, resulting in the overconsumption of primary producers.

trophic level
Classification of organisms based on their source of food; for example, primary producers, herbivores, and carnivores.

trophic level efficiency
The fraction of energy that the organisms in one trophic level store as biomass and make available to the next trophic level.

tropical rain forest
An area in which plentiful rainfall (greater than 2,000 mm per year) and warm climate support an enormous diversity of plants and animals.

tropical savanna
An ecosystem of open grasslands and a few scattered trees, which occurs where rainfall is highly seasonal and drought conditions generally persist for more than half of the year.

tropical seasonal forest
Tropical forests and woodlands dominated by trees that lose their leaves during seasonal dry periods.

tropical zone
The climate zone that straddles the equator and extends to about 25° latitude north and south, corresponding to the Hadley wind cells in the troposphere.

troposphere
The lowest layer of the atmosphere based on differences in temperature and chemical properties.

tundra
Treeless landscapes dominated by grasses, herbs, and low shrubs in polar regions with short growing seasons.

U

umbrella species
Species whose ecological requirements are highly correlated with the needs of many other species.

unconfined aquifer
A zone of water-saturated rock or sediment that is open to water recharge from the rock and sediment layers above it.

unconventional reservoir
Oil or natural gas found in atypical geological settings such as shale or tar sands.

uneven-aged management
Maintaining biodiversity and ecosystem services through less intensive methods of harvest in which trees, dead snags, and woody debris are retained.

uranium
An element used to produce fuel rods for nuclear reactors. The decay of uranium isotope ^{235}U releases large amounts of energy.

urban
An area in which the average population density is at least 500 people per square mile.

urban decay
Decline in economic and community activity and deterioration of infrastructure that has occurred in many cities.

urban ecosystem
A region of Earth that supports the commerce, industry, and community interactions of a large number and high density of humans.

urban growth boundary (UGB)
A form of regional zoning that delineates areas suitable for high-density urban development from those suitable for lower density development.

urban heat island
The tendency for cities to be warmer than surrounding rural areas due to urban features that affect the absorption of solar energy, the rate of heat loss, and patterns of air movement.

urban planning
The profession that unites disciplines such as engineering, architecture, ecology, and sociology to design and build environments and transportation systems for urban communities.

urban sprawl
A widespread growth in the amount of land included in metropolitan areas.

urbanization
The increasing concentration of human populations in densely populated cities.

utilitarianism
An example of consequence-based ethics that defines right actions as those that deliver the greatest good to the greatest number.

V

vapor pressure
The fraction of atmospheric pressure that is a consequence of water molecules.

variables
The properties being measured in scientific observations and experiments.

vector
Intermediary organism that transmits a parasite from host to host.

vermicomposting
The use of earthworms to break down organic waste into worm castings, which can be used as a rich fertilizer.

virtue ethics
A system of values based on the belief that an action is right if it is motivated by virtues, such as kindness, honesty, loyalty, and justice.

virulence
The extent to which a pathogen impairs the growth and reproduction of its host.

volatile organic compounds (VOCs)
Diverse array of organic chemicals that can vaporize into the air; many VOCs can impair the health of organisms, including humans.

volts
An electrical potential energy measurement; the greater the difference in electrical potential between the terminals, the greater the voltage of the current.

volunteerism
Work performed freely on behalf of a community.

W

waste-to-energy facility (WTE)
Incinerators that use the heat produced by burning waste to generate electricity.

water table
The underground depth where rock and sediment are completely saturated with water.

watershed
The area of land from which rainfall drains into a river or lake; also called drainage basins.

watt-hour (Wh)
The energy unit most commonly used to measure the everyday use of electricity.

wave power
The kinetic energy of ocean waves is used to generate electricity.

weather
Short-term variations in local atmospheric conditions.

wet deposition
Trace gases and particles are captured and removed from the atmosphere in raindrops, snowflakes, or droplets of fog.

wind farms
Clusters of turbines built close together in areas with the strongest and most regular winds.

wind power
Using wind to generate electricity.

wind turbines
Mechanical assemblies that are designed to transform air movement into electricity.

windbreaks
Trees planted along the edge of an agricultural field to slow the wind and reduce soil erosion.

work
The force applied to an object over a distance.

X

xylem
Specialized plant tissue that transports water and mineral nutrients up the stem.

Z

zoning
The designation of specific areas for different categories of land use.

zoonosis
A disease that is shared by humans and other animals.

zooplankton
Diverse array of tiny protists and animals suspended in pelagic waters.

zygote
A new, genetically unique cell formed by the union of an egg and a sperm.

PHOTO CREDITS

Chapter 1

Chapter Opener: Yuriy Poznukhov/123RF; **1.1** Chris Schmid/Aurora Open/Corbis; **1.2** NASA; **1.3** Steve C.Mitchel/epa/Corbis; **1.4a** Lynn Johnson/National Geographic Society/Alamy; **1.4b** Ashley Whitworth/Fotolia; **1.5** SuperStock; **1.6** Steve Terrill/Corbis, Steve Prezant/Image Source/Corbis, Steve Glass/Aurora Open/Corbis, Andre Jenny/Alamy, Dainis Derics/Shutterstock; **1.7** Reuters; **1.8** NASA; **1.10a** Lissa Leege; **1.10b** Lissa Leege; **1.10c** Lissa Leege; **1.11** Lost Horizon Images/Corbis; **1.12** Lost Horizon Images/Corbis; **1.13a** Donald R. Swartz/Shutterstock; **1.13b** JTB Media Creation, Inc.; **1.14** NASA; **1.15** Eryk Jaegermann/Design Pics/AGE Fotostock; **1.16** Ollirg/Shutterstock; **1.17** Rabh images/Alamy; **1.18** Fotog/Tetra Images/AGE Fotostock; **1.19** Dave Rudkin/DK Images; **1.20** WideAwake/Fotolia; **1.21** Louise Murray/Robert Harding Picture Library Ltd/Alamy; **1.23** Rolf Peterson/Michigan Tech University/AP Images, Rolf Peterson/Minneapolis Star Tribune/MCT/Newscom; **1.24** Tatiyana Zenkovich/epa/Corbis; **1.26** Max Alexander/DK Images; **1.27** Philippe Michel/AGE Fotostock; Agents of Change: **1.1** Rachel Barge; **1.2** Werner Dieterich/Alamy; **1.3** Rich Pedroncelli/AP Images

Chapter 2

Chapter Opener: Christoph Martin/Getty Images; **2.1** Wadsworth Atheneum Museum of Art/Art Resource, NY; **2.2** Library of Congress Prints and Photograph Division; **2.3** Dan Schreiber/Shutterstock; **2.4** Holton Collection/Superstock; **2.5** McPhoto/Blickwinkel/AGE Fotostock; **2.6** Library of Congress Prints and Photographs Division Washington, D.C. [LC-D41-12683]; **2.7** Corbis, Nelson Sirlin/Shutterstock; **2.8** John McConnico/AP Images; **2.9** David Grossman/Alamy, Mike Simons/Tulsa World/AP Images; **2.10** Frank Greenaway/DK Images, David Peart/DK Images, Juniors Bildarchiv GmbH/Alamy, Sharon Morris/Shutterstock; **2.11** Simon Maina//AFP/Getty Images/Newscom; **2.12** Mary Ann Chastain/AP Images; **2.13** Jacques Jangoux/Alamy; **2.14** Zheng Xianzhang/Alamy; **2.16a** Alan Hills/DK Images; **2.16b** AGE Fotostock; **2.17** Sheri Blaney/Getty Images; **2.18** Krapels/Fotolia; **2.19** NASA; **2.20** Jaime Florez/World Wildlife Fund; **2.21** Fax Sinclair; **2.22** Edward Parker/Alamy; **2.24** Mike Segar/Reuters/Corbis; **2.25** Robert Tinker/Design Pics Inc/Alamy; **2.26** Photoshot/Balance/AGE Fotostock; **2.27** Wade H. Massie/Shutterstock; **2.29** PureStock/Alamy; **2.30** Jan kranendonk/Shutterstock; **2.31** United Nations Photo Library; **2.34** Frans lemmens/Alamy; Agents of Change: **2.1** Sol Weiner; **2.2** Shawn Rocco/The News & Observer/AP Images

Chapter 3

Chapter Opener: Lauracrane/Fotolia; **3.1** NASA; **3.2** NASA; **3.3** Luc Novovitch/Alamy; **3.7** David Murray/DK Images; **3.8** Karl Shone/DK Images; **3.9** Volodymyr Goinyk/Shutterstock; **3.10** Andy Crawford/DK Images, Clive Streeter/DK Images, Design56/Shutterstock; **3.16** Ragnar Schmuck/fStop/Alamy; **3.17** Ugurhan/E+/Getty Images; **3.19** Dave King/DK Images; **3.20** Roman Sigaev/Shutterstock; **3.21** Jeffrey Greenberg/Science Source; **3.22** U.S. Department of Energy; **3.23** NASA; **3.28** Kip Evans/Alamy; **3.31** Jiri Foltyn/Fotolia; **3.32** Demetrio Carrasco/DK Images; **3.33** Santiago Urquijo/Moment/Getty Images; **3.36a** NASA; **3.36b** NASA; **3.38** Jim Henderson/Alamy; **3.42** Lloyd Park/DK Images; **3.43** Tyler Olson/Age Fotostock

Chapter 4

Chapter Opener: Jean-Paul Chatagnon/Biospohoto/Science Source; **4.1** Everett Collection; **4.2** Embassy of Australia; **4.3** Julie Dermansky/Corbis; **4.5** Peter Anderson/DK Images; **4.6** NOAA National Geophysical Data Center; **4.7** Dave King/DK Images; **4.8** Steve Gschmeissner/Science Photo Library/Alamy, Monty Rakusen/Cultura/AGE Fotostock; **4.9a** Michael Abbey/Science Source; **4.9b** Michael Abbey/Science Source; **4.9c** Michael Abbey/Science Source; **4.9d** Michael Abbey/Science Source; **4.10a** Science Source; **4.10b** Phil Degginger/AGE Fotostock; **4.10c** Alan & Linda Detrick/Science Source; **4.11** Aleksandar Zoric/Shutterstock; **4.12** DK Images, Dave King/DK Images, Jerry Young/DK Images, Dave King/DK Images, M.I. Walker/DK Images, DK Images; **4.14** Dieter Hopf/Imagebroker/Alamy; Focus on Science: DeAgostini/SuperStock; **4.17** Gregory Erickson; **4.18** Gregory Erickson; **4.23** Tom Reichner/Shutterstock, Devist/Fotolia; **4.24a** FLPA/Andrew Parkinson / AGE Fotostock; **4.24b** Arco/Robiller F./Glow Images; **4.25** Alexander/Fotolia; **4.26** Darwin notebook drawing supplied by N. Christensen; **4.27a** Tierbild Okapia/Science Source; **4.27b** Celia Mannings/Alamy; **4.27c** Gerald & Buff Corsi/Visuals Unlimited/Getty Images; **4.29** Wayne Hughes/Alamy; **4.34** Maresa Pryor/Danita Delimont/Alamy, Accent Alaska/Accent Alaska.com/Alamy, Bob Gibbons/Alamy; **4.37** Christian Charisius/Reuters; **4.38** Shattil & Rozinski/Nature Picture Library; **4.39** Robin Hunter/Queen/DK Images; Matthew Ward/DK Images, M.I. Walker/DK Images, Dr Julian Thorpe/DK Images, Eye of Science/Science Source, DK Images, Jerry Young/DK Images, Laura Wickenden/DK Images, Karl Shone/DK Images, Dave King/DK Images, Geoff Brightling/DK Images, Jerry Young/DK Images, Dave King/Jeremy Hunt/DK Images, Geoff Dann/DK Images, Dave King/DK Images, Jerry Young/DK Images, Steve Gorton/DK Images, Kim Taylo/DK Images, Frank Greenaway/DK Nature Encyclopedia/DK, David Peart/DK Images, Jerry Young/DK Images, Sergey Goruppa/Shutterstock, D. Kucharski K. Kucharska/Shutterstock

Chapter 5

Chapter Opener: Kazuyoshi Nomachi/Terra/Corbis; **5.4a** Kim Walker/Robert Harding World Imagery/Alamy; **5.4b** Nmelnychuk/Fotolia; **5.4c** Andy Crawford/DK Images; **5.4d** DK Images; **5.4e** Stephen Mcsweeny/Shutterstock; **5.6** Daniel Berehulak/Getty Images; **5.7** Tibor Bognar/Alamy; **5.9b** E.R. Degginger/Alamy; **5.14** Szefei/Shutterstock; **5.18** Tariq Mahmood/AFP/Getty Images/Newscom; **5.21** Adrian Raftery; **5.24** Steve Gortin/DK Images; **5.25 inset** Kazuyoshi Nomachi/Terra/Corbis; **5.25** Jon Hicks/Corbis; **5.27** Jon Hicks/Corbis; **5.27a** Ivan_Sabo/Shutterstock; **5.27b** David R. Frazier/Photolibrary, Inc/Alamy; **5.27c** Lisa S/Shutterstock; **5.28a** Peter Menzel/menzelphoto.com; **5.28b** Peter Menzel Photography; **5.29** NASA Images; **5.30** Lambert/Contributor/Archive Photos/Getty Images; **5.31** David Turnley/Corbis; **5.32** Barry Lewis/Alamy; **5.33** Dinodia Photo/AGE Fotostock; **5.34** Giotto/Demotix/Demotix/Corbis

Chapter 6

Chapter Opener: Majk11/Fotolia; **6.1** Radius/Superstock; **6.2a** Biophoto Associates/Science Source, Dani Vincek/Fotolia; **6.2b** Richard Alan Wood/Age Fotostock; **6.2c** Jared Hobbs/All Canada Photos/Corbis; **6.3** Anneka/Shutterstock; **6.4a** Eric V. Grave/Science Source; **6.4b** Michael Abbey/Science Source; **6.7** Chris Curtis/Shutterstock; **6.8** Anup Shah/Nature picture library; **6.9** Biophoto Associates/Science Source; **6.10a** James Laurie/Shutterstock; **6.10b** Mint Images - Frans Lanting/Mint Images/Getty Images, Klaas Lingbeek- van Kranen/Getty Images; **6.10a** Alexander_photo/Fotolia; **6.10b** Keith Douglas/Alamy; **6.10c** Don Johnston/AGE Fotostock; **6.11** Paul Souders/Digital Vision/Getty Images; **6.12** Brian Lasenby/Fotolia; **6.14a** The Fight Between Carnival and Lent/Peter Bruegel; **6.14b** John Downer/Oxford Scientific/Getty Images; **6.16** Gregory G. Dimijian, M.D./Science Source; **6.17** Martin Shields/Science Source; **6.18** Frank Greenaway/DK Images; **6.19a** Cynthia Kidwell/Shutterstock; **6.19c** Artcasta/Shutterstock; **6.19d** Africa Studio/Fotolia; **6.19e** Martin Pateman/Shutterstock; **6.20a** Geoff Dann/DK Images; **6.20b** Steve Shott/DK Image; **6.21a** Cynthia Kidwell/Shutterstock; **6.21b** GVictoria/Shutterstock; **6.21c** Artcasta/Shutterstock; **6.21d** Africa Studio/Fotolia; **6.21e** Martin Pateman/Shutterstock; **6.22a** Roger Phillips/DK Images; **6.22b** Steve Gorton/DK Images; **6.22c** Harbour Seal/DK Images; **6.22d** Dave King/DK Images; **6.22e** Sobur/Shutterstock; **6.22f** Cyril Laubscher/DK Images; **6.22g** Philip Dowell/DK Images; **6.22h** Frank Greenaway/DK Images; **6.22i** Model of Fiddler Crab/DK Images; **6.22j** Clive Streeter and Patrick McLeavy,Dorling Kindersley; **6.22k** Clive Streeter and Patrick McLeavy/DK Images; **6.22l** Frank Greenaway/DK Images; **6.22m** Dave King/JE962/DK Images; **6.22n** Biology Pics/Science Source; **6.22o** Felix MAckel/E+/Getty Images; **6.22p** Airphoto.gr/Shutterstock; **6.23a** George Burba/Shutterstock; **6.23b** hvoya/Shutterstock; **6.23c** Steve Byland/Shutterstock; **6.23d** Mike Truchon/Shutterstock; **6.23e** Ornitolog82/Fotolia; **6.23f** Steve Byland/Shutterstock; **6.23g** fivespots/Shutterstock; **6.23h** Jane Burton/DK Images; **6.23i** Kim Taylor/DK Images; **6.23j** Crepesoles/Shutterstock; **6.23k** Peter Anderson/DK Images; **6.23l** DK Images; **6.23m** Mediscan/Medical-on-Line/Alamy; **6.24** Norm Christensen; **6.25a** Alamy; **6.25b** Ronnie Howard/Shutterstock; **6.26** Jon Chase; **6.27** Stephen How/Alamy; **6.28** LianeM/Shutterstock; **6.34** Kathie Thomas/Dreamstime; **6.35a** Tnymand/Fotolia; **6.35b** Wildlife GmbH/Alamy; **6.35c** Joe Klune/Shutterstock; **6.35d** Design Pics/Superstock; **6.36** Norm Christensen; **6.37a** Norm Christensen; **6.37b** Norm Christensen; **6.37c** Norm Christensen; **6.39** Peter Anderson/DK Images; **6.40** Newscom; **6.41** Mark Newman/Frank Lane Picture Agency; Agents of Change: **6.2–6.3** Elizabeth Brajevich

Chapter 7

Chapter Opener: Paul Banton/Shutterstock; **7.1a** G Kopp/Blickwinkel/Age Fotostock; **7.1b** Carlos S. Pereyra/easyFotostock/Age Fotostock; **7.1c** J. Norman Reid/Shutterstock; **7.1d** Andoni Canela/Age fotostock; **7.1e** Janis Lacis/Shutterstock; **7.1f** John E Marriott/All Canada Photos/Age Fotostock; **7.2** Norm Christensen; **7.3a** Kenneth M. Highfill/Science Source; **7.3b** Louis M Landry; **7.3c** Suzanne L. & Joseph T. Collins/Science Source; **7.4a** Blickwinkel/Alamy; **7.4b** Imagebroker/Alamy; **7.6a** Inga Nielsen/Shutterstock; **7.6b** PiLensPhoto/Fotolia; **7.6d** Les Cunliffe/Fotolia; **7.6e** Littlesam/Shutterstock; **7.6f** Patrick Poendl/Shutterstock; **7.6g** Christian Heinrich/image Broker/AGE Fotostock; **7.6h** Dmitry Pichugin/Fotolia; **7.7** George Grall/National Geographic/Corbis; **7.8** Wolfgang Lucht; **7.10** Superstock; **7.12** Erin Eldermire; **7.13** Terry Whittaker/Alamy; **7.14** Norm Christensen; **7.15** Gerry Ellis/Minden Picture/Corbis; **7.16** Sborisov/Fotolia; **7.18** Norm Christensen; **7.19** Norm Christensen; **7.20a** Norm Christensen; **7.20b** Norm Christensen; **7.21** Andrea Jones/Alamy; **7.22** Inga Spence/Science Source; **7.23** Michael Snell/Alamy; **7.24** Norm Christensen; **7.25** Radius Images/Corbis; **7.26** Gunter Marx/Alamy; **7.27** Corbis/AGE Fotostock; **7.28** Edward Schuur/University of Florida/AP images; **7.29a** Eric Carr/Alamy; **7.29b** Stefan Auth/image Broker/AGE Fotostock; **7.29c** Kenny Tong/Shutterstock; **7.30** Thomas Barrat/Shutterstock; **7.31** Rick & Nora Bowers/Alamy; **7.34** Ron Niebrugge/Alamy; **7.35** Kevin Schafer/Alamy; **7.36** Jack Hobhouse/Alamy; **7.37a** Blickwinkel/Hecker/Alamy; **7.37b** John Elk III/Alamy; **7.39** Panoramic Stock Images/National Geographic Creative; **7.40a** Marevision/AGE Fotostock; **7.40b** Marko K?nig/Image Brokr/AGE Fotostock; **7.40c** Maximilian Weinzierl/Alamy; **7.42** Laguna Design/Science Source; **7.43a** Bill Draker/Image Broker/AGE Fotostock; **7.43b** Jeremy Woodhouse/Exactostock/Superstock; **7.44** Beverly Joubert/National Geographic Creative; **7.45** Michael Melford/National Geographic Creative; **7.46** Warren Price Photography/Shutterstock; **7.47** James Steinberg/Science Source; **7.48** Andrew McLachlan/All Canada Photos/AGE Fotostock; **7.49** Richard Broadwell/Alamy; **7.50** NASA;

7.51 Willard R. Culver/National Geographic/Getty Images; **7.52a** FLPA/Alamy; **7.53** George Grall/Getty Images; **7.54** Tim Graham/Latitude/Corbis; **7.56** Jeff Rotman/Science Source; **7.57** WaterFrame_jdo/Alamy; **7.59** Mark Doherty/Shutterstock; **7.60** Emory Kristof/National Geographic

Chapter 8

Chapter Opener: Andre Seale/AGE Fotostock; **8.1** Paul Baker/KRT/Newscom; **8.2** Bill Waldman/Alamy; **8.3** Douglas Peebles Photography/Alamy; **8.4** Jared Hobbs/All Canada Photos/Corbis; **8.5** Don Johnston/AGE Fotostock; **8.6** Martin Shields/Alamy; **8.7** Norm Christensen; **8.8a** Geoff Dann/DK Images; **8.8b** Colin Keates/DK Images; **8.9a** Neil Fletcher/DK Images; **8.9b** Nigel Hicks/DK Images; **8.10a** John Anderson/iStock/360/Getty Images; **8.10b** David Peart/DK Images; **8.11** Rene Frederic/Pixtal/AGE Fotostock; **8.12** Rieger Bertrand/hemis/AGE Fotostock; **8.13** Norm Christensen; **8.14a** Georgie Holland/AGE Fotostock; **8.14b** David Courtenay/Oxford Scientific/Getty Images; **8.14c** Mirko Zanni/WaterFrame/Getty Images; **8.14d** Pi-Lens/Shutterstock; **8.16** Ralph Lee Hopkins/National Geographic Image Collection/Alamy; **8.18a** Heeb Christian/Prisma/AGE Fotostock; **8.18b** Nomad/SuperStock; **8.20** Norm Christensen; **8.21** WaterFrame/Alamy; **8.22** Daryl Balfour/Getty Images; **8.23** David Rudkin/DK Images; **8.24a** Tom Uhlman/Alamy; **8.24b** Clint Farlinger/Alamy; **8.26** B.A.E. Inc./Alamy; **8.27** Jason Edwards/National Geographic Image Collection/Alamy; **8.28a** Derrick Hamrick/Imagebroker/AGE Fotostock; **8.28b** Clifton/Imagebroker/AGE Fotostock; **8.28c** Gilbert S. Grant/Science Source; **8.29** David Peart/DK Images; **8.30** William Leaman/Alamy; **8.32a** Peter Sutter/Arco Images/Imagebroker/AGE Fotostock; **8.32b** Alamy; **8.33** Bildagentur Zoonar GmbH/Shutterstock; **8.34** Frank Greenaway/DK Images; **8.35a** H. Mark Weidman Photography/Alamy; **8.35b** Harrison Shull/Aurora Photos/Alamy; **8.36** Pablo Galán Cela/AGE Fotostock; **8.37** William Leaman/Alamy; **8.38** Lars Johansson/Alamy; **8.39** Mike Veitch/Alamy; **8.40** Paulo Santos/Interfoto/AP Images; **8.41** NASA; **8.42** Gary K Smith/Frank Lane Picture Agency Limited; **8.43** Dave King/DK Images; **8.44** STRDEL/AFP/Getty Images/Newscom; **8.45** Nick Haddad; **8.46** Ellen Damschen/University of Wisconsin; **8.47** Q-Images/Alamy; **8.48** Demetrio Carrasco/DK Images; **8.49** GSK/Shutterstock; **8.50** Henk Meijer/Alamy; **8.52** N.A.S./Science Source; **8.54** FloridaStock/Shutterstock; **8.55** Christie's Images/Superstock; **8.56a** Photoshot Holdings Ltd/Alamy; **8.56b** Piotr Gatlik/Shutterstock; **8.57** Peter Luethi/AP images; **8.59** Eleanor Briggs; Agents of Change: **8.1** Rikshaw Films; **8.2** Rikshaw Films; **8.3** Prisma/Superstock; **8.4** Rikshaw Films

Chapter 9

Chapter Opener: Anna Henly/Digital Vision/Getty Images; **9.1** Justin Lewis/Getty Images; **9.1** Alamy; **9.3** Ashley Cooper/Alamy; **9.4** NASA; **9.7** Oregon state university; **9.7** NASA; **9.8** Bruno Barbier/Getty Images; **9.9** Werner Forman/Getty Images; **9.10** David Watkins/Fotolia; **9.11** CuboImages srl/Alamy; **9.12a** NOAA; **9.12b** NOAA; **9.16** Philippe Bourseiller/Getty Images; **9.18** Historic Map Works LLC/Getty Images, Miro Vrlik Photography/Alamy; **9.21** Thomas Kitchin & Victoria Hurst/All Canada Photos/Corbis, Fotog/Tetra Images/Corbis, Michael R. Smith, Josu Altzelai/Age Fotostock Spain, S.L./Alamy; **9.24a** Fotolia; **9.24b** Eddie J. Rodriquez/Shutterstock; **9.27a** NASA; **9.27b** NASA; **9.27c** NASA; **9.29a** J.W.Alker/Imagebroker/Age Fotostock; **9.29b** Federica Grassi/Getty Images; **9.30** Blickwinkel/Alamy; **9.31** Design Pics Inc/Alamy; **9.36** Michael DeYoung/Design Pics/Corbis; **9.38** Jeff Greenberg/Alamy; **9.40** Michael Betts/Getty Images; **9.41** Ron Chapple Stock/Alamy; **9.42a** Risto Bozovic/AP Images; **9.42b** Paul Spinelli/AP images; **9.43** Bob Neimy/Agstockusa/AP Images; **9.44** Dirk Visbach/AGE Fotostock; **9.45** Katie Arkema; **9.48** United Nations Department of Public Information; **9.49** Tomas Abad/AGE Fotostock; **9.50a** M. Frustino/AP Images; **9.50b** Kyodo/AP Images; **9.50c** Ministry of Foreign Affairs of D/Xinhua/Photoshot/Newscom; Agents of Change: **9.1** Cravey, Jacob; **9.2** Cravey, Jacob; **9.3** Arkorn/Shutterstock

Chapter 10

Chapter Opener: Greg Baker/AP Images; **10.1** Anonymous/ap Images; **10.2** Bettmann/Corbis; **10.3** Dawson, John; **10.4** Eyepress/AP Images; **10.7** Stan Pritchard/Alamy; **10.8** Design Pics/AGE Fotostock; **10.9** AGE Fotostock; **10.11a** NASA Goddard Space Flight Center; **10.11b** NASA Goddard Space Flight Center; **10.11c** NASA Goddard Space Flight Center; **10.11d** Robert Simmon/NASA; **10.12a** Lee Celano/Reuters; **10.12b** Agencia el Universal/Newscom; **10.12c** Andreas Manolis/Reuters; **10.13** Javarman/Shutterstock; **10.14** Murray Lee/AGE Fotostock; **10.17** Kevin T Brett/Shutterstock; **10.18a** Ahsgan/Corbis Wire/Corbis; **10.18b** Jorge Dan Lopez/Reuters/Corbis; **10.18c** Alan Copson/Alloy/Corbis; **10.23** Oliver Knight/Alamy; **10.25** AGE Fotostock; **10.26** Banzaert, Amy; **10.27** MarcelClemens/Shutterstock; **10.29** AP Images; **10.30** Bettmann/Corbis; **10.31** Konstantin Sutyagin/Shutterstock.com; **10.33** Diana Hirsch/Fundamental Photos; Agents of Change: **10.1** Marisol Becerra; **10.2** Don Despain/Alamy

Chapter 11

Chapter Opener: David Keith Jones /Images of Africa Photobank/Alamy; **11.2a** Shashank BengaliMCT/Newscom; **11.2b** Shashank Bengali/MCT/Newscom; **11.2c** John Scofield/National Geographic/Getty Images; **11.2d** Kevin O'Hara/Age Fotostock; **11.4** Thumb/Shutterstock; **11.6** Peter Johnson/Corbis; **11.7** Jose Pedro Fernandes/Alamy; **11.14** Jon Spaull/DK Images; **11.16c** JuneJ/Shutterstock; **11.17** Tim Boyle/Getty Images; **11.18** NASA; **11.21a** USGS; **11.21b** John Raoux/AP Images; **11.22** AFP/Getty Images/Newscom; **11.26a** Alan Spencer Photography/Alamy; **11.26b** Science Source; **11.27** NASA Images; **11.28b** Justin B. Ries, Ph.D.; **11.29** Rosanne Tackaberry/Alamy; **11.30** Newscom; **11.31** Ron Chapple/Corbis; **11.32** P. Schuetz/AGE Fotostock; **11.33** U.S. Coast Guard; **11.35** Rich Reid/National Geographic Image Collection/Alamy; **11.37a** Ballantine, Katherine; **11.37b** Schneider, Rebecca; **11.39a** Jim Parkin/Alamy; **11.39b** videowokart /123RF; **11.40** Jeff Greenberg "0 people images"/Alamy; **11.41** Haider Y. Abdulla/Fotolia; **11.43** Patrick Cloutier; **11.45** John Todd Ecological Design; **11.46** John Todd Ecological Design; **11.48** Andrey Volokhatiuk/Fotolia; **11.50** Colin Weston / LatitudeStock /Alamy; **11.52** Leonard F Wilcox/Alamy; **11.53** Anonymous/AP Images; Agents of Change: **11.1–11.2** Martin Figueroa

Chapter 12

Chapter Opener: AGE Fotostock America; **12.1** Sam Fried/Science Source; **12.2** NOAA/MCT/Newscom; **12.3** Stephen Mcsweeny/Panther Media/AGE Fotostock; **12.4** Nigel Pavitt/John Warburton-Lee Photography/Alamy; **12.5** worker/Shutterstock, Dave King/DK Images, Roger Dixon/DK Images, Sian Irvine/DK Images, Roger Phillips/DK Images, Geoff Dann/DK Images, Lan O'Leary/DK Images, Max Alexander/DK Images, Roger Phillips/DK Images, Roger Phillips/DK Images, Frank Greenaway/DK Images, Roger Dixon/DK Images, Peter Anderson/DK Images, Philip Dowell/DK Images, Dave King/DK Images, Andy Crawford/DK Images, Paul Bricknell/DK Images, Roger Phillips/DK Images, Roger Phillips/DK Images, Philip Dowell/DK Images, DK Images, Pixtal/Dinodia/AGE Fotostock; **12.6** Andrew McRobb/DK Images; **12.7** Gonzalo Azumendi/age fotostock Spain, S.L./Alamy; **12.8** Vasiliy Koval/Shutterstock, Ann Ronan Pictures/Heritage Images/Glow Images, Zev Radovan/BibleLandPictures/Alamy, Purestock/Alamy, Stephen Finn/Shutterstock, Joy Brown/Shutterstock, Max Alexander/DK Images, Roger Phillips/DK Images, Andy Crawford/DK Images, Peter Anderson/DK Images, Dorling Kindersley Limited, Philip Dowell/DK Images, Erich Lessing/Art Resource, Arogan/Shutterstock; **12.9** inga spence/Alamy; **12.10** Sandra Cunningham/Shutterstock, Maridav/Shutterstock, Geoff Brightling/DK Images; **12.14** Georg Gerster/Science Source; **12.15** Henk Bentlage/Shutterstock; **12.16** Anita P Peppers/Fotolia; **12.17** Casther/Fotolia; **12.18a** Christian Heinrich/image Broker/AGE Fotostock; **12.18b** Inga Nielsen/Shutterstock; **12.19a** Eric Issel/Fotolia; **12.19b** Jenoche/Fotolia; **12.20** Schankz/Shutterstock; **12.21a** Colin Keates/Courtesy of the Natural History **12.21b** Dave King/DK Images; Museum, London/DK Images; **12.21c** Dave King/DK Images; **12.24a** David R. Frazier Photolibrary, Inc./Alamy; **12.24b** Marko5/Shutterstock; **12.24c** Photo by Erwin Cole, USDA Natural Resources **12.24d** Bill Barksdale/Design Pics/Corbis; Conservation Service; **12.24e** Anne Gilbert/Alamy; **12.25** Jim Richardson Photography; **12.27** Frank Greenaway/DK Images; **12.28** Mark Wagner Aviation-Images/Alamy; **12.29** Holger Weitzel/Imagebroker/Alamy; **12.30a** USGS; **12.30b** EcoView/Fotolia; **12.31** Vallefrias/Fotolia; **12.34a** Frank Lane Picture Agency Limited; **12.34b** Muzsy/Shutterstock; **12.35** Lou Linwei/Alamy; **12.36** Bol/K Thomas/AGE Fotostock; **12.37** Emilio Ereza/Pixtal/AGE Fotostock; **12.37** Madlen/Shutterstock; **12.38** James King-Holmes/Science Source; **12.39a** Tertman/Shutterstock; **12.39b** Keith J Smith./Alamy; **12.41a** Carpenter/Fotolia; **12.41b** Zcw/Shutterstock; **12.42** Evan Golub/Demotix/Corbis; **12.43a** Jose Acosta/AP Images; **12.43b** Nigel Cattlin/Alamy; **12.43c** Photo Fun/Shutterstock; **12.44** Juan Carlos Muñoz/AGE Fotostock, DP Wilson/FLPA/AGE Fotostock, blickwinkel/Alamy, Crisod/Fotolia; **12.45** Robert Reich/Fotolia; **12.46** Bill Barksdale/Agstockusa/AGE Fotostock; **12.47** Raymond Klass/Alamy; **12.48** Nigel Cattlin/Alamy, David R. Frazier/Science Source, Age Fotostock/Superstock, Flirt/Superstock, Stoonn/Shutterstock; **12.49a** Kristian Cabanis/AGE Fotostock; **12.49b** Pictorium/Alamy; **12.50** Jeff Carroll/Agstockusa/AGE Fotostock; **12.52** Photology1971/Fotolia, lnzyx/Fotolia, Maxim Golubchikov/Shutterstock; **12.53** Oscar Durand/The New York Times; **12.54** Goran Bogicevic/Shutterstock; **12.55** Adam Jones/Science Source; **12.56** The Guardian/Alamy; **12.57** Sean Sprague/Alamy; **12.60** Angela Waite/Alamy; Agents of Change: **12.1** Kathy Shea; **12.2** Hillary King

Chapter 13

Chapter Opener: Harald Woeste/AGE Fotostock; **13.1** Stevens Frederic/Sipa/Newscom; **13.2** James P. Blair/Getty Images; **13.3** Norm Christensen; **13.4** Norm Christensen; **13.5** Kip Evans/Alamy; **13.7** Isifa Image Service s.r.o/Alamy; **13.8** Angelo Cavalli/AGE Fotostock; **13.9a** Shibo Yang/Alamy; **13.9b** JG Photography/Alamy; **13.9c** Alex Kosev/Shutterstock; **13.9d** David Hoffman/Alamy; **13.10** EyePress News EyePress/Newscom; **13.11** Redbrickstock.com/Alamy, Image Source/AGE Fotostock,Catherine Karnow/National Geographic Image Collection/Alamy, Blickwinkel/McPHOTO/HRM/Alamy, Leungchopan/Fotolia, Ainars Aunins/Shutterstock, Paul Dronsfield/Alamy; **13.12** Stefan Jannides/Alamy; **13.13** AlessandroZocc /Shutterstock; **13.14a** Vivian A Thode/Shutterstock.com; **13.14b** vrabelpeter1/Fotolia; **13.15** Don Mason/Palladium/AGE Fotostock; **13.16a** Norm Christensen; **13.16b** Norm Christensen; **13.16c** Norm Christensen; **13.16d** Norm Christensen; **13.17** Arnold John Labrentz/Shutterstock; **13.18** Jeff Barnard/AP Images; **13.20** Imagebroker.net/Superstock; **13.21** John Baran/Alamy; **13.22** Marcot, Bruce G.; **13.23** NASA; **13.23** NASA; **13.24** NASA; **13.25** John Winings; **13.26** John Winings; **13.27a** US National Forest Service, Coconino National Forest; **13.27b** US National Forest Service/Coconino National Forest; **13.27c** US National Forest Service/Coconino National

Forest; **13.28** NASA; **13.29** Artlosk/Fotolia; **13.30** Motoring Picture Library/Alamy; **13.31a** Calvin Larsen/Science Source; **13.31b** Technology And Industry Concepts/Alamy; Aleksander Bolbot/Alamy; **13.32a** Againstar/Shutterstock; **13.33** Charlie Borland/Fogstock/AGE Fotostock; **13.34** Norm Christensen; Agents of Change: **13.1** Bryan Robin; 13.2 SuperStock; **13.3** Dave Reede/All Canada Photos/Superstock

Chapter 14

Chapter Opener: Huyangshu/Shutterstock; **14.2** Kern County Museum; **14.3** Jonathan Blair/Documentary Value/Corbis; **14.4** Stephen Oliver/DK Images, Corey Hochachka/Design Pics/AGE Fotostock, Corbis/AGE Fotostock; **14.4** Ray Moller/DK Images; **14.5** David Frazier/Corbis; **14.7** Zack Seckler/Corbis; **14.8** Peter & Georgina Bowater Stock Connection Worldwide/Newscom; **14.11** RichardBakerWork/Alamy; **14.12** Pi-Lens/Shutterstock; **14.13** Philip Scalia/Alamy; **14.14** Tim Parmenter/Courtesy of the Natural History Museum, London/DK Images, Gary Ombler/DK Images; Stephen Oliver/DK Images; **14.16** AGE Fotostock America; **14.17** Daniel Shea/Galeries/Corbis; **14.18** Hans/Imagebroker/AGE Fotostock; **14.19** Thomas R. Fletcher/Alamy; **14.20** Charles E. Rotkin/Corbis; **14.21** U.S. Department of Energy; **14.23** Ron Nickel/Design Pics/AGE Fotostock; **14.24** Horizon/Horizon Stock Images/AGE Fotostock; **14.26** Dan Barnes/E+/Getty Images; **14.27** Corbis/AGE Fotostock; **14.28** Roy Morsch/Corbis; **14.29** Tom & Pat Leeson/Science Source; **14.30** Matthew D White/Photolibrary/Getty Images,; Anonymous/AP Images; **14.31** Alberto Velando; **14.32** Pi-Lens/Shutterstock; **14.34** Radius/Superstock; **14.35** Alamy; **14.35a** Paolo Koch/Science Source; **14.35b** RIA Novosti/Alamy; **14.35c** Clive Streeter/Courtesy of The Science Museum, London/DK Images; **14.35d** RIA Novosti/Alamy; **14.37** RIA Novosti/Alamy; **14.38** Mainichi Newspaper/Corbis News Premium/Corbis; **14.39** Nomad/Superstock,; Nuclear Regulatory Commission

Chapter 15

Chapter Opener: David Keith Jones/Alamy; **15.2** EschCollection/Photonica/Getty Images; **15.3** Jim Lopes/Shutterstock,; Sailorr/Shutterstock,; NASA Images; **15.6** Jin Linpeng/Xinhua Press/Corbis Wire/Corbis; **15.7** Topher Donahue/Aurora Open/Alloy/Corbis; **15.9** Tom Grundy/Shutterstock; **15.10** Sean Hunter/DK Images; **15.12** Fotosearch Value/Getty Images; **15.13** Mike Kahn/

Green Stock Media(Agent)/Alamy; **15.14** Terrance Emerson/Shutterstock; **15.15a** Sergwsq/123RF; **15.15b** Nigel Hicks/DK Images; **15.16** Orlin Wagner/AP Images; **15.17** Ron Tarver/MCT/Newscom; **15.18** NASA Images; **15.20** Siepmann/Image Broker/Alamy; **15.22** Dave King/DK Images; **15.24** Geoffrey Kuchera/Shutterstock; **15.28a** NASA; **15.28b** NASA; **15.29** Flpa/Mark/AGE Fotostock; **15.31** M. Balan/DK Images; **15.32** David Moir/Reuters/Corbis; **15.34** Dan Breckwoldt/Shutterstock; **15.39** Dave Rudkin/Shutterstock,; Satina/123RF; **15.40** Danil Roudenko/123RF; **15.43** Public domain; Agents of Change: **15.1** Jan Kelso and Amber White; **15.2–15.3** Jan Kelso and Amber White

Chapter 16

Chapter Opener: Rabbit75_fot/Fotolia; **16.1** Alamy; **16.2** NASA; **16.3** Photo provided by www.cyclingsolution.blogspot.com; **16.4** Stephen Finn/Shutterstock; **16.5** Wesley Bocxe/Science Source; **16.6** Zhu Difeng/Fotolia; **16.7** Mardis Coers/Getty Images; **16.8a** Jian Chen Stock Connection Worldwide/Newscom; **16.8b** Andrew Burns/Shutterstock; **16.8c** Fincher Lindsay; **16.10** Rudi1976/Fotolia; **16.11** NASA; **16.13** Ken Cavanagh/Alamy; **16.14** Indianapolis Cultural Trail, Inc.; **16.15** NASA; **16.17** Jeff Haynes/AFP/Getty Images; **16.18** Ric Francis/AP Images; **16.19a** Trekandshoot/Shutterstock; **16.19b** Stock Connection Blue/Alamy; **16.20a** Alamy; **16.20b** NASA; **16.20c** NASA; **16.21** Larry Roberg/Fotolia; **16.22** Imagebroker.net/Superstock; **16.23** David trevor/Alamy; **16.24** Chung, Payton; **16.25** David R. Frazier Photolibrary, Inc./Alamy; **16.26** Rich Addicks/The New York Times/Redux; **16.28** U.S. Environmental Protection Agency; **16.32** Jimmy Collins/Corbis; **16.35** Jeremy Pembrey/Alamy; **16.36** Wanderlust Photos; **16.37** Scott Hortop/Alamy; **16.38** A. Jagel/Blickwinkel/AGE Fotostock; **16.39** Javier Castro/Fotolia; **16.41** Patricia Leege; **16.42** Ella Milroy/Departure Lounge/DK Images; **16.43** Jenny Hager/The Image Works; **16.44** Terese Loeb Kreuzer/Alamy; **16.45** Dave King/DK Images; **16.46** Paul Franklin/DK Images; **16.47** John W. Dover; **16.49a** Mark Newman/Frank Lane Picture Agency Limited; **16.49b** DK Images; **16.50** Tyrone Turner/National Geographic Image Collection/Alamy; **16.50a** Alex Farnsworth/The Image Works; **16.50b** Alamy; **16.50c** Metropolitan Planning Council Metropolitan Planning Council/Chicago U.S.A; **16.50d** Robert E. Klein/AP Images; Agents of Change: **16.1–16.4** New Jersey Tree Foundation

Chapter 17

Chapter Opener: Tom & Dee Ann McCarthy/Corbis; **17.1a** Georg Gerster/Science Source; **17.1b** Walter Bibikow/Danita Delimont.com "Danita Delimont Photography"/Newscom; **17.1c** RIA Novosti/Alamy; **17.1d** Jeff Greenberg/Alamy; **17.4** Raga Jose Fuste/Prisma Bildagentur AG/Alamy; **17.6** Shutterstock; **17.8** David R.Frazier/The Image Works; **17.9** Accent Alaska/Alamy; **17.10** Steve Shepard/E+/Getty Images; **17.11** Sascha Schuermann/Getty Images; **17.12** Stringer Shanghai/Reuters; **17.13** Science Source; **17.15** Matt Miller/Topham/The Image Works; **17.17a** Mavar/Shutterstock; **17.17b** Chris Knapton/Alamy; **17.17c** Norman Chan/Shutterstock; **17.18a** Dave Rudkin/DK Images; **17.18b** Angela Hampton Picture Library/Alamy; **17.18c** Charles O. Cecil/Alamy; **17.19** Ray Pfortner/Photolibrary/Getty Images; **17.21** Remi Trudel; **17.24** Christopher Gable/Sally Gable/DK Images; **17.25a** Royce Bair/Moment/Getty Images; **17.25b** Jetta Productions/Iconica/Getty Images; **17.25c** Kurhan/Shutterstock; **17.25d** Robert Daly/OJO Images/Getty Images; **17.25e** Sergey Yechikov/Shutterstock; **17.25f** Lya_Cattel/E+/Getty Images; **17.26** Aaron Engel-Hall; **17.27** Sunsetman/Shutterstock; **17.28** USGS National Center; **17.29** People and Politics/Alamy; **17.30** Diane Collins and Jordan Hollender/Photodisc/Getty Images; Agents of Change: **17.1–17.3** Alex Freid

Chapter 18

Chapter Opener: Carson Ganci/Design Pics/AGE Fotostock; **18.3** Kletr/Shutterstock; **18.4** Andy Crump,TDR, World Health Organization/Science Source; **18.7a** Elwynn/Shutterstock; **18.7b** Erik Nguyen/Eureka Premium/Corbis; **18.7c** Mediacolor's/Alamy; **18.7d** Alex Segre/Alamy; **18.8** J.R. Bale/Alamy; **18.9** Chine Nouvelle/Sipa/Newscom; **18.11a** Ryan Remiorz/The Canadian Press/AP Images; **18.11b** Denis Poroy/AP Images; **18.13** US Navy/Handout/Digital/Corbis News/Corbis; **18.15a** NASA/Terra Deutsche Presse-Agentur/Newscom; **18.15b** A9999 Db NASA/Terra Deutsche Presse-Agentur/Newscom; **18.16** A.J. Sisco/Upi/Newscom; **18.17** Franck PrevelAP Images; **18.18** Newscom; **18.21** Inga Spence/Science Source; **18.23** Eye of Science/Science Source; **18.25** Giuseppe_R/Shutterstock; **18.26a** Steamroller_blues/Shutterstock; **18.26b** Tischenko Irina/Shutterstock; **18.27** Ames Bruce; **18.28** Bob Gibbons/Alamy; **18.29a** Fatih Kocyildir/Shutterstock; **18.29b** James Breeden/PacificCoastNews/Newscom; **18.30a** Corbis/

Superstock; **18.30b** Scott Camazine/Alamy; **18.31a** GoldSmith/Bsip/AGE Fotostock; **18.31b** Javier Larrea/AGE Fotostock; **18.33** Cobus Bodenstein/AP Images; **18.34a** Ulrich Doering/Alamy; **18.34b** Frank Greenaway/DK Images; **18.35** Diez,Ottmar/Bon Appetit/Alamy; **18.36** Eye of Science/Science Source; **18.38a** Ivan Kuzmin/Shutterstock; **18.38b** Cynthia Goldsmith/CDC; **18.40a** Sinclair Stammers/Science Source; **18.40b** DK images; **18.41** Christine Osborne/Alamy; **18.42** Wes Pope/MCT/Newscom; **18.43** USDA/ARS; **18.45** Chris Bjornberg/Science Source; Agents of Change: **18.1** Perez Will; **18.2** Sol Weiner, Brianafrica/Alamy; **18.3** Norm Christensen

Chapter 19

Chapter Opener: E.V Binstock/Corbis; **19.1** Thor Swift/The Image Works, SuperStock/SuperStock; **19.2** Jeffrey Hamilton/Exactostock/Superstock, Michael DeYoung/Design Pics/Corbis, Charlie Borland/Fogstock/AGE Fotostock, Tomas Abad/AGE Fotostock; **19.3** Gary K Smith/Frank Lane Picture Agency Limited, Vallefrias/Fotolia, Michael Melford/National Geographic Creative; Ben Curtis/AP Images; **19.4** Rich Reid/National Geographic Creative, Jeff Greenberg "0 people images"/Alamy, R. Hamilton Smith/AgStock Images, Inc./Alamy, Nyo09/Fotolia; **19.5** Ron Tarver/MCT/Newscom, Chris A Selby/Alamy, John Zellmer/Getty Images, Qingqing/Shutterstock, Danil Roudenko/123RF; **19.6** Edward Parker/Alamy, Raphael Gaillarde/Getty Images, Simon Maina/AFP/Getty Images/Newscom, Angelo Cavalli/AGE Fotostock, Harald Woeste/AGE Fotostock; **19.7** Mazzzur/Shutterstock, Steve Gortin/DK Images, Norm Christensen, Dinodia Photo/AGE Fotostock, Ivan_Sabo/Shutterstock, David R. Frazier/Photolibrary, Inc/Alamy, Lisa S/Shutterstock, Jon Hicks/Corbis; **19.8** Ornitolog82/Fotolia, George Burba/Shutterstock, hvoya/Shutterstock, Steve Byland/Shutterstock, Mike Truchon/Shutterstock, Steve Byland/Shutterstock, fivespots/Shutterstock, Kim Taylor/DK Images, Jane Burton/DK Images, DK Images, Mediscan/Medical-on-Line/Alamy, Georgia Southern University; **19.9** Anna Henly/Digital Vision/Getty Images; **19.11** Deshakalyan Chowdhury/AFP/Getty Images/Newscom; **19.12** Lissa Leege; **19.13** Marmaduke St. John/Alamy, Picturebank/Alamy, Aurelia Ventura/La Opinio/Newscom, Oberlin College, Wig Worland/Alamy; **19.14** ZUMA Press/Newscom; **19.15** Duke University; **19.16** Marisol Becerra; Agents of Change: **19.1–19.3** Andrew Sartain

TEXT CREDITS

1.25 Data from: Blus, L.J., et al. 1971. Eggshell thinning in the brown pelican: implication of DDE. BioScience 21: 1213.1215. **p. 8** Source: from the World Commission on Environment and Development's (the Brundtland Commission) report Our

Common Future (Oxford: Oxford University Press, 1987). **p. 20** Source: Daniel J Boorstin, The Discoverers: A history of mans' search to know his world and himself. New York, NY: Random House, 1983. **2.19** Data from Constanza, R. et al. 2014, "Changes in the

global value of ecosystem services" Global Environmental Change 26: 152.158. **2.33** Based on http://policymix.nina.no/Portals/policymix/Images/Case%20studies/Costa%20Rica/CR_biological%20corridor%20map.jpg. **p. 32** Source: Gifford Pinchot, quoted in,

John W.Simpson, Dam!: Water, Power, Politics and Preservation (New York,Pantheon Books, 2005), 324. **pp. 34–35** Source: Bible, Genesis 1:28, King James Version. **p. 37** Source: Leopold, A. 1949. A Sand County Almanac. Oxford University press, Oxford, UK.

p. 39 Source: Leopold, A. 1949. A Sand County Almanac. Oxford University Press, Oxford, UK. **p. 47** Source: Based on Ricketts, T.H., G.C. Daily, P.R. Ehrlich and C.D. Michener. 2004. Economic value of tropical forest to coffee production. Proceedings of the National Academy of Sciences 101:12579–12582. **p. 54** Source: Tenth Amendment to the United States Constitution. **p. 54** Source: Article IV of the US Constitution. **pp. 58–59** © Pearson Education, Inc. **p. 36** Source: Henry David Thoreau, Walden, or, Life in the woods. Boston, Ticknor and Fields, 1854. **3.18** Copyright DK LTD. **3.37** Copyright DK Ltd. **3.40** U.S. National Oceanographic and Atmospheric Administration, National Weather Service 2014. **Box 4.1** Based on Erickson, G.M, P. J. Currie, B. D. Inouye, A. A. Winn. 2006. Tyrannosaur Life Tables: An Example of Nonavian Dinosaur Population Biology. Science 313:213–217. **4.28** Grant, B.R. and P.R. Grant. 1989. Natural selection in a population of Darwin's finches. American Naturalist 133:377–393. **4.30** Grant, B.R., and P.R. Grant, 1989. Natural selection in a population of Darwin's finches. American Naturalist 133: 377–393. **p. 112** Source: Darwin, Charles (1845), Journal of researches into the natural history and geology of the countries visited during the voyage of H.M.S. Beagle round the world, under the Command of Capt. Fitz Roy, R.N (2nd. ed.), London: John Murray. **5.1** Inspired by Roberts 2011, Science 333 (29)540–543. **5.2** Cavalli-Sforza, L.L. and M.W. Feldman. 2003. The application of molecular genetic approaches to the study of human evolution. Nature Genetics 33: 266–275. **5.3** U.S. Central Intelligence Agency World Factbook 2013. **5.6** United Nations World Population Projection Revision 2012. **5.7** Based on United Nations World Population Projection Revision 2012. **5.8a** Data from CIA Factbook 2013. http://www.indexmundi.com/map/?v=25. **5.8b** Data from, http://data.worldbank.org/indicator/NY.GDP.PCAP.CD/countries/all?display=default. **5.9** Data from: United Nations World Population Prospects: The 2012 Revision Population Database. **5.10** Data from: United Nations World Population Prospects: The 2012 Revision Population Database. **5.11** Data from: United Nations World Population Prospects: The 2012 Revision Population Database. **5.12** Data from: United Nations World Population Prospects: The 2012 Revision Population Database. **5.13** Data from: United Nations World Population Prospects: The 2012 Revision Population Database. **5.14** Data from: United Nations World Population Prospects: The 2012 Revision Population Database. **5.15** Data from: CIA World Factbook 2013. **5.16** Data from: United Nations World Population Prospects: The 2012 Revision Population Database. **5.17** Data from: United Nations World Population Prospects: The 2012 Revision Population Database. **5.19** Data from: United Nations World Population Prospects: The 2012 Revision Population Database. **5.20** Data from: United Nations World Population Prospects: The 2012 Revision Population Database. **5.22** From Raftery, A.E. et al. 2012. Bayesian probabilistic population projections for all countries: supporting information. Proceedings of the National Academy of Sciences. http://www.pnas.org/cgi/doi/10.1073/pnas.1211452109. **5.24** Data from: Ecological Footprint Atlas 2010. **5.25** Data from: Global Footprint Network, www.fooprintnetwork.org. **5.26** Global Footprint Network, www.footprintnetwork.org. **5.32** Data from United Nations World Population Prospects: The 2012 Revision Population Database. **p. 147** Source: United Nations Fund for Population Activities (UNFPA). **p. 147** Source: Statement of Thoraya Ahmed Obaid, UNFPA Executive Director Address to the Board. **p. 147** Source: UNFPA Dispatch, 30 October 2013 "First School, Then Marriage and Babies". **6.4** Based on Gause, G.F. 1932. Experimental studies on the struggle for existence. Journal of Experimental Biology 9: 389–402. **6.5** Based on, MacArthur, R.H. 1958. Population ecology of some warblers of northeastern coniferous forests. Ecology 39: 599–619. **6.13** Data from: MacLulich, D. A. (1937). 'Fluctuations in numbers of the varying hare (Lepus americanus).' Univ. Toronto Studies, Biol. Ser. no. 43: 1–136. **6.38** Johnston, D.W. and E.P. Odum. 1956. Breeding bird populations in relation to plant succession on the Georgia Piedmont. Ecology 37: 50–62. **6.31** "Net Primary Productivity" from ATLAS OF BIOSPHERE from http://www.sage.wise.edu/atlas/maps/npp/atl_npp.jpg. Reprinted by permission of The Nelson Institute Center for Sustainability and the Global Environment at the University of Wisconsin-Madison. **7.6** Whitaker, R.H., 1975. Communities and ecosystems. New York: Macmillan. **7.9** Fischlin, A., G.F. Midgley, J.T. Price, R. Leemans, B. Gopal, C. Turley, M.D.A. Rounsevell, O.P. Dube, J. Tarazona, A.A. Velichko et al., 2007. Ecosystems, their properties, goods, and services. Climate Change 2007: Impacts, Adaptation and Vulnerability. Contribution of Working Group II to the Fourth Assessment Report of the Intergovernmental Panel on Climate Change, M.L. Parry, O.F. Canziani, J.P. Palutikof, P.J. van der Linden, and C.E. Hanson, eds. Cambridge: Cambridge University Press, 211–272. **7.17a** Source of 1620 and 1920 maps: Greeley, W.B. 1925. The relation of geography to timber supply. Economic Geography 1:1–11. **7.17b** Source of 2010 map: http://fia.fs.fed.us/slides/current-data.pdf. **7.50** Chesapeake Bay Foundation. http://www.cbf.org/page.aspx?pid=943. Last accessed 8.9.11. **7.58** Data from, United Nations Environmental Program World Conservation Monitoring Centre. URL: http://data.unep-wcmc.org/datasets/13. **8.5** Whittaker, R.H. 1956. The vegetation of the great smoky mountains. Ecological Monographs 26: 1–80. **Box 8.1** Based on Mora, C., D.P. Tittensor, S. Adl, A.G.B. Simpson, B. Worm. 2011. How many species are there on Earth and in the Ocean? PLoS Biology, 2011; 9 (8): e1001127 DOI:10.1371/journal.pbio.1001127. **8.14** Data from: Tilman, D. and J.A. Downing, 1994. Biodiversity and stability in grasslands. Nature 367: 363–365. **8.18** Bartblott, W. et al. 1999. Terminological and methodological aspects of the mapping and analysis of global biodiversity. Acta Botanica Fennica162: 103–110. **8.25** Based on MacArthur, R.H. and E.O. Wilson (1967) The Theory of Island Biogeography, Princeton University Press, Figures 8 and 11. **8.31** Based on, United Nations Food and Agriculture Organization, 2012. State of the World's Fisheries and Aquaculture Report. Rome, Italy: FAO. Figure 18, Page 56 of report. **p. 235** Source: United Nations Charter for Nature. **p. 256** Source: Endangered Species Act, Signed into law by President Richard Nixon on December 28, 1973. **9.7** Based on, Petit, J.R., et al., 1999. Climate and atmospheric history of the past 420,000 years from the Vostok Ice Core, Antarctica. Nature 399: 429–436. **9.13** U.S. National Oceanographic and Atmospheric Administration. **9.14** U.S. National Aeronautics and Space Administration. **9.18** Based on, Boden, T.A., G. Marland, and R.J. Andres (2010). Global, Regional, and National Fossil-Fuel CO_2 Emissions. Carbon Dioxide Information Analysis Center, Oak Ridge National Laboratory, U.S. Department of Energy, Oak Ridge, Tenn., U.S.A. doi 10.3334/CDIAC/00001_V2010. **9.19** Data from, NOAA, 2014. Mauna Loa CO_2 monthly mean data. http://www.esrl.noaa.gov/gmd/ccgg/trends/#mlo_data. **9.20** NOAA, 2013. Annual greenhouse gas index (AGGI). http://www.esrl.noaa.gov/gmd/aggi/. **9.23** United Nations World Development Indicators (WDI). http://data.worldbank.org/data-catalog/world-development-indicators/wdi-2012. **9.25** Data from The Rutgers University Snow Lab. URL = http://climate.rutgers.edu/snowcover/table_area.php?ui_set=2. **9.28** Based on University of Colorado Sea Level Lab.http://sealevel.colorado.edu. **9.32** IPCC, 2013. Climate change 2013: The physical basis: summary for policy makers. http://www.ipcc.ch/report/ar5/wg1/#.Uq8GqLTWuJI. **9.33** IPCC. 2014. Climate change 2014: impacts, adaptation, and vulnerability. Summary for policymakers. http://ipcc-wg2.gov/AR5/images/uploads/IPCC_WG2AR5_SPM_Approved.pdf (last reviewed 7/21/2014). **9.34** IPCC. 2007. Climate change 2007: Synthesis report. New York: Cambridge University Press. **9.35** IPCC. 2007. Climate change 2007: Synthesis report. New York: Cambridge University Press. **9.37** Based on, Pacala, S., and R. Socolow, 2004. Stabilization wedges: Solving the climate problem for the next 50 years with current technologies. Science 305: 968–972. **9.46** Based on, Arkema et al. Nature Climate Change 3, 913–918 (2013) doi:10.1038/nclimate1944 Figure 1. **9.47** Based on, Arkema K.K., 2013. Coastal habitats shield people and property from sea-level rise and storms. Nature: Climate Change 3:913–918. Figure 3. **9.51** Data from, UN World Development Index. URL=http://data.worldbank.org/data-catalog/world-developmentindicators/wdi-2012. **p. 268** Source: Sheila Watt-Cloutier, past chair of the Inuit Circumpolar Conference, By The Progressive on October 21, 2005. **p. 269** Source: Inuit Circumpolar Conference. **Box 10.1** Based on, Estrada, F., P. Perron, and B. Martínez-López. 2013. Statistically derived contributions of diverse human influences to twentieth-century temperature changes. Nature Geoscience 6:1050-1055. **10.15** National Atmospheric Deposition Program 2014 National Trends Network http://nadp.sws.uiuc.edu/ntn. **10.17** Based on, http://www.epa.gov/air/airtrends/lead.html#pbnat. **10.19** Data from EPA (2014) National trends in lead levels www.epa.gov/air/airtrends/lead.html. **10.21** http://www.airnow.gov/?action=aqibasics.aqi. **10.22** Based on "Deaths From Indoor Smoke from Solid Fuels" from http://www.who.int/heli/risks/indoorair/indoorair/en/index.html. World Health Organization Press, Geneva, Switzerland. **10.28** http://energy.cr.usgs.gov/radon/usrnpot.gif. **10.32** US EPA. http://www.epa.gov/air/airtrends/sulfur.html. Last accessed 1-16.2014. **10.33** Based on, World Meteorlogical Organization. 2011. Scientific Assessment of Ozone Depletion: 2010. Global Ozone Research and Monitoring Report No. 52, 516 pp. **AOC 10.3** Google Maps. **p. 327** Source: The Air Pollution Control Act of 1955 enacted by Congress. **11.1** http://earthobservatory.nasa.gov/IOTD/view.php?id=82235. **11.3** Adapted from Vital Water, www.unep.org; sourced from satellite images provided by NASA Goddard Space Flight Center. **11.7b** "Large Scale" from www.watersheds.org. Reprinted by permission. **11.8** Data from: Pacific Institute. **11.9** Based on, Igor Shiklomanov's chapter "World fresh water resources" in Peter H. Gleick (editor), 1993. Water in Crisis: A Guide to the World's Fresh Water Resources. **11.10** Based on, WRI. Aqueduct Metatdata Document: Aqueduct Global Map. 2.0 Francis Gassert, Matt Landis, Matt Luck, Paul Reig, and Tien Shiao. **11.15** Source: http://kornfeldblog-environmentallawblog.blogspot.com/2011/09/elements-of-water-cycle-storage-inice.htm. **11.16a** Data from USGS surface-water daily statistics for the nation USGS 01649500 Northeast Branch Anacostia River at Riverdale, Maryland. **11.16b** Data from USGS surface-water daily statistics for the nation USGS 01649500 Northeast Branch Anacostia River at Riverdale, Maryland. **11.19** Map based on National Geographic: Change the Course: Colorado River Map.Data Sources: Flow volumes: US Geological Survey; Diversion flow volumes: US Bureau of Reclamation; Dams, canals, pipelines, and aqueducts: USGS and National Geographic. **11.20** Based on Steward et al. (2013) "Tapping unsustainable groundwater stores for agricultural production in the High Plains Aquifer of Kansas, projections to 2110." Proceedings of the National Academy of Sciences 110(37): E3477–E3486. **11.24** Data reported by states to the EPA for 2012 under Sections 305(b) and 303(d) of the Clean Water Act. **11.28** Modified after R.A. Feely, Bulletin of the American Meteorological Society, July 2008. **11.34** Dahl, T. E. 2006. Status and trends of wetlands in the conterminous United States 1998 to 2004. Washington, D.C.: U.S. Fish and Wildlife Service. **11.36** Nichols, F. H., 2007. The San Francisco Bay and Delta–An Estuary Undergoing Change. U.S. Geological

Survey. http://sfbay.wr.usgs.gov/general_factsheets/change.html. **11.38** Ballantine, K. and R. Schneider. 2009. Fifty-five years of soil development in restored freshwater depressional wetlands. Ecological Applications 19: 1467–1480. **11.51** Based on, "MWRA Water System Demand 1986–2009." **AOC 11.3** Badger Meter. **p. 364** Source: Reprinted by permission of John Todd, John Todd Ecological Design, Inc. **p. 369** Source: Clean Water Act (CWA) of 1972, Enacted by the 92nd United States Congress. **12.13** United Nations. 2007. Millennium Ecosystem Assessment—latest available data. **12.18** From "Figure 4.10" from COMMUNITIES AND ECOSYSTEMS, 1st Edition, by R.H. Whittaker. Copyright (c) 1900. Adapted by permission of Pearson Education, Inc., Upper Saddle River, NJ. **12.26** L. DeHaan. Spring 2013. To select a needle in a haystack. The Land Report 105: 7. **12.40** Data from USDA, Economic Research Service and National Agricultural Statistics Service. **12.59** http://nationalatlas.gov/agriculture.html. **p. 376** Source: Slogan of W.P. Soash, a real estate man from The worst hard time: the untold story of those who survived The Great American dust bowl. Author: Timothy Egan. Publisher: Houghton Mifflin Harcourt, 2005. **p. 392** Source: The National Resource Conservation Service. **Box 13.10** Based on, McCarthy, H.R., R. Oren, K.H. Johnsen, et al., 2010. Re-assessment of plant carbon dynamics at the Duke free-air CO_2 enrichment site: interactions of atmospheric CO_2 with nitrogen and water availability over stand development. New Phytologist 185: 514–528. **13.6** Data from, Food and Agriculture Organization of the United Nations (FAO), 2010.Global Forest Resources Assessment 2010: Main Report.FAO Forestry Paper Rome: FAO, 378 pp. **13.17** Data from, Romme, W.H. Romme and D.G. Despain, 1989. Historical perspective on the Yellowstone fires of 1988.BioScience 39: 695–699. **13.24** Hansen, M.C., et al. 2013 High-resolution global maps of 21st century forest cover change. Science 342: 850–583. **p. 433** Source: Deforestation, United Nations Food and Agriculture Organization quoted in Generating knowledge through National Forest Assessments towards improved forest, land use and livelihood policies. **Box 14.1** Based on, Perez, C., A. Velando, I. Munilla, M. Lopez-Alonso and D. Oro, 2008. Monitoring polycyclic aromatic hydrocarbon pollution in the marine environment after the Prestige oil spill by means of seabird blood analysis. Environment Science & Technology 42: 707–713. **14.6** Based on, International Energy Agency (IEA). 2012. Key World Energy Statistics 2012. http://www.iea.org/publications/freepublications/publication/kwes.pdf. **14.15** Data from: British Petroleum, 2013. Statistical review of world energy June 2013. URL http://www.bp.com/content/dam/bp/pdf/statistical-review/statistical_review_of_world_energy_2013.pdf. Last accessed 1-30-14. **14.22** "Typical Oil Gas Reservoir" from GLOBAL CLIMATE CHANGE AND ENERGY: CARBON DIOXIDE CAPTURE AND STORAGE, http://www.planetseed.com/node/1520. Copyright (c) 2011 by Schlumberger Excellence in Education Development (SEED) Inc. All rights reserved. Reprinted by permission of SEED. For more information, visit our website at www.planetseed.com. **14.23** Data from: British Petroleum, 2013. Statistical review of world energy June 2013. www.bp.com/statistical review, p. 6. Last accessed February 7, 2014. **14.24** Data from: British Petroleum, 2013. Statistical review of world energy June. http://www.bp.com/en/global/corporate/about-bp/energy-economics/statistical-review-of-world-energy-2013.html. Last accessed February 30, 2014. **14.33** Data from: Uranium 2013: Resources, Production and Demand. A Joint Report by the OECD Nuclear Energy Agencyand the International Atomic Energy Agency. **p. 468** Source: Sheikh Zaki Yamani, quoted in "The End of the Oil Age," The Economist, October 23, 2003. **15.4** Data from: Renewables 2013: Global Status Report: 2013 Update. http://www.ren21.net/REN21Activities/GlobalStatusReport.aspx. Last accessed 2.28.2014. **15.5** Data from Renewables 2013: Global Status Report Panel Price data from http://www.renewableenergyworld.com/rea/news/article/2013/03/solarpv-profits-last-stand. **15.11** U.S. Department of Energy, National Renewable Energy Laboratory. URL: http://www.nrel.gov/csp/troughnet/images/map_normal_radiation.gif. Last accessed 2/14/2014. **15.16** U.S. Department of Energy, National Renewable Energy Laboratory. URL: http://www.nrel.gov/csp/troughnet/images/map_normal_radiation.gif. Last accessed 2/14/2014. **15.23** U.S. Department of Energy, National Renewable Energy Laboratory. URL: http://windeis.anl.gov/guide/maps/map2.html. Last accessed 10/10/2011. **15.27** Data from: British Petroleum, 2013. Statistical review of world energy June 2013. URL http://www.bp.com/content/dam/bp/pdf/statistical-review/statistical_review_of_world_energy_2013.pdf. Last accessed 1-30-14. **5.30** Data from: Andrews, E.D. and L.A. Pizzi. 2000. Origin of the Colorado River experimental flood in Grand Canyon. Hydrological Sciences 45: 607–627. **15.36** U.S. Department of Energy, National Renewable Energy Laboratory. http://www1.eere.energy.gov/tribalenergy/guide/images/photo_geomap.jpg. Last accessed 2/14/2014. **15.37** Map from BP Statistical Review of World Energy, 2013. URL www.bp.com/statistical review, p. 6. Last accessed 2.7.14; Hawkins, T.R., B. Singh, G. Majeau-Bettz, and A.H. Strømman. 2012. Comparative environmental life cycle assessment of conventional and electric vehicles. Journal of Industrial Ecology 17: 53–64. **15.38** U.S. Environmental Protection Agency. 2007. Energy trends in selected manufacturing sectors: Opportunities and challenges for environmentally preferable energy outcomes. URL http://www.epa.gov/sectors/pdf/energy/ch2.pdf. pages 2.4. Last accessed 1-30-14; Data from, U.S. Energy Information Administration: International Energy Statistics. **15.39** Based on, U.S. Bureau of Transportation Statistics 2012 Transportation Statistics Annual Report. URL: http://http://www.rita.dot.gov/bts/sites/rita.dot.gov.bts/files/tsar_2012.pdf.pdfLast accessed 2.25.2014. **15.41** Figure drafted by Christensen using data from Hawkins, T.R., B. Singh, G. Majeau-Bettz, and A.H. Strømman. 2012. Comparative environmental life cycle assessment of conventional and electric vehicles. Journal of Industrial Ecology 17:53.64. This figure was not published in a journal. **16.12** Adapted from: Figure 2, p. 157 of Lemmen, D.S. and F.J. Warren (eds.).2004. Climate Change Impacts and Adaptation: A Canadian Perspective. Climate Change Impacts and Adaptation Program, Natural Resources Canada, Ottawa, ON, 174 pp. http://www.nrcan.gc.ca/files/earthsciences/pdf/perspective/pdf/report_e.pdf.Last accessed July 18, 2014. **16.16** Based on Hobbs, Frank and Nicole Stoops, U.S. Census Bureau, Census 2000 Special Reports, Series CENSR-4, Demographic Trends in the 20th Century, U.S. Government Printing Office, Washington, DC, 2002 URL:http://www.census.gov/prod/2002pubs/censr-4.pdfLast accessed 10.1-2011. **16.31** Thomas, Randall ed. Max Fordham LLP. 2006. Environmental Design: An Introduction for Architects and Engineers, 3rd edition. p. 72. **Table 16.31** Data from table 6.2 Broad Comparative Energy Requirements of Building Materials, p 72. Environmental Design, An Introduction for Architects and Engineers, 3rd Edition. Edited by Randall Thomas Max Fordham LLP. Taylor & Francis, 2006. **16.33** Data from Newman, P.W.G., and J.R. Kenworthy, 1989. Gasoline consumption and cities. Journal of the American Planning Association 55: 24–37. **16.40** Davis, B.N.K. 1979. The ground arthropods of London Gardens. London Naturalist 58: 15–24. London Naturalist is published by the London Natural History Society http://lnhs.org.uk/. **16.48** Data from: Caroline Chiquet, John W. Dover & Paul Mitchell. 2013. Birds and the urban environment: the value of green walls. Urban Ecosyst 16:453–462. **p. 539** Source: Christine Todd Whitman (2006). **p. 547** Source: Aldo Leopold, A Sand County Almanac. 1949. Oxford University Press. **17.2** U.S. EPA. 2013. Municipal solid waste (MSW) in the United States: Facts and figures. URL:http://www.epa.gov/osw/nonhaz/municipal/. Last accessed 4.2.2014. **17.3** U.S. EPA. 2012. Municipal solid waste (MSW) in the United States: Facts and figures. URL: http://www.epa.gov/osw/nonhaz/municipal/. Last accessed 4.2.2014. **17.16** U.S. EPA. 2013. Municipal solid waste (MSW) in the United States: Facts and figures. URL: http://www.epa.gov/osw/nonhaz/municipal/. Last accessed 4.2.2014. **17.20** U.S. EPA. 2013 Municipal solid waste (MSW) in the United States: Facts and figures. URL: http://www.epa.gov/osw/nonhaz/municipal/. Last accessed 4.2.2014. **17.22** Trudel, R. and J.J. Argo. 2013. The effect of product size and form distortion on consumer recycling behavior. Journal of Consumer Research 40. DOI: 10.1086/671475. **17.23** Trudel, R. and J.J. Argo. 2013. The effect of product size and form distortion on consumer recycling behavior. Journal of Consumer Research 40. DOI: 10.1086/671475. **17.25** Sony Ericsson Communications. 2009. Sustainability Report. http://blogs.sonymobile.com/wp-content/uploads/2012/09/SESustainabilityReport2008.pdf. Last accessed April 2, 2014. **p. 573** Source: On September 20, 2012, EPA Administrator Lisa P. Jackson launched the SMM Electronics **Box 18.1** Based on, Ames, B.N., M. Profet and L.W. Gold. 1990. Nature's chemicals and synthetic chemicals: Comparative toxicology. Proceedings of the National Academy of Sciences 87:7782–7786. **18.2** Based on, Utzinger, J., Tozan, Y. and Singer, BH. 2001. Efficacy and cost effectiveness of environmental management for malaria control. Tropical Medicine and International Health 6, 677–687. **18.5** Based on, World Health Organization. 2013. The top 10 causes of death. Fact Sheet no. 310. URL:http://www.who.int/mediacentre/factsheets/fs310/en/index2.html#. Last accessed 4.15.1014. **18.6** Based on, World Health Organization. 2013. The top 10 causes of death. Fact Sheet no. 310. URL:http://www.who.int/mediacentre/factsheets/fs310/en/index2.html#. Last accessed 4.15.2014. **18.10** Global Seismic Hazard Assessment Program. URL: http://www.seismo.ethz.ch/static/gshap/. Last accessed 4/16/2014. **18.12** Change in tsunami wave shape as it approaches land from http://www.walkatoregion.govt.nz/services/Regional-services/Regional-hazards-and-emergency-mangement/coastal-hazards/Tsunami/. Reprinted by permission of Walkato Regional Council. **18.14** National Oceanic and Atmospheric Administration. URL: http://csc.noaa.gov/hurricanes. **18.24** Based on, Swan, S.H., E.P. Elkin and L. Fenster. 2000. The question of declining sperm density revisited: An analysis of 101 studies published 1934.1996. Environmental Health Perspectives 108: 961–966. **18.32** Pascual, M., X., Rodó, S.P. Ellner, et al. 2000. Cholera dynamics and El Nino-Southern Oscillation. Science 289: 1766–1769. **18.39** Based on, U.S. Centers for Disease Control. URL=http://www.cdc.gov/asthma/nhis/default.htm. Last accessed 4/16/2014. **18.43** U.S. Centers for Disease Control. URL: http://www.cdc.gov/mmwr/preview/mmwrhtml/figures/R807A2F1.GIF. Last accessed 4/16/2014. **18.44** U.S. Centers for Disease Control. http://www.cdc.gov/lyme/stats/chartstables/reportedcases_statelocality.html. Last accessed April 16, 2014. **18.45** Data from, Pan AmericanHealth Organization. URL=http://www.paho.org/hq/index.php?option=com_content&view=article&id=264&Itemid=40170&lang=en. Last accessed 4/16/2014. **p. 581** Source: In 1947, the World Health Organization's Expert Committee on Malaria. **p. 591** Source: Deishmann, W., & Henschler, D. (1986). Where is there that is not poison? A study of the Third Defense by Paracelsus. Archives of Toxicology, 58(4), 207–213. **p. 597** Source: Silent Spring, Rachel Carson, 1962. **19.11** Data from, Helliwell, J.F., R. Layard, and J.D. Sachs. 2013. The World Happiness Report 2013. Sustainable Development Solutions Network. New York, NY. **p. 625** Navajivan Trust for the Estate of Mahatma Gandhi. **p. 629** Edmund Burke.

Q

R